THOUGHT-PROVOKING APPROACH

This text encourages students to think about data and their analyses by asking and answering interesting statistical questions.

- **Figure Caption Questions** prompt students to explore graphical displays of data.
- **Questions to Explore** in worked-out examples help students to isolate the question being asked.
- **Exercises** encourage students to explain their responses rather than simply providing a numerical answer.

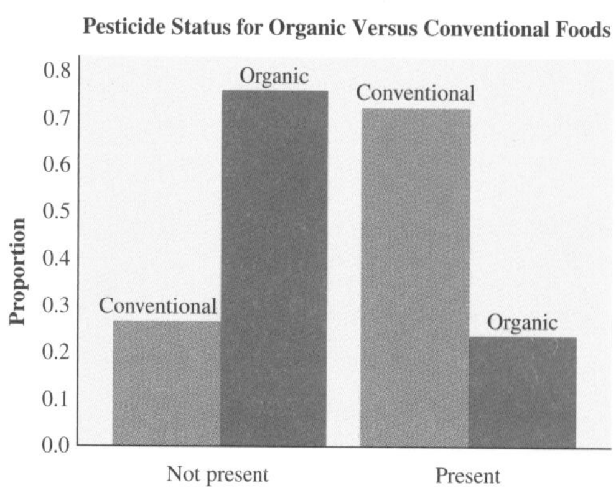

▲ **Figure 3.2 Conditional Proportions on Pesticide Status, Given the Food Type.** For a particular pesticide status category, the side-by-side bars compare the two food types. **Question** Comparing the bars, how would you describe the difference between organic and conventionally grown foods in the conditional proportion with pesticide residues present?

Page 95

REAL-WORLD APPLICATIONS

To help students understand the difference between academic statistics and what is done outside the classroom, **In Practice** boxes explain how statisticians analyze data in the real world.

In Practice Conditional Probabilities in the Media

When you read or hear a news report that uses a probability statement, be careful to distinguish whether it is reporting a conditional probability. Most statements are conditional on some event and must be interpreted in that context. For instance, probabilities reported by opinion polls are often conditional on a person's gender, race, or age group.

Page 234

EXTENSIVE PEDAGOGY

Key definitions, guidelines, procedures, and summaries are highlighted in boxes, creating at-a-glance learning aids to help in comprehending the statistical concepts. Three types of margin notes guide student learning:

- **Did You Know** helps readers understand the context of the statistical question.
- **Caution** alerts students to areas to which they need to pay special attention.
- **Recall** directs students to previous text discussions to review and reinforce concepts and methods.

Did You Know?

The **law of large numbers** is what helps keep casinos in business. For a short time, a gambler may be lucky and make money, but the casino knows that, in the long run, ahead. ◄

Page 213

Caution

Although technology will report the endpoints of a confidence interval for a difference of proportions to several decimal places, it is simpler to report and interpret these confidence inte decimal places for th

Page 469

Recall

We use *z*-scores to standardize the unit of measurement for a normal distribution. Once the units are standardized, we have a standard normal distribution with mean 0 and standard deviation of 1. ◄

Page 339

A Guide to Learning From the Art in This Text

We use color to help distinguish between the different shapes that graphs may take:

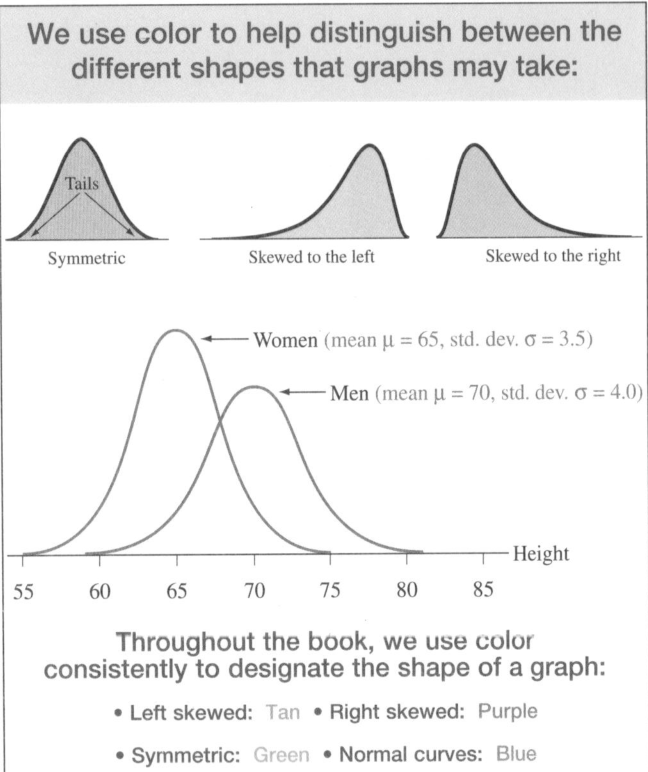

Throughout the book, we use color consistently to designate the shape of a graph:

- **Left skewed:** Tan • **Right skewed:** Purple
- **Symmetric:** Green • **Normal curves:** Blue

And between important measures such as the sample median and sample mean

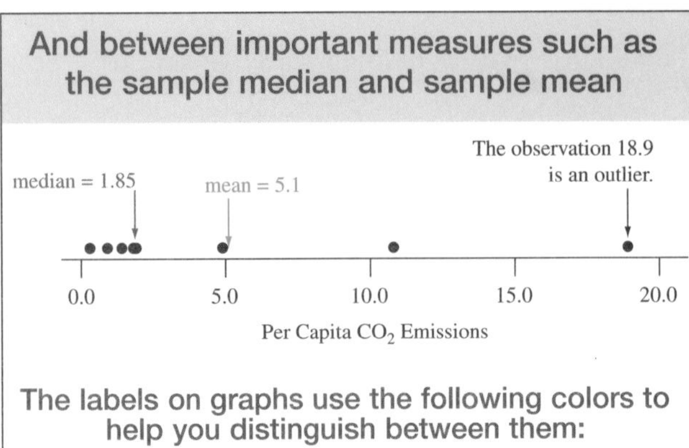

The labels on graphs use the following colors to help you distinguish between them:

- **Sample Median (Q2):** Green
- **Sample Mean $\bar{x}$:** Red

and between some of the most important statistics and parameters

	Sample Statistic	Population Parameter
Mean	$\bar{x}$	μ
Standard Deviation	s	σ
Proportion	$\hat{p}$	p

We show **Sampling Distributions of Sample Proportions** in blue because the normal distribution is used to describe the sampling distribution of $\hat{p}$.

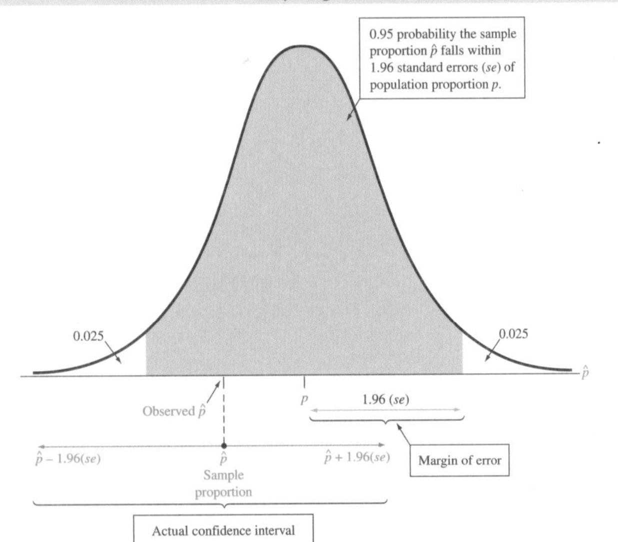

▲ **Figure 8.3 Sampling Distribution of Sample Proportion $\hat{p}$.** For large random samples, the sampling distribution is normal around the population proportion p, so $\hat{p}$ has probability 0.95 of falling within 1.96(*se*) of p. As a consequence, $\hat{p} \pm 1.96(se)$ is a 95% confidence interval for p. **Question** Why is the confidence interval $\hat{p} \pm 1.96(se)$ instead of $p \pm 1.96(se)$?

We show **Sampling Distributions of Sample Means** in green because the symmetric *t* distribution is used to describe the sampling distribution of $\bar{x}$.

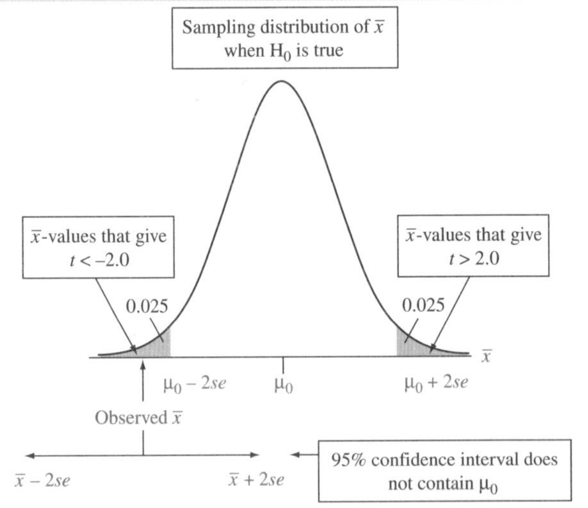

▲ **Figure 9.10 Relation between Confidence Interval and Significance Test.** With large samples, if the sample mean falls more than about two standard errors from μ_0, then μ_0 does not fall in the 95% confidence interval and also μ_0 is rejected in a test at the 0.05 significance level. **Question** Inference about proportions does not have an *exact* equivalence between confidence intervals and tests. Why? (*Hint:* Are the same standard error values used in the two methods?)

Read and think about the questions that appear in selected figures. The answers are given at the beginning of each Chapter Review section.

Statistics

The Art and Science of Learning from Data

Statistics

The Art and Science of Learning from Data

Third Edition

Alan Agresti
University of Florida

Christine Franklin
University of Georgia

PEARSON

Boston Columbus Indianapolis New York San Francisco Upper Saddle River
Amsterdam Cape Town Dubai London Madrid Milan Munich Paris Montréal Toronto
Delhi Mexico City São Paulo Sydney Hong Kong Seoul Singapore Taipei Tokyo

Editor in Chief: Deirdre Lynch
Acquisitions Editor: Marianne Stepanian
Senior Development Editor: Elaine Page
Executive Content Editor: Christine O'Brien
Senior Content Editor: Chere Bemelmans
Associate Content Editor: Dana Bettez
Editorial Assistant: Sonia Ashraf
Senior Managing Editor: Karen Wernholm
Senior Production Project Manager: Beth Houston
Digital Assets Manager: Marianne Groth
Supplements Production Coordinator: Katherine Roz
Manager, Multimedia Production: Christine Stavrou
Executive Marketing Manager: Roxanne McCarley
Marketing Manager: Erin Lane
Marketing Coordinator: Kathleen DeChavez
Rights and Permissions Advisor: Michael Joyce
Image Manager: Rachel Youdelman
Senior Manufacturing Buyer: Debbie Rossi
Senior Media Buyer: Ginny Michaud
Design Manager: Andrea Nix
Senior Designer: Heather Scott
Text Design: Ellen Pettengell
Production Coordination, Composition, and Illustrations: Integra
Cover Image: Robyn Mackenzie/iStockphoto

Library of Congress Cataloging-in-Publications Data

Agresti, Alan
Statistics: the art and science of learning from data / Alan Agresti, Christine Franklin.—3rd ed.
p. cm.
Includes Index
ISBN 0-321-75594-4
1. Statistic-Textbooks. I. Franklin, Christine A. II. Title.
QA276.12.A35 2013
519. 5—dc22 2011010804

1 2 3 4 5 6 7 8 9 10—Quad—15 14 13 12 11

www.pearsonhighered.com

ISBN-10: 0-321-75594-4
ISBN-13: 978-0-321-75594-0

Dedication

To my wife Jacki for her extraordinary support, including making numerous suggestions and putting up with the evenings and weekends I was working on this book.

ALAN AGRESTI

To Corey and Cody, who have shown me the joys of motherhood, and to my husband, Dale, for being a dear friend and a dedicated father to our boys.

CHRIS FRANKLIN

Agresti/Franklin CD Contents

Data sets are provided in a number of formats

1. .csv
2. TI-83/84 Plus
3. .txt

Look at the back endpapers of the book for a complete list of data files. Information is given on which examples, exercises, and activities require or reference the data files and applets.

Applets

1. Sample from a population
2. Sampling distributions
3. Random numbers
4. Long-run probability demonstrations:
   a. Simulating the probability of rolling a 6
   b. Simulating the probability of rolling a 3 or 4
   c. Simulating the probability of head with a fair coin
   d. Simulating the probability of head with an unfair coin (P(H) = 0.2)
   e. Simulating the probability of head with an unfair coin (P(H) = 0.8)
   f. Simulating the stock market
5. Mean versus median
6. Standard deviation
7. Confidence intervals for a proportion
8. Confidence intervals for a mean (for studying the impact of confidence level and the impact of not knowing the standard deviation)
9. Hypothesis tests for a proportion
10. Hypothesis tests for a mean
11. Correlation by eye
12. Regression by eye
13. Binomial distribution

A more detailed description of the Applets appears on page x.

Contents

An Introduction to the Applets

The applets on the CD-ROM that is bound inside all new copies of this text are designed to help students understand a wide range of introductory statistics topics.

- The **sample from a population** applet lets the user select samples of various sizes from a wide range of population shapes including uniform, bell-shaped, skewed, and binary populations (including a range of values for the population proportion, p). In addition, one can alter any of the default populations to create a custom distribution by dragging the mouse over the population or by going to Custom binary and typing in the desired population proportion. Small samples are drawn in an animated fashion to help students understand the basic idea of sampling. Larger samples are drawn in an unanimated fashion so that characteristics of larger samples can be quickly compared to population characteristics.
- The **sampling distributions** applet builds off the previous applet by adding the values of user-selected statistics for each sample. Students can study the resulting sampling distribution and see how characteristics of the sampling distribution, such as center and spread, are affected by sample size and population shape. Students can also compare sampling distributions of different statistics such as the sample mean and the sample median.
- The **random numbers** applet lets students select a random sample from a range of user-defined integer values. Students can use the applet to study basic probability by considering the relative frequency of particular outcomes among the samples. They can also select samples from a list of values for a hands-on sampling activity.
- Six **long-run probability demonstration** applets simulate rolling a die, flipping a coin, and fluctuation of the stock market. Students can select the number of times a simulation occurs, and whether they would like it animated. The relative frequency of an event of interest is plotted versus the number of simulations. As the number of simulations increases, the convergence of the relative frequency to the true probability of the event will be evident.
- The **mean versus median** applet lets students construct a data set interactively by clicking on a graphic that displays the mean and median of the data. Using the applet lets students study the effects of shape and outliers on the mean and the median.
- The **standard deviation** applet provides a similar type of exploration. This applet is offered in a stacked form so that data sets with different standard deviations can be compared easily.
- Three applets help students better understand confidence intervals. The **confidence intervals for a proportion** applet lets students simulate 95% and 99% confidence intervals for a population proportion and gain an understanding of how to interpret a 95% and 99% confidence level. The confidence intervals are plotted illustrating their relationship in terms of width and their random nature. The sample size and the true underlying proportion are specified by the user. Two applets lets students study **confidence intervals for a mean** in a similar manner. The first can be used to show how sample size and distributional shape affect the performance of classic t intervals for the mean. The second lets students compare the performance of z and t intervals for different distributional shapes and samples sizes.
- The applets for **hypothesis tests for a proportion** and **hypothesis tests for a mean** help students understand how the underlying assumptions affect the performance of hypothesis tests. These applets plot test statistics and corresponding P-values for data generated under different user-supplied conditions. Tabled rejection proportions allow students to determine how the conditions specified affect the true level of significance (Type I error probability) for the tests. The concepts of power and Type II error can also be explored with these applets.
- The **correlation by eye** applet helps students guess the value of the correlation coefficient based on a scatterplot of simulated data. In addition, students can see how adding and deleting points affects the correlation coefficient. Likewise, the **regression by eye** applet lets students attempt interactively determining the regression line for simulated data.
- The **binomial distribution** applet generates samples from the binomial distribution at user-specified parameter values. By varying the parameters, students can develop an understanding of how these parameters affect the binomial distribution.

Preface

We have each taught introductory statistics for more than 30 years, and we have witnessed the welcome evolution from the traditional formula-driven mathematical statistics course to a concept-driven approach. This concept-driven approach places more emphasis on why statistics is important in the real world and places less emphasis on probability. One of our goals in writing this book was to help make the conceptual approach more interesting and more readily accessible to college students. At the end of the course, we want students to look back at their statistics course and realize that they learned practical concepts that will serve them well for the rest of their lives.

We also want students to come to appreciate that in practice, assumptions are not perfectly satisfied, models are not exactly correct, distributions are not exactly normally distributed, and all sorts of factors should be considered in conducting a statistical analysis. The title of our book reflects the experience of data analysts, who soon realize that statistics is an art as well as a science.

What's New in This Edition

Our goal in writing the third edition of our textbook was to improve the student and instructor user experience. We have:

- Clarified terminology and streamlined writing throughout the text to improve ease of reading and facilitate comprehension.
- Modified the design to clearly show pedagogical hierarchy and distinguishing features.
- Added concept tags to all examples, which makes it easy for students and instructors to identify what is being demonstrated in the example.
- Added margin Caution boxes to alert students to areas where they need to pay special attention, such as common mistakes to avoid.
- Updated or replaced at least 25 percent of the exercises and examples. In addition, we have updated all General Social Services (GSS) data with the most current data available.
- Significantly rewritten Chapter 7: Sampling Distribution. In this chapter we now emphasize simulation to develop the concepts of sampling distributions, with less emphasis on the more traditional mathematical approach. We have reorganized the chapter to better distinguish a population, data, and sampling distribution. We now introduce standard error terminology in Chapter 8, where in practice we use the sample proportion and sample standard deviation to estimate the standard deviation of a sampling distribution. We believe this will result in less confusion for the student and emphasize that in practice, when we use the term *standard error*, we most often are referencing the estimated standard deviation of a sampling distribution, not the theoretical standard deviation.
- Added Learning Objectives for each chapter to the Instructor's Edition, which helps when preparing lectures.

Our Approach

In 2005, the American Statistical Association (ASA) endorsed guidelines and recommendations for the introductory statistics course as described in the report, "Guidelines for Assessment and Instruction in Statistics Education (GAISE) for the College Introductory Course" (www.amstat.org/education/gaise). The report states that the overreaching goal of all introductory statistics courses is to produce statistically educated students, which means that students should develop statistical literacy and the ability to think statistically. The report gives six key recommendations for the college introductory course:

- Emphasize statistical literacy and develop statistical thinking.
- Use real data.
- Stress conceptual understanding rather than mere knowledge of procedures.
- Foster active learning in the classroom.
- Use technology for developing concepts and analyzing data.
- Use assessment to evaluate and improve student learning.

We wholeheartedly endorse these recommendations, and our textbook takes every opportunity to support these guidelines.

Ask and Answer Interesting Questions

In presenting concepts and methods, we encourage students to think about the data and the appropriate analyses by posing questions. Our approach, learning by framing questions, is carried out in various ways, including (1) presenting a structured approach to examples that separates the question and the analysis from the scenario presented, (2) providing homework problems that encourage students to think and write, and (3) asking questions in the figure captions that are answered in the Chapter Review.

Present Concepts Clearly

Students have told us that this book is more "readable" and interesting than other introductory statistics texts because of the wide variety of intriguing real data examples and exercises. We have simplified our prose wherever possible, without sacrificing any of the accuracy that instructors expect in a textbook.

A serious source of confusion for students is the multitude of inference methods that derive from the many combinations of confidence intervals and tests, means and proportions, large sample and small sample, variance known and unknown, two-sided and one-sided inference, independent and dependent samples, and so on. We emphasize the most important cases for practical application of inference: large sample, variance unknown, two-sided inference, and independent samples. The many other cases are also covered (except for known variances), but more briefly, with the exercises focusing mainly on the way inference is commonly conducted in practice.

Connect Statistics to the Real World

We believe it's important for students to be comfortable with analyzing a balance of both quantitative and categorical data so students can work with the data they most often see in the world around them. Every day in the media, we see and hear percentages and rates used to summarize results of opinion polls, outcomes of medical studies, and economic reports. As a result, we have increased the attention paid to the analysis of proportions. For example, we use contingency tables early in the text to illustrate the concept of association between two categorical variables and to show the potential influence of a lurking variable.

Organization of the Book

The statistical investigative process has the following components: (1) asking a statistical question; (2) designing an appropriate study to collect data; (3) analyzing the data; and (4) interpreting the data and making conclusions to answer the statistical questions. With this in mind, the book is organized into four parts.

Part 1 focuses on gathering and exploring data. This equates to components 1, 2, and 3, when the data is analyzed descriptively (both for one variable and the association between two variables).

Part 2 covers probability, probability distributions, and the sampling distribution. This equates to component 3, when the student learns the underlying probability necessary to make the step from analyzing the data descriptively to analyzing the data inferentially (for example, understanding sampling distributions to develop the concept of a margin of error and a P-value).

Part 3 covers inferential statistics. This equates to components 3 and 4 of the statistical investigative process. The students learn how to form confidence intervals and conduct significance tests and then make appropriate conclusions answering the statistical question of interest.

Part 4 covers analyzing associations (inferentially) and looks at extended statistical methods.

The chapters are written in such a way that instructors can teach out of order. For example, after Chapter 1, an instructor could easily teach Chapter 4, Chapter 2, and Chapter 3. Alternatively, an instructor may teach Chapters 5, 6, and 7 after Chapters 1 and 4.

Features of the Third Edition

Promoting Student Learning

To motivate students to think about the material, ask appropriate questions, and develop good problem-solving skills, we have created special features that distinguish this text.

Student Support

To draw students to important material we highlight key definitions, guidelines, procedures, "In Practice" remarks, and other summaries in boxes throughout the text. In addition, we have four types of margin notes:

- **In Words:** This feature explains, in plain language, the definitions and symbolic notation found in the body of the text (which, for technical accuracy, must be more formal).
- **Caution:** These margin boxes alert students to areas to which they need to pay special attention, particularly where they are prone to make mistakes or incorrect assumptions.
- **Recall:** As the student progresses through the book, concepts are presented that depend on information learned in previous chapters. The Recall margin boxes direct the reader back to a previous presentation in the text to review and reinforce concepts and methods already covered.
- **Did You Know:** These margin boxes provide information that helps with the contextual understanding of the statistical question under consideration.

Graphical Approach

Because many students are visual learners, we have taken extra care to make the **text figures** informative. We've annotated many of the figures with labels that clearly identify the noteworthy aspects of the illustration. Further, most figure captions include a question (answered in the Chapter Review) designed to challenge the student to interpret and think about the information being communicated by the graphic. The graphics also feature a pedagogical use of color to help students recognize patterns and distinguish between statistics and parameters. The use of color is explained in the very front of the book for easy reference.

Hands-On Activities and Simulations

Chapters 1 through 12 include at least one **activity** each. The instructor can elect to carry out the activities in class, outside of class, or a combination of both. The activity often involves simulation, commonly using an applet available on the companion CD-ROM and within MyStatLab™. These hands-on activities and simulations encourage students to learn by doing.

Connection to History: On the Shoulders of...

We believe that knowledge pertaining to the evolution and history of the statistics discipline is relevant to understanding the methods we use for designing studies and analyzing data. Throughout the text, several chapters feature a spotlight on people who have made major contributions to the statistics discipline. These spotlights are titled **On the Shoulders of...**

Real World Connections

Chapter-Opening Example

Each chapter begins with a **high-interest example** that raises key questions and establishes themes that are woven throughout the chapter. Illustrated with engaging photographs, this example is designed to grab students' attention and draw them into the chapter. The issues discussed in the chapter's opening example are referred to and revisited in examples within the chapter. All chapter-opening examples use real data from a variety of applications.

Statistics: In Practice

We realize that there is a difference between proper "academic" statistics and what is actually done in practice. Data analysis in practice is an art as well as a science. Although statistical theory has foundations based on precise assumptions and conditions, in practice the real world is not so simple. **In Practice** boxes and text references alert students to the way statisticians actually analyze data in practice. These comments are based on our extensive consulting experience and research and by observing what well-trained statisticians do in practice.

Exercises and Examples

Innovative Example Format

Recognizing that the worked examples are the major vehicle for engaging and teaching students, we have developed a unique structure to help students learn to model the question-posing and investigative thought process required to examine issues intelligently using statistics. The five components are as follows:

- **Picture the Scenario** presents background information so students can visualize the situation. This step places the data to be investigated in context and often provides a link to previous examples.
- **Questions to Explore** reference the information from the scenario and pose questions to help students focus on what is to be learned from the example and what types of questions are useful to ask about the data.

- **Think It Through** is the heart of each example. Here, the questions posed are investigated and answered using appropriate statistical methods. Each solution is clearly matched to the question so students can easily find the response to each Question to Explore.
- **Insight** clarifies the central ideas investigated in the example and places them in a broader context that often states the conclusions in less technical terms. Many of the Insights also provide connections between seemingly disparate topics in the text by referring to concepts learned previously and/or foreshadowing techniques and ideas to come.
- **Try Exercise:** Each example concludes by directing students to an end-of-section exercise that allows immediate practice of the concept or technique within the example.

Concept tags are included with each example so that students can easily identify the concept demonstrated in the example.

Relevant and Engaging Exercises

The text contains a strong emphasis on real data in both the examples and exercises. We have updated the exercise sets in the third edition to ensure that students have ample opportunity to practice techniques and apply the concepts. Nearly all of the chapters contain more than 100 exercises, and more than 25 percent of the exercises are new to this edition or have been updated with current data. These exercises are realistic and ask students to provide interpretations of the data or scenario rather than merely to find a numerical solution. We show how statistics addresses a wide array of applications including opinion polls, market research, the environment, and health and human behavior. Because we believe that most students benefit more from focusing on the underlying concepts and interpretations of data analyses rather from the actual calculations, the exercises often show summary statistics and printouts and ask what can be learned from them.

We have exercises in three places:

- **At the end of each section.** These exercises provide immediate reinforcement and are drawn from concepts within the section.
- **At the end of each chapter.** This more comprehensive set of exercises draws from all concepts across all sections within the chapter.
- **In the Part Reviews.** These exercises draw from across all chapters in the part.

Each exercise has a descriptive label. Exercises for which technology is recommended are indicated with the icon ■. Larger data sets used in examples and exercises are referenced in the text, listed in the back endpapers, and made available on the companion CD-ROM. The exercises are divided into the following three categories:

- **Practicing the Basics** are the section exercises and the first group of end-of-chapter exercises; they reinforce basic application of the methods.
- **Concepts and Investigations** exercises require the student to explore real data sets and carry out investigations for mini-projects. They may ask students to explore concepts and related theory, or be extensions of the chapter's methods. This section contains some multiple-choice and true-false exercises to help students check their understanding of the basic concepts and prepare for tests. A few more difficult, optional exercises (highlighted with the ♦♦ icon) are included to present some additional concepts and methods. Concepts and Investigations exercises are found in the end-of-chapter exercises and the Part Reviews.
- **Student Activities** are designed for group work based on investigations done by each of the students on a team. Student Activities are found in the end-of-chapter exercises, and additional activities may be found within chapters as well.

Technology Integration

Up-to-Date Use of Technology

The availability of technology enables instruction that is less calculation-based and more concept-oriented. Output from software applications and calculators is displayed throughout the textbook, and discussion focuses on interpretation of the output, rather than on the keystrokes needed to create the output. Although most of our output is from Minitab® and the TI-83+/84, we also show screen captures from IBM® SPSS® and Microsoft Excel® as appropriate. Technology-specific manuals containing keystroke information are available with this text. See the supplements listing for more information.

Applets

Applets referred to in the text are found on the companion CD-ROM or within MyStatLab. Applets have great value because they demonstrate concepts to students visually. For example, creating a sampling distribution is accomplished more readily with applets than with a static text figure. The applets are presented as optional explorations in the text. (Description of the applets may be found on page x.)

Data Sets

We use a wealth of real data sets throughout the textbook. These data sets are available on the companion CD-ROM and on the website www.pearsonhighered.com/mathstatsresources/. The same data set is often used in several chapters, helping reinforce the four components of the statistical investigative process and allowing the students to see the big picture of statistical reasoning. Exercises using data sets are noted with this icon:

An Invitation Rather Than a Conclusion

We hope that students using this textbook will gain a lasting appreciation for the vital role the art and science of statistics plays in analyzing data and helping us make decisions in our lives. Our major goals for this textbook are that students learn how to:

- Produce data that can provide answers to properly posed questions.
- Appreciate how probability helps us understand randomness in our lives, as well as grasp the crucial concept of a sampling distribution and how it relates to inference methods.
- Choose appropriate descriptive and inferential methods for examining and analyzing data and drawing conclusions.
- Communicate the conclusions of statistical analyses clearly and effectively.
- Understand the limitations of most research, either because it was based on an observational study rather than a randomized experiment or survey, or because a certain lurking variable was not measured that could have explained the observed associations.

We are excited about sharing the insights that we have learned from our experience as teachers and from our students through this text. Many students still enter statistics classes on the first day with dread because of its reputation as a dry, sometimes difficult, course. It is our goal to inspire a classroom environment that is filled with creativity, openness, realistic applications, and learning that students find inviting and rewarding. We hope that this textbook will help the instructor and the students experience a rewarding introductory course in statistics.

Supplements

For the Student

Student's Solutions Manual, by Sarah Streett, contains fully worked solutions to odd-numbered exercises. (ISBN-10: 0-321-75619-3; ISBN-13: 978-0-321-75619-0)

Video Resources on DVD contain example-level videos that explain how to work examples from the text. The videos provide excellent support for students who require additional assistance or want reinforcement on topics and concepts learned in class. (ISBN-10: 0-321-78051-5; ISBN-13: 978-0-321-78051-5)

Excel® Manual (download only), by Jack Morse (University of Georgia), provides detailed tutorial instructions and worked-out examples and exercises for Excel. Available for download from www.pearsonhighered.com/mathstatsresources or within MyStatLab.

Graphing Calculator Manual (download only), by Peter Flanagan-Hyde (Phoenix Country Day School), provides detailed tutorial instructions and worked-out examples and exercises for the TI-83/84 Plus. Available for download from www.pearsonhighered.com/mathstatsresources or within MyStatLab.

MINITAB® Manual (download only), by Linda Dawson (University of Washington, Tacoma), provides detailed tutorial instructions and worked-out examples and exercises for MINITAB. Available for download from www.pearsonhighered.com/mathstatsresources or within MyStatlab.

Student Laboratory Workbook, by Megan Mocko (University of Florida) and Maria Ripol (University of Florida), is a study tool for the first ten chapters of the text. This workbook provides section-by-section review and practice and additional activities that cover fundamental statistical topics. (ISBN-10: 0-321-78342-5; ISBN-13: 978-0-321-78342-4)

Study Cards for Statistics Software This series of study cards, available for Excel®, MINITAB®, JMP®, SPSS®, R®, StatCrunch®, and the TI-83/84® graphing calculators provides students with easy, step-by-step guides to the most common statistics software. Visit www.myPearsonStore.com for more information.

For the Instructor

Instructor's Edition (IE) contains comprehensive Instructor's Notes for each chapter. Broken down by section, they offer a valuable introduction to each chapter by presenting learning objectives (new to this edition), the author's rationale for content and presentation decisions made in the chapter, tips for introducing complex material, common pitfalls students encounter, additional examples and activities to use in class, and suggestions for how to integrate applets and activities effectively. Short answers to all of the exercises are given in the Answer Appendix. Full solutions to all of the exercises are in the Instructor's Solutions Manual. (ISBN-10: 0-321-75610-X; ISBN-13: 978-0-321-75610-7)

Instructor to Instructor Videos provide an opportunity for adjuncts, part-timers, TAs, or other instructors who are new to teaching from this text or have limited class prep time to learn about the book's approach and coverage directly from Chris Franklin. The videos focus on those topics that have proven to be most challenging to students. Chris offers suggestions, pointers, and ideas about how to present these topics and concepts effectively based on her many years of teaching introductory statistics. She also shares insights on how to help students use the textbook in the most effective way to realize success in the course. The videos are available for download from Pearson's online catalog at www.pearsonhighered.com/irc and through MyStatLab.

Instructor's Solutions Manual, by Sarah Streett, contains fully worked solutions to every textbook exercise. Available for download from Pearson's online catalog at www.pearsonhighered.com/irc and through MyStatLab.

Answers to the Student Laboratory Manual is available for download from Pearson's online catalog at www.pearsonhighered.com/irc and through MyStatLab.

PowerPoint Lecture Slides are fully editable and printable slides that follow the textbook. These slides can be used during lectures or posted to a Web site in an online course. The PowerPoint Lecture Slides are available from Pearson's online catalog at www.pearsonhighered.com/irc and through MyStatLab.

Active Learning Questions are prepared in PowerPoint® and intended for use with classroom response systems. Several multiple-choice questions are available for each chapter of the book, allowing instructors to quickly assess mastery of material in class. The Active Learning Questions are available from Pearson's online catalog at www.pearsonhighered.com/irc and through MyStatLab.

TestGen® (www.pearsoned.com/testgen) enables instructors to build, edit, print, and administer tests using a computerized bank of questions developed to cover all the objectives of the text. TestGen is algorithmically based, allowing instructors to create multiple but equivalent versions of the same question or test with the click of a button. Instructors can also modify test bank questions or add new questions. The software and test bank are available for download from Pearson's online catalog at www.pearsonhighered.com/irc and through MyStatLab.

The ***Online Test Bank*** is a test bank derived from TestGen®. It includes multiple choice and short answer questions for each section of the text, along with the answer keys. Available for download from Pearson's online catalog at www.pearsonhighered.com/irc and through MyStatLab.

Technology Resources

Companion CD-ROM

Each new copy of the text comes with a companion CD-ROM containing data sets (.csv, TI-83/84, and .txt files) and applets referenced in the text, which are useful for illustrating statistical concepts.

MyStatLab™ Online Course (access code required)

MyStatLab is a course management system that delivers **proven results** in helping individual students succeed.

- MyStatLab can be successfully implemented in any environment—lab-based, hybrid, fully online, traditional—and demonstrates the quantifiable difference that integrated usage has on student retention, subsequent success, and overall achievement.
- MyStatLab's comprehensive online gradebook automatically tracks students' results on tests, quizzes, homework, and in the study plan. Instructors can use the gradebook to intervene if students have trouble or to provide positive feedback. Data can be easily exported to a variety of spreadsheet programs, such as Microsoft Excel.

MyStatLab provides **engaging experiences** that personalize, stimulate, and measure learning for each student.

- **Tutorial Exercises with Multimedia Learning Aids:** The homework and practice exercises in MyStatLab align with the exercises in the textbook, and they regenerate algorithmically to give students unlimited opportunity for practice and mastery. Exercises offer immediate helpful feedback, guided solutions, sample problems, animations, videos, and eText clips for extra help at point-of-use.

- **Getting Ready for Statistics:** A library of questions now appears within each MyStatLab course to offer the developmental math topics students need for the course. These can be assigned as a prerequisite to other assignments, if desired.
- **Conceptual Question Library:** In addition to algorithmically regenerated questions that are aligned with your textbook, there is a library of 1,000 Conceptual Questions available in the assessment managers that require students to apply their statistical understanding.
- **StatCrunch:** MyStatLab includes a web-based statistical software, StatCrunch, within the online assessment platform so that students can easily analyze data sets from exercises and the text. In addition, MyStatLab includes access to **www.StatCrunch.com**, a web site where users can access more than 13,000 shared data sets, conduct online surveys, perform complex analyses using the powerful statistical software, and generate compelling reports.
- **Integration of Statistical Software:** Knowing that students often use external statistical software, we make it easy to copy our data sets, both from the ebook and MyStatLab questions, into software like StatCrunch, Minitab, Excel and more. Students have access to a variety of support—Technology Instruction Videos, Technology Study Cards, and Manuals—to learn how to effectively use statistical software.
- **Expert Tutoring:** Although many students describe the whole of MyStatLab as "like having your own personal tutor," students also have access to live tutoring from Pearson. Qualified statistics instructors provide tutoring sessions for students via MyStatLab.

And, MyStatLab comes from a **trusted partner** with educational expertise and an eye on the future.

Knowing that you are using a Pearson product means knowing that you are using quality content. That means that our eTexts are accurate, that our assessment tools work, and that our questions are error-free. And whether you are just getting started with MyStatLab, or have a question along the way, we're here to help you learn about our technologies and how to incorporate them into your course.

To learn more about how MyStatLab combines proven learning applications with powerful assessment, visit **www.mystatlab.com** or contact your Pearson representative.

MathXL® for Statistics Online Course (access code required)

MathXL® is the homework and assessment engine that runs MyStatLab. (MyStatLab is MathXL plus a learning management system.) With MathXL for Statistics, instructors can:

- Create, edit, and assign online homework and tests using algorithmically generated exercises correlated at the objective level to the textbook.
- Create and assign their own online exercises and import TestGen tests for added flexibility.
- Maintain records of all student work, tracked in MathXL's online gradebook.

With MathXL for Statistics, students can:

- Take chapter tests in MathXL and receive personalized study plans and/or personalized homework assignments based on their test results.
- Use the study plan and/or the homework to link directly to tutorial exercises for the objectives they need to study.

- Students can also access supplemental animations and video clips directly from selected exercises.
- Knowing that students often use external statistical software, we make it easy to copy our data sets, both from the eText and the MyStatLab questions, into software like StatCrunch, Minitab, Excel and more.

MathXL for Statistics is available to qualified adopters. For more information, visit our website at www.mathxl.com, or contact your Pearson representative.

StatCrunch®

StatCrunch® is powerful web-based statistical software that allows users to perform complex analyses, share data sets, and generate compelling reports of their data. The vibrant online community offers more than 13,000 data sets for students to analyze.

- **Collect.** Users can upload their own data to StatCrunch or search a large library of publicly shared data sets, spanning almost any topic of interest. Also, an online survey tool allows users to quickly collect data via web-based surveys.
- **Crunch.** A full range of numerical and graphical methods allow users to analyze and gain insights from any data set. Interactive graphics help users understand statistical concepts, and are available for export to enrich reports with visual representations of data.
- **Communicate.** Reporting options help users create a wide variety of visually-appealing representations of their data.

Full access to StatCrunch is available with a MyStatLab kit, and StatCrunch is available by itself to qualified adopters. For more information, visit our website at www.statcrunch.com, or contact your Pearson representative.

The Student Edition of MINITAB® (CD Only)

The Student Edition of MINITAB is a condensed version of the Professional Release of MINITAB statistical software. It offers the full range of statistical methods and graphical capabilities, along with worksheets that can include up to 10,000 data points. Only available for bundling with the text. (ISBN-10: 0-321-11313-6; ISBN-13: 978-0-321-11313-9)

JMP® Student Edition

JMP Student Edition is easy-to-use, streamlined version of JMP desktop statistical discovery software from SAS Institute Inc. and is only available for bundling with the text. (ISBN-10: 0-321-67212-7; ISBN-13: 978-0-321-67212-4)

IBM® SPSS® Statistics Student Version

SPSS, a statistical and data management software package, is also available for bundling with the text. (ISBN-10: 0-321-67537-1; ISBN-13: 978-0-321-67537-8)

XLSTAT for Pearson

Used by leading businesses and universities, XLSTAT is an Excel® add-in that offers a wide variety of functions to enhance the analytical capabilities of Microsoft Excel, making it the ideal tool for your everyday data analysis and statistics requirements. XLSTAT is compatible with all Excel versions (except Mac 2008). Available for bundling with the text. (ISBN-10: 0-321-75932-X; ISBN-13: 978-0-321-75932-0).

For more information, please contact your local Pearson Education Sales Representative.

Acknowledgments

We are indebted to the following individuals, who provided valuable feedback for the third edition:

Larry Ammann, *University of Texas, Dallas*
Ellen Breazel, *Clemson University*
Dagmar Budikova, *Illinois State University*
Richard Cleary, *Bentley University*
Winston Crawley, *Shippensburg University*
Jonathan Duggins, *Virginia Tech*
Brian Karl Finch, *San Diego State University*
Kim Gilbert, *University of Georgia*
Hasan Hamdan, *James Madison University*
John Holcomb, *Cleveland State University*
Nusrat Jahan, *James Madison University*
Martin Jones, *College of Charleston*
Gary Kader, *Appalachian State University*
Jackie Miller, *The Ohio State University*
Megan Mocko, *University of Florida*
June Morita, *University of Washington*
Sister Marcella Louise Wallowicz, *Holy Family University*
Peihua Qui, *University of Minnesota*

We are also indebted to the many reviewers, class testers, and students who gave us invaluable feedback and advice on how to improve the quality of the book.

ARIZONA Russel Carlson, University of Arizona; Peter Flanagan-Hyde, Phoenix Country Day School ■ **CALIFORINIA** James Curl, Modesto Junior College; Christine Drake, University of California at Davis; Mahtash Esfandiari, UCLA; Dawn Holmes, University of California Santa Barbara; Rob Gould, UCLA; Rebecca Head, Bakersfield College; Susan Herring, Sonoma State University; Colleen Kelly, San Diego State University; Marke Mavis, Butte Community College; Elaine McDonald, Sonoma State University; Corey Manchester, San Diego State University; Amy McElroy, San Diego State University; Helen Noble, San Diego State University; Calvin Schmall, Solano Community College ■ **COLORADO** David Most, Colorado State University ■ **CONNECTICUT** Paul Bugl, University of Hartford; Anne Doyle, University of Connecticut; Pete Johnson, Eastern Connecticut State University; Dan Miller, Central Connecticut State University; Kathleen Mclaughlin, University of Connecticut; Nalini Ravishanker, University of Connecticut; John Vangar, Fairfield University; Stephen Sawin, Fairfield University ■ **DISTRICT OF COLUMBIA** Hans Engler, Georgetown University; Mary W. Gray, American University; Monica Jackson, American University ■ **FLORIDA** Nazanin Azarnia, Santa Fe Community College; Brett Holbrook; James Lang, Valencia Community College; Karen Kinard, Tallahassee Community College; Maria Ripol, University of Florida; James Smart, Tallahassee Community College; Latricia Williams, St. Petersburg Junior College, Clearwater; Doug Zahn, Florida State University ■ **GEORGIA** Carrie Chmielarski, University of Georgia; Ouida Dillon, Oconee County High School; Katherine Hawks, Meadowcreek High School; Todd Hendricks, Georgia Perimeter College; Charles LeMarsh, Lakeside High School; Steve Messig, Oconee County High School; Broderick Oluyede, Georgia Southern University; Chandler Pike, University of Georgia; Kim Robinson, Clayton State University; Jill Smith,

University of Georgia; John Seppala, Valdosta State University; Joseph Walker, Georgia State University ▪ **IOWA** John Cryer, University of Iowa; Kathy Rogotzke, North Iowa Community College; R. P. Russo, University of Iowa; William Duckworth, Iowa State University ▪ **ILLINOIS** Linda Brant Collins, University of Chicago; Ellen Fireman, University of Illinois; Jinadasa Gamage, Illinois State University; Richard Maher, Loyola University Chicago; Cathy Poliak, Northern Illinois University; Daniel Rowe, Heartland Community College ▪ **KANSAS** James Higgins, Kansas State University; Michael Mosier, Washburn University ▪ **KENTUCKY** Lisa Kay, Eastern Kentucky University ▪ **MASSACHUSETTS** Katherine Halvorsen, Smith College; Xiaoli Meng, Harvard University; Daniel Weiner, Boston University ▪ **MICHIGAN** Kirk Anderson, Grand Valley State University; Phyllis Curtiss, Grand Valley State University; Roy Erickson, Michigan State University; Jann-Huei Jinn, Grand Valley State University; Sango Otieno, Grand Valley State University; Alla Sikorskii, Michigan State University; Mark Stevenson, Oakland Community College; Todd Swanson, Hope College; Nathan Tintle, Hope College ▪ **MINNESOTA** Bob Dobrow, Carleton College; German J. Pliego, University of St.Thomas; Engin A. Sungur, University of Minnesota–Morris ▪ **MISSOURI** Lynda Hollingsworth, Northwest Missouri State University; Larry Ries, University of Missouri–Columbia; Suzanne Tourville, Columbia College ▪ **MONTANA** Jeff Banfield, Montana State University ▪ **NEW JERSEY** Harold Sackrowitz, Rutgers, The State University of New Jersey; Linda Tappan, Montclair State University ▪ **NEW MEXICO** David Daniel, New Mexico State University ▪ **NEW YORK** Brooke Fridley, Mohawk Valley Community College; Martin Lindquist, Columbia University; Debby Lurie, St. John's University; David Mathiason, Rochester Institute of Technology; Steve Stehman, SUNY ESF; Tian Zheng, Columbia University ▪ **NEVADA:** Alison Davis, University of Nevada-Reno ▪ **NORTH CAROLINA** Pamela Arroway, North Carolina State University; E. Jacquelin Dietz, North Carolina State University; Alan Gelfand, Duke University; Scott Richter, UNC Greensboro; Roger Woodard, North Carolina State University ▪ **NEBRASKA** Linda Young, University of Nebraska ▪ **OHIO** Jim Albert, Bowling Green State University; Stephan Pelikan, University of Cincinnati; Teri Rysz, University of Cincinnati; Deborah Rumsey, The Ohio State University; Kevin Robinson, University of Akron ▪ **OREGON** Michael Marciniak, Portland Community College; Henry Mesa, Portland Community College, Rock Creek; Qi-Man Shao, University of Oregon; Daming Xu, University of Oregon ▪ **PENNSYLVANIA** Douglas Frank, Indiana University of Pennsylvania; Steven Gendler, Clarion University; Bonnie A. Green, East Stroudsburg University; Paul Lupinacci, Villanova University; Deborah Lurie, Saint Joseph's University; Linda Myers, Harrisburg Area Community College; Tom Short, Villanova University; Kay Somers, Moravian College ▪ **SOUTH CAROLINA** Beverly Diamond, College of Charleston; Murray Siegel, The South Carolina Governor's School for Science and Mathematics; ▪ **SOUTH DAKOTA** Richard Gayle, Black Hills State University; Daluss Siewert, Black Hills State University; Stanley Smith, Black Hills State University ▪ **TENNESSEE** Bonnie Daves, Christian Academy of Knoxville; T. Henry Jablonski, Jr., East Tennessee State University; Robert Price, East Tennessee State University; Ginger Rowell, Middle Tennessee State University; Edith Seier, East Tennessee State University ▪ **TEXAS** Tom Bratcher, Baylor University; Jianguo Liu, University of North Texas; Mary Parker, Austin Community College; Robert Paige, Texas Tech University; Walter M. Potter, Southwestern University; Therese Shelton, Southwestern University; James Surles, Texas Tech University; Diane Resnick, University of Houston-Downtown ▪ **UTAH** Patti Collings, Brigham Young University; Carolyn Cuff, Westminster College; Lajos Horvath, University of Utah; P. Lynne Nielsen, Brigham Young University ▪ **VIRGINIA** David Bauer, Virginia Commonwealth University; Ching-Yuan Chiang, James Madison University; Steven Garren, James Madison University; Debra Hydorn, Mary Washington College; D'Arcy Mays, Virginia Commonwealth University;

Stephanie Pickle, Virginia Polytechnic Institute and State University ▪ **WASHINGTON** Rich Alldredge, Washington State University; Brian T. Gill, Seattle Pacific University ▪ **WISCONSIN** Brooke Fridley, University of Wisconsin–LaCrosse; Loretta Robb Thielman, University of Wisconsin–Stoutt. ▪ **WYOMING** Burke Grandjean, University of Wyoming ▪ **CANADA** Mike Kowalski, University of Alberta; David Loewen, University of Manitoba

We thank the following individuals, who made invaluable contributions to the third edition:

Ellen Breazel, *Clemson University*

Linda Dawson, *Washington State University, Tacoma*

Bernadette Lanciaux, *Rochester Institute of Technology*

Scott Nickleach, *Sonoma State University*

The detailed assessment of the text fell to our accuracy checkers, Ann Cannon, Cornell College; Dave Bregenzer, Utah State University; Stan Seltzer, Ithaca College; Sarah Streett; and the Pearson math tutors Alice Armstrong and Abdellah Dakhama, who checked the manuscript in both the preliminary and final versions.

Thank you to Sarah Streett, who took on the task of revising the solutions manuals to reflect the many changes to the third edition. We also want to thank Jackie Miller (The Ohio State University) for her contributions to the Instructor's Notes, Webster West (Texas A & M) for his work in producing the applets, and our student technology manual and workbook authors, Jack Morse (University of Georgia), Linda Dawson (University of Washington, Tacoma), Peter Flanagan-Hyde (Phoenix Country Day School), Megan Mocko (University of Florida), and Maria Ripol (University of Florida).

We would like to thank the Pearson team who has given countless hours in developing this text; without their guidance and assistance, the text would not have come to completion. We thank Marianne Stepanian, Chere Bemelmans, Dana Bettez, Sonia Ashraf, Beth Houston, Erin Lane, Kathleen DeChavez, and Christine Stavrou. We also thank Allison Campbell, Senior Project Manager at Integra-Chicago, for keeping this book on track throughout production. And we extend a very special note of appreciation to Elaine Page, our development editor.

Alan Agresti would like to thank those who have helped us in some way, often by suggesting data sets or examples. These include Anna Gottard, Wolfgang Jank, Bernhard Klingenberg, René Lee-Pack, Jacalyn Levine, Megan Lewis, Megan Meece, Dan Nettleton, Yongyi Min, and Euijung Ryu. Many thanks also to Tom Piazza for his help with the General Social Survey. Finally, Alan Agresti would like to thank his wife Jacki Levine for her extraordinary support throughout the writing of this book. Besides putting up with the evenings and weekends he was working on this book, she offered numerous helpful suggestions for examples and for improving the writing.

Chris Franklin gives a special thank you to her husband and sons, Dale, Corey, and Cody Green. They have patiently sacrificed spending many hours with their spouse and mom as she has worked on this book through three editions. A special thank you also to her parents Grady and Helen Franklin and her two brothers, Grady and Mark, who have always been there for their daughter and sister. Chris also appreciates the encouragement and support of her colleagues and her many students who used the book, offering practical suggestions for improvement. Chris appreciates the support of teachers who have used the previous editions of the book. Finally, Chris thanks her coauthor, Alan Agresti, for making this book a reality, a book they began discussing oh so many years ago.

ALAN AGRESTI, *Gainesville, Florida*

CHRIS FRANKLIN, *Athens, Georgia*

About the Authors

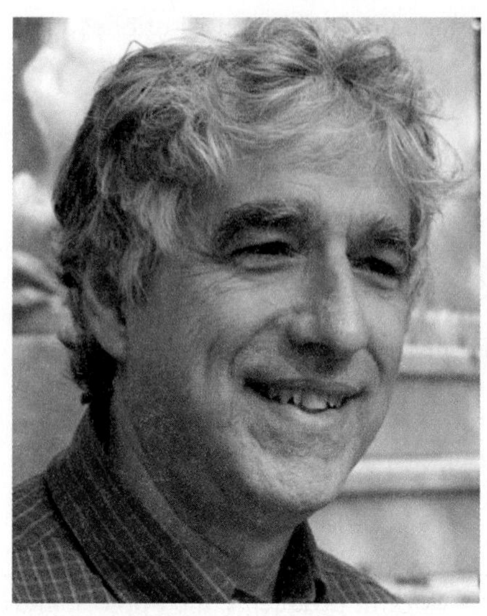

Alan Agresti is Distinguished Professor Emeritus in the Department of Statistics at the University of Florida. He taught statistics there for 38 years, including the development of three courses in statistical methods for social science students and three courses in categorical data analysis. He is author of more than 100 refereed articles and five texts including *Statistical Methods for the Social Sciences* (with Barbara Finlay, Prentice Hall, 4th edition, 2009) and *Categorical Data Analysis* (Wiley, 2nd edition, 2002). He is a Fellow of the America Statistical Association and recipient of an Honorary Doctor of Science from De Montfort University in the UK. In 2003 he was named Statistician of the Year by the Chicago chapter of the American Statistical Association, and in 2004 he was the first honoree of the Herman Callaert Leadership Award in Biostatistical Education and Dissemination, awarded by the University of Limburgs, Belgium. He has held visiting positions at Harvard University, Boston University, the London School of Economics, and Imperial College and has taught courses or short courses for universities and companies in about 30 countries worldwide. He has also received teaching awards from the University of Florida and an excellence in writing award from John Wiley & Sons.

Christine Franklin is a Senior Lecturer and Lothar Tresp Honoratus Honors Professor in the Department of Statistics at the University of Georgia. She has been teaching statistics for more than 30 years at the college level. Chris has been actively involved at the national and state level with promoting statistical education at Pre-K–16 since the 1980s. She is a past Chief Reader for AP Statistic. She has developed three graduate level courses at the University of Georgia in statistics for elementary, middle, and secondary teachers. Chris served as the lead writer for the ASA-endorsed Guidelines for Assessment and Instruction in Statistics Education (GAISE) Report: A Pre- K–12 Curriculum Framework. Chris has been honored by her selection as a Fellow of the American Statistical Association, the 2006 Mu Sigma Rho National Statistical Education Award recipient for her teaching and lifetime devotion to statistics education, and numerous teaching and advising awards at the University of Georgia including election to the UGA Teaching Academy. Chris has written more than 50 journal articles and resource materials for textbooks. Most important for Chris is her family. Her most recent memorable experience was a mission trip to Mexico with her husband and sons. In June 2012, Chris will be returning to Philmont, New Mexico, for her third trip backpacking for 11 days with both her sons and other Boy Scouts.

Part

1

Gathering and Exploring Data

1 Statistics: The Art and Science of Learning from Data

1.1 Using Data to Answer Statistical Questions

1.2 Sample Versus Population

1.3 Using Calculators and Computers

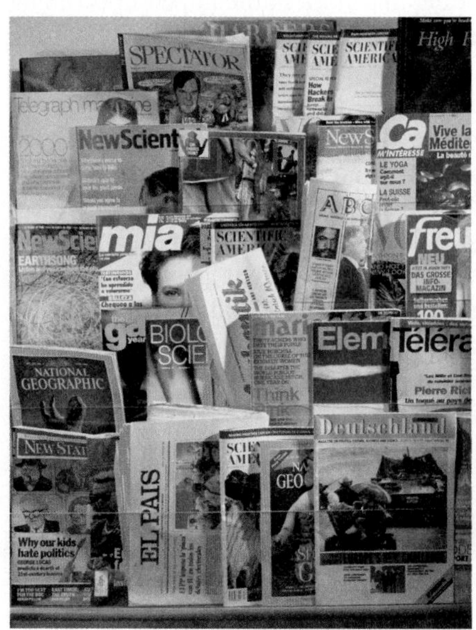

Example 1

How Statistics Helps Us Learn about the World

Picture the Scenario

In this book, you will explore a wide variety of everyday scenarios. For example, you will evaluate media reports about opinion surveys, medical research studies, the state of the economy, and environmental issues. You'll face financial decisions such as choosing between an investment with a sure return and one that could make you more money but could possibly cost you your entire investment. You'll learn how to analyze the available information to answer necessary questions in such scenarios. One purpose of this book is to show you why an understanding of statistics is essential for making good decisions in an uncertain world.

Questions to Explore

This book will show you how to collect appropriate information and how to apply statistical methods so you can better evaluate that information and answer the questions posed. Here are some examples of questions we'll investigate in coming chapters:

- How can you evaluate evidence about global warming?
- Are cell phones dangerous to your health?
- What's the chance your tax return will be audited?
- How likely are you to win the lottery?
- Is there bias against women in appointing managers?
- What "hot streaks" should you expect in basketball?
- How can you analyze whether a diet really works?
- How can you predict the selling price of a house?

Thinking Ahead

Each chapter uses questions like these to introduce a topic and then introduces tools for making sense of the available information. We'll see that **statistics** is the art and science of designing studies and analyzing the information that those studies produce.

In the business world, managers use statistics to analyze results of marketing studies about new products, to help predict sales, and to measure employee performance. In finance, statistics is used to study stock returns and investment opportunities. Medical studies use statistics to evaluate whether new ways to treat disease are better than existing ways. In fact, most professional occupations today rely heavily on statistical methods. In a competitive job market, understanding statistics provides an important advantage.

But it's important to understand statistics even if you will never use it in your job. Understanding statistics can help you make better choices. Why? Because every day you are bombarded with statistical information from news reports, advertisements, political campaigns, and surveys. How do you know what to heed and what to ignore? An understanding of the statistical reasoning—and in some cases statistical misconceptions—underlying these pronouncements will help. For instance, this book will enable you to evaluate claims about medical research studies more effectively so that you know when you should be skeptical. For example, does taking an aspirin daily truly lessen the chance of cancer?

We realize that you are probably not reading this book in the hope of becoming a statistician. (That's too bad, because there's a severe shortage of statisticians—more jobs than trained people. And with the ever-increasing ways in which statistics is being applied, it's an exciting time to be a statistician.) You may even suffer from math phobia. Please be assured that to learn the main concepts of statistics, logical thinking and perseverance are more important than high-powered math skills. Don't be frustrated if learning comes slowly and you need to read about a topic a few times before it starts to make sense. Just as you would not expect to sit through a single foreign language class session and be able to speak that language fluently, the same is true with the language of statistics. It takes time and practice. But we promise that your hard work will be rewarded. Once you have completed even part of this text, you will understand much better how to make sense of statistical information, and hence the world around you.

1.1 Using Data to Answer Statistical Questions

Does a low-carbohydrate diet result in significant weight loss? Are people more likely to stop at a Starbucks if they've seen a recent Starbucks TV commercial? Information gathering is at the heart of investigating answers to such questions. The information we gather with experiments and surveys is collectively called **data**.

For instance, consider an experiment designed to evaluate the effectiveness of a low-carbohydrate diet. The data might consist of the following measurements for the people participating in the study: weight at the beginning of the study, weight at the end of the study, number of calories of food eaten per day, carbohydrate intake per day, body-mass index (BMI) at the start of the study, and gender. A marketing survey about the effectiveness of a TV ad for Starbucks could collect data on the percentage of people who went to a Starbucks since the ad aired and analyze how it compares for those who saw the ad and those who did not see it.

Defining Statistics

You already have a sense of what the word *statistics* means. You hear statistics quoted about sports events (number of points scored by each player on a basketball team), statistics about the economy (median income, unemployment rate), and statistics about opinions, beliefs, and behaviors (percentage of students who indulge in binge drinking). In this sense, a statistic is merely a number calculated from data. But statistics as a field can be broadly viewed as a way of thinking about data and quantifying uncertainty, not a maze of numbers and messy formulas.

> **Statistics**
>
> **Statistics** is the art and science of designing studies and analyzing the data that those studies produce. Its ultimate goal is translating data into knowledge and understanding of the world around us. In short, *statistics is the art and science of learning from data*.

Statistical methods help us investigate questions in an objective manner. Statistical problem solving is an investigative process that involves four components: (1) formulate a statistical question, (2) collect data, (3) analyze data, and (4) interpret results. The following examples ask questions that we'll learn how to answer using statistical investigations.

Scenario 1: Predicting an Election Using an Exit Poll In elections, television networks often declare the winner well before all the votes have been counted. They do this using exit polling, interviewing voters after they leave the voting booth. Using an exit poll, a network can often predict the winner after learning how several thousand people voted, out of possibly millions of voters.

The 2010 California gubernatorial race pitted Democratic candidate Jerry Brown against Republican candidate Meg Whitman. A TV exit poll used to project the outcome reported that 53.1% of a sample of 3889 voters said they had voted for Jerry Brown.[1] Was this sufficient evidence to project Brown as the winner, even though information was available from such a small portion of the more than 9.5 million voters in California? We'll learn how to answer that question in this book.

Scenario 2: Making Conclusions in Medical Research Studies Statistical reasoning is at the foundation of the analyses conducted in most medical research studies. Let's consider three examples of how statistics can be relevant.

Heart disease is the most common cause of death in industrialized nations. In the United States and Canada, nearly 30% of deaths yearly are due to heart disease, mainly heart attacks. Does regular aspirin intake reduce deaths from heart attacks? Harvard Medical School conducted a landmark study to investigate. The people participating in the study regularly took either an aspirin or a placebo (a pill with no active ingredient). Of those who took aspirin, 0.9% had heart attacks during the study. Of those who took the placebo, 1.7% had heart attacks, nearly twice as many.

Can you conclude that it's beneficial for people to take aspirin regularly? Or, could the observed difference be explained by how it was decided which people would receive aspirin and which would receive the placebo? For instance, might those who took aspirin have had better results merely because they were healthier, on average, than those who took the placebo? Or, did those taking aspirin have a better diet or exercise more regularly, on average?

For years there has been controversy about whether regular intake of large doses of vitamin C is beneficial. Some studies have suggested that it is. But some scientists have criticized those studies' designs, claiming that the subsequent statistical analysis was meaningless. How do we know when we can trust the statistical results in a medical study that is reported in the media?

Suppose you wanted to investigate whether, as some have suggested, heavy use of cell phones makes you more likely to get brain cancer. You could pick half the students from your school and tell them to use a cell phone each day for the next 50 years, and tell the other half never to use a cell phone. Fifty years from now you could see whether more users than nonusers of cell phones got brain cancer. Obviously it would be impractical to carry out such a study. And who wants to wait 50 years to get the answer? Years ago, a British statistician figured out how to study whether a particular type of behavior has an effect on cancer, using already available data. He did this to answer a then controversial question: Does smoking cause lung cancer? How did he do this?

This book will show you how to answer questions like these. You'll learn when you can trust the results from studies reported in the media and when you should be skeptical.

Scenario 3: Using a Survey to Investigate People's Beliefs How similar are your opinions and lifestyle to those of others? It's easy to find out. Every other year, the National Opinion Research Center at the University of Chicago conducts the General Social Survey (GSS). This survey of a few thousand adult Americans provides data about the opinions and behaviors of the American public. You can use it to investigate how adult Americans answer a wide diversity of questions, such as, "Do you believe in life after death?" "Would you be willing to pay higher prices in order to protect the environment?" "How much TV do you watch per day?" and "How many sexual partners have you had in the past

[1]*Source:* Data from www.cnn.com/ELECTION/2010/results/polls/.

year?" Similar surveys occur in other countries, such as the Eurobarometer survey within the European Union. We'll use data from such surveys to illustrate the proper application of statistical methods.

Reasons for Using Statistical Methods

The scenarios just presented illustrate the three main components of statistics for answering a statistical question:

- **Design:** Planning how to obtain data to answer the questions of interest
- **Description:** Summarizing and analyzing the data that are obtained
- **Inference:** Making decisions and predictions based on the data for answering the statistical question

Design refers to planning how to obtain data that will efficiently shed light on the problem of interest. How could you conduct an experiment to determine reliably whether regular large doses of vitamin C are beneficial? In marketing, how do you select the people to survey so you'll get data that provide good predictions about future sales?

Description means exploring and summarizing patterns in the data. Files of raw data are often huge. For example, over time the General Social Survey has collected data about hundreds of characteristics on many thousands of people. Such raw data are not easy to assess—we simply get bogged down in numbers. It is more informative to use a few numbers or a graph to summarize the data, such as an average amount of TV watched or a graph displaying how number of hours of TV watched per day relates to number of hours per week exercising.

In Words

The verb **infer** means to arrive at a decision or prediction by reasoning from known evidence.

Inference means making decisions or predictions based on the data. Usually the decision or prediction refers to a larger group of people, not merely those in the study. For instance, in the exit poll described in Scenario 1, of 3889 voters sampled, 53.1% said they voted for Jerry Brown. Using these data, we can predict (infer) that a majority of the 9.5 million voters voted for him. Stating the percentages for the sample of 3889 voters is *description*, while predicting the outcome for all 9.5 million voters is *inference*.

Statistical description and inference are complementary ways of analyzing data. Statistical description provides useful summaries and helps you find patterns in the data, while inference helps you make predictions and decide whether observed patterns are meaningful. You can use both to investigate questions that are important to society. For instance, "Has there been global warming over the past decade?" "Is having the death penalty available for punishment associated with a reduction in violent crime?" "Does student performance in school depend on the amount of money spent per student, the size of the classes, or the teachers' salaries?"

Long before we analyze data, we need to give careful thought to posing the questions to be answered by that analysis. The nature of these questions has an impact on all stages—design, description, and inference. For example, in an exit poll, do we just want to predict which candidate won, or do we want to investigate *why* by analyzing how voters' opinions about certain issues related to how they voted? We'll learn how questions such as these and the ones posed in the previous paragraph can be phrased in terms of statistical summaries (such as percentages and means) so that we can use data to investigate their answers.

In Words

We'll see in Activity 1 that the term **variable** refers to the characteristic being measured, such as number of hours per day that you watch TV.

Finally, a topic that we have not mentioned yet but that is fundamental for statistical inference is **probability**, which is a framework for quantifying how likely various possible outcomes are. We'll study probability because it will help us to answer questions such as, "If Brown were actually going to lose the election (that is, if he were supported by less than half of all voters), what's the chance that an exit poll of 3889 voters would show support by 53.1% of the voters?" If the chance were extremely small, we'd feel comfortable making the inference that his reelection was supported by the majority of all 9.5 million voters.

Activity 1

Downloading Data from the Internet

It is simple to get descriptive summaries of data from the General Social Survey (GSS). We'll demonstrate, using one question asked in recent surveys, "On a typical day, about how many hours do you personally watch television?"

- Go to the Web site sda.berkeley.edu/GSS.
- Click on GSS—with *No Weight* as the default weight selection.
- The GSS name for the number of hours of TV watching is TVHOURS. Type TVHOURS as the row variable name.
- In the Weight menu, make sure that *No Weight* is selected. Click on *Run the Table.*

Now you'll see a table that shows the number of people and, in bold, the percentage who made each of the possible responses. For all the years combined in which this question was asked, the most common response was 2 hours of TV a day (about 27% made this response).

What percentage of the people surveyed reported watching 0 hours of TV a day? How many people reported watching TV 24 hours a day?

Another question asked in the GSS is, "Taken all together, would you say that you are very happy, pretty happy, or not too happy?" The GSS name for this item is HAPPY. What percentage of people reported being very happy?

You might use the GSS to investigate what sorts of people are more likely to be very happy. Those who are happily married? Those who are in good health? Those who have lots of friends? We'll see how to find out in this book.

*If this doesn't work, your computer's firewall settings may be restricting access.

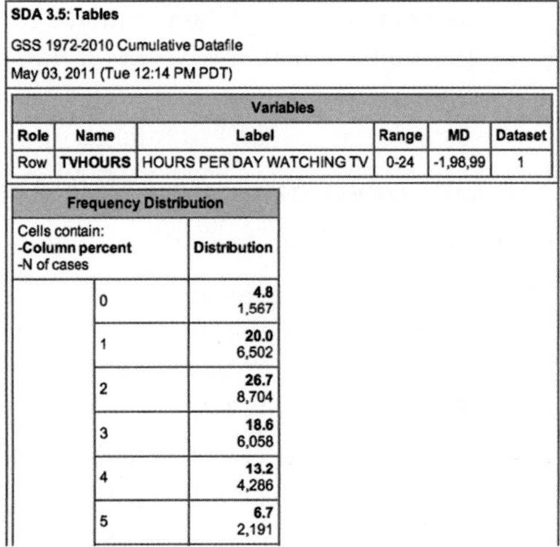

SDA 3.5: Tables

GSS 1972-2010 Cumulative Datafile

May 03, 2011 (Tue 12:14 PM PDT)

Variables

Role	Name	Label	Range	MD	Dataset
Row	**TVHOURS**	HOURS PER DAY WATCHING TV	0-24	-1,98,99	1

Frequency Distribution

Cells contain: -**Column percent** -N of cases		**Distribution**
	0	**4.8** 1,567
	1	**20.0** 6,502
	2	**26.7** 8,704
	3	**18.6** 6,058
	4	**13.2** 4,286
	5	**6.7** 2,191

Try Exercises 1.3 and 1.4

1.1 Practicing the Basics

1.1 Aspirin and heart attacks The Harvard Medical School study mentioned in Scenario 2 included about 22,000 male physicians. Whether a given individual would be assigned to take aspirin or the placebo was determined by flipping a coin. As a result, about 11,000 physicians were assigned to take aspirin and about 11,000 to take the placebo. The researchers summarized the results of the experiment using percentages. Of the physicians taking aspirin, 0.9% had a heart attack, compared to 1.7% of those taking the placebo. Based on the observed results, the study authors concluded that taking aspirin reduces the risk of having a heart attack. Specify the aspect of this study that pertains to (a) design, (b) description, and (c) inference.

1.2 Poverty and race The Current Population Survey (CPS) is a survey conducted by the U.S. Census Bureau for the Bureau of Labor Statistics. It provides a comprehensive body of data on the labor force, unemployment, wealth, poverty, etc. The data can be found online at www.census.gov/hhes/www/cpstc/cps_table_creator.html. A report from the 2009 CPS focused on a sample of about 50,000 households, each consisting of at least one related person under the age of 18. The report indicated that 14.7% of white households, 30.4% of black households, and 11.1% of Asian households had annual incomes below the poverty line. Based on these results, the study authors concluded that the percentage of *all* such black households with annual incomes below the poverty line is between 28.6% and 32.2%. Specify the aspect of this study that pertains to (a) description and (b) inference.

1.3 **GSS and heaven** Go to the General Social Survey Web site, http://sda.berkeley.edu/GSS. Enter HEAVEN as the row variable and then click *Run the Table*. When asked whether or not they believed in heaven, what percentage of those surveyed said yes, definitely; yes, probably; no, probably not; and no, definitely not? (Data from CSM, UC Berkeley.)

TRY

1.4 **GSS and heaven and hell** Refer to the previous exercise. You can obtain data for a particular survey year such as 2008 by entering YEAR(2008) in the Selection Filter option box before you click on *Run the Table*.

TRY

a. Do this for HEAVEN in 2008, giving the percentages for the four possible outcomes.

b. Summarize opinions in 2008 about belief in hell (row variable HELL). Was the percentage of "yes, definitely" responses higher for belief in heaven or in hell?

1.5 **GSS for subject you pick** At the GSS Web site, click on *Standard Codebook* under Codebooks and then on *Sequential Variable List*. Find a subject that interests you and look up a relevant GSS code name to enter as the row variable. Summarize the results that you obtain.

1.2 Sample Versus Population

We've seen that statistics consists of methods for **designing** investigative studies, **describing** (summarizing) data obtained for those studies, and making **inferences** (decisions and predictions) based on those data to answer a statistical question of interest.

We Observe Samples But Are Interested in Populations

The entities that we measure in a study are called the **subjects**. Usually subjects are people, such as the individuals interviewed in a General Social Survey. But they need not be. For instance, subjects could be schools, countries, or days. We might measure characteristics such as the following:

- For each school: the per-student expenditure, the average class size, the average score of students on an achievement test
- For each country: the percentage of residents living in poverty, the birth rate, the percentage unemployed, the percentage who are computer literate
- For each day in an Internet café: the amount spent on coffee, the amount spent on food, the amount spent on Internet access

The **population** is the set of all the subjects of interest. In practice, we usually have data for only *some* of the subjects who belong to that population. These subjects are called a **sample**.

Population and Sample

The **population** is the total set of subjects in which we are interested. A **sample** is the subset of the population for whom we have (or plan to have) data, often randomly selected.

In the 2008 General Social Survey (GSS), the sample was the 2023 people who participated in this survey. The population was the set of all adult Americans at that time—more than 200 million people.

Sample and population

Example 2

An Exit Poll

Picture the Scenario

Scenario 1 in the previous section discussed an exit poll. The purpose was to predict the outcome of the 2010 gubernatorial election in California. The exit poll sampled 3889 of the 9.5 million people who voted.

Question to Explore

For this exit poll, what was the population and what was the sample?

Think It Through

The population was the total set of subjects of interest, namely, the 9.5 million people who voted in this election. The sample was the 3889 voters who were interviewed in the exit poll. These are the people from whom the poll obtained data about their votes.

Insight

The ultimate goal of most studies is to learn about the *population*. For example, the sponsors of this exit poll wanted to make an inference (prediction) about *all* voters, not just the 3889 voters sampled by the poll.

Try Exercises 1.9 and 1.10

Did You Know?

Examples in this book use the five parts shown in this example: **Picture the Scenario** introduces the context. **Question to Explore** states the question addressed. **Think It Through** shows the reasoning used to answer that question. **Insight** gives follow-up comments related to the example. **Try Exercises** direct you to a similar "Practicing the Basics" exercise at the end of the section. Also, each example title is preceded by a label highlighting the example's **concept**. In this example, the concept label is "Sample and population." ◀

Occasionally data are available from an entire population. For instance, every ten years the U.S. Census Bureau gathers data from the entire U.S. population (or nearly all). But the census is an exception. Usually, it is too costly and time-consuming to obtain data from an entire population. It is more practical to get data for a sample. The General Social Survey and polling organizations such as the Gallup poll usually select samples of about 1000 to 2500 Americans to learn about opinions and beliefs of the population of *all* Americans. The same is true for surveys in other parts of the world, such as the Eurobarometer in Europe.

Descriptive Statistics and Inferential Statistics

Using the distinction between samples and populations, we can now tell you more about the use of **description** and **inference** in statistical analyses.

Description in Statistical Analyses

Descriptive statistics refers to methods for summarizing the collected data (where the data constitutes either a sample or a population). The summaries usually consist of graphs and numbers such as averages and percentages.

A descriptive statistical analysis usually combines graphical and numerical summaries. For instance, Figure 1.1 is a **bar graph** that shows the percentages of various types of U.S. households in 2005. It summarizes a survey of 50,000 American households by the U.S. Census Bureau. The main purpose of descriptive statistics is to reduce the data to simple summaries without distorting or losing much information. Graphs and numbers such as percentages

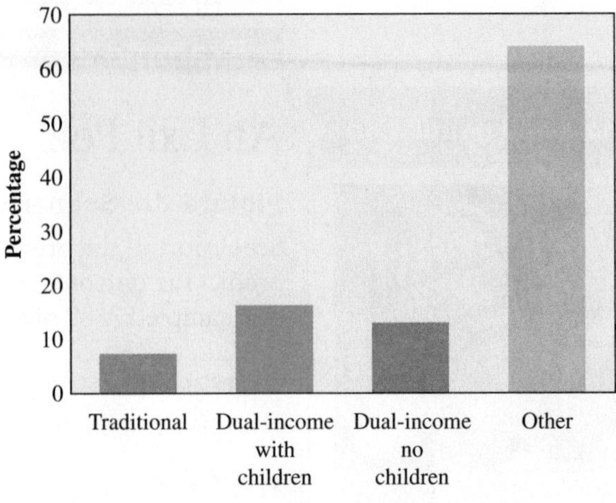

▲ **Figure 1.1** **Types of U.S. Households, Based on a Sample of 50,000 Households in the 2005 Current Population Survey.** (*Source:* Data from United States Census Bureau.)

and averages are easier to comprehend than the entire set of data. It's much easier to get a sense of the data by looking at Figure 1.1 than by reading through the questionnaires filled out by the 50,000 sampled households. From this graph, it's readily apparent that the "traditional" household, defined as being a married man and woman with children in which only the husband is in the labor force, is no longer very common in the United States. In fact, "Other" households, which include female-headed households and households headed by young adults or older Americans who do not reside with spouses are most common.

Descriptive statistics are also useful when data are available for the entire population, such as in a census. By contrast, inferential statistics are used when data are available for a sample only, but we want to make a decision or prediction about the entire population.

Inference in Statistical Analyses

Inferential statistics refers to methods of making decisions or predictions about a population, based on data obtained from a sample of that population.

In most surveys, we have data for a sample, not for the entire population. We use descriptive statistics to summarize the sample data and inferential statistics to make predictions about the population.

Descriptive and inferential statistics

Example 3

Polling Opinions on Handgun Control

Picture the Scenario

Suppose we'd like to know what people think about controls over the sales of handguns. Let's consider how people feel in Florida, a state with a relatively high violent crime rate. The population of interest is the set of more than 10 million adult residents of Florida.

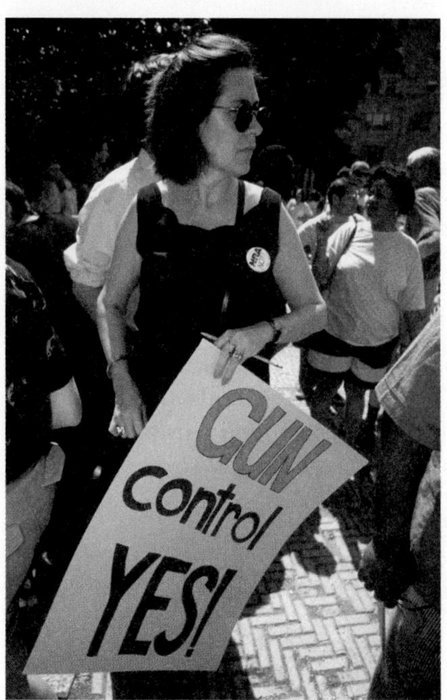

Since it is impossible to discuss the issue with all these people, we can study results from a recent poll of 834 Florida residents conducted by the Institute for Public Opinion Research at Florida International University. In that poll, 54.0% of the sampled subjects said they favored controls over the sales of handguns. A newspaper article about the poll reports that the "margin of error" for how close this number falls to the population percentage is 3.4%. We'll see (later in the textbook) that this means we can predict with high confidence (about 95% certainty) that the percentage of *all* adult Floridians favoring control over sales of handguns falls within 3.4% of the survey's value of 54.0%, that is, between 50.6% and 57.4%.

Question to Explore

In this analysis, what is the descriptive statistical analysis and what is the inferential statistical analysis?

Think It Through

The results for the sample of 834 Florida residents are summarized by the percentage, 54.0%, who favored handgun control. This is a descriptive statistical analysis. We're interested, however, not just in those 834 people but in the *population of all* adult Florida residents. The prediction that the percentage of *all* adult Floridians who favor handgun control falls between 50.6% and 57.4% is an inferential statistical analysis. In summary, we *describe* the *sample*, and we make *inferences* about the *population*.

Insight

The sample size of 834 was small compared to the population size of more than 10 million. However, because the values between 50.6% and 57.4% are all above 50%, the study concluded that a slim majority of Florida residents favored handgun control.

Try Exercises 1.11, part a, and 1.12, parts a–c

An important aspect of statistical inference involves reporting the likely *precision* of a prediction. How close is the *sample* value of 54% likely to be to the true (unknown) percentage of the *population* favoring gun control? We'll see (in Chapters 4 and 6) why a well-designed sample of 834 people yields a sample percentage value that is very likely to fall within about 3–4% (the so-called *margin of error*) of the population value. In fact, we'll see that inferential statistical analyses can predict characteristics of entire populations quite well by selecting samples that are small relative to the population size. Surprisingly, the absolute size of the sample matters much more than the size relative to the population total. For example, the population of China is about four times that of the United States, but a random sample of 1000 people from the Chinese population and a random sample of 1000 people from the U.S. population would achieve similar levels of accuracy. That's why most polls take samples of only about a thousand people, even if the population has millions of people. In this book, we'll see why this works.

Sample Statistics and Population Parameters

In Example 3, the percentage of the sample favoring handgun control is an example of a **sample statistic**. It is crucial to distinguish between sample statistics and the corresponding values for the population. The term **parameter** is used for a numerical summary of the population.

Recall

A **population** is the total group of individuals about whom you want to make conclusions. A **sample** is a subset of the population for whom you actually have data. ◀

Parameter and Statistic

A **parameter** is a numerical summary of the population. A **statistic** is a numerical summary of a sample taken from the population.

For example, the percentage of the population of all adult Florida residents favoring handgun control is a parameter. We hope to learn about parameters so that we can better understand the population, but the true parameter values are almost always unkown. Thus, we use sample statistics to estimate the parameter values.

Randomness and Variability

Random is often thought to mean chaotic or haphazard, but randomness is an extremely powerful tool for obtaining good samples and conducting experiments. A sample tends to be a good reflection of a population when each subject in the population has the same chance of being included in that sample. That's the basis of **random sampling**, which is designed to make the sample representative of the population. A simple example of random sampling is when a teacher puts each student's name on a slip of paper, places it in a hat, and then draws names from the hat without looking.

- Random sampling allows us to make powerful inferences about populations.
- Randomness is also crucial to performing experiments well.

If, as in Scenario 2 on page 5, we want to compare aspirin to a placebo in terms of the percentage of people who later have a heart attack, it's best to randomly select those in the sample who use aspirin and those who use placebo. This approach tends to keep the groups balanced on other factors that could affect the results. For example, suppose we allowed people to choose whether or not to use aspirin (instead of randomizing whether the person receives aspirin or the placebo). Then, the people who decided to use aspirin might have tended to be healthier than those who didn't, which could produce misleading results.

People are different from each other, so, not surprisingly, the measurements we make on them *vary* from person to person. For the GSS question about TV watching in Activity 1 on page 7, different people reported different amounts of TV watching. In the exit poll of Example 1, not all people voted the same way. If subjects did not vary, we'd need to sample only one of them. We learn more about this variability by sampling more people. If we want to predict the outcome of an election, we're better off sampling 100 voters than one voter, and our prediction will be even more reliable if we sample 1000 voters.

- Just as people vary, so do samples vary.

Suppose you take an exit poll of 1000 voters to predict the outcome of an election. Suppose the Gallup organization also takes an exit poll of 1000 voters. Your sample will have different people than Gallup's. Consequently, the predictions will also differ. Perhaps your exit poll of 1000 voters has 480 voting for the Republican candidate, so you predict that 48% of all voters voted for that person. Perhaps Gallup's exit poll of 1000 voters has 440 voting for the Republican candidate, so they predict that 44% of all voters voted for that person. Activity 2 at the end of the chapter shows that with random sampling, the amount of variability from sample to sample is actually quite predictable. Both of your predictions are likely to fall within 5% of the actual population percentage who voted Republican, assuming the samples are random. If, on the other hand, Republicans are more likely than Democrats to refuse to participate in the exit poll, then we would need to account for this. In the 2004 U.S. presidential election, much controversy arose when George W. Bush won several states in which exit polling predicted that John Kerry had won. Is it likely that the way the exit polls were conducted led to these incorrect predictions?

The Basic Ideas of Statistics

Here is a summary of the key concepts of statistics that you'll learn about in this book:

Chapter 2: Exploring Data with Graphs and Numerical Summaries How can you present simple summaries of data? You replace lots of numbers with simple graphs and numerical summaries.

Chapter 3: Association: Contingency, Correlation, and Regression How does annual income ten years after graduation correlate with college GPA? You can find out by studying the association between those characteristics.

Chapter 4: Gathering Data How can you design an experiment or conduct a survey to get data that will help you answer questions? You'll see why results may be misleading if you don't use randomization.

Chapter 5: Probability in Our Daily Lives How can you determine the chance of some outcome, such as winning a lottery? Probability, the basic tool for evaluating chances, is also a key foundation for inference.

Chapter 6: Probability Distributions You've probably heard of the normal distribution or "bell-shaped curve" that describes people's heights or IQs or test scores. What is the normal distribution, and how can we use it to find probabilities?

Chapter 7: Sampling Distributions Why is the normal distribution so important? You'll see why, and you'll learn its key role in statistical inference.

Chapter 8: Statistical Inference: Confidence Intervals How can an exit poll of 3889 voters possibly predict the results for millions of voters? You'll find out by applying the probability concepts of Chapters 5, 6, and 7 to make statistical inferences that show how closely you can predict summaries such as population percentages.

Chapter 9: Statistical Inference: Significance Tests about Hypotheses How can a medical study make a decision about whether a new drug is better than a placebo? You'll see how you can control the chance that a statistical inference makes a correct decision about what works best.

Chapters 10–15: Applying Descriptive and Inferential Statistics to Many Kinds of Data After Chapters 2–9 introduce you to the key concepts of statistics, the rest of the book shows you how to apply them in lots of situations. For instance, Chapter 10 shows how to compare two groups, such as using a sample of students from your university to make an inference about whether male and female students have different rates of binge drinking.

1.2 Practicing the Basics

1.6 Description and inference

a. Distinguish between *description* and *inference* as reasons for using statistics. Illustrate the distinction using an example.

b. You have data for a population, such as obtained in a census. Explain why descriptive statistics are helpful but inferential statistics are not needed.

1.7 Number of good friends One year the General Social Survey asked, "About how many good friends do you have?" Of the 840 people who responded, 6.1% reported having only one good friend. Identify (a) the sample, (b) the population, and (c) the statistic reported. (*Source:* Data from CSM, Berkeley.)

1.8 Concerned about global warming? The Institute for Public Opinion Research at Florida International University has conducted the FIU/Florida Poll (www2.fiu.edu/orgs/ipor/globwarm2.htm) of about 1200 Floridians annually since 1988 to track opinions on a wide variety of issues. In 2006 the poll asked, "How concerned are you about the problem of global warming?" The possible responses were very concerned, somewhat concerned, not very concerned, and haven't heard about it. The poll reported percentages (44, 30, 21, 6) in these categories.

a. Identify the sample and the population.

b. Are the percentages quoted statistics or parameters? Why?

1.9 EPA TRY The Environmental Protection Agency (EPA) uses a few new automobiles of each model every year to collect data on pollution emissions and gasoline mileage performance. For the Honda Accord model, identify what's meant by the (a) subject, (b) sample, and (c) population.

1.10 Babies and social preference TRY A recent study at Yale University's Infant Cognition Center, published in the journal *Nature*, investigated whether babies develop social preferences at an early age. As part of the study, 16 six-month-old infants were each shown a sequence of videos. One video focused on a figure whose actions toward others were helpful, while the other focused on a figure whose actions were hurtful. After viewing the videos, each infant was presented with the two figures and allowed to choose one to play with. Of the 16 infants in the study, 14 chose to play with the helper object. The researchers concluded that six-month-old infants have both the ability to recognize and the preference to align themselves with the helpful figure. Identify (a) the sample, (b) the population, and (c) the inference being drawn.

1.11 Graduating seniors' salaries TRY The job placement center at your school surveys *all* graduating seniors at the school. Their report about the survey provides numerical summaries such as the average starting salary and the percentage of students earning more than $30,000 a year.

a. Are these statistical analyses descriptive or inferential? Explain.

b. Are these numerical summaries better characterized as statistics or as parameters?

1.12 At what age did women marry? TRY A historian wants to estimate the average age at marriage of women in New England in the early 19th century. Within her state archives she finds marriage records for the years 1800–1820, which she treats as a sample of all marriage records from the early 19th century. The average age of the women in the records is 24.1 years. Using the appropriate statistical method, she estimates that the average age of brides in early 19th-century New England was between 23.5 and 24.7.

a. Which part of this example gives a descriptive summary of the data?

b. Which part of this example draws an inference about a population?

c. To what population does the inference in part b refer?

d. The average age of the sample was 24.1 years. Is 24.1 a statistic or a parameter?

1.13 Age pyramids as descriptive statistics The figure shown is a graph published by Statistics Sweden. It compares Swedish society in 1750 and in 2010 on the numbers of men and women of various ages, using "age pyramids." Explain how this indicates that

a. In 1750, few Swedish people were old.

b. In 2010, Sweden had many more people than in 1750.

c. In 2010, of those who were very old, more were female than male.

d. In 2010, the largest five-year group included people born during the era of first manned space flight.

Age pyramids, 1750 and 2010

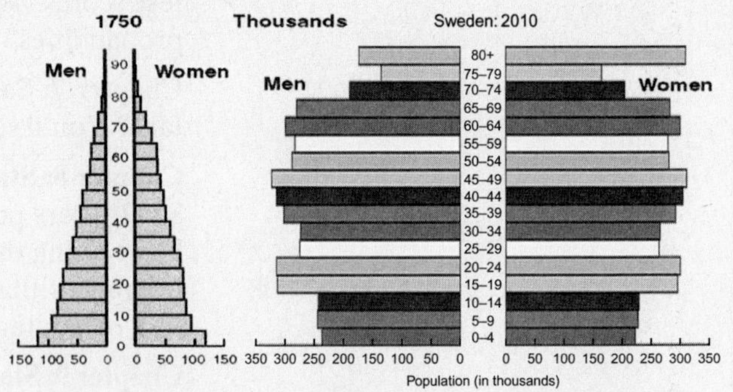

Graphs of number of men and women of various ages, in 1750 and in 2010. (*Source:* From Statistics Sweden.)

1.14 Gallup polls Go to the Web site www.galluppoll.com for the Gallup poll. From reports listed on their home page, give an example of (a) a descriptive statistical analysis and (b) an inferential statistical analysis.

1.15 National service Consider the population of all students at your school. A certain proportion support mandatory national service (MNS) following high school. Your friend randomly samples 20 students from the school, and uses the sample proportion who support MNS to predict the population proportion at the school. You take your own, separate random sample of 20 students, and find the sample proportion that supports MNS.

a. For the two studies, are the populations the same?

b. For the two studies, are the sample proportions necessarily the same? Explain.

1.16 Samples vary less with more data We'll see that the amount by which statistics vary from sample to sample always depends on the sample size. This important fact can be illustrated by thinking about what would happen in repeated flips of a fair coin.

a. Which case would you find more surprising—flipping the coin five times and observing all heads or flipping the coin 500 times and observing all heads?

b. Imagine flipping the coin 500 times, recording the proportion of heads observed, and repeating this experiment many times to get an idea of how much the proportion tends to vary from one sequence to another. Different sequences of 500 flips tend to result in proportions of heads observed which are less variable than the proportion of heads observed in sequences of only five flips each. Using part a, explain why you would expect this to be true.

1.3 Using Calculators and Computers

Today's researchers (and students) are lucky: Unlike those in the previous generation, they don't have to do complex statistical calculations by hand. Powerful user-friendly computing software and calculators are now readily available for statistical analyses. This makes it possible to do calculations that would be extremely tedious or even impossible by hand, and it frees time for interpreting and communicating the results.

Using (and Misusing) Statistics Software and Calculators

MINITAB and **SPSS** are two popular statistical software packages on college campuses. The **TI-83+** and **TI-84** graphing calculators,[2] which have similar output, are useful as portable tools for generating simple statistics and graphs. The **Microsoft Excel** software can conduct some statistical methods, sorting and analyzing data with its spreadsheet program, but its capabilities are limited. Throughout this text, we'll show examples of MINITAB and TI-83+/84 output. Occasional exercises use SPSS or Excel. The emphasis in this book is on how to interpret the output, not on the details of using such software.

In Practice Selecting Valid Analyses

Given the current software capabilities, why do you still have to learn about statistical methods? Can't computers do all this analysis for you? The problem is that a computer will perform the statistical analysis you request whether or not its use is valid for the given situation. Just knowing how to use software does not guarantee a proper analysis. You'll need a good background in statistics to understand which statistical method to use, which options to choose with that method, and how to interpret and make valid conclusions from the computer output. This text helps give you this background.

[2] We will use the shorthand, TI-83+/84, to indicate output from a graphing calculator.

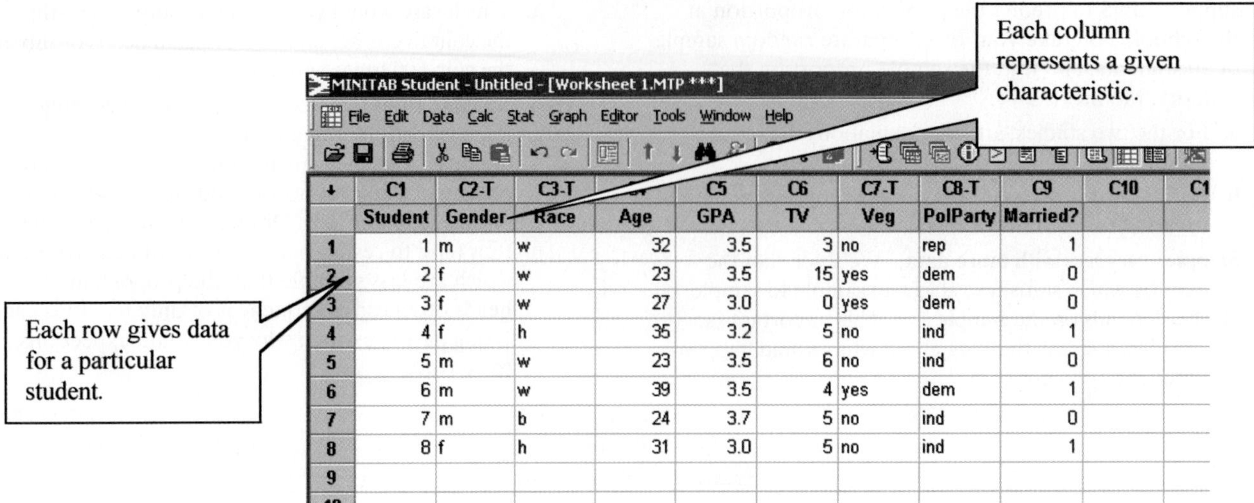

	C1	C2-T	C3-T		C5	C6	C7-T	C8-T	C9	C10	C1
	Student	Gender	Race	Age	GPA	TV	Veg	PolParty	Married?		
1	1	m	w	32	3.5	3	no	rep	1		
2	2	f	w	23	3.5	15	yes	dem	0		
3	3	f	w	27	3.0	0	yes	dem	0		
4	4	f	h	35	3.2	5	no	ind	1		
5	5	m	w	23	3.5	6	no	ind	0		
6	6	m	w	39	3.5	4	yes	dem	1		
7	7	m	b	24	3.7	5	no	ind	0		
8	8	f	h	31	3.0	5	no	ind	1		
9											

▲ **Figure 1.2** Part of a MINITAB Data File.

Data Files

To make statistical analysis easier, large sets of data are organized in a **data file**. This file usually has the form of a spreadsheet. It is the way statistical software receives the data.

Figure 1.2 is an example of part of a data file. It shows how a data file looks in MINITAB. The file shows data for eight students on the following characteristics:

- Gender (f = female, m = male)
- Racial–ethnic group (b = black, h = Hispanic, w = white)
- Age (in years)
- College GPA (scale 0 to 4)
- Average number of hours per week watching TV
- Whether a vegetarian (yes, no)
- Political party (dem = Democrat, rep = Republican, ind = independent)
- Marital status (1 = married, 0 = unmarried)

Figure 1.2 shows the two basic rules for constructing a data file:

- Any one row contains measurements for a particular subject (for instance, person).
- Any one column contains measurements for a particular characteristic.

Some characteristics have numerical data, such as the values for hours of TV watching. Some characteristics have data that consist of categories or labels, such as the categories (yes, no) for vegetarians or the labels (dem, rep, ind) for political affiliation.

Figure 1.2 resembles a larger data file from a questionnaire administered to a sample of students at the University of Florida. That data file, which includes other characteristics as well, is called "FL student survey" and is on the CD that comes with this text. (You may want to find this file on the CD now, to practice accessing data files used in some homework exercises.)

To construct a similar data file for your class, try Activity 3 in the Student Activities at the end of the chapter.

Data files

Example 4

Ads on Facebook

Picture the Scenario

You are the manager of a digital media store (e.g., DVDs, video games). Your sales have been shrinking because of Internet competition such as Amazon .com, so you decide to try advertising and selling your products online. After launching your new Web site, you enter an advertising agreement with Facebook, the terms of which include your ad being displayed to 1000 Facebook users.

Anyone who clicks on the ad will go to your Web site and a display of 10 featured items. For Facebook advertising to be profitable, the 1000 users to whom your ad is shown need to spend, on average, \$0.75 or more on the 10 featured items. The first person shown your ad doesn't click on it and orders nothing. The second person visits your site but does not purchase anything. The third person orders one copy of item 3 at a cost of \$8 and two copies of item 5 at a cost of \$6 each. At the end of the advertising period, you create a data file containing the orders (in dollars) of the 10 featured items for each of the 1000 potential customers.

Questions to Explore

a. How do you record the responses from the first three people in the data file for the 10 featured items?

b. How many rows of data will your file contain?

Think It Through

a. Each row refers to a particular person to whom your ad was shown. Each column refers to one of the featured items. Each entry of data is the dollar amount spent on a given featured item by a particular person. So, the first three rows of the data file are

Person	item1	item2	item3	item4	item5	item6	item7	item8	item9	item10
1	0	0	0	0	0	0	0	0	0	0
2	0	0	0	0	0	0	0	0	0	0
3	0	0	8	0	12	0	0	0	0	0

b. The entire data file would consist of 1000 lines of data, one line for each potential customer.

Insight

From the entire data file, suppose you calculate that the average sales per person equals \$0.90. What could you conclude about the average sales if you decide to enter a new agreement to display your ad to 100,000 Facebook users? Is it likely to be at least \$0.75? As we'll see (in Chapters 4 and 9), you can use additional information about the customers and about the method of sampling to help answer this question. If you decide to advertise online, consider how statistics can help you use information gathered about customers' demographics and past purchases to develop and maintain a promising list of potential customers.

Try Exercise 1.18

Databases

Most studies design experiments or surveys to collect data to answer the questions of interest. Often, though, it is adequate to take advantage of existing archived collections of data files, called **databases**. Many Internet databases are available. By browsing various Web sites, you can obtain information on many topics.

The General Social Survey, discussed in Scenario 3 on page 5 and Activity 1 on page 7, is one such database that is accessed from the Internet. The CD that comes with this text contains several data files that you'll use in some exercises. Here are some other databases that can be fun and informative to browse:

- Are you interested in pro baseball or basketball or football? Click on the statistics links at www.mlb.com or www.nba.com or www.nfl.com.
- Are you interested in what people believe around the world? Check out www.globalbarometer.net or www.europa.eu.int/comm/public_opinion or www.latinobarometro.org.
- Are you interested in results of Gallup polls about people's beliefs? See www.galluppoll.com.
- Are you interested in population growth, unemployment rates, or the spread of sexually transmitted diseases? See www.google.com/publicdata for these and many other data sets, along with visualization tools.

Also very useful are search engines such as Google. Type in "U.S. Census data" in the search window at www.google.com, and it lists databases such as those maintained by the U.S. Census Bureau. Type in "Statistics Canada data" for census databases from Canada. A search engine also is helpful for finding data or descriptive statistics on a particular topic. For instance, if you want to find summaries of Canadians' opinions about the legalization of marijuana for medical treatment, try typing

poll legalize marijuana Canada medicine

in the search window.

In Practice Always Check Sources

Not all databases or reported data summaries give reliable information. Before you give credence to such data, verify that the data are from a trustworthy source and that the source provides information about the way the study was conducted.

For example, many news organizations and search Web sites (such as cnn.com) ask you to participate in a "topic of the day" poll on their home page. Their summaries of how people responded are not reliable indications of how the entire population feels. They are biased because it is unlikely that the people who respond are representative of the population. We'll explain why in Chapter 4.

Applets

Just like riding a bike, it's easier to learn statistics if you are actively involved, learning by doing. One way is to practice, by using software or calculators with data files or data summaries that we'll provide. Another way is to perform activities that illustrate the ideas of statistics by using **applets**. An applet is a small *application* program for performing a specific task. We'll use them throughout the text. Using an applet, you can take samples from artificial populations and analyze them to discover properties of statistical methods applied to those samples. This is a type of **simulation**—using a computer to mimic what would actually happen if you selected a sample and used statistics in real life. So, let's get started with your active involvement.

Activity 2

Simulating Randomness and Variability*

To get a feel for randomness and variability, let's simulate taking an exit poll of voters using the *sample from a population* applet on the text CD. This applet generates samples resembling those we'd get with random sampling. Select from the menu for the population, Binary: p = 0.5. A graph will appear with a bar at 0 and a bar at 1, representing two possible outcomes. Each observation from an individual in the population has two possible outcomes, as in sampling a person who voted in an election with two candidates.

Let's see what would happen with exit polling when 50% of the entire population (a proportion of 0.50) voted for each candidate. We'll use a small poll, only 10 voters. First, represent this by taking a coin and flipping it ten times. Let the number of heads represent the number who voted for the Democrat. What proportion in your sample of size 10 voted for the Democrat?

Now let's do this with the applet. We'll regard outcome 1 as voting for the Democratic candidate and outcome 0 as voting for the Republican candidate. You should already have the applet set at Binary: p = 0.5. To take a sample of size 10, go to the Sample Size menu, select $n = 10$, and then click on *Sample*. You'll see that a certain number of outcomes occurred for each type. For example, when we did this, we got outcome 1 four times and outcome 0 six times. This simulates sampling 10 voters in the exit poll, in which 4 said they voted for the Democrat and 6 for the Republican. It corresponds to a sample proportion of 0.40 voting for the Democrat.

To illustrate how samples vary, take your own sample of size 10 using this applet. When you do this, you will probably get a proportion different than 0.40 voting for the Democrat because the process is random. What did you get? Now collect another sample of size 10 and find the sample proportion. What did you get? Repeat taking samples of size 10 at least five times. Note the sample proportion for each sample and how the sample proportions compare to the population proportion, 0.50. Are they always close?

Now repeat this for a larger exit poll, taking a sample of 1000 voters instead of 10 (set *Sample* $n = 1000$ on the applet menu). What proportion voted for the Democrat? Repeat taking samples of size 1000 at least five times. Note the sample proportion for each sample and how these sample proportions compare to the population proportion, 0.50. Do the sample proportions tend to be close to 0.50? We predict that all of your sample proportions will fall between 0.45 and 0.55. Are we correct?

We would expect that some sample proportions generated using a sample size of 10 fell much farther from 0.50 than the sample proportions generated using a sample size of 1000. This illustrates by simulation that sample proportions tend to be closer to the population proportion when the sample size is larger. In fact, we will discover as we move forward in the textbook that we do much better in making inferences about the population with larger sample sizes.

*For more information about the applets see page x and the back left endpaper of this book.

Try Exercises 1.22, 1.23, and 1.35

1.3 Practicing the Basics

1.17 Data file for friends Construct (by hand) a data file of the form of Figure 1.2, for two characteristics with a sample of four of your friends. One characteristic should take numerical values, and the other should take values that are categories.

1.18 Shopping sales data file Construct a data file describing the purchasing behavior of the five people, described below, who visit a shopping mall. Enter purchase amounts each spent on clothes, sporting goods, books, and music CDs as the data. Customer 1 spent $49 on clothes and $16 on music CDs, customer 4 spent $92 on books, and the other three customers did not buy anything.

1.19 Sample with caution Individuals with children who read the Ann Landers column were asked if they had it to do all over again, whether they would want to have children. Of the nearly 10,000 readers who responded, only 30% responded by saying yes. Why is it not safe to infer anything from this survey about the proportion of the general population who would still want to have children if given the opportunity to do things over?

1.20 Create a data file with software Your instructor will show you how to create data files using the software for your course. Use it to create the data file you constructed by hand in Exercise 1.17 or 1.18.

1.21 Use a data file with software You may need to learn how to open a data file from the text CD or download one from the Web for use with the software for your course. Do this for the "FL student survey" data file on the text CD, from the survey mentioned following Figure 1.2.

1.22 Simulate with the sample from the Population applet Refer to Activity 2 on page 19.

a. Repeat the activity using a population proportion 0.60: Take at least five samples of size 10 each, and observe how the sample proportions of the one outcome vary around 0.60, and then do the same thing with at least five samples of size 1000 each.

b. In part a, what seems to be the effect of the sample size on the amount by which sample proportions tend to vary around the population proportion, 0.60?

c. What is the practical implication of the effect of the sample size summarized in part b with respect to making inferences about the population proportion when you collect data and observe only the sample proportion?

1.23 Is a sample unusual? Suppose 70% of all voters voted for the Republican candidate. If an exit poll of 50 people were chosen randomly, would it be surprising if less than half who were sampled said they voted Republican? To reason an answer, use the applet described in Activity 2 to conduct at least ten simulations of taking samples of size 50 from a population with proportion 0.70. Note the sample proportions. Do you observe any sample proportion less than 0.50? What does this suggest?

Chapter Review

CHAPTER SUMMARY

- Statistics consists of methods for conducting research studies and for analyzing and interpreting the data produced by those studies. **Statistics is the art and science of learning from data**.
- The first part of the statistical process for answering a statistical question involves **design**—planning an investigative study to obtain relevant data for answering the statistical question. The design often involves taking a **sample** from a **population** where the population contains *all* the **subjects** (usually, people) of interest. After we've collected the data, there are two types of statistical analyses:

  Descriptive statistics summarize the sample data with numbers and graphs.

  Inferential statistics make decisions and predictions about the entire population, based on the information in the sample data.
- With **random sampling**, each subject in the population has the same chance of being in the sample. This is desirable because then the sample tends to be a good reflection of the population. Randomization is also important for good experimental design, for example, randomly assigning who gets the medicine and who gets the placebo in a medical study.
- The measurements we make of a characteristic **vary** from individual to individual. Likewise, results of descriptive and inferential statistics **vary**, depending on the sample chosen. We'll see that the study of **variability** is a key part of statistics. **Simulation** investigations generate many samples randomly, often using an **applet**. They provide a way of learning about the impact of randomness and variability from sample to sample.
- The calculations for data analysis can use computer software. The data are organized in a **data file**. This file has a separate row of data for each subject and a separate column for each characteristic. However, you'll need a good background in statistics to understand which statistical method to use and how to interpret and make valid conclusions from the computer output.

CHAPTER PROBLEMS

Practicing the Basics

1.24 UW Student survey In a University of Wisconsin (UW) study about alcohol abuse among students, 100 of the 40,858 members of the student body in Madison were sampled and asked to complete a questionnaire. One question asked was, "On how many days in the past week did you consume at least one alcoholic drink?"

a. Identify the population and the sample.

b. For the 40,858 students at UW, one characteristic of interest was the percentage who would respond "zero" to this question. For the 100 students sampled, suppose 29% gave this response. Does this mean that 29% of the entire population of UW students would make this response? Explain.

c. Is the numerical summary of 29% a sample statistic, or a population parameter?

1.25 ESP For several years, the General Social Survey asked subjects, "How often have you felt as though you were in touch with someone when they were far away from you?" Of 3887 sampled subjects who had an opinion, 1407 said never and 2480 said at least once. The proportion who had at least one such experience was 2480/3887 = 0.638.

a. Describe the population of interest.

b. Explain how the sample data are summarized using descriptive statistics.

c. For what population parameter might we want to make an inference?

1.26 Presidential popularity Each month the Gallup organization conducts a poll for CNN and *USA Today* of the U.S. president's current popularity rating (see www.pollingreport.com). For a poll conducted February 2–3, 2011, of approximately 1500 Americans, it was reported, "45% of people polled said they approve of how Obama is handling the presidency. The margin of error is plus or minus 3 percentage points." Explain how this margin of error provides an *inferential* statistical analysis.

1.27 Breaking down Brown versus Whitman Example 2 of this chapter discusses an exit poll taken during the 2010 California gubernatorial election. The administrators of the poll also collected demographic data, which allows for further breakdown of the 3889 voters from whom information was collected. Of the 1633 voters registered as Democrats, 91% voted for Brown, with a margin of error of 1.4%. Of the 1206 voters registered as Republicans, 10% voted for Brown, with a margin of error of 1.7%. And of the 1050 Independent voters, 42% voted for Brown, with a margin of error of 3.0%.

a. Do these results summarize sample data or population data?

b. Identify a descriptive aspect of the results.

c. Identify an inferential aspect of the results.

1.28 Reducing stress Your school wants to make an inference about the percentage of students at the school who prefer having a several-day period between the end of classes and the start of final exams to help reduce the level of stress as students prepare for exams. A survey is taken of 100 students.

a. Identify the sample and the population.

b. For the study, explain the purpose of using (i) descriptive statistics, and (ii) inferential statistics.

1.29 Marketing study For the marketing study about sales in Example 4, identify the (a) sample and population, and (b) descriptive and inferential aspects.

1.30 Multiple choice: Believe in reincarnation? In a survey of 750 Americans conducted by the Gallup organization, 24% indicated a belief in reincarnation. A method presented later in this book allows us to predict that for *all* adult Americans, the percentage believing in reincarnation falls between 21% and 27%. This prediction is an example of

a. descriptive statistics

b. inferential statistics

c. a data file

d. designing a study

1.31 Multiple choice: Use of inferential statistics? Inferential statistics are used

a. to describe whether a sample has more females or males.

b. to reduce a data file to easily understood summaries.

c. to make predictions about populations using sample data.

d. when we can't use statistical software to analyze data.

e. to predict the sample data we will get when we know the population.

1.32 True or false? In a particular study, you could use descriptive statistics, or you could use inferential statistics, but you would rarely need to use both.

Concepts and Investigations

1.33 Statistics in the news Pick up a recent issue of a national newspaper, such as *The New York Times* or *USA Today*, or consult a news Web site, such as msnbc.com or cnn.com. Identify an article that used statistical methods. Did it use descriptive statistics, or inferential statistics, or both? Explain.

1.34 What is statistics? On a final exam that one of us recently gave, students were asked, "How would you define 'statistics' to someone who has never taken a statistics course?" One student wrote, "You want to know the answer to some question. There's no answer in the back of a book. You collect some data. Statistics is the body of procedures that helps you analyze the data to figure out the answer and how sure you can be about it." Pick a question that interests you, and explain how you might be able to use statistics to investigate the answer.

1.35 Surprising ESP data? In Exercise 1.25, of 3887 sampled subjects, 63.8% said that at least once they felt as though they were in touch with someone when they were far away. That is, of 3887 sampled subjects who had an opinion, 1407 said never and 2480 said at least once. Suppose that only 20% of the entire population would report at least one such experience. If the sample of 3887 people were a random sample, would this sample proportion result of 0.638 be surprising? Investigate, using the applet described in Activity 2. Do this by simulating samples from the population with a proportion of 0.20. Use the sample size of 4000 as an approximation to the actual sample size of 3887.

TRY

1.36 Create a data file Using the statistical software that your instructor has assigned for the course, find out how to enter a data file. Create a data file using the data in Figure 1.2.

Student Activities

1.37 Activity 3 Your instructor will help the class create a data file consisting of the values for class members of characteristics based on responses to a questionnaire like the one that follows. Alternatively, your instructor may ask you to use a data file of this type already prepared with a class of students at the University of Florida, the "FL student survey" data file on the text CD. Using a spreadsheet program or the statistical software the instructor has chosen for your course, create a data file containing this information. What are some questions you might ask about these data? Homework exercises in each chapter will use these data.

GETTING TO KNOW THE CLASS

Please answer the following questions. Do not put your name on this sheet. Skip any question that you feel uncomfortable answering. These data are being collected to learn more about you and your classmates and to form a database for the class to analyze.

1. What is your height (recorded in inches)? _____
2. What is your gender (M = Male, F = Female) _____
3. How much did you spend on your last haircut? _____
4. Do you have a paying job during the school year at which you work on average at least 10 hours a week (y = yes, n = no) _____
5. Aside from class time, how many hours a week, on average, do you expect to spend studying and completing assignments for this course? _____
6. Do you smoke cigarettes (y = yes, n = no)? _____
7. How many different people have you dated in the past 30 days? _____
8. What was (is) your high school GPA (based upon a 4.0 scale)? _____
9. What is your current college GPA? _____
10. What is the distance (in miles) between your current residence and this class? _____
11. How many minutes each day, on average, do you spend browsing the Internet? _____
12. How many minutes each day, on average, do you watch TV? _____
13. How many hours each week, on average, do you participate in sports or have other physical exercise? _____
14. How many times a week, on average, do you read a daily newspaper? _____
15. Do you consider yourself a vegetarian? (y = yes, n = no) _____
16. How would you rate yourself politically? (1 = very liberal, 2 = liberal, 3 = slightly liberal, 4 = moderate, 5 = slightly conservative, 6 = conservative, 7 = very conservative) _____
17. What is your political affiliation? (D = Democrat, R = Republican, I = Independent) _____

Exploring Data with Graphs and Numerical Summaries

2

Example 1

Statistics Informs About Threats to Our Environment

Picture the Scenario

Air pollution concerns many people, and the consumption of energy from fossil fuels affects air pollution. Most technologically advanced nations, such as the United States, use vast quantities of a diminishing supply of nonrenewable fossil fuels. Countries fast becoming technologically advanced, such as China and India, are greatly increasing their energy use. The result is increased emissions of carbon dioxide and other pollutants into the atmosphere. Emissions of carbon dioxide may also contribute to global warming. Scientists use descriptive statistics to explore energy use, to learn whether air pollution has an effect on climate, to compare different countries in how they contribute to the worldwide problems of air pollution and climate change, and to measure the impact of global warming on the environment.

Questions to Explore

- How can we investigate which countries emit the highest amounts of carbon dioxide or use the most nonrenewable energy?
- Is climate change occurring, and how serious is it?

Thinking Ahead

Chapter 1 distinguished between the ideas of **descriptive statistics** (summarizing data) and **inferential statistics** (making a prediction or decision about a population using sample data). In this chapter, we discuss both graphical and numerical summaries that are the key elements of descriptive statistics. And, we will use these summaries to investigate questions like the ones asked above. In Example 3, we'll examine data on the dependence of the United States and Canada on fossil fuels. Examples 11 and 18 explore carbon dioxide emissions for the world's largest nations and for nations in the European Union. Example 9 uses descriptive statistics to analyze data about temperature change over time.

Before we begin our journey into the art and science of analyzing data using descriptive statistics, we need to learn a few new terms. We'll then study ways of using graphs to describe data, followed by ways of summarizing the data numerically.

2.1 Different Types of Data

Life would be uninteresting if everyone looked the same, ate the same food, and had the same thoughts. Fortunately, there is **variability** everywhere, and statistical methods provide ways to measure and understand variability. For some characteristics in a study, we often see variation among the subjects. For example, there's variability among your classmates in weight, major, GPA, favorite sport, and religious affiliation. Other characteristics may vary both by subject and across time. For instance, the amount of time spent studying in a day can vary both by student and by day.

Variables

Variables are the characteristics observed in a study. In the previous paragraph, weight and major are two variables we might want to study.

Variable

A **variable** is any characteristic observed in a study.

The term *variable* highlights that data values *vary*. For example, to investigate whether global warming has occurred where you live, you could gather data on the high temperature each day over the past century at the nearest weather station. The variable is the high temperature. Examples of other variables for each day are the low temperature, whether it rained that day, and the number of centimeters of precipitation.

Variables Can Be Quantitative (Numerical) or Categorical (in Categories)

The data values that we observe for a variable are called **observations**. Each observation can be a **number** such as the number of centimeters of precipitation in a day. Or each observation belongs to a **category**, such as "yes" or "no" for whether it rained.

In Words

Consider your classmates. From person to person, there is variability in age, GPA, major, and whether he or she is dating someone. These are **variables.** Age and GPA are **quantitative**, as their values are numerical. Major and dating status are **categorical** since their values are categories, such as (psychology, business, history) for major and (yes, no) for dating status.

Categorical and Quantitative Variables

A variable is called **categorical** if each observation belongs to one of a set of categories. A variable is called **quantitative** if observations on it take numerical values that represent different magnitudes of the variable.

The daily high temperature and the amount of precipitation are quantitative variables. Other examples of quantitative variables are age, number of siblings, annual income, and number of years of education completed. For human subjects, examples of categorical variables include gender (with categories male and female), religious affiliation (with categories such as Catholic, Jewish, Muslim, Protestant, Other, None), type of residence (house, condominium, apartment, dormitory, other), and belief in life after death (yes, no).

In the definition of a quantitative variable, why do we say that numerical values must *represent different magnitudes*? Quantitative variables measure "how much" of something (that is, *quantity* or *magnitude*). With quantitative variables, we can find arithmetic summaries such as averages. However, some numerical variables, such as area codes, are not considered quantitative variables because they do not vary in quantity. For example, a bank might be interested in the average of the sizes of loans made to its customers, but an "average" area code does not make sense.

Graphs and numerical summaries describe the main features of a variable:

- For **quantitative** variables, key features to describe are the **center** and the **variability** (sometimes referred to as "**spread**") of the data. For instance, what's a typical annual amount of precipitation? Is there much variation from year to year?
- For **categorical** variables, a key feature to describe is the relative number of observations in the various categories. For example, what percentage of students at a certain college are Democrats?

Quantitative Variables Are Discrete or Continuous

For a quantitative variable, each value it can take is a number, and we classify quantitative variables as being either **discrete** or **continuous**.

In Words

A **discrete** variable is usually a count ("the number of ..."). A **continuous** variable has a continuum of infinitely many possible values (such as time, distance, or physical measurements such as weight and height).

Discrete and Continuous Variables

A quantitative variable is **discrete** if its possible values form a set of separate numbers, such as 0, 1, 2, 3,.... A quantitative variable is **continuous** if its possible values form an interval.

Examples of discrete variables are the number of pets in a household, the number of children in a family, and the number of foreign languages in which a person is fluent. Any variable phrased as "the number of..." is discrete. The possible values are separate numbers such as {0, 1, 2, 3, 4,...}. The outcome of the variable is a count. *Any variable with a finite number of possible values is discrete.*

Examples of continuous variables are height, weight, age, and the amount of time it takes to complete an assignment. The collection of all the possible values of a continuous variable does not consist of a set of separate numbers, but rather an infinite region of values. The amount of time needed to complete an assignment, for example, could take the value 2.496631...hours. *Continuous variables have an infinite continuum of possible values.*

In Practice Data Analysis Depends on Type of Variable

Why do we care whether a variable is *quantitative* or *categorical*, or whether a quantitative variable is *discrete* or *continuous*? We'll see that the method used to analyze a data set will depend on the type of variable the data represent.

Frequency Tables

The first step in numerically summarizing data about a variable is to look at the possible values and count how often each occurs. For a categorical variable, each observation falls in one of the categories. The category with the highest frequency is called the **modal category**. For a quantitative variable, the numerical value that occurs most frequently is the **mode**. We can use proportions or percentages to summarize the numbers of observations in the various categories.

Proportion and Percentage (Relative Frequencies)

The **proportion** of the observations that fall in a certain category is the frequency (count) of observations in that category divided by the total number of observations. The **percentage** is the proportion multiplied by 100. Proportions and percentages are also called **relative frequencies** and serve as a way to summarize the measurements in categories of a categorical variable.

Frequency Table

A **frequency table** is a listing of possible values for a variable, together with the number of observations for each value.

Example 2

Categorical variable

Shark Attacks

Picture the Scenario

In the United States, shark attacks most commonly occur in Florida. There were 268 reported shark attacks in Florida between 2000 and 2010. Which regions of the world besides Florida have experienced shark attacks, and

how many occur in each region? The International Shark Attack File (ISAF) provides data on shark attacks. Table 2.1 classifies 715 shark attacks reported from 2000 through 2010, listing countries where attacks occurred as well as U.S. states where shark attacks were most common. The number of reported shark attacks for a particular region is a **frequency**, or count. The proportion is found by dividing the frequency by the total count of 715. The percentage equals the proportion multiplied by 100.

Questions to Explore

a. What is the variable? Is it categorical or quantitative?

b. Show how to find the proportion and percentage for Florida, reported in Table 2.1.

c. Identify the modal category for these data.

Table 2.1 Frequency of Shark Attacks in Various Regions for 2000–2010*

Region	Frequency	Proportion	Percentage
Florida	268	0.375	37.5
Hawaii	41	0.057	5.7
California	32	0.045	4.5
South Carolina	32	0.045	4.5
North Carolina	28	0.039	3.9
Australia	120	0.168	16.8
South Africa	41	0.057	5.7
Brazil	20	0.028	2.8
Bahamas	12	0.017	1.7
Other	121	0.169	16.9
Total	**715**	**1.000**	**100.0**

*Data current as of January 2011.
Source: Data from www.flmnh.ufl.edu/fish/sharks/statistics/statsw.htm.

Think It Through

a. Each observation (a shark attack) identifies a region where the attack occurred. Region is the variable. It is categorical, with categories shown in the first column of Table 2.1.

b. Of the 715 reported shark attacks, 268 were in Florida. This is a proportion of nearly 4 out of every 10 attacks. The percentage is $100(0.375) = 37.5\%$.

c. For the regions listed, the greatest number of attacks was in Florida, with 37.5% of the reported worldwide attacks. Florida is the modal category because it has the greatest frequency of attacks.

Insight

Don't mistake the frequencies as values of a variable. They are merely a summary of how many times the observation (a shark attack) occurred in each category (region). The variable summarized here is the region in which it took place. In tables that summarize frequencies, *the total proportion is 1.0 and the total percentage is 100%*. In practice, the separate values may sum to a slightly different number (such as 99.9% or 100.1%) because of rounding.

Try Exercise 2.8

Frequency Table: Daily TV Watching

No. Hours	Frequency	Percent
0–1	360	27.2
2–3	569	43.0
4–5	247	18.7
6–7	69	5.2
8 or more	79	6.0
Total	**1,324**	**100.1**

Source: Data from CSM, UC Berkeley.

In example 2, Table 2.1 was a frequency table for a categorical variable. For a quantitative variable, a frequency table usually divides the possible values into a set of intervals and displays the number of observations in each interval. For example, in the next section, we'll analyze General Social Survey data on responses to the question, "On an average day, about how many hours do you personally watch TV?" This was measured as a discrete variable, with values 0, 1, 2, and so on. The table shown in the margin is one way of constructing a frequency table for the 1324 subjects who responded. A larger table could list each distinct value for the variable.

2.1 Practicing the Basics

2.1 Categorical/quantitative difference

a. Explain the difference between categorical and quantitative variables.

b. Give an example of each.

2.2 U.S. married-couple households According to a recent Current Population Survey of U.S. married-couple households, 13% are traditional (with children and with only the husband in the labor force), 31% are dual-income with children, 25% are dual-income with no children, and 31% are other (such as older married couples whose children no longer reside in the household). Is the variable "household type" categorical or quantitative? Explain.

2.3 Identify the variable type Identify each of the following variables as categorical or quantitative.

a. Number of pets in family

b. County of residence

c. Choice of auto to buy (domestic or import)

d. Distance (in kilometers) of commute to work

2.4 Categorical or quantitative? Identify each of the following variables as either categorical or quantitative.

a. Choice of diet (vegetarian, nonvegetarian)

b. Time spent in previous month attending a place of religious worship

c. Ownership of a personal computer (yes, no)

d. Number of people you have known who have been elected to a political office

2.5 Discrete/continuous

a. Explain the difference between a discrete variable and a continuous variable.

b. Give an example of each type.

2.6 Discrete or continuous? Identify each of the following variables as continuous or discrete.

a. The length of time to run a marathon

b. The number of people in line at a box office to purchase theater tickets

c. The weight of a dog

d. The number of people you have dated in the past month

2.7 Discrete or continuous 2 Repeat the previous exercise for the following:

a. The total playing time of a CD

b. The number of courses for which a student has received credit

c. The amount of money in your pocket (*Hint:* You could regard a number such as $12.75 as 1275 in terms of "the number of cents.")

d. The distance between where you live and your statistics classroom, when you measure it precisely with values such as 0.5 miles, 2.4 miles, 5.38 miles.

2.8 Number of children In the 2008 General Social Survey, 2020 respondents answered the question, "How many children have you ever had?" The results were
TRY

No. children	0	1	2	3	4	5	6	7	8+	Total
Count	521	323	524	344	160	77	30	19	22	2020

a. Is the variable, number of children, categorical or quantitative?

b. Is the variable, number of children, discrete or continuous?

c. Add proportions and percentages to this frequency table.

d. Which response is the mode?

2.2 Graphical Summaries of Data

Looking at a graph often gives you more of a feel for a data set than looking at the raw data or a frequency table. In this section, we'll learn about graphs for categorical variables and then graphs for quantitative variables. We'll find out what we should look for in a graph to help us understand the data better.

Graphs for Categorical Variables

The two primary graphical displays for summarizing a categorical variable are the **pie chart** and the **bar graph**.

- A **pie chart** is a circle having a "slice of the pie" for each category. The size of a slice corresponds to the percentage of observations in the category.
- A **bar graph** displays a vertical bar for each category. The height of the bar is the percentage of observations in the category. Typically, the vertical bars for each category are apart, not side by side.

Example 3

Categorical graphical summaries

Renewable Electricity

Picture the Scenario

In the United States, most electricity comes from coal and natural gas. As the demand for energy increases, the availability of these nonrenewable sources diminishes. Each source is also a significant contributor to the emission of carbon dioxide, a factor which may lead to climate change. Consequently, much attention has been focused on developing alternative sources of energy. *Renewable* sources such as hydropower, solar power, and wind power can generate electricity without depleting natural resource stockpiles and without the potential damage to the environment associated with fossil fuels. Table 2.2 displays a percentage breakdown by source of all electricity generated in the United States and Canada in 2009.

Table 2.2 Sources of Electricity in the United States and Canada, 2009

Source	U.S. Percentage	Canada Percentage
Coal	45	17
Natural gas	23	7
Nuclear	20	15
Hydropower	7	59
Other Renewables	4	1
Petroleum	1	1
Total	100	100

Source: Data from U.S. Energy Information Administration, Canadian Electricity Association.

Questions to Explore

a. Display the U.S. information from Table 2.2 in a pie chart and a bar graph.

b. What percentage of electricity was generated from renewable sources (hydropower and other renewables) in the United States? In Canada?

Think It Through

a. Source of electricity is a categorical variable. Figure 2.1 is a pie chart of the U.S. data from Table 2.2. Sources of electricity with larger percentages have larger slices of the pie. The percentages are also included in the labels for each slice of the pie. Figure 2.2 shows the bar graph. The categories with larger percentages have higher bars. The scale for the percentages is shown on the vertical axis. The width is the same for each bar.

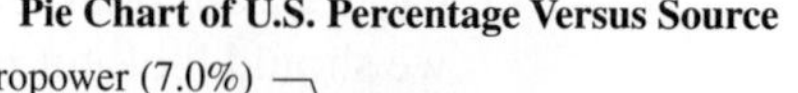

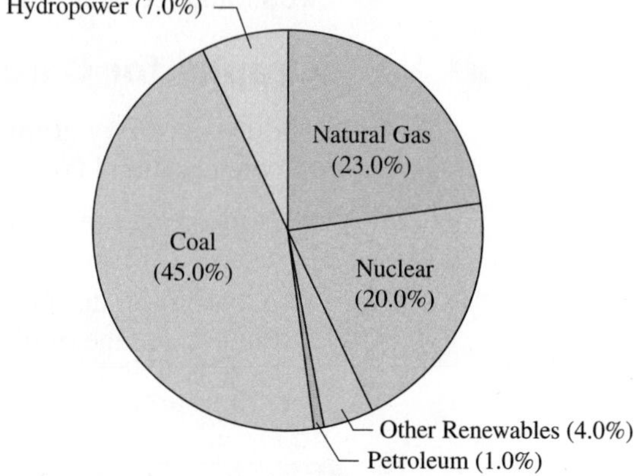

▲ **Figure 2.1 Pie Chart of Electricity Sources in the United States.** The label for each slice of the pie gives the category and the percentage of electricity generated from that source. The slice that represents the percentage generated by coal is 45% of the total area of the pie. **Question** Why is it beneficial to label the pie wedges with the percent? (Hint: Is it always clear which of two slices is larger and what percent a slice represents?)

Chart of U.S. Percentage Versus Source

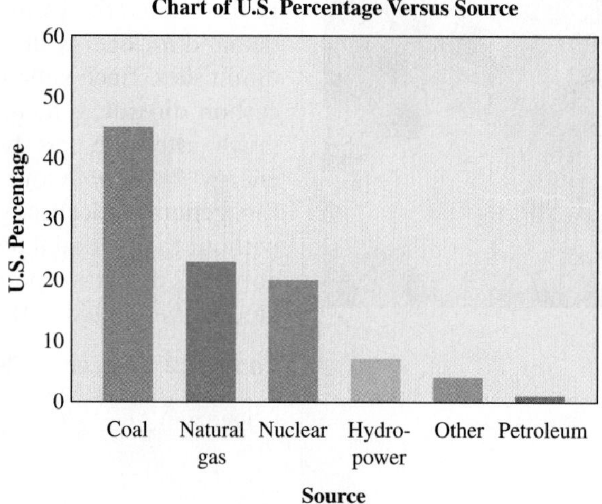

▲ **Figure 2.2 Bar Graph of Electricity Sources in the United States.** The bars are ordered from largest to smallest based on the percentage use.

b. In the United States, the renewable sources (hydropower and other renewables) provide only about 11% (7 + 4 = 11) of the electricity. It is easy to see from either the pie chart or bar graph that the top source (the modal category) is coal. By contrast, Table 2.2 tells us that renewable sources provide about 60% of the electricity in Canada and the modal category is hydropower.

Insight

The pie chart and bar graph are both simple to construct using software. The bar graph is generally easier to read and more flexible. With a pie chart, when two slices are about the same size, it's often unclear which value is larger. This distinction is clearer when comparing heights of bars in a bar graph. We'll see that the bar graph can easily summarize how results compare for different groups (for instance, if we wanted to compare the United States and Canada for Table 2.2). Also, the bar graph is a better visual display when there are many categories.

Try Exercise 2.9

The bar graph in Figure 2.2 displays the categories in decreasing order of the category percentages. This order makes it easy to separate the categories with high percentages visually. In some applications, it is more natural to display them according to their alphabetical order or some other criterion. For instance, if the categories have a natural order, such as summarizing the percentages of grades (A, B, C, D, F) for students in a course, we'd use that order in listing the categories on the graph.

Pareto Charts

Figure 2.2 is a special type of bar graph called a **Pareto chart**. Named after Italian economist Vilfredo Pareto (1848–1923) who advocated its use, it is a bar graph with categories ordered by their frequency, from the tallest bar to the shortest bar. The Pareto chart is often used in business applications to identify the most common outcomes, such as identifying products with the highest sales or identifying the most common types of complaints that a customer service center receives. The chart helps to portray the **Pareto principle**, which states that a small subset of categories often contains most of the observations. For example, Figure 2.2 shows that three categories (coal, nuclear, and natural gas) were responsible for about 88% of U.S. electricity sources.

Graphs for Quantitative Variables

Now let's explore how to summarize *quantitative* variables graphically. We'll look at three types of displays—the dot plot, stem-and-leaf plot, and histogram—and illustrate the graphs by analyzing data for a common daily food (cereal).

Dot Plots

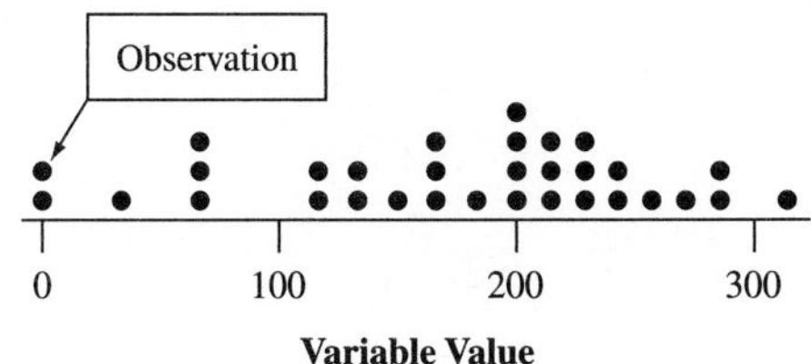

A **dot plot** shows a dot for each observation, placed just above the value on the number line for that observation. To construct a dot plot,

- Draw a horizontal line. Label it with the name of the variable and mark regular values of the variable on it.
- For each observation, place a dot above its value on the number line.

Example 4

Health Value of Cereals

Picture the Scenario

Let's investigate the amount of sugar and salt (sodium) in breakfast cereals. Table 2.3 lists 20 popular cereals and the amounts of sodium and sugar contained in a single serving. The sodium and sugar amounts are both quantitative variables. The variables are continuous because they measure amounts that can take any positive real number value. In this table, the amounts are rounded to the nearest number of grams for sugar and milligrams for sodium.

Questions to Explore

a. Construct a dot plot for the sodium values of the 20 breakfast cereals. (We'll consider sugar amounts in the exercises.)

b. What does the dot plot tell us about the data?

Did You Know?

Nutritionists recommend that daily consumption should not exceed 2400 milligrams (mg) for sodium and 50 grams for sugar (on a 2000-calorie-a-day diet). ◀

Caution

Continuous data is often recorded to the nearest whole number. Although the data may now appear as discrete, the data are still analyzed and interpreted as continuous data. ◀

Table 2.3 Sodium and Sugar Amounts in 20 Breakfast Cereals

The amounts refer to one National Labeling and Education Act (NLEA) serving. A third variable, Type, classifies the cereal as being popular for adults (Type A) or children (Type C).

Cereal	Sodium (mg)	Sugar (g)	Type
Frosted Mini Wheats	0	11	A
Raisin Bran	340	18	A
All Bran	70	5	A
Apple Jacks	140	14	C
Cap'n Crunch	200	12	C
Cheerios	180	1	C
Cinnamon Toast Crunch	210	10	C
Crackling Oat Bran	150	16	A
Fiber One	100	0	A
Frosted Flakes	130	12	C
Froot Loops	140	14	C
Honey Bunches of Oats	180	7	A
Honey Nut Cheerios	190	9	C
Life	160	6	C
Rice Krispies	290	3	C
Honey Smacks	50	15	A
Special K	220	4	A
Wheaties	180	4	A
Corn Flakes	200	3	A
Honeycomb	210	11	C

Source: www.weightchart.com/nutrition.

Think It Through

a. Figure 2.3 shows a dot plot. Each cereal sodium value is represented with a dot above the number line. For instance, the labeled dot above 0 represents the sodium value of 0 mg for Frosted Mini Wheats.

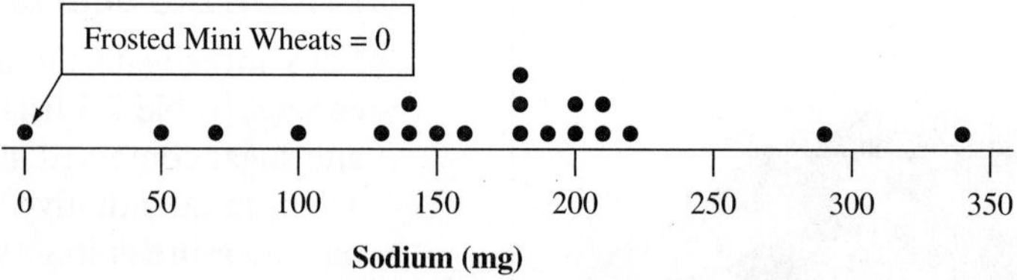

▲ **Figure 2.3 Dot Plot for Sodium Content of 20 Breakfast Cereals.** The sodium value for each cereal is represented with a dot above the number line. **Question** What does it mean when more than one dot appears above a value?

b. The dot plot gives us an overview of all the data. We see clearly that the sodium values fall between 0 and 340 mg, with most cereals falling between 125 and 250 mg.

Insight

The dot plot displays the individual observations. The number of dots above a value on the number line represents the frequency of occurrence of that value. From a dot plot, we can reconstruct (at least approximately) all the data in the sample.

Try Exercises 2.14 and 2.16, part b

Stem-and-Leaf Plots

Another type of graph, called a **stem-and-leaf plot**, is similar to the dot plot in that it displays individual observations.

- Each observation is represented by a **stem** and a **leaf**. Usually the stem consists of all the digits except for the final one, which is the leaf.
- Sort the data in order from smallest to largest. Place the stems in a column, starting with the smallest. Place a vertical line to their right. On the right side of the vertical line, indicate each leaf (final digit) that has a particular stem. List the leaves in increasing order.

Stems	Leaves
7	69
8	00125699
9	12446

Observation = 92, in a sample of 15 test scores

Stem-and-leaf plot

Example 5

Health Value of Cereals

Picture the Scenario

Let's reexamine the sodium values for the 20 breakfast cereals, shown again in the margin of the next page.

Questions to Explore

a. Construct a stem-and-leaf plot of the 20 sodium values.

b. How does the stem-and-leaf plot compare to the dot plot?

Think It Through

a. In the stem-and-leaf plot, we'll let the final digit of a sodium value form the leaf and the other digits form the stem. For instance, the sodium value for Honey Smacks is 50. The stem is 5, and the leaf is 0. Each stem is placed to the left of a vertical bar. Each leaf is placed to the right of the bar. Figure 2.4 shows the plot. Notice that a leaf has only one digit, but a stem can have one or more digits.

The Honey Smacks observation of 50 is labeled on the graph. Two observations have a stem of 20 and a leaf of 0. These are the sodium values of 200 for Cap'n Crunch and Corn Flakes.

b. The stem-and-leaf plot looks like the dot plot turned on its side, with the leaves taking the place of the dots. Often, with a stem-and-leaf plot, it is easier to read the actual observation value. In summary, we generally get the same information from a stem-and-leaf plot as from a dot plot.

Cereal	Sodium
Frosted Mini Wheats	0
Raisin Bran	340
All Bran	70
Apple Jacks	140
Cap'n Crunch	200
Cheerios	180
Cinnamon Toast Crunch	210
Crackling Oat Bran	150
Fiber One	100
Frosted Flakes	130
Froot Loops	140
Honey Bunches of Oats	180
Honey Nut Cheerios	190
Life	160
Rice Krispies	290
Honey Smacks	50
Special K	220
Wheaties	180
Corn Flakes	200
Honeycomb	210

Insight

A stem is shown for each possible value between the minimum and maximum even if no observation occurs for that stem. A stem has no leaf if there is no observation at that value. These are the values at which the dot plot has no dots.

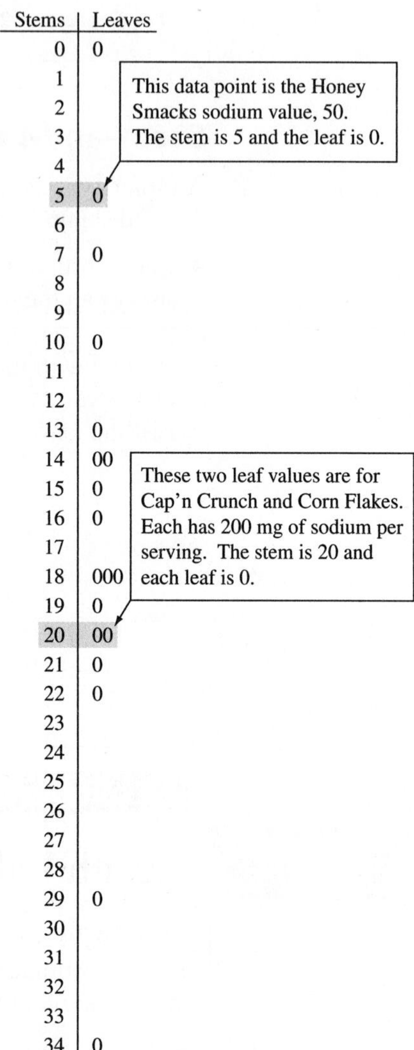

▲ **Figure 2.4 Stem-and-Leaf Plot for Cereal Sodium Values.** The final digit of a sodium value forms the leaf, and the other digits form the stem. **Question** Why do some stems not have a leaf?

Try Exercise 2.15

Truncated Data	
Frosted Mini Wheats	0
Raisin Bran	34
All Bran	7
Apple Jacks	14
Cap'n Crunch	20
Cheerios	18
Cinnamon Toast Crunch	21
Crackling Oat Bran	15
Fiber One	10
Frosted Flakes	13
Froot Loops	14
Honey Bunches of Oats	18
Honey Nut Cheerios	19
Life	16
Rice Krispies	29
Honey Smacks	5
Special K	22
Wheaties	18
Corn Flakes	20
Honeycomb	21

To make a stem-and-leaf plot more compact, we can **truncate** these data values: Cut off the final digit (it's not necessary to round it), as shown in the margin, and plot the data as 0, 34, 7, 14, 20, and so on, instead of 0, 340, 70, 140, 200,.... Arranging the leaves in increasing order on each line, we then get the stem-and-leaf plot

0	057
1	0344568889
2	001129
3	4

This is a bit *too* compact, as it does not portray where the data fall as clearly as Figure 2.4 or the dot plot. We could instead list each stem twice, putting leaves from 0 to 4 on the first stem and from 5 to 9 on the second stem. We then get

0	0
0	57
1	0344
1	568889
2	00112
2	9
3	4

Like the dot plot, this gives us the sense that most sodium values are relatively high, with a couple of cereals being considerably lower than the others.

Histograms

With a dot plot or a stem-and-leaf plot, it's easy to reconstruct the original data set because the plot shows the individual observations. This becomes unwieldy for large data sets. A more versatile way to graph the data, useful for very large data sets, uses bars to summarize frequencies of outcomes.

Histogram

A **histogram** is a graph that uses bars to portray the frequencies or the relative frequencies of the possible outcomes for a quantitative variable.

Example 6

Histogram

TV Watching

Picture the Scenario

The 2004 General Social Survey asked, "On an average day, about how many hours do you personally watch television?" Figure 2.5 shows the histogram of the 899 responses.

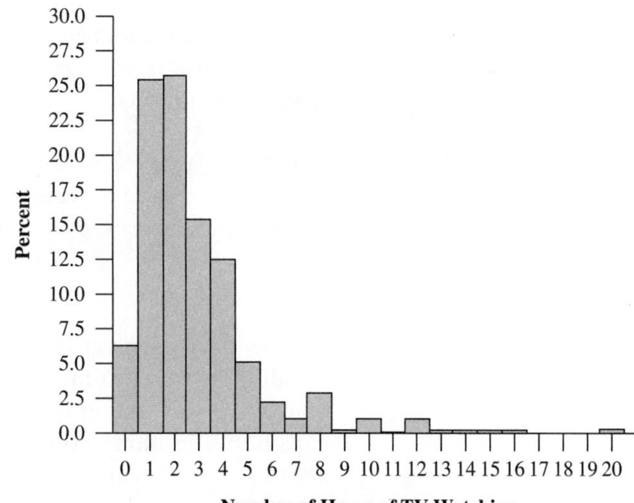

▲ **Figure 2.5** Histogram of GSS Responses about Number of Hours Spent Watching TV on an Average Day. *Source:* Data from CSM, UC Berkeley.

Questions to Explore

a. What was the most common outcome?

b. What percentage of people reported watching TV no more than 2 hours per day?

Think It Through

a. The most common outcome (the mode) is the value with the highest bar. This is 2 hours of TV watching.

b. To find the percentage for "no more than 2 hours per day," we need to look at the percentages for 0, 1, and 2 hours per day. They seem to be about 6, 25, and 26. Adding these percentages together tells us that about 57% of the respondents reported watching no more than 2 hours of TV per day.

Insight

In theory, TV watching is a continuous variable. However, the possible responses subjects were able to make here were 0, 1, 2,..., so it was measured as a discrete variable. Figure 2.5 is an example of a histogram of a discrete variable. Note that since the variable is treated as discrete, the histogram in Figure 2.5 could have been constructed with the bars apart instead of beside each other.

Try Exercise 2.25, parts a and b

Caution

The term *histogram* is used for a graph with bars representing a quantitative variable. The term *bar graph* is used for a graph with bars representing a categorical variable. ◀

For a discrete variable, a histogram usually has a separate bar for each possible value. For a continuous variable, you need to divide the interval of possible values into smaller intervals formed with values grouped together. You can also do this when a discrete variable has a large number of possible values, such as a score on an exam. In such cases, you can form a frequency table for the intervals and graph the frequencies or percentages for those intervals. Typically, the intervals should each have the same width.

SUMMARY: Steps for Constructing a Histogram

- Divide the range of the data into intervals of equal width. For a discrete variable with few values, use the actual possible values.
- Count the number of observations (the frequency) in each interval, forming a frequency table.
- On the horizontal axis, label the values or the endpoints of the intervals. Draw a bar over each value or interval with height equal to its frequency (or percentage), values of which are marked on the vertical axis.

How do you select the intervals? If you use too few intervals, the graph is usually too crude. It may consist mainly of a couple of tall bars. If you use too many intervals, the graph may be irregular, with many very short bars and/or gaps between bars. You can then lose information about the shape. Usually about 5–10 intervals are adequate, with perhaps additional intervals when the sample size is quite large. There is no one right way to select the intervals. Software can select them for you, find the counts and percentages, and construct the histogram.

Histogram

Cereal	Sodium
Frosted Mini Wheats	0
Raisin Bran	340
All Bran	70
Apple Jacks	140
Cap'n Crunch	200
Cheerios	180
Cinnamon Toast Crunch	210
Crackling Oat Bran	150
Fiber One	100
Frosted Flakes	130
Froot Loops	140
Honey Bunches of Oats	180
Honey Nut Cheerios	190
Life	160
Rice Krispies	290
Honey Smacks	50
Special K	220
Wheaties	180
Corn Flakes	200
Honeycomb	210

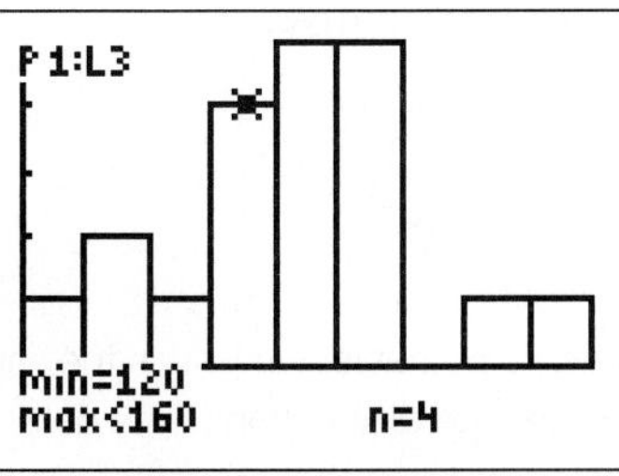

TI-83+/84 output

Example 7

Health Value of Cereals

Picture the Scenario

Let's reexamine the sodium values of the 20 breakfast cereals. Those values are shown again in the margin.

Questions to Explore

a. Construct a frequency table.

b. Construct a histogram.

c. What information does the histogram not show that you can get from a dot plot or a stem-and-leaf plot?

Think It Through

a. To construct a frequency table, we divide the possible sodium values into separate intervals and count the number of cereals in each. The sodium values range from 0 to 340. We created Table 2.4 using eight intervals, each with a width of 40. With the interval labels shown in the table, for a continuous variable 0 to 39 actually represents 0 to 39.999999…, that is, 0 up to every number *below* 40. So, 0 to 39 is then shorthand for "0 to less than 40." Sometimes you will see the intervals written as 0 to 40, 40 to 80, 80 to 120, and so on. However, for an observation that falls at an interval endpoint, then it's not clear in which interval it goes. When reading the histogram, we generally use a left endpoint convention where if an observation is an endpoint, it belongs to the interval with the observation as the left endpoint.

Table 2.4 Frequency Table for Sodium in 20 Breakfast Cereals

The table summarizes the sodium values using eight intervals and lists the number of observations in each, as well as the proportions and percentages.

Interval	Frequency	Proportion	Percentage
0 to 39	1	0.05	5%
40 to 79	2	0.10	10%
80 to 119	1	0.05	5%
120 to 159	4	0.20	20%
160 to 199	5	0.25	25%
200 to 239	5	0.25	25%
240 to 279	0	0.00	0%
280 to 319	1	0.05	5%
320 to 359	1	0.05	5%

b. Figure 2.6 shows the histogram for this frequency table. A bar is drawn over each interval of values, with the height of each bar equal to its corresponding frequency. The histogram created using the TI-83+/84 calculator is in the margin.

c. The histogram does not show the actual numerical values. For instance, we know that one observation falls below 40, but we do not know its actual value. In summary, with a histogram, we may lose the actual

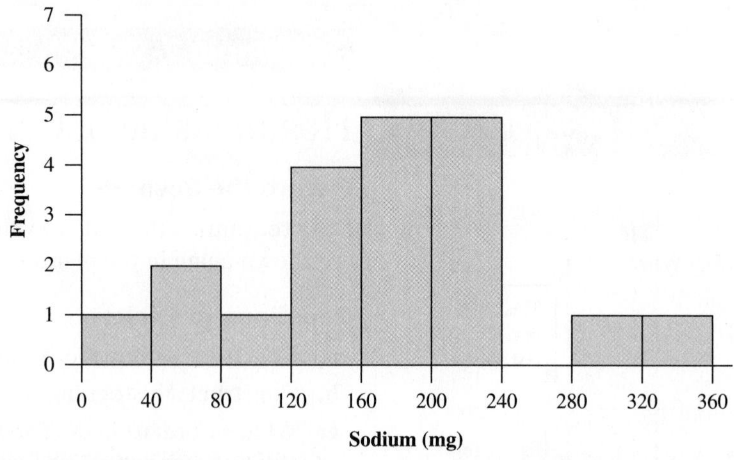

▲ **Figure 2.6 Histogram of Breakfast Cereal Sodium Values.** The rectangular bar over an interval has height equal to the number of observations in the interval.

numerical values of individual observations, unlike with a dot plot or a stem-and-leaf plot.

Insight

The histogram in Figure 2.6 labels the vertical axis with the frequencies. If it were labeled with proportions or percentages, the heights of the bars would be the same, just the vertical axis labels would change. The histogram using proportions would be identical in appearance to the one using percentages except for the vertical axis labels, and we would have the same graphical information about the sodium values. Any of these histograms would suggest that the cereal we choose to eat *does* matter if we wish to monitor our sodium intake.

Try Exercise 2.21

Choosing a Graph Type

We've now studied three graphs for quantitative variables—the dot plot, stem-and-leaf plot, and histogram. How do we decide which to use? Here are some guidelines:

- The dot plot and stem-and-leaf plot are more useful for small data sets, since they portray each individual observation. With large data sets, histograms usually work better, being more compact than the other displays.
- More flexibility is possible in defining the intervals with a histogram than in defining the stems with a stem-and-leaf plot.
- Data values are retained with the stem-and-leaf plot and dot plot but not with the histogram.

Unless the data set is small (say, about 50 or fewer observations), the histogram is usually preferred. When in doubt, create a histogram and the dot plot or stem-and-leaf plot, and then use whichever is clearer and more informative.

In Practice Technology Can Construct Graphical Displays

Graphical displays are easily constructed using statistical software or graphing calculators. They choose the intervals for you when they construct histograms. In practice, you won't have to draw graphs yourself. But it is important to understand *how* they are drawn and *how to interpret* them.

The Shape of a Distribution

Distribution

A graph for a data set describes the **distribution** of the data, that is, the values the variable takes and the frequency of occurrence of each value. The distribution of the data (or so-called data distribution) can also be described by a frequency table.

Here are some things to look for in a data distribution of a quantitative variable, whether portrayed by a frequency table or by a graph such as a histogram, stem-and-leaf plot, or dot plot:

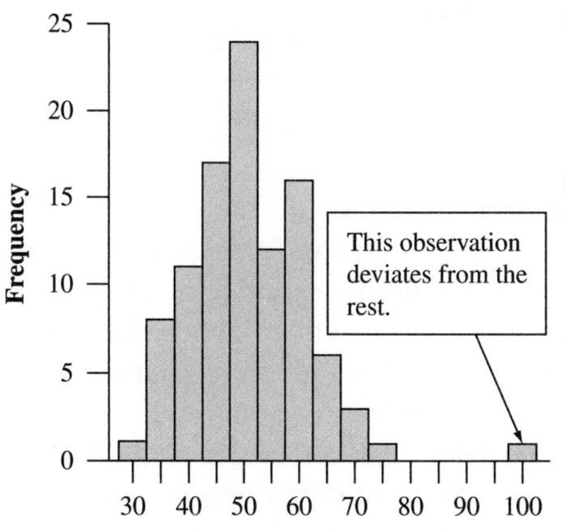

- Look for the **overall pattern**. Do the data cluster together, or is there a **gap** such that one or more observations noticeably deviate from the rest, as in the histogram in the margin? We'll discuss such "outlier" observations later in the chapter.
- Do the data have a single mound? A distribution of such data is called **unimodal**. The highest point is at the **mode**. A distribution with *two* distinct mounds is called **bimodal**. A bimodal distribution can result, for example, when a population is polarized on a controversial issue. Suppose each subject is presented with ten scenarios in which a person found guilty of murder may be given the death penalty. If we count the number of those scenarios in which subjects feel the death penalty would be just, many responses would be close to 0 (for subjects who oppose the death penalty generally) and many would be close to 10 (for subjects who think it's always or usually warranted for murder).

Did You Know?

At the very front of the text, you will find a guide showing how the art in the book aids learning. Check out this guide to learn how color is used in the graphs. ◀

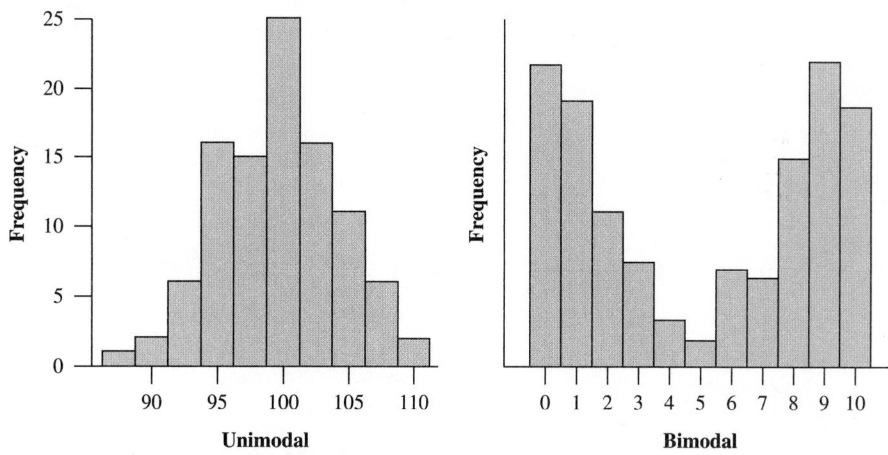

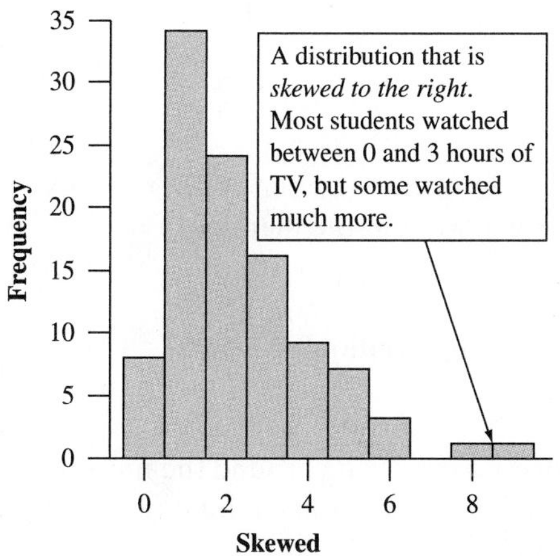

- The **shape** of the distribution is often described as **symmetric** or **skewed.** A distribution is *symmetric* if the side of the distribution below a central value is a mirror image of the side above that central value. The distribution is *skewed* if one side of the distribution stretches out longer than the other side. For instance, the skewed distribution in the margin resulted from students being asked how many hours they watched TV on the previous day.

In picturing features such as skew and symmetry, it's common to use smooth curves as in Figure 2.7 to summarize the shape of a histogram. You can think of this as what can happen when you choose more and more intervals (making each interval narrower) and collect more data, so the histogram gets "smoother." The parts of the curve for the lowest values and for the highest values are called the **tails** of the distribution.

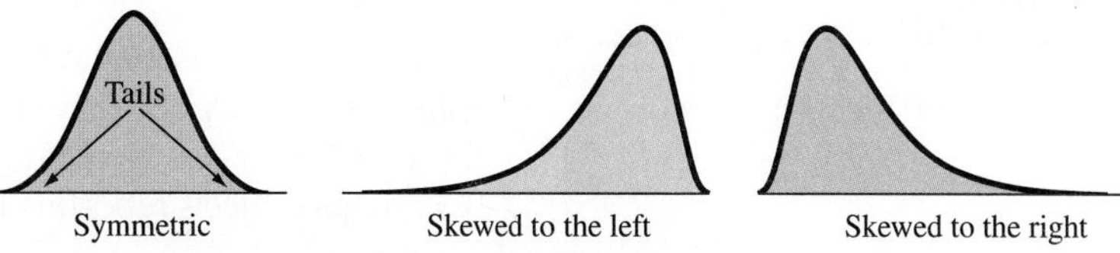

▲ **Figure 2.7 Curves for Data Distributions Illustrating Symmetry and Skew.**
Question What does the longer tail indicate about the direction of skew?

Skewed Distribution

To **skew** means to pull in one direction.
A distribution is **skewed to the left** if the left tail is longer than the right tail.
A distribution is **skewed to the right** if the right tail is longer than the left tail.

Identifying Skew

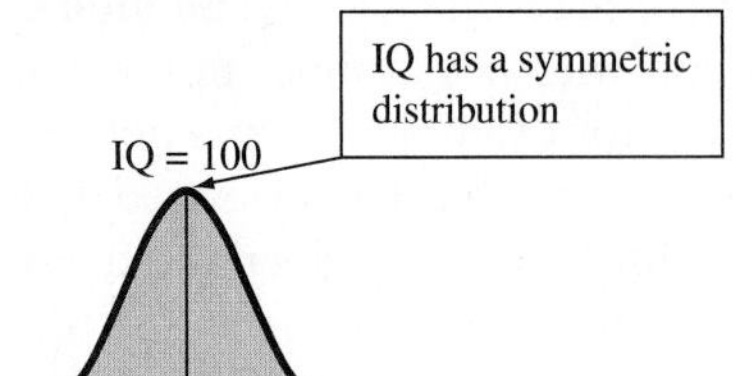

Let's consider some variables and think about what shape their distributions would have. How about IQ? Values cluster around 100 and tail off in a similar fashion in both directions. The appearance of the distribution on one side of 100 is roughly a mirror image of the other side, with tails of similar length. The distribution is approximately symmetric (see the green shaded graph in the margin).

How about life span for humans? Most people in advanced societies live to be at least about 60 years old, but some die at a very young age, so the distribution of life span would probably be skewed to the left (see the tan shaded graph below).

What shape would we expect for the distribution of annual incomes of adults? There would probably be a long right tail, with some people having incomes much higher than the overwhelming majority of people. This suggests that the distribution would be skewed to the right (see the purple shaded graph below).

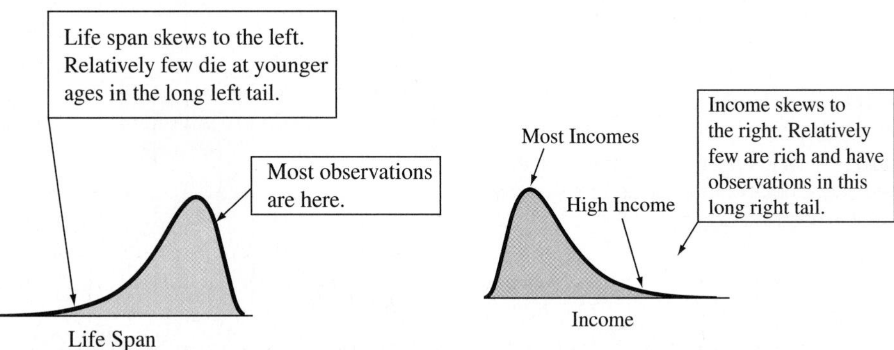

Shape of the distribution

Example 8

TV Watching

Picture the Scenario

In Example 6, we constructed a histogram of the number of hours of TV watching reported in the GSS. It is shown again in the margin.

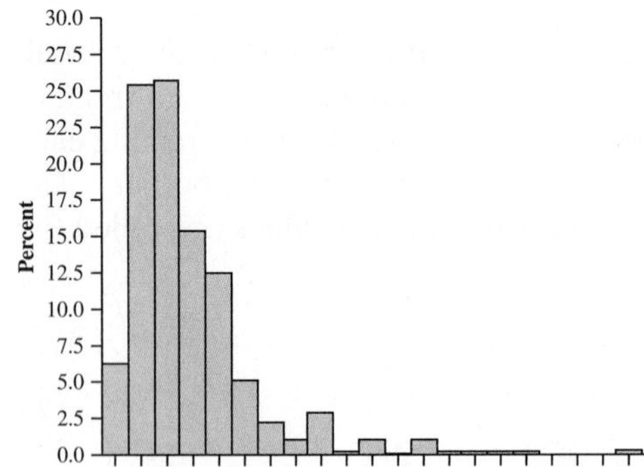

Question to Explore

How would you describe the shape of the distribution?

Think It Through

There appears to be a single mound of data clustering around the mode of 2. The distribution is unimodal. There also appears to be a long right tail, so the distribution is skewed to the right.

Insight

In surveys, the observation that a subject reports is not necessarily the true value. Often they either round or don't remember exactly and just guess. In this distribution, the percentage is quite a bit higher for 8 hours than for 7 or 9. Perhaps subjects reporting high values tend to pick even numbers.

Try Exercises 2.23 and 2.25, part c

Time Plots: Displaying Data over Time

For some variables, observations occur over time. Examples include the daily closing price of a stock and the population of a country measured every decade in a census. A data set collected over time is called a **time series**.

We can display time-series data graphically using a **time plot**. This charts each observation, on the vertical scale, against the time it was measured, on the horizontal scale. A common pattern to look for is a **trend** over time, indicating a tendency of the data to either rise or fall. To see a trend more clearly, it is beneficial to connect the data points in their time sequence.

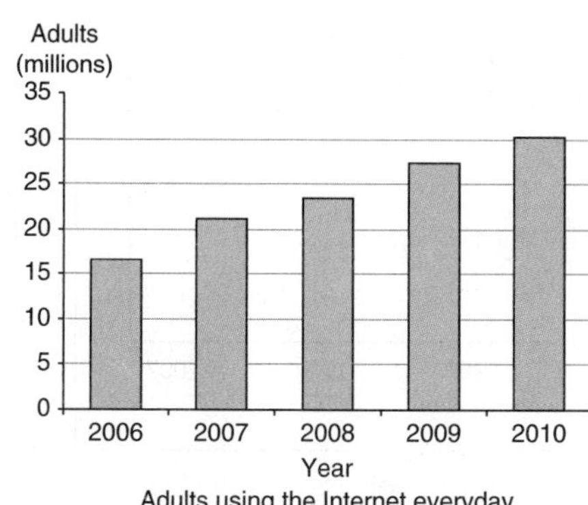
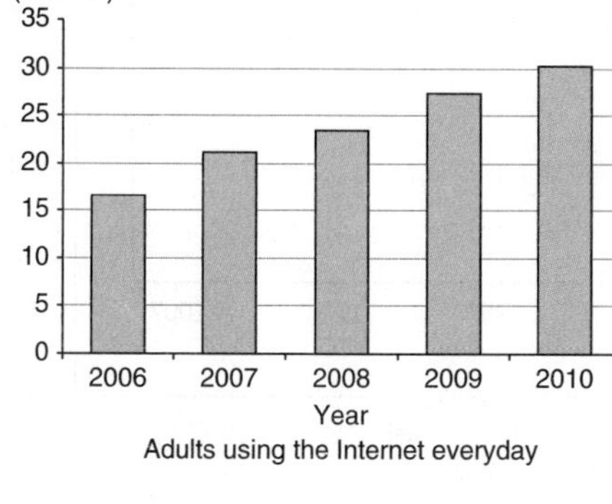

Adults using the Internet everyday

Another way time series data is displayed is with a type of bar graph. The figure in the margin is such a graph displaying the number of people in the United Kingdom between 2006 and 2010 who used the Internet[1] (in millions). In 2006, it is estimated 16.5 million were using the Internet. By 2010, nearly double (30 million) were doing so. There is a clear increasing trend over time. In practice, there's often not such a clear trend, as the next example using a time plot illustrates.

With a computer, it is also possible to make animated plots, unleashing time from the horizontal axis. The site www.gapminder.org/ has many examples and tools for dynamic time plots.

Example 9

Time plot

Warming Trend in New York City

Picture the Scenario

In a given year, the annual average temperature is the average of the daily average temperatures for that year. Let's analyze data on annual average temperature (in degrees Fahrenheit) in Central Park, New York City, from 1869 to 2010. This is a continuous, quantitative variable. The data are in the Central Park Yearly Temps data file on the text CD.

Question to Explore

What do we learn from a time plot of these annual average temperatures? Is there a trend toward warming in New York City?

Think It Through

Figure 2.8 shows a time plot, constructed using MINITAB software. The time plot constructed using the TI-83+/84 calculator is in the margin. The observations fluctuate considerably, but the figure does suggest an increasing trend in the annual average temperatures in New York City.

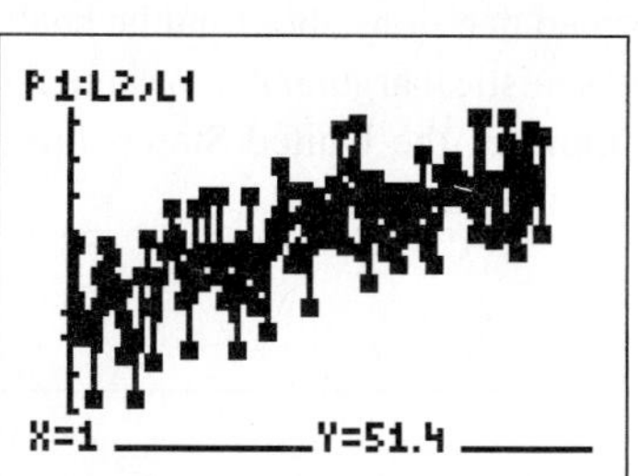

TI-83+/84 output

Insight

The short-term fluctuations in a time plot can mask an overall trend. It's possible to get a clearer picture by "smoothing" the data. This is beyond our scope here. MINITAB presents the option of smoothing the data to portray a general trend (*click* on *data view* and choose the Lowess option under Smoother). This is the smooth curve passing through the data points in Figure 2.8. This curve goes from a level of 51.4 degrees in 1869 to 56.7 degrees in 2010.

[1]*Source:* www.statistics.gov.uk/cci/nugget.asp?id=8. Copyright © 2000–2011, Miniwatts Marketing Group. All rights reserved.

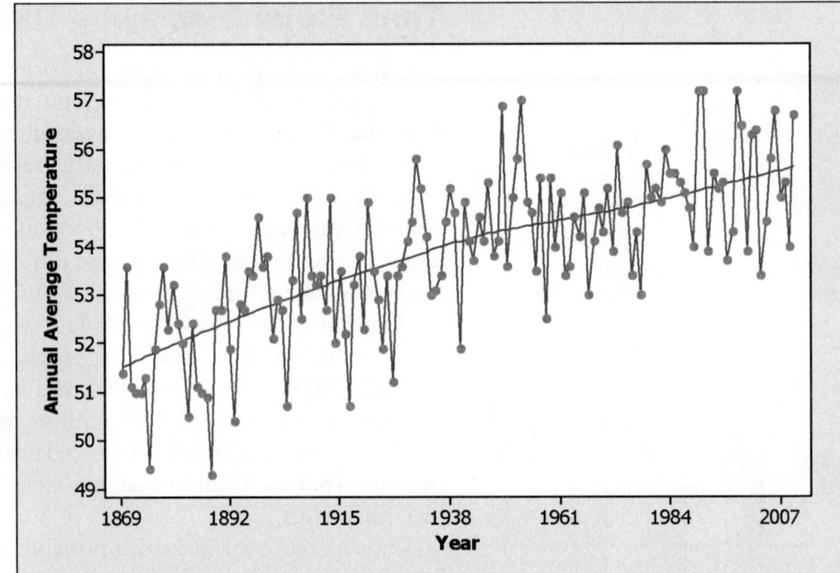

▲ **Figure 2.8** **MINITAB Output for a Time Plot of Central Park, New York City, Average Annual Temperatures.** The annual average temperatures are plotted against the year from 1869 to 2010. A smoothing curve is superimposed. **Question:** Are the annual average temperatures tending to increase, decrease, or stay the same over time?

The data reported in Figure 2.8 refer to one location in the United States. In a study of climate change, it would be important to explore other locations around the world to see if similar trends are evident.

Try Exercises 2.28 and 2.29

On the Shoulders of... Florence Nightingale

Graphical Displays Showing Deaths From Disease Versus Military Combat

During the Crimean War in 1854, the British nurse Florence Nightingale (1820–1910) gathered data on the number of soldiers who died from various causes. She prepared graphical displays such as time plots and pie charts for policy makers. The graphs showed that more soldiers were dying from contagious diseases than from war-related wounds. The plots were revolutionary for her time. They helped to promote her national cause of improving hospital conditions.

After implementing sanitary methods, Nightingale showed with time plots that the relative frequency of soldiers' deaths from contagious disease decreased sharply and no longer exceeded that of deaths from wounds.

Throughout the rest of her life, Nightingale promoted the use of data for making informed decisions about public health policy. For example, she used statistical arguments to campaign for improved medical conditions in the United States during the Civil War in the 1860s (Franklin, 2002).

2.2 Practicing the Basics

2.9 Federal spending on financial aid A 2011 Roper Center survey asked: "If you were making up the budget for the federal government this year (2011), would you increase spending, decrease spending, or keep spending the same for financial aid for college students?" Of those surveyed, 44% said to increase spending, 16% said to decrease spending, 37% said to keep spending the same, and 3% either had no opinion or refused to answer.

TRY

a. Sketch a bar chart to display the survey results.

b. Which is easier to sketch relatively accurately, a pie chart or a bar chart?

c. What is the advantage of using a graph to summarize the results instead of merely stating the percentages for each response?

2.10 What do alligators eat? The bar chart (constructed using MINITAB) is from a study[2] investigating the factors that influence alligators' choice of food. For 219 alligators captured in four Florida lakes, researchers classified the primary food choice (in volume) found in the alligator's stomach in one of the categories—fish, invertebrate (snails, insects, crayfish), reptile (turtles, baby alligators), bird, or other (amphibian, mammal, plants).

a. Is primary food choice categorical or quantitative?

b. Which is the modal category for primary food choice?

c. About what percentage of alligators had fish as the primary food choice?

d. This type of bar chart, with categories listed in order of frequency, has a special name. What is it?

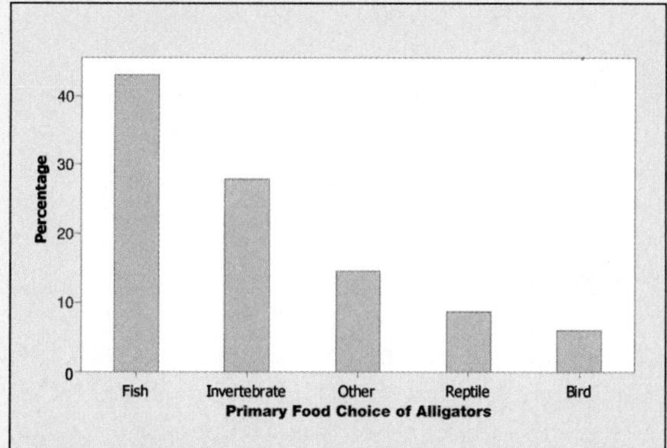

2.11 Weather stations The pie chart (constructed using EXCEL) shown portrays the regional distribution of weather stations in the United States.

a. Do the slices of the pie portray (i) variables or (ii) categories of a variable?

b. Identify what the two numbers mean that are shown for each slice of the pie.

c. Without inspecting the numbers, would it be easier to identify the modal category using this graph or using the corresponding bar graph? Why?

[2]Data courtesy of Clint Moore.

Regional Distribution of Weather Stations

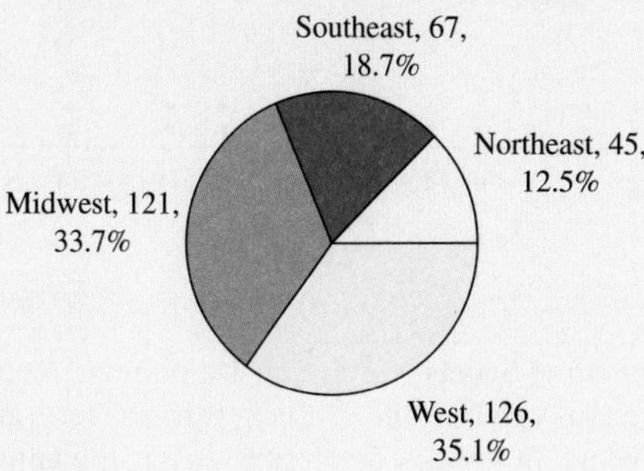

2.12 France is most popular holiday spot Which countries are most frequently visited by tourists from other countries? The table shows results according to *Travel and Leisure* magazine (2005).

a. Is country visited a categorical or a quantitative variable?

b. In creating a bar graph of these data, would it be most sensible to list the countries alphabetically or in the form of a Pareto chart? Explain.

c. Does either a dot plot or stem-and-leaf plot make sense for these data? Explain.

Most Visited Countries, 2005

Country	Number of Visits (millions)
France	77.0
China	53.4
Spain	51.8
United States	41.9
Italy	39.8
United Kingdom	24.2
Canada	20.1
Mexico	19.7

Source: Data from *Travel and Leisure* magazine, 2005.

2.13 Shark attacks worldwide Table 2.1, part of which is shown again below, summarized shark attacks for different regions of the world. Using software or sketching, construct a bar graph, ordering the regions (i) alphabetically, and (ii) as in a Pareto chart. Which do you prefer? Why?

Region	Frequency
Florida	268
Hawaii	41
California	32
South Carolina	32
North Carolina	28
Australia	120
South Africa	41
Brazil	20
Bahamas	12
Other	121

2.14 TRY **Sugar dot plot** For the breakfast cereal data given in Table 2.3, a dot plot for the sugar values (in grams) is shown:

a. Identify the minimum and maximum sugar values.

b. Which sugar outcomes occur most frequently? What are these values called?

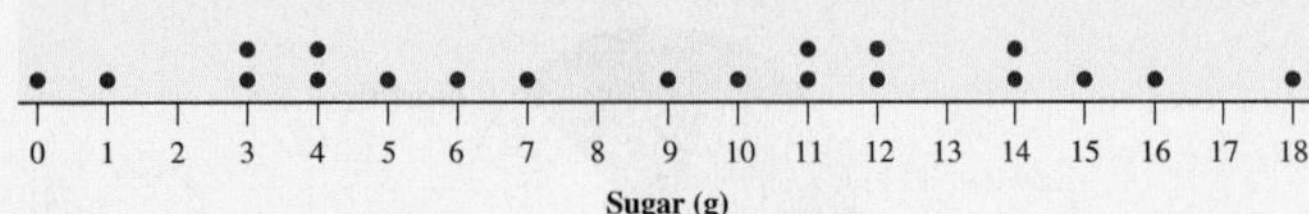

2.15 TRY **Super Bowl tickets** StubHub is a popular Web site where fans can buy and sell tickets to concerts and sporting events. Below are data representing the amounts (in dollars) that buyers using StubHub spent on Super Bowl XLV tickets.

2275 3050 2800 4200 7500 3500 2400 2575

2890 2395 5000 3300 2475 2195 2999 3650

a. Construct a stem-and-leaf plot. Truncate the data to the first two digits for purposes of constructing the plot. For example, 2275 becomes 22.

b. Summarize what this plot tells you about the data.

c. Reconstruct the stem-and-leaf plot in part a using split stems; that is, two stems of 2, two stems of 3, etc. Compare the two stem-and-leaf plots. Explain how one may be more informative than the other.

2.16 TRY **Graphing exam scores** A teacher shows her class the scores on the midterm exam in the stem-and-leaf plot shown:

6	588
7	01136779
8	1223334677789
9	011234458

a. Identify the number of students and their minimum and maximum scores.

b. Sketch how the data could be displayed in a dot plot.

c. Sketch how the data could be displayed in a histogram with four intervals.

2.17 **Fertility rates** The fertility rate for a nation is the average number of children per adult woman. The table below part c shows results for western European nations, the United States, Canada, and Mexico, as reported by the United Nations in 2005.

a. Construct a stem-and-leaf plot using stems 1 and 2 and the decimal parts of the numbers for the leaves. What is a disadvantage of this plot?

b. Construct a stem-and-leaf plot using split stems (Take the first stem for 1 to have leaves 0 through 4, the second stem for 1 to have leaves 5 through 9, and the stem 2 to have leaves 0 through 4.)

c. Construct a histogram using the intervals 1.1–1.3, 1.4–1.6, 1.7–1.9, 2.0–2.2, 2.3–2.5.

Country	Fertility	Country	Fertility
Austria	1.4	Netherlands	1.7
Belgium	1.7	Norway	1.8
Denmark	1.8	Spain	1.3
Finland	1.7	Sweden	1.6
France	1.9	Switzerland	1.4
Germany	1.3	United Kingdom	1.7
Greece	1.3	United States	2.0
Ireland	1.9	Canada	1.5
Italy	1.3	Mexico	2.4

2.18 **Fertility plotted** For the fertility data in the previous exercise, MINITAB reports the stem-and-leaf plot shown below. (You can ignore the cumulative counts in the left column if your instructor has not explained this feature of a MINITAB stem-and-leaf plot.)

a. Explain how this plot was formed using the data in the table. (*Hint:* What does "Leaf Unit = 0.010" indicate about how to read the values from the plot?)

b. The plot shows one observation that stands out from the others. Identify it.

c. Sketch a dot plot. Identify the observation that stands out from the others.

d. Sketch the histogram that has a bar at each value that has a stem in this stem-and-leaf plot (i.e., bars at 1.3, 1.4, 1.5, …, 2.4). Explain how it also highlights the observation that stands out.

Stem-and-leaf of Fertility Rate N = 18
Leaf Unit = 0.010

4	13	0000
6	14	00
7	15	0
8	16	0
(4)	17	0000
6	18	00
4	19	00
2	20	0
1	21	
1	22	
1	23	
1	24	0

2.19 **Leaf unit** When the observations are large numbers, their final digits are not shown in a stem-and-leaf plot. The plot specifies a **leaf unit** by which to multiply each observation. For instance, for the cereal sugar data from Table 2.3 expressed in milligrams (an excerpt of which is shown), MINITAB software reports the figure shown here, indicating that "Leaf Unit = 1000." For instance, the observation of 18 in the final row of the plot represent observation 18,000.

Cereal	Sugar (mg)
Frosted Mini Wheats	11,000
Raisin Bran	18,000
All Bran	5,000

Stem and Leaf Plot
with Leaf Unit = 1000

0	01
0	33
0	445
0	67
0	9
1	011
1	22
1	445
1	6
1	8

a. In milligrams, what values are represented on the first line of the plot?

b. Which sugar outcomes occur most frequently?

2.20 Truncated data and split stems The figure below shows the stem-and-leaf plot for the cereal sodium values constructed after Example 5 using *split stems*, with leaves from 0 to 4 on the first split stem and leaves from 5 to 9 on the second.

Split Stems for Truncated Cereal Sodium Values

0	0
0	57
1	0344
1	568889
2	00112
2	9
3	4

a. Explain why the truncated data shown here go from 0 to 34.

b. Identify on the plot the observation of 220 for Special K.

c. What features of the data shown in a nontruncated plot are lost in this figure? (*Hint*: Is the large gap between 220 and 290 readily visible?)

2.21 Histogram for sugar For the breakfast cereal data, the figure at the top of next column shows a histogram (constructed using MINITAB) for the sugar values, in grams.

a. Identify the intervals of sugar values used for the plot.

b. Describe the shape of the distribution. What do you think might account for this unusual shape? (*Hint:* How else are the cereals classified in Table 2.3?)

c. What information can you get from the dot plot or stem-and-leaf plot of these data shown in Exercises 2.14 and 2.19 that you cannot get from this plot?

d. This histogram shows frequencies. If you were to construct a histogram using the *percentages* for each interval, how (if at all) would the shape of this histogram change?

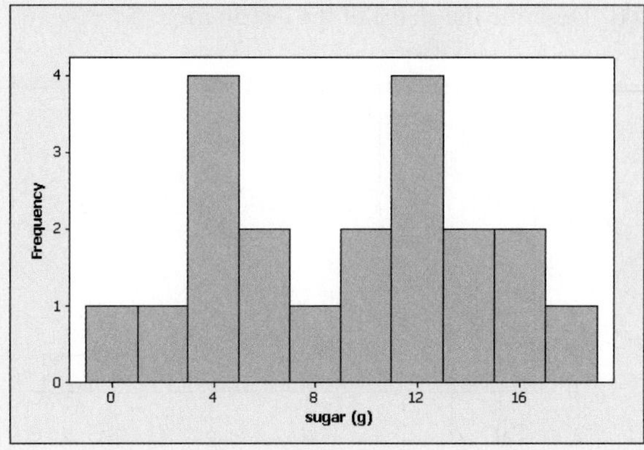

2.22 Sugar plots Using software with the Cereal data set on the text CD, construct (a) a dot plot, (b) a stem-and-leaf plot, and (c) a histogram. Explain how to interpret each plot.

2.23 Shape of the histogram For each of the following variables, indicate whether you would expect its histogram to be symmetric, skewed to the right, or skewed to the left. Explain why.

a. Assessed value of houses in a large city (*Hint:* Would the relatively few homes with extremely high assessed value result in a long right tail or a long left tail?)

b. Number of times checking account overdrawn in the past year for the faculty in your school

c. IQ for the general population

d. The height of female college students

2.24 More shapes of histograms Repeat the preceding exercise for

a. The scores of students (out of 100 points) on a very easy exam in which most score perfectly or nearly so, but a few score very poorly

b. The weekly church contribution for all members of a congregation, in which the three wealthiest members contribute generously each week

c. Time needed to complete a difficult exam (maximum time is 1 hour)

d. Number of music CDs (compact discs) owned, for each student in your school.

2.25 How often do students read the newspaper? Question 14 on the class survey (Activity 3 in Chapter 1 on pages 22–23) asked, "Estimate the number of times a week, on average, that you read a daily newspaper."

a. Is this variable continuous, or discrete? Explain.

b. The histogram shown gives results of this variable when this survey was administered to a class of 36 University of Georgia students. Report the (i) minimum response, (ii) maximum response, (iii) number of students who did not read the newspaper at all, and (iv) mode.

c. Describe the shape of the distribution.

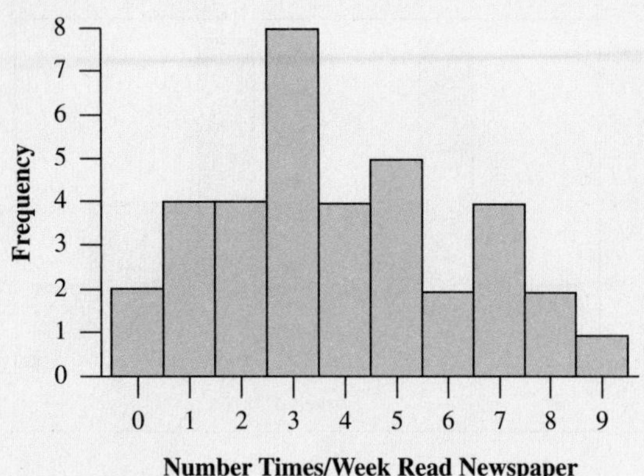

2.26 Blossom widths A data set analyzed by the famous statistician R. A. Fisher consisted of measurements of different varieties of iris blossoms. Below is a histogram representing the widths of the petals of *Iris setosa.*

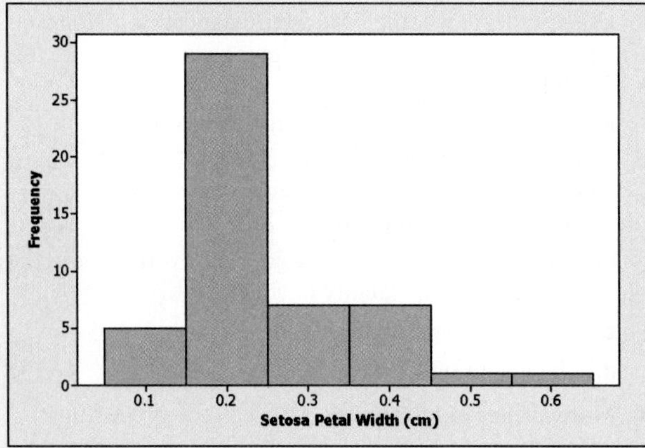

a. Describe the shape of the distribution of *setosa* petal widths.

b. Of the 50 *setosa* blossoms in the data set, approximately what percentage has a petal width of more than 0.25 cm?

c. Is it possible to accurately determine the percentage with a width of more than 0.3 cm? Why or why not?

2.27 Central Park temperatures The first figure shows a histogram of the Central Park, New York, annual average temperatures from 1869–2010.

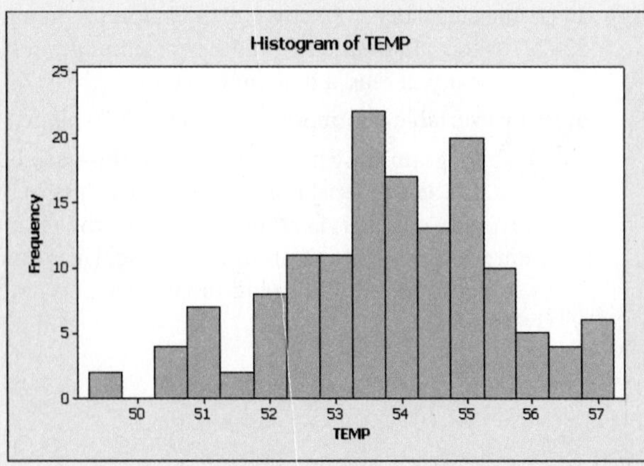

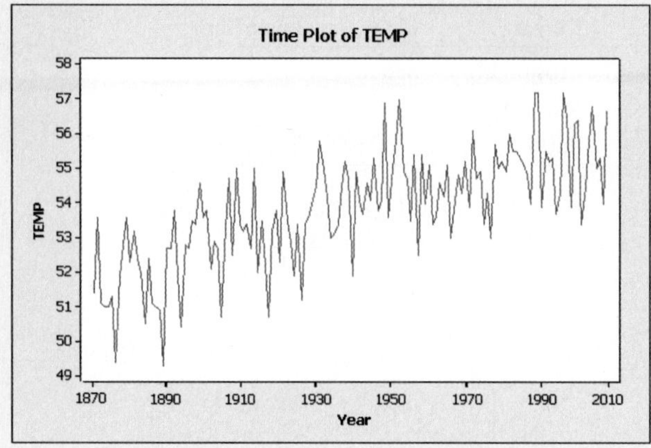

a. Describe the shape of the distribution.

b. What information can the time plot above show that a histogram cannot provide?

c. What information does the histogram show that a time plot does not provide?

2.28 Is whooping cough close to being eradicated? In the first half of the 20th century, whooping cough was a frequently occurring bacterial infection that often resulted in death, especially among young children. A vaccination for whooping cough was developed in the 1940s. How effective has the vaccination been in eradicating whooping cough? One measure to consider is the **incidence rate** (number of infected individuals per 100,000 population) in the United States. The table shows incidence rates from 1925–1970.

TRY

Incidence Rates for Whooping Cough, 1925–1970

Year	Rate per 100,000
1925	131.2
1930	135.6
1935	141.9
1940	139.6
1945	101.0
1950	80.1
1955	38.2
1960	8.3
1965	3.5
1970	2.1

Source: Data from Historical Statistics of the United States, Colonial Times to 1970, U.S. Department of Commerce, p. 77.

a. Sketch a time plot. What type of trend do you observe? Based on the trend from 1945–1970, was the whooping cough vaccination proving effective in reducing the incidence of whooping cough?

b. The figure shown is a time plot of the data from 1980–1999, reported in the Morbidity and Mortality Weekly Reports (MMWR).[3] What has the incidence rate been since about 1993? How does the incidence rate since 1993 compare with the incidence rate in 1970? Is the United States close to eradicating whooping cough? (Note: In the years 2006–2008, the incidence rates were 5.27, 3.49, 4.40 respectively.)

[3]*Source:* Data from www.cdr.gov/mmwr/.

c. Would a histogram of the incidence rates since 1935 address the question about the success of the vaccination for whooping cough? Why or why not?

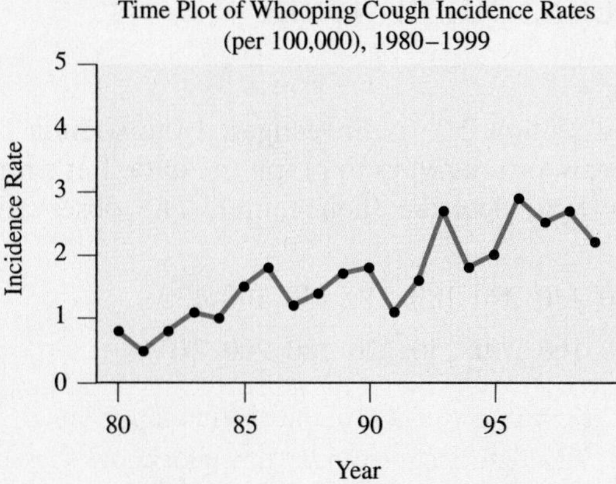

2.29 Warming in Newnan, Georgia? Access the Newnan, GA Temps file on the text CD, which reports the average annual temperatures during the 20th century for Newnan, Georgia. Construct a time plot to investigate a possible trend over time. Is there evidence of climate change?

2.3 Measuring the Center of Quantitative Data

Section 2.2 introduced **graphical summaries** that display data for both categorical and quantitative variables. For a categorical variable, a graph shows the proportion of observations in each category and the modal category. For a quantitative variable, a graph shows the important feature of **shape.** It's always a good idea to look at the data first with a graph, to get a "feel" for the data. You can then consider **numerical summaries of the sample (statistics).** For quantitative variables, we'll attempt to answer the questions, "What is a representative observation like?" and "Do the observations take similar values, or do they vary quite a bit?" Statistics that answer the first question describe the **center** of the distribution. Statistics that answer the second question describe the **variability (or spread)** of the distribution. We'll also study how the shape of the distribution influences the statistics and our choice of which are suitable for a given data set.

Recall

A statistic is a numerical summary of a sample. A parameter is a numerical summary of the population. ◄

Describing the Center: The Mean and the Median

The best-known and most frequently used measure of the center of a distribution of a quantitative variable is the **mean**. It is found by averaging the observations.

In Words

The **mean** refers to averaging, that is, adding up the data points and dividing by how many there are. The **median** is the point that splits the data in two, half the data below it and half above it (much like the median on a highway splits a road into two equal parts).

Mean

The **mean** is the sum of the observations divided by the number of observations. It is interpreted as the balance point of the distribution.

Another popular measure is the **median**. Half the observations are smaller than it, and half are larger.

Median

The **median** is the middle value of the observations when the observations are ordered from the smallest to the largest (or from the largest to the smallest).

Example 10

Mean and median

Center of the Cereal Sodium Data

Picture the Scenario

In Examples 4, 5, and 7 in Section 2.2, we investigated the sodium level in 20 breakfast cereals and saw various ways to graph the data. Let's return to those data and learn how to describe their center. The observations (in mg) are

0 340 70 140 200 180 210 150 100 130
140 180 190 160 290 50 220 180 200 210

Questions to Explore

a. Find the mean.
b. Find the median.

Think It Through

a. We find the mean by adding all the observations and then dividing this sum by the number of observations, which is 20:

$$\text{Mean} = (0 + 340 + 70 + \ldots + 210)/20 = 3340/20 = 167.$$

b. To find the median, we arrange the data from the smallest to the largest observation.

0 50 70 100 130 140 140 150 160 180
180 180 190 200 200 210 210 220 290 340

For the 20 observations, the smaller 10 (on the first line) range from 0 to 180, and the larger 10 (on the second line) range from 180 to 340. The median is 180, which is the average of the two middle values, the tenth and eleventh observations, (180 + 180)/2.

Insight

The mean and median take different values. Why? The median measures the center by dividing the data into two equal parts, regardless of the actual numerical values above that point or below that point. The mean takes into account the actual numerical values of all the observations.

Try Exercise 2.31

In this example, what if the smaller 10 observations go from 0 to 180, and the larger ten go from 190 to 340? Then, the median is the average of the two middle observations, which is (180 + 190)/2 = 185.

SUMMARY: How to Determine the Median

- Put the n observations in order of their size.
- When the number of observations n is odd, the median is the middle observation in the ordered sample.
- When the number of observations n is even, two observations from the ordered sample fall in the middle, and the median is their average.

A Closer Look at the Mean and the Median

Notation for the mean is used both in formulas and as a shorthand in text.

In Words

The formula

$$\bar{x} = \frac{\Sigma x}{n}$$

is short for "sum the values on the variable x and divide by the sample size."

Notation for a Variable and Its Mean

Variables are symbolized by letters near the end of the alphabet, most commonly x and y. The sample size is denoted by n. For a sample of n observations on a variable x, the mean is denoted by $\bar{x}$. Using the mathematical symbol Σ for "sum," the mean has the formula

$$\bar{x} = \frac{\Sigma x}{n}.$$

For instance, the cereal data set has $n = 20$ observations. As we saw in Example 10,

$$\bar{x} = (\Sigma x)/n = (0 + 340 + 70 + \ldots + 210)/20 = 3340/20 = 167.$$

Did You Know?

The mean is also interpreted as the "fair share" value; for example, when considering these 20 cereals, the mean of 167 mg can be interpreted as the amount of sodium each cereal serving contains if all 20 cereals have the same amount. ◀

Here are some basic **properties of the mean**:

- The mean is the *balance point* of the data: If we were to place identical weights on a line representing where the observations occur, then the line would balance by placing a fulcrum at the mean.

The Fulcrum Shows the Mean of the Cereal Sodium Data

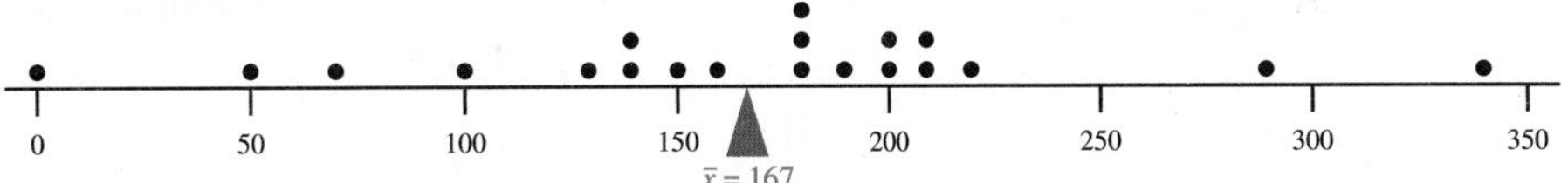

- For a skewed distribution, the mean is pulled in the direction of the longer tail, relative to the median. The next example illustrates this idea.
- The mean can be highly influenced by an **outlier,** which is an unusually small or unusually large observation.

Outlier

An **outlier** is an observation that falls well above or well below the overall bulk of the data.

Outliers typically call for further investigation to see, for example, whether they resulted from an error in data entry or from some surprising or unusual occurence.

Example 11

Mean, median, and outliers

CO_2 Pollution

Picture the Scenario

The Pew Center on Global Climate Change[4] reports that global warming is largely a result of human activity that produces carbon dioxide (CO_2) emissions and other greenhouse gases. The CO_2 emissions from fossil fuel combustion are the result of electricity, heating, industrial processes, and gas

[4] *Source:* www.pewclimate.org/global-warming-basics/facts_and_figures/.

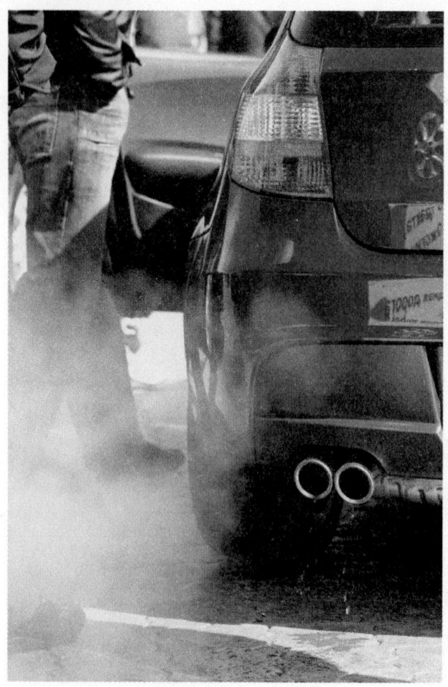

Did You Know?

A metric ton is 1000 kilograms, which is about 2200 pounds. ◀

consumption in automobiles. The International Energy Agency[5] reported the per capita CO_2 emissions by country (that is, the total CO_2 emissions for the country divided by the population size of that country) for 2007. For the eight largest countries in population size (which make up more than half the world's population), the values were, in metric tons per person:

China 4.9	Brazil 1.9
India 1.4	Pakistan 0.9
United States 18.9	Russia 10.8
Indonesia 1.8	Bangladesh 0.3

Questions to Explore

a. For these eight values, the mean is 5.1. What is the median?

b. Is any observation a potential outlier? Discuss its impact on how the mean compares to the median.

c. Using this data set, explain the effect an outlier can have on the mean.

Think It Through

a. The CO_2 values have $n = 8$ observations. The ordered values are

$$0.3,\ 0.9,\ 1.4,\ 1.8,\ 1.9,\ 4.9,\ 10.8,\ 18.9$$

Since n is even, two observations are in the middle, the fourth and fifth ones in the ordered sample. These are 1.8 and 1.9. The median is their average, 1.85.

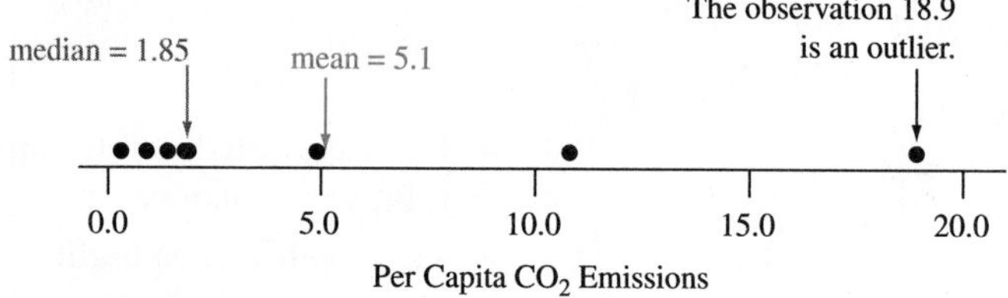

b. Let's consider a dot plot, as shown. The relatively high value of 18.9 falls well above the rest of the data. It is an outlier. This value, as well as the value at 10.8, causes the mean, which is 5.1, to fall well above the median, which is 1.85.

c. The size of the outlier affects the calculation of the mean but not the median. If the observation of 18.9 for the United States had been the same as for China (4.9), the eight observations would have had a mean of 3.4, instead of 5.1. The median would still have been 1.85. In summary, a single outlier can have a large impact on the value of the mean.

Insight

The mean may not be representative of where the bulk of the observations fall. This is fairly common with small samples when one observation is much larger or much smaller than the others. It's not surprising that the United States is an outlier, as the other nations are not nearly as economically advanced. Later in the chapter (Example 18), we'll compare carbon dioxide emissions for the United States with those of European nations.

Try Exercise 2.32

[5] *Source*: www.iea.org/.

Comparing the Mean and Median

The shape of a distribution influences whether the mean is larger or smaller than the median. For instance, an extremely large value out in the right-hand tail pulls the mean to the right. The mean then usually falls above the median, as we observed with the CO_2 data in Example 11.

Generally, if the shape is

- perfectly symmetric, the mean equals the median.
- skewed to the right, the mean is larger than the median.
- skewed to the left, the mean is smaller than the median.

As Figure 2.9 illustrates, the mean is drawn in the direction of the longer tail.

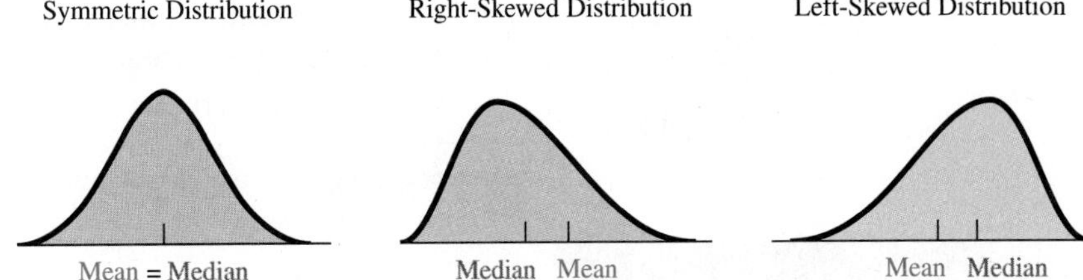

▲ **Figure 2.9 Relationship Between the Mean and Median. Question** For skewed distributions, what causes the mean and median to differ?

When a distribution is close to symmetric, the tails will be of similar length, and therefore the median and mean are similar. For skewed distributions, the mean lies toward the direction of skew (the longer tail) relative to the median, as Figure 2.9 shows. This is because extreme observations in a tail affect the balance point for the distribution, which is the mean. The more highly skewed the distribution, the more the mean and median tend to differ.

Example 11 illustrated this property. The dot plot of CO_2 emissions shown there is skewed to the right. As expected, the mean of 5.1 falls in the direction of skew, above the median of 1.5. Another example is given by mean household income in the United States. In 2009, the mean was about $67,976 and the median was about $49,777 (U.S. Bureau of the Census). This suggests that the distribution of household incomes in the United States is skewed to the right.

Why is the median not affected by an outlier? How far an outlier falls from the middle of the distribution does not influence the median. The median is determined solely by having an equal number of observations above it and below it.

For the CO_2 data, for instance, if the value 18.9 for the United States were changed to 90, as shown below, the median would still equal 1.85. However, the calculation of the mean uses *all* the numerical values. So, unlike the median, it depends on how far observations fall from the middle. Because the mean is the balance point, an extreme value on the right side pulls the mean toward the right tail. Because the median is not affected, it is said to be **resistant** to the effect of extreme observations.

The Median Is Resistant to Outliers

1. Change 18.9 to 90 for United States.

   0.3, 0.9, 1.4, 1.8, 1.9, 4.9, 10.8, 90

   median = (1.8 + 1.9)/2 = 1.85

2. Find the mean using the value of 90.

   0.3, 0.9, 1.4, 1.8, 1.9, 4.9, 10.8, 90

   mean = 14

Resistant

A numerical summary of the observations is called **resistant** if extreme observations have little, if any, influence on its value.

The median is resistant. The mean is not.

From these properties, you might think that it's always better to use the median rather than the mean. That's not true. The mean has other useful properties that we'll learn about and take advantage of in later chapters. Also, there are advantages to having a measure use the numerical values of *all* the data. For instance, *for discrete data that take only a few values, quite different patterns of data can give the same result for the median*. It is then *too* resistant.

The following example illustrates this situation and also shows how to find the mean and median from a frequency table.

Example 12

Mean and median

Marriage Statistics

Picture the Scenario

A Census Bureau report[6] gave data on the number of times U.S. residents had been married, for subjects of various ages. Table 2.5 summarizes responses for subjects of age 20–24. The frequencies are actually *thousands* of people, for instance 8,418,000 men never married, but this does not affect calculations about the mean or median.

Table 2.5 Number of Times Married, for Subjects of Age 20–24

Number Times Married	Frequency	
	Women	**Men**
0	7350	8418
1	2587	1594
2	80	10
Total	**10,017**	**10,022**

Questions to Explore

a. Find the median and the mean for each gender.

b. Why is the median not particularly informative?

Think It Through

a. For the 10,017 women, the ordered sample would be

000000.....0 111111...1 222222...2

(7350 0s) (2587 1s) (80 2s)

Since $n = 10{,}017$ the observations would be listed from 1st to 10,017th, and the middle observation would be the $(1 + 10{,}017)/2 = 5009$th observation. However, only three distinct responses occur, and more than 5009 of them are 0. So, the median is 0. Likewise, the median is 0 for men.

[6]*Source:* www.census.gov/hhes/socdemo/marriage/data/sipp/2004/tables.html.

To calculate the mean for women, it is unnecessary to add the 10,017 separate observations to obtain the numerator of $\bar{x}$, because each value occurred many times. We can find the sum of the 10,017 observations by multiplying each possible value by its frequency and then adding:

$$\Sigma x = 7350(0) + 2587(1) + 80(2) = 2747.$$

Since the sample size is 10,017, the mean is 2747/10,017 = 0.274. Likewise, you can show that the mean for men is 0.161.

b. The median is 0 for women and for men. This makes it seem as if there's no difference between men and women on this variable. By contrast, the mean uses the numerical values of all the observations, not just the ordering. In summary, in this age group we learn that, on average, women have been married more often than men.

Caution

The mean is not necessarily a possible value for a discrete variable. The discrete variable "number of times married" takes on values 0,1, and 2. However, the mean number of times married is 0.274 for women and 0.161 for men. ◀

Insight

The median ignores too much information when the data are highly discrete—that is, when the data take only a few values. An extreme case occurs for **binary data**, which take only two values, 0 and 1. The median equals the more common outcome but gives no information about the relative number of observations at the two levels.

For instance, consider a sample of size 5 for the variable, number of times married. The observations (1, 1, 1, 1, 1) and the observations (0, 0, 1, 1, 1) both have a median of 1. The mean is 1 for (1, 1, 1, 1, 1) and 3/5 for (0, 0, 1, 1, 1). When observations take values of only 0 or 1, the mean equals the *proportion* of observations that equal 1. It is much more informative than the median. When the data are highly discrete but have more than two categories, such as in Table 2.5, it is more informative to report the proportions (or percentages) for the possible outcomes than to report the median *or* the mean.

Try Exercise 2.45

If you want to summarize the emission data of Example 11, the median may be more relevant because of the skew resulting from the extremely large value for the United States. But for the marriage data in Example 12, the median discards too much information and the mean is more informative. In practice, both the mean and median are useful, and frequently *both* values are reported.

In Practice Effect of Shape on Choice of Mean or Median

- If a distribution is highly skewed, the median is usually preferred over the mean because it better represents what is typical.
- If the distribution is close to symmetric or only mildly skewed, or if it is discrete with few distinct values, the mean is usually preferred because it uses the numerical values of all the observations.

The Mode

We've seen that the **mode** is the value that occurs most frequently. It describes a typical observation in terms of the most common outcome. The concept of the mode is most often used to describe the category of a categorical variable that has the highest frequency (the modal category). With quantitative variables, the mode is most useful with discrete variables taking a small number of possible values. For the marriage data of Table 2.5, for example, the mode is 0 for each gender.

Caution

The mode need not be near the center of the distribution. It may be the largest or the smallest value, as in the marriage example. Thus, it is somewhat inaccurate to call the mode a measure of center, but often it is useful to report the most common outcome. ◀

Activity 1

Using an Applet to Simulate the Relationship between the Mean and Median*

The Mean Versus Median applet on the text CD allows you to add and delete data points from a sample. Different colored arrows indicate the mean and median after each new point is added. Open the applet and set the number line to have a range of values from 0 to 20 and click on *Update.*

- In Example 11 on CO_2, use the per capita values (metric tons) for the eight largest nations (0.3, 0.9, 1.4, 1.8, 1.9, 4.9, 10.8, 18.9). Create this sample using the applet. Now, update the line by changing the upper limit to 50 and clicking "Update." Investigate what happens to the mean and median as you move the highest observation of 18.9 even higher, up to 50, or move it lower, down to 10. How does the outlier influence the relationship between the mean and median?
- Consider the effect of the sample size on the relationship between the mean and median, by adding to this distribution 20 values that are like the nonoutliers. Does the outlier of 18.9 have as much effect on the mean when the sample size is larger?

*For more information about the applets see page x and the back left endpaper of this book.

Try Exercises 2.38 and 2.148

2.3 Practicing the Basics

2.30 Median versus mean The mean and median describe the center.

a. Why is the median sometimes preferred? Give an example.

b. Why is the mean sometimes preferred? Give an example.

2.31 More on CO_2 emissions The Energy Information Agency
TRY reported the CO_2 emissions from fossil fuel combustion for the seven countries in 2008 with the highest emissions. These values, reported as million metric tons of carbon equivalent, are 6534 (China), 5833 (United States), 1729 (Russia), 1495 (India), 1214 (Japan), 829 (Germany), and 574 (Canada).

a. Find the mean and median.

b. The totals reported here do not take into account a nation's population size. Explain why it may be more sensible to analyze *per capita* values, as was done in Example 11.

2.32 Resistance to an outlier Consider the following three
TRY sets of observations:

Set 1: 8, 9, 10, 11, 12

Set 2: 8, 9, 10, 11, 100

Set 3: 8, 9, 10, 11, 1000

a. Find the median for each data set.

b. Find the mean for each data set.

c. What do these data sets illustrate about the resistance of the median and mean?

2.33 Income and race According to the U.S. Bureau of the Census, *Current Population Reports*, in 2009 the median household income was \$51,861 for whites and \$32,750 for blacks, whereas the mean was \$70,544 for whites and \$46,280 for blacks. Does this suggest that the distribution of household income for each race is symmetric, skewed to the right, or skewed to the left? Explain.

2.34 Labor dispute The workers and the management of a company are having a labor dispute. Explain why the workers might use the median income of all the employees to justify a raise but management might use the mean income to argue that a raise is not needed.

2.35 Cereal sodium The dot plot shows the cereal sodium values from Example 4. What aspect of the distribution causes the mean to be less than the median?

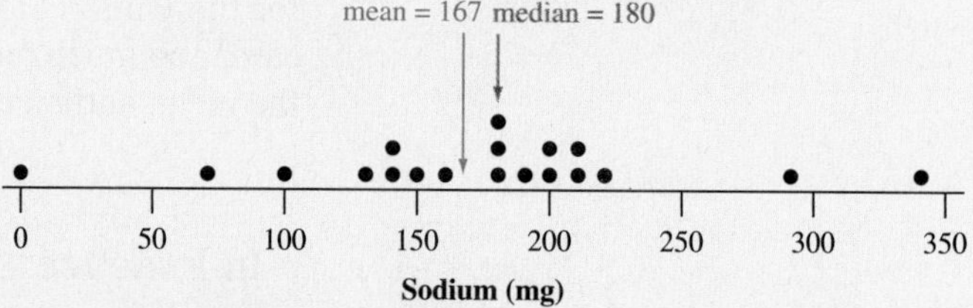

2.36 Center of plots The figure shows dot plots for three sample data sets.

a. For which, if any, data sets would you expect the mean and the median to be the same? Explain why.

b. For which, if any, data sets would you expect the mean and the median to differ? Which would be larger, the mean or the median? Why?

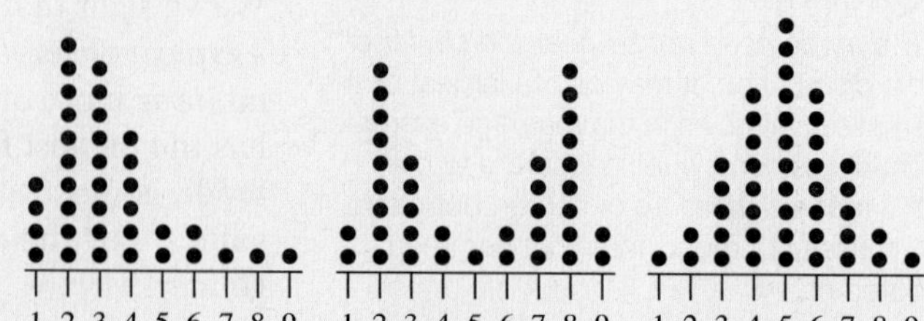

2.37 Public transportation—center The owner of a company in downtown Atlanta is concerned about the large use of gasoline by her employees due to urban sprawl, traffic congestion, and the use of energy inefficient vehicles such as SUVs. She'd like to promote the use of public transportation. She decides to investigate how many miles her employees travel on public transportation during a typical day. The values for her 10 employees (recorded to the closest mile) are

0 0 4 0 0 0 10 0 6 0

a. Find and interpret the mean, median, and mode.

b. She has just hired an additional employee. He lives in a different city and travels 90 miles a day on public transport. Recompute the mean and median. Describe the effect of this outlier.

2.38 Public transportation—outlier Refer to the previous exercise.
TRY

a. Use the Mean Versus Median applet to investigate what effect adding the outlier of 90 to the data set has on the mean and median.

b. Now add 10 more data values that are near the mean of 2 for the original 10 observations. Does the outlier of 90 still have such a strong effect on the mean?

2.39 Student hospital costs If you had data for all students in your school on the amount of money spent in the previous year on overnight stays in a hospital, probably the median and mode would be 0 but the mean would be positive.

a. Explain why.

b. Give an example of another variable that would have this property.

2.40 Net worth by degree The *Statistical Abstract of the United States* reported that in 2004 for those with a college education, the median net worth was $226,100 and the mean net worth was $851,300. For those with a high school diploma only, the values were $68,700 and $196,800.

a. Explain how the mean and median could be so different for each group.

b. Which measure do you think gives a more realistic measure of a typical net worth, the mean or the median. Why?

2.41 Canadian income According to Statistics Canada, in 2004 the median household income in Canada was $58,100 and the mean was $76,100. What would you predict about the shape of the distribution? Why?

2.42 Baseball salaries The players on the New York Yankees baseball team in 2010 had a mean salary of $7,935,531 and a median salary of $4,525,000.[7] What do you think causes these two values to be so different?

2.43 European fertility The European fertility rates (mean number of children per adult woman) from Exercise 2.17 are shown again in the table.

a. Find the median of the fertility rates. Interpret.

b. Find the mean of the fertility rates. Interpret.

c. For each woman, the number of children is a whole number, such as 2 or 3. Explain why it makes sense to measure a *mean* number of children per adult woman (which is not a whole number) to compare fertility levels, such as the fertility levels of 1.5 in Canada and 2.4 in Mexico.

Country	Fertility	Country	Fertility
Austria	1.4	Netherlands	1.7
Belgium	1.7	Norway	1.8
Denmark	1.8	Spain	1.3
Finland	1.7	Sweden	1.6
France	1.9	Switzerland	1.4
Germany	1.3	United Kingdom	1.7
Greece	1.3	United States	2.0
Ireland	1.9	Canada	1.5
Italy	1.3	Mexico	2.4

2.44 Sex partners A recent General Social Survey asked female respondents, "How many sex partners have you had in the last 12 months?" Of the 365 respondents, 102 said 0 partners, 233 said 1 partner, 18 said 2 partners, 9 said 3 partners, 2 said 4 partners, and 1 said 5 partners. (*Source:* Data from CSM, UC Berkeley.)

a. Display the data in a table. Explain why the median is 1.

b. Show that the mean is 0.85.

c. Suppose the 102 women who answered 0 instead had answered 5 partners. Show that the median would still be 1. (The mean would increase to 2.2. The mean uses the numerical values of the observations, not just their ordering.)

TRY **Knowing homicide victims** The table summarizes responses of 4383 subjects in a recent General Social Survey to the question, "Within the past month, how many people have you known personally that were victims of homicide?"

Number of People You Have Known Who Were Victims of Homicide

Number of Victims	Frequency
0	3944
1	279
2	97
3	40
4 or more	23
Total	**4383**

(*Source:* Data from CSM, UC Berkeley.)

a. To find the mean, it is necessary to give a score to the "4 or more" category. Find it, using the score 4.5. (In practice, you might try a few different scores, such as 4, 4.5, 5, 6, to make sure the resulting mean is not highly sensitive to that choice.)

b. Find the median. Note that the "4 or more" category is not problematic for it.

c. If 1744 observations shift from 0 to 4 or more, how do the mean and median change?

d. Why is the median the same for parts b and c, even though the data are so different?

2.46 Accidents One variable in a study measures how many serious motor vehicle accidents a subject has had in the past year. Explain why the mean would likely be more useful than the median for summarizing the responses of the 60 subjects.

---

[7]espn.go.com/mlb/team/salaries/_/name/nyy/new-york-yankees.

2.4 Measuring the Variability of Quantitative Data

A measure of the center is not enough to adequately describe a distribution for a quantitative variable. It tells us nothing about the variability of the data. With the cereal sodium data, if we report the mean of 167 mg to describe the center, would the value of 210 mg for Honeycomb be considered quite high, or are most of the data even farther from the mean? To answer this question, we need numerical summaries of the variability of the distribution.

Measuring Variability: The Range

To see why a measure of the center is not enough, let's consider Figure 2.10. This figure compares hypothetical income distributions of music teachers in public schools in Denmark and in the United States. Both distributions are symmetric and have a mean of about $40,000. However, the annual incomes in Denmark (converted to U.S. dollars) go from $35,000 to $45,000, whereas those in the United States go from $20,000 to $60,000. Incomes are more similar in Denmark and vary more in the United States. A simple way to describe this is with the **range**.

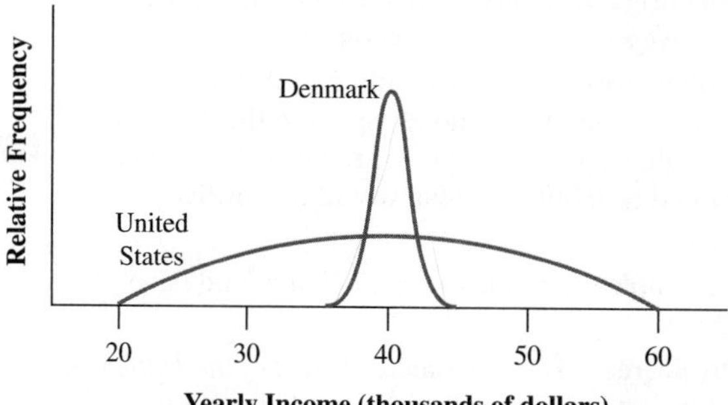

▲ **Figure 2.10** Income Distributions for Music Teachers in Denmark and in the United States. The distributions have the same mean, but the one for the United States varies more from the mean. **Question** How would the range for Denmark change if one teacher earned $100,000?

> **Range**
>
> The **range** is the difference between the largest and the smallest observations.

In Denmark the range is $45,000 − $35,000 = $10,000. In the United States the range is $60,000 − $20,000 = $40,000. The range is a larger value when the data vary more in the distribution.

The range is simple to compute and easy to understand, but it uses only the extreme values and ignores the other values. Therefore, it's affected severely by outliers. For example, if one teacher in Denmark made $100,000, the range would change from $45,000 − $35,000 = $10,000 to $100,000 − $35,000 = $65,000. The range is not a resistant statistic. It shares the worst property of the mean, not being resistant, and the worst property of the median, ignoring the numerical values of nearly all the data.

Measuring Variability: The Standard Deviation

In practice we don't use the range much since it only utilizes the largest and smallest observations. A much better summary of variability uses *all* the data, and it describes a typical distance of how far the data falls from the mean. It does this by summarizing **deviations** from the mean.

- The **deviation** of an observation x from the mean $\bar{x}$ is $(x - \bar{x})$, the difference between the observation and the sample mean.

For the cereal sodium values, the mean is $\bar{x} = 167$. The observation of 210 for Honeycomb has a deviation of $210 - 167 = 43$. The observation of 50 for Honey Smacks has a deviation of $50 - 167 = -117$. Figure 2.11 shows these deviations.

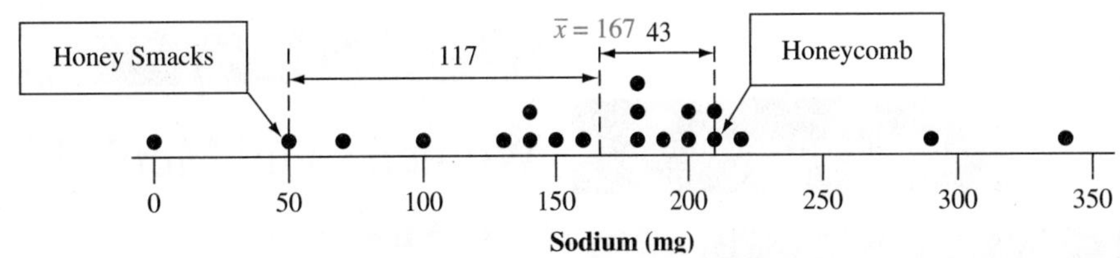

▲ **Figure 2.11** Dot Plot for Cereal Sodium Data, Showing Deviations for Two Observations. **Question** When is a deviation positive and when is it negative?

- Each observation has a deviation from the mean.
- A deviation $x - \bar{x}$ is *positive* when the observation falls *above* the mean. A deviation is *negative* when the observation falls *below* the mean.
- The interpretation of the mean as the balance point implies that the positive deviations counterbalance the negative deviations. Because of this, *the sum of the deviations always equals zero.* Summary measures of variability from the mean use either the squared deviations or their absolute values.
- The average of the squared deviations is called the **variance**. Because the variance uses the *square* of the units of measurement for the original data, its square root is easier to interpret. This is called the **standard deviation.**
- The symbol $\Sigma(x - \bar{x})^2$ is called a **sum of squares**. It represents finding the deviation for each observation, squaring each deviation, and then adding them.

Did You Know?

Another measure of the typical or average distance of observations from the mean is the Mean Absolute Deviation (MAD). The absolute value of the deviation is found instead of the square. Exercise 2.145 explores the MAD. ◀

The Standard Deviation s

The **standard deviation s** of n observations is

$$s = \sqrt{\frac{\Sigma(x - \bar{x})^2}{n - 1}} = \sqrt{\frac{\text{sum of squared deviations}}{\text{sample size} - 1}}.$$

This is the square root of the **variance s^2**, which is an average of the squares of the deviations from their mean,

$$s^2 = \frac{\Sigma(x - \bar{x})^2}{n - 1}.$$

A calculator can compute the standard deviation s easily. Its interpretation is quite simple: *Roughly, the standard deviation s represents a typical distance or a type of average distance of an observation from the mean.* The most basic property of the standard deviation is this:

- The larger the standard deviation s, the greater the variability of the data.

A small technical point: You may wonder why the denominators of the variance and the standard deviation use $n - 1$ instead of n. We said that the variance was an *average* of the n squared deviations, so should we not divide by n? Basically it is because the deviations have only $n - 1$ pieces of information about variability: That is, $n - 1$ of the deviations determine the last one, because the deviations sum to 0. For example, suppose we have $n = 2$ observations and the first observation has deviation $(x - \bar{x}) = 5$. Then the second observation must have deviation $(x - \bar{x}) = -5$ because the deviations must add to 0. With $n = 2$, there's only $n - 1 = 1$ nonredundant piece of information about variability. And with $n = 1$, the standard deviation is undefined because with only one observation, it's impossible to get a sense of how much the data vary.

Standard deviation

Example 13

Women's and Men's Ideal Number of Children

Picture the Scenario

Students in a class were asked on a questionnaire at the beginning of the course, "How many children do you think is ideal for a family?" The observations, classified by student's gender, were

Men: 0, 0, 0, 2, 4, 4, 4

Women: 0, 2, 2, 2, 2, 2, 4

Question to Explore

Both men and women have a mean of 2 and a range of 4. Do the distributions of data have the same amount of variability around the mean? If not, which distribution has more variability?

Think It Through

Let's check dot plots for the data.

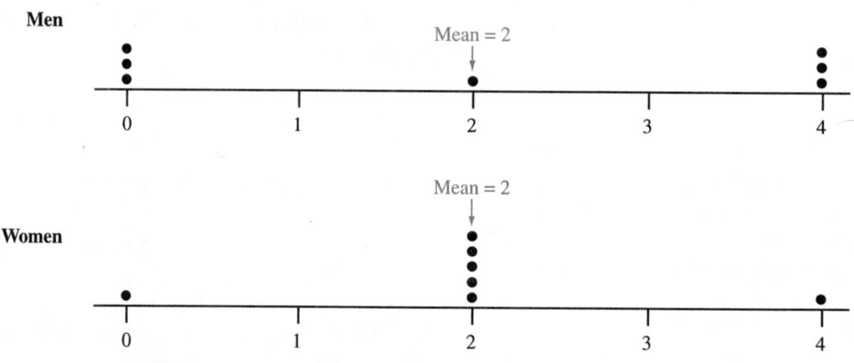

The typical deviation from the mean for the male observations appears to be about 2. The observations for females mostly fall right at the mean, so their typical deviation is smaller.

Let's calculate the standard deviation for men. Their observations are 0, 0, 0, 2, 4, 4, 4. The deviations and squared deviations about their mean of 2 are

Value	Deviation	Squared Deviation
0	$(0 - 2) = -2$	4
0	$(0 - 2) = -2$	4
0	$(0 - 2) = -2$	4
2	$(2 - 2) = 0$	0
4	$(4 - 2) = 2$	4
4	$(4 - 2) = 2$	4
4	$(4 - 2) = 2$	4

The sum of squared deviations equals

$$\Sigma(x - \bar{x})^2 = 4 + 4 + 4 + 0 + 4 + 4 + 4 = 24.$$

In Practice Rounding

Statistical software and calculators can find the standard deviation s for you. Try calculating s for a couple of small data sets to help you understand what it represents. After that, rely on software or a calculator. To ensure accurate results, don't round off while doing a calculation. (For example, use a calculator's memory to store intermediate results.) When presenting the solution, however, round off to two or three significant digits. In calculating s for women, you get $s = \sqrt{1.3333\ldots} = 1.1547005\ldots$. Present the value $s = 1.2$ or $s = 1.15$ to make it easier for a reader to comprehend.

The standard deviation of these $n = 7$ observations equals

$$s = \sqrt{\frac{\Sigma(x - \bar{x})^2}{n - 1}} = \sqrt{\frac{24}{6}} = \sqrt{4} = 2.0.$$

This indicates that for men a typical distance of an observation from the mean is 2.0. By contrast, you can check that the standard deviation for women is $s = 1.2$. The observations for males tended to be farther from the mean than those for females, as indicated by $s = 2.0 > s = 1.2$. In summary, the men's observations varied more around the mean.

Insight

The standard deviation is more informative than the range. For these data, the standard deviation detects that the women were more consistent than the men in their viewpoints about the ideal number of children. The range does not detect the difference, as it equals 4 for each gender.

Try Exercise 2.47

Standard deviation

Example 14

Exam Scores

Picture the Scenario

The first exam in your statistics course is graded on a scale of 0 to 100. Suppose that the mean score in your class is 80.

Question to Explore

Which value is most plausible for the standard deviation s: 0, 10, or 50?

Think It Through

The standard deviation s is a *typical distance* of an observation from the mean. A value of $s = 0$ seems unlikely. For that to happen, every deviation would have to be 0. This means that every student must then score 80, the mean. A value of $s = 50$ is implausibly large since 50 or -50 would not be a typical distance of a score from the mean of 80. For instance, you can't score 130 on the exam. We would instead expect to see a value of s such as 10. With $s = 10$, a typical distance is 10, as occurs with the scores of 70 and 90.

Insight

In summary, we've learned that s is a typical distance of observations from the mean, larger values of s represent greater variability, and $s = 0$ means that all observations take the same value.

Try Exercises 2.50 and 2.51

SUMMARY: Properties of the Standard Deviation, s

- The greater the variability from the mean of the data, the larger is the value of s.
- $s = 0$ only when all observations take the same value. For instance, if the reported ideal number of children for seven people is 2, 2, 2, 2, 2, 2, 2, then the mean equals 2, each of the seven deviations equals 0, and $s = 0$. This is the minimum possible variability for a sample.

- s can be influenced by outliers. It uses the mean, which we know can be influenced by outliers. Also, outliers have large deviations, and so they tend to have *extremely* large squared deviations. We'll see (for example, in Exercise 2.47 and the beginning of Section 2.6) that these can inflate the value of s and make it sensitive to outliers.

Bell-shaped Distribution

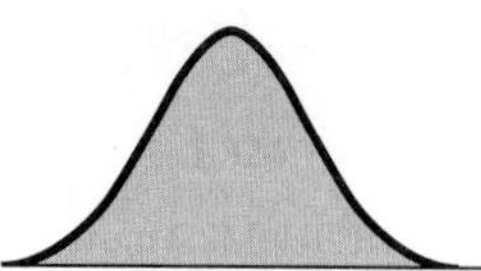

Interpreting the Magnitude of s: The Empirical Rule

Suppose that a distribution is unimodal and approximately symmetric with a **bell shape**, as in the margin figure. The value of s then has a more precise interpretation. Using the mean and standard deviation, we can form intervals that contain certain percentages (approximately) of the data.

In Words

- $\bar{x} - s$ denotes the value 1 standard deviation below the mean.
- $\bar{x} + s$ denotes the value 1 standard deviation above the mean.
- $\bar{x} \pm s$ denotes the values that are 1 standard deviation from the mean in either direction.

Empirical Rule

If a distribution of data is bell shaped, then approximately

- 68% of the observations fall within 1 standard deviation of the mean, that is, between the values of $\bar{x} - s$ and $\bar{x} + s$ (denoted $\bar{x} \pm s$).
- 95% of the observations fall within 2 standard deviations of the mean ($\bar{x} \pm 2s$).
- All or nearly all observations fall within 3 standard deviations of the mean ($\bar{x} \pm 3s$).

Figure 2.12 is a graphical portrayal of the empirical rule.

Did You Know?

The **empirical rule** has this name because many distributions of data observed in practice (*empirically*) are approximately bell shaped. ◀

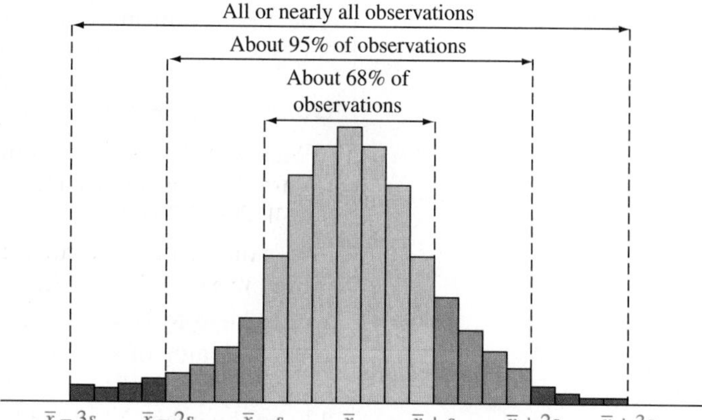

▲ **Figure 2.12 The Empirical Rule.** For bell-shaped distributions, this tells us approximately how much of the data fall within 1, 2, and 3 standard deviations of the mean. **Question** About what percentage would fall *more than* 2 standard deviations from the mean?

Empirical rule

Example 15

Female Student Heights

Picture the Scenario

Many human physical characteristics have bell-shaped distributions. Let's explore height. Question 1 on the student survey in Activity 3 of Chapter 1 asked for the student's height. Figure 2.13 shows a histogram of the heights from responses to this survey by 261 female students at the University of Georgia. (The data are in the Heights data file on the text CD. Note that the height of

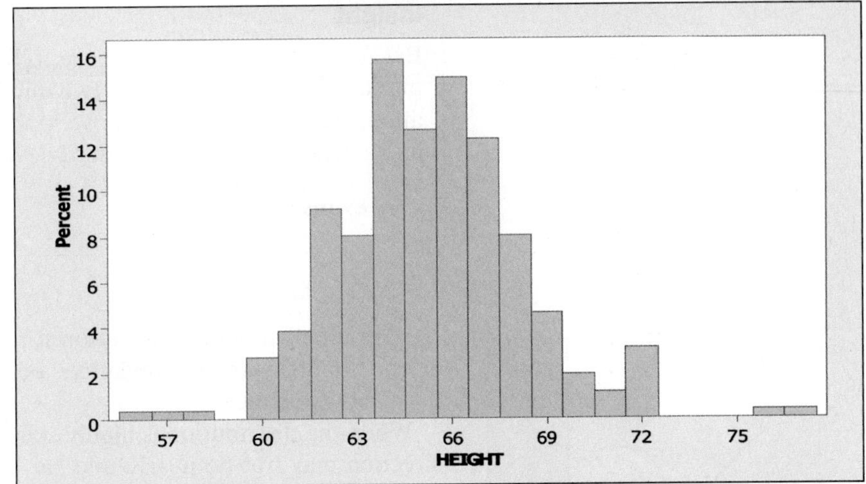

▲ **Figure 2.13 Histogram of Female Student Height Data.** This summarizes heights, in inches, of 261 female college students. **Question** How would you describe the shape, center, and variability of the distribution?

92 inches was omitted from the analysis.) Table 2.6 presents some descriptive statistics using MINITAB.

Question to Explore

Can we use the empirical rule to describe the variability from the mean of these data? If so, how?

Table 2.6 MINITAB Output for Descriptive Statistics of Student Height Data

Variable	N	Mean	Median	StDev	Minimum	Maximum
HEIGHT	261	65.284	65.000	2.953	56.000	77.000

Think It Through

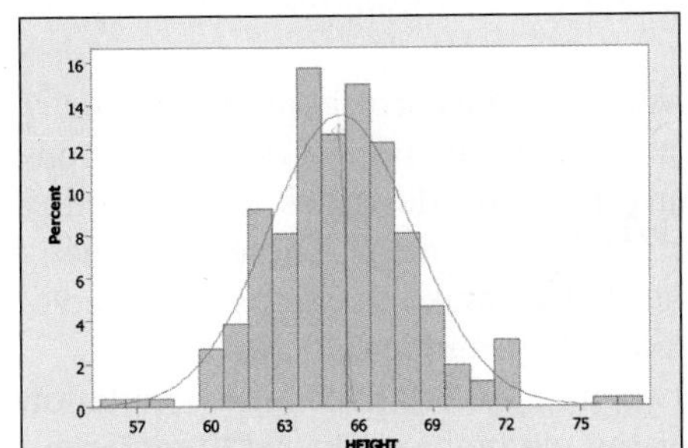

Figure 2.13 has approximately a bell shape. The figure in the margin shows a bell-shaped curve that approximates the histogram. From Table 2.6, the mean and median are close, about 65 inches, which reflects an approximately symmetric distribution. The empirical rule is applicable.

From Table 2.6 the mean is 65.3 inches and the standard deviation (labeled StDev) is 3.0 inches (rounded). By the empirical rule, approximately

- 68% of the observations fall between

  $$\bar{x} - s = 65.3 - 3.0 = 62.3 \text{ and } \bar{x} + s = 65.3 + 3.0 = 68.3,$$

  that is, within the interval (62.3, 68.3), about 62 to 68 inches.

- 95% of the observations fall within

  $\bar{x} \pm 2s$, which is $65.3 \pm 2(3.0)$, or (59.3, 71.3), about 59 to 71 inches.

- All or nearly all observations fall within $\bar{x} \pm 3s$, or (56.3, 74.3).

Of the 261 observations, by actually counting we find that

- 187 observations, 72%, fall within (62.3, 68.3).
- 248 observations, 95%, fall within (59.3, 71.3).
- 258 observations, 99%, fall within (56.3, 74.3).

In summary, the percentages predicted by the empirical rule are near the actual ones.

Caution

The empirical rule (which applies only to bell-shaped distributions) is not a general interpretation for what the standard deviation measures. The general interpretation is that the standard deviation measures the typical distance of observations from mean (this is for all distributions). ◀

Insight

Because the distribution is close to bell shaped, we can predict simple summaries effectively using only two numbers—the mean and the standard deviation. We can do the same if we look at the data only for the 117 males, who had a mean of 70.9 and standard deviation of 2.9.

Try Exercise 2.52

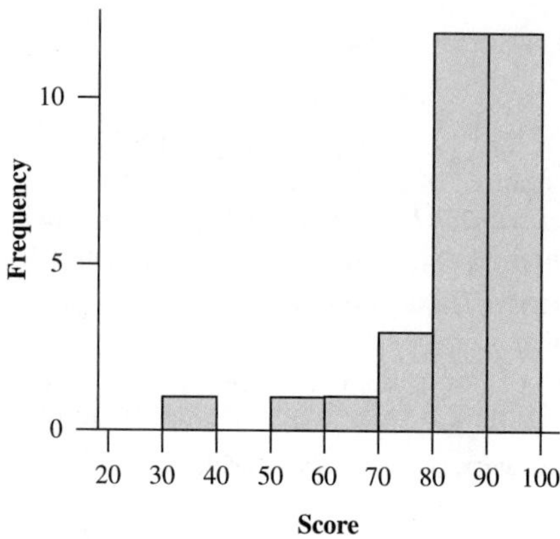

With a bell-shaped distribution for a large data set, the observations usually extend about 3 standard deviations below the mean and about 3 standard deviations above the mean.

When the distribution is highly skewed, the most extreme observation in one direction may not be nearly that far from the mean. For instance, on a recent exam, the scores ranged between 30 and 100, with median = 88, $\bar{x} = 84$, and $s = 16$. The maximum score of 100 was only 1 standard deviation above the mean (that is, $100 = \bar{x} + s = 84 + 16$). By contrast the minimum score of 30 was more than 3 standard deviations below the mean. This happened because the distribution of scores was highly skewed to the left. (See the margin figure.)

Remember that the empirical rule is only for bell-shaped distributions. In Chapter 6, we'll see why the empirical rule "works." For a special family of smooth, bell-shaped curves (called the **normal distribution**), we'll see how to find the percentage of the distribution in *any* particular region by knowing only the mean and the standard deviation.

Caution: Using the Empirical Rule

The empirical rule may approximate the actual percentages falling within 1, 2, and 3 standard deviations of the mean poorly if the data are highly skewed or highly discrete (the variable taking relatively few values). See Exercise 2.57. ◀

Recall

From Section 1.2, a **population** is the total group about whom you want to make conclusions. A **sample** is a subset of the population for whom you actually have data. A **parameter** is a numerical summary of the population, and a **statistic** is a numerical summary of a sample. ◀

Sample Statistics and Population Parameters

Of the numerical summaries introduced so far, the mean $\bar{x}$ and the standard deviation s are the most commonly used in practice. We'll use them frequently in the rest of the text. The formulas that define $\bar{x}$ and s refer to *sample* data. They are *sample statistics.*

We will distinguish between sample statistics and the corresponding *parameter* values for the population. The population mean is the average of all observations in the population. The population standard deviation describes the variability of the population observations about the population mean. These are usually unknown. Inferential statistical methods help us to make decisions and predictions about the population parameters based on the sample statistics.

In later chapters, to help distinguish between sample statistics and population parameters, we'll use different notation for the parameter values. Often, Greek letters are used to denote the parameters. For instance, we'll use μ (the Greek letter, mu) to denote the population mean and σ (the Greek letter lowercase sigma) to denote the population standard deviation.

2.4 Practicing the Basics

2.47 Sick leave A company decides to investigate the amount of sick leave taken by its employees. A sample of eight employees yields the following numbers of days of sick leave taken in the past year:
TRY

0 0 4 0 0 0 6 0

a. Find and interpret the range.

b. Find and interpret the standard deviation s.

c. Suppose the 6 was incorrectly recorded and is supposed to be 60. Redo parts a and b with the correct data and describe the effect of this outlier.

2.48 Life expectancy The *Human Development Report 2006*, published by the United Nations, showed life expectancies by country. For Western Europe, the values reported were

Denmark 77, Portugal 77, Netherlands 78, Finland 78, Greece 78, Ireland 78, UK 78, Belgium 79, France 79,

Germany 79, Norway 79, Italy 80, Spain 80, Sweden 80, Switzerland 80.

For Africa, the values reported (many of which were substantially lower than five years earlier because of the prevalence of AIDS) were

Botswana 37, Zambia 37, Zimbabwe 37, Malawi 40, Angola 41, Nigeria 43, Rwanda 44, Uganda 47, Kenya 47, Mali 48, South Africa 49, Congo 52, Madagascar 55, Senegal 56, Sudan 56, Ghana 57.

a. Which group (Western Europe or Africa) of life expectancies do you think has the larger standard deviation? Why?

b. Find the standard deviation for each group. Compare them to illustrate that s is larger for the group that shows more variability from the mean.

2.49 Life expectancy including Russia For Russia, the United Nations reported a life expectancy of 65 (down from 75 years earlier). Suppose we add this observation to the data set for Western Europe in the previous exercise. Would you expect the standard deviation to be larger, or smaller, than the value for the Western European countries alone? Why?

2.50 Shape of home prices? TRY According to the National Association of Home Builders, the median selling price of new homes in the United States in January 2007 was \$239,800. Which of the following is the most plausible value for the standard deviation: −\$15,000, \$1000, \$60,000, or \$1,000,000? Why? Explain what's unrealistic about each of the other values.

2.51 Exam standard deviation TRY For an exam given to a class, the students' scores ranged from 35 to 98, with a mean of 74. Which of the following is the most realistic value for the standard deviation: −10, 0, 3, 12, 63? Clearly explain what's unrealistic about each of the other values.

2.52 Heights TRY For the sample heights of Georgia college students in Example 15, the males had $\bar{x} = 71$ and $s = 3$ and the females had $\bar{x} = 65$ and $s = 3$.

a. Use the empirical rule to describe the distribution of heights for males.

b. The standard deviation for the overall distribution (combining females and males) was 4. Why would you expect it to be larger than the standard deviations for the separate male and female height distributions? Would you expect the overall distribution to still be unimodal?

2.53 Histograms and standard deviation The figure shows histograms for three different samples, each with sample size $n = 100$.

a. Which sample has the (i) largest and (ii) smallest standard deviation?

b. To which sample(s) is the empirical rule relevant? Why?

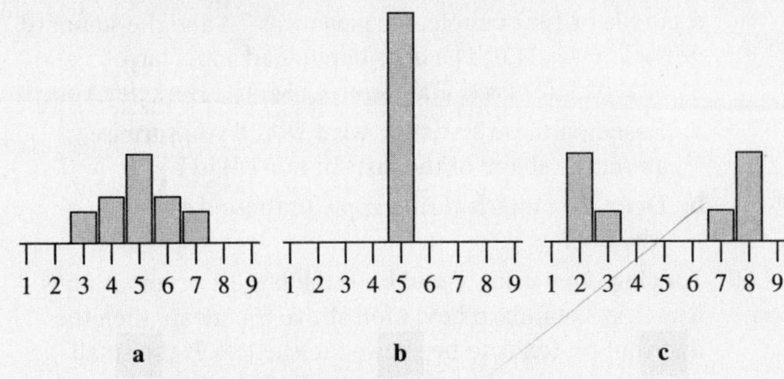

Histograms and relative sizes of standard deviations.

2.54 Female strength The High School Female Athletes data file on the text CD has data for 57 female high school athletes on the maximum number of pounds they were able to bench press. The data are roughly bell shaped, with $\bar{x} = 79.9$ and $s = 13.3$. Use the empirical rule to describe the distribution.

2.55 Female body weight The College Athletes data file on the text CD has data for 64 female college athletes. The data on weight (in pounds) are roughly bell shaped with $\bar{x} = 133$ and $s = 17$.

a. Give an interval within which about 95% of the weights fall.

b. Identify the weight of an athlete who is three standard deviations above the mean in this sample. Would this be a rather unusual observation? Why?

2.56 Shape of cigarette taxes A recent summary for the distribution of cigarette taxes (in cents) among the 50 states and Washington, D.C. in the United States reported $\bar{x} = 73$ and $s = 48$. Based on these values, do you think that this distribution is bell shaped? If so, why? If not, why not, and what shape would you expect?

2.57 Empirical rule and skewed, highly discrete distribution Example 12 gave data on the number of times married. For the observations for men, shown below, $\bar{x} = 0.16$ and $s = 0.37$.

No. Times	Count	Percentage
0	8418	84.0
1	1594	15.9
2	10	0.1
Total	**10022**	**100.0**

a. Find the actual percentages of observations within 1, 2, and 3 standard deviations of the mean. How do these compare to the percentages predicted by the empirical rule?

b. How do you explain the results in part a?

2.58 How much TV? The 2008 General Social Survey asked, "On the average day, about how many hours do you personally watch television?" Of 1,324 responses, the mode was 2, the median was 2, the mean was 2.98, and the standard deviation was 2.66. Based on these statistics, what would you surmise about the shape of the distribution? Why? (*Source:* Data from CSM, UC Berkeley.)

2.59 How many friends? A recent General Social Survey asked respondents how many close friends they had. For a sample of 1467 people, the mean was 7.4 and the standard deviation was 11.0. The distribution had a median of 5 and a mode of 4. (*Source:* Data from CSM, UC Berkeley.)

a. Based on these statistics, what would you surmise about the shape of the distribution? Why?

b. Does the empirical rule apply to these data? Why or why not?

2.60 Judging skew using $\bar{x}$ and s If the largest observation is less than 1 standard deviation above the mean, then the distribution tends to be skewed to the left. If the smallest observation is less than 1 standard deviation below the mean, then the distribution tends to be skewed to the right. A professor examined the results of the first exam given in her statistics class. The scores were

35 59 70 73 75 81 84 86

The mean and standard deviation are 70.4 and 16.7. Using these, determine if the distribution is either left or right skewed. Construct a dot plot to check.

2.61 EU data file The European Union Unemployment data file on the text CD contains unemployment rates in December 2003 for the 25 countries that were in the European Union in 2004. Using software,

a. Construct a graph to describe these values.

b. Find the standard deviation. Interpret.

2.62 Create data with a standard deviation Use the Standard Deviation applet on the text CD to investigate how the standard deviation changes as the data change.

a. Create 10 observations that have a mean of 5 and a standard deviation of about 2.

b. Create 10 observations that have a mean of 5 and a standard deviation of about 4.

c. Placing 10 values between 0 and 10, what is the largest standard deviation you can get? What are the values that have that standard deviation value?

2.5 Using Measures of Position to Describe Variability

The mean and median describe the center of a distribution. The range and the standard deviation describe the variability of the distribution. We'll now learn about some other ways of describing a distribution using measures of **position.**

One type of measure of position tells us the point where a certain percentage of the data fall above or fall below that point. The median is an example. It specifies a location such that half the data fall below it and half fall above it. The range uses two other measures of position, the maximum value and the minimum value. Another type of measure of position tells us *how far* an observation falls from a particular point, such as the number of standard deviations an observation falls from the mean.

Measures of Position: The Quartiles and Other Percentiles

The median is a special case of a more general set of measures of position called **percentiles.**

p%

pth Percentile

Percentile

The **pth percentile** is a value such that *p* percent of the observations fall below or at that value.

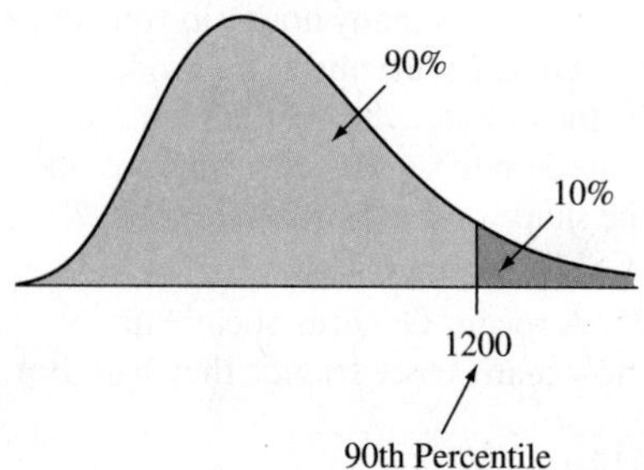

Suppose you're informed that your score of 1200 (out of 1600) on the SAT college entrance exam falls at the 90th percentile. Set $p = 90$ in this definition. Then, 90% of those who took the exam scored between the minimum score and 1200. Only 10% of the scores were higher than yours.

Substituting $p = 50$ in this definition gives the 50th percentile. For it, 50% of the observations fall below or at it and 50% above it. But this is simply the median. *The 50th percentile is usually referred to as the median.*

Three useful percentiles are the **quartiles.** The **first quartile** has $p = 25$, so it is the 25th percentile. The lowest 25% of the data fall below it. The **second quartile** has $p = 50$, so it is the 50th percentile, which is the median. The **third quartile** has $p = 75$, so it is the 75th percentile. The highest 25% of the data fall above it. *The quartiles split the distribution into four parts, each containing one quarter (25%) of the observations*. See Figure 2.14.

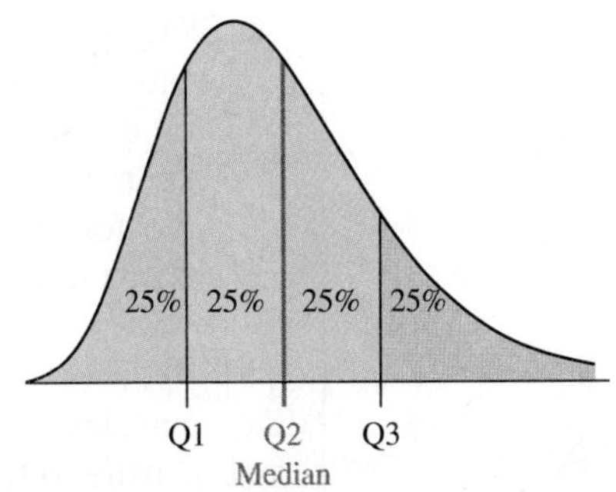

▲ **Figure 2.14 The Quartiles Split the Distribution Into Four Parts.** 25% is below the first quartile (Q1), 25% is between the first quartile and the second quartile (the median, Q2), 25% is between the second quartile and the third quartile (Q3), and 25% is above the third quartile. **Question** Why is the second quartile also the median?

The quartiles are denoted by Q1 for the first quartile, Q2 for the second quartile, and Q3 for the third quartile. Notice that one quarter of the observations fall below Q1, two quarters (one half) fall below Q2 (the median), and three quarters fall below Q3.

SUMMARY: Finding Quartiles

- Arrange the data in order.
- Consider the median. This is the **second quartile, Q2.**
- Consider the lower half of the observations (excluding the median itself if n is odd). The median of these observations is the **first quartile, Q1.**
- Consider the upper half of the observations (excluding the median itself if n is odd). Their median is the **third quartile, Q3.**

Example 16

The quartiles

Cereal Sodium Data

Picture the Scenario

Let's again consider the sodium values for the 20 breakfast cereals.

Questions to Explore

a. What are the quartiles for the 20 cereal sodium values?

b. Interpret the quartiles in the context of the cereal data.

Think it Through

a. From Table 2.3 the sodium values, arranged in ascending order, are

Q1 = 135

0	50	70	100	**130**	**140**	140	150	160	**180**
180	180	190	200	**200**	**210**	210	220	290	340

Q3 = 205

- The median of the 20 values is the average of the 10th and 11th observations, 180 and 180, which is **Q2 = 180 mg.**
- The first quartile Q1 is the median of the 10 smallest observations (in the top row), which is the average of 130 and 140, **Q1 = 135 mg.**
- The third quartile Q3 is the median of the 10 largest observations (in the bottom row), which is the average of 200 and 210, **Q3 = 205 mg.**

b. The quartiles tell you how the data split into four parts. The sodium values range from 0 mg to 135 mg for the first quarter, 135 mg to 180 mg for the second quarter, 180 mg to 205 mg for the third quarter, and 205 mg to 340 mg for the fourth quarter. By dividing the distribution into four quarters, we see that some quarters of the distribution vary more than others.

Insight

The quartiles also give information about shape. The distance of 45 from the first quartile to the median exceeds the distance of 25 from the median to the third quartile. This commonly happens when the distribution is skewed to the left, as shown in the margin figure. Although each quarter of a distribution may span different lengths (indicating different amounts of variability within the quarters of the distribution), each quarter contains the same number (25%) of observations.

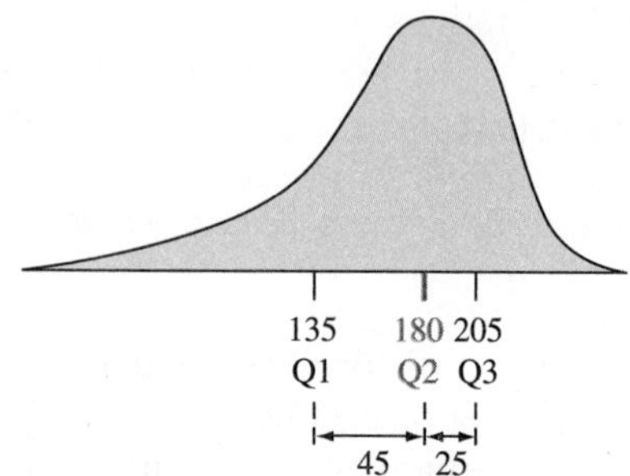

Try Exercise 2.63

In Practice Finding Percentiles Using Technology

Percentiles other than the quartiles are reported only for large data sets. Software can do the computational work for you, and we won't go through the details. Precise algorithms for the calculations use interpolations, and different software often uses slightly different rules. This is true even for the quartiles Q1 and Q3: Most software, but not all, does not use the median observation itself in the calculation when the number of observations *n* is odd.

Q1 Q2 Q3

IQR

Measuring Variability: The Interquartile Range

The quartiles are also used to define a measure of variability that is more resistant than the range and the standard deviation. This measure summarizes the range for the *middle half* of the data. The middle 50% of the observations fall between the first quartile and the third quartile—25% from Q1 to Q2 and 25% from Q2 to Q3. The distance from Q1 to Q3 is called the **interquartile range**, denoted by **IQR**.

In Words

If the **interquartile range** of U.S. music teacher salaries equals $16,000, this means that for the middle 50% of the distribution of salaries, $16,000 is the distance between the largest and smallest salaries.

Interquartile Range (IQR)

The **interquartile range** is the distance between the third and first quartiles,

IQR = Q3 − QI

For instance, for the breakfast cereal sodium data, we just saw in Example 16 that

- Minimum value = 0
- First quartile Q1 = 135
- Median = 180
- Third quartile Q3 = 205
- Maximum value = 340

The range is 340 − 0 = 340. The interquartile range is Q3 − Q1 = 205 − 135 = 70.

As with the range and standard deviation s, the more varied the data, the larger the IQR tends to be. But unlike those measures, the IQR is not affected by any observations below the first quartile or above the third quartile. In other words, it is not affected by outliers. In contrast, the range depends solely on the minimum and the maximum values, the most extreme values, so the range changes as either extreme value changes. For example, if the highest sodium value were 1000 instead of 340, the range would change dramatically from 340 to 1000, but the IQR would not change. So, it's often better to use the IQR instead of the range or standard deviation to compare the variability for distributions that are very highly skewed or that have severe outliers.

Detecting Potential Outliers

Examining the data for unusual observations, such as outliers, is important in any statistical analysis. Is there a formula for flagging an observation as potentially being an outlier? One way uses the interquartile range.

> **The 1.5 × IQR Criterion for Identifying Potential Outliers**
>
> An observation is a potential outlier if it falls more than 1.5 × IQR below the first quartile or more than 1.5 × IQR above the third quartile.

Recall

Ordered Sodium Values

0 50 . . . 340

From Example 16, the breakfast cereal sodium data has Q1 = 135 and Q3 = 205. So, IQR = Q3 − Q1 = 205 − 135 = 70. For those data

1.5 × IQR = 1.5 × 70 = 105.

Q1 − 1.5 × IQR = 135 − 105 = 30 (lower boundary, potential outliers below),

and

Q3 + 1.5 × IQR = 205 + 105 = 310 (upper boundary, potential outliers above).

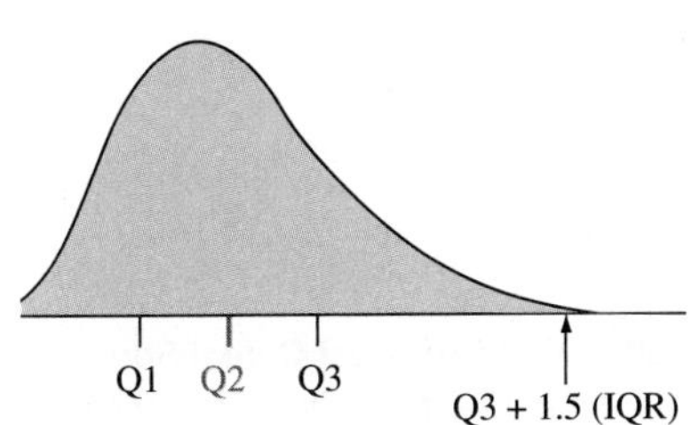

By the 1.5 × IQR criterion, observations below 30 or above 310 are potential outliers. The only observations below 30 or above 310 are the sodium values of 0 mg for Frosted Mini Wheats and 340 mg for Raisin Bran. These are the only potential outliers.

Why do we identify an observation as a *potential* outlier rather than calling it a *definite* outlier? When a distribution has a long tail, some observations may be more than 1.5 × IQR below the first quartile or above the third quartile even if they are not outliers, in the sense that they are not separated far from the bulk of the data. For instance, in a long right tail there need not be a long gap between the largest observation and the rest of the data.

The Box Plot: Graphing a Five-Number Summary of Positions

The quartiles and the maximum and minimum values are five numbers often used as a set to summarize positions that help to describe center and variability of a distribution.

The **five-number summary** is the basis of a graphical display called the **box plot**. The **box** of a box plot contains the central 50% of the distribution, from the

> **The Five-Number Summary of Positions**
>
> The five-number summary of a data set is the minimum value, first quartile Q1, median, third quartile Q3, and the maximum value.

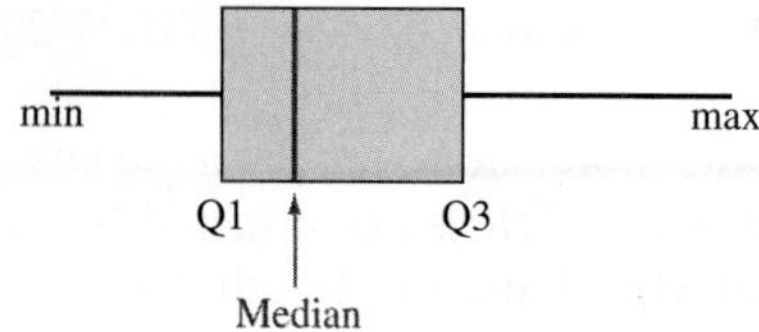

first quartile to the third quartile (see the margin figure). A line inside the box marks the median. The lines extending from the box are called **whiskers**. These extend to encompass the rest of the data, except for potential outliers, which are shown separately.

SUMMARY: Constructing a Box Plot

- A **box** goes from the lower quartile Q1 to the upper quartile Q3.
- A line is drawn inside the box at the median.
- A line goes from the lower end of the box to the smallest observation that is not a potential outlier. A separate line goes from the upper end of the box to the largest observation that is not a potential outlier. These lines are called **whiskers**. The potential outliers (more than 1.5 IQR below the first quartile or above the third quartile) are shown separately.

Box plot

Example 17

Cereal Sodium Data

Picture the Scenario

Example 7 constructed a histogram for the cereal sodium values. That figure is shown again in the margin. Figure 2.15 shows a box plot for the sodium values. Labels are also given for the five-number summary of positions.

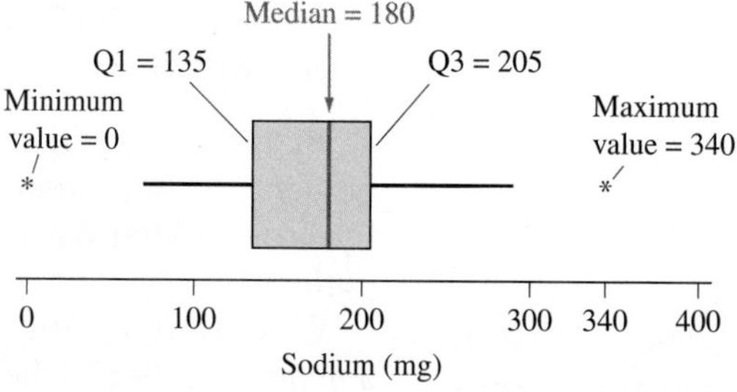

▲ **Figure 2.15 Box Plot and Five-Number Summary for 20 Breakfast Cereal Sodium Values.** The central box contains the middle 50% of the data. The line in the box marks the median. Whiskers extend from the box to the smallest and largest observations, which are not identified as potential outliers. Potential outliers are marked separately. **Question** Why is the left whisker drawn down only to 50 rather than to 0?

Questions to Explore

a. Which, if any, values are considered outliers?

b. Explain how the box plot in Figure 2.15 was constructed and how to interpret it.

Think It Through

a. We can identify outliers using the 1.5*IQR criterion. We know that Q1 = 135 mg and Q3 = 205 mg. Thus IQR = 205 − 135 = 70 mg, and the lower and upper boundaries are 135 − 1.5*70 = 30 mg and 205 + 1.5*70 = 310 mg, respectively. The sodium value of 0 mg for Frosted Mini Wheats is less than 30 mg and the value of 340 mg for

Raisin Bran is greater than 310 mg, so each of these values is considered an outlier.

b. The five-number summary of sodium values shown on the box plot is minimum = 0, Q1 = 135, median = 180, Q3 = 205, and maximum = 340. The middle 50% of the distribution of sodium values range from Q1 = 135 mg to Q3 = 205 mg, which are the two outer lines of the box. The median of 180 mg is indicated by the center line through the box. As we saw in part a, the 1.5 * IQR criterion flags the sodium values of 0 mg for Frosted Mini Wheats and 340 mg for Raisin Bran as outliers. These values are represented on the box plot as asterisks. The whisker extending from Q1 is drawn down to 50, which is the smallest observation that is not below the lower boundary of 30. The whisker extending from Q3 is drawn up to 290, which is the largest observation that is not above the upper boundary of 310.

Insight

Most software identifies observations that are more than 1.5 IQR from the quartiles by a symbol, such as *. (Some software uses a second symbol for observations more than 3 IQR from the quartiles.) The TI-83+/84 output for the box plot is shown in the margin. Why show potential outliers separately? One reason is to identify them for further study. Was the observation incorrectly recorded? Was that subject fundamentally different from the others in some way? Often it makes sense to repeat a statistical analysis without an outlier, to make sure the results are not overly sensitive to a single observation. For the cereal data, the sodium value of 0 (for Frosted Mini Wheats) was not incorrectly recorded. It is merely an unusually small value, relative to sodium levels for the other cereals. The sodium value of 340 for Raisin Bran was merely an unusually high value.

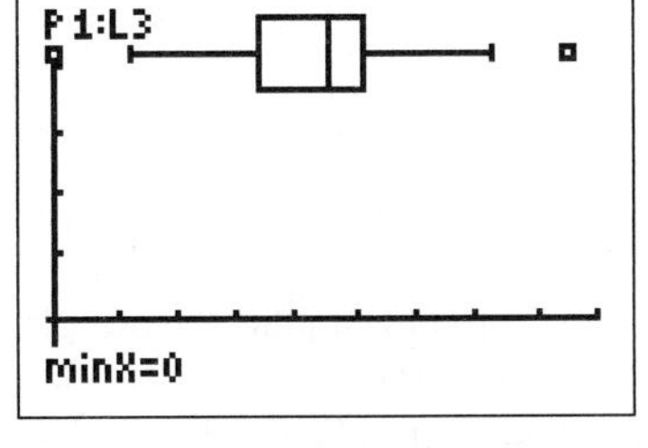

TI-83+/84 output

Try Exercises 2.72 and 2.74

Another reason for keeping outliers separate in a box plot is that they do not provide much information about the shape of the distribution, especially for large data sets. Some software can also provide a box plot that extends the whiskers to the minimum and maximum, even if outliers exist. However, an extreme value can then cause the box plot to give the impression of severe skew when actually the remaining observations are not at all skewed.

1000 1100 1200

The Box Plot Compared with the Histogram

A box plot does not portray certain features of a distribution, such as distinct mounds and possible gaps, as clearly as does a histogram. For example, the histogram and box plot in the margin refer to the same data. The histogram suggests that the distribution is bimodal (two distinct mounds), but we could not learn this from the box plot. This comparison shows that more than one type of graph may be needed to summarize a data set well.

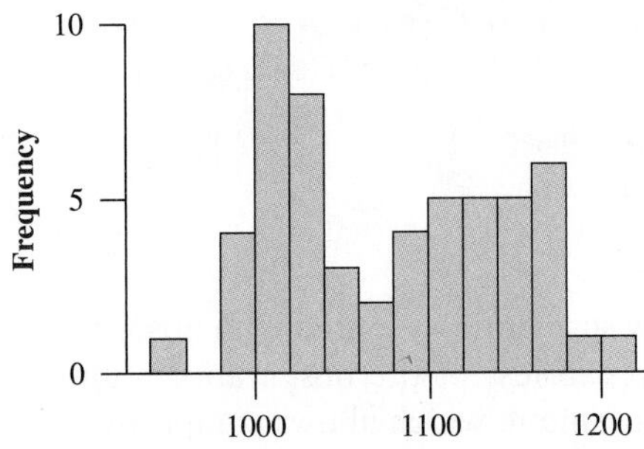

A box plot does indicate skew from the relative lengths of the whiskers and the two parts of the box. The side with the larger part of the box and the longer whisker usually has skew in that direction. However, the box plot will not show us if there is a large gap in the distribution contributing to the skew. And as we've seen, **box plots are useful for identifying potential outliers**. We'll see next that they're also useful for graphical comparisons of distributions.

In Practice

The histogram works best for larger data sets. For smaller data sets (those easily constructed by hand), the dot plot is a useful graph of the actual data to construct along with the box plot.

Side-by-Side Box Plots Help to Compare Groups

In Example 15 we looked at female college student heights from the Heights data file on the text CD. To compare heights for females and males, we could look at side-by-side box plots, as shown in Figure 2.16.

The box plots suggest that both distributions are approximately symmetric. The median (the center line in a box) is approximately 71 inches for the males and 65 inches for the females. Although the centers differ, the variability of the middle 50% of the distribution is similar, as indicated by the width of the boxes (which is the IQR) being similar. Both samples have heights that are unusually short or tall, flagged as potential outliers. The upper 75% of the male heights are higher than the lower 75% of female heights. That is, 75% of the female heights fall below their third quartile, about 67 inches, whereas 75% of the male heights fall above their first quartile, about 69 inches.

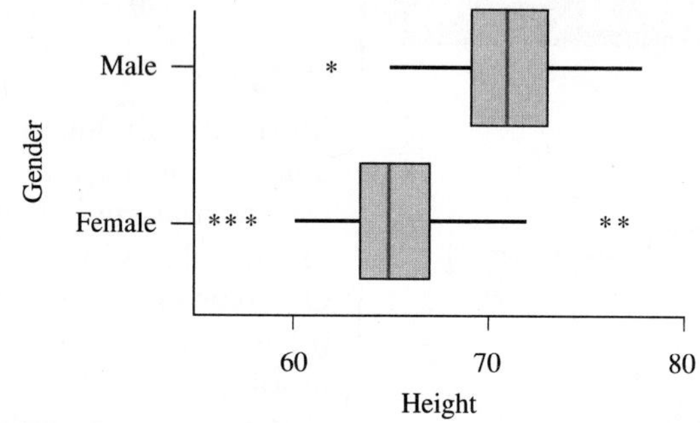

▲ **Figure 2.16 Box Plots of Male and Female College Student Heights.** The box plots use the same scale for height. **Question** What are approximate values of the quartiles for the two groups?

The *z*-Score Also Identifies Position and Potential Outliers

The empirical rule tells us that for a bell-shaped distribution, it is unusual for an observation to fall more than 3 standard deviations from the mean. An alternative criterion for identifying potential outliers uses the standard deviation.

- An observation in a bell-shaped distribution is regarded as a potential outlier if it falls more than 3 standard deviations from the mean.

How do we know the number of standard deviations that an observation falls from the mean? When $\bar{x} = 84$ and $s = 16$, a value of 100 is 1 standard deviation above the mean, since $(100 - 84) = 16$. Alternatively, $(100 - 84)/16 = 1$.

Taking the difference between an observation and the mean and dividing by the standard deviation tells us the number of standard deviations that the observation falls from the mean. This number is called the ***z*-score**.

z-score

The ***z*-score** for an observation is the number of standard deviations that it falls from the mean. A positive *z*-score indicates the observation is above the mean. A negative *z*-score indicates the observation is below the mean. For sample data, the *z*-score is calculated as

$$z = \frac{\text{observation} - \text{mean}}{\text{standard deviation}}.$$

The *z*-score allows us to quickly tell how surprising or extreme an observation is. The *z*-score converts an observation (regardless of the observation's unit of measurement) to a common scale of measurement, which allows comparisons.

Example 18

z-scores

Pollution Outliers

Picture the Scenario

Let's consider air pollution data for the European Union (EU). The Energy-EU data file[8] on the text CD contains data on per capita carbon-dioxide (CO_2) emissions, in metric tons, for the 27 nations in the EU. The mean was 8.3 and the standard deviation was 3.6. Figure 2.17 shows a box plot of the data. The maximum of 21.3, representing Luxembourg, is highlighted as a potential outlier.

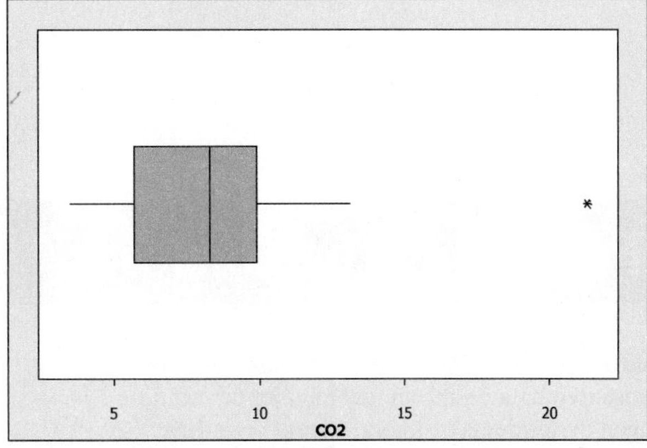

▲ **Figure 2.17** **MINITAB Box Plot of Carbon Dioxide Emissions for European Union Nations.** Question Can you use this plot to approximate the five-number summary?

Questions to Explore

a. How many standard deviations from the mean was the CO_2 value of 21.3 for Luxembourg?

b. The CO_2 value for the United States was 18.9. According to the three standard deviation criterion, is the United States an outlier on carbon dioxide emissions relative to the EU?

Think It Through

a. Since $\bar{x} = 8.3$ and $s = 3.6$ inches, the z-score for the observation of 21.3 is

$$z = \frac{\text{observation} - \text{mean}}{\text{standard deviation}} = \frac{21.3 - 8.3}{3.6} = \frac{13}{3.6} = 3.6.$$

The carbon dioxide emission (per capita) for Luxembourg is 3.6 standard deviations above the mean. By the 3 standard deviation criterion, this is a potential outlier. Since it's well removed from the rest of the data, we'd regard it as an actual outlier. However, Luxembourg has only 350,000 people, so in terms of the *amount* of pollution it is not a major polluter in the EU.

b. The z-score for the CO_2 value of the United States is $z = (18.9 - 8.3)/3.6 = 2.9$. Although the 3 standard deviation rule fails to flag the United States as an outlier relative to EU nations, the value of 2.9 is close enough to 3 to garner some attention. Furthermore, because of

[8] *Source:* www.iea.org/stats/index.asp.

the relatively large size of the U.S. population, a z-score this close to 3 indicates that the U.S. is a significant contributor to overall CO_2 emission.

Insight

The z-scores of 3.6 and 2.9 are positive. This indicates that the observations are *above* the mean, because an observation above the mean has a positive z-score. In fact, these large positive z-scores tell us that Luxembourg and the United States have very high CO_2 emissions compared to the other nations. The z-score is negative when the observation is *below* the mean. For instance, France has a CO_2 value of 5.7, which is below the mean of 8.3 and has a z-score of −0.7.

Try Exercises 2.76 and 2.77

2.5 Practicing the Basics

2.63 Vacation days *National Geographic Traveler* magazine recently presented data on the annual number of vacation days averaged by residents of eight different countries. They reported 42 days for Italy, 37 for France, 35 for Germany, 34 for Brazil, 28 for Britain, 26 for Canada, 25 for Japan, and 13 for the United States.

TRY

a. Report the median.

b. By finding the median of the four values below the median, report the first quartile.

c. Find the third quartile.

d. Interpret the values found in parts a–c in the context of these data.

2.64 European unemployment In recent years, many European nations have suffered from relatively high unemployment. For the 15 nations that made up the European Union in 2003, the table shows the unemployment rates reported by Eurostat as of January 2007.

a. Find and interpret the median.

b. Find the first quartile (Q1) and the third quartile (Q3).

c. Find and interpret the mean.

European Union 2007 Unemployment Rates (www.europa.eu.int/comm/eurostat)

Belgium 7.8	France 8.4	Italy 6.7
Denmark 3.2	Portugal 7.2	Finland 7.0
Germany 7.7	Netherlands 3.6	Austria 4.5
Greece 8.7	Luxembourg 5.0	Sweden 6.0
Spain 8.6	Ireland 4.4	U.K. 5.4

Source: © European Union, 1995–2011.

2.65 Female strength The High School Female Athletes data file on the text CD has data for 57 high school female athletes on the maximum number of pounds they were able to bench press, which is a measure of strength. For these data, $\bar{x} = 79.9$, Q1 = 70, median = 80, Q3 = 90.

a. Interpret the quartiles.

b. Would you guess that the distribution is skewed or roughly symmetric? Why?

2.66 Female body weight The College Athletes data file on the text CD has data for 64 college female athletes. The data on weight (in pounds) has $\bar{x} = 133$, Q1 = 119, median = 131.5, Q3 = 144.

a. Interpret the quartiles.

b. Would you guess that the distribution is skewed, or roughly symmetric? Why?

2.67 Ways to measure variability The standard deviation, the range, and the interquartile range (IQR) summarize the variability of the data.

a. Why is the standard deviation s usually preferred over the range?

b. Why is the IQR sometimes preferred to s?

c. What is an advantage of s over the IQR?

2.68 Variability of cigarette taxes Here's the five-number summary for the distribution of cigarette taxes (in cents) among the 50 states and Washington, D.C. in the United States.

Minimum = 2.5, Q1=36, Median = 60,
Q3 = 100, Maximum = 205

a. About what proportion of the states have cigarette taxes (i) greater than 36 cents and (ii) greater than $1?

b. Between which two values are the middle 50% of the observations found?

c. Find the interquartile range. Interpret it.

d. Based on the summary, do you think that this distribution was bell shaped? If so, why? If not, why not, and what shape would you expect?

2.69 Sick leave Exercise 2.47 showed data for a company that investigated the annual number of days of sick leave taken by its employees. The data are

0 0 4 0 0 0 6 0

a. The standard deviation is 2.4. Find and interpret the range.

b. The quartiles are Q1 = 0, median = 0, Q3 = 2. Find the interquartile range.

c. Suppose the 6 was incorrectly recorded and is supposed to be 60. The standard deviation is then 21.1 but the quartiles do not change. Redo parts a–c with the correct data and describe the effect of this outlier. Which measure of variability, the range, IQR, or standard deviation, is least affected by the outlier? Why?

2.70 Infant mortality Africa The *Human Development Report 2006,* published by the United Nations, showed infant mortality rates (number of infant deaths per 1000 live births) by country. For Africa, some of the values reported were:

South Africa 54, Sudan 63, Ghana 68, Madagascar 76, Senegal 78, Zimbabwe 79, Uganda 80, Congo 81, Botswana 84, Kenya 96, Nigeria 101, Malawi 110, Mali 121, Angola 154.

a. Find the first quartile (Q1) and the third quartile (Q3).

b. Find the interquartile range (IQR). Interpret it.

2.71 Infant mortality Europe For Western Europe, the infant mortality rates reported by the *Human Development Report 2006* were

Sweden 3, Finland 3, Spain 3, Belgium 4, Denmark 4, France 4, Germany 4, Greece 4, Italy 4, Norway 4, Portugal 4, Netherlands 5, Switzerland 5, UK 5.

Show that Q1 = Q2 = Q3 = 4. (The quartiles, like the median, are less useful when the data are highly discrete.)

2.72 Computer use TRY During a recent semester at the University of Florida, students having accounts on a mainframe computer had storage space use (in kilobytes) described by the five-number summary, minimum = 4, Q1 = 256, median = 530, Q3 = 1105, and maximum = 320,000.

a. Would you expect this distribution to be symmetric, skewed to the right, or skewed to the left? Explain.

b. Use the 1.5 × IQR criterion to determine if any potential outliers are present.

2.73 Central Park temperature distribution revisited Exercise 2.27 showed a histogram for the distribution of Central Park annual average temperatures for the 20th century. The box plot for these data is shown here.

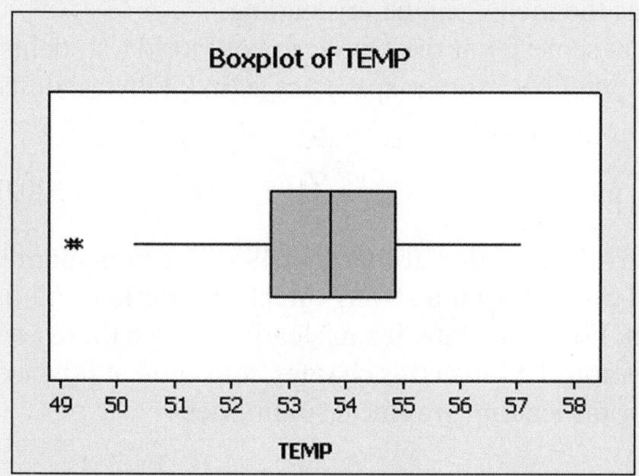

a. If this distribution is skewed, would you expect it to be skewed to the right or to the left? Explain.

b. Approximate each component of the five-number summary, and interpret.

2.74 Box plot for exam TRY The scores on an exam have mean = 88, standard deviation = 10, minimum = 65, Q1 = 77, median = 85, Q3 = 91, maximum = 100. Sketch a box plot, labeling which of these values are used in the plot.

2.75 Public transportation Exercise 2.37 described a survey about how many miles per day employees of a company use public transportation. The sample values were

0 0 4 0 0 0 10 0 6 0

a. Identify the five-number summary, and sketch a box plot.

b. Explain why Q1 and the median share the same line in the box.

c. Why does the box plot not have a left whisker?

2.76 Energy statistics TRY The Energy Information Administration records per capita consumption of energy by country. The 2006 data for the 27 nations that now make up the European Union are used to create the boxplot below. The energy values (in millions of BTUs) have a mean of 167.8 and a standard deviation of 72.8, and are roughly bell shaped, except for the value of 424.1 for Luxembourg.

a. Using the MINITAB box plot shown, give approximate values for the five-number summary and indicate whether any countries were judged to be potential outliers according to that plot.

b. Italy had a value of 138.7. How many standard deviations from the mean was it?

c. The United States is not in the data used below, but its value was 334. Relative to the distribution for the EU nations, how many standard deviations from the mean was it?

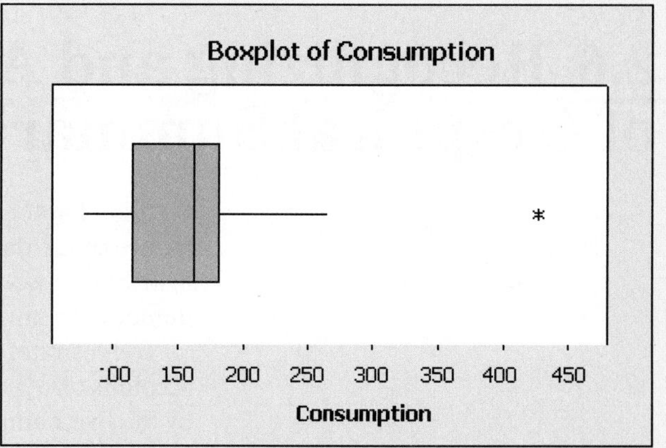

2.77 European Union unemployment rates TRY The 2007 unemployment rates of countries in the European Union shown in Exercise 2.64 ranged from 3.2 to 8.7, with Q1 = 4.5, median = 6.7, Q3 = 7.8, a mean of 6.3, and standard deviation of 1.8.

a. In a box plot, what would be the values at the outer edges of the box, and what would be the values to which the whiskers extend?

b. Greece had the highest unemployment rate of 8.7. Is it an outlier according to the 3 standard deviation criterion? Explain.

c. What unemployment value for a country would have a *z*-score equal to 0?

2.78 Air pollution Example 18 discussed EU carbon dioxide emissions, which had a mean of 8.3 and standard deviation of 3.6.

a. Canada's observation was 16.5. Find its *z*-score relative to the distribution of values for the EU nations, and interpret.

b. Sweden's observation was 5.0. Find its *z*-score, and interpret.

2.79 Female heights For the 261 female heights shown in the box plot in Figure 2.16, the mean was 65.3 inches and the standard deviation was 3.0 inches. The shortest person in this sample had a height of 56 inches.

a. Find the *z*-score for the height of 56 inches.

b. What does the negative sign for the *z*-score represent?

c. Is this observation a potential outlier according to the 3 standard deviation distance criterion? Explain.

2.80 Hamburger sales The manager of a fast-food restaurant records each day for a year the amount of money received from sales of food that day. Using software, he finds a bell-shaped histogram with a mean of $1165 and a standard deviation of $220. Today the sales equaled $2000. Is this an unusually good day? Answer by providing statistical justification.

2.81 Florida students again Refer to the FL Student Survey data set on the text CD and the data on weekly hours of TV watching.

a. Use software to construct a box plot. Interpret the information on the plot, and use it to describe the shape of the distribution.

b. Using a criterion for outliers, investigate whether there are any potential outliers.

2.82 Females or males watch more TV? Refer to the previous exercise. Suppose you wanted to compare TV watching of males and females. Construct a side-by-side box plot to do this. Interpret.

2.83 CO_2 comparison The MINITAB vertical side-by-side box plots shown below compare the values reported by the UN of per capita carbon dioxide emissions for nations in the European Union and in South America, in 2003.

a. Give the approximate value of carbon dioxide emissions for the outlier shown.

b. What shape would you predict for the distribution in South America? Why?

c. Summarize how the carbon dioxide emissions compare in Europe and South America.

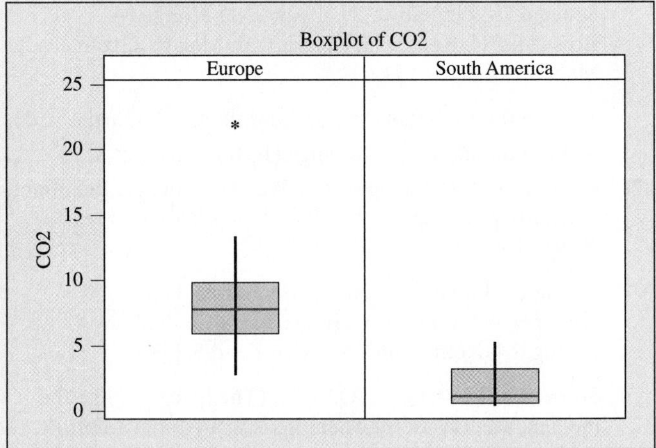

2.6 Recognizing and Avoiding Misuses of Graphical Summaries

In this chapter, we've learned how to describe data. We've learned that the nature of the data affects how we can effectively describe them. For example, if a distribution is very highly skewed or has extreme outliers, we've seen that some numerical summaries, such as the mean, can be misleading.

To illustrate, during a recent semester at the University of Florida, students with accounts on a mainframe computer had storage space usage (in kilobytes) described by the five-number summary:

minimum = 4, Q1 = 256, median = 530, Q3 = 1105, maximum = 320,000

and by the mean of 1921 and standard deviation of 11,495. The maximum value was an extreme outlier. You can see what a strong effect that outlier had on the mean and standard deviation. For these data, it's misleading to use these two values to summarize the distribution. To finish this chapter, let's look at how we also need to be on the lookout for misleading graphical summaries.

Beware of Poor Graphs

With modern computer-graphic capabilities, Web sites, newspapers, and other periodicals use graphs in an increasing variety of ways to portray information. The graphs are not always well designed, however, and you should look at them skeptically.

Example 19

Misleading graphical summaries

Recruiting STEM Majors

Picture the Scenario

Look at Figure 2.18. According to the title and the two-sentence caption, the graph is intended to display how total enrollment has risen at a United States (U.S.) university in recent years while the number of STEM (science, technology, engineering, and mathematics) students has "dipped just about every year." A graphic designer used a software program to construct a graph for use in a local newspaper. The Sunday headline story was about the decline of STEM majors. The graph is a time plot showing the enrollment between 2004 and 2012, using outlined human figures to portray total enrollment and blue human figures to portray STEM enrollment.

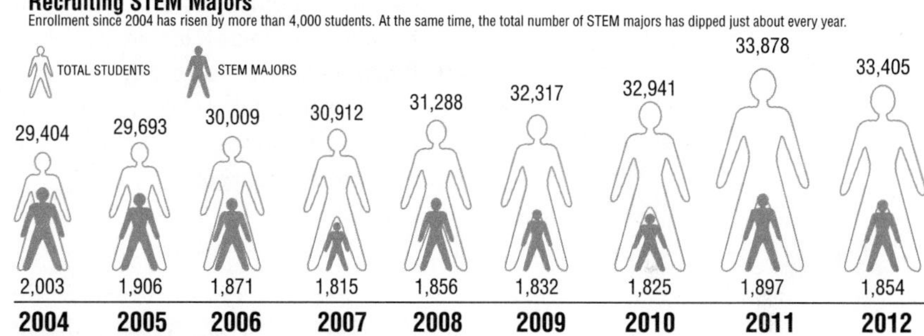

▲ **Figure 2.18 An Example of a Poor Graph. Question** What's misleading about the way the data are presented?

Questions to Explore

a. Do the heights of the human figures accurately represent the counts? Are the areas of the figures in accurate proportions to each other? Answer by first comparing the two observations for 2004 and then by comparing those for 2004 to those for 2012.

b. Is an axis shown and labeled as a reference line for the enrollments displayed?

c. What other design problems do you see?

Think It Through

a. In the year 2004, the number of STEM majors (2003) is about 7% of the total of 29,404. However, the height of the number of STEM majors figure is about 2/3 the height of the total enrollment figure. The graphs also mislead by using different widths for the figures. Compare, for instance, the figures for total enrollment in 2004 and 2012. The total enrollments are not nearly as different as the areas taken up by the figures make them appear.

b. Figure 2.18 does not provide a vertical axis, much less a labeling of that axis. It is not clear whether the counts are supposed to be represented by the heights or by the areas of the human figures.

c. It is better to put the figures next to each other, rather than to overlay the two sets of figures as was done here. The use of solid blue figures and outlined figures can easily distort your perception of the graph. While it may not have been deliberate, the design obscures the data.

Insight

The intent of Figure 2.18 was to provide an intriguing visual display that quickly tells the story of the newspaper article. However, deviating from standard graphs gave a misleading representation of the data.

Try Exercise 2.87

SUMMARY: Guidelines For Constructing Effective Graphs

- Label both axes and provide a heading to make clear what the graph is intended to portray. (Figure 2.18 does not even provide a vertical axis.)
- To help our eyes visually compare relative sizes accurately, the vertical axis should usually start at 0.
- Be cautious in using figures, such as people, in place of the usual bars or points. It can make a graph more attractive, but it is easy to get the relative percentages that the figures represent incorrect.
- It can be difficult to portray more than one group on a single graph when the variable values differ greatly. (In Figure 2.18, one frequency is very small compared to the other.) Consider instead using separate graphs, or plotting relative sizes such as ratios or percentages.

Example 20

Improving graphical summaries

Recruiting STEM Majors

Picture the Scenario

Let's look at Figure 2.18 again—because it's such a good example of how things can go wrong. Suppose the newspaper asks you to repair the graph, providing an accurate representation of the data.

Questions to Explore

a. Show a better way to graphically portray the same information about the enrollment counts in Figure 2.18.

b. Describe two alternative ways to portray the data that more clearly show the number of STEM majors relative to the size of the student body over time.

Think It Through

a. Since the caption of Figure 2.18 indicates that counts are the desired data display, we can construct a time plot showing total and STEM major enrollments on the same scale. Figure 2.19 shows this revised

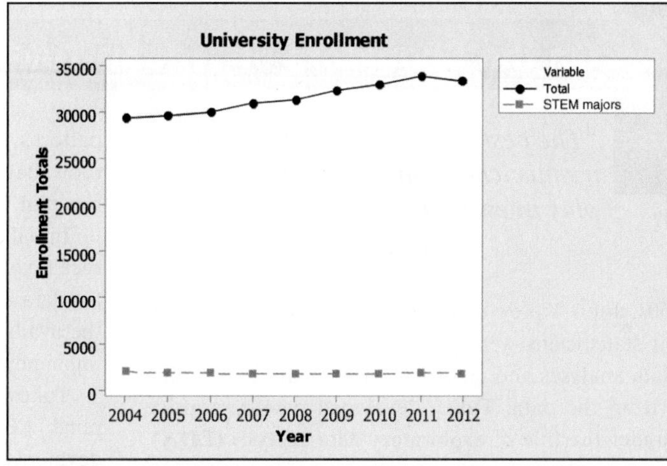

▲ **Figure 2.19** A Better Graph for the Data in Figure 2.18. **Question** What trends do you see in the enrollments from 2004 to 2012?

graph. This time plot clearly labels the horizontal and vertical axes. There is the natural reference starting point of 0 for the vertical axis, allowing for better relative comparison of the two groups. We can see that the STEM major enrollment is a small part of the total enrollment. Between 2004 and 2012 there's been some tendency upward in the total enrollment, with relatively little change in the STEM major enrollment.

b. Figure 2.19 is better than Figure 2.18 but is not ideal. Any trend that may be there for STEM majors is not clear because their counts are small compared to the total enrollment counts. Given that both the total and the STEM major counts vary from year to year, it's more meaningful to look at the STEM major *percentage* of the student body and plot that over time. See Exercise 2.86, which shows that this percentage went down from 2004 to 2012.

Even if we plot the counts, as in Figure 2.19, it may be better to show a separate plot on its own scale for the STEM major counts. Or, in a single graph, rather than comparing the STEM major counts to the *total* enrollment (which contains the STEM majors group), it may be more meaningful to compare them to the *rest* of the enrollment, that is, the total enrollment minus the number of STEM majors.

Insight

In constructing graphs, strive for clarity and simplicity. Many books have been written in recent years showing innovative ways to portray information clearly in graphs. Examples are *The Visual Display of Quantitative Information* and *Envisioning Information*, by Edward Tufte (Graphics Press).

Try Exercise 2.86

When you plan to summarize data by graphical and numerical descriptive statistics, think about questions such as the following: What story are you attempting to convey with the data? Which graphical display and numerical summaries will most clearly and accurately convey this story?

John Tukey

On the Shoulders of...John Tukey

"The best thing about being a statistician is that you get to play in everyone's backyard."

—John Tukey (1915–2000)

In the 1960s, John Tukey of Princeton University was concerned that statisticians were putting too much emphasis on complex data analyses and ignoring simpler ways to examine and learn from the data. Tukey developed new descriptive methods, under the title of **exploratory data analysis (EDA)**. These methods make few assumptions about the structure of the data and emphasize data display and ways of searching for patterns and deviations from those patterns. Two graphical tools that Tukey invented were the stem-and-leaf plot and the box plot.

Initially, few statisticians promoted EDA. However, in recent years, some EDA methods have become common tools for data analysis. Part of this acceptance has been inspired by the availability of computer software and calculators that can implement some of Tukey's methods more easily.

Tukey's work illustrates that statistics is an evolving discipline. Almost all of the statistical methods used today were developed in the past century, and new methods continue to be created, largely because of increasing computer power.

2.6 Practicing the Basics

2.84 Great pay (on the average) The six full-time employees of Linda's Tanning Salon near campus had annual incomes last year of \$8900, \$9200, \$9200, \$9300, \$9500, \$9800. Linda herself made \$250,000.

a. For the seven annual incomes at Linda's Salon, report the mean and median.

b. Why is it misleading for Linda to boast to her friends that the average salary at her business is more than \$40,000?

2.85 Market share for food sales The pie chart shown was displayed in an article in *The Scotsman* newspaper (January 15, 2005) to show the market share of different supermarkets in Scotland.

a. Pie charts can be tricky to draw correctly. Identify two problems with this chart.

b. From looking at the graph without inspecting the percentages, would it be easier to identify the mode using this graph or using a bar graph? Why?

Market Share of Scottish Supermarket Chain

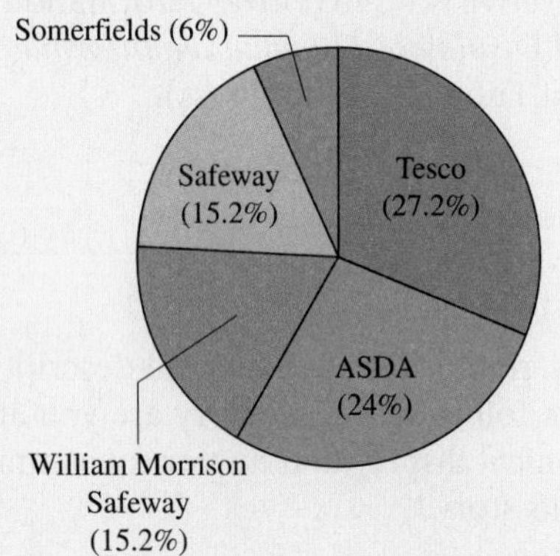

Source: Copyright The Scotsman Publications, Ltd.

2.86 Enrollment trends TRY Examples 19 and 20 presented graphs showing the total student enrollment at a U.S. university between 2004 and 2012 and the data for STEM major enrollment during that same time period. The data are repeated here.

Enrollment at the U.S. University

Year	Total Students	STEM Majors
2004	29,404	2,003
2005	29,693	1,906
2006	30,009	1,871
2007	30,912	1,815
2008	31,288	1,856
2009	32,317	1,832
2010	32,941	1,825
2011	33,878	1,897
2012	33,405	1,854

a. Construct a graph for only STEM major enrollments over this period. Describe the trend in these enrollment counts.

b. Find the percentage of enrolled students each year who are STEM majors, and construct a time plot of the percentages.

c. Summarize what the graphs constructed in parts a and b tell you that you could not learn from Figures 2.18 and 2.19 in Examples 19 and 20.

2.87 Terrorism and war in Iraq TRY In 2004, a college newspaper reported results of a survey of students taken on campus. One question asked was, "Do you think going to war with Iraq has made Americans safer from terrorism, or not?" The figure shows the way the magazine reported results.

a. Explain what's wrong with the way this bar chart was constructed.

b. Explain why you would not see this error made with a pie chart.

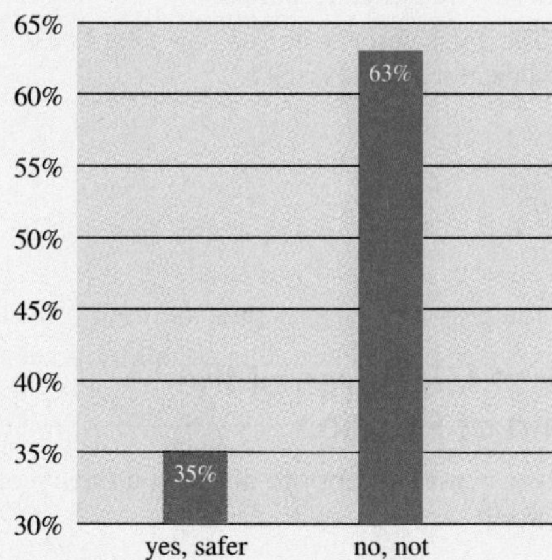

2.88 BBC license fee Explain what is wrong with the time plot shown of the annual license fee paid by British subjects for watching BBC programs. *(Source: Evening Standard*, November 29, 2006.)

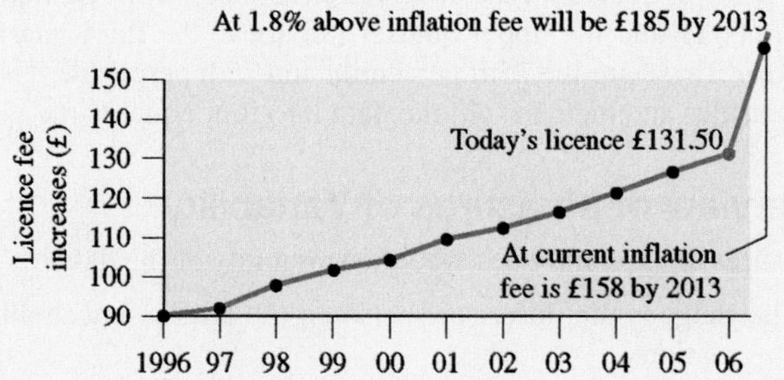

2.89 Federal government spending Explain what is wrong with the following pie chart, which depicts the federal government breakdown by category for 2010.

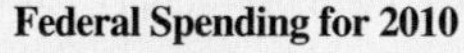

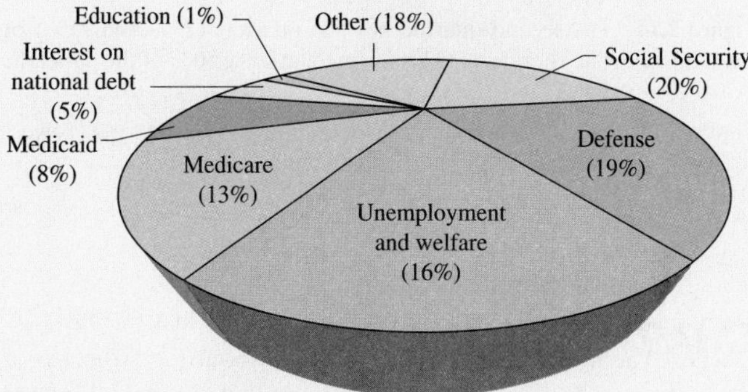

Source: www.gpoaccess.gov/usbudget/fy10.

2.90 Bad graph Search some publications and find an example of a graph that violates at least one of the principles for constructing good graphs. Summarize what's wrong with the graph and explain how it could be improved.

Chapter Review

ANSWERS TO THE CHAPTER FIGURE QUESTIONS

Figure 2.1 It is not always clear exactly what portion or percentage of a circle the slice represents, especially when a pie has many slices. Also, some slices may be representing close to the same percentage of the circle; thus, it is not clear which is larger or smaller.

Figure 2.3 The number of dots above a value on the number line represents the frequency of occurrence of that value.

Figure 2.4 There's no leaf if there's no observation having that stem.

Figure 2.7 The direction of the longer tail indicates the direction of the skew.

Figure 2.8 The annual average temperatures are tending to increase over time.

Figure 2.9 The mean is pulled in the direction of the longer tail since extreme observations affect the balance point of the distribution. The median is not affected by the size of the observation.

Figure 2.10 The range would change from \$10,000 to \$65,000.

Figure 2.11 A deviation is positive when the observation falls above the mean and negative when the observation falls below the mean.

Figure 2.12 Since about 95% of the data fall within 2 standard deviations of the mean, about 100% − 95% = 5% fall more than 2 standard deviations from the mean.

Figure 2.13 The shape is approximately bell shaped. The center as measured by the mean and median is about 65 inches. The heights range from 56 to 77 inches.

Figure 2.14 The second quartile has two quarters (25% and 25%) of the data below it. Therefore, Q2 is the median since 50% of the data falls below the median.

Figure 2.15 The left whisker is drawn only to 50 since the next lowest sodium value, which is 0, is identified as a potential outlier.

Figure 2.16 The quartiles are about Q1 = 69, Q2 = 71, and Q3 = 73 for males and Q1 = 63, Q2 = 65, and Q3 = 67 for females.

Figure 2.17 Yes. The left whisker extends to the minimum (about 2), the left side of the box is at Q1 (about 6), the line inside the box is at the median (about 8), the right side of the box is at Q3 (about 10), and the outlier identified with a star is at the maximum (about 21).

Figure 2.18 Neither the height nor the area of the human figures accurately represents the frequencies reported.

Figure 2.19 The total enrollment goes up until 2003. The African American enrollment seems fairly stable.

CHAPTER SUMMARY

This chapter introduced **descriptive statistics**—graphical and numerical ways of **describing** data. The characteristics of interest that we measure are called **variables**, because values on them vary from subject to subject.

- A **categorical variable** has observations that fall into one of a set of categories.
- A **quantitative variable** takes numerical values that represent different magnitudes of the variable.
- A quantitative variable is **discrete** if it has separate possible values, such as the integers 0, 1, 2,... for a variable expressed as "the number of..." It is **continuous** if its possible values form an interval.

When we explore quantitative data, key features to describe are the **shape**, using graphical displays, and the **center** and **variability**, using numerical summaries. When we explore categorical data, key features to describe are the frequency (or percentage) of observations for each category and the modal category.

Overview of Graphical Methods

- For categorical variables, data are displayed using **pie charts** and **bar graphs**. Bar graphs provide more flexibility and make it easier to compare categories having similar percentages.
- For quantitative variables, a **histogram** is a graph of a frequency table. It displays bars that specify frequencies or relative frequencies (percentages) for possible values or intervals of possible values. The **stem-and-leaf plot** (a vertical line dividing the final digit, the leaf, from the stem) and **dot plot** (dots above the number line) show the individual observations. They are useful for small data sets. These three graphs all show shape, such as whether the distribution is approximately bell shaped, skewed to the right (longer tail pointing to the right), or skewed to the left.
- The **box plot** has a box drawn between the first quartile and third quartile, with a line drawn in the box at the median. It has whiskers that extend to the minimum and maximum values, except for potential **outliers**. An **outlier** is an extreme value falling far below or above the bulk of the data.
- A **time plot** graphically displays observations for a variable measured over time. This plot can visually show **trends** over time.

Overview of Measures of the Center and of Position

Measures of the center attempt to describe a typical or representative observation.

- The **mean** is the sum of the observations divided by the number of observations. It is the balance point of the data.
- The **median** divides the ordered data set into two parts of equal numbers of observations, half below and half above that point. The median is the 50th percentile (second quartile). It is a more representative summary than the mean when the data are highly skewed.
- The lower quarter of the observations fall below the **first quartile (Q1)**, and the upper quarter fall above the **third quartile (Q3)**. These are the 25th percentile and 75th percentile. These quartiles and median split the data into four equal parts.

Overview of Measures of Variability

Measures of variability describe the variability of the data.

- The **range** is the difference between the largest and smallest observations.
- The **deviation** of an observation x from the mean is $x - \bar{x}$. The **variance** is an average of the squared deviations. Its square root, the **standard deviation**, is more useful, describing a typical distance from the mean.
- The **empirical rule** states that for a bell-shaped distribution:

About 68% of the data fall within 1 standard deviation of the mean, $\bar{x} \pm s$.

About 95% of the data fall within 2 standard deviations, $\bar{x} \pm 2s$.

Nearly all the data fall within 3 standard deviations, $\bar{x} \pm 3s$.

- The **interquartile range (IQR)** is the difference between the third and first quartiles, which span the middle 50% of the data in a distribution. It is more **resistant** than the range and standard deviation, being unaffected by extreme observations.
- An observation is a potential outlier if it falls (a) more than 1.5 × IQR below Q1 or more than 1.5 × IQR above Q3, or (b) more than 3 standard deviations from the mean. The ***z*-score** is the number of standard deviations that an observation falls from the mean.

SUMMARY OF NOTATION

Mean $\bar{x} = \frac{\Sigma x}{n}$ where x denotes the variable, n is the sample size, and Σ indicates to sum

Standard deviation $s = \sqrt{\frac{\Sigma(x - \bar{x})^2}{n - 1}}$

$z\text{-score} = \frac{\text{observed value} - \text{mean}}{\text{standard deviation}}$

CHAPTER PROBLEMS

Practicing the Basics

2.91 Categorical or quantitative? Identify each of the following variables as categorical or quantitative.

a. Number of children in family

b. Amount of time in football game before first points scored

c. College major (English, history, chemistry,...)

d. Type of music (rock, jazz, classical, folk, other)

2.92 Continuous or discrete? Which of the following variables are continuous, when the measurements are as precise as possible?

a. Age of mother

b. Number of children in a family

c. Cooking time for preparing dinner

d. Latitude and longitude of a city

e. Population size of a city

2.93 Immigration into United States The table shows the number (in millions) of the foreign-born population of the United States in 2004, by place of birth.

Foreign-Born Population in the United States

Place of Birth	Number (in Millions)
Europe	4.7
Caribbean	3.3
Central America	12.9
South America	2.1
Asia	8.7
Other	2.6
Total	**34.3**

Source: U.S. Statistical Abstract, 2006.

a. Is Place of Birth quantitative or categorical? Show how to summarize results by adding a column of percentages to the table.

b. Which of the following is a sensible numerical summary for these data: Mode (or modal category), mean, median? Explain, and report whichever is/are sensible.

c. How would you order the Place of Birth categories for a Pareto chart? What's its advantage over the ordinary bar graph?

2.94 Cool in China A recent survey[9] asked 1200 university students in China to pick the personality trait that most defines a person as "cool." The possible responses allowed, and the percentage making each, were individualistic and innovative (47%), stylish (13.5%), dynamic and capable (9.5%), easygoing and relaxed (7.5%), other (22.5%).

a. Identify the variable being measured.

b. Classify the variable as categorical or quantitative.

c. Which of the following methods could you use to describe these data?: (i) bar chart, (ii) dot plot, (iii) box plot, (iv) median, (v) mean, (vi) mode (or modal category), (vii) IQR, (viii) standard deviation.

2.95 Chad voting problems The 2000 U.S. presidential election had various problems in Florida. One was "overvotes"—people mistakenly voting for more than one presidential candidate. (There were multiple minor-party candidates.) There were 110,000 overvote ballots, with Al Gore marked on 84,197 and George W. Bush on 37,731. These ballots were disqualified. Was overvoting related to the design of the ballot? The figure shows MINITAB dot plots of the overvote percentages for 65 Florida counties organized by the type of voting machine—optical scanning, Votomatic (voters manually punch out chads), and Datavote (voter presses a lever that punches out the chad mechanically)—and the number of columns on the ballot (1 or 2).

a. The overvote was highest (11.6%) in Gadsden County. Identify the number of columns and the method of registering the vote for that county.

b. Of the six ballot type and method combinations, which two seemed to perform best in terms of having relatively low percentages of overvotes?

c. How might these data be summarized further by a bar graph with six bars?

Overvote Percentages by Number of Columns on Ballot and Method of Voting

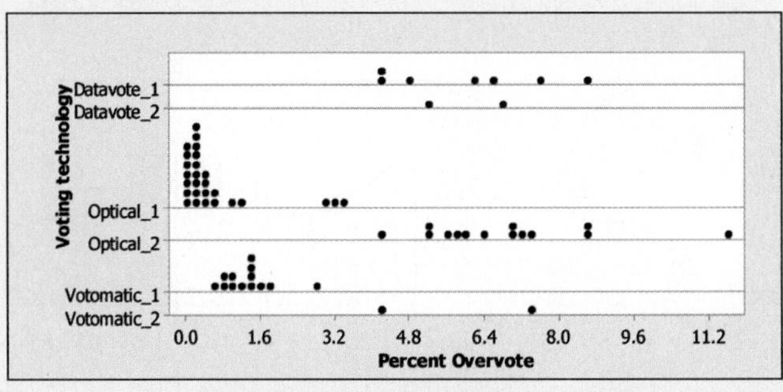

Source: A. Agresti and B. Presnell, *Statistical Science*, vol. 17, pp. 1–5, 2002.

[9] *Source:* Public relations firm Hill & Knowlton, as reported by *Time magazine*.

2.96 Number of children For the question "How many children have you ever had?" in the 2008 General Social Survey, the results were

No. children	0	1	2	3	4	5	6	7	8+
Count	521	323	524	344	160	77	30	19	22

a. Which is the most appropriate graph to display the data—dot plot, stem-and-leaf plot, or histogram? Why?

b. Based on sketching or using software to construct the graph, characterize this distribution as skewed to the left, skewed to the right, or symmetric. Explain.

2.97 Newspaper reading Exercise 2.25 gave results for the number of times a week a person reads a daily newspaper for a sample of 36 students at the University of Georgia. The frequency table is shown below.

a. Construct a dot plot of the data.

b. Construct a stem-and-leaf plot of the data. Identify the stems and the leaves.

c. The mean is 3.94. Find the median.

d. Is the distribution skewed to the left, skewed to the right, or symmetric? Explain.

No. Times	Frequency
0	2
1	4
2	4
3	8
4	4
5	5
6	2
7	4
8	2
9	1

2.98 Match the histogram Match each lettered histogram with one of the following descriptions: Skewed to the left, symmetric and bimodal, symmetric and unimodal, skewed to the right.

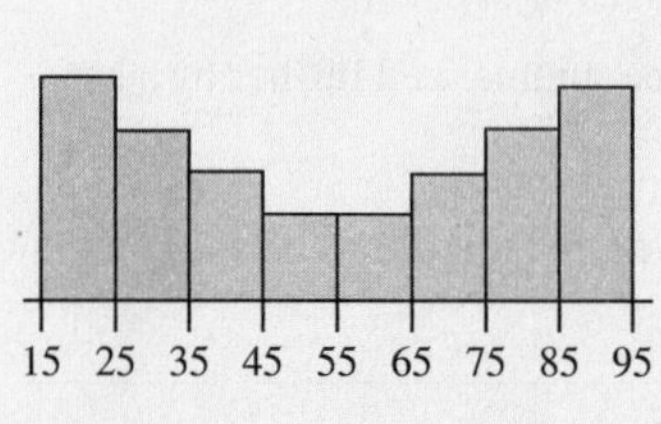

a

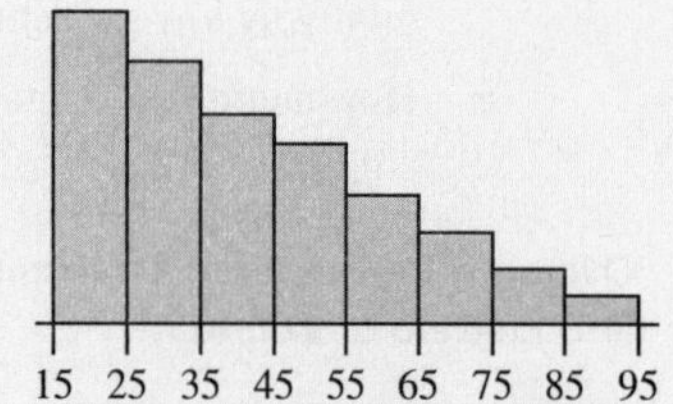

b

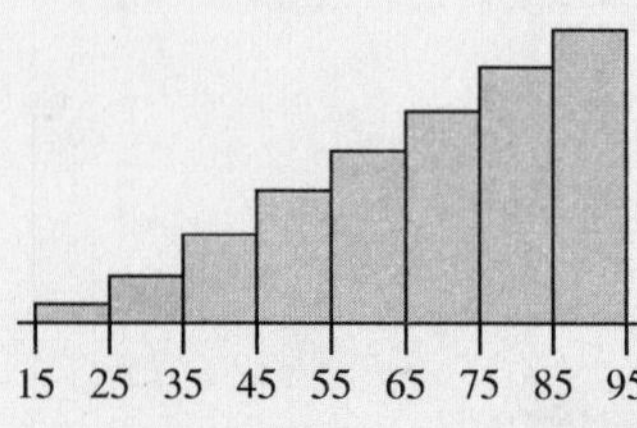

c

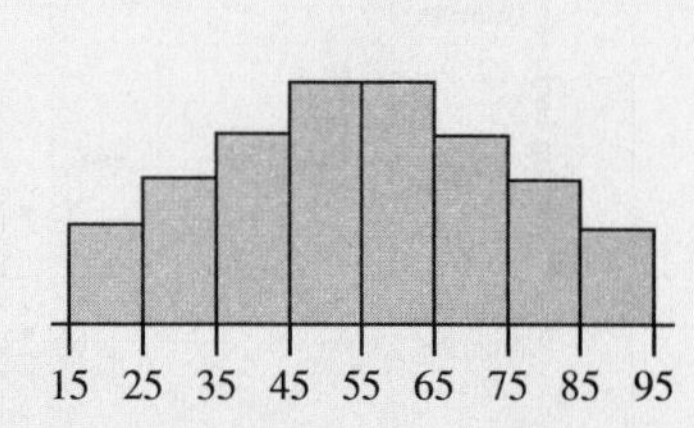

d

2.99 Sandwiches and protein Listed in the table below are the prices of six-inch Subway sandwiches at a particular franchise and the number of grams of protein contained in each sandwich.

Sandwich	Cost($)	Protein(g)
BLT	$2.99	17
Ham (Black Forest, without cheese)	$2.99	18
Oven Roasted Chicken	$3.49	23
Roast Beef	$3.69	26
Subway Club®	$3.89	26
Sweet Onion Chicken Teriyaki	$3.89	26
Turkey Breast	$3.49	18
Turkey Breast & Ham	$3.49	19
Veggie Delite®	$2.49	8
Cold Cut Combo	$2.99	21
Tuna	$3.10	21

a. Construct a stem-and-leaf plot of the protein amounts in the various sandwiches.

b. What is the advantage(s) of using the stem-and-leaf plot instead of a histogram?

c. Summarize your findings from these graphs.

2.100 Sandwiches and cost Refer to the previous exercise. Repeat parts a-c for the cost of the sandwiches. Summarize your findings.

2.101 What shape do you expect? For the following variables, indicate whether you would expect its histogram to be bell shaped, skewed to the right, or skewed to the left. Explain why.

a. Number of times arrested in past year

b. Time needed to complete difficult exam (maximum time is 1 hour)

c. Assessed value of home

d. Age at death

2.102 Sketch plots For each of the following, sketch roughly what you expect a histogram to look like, and explain whether the mean or the median would be greater.

a. The selling price of new homes in 2010.

b. The number of children ever born per woman age 40 or over

c. The score on an easy exam (mean = 88, standard deviation = 10, maximum = 100, minimum = 50)

d. Number of months in which subject drove a car last year

2.103 Median versus mean sales price of new homes In December 2010, the U.S. Census Bureau reported that the median U.S. sales price of new homes was $241,500. Would you expect the mean sales price to have been higher or lower? Explain.

2.104 Household net worth A study reported that in 2007 the mean and median net worth of American families were $556,300 and $120,300, respectively.

a. Is the distribution of net worth for these families likely to be symmetric, skewed to the right, or skewed to the left? Explain.

b. During the Great Recession of 2008, many Americans lost wealth due to the large decline in values of assets such as homes and retirement savings. In 2009, reported mean and median net worth were reported as \$434,782 and \$91,304. Why do you think the difference in decline from 2007 to 2009 was larger for the mean than the median?

2.105 Golfers' gains During the 2010 Professional Golfers Association (PGA) season, 90 golfers earned at least \$1 million in tournament prize money. Of those, 5 earned at least \$4 million, 11 earned between \$3 million and \$4 million, 21 earned between \$2 million and \$3 million, and 53 earned between \$1 million and \$2 million.

a. Would the data for all 90 golfers be symmetric, skewed to the left, or skewed to the right?

b. Two measures of central tendency of the golfers' winnings were \$2,090,012 and \$1,646,853. Which do you think is the mean and which is the median?

2.106 Hiking In a guidebook about interesting hikes to take in national parks, each hike is classified as easy, medium, or hard and by the length of the hike (in miles). Which classification is quantitative and which is categorical?

2.107 Lengths of hikes Refer to the previous exercise.

a. Give an example of five hike lengths such that the mean and median are equal.

b. Give an example of five hike lengths such that the mode is 2, the median is 3, and the mean is larger than the median.

2.108 Central Park monthly temperatures The MINITAB graph below uses dot plots to compare the distributions of the Central Park temperatures from 1869–2010 for the months of January and July.

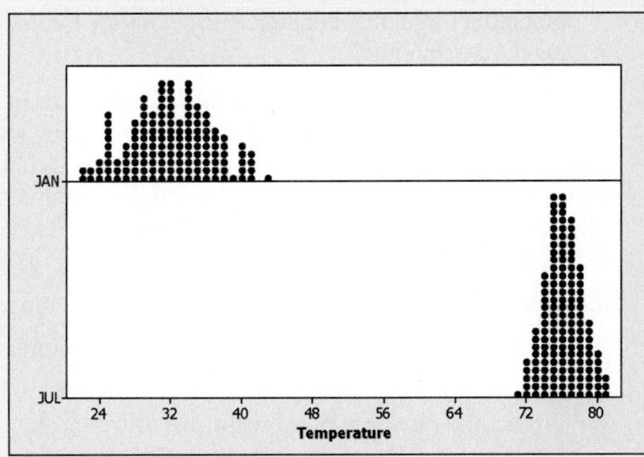

a. Describe the shape of each of the two distributions.

b. Estimate the balance point for each of the two distributions and compare.

c. Compare the variability of the two distributions. Estimate from the dot plots how the range and standard deviation for the January temperature distribution compare to the July temperature distribution. Are you surprised by your results for these two months?

2.109 What does s equal?

a. For an exam given to a class, the students' scores ranged from 35 to 98, with a mean of 74. Which of the following is the most realistic value for the standard deviation: –10, 1, 12, 60? Clearly explain what is unrealistic about the other values.

b. The sample mean for a data set equals 80. Which of the following is an impossible value for the standard deviation? 200, 0, –20? Why?

2.110 Female heights According to a recent report from the U.S. National Center for Health Statistics, females between 25 and 34 years of age have a bell-shaped distribution for height, with mean of 65 inches and standard deviation of 3.5 inches.

a. Give an interval within which about 95% of the heights fall.

b. What is the height for a female who is 3 standard deviations below the mean? Would this be a rather unusual height? Why?

2.111 Energy and water consumption In parts a and b, what shape do you expect for the distributions of electricity use and water use in a recent month in Gainesville, Florida? Why? (Data supplied by N. T. Kamhoot, Gainesville Regional Utilities.)

a. Residential electricity used had mean = 780 and standard deviation = 506 kilowatt hours (Kwh). The minimum usage was 3 Kwh and the maximum was 9390 Kwh.

b. Water consumption had mean = 7100 and standard deviation = 6200 (gallons).

2.112 Mean versus median and income A U.S. Federal Reserve study calculated the mean and median incomes for 2007 for each of the different income groups represented in the table.

Income percentile	Mean	Median
Below 20%	\$10,520	\$8,100
20 to 39.9%	\$134,900	\$37,900
40 to 59.9%	\$209,900	\$88,100
60 to 79.9%	\$375,100	\$204,900
80 to 89.9%	\$606,300	\$356,200
90 to 100%	\$3,306,000	\$1,119,000

Why does the disparity between mean and median income get larger as income gets larger?

2.113 Student heights The Heights data file on the text CD has heights for female and male students (in inches). For males, the mean is 70.9 and the standard deviation is 2.9. For females, the mean is 65.3 and the standard deviation is 3.0.

a. Interpret the male heights using the empirical rule.

b. Compare the center and variability of the height distributions for females and males.

c. The lowest observation for males was 62. How many standard deviations below the mean is it?

2.114 Cigarette tax How do cigarette taxes per pack vary from one state to the next? The data set of 2003 cigarette taxes for all 50 states and Washington, D.C. is in the Cigarette Tax data file on the text CD.

a. Use software to construct a histogram. Write a short description of the distribution, noting shape and possible outliers.

b. Find the mean and median for the cigarette taxes. Which is larger, and why would you have expected that from the histogram?

c. Find the standard deviation, and interpret it.

2.115 Cereal sugar values Revisit the sugar data for breakfast cereals that are given in the table.

a. Interpret the box plot in the figure (MINITAB output) by giving approximate values for the five-number summary.

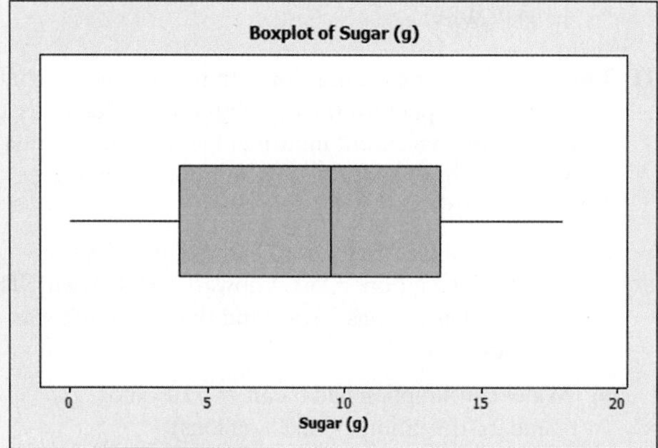

b. What does the box of the box plot suggest about possible skew?

c. The mean is 8.75 and the standard deviation is 5.32. Find the z-score associated with the minimum sugar value of 0. Interpret.

Cereal	Sugar (g)
Frosted Mini Wheats	11
Raisin Bran	18
All Bran	5
Apple Jacks	14
Cap'n Crunch	12
Cheerios	1
Cinnamon Toast Crunch	10
Crackling Oat Bran	16
Fiber One	0
Frosted Flakes	12
Froot Loops	14
Honey Bunches of Oats	7
Honey Nut Cheerios	9
Life	6
Rice Krispies	3
Honey Smacks	15
Special K	4
Wheaties	4
Corn Flakes	3
Honeycomb	11

2.116 Stock prices positions The data values below represent the prices per share of the 20 most actively traded stocks on the New York Stock Exchange (rounded to the nearest dollar) on February 18, 2011.

5	15	2	16	5	5	21	33	19	9
7	9	48	39	52	17	85	13	35	10

a. Sketch a dot plot or construct a stem-and-leaf plot.

b. Find the median, the first quartile, and the third quartile.

c. Sketch a box plot. What feature of the distribution displayed in the plot in part a is not obvious in the box plot? (*Hint*: Are there any gaps in the data?)

2.117 Temperatures in Central Park Access the Central Park temps data file on the text CD.

a. Using software, construct a histogram of average March temperatures and interpret, noting shape, center, and variability.

b. Find and interpret the mean and standard deviation of March temperatures.

c. Construct a histogram of average November temperatures. Using the two histograms, make comparative statements regarding the March and November temperatures.

d. Make a side-by-side box plot of March and November temperatures. Again make some comparative statements regarding the two. How is the side-by-side box plot more useful than multiple histograms for comparing the two months?

2.118 Teachers' salaries According to *Statistical Abstract of the United States, 2006,* average salary (in dollars) of secondary school classroom teachers in 2004 in the United States varied among states with a five-number summary of: minimum = 33,100, Q1 = 39,250, Median = 42,700, Q3 = 48,850, Maximum = 61,800.

a. Find and interpret the range and interquartile range.

b. Sketch a box plot, marking the five-number summary on it.

c. Predict the direction of skew for this distribution. Explain.

d. If the distribution, although skewed, is approximately bell shaped, which of the following would be the most realistic value for the standard deviation: (i) 100, (ii) 1000, (iii) 6000, or (iv) 25,000? Explain your reasoning.

2.119 Health insurance In 2004, the five-number summary of positions for the distribution of statewide percentage of people without health insurance had a minimum of 8.9% (Minnesota), Q1 = 11.6, Median = 14.2, Q3 = 17.0, and maximum of 25.0% (Texas) *(Statistical Abstract of the United States, 2006).*

a. Do you think the distribution is symmetric, skewed right, or skewed left? Why?

b. Which is most plausible for the standard deviation: − 16, 0, 4, 15, or 25? Why? Explain what is unrealistic about the other values.

2.120 What box plot do you expect? For each of the following variables, sketch a box plot that would be plausible.

a. Exam score (min = 0 max = 100, mean = 87, standard deviation = 10)

b. IQ mean = 100 and standard deviation = 16

c. Weekly religious contribution median = $10 and mean = $17)

2.121 High school graduation rates The distribution of high school graduation rates in the United States in 2004 had a minimum value of 78.3 (Texas), first quartile of 83.6, median of 87.2, third quartile of 88.8, and maximum value of 92.3 (Minnesota) *(Statistical Abstract of the United States, 2006).*

a. Report the range and the interquartile range.

b. Would a box plot show any potential outliers? Explain.

2.122 SAT scores revisited The U.S. statewide average total SAT scores math + reading + writing for 2010 are summarized in the box plot. These SAT scores are out of a possible 2400.

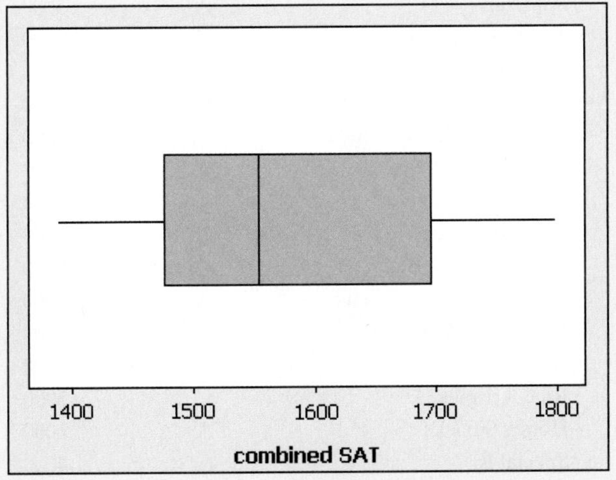

a. Explain how the box plot gives you information about the distribution shape.

b. Using the box plot, give the approximate value for each component of the five-number summary. Interpret each.

c. A histogram for these data is also shown. What feature of the distribution would be missed by using only the box plot?

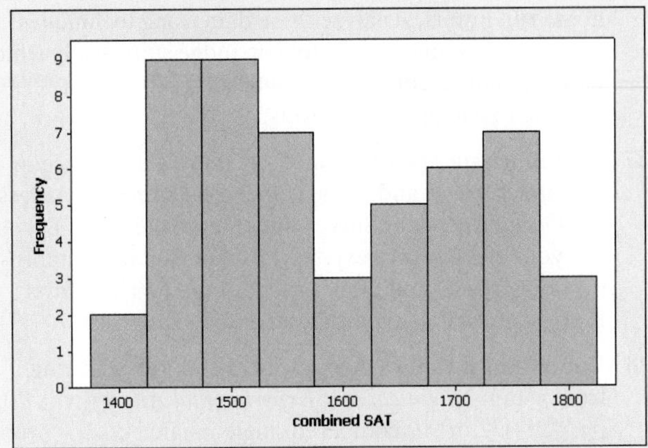

2.123 Blood pressure A World Health Organization study (the MONICA project) of health in various countries reported that in Canada, systolic blood pressure readings have a mean of 121 and a standard deviation of 16. A reading above 140 is considered to be high blood pressure.

a. What is the z-score for a blood pressure reading of 140? How is this z-score interpreted?

b. The systolic blood pressure values have a bell-shaped distribution. Report an interval within which about 95% of the systolic blood pressure values fall.

2.124 No cereal sodium The cereal sodium values have a mean of 167 and a standard deviation of 77.3. Find the z-score for the cereal that has a sodium value of 0. Interpret.

2.125 Who was Roger Maris? Roger Maris, who spent most of his professional baseball career with the New York Yankees, held the record for the most home runs in one season (61) from 1961 until 1998, when the record was broken by Mark McGwire. Maris played in the major leagues from 1957 to 1968. The number of home runs he hit in each year that he played is summarized in MINITAB output as shown.

```
Variable  N        Mean     Median  StDev
RMHR      12       22.92     19.50  15.98
Variable  Minimum  Maximum      Q1     Q3
RMHR      5.00     61.00     10.00  31.75
```

a. Use the 3 standard deviation criterion to determine if any potential outliers are present.

b. The criterion in (part a) requires the distribution to be approximately bell-shaped. Is there any evidence here to contradict this? Explain.

c. A sports writer commented that Roger Maris hit *only* 13 home runs in 1966. Was this unusual for Maris? Comment, using statistical justification.

Concepts and Investigations

2.126 Baseball's great home run hitters The file Baseball's HR Hitters on the text CD contains data on the number of home runs hit each season by some of baseball's great home run hitters. Analyze these data using techniques introduced in this chapter to help judge statistically which player might be considered the best. Specify the criterion you use to compare the players.

2.127 How much spent on haircuts? Is there a difference in how much males and females spend on haircuts? Access the Georgia Student Survey data file on the text CD or use your class data to explore this question using appropriate graphical and numerical methods. Write a brief report describing your analyses and conclusions.

2.128 Controlling asthma A study of 13 children suffering from asthma (*Clinical and Experimental Allergy,* vol. 20, pp. 429–432, 1990) compared single inhaled doses of formoterol (F) and salbutamol (S). Each child was evaluated using both medications. The outcome measured was the child's peak expiratory flow (PEF) eight hours following treament. Is there a difference in the PEF level for the two medications? The data on PEF follow:

Child	F	S
1	310	270
2	385	370
3	400	310
4	310	260
5	410	380
6	370	300
7	410	390
8	320	290
9	330	365
10	250	210
11	380	350
12	340	260
13	220	90

a. Construct plots to compare formoterol and salbutamol. Write a short summary comparing the two distributions of the peak expiratory flow.

b. Consider the distribution of differences between the PEF levels of the two medications. Find the 13 differences and construct and interpret a plot of the differences. If on the average there is no difference between the PEF level for the two brands, where would you expect the differences to be centered?

2.129 Google trend Go to www.google.com/trends. Look up a subject of interest to you, and create a time plot to describe a trend over time. Interpret the plot.

2.130 Google again Click on one of the subjects listed at www.google.com/trends. Explain how the graph Google presents could be improved.

2.131 Back-to-back stem-and-leaf plot To compare two groups graphically, one can use the same stems for each and put leaves for one group on one side and for the other group on the other side. This is called a **back-to-back stem-and-leaf plot**. The figure shown compares sugar amounts (expressed in milligrams) for cereals listed in the table according to whether they were intended for adults or children.

Back-to-Back Stem-and-Leaf Plot of Sugar for Adult and Children Cereals

Adult		Children
0	000	
	100	0
	200	
0	300	0
0 0	400	
0	500	
	600	0
0	700	
	800	
	900	0
	1000	0
0	1100	0
	1200	0 0
	1300	
	1400	0 0
0	1500	
0	1600	
	1700	
0	1800	

Cereal	Type	Sugar (mg)
Frosted Mini Wheats	A	11,000
Raisin Bran	A	18,000
All Bran	A	5000
Apple Jacks	C	14,000
Cap'n Crunch	C	12,000
Cheerios	C	1000
Cinnamon Toast Crunch	C	10,000
Crackling Oat Bran	A	16,000
Fiber One	A	0
Frosted Flakes	C	12,000
Froot Loops	C	14,000
Honey Bunches of Oats	A	7000
Honey Nut Cheerios	C	9000
Life	C	6000
Rice Krispies	C	3000
Honey Smacks	A	15,000
Special K	A	4000
Wheaties	A	4000
Corn Flakes	A	3000
Honeycomb	C	11,000

a. Is the distribution of sugar values similar or different for the two cereal types? If they are different, describe the difference.

b. Construct a back-to-back stem-and-leaf plot for the sodium values, listed in Table 2.3. Is the distribution of sodium values similar, or different, for the two cereal types?

2.132 You give examples Give an example of a variable that you'd expect to have a distribution that is

a. Approximately symmetric

b. Skewed to the right

c. Skewed to the left

d. Bimodal

e. Skewed to the right, with a mode and median of 0 but a positive mean

2.133 Political conservatism and liberalism Where do Americans tend to fall on the conservative–liberal political spectrum? The General Social Survey asks, "I'm going to show you a seven-point scale on which the political views that people might hold are arranged from extremely liberal, point 1, to extremely conservative, point 7. Where would you place yourself on this scale?" The table shows the seven-point scale and the distribution of 1933 responses for a recent survey (2008).

Score	Category	Frequency
1.	Extremely liberal	69
2.	Liberal	240
3.	Slightly liberal	221
4.	Moderate	740
5.	Slightly conservative	268
6.	Conservative	327
7.	Extremely conservative	68

(*Source:* Data from CSM, UC Berkeley.)

This is a categorical scale with ordered categories, called an **ordinal scale**. Ordinal scales are often treated in a quantitative manner by assigning scores to the categories and then using numerical summaries such as the mean and standard deviation.

a. Using the scores shown in the table, the mean for these data equals 4.11. Using the reasoning from Example 12, set up the way this would be calculated.

b. Identify the mode (or modal category).

c. In which category does the median fall? Why?

2.134 Mode but not median and mean The previous exercise showed how to find the mean and median when a categorical variable has ordered categories. A categorical scale that does *not* have ordered categories (such as choice of religious affiliation or choice of major in college) is called a **nominal scale**. For such a variable, the mode (or modal category) applies, but not the mean or median. Explain why.

2.135 Multiple choice: GRE scores In a study of graduate students who took the Graduate Record Exam (GRE), the Educational Testing Service reported that for the quantitative exam, U.S. citizens had a mean of 529 and standard deviation of 127, whereas the non-U.S. citizens had a mean of 649 and standard deviation of 129. Which of the following is true?

a. Both groups had about the same amount of variability in their scores, but non-U.S. citizens performed better, on the average, than U.S. citizens.

b. If the distribution of scores was approximately bell shaped, then almost no U.S. citizens scored below 400.

c. If the scores range between 200 and 800, then probably the scores for non-U.S. citizens were symmetric and bell shaped.

d. A non-U.S. citizen who scored 3 standard deviations below the mean had a score of 200.

2.136 Multiple choice: Fact about *s* Which statement about the standard deviation s is false?

a. s can never be negative.

b. s can never be zero.

c. For bell-shaped distributions, about 95% of the data fall within $\bar{x} \pm 2s$.

d. s is a nonresistant (sensitive to outliers) measure of variability, as is the range.

2.137 Multiple choice: Relative GPA The mean GPA for all students at a community college in the fall semester was 2.77. A student with a GPA of 2.0 wants to know her relative standing in relation to the mean GPA. A numerical summary that would be useful for this purpose is the

a. standard deviation

b. median

c. interquartile range

d. number of students at the community college

2.138 True or false:

a. The mean, median, and mode can never all be the same.

b. The mean is always one of the data points.

c. When n is odd, the median is one of the data points.

d. The median is the same as the second quartile and the 50th percentile.

2.139 Bad statistic A teacher summarizes grades on an exam by Min = 26, Q1 = 67, Q2 = 80, Q3 = 87, Max = 100, Mean = 76, Mode = 100, Standard deviation = 76, IQR = 20

She incorrectly recorded one of these. Which one do you think it was? Why?

2.140 True or false: Soccer According to a story in the *Guardian* newspaper (football.guardian.co.uk), in the United Kingdom the mean wage for a Premiership player in 2006 was £676,000. True or false: If the income distribution is skewed to the right, then the median salary was even larger than £676,000.

2.141 ◆◆ **Mean for grouped data** Refer to the calculation of the mean in Example 12 or in Exercise 2.133. Explain why the mean for grouped data can be expressed as a sum, taking each possible outcome times the *proportion* of times it occurred.

2.142 ◆◆ **Male heights** According to a recent report from the U.S. National Center for Health Statistics, for males aged 25–34 years, 2% of their heights are 64 inches or less, 8% are 66 inches or less, 27% are 68 inches or less, 39% are 69 inches or less, 54% are 70 inches or less, 68% are 71 inches or less, 80% are 72 inches or less, 93% are

74 inches or less, and 98% are 76 inches or less. These are called **cumulative percentages**.

a. Which category has the median height? Explain why.

b. Nearly all the heights fall between 60 and 80 inches, with fewer than 1% falling outside that range. If the heights are approximately bell-shaped, give a rough approximation for the standard deviation of the heights. Explain your reasoning.

2.143 ◆◆ **Range and standard deviation approximation** Use the empirical rule to explain why the standard deviation of a bell-shaped distribution for a large data set is often roughly related to the range by evaluating Range $\approx 6s$. (For small data sets, one may not get any extremely large or small observations, and the range may be smaller, for instance about 4 standard deviations.)

2.144 ◆◆ **Range the least resistant** We've seen that measures such as the mean, the range, and the standard deviation can be highly influenced by outliers. Explain why the range is worst in this sense. (*Hint:* As the sample size increases, explain how a single extreme outlier has less effect on the mean and standard deviation, but can still have a large effect on the range.)

2.145 ◆◆ **Using MAD to measure variability** The standard deviation is the most popular measure of variability from the mean. It uses squared deviations, since the ordinary deviations sum to zero. An alternative measure is the **mean absolute deviation,** $\Sigma|x - \bar{x}|/n$.

a. Explain why greater variability tends to result in larger values of this measure.

b. Would the MAD be more, or less, resistant than the standard deviation? Explain.

2.146 ◆◆ **Rescale the data** The mean and standard deviation of a sample may change if data are rescaled (for instance, temperature changed from Fahrenheit to Celsius). For a sample with mean $\bar{x}$, adding a constant c to each observation changes the mean to $\bar{x} + c$, and the standard deviation s is unchanged. Multiplying each observation by $c > 0$ changes the mean to $c\,\bar{x}$ and the standard deviation to cs.

a. Scores on a difficult exam have a mean of 57 and a standard deviation of 20. The teacher boosts all the scores by 20 points before awarding grades. Report the mean and standard deviation of the boosted scores. Explain which rule you used, and identify c.

b. Suppose that annual income for some group has a mean of $39,000 and a standard deviation of $15,000. Values are converted to British pounds for presentation to a British audience. If one British pound equals $2.00, report the mean and standard deviation in British currency. Explain which rule above you used, and identify c.

c. Adding a constant and/or multiplying by a constant is called a **linear transformation** of the data. Do linear transformations change the *shape* of the distribution? Explain your reasoning.

Student Activities

2.147 **The average student** Refer to the data file you created in Activity 3 in Chapter 1. For variables chosen by your instructor, describe the "average student" in your class. Prepare a one-page summary report. In class, students will compare their analyses of what makes up an average student.

2.148 TRY **Create own data** For the Mean Versus Median applet, your instructor will give you a data set to illustrate the effect of extreme observations on the mean and median. Write a short summary of your observations.

2.149 **GSS** Access the General Social Survey at sda.berkeley.edu/GSS.

a. Find the frequency table and histogram for Example 6 on TV watching. (*Hint:* Enter TVHOURS as the row variable, YEAR(2008) as the selection filter, choose bar chart for Type of Chart, and click on *Run the Table.*)

b. Your instructor will have you obtain graphical and numerical summaries for another variable from the GSS. Students will compare results in class.

BIBLIOGRAPHY

Franklin, Christine A. (2002). "The Other Life of Florence Nightingale," *Mathematics Teaching in the Middle School* 7(6): 337–339.

Tukey, J. W. (1977). *Exploratory Data Analysis.* Reading, MA: Addison-Wesley.

Association: Contingency, Correlation, and Regression

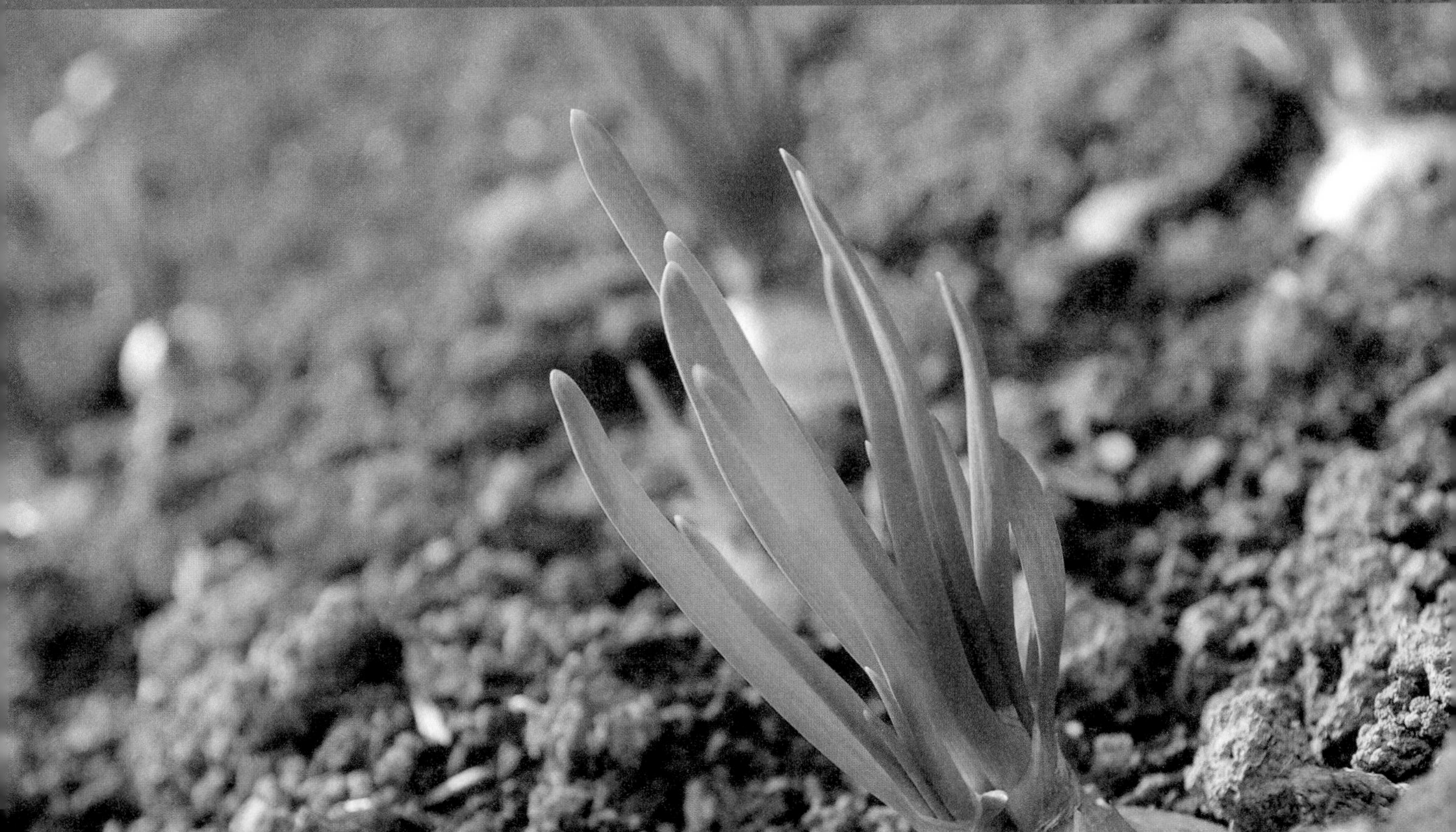

Example 1

Smoking and Your Health

Picture the Scenario

While numerous studies have concluded that cigarette smoking is harmful to your health in many ways, one study found conflicting evidence. As part of the study, 1314 women in the United Kingdom were asked if they smoked.[1] Twenty years later, a follow-up survey observed whether each woman was deceased or still alive. The researchers studied the possible link between whether a woman smoked and whether she survived the 20-year study period. During that period, 24% of the smokers died and 31% of the nonsmokers died. The higher death rate for the nonsmokers is surprising.

Questions to Explore

- Is smoking actually beneficial to your health, since a smaller percentage of smokers died?
- What descriptive statistical methods can we use in exploring the data?
- If we observe a link between smoking status and survival status, is there something that could explain how this happened?

Thinking Ahead

In Chapter 2 we distinguished between categorical and quantitative variables. In this study, we can identify two categorical variables. One is smoking status—whether a woman was a smoker (yes or no). The other is survival status—whether a woman survived the 20-year study period (yes or no). In practice, research investigations almost always need to analyze more than one variable. The link between two variables is often the primary focus.

This chapter presents descriptive statistics for examining data on two variables. We'll analyze what these data suggest about the link between smoking and cancer. We'll learn that a third variable can influence the results. In this case, we'll see that the age of the woman is important. We'll revisit this example in Example 16 and analyze the data while taking age into account as well.

Example 1 has two categorical variables, but we'll also learn how to examine links between pairs of quantitative variables. For instance, we might want to answer questions such as, "What's the relationship between the daily amount of gasoline use by automobiles and the amount of air pollution?" or "Do high schools with higher per-student funding tend to have higher mean SAT scores for their students?"

Response Variables and Explanatory Variables

When we analyze data on two variables, our first step is to distinguish between the **response variable** and the **explanatory variable**.

In Words

The data analysis examines how the outcome on the **response** variable *depends on* or is *explained by* the value of the **explanatory** variable.

Response Variable and Explanatory Variable

The **response variable** is the outcome variable on which comparisons are made. When the **explanatory variable** is categorical, it defines the groups to be compared with respect to values for the response variable. When the explanatory variable is quantitative, it defines the change in different numerical values to be compared with respect to values for the response variable.

[1]Described by D. R. Appleton et al., *American Statistician*, vol. 50, pp. 340–341 (1996).

In Example 1, survival status (whether a woman is alive after 20 years) is the response variable. Smoking status (whether the woman was a smoker) is the explanatory variable. In a study of air pollution in several countries, the carbon dioxide (CO_2) level in a country's atmosphere might be a response variable, and the explanatory variable could be the country's amount of gasoline use for automobiles. In a study of achievement for a sample of college students, college GPA might be a response variable, and the explanatory variable could be the number of hours a week spent studying. In each of these studies, there is a natural explanatory/response relationship between these quantitative variables.

Some studies regard *either* or *both* variables as response variables. For example, this might be the case if we were analyzing data on smoking status and on alcohol status (whether the subject has at least one alcoholic drink per week). There is no clear distinction as to which variable would be explanatory for the other.

The main purpose of a data analysis with two variables is to investigate whether there is an **association** and to describe the nature of that association.

In Words

When there's an **association**, the likelihood of a particular value for one variable depends on the value of the other variable. It is more likely that someone with a high school GPA of 4 will have a college GPA above 3.5 than a student who has a high school GPA of 3. So high school GPA and college GPA have an association.

Association Between Two Variables

An **association** exists between two variables if a particular value for one variable is more likely to occur with certain values of the other variable.

In Example 1, surviving the study period was more likely for smokers than for nonsmokers. So, there is an association between survival status and smoking status. For higher levels of energy use, does the CO_2 level in the atmosphere tend to be higher? If so, then there is an association between energy use and CO_2 level.

This chapter presents methods for studying whether associations exist and for describing how strong they are. We explore associations between categorical variables in Section 3.1 and between quantitative variables in Sections 3.2 and 3.3.

3.1 The Association Between Two Categorical Variables

How would you respond to the question, "Taken all together, would you say that you are very happy, pretty happy, or not too happy?" We could summarize the percentage of people who have each of the three possible outcomes for this categorical variable using a table, a bar graph, or a pie chart. However, we'd probably want to know how the percentages depend on the values for other variables, such as a person's marital status (married, unmarried). Is the percentage who report being very happy higher for married people than for unmarried people?

We'll now look at ways to summarize the association between two categorical variables. You can practice these concepts with data on personal happiness in Exercises 3.3, 3.10, and 3.64.

Categorical explanatory and response variables

Example 2

Pesticides in Organic Foods

Picture the Scenario

One appeal of eating organic foods is the belief that they are pesticide-free and thus healthier. However, little fruit and vegetable acreage is organic (only 2% in the United States), and consumers pay a premium for organic food.

How can we investigate how the percentage of foods carrying pesticide residues compares for organic and conventionally grown foods? The Consumers Union led a study based on sampling carried out by the U.S. Department of Agriculture (USDA) and the state of California.[2] The sampling was part of regulatory monitoring of foods for pesticide residues. For this study, Table 3.1 displays the frequencies of foods for all possible category combinations of the two variables, food type and pesticide status.

Table 3.1 Frequencies for Food Type and Pesticide Status

The row totals and the column totals are the frequencies for the categories of each variable. The counts inside the table give information about the association.

	Pesticide Status		
Food Type	**Present**	**Not Present**	**Total**
Organic	29	98	127
Conventional	19,485	7,086	26,571
Total	**19,514**	**7,184**	**26,698**

Questions to Explore

a. What is the response variable, and what is the explanatory variable?
b. How does the proportion of foods with pesticides present compare for the two food types?
c. What proportion of all sampled produce contained pesticide residues?

Think It Through

a. Pesticide status, namely whether pesticide residues are present, is the outcome of interest. The food type, organic or conventionally grown, is the variable that defines two groups to be compared on their pesticide status. So, pesticide status is the response variable and food type is the explanatory variable.
b. From Table 3.1, 29 out of 127 organic foods contained pesticide residues. The proportion with pesticides is $29/127 = 0.228$. Likewise, 19,485 out of 26,571 conventionally grown foods contained pesticide residues. The proportion is $19{,}485/26{,}571 = 0.733$, much higher than for organic foods. Since $0.733/0.228 = 3.2$, the relative occurrence of pesticide residues for conventionally grown produce is approximately three times that for organically grown produce.
c. The overall proportion of sampled produce with pesticide residues is the total number with pesticide residues out of the total number of food items, or

$$(29 + 19{,}485)/(127 + 26{,}571) = 19{,}514/26{,}698 = 0.731.$$

We can find this result using the column totals of Table 3.1.

Insight

The value of 0.731 for the overall proportion with pesticide residues is close to the proportion for conventionally grown foods alone, which we found to be 0.733. This is because conventionally grown foods make up a

Did You Know?

Should we be alarmed that about 73% of the food tested contained pesticide residues? The study stated that the level is usually far below the limits set by the Environmental Protection Agency. Scientists did find that the pesticide residues on organic produce were generally less toxic. ◀

[2] *Source: Food Additives and Contaminants* 2002, vol.19 no. 5, 427–446.

high percentage of the sample. (From the row totals, the proportion of the sampled items that were conventionally grown was 26,571/26,698 = 0.995.) In summary, pesticide residues occurred in more than 73% of the sampled items, and they were much more common (about three times as common) for conventionally grown than organic foods.

Try Exercise 3.3

Contingency Tables

Table 3.1 has two categorical variables: food type and pesticide status. We can analyze the categorical variables separately, through the column totals for pesticide status and the row totals for food type. These totals are the category counts for the separate variables, for instance (19,514, 7,184) for the (present, not present) categories of pesticide status. We can also study the association between them, as we did by using the counts for the category combinations to find proportions, in Example 2, part b. Table 3.1 is an example of a **contingency table**.

Contingency Table

A **contingency table** is a display for two categorical variables. Its rows list the categories of one variable and its columns list the categories of the other variable. Each entry in the table is the number of observations in the sample at a particular combination of categories of the two categorical variables.

	Pesticides	
Food Type	**Yes**	**No**
Organic	29	98
Conventional	19,485	7,086
	cell ↑	

Each row and column combination in a contingency table is called a **cell.** For instance, the first cell in the second row of Table 3.1 (shown again in the margin) has the frequency 19,485, the number of observations in the conventional category of food type with pesticides present. The process of taking a data file and finding the frequencies for the cells of a contingency table is referred to as **cross-tabulation** of the data. Table 3.1 is formed by cross-tabulation of food type and pesticide status for the 26,698 sampled food items.

Conditional Proportions

Consider the question, "Do organic and conventionally grown foods differ in the proportion of food items with pesticide residues?" To answer, we find the proportions on pesticide status within each category of food type and compare them. From Example 2, the proportions are 29/127 = 0.228 for organic foods and 19,485/26,571 = 0.733 for conventionally grown foods.

These proportions are called **conditional proportions** because their formation is **conditional** on (in this example) food type. Restricting our attention to organic foods, the proportion of food items with pesticides present equals 0.23. Table 3.2 shows the conditional proportions.

Recall

Using Table 3.1, shown again below, we obtain 0.23 from 29/127 and 0.77 from 98/127, the cell counts divided by the first row total. ◀

	Pesticide Status		
Food Type	**Present**	**Not Present**	**Total**
Organic	29	98	**127**
Conven.	19,485	7,086	**26,571**
Total	**19,514**	**7,184**	**26,698**

Table 3.2 Conditional Proportions on Pesticide Status, for Two Food Types

These conditional proportions (using two decimal places) treat pesticide status as the response variable. The sample size n in a row shows the total on which the conditional proportions in that row were based.

	Pesticide Status			
Food Type	**Present**	**Not Present**	**Total**	n
Organic	0.23	0.77	**1.00**	127
Conventional	0.73	0.27	**1.00**	26,571

The conditional proportions in each row sum to 1.0. The sample size n for each set of conditional proportions is listed so you can determine the frequencies on which the conditional proportions were based. *Whenever we distinguish between a response variable and an explanatory variable, it is natural to form conditional proportions (based on the explanatory variable) for categories of the response variable.*

By contrast, the proportion of *all* sampled produce items with pesticide residues, which we found in part c of Example 2, is not a conditional proportion. It is not found for a particular food type. We ignored the information on food type and used the counts in the bottom margin of the table to form the proportion $19{,}514/26{,}698 = 0.731$. Such a proportion is called a **marginal proportion**. It is found using counts in the *margin* of the table.

Example 3

Graphing the data

Comparing Pesticide Residues for the Food Types

Picture the Scenario

For the food type and pesticide status, we've now seen two ways to display the data. Table 3.1 showed cell frequencies in a contingency table, and Table 3.2 showed conditional proportions.

Questions to Explore

a. How can we use a single graph to show the relationship?
b. What does the graph tell us?

Think It Through

a. We've seen that to compare the two food types on pesticide status, the response variable, we can find conditional proportions for the pesticide status categories (shown in Table 3.2). Figure 3.1 shows bar graphs of the conditional proportions, one for organic foods and another for conventionally grown foods. More efficiently, we can construct a single bar graph that shows **side-by-side bars** to compare the conditional proportion of pesticide residues in conventionally grown and organic foods.

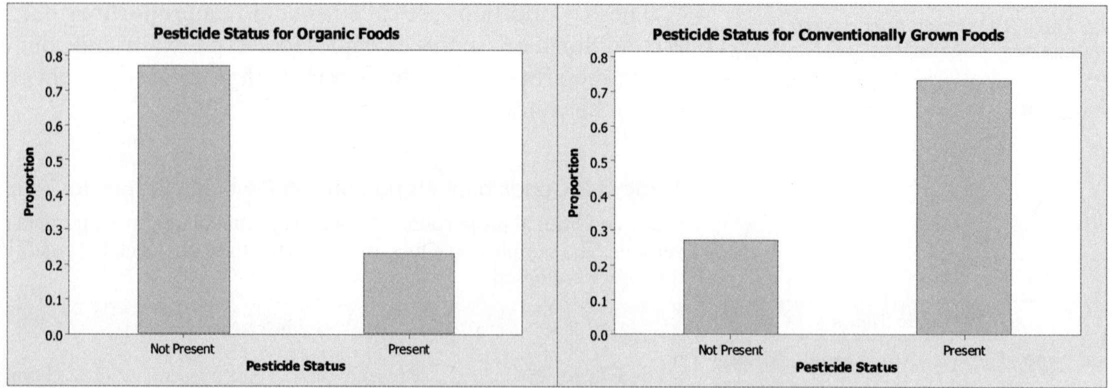

▲ **Figure 3.1 MINITAB Output of Conditional Proportions on Pesticide Status.** In this graph, the conditional proportions are shown separately for each food type. **Question** Can you think of a way to display the results that makes it easier to compare the food types on the pesticide status?

Figure 3.2 shows this graph, which conveys the same information as the two separate bar graphs in Figure 3.1. The advantage of the side-by-side bar graph is that it allows for easy comparison of the two food types.

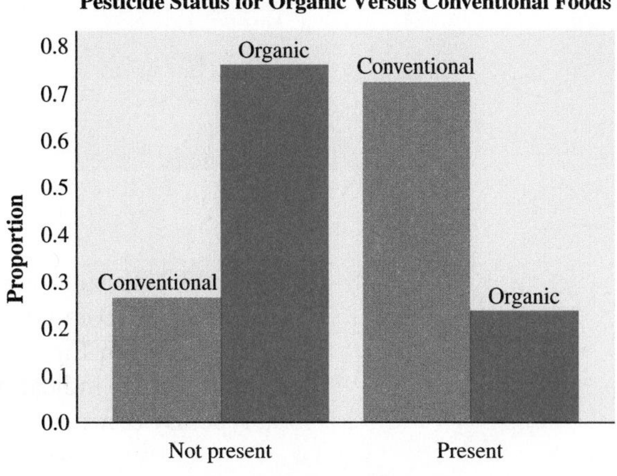

▲ **Figure 3.2** Conditional Proportions on Pesticide Status, Given the Food Type. For a particular pesticide status category, the side-by-side bars compare the two food types. **Question** Comparing the bars, how would you describe the difference between organic and conventionally grown foods in the conditional proportion with pesticide residues present?

b. Figure 3.2 clearly shows that the proportion of foods with pesticides present is much higher for conventionally grown food than for organically grown food.

Insight

Chapter 2 used bar graphs to display proportions for a single variable. With two variables, bar graphs usually display conditional proportions, as in Figure 3.2. This display is useful for making comparisons, such as the way Figure 3.2 compares organic and conventionally grown foods in terms of the proportion of pesticide residues.

Try Exercise 3.4

Looking for an Association

When forming a contingency table, determine if one variable should be the response variable. If there is a clear explanatory/response relationship, that dictates which way we compute the conditional proportions. In some cases, either variable could be the response variable, such as in cross-tabulating belief in heaven (yes, no) with belief in hell (yes, no). Then you can form conditional proportions in either or both directions. Studying the conditional proportions helps you judge whether there is an *association* between the variables.

Table 3.2 suggests that there is a reasonably strong association between food type and pesticide status because the proportion of food items with pesticides present differs considerably (0.23 versus 0.73) between the two food types. There would be *no association* if, instead, the proportion with pesticides present had been the same for each food type. For instance, *suppose* that for each food type, 60% had pesticides present and 40% did not have pesticides present, as shown in Table 3.3; the food types would have the same pesticide status distribution. We then say that pesticide status is **independent** of food type.

Table 3.3 Hypothetical Conditional Proportions on Pesticide Status for Each Food Type, Showing No Association

The conditional proportions for the response variable (pesticide status) categories are the same for each food type.

	Pesticide Status		
Food Type	**Present**	**Not Present**	**Total**
Organic	0.40	0.60	**127**
Conventional	0.40	0.60	**26,571**

Figure 3.3 is a side-by-side display of the hypothetical conditional proportions from Table 3.3. Compare Figure 3.3, no association, with Figure 3.2, which shows an association. In Figure 3.3, for any particular response category, the bars are the same height, indicating no association: Pesticide status does not depend on food type. By contrast, in Figure 3.2, the bars have quite different heights.

In Practice Comparing Population and Sample Conditional Proportions

When sampling from a population, even if there is *no* association between the two variables in the *population*, you can't expect the *sample* conditional proportions to be *exactly* the same because of ordinary random variation from sample to sample. Later in the text, we'll present inferential methods to determine if observed sample differences between conditional proportions are large enough to indicate that the variables are associated in the population.

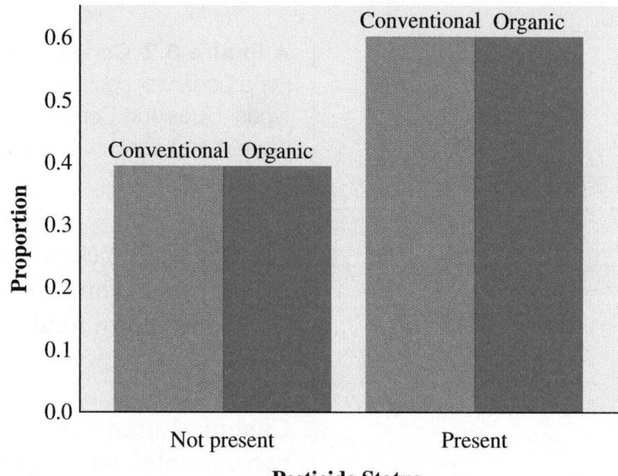

▲ **Figure 3.3 Hypothetical Conditional Proportions on Pesticide Status, Given Food Type, Showing No Association.** Question What's the difference between Figures 3.2 and 3.3 in the pattern shown by the bars in the graph?

In Chapter 5, we will discuss the expected relationship of a marginal proportion to a conditional proportion when two variables are independent (not associated).

3.1 Practicing the Basics

3.1 Which is the response/explanatory variable? For the following pairs of variables, which more naturally is the response variable and which is the explanatory variable?

a. College grade point average (GPA) and high school GPA

b. Number of children and mother's religion

c. Happiness (not too happy, pretty happy, very happy) and whether married (yes, no)

3.2 Sales and advertising Each month, the owner of Fay's Tanning Salon records in a data file the monthly total sales receipts and the amount spent that month on advertising.

a. Identify the two variables.

b. For each variable, indicate whether it is quantitative or categorical.

c. Identify the response variable and the explanatory variable.

3.3 Does higher income make you happy? Every
TRY General Social Survey includes the question, "Taken all together, would you say that you are very happy, pretty happy, or not too happy?" The table uses the 2008 survey to cross-tabulate happiness with family income, measured as the response to the question, "Compared with American families in general, would you say that your family income is below average, average, or above average?"

Happiness and Family Income, from General Social Survey

	Happiness			
Income	**Not Too Happy**	**Pretty Happy**	**Very Happy**	**Total**
Above average	26	233	164	**423**
Average	117	473	293	**883**
Below average	172	383	132	**687**
Total	**315**	**1,089**	**589**	**1,993**

a. Identify the response variable and the explanatory variable.

b. Construct the conditional proportions on happiness at each level of income. Interpret and summarize the association between these variables.

c. Overall, what proportion of people reported being very happy?

3.4 Religious activities In a recent General Social Survey,
TRY respondents answered the question, "In the past month, about how many hours have you spent praying, meditating, reading religious books, listening to religious broadcasts, etc.?" The responses on this variable were cross-tabulated with the respondent's gender. The table shows the results.

Religious Activity by Gender

	Number of Hours of Religious Activity					
Gender	**0**	**1–9**	**10–19**	**20–39**	**40 or more**	**Total**
Female	229	297	88	103	49	**766**
Male	276	243	59	40	16	**634**

a. Identify the response variable and the explanatory variable.

b. Find the conditional proportions for categories of the response variable, given gender, and interpret them.

c. Using software, create a side-by-side bar graph that compares males and females on the response variable. Summarize results.

d. What would be a disadvantage of using two pie charts to make this comparison?

3.5 Alcohol and college students The Harvard School of Public Health, in its College Alcohol Study Survey, surveyed college students in about 200 colleges in 1993, 1997, 1999, and 2001. The survey asked students questions about their drinking habits. Binge drinking was defined as five drinks in a row for males and four drinks in a row for females. The table shows results from the 2001 study, cross-tabulating subjects' gender by whether they have participated in binge drinking.

Binge Drinking by Gender

	Binge Drinking Status		
Gender	**Binge Drinker**	**Non-Binge Drinker**	**Total**
Male	1,908	2,017	**3,925**
Female	2,854	4,125	**6,979**
Total	**4,762**	**6,142**	**10,904**

a. Identify the response variable and the explanatory variable.

b. Report the cell counts of subjects who were (i) male and a binge drinker, (ii) female and a binge drinker.

c. Can you compare the counts in part b to answer the question, "Is there a difference between male and female students in the proportion who binge drink?" Explain.

d. Construct a contingency table that shows the conditional proportions of sampled students who do or do not binge drink, given gender. Interpret.

e. Based on part d, does it seem that there is an association between binge drinking and gender? Explain.

3.6 How to fight terrorism? A survey of 1000 adult Americans (*Rasmussen Reports*, April 15, 2004) asked each whether the best way to fight terrorism is to let the terrorists know we will fight back aggressively or to work with other nations to find an international solution. The first option was picked by 53% of the men but by only 36% of the women in the sample. Assume there were 600 men and 400 women in the sample.

a. Identify the response variable and the explanatory variable, and their categories.

b. Construct a contingency table (similar to Table 3.1) that shows the counts for the different combination of categories.

c. Use a contingency table to display the percentages for the two options, separately for females and for males.

d. Explain why the percentages reported here are *conditional* percentages.

e. Give an example of how results would have to differ from these for you to conclude that there's *no* evidence of association between these variables.

3.7 Heaven and hell A General Social Survey question asked, "Do you believe in heaven?" and "Do you believe in hell?" Explain how we could regard either variable (opinion about heaven, opinion about hell) as a response variable.

3.8 Heaven and hell data The last time the questions in the previous exercise were asked in the GSS, 955 subjects answered "yes" to both questions, 188 answered "no" to both, 162 answered "yes" to heaven but "no" to hell, and 9 answered "no" to heaven but "yes" to hell.

a. Display the data as a contingency table, labeling the variables and the categories.

b. Find the conditional proportions that treat opinion about heaven as the response variable and opinion about hell as the explanatory variable. Interpret.

c. Find the conditional proportions that treat opinion about hell as the response variable and opinion about heaven as the explanatory variable. Interpret.

d. Find the marginal proportion who (i) believe in heaven, (ii) believe in hell.

3.9 Gender gap in party ID In recent election years, political scientists have analyzed whether a gender gap exists in political beliefs and party identification. The table shows data collected from the 2004 General Social Survey on gender and party identification (ID).

Party ID by Gender

	Party Identification			
Gender	**Democrat**	**Independent**	**Republican**	**Total**
Male	299	365	232	**896**
Female	422	381	273	**1,076**
Total	**721**	**746**	**505**	**1,972**

a. Identify the response and explanatory variables.

b. What proportion of sampled individuals is (i) male and Republican, (ii) female and Republican?

c. What proportion of the overall sample is (i) male, (ii) Republican?

d. The figure (MINITAB output) displays the proportion of individuals identifying with each political party, given gender. What are these proportions called? Is there a difference between males and females in the proportions that identify with a particular party? Summarize whatever gender gap you observe.

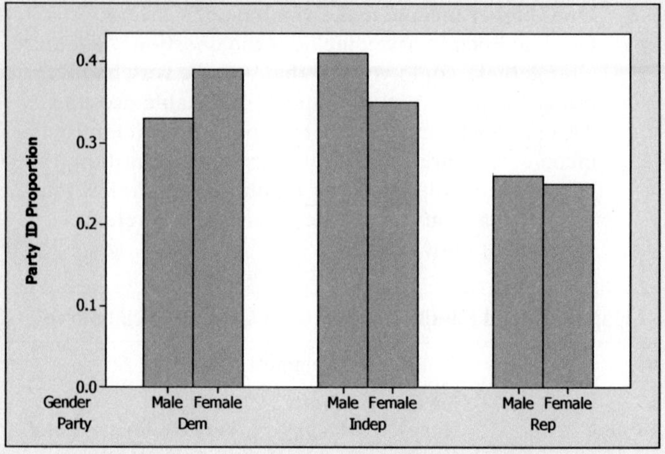

3.10 Use the GSS Go to the GSS Web site sda.berkeley.edu/GSS, click on GSS, with *No Weight* as the default weight selection, type SEX for the row variable and HAPPY for the column variable, put a check in the row box only for percentaging in the table options, and click on *Run the Table.*

a. Report the contingency table of counts.

b. Report the conditional proportions to compare the genders on reported happiness.

c. Are females and males similar, or quite different, in their reported happiness?

3.2 The Association Between Two Quantitative Variables

In practice, when we investigate the association between two variables, there are three types of cases:

- The variables could be *categorical* as food type and pesticide status are. In this case, as we have already seen, the data are displayed in a contingency table, and we can explore the association by comparing conditional proportions.
- One variable could be *quantitative* and one could be *categorical* as income and gender. As we saw in Chapter 2, we can compare the categories (such as females and males) using summaries of center and variability for the quantitative variable (such as the mean and standard deviation of income) and graphics such as side-by-side box plots.
- Both variables could be *quantitative*. In this case, we analyze how the outcome on the response variable tends to change as the value of the explanatory variable changes. The rest of the chapter considers this case.

Recall

Figure 2.16 in Chapter 2 used **side-by-side box plots** to compare heights for females and males. ◀

In exploring the relationship between two quantitative variables, we'll use the principles introduced in Chapter 2 for exploring the data of a single variable. We first use graphics to look for an overall pattern. We follow up with numerical summaries and check also for unusual observations that deviate from the overall pattern and may affect results.

Example 4

Numerical and graphical summaries

Worldwide Internet and Facebook Use

Picture the Scenario

The number of worldwide Internet users and the number of users of social networking sites such as Facebook have grown significantly over the past decade. This growth though has not been distributed evenly throughout the world. Countries such as Australia, Sweden, and the Netherlands have achieved an Internet penetration of more than 80%, while only 7.1% of India's population uses the Internet. The story with Facebook is similar. More than 40% of the populations of countries such as the United States and Australia use Facebook, compared to fewer than 3% of the populations of countries such as China, India, and Russia.

The Internet Use data file on the text CD contains recent data for 33 countries on Internet penetration, Facebook penetration, broadband subscription percentage, and other variables related to Internet use. In this example, we'll investigate the relationship between Internet penetration and Facebook penetration. Note that we will often say "use" instead of "penetration" in these two variable names. Table 3.4 displays the values of these two variables for each of the 33 countries.

Table 3.4 Internet and Facebook Penetration Rates For 33 Countries

Country	Internet Penetration	Facebook Penetration
Argentina	49.40%	30.53%
Australia	80.60%	46.01%
Belgium	67.30%	36.98%
Brazil	37.76%	4.39%
Canada	72.30%	52.08%
Chile	50.90%	46.14%
China	22.40%	0.05%
Colombia	38.80%	25.90%
Egypt	12.90%	5.68%
France	65.70%	32.91%
Germany	67.00%	14.07%
Hong Kong	69.50%	52.33%
India	7.10%	1.52%
Indonesia	10.50%	13.49%
Italy	48.80%	30.62%
Japan	73.80%	2.00%
Malaysia	62.80%	37.77%
Mexico	24.90%	16.80%
Netherlands	82.90%	20.54%

(Continued)

Country	Internet Penetration	Facebook Penetration
Peru	26.20%	13.34%
Philippines	21.50%	19.68%
Poland	52.00%	11.79%
Russia	27.00%	2.99%
Saudi Arabia	22.70%	11.65%
South Africa	10.50%	7.83%
Spain	66.80%	30.24%
Sweden	80.70%	44.72%
Taiwan	66.10%	38.21%
Thailand	20.50%	10.29%
Turkey	35.00%	31.91%
USA	77.33%	46.98%
UK	70.18%	45.97%
Venezuela	25.50%	28.64%

Source: Data from www.internetworldstats.com and www.checkfacebook.com.

Question to Explore

Use numerical and graphical summaries to describe the shape, center, and variability of the distributions of Internet penetration and Facebook penetration.

Think It Through

Using MINITAB, we obtain the following numerical measures of center and variability:

Variable	N	Mean	StDev	Minimum	Q1	Median	Q3	Maximum
Internet Use	33	47.00	24.40	7.00	24.00	49.00	68.50	83.00
Facebook Use	33	24.73	16.49	0.00	11.00	26.00	38.00	52.00

Figure 3.4 portrays the distributions using histograms. We observe that the distribution of Internet use is unimodal and skewed to the left. The distribution of Facebook use can be characterized as roughly symmetric with multiple modes.

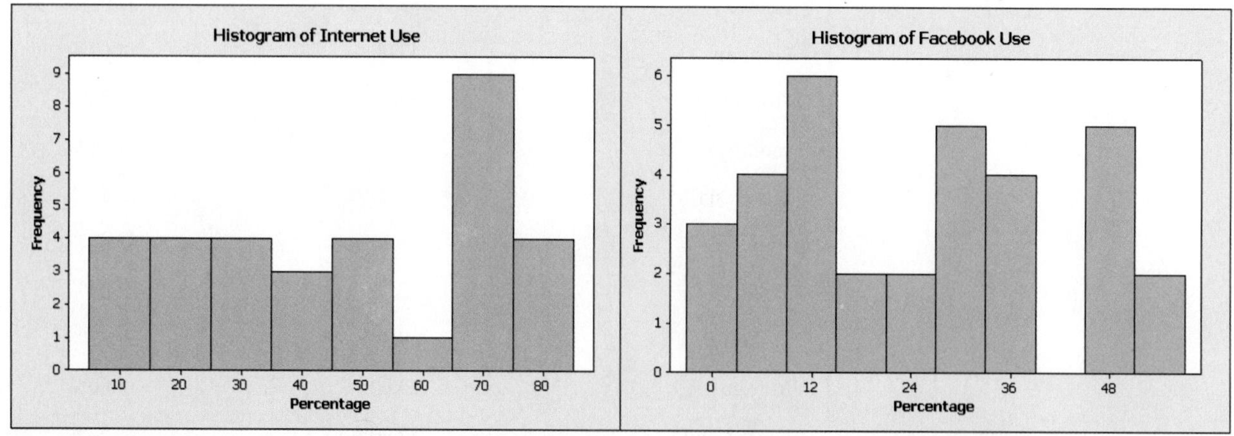

▲ **Figure 3.4 MINITAB Histograms of Internet Use and Facebook Use for the 33 Countries.** Question Which nations, if any, might be outliers in terms of Internet use? Facebook use? Which graphical display would more clearly identify potential outliers?

Insight

The histograms portray each variable separately. How can we portray the association between Internet use and Facebook use on a single display? We'll study how to do that next.

Try Exercise 3.13, part a

Looking for a Trend: The Scatterplot

With two quantitative variables, it is common to denote the response variable y and the explanatory variable x. We use this notation because graphical plots for examining the association use the y-axis for values of the response variable and the x-axis for values of the explanatory variable. This graphical plot is called a ***scatterplot***.

Scatterplot

A *scatterplot* is a graphical display for two quantitative variables using the horizontal (x) axis for the explanatory variable x and the vertical (y) axis for the response variable y. The values of x and y for a subject are represented by a point relative to the two axes. The observations for the n subjects are n points on the scatterplot.

Example 5

Scatterplots

Internet and Facebook Use

Picture the Scenario

We return to the data from Example 4 for 33 countries on Internet and Facebook use.

Questions to Explore

a. Display the relationship between Internet use and Facebook use with a scatterplot.

b. What can we learn about the association by inspecting the scatterplot?

Think It Through

a. The first step is to identify the response variable and the explanatory variable. We'll study how Facebook use depends on Internet use. A temporal relationship exists between when individuals become Internet users and when they become Facebook users; the former precedes the latter. We will treat Internet use as the explanatory variable and Facebook use as the response variable. Thus, we use x to denote Internet use and y to denote Facebook use. We plot Internet use on the horizontal axis and Facebook use on the vertical axis. Any statistical software package can create a scatterplot. Using data such as that in Table 3.4, place Internet use in one column and Facebook use in another. Select the variable that plays the role of x and the variable that plays the role of y. Figure 3.5 shows the scatterplot created with MINITAB.

Consider the observation for the United States. It has Internet use $x = 77$ and Facebook use $y = 47$. *Find its point circled in black.*

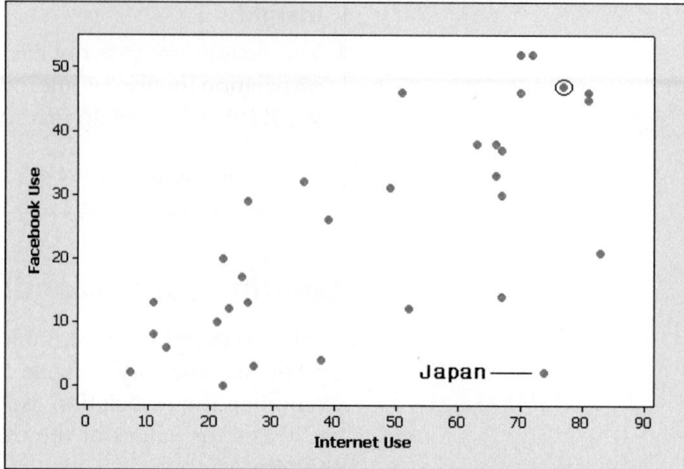

▲ **Figure 3.5 MINITAB Scatterplot for Internet Use and Facebook Use for 33 Countries.** The point for Japan is labeled and has coordinates $x = 74$ and $y = 2$. **Question** Is there any point that you would identify as standing out in some way? Which country does it represent, and how is it unusual in the context of these variables?

b. Here are some things we learn by inspecting the scatterplot:

- There is a clear trend. Nations with larger percentages of Internet use generally have larger percentages of Facebook use.
- For countries with relatively low Internet use (below 20%), there is little variability in Facebook use.
- Facebook use ranges from about 2% to 13% for each such country.
- For countries with high Internet use (above 20%), there is high variability in Facebook use. Facebook use ranges from about 2% to 52% for these countries.
- The point for Japan seems unusual. Its Internet use is among the highest of all countries (74%), while its Facebook use is among the lowest (2%). Based on values for other countries with similarly high Internet use, we might expect Facebook use to be between 25% and 50% rather than 2%. Although not as unusual as Japan, Facebook use for the Netherlands (21%) is a little lower than we'd expect for a country with such high Internet use (83%). Can you identify the point for the Netherlands on the scatterplot?

Insight

Although the points for Japan and, to a lesser extent, the Netherlands, can be considered atypical, there is a clear overall association. The countries with lower Internet use tend to have lower Facebook use, and the countries with high Internet use tend to have high Facebook use.

Try Exercise 3.12, parts a and b

How to Examine a Scatterplot

We examine a scatterplot to study **association**. How do values on the response variable change as values of the explanatory variable change? As Internet use gets higher, for instance, we see that Facebook use gets higher. When there's a trend in a scatterplot, what's the direction? Is the association **positive** or **negative?**

Positive Association and Negative Association

Two quantitative variables *x* and *y* are said to have a positive association when high values of *x* tend to occur with high values of *y*, and when low values of *x* tend to occur with low values of *y*. **Positive association:** As *x* goes up, *y* tends to go up.

Two quantitative variables have a negative association when high values of one variable tend to pair with low values of the other variable, and low values of one pair with high values of the other. **Negative association:** As *x* goes up, *y* tends to go down.

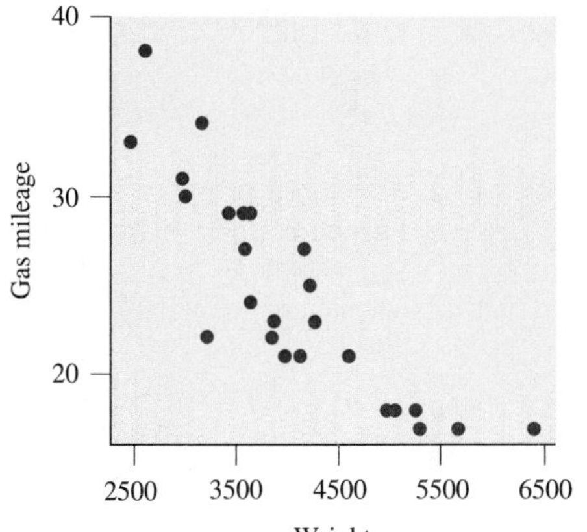

Figure 3.5 displays a positive association, because high (low) values of Internet use tend to occur with high (low) values of Facebook use. If we were to study the association between x = weight of car and y = gas mileage (in miles per gallon) we'd expect a negative association: Heavier cars would tend to get poorer gas mileage. The figure in the margin illustrates this idea, using data from the Car Weight and Mileage data file on the text CD.

Here are some questions to explore when you examine a scatterplot:

- Does there seem to be a positive association, a negative association, or no association?
- Can the trend in the data points be approximated reasonably well by a straight line? In that case, do the data points fall close to the line, or do they tend to scatter quite a bit?
- Are some observations unusual, falling well apart from the overall trend of the data points? What do the unusual points tell us?

Examining scatterplots

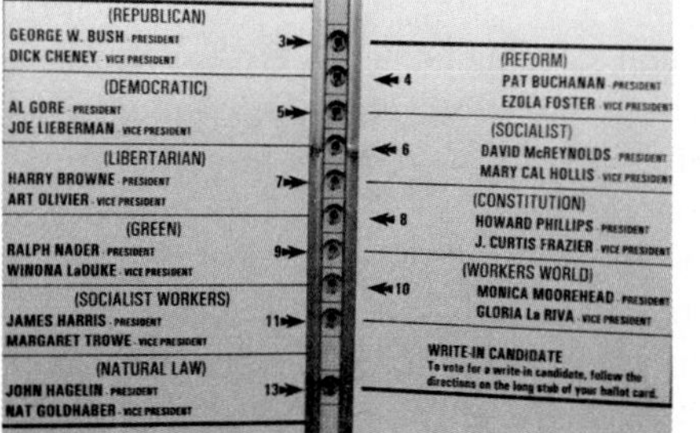

Example 6

The Butterfly Ballot and the 2000 U.S. Presidential Election

Picture the Scenario

Al Gore and George W. Bush were the Democatric and Republican candidates in the 2000 U.S. presidential election. In Palm Beach County, Florida, initial election returns reported 3407 votes for the Reform party candidate, Pat Buchanan. Political analysts thought this total seemed surprisingly large. They felt that most of these votes may have actually been intended for Gore (whose name was next to Buchanan's on the ballot) but were wrongly cast for Buchanan because of the design of the "butterfly ballot" used in that county, which some voters found confusing. On the butterfly ballot, Bush appeared first in the left column, followed by Buchanan in the right column, and Gore in the left column (see the photo in the margin).

Question to Explore

For each of the 67 counties in Florida, the Buchanan and the Butterfly Ballot data file on the text CD includes the Buchanan vote and the vote for the Reform party candidate in 1996 (Ross Perot). How can we explore graphically whether the Buchanan vote in Palm Beach County in 2000 was in fact surprisingly high, given the Reform party voting totals in 1996?

Think It Through

Figure 3.6 is a scatterplot of the countywide vote for the Reform party candidates in 2000 (Buchanan) and in 1996 (Perot). Each point represents a county. This figure shows a strong positive association statewide: Counties with a high Perot vote in 1996 tended to have a high Buchanan vote in 2000, and counties with a low Perot vote in 1996 tended to have a low Buchanan vote in 2000. The Buchanan vote in 2000 was roughly only 3% of the Perot vote in 1996.

In Figure 3.6, one point falls well above the others. This severe outlier is the observation for Palm Beach County, the county that had the butterfly ballot. It is far removed from the overall trend for the other 66 data points, which follow an approximately straight line.

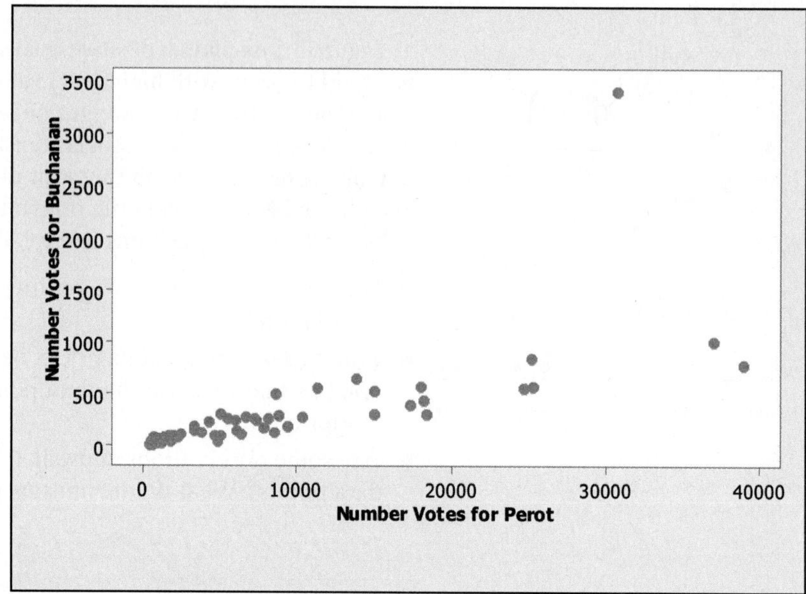

▲ **Figure 3.6** MINITAB Scatterplot of Florida Countywide Vote for Reform Party Candidates Pat Buchanan in 2000 and Ross Perot in 1996. Question Why is the top point, but not each of the two rightmost points, considered an outlier relative to the overall trend of the data points?

Insight

Alternatively, you could plot the Buchanan vote against the Gore vote or against the Bush vote (Exercise 3.23). These and other analyses conducted by statisticians[3] predicted that fewer than 900 votes were truly intended for Buchanan in Palm Beach County, compared to the 3407 votes he actually received. Bush won the state by 537 votes and, with it, the Electoral College and the election. So, this vote may have been a pivotal factor in determining the outcome of that election. Other factors that played a role were 110,000 disqualified overvote ballots in which people mistakenly voted for more than one presidential candidate (with Gore marked on 84,197 ballots and Bush on 37,731), often because of confusion from names being listed on more than one page of the ballot, and 61,000 undervotes caused by factors such as "hanging chads" from manual punch-card machines in some counties.

Try Exercise 3.23

In practice, data points in a scatterplot sometimes fall close to a straight line trend, as we saw for all data except Palm Beach County in Figure 3.6. This pattern represents a strong association in the sense that we can predict the y-value quite well from knowing the x-value. The next step in analyzing a relationship is summarizing the strength of the association.

[3]For further discussion of these and related data, see Exercise 2.95 and the article by A. Agresti and B. Presnell, *Statistical Science,* vol. 17, pp. 1–5, 2002.

Summarizing the Strength of Association: The Correlation

When the data points follow a roughly straight-line trend, the variables are said to have an approximately **linear** relationship. In some cases, the data points fall close to a straight line, but more often there is quite a bit of variability of the points around the straight-line trend. A summary measure called the **correlation** describes the strength of the linear association.

Correlation

The **correlation** summarizes the direction of the association between two quantitative variables and the strength of its linear (straight-line) trend. Denoted by r, it takes values between -1 and $+1$.

- A positive value for r indicates a positive association and a negative value for r indicates a negative association.
- The closer r is to ± 1 the closer the data points fall to a straight line, and the stronger the linear association is. The closer r is to 0, the weaker the linear association is.

Let's get a feel for the correlation r by looking at its values for the scatterplots shown in Figure 3.7:

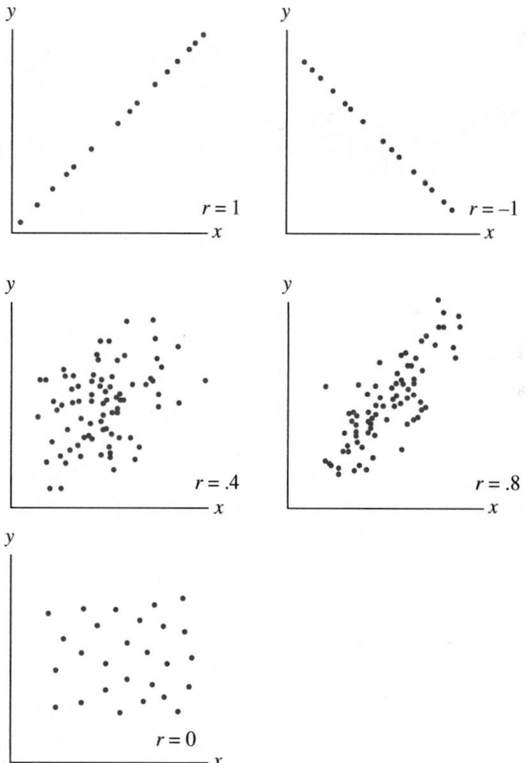

▲ **Figure 3.7 Some Scatterplots and Their Correlations.** The correlation gets closer to ± 1 when the data points fall closer to a straight line. **Question** Why are the cases in which the data points are closer to a straight line considered to represent stronger association?

The correlation r takes the extreme values of $+1$ and -1 only when the data points follow a straight line pattern *perfectly* as seen in the top two graphs in Figure 3.7. When $r = +1$ occurs, the line slopes upward. The association is positive since higher values of x tend to occur with higher values of y. The value $r = -1$ occurs when the line slopes downward, corresponding to a negative association.

In practice, don't expect the data points to fall perfectly on a straight line. However, the closer they come to that ideal, the closer the correlation is to 1 or -1. For instance, the scatterplot in Figure 3.7 with correlation $r = 0.8$ shows a stronger association than the one with correlation $r = 0.4$, for which the data points fall farther from a straight line.

Properties of the Correlation

Recall

The **absolute value** of a number gives the distance the number falls from zero on the number line. The correlation values of -0.9 and 0.9 both have an absolute value of 0.9. They both represent a stronger association than correlation values of -0.6 and 0.6, for example. ◀

- The correlation r always **falls between -1 and $+1$**. The closer the value to 1 in absolute value (see the margin comments), the stronger the **linear (straight-line) association**, as the data points fall nearer to a straight line.
- A **positive correlation** indicates a **positive association**, and a **negative correlation** indicates a **negative association**.
- The value of the correlation **does not depend on the variables' units**. For example, suppose one variable is the income of a subject, in dollars. If we change the observations to units of euros or to units of thousands of dollars, we'll get the same correlation.
- **Two variables have the same correlation no matter which is treated as the response variable and which is treated as the explanatory variable.**

Finding and interpreting the correlation value

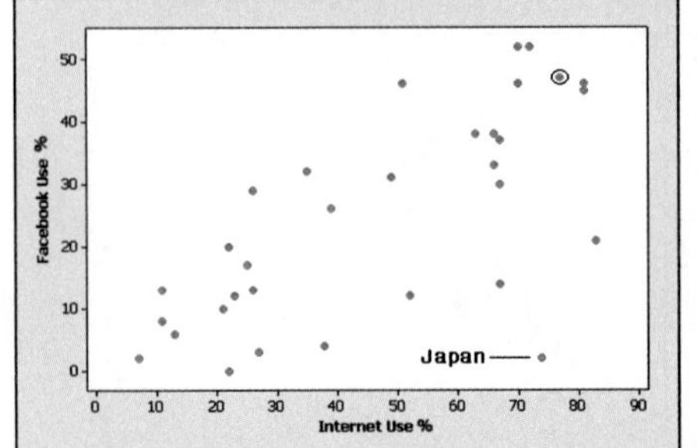

Example 7

Internet Use and Facebook Use

Picture the Scenario

Example 5 displayed a scatterplot for Internet use and Facebook use for 33 countries, shown again in the margin. We observed a positive association.

Questions to Explore

a. What value does software give for the correlation?

b. How can we interpret the correlation value?

Think It Through

Since the association is positive, we expect to find $r > 0$. If we input the columns of Internet use and Facebook use into MINITAB and request the correlation from the Basic Statistics menu, we get

Correlations: Internet Use, Facebook Use
Pearson correlation of Internet Use and Facebook Use = 0.682.

The correlation of $r = 0.682$ is positive. This result confirms the positive linear association we observed in the scatterplot. In summary, a country's extent of Facebook use is moderately associated with its Internet use, with higher Internet use tending to correspond to higher Facebook use.

Insight

The identifier *Pearson* for the correlation in the MINITAB output refers to the British statistician, Karl Pearson. In 1896 he provided the formula used to compute the correlation value from sample data. This formula is shown next.

Try Exercises 3.14 and 3.15

Recall

From Section 2.5 the **z-score** for an observation indicates the number of standard deviations and the direction (below or above) that the observation falls from the mean. ◀

Formula for the Correlation Value

Although software can compute the correlation for us, it helps to understand it if you see its formula. For an observation x on the explanatory variable, let z_x denote the z-score that represents the number of standard deviations and direction that x falls from the mean of x. That is,

$$z_x = \frac{\text{observed value} - \text{mean}}{\text{standard deviation}} = \frac{(x - \bar{x})}{s_x},$$

where s_x denotes the standard deviation of the x-values. Similarly, let z_y denote the number of standard deviations and direction that an observation y on the response variable falls from the mean of y. To obtain r, you calculate the product $z_x z_y$ for each observation, then find a typical value (a type of average) of those products.

Calculating the Correlation r

$$r = \frac{1}{n-1}\Sigma z_x z_y = \frac{1}{n-1}\Sigma\left(\frac{x-\bar{x}}{s_x}\right)\left(\frac{y-\bar{y}}{s_y}\right)$$

where n is the number of points, $\bar{x}$ and $\bar{y}$ are means, and s_x and s_y are standard deviations for x and y. The sum is taken over all n observations.

In Practice Using Technology to Calculate r

Hand **calculation of the correlation** r is tedious. You should rely on software or a calculator. It's more important to understand how the correlation describes association in terms of how it reflects the relative numbers of points in the four quadrants.

For x = Internet use and y = Facebook use, Example 7 found the correlation $r = 0.682$, using statistical software. To visualize how the formula works, let's revisit the scatterplot, reproduced in Figure 3.8, with a vertical line at the mean of x and a horizontal line at the mean of y. These lines divide the scatterplot into four **quadrants**. The summary statistics are

$$\bar{x} = 47.0 \qquad \bar{y} = 24.7$$
$$s_x = 24.4 \qquad s_y = 16.5.$$

The point for Japan ($x = 74$, $y = 2$) has as its z-scores $z_x = 1.11$, $z_y = -1.38$. This point is labeled in Figure 3.8. Since $x = 74$ is to the right of the mean for x and $y = 2$ is below the mean of y, it falls in the lower-right quadrant. This makes Japan somewhat atypical in the sense that all but 7 of the 33 countries have points that fall in the upper-right and lower-left quadrants.

In Words

A **quadrant** is any of the four regions created when a plane is divided by a horizontal line and a vertical line.

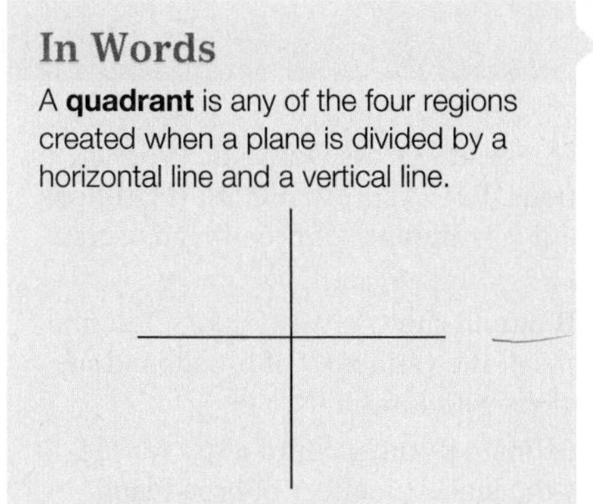

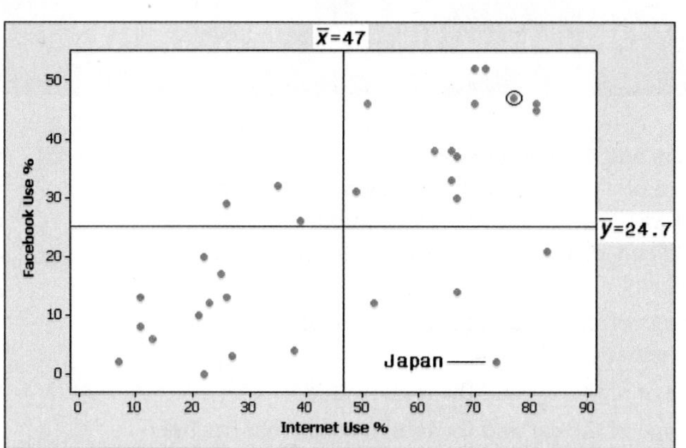

▲ **Figure 3.8 MINITAB Scatterplot of Internet Use and Facebook Use Divided Into Quadrants at ($\bar{x}$, $\bar{y}$)** Of the 33 data points, 26 lie in the upper-right quadrant (above the mean on each variable) or the lower-left quadrant (below the mean on each variable). **Question** Do the points in these two quadrants make a positive or a negative contribution to the correlation value? (***Hint:*** Is the product of z-scores for these points positive or negative?)

SUMMARY: Product of z-scores and correlation

- The product of the z-scores for any point in the upper-right quadrant is positive. The product is also positive for each point in the lower-left quadrant. Such points contribute to a positive correlation.
- The product of the z-scores for any point in the upper-left and lower-right quadrants is negative. Such points contribute to a negative correlation.

The overall correlation reflects the number of points in the various quadrants and how far they fall from the means. For example, if all points fall in the upper-right and lower-left quadrants, the correlation must be positive.

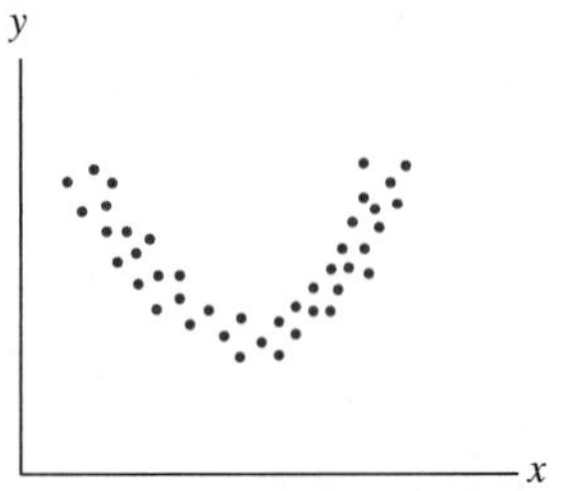

▲ **Figure 3.9 The Correlation Poorly Describes the Association When the Relationship Is Curved.** For this U-shaped relationship, the correlation is 0 (or close to 0), even though the variables are strongly associated. **Question** Can you use the formula for r, in terms of how points fall in the quadrants, to reason why the correlation would be close to 0?

Graph Data to See If the Correlation Is Appropriate

The correlation is an efficient way to summarize the association shown by lots of data points with a single number. But be careful to use it only when it is appropriate. Figure 3.9 illustrates why. It shows a scatterplot in which the data points follow a U-shaped curve. There is an association because as x increases, y first tends to decrease and then it tends to increase. For example, this might happen if x = age of person and y = annual medical expenses. Medical expenses tend to be high for newly born and young children, then they tend to be low until the person gets old when they become high again. However, $r = 0$ for the data in Figure 3.9.

The correlation is designed for straight-line relationships. For Figure 3.9, $r = 0$, and it fails to detect the association. The correlation is not valid for describing association when the points cluster around a curve rather than around a straight line.

This figure highlights an important point to remember about *any* data analysis:

- **Always plot the data.**

If we merely used software to calculate the correlation for the data in Figure 3.9 but did not plot the data, we might mistakenly conclude that the variables have no association. They *do* have one, but it is not a straight-line association.

In Practice Always Construct a Scatterplot

Always **construct a scatterplot** to display a relationship between two quantitative variables. The correlation indicates the direction and strength only of an approximate *straight-line* relationship.

3.2 Practicing the Basics

3.11 Used cars and direction of association For the 100 cars on the lot of a used-car dealership, would you expect a positive association, negative association, or no association between each of the following pairs of variables? Explain why.

a. The age of the car and the number of miles on the odometer

b. The age of the car and the resale value

c. The age of the car and the total amount that has been spent on repairs

d. The weight of the car and the number of miles it travels on a gallon of gas

3.12 Broadband and GDP The Internet Use data file on the text CD contains data on the number of individuals with broadband access and Gross Domestic Product (GDP) for 33 nations. Let x represent GDP (in billions of U.S. dollars) and y = number of broadband users (in thousands).
TRY

a. The MINITAB output shows a scatterplot. Describe this plot in terms of the variability of broadband subscribers for nations with low GDP.

b. Give the approximate x- and y-coordinates for the nation that has the highest number of broadband subscribers.

c. Use software to calculate the correlation coefficient between the two variables. What is the sign of the coefficient? Explain what the sign means in the context of the problem.

d. Identify one nation that appears to have fewer broadband subscribers than you might expect based on that nation's GDP, and one that appears to have more.

e. If you recalculated the correlation coefficient after removing the two observations you identified in part d, how would you expect the resulting coefficient to compare to the one obtained in part c?

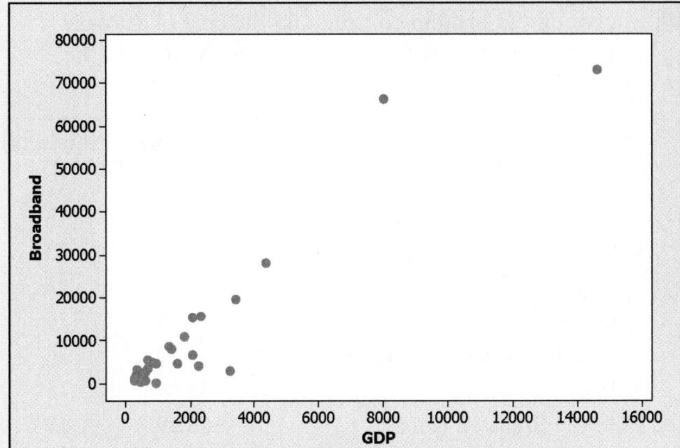

3.13 Economic development based on GDP The previous problem discusses GDP, which is a commonly used measure of the overall economic activity of a nation. For this group of nations, the GDP data have a mean of 1771 and a standard deviation of 2781 (in billions of U.S. dollars).

TRY

a. The five-number summary of GDP is minimum = 245, Q1 = 396, median = 731, Q3 = 2033, and maximum = 14,581. Sketch a box plot.

b. Based on your graph in part a, as well as the mean and standard deviation, describe the shape of the distribution of GDP values.

c. The data set also contains per capita GDP, or the overall GDP divided by the nation's population size. Suppose we were to construct a scatterplot of per capita GDP and GDP. What trend might you expect to see?

d. Your friend, Joe, argues that the correlation between the two variables must be 1 since they are both measuring the same thing. In reality, the actual correlation between per capita GDP and GDP is only 0.247. Identify the flaw in Joe's reasoning.

3.14 Politics and newspaper reading For the FL Student Survey data file on the text CD, the correlation between y = political ideology (scored 1 = very liberal to 7 = very conservative) and x = number of times a week reading a newspaper is −0.07.

TRY

a. Would you call this association strong or weak? Explain.

b. The correlation between political ideology and religiosity (how often attend religious services) is 0.58. For this sample, which explanatory variable, newspaper reading or religiosity, seems to have a stronger association with y? Explain.

3.15 Internet use correlations For the 33 nations in the Internet Use data file on the text CD, consider the following correlations:

TRY

Variable 1	Variable 2	Correlation
Internet users	Facebook users	0.507
Internet users	Broadband subscribers	0.949
Internet users	Population	0.744
Facebook users	Broadband subscribers	0.619
Facebook users	Population	0.097
Broadband subscribers	Population	0.533

a. Which pair of variables exhibits the *strongest* linear relationship?

b. Which pair of variables exhibits the *weakest* linear relationship?

c. In Example 7, we found the correlation between Internet use and Facebook use (measured in percentages of the population) to be 0.682. Why does the correlation between total number of Internet users and Facebook users differ from that of Internet use and Facebook use?

3.16 Match the scatterplot with r Match the scatterplots below with the correlation values.

1. $r = -0.9$
2. $r = -0.5$
3. $r = 0$
4. $r = 0.6$

(a)
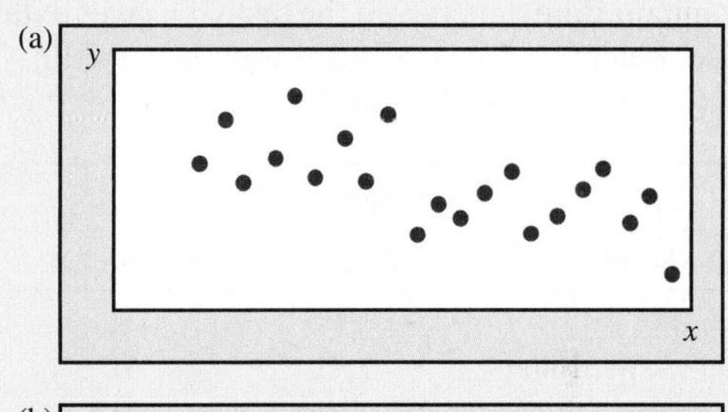

(b)
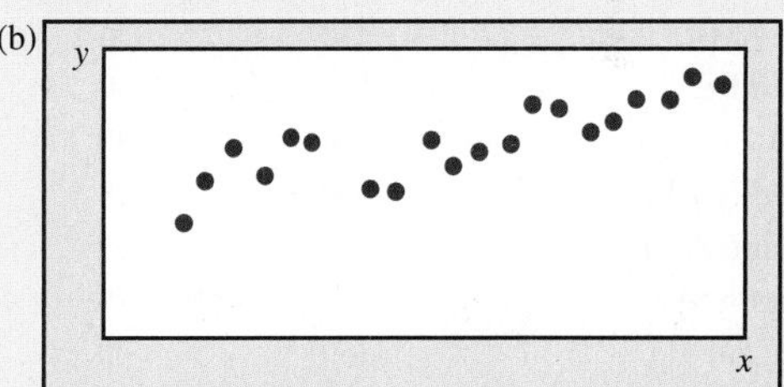

(c)
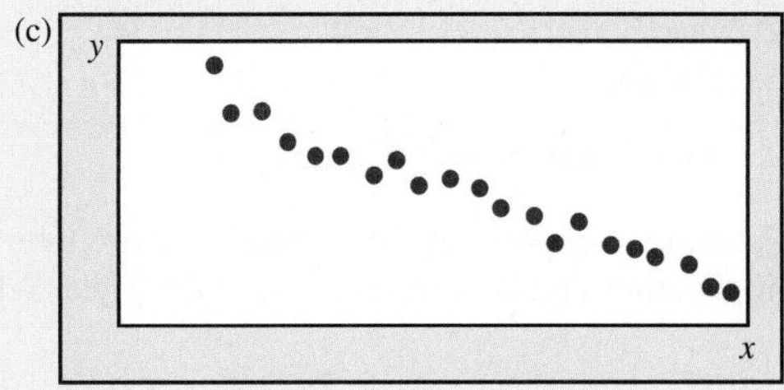

(d)
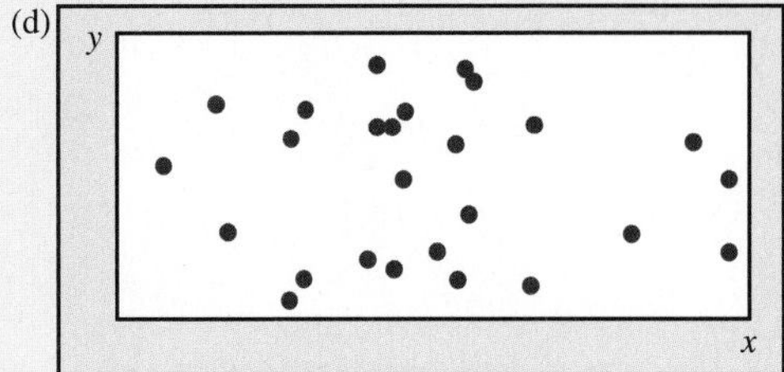

3.17 What makes $r = 1$? Consider the data:

x	3	4	5	6	7
y	8	13	12	14	16

a. Sketch a scatterplot.
b. If one pair of (x, y) values is removed, the correlation for the remaining four pairs equals 1. Which pair is it?
c. If one y value is changed, the correlation for the five pairs equals 1. Identify the y value and how it must be changed for this to happen.

3.18 z-scores for $r = 1$ Use the points from the previous exercise with $x = 3, 5, 6, 7$.

a. Find the z-scores on x and on y for each point. Comment on how z_x and z_y relate to each other for each point.
b. Compute r using the z-scores from part a for the four observations. Is this the value you expected to get for r? Why?

3.19 $r = 0$ Sketch a scatterplot for which $r > 0$, but $r = 0$ after one of the points is deleted.

3.20 Correlation inappropriate Describe a situation in which it is inappropriate to use the correlation to measure the association between two quantitative variables.

3.21 Which mountain bike to buy? Is there a relationship between the weight of a mountain bike and its price? A lighter bike is often preferred, but do lighter bikes tend to be more expensive? The following table, from the Mountain Bike data file on the text CD, gives data on price, weight, and type of suspension (FU = full, FE = front end) for 12 brands.

Mountain Bikes

Brand and Model	Price($)	Weight(LB)	Type
Trek VRX 200	1000	32	FU
Cannondale Super V400	1100	31	FU
GT XCR-4000	940	34	FU
Specialized FSR	1100	30	FU
Trek 6500	700	29	FE
Specialized Rockhop	600	28	FE
Haro Escape A7.1	440	29	FE
Giant Yukon SE	450	29	FE
Mongoose SX 6.5	550	30	FE
Diamondback Sorrento	340	33	FE
Motiv Rockridge	180	34	FE
Huffy Anorak 36789	140	37	FE

Source: Data from *Consumer Reports,* June 1999.

a. You are shopping for a new bike. You are interested in whether and how weight affects the price. Which variable is the logical choice for the (i) explanatory variable, (ii) response variable?
b. Construct a scatterplot of price and weight. Does the relationship seem to be approximately linear? In what way does it deviate from linearity?
c. Use your software to verify that the correlation equals −0.32. Interpret it in context. Does weight appear to affect the price strongly in a linear manner?

3.22 Prices and protein revisited Is there a relationship between the protein content and the cost of Subway sandwiches? Use software to analyze the data in the following table:

Sandwich	Cost ($)	Protein (g)
BLT	$2.99	17
Ham (Black Forest, without cheese)	$2.99	18
Oven Roasted Chicken	$3.49	23
Roast Beef	$3.69	26
Subway Club®	$3.89	26
Sweet Onion Chicken Teriyaki	$3.89	26
Turkey Breast	$3.49	18
Turkey Breast & Ham	$3.49	19
Veggie Delite®	$2.49	8
Cold Cut Combo	$2.99	21
Tuna	$3.10	21

a. Construct a scatterplot to show how protein depends on cost. Is the association positive or negative? Do you notice any unusual observations?
b. What might explain the gap observed in the scatterplot? (Hint: Are vegetables generally high or low in protein relative to meat and poultry products?)
c. Obtain the correlation between cost and protein, r. Interpret this value in context.

3.23 Buchanan vote TRY Refer to Example 6 and the Buchanan and the Butterfly Ballot data file on the text CD. Let y = Buchanan vote and x = Gore vote.

a. Construct a box plot for each variable. Summarize what you learn.
b. Construct a scatterplot. Identify any unusual points. What can you learn from a scatterplot that you cannot learn from box plots?
c. For the county represented by the most outlying observation, about how many votes would you have expected Buchanan to get if the point followed the same pattern as the rest of the data?
d. Repeat parts a and b using y = Buchanan vote and x = Bush vote.

3.3 Predicting the Outcome of a Variable

Recall

The correlation does not require one variable to be designated the response and the other variable the explanatory. ◀

We've seen how to explore the relationship between two quantitative variables graphically with a scatterplot. When the relationship has a straight-line pattern, the correlation describes it numerically. We can analyze the data further by finding an equation for the straight line that best describes that pattern. This equation predicts the value of the variable designated as the response variable from the value of the variable designated as the explanatory variable.

In Words

The **symbol $\hat{y}$**, which denotes the predicted value of *y*, is pronounced *y-hat.*

Regression Line: An Equation for Predicting the Response Outcome

The **regression line** predicts the value for the response variable *y* as a straight-line function of the value *x* of the explanatory variable. Let $\hat{y}$ denote the **predicted value** of *y*. The equation for the regression line has the form

$$\hat{y} = a + bx.$$

In this formula, *a* denotes the **y-intercept** and *b* denotes the **slope.**

Example 8

Predict an outcome

Height Based on Human Remains

Picture the Scenario

Anthropologists can reconstruct information using partial human remains at burial sites. For instance, after finding a femur (thighbone), they can predict how tall an individual was. They use the regression line, $\hat{y} = 61.4 + 2.4x$, where $\hat{y}$ is the predicted height and x is the length of the femur, both in centimeters.

Questions to Explore

How can we graph the line that depicts how the predicted height depends on the femur length? A femur found at a particular site has a length of 50 cm. What is the predicted height of the person who had that femur?

Think It Through

The formula $\hat{y} = 61.4 + 2.4x$ has *y*-intercept 61.4 and slope 2.4. It has the straight-line form $\hat{y} = a + bx$ with $a = 61.4$ and $b = 2.4$.

Each number x, when substituted into the formula $\hat{y} = 61.4 + 2.4x$, yields a value for $\hat{y}$. For simplicity in plotting the line, we start with $x = 0$, although in practice this would not be an observed femur length. The value $x = 0$ has $\hat{y} = 61.4 + 2.4(0) = 61.4$. Now, all points on the *y*-axis have $x = 0$, so the line has height 61.4 at the point of its intersection with the *y*-axis. Because of this placement, the constant 61.4 in the equation is called the ***y*-intercept**. The line intersects the *y*-axis at the point with (x, y) coordinates (0, 61.4), which is 61.4 units up the *y*-axis.

The value $x = 50$ has $\hat{y} = 61.4 + 2.4(50) = 181.4$. When the femur length is 50 cm, the predicted height of the person is 181.4 cm. The coordinates are (50, 181.4). We can plot the line by connecting the points (0, 61.4) and (50, 181.4). Figure 3.10 plots the straight line for x between 0 and 50. In summary, the predicted height $\hat{y}$ increases from 61.4 to 181.4 as x increases from 0 to 50.

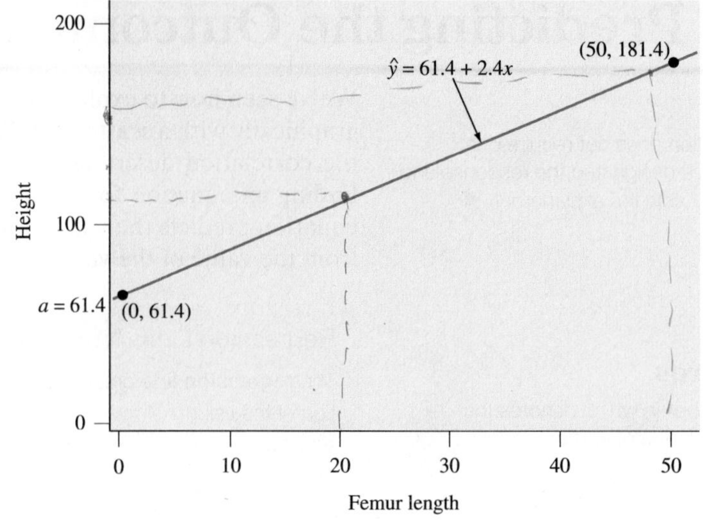

▲ **Figure 3.10** Graph of the Regression Line for x = Femur Length and y = Height of Person. Questions At what point does the line cross the y-axis? How can you interpret the slope of 2.4?

In Practice Notation for the Regression Line

The formula $\hat{y} = a + bx$ uses slightly different notation from the traditional formula, which is $y = mx + b$. In that equation, m = *the slope* (the coefficient of x) and b = y-intercept. Regardless of the notation, the interpretation of the y-intercept and slope are the same.

Insight

A regression equation is often called a **prediction equation** since it predicts the value of the response variable y at any value of x. Sadly, this particular prediction equation had to be applied to bones found in mass graves in Kosovo, to help identify Albanians who had been executed by Serbians in 1998.[4]

Try Exercises 3.25, part a, and 3.26, part a

Recall

Math facts:

***y*-intercept** is the value of the line in the y direction when $x = 0$.

Slope = rise/run = change in y/change in x ◀

Interpreting the *y*-Intercept and Slope

The ***y*-intercept** is the predicted value of y when $x = 0$. This fact helps us plot the line, but it may not have any interpretative value if no observations had x values near 0. It does not make sense for femur length to be 0 cm, so the y-intercept for the equation $\hat{y} = 61.4 + 2.4x$ is not a relevant predicted height.

The **slope** b in the equation $\hat{y} = a + bx$ equals the amount that $\hat{y}$ changes when x increases by one unit. For two x values that differ by 1.0, the $\hat{y}$ values differ by b. For the line $\hat{y} = 61.4 + 2.4x$, we've seen that $\hat{y} = 181.4$ at $x = 50$. If x increases by 1.0 to $x = 51$, we get $\hat{y} = 61.4 + 2.4(51) = 183.8$ The increase in $\hat{y}$ is from 181.4 to 183.8, which is 2.4, the slope value. For each 1-cm increase in femur length, height is predicted to increase by 2.4 cm. Figure 3.11 portrays this interpretation.

When the slope is negative, the predicted value $\hat{y}$ *decreases* as x increases. The straight line then goes downward, and the association is *negative.*

When the slope = 0, the regression line is horizontal (parallel to the x-axis). The predicted value $\hat{y}$ of y stays constant at the y-intercept for any value of x. Then the predicted value $\hat{y}$ does not change as x changes, and the variables do not exhibit an association. Figure 3.12 illustrates the three possibilities for the sign of the slope.

The absolute value of the slope describes the *magnitude* of the change in $\hat{y}$ for a 1-unit change in x. The larger the absolute value, the steeper the regression line.

[4] "The Forensics of War," by Sebastian Junger in *Vanity Fair*, October 1999.

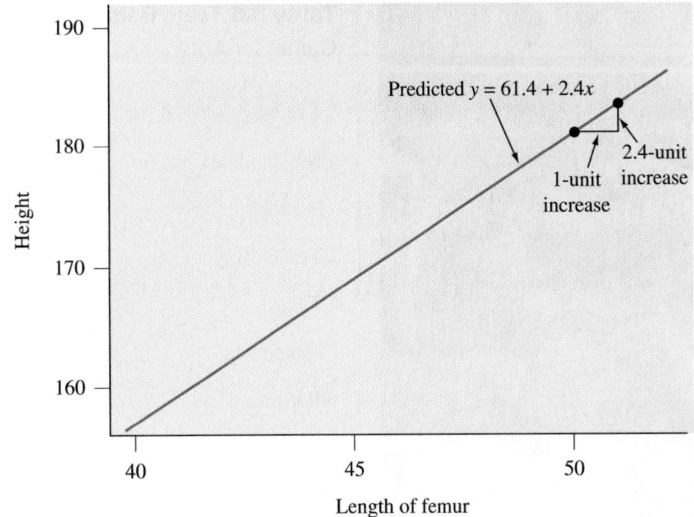

▲ **Figure 3.11 The Slope of a Straight Line.** The slope is the change in the predicted value $\hat{y}$ of the response variable for a 1-unit increase in the explanatory variable x. For an increase in femur length from 50 cm to 51 cm, the predicted height increases by 2.4 cm. **Question** What does it signify if the slope equals 0?

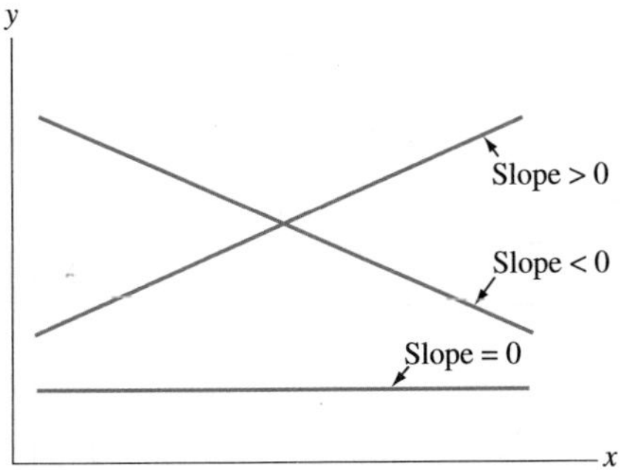

▲ **Figure 3.12 Three Regression Lines Showing Positive Association (slope $>$ 0), Negative Association (slope $<$ 0) and No Association (slope $=$ 0).** **Question** Would you expect a positive or negative slope when y = annual income and x = number of years of education?

A line with $b = 4.2$, such as $\hat{y} = 61.4 + 4.2x$, is steeper than one with $b = 2.4$. A line with $b = -0.07$ is steeper than one with $b = -0.04$.

Depending on the units of measurement, a 1-unit increase in a predictor x could be a trivial amount, or it could be huge. We will gain a better feel for how the slope works in context as we explore upcoming examples.

Finding a Regression Equation

How can we use the data to find the regression equation? We should first construct a scatterplot to make sure that the relationship has a roughly straight line trend. If so, then software or calculators can easily find the straight line that best fits the data.

Regression equation

Example 9

Predicting Baseball Scoring Using Batting Average

Picture the Scenario

In baseball, two summaries of a team's offensive ability are the team batting average (the proportion of times the team's players get a hit, out of the times they are officially at bat) and team scoring (the team's mean number of runs scored per game). Table 3.5 shows the 2010 statistics for the American League teams, from the AL Team Statistics data file on the text CD.

Scoring runs is a result of hitting, so team scoring is the response variable y and team batting average is the explanatory variable x. Figure 3.13 is the scatterplot. There is a trend summarized by a positive correlation, $r = 0.568$.

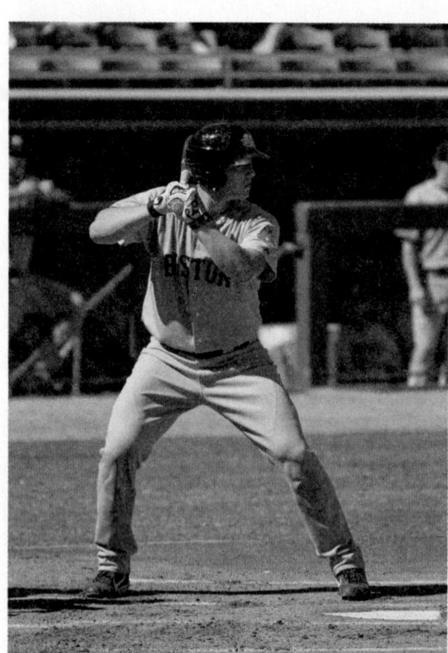

Table 3.5 Team Batting Average and Team Scoring (Mean Number of Runs per Game) for American League Teams in 2010[5]

Team	Batting Average	Team Scoring
NY Yankees	0.267	5.30
Boston	0.268	5.05
Tampa Bay	0.247	4.95
Texas	0.276	4.86
Minnesota	0.273	4.82
Toronto	0.248	4.66
Chicago Sox	0.268	4.64
Detroit	0.268	4.64
LA Angels	0.248	4.20
Kansas City	0.274	4.17
Oakland	0.256	4.09
Cleveland	0.248	3.99
Baltimore	0.259	3.78
Seattle	0.236	3.17

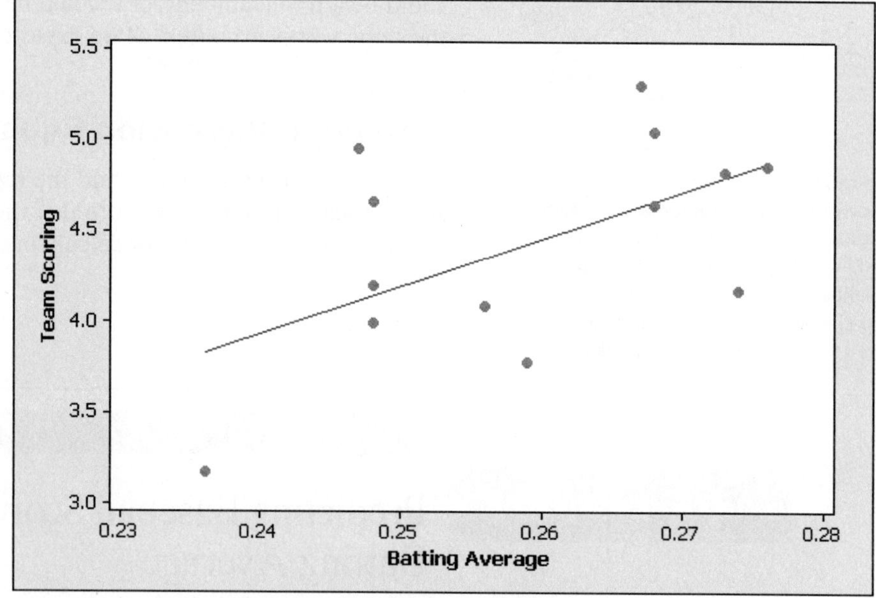

▲ **Figure 3.13** MINITAB Output for Scatterplot of Team Batting Average and Team Scoring, with Regression Line Superimposed. Question How can you find the prediction error that results when you use the regression line to predict team scoring for a team?

Questions to Explore

a. According to software, what is the regression equation?

b. If a team has a batting average of 0.275 next year, what is their predicted mean number of runs per game?

c. How do you interpret the slope in this context?

---

[5] *Source:* Data from espn.go.com/mlb/stats.

Think It Through

a. With the software package MINITAB, when we choose the regression option in the Regression part of the Statistics menu, part of the output tells us:

The regression equation is

Team Scoring = −2.32 + 26.1 Batting Average

The y-intercept is $a = -2.32$. The slope is the coefficient of the explanatory variable, denoted in the data file by BAT_AVG. It equals $b = 26.1$. The regression equation is

$$\hat{y} = a + bx = -2.32 + 26.1x.$$

The TI-83+/84 calculator provides the output shown in the margin.

b. We predict that an American League team with a team batting average of 0.275 will score an average of $\hat{y} = -2.32 + 26.1(0.275) = 4.86$ runs per game.

c. Since the slope $b = 26.1$ is positive, the association is positive: The predicted team scoring increases as team batting average increases. The slope refers to the change in $\hat{y}$ for a 1-unit change in x. However, x = team batting average is a proportion. In Table 3.5, the team batting averages fall between about 0.23 and 0.28, a range of 0.05. An increase of 0.05 in x corresponds to an increase of (0.05)26.1 = 1.3 in predicted team scoring. The mean number of runs scored per game is predicted to be about 1.3 higher for the best hitting teams than for the worst hitting teams.

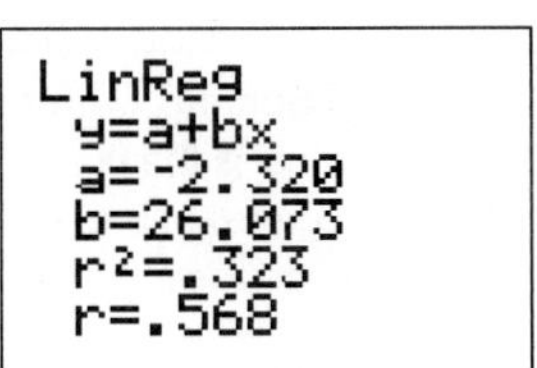

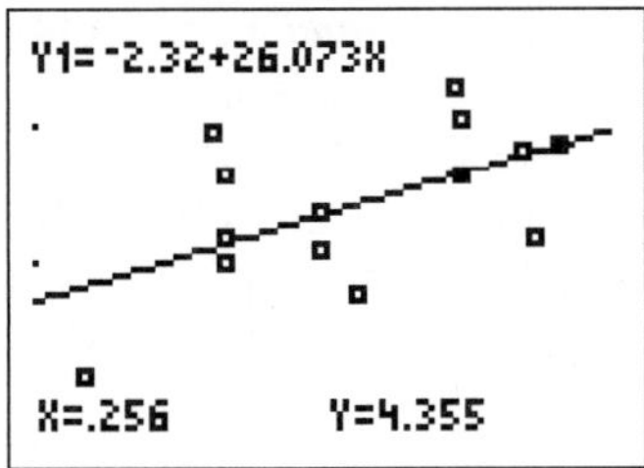

TI-83+/84 output

Insight

Figure 3.13 shows the regression line superimposed over the scatterplot. It applies only over the range of observed batting averages. For instance, it's not sensible to predict that a team with batting average of 0.0 will average −2.32 runs per game.

Try Exercises 3.25, part b, and 3.32 to get a feel for fitting a regression line; use the Regression by Eye applet discussed in Exercise 3.115

Residuals Measure the Size of Prediction Errors

The regression equation $\hat{y} = -2.32 + 26.1x$ predicts team scoring for a given level of x = team batting average. Once we have used the regression equation, we can compare the predicted values to the actual team scoring to check the accuracy of those predictions.

For example, New York had $y = 5.3$ and $x = 0.267$. The prediction for y = mean number of runs per game at 0.267 is $-2.32 + 26.1x = -2.32 + 2.61(0.267) = 4.65$. The prediction error is the difference between the actual y value of 5.3 and the predicted value of 4.65, which is $y - \hat{y} = 5.3 - 4.65 = 0.65$. For New York, the regression equation under predicts y by 0.65 runs per game. For Seattle, $x = 0.236$ and $y = 3.17$. The regression line yields a predicted value of $-2.32 + 26.1(0.236) = 3.84$, so the prediction is too high. The prediction error is $3.17 - 3.84 = -0.67$. These prediction errors are called **residuals**.

Each observation has a residual. As we just saw, some are positive and some are negative. A positive residual occurs when the actual y is larger than the predicted value $\hat{y}$, so that $y - \hat{y} > 0$. A negative residual results when the actual y is smaller than the predicted value $\hat{y}$. The smaller the absolute value of a residual, the closer the predicted value is to the actual value, so the better is the prediction. If the predicted value is the same as the actual value, the residual is zero.

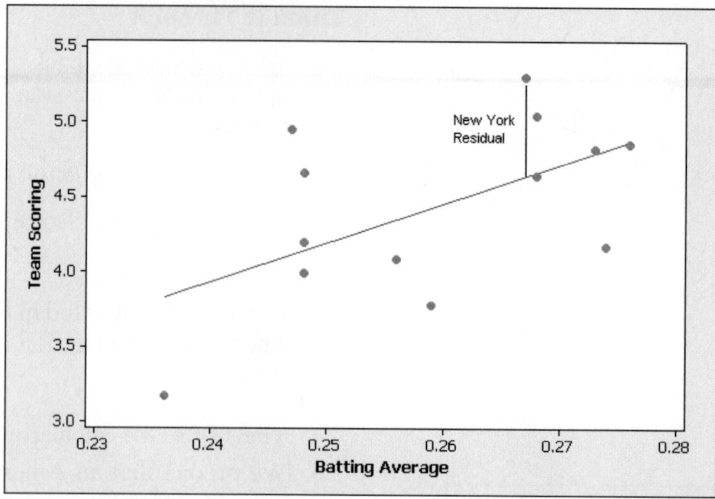

▲ **Figure 3.14 Scatterplot of Team Batting Average and Team Scoring, with the Residual for New York at the Point ($x = 0.267$, $y = 5.3$).** The residual is the prediction error, which is represented by the vertical distance of the point from the regression line. **Question** Why is a residual represented by a *vertical* distance from the regression line?

Graphically in the scatterplot, *for an observation, the vertical distance between the point and the regression line is the absolute value of the residual.* Figure 3.14 illustrates this fact for the positive residual for New York. The residuals are vertical distances because the regression equation predicts y, the variable on the vertical axis, at a given value of x. Notice the parallel with analyzing contingency tables by studying values of the response variable, *conditional* on (given) values of the explanatory variable.

Residual

In a scatterplot, the vertical distance between the point and the regression line is the absolute value of the residual. The residual is denoted as the difference $y - \hat{y}$ between the actual value and the predicted value of the response variable.

Residuals

Example 10

Detecting an Unusual Vote Total

Picture the Scenario

Example 6 investigated whether the vote total in Palm Beach County, Florida in the 2000 presidential election was unusually high for Pat Buchanan, the Reform party candidate. We did this by plotting, for all 67 counties in Florida, Buchanan's vote against the Reform party candidate (Ross Perot) vote in the 1996 election.

Question to Explore

If we fit a regression line to y = Buchanan vote and x = Perot vote for the 67 counties, would the residuals help us detect any unusual vote totals for Buchanan?

Think It Through

For a county with a large residual, the predicted vote is far from the actual vote, which would indicate an unusual vote total. We can easily have software find the regression line and the residual for each of the 67 counties. Then we can

quickly see whether some residuals are particularly large by constructing a histogram of the residuals. Figure 3.15 shows this histogram for the vote data. The residuals cluster around 0, but one is very large, greater than 2000. Inspection of the results shows that this residual applies to Palm Beach County, for which the actual Buchanan vote was $y = 3407$ and the predicted vote was $\hat{y} = 1100$. Its residual is $y - \hat{y} = 3407 - 1100 = 2307$. In summary, in Palm Beach County Buchanan's vote was much higher than predicted.[6]

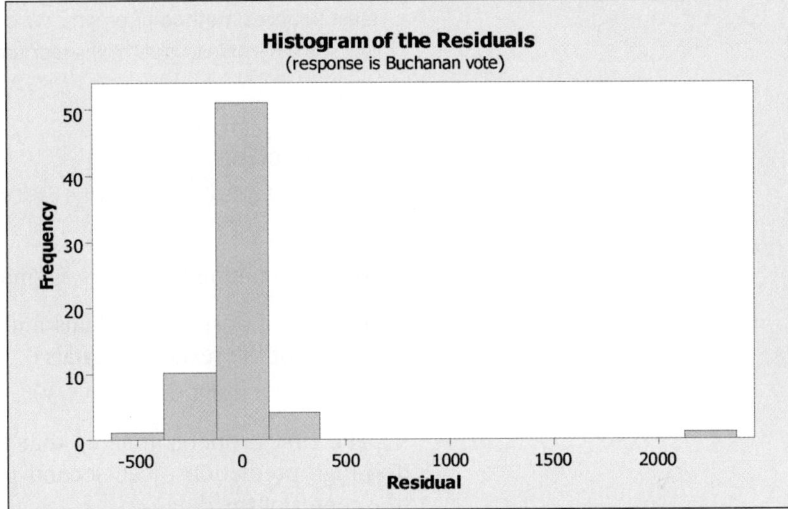

▲ **Figure 3.15** MINITAB Output of Residuals for Predicting 2000 Buchanan Presidential Vote in Florida Counties Using 1996 Perot Vote. Question What does the rightmost bar represent?

Insight

As we'll discuss in the next section, an extreme outlier can pull the regression line toward it. Because of this, it's a good idea to fit the data *without* the Palm Beach County observation and see how well that line predicts results for Palm Beach County. We'd then get the regression equation $\hat{y} = 45.7 + 0.02414x$. Since the Perot vote in Palm Beach County was 30,739, this line would predict a Buchanan vote there of $\hat{y} = 45.7 + 0.02414(30.739) = 788$. This compares to the actual Buchanan vote of 3407 in Palm Beach County, for a residual of $3407 - 788 = 2619$. Again, the actual Buchanan vote seems surprisingly high.

Try Exercise 3.33

The Method of Least Squares Yields the Regression Line

We've seen that software finds the regression line. It chooses the optimal line to fit through the data points by making the residuals as small as possible. This process involves compromise because a line can perfectly predict one point (resulting in a residual of 0) but poorly predict many other points (resulting in larger residuals). The actual summary measure used to evaluate regression lines is

$$\text{Residual sum of squares} = \Sigma(\text{residual})^2 = \Sigma(y - \hat{y})^2.$$

This formula squares each vertical distance between a point and the line and then adds them up. The better the line, the smaller the residuals tend to be, and

[6]A more complex analysis accounts for larger counts tending to vary more. However, our analysis is adequate for highlighting data points that fall far from the linear trend.

the smaller the residual sum of squares tends to be. Each potential line has a set of predicted values, a set of residuals, and a residual sum of squares. The line that software reports is the one having the *minimum* residual sum of squares. This way of selecting a line is called the **least squares method**.

Least Squares Method

Among the many possible lines that could be drawn through data points in a scatterplot, the **least squares method** gives what we call the regression line. This method produces the line that has the smallest value for the residual sum of squares using $\hat{y} = a + bx$ to predict y.

In Practice

It's simple for software to use the least squares method to find the regression line for us.

Besides making the errors as small as possible, this regression line

- Has some positive residuals and some negative residuals, and the sum (and mean) of the residuals equals 0.
- Passes through the point $(\bar{x}, \bar{y})$.

The first property tells us that the too-low predictions are balanced by the too-high predictions. The second property tells us that the line passes through the center of the data.

Even though we usually rely on technology to compute the regression line, the method of least squares does provide formulas for the y-intercept and slope, based on summary statistics for the sample data. Let $\bar{x}$ denote the mean of x, $\bar{y}$ the mean of y, s_x the standard deviation of the x values and s_y the standard deviation of the y values.

Regression Formulas for y-Intercept and Slope

The **slope** equals $b = r\left(\frac{s_y}{s_x}\right)$.

The **y-intercept** equals $a = \bar{y} - b(\bar{x})$.

Notice that the slope b is directly related to the correlation r, and the y-intercept depends on the slope. Let's return to the baseball data in Example 9 to illustrate the calculations. For that data set we have $\bar{x} = 0.2597$ for batting average, $\bar{y} = 4.45$ for team scoring, $s_x = 0.01257$, and $s_y = 0.577$. The correlation is $r = 0.568$, so

$$b = r\left(\frac{s_y}{s_x}\right) = 0.568(0.577/0.01257) = 26.07$$

$$\text{and } a = \bar{y} - b(\bar{x}) = 4.45 - 26.07(0.2597) = -2.32$$

The regression line to predict team scoring from batting average is $\hat{y} = -2.32 + 26.1x$.

In Practice

The formulas for a and b help us to interpret the regression line $\hat{y} = a + bx$ (for example, see Exercises 3.110 and 3.111), but you should rely on software[7] for calculations. Software doesn't round during the different steps of the calculation, so it will give more accurate results.

---

[7]In fact, software finds the slope without first finding the correlation, using the formula

$$b = \frac{1}{(n-1)s_x^2}\Sigma(x - \bar{x})(y - \bar{y}).$$

The Slope, the Correlation, and the Units of the Variables

We've used the correlation to describe the strength of the association. Why can't we use the *slope* to do this, with bigger slopes representing stronger associations? The reason is that the numerical value of the slope depends on the units for the variables.

For example, considering the 33 countries described in Table 3.4 and data for these countries found in the Internet Use file, the regression line between y = Internet penetration and x = GDP is $\hat{y} = 12.4 + 1.6x$. GDP was measured in *thousands* of U.S. dollars (per capita). Suppose we instead measure GDP in dollars, such as $x = 38{,}100$ for Australia instead of $x = 38.1$. A one-unit increase in GDP then refers to a single dollar (per capita). This is only 1/1000 as much as one thousand-dollar increase, so the change in the predicted value of y would be 1/1000 as much, or $(1/1000)1.6 = 0.0016$. Thus, if x = GDP in dollars, the slope of the regression equation is 0.0016 instead of 1.6. The strength of the association is the same in each case, since the variables and data base are identical. Only the units of measurement for one variable changed.

So the slope b doesn't tell us whether the association is strong or weak since we can make b as large or as small as we want by changing the units. By contrast, *the correlation does not change when the units of measurement change.* It is 0.903 between Internet penetration and GDP, whether we measure GDP in dollars or in thousands of dollars.

In summary, we've learned that the correlation describes the strength of the linear association. We've also seen how the regression line predicts the response variable y using the explanatory variable x. Although correlation and regression methods serve different purposes, there are strong connections between them:

- They are both appropriate when the relationship between two quantitative variables can be approximated by a straight line.
- The correlation and the slope of the regression line have the same sign. If one is positive, so is the other one. If one is negative, so is the other one. If one is zero, the other is also zero.

However, there are some differences between correlation and regression methods. With regression, we must identify response and explanatory variables. We get a different line if we use x to predict y than if we use y to predict x. By contrast, correlation does not make this distinction. We get the same correlation either way. Also, the values for the y-intercept and slope of the regression line depend on the units, whereas the correlation does not. Finally, the correlation falls between -1 and $+1$, whereas the regression slope can equal any real number.

r-Squared (r^2)

Recall

Using the baseball data, following Example 9 we noted that the prediction error using the regression equation is $y - \hat{y}$, called the residual. ◀

When we predict a value of y, why should we use the regression line? We could instead predict y using the center of its distribution, such as the sample mean, $\bar{y}$. The reason for using the regression line is that if x and y have an association, then we can predict most y values more accurately by substituting x values into the regression equation, $\hat{y} = a + bx$, than by using the sample mean $\bar{y}$ for prediction.

Another way to describe the strength of association refers to how much more accurately you can predict values of y using the regression equation instead of the sample mean $\bar{y}$. For a particular observation, the prediction error from using the regression equation is $y - \hat{y}$, the residual. The prediction error from merely using the sample mean as the prediction is $y - \bar{y}$. The square of the correlation summarizes how much less prediction error there is when you use the regression line to predict y, compared to using $\bar{y}$ to predict y.

For the quantitative variables Internet use and Facebook use in Example 5, $r = 0.682$, so $r^2 = (0.682)^2 = 0.465$. This means that in an average sense for all the observations, the prediction error using the regression line to predict y is

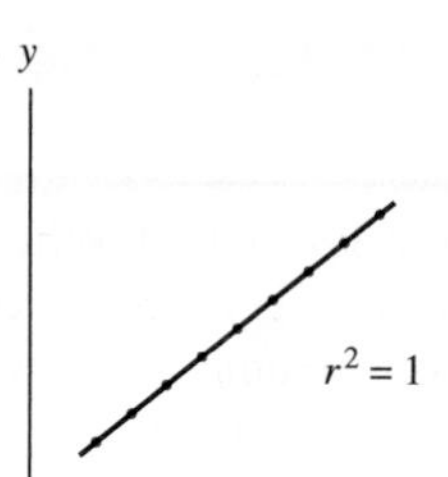

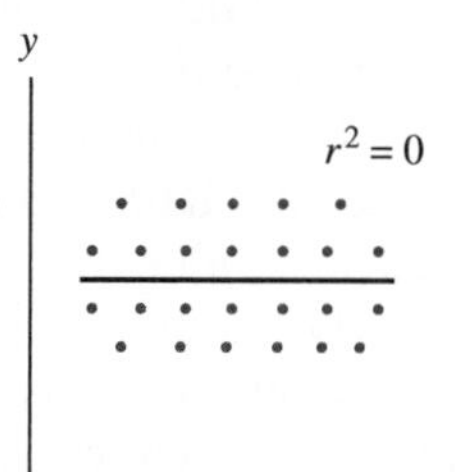

46.5% smaller than the prediction error using $\bar{y}$ to predict y. Chapter 12 will show the details about how to summarize the prediction error for all the observations using a variance type measure for each of the two predictors.

The typical way to interpret r^2 is as *the proportion of the variation in the y-values that is accounted for by the linear relationship of y with x.* Specifically, $r^2 = 0.465$ means that the variation of the $\hat{y}$-values from the linear prediction equation is 46.5% of the variation of the observed y-values for the sample. When the data points fall exactly in a straight-line pattern, as in the scatterplot in the margin, y and $\hat{y}$ are identical for all points. Then $r^2 = 1$ and the variation of the $\hat{y}$-values is 100% of the variation of the observed y-values for the sample. When $r = 0$, then the slope b of the regression equation is 0. So, all $\hat{y}$-values are identical, and $r^2 = 0.0$ means that the variation of the $\hat{y}$-values is 0% of the variation of the observed y-values. The stronger the association, the closer the variation of the $\hat{y}$-values is to the variation of the y-values.

Again, we'll give a more detailed explanation of r^2 later in the text. We mention r^2 here because you'll see it listed on regression software output, and we want you to have a rough idea of what it represents.

Associations with Quantitative and Categorical Variables

In this chapter, we've learned how to explore an association between categorical variables and between quantitative variables. It's also possible to mix the variable types or add other variables. For example, with two quantitative variables, we can identify points in the scatterplot according to their values on a relevant categorical variable. This is done by using different symbols or colors on the scatterplot to portray the different categories.

Example 11

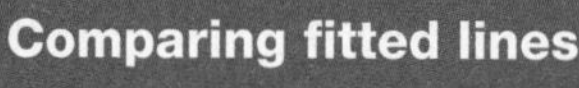

The Gender Difference in Winning Olympic High Jumps

Picture the Scenario

The summer Olympic Games occur every four years, and one of the track and field events is the high jump. Men have competed in the high jump since 1896 and women since 1928. The High Jump data file on the text CD contains the winning heights (in meters) for each year.[8]

Questions to Explore

a. How can we display the data on these two quantitative variables (winning height, year) and the categorical variable (gender) graphically?

b. How have the winning heights changed over time? How different are the winning heights for men and women in a typical year?

Think It Through

a. Figure 3.16 shows a scatterplot with x = year and y = winning height. The data points are displayed with a black circle for men and a red square for women. There were no Olympic Games during World War II, so no observations appear for 1940 or 1944.

[8]From www.olympic.org/medallists-results.

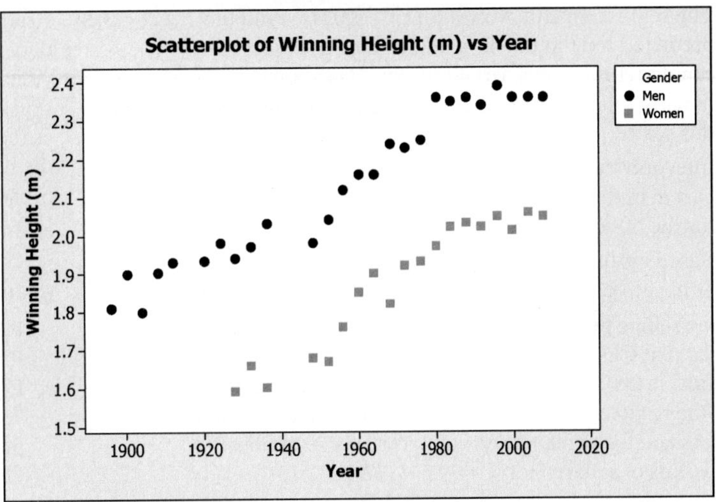

▲ **Figure 3.16** MINITAB Scatterplot for the Winning High Jumps (in Meters) in the Olympics. The black dots represent men and the red squares represent women. **Question** In a typical year, what is the approximate difference between the winning heights for men and for women?

b. The scatterplot shows that for each gender the winning heights have an increasing trend over time. Men have consistently jumped higher than women, between about 0.3 and 0.4 meters in a given year. The women's winning heights are similar to those for the men about 60 years earlier—for instance, about 2.0 meters in 1990–2008 for women and in 1930–1940 for men.

Insight

We could describe these trends by fitting a regression line to the points for men and a separate regression line to the points for women. However, note that in recent Olympics the winning distances have leveled off somewhat. We should be cautious in using regression lines to predict future winning heights.

Try Exercise 3.42

3.3 Practicing the Basics

3.24 Sketch plots of lines Identify the values of the y-intercept a and the slope b, and sketch the following regression lines, for values of x between 0 and 10.

a. $\hat{y} = 7 + 0.5x$

b. $\hat{y} = 7 + x$

c. $\hat{y} = 7 - x$

d. $\hat{y} = 7$

3.25 Sit-ups and the 40-yard dash Is there a relationship between how many sit-ups you can do and how fast you can run 40 yards? The EXCEL output shows the relationship between these variables for a study of female athletes to be discussed in Chapter 12.

TRY

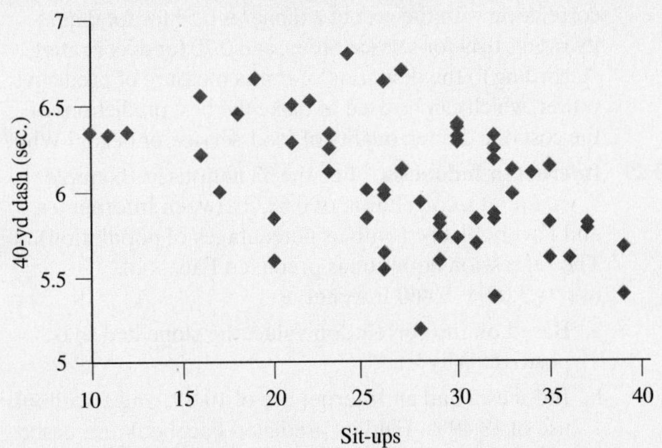

Excel scatterplot of time to run 40-yard dash by number of sit-ups.

a. The regression equation is $6.71\hat{y} - 0.024x$. Find the predicted time in the 40-yard dash for a subject who can do (i) 10 sit-ups, (ii) 40 sit-ups. Based on these times, explain how to sketch the regression line over this scatterplot.

b. Interpret the y-intercept and slope of the equation in part a, in the context of the number of sit-ups and time for the 40-yard dash.

c. Based on the slope in part a, is the correlation positive, or negative? Explain.

3.26 Home selling prices The House Selling Prices FL data file on the text CD lists selling prices of homes in Gainesville, Florida, in 2003 and some predictors for the selling price. For the response variable y = selling price in thousands of dollars and the explanatory variable x = size of house in thousands of square feet, $\hat{y} = 9.2 + 77.0x$.

TRY

a. How much do you predict a house would sell for if it has (i) 2000 square feet, (ii) 3000 square feet?

b. Using results in part a, explain how to interpret the slope.

c. Is the correlation between these variables positive or negative? Why?

d. One home that is 3000 square feet sold for $300,000. Find the residual, and interpret.

3.27 Rating restaurants Zagat restaurant guides publish ratings of restaurants for many large cities around the world (see www.zagat.com). The review for each restaurant gives a verbal summary as well as a 0- to 30-point rating of the quality of food, décor, service, and the cost of a dinner with one drink and tip. For Italian restaurants in London in 2007, the food ratings had a mean of 20.46 and standard deviation of 2.70. The cost of a dinner (in U.S. dollars) had a mean of $79.76 and standard deviation of $20.54. The equation that predicts the cost of a dinner using the rating of the quality of food is $\hat{y} = 2.5 + 4.0x$. The correlation between these two variables is 0.53.

a. Predict the cost of a dinner in a restaurant that gets the (i) lowest possible food quality rating of 0, (ii) highest possible food quality rating of 30.

b. Interpret the slope.

c. Interpret the correlation.

d. Show how the slope can be obtained from the correlation and other information given.

3.28 Predicting cost of meal Refer to the previous exercise. The correlation with the cost of a dinner is 0.53 for food quality rating, 0.49 for service rating, and 0.70 for décor rating. According to the definition of r^2 as a measure of predictive power, which can be used to make the best predictions of the cost of a dinner: quality of food, service, or décor? Why?

3.29 Internet in Indonesia For the 33 nations in Example 7, we found a correlation of 0.682 between Internet use and Facebook use (both as percentages of population). The regression equation is predicted Facebook use = 3.09 + 0.460 Internet use

a. Based on the correlation value, the slope had to be positive. Why?

b. Indonesia had an Internet use of 10.5% and Facebook use of 13.49%. Find its predicted Facebook use based on the regression equation.

c. Find the residual for Indonesia. Interpret.

3.30 Broadband subscribers and population The Internet Use data file on the text CD contains data on the number of individuals in a country with broadband access and the population size for each of 33 nations. When using population size as the explanatory variable, x, and broadband subscribers as the response variable, y, the regression equations is predicted broadband subscribers = 4,981,673 + 0.0308 population

a. Interpret the slope of the regression equation. Is the association positive or negative? Explain what this means.

b. Predict broadband subscribers at the (i) minimum population size x value of 7,019,731, (ii) at the maximum population size x value of 1,330,257,143.

c. For the United States, broadband subscribers = 73,206,000, and population = 310,232,863. Find the predicted broadband use and the residual for the United States. Interpret the value of this residual.

3.31 SAT reading and math scores The SAT2010 data file on the text CD contains average reading and math SAT scores for each of the 50 states and Washington D.C. Let the explanatory variable x = reading and the response variable y = math. The regression equation is $\hat{y} = 18.1 + 0.975x$.

a. California had an average reading score of 501. Use the regression equation to predict the average math score for California.

b. Calculate the residual associated with California and comment on its value in the context of the problem.

c. Does it appear that a state's average reading score is a reliable predictor of its average math score?

3.32 How much do seat belts help? A study in 2000 by the National Highway Traffic Safety Administration estimated that failure to wear seat belts led to 9200 deaths in the previous year, and that the number of deaths would decrease by 270 for every 1 percentage point gain in seat belt usage. Let $\hat{y}$ = predicted number of deaths in a year and x = percentage of people who wear seat belts.

TRY

a. Report the slope b for the equation $\hat{y} = a + bx$.

b. If the y-intercept equals 28,910, then predict the number of deaths in a year if (i) no one wears seat belts, (ii) 73% of people wear seat belts (the value in 2000), (iii) 100% of people wear seat belts.

3.33 Regression between cereal sodium and sugar The figure shows the result of a MINITAB regression analysis of the explanatory variable x = sugar and the response variable y = sodium for the breakfast cereal data set discussed in Chapter 2 (the Cereal data file on the text CD).

TRY

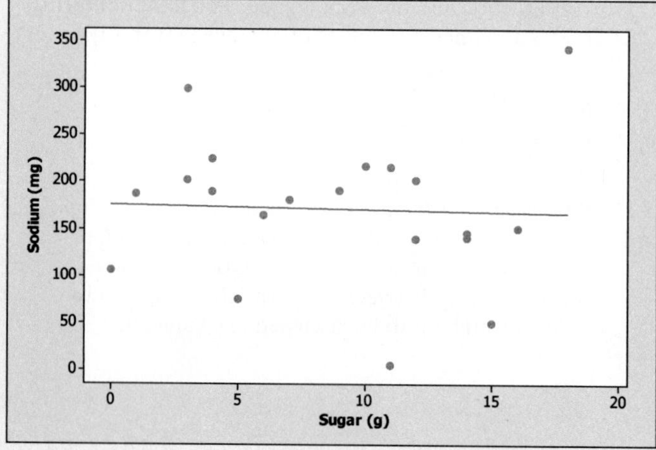

a. Suppose you had fit a line to the scatterplot by eye-balling. In what sense would the line calculated by MINITAB be better than your line?

b. Now let's look at a histogram of the residuals. Explain what the two short bars on the far right of the histogram mean in the context of the problem. Which two brands of cereal do they represent?

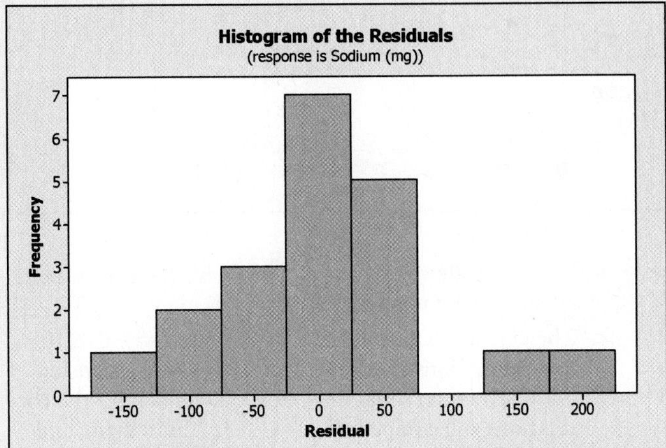

c. In general, how reliable would you say amount of sugar is as a predictor of the amount of sodium?

3.34 Regression and correlation between cereal sodium and sugar Refer to the previous exercise. Show the algebraic relationship between the correlation of −0.017 and the slope of the regression equation $b = -0.25$, using the fact that the standard deviations are 5.32 g for sugar and 77.3 mg for sodium. (Hint: Recall that $b = r\left(\frac{s_y}{s_x}\right)$.)

3.35 Advertising and sales Each month, the owner of Fay's Tanning Salon records in a data file y = monthly total sales receipts and x = amount spent that month on advertising, both in thousands of dollars. For the first three months of operation, the observations are as shown in the table.

Advertising	Sales
0	4
1	6
2	8

a. Sketch a scatterplot.

b. From inspection of the scatterplot, state the correlation and the regression line. (*Note*: You should be able to figure them out without using software or formulas.)

c. Find the mean and standard deviation for each variable.

d. Using part c, find the regression line, using the formulas for the slope and the y-intercept. Interpret the y-intercept and the slope.

3.36 Midterm–final correlation For students who take Statistics 101 at Lake Wobegon College in Minnesota, both the midterm and final exams have mean = 75 and standard deviation = 10. The professor explores using the midterm exam score to predict the final exam score. The regression equation relating y = final exam score to x = midterm exam score is $\hat{y} = 30 + 0.60x$.

a. Find the predicted final exam score for a student who has (i) midterm score = 100, (ii) midterm score = 50. Note that in each case the predicted final exam score *regresses toward the mean* of 75. (This is a property of the regression equation that is the origin of its name, as Chapter 12 will explain.)

b. Show that the correlation equals 0.60, and interpret it. (*Hint:* Use the relation between the slope and correlation.)

3.37 Predict final exam from midterm In an introductory statistics course, x = midterm exam score and y = final exam score. Both have mean = 80 and standard deviation = 10. The correlation between the exam scores is 0.70.

a. Find the regression equation.

b. Find the predicted final exam score for a student with midterm exam score = 80.

3.38 NL baseball Example 9 related y = team scoring (per game) and x = team batting average for American League teams. For National League teams in 2010, $\hat{y} = -6.25 + 41.5x$.

a. The team batting averages fell between 0.242 and 0.272. Explain how to interpret the slope in context.

b. The standard deviations were 0.00782 for team batting average and 0.3604 for team scoring. The correlation between these variables was 0.900. Show how the correlation and slope of 41.5 relate, in terms of these standard deviations.

c. Software reports $r^2 = 0.81$. Explain how to interpret this measure.

3.39 Study time and college GPA A graduate teaching assistant (Euijung Ryu) for Introduction to Statistics (STA 2023) at the University of Florida collected data from one of her classes in spring 2007 to investigate the relationship between using the explanatory variable x = study time per week (average number of hours) to predict the response variable y = college GPA. For the 21 females in her class, the correlation was 0.42. For the eight males in her class, the data were as shown in the table.

a. Create a data file and use it to construct a scatterplot. Interpret.

b. Find and interpret the correlation.

c. Find and interpret the prediction equation by reporting the predicted GPA for a student who studies (i) 5 hours per week, (ii) 25 hours per week.

Student	Study Time	GPA
1	14	2.8
2	25	3.6
3	15	3.4
4	5	3.0
5	10	3.1
6	12	3.3
7	5	2.7
8	21	3.8

3.40 Oil and GDP An article in the September 16, 2006, issue of *The Economist* showed a scatterplot for many nations relating the response variable y = annual oil consumption per person (in barrels) and the explanatory variable x = gross domestic product (GDP, per person, in thousands of dollars). The values shown on the plot were approximately as shown in the table.

a. Create a data file and use it to construct a scatterplot. Interpret.

b. Find and interpret the prediction equation.

c. Find and interpret the correlation.

d. Find and interpret the residual for Canada.

Nation	GDP	Oil Consumption
India	3	1
China	8	2
Brazil	9	4
Mexico	10	7
Russia	11	8
S. Korea	20	18
Italy	29	12
France	30	13
Britain	31	11
Germany	31	12
Japan	31	16
Canada	34	26
U.S.	41	26

3.41 Mountain bikes revisited Is there a relationship between the weight and price of a mountain bike? This question was considered in Exercise 3.21. We will analyze the Mountain Bike data file on the text CD. (The data also were shown in Exercise 3.21.)

a. Construct a scatterplot. Interpret.

b. Find the regression equation. Interpret the slope in context. Does the y-intercept have contextual meaning?

c. You decide to purchase a mountain bike that weighs 30 pounds. What is the predicted price for the bike?

3.42 Mountain bike and suspension type Refer to the previous exercise. The data file contains price, weight, and type of suspension system (FU = full, FE = front-end in the scatterplot shown).

a. Do you observe a linear relationship? Is the single regression line, which is $\hat{y} = 1896 - 40.45x$, the best way to fit the data? How would you suggest fitting the data?

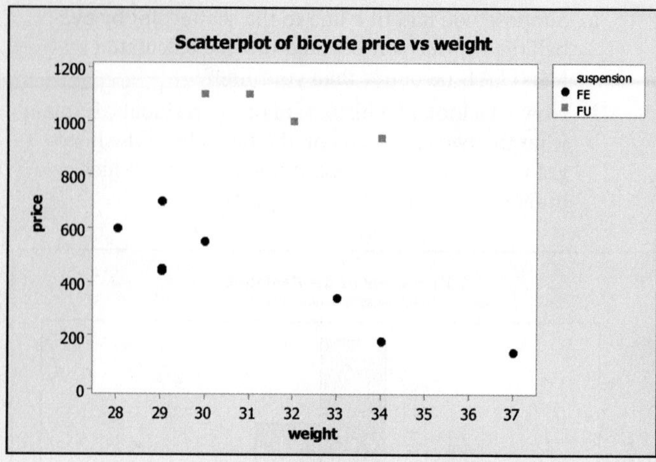

b. Find separate regression equations for the two suspension types. Summarize your findings.

c. The correlation for all 12 data points is $r = -0.32$. If the correlations for the full and front-end suspension bikes are found separately, how do you believe the correlations will compare to $r = -0.32$? Find them, and interpret.

d. You see a mountain bike advertised for $700 that weighs 28.5 lb. The type of suspension is not given. Would you predict that this bike has a full or a front-end suspension? Statistically justify your answer.

3.43 SAT participation The SAT2010 data file on the text CD contains combined average SAT scores for each of the 50 states and Washington D.C., and also the corresponding participation rate of each state. Let's consider using the explanatory variable x = participation rate (in %) to predict the response variable y = combined average score. The regression equation is $\hat{y} = 1718 - 3.36x$.

a. Use the regression equation to predict the combined average score for the states with the smallest and largest participation rates.

b. West Virginia has a participation rate of 16% and a combined average score of 1522. Calculate the residual associated with West Virginia and comment on its value in the context of the problem.

c. The correlation between combined average SAT scores and participation rate is −0.877. What factor(s) or characteristic(s) of the students taking the SAT might contribute to this very strong negative relationship?

3.4 Cautions in Analyzing Associations

This chapter has introduced ways to explore **associations** between variables. When using these methods, you need to be cautious about certain potential pitfalls.

Extrapolation Is Dangerous

Extrapolation refers to using a regression line to predict y values for x values outside the observed range of data. This is riskier as we move farther from that range. If the trend changes, extrapolation gives poor predictions. For example, regression analysis is often applied to observations of a quantitative response variable over time. The regression line describes the time trend. But it is dangerous to use the line to make predictions far into the future.

Example 12

Extrapolation

Forecasting Future Global Warming

Picture the Scenario

In Chapter 2, we explored trends in temperatures over time using time plots. Let's use regression with the Central Park Yearly Temps data file on the text CD to describe the trend over time of the annual mean temperatures from 1869–2010 for Central Park, New York City.[9]

Questions to Explore

a. What does a regression line tell us about the trend over the 20th century?

b. What does it predict about the annual mean temperature in the year (i) 2015, (ii) 3000? Are these extrapolations sensible?

Think It Through

a. Figure 3.17 shows a time plot of the Central Park annual mean temperatures. For a regression analysis, the mean annual temperature is the response variable. Time (the year) is the explanatory variable. Software tells us that the regression line is $\hat{y} = -1.636 + 0.029x$. Figure 3.17 superimposes the regression trend line over the time plot.

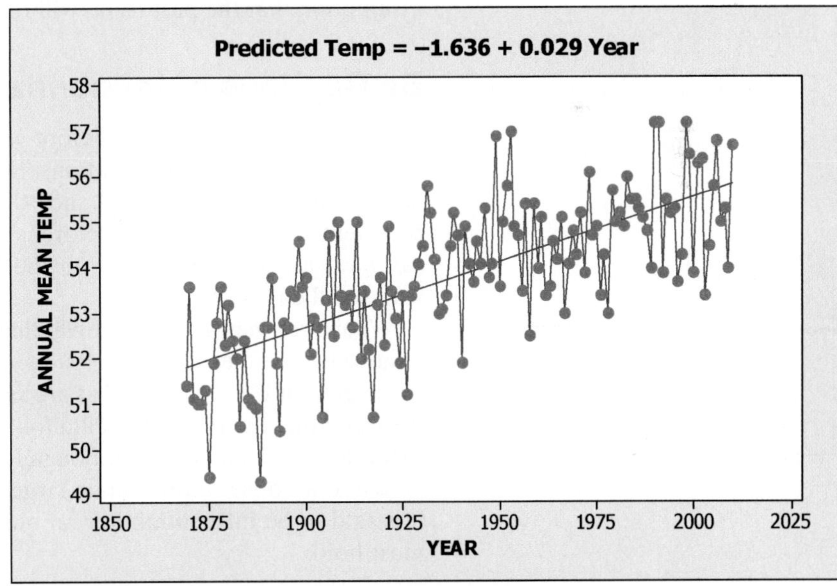

▲ **Figure 3.17 MINITAB Time Plot of Central Park Annual Mean Temperature Versus Time, Showing Fitted Regression Line. Question** If the present trend continues, what would you predict for the annual mean temperature for 2015?

The positive slope reflects an increasing trend: The annual mean temperature tended upward over the century. *The slope $b = 0.029$*

[9]*Source:* www.erh.noaa.gov/okx/climate/records/monthannualtemp.html.

indicates that for each one-year increase, the predicted annual mean temperature increases by 0.029 degrees Fahrenheit. The slope value of 0.029 seems close to 0, indicating little warming. However, 0.029 is the predicted change *per year.* Over a century, the predicted change is $100(0.029) = 2.9$ degrees, which is quite significant.

b. Using the trend line, the predicted annual mean temperature for the year 2015 is

$$\hat{y} = -1.636 + 0.029(2015) = 56.8 \text{ degrees Fahrenheit.}$$

Farther into the future we see more dramatic increases. At the next millennium for the year 3000, the forecast is $\hat{y} = -1.636 + 0.029(3000) = 85.4$. If this is accurate, it would be exceedingly uncomfortable to live in New York!

Insight

It is dangerous to extrapolate far outside the range of observed x values. There's no guarantee that the relationship will have the same trend outside that range. It seems reasonable to predict for 2015. That's not looking too far into the future from the last observation in 2010. However, it is foolhardy to predict for 3000. It's not sensible to assume that the same straight-line trend will continue for the next 990 years. As time moves forward, the annual mean temperatures may increase even faster or level off or even decrease.

Try Exercise 3.45

Predictions about the future using time series data are called **forecasts**. When we use a regression line to forecast a trend for future years, we must make the assumption that the past trend will remain the same in the future. This is risky.

Be Cautious of Influential Outliers

One reason to plot the data before you do a correlation or regression analysis is to check for unusual observations. Such an observation can tell you something interesting, as in Examples 6 and 10 about the Buchanan vote in the 2000 U.S. presidential election. Furthermore, a data point that is an outlier on a scatterplot can have a substantial effect on the regression line and correlation, especially with small data sets.

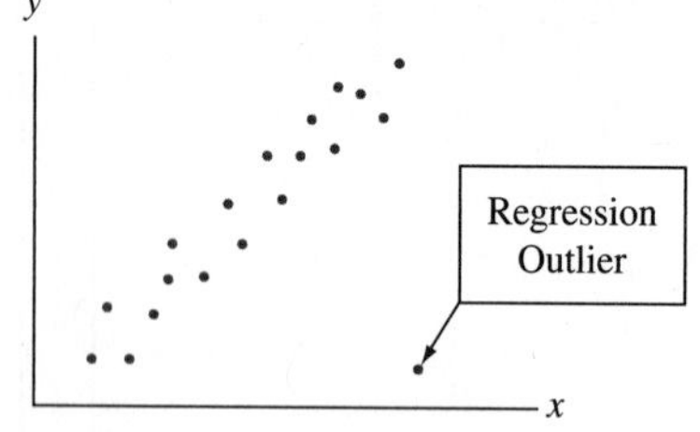

What's relevant here is not whether an observation is an outlier in its x value, relative to the other x values, or in its y value, relative to the other y values. Instead, we search for observations that are **regression outliers**, being well removed from the trend that the rest of the data follow. The margin figure shows an observation that is a regression outlier, although it is not an outlier on x alone or on y alone.

When an observation has a large effect on results of a regression analysis, it is said to be **influential**. For an observation to be influential, two conditions must hold:

- Its x value is relatively low or high compared to the rest of the data.
- The observation is a regression outlier, falling quite far from the trend that the rest of the data follow.

When both of these happen, the line tends to be pulled toward that data point and away from the trend of the rest of the points.

Figure 3.18 shows two regression outliers. The correlation without these two points equals 0.00. The first regression outlier is near the middle of the range of x. It does not have much potential for tilting the line up or down. It has little influence on the slope or the correlation. The correlation changes only to 0.03 when we add it to the data set. The second regression outlier is at the high end of the range of x-values. It is influential. The correlation changes to 0.47 when we add it to the data set.

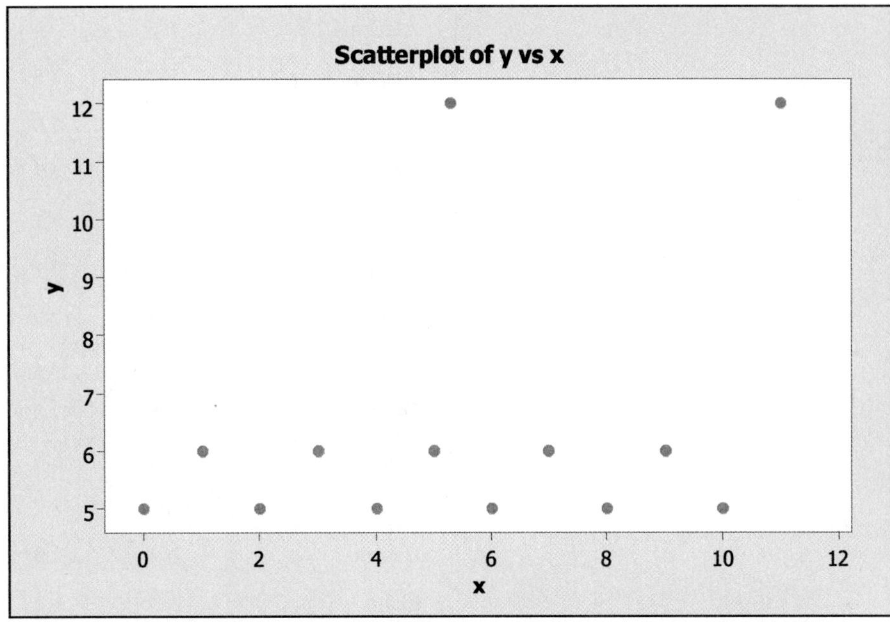

▲ **Figure 3.18 An Observation Is a Regression Outlier if it is Far Removed from the Trend that the Rest of the Data Follow.** The top two points are regression outliers. Not all regression outliers are influential in affecting the correlation or slope. **Question** Which regression outlier in this figure is influential?

Influential outliers

Example 13

Higher Education and Higher Murder Rates

Picture the Scenario

Table 3.6 shows data[10] for the 50 states and the District of Columbia on

Violent crime rate: The annual number of murders, forcible rapes, robberies, and aggravated assaults per 100,000 people in the population.

Murder rate: The annual number of murders per 100,000 people in the population.

Poverty: Percentage of the residents with income below the poverty level.

High school: Percentage of the adult residents who have at least a high school education.

College: Percentage of the adult residents who have a college education.

Single parent: Percentage of families headed by a single parent.

The data are in the U.S. Statewide Crime data file on the text CD. Let's look at the relationship between y = murder rate and x = college. We'll look at other variables in the exercises.

[10]From *Statistical Abstract of the United States*, 2003.

Table 3.6 Statewide Data on Several Variables

State	Violent Crime	Murder Rate	Poverty	High School	College	Single Parent
Alabama	486	7.4	14.7	77.5	20.4	26.0
Alaska	567	4.3	8.4	90.4	28.1	23.2
Arizona	532	7.0	13.5	85.1	24.6	23.5
Arkansas	445	6.3	15.8	81.7	18.4	24.7
California	622	6.1	14.0	81.2	27.5	21.8
Colorado	334	3.1	8.5	89.7	34.6	20.8
Connecticut	325	2.9	7.7	88.2	31.6	22.9
Delaware	684	3.2	9.9	86.1	24.0	25.6
District of Columbia	1508	41.8	17.4	83.2	38.3	44.7
Florida	812	5.6	12.0	84.0	22.8	26.5
Georgia	505	8.0	12.5	82.6	23.1	25.5
Hawaii	244	2.9	10.6	87.4	26.3	19.1
Idaho	253	1.2	13.3	86.2	20.0	17.7
Illinois	657	7.2	10.5	85.5	27.1	21.9
Indiana	349	5.8	8.3	84.6	17.1	22.8
Iowa	266	1.6	7.9	89.7	25.5	19.8
Kansas	389	6.3	10.5	88.1	27.3	20.2
Kentucky	295	4.8	12.5	78.7	20.5	23.2
Louisiana	681	12.5	18.5	80.8	22.5	29.3
Maine	110	1.2	9.8	89.3	24.1	23.7
Maryland	787	8.1	7.3	85.7	32.3	24.5
Massachusetts	476	2.0	10.2	85.1	32.7	22.8
Michigan	555	6.7	10.2	86.2	23.0	24.5
Minnesota	281	3.1	7.9	90.8	31.2	19.6
Mississippi	361	9.0	15.5	80.3	18.7	30.0
Missouri	490	6.2	9.8	86.6	26.2	24.3
Montana	241	1.8	16.0	89.6	23.8	21.4
Nebraska	328	3.7	10.7	90.4	24.6	19.6
Nevada	524	6.5	10.1	82.8	19.3	24.2
New Hampshire	175	1.8	7.6	88.1	30.1	20.0
New Jersey	384	3.4	8.1	87.3	30.1	20.2
New Mexico	758	7.4	19.3	82.2	23.6	26.6
New York	554	5.0	14.7	82.5	28.7	26.0
North Carolina	498	7.0	13.2	79.2	23.2	24.3
North Dakota	81	0.6	12.8	85.5	22.6	19.1

State	Violent Crime	Murder Rate	Poverty	High School	College	Single Parent
Ohio	334	3.7	11.1	87.0	24.6	24.6
Oklahoma	496	5.3	14.1	86.1	22.5	23.5
Oregon	351	2.0	12.9	88.1	27.2	22.5
Pennsylvania	420	4.9	9.8	85.7	24.3	22.8
Rhode Island	298	4.3	10.2	81.3	26.4	27.4
South Carolina	805	5.8	12.0	83.0	19.0	27.1
South Dakota	167	0.9	9.4	91.8	25.7	20.7
Tennessee	707	7.2	13.4	79.9	22.0	27.9
Texas	545	5.9	14.9	79.2	23.9	21.5
Utah	256	1.9	8.1	90.7	26.4	13.6
Vermont	114	1.5	10.3	90.0	28.8	22.5
Virginia	282	5.7	8.1	86.6	31.9	22.2
Washington	370	3.3	9.5	91.8	28.6	22.1
West Virginia	317	2.5	15.8	77.1	15.3	22.3
Wisconsin	237	3.2	9.0	86.7	23.8	21.7
Wyoming	267	2.4	11.1	90.0	20.6	20.8

Questions to Explore

a. Construct the scatterplot between y = murder rate and x = college. Does any observation look like it could be influential in its effect on the regression line?

b. Use software to find the regression line. Check whether the observation identified in part a actually is influential by finding the line again without that observation.

Think It Through

a. Figure 3.19 shows the scatterplot. The observation out by itself is D.C. with $x = 38.3$ and $y = 41.8$, which is the largest observation on both these variables. It satisfies both conditions for an observation to be influential: It has a relatively extreme value on the explanatory variable (college), and it is a regression outlier, falling well away from the linear trend of the other points.

b. Using software, the regression line fitted to all 51 observations, including D.C., equals $\hat{y} = -3.1 + 0.33x$. The slope is *positive*, as shown in the first plot in Figure 3.20. You can check that the predicted murder rates *increase* from 1.9 to 10.1 as the percentage with a college education increases from $x = 15\%$ to $x = 40\%$, roughly the range of observed x values. By contrast, when we fit the regression line only to the 50 states, excluding the observation for D.C., $\hat{y} = 8.0 - 0.14x$. The slope of -0.14 reflects a *negative* trend, as shown in the second plot in Figure 3.20. Now, the predicted murder rate *decreases* from 5.9 to 2.4 as the percentage with a college education increases from 15% to 40%.

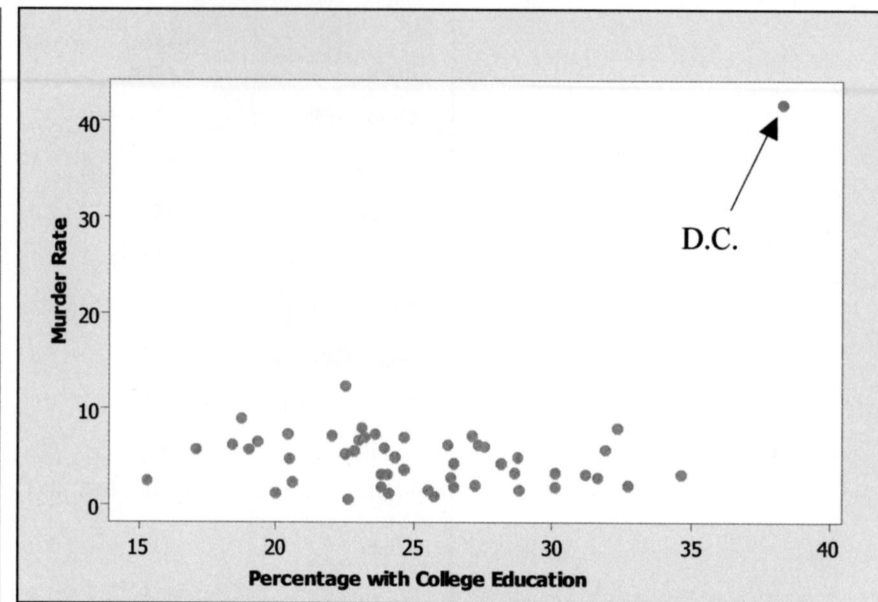

▲ **Figure 3.19** MINITAB Scatterplot Relating Murder Rate to Percentage with College Education. Question How would you expect the slope to change if D.C. is excluded from the regression analysis?

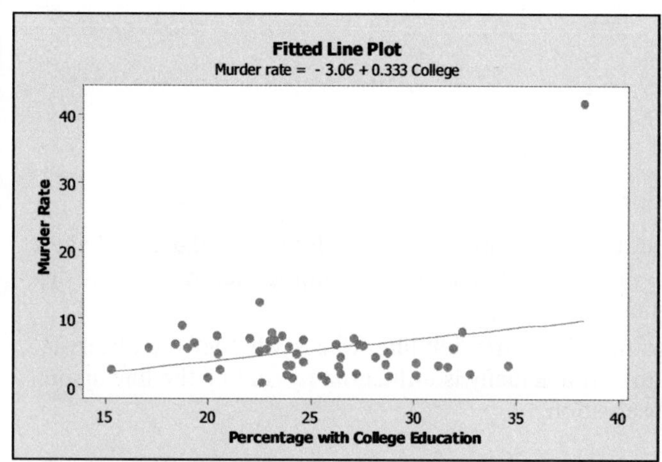

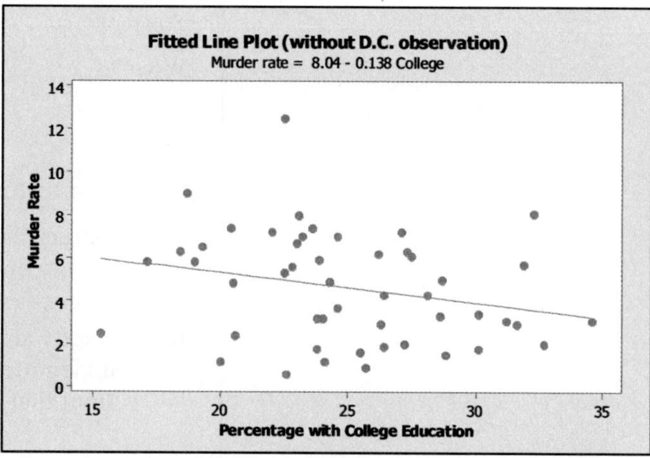

▲ **Figure 3.20** MINITAB Scatterplots Relating Murder Rate to Percentage with College Education, With and Without Observation for D.C. Question Which line better describes the trend for the 50 states?

Insight

Including D.C. in the regression analysis has the effect of pulling the slope of the regression line upward. The regression line then makes it seem, misleadingly, as if the murder rate *increases* when the percentage with a college education increases. In fact, for the rest of the data, the predicted murder rate *decreases.* The regression line including D.C. distorts the relationship for the other 50 states. In summary, the D.C. data point is highly influential. The regression line for the 50 states alone better represents the overall negative trend. In reporting these results, we should show this line and then note that D.C. is a regression outlier that falls well outside this trend.

Try Exercise 3.47

Caution

Always construct a scatterplot before finding a correlation coefficient or fitting a regression line. ◀

This example shows the correlation and the regression line are **nonresistant**: They are prone to distortion by outliers. *Investigate any regression outlier.* Was the observation recorded incorrectly, or is it merely different from the rest of the data in some way? It is often a good idea to refit the regression line without it to see if it has a large effect, as we did in this last example.

Correlation Does Not Imply Causation

In a regression analysis, suppose that as x goes up, y also tends to go up (or go down). Can we conclude that there's a *causal* connection, with changes in x causing changes in y?

The concept of causality is central to science. We are all familiar with it, at least in an informal sense. We know, for example, that exposure to a virus can cause the flu. But just observing an association between two variables is not enough to imply a causal connection. There may be some alternative explanation for the association.

Example 14

Correlation and causation

Education and Crime

Picture the Scenario

Figure 3.21 shows recent data on y = crime rate and x = education for Florida counties, from the FL Crime data file on the text CD. Education was measured as the percentage of residents aged at least 25 in the county who had at least a high school degree. Crime rate was measured as the number of crimes in that county in the past year per 1000 residents.

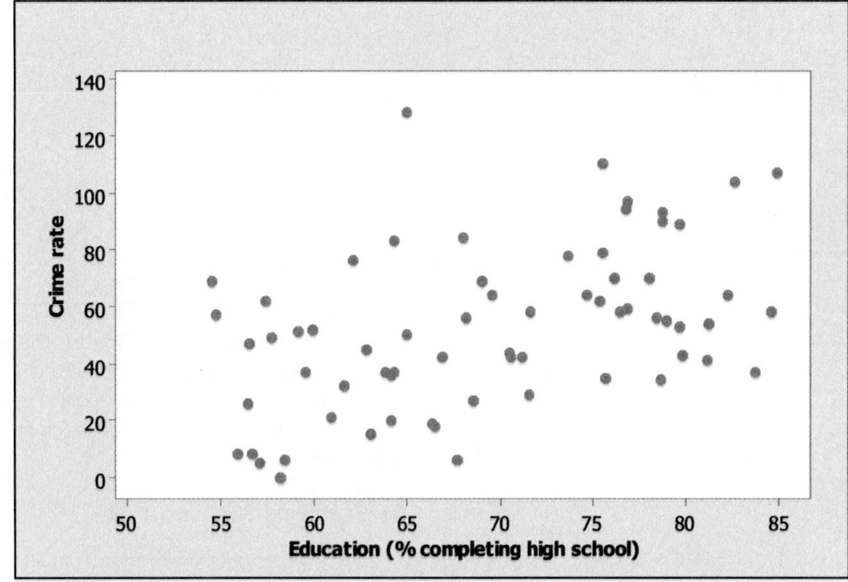

▲ **Figure 3.21 MINITAB Scatterplot of Crime Rate and Percentage With at Least a High School Education.** There is a moderate positive association ($r = 0.47$). **Question** Does more education cause more crime, or does more crime cause more education, or possibly neither?

As the figure shows, these variables have a positive association. The correlation is $r = 0.47$. Unlike the previous example, there is no obviously influential observation causing the positive correlation between education and higher crime rate. But, another variable measured for these counties is urbanization, measured as the percentage of the residents who live in metropolitan areas. It has a correlation of 0.68 with crime rate and 0.79 with education.

Question to Explore

From the positive correlation between crime rate and education, can we conclude that having a more highly educated populace causes the crime rate to go up?

Think It Through

The strong correlation of 0.79 between urbanization and education tells us that highly urban counties tend to have higher education levels. The moderately strong correlation of 0.68 between urbanization and crime rate tells us that highly urban counties also tend to have higher crime. So, perhaps the reason for the positive correlation between education and crime rate is that education tends to be greater in more highly urbanized counties, but crime rates also tend to be higher in such counties. In summary, a correlation could occur without any causal connection.

Insight

For counties with similar levels of urbanization, the association between crime rate and education may look quite different. You may then see a *negative* correlation. Figure 3.22 portrays how this could happen. It shows a negative trend between crime rate and education for counties having urbanization = 0 (none of the residents living in a metropolitan area), a separate negative trend for counties having urbanization = 50, and a separate negative trend for counties having urbanization = 100. If we ignore the urbanization values and look at all the points, however, we see a positive trend—higher crime rate tending to occur with higher education levels, as reflected by the overall positive correlation.

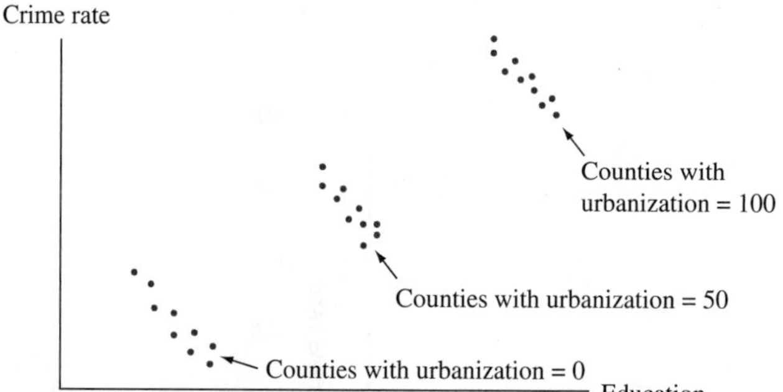

▲ **Figure 3.22 Hypothetical Scatter Diagram Relating Crime Rate and Education.** The points are also labeled by whether urbanization = 0, 50, or 100. **Question** Sketch lines that represent (a) the overall positive relationship between crime rate and education, and (b) the negative relationship between crime rate and education for counties having urbanization = 0.

Try Exercises 3.53 and 3.57

Whenever two variables are associated, other variables may have influenced that association. In Example 14, urbanization influenced the association between crime rate and education. This illustrates an important point: **Correlation does not imply causation.**

In Example 14, crime rate and education were positively correlated, but that does not mean that having a high level of education causes a county's crime rate to be high. Whenever we observe a correlation between variables x and y, there may be a third variable correlated with both x and y that is responsible for their association. Let's look at another example to illustrate this point.

Example 15

Lurking variable

Ice Cream and Drowning

Picture the Scenario

The Gold Coast of Australia, south of Brisbane, is famous for its beaches. Because of strong rip tides, however, each year many people drown. Data collected monthly show a positive correlation between y = number of people who drowned in that month and x = number of gallons of ice cream sold in refreshment stands along the beach in that month.

Question to Explore

Identify another variable that could be responsible for this association.

Think It Through

In the summer in Australia (especially January and February), the weather is hot. People tend to buy more ice cream in those months. They also tend to go to the beach and swim more in those months, and more people drown. In the winter, it is cooler. People buy less ice cream, fewer people go to the beach, and fewer people drown. So, the mean temperature in the month is a variable that could be responsible for the correlation. As mean temperature goes up, so does ice cream sales and so does the number of people who drown.

Insight

If we looked only at months having similar mean temperatures, probably we would not observe any association between ice cream sales and the number of people who drown.

Try Exercises 3.54

A third variable that is not measured in a study (or perhaps even known about to the researchers) but that influences the association between the response variable and the explanatory variable is referred to as a **lurking variable**.

Lurking Variable

A **lurking variable** is a variable, usually unobserved, that influences the association between the variables of primary interest.

In interpreting the positive correlation between crime rate and education for Florida counties, we'd be remiss if we failed to recognize that the correlation could be due to a lurking variable. This could happen if we observed those two variables but not urbanization, which would then be a lurking variable. Likewise, if we got excited by the positive correlation between ice cream sales and the number drowned in a month, we'd fail to recognize that the monthly mean temperature is a lurking variable.

Simpson's Paradox

We can express the statement that *correlation does not imply causation* more generally as **association does not imply causation.** This warning holds whether we analyze associations between quantitative variables or between categorical variables.

The direction of an association between two variables can change after we include a third variable and analyze the data at separate levels of that variable; this is known as **Simpson's paradox.**[11] We observed Simpson's paradox in Figure 3.22 in Example 14, in which a positive correlation between crime rate and education changed to a negative correlation when data were considered at separate levels of urbanization. This example serves as a warning: *Be cautious about interpreting an association.* Always be wary of lurking variables that may influence the association.

Let's illustrate by revisiting Example 1, which presented a study indicating that smoking could apparently be beneficial to your health. Could a lurking variable be responsible for this association?

Example 16

Reversal in direction of association

Smoking and Health

Picture the Scenario

Example 1 mentioned a survey[12] of 1314 women in the United Kingdom that asked each woman whether she was a smoker. Twenty years later, a follow-up survey observed whether each woman was dead or still alive. Table 3.7 is a contingency table of the results. The response variable is survival status after 20 years. We find that 139/582, which is 24%, of the smokers died, and 230/732, or 31%, of the nonsmokers died. There was a greater survival rate for the smokers.

Table 3.7 Smoking Status and 20-Year Survival in Women

	Survival Status		
Smoker	**Dead**	**Alive**	**Total**
Yes	139	443	582
No	230	502	732
Total	**369**	**945**	**1,314**

Could the age of the woman at the beginning of the study explain the association? Presumably it could if the older women were less likely than the younger women to be smokers. Table 3.8 shows separate contingency tables relating smoking status and survival status for these 1314 women separated into four age groups.

[11]The paradox is named after a British statistician who in 1951 investigated conditions under which this flip-flopping of association can happen.

[12]Described in article by D. R. Appleton et al., *American Statistician*, 50, 1996, 340–341.

Table 3.8 Smoking Status and 20-Year Survival, for Four Age Groups

	Age Group							
	18–34 Survival?		35–54 Survival?		55–64 Survival?		65+ Survival?	
Smoker	**Dead**	**Alive**	**Dead**	**Alive**	**Dead**	**Alive**	**Dead**	**Alive**
Yes	5	174	41	198	51	64	42	7
No	6	213	19	180	40	81	165	28

Questions to Explore

a. Show that the counts in Table 3.8 are consistent with those in Table 3.7.

b. Use conditional percentages to describe the association in Table 3.8 between smoking status and survival status for each age group.

c. How can you explain the association in Table 3.7, whereby smoking seems to help women live a longer life? How can this association be so different from the one shown in Table 3.8?

Think It Though

a. If you add the counts in the four separate parts of Table 3.8, you'll get the counts in Table 3.7. For instance, from Table 3.7, we see that 139 of the women who smoked died. From Table 3.8, we get this from $5 + 41 + 51 + 42 = 139$, taking the number of smokers who died from each age group. Doing this for each cell, we see that the counts in the two tables are consistent with each other.

b. Table 3.9 shows the conditional percentages who died, for smokers and nonsmokers in each age group. Figure 3.23 plots them. For each age group, a higher percentage of smokers than nonsmokers died.

Table 3.9 Conditional Percentages of Deaths for Smokers and Nonsmokers, by Age.

For instance, for smokers of age 18–34, from Table 3.8 the proportion who died was $5/(5 + 174) = 0.028$, or 2.8%.

	Age Group			
Smoker	**18–34**	**35–54**	**55–64**	**65+**
Yes	2.8%	17.2%	44.3%	85.7%
No	2.7%	9.5%	33.1%	85.5%

c. Could age explain the association? From Table 3.8, the proportion of women who smoked at the beginning of the study tended to be higher for the younger women. The percentage of smokers was 45% in the 18–34 age group (that is, $(5 + 174)/(5 + 174 + 6 + 213) = 0.45$), but only 20% in the 65+ age group. At the same time, younger women were less likely to die during the 20-year study period. For instance, the proportion who died was $(5 + 6)/(5 + 174 + 6 + 213) = 0.03$ in the 18–34 age group but 0.86 in the 65+ age group. In summary, the overall association in Table 3.7 could merely reflect younger women being more likely to be smokers, while the younger women were also less likely to die during this time frame.

Table 3.7 indicated that smokers had *higher* survival rates than nonsmokers. When we looked at the data separately for each age group in Table 3.8, we saw the reverse: Smokers had *lower* survival rates

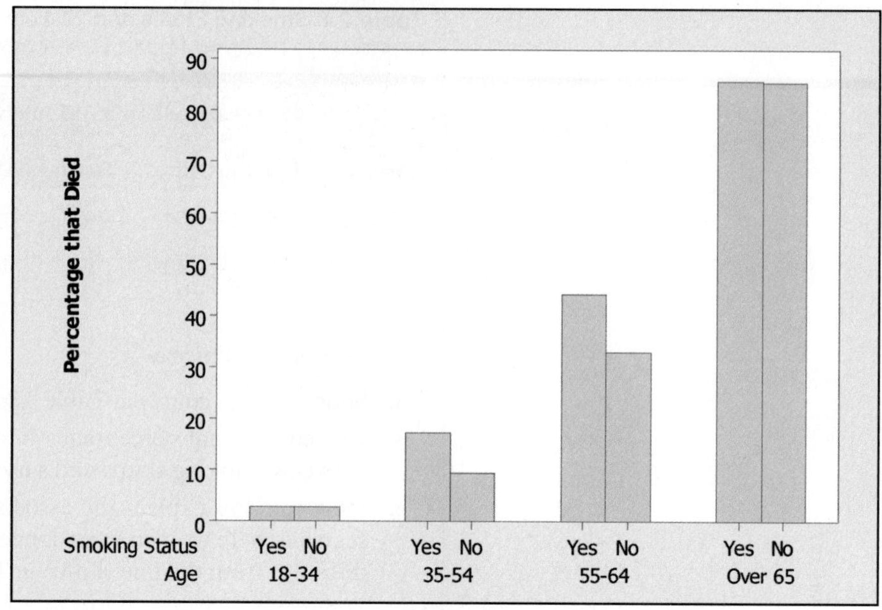

▲ **Figure 3.23 MINITAB Bar Graph Comparing Percentage of Deaths for Smokers and Nonsmokers, by Age.** This side-by-side bar graph shows the conditional percentages from Table 3.9.

than nonsmokers. The analysis using Table 3.7 did not account for age, which strongly influences the association.

Insight

Because of the reversal in the association after taking age into account, the researchers did *not* conclude that smoking is beneficial to your health. This example illustrates the dramatic influence that a lurking variable can have, which would be unknown to researchers if they fail to include a certain variable in their study. An association can look quite different after adjusting for the effect of a third variable by grouping the data according to its values.

Try Exercise 3.58

The Effect of Lurking Variables on Associations

Lurking variables can affect associations in many ways. For instance, a lurking variable may be a **common cause** of both the explanatory and response variable. In Example 15, the mean temperature in the month is a common cause of both ice cream sales and the number of people who drown.

In practice, there's usually not a single variable that causally explains a response variable or the association between two variables. More commonly, there are **multiple causes**. When there are multiple causes, the association among them makes it difficult to study the effect of any single variable. For example, suppose someone claims that growing up in poverty causes crime. Realistically, probably lots of things contribute to crime. Many variables that you might think of as possible causes, such as a person's educational level, whether the person grew up in a stable family, and the quality of the neighborhood in which the person lives, are themselves likely to be associated with whether a person grew up in poverty. Perhaps people growing up in poverty tend to have poorly educated parents, grow up in high-crime neighborhoods, and achieve low levels of education. Perhaps all these factors make a person more likely to become a criminal. Growing up in poverty may have a direct effect on crime but also an indirect effect through these other variables.

It's especially tricky to study cause and effect when two variables are measured over time. The variables may be associated merely because they both have a *time trend.* Suppose that both the divorce rate and the crime rate have an increasing trend over a 10-year period. They will then have a positive correlation: Higher crime rates occur in years that have higher divorce rates. Does this imply that an increasing divorce rate *causes* the crime rate to increase? Absolutely not. They would also be positively correlated with all other variables that have a positive time trend, such as annual average house price and the annual use of cell phones. There are likely to be other variables that are themselves changing over time and have causal influences on the divorce rate and the crime rate.

Confounding

When two explanatory variables are both associated with a response variable but are also associated with each other, **confounding** occurs. It is difficult to determine whether either of them truly causes the response because a variable's effect could be at least partly due to its association with the other variable. Example 16 illustrates a study with confounding. Over the 20-year study period, smokers had a greater survival rate than nonsmokers. However, age was a confounding variable. Older subjects were less likely to be smokers, and older subjects were more likely to die. Within each age group, smokers had a *lower* survival rate than nonsmokers. Age had a dramatic influence on the association between smoking and survival status.

What's the difference between a confounding variable and a lurking variable? A lurking variable is not measured in the study. It has the *potential* for confounding. If it were included in the study and if it were associated both with the response variable and the explanatory variable, it would become a confounding variable. It would affect the relationship between the response and explanatory variables.

The potential for lurking variables to affect associations is the main reason it is difficult to study many issues of importance, whether it be medical issues such as the effect of smoking on cancer or social issues such as what causes crime, what causes the economy to improve, or what causes students to succeed in school. It's not impossible—statistical methods can analyze the effect of an explanatory variable after adjusting for confounding variables (as we adjusted for age in Example 16)—but there's always the chance that an important variable was not included in the study.

3.4 Practicing the Basics

3.44 Extrapolating murder The SPSS figure shows the data and regression line for the 50 states in Table 3.6 relating

x = percentage of single-parent families to

y = annual murder rate (number of murders per 100,000 people in the population).

a. The lowest x value was for Utah and the highest was for Mississippi. Using the figure, approximate those x-values.

b. Using the regression equation stated above the figure, find the predicted murder rate at $x = 0$. Why is it not sensible to make a prediction at $x = 0$ based on these data?

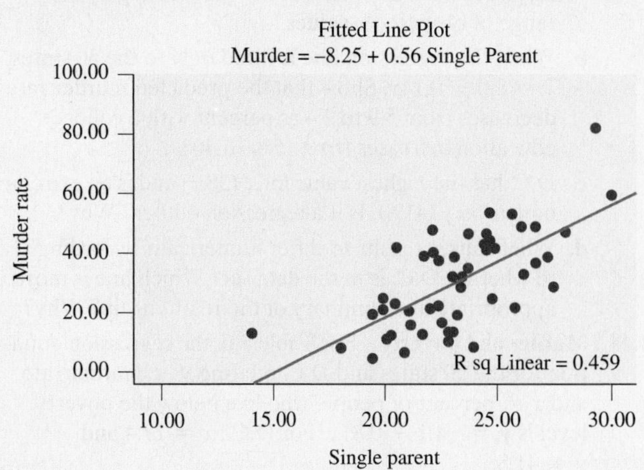

3.45 Men's Olympic long jumps The Olympic winning men's long jump distances (in meters) from 1896 to 2008 and the fitted regression line for predicting them using x = year are displayed in the MINITAB output below.

TRY

a. Identify an observation that may influence the fit of the regression line. Why did you identify this observation?

b. Which do you think is a better prediction for the year 2012—the sample mean of the y values in this plot or the value obtained by plugging 2012 into the fitted regression equation?

c. Would you feel comfortable using the regression line shown to predict the winning long jump for men in the year 3000? Why or why not?

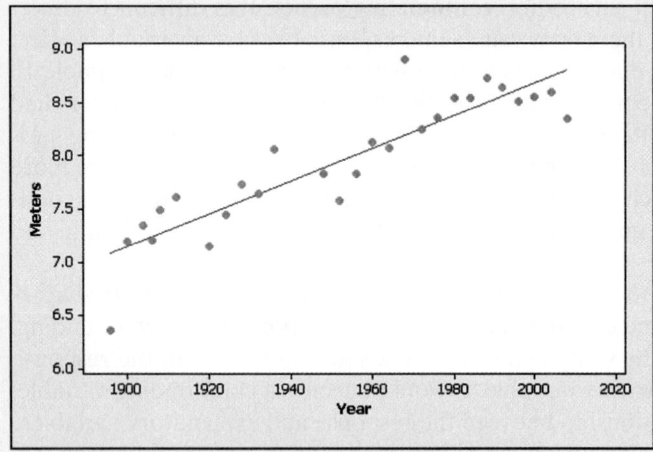

3.46 U.S. average annual temperatures Use the U.S. Temperatures data file on the text CD.

a. Fit a trend line, and interpret the slope.

b. Predict the annual mean U.S. temperature for the year (i) 2010 and (ii) 3000.

c. In which prediction in part b do you have more faith? Why?

3.47 Murder and education Example 13 found the regression line $\hat{y} = -3.1 + 0.33x$ for all 51 observations on y = murder rate and x = percent with a college education.

TRY

a. Show that the predicted murder rates increase from 1.85 to 10.1 as percent with a college education increases from $x = 15\%$ to $x = 40\%$, roughly the range of observed x values.

b. When the regression line is fitted only to the 50 states, $\hat{y} = 8.0 - 0.14x$. Show that the predicted murder rate decreases from 5.9 to 2.4 as percent with a college education increases from 15% to 40%.

c. D.C. has the highest value for x (38.3) and is an extreme outlier on y (41.8). Is it a regression outlier? Why?

d. What causes results to differ numerically according to whether D.C. is in the data set? Which line is more appropriate as a summary of the relationship? Why?

3.48 Murder and poverty For Table 3.6, the regression equation for the 50 states and D.C. relating y = murder rate and x = percent of people who live below the poverty level is $\hat{y} = -4.1 + 0.81x$. For D.C., $x = 17.4$ and $y = 41.8$.

a. When the observation for D.C. is removed from the data set, $\hat{y} = 0.4 + 0.36x$. Does D.C. have much influence on this regression analysis? Explain.

b. If you were to look at a scatterplot, based on the information given, do you think that the poverty value for D.C. would be relatively large, or relatively small? Explain.

3.49 TV watching and the birth rate The figure shows recent data on x = the number of televisions per 100 people and y = the birth rate (number of births per 1000 people) for six African and Asian nations. The regression line, $\hat{y} = 29.8 - 0.024x$ applies to the data for these six countries. For illustration, another point is added at (81, 15.2) which is the observation for the United States. The regression line for all seven points is $\hat{y} = 31.2 - 0.195x$. The figure shows this line and the one without the U.S. observation.

a. Does the U.S. observation appear to be (i) an outlier on x, (ii) an outlier on y, or (iii) a regression outlier relative to the regression line for the other six observations?

b. State the two conditions under which a single point can have a dramatic effect on the slope, and show that they apply here.

c. This one point also drastically affects the correlation, which is $r = -0.051$ without the United States but $r = -0.935$ with the United States. Explain why you would conclude that the association between birth rate and number of televisions is (i) very weak without the U.S. point and (ii) very strong with the U.S. point.

d. Explain why the U.S. residual for the line fitted using that point is very small. This shows that a point can be influential even if its residual is not large.

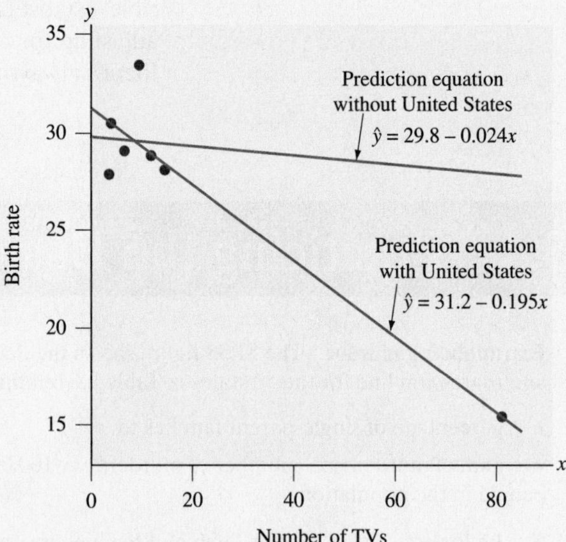

3.50 Looking for outliers Using software, analyze the relationship between x = college education and y = percentage single-parent families, for the data in Table 3.6, which are in the U.S. Statewide Crime data file on the text CD.

a. Construct a scatterplot. Based on your plot, identify two observations that seem quite different from the others, having a y value relatively large in one case and somewhat small in the other case.

b. Find the regression equation (i) for the entire data set, (ii) deleting only the first of the two outlying observations, and (iii) deleting only the second of the two outlying observations.

c. Is either deleted observation influential on the slope? Summarize the influence.

d. Including D.C., $\hat{y} = 21.2 + 0.089x$, whereas deleting that observation, $\hat{y} = 28.1 - 0.206x$. Find $\hat{y}$ for D.C. in the two cases. Does the predicted value for D.C. depend much on which regression equation is used?

3.51 Regression between cereal sodium and sugar Let x = sodium and y = sugar for the breakfast cereal data in the Cereal data file on the text CD and in Table 2.3 in Chapter 2.

a. Construct a scatterplot. Do any points satisfy the two criteria for a point to be potentially influential on the regression? Explain.

b. Find the regression line and correlation using all the data points, and then using all except a potentially influential observation. Summarize the influence of that point.

3.52 TV in Europe An article (by M. Dupagne and D. Waterman, *Journal of Broadcasting and Electronic Media*, vol. 42, 1998 pp. 208–220), studied variables relating to the percentage of TV programs in 17 Western European countries that consisted of fiction programs imported from the United States. One explanatory variable considered was the percentage of stations in the country that are private. The data are in the TV Europe data file on the text CD.

a. Construct a scatterplot. Describe the direction of the association.

b. Find the correlation and the regression line. Draw the line on the scatterplot.

c. Do you observe an outlier? If so, what makes the observation unusual? Would you expect it to be an influential observation?

d. Delete the observation identified in part c from the data set. Find the regression line and the correlation. Compare your results to those in part b. Was the observation influential? Explain.

3.53 Height and vocabulary TRY Do tall students tend to have better vocabulary skills than short students? We might think so looking at a sample of students from grades 1, 6, and 12 of Lake Wobegon school district. The correlation was 0.81 between their height and their vocabulary test score: Taller students tended to have higher vocabulary test scores.

a. Is there a causal relationship, whereby being taller gives you a better vocabulary?

b. Explain how a student's age could be a lurking variable that might be responsible for this association, being a common cause of both height and vocabulary test score.

c. Sketch a hypothetical scatterplot (as we did in Figure 3.22 for the example on crime and education), labeling points by the child's grade (1, 6, and 12), such that overall there is a positive trend, but the slope would be about 0 when we consider only children in a given grade. How does the child's age play a role in the association between height and test score?

3.54 More firefighters cause worse fires? TRY Data are available for all fires in Chicago last year on x = number of firefighters at the fire and y = cost of damages due to the fire.

a. If the correlation is positive, does this mean that having more firefighters at a fire causes the damage to be worse? Explain.

b. Identify a third variable that could be a common cause of x and y. Construct a hypothetical scatterplot (like Figure 3.22 for crime and education), identifying points according to their value on the third variable, to illustrate your argument.

3.55 Antidrug campaigns An Associated Press story (June 13, 2002) reported, "A survey of teens conducted for the Partnership for a Drug Free America found kids who see or hear antidrug ads at least once a day are less likely to do drugs than youngsters who don't see or hear ads frequently. When asked about marijuana, kids who said they saw the ads regularly were nearly 15 percent less likely to smoke pot."

a. Discuss at least one lurking variable that could affect these results.

b. Explain how multiple causes could affect whether a teenager smokes pot.

3.56 What's wrong with regression? Explain what's wrong with the way regression is used in each of the following examples:

a. Winning times in the Boston marathon (at www.bostonmarathon.org) have followed a straight line decreasing trend from 160 minutes in 1927 (when the race was first run at the Olympic distance of about 26 miles) to 130 minutes in 2004. After fitting a regression line to the winning distances, you use the equation to predict that the winning time in the year 2300 will be about 13 minutes.

b. Using data for several cities on x = % of residents with a college education and y = median price of home, you get a strong positive correlation. You conclude that having a college education causes you to be more likely to buy an expensive house.

c. A regression between x = number of years of education and y = annual income for 100 people shows a modest positive trend, except for one person who dropped out after 10th grade but is now a multimillionaire. It's wrong to ignore any of the data, so we should report all results including this point. For this data, the correlation $r = -0.28$.

3.57 Education causes crime? TRY The table shows a small data set that has a pattern somewhat like that in Figure 3.22 in Example 14. As in that example, education is measured as the percentage of adult residents who have at least a high school degree. Using software,

a. Construct a data file, with columns for education, crime rate, and a rural/urban label.

b. Construct a scatterplot between y = crime rate and x = education, labeling each point as rural or urban.

c. Find the overall correlation between crime rate and education for all eight data points. Interpret.

d. Find the correlation between crime rate and education for the (i) urban counties alone and (ii) rural counties alone. Why are these correlations so different from the correlation in part c?

Urban Counties		Rural Counties	
Education	**Crime Rate**	**Education**	**Crime Rate**
70	140	55	50
75	120	58	40
80	110	60	30
85	105	65	25

3.58 Death penalty and race The table shows results of whether the death penalty was imposed in murder trials in Florida between 1976 and 1987. For instance, the death penalty was given in 53 out of 467 cases in which a white defendant had a white victim.

TRY

Death Penalty, by Defendant's Race and Victim's Race

Victim's Race	Defendant's Race	Death Penalty		
		Yes	No	Total
White	White	53	414	**467**
	Black	11	37	**48**
Black	White	0	16	**16**
	Black	4	139	**143**

Source: Originally published in *Florida Law Review*. Michael Radelet and Glenn L. Pierce, Choosing Those Who Will Die: Race and the Death Penalty in Florida, vol. 43, *Florida Law Review* 1 (1991).

a. First, consider only those cases in which the victim was white. Find the conditional proportions that got the death penalty when the defendant was white and when the defendant was black. Describe the association.

b. Repeat part a for cases in which the victim was black. Interpret.

c. Now add these two tables together to get a summary contingency table that describes the association between the death penalty verdict and defendant's race, ignoring the information about the victim's race. Find the conditional proportions. Describe the association, and compare to parts a and b.

d. Explain how these data satisfy Simpson's paradox. How would you explain what is responsible for this result to someone who has not taken a statistics course?

e. In studying the effect of defendant's race on the death penalty verdict, would you call victim's race a confounding variable? What does this mean?

3.59 NAEP scores Eighth-grade math scores on the National Assessment of Educational Progress had a mean of 277 in Nebraska compared to a mean of 271 in New Jersey (H. Wainer and L. Brown, *American Statistician*, vol. 58, 2004, p. 119).

a. Identify the response variable and the explanatory variable.

b. For white students, the means were 281 in Nebraska and 283 in New Jersey. For black students, the means were 236 in Nebraska and 242 in New Jersey. For other nonwhite students, the means were 259 in Nebraska and 260 in New Jersey. Identify the third variable given here. Explain how it is possible for New Jersey to have the higher mean for each race, yet for Nebraska to have the higher mean when the data are combined. (This is a case of Simpson's paradox for a quantitative response.)

3.60 Age a confounder? A study observes that the subjects in the study who say they exercise regularly reported only half as many serious illnesses per year, on the average, as those who say they do not exercise regularly. One paragraph in the results section of an article about the study starts out, "We next analyzed whether age was a confounding variable in studying this association."

a. Explain what this sentence means and how age could potentially explain the association between exercising and illnesses.

b. If age was not actually measured in the study and the researchers failed to consider its effects, could it be a confounding variable or a lurking variable? Explain the difference between a lurking variable and a confounding variable.

Chapter Review

ANSWERS TO THE CHAPTER FIGURE QUESTIONS

Figure 3.1 Construct a single graph with side-by-side bars (as in Figure 3.2) to compare the conditional proportions of pesticide residues in the two types of food.

Figure 3.2 The proportion of foods having pesticides present is much higher for conventionally grown food than for organically grown food.

Figure 3.3 For a particular response, the bars in Figure 3.2 have different heights, in contrast to Figure 3.3 for which the bars have the same height.

Figure 3.4 There are no apparent outliers for Internet use. Even though there is a small gap for the distribution of Facebook use, it is unclear if any outliers are present. A modified box plot would more clearly identify if any potential outliers exist for Facebook use.

Figure 3.5 The point labeled Japan appears atypical. All of the other countries with comparable Internet use have at least 15% Facebook use, whereas Japan has only about 2% Facebook use.

Figure 3.6 The top point is far above the overall straight-line trend of the other 66 points on the graph. The two rightmost points fall in the overall increasing trend.

Figure 3.7 If data points are closer to a straight line, there is a smaller amount of variability around the line; thus, we can more accurately predict one variable knowing the other variable, indicating a strong association.

Figure 3.8 These 36 points make a positive contribution to the correlation, since the product of their z-scores will be positive.

Figure 3.9 Yes, the points fall in a balanced way in the quadrants. The positive cross products in the upper-right quadrant are roughly counterbalanced by the negative cross products in the upper-left quadrant. The positive cross products in the lower-left quadrant are roughly counterbalanced by the negative cross products in the lower-right quadrant.

Figure 3.10 The line crosses the y-axis at 61.4. For each femur length increase of 1 cm, the height is predicted to increase 2.4 cm.

Figure 3.11 The height is predicted to be the same for each femur length.

Figure 3.12 We would expect a positive slope.

Figure 3.13 The error from predicting with the regression line is represented by a vertical line from the regression line to the observation. The distance between the point and line is the prediction error.

Figure 3.14 The absolute values of the residuals are vertical distances because the regression equation predicts y, the variable on the vertical axis.

Figure 3.15 The rightmost bar represents a county with a large positive residual. For it, the actual vote for Buchanan was much higher than what was predicted.

Figure 3.16 The difference in winning heights for men and women is typically about 0.3 to 0.4 meters.

Figure 3.17 $\hat{y} = a + bx = -1.636 + 0.029(2015) = 56.8$

Figure 3.18 The observation in the upper-right corner will influence the tilt of the straight-line fit.

Figure 3.19 The slope will decrease if D.C. is excluded from the regression analysis.

Figure 3.20 The fitted line to the data that does not include D.C.

Figure 3.21 Possibly neither. There may be another variable influencing the positive association between crime rate and amount of education.

Figure 3.22 Line (a) will pass through all the points, with a positive slope. In case (b), the lines passes only through the left set of points with *urbanization* = 0. It has the same negative slope that those points show.

CHAPTER SUMMARY

This chapter introduced descriptive statistics for studying the **association** between two variables. We explored how the value of a **response variable** (the outcome of interest) is related to the value of an **explanatory variable.**

- For two *categorical variables*, **contingency tables** summarize the counts of observations at the various combinations of categories of the two variables. Bar graphs can plot **conditional proportions** or percentages for the response variable, at different given values of the explanatory variable.
- For two *quantitative variables*, **scatterplots** display the relationship and show whether there is a **positive association** (upward trend) or a **negative association** (downward trend). The **correlation,** r, describes this direction and the strength of linear (straight-line) association. It satisfies $-1 \le r \le 1$. The closer r is to -1 or 1, the stronger the linear association.
- When a relationship between two quantitative variables approximately follows a straight line, it can be described by a **regression line,** $\hat{y} = a + bx$. The slope b describes the direction of the association (positive or negative, like the correlation) and gives the effect on $\hat{y}$ of a one-unit increase in x.
- The correlation and the regression equation can be strongly affected by an **influential observation.** This observation takes a relatively small or large value on x and is a **regression outlier,** falling away from the straight-line trend of the rest of the data points. Be cautious of **extrapolating** a regression line to predict y at values of x that are far above or below the observed x-values.
- **Association does not imply causation.** A **lurking variable** may influence the association. It is even possible for the association to reverse in direction after we adjust for a third variable. This phenomenon is called **Simpson's paradox.**

SUMMARY OF NOTATION

$$\text{Correlation } r = \frac{1}{n-1}\Sigma z_x z_y = \frac{1}{n-1}\Sigma\left(\frac{x-\bar{x}}{s_x}\right)\left(\frac{y-\bar{y}}{s_y}\right)$$

where n is the number of points, $\bar{x}$ and $\bar{y}$ are means, s_x and s_y are standard deviations for x and y.

Regression equation: $\hat{y} = a + bx$, for predicted response $\hat{y}$, y-intercept a, slope b.

Formulas for the slope and y-intercept are

$$b = r\left(\frac{s_y}{s_x}\right) \text{ and } a = \bar{y} - b(\bar{x}).$$

CHAPTER PROBLEMS

Practicing the Basics

3.61 Choose explanatory and response For the following pairs of variables, identify the response variable and the explanatory variable.

a. Number of square feet in a house and assessed value of the house.

b. Political party preference (Democrat, Independent, Republican) and gender.

c. Annual income and number of years of education.

d. Number of pounds lost on a diet and type of diet (low-fat, low-carbohydrate).

3.62 Graphing data For each case in the previous exercise,

a. Indicate whether each variable is quantitative or categorical.

b. Describe the type of graph that could best be used to portray the results.

3.63 Life after death for males and females In a recent General Social Survey, respondents answered the question, "Do you believe in a life after death?" The table shows the responses cross-tabulated with gender.

Opinion About Life After Death by Gender

	Opinion About Life After Death	
Gender	**Yes**	**No**
Male	621	187
Female	834	145

a. Construct a table of conditional proportions.

b. Summarize results. Is there much difference between responses of males and females?

3.64 God and happiness Go to the GSS Web site sda.berkeley .edu/GSS, click on GSS, with *No Weight* as the default weight selection, type GOD for the row variable and HAPPY for the column variable, and click on *Run the Table*.

a. Report the contingency table of counts.

b. Treating reported happiness as the response variable, find the conditional proportions. For which opinion about God are subjects most likely to be very happy?

c. To analyze the association, is it more informative to view the proportions in part b or the frequencies in part a? Why?

3.65 Degrees and income The mean annual salaries earned in 2005 by year-round workers with various educational degrees are given in the table:

Degree	**Mean Salary**
No diploma	$19,964
High school diploma	$29,448
Bachelor's degree	$54,689
Master's degree	$67,898
Doctoral degree	$92,863
Professional degree	$119,009

Source: U.S. Census Bureau.

a. Identify the response variable. Is it quantitative or categorical?

b. Identify the explanatory variable. Is it quantitative or categorical?

c. Explain how a bar graph could summarize the data.

3.66 Whooping cough Consider the Whooping Cough data file on the text CD.

a. Identify the response variable and the explanatory variable.

b. Construct a graph using bars to display the incidence rate of whooping cough contingent on year. Interpret.

3.67 Women managers in the work force The following side-by-side bar graph appeared in a 2003 issue of the *Monthly Labor Review* about women as managers in the work force. The graph summarized the percentage of managers in different occupations who were women, for the years 1972 and 2002.

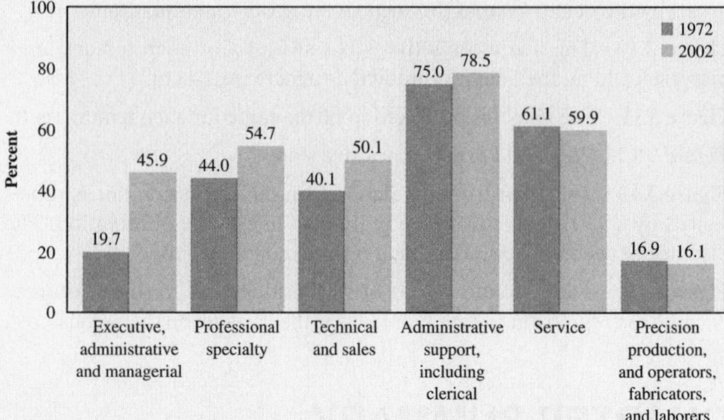

Source: Monthly Labor Review, vol. 126, no. 10, 2003, p. 48. Bureau of Labor Statistics.

a. Consider the first two bars in this graph. Identify the response variable and explanatory variable.

b. Express the information from the first two bars in the form of conditional proportions in a contingency table for two categorical variables.

c. Based on part b, does it seem as if there's an association between these variables? Explain.

d. The entire graph shows two explanatory variables. What are they?

3.68 RateMyProfessor.com The Web site RateMyProfessors .com[13] reported a correlation of 0.62 between the quality rating of the professor (on a simple 1 to 5 scale with higher values representing higher quality) and the rating of how easy a grader the professor is. This correlation is based on ratings of nearly 7000 professors.

a. How would you interpret this correlation?

b. What would you expect this correlation to equal if the quality rating did not tend to depend on the easiness rating of the professor?

3.69 Women in government and economic life The OECD (Organization for Economic Cooperation and Development) consists of advanced, industrialized countries that accept the principles of representative democracy and a free market economy. For the nations outside of Europe that are in the OECD, the table shows UN data from 2007 on the percentage of seats in parliament held by women and female economic activity as a percentage of the male rate.

a. Treating women in parliament as the response variable, prepare a scatterplot and find the correlation. Explain how the correlation relates to the trend shown in the scatterplot.

[13] See insidehighered.com/news/2006/05/08/rateprof.

b. Use software or a calculator to find the regression equation. Explain why the y-intercept is not meaningful.

c. Find the predicted value and residual for the United States. Interpret the residual.

d. With UN data for all 23 OECD nations, the correlation between these variables is 0.56. For women in parliament, the mean is 26.5% and the standard deviation is 9.8%. For female economic activity, the mean is 76.8 and the standard deviation is 7.7. Find the prediction equation, treating women in parliament as the response variable.

Nation	Women in Parliament (%)	Female Economic Activity
Iceland	33.3	87
Australia	28.3	79
Canada	24.3	83
Japan	10.7	65
United States	15.0	81
New Zealand	32.2	81

3.70 African droughts and dust Is there a relationship between the amount of dust carried over large areas of the Atlantic and the Caribbean and the amount of rainfall in African regions? In an article (by J. M. Prospero and P. J. Lamb, *Science*, vol. 302, 2003, pp. 1024–1027) the following scatterplots were given along with corresponding regression equations and correlations. The precipitation index is a measure of rainfall.

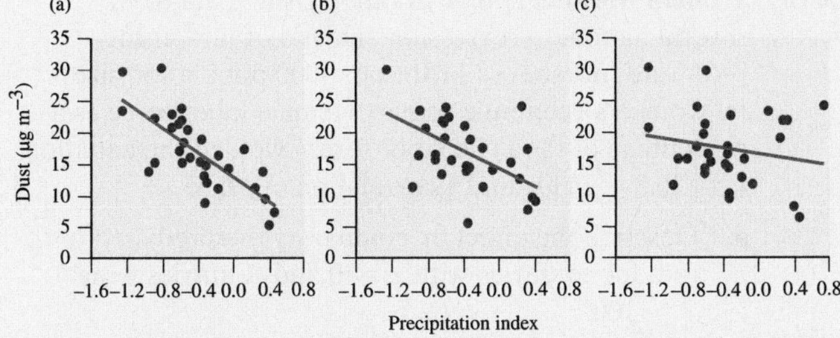

Source: J. M. Prospero and P. J. Lamb, *Science,* vol. 302, 2003, pp. 1024–1027.

a. Match the following regression equations and correlations with the appropriate graph.

(i) $\hat{y} = 14.05 - 7.18x;\quad r = -0.75$

(ii) $\hat{y} = 16.00 - 2.36x;\quad r = -0.44$

(iii) $\hat{y} = 12.80 - 9.77x;\quad r = -0.87$

b. Based on the scatterplots and information in part a, what would you conclude about the relationship between dust amount and rainfall amounts?

3.71 Crime rate and urbanization For the data in Example 14 on crime in Florida, the regression line between y = crime rate (number of crimes per 1000 people) and x = percentage living in an urban environment is $\hat{y} = 24.5 + 0.56x$.

a. Using the slope, find the difference in predicted crime rates between counties that are 100% urban and counties that are 0% urban. Interpret.

b. Interpret the correlation of 0.67 between these variables.

c. Show the connection between the correlation and the slope, using the standard deviations of 28.3 for crime rate and 34.0 for percentage urban.

3.72 Predict crime using poverty A recent analysis of data for the 50 U.S. states on y = violent crime rate (measured as number of violent crimes per 100,000 people in the state) and x = poverty rate (percent of people in the state living at or below the poverty level) yielded the regression equation, $\hat{y} = 209.9 + 25.5x$.

a. Interpret the slope.

b. The state poverty rates ranged from 8.0 (for Hawaii) to 24.7 (for Mississippi). Over this range, find the range of predicted values for the violent crime rate.

c. Would the correlation between these variables be positive or negative? Why?

3.73 Height and paycheck The headline of an article in the *Gainesville Sun* (October 17, 2003) stated, "Height can yield a taller paycheck." It described an analysis of four large studies in the United States and Britain by a University of Florida professor on subjects' height and salaries. The article reported that for each gender, "an inch is worth about $789 a year in salary. So, a person who is 6 feet tall will earn about $5,523 more a year than a person who is 5 foot 5."

a. For the interpretation in quotes, identify the response variable and explanatory variable.

b. State the slope of the regression equation, when height is measured in inches and income in dollars.

c. Explain how the value $5,523 relates to the slope.

3.74 Predicting college GPA An admissions officer claims that at his college the regression equation $\hat{y} = 0.5 + 7x$ approximates the relationship between y = college GPA and x = high school GPA, both measured on a four-point scale.

a. Sketch this equation between $x = 0$ and 4, labeling the x- and y-axes. Is this equation realistic? Why or why not?

b. Suppose that actually $\hat{y} = 0.5 + 0.7x$. Predict the GPA for two students having GPAs of 3.0 and 4.0. Interpret, and explain how the difference between these two predictions relates to the slope.

3.75 College GPA = high school GPA Refer to the previous exercise. Suppose the regression equation is $\hat{y} = x$. Identify the y-intercept and slope. Interpret the line in context.

3.76 What's a college degree worth? In 2002, a Census Bureau survey reported that the mean total earnings that a full-time worker in the United States can expect to earn between ages 25 and 64 is $1.2 million for those with only a high-school education and $2.1 million for those with a college degree but no advanced degree.

a. Assuming four years for a college degree and a straight-line regression of y = total earnings on x = number years of education, what is the slope? Interpret it.

b. If y instead measures earnings per year (rather than for 40 years), then what is the slope? Interpret.

3.77 Car weight and gas hogs: The table shows a short excerpt from the Car Weight and Mileage data file on the text CD. That file lists several 2004 model cars with automatic transmission and their x = weight (in pounds) and y = mileage (miles per gallon of gas). The prediction equation is $\hat{y} = 47.32 - 0.0052x$.

Automobile Brand	Weight	Mileage
Honda Accord Sedan LX	3,164	34
Toyota Corolla	2,590	38
Dodge Dakota Club Cab	3,838	22
Jeep Grand Cherokee Laredo	3,970	21
Hummer H2	6,400	17

Sources: auto.consumerguide.com, honda.com, toyota.com, landrover.com, ford.com.

a. Interpret the slope in terms of a 1000 pound increase in the vehicle weight.

b. Find the predicted mileage and residual for a Hummer H2. Interpret.

3.78 Predicting Internet use from cell phone use We now use data from the Human Development data file on cell phone use and Internet use for 39 countries.

a. The MINITAB output below shows a scatterplot. Describe it in terms of (i) identifying the response variable and the explanatory variable, (ii) indicating whether it shows a positive or a negative association, and (iii) describing the variability of Internet use values for nations that are close to 0 on cell phone use.

b. Identify the approximate x- and y-coordinates for a nation that has less Internet use than you would expect, given its level of cell phone use.

c. The prediction equation is $\hat{y} = 1.27 + 0.475x$. Describe the relationship by noting how $\hat{y}$ changes as x increases from 0 to 90, which are roughly its minimum and maximum.

d. For the United States, $x = 45.1$ and $y = 50.15$. Find its predicted Internet use and residual. Interpret the large positive residual.

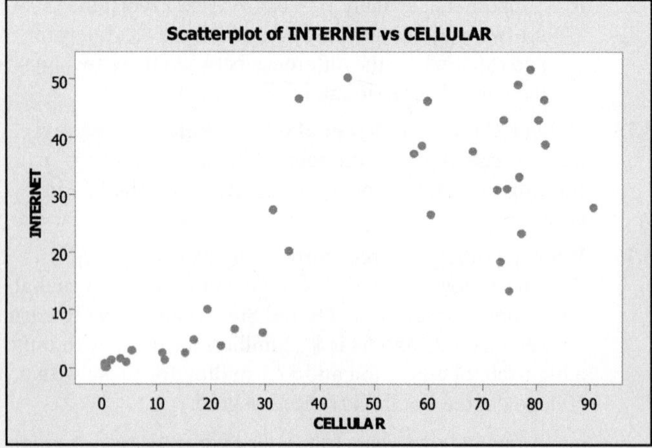

3.79 Income depends on education? For a study of counties in Florida, the table shows part of a printout for the regression analysis relating y = median income (thousands of dollars) to x = percent of residents with at least a high school education.

a. County A has 10% more of its residents than County B with at least a high school education. Find their difference in predicted median incomes.

b. Find the correlation. (*Hint:* Use the relation between the correlation and the slope of the regression line.) Interpret the (i) sign and (ii) strength of association.

Variable	Mean	Std Dev
Income	24.51	4.69
Education	69.49	8.86

The regression equation is income = −4.63 + 0.42 education

3.80 Fertility and GDP Refer to the Human Development data file on the text CD. Use x = GDP and y = fertility (mean number of children per adult woman).

a. Construct a scatterplot, and indicate whether regression seems appropriate.

b. Find the correlation and the regression equation.

c. With x = percent using contraception, $\hat{y} = 6.7 - 0.065x$. Can you compare the slope of this regression equation with the slope of the equation with GDP as a predictor to determine which has the stronger association with y? Explain.

d. Contraception has a correlation of −0.887 with fertility. Which variable has a stronger association with fertility: GDP or contraception?

3.81 Women working and birth rate Using data from several nations, a regression analysis of y = crude birth rate (number of births per 1000 population size) on women's economic activity (female labor force as a percentage of the male labor force) yielded the equation $\hat{y} = 36.3 - 0.30x$ and a correlation of −0.55.

a. Describe the effect by comparing the predicted birth rate for countries with $x = 0$ and countries with $x = 100$.

b. Suppose that the correlation between the crude birth rate and the nation's GNP equals −0.35. Which variable, GNP or women's economic activity, seems to have the stronger association with birth rate? Explain.

3.82 Education and income The regression equation for a sample of 100 people relating x = years of education and y = annual income (in dollars) is $\hat{y} = -20{,}000 + 4000x$, and the correlation equals 0.50. The standard deviations were 2.0 for education and 16,000 for annual income.

a. Show how to find the slope in the regression equation from the correlation.

b. Suppose that now we let x = annual income and y = years of education. Will the correlation or the slope change in value? If so, show how.

3.83 Income in euros Refer to the previous exercise. Results in the regression equation $\hat{y} = -20{,}000 + 4000x$ for y = annual income were translated to units of euros, at a time when the exchange rate was $1.25 per euro.

a. Find the intercept of the regression equation. (*Hint*: What does 20,000 dollars equal in euros?)

b. Find the slope of the regression equation.

c. What is the correlation when annual income is measured in euros? Why?

3.84 Changing units for cereal data Refer to the Cereal data file on the text CD, with x = sugar (g) and y = sodium (mg), for which $\hat{y} = 169 - 0.25x$.

a. Convert the sugar measurements to mg and calculate the line obtained from regressing sodium (mg) on sugar (mg). Which statistics change and which remain the same? Clearly interpret the slope coefficient.

b. Suppose we instead convert the sugar measurements to ounces. How would this effect the slope of the regression line? Can you determine the new slope just from knowing that 1 ounce equals rougth 28.35 grams?

3.85 Murder and single-parent families For Table 3.6 on the 50 states and D.C., the figure below shows the relationship between the murder rate and the percentage of single-parent families.

a. For D.C., the percentage of single-parent families = 44.7 and the murder rate = 41.8. Identify D.C. on the scatterplot, and explain the effect you would expect it to have on a regression analysis.

b. The regression line fitted to all 51 observations is $\hat{y} = -21.4 + 1.14x$. The regression line fitted only to the 50 states is $\hat{y} = -8.2 + 0.56x$. Summarize the effect of including D.C. in the analysis.

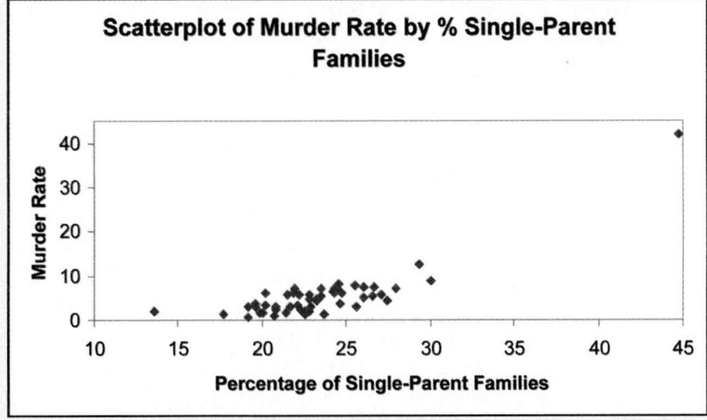

3.86 Violent crime and college education For the U.S. Statewide Crime data file on the text CD, let y = violent crime rate and x = percent with a college education.

a. Construct a scatterplot. Identify any points that you think may be influential in a regression analysis.

b. Fit the regression line using all 51 observations. Interpret the slope.

c. Fit the regression line after deleting the observation identified in part a. Interpret the slope, and compare results to part b.

3.87 Violent crime and high school education Repeat the previous exercise using x = percent with at least a high school education. This shows that an outlier is not especially influential if its x-value is not relatively large or small.

3.88 Crime and urbanization For the U.S. Statewide Crime data file on the text CD, using MINITAB to analyze y = violent crime rate and x = urbanization (percentage of the residents living in metropolitan areas) gives the results shown:

Variable	N	Mean	StDev	Minimum	Q1	Median	Q3	Maximum
violent	51	441.6	241.4	81.0	281.0	384.0	554.0	1508.0
urban	51	68.36	20.85	27.90	49.00	70.30	84.50	100.00

The regression equation is violent = 36.0 + 5.93 urban

a. Using the five-number summary of positions, sketch a box plot for y. What does your graph and the reported mean and standard deviation of y tell you about the shape of the distribution of violent crime rate?

b. Construct a scatterplot. Does it show any potentially influential observations? Would you predict that the slope would be larger, or smaller, if you delete that observation? Why?

c. Fit the regression without the observation highlighted in part b. Describe the effect on the slope.

3.89 High school graduation rates and health insurance Access the HS Graduation Rates file on the text CD, which contains statewide data on x = high school graduation rate and y = percentage of individuals without health insurance.

a. Construct a scatterplot. Describe the relationship.

b. Find the correlation. Interpret.

c. Find the regression equation for the data. Interpret the slope, and summarize the relationship.

3.90 Women's Olympic high jumps Example 11 discussed how the winning height in the Olympic high jump changed over time. Using the High Jump data file on the text CD, MINITAB reports

Women_Meters = −10.9 + 0.00650 Year_Women

for predicting the women's winning height (in meters) using the year number.

a. Predict the winning Olympic high jump distance for women in (i) 2016 and (ii) 3000.

b. Do you feel comfortable making either prediction in part a? Explain.

3.91 Income and height A survey of adults revealed a positive correlation between the height of the subjects and their income in the previous year.

a. Explain how gender could be a potential lurking variable that could be responsible for this association.

b. If gender had actually been one of the variables measured in the study, would it be a lurking variable or a confounding variable? Explain.

3.92 More TV watching goes with fewer babies? For United Nations data from several countries, there is a strong negative correlation between the birth rate and the per capita television ownership.

a. Does this imply that having higher television ownership causes a country to have a lower birth rate?

b. Identify a lurking variable that could provide an explanation for this association.

3.93 More sleep causes death? An Associated Press story (February 15, 2002) quoted a study at the University of

California at San Diego that reported, based on a nationwide survey, that those who averaged at least 8 hours sleep a night were 12% more likely to die within six years than those who averaged 6.5 to 7.5 hours of sleep a night.

a. Explain how the subject's age could be positively associated both with time spent sleeping and with an increased death rate, and hence could be a lurking variable responsible for the observed association between sleeping and the death rate.

b. Explain how the subject's age could be a common cause of both variables.

3.94 Ask Marilyn Marilyn vos Savant writes a column for *Parade* magazine to which readers send questions, often puzzlers or questions with a twist. In the April 28, 1996, column, a reader asked, "A company decided to expand, so it opened a factory generating 455 jobs. For the 70 white-collar positions, 200 males and 200 females applied. Of the people who applied, 20% of the females and only 15% of the males were hired. Of the 400 males applying for the blue-collar positions, 75% were hired. Of the 100 females applying, 85% of were hired. A federal Equal Employment Opportunity Commission (EEOC) enforcement official noted that many more males were hired than females, and decided to investigate. Responding to charges of irregularities in hiring, the company president denied any discrimination, pointing out that in both the white-collar and blue-collar fields, the percentage of female applicants hired was greater than it was for males. But the government official produced his own statistics, which showed that a female applying for a job had a 58% chance of being denied employment while male applicants had only a 45% denial rate. As the current law is written, this constituted a violation. ... Can you explain how two opposing statistical outcomes are reached from the same raw data?" (Copyright 1996 Marilyn vos Savant. Initially published in *Parade* Magazine. All rights reserved.)

a. Construct two contingency tables giving counts relating gender to whether hired (yes or no), one table for white-collar jobs and one table for blue-collar jobs.

b. Construct a single contingency table for gender and whether hired, combining all 900 applicants into one table. Verify that the percentages not hired are as quoted above by the government official.

c. Comparing the data in the tables constructed in parts a and b, explain why this is an example of Simpson's paradox.

Concepts and Investigations

3.95 NL baseball team ERA and number of wins Is a baseball team's earned run average (ERA = the average number of earned runs they give up per game) a good predictor of the number of wins that a team has for a season? The data for the National League teams in 2010 are available in the NL Team Statistics file on the text CD. Conduct a correlation and regression analysis, including graphical and numerical descriptive statistics. Summarize results in a two-page report.

3.96 Time studying and GPA Is there a relationship between the amount of time a student studies and a student's GPA? Access the Georgia Student Survey file on the text CD or use your class data to explore this question using appropriate graphical and numerical methods. Write a brief report summarizing results.

3.97 Warming in Newnan, Georgia Access the Newnan GA Temp's file on the text CD, which contains data on average annual temperatures for Newnan, Georgia, during the 20th century. Fit a regression line to these temperatures and interpret the trend. Compare the trend to the trend found in Example 12 for Central Park, New York, temperatures.

3.98 Regression for dummies You have done a regression analysis for the catalog sales company you work for, using monthly data for the last year on y = total sales in the month and x = number of catalogs mailed in preceding month. You are asked to prepare a 200-word summary of what regression does under the heading "Regression for Dummies," to give to fellow employees who have never taken a statistics course. Write the summary, being careful not to use any technical jargon with which the employees may not be familiar.

3.99 Fluoride and AIDS An Associated Press story (August 25, 1998) about the lack of fluoride in most of the water supply in Utah quoted antifluoride activist Norma Sommer as claiming that fluoride may be responsible for AIDS, since the water supply in San Francisco is heavily fluoridated and that city has an unusually high incidence of AIDS. Describe how you could use this story to explain to someone who has never studied statistics that association need not imply causation.

3.100 Fish fights Alzheimer's An AP story (July 22, 2003) described a study conducted over four years by Dr. Martha Morris and others from Chicago's Rush-Presbyterian-St. Luke's Medical Center involving 815 Chicago residents aged 65 and older (*Archives of Neurology*, July 21, 2003). Those who reported eating fish at least once a week had a 60% lower risk of Alzheimer's than those who never or rarely ate fish. However, the story also quoted Dr. Rachelle Doody of Baylor College of Medicine as warning, "Articles like this raise expectations and confuse people. Researchers can show an association, but they can't show cause and effect." She said it is not known whether those people who had a reduced risk had eaten fish most of their lives, and whether other dietary habits had an influence. Using this example, describe how you would explain to someone who has never taken statistics (a) what a lurking variable is, (b) how there can be multiple causes for any particular response variable, and (c) why they need to be skeptical when they read new research results.

3.101 Dogs make you healthier A study published in the *British Journal of Health Psychology* (D. Wells, vol.12, 2007, pp. 145–156) found that dog owners are physically healthier than cat owners. The author of the study was quoted as saying, "It is possible that dogs can directly promote our well being by buffering us from stress. The ownership of a dog can also lead to increases in physical activity and facilitate the development of social contacts, which may enhance physiological and psychological human health in a more indirect manner." Identify lurking variables in this explanation, and use this quote to explain how lurking variables can be responsible for an association.

3.102 Multiple choice: Correlate GPA and GRE In a study of graduate students who took the Graduate Record Exam (GRE), the Educational Testing Service reported a correlation of 0.37 between undergraduate grade point average (GPA) and the graduate first year GPA.[14] This means that

a. As undergraduate GPA increases by one unit, graduate first-year GPA increases by 0.37 unit.

b. Since the correlation is not 0, we can predict a person's graduate first-year GPA perfectly if we know their undergraduate GPA.

c. The relationship between undergraduate GPA and graduate first-year GPA follows a curve rather than a straight line.

d. As one of these variables increases, there is a weak tendency for the other variable to increase also.

3.103 Multiple choice: Properties of *r* Which of the following is *not* a property of r?

a. r is always between -1 and 1.

b. r depends on which of the two variables is designated as the response variable.

c. r measures the strength of the linear relationship between x and y.

d. r does not depend on the units of y or x.

e. r has the same sign as the slope of the regression equation.

3.104 Multiple choice: Interpreting *r* One can interpret $r = 0.30$ as

a. a weak, positive association

b. 30% of the time $\hat{y} = y$

c. $\hat{y}$ changes 0.30 units for every one-unit increase in x

d. a stronger association than two variables with $r = -0.70$

3.105 Multiple choice: Correct statement about *r* Which one of the following statements is correct?

a. The correlation is always the same as the slope of the regression line.

b. The mean of the residuals from the least-squares regression line is 0 only when $r = 0$.

c. The correlation is the percentage of points that lie in the quadrants where x and y are both above the mean or both below the mean.

d. The correlation is inappropriate if a U-shaped relationship exists between x and y.

3.106 Multiple choice: Describing association between categorical variables You can summarize the data for two categorical variables x and y by

a. drawing a scatterplot of the x- and y-values.

b. constructing a contingency table for the x- and y-values.

c. calculating the correlation between x and y.

d. constructing a box plot for each variable.

3.107 Multiple choice: Slope and correlation The slope of the regression equation and the correlation are similar in the sense that

a. they do not depend on the units of measurement.

b. they both must fall between -1 and $+1$.

c. they both have the same sign.

d. neither can be affected by severe regression outliers.

3.108 True or false The variables y = annual income (thousands of dollars), x_1 = number of years of education, and x_2 = number of years experience in job are measured for all the employees having city-funded jobs in Knoxville, Tennessee. Suppose that the following regression equations and correlations apply:

(i) $\hat{y} = 10 + 1.0x_1, \quad r = 0.30.$

(ii) $\hat{y} = 14 + 0.4x_2, \quad r = 0.60.$

The correlation is -0.40 between x_1 and x_2. Which of the following statements are true?

a. The weakest association is between x_1 and x_2.

b. The regression equation using x_2 to predict x_1 has negative slope.

c. Each additional year on the job corresponds to a $400 increase in predicted income.

d. The predicted mean income for employees having 20 years of experience is $4000 higher than the predicted mean income for employees having 10 years of experience.

3.109 Correlation doesn't depend on units ◆◆ Suppose you convert y = income from British pounds to dollars, and suppose a pound equals 2.00 dollars.

a. Explain why the y values double, the mean of y doubles, the deviations $(y - \bar{y})$ double, and the standard deviation s_y doubles.

b. Using the formula for calculating the correlation, explain why the correlation would not change value.

3.110 When correlation = slope ◆◆ Consider the formula $b = r(s_y/s_x)$ that expresses the slope in terms of the correlation. Suppose the data are equally spread out for each variable. That is, suppose the data satisfy $s_x = s_y$. Show that the correlation and the slope are the same. (In practice, the standard deviations are not usually identical. However, this provides an interpretation for the correlation as representing what we would get for the slope of the regression line if the two variables were equally spread out.)

3.111 Center of the data ◆◆ Consider the formula $a = \bar{y} - b\bar{x}$ for the y-intercept.

a. Show that $\bar{y} = a + b\bar{x}$. Explain why this means that the predicted value of the response variable is $\hat{y} = \bar{y}$ when $x = \bar{x}$.

b. Show that an alternative way of expressing the regression model is as $(\hat{y} - \bar{y}) = b(x - \bar{x})$. Interpret this formula.

[14] *Source:* GRE Guide to the Use of Scores, 1998–1999, www.gre.org.

3.112 Final exam "regresses toward mean" of midterm Let
◆◆ y = final exam score and x = midterm exam score. Suppose that the correlation is 0.70 and that the standard deviation is the same for each set of scores.

a. Using part b of the previous exercise and the relation between the slope and correlation, show that $(\hat{y} - \bar{y}) = 0.70(x - \bar{x})$.

b. Explain why this indicates that the predicted difference between your final exam grade and the class mean is 70% of the difference between your midterm exam score and the class mean. Your score is predicted to "regress toward the mean." (The concept of *regression toward the mean*, which is responsible for the name of regression analysis, will be discussed in Section 12.2.)

Student Activities

3.113 Analyze your data Refer to the data file the class created in Activity 3 in Chapter 1. For variables chosen by your instructor, conduct a regression and correlation analysis. Prepare a one-page report summarizing your analyses, interpreting your findings.

3.114 Activity: Effect of moving a point The Regression by Eye applet on the text CD lets you add and delete points on a scatterplot. The regression line is automatically calculated for the points you provide.

a. Your instructor will give you five data points that have an approximate linear relation between x and y but a slope near 0. Plot these in a scatterplot.

b. Add a sixth observation that is influential. What did you have to do to get the slope to change noticeably from 0?

c. Now consider the effect of sample size on the existence of influential points. Start by creating 20 points with a similar pattern as in part a that have a slope near 0. Is it now harder to add a single point that makes the slope change noticeably from 0? Why? Compare your results in part c to the results from part b. Your class will discuss the results.

d. Try this activity, with the Correlation by Eye applet, looking at the effect of an influential point on the correlation coefficient, r. Repeat parts a–c in terms of the correlation instead of the slope.

3.115 Activity: Guess the correlation and regression The
TRY Regression by Eye applet and the Correlation by Eye applet on the text CD allow you to guess the correlation r and the regression line for a scatterplot with randomly generated data points. Your instructor will give you instructions on how to use these applets to practice matching the correct value of r to a scatterplot and to estimate the correct regression line.

Gathering Data

4

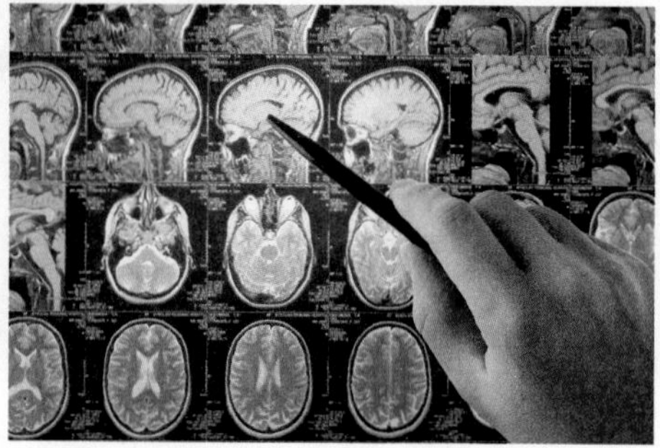

Example 1

Cell Phones and Your Health

Picture the Scenario

How safe are cell phones to use? Cell phones emit electromagnetic radiation, and a cell phone's antenna is the main source of this energy. The closer the antenna is to the user's head, the greater the exposure to radiation. With the increased use of cell phones, there has been a growing concern over the potential health risks. Several studies have explored the possibility of such risks:

Study 1 A German study (Stang et al., 2001) compared 118 patients with a rare form of eye cancer called uveal melanoma to 475 healthy patients who did not have this eye cancer. The patients' cell phone use was measured using a questionnaire. On average, the eye cancer patients used cell phones more often.

Study 2 A British study (Hepworth et al., 2006) compared 966 patients with brain cancer to 1716 patients without brain cancer. The patients' cell phone use was measured using a questionnaire. Cell phone use for the two groups was similar.

Study 3 A study published in *The Journal of the American Medical Association* (Volkow et. al., 2011) indicates that cell phone use speeds up activity in the brain. As part of a randomized crossover study, 47 participants were fitted with a cell phone device on each ear and then underwent two positron emission topography, or PET, scans to measure a specific type of brain activity. During one scan, the cell phones were both turned off. During the other scan, an automated 50-minute muted call was made to the phone on the right ear. The order of when the call was received (for the first or second scan) was randomized. Comparison of the PET scans showed a significant increase in activity in the part of the brain closest to the antenna during the transmission of the automated call.

Studies 1 and 3 found potential physiological responses to cell phone use. Study 2 did not.

Question to Explore

- Why do results of different medical studies sometimes disagree?
- What is the best study design for gathering data to explore whether cell phone use is associated with various types of physiological activities in our bodies? Can such a study design establish a direct causal link between cell phone use and potential health risks? Is this study design ethically feasible to conduct?

Thinking Ahead

A knowledge of different **study designs for gathering data** helps us understand how contradictory results can happen in scientific research studies and determine which studies deserve our trust. In this chapter, we'll learn that the study design can have a major impact on its results. Unless the study is well designed and implemented, the results may be meaningless, or worse yet, misleading. Throughout Chapter 4, we will return to the cell phone studies introduced in Example 1 to ask more questions and understand how study design can affect what are appropriate conclusions to infer from an analysis of the data.

Chapters 2 and 3 introduced graphical and numerical ways of describing data. We described shape, center, and variability, looked for patterns and unusual observations, and explored associations between variables. For these and other statistical analyses to be useful, we must have "good data." But what's the best way to gather data to ensure they are "good"? The studies described in Example 1 used two different methods. The first two studies merely *observed* subjects, whereas the third study *conducted an experiment* with them. We'll now learn about such methods, study their pros and cons, and see both good and bad ways to use them. This will help us understand when we can trust the conclusions of a study and when we need to be skeptical.

4.1 Experimental and Observational Studies

We use statistics to learn about a **population.** The German and British studies in Example 1 examined subjects' cell phone use and whether or not the subjects had brain or eye cancer. Since it would be too costly and time-consuming to seek out *all* individuals in the populations of Germany or Great Britain, the researchers used **samples** to gather data. These studies, like many, have two variables of primary interest—a **response variable** and an **explanatory variable.** The response variable is whether or not a subject has cancer (eye or brain). The explanatory variable is the amount of cell phone use.

Recall

From Section 1.2, the **population** consists of *all* the subjects of interest. The **sample** is the subset of the population for whom the study collects data.

From the beginning of Chapter 3, the **response variable** measures the outcome of interest. Studies investigate how that outcome depends on the **explanatory variable**. ◀

Types of Studies: Experimental and Observational

Some studies, such as Study 3 in Example 1, perform an **experiment.** Subjects are assigned to experimental conditions, such as both cell phones are deactivated ("off" condition) or the right cell phone is activated ("on" condition), that we want to compare on the response outcome. These experimental conditions are called **treatments.** They correspond to different values of the explanatory variable.

Experimental Study

A researcher conducts an experimental study, or more simply, an **experiment,** by assigning subjects to certain experimental conditions and then observing outcomes on the response variable. The experimental conditions, which correspond to assigned values of the explanatory variable, are called **treatments.**

For example, Study 3 used 47 participants as the subjects. In this experimental study, each participant was assigned to both treatments. These are the categories of the explanatory variable, whether or not the cell phone is in the "off" condition or the "on" condition. The purpose of the experiment was to examine the association between this variable and the response variable—whether or not the amount of brain activity changes when the cell phone is in use.

An experiment assigns each subject to a treatment (or to both treatments as in Study 3) and then observes the response. By contrast, many studies merely *observe* the values on the response variable and the explanatory variable for the sampled subjects without doing anything to them. Such studies are called **observational studies.** They are **nonexperimental.**

Observational Study

In an **observational study**, the researcher observes values of the response variable and explanatory variables for the sampled subjects, without anything being done to the subjects (such as imposing a treatment).

In short, an **observational study** merely *observes* rather than *experiments* with the study subjects. An **experimental study** assigns to each subject a treatment and then observes the outcome on the response variable.

Experimental and observational studies

Example 2

Cell Phone Use

Picture the Scenario

Example 1 described three studies about whether relationships might exist between cell phone use and physiological activity in the human body. We've seen that Study 3 was an experiment. Studies 1 and 2 both observed the amount of cell phone use for cancer patients and for noncancer patients using a questionnaire.

Questions to Explore

a. Were Studies 1 and 2 experimental or observational studies?

b. How were Studies 1 and 2 fundamentally different from Study 3?

Think It Through

a. In Studies 1 and 2, information on the amount of cell phone use was gathered by giving a questionnaire to the sampled subjects. The subjects (people) decided how much they would use a cell phone and thus determined their amount of radiation exposure. The studies merely observed this exposure. No experiment was performed. So, Studies 1 and 2 were *observational* studies.

b. In Study 3, each subject was given both treatments (the "off" condition and the "on" condition). The researchers did not merely observe the subjects for the amount of brain activity but exposed the subjects to both treatments and determined which treatment each would receive first: some received the call during the first PET scan and some during the second scan. The researchers imposed the treatments on the subjects. So, Study 3 was an *experimental* study.

Insight

One reason that results of different medical studies sometimes disagree is that they are not the same *type* of study. Experimental studies are generally not directly comparable to observational studies. As we'll see, there are different types of both observational and experimental studies, and some are more trustworthy than others.

Try Exercises 4.1 and 4.3

Let's consider another study. From its description, we'll try to determine if it is an experiment or an observational study.

Example 3

Experiment or observational study

Drug Testing and Student Drug Use

Picture the Scenario

"Student Drug Testing Not Effective in Reducing Drug Use" was the headline in a news release from the University of Michigan. It reported results from a study of 76,000 students in 497 high schools and 225 middle schools nationwide.[1] Each student in the study filled out a questionnaire. One question asked if the student used drugs. The study found that drug use was similar in schools that tested for drugs and in schools that did not test for drugs. For instance, the table in the margin shows the conditional proportions on drug use for sampled twelfth graders from the two types of schools.

Conditional Proportions on Drug Use

	Drug Use		
Drug Tests?	**Yes**	**No**	*n*
Yes	0.37	0.63	5,653
No	0.36	0.64	17,437

Questions to Explore

a. What were the response and explanatory variables?

b. Was this an observational study or an experiment?

Think It Through

a. The study compared the percentage of students who used drugs in schools that tested for drugs and in schools that did not test for drugs. Whether the student used drugs was the response variable. The explanatory variable was whether the student's school tested for drugs. Both variables were categorical, with categories "yes" and "no." For each grade, the data were summarized in a contingency table, as shown in the margin for twelfth graders.

b. For each student, the study merely observed whether his or her school tested for drugs and whether he or she used drugs. So this was an observational study.

Insight

An experiment would have assigned schools to use or not use drug testing, rather than leaving the decision to the schools. For instance, the study could have randomly selected half the schools to perform drug testing and half not to perform it, and then a year later measured the student drug use in each school.

Try Exercise 4.7

As we will soon discuss, an experimental study gives the researcher more control over outside influences. This control can allow more accuracy in studying the association.

Advantage of Experiments Over Observational Studies

Recall

From Section 3.4, a **lurking variable** is a variable not observed in the study that influences the association between the response and explanatory variables due to its own association with each of those variables. ◀

In an observational study, lurking variables can affect the results. Study 1 in Example 1 found an association between cell phone use and eye cancer. However, there could be lifestyle, genetic, or health differences between the subjects with eye cancer and those without it, and between those who use cell phones a lot and those who do not. A lurking variable could affect the association. For example,

[1]Study by R. Yamaguchi et al., reported in *Journal of School Health*, vol. 73, pp. 159–164, 2003.

a possible lifestyle lurking variable is computer use. Perhaps those who use cell phones often also use computers frequently. Perhaps high exposure to computer screens increases the chance of eye cancer. In that case, the higher prevalence of eye cancer for heavier users of cell phones could be due to their higher use of computers, not their higher use of cell phones.

By contrast, an experiment reduces the potential for lurking variables to affect the results. Why? We'll see that with a type of "random" selection to determine which subjects receive each treatment, we are attempting to form and balance groups of subjects receiving the different treatments; that is, the groups have similar distributions on other variables, such as lifestyle, genetic, or health characteristics. For instance, suppose the researchers in cell phone Study 3 decided to have 24 subjects receive the "on" condition and 23 subjects receive the "off" condition instead of having each subject receive both treatments. The health of the subjects could be a factor in how the brain responds to the treatments and the amount of brain activity that appears on the scans. If we randomly determine which subjects receive the "on" condition and which do not, the two groups of subjects are expected to have similar distributions on health. One group will not be much healthier, on average, than the other group. When the groups are balanced on a lurking variable, there is no association between the lurking variable and the explanatory variable (for instance, between health of the subject and whether the subject receives the "on" condition). Then, the lurking variable will not affect how the explanatory variable is associated with the response variable. One treatment group will not tend to have more brain activity because of health differences, on the average.

In Study 3 described in Example 1, by using the same subjects for both treatments, the researchers were attempting to further control the variability from the subject's health. In this experimental study, it was important to randomize the order in which the subject received the treatment.

Establishing **cause and effect** is central to science. But it's not possible to establish cause and effect definitively with observational studies. There's always the possibility that some lurking variable could be responsible for the association. If people who make greater use of cell phones have a higher rate of eye cancer, it may be because of some variable that we failed to measure in our study, such as computer use. As we learned in Section 3.4, **association does not imply causation.**

Because it's easier to adjust for lurking variables in an experiment than in an observational study, *we can study the effect of an explanatory variable on a response variable more accurately with an experiment than with an observational study*. With an experiment, the researcher has control over which treatment each subject receives. If we find an association between the explanatory variable and the response variable, we can be more sure that we've discovered a causal relationship than if we merely find an association in an observational study. Consequently, the best method for determining causality is to conduct an experiment.

Determining Which Type of Study Is Possible

If experiments are preferable, why ever conduct an observational study? Why bother to measure whether cell phone usage was greater for those with cancer than for those not having it? Why not instead conduct an experiment, such as the following: Pick half the students from your school at random and tell them to use a cell phone each day for the next 50 years. Tell the other half of the student body not to ever use cell phones. Fifty years from now, analyze whether cancer was more common for those who used cell phones.

There are obvious difficulties with such an experiment:

- It's not ethical for a study to expose over a long period of time some of the subjects to something (such as cell phone radiation) that we suspect may be harmful.

- It's difficult in practice to make sure that the subjects behave as told. How can you monitor them to ensure that they adhere to their treatment assignment over the 50-year experimental period?
- Who wants to wait 50 years to get an answer?

For these reasons, medical experiments are often performed over a short time period or often with exposure to treatments that would not be out of the ordinary (such as a one-time exposure to cell phone activity). To measure the effects on people of longer exposure to potentially harmful treatments, often the experiment uses animals such as mice instead of people. Because inferences about human populations are more trustworthy when we use samples of human subjects than when we use samples of animal subjects, scientists often resort to observational studies. Then we are not deliberately placing people in harm's way, but observing people who have chosen on their own to take part in the activity of interest. We'll study methods for designing a good observational study in Section 4.4. This can yield useful information when an experiment is not practical.

Finally, another reason nonexperimental studies are common is that many questions of interest do not involve trying to assess causality. For instance, if we want to gauge the public's opinion on some issue, or if we want to conduct a marketing study to see how people rate a new product, it's completely adequate to use a study that samples (appropriately) the population of interest.

Using Data Already Available

Of course, you will not conduct a study every time you want to answer some question, such as whether cell phone use is dangerous. It's human nature to rely instead on already available data. The most readily available data come from your personal observations. Perhaps a friend recently diagnosed with brain cancer was a frequent user of cell phones. Is this strong evidence that frequent cell phone use increases the likelihood of getting brain cancer?

Informal observations of this type are called **anecdotal evidence.** Unfortunately, there is no way to tell if they are representative of what happens for an entire population. Sometimes you hear people give anecdotal evidence to attempt to disprove causal relationships. "My Uncle Geoffrey is 85 years old and he's smoked a pack of cigarettes a day for his entire adult life, yet he's as healthy as a horse." An association does not need to be perfect, however, to be causal. Not all people who smoke a pack of cigarettes each day will get lung cancer, but a much higher proportion of them will do so than people who are nonsmokers. Perhaps Uncle Geoffrey is lucky to be in good health, but that should not encourage you to smoke regularly.

Did You Know?

Examples of sources for available data:

- General Social Survey (GSS)
- www.fedstats.gov (for United States)
- www.statcan.ca (for Canada)
- www.inegi.gob.mx (for Mexico)
- www.statistics.gov.uk (for United Kingdom)
- www.abs.gov.au (for Australia) ◀

Instead of using anecdotal evidence to draw conclusions, you should rely on data from reputable research studies. You can find research results on topics of interest by entering keywords in Internet search engines. This search directs you to published results, such as in medical journals for medical studies.[2] Results from well-designed studies are more trustworthy than anecdotal evidence. Excellent sources of available data are listed in the margin.

The Census and Other Sample Surveys

The General Social Survey (GSS) is an example of a **sample survey.** It gathers information by interviewing a sample of subjects from the U.S. adult population to provide a snapshot of that population. The study in Example 3 on student drug use in schools also used a sample survey.

---

[2]Most surveys of research about cell phones suggest that there is no convincing evidence yet of adverse radiation effects (e.g., D. R. Cox, *J. Roy. Statist. Soc., Ser. A*, vol. 166, pp. 241–246, 2003). The primary danger appears to be people using cell phones while driving!

> **Sample Survey**
>
> A **sample survey** selects a sample of subjects from a population and collects data from them.

A sample survey is a type of nonexperimental study. The subjects provide data on the variables measured. There is no assignment of subjects to different treatments.

Most countries conduct a regular **census**. This is a survey that attempts to count the number of people in the population and to measure certain characteristics about them. It is different from most surveys, which sample only a small part of the entire population.

In Article 1, Section 2, the U.S. Constitution states that a complete counting of the U.S. population is to be done every 10 years. The first census was taken in 1790. It counted 3.9 million people, while today the U.S. population is estimated to be more than 300 million (308,745,538 by the 2010 U.S. census). Other than counting the population size, here are three key reasons for conducting the U.S. census:

- The Constitution mandates that seats in the House of Representatives be apportioned to states based on their portion of the population measured by the census. When the 1910 census was completed, Congress fixed the number of seats at 435. With each new census, states may gain or lose seats depending on how their population size compares with other states.
- Census data are used in the drawing of boundaries for electoral districts.
- Census data are used to determine the distribution of federal dollars to states and local communities.

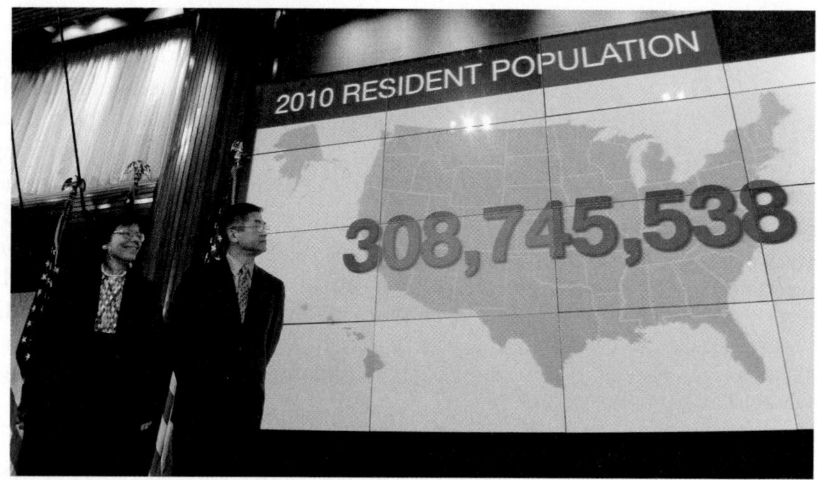

U.S. Commerce Secretary Gary Locke stands beside a screen showing the country's resident population during the 2010 Census presentation at the National Press Club in Washington.

Methods for taking the U.S. census have varied since horseback riders served as enumerators (the people who make the counts) in the first censuses. From the 1930s to the 1960s, women who were homemakers served as the enumerators. But as women increasingly took jobs outside the home, too few enumerators were available, so in 1970 the Census Bureau began to mail forms to households. The mail-back rate for the forms was 90%. However, this decreased to 74% for the 2010 census. The 26% who did not respond required follow-ups. The Census Bureau sends a reminder requesting that the form be returned. If this is unsuccessful, enumerators make personal visits to the household address. It should be noted that in the year 2000, census forms could be completed using the Internet.

Although it's the intention of a census to sample *everyone* in a population, in practice this is not possible. Some people do not have known addresses for the

Census Bureau to send a census form. Some people are homeless or transient. Because data are needed regularly on economic variables, such as the unemployment rate, the U.S. Census Bureau continually takes smaller samples of the population rather than relying solely on the complete census. An example is the monthly Current Population Survey of about 50,000 households.

It is usually more practical to take a sample rather than to try to measure everyone in a population. In fact, in the late 1990s, statisticians at the Census Bureau proposed using sampling methods to estimate numbers of people missed by the ordinary census. Their proposal became a political battleground between Republicans (who did not want sampling used) and Democrats (who supported sampling). In 1999, the Supreme Court ruled 5 to 4 that sampling could not be used to adjust the counts for apportioning seats in Congress but could be used to adjust counts for federal funding.

For a sample survey to be informative, it is important that the sample reflect the population well. As we'll discuss next, random selection—letting chance determine which subjects are in the sample—is the key to getting a good sample.

4.1 Practicing the Basics

4.1 **Cell phones** Consider the cell phone Study 3 described in Example 1.
TRY

a. Identify the response variable and the explanatory variable.

b. Was this an observational study or an experiment? Explain why.

4.2 **High blood pressure and binge drinking** Many studies have demonstrated that high blood pressure increases the risk of developing heart disease or having a stroke. It is also safe to say that the health risks associated with binge drinking far outweigh any benefits. A study published in *Heath Magazine* in 2010 suggested that a combination of the two could be a lethal mix. As part of the study that followed 6100 South Korean men aged 55 and over for two decades, men with high blood pressure who binge drank even occasionally had double the risk of dying from a stroke or heart attack when compared to teetotalers with normal blood pressure.

a. Is this an observational or experimental study?

b. Identify the explanatory and response variable(s).

c. Does the study prove that a combination of high blood pressure and binge drinking causes an increased risk of death by heart attack or stroke? Why or why not?

4.3 **Chocolate good for you?** A few studies published in 2007 claimed health benefits from eating dark chocolate or drinking cocoa. One study found much lower rates of death due to cancer among the Kuna Indians, who live in the San Blas islands of Panama and drink cocoa as their main beverage, compared to residents of mainland Panama (V. Bayard et al., *Int. J. Med. Sci.*, 2007; 4: 53–58).
TRY

a. Identify the response variable and the explanatory variable.

b. Was this study an observational study or an experimental study? Explain.

c. Did the design of this study take into account potential lurking variables? If so, explain how. If not, identify a lurking variable that could affect the results.

4.4 **Experiments versus observational studies** When either type of study is feasible, an experiment is usually preferred over an observational study. Explain why, using an example to illustrate. Also explain why it is not always possible for researchers to carry out a study in an experimental framework. Give an example of such a situation.

4.5 **School testing for drugs** Example 3 discussed a study comparing high schools that tested for drugs with high schools that did not test for drugs, finding similar levels of student drug use in each. State a potential lurking variable that could affect the results of such a study. Describe what the effect could be.

4.6 **Hormone therapy and heart disease** Since 1976 the Nurses' Health Study has followed more than 100,000 nurses. Every two years, the nurses fill out a questionnaire about their habits and their health. Results from this study indicated that postmenopausal women have a reduced risk of heart disease if they take a hormone replacement drug.

a. Suppose the hormone-replacement drug actually has no effect. Identify a potential lurking variable that could explain the results of the observational study. (*Hint*: Suppose that the women who took the drug tended to be more conscientious about their personal health than those who did not take it.)

b. Recently a randomized experiment called the Women's Health Initiative was conducted by the National Institutes of Health to see if hormone therapy is truly helpful. The study, planned to last for eight years, was stopped after five years when analyses showed that women who took hormones had 30% more heart attacks. This study suggested that rather than reducing the risk of heart attacks, hormone replacement drugs actually increase the risk.[3] How is it that two studies could reach such different conclusions? (For attempts to reconcile the studies, see a story by Gina Kolata in *The New York Times*, April 21, 2003.)

c. Explain why randomized experiments, when feasible, are preferable to observational studies.

4.7 Hairdressers at risk In a study by Swedish researchers (*Occupational and Environmental Medicine* 2002, 59:517–522), 2410 women who had worked as hairdressers and given birth to children were compared to 3462 women from the general population who had given birth. The hairdressers had a slightly higher percentage of infants with a birth defect.

TRY

a. Identify the response variable and the explanatory variable.

b. Is this study an observational study or an experiment? Explain.

c. Can we conclude that there's something connected with being a hairdresser that causes higher rates of birth defects? Explain.

4.8 Breast-cancer screening A study published in 2010 in the *New England Journal of Medicine* discusses a breast-cancer screening program that began in Norway in 1996 and was expanded geographically through 2005. Women in the study were offered mammography screening every two years. The goal of the study was to compare incidence-based rates of death from breast cancer across four groups:

1. Women who from 1996 through 2005 were living in countries with screening.
2. Women who from 1996 through 2005 were living in countries without screening.
3. A historical-comparison group who lived in screening countries from 1986 through 1995.
4. A historical-comparison group who lived in nonscreening countries from 1986 through 1995.

Data were analyzed for 40,075 women. Rates of death were reduced in the screening group as compared to the historical screening group, and also in the nonscreening group as compared to the historical nonscreening group.

a. Is this an observational or experimental study?

b. Identify the explanatory and response variable(s).

c. Does the study prove that being offered mammography screening causes a reduction in death rates associated with breast cancer? Why or why not?

[3]See article by H. N. Hodis et al., *New England Journal of Medicine*, August 7, 2003.

4.9 Experiment or observe? Explain whether an experiment or an observational study would be more appropriate to investigate the following:

a. Whether or not smoking has an effect on coronary heart disease

b. Whether or not higher SAT scores tend to be positively associated with higher college GPAs

c. Whether or not a special coupon attached to the outside of a catalog makes recipients more likely to order products from a mail-order company

4.10 Baseball under a full moon During a baseball game between the Boston Brouhahas and the Minnesota Meddlers, the broadcaster mentions that the away team has won "13 consecutive meetings between the two teams played on nights with a full moon."

a. Is the broadcaster's comment based on observational or experimental data?

b. The current game is being played in Boston. Should the Boston Brouhahas be concerned about the recent full moon trend?

4.11 Seat belt anecdote Andy once heard about a car crash victim who died because he was pinned in the wreckage by a seat belt he could not undo. As a result, Andy refuses to wear a seat belt when he rides in a car. How would you explain to Andy the fallacy behind relying on this anecdotal evidence?

4.12 Poker as a profession? Tony's mother is extremely proud that her son will graduate college in a few months. She expresses concern, however, when Tony tells her that following graduation, he plans to move to Las Vegas to become a professional poker player. He mentions that his friend Nick did so and is now earning more than a million dollars per year. Should Tony's anecdotal evidence about Nick soothe his mother's concern?

4.13 Census every 10 years? A nationwide census is conducted in the United States every 10 years.

a. Give at least two reasons why the United States takes a census only every 10 years.

b. What are reasons for taking the census at all?

c. The most commonly discussed characteristic learned from a census is the size of the population. However, other characteristics of the population are measured during each census. Using the Internet, report two such characteristics recorded during the 2010 U.S. census. (*Hint:* Visit the following Web site: 2010.census.gov/2010census)

4.2 Good and Poor Ways to Sample

The sample survey is a common type of nonexperimental study. The first step of a sample survey is to define the population targeted by the study. For instance, the Gallup organization (www.gallup.com) conducts a monthly survey of about 1000 adult Americans to report the percentage of those sampled who respond "approve" when asked, "Do you approve or disapprove of the way [the current president of the United States] is handling his job as president?" The population consists of all adults living in the United States.

Sampling Frame and Sampling Design

Once you've identified the population, the second step is to compile a list of subjects, so you can sample from it. This list is called the **sampling frame.**

> **Sampling Frame**
>
> The **sampling frame** is the list of subjects in the population from which the sample is taken.

Ideally, the sampling frame lists the entire population of interest. In practice, as in a census, it's usually hard to identify every subject in the population.

Suppose you plan to sample students at your school, asking students to complete a questionnaire about various issues. The population is all students at the school. One possible sampling frame is the student directory. Another one is a list of registered students.

Once you have a sampling frame, you need to specify a method for selecting subjects from it. The method used is called the **sampling design.** Here's one possible sampling design for sampling students at your school: You sample all students in your statistics class. Do you think your class is necessarily reflective of the entire student body? Do you have a representative mixture of freshmen through seniors, males and females, athletes and nonathletes, working and nonworking students, political party affiliations, and so on? With this sampling design, it's doubtful.

When you pick a sample merely by convenience, the results may not be representative of the population. Some response outcomes may occur much more frequently than in the population, and some may occur much less. Consider a survey question about the number of hours a week that you work at an outside job. If your class is primarily juniors and seniors, they may be more likely to work than freshmen or sophomores. The mean response in the sample may be much larger than in the overall student population. Information from the sample may be misleading.

Simple Random Sampling

You're more likely to obtain a representative sample if you let *chance*, rather than *convenience*, determine the sample. The sampling design should give each student an equal chance to be in the sample. It should also enable the data analyst to figure out how likely it is that descriptive statistics (such as sample means) fall close to corresponding values we'd like to make inferences about for the entire population. These are reasons for using **random sampling.**

You may have been part of a selection process in which your name was put in a box or hat along with many other names, and someone blindly picked a name for a prize. If the names were thoroughly mixed before the selection, this emulates a *random* type of sampling called a **simple random sample.**

Recall

As in Chapters 2 and 3, n denotes the *number* of observations in the sample, called the **sample size.** ◀

> **Simple Random Sample**
>
> A **simple random sample** of n subjects from a population is one in which each possible sample of that size has the same chance of being selected.

A *simple random sample* is often just called a **random sample.** The "simple" adjective distinguishes this type of sampling from more complex random sampling designs presented in Section 4.4.

Simple random samples

Example 4

Drawing Prize Winners

Picture the Scenario

A campus club decides to raise money for a local charity by selling tickets to a dinner banquet and raffle. The Athletic Department has donated two pairs of football season tickets as prizes. The group of 75 individuals who purchased tickets to the banquet comprises the population. The winners will be drawn randomly from everyone who purchased a ticket and is actually in attendance at the banquet. The group of individuals in attendance makes up the sampling frame. To select the two winners, the organizers choose a simple random sample of size $n = 2$ from the 60 individuals in attendance. Each individual at the banquet receives an entry with a number between 1 and 60. Duplicate entries numbered 1 to 60 are placed in a container and mixed, after which the two winning entries are chosen at random. Whoever holds the entries containing the same numbers as those drawn at random wins the football tickets.

Questions to Explore

a. What are the possible samples?

b. What is the chance that a particular sample of size 2 will be drawn?

c. Professor Shaffer is in attendance at the banquet and holds entry number 1. What is the chance that her entry will be chosen?

Think It Through

a. The entire collection of possible samples is too long to list, but we can visualize the possibilities by listing certain pairs:

(1,2), (1,3), (1,4),. . ., (1,58), (1,59), (1,60)
(2,3), (2,4),. . ., (2,58), (2,59), (2,60)

.

.

.

(57,58), (57,59), (57,60)
(58,59), (58,60)
(59,60)

b. The first row above contains 59 possible samples, the second row contains 58, and so forth. There are thus 59 + 58 + . . . + 2 + 1, which (using a calculator!) equals 1770 different possible samples. The process of randomly selecting two entries ensures that each possible sample of size 2 has an equal chance of occurring. Since there are 1770 possible samples, the chance of any one sample being selected is 1 out of 1770.

c. Looking at the top row in part a, we see that 1 shows up in 59 of the samples (all of the samples listed in the top row). So the chance that Professor Shaffer wins season tickets is 59 out of 1770.

Insight

The chance of 59/1770 for Professor Shaffer is the same as that of anyone else in attendance. In practice, especially with larger populations, it is difficult to mix the entrants' numbers in a container so that each sample truly has an equal chance of selection. There are better ways of selecting simple random samples, as we'll learn next.

Try Exercise 4.14

Did You Know?

The Random Numbers applet on the text CD can also be used to generate random numbers. ◀

Selecting a Simple Random Sample

What's a better way to take a simple random sample than blindly drawing slips of paper out of a hat? You first number the subjects in the sampling frame. You then generate a set of those numbers randomly. Finally, you sample the subjects whose numbers were generated.

You can generate numbers randomly with a **random number table,** with software, or with a statistical calculator. A random number table contains a sequence of digits such that any particular digit is equally likely to be any of the numbers $0, 1, 2, \ldots, 9$ and does not depend on the other numbers generated. If a particular digit is a 6, for instance, the next digit is just as likely to be a 6 (or a 0) as any other number. The numbers fluctuate according to no set pattern. Table 4.1 shows part of a random number table.

Table 4.1 A Portion of a Table of Random Numbers

Line/Col.	(1)	(2)	(3)	(4)	(5)	(6)	(7)	(8)
1	10480	15011	01536	02011	81647	91646	69179	14194
2	22368	46573	25595	85393	30995	89198	27982	53402
3	24130	48360	22527	97265	76393	64809	15179	24830
4	42167	93093	06243	61680	07856	16376	39440	53537
5	37570	39975	81837	16656	06121	91782	60468	81305

SUMMARY: Using Random Numbers to Select a Simple Random Sample

To select a simple random sample,

- Number the subjects in the sampling frame, using numbers of the same length (number of digits).
- Select numbers of that length from a table of random numbers or using software or a calculator with a random number generator.
- Include in the sample those subjects having numbers equal to the random numbers selected.

Simple random samples

Example 5

Auditing a School District

Picture the Scenario

Local school districts must be prepared for annual visits from state auditors whose job is to verify the actual dollar amount of the accounts within the school district and determine if the money is being spent appropriately. It is too time-consuming and expensive for auditors to review all accounts, so they typically review only some of them. So that a school district cannot anticipate which accounts will be reviewed, the auditors often take a simple random sample of the accounts.

Questions to Explore

a. How can the auditors use random numbers to select 10 accounts to audit in a school district that has 60 accounts?

b. Why is it important for the auditors not to use personal judgment in selecting the accounts to audit?

Think It Through

a. The sampling frame consists of the 60 accounts. We first number the accounts 01 through 60. In a random number table, we select two digits at a time until we have 10 unique two-digit numbers between 01 and 60. (We number the accounts 01 through 60 rather than 1 through 60 because we select two-digit numbers from the table.) Any pair of random digits has the same chance of selection. If digits 61–99 are selected, they are discarded.

Choose a random starting place in the table. We illustrate by starting with row 1, column 1 of Table 4.1. The random numbers in this row are

10480 15011 01536 02011 81647 91646 69179 14194.

The first 10 two-digit numbers are

10 48 01 50 11 01 53 60 20 11.

The first account chosen is the one numbered 10. The second chosen is the one numbered 48, and so forth. We observe that 01 and 11 occur twice. We discard the two repeats, and we need two more numbers between 01 and 60. The next 10 two-digit numbers are

81 64 79 16 46 69 17 91 41 94.

We skip the numbers 81, 64, 79 because no account in the sampling frame of 60 accounts has an assigned number that large. The last two accounts sampled are 16 and 46. In summary, the auditors should audit the accounts numbered 01, 10, 11, 16, 20, 46, 48, 50, 53, and 60.

b. By using simple random sampling, the auditors hold the school district responsible for all accounts. If the auditors personally chose the accounts to audit, they might select certain accounts each year. A school district would soon learn which accounts they need to have in order and which accounts can be given less attention.

Insight

Likewise, the Internal Revenue Service (IRS) randomly selects some tax returns for audit, so taxpayers cannot predict ahead of time whether they will be audited.

Try Exercise 4.15

In Practice Using Technology to Generate Random Numbers

You can obtain **random numbers** on the Internet (see *random.org*), the Random Numbers applet on the text CD, or using a random integer function with a statistical calculator or software to generate random numbers. You can instruct MINITAB to generate randomly any number of integers between a minimum of 01 and maximum of 60, as shown in Figure 4.1, or other ranges depending on the numbers you've assigned to subjects in the sampling frame.

Methods of Collecting Data in Sample Surveys

In Example 5, the subjects sampled were accounts. More commonly, they are people. For any survey, it is difficult to get a good sampling frame since a list of all adults in a population typically does not exist. So, it's necessary to pick a place where almost all people can be found—a person's place of residence.

Once we identify the desired sample for a survey, how do we contact the people to collect the data? The three most common methods are personal interview, telephone interview, and self-administered questionnaire.

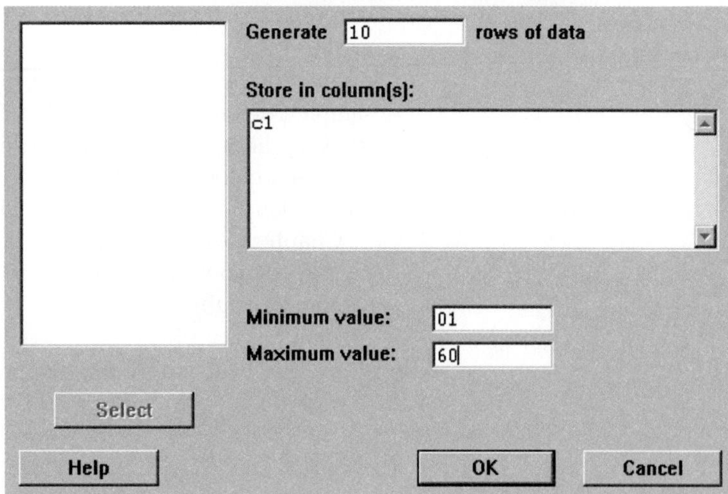

▲ **Figure 4.1 This MINITAB screen is used to randomly generate 10 integers between 01 and 60, inclusively.** The 10 random numbers will be stored in Column 1 of the MINITAB spreadsheet. Because some integers may be repeats, it is best to generate more than the desired sample size (for example, with the school audit, generate 20 integers instead of the desired 10).

Personal interview In a personal (face-to-face) interview, an interviewer asks prepared questions and records the subject's responses. An advantage is that subjects are more likely to participate. A disadvantage is the cost. Also, some subjects may not answer sensitive questions pertaining to opinions or lifestyle, such as alcohol and drug use, that they might answer on a printed questionnaire.

Telephone interview A telephone interview is like a personal interview but conducted over the phone. A main advantage is lower cost, since no travel is involved. A disadvantage is that the interview might have to be short. Subjects aren't as patient on the phone and may hang up before starting or completing the interview.

Self-administered questionnaire Subjects are requested to fill out a questionnaire mailed to them by post or e-mail. An advantage is that it is cheaper than a personal interview. A disadvantage is that more subjects may fail to participate.

So which method is most used? Most major national polls that survey regularly to measure public opinion use the telephone interview. An example is the polls conducted by the Gallup organization. The General Social Survey is an exception, using personal interviews for their questionnaire, which is quite long.

For telephone interviews, since the mid-1980s, many sample surveys have used **random digit dialing** to select households. The survey can then obtain a sample without having a sampling frame. In the United States, typically the area code and the 3-digit exchange are randomly selected from the list of all such codes and exchanges. Then, the last four digits are dialed randomly, and an adult is selected randomly within the household. Although this sampling design incorporates randomness, it is not a simple random sample because each sample is not equally likely to be chosen. (Do you see why not? Does everyone have a telephone, or exactly one telephone?)

Did You Know?

Besides Gallup, there are other polling groups that conduct national surveys, such as the Pew Research Center (www.people-press.org) and Zogby International (www.zogby.com). TV stations and newspapers commonly use such polling organizations to conduct polls. ◀

Accuracy of the Results from Surveys with Random Sampling

The most common use of data from sample surveys is to estimate population percentages. For instance, a Gallup poll recently reported that 30% of Americans worried that they might not be able to pay health-care costs during the next 12 months. How good is such a sample estimate of the population percentage? When you read results of surveys, you'll often see a statement such as, "The **margin of error** is plus or minus 3 percentage points." This means that it's very likely that the population percentage is no more than 3% lower or 3% higher than the reported sample percentage. So if Gallup reports that 30% worry about health-care costs, it's very likely that in the entire population, the percentage that worry about health-care costs is between about 27% and 33% (that is, within 3% of 30%). "Very likely" means that about 95 times out of 100 such statements are correct.

Chapters 7 and 8 will show details about margin of error and how to calculate it. For now, we'll use a rough approximation. When using a simple random sample of n subjects,

$$\text{approximate margin of error} = \frac{1}{\sqrt{n}} \times 100\%.$$

Margin of error

Example 6

Gallup Poll

Picture the Scenario

On April 20, 2010, one of the worst environmental disasters took place in the Gulf of Mexico. As a result of an explosion on an oil drilling platform, oil flowed freely into the Gulf of Mexico for nearly three months until it was finally capped on July 15, 2010. It is estimated that more than 200 million gallons of crude oil spilled, causing extensive damage to marine and wildlife habitats and crippling the Gulf's fishing and tourism industries. In response to the spill, many activists called for an end to deepwater drilling off the U.S. coast and for increased efforts to eliminate our dependence on oil.

Meanwhile, approximately nine months after the Gulf disaster, political turbulence gripped the Middle East, causing the price of gasoline in the United States to approach an all-time high. Between March 3 and 6, 2011, Gallup's annual environmental survey[4] reported that 60% of Americans favored offshore drilling as a means to reduce U.S. dependence on foreign oil, 37% opposed offshore drilling, and the remaining 3% had no opinion. The poll was based on interviews conducted with a random sample of 1021 adults, aged 18 and older, living in the continental United States, selected using random digit dialing.

Questions to Explore

a. Find an approximate margin of error for these results reported in the environmental survey report.

b. How is the margin of error interpreted?

Think It Through

a. The sample size was $n = 1021$ U.S. adults. The Gallup poll was a random sample but not a simple random sample. It used random digit dialing, which can give results nearly as accurate as with a simple random sample. We estimate the margin of error as approximately

$$\frac{1}{\sqrt{n}} \times 100\% = \frac{1}{\sqrt{1021}} \times 100\% = 0.03 \times 100\% = 3\%.$$

b. Gallup reported that 60% of Americans support offshore drilling. This percentage refers to the sample of 1021 U.S. adults. The margin of

[4]www.gallup.com/poll/146615/Oil-Drilling-Gains-Favor-Americans.aspx

error of 3% suggests that in the population of adult Americans, it's likely that between about 57% and 63% support offshore drilling.

Insight

You may be surprised or skeptical that a sample of only 1021 people out of a huge population (such as 200 million adult Americans) can provide such a precise inference. That's the power of random sampling. We'll see why this works in Chapter 7. The more precise formula shown there will give a somewhat smaller margin of error value when the sample percentage is far from 50%.

Try Exercises 4.20 and 4.21 and try using the Sample from a Population applet as described in Exercise 4.115

Sources of Potential Bias in Sample Surveys

A variety of problems can cause responses from a sample to favor some parts of the population over others. Then, results from the sample are not representative of the population and are said to exhibit **bias.**

In Words

There is **bias** if, because of the way the study was designed, certain outcomes will occur more often in the sample than they do in the population. For example, consider the population of adults in your home town. The results of an opinion survey asking about raising the sales tax to support public schools may be biased in the direction of approve if you sample only educators or biased in the direction of disapprove if you sample only business owners.

Sampling bias Bias may result from the sampling method. The main way this occurs is if the sample is not random. Another way it can occur is due to **undercoverage**—having a sampling frame that lacks representation from parts of the population. A telephone survey will not reach homeless people, prison inmates, or people who don't have telephones. If its sampling frame consists of the names in a telephone directory, it will not reach those with unlisted numbers. Most major polling agencies use randomized dialing. Responses by those who are not in the sampling frame might be quite different from those who are in the frame. Bias resulting from the sampling method, such as nonrandom sampling or undercoverage, is called **sampling bias.**

Nonresponse bias A second type of bias occurs when some sampled subjects cannot be reached or refuse to participate. This is called **nonresponse bias.** The subjects who are willing to participate may be different from the overall sample in some way, perhaps having strong emotional convictions about the issues being surveyed. Even those who do participate may not respond to some questions, resulting in nonresponse bias due to **missing data.** All major surveys suffer from some nonresponse bias. The General Social Survey has a nonresponse rate of about 20–30%. The nonresponse rate is much higher for many telephone surveys. By contrast, government-conducted surveys often have lower nonresponse rates. The Current Population Survey, which measures factors related to employment, has a nonresponse rate of only about 7%. To reduce nonresponse bias, investigators try to make follow-up contact with the subjects who do not return questionnaires.

Results have dubious worth when there is substantial nonresponse. For instance, in her best-selling and controversial book *Women and Love* (1987), author and feminist Shere Hite presented results of a survey of adult women in the United States. One of her conclusions was that 70% of women who had been married at least five years have extramarital affairs. She based this conclusion on responses to questionnaires returned from a sample of 4500 women. This sounds impressively large. However, the questionnaire was mailed to about 100,000 women. We cannot know whether this sample of 4.5% of the women who responded is representative of the 100,000 who received the questionnaire, much less the entire population of adult American women. In fact, Hite was criticized for her research methodology of using nonrandom samples and the large nonresponse rate.

Response bias A third type of potential bias is in the actual responses made. This is called **response bias.** An interviewer might ask the questions in a leading way, such that subjects are more likely to respond a certain way. Or, subjects may lie because they think their response is socially unacceptable, or they may give the response that they think the interviewer prefers.

If you design an interview or questionnaire, you should strive to construct questions that are clear and understandable. *Avoid questions that are confusing, long, or leading.* The wording of a question can greatly affect the responses. A Roper poll was designed to determine the percentage of Americans who express some doubt that the Holocaust occurred in World War II. In response to the question, "Does it seem possible or does it seem impossible to you that the Nazi extermination of the Jews never happened?" 22% said it was possible the Holocaust never happened. The Roper organization later admitted that the question was worded in a confusing manner. When they asked, "Does it seem possible to you that the Nazi extermination of the Jews never happened, or do you feel certain that it happened?" only 1% said it was possible it never happened.[5]

In Practice Be Alert for Potential Bias

A sound sampling design can prevent sampling bias. It cannot prevent nonresponse bias and response bias. With any sample survey, carefully scrutinize the results. Look for information about how the sample was selected, how large the sample size was, the nonresponse rate, how the questions were worded, and who sponsored the study. The less you know about these details, the less you should trust the results.

Even the order in which questions are asked can dramatically influence results. One study[6] asked, during the Cold War, "Do you think the U.S. should let Russian newspaper reporters come here and send back whatever they want?" and "Do you think Russia should let American newspaper reporters come in and send back whatever they want?" For the first question, the percentage of yes responses was 36% when it was asked first and 73% when it was asked second.

SUMMARY: Types of Bias in Sample Surveys

- **Sampling bias** occurs from using nonrandom samples or having undercoverage.
- **Nonresponse bias** occurs when some sampled subjects cannot be reached or refuse to participate or fail to answer some questions.
- **Response bias** occurs when the subject gives an incorrect response (perhaps lying) or the way the interviewer asks the questions (or wording of a question in print) is confusing or misleading.

Poor Ways to Sample

How the sample is obtained can also result in bias. Two sampling methods that may be necessary but are not ideal are a convenience sample and a volunteer sample.

Convenience samples Have you even been stopped on the street or at a shopping mall to participate in a survey? Such a survey is not a random sample but rather a **convenience sample.** It is easy for the interviewer to obtain data relatively cheaply. But the sample may poorly represent the population. Biases may result because of the time and location of the interview and the judgment of the interviewer about whom to interview. For example, working people might be underrepresented if the interviews are conducted on workdays between 9 A.M. and 5 P.M. Poor people may be underrepresented if the interviewer conducts interviews at an upscale shopping mall.

Volunteer samples Have you ever answered a survey you've seen posted on the Internet, such as at the home page for a news organization? A sample of this type, called a **volunteer sample,** is the most common type of convenience sample. As the name implies, subjects *volunteer* to be in the sample. One segment

[5] *Newsweek*, July 25, 1994.

[6] Described in Crossen (1994).

of the population may be more likely to volunteer than other segments because they have stronger opinions about the issue or are more likely to visit that Internet site. This results in sampling bias. For instance, a survey by the Pew Research Center (January 6, 2003) estimated that 46% of Republicans said they like to register their opinions in online surveys, compared with only 28% of Democrats. (Why? One possible lurking variable is personal income: On average, Republicans are wealthier than Democrats, and wealthier people are more likely to have access to and use the Internet.) Thus, results of online surveys may be more weighted in the direction of Republicans' beliefs than the general population.

Convenience samples are not ideal. Sometimes, however, they are necessary, both in observational studies and in experiments. This is often true in medical studies. Suppose we want to investigate how well a new drug performs compared to a standard drug, for subjects who suffer from high blood pressure. We're not going to find a sampling frame of all who suffer from high blood pressure and take a simple random sample of them. We may, however, be able to sample such subjects at medical centers or using volunteers. Even then, randomization should be used wherever possible. For the study patients, we can randomly select who is given the new drug and who is given the standard one.

Some samples are poor not only because of their convenience but also because they use an inappropriate sampling frame. An example of this was a poll in 1936 to predict the result of a presidential election.

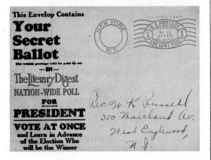

Example 7

A poor sample

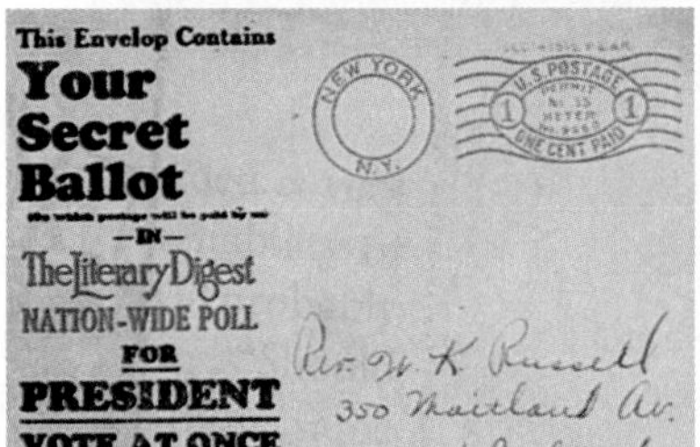

The *Literary Digest* Poll

Picture the Scenario

The *Literary Digest* magazine conducted a poll to predict the result of the 1936 presidential election between Franklin Roosevelt (Democrat and incumbent) and Alf Landon (Republican). At the time, the magazine's polls were famous because they had correctly predicted three successive elections. In 1936, *Literary Digest* mailed questionnaires to 10 million people and asked how they planned to vote. The sampling frame was constructed from telephone directories, country club memberships, and automobile registrations. Approximately 2.3 million people returned the questionnaire. The *Digest* predicted that Landon would win, getting 57% of the vote. Instead, Landon actually got only 36%, and Roosevelt won in a landslide.

Questions to Explore

a. What was the population?

b. How could the *Literary Digest* poll make such a large error, especially with such a huge sample size? What type of bias could have occurred in this poll?

Think It Through

a. The population was all registered voters in the United States in 1936.

b. This survey had two severe problems:

- Sampling bias due to undercoverage of the sampling frame and a nonrandom sample: In 1936, the United States was in the Great Depression. Those who had cars and country club memberships and thus received questionnaires tended to be wealthy. The wealthy tended to be primarily Republican, the political party of Landon.

Many potential voters were not on the lists used for the sampling frame. There was also no guarantee that a subject in the sampling frame was a registered voter.
- Nonresponse bias: Of the 10 million people who received questionnaires, 7.7 million did not respond. As might be expected, those individuals who were unhappy with the incumbent (Roosevelt) were more likely to respond.

Insight

For this same election, a pollster who was getting his new polling agency off the ground surveyed 50,000 people and predicted that Roosevelt would win. Who was this pollster? George Gallup. The Gallup organization is still with us today. However, the *Literary Digest* went out of business soon after the 1936 election.[7]

Try Exercise 4.27

A Large Sample Size Does Not Guarantee an Unbiased Sample

Many people think that as long as the sample size is large, it doesn't matter how the sample was selected. This is incorrect, as illustrated by the *Literary Digest* poll. A sample size of 2.3 million did not prevent poor results. The sample was not representative of the population, and the sample percentage of 57% who said they would vote for Landon was far from the actual population percentage of 36% who voted for him. Many Internet surveys have thousands of respondents, but a volunteer sample of thousands is not as good as a random sample, even if that random sample is much smaller. *We're almost always better off with a simple random sample of 100 people than with a volunteer sample of thousands of people.*

SUMMARY: Key Parts of a Sample Survey

- Identify the **population** of all the subjects of interest.
- Construct a **sampling frame,** which attempts to list all the subjects in the population.
- Use a **random sampling design**, implemented using random numbers, to select n subjects from the sampling frame.
- Be cautious about **sampling bias** due to nonrandom samples (such as volunteer samples) and sample undercoverage, **response bias** from subjects not giving their true response or from poorly worded questions, and **nonresponse bias** from refusal of subjects to participate.

In Section 4.1 we learned that *experimental* studies are preferable to *nonexperimental* studies but are not always possible. Some types of nonexperimental studies have fewer potential pitfalls than others. For a sample survey with random sampling, we can make inferences about the population of interest. By contrast, with a study using a convenience sample, results apply only to those subjects actually observed. For this reason, some researchers use the term *observational study* to refer only to studies that use available subjects (such as a convenience sample) and not to sample surveys that randomly select their sample.

[7]Bryson, M. C. (1976), *American Statistician*, vol. 30, pp. 184–185.

4.2 Practicing the Basics

4.14 Choosing officers A campus club consists of five officers: president (P), vice president (V), secretary (S), treasurer (T), and activity coordinator (A). The club can select two officers to travel to New Orleans for a conference; for fairness, they decide to make the selection at random. In essence, they are choosing a simple random sample of size $n = 2$.
TRY

a. What are the possible samples of two officers?

b. What is the chance that a particular sample of size 2 will be drawn?

c. What is the chance that the activity coordinator will be chosen?

4.15 Simple random sample of students In Example 4, a random drawing was held to select the winners of the football tickets. Organizers randomly chose two numbers from a collection of slips of paper numbered 1 through 60. Using the Table of Random Digits, begin the selection process at line 10. Which individuals did you choose? Repeat the procedure, this time beginning at line 18. (*Note:* In practice, a statistician would use a random starting place in the table of random numbers. A row was specified for convenience so everyone would have the same random numbers.)
TRY

4.16 Random numbers applet Use the Random Numbers applet on the text CD or use software to (a) answer the previous exercise and (b) select a simple random sample of three students out of a class of 500 students.

4.17 Auditing accounts—applet Use the Random Numbers applet on the text CD to select 10 of the 60 school district accounts described in Example 5. Explain how you did this, and identify the accounts to be sampled.

4.18 Auditing accounts—software Use either software or a statistical calculator to select 10 of the 60 school district accounts described in Example 5. Explain how you did this, and identify the accounts to be sampled.

4.19 Sampling from a directory A local telephone directory has 50,000 names, 100 per page for 500 pages. Explaining how you found and used random numbers, select 10 numbers to identify subjects for a simple random sample of 10 names.

4.20 Comparing polls The following table shows the result of the 2008 presidential election along with the vote predicted by several organizations in the days before the election. The sample sizes were typically about 1000 to 2000 people. The percentages for each poll do not sum to 100 because of voters who indicated they were undecided or preferred another candidate.
TRY

a. Treating the sample sizes as 1000 each, find the approximate margin of error.

b. Do most of the predictions fall within the margin of error of the actual vote percentages? Considering the relative sizes of the sample and the population and the undecided factor, would you say that these polls had good accuracy?

Predicted Vote

Poll	Obama	McCain
Gallup/USA Today	55	44
Harris	52	44
ABC/Wash Post	53	44
CBS	51	42
NBC/WSJ	51	43
Pew Research	52	46
Actual vote	52.7	46.0

Source: www.ncpp.org/files/08FNLncppNatlPolls_010809.pdf. Copyright 2006 by the National Council on Public Polls.

4.21 Margin of error and n The Gallup poll in Example 6 reported that during March 2011, 60% of Americans favored offshore drilling as a means of reducing U.S. dependence on foreign oil. The poll was based on the responses of $n = 1021$ individuals, and resulted in a margin of error of approximately 3%. Find the approximate margin of error had the poll been based on a sample of size (a) $n = 100$, (b) $n = 400$, and (c) $n = 1600$. Explain how the margin of error changes as n increases.
TRY

4.22 Bias due to perceived race A political scientist at the University of Chicago studied the effect of the race of the interviewer.[8] Following a phone interview, respondents were asked whether they thought the interviewer was black or white (all were actually black). Perceiving a white interviewer resulted in more conservative opinions. For example, 14% agreed that "American society is fair to everyone" when they thought the interviewer was black, but 31% agreed to this statement when posed by an interviewer that the respondent thought was white. Which type of bias does this illustrate: Sampling bias, nonresponse bias, or response bias? Explain.

4.23 Confederates Some southern states in the United States have wrestled with the issue of a state flag that is sensitive to African Americans and not divisive. Suppose a survey asks, "Do you oppose the present state flag that contains the Confederate symbol, a symbol of past slavery in the South and a flag supported by extremist groups?"

a. Explain why this is an example of a leading question.

b. Explain why a better way to ask this question would be, "Do you favor or oppose the current state flag containing the Confederate symbol?"

4.24 Instructor ratings The Web site www.ratemyprofessors.com provides students an opportunity to view ratings for instructors at their universities. A group of students planning to register for a statistics course in the upcoming semester are trying to identify the instructors who receive the highest ratings on the site. One student decides to register for Professor Smith's course because she has the best ratings of all statistics instructors. Another student comments:

a. The Web site ratings are unreliable because the ratings are from students who voluntarily visit the site to rate their instructors.

[8]Study by Lynn Sanders, as reported by the *Washington Post*, June 26, 1995.

b. To obtain reliable information about Professor Smith, they would need to take a simple random sample of the 78 ratings left by students on the site, and compile new overall ratings based on those in the random sample.

Which, if either, of the student's comments are valid?

4.25 Job market for MBA students A February 2, 2003, *Atlanta Journal Constitution* article about the bleak job market for graduating MBA students described an opinion survey conducted by a graduate student at a major state university. The student polled 1500 executive recruiters, asking their opinions on the industries most likely to hire. He received back questionnaires from 97 recruiters, of whom 54 indicated that health care was the industry most likely to see job growth.

a. What is the population for this survey?

b. What was the intended sample size? What was the sample size actually observed? What was the percentage of nonresponse?

c. Describe two potential sources of bias with this survey.

4.26 Gun control More than 75% of Americans answer yes when asked, "Do you favor cracking down against illegal gun sales?" but more than 75% say no when asked, "Would you favor a law giving police the power to decide who may own a firearm?"

a. Which statistic would someone who opposes gun control prefer to quote?

b. Explain what is wrong with the wording of each of these statements.

4.27 Stock market associated with poor mental health An Internet survey of 545 Hong Kong residents suggested that close daily monitoring of volatile financial affairs may not be good for your mental health (*J. Social and Clinical Psychology* 2002: 21: 116–128). Subjects who felt that their financial future was out of control had the poorest overall mental health, whereas those who felt in control of their financial future had the best mental health.

TRY

a. What is the population of interest for this survey?

b. Describe why this is an observational study.

c. Briefly discuss the potential problems with the sampling method used and how these problems could affect the survey results.

4.28 "What rots beneath" This was the headline of a *New York Times* article (May 19, 2003) about the Gowanus Canal in Brooklyn, which had become infamous for its contamination from sewage and industrial waste. Scientists and Army Corps of Engineers technicians used augurs, drills, and a split-spoon, which sucks up the muck of the canal bottom, to analyze what is "living" in the canal.

a. Describe the population under study.

b. Explain why a census is not practical for this study. What advantages does sampling offer?

4.29 Teens buying alcohol over Internet In August 2006, a trade group for liquor retailers put out a press release with the headline, "Millions of Kids Buy Internet Alcohol, Landmark Survey Reveals." Further details revealed that in an Internet survey of 1001 teenagers, 2.1% reported that they had bought alcohol online. In such a study, explain how there could be

a. Sampling bias. (*Hint:* Are all teenagers equally likely to respond to the survey?)

b. Nonresponse bias, if some teenagers refuse to participate.

c. Response bias, if some teenagers who participate are not truthful.

4.30 Online dating A story titled "Personals, Sex Sites Changing the Rules of Love" at www.msnbc.msn.com reported results of a study about online dating by the MSNBC network. The study used online responses of 15,246 people. Of those who responded, three fourths were men and about two thirds had at least a bachelor's degree. One reported finding was that "29% of men go online intending to cheat." Identify the potential bias in this study that results from

a. Sampling bias due to undercoverage

b. Sampling bias due to the sampling design

c. Response bias

4.31 Identify the bias A newspaper designs a survey to estimate the proportion of the population willing to invest in the stock market. It takes a list of the 1000 people who have subscribed to the paper the longest and sends each of them a questionnaire that asks, "Given the extremely volatile performance of the stock market as of late, are you willing to invest in stocks to save for retirement?" After analyzing result from the 50 people who reply, they report that only 10% of the local citizens are willing to invest in stocks for retirement. Identify the bias that results from the following:

a. Sampling bias due to undercoverage

b. Sampling bias due to the sampling design

c. Nonresponse bias

d. Response bias due to the way the question was asked

4.32 Types of bias Give an example of a survey that would suffer from

a. Sampling bias due to the sampling design

b. Sampling bias due to undercoverage

c. Response bias

d. Nonresponse bias

4.3 Good and Poor Ways to Experiment

Just as there are good and poor ways to gather a sample in an observational survey, there are good and poor ways to conduct an experiment. First, let's recall the definition of an experimental study from Section 4.1: We assign each subject to an experimental condition, called a **treatment.** We then observe the outcome on the response variable. The goal of the experiment is to investigate the association—how the treatment affects the response. An advantage of an experimental study over a nonexperimental study is that it provides stronger evidence for causation.

In an experiment, subjects are often referred to as **experimental units.** This name emphasizes that the objects measured need not be human beings. They could, for example, be schools, stores, mice, or computer chips.

The Elements of a Good Experiment

Let's consider another example to help us learn what makes a good experiment. It is common knowledge that smoking is a difficult habit to break. Studies have reported that regardless of what smokers do to quit, most relapse within a year. Some scientists have suggested that smokers are less likely to relapse if they take an antidepressant regularly after they quit. How can you design an experiment to study whether antidepressants help smokers to quit?

For this type of study, as in most medical experiments, it is not feasible to randomly sample the population (all smokers who would like to quit). We need to use a convenience sample. For instance, a medical center might advertise to attract volunteers from the smokers who would like to quit.

Control comparison group Suppose you have 400 volunteers who would like to quit smoking. You could ask them to quit, starting today. You could have each start taking an antidepressant, and then a year from now check how many have relapsed. Perhaps 42% of them would relapse. But this is not enough information.[9] You need to be able to compare this result to the percentage who would relapse if they were *not* taking the antidepressant.

An experiment normally has a primary treatment of interest, such as receiving an antidepressant. But it should also have a second treatment for comparison to help you analyze the effectiveness of the primary treatment. So the volunteers should be split into two groups: One group receives the antidepressant, and the other group does not. You could give the second group a pill that looks like the antidepressant but that does not have any active ingredient—a placebo. This second group, using the placebo treatment, is called the **control group.** After subjects are assigned to the treatments and observed for a certain period of time, the relapse rates are compared.

Why bother to give the placebo to the control group, if the pill doesn't contain any active ingredient? This is partly so that the two treatments appear identical to the subjects. (As we'll discuss, subjects should not know which treatment they are receiving.) This is also because people who take a placebo tend to respond better than those who receive nothing, perhaps for psychological reasons. This is called the **placebo effect.** For instance, of the subjects not receiving the antidepressant, perhaps 75% would relapse within a year, but if they received a placebo pill perhaps 55% of them would relapse. If the relapse rate was 42% for subjects who received antidepressants, then in comparing relapse rates for the antidepressant and control groups it makes a big difference how you define the control group.

---

[9]Chapter 12 explains another reason a control group is needed. The regression effect implies that, over time, poor subjects tend to improve and good subjects tend to get worse, in relative terms.

In some experiments, *a control group may receive an existing treatment rather than a placebo.* For instance, a smoking cessation study might analyze whether an antidepressant works better than a nicotine patch in helping smokers to quit. It may not be necessary to include a placebo group if the nicotine patch has already been shown in previous studies to be more effective than a placebo. Or the experiment could compare all three treatments: antidepressant, nicotine patch, and a placebo.

Randomization In a smoking cessation experiment, how should the 400 study subjects be assigned to the treatment groups? Should you personally decide which treatment each subject receives? This could result in bias. If you are conducting the study to show that the antidepressant is effective, you might consciously or subconsciously place smokers you believe will be more likely to succeed into the group that receives the antidepressant.

It is better to use **randomization** to assign the subjects: Randomly assign 200 of the 400 subjects to receive the antidepressant and the other 200 subjects to form the control group. Randomization helps to prevent bias from one treatment group tending to be different from the other in some way, such as having better health or being younger. In using randomization, we attempt to *balance the treatment groups* by making them similar with respect to their distribution on potential lurking variables. This enables us to attribute any difference in their relapse rates to the treatments they are using, not to lurking variables or to researcher bias.

You might think you can do better than randomization by using your own judgment to assign subjects to the treatment groups. For instance, when you identify a potential lurking variable, you could try to balance the groups according to its values. If age is that variable, every time you put someone of a particular age in one treatment group, you could put someone of the same age in the other treatment group. There are two problems with this: (1) *many* variables are likely to be important, and it is difficult to balance groups on *all* of them at once, and (2), you may not have thought of other relevant lurking variables. Even if you can balance the groups on the variables you identified, the groups could be unbalanced on these other variables, causing the overall results to be biased in favor of one treatment.

You can feel more confident about the worthiness of new research findings if they come from a randomized experiment with a control group rather than from an experiment without a control group, from an experiment that did not use randomization to assign subjects to the treatments, or from an observational study. An analysis[10] of published medical studies about treatments for heart attacks indicated that the new therapy provided improved treatment 58% of the time in studies without randomization and control groups but only 9% of the time in studies having randomization and control groups. Although this does not prove anything (as this analysis itself was an observational study), it does suggest that studies conducted without randomization or without other ways to reduce bias may produce results that tend to be overly optimistic.

SUMMARY: The Role of Randomization

Use randomization for assigning subjects to the treatments

- To eliminate bias that may result if you (the researchers) assign the subjects
- To balance the groups on variables that you know affect the response
- To balance the groups on lurking variables that may be unknown to you

[10]See Crossen (1994), p. 168.

Blinding the study It is important that the treatment groups be treated as equally as possible. Ideally, the subjects are **blind** to the treatment to which they are assigned. In the smoking cessation study, the subjects should not know whether they are taking the antidepressant or a placebo. Whoever has contact with the subjects during the experiment, including the data collectors who record the subjects' response outcomes, should also be blind to the treatment information. Otherwise they could intentionally or unintentionally provide extra support to one group. When neither the subject nor those having contact with the subject know the treatment assignment, the study is called **double-blind.** That's ideal.

Example 8

Study design

Antidepressants for Quitting Smoking

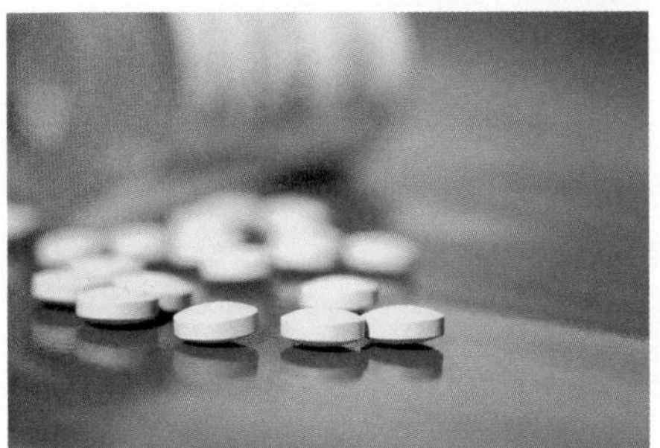

Picture the Scenario

To investigate whether antidepressants help people quit smoking, one study[11] used 429 men and women who were 18 or older and had smoked 15 cigarettes or more per day for the previous year. The subjects were highly motivated to quit and in good health. They were assigned to one of two groups: One group took 300 mg daily of an antidepressant that has the brand name bupropion. The other group did not take an antidepressant. At the end of a year, the study observed whether each subject had successfully abstained from smoking or had relapsed.

Questions to Explore

a. Identify the response and explanatory variables, the treatments, and the experimental units.

b. How should the researchers assign the subjects to the two treatment groups?

c. Without knowing more about this study, what would you identify as a potential problem with the study design?

Think It Through

a. This experiment has

Response variable: Whether the subject abstains from smoking for one year

Explanatory variable: Whether the subject received bupropion (yes or no)

Treatments: bupropion, no bupropion

Experimental units: The 429 volunteers who are the study subjects

b. The researchers should randomize to assign subjects to the two treatments. They could use random numbers to randomly assign half (215) of the subjects to form the group that uses bupropion and 214 subjects to the group who does not receive bupropion (the control group). The procedure would be as follows:

- Number the study subjects from 001 to 429.
- Pick a three-digit random number between 001 and 429. If the number is 392, then the subject numbered 392 is put in the bupropion group.

[11] *Annals of Internal Medicine* 2001; 135:423–433.

- Continue to pick three-digit numbers until you've picked 215 distinct values between 001 and 429. This determines the 215 subjects who will receive bupropion. The remaining 214 subjects will form the control group.

c. The description of the experiment in Picture the Scenario did not say whether the subjects who did *not* receive bupropion were given a placebo or whether the study was blinded. If not, these would be potential sources of bias.

Insight

In the actual reported study, the subjects were randomized to receive bupropion or a placebo for 45 weeks. The study *was* double-blinded. This experimental study was well designed. At the end of one year, 55.1% of the subjects receiving bupropion were not smoking, compared with 42.3% in the placebo group. After 18 months, 47.7% of the bupropion subjects were not smoking compared to 37.7% for the placebo subjects. However, after two years, the percentage of nonsmokers was similar for the two groups, 41.6% versus 40%.

Try Exercise 4.34

Did You Know?

A randomized experiment comparing medical treatments is often referred to as a **clinical trial**. ◀

In Practice The Randomized Experiment in Medicine

In medicine, the randomized experiment (clinical trial) has become the gold standard for evaluating new medical treatments. The Cochrane Collaboration (www.cochrane.org) is an organization devoted to synthesizing evidence from medical studies all over the world. According to this organization, there have been hundreds of thousands of randomized experiments comparing medical treatments. In most countries, a company cannot get a new drug approved for sale unless it has been tested in a randomized experiment.

Source: Background information from S. Senn (2003), *Dicing with Death.* Cambridge University Press, p. 68.

Sample Size and Statistical Significance

Now suppose you've conducted an experiment, and it's a year later. You find that 44% of the subjects taking bupropion have relapsed, whereas 55% of the control group relapsed. Can you conclude that bupropion was effective in helping smokers quit?

Not quite yet. You must convince yourself that this difference between 44% and 55% cannot be explained by the variation that occurs naturally just by chance. Even if the effect of bupropion is no different from the effect of placebo, the sample relapse rates would not be *exactly* the same for the two groups. Just by ordinary chance, in the random assignment of subjects to treatment groups, on average, one group may be slightly more committed to quitting smoking than the other group.

Recall

For simple random sampling, the margin of error is approximately $\frac{1}{\sqrt{n}} \times 100\%$, which is 10% for $n = 100$, 7% for $n = 200$, and 3% for $n = 1000$. ◀

In a randomized experiment, the variation that could be expected to occur just by chance alone is roughly like the margin of error with simple random sampling. So if a treatment has $n = 215$ observations, this is about $(1/\sqrt{215}) \times 100\%$ or about 7% for a percentage. If the population percentage of people who relapse is 50%, the sample percentage of 215 subjects who relapse is very likely to fall between 43% and 57% (that is, within 7% of 50%). The difference in a study between 44% relapsing and 55% relapsing could be explained by the ordinary variation expected for this size of sample.

By contrast, if each treatment had $n = 1000$ observations, the ordinary variation we'd expect due to chance is only about 3% for each percentage. Then, the difference between 44% and 55% could not be explained by ordinary variation: There is no plausible common population percentage for the two treatments such that 44% and 55% are both within 3% of its value. Bupropion would truly seem to be better than placebo. The difference expected due to ordinary variation is smaller with larger samples. You can be more confident that sample results reflect a true effect when the sample size is large than when it is small. Obviously, we cannot learn much by using only $n = 1$ subject for each treatment. The process of assigning *several* experimental units to each treatment is called **replication.**

When the difference between the results for the two treatments is so large that it would be rare to see such a difference by ordinary random variation, we say that the results are **statistically significant.** For example, the difference between 55% relapse and 44% relapse when $n = 1000$ for each group is statistically significant. We can then conclude that the observed difference is likely due to the effect of the treatments, rather than merely due to ordinary random variation. Thus, in the population, we infer that the response truly depends on the treatment. How to determine this and how it depends on the sample size are topics we'll study in later chapters.[12] For now, suffice it to say the larger the sample size, the better.

Caution

The term *replication* carries two meanings in experimental studies: (1) assigning many experimental units to a treatment and (2) repeating similar experiments. Carefully communicate the context in which you are using the term *replication.* ◀

When a study has statistically significant results, it's still possible that the observed effect was merely due to chance. Even if the treatments are identical, one may seem better just because, by sheer luck, random assignment gave it many subjects who tend to respond better, perhaps being healthier on average. For this reason, another type of replication is also important. You can feel more confident if other researchers perform similar experiments and get similar results.

Generalizing Results to Broader Populations

We've seen that random samples are preferable to convenience samples, yet convenience samples are more feasible in many experiments. When an experiment uses a convenience sample, be cautious about the extent to which results generalize to a larger population. Look carefully at the characteristics of the sample. Do they seem representative of the overall population? If not, the results have dubious worth.

Many medical studies use volunteers at *several* medical centers to try to obtain a broad cross section of subjects. But some studies mistakenly try to generalize to a broader population than the one from which the sample was taken. A psychologist may conduct an experiment using a sample of students from an introductory psychology course. For the results to be of wider interest, however, the psychologist might claim that the conclusions generalize to *all* college students, to all young adults, or even to all adults. Such generalizations may be wrong since the sample may differ from those populations in fundamental ways such as in average socioeconomic status, race, or gender.

SUMMARY: Key Parts of a Good Experiment

- A good experiment has a **control comparison group**, **randomization** in assigning experimental units to treatments, **blinding**, and **replication**.
- The **experimental units** are the subjects—the people, animals, or other objects to which the treatments are applied.
- The **treatments** are the experimental conditions imposed on the experimental units. One of these may be a **control** (for instance, either a placebo or an existing treatment) that provides a basis for determining if a particular treatment is effective. The treatments correspond to values of an explanatory variable.
- **Randomize** in assigning the experimental units to the treatments. This tends to balance the comparison groups with respect to lurking variables.
- **Replicating** the treatments on many experimental units helps, so that observed effects are not due to ordinary variability but instead are due to the treatment. Repeat studies to increase confidence in the conclusions.

---

[12] It's a bit more complicated than the reasoning shown here. The margin of error for comparing two percentages differs from the one for estimating a single percentage.

Carefully assess the scope of conclusions in research articles, mass media, and advertisements. Evaluate critically the basis for the conclusions by noting the experimental design or the sampling design upon which the conclusions are based.

4.3 Practicing the Basics

4.33 Smoking affects lung cancer? You would like to investigate whether smokers are more likely than nonsmokers to get lung cancer. From the students in your class, you pick half at random to smoke a pack of cigarettes each day and half not to ever smoke. Fifty years from now, you will analyze whether more smokers than nonsmokers got lung cancer.

a. Is this an experiment or an observational study? Why?

b. Summarize at least three practical difficulties with this planned study.

4.34 Never leave home without duct tape There have been anecdotal reports of the ability of duct tape to remove warts. In an experiment conducted at the Madigan Army Medical Center in the state of Washington (*Archives of Pediatric and Adolescent Medicine* 2002; 156: 971–974), 51 patients between the ages of 3 and 22 were randomly assigned to receive either duct-tape therapy (covering the wart with a piece of duct tape) or cryotherapy (freezing a wart by applying a quick, narrow blast of liquid nitrogen). After two months, the percentage successfully treated was 85% in the duct tape group and 60% in the cryotherapy group.

TRY

a. Identify the response variable, the explanatory variable, the experimental units, and the treatments.

b. Describe the steps of how you could randomize in assigning the 51 patients to the treatment groups.

4.35 More duct tape In a follow-up study, 103 patients in the Netherlands having warts were randomly assigned to use duct tape or a placebo, which was a ring covered by tape so that the wart itself was kept clear (*Arch. Pediat. Adoles. Med.* 2006; 160: 1121–1125).

a. Identify the response variable, the explanatory variable, the experimental units, and the treatments.

b. After six weeks, the warts had disappeared for 16% of the duct tape group and 6% of the placebo group. However, the difference was declared to be "not statistically significant." Explain what this means.

4.36 Vitamin B A *New York Times* article (March 12, 2006) described two studies in which subjects who had recently had a heart attack were randomly assigned to one of four treatments: placebo and three different doses of vitamin B. In each study, after years of study, the differences among the proportions having a heart attack were judged to be not statistically significant. Identify the (a) response variable, (b) explanatory variable, (c) experimental units, (d) treatments, and (e) explain what it means to say that differences "were judged to be not statistically significant."

4.37 Smoking cessation A study published in 2010 in *The New England Journal of Medicine* investigated the effect of financial incentives on smoking cessation. As part of the study, 878 employees of a company, all of whom were smokers, were randomly assigned to one of two treatment groups. One group (442 employees) was to receive information about smoking cessation programs, while the other (436 employees) was to receive that same information, as well as a financial incentive to quit smoking. The primary endpoint of the study was smoking cessation status six months after the initial cessation was reported. After implementation of the program, 14.7% of individuals in the financial incentive group reported cessation six months after the initial report, compared to 5.0% of the information-only group.

a. For this study, identify the experimental units, explanatory and response variable(s), and treatments.

b. Assuming the observed difference in cessation rates between the groups (14.7% − 5.0% = 9.7%) is statistically significant, is this convincing evidence that the difference was due to the effect of the financial incentive and not due to ordinary random variation?

4.38 No statistical significance A randomized experiment investigates whether an herbal treatment is better than a placebo in treating subjects suffering from depression. Unknown to the researchers, the herbal treatment has no effect: Subjects have the same score on a rating scale for depression (for which higher scores represent worse depression) no matter which treatment they take.

a. The study will use eight subjects, numbered 1 to 8. Using random numbers, pick the four subjects who will take the herbal treatment. Identify the four who will take the placebo.

b. After taking the assigned treatment for three months, subjects' results on the depression scale are as follows:

Subject	Response	Subject	Response	Subject	Response	Subject	Response
1	85	2	60	3	44	4	95
5	69	6	78	7	50	8	75

Based on the treatment assignment in part a, find the sample mean response for those who took the herbal supplement and for those who took the placebo.

c. Using the means in part b, explain how (i) sample means can be different even when there is "no effect" in a population of interest, and (ii) a difference between two sample means may not be "statistically significant" even though those sample means are not equal.

4.39 Pain reduction medication Consider an experiment being designed to study the effectiveness of an experimental pain reduction medication. The plan includes recruiting 100 individuals suffering from moderate to severe pain to participate. One half of the group will be assigned to take the actual experimental drug, and the other half will be assigned a placebo. The study will be blind in the sense that the individuals will not know which treatment they are receiving. At the end of the study, individuals will be asked to record using a standardized scale how much pain relief they experienced. Why is it important to use a placebo in such a study?

4.40 Pain reduction medication, continued Consider the same setting as that of Exercise 4.39. Of the 100 participants, 45 are male and 55 are female. During the design of the study, one member of the research team suggests that all males be given the active drug and all females be given the placebo. Another member of the team wants to randomly assign each of the total group of 100 participants to one of the two treatments.

a. Which researcher's plan is the best experimental design for measuring the effectiveness of the medication if the results of the experiment are to be generalized to the entire population, which consists of both females and males?

b. Does the fact that the participants of the study were recruited rather than selected at random prohibit generalization of the results?

4.41 Pain reduction medication, continued Revisit the setting of Exercise 4.39. Suppose that in addition to the participants being blinded, the researchers responsible for recording the results of the study are also blinded. Why does it matter whether the researchers know which participants receive which treatment?

4.42 Colds and vitamin C For some time there has been debate about whether regular large doses of vitamin C reduce the chance of getting a common cold.

a. Explain how you could design an experiment to test this. Describe all parts of the experiment, including (i) what the treatments are, (ii) how you assign subjects to the treatments, and (iii) how you could make the study double-blind.

b. An observational study indicates that people who take vitamin C regularly get fewer colds, on the average. Explain why these results could be misleading.

4.43 Reducing high blood pressure A pharmaceutical company has developed a new drug for treating high blood pressure. They would like to compare its effects to those of the most popular drug currently on the market. Two hundred volunteers with a history of high blood pressure and who are currently not on medication are recruited to participate in a study.

a. Explain how the researchers could conduct a randomized experiment. Indicate the experimental units, the response and explanatory variables, and the treatments.

b. Explain what would have to be done to make this study double-blind.

4.4 Other Ways to Conduct Experimental and Nonexperimental Studies

In this chapter, we've learned the basics of good ways to conduct nonexperimental and experimental studies. This final section shows ways these methods are often extended in practice so they are even more powerful. We'll first learn about alternatives to simple random sampling in sample surveys, and then about types of observational studies that enable us to study questions for which experiments with humans are not feasible. Finally, we'll learn how to study the effects of two or more explanatory variables with a single experiment.

Sample Surveys: Other Random Sampling Designs Useful in Practice

We've seen that an experiment can better investigate cause and effect than an observational study. With an observational study, it's possible that an association is due to some lurking variable. However, observational studies do have advantages. It's often not practical to conduct an experiment. For instance, with human subjects it's not ethical to design an experiment to study the long-term effect of cell phone use on getting cancer. Nevertheless, it is possible to design an observational study that controls for identified lurking variables.

A sample survey that selects subjects randomly is another example of a well-designed and informative study that is not experimental. Simple random sampling gives every possible sample the same chance of selection. In practice, more complex random sampling designs are often easier to implement. Sometimes they are even preferable to simple random sampling.

Cluster random sampling To use simple random sampling, we need a sampling frame—the list of all, or nearly all, subjects in the population. Unfortunately, this information is often not available. It may be easier to identify **clusters** of subjects. A study of residents of a country can identify counties or census tracts. A study of students can identify schools. A study of the elderly living in institutions can identify nursing homes. We can obtain a sample by randomly selecting the clusters and observing each subject in the clusters chosen.

For instance, suppose you would like to sample about 1% of the families in your city. You could use city blocks as clusters. Using a map to label and number city blocks, you could select a simple random sample of 1% of the blocks and then select every family on each of those blocks for your observations.

Cluster Random Sample

Divide the population into a large number of **clusters**, such as city blocks. Select a simple random sample of the clusters. Use the subjects in those clusters as the sample. This is a **cluster random sample**.

For personal interviews, when the subjects within a cluster are close geographically, cluster sampling is less expensive per observation than simple random sampling. By interviewing every family in a particular city block, you can obtain many observations quickly and with little travel.

SUMMARY: Advantages and Disadvantages of a Cluster Random Sample

Cluster random sampling is a preferable sampling design if

- A reliable sampling frame is not available, or
- The cost of selecting a simple random sample is excessive.

A disadvantage is that we usually need a larger sample size with a cluster random sample than with a simple random sample in order to achieve a particular margin of error.

Stratified random sampling To estimate the mean number of hours a week that students in your school work on outside employment, use the student directory to take a simple random sample of $n = 40$ students. You plan to analyze how the mean compares for freshmen, sophomores, juniors, and seniors. Merely by chance, you may get only a few observations from one of the classes, making it hard to estimate the mean well for that class. If the student directory also identifies students by their class, you could amend the sampling scheme to take a simple random sample of 10 students from each class, still having an overall $n = 40$. The four classes are called **strata** of the population. This type of sample is called a **stratified random sample.**

Stratified Random Sample

A **stratified random sample** divides the population into separate groups, called **strata**, and then selects a simple random sample from each stratum.

A limitation of stratification is that you must have a sampling frame and know the stratum into which each subject in the sampling frame belongs. You might want to stratify students in your school by whether the student has a job, to make sure you get enough students who work and who don't work. However, this information may not be available for each student.

SUMMARY: Advantages and Disadvantages of a Stratified Random Sample

Stratified random sampling has the

- Advantage that you can include in your sample enough subjects in each group (stratum) you want to evaluate.
- Disadvantage that you must have a sampling frame and know the stratum into which each subject belongs.

What's the difference between a stratified sample and a cluster sample? A stratified sample *uses every stratum.* By contrast, a cluster sample *uses a sample of the clusters*, rather than all of them. Figure 4.2 illustrates the distinction among sampling subjects (simple random sample), sampling clusters of subjects (cluster random sample), and sampling subjects from within strata (stratified random sample).

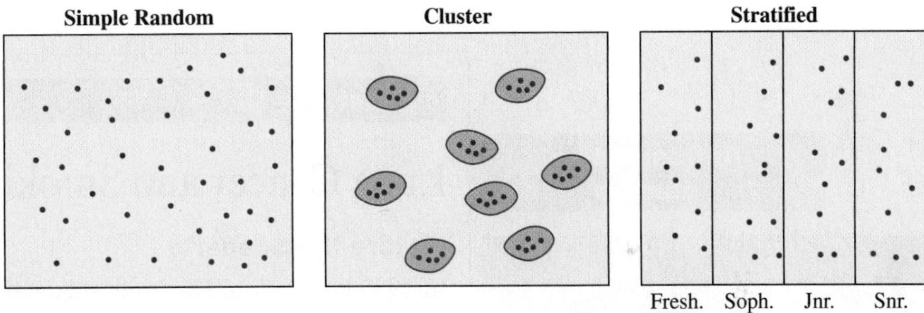

▲ **Figure 4.2 Ways of Randomly Sampling 40 Students.** The figure is a schematic for a simple random sample, a cluster random sample of 8 clusters of students who live together, and a stratified random sample of 10 students from each class (Fresh., Soph., Jnr., Snr.). **Question** What's the difference between clustering and stratifying?

Comparison of different random sampling methods A good sampling design ensures that each subject in a population has an opportunity to be selected. The design should incorporate randomness. Table 4.2 summarizes the random sampling methods we've presented.

Table 4.2 Summary of Random Sampling Methods

Method	Description	Advantages
Simple random sample	Each possible sample is equally likely	Sample tends to be a good reflection of the population
Cluster random sample	Identify clusters of subjects, take simple random sample of the clusters	Do not need a sampling frame of subjects, less expensive to implement
Stratified random sample	Divide population into groups (strata), take simple random sample from each stratum	Ensures enough subjects in each group that you want to compare

In practice, sampling designs often have two or more stages. When carrying out a large survey for predicting national elections, the Gallup organization often (1) identifies election districts as clusters and takes a simple random sample of them, and (2) takes a simple random sample of households within each selected election district, rather than sampling every household in a district, which might be infeasible.

Marketing companies also commonly use *two-stage cluster sampling*, first randomly selecting test market cities (the clusters) and then randomly selecting consumers within each test market city. Many major surveys, such as the General Social Survey, Gallup polls, and the Current Population Survey, incorporate both stratification and clustering.

In future chapters, when we use the term *random sampling*, we'll mean *simple random sampling.* The formulas for most statistical methods assume simple random sampling. Similar formulas exist for other types of random sampling, but they are complex and beyond the scope of this text.

Retrospective and Prospective Observational Studies

Rather than taking a cross section of a population at some time, such as with a sample survey, some studies are *backward looking* **(retrospective)** or *forward looking* **(prospective).** Observational studies in medicine are often retrospective. How can we study whether there is an association between long-term cell phone use and brain cancer if we cannot perform an experiment? We can form a sample of subjects who have brain cancer and a similar sample of subjects who do not and then compare the past use of cell phones for the two groups. This approach was first applied to study smoking and lung cancer.

Example 9

Retrospective studies

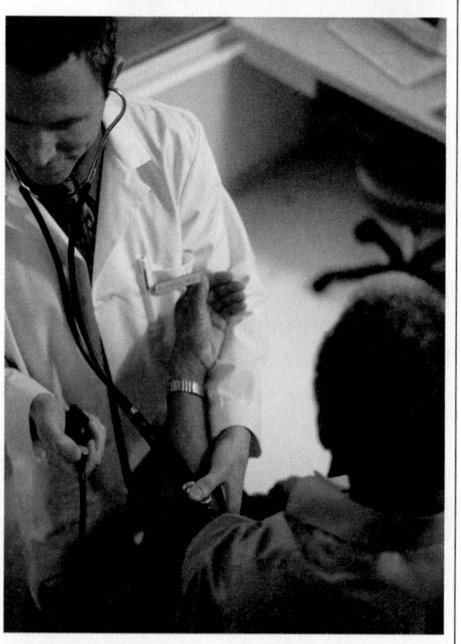

Lung Cancer and Smoking

Picture the Scenario

In 1950 in London, England, medical statisticians Austin Bradford Hill and Richard Doll conducted one of the first studies linking smoking and lung cancer. In 20 hospitals, they matched 709 patients admitted with lung cancer in the preceding year with 709 noncancer patients at the same hospital of the same gender and within the same five-year grouping on age. All patients were queried about their smoking behavior. A smoker was defined as a person who had smoked at least one cigarette a day for at least a year. The study used a *retrospective* design to "look into the past" in measuring the patients' smoking behavior.

Table 4.3 shows the results. The 709 *cases* in the first column of the table were the patients with lung cancer. The 709 *controls* in the second column were the matched patients without lung cancer.

Table 4.3 Results of Retrospective Study of Smoking and Lung Cancer

The cases had lung cancer. The controls did not. The *retrospective* aspect refers to studying whether subjects had been smokers in the past.

	Lung Cancer	
Smoker	**Cases**	**Controls**
Yes	688	650
No	21	59
Total	**709**	**709**

Question to Explore

Compare the proportions of smokers for the lung cancer cases and the controls. Interpret.

Think It Through

For the lung cancer cases, the proportion who were smokers was 688/709 = 0.970, or 97%. For the controls (not having lung cancer), the proportion who were smokers was 650/709 = 0.917, or about 92%. The lung cancer cases were more likely than the controls to have been smokers.

Insight

An inferential analysis showed that these results were statistically significant. This suggested that an association exists between smoking and lung cancer.

Try Exercise 4.48

Case-control studies The type of retrospective study used in Example 9 to study smoking and lung cancer is called a **case-control study**.

In Words

In Example 9, the response outcome of interest was having lung cancer. The cases had lung cancer, the controls did not have lung cancer, and they were compared on the explanatory variable—whether the subject had been a smoker (yes or no).

Case-Control Study

A **case-control study** is a retrospective observational study in which subjects who have a response outcome of interest (the cases) and subjects who have the other response outcome (the controls) are compared on an explanatory variable.

This is a popular design for medical studies in which it is not practical or ethical to perform an experiment. We can't randomly assign subjects into a smoking group and a nonsmoking group—this would involve asking some subjects to start smoking. Since we can't use randomization to balance effects of potential lurking variables, usually the cases and controls are matched on such variables. In Example 9, cases and controls were matched on their age, gender, and hospital.

In a case-control study, the number of cases and the number of controls is fixed. The random part is observing the outcome for the explanatory variable. For instance, in Example 9, for the 709 patients of each type, we looked back in time to see whether they smoked. We found percentages for the categories of the explanatory variable (smoker, nonsmoker) given lung cancer status (cases or control). It is not meaningful to form a percentage for a category of the response variable. For example, we cannot estimate the population percentage of subjects who have lung cancer, for smokers or nonsmokers. By the study design, *half* the subjects had lung cancer. This does not mean that half the population had lung cancer.

Example 10

Case-control studies

Cell Phone Use

Picture the Scenario

Studies 1 and 2 about cell phone use in Example 1 were both case-control studies. In Study 2, the cases were brain cancer patients. The controls were randomly sampled from general practitioner lists and were matched with the cases on age, sex, and place of residence. In Study 1, the cases had a type of eye cancer. In forming a sample of controls, Study 1 did not attempt to match subjects with the cases.

Question to Explore

Why might researchers decide to match cases and controls on characteristics such as age?

Think It Through

We've seen that one way to balance groups on potential lurking variables such as age is to randomize in assigning subjects to groups. However, these were observational studies, not experiments. So it was not possible for the researchers to use randomization to balance treatments on potential lurking variables. Matching is an attempt to achieve the balance that randomization provides.

When researchers fail to use relevant variables to match cases with controls, those variables could influence the results. They could mask the true relationship. The results could suggest an association when actually there is not one, or the reverse. Without matching, results are more susceptible to effects of lurking variables.

Insight

The lack of matching in Study 1 may be one reason that the results from the two studies differed, with Study 2 not finding an association and Study 1 finding one. For example, in Study 1 suppose the eye cancer patients tended to be older than the controls, and suppose older people tend to be heavier users of cell phones. Then, age could be responsible for the observed association between cellphone use and eye cancer.

Try Exercise 4.49

Prospective studies A *retrospective* study, such as a case-control study, looks into the past. By contrast, a *prospective* study follows its subjects into the future.

Example 11

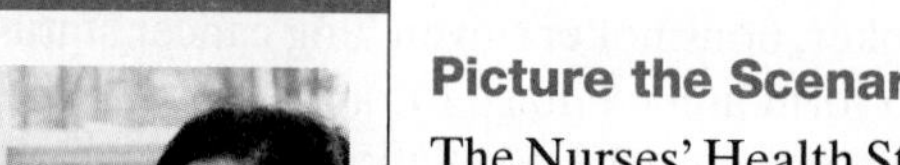

Nurses' Health

Picture the Scenario

The Nurses' Health Study, conducted by researchers at Harvard University,[13] began in 1976 with 121,700 female nurses age 30 to 55. The purpose of the study was to explore relationships among diet, hormonal factors, smoking habits, and exercise habits, and the risk of coronary heart disease, pulmonary disease, and stroke. Since the initial survey in 1976, the nurses have filled out a questionnaire every two years.

Question to Explore

What does it mean for this observational study to be called *prospective*?

Think It Through

The retrospective smoking study in Example 9 looked to the past to learn if its lung cancer subjects had been smokers. By contrast, at the start of the Nurses' Health Study, it was not known whether a particular nurse would eventually

[13] More information on this study can be found at clinicaltrials.gov/show/NCT00005152 and at www.channing.harvard.edu/nhs.

have an outcome such as lung cancer. The study followed each nurse into the future to see whether she developed that outcome and to analyze whether certain explanatory variables (such as smoking) were associated with it.

Insight

Over the years, several analyses have been conducted using data from the Nurses' Health Study. One finding (reported in *The New York Times*, February 11, 2003) was that nurses in this study who were highly overweight 18-year-olds were five times as likely as young women of normal weight to need hip replacement later in life.

Try Exercise 4.50

SUMMARY: Types of Nonexperimental Studies

- A **retrospective** study looks into the past.
- A **prospective** study identifies a group (cohort) of people and observes them in the future. In research literature, prospective studies are often referred to as cohort studies.
- A **sample survey** takes a **cross section** of a population at the current time. Sample surveys are sometimes called cross-sectional studies.

Observational Studies and Causation

Can we ever *definitively* establish causation with an observational study? No, we cannot. For example, because the smoking and lung cancer study in Example 9 was observational, cigarette companies argued that a lurking variable could have caused this association. So why are doctors so confident in declaring that smoking causes lung cancer? For a combination of reasons: (1) Experiments conducted using animals have shown an association, (2) in many countries, over time female smoking has increased relative to male smoking, and (3) the incidence of lung cancer has increased in women compared to men. Other studies, both retrospective and prospective, have added more evidence. For example, when a prospective study begun in 1951 with 35,000 doctors ended in 2001, researchers[14] estimated that cigarettes took an average of 10 years off the lives of smokers who never quit. This study estimated that at least half the people who smoke from youth are eventually killed by their habit.

Most importantly, studies carried out on different populations of people have *consistently* concluded that smoking is associated with lung cancer, even after adjusting for all potentially confounding variables that researchers have suggested. As more studies are done that adjust for confounding variables, the chance that a lurking variable remains that can explain the association is reduced. As a consequence, although we cannot definitively conclude that smoking causes lung cancer, physicians will not hesitate to tell you that they believe it does.

In Words

A **factor** is a categorical explanatory variable (such as whether the subject takes an antidepressant) having as categories the experimental conditions (the **treatments**, such as bupropion or no bupropion).

Multifactor Experiments

Now let's learn about other ways of performing experiments. First, we'll learn how a single experiment can help us analyze effects of two explanatory variables at once.

Categorical explanatory variables in an experiment are often referred to as **factors.** For example, consider the experiment described in Example 8 to study whether taking an antidepressant can help a smoker stop smoking. The factor

[14]See article by R. Doll et al., *British Med. J.*, June 26, 2004, p. 1519.

measured whether a subject used the antidepressant bupropion (yes or no). Suppose the researchers also wanted to study a second factor, using a nicotine patch versus not using one. The experiment could then have four treatment groups that result from cross-classifying the two factors. See Figure 4.3. The four treatments are bupropion alone, nicotine patch alone, bupropion and nicotine patch, neither bupropion nor nicotine patch.

		Nicotine patch (factor 1)	
		Yes	No
Bupropion (factor 2)	Yes	1	2
	No	3	4

▲ **Figure 4.3 An Experiment Can Use Two (or More) Factors at Once.** The treatments are the combinations of categories of the factors, as numbered in the boxes.

Why use both factors at once in an experiment? Why not do one experiment about bupropion and a separate experiment about the nicotine patch? The reason is that we can learn more from a two-factor experiment. For instance, the combination of using both a nicotine patch and bupropion may be more effective than using either method alone.

Example 12

Multifactor experiments

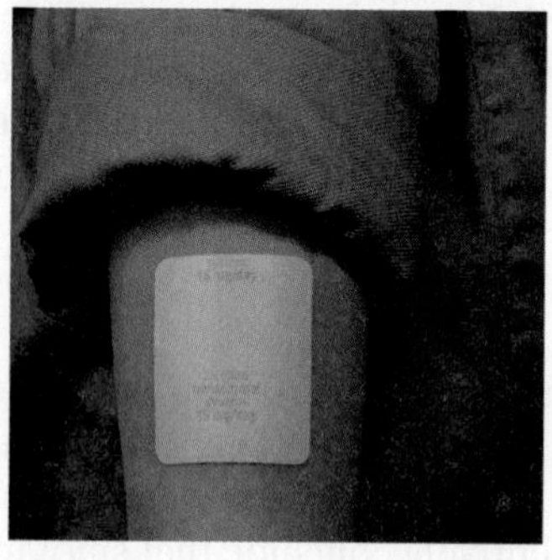

Antidepressants and/or Nicotine Patches

Picture the Scenario

Example 8 analyzed a study about whether bupropion helps a smoker quit cigarettes. Let's now also consider nicotine patches as another possible cessation aid.

Question to Explore

How can you design a single study to investigate the effects of nicotine patches and bupropion on whether a subject relapses into smoking?

Think It Through

You could use the two factors with four treatment groups shown in Figure 4.3. Figure 4.4 on the next page portrays the design of a randomized experiment, with placebo alternatives to bupropion and to the nicotine patch.

The study should be double-blind. After a fixed length of time, you compare the four treatments as to the percentages who have relapsed.

Insight

In fact, a two-factor experiment *has* been conducted to study both these factors.[15] The most effective treatments were bupropion alone or in combination with a nicotine patch. We'll explore the results in Exercise 4.75. If we included a third factor of whether or not the subject receives counseling, we would have three factors with $2 \times 2 \times 2 = 8$ treatments to compare.

Recall

This study is double-blind if both the subjects and the evaluator(s) of the subjects do not know whether or not the pill is active and whether or not the patch is active. ◀

[15] Jorenby, D. et al. (1999), *New England Journal of Medicine*, 340(9): 685–691.

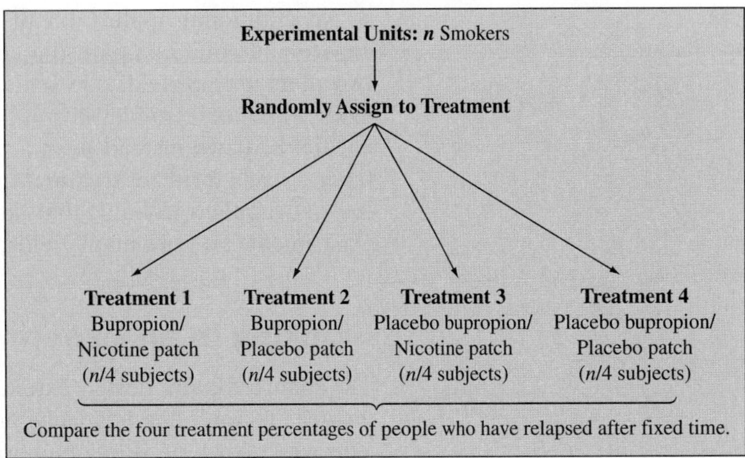

▲ **Figure 4.4** Diagram of a Randomized Experiment with Two Factors (Whether Use Bupropion and Whether Use a Nicotine Patch). Question A three-factor design could also incorporate whether or not a subject receives counseling to discourage smoking. How many treatments would such a design have?

Try Exercise 4.52

Matched Pairs Designs

The experiments described in Examples 8 and 12 used **completely randomized designs:** The subjects were randomly assigned to one of the treatments. Sometimes we can use an alternative experimental design in which *each treatment is observed for each subject.* Medical experiments often do this to compare treatments for chronic conditions that do not have a permanent cure.

To illustrate, suppose a study plans to compare an oral drug and a placebo for treating migraine headaches. The subjects could take one treatment the first time they get a migraine headache and the other treatment the second time they get a migraine headache. The response is whether the subject's pain is relieved. The first three subjects might contribute the results that follow to the data file.

Subject	Drug	Placebo	
1	Relief	No relief	← first *matched pair*
2	Relief	Relief	
3	No relief	No relief	

For the entire sample, we would compare the percentage of subjects who have pain relief with the drug to the percentage who have pain relief with the placebo. The two observations for a particular subject are called a **matched pair** because they both come from the same person.

A matched-pairs design in which subjects cross over during the experiment from using one treatment to using another treatment is called a **crossover design.** Crossover design helps remove certain sources of potential bias. Using the same subjects for each treatment keeps potential lurking variables from affecting the results because those variables take the same values for each treatment. For instance, any difference observed between the drug and the placebo responses is not because subjects taking the drug had better overall health. Another example of a matched pair, crossover design is Study 3 discussed in Example 1.

An additional application of a matched-pairs design occurs in agricultural experiments that compare the crop yield for two fertilizer types. In each field, two plots are marked, one for each fertilizer. Two plots in the same field are likely to be more similar in soil quality and moisture than two plots from different fields. If we instead used a completely randomized design to assign entire fields to each fertilizer treatment, there's the possibility that by chance one fertilizer gets applied to fields that have better soil. This is especially true when the experiment can't use many fields.

Blocking in an Experiment

The matched-pairs design extends to the comparison of more than two treatments. The migraine headache crossover study could investigate relief for subjects who take a low dose of an oral drug, a high dose of that drug, and a placebo. The crop yield study could have several fertilizer types.

In experiments with matching, a set of matched experimental units is referred to as a **block.** A block design with random assignment of treatments to units within blocks is called a **randomized block** design. In the migraine headache study, each person is a block. In the crop yield study, each field is a block. A purpose of using blocks is to provide similar experimental units for the treatments, so they're on an equal footing in any subsequent comparisons.

To reduce possible bias, treatments are usually randomly assigned within a block. In the migraine headache study, the order in which the treatments are taken would be randomized. Bias could occur if all subjects received one treatment before another. An example of potential bias in a crossover study is a positive carry-over effect if the first drug taken has lingering effects that can help improve results for the second drug taken. Likewise, in a crop yield study, the fertilizers are randomly assigned to the plots in a field.

On the Shoulders of... Austin Bradford Hill (1897–1991) and Richard Doll (1912–2005)

How did statisticians become pioneers in examining the effects of smoking on lung cancer?

In the mid-20th century, Austin Bradford Hill was Britain's leading medical statistician. In 1946, to assess the effect of streptomycin in treating pulmonary tuberculosis, he designed what is now regarded as the first controlled randomized medical experiment. About that time, doctors noticed the alarming increase in the number of lung cancer cases. The cause was unknown but suspected to be atmospheric pollution, particularly coal smoke. In 1950, Bradford Hill and medical doctor and epidemiologist Richard Doll published results of the case-control study (Example 9) that identified smoking as a potentially important culprit. They were also pioneers in developing a sound theoretical basis for the case-control method. They followed their initial study with other case-control studies and with a long-term prospective study that showed that smoking was not only the predominant risk factor for lung cancer but also correlated with other diseases.

At first, Bradford Hill was cautious about claiming that smoking causes lung cancer. As the evidence mounted from various studies, he accepted the causal link as being overwhelmingly the most likely explanation. An influential article written by Bradford Hill in 1965 proposed criteria that should be satisfied before you conclude that an association reflects a causal link. For his work, in 1961 Austin Bradford Hill was knighted by Queen Elizabeth. Richard Doll was knighted in 1969.

4.4 Practicing the Basics

4.44 Student loan debt A researcher wants to compare student loan debt for students who attend four-year public universities with those who attend four-year private universities. She plans to take a random sample of 100 recent graduates of public universities and 100 recent graduates of private universities. Which type of random sampling is utilized in her study design?

4.45 Club officers again In Exercise 4.14, two officers were to be selected to attend a conference in New Orleans. Three of the officers are female and two are male. It is decided to send one female and one male to the convention.

a. Labeling the officers as 1, 2, 3, 4, 5, where 4 and 5 are male, draw a stratified random sample using random numbers. Explain how you did this.

b. Explain why this sampling design is not a simple random sample.

c. If the activity coordinator is female, what are her chances of being chosen? If male?

4.46 Security awareness training Of 400 employees at a company, 25% work in production, 40% work in sales and marketing, and 35% work in new product development. As part of a security awareness training program, the group overseeing implementation of the program will randomly choose a sample of 20 employees to begin the training; the percentages of workers from each department in the sample are to align with the percentages throughout the company.

a. What type of sampling could be used to achieve this goal?

b. Explain how to carry out the sampling using a table of random digits.

4.47 Teaching and learning model A school district comprises 24 schools. The numbers of students in each of the schools are as follows:

455 423 399 388 348 344 308 299 297 272 266 260
252 244 222 209 175 161 151 148 128 109 101 98

The district wants to implement an experimental teaching and learning model for approximately 20% of the 6057 students in the district. Administrators want to choose the 20% randomly, but they will not be able to use simple random sampling throughout the entire district because the new model can be implemented only at an entire school, not just for a select group of students at each school. The schools not selected will continue to use the current teaching and learning model.

a. Explain how one could use cluster random sampling to achieve the goal of choosing approximately 20% of the students in the district for the experimental model.

b. Use cluster random sampling and a table of random digits to decide which schools will use the new model. Begin your sampling at line 10 of the table. How many schools are in your sample? How many students are in your sample?

c. Repeat part b, this time beginning at line 15. Did you obtain the same number of schools?

d. Would it be possible for the school district to implement stratified random sampling? Explain.

4.48 German mobile study TRY The contingency table shows results from the German study about whether there was an association between mobile phone use and eye cancer (Stang et al., 2001).

a. The study was retrospective. Explain what this means.

b. Explain what is meant by cases and controls in the headings of the table.

c. What proportion had used mobile phones, of those in the study who (i) had eye cancer and (ii) did not have eye cancer?

Eye Cancer and Use of Mobile Phones

Mobile Phones	Cases	Controls
Yes	16	46
No	102	429
Total	**118**	**475**

4.49 Smoking and lung cancer TRY Refer to the smoking case-control study in Example 9. Since subjects were not matched according to *all* possible lurking variables, a cigarette company can argue that this study does not prove a causal link between smoking and lung cancer. Explain this logic, using diet as the lurking variable.

4.50 Smoking and death TRY Example 1 in Chapter 3 described a survey of 1314 women during 1972–1974, in which each woman was asked whether she was a smoker. Twenty years later, a follow-up survey observed whether each woman was deceased or still alive. Was this study a retrospective study, or a prospective study? Explain.

4.51 Baseball under a full moon Exercise 4.10 mentioned that the away team has won 13 consecutive games played between the Boston Brouhahas and Minnesota Meddlers during full moons. This is a statement based on retrospective observational data.

a. Many databases are huge, including those containing sports statistics. If you had access to the database, do you think you could uncover more surprising trends?

b. Would you be more convinced that the phase of the moon has predictive power if the away team were to win the *next* 13 games played under a full moon between Boston and Minnesota?

c. The results of which type of observational study are generally more reliable, retrospective or prospective?

4.52 Two factors helpful? TRY A two-factor experiment designed to compare two diets and to analyze whether results depend on gender randomly assigns 20 men and 20 women to the two diets, 10 of each to each diet. After three months the sample mean weight losses are as shown in the table.

Sample Mean Weight Loss by Diet Type and Gender

	Gender		
Diet	**Female**	**Male**	**Overall**
Low-carb	12	0	6
Low-fat	0	12	6
Overall	6	6	6

a. Identify the two factors (explanatory variables) and the response variable.

b. What would the study have concluded if it did only a one-factor analysis on diet, looking at the overall results and not having the information about gender?

c. What could the study learn from the two-factor study that it would have missed by doing a one-factor study on diet alone or a one-factor study on gender alone?

4.53 Caffeine jolt A study (*Psychosomatic Medicine* 2002; 64: 593–603) claimed that people who consume caffeine regularly may experience higher stress and higher blood pressure. In the experiment, 47 regular coffee drinkers consumed 500 milligrams of caffeine in a pill form (equivalent to four 8-oz cups) during one workday, and a placebo pill during another workday. The researchers monitored the subjects' blood pressure and heart rate, and the subjects recorded how stressed they felt.

a. Identify the response variable(s), explanatory variable, experimental units, and the treatments.

b. Is this an example of a completely randomized design, or a crossover design? Explain.

4.54 Allergy relief An experiment is being designed to compare relief from hay fever symptoms given by a low dose of a drug, a high dose of the drug, and a placebo. Each subject who suffers from hay fever and volunteers for the study is observed on three separate days, with a different treatment used each day. There are two days between treatments, so that a treatment does not have a carry-over effect for the next treatment assigned.

a. What are the blocks in this block design? What is this type of block design called?

b. Suppose the study is conducted as a double-blind study. Explain what this means.

c. Explain how randomization within blocks could be incorporated into the study.

4.55 Effect of partner smoking in smoking cessation study Smokers may have a more difficult time quitting smoking if they live with another smoker. How can an experiment explore this possibility in a study to compare bupropion with placebo? Suppose the researchers split the subjects into two groups: those who live with another smoker, and those who do not live with smokers. Within each group, the subjects are randomly assigned to take bupropion or a placebo. The figure shows a flow chart of this design, when 250 of the 429 study subjects live with nonsmokers and 179 live with another smoker.

a. Is this design a *completely* randomized design? Why or why not? (*Hint*: Is the smoking status of the person a subject lives with randomly determined?)

b. Does this experiment have blocks? If so, identify them.

c. Is this design a randomized block design? Explain why or why not.

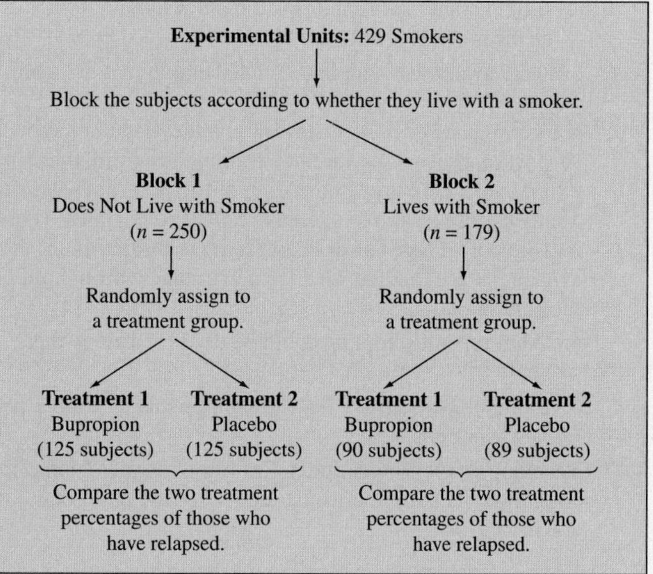

Diagram of an experiment to account for smoking status of partner

Chapter Review

ANSWERS TO THE CHAPTER FIGURE QUESTIONS

Figure 4.2 In a stratified sample, every stratum is used. A simple random sample is taken within each stratum. A cluster sample takes a simple random sample of the clusters. All the clusters are *not* used.

Figure 4.4 The design would have $2 \times 2 \times 2 = 8$ treatments.

CHAPTER SUMMARY

This chapter introduced methods for gathering data.

- An **experiment** assigns subjects to experimental conditions (such as drug or placebo) called **treatments**. These are categories of the explanatory variable. The outcome on the response variable is then observed for each subject.
- An **observational study** is a type of nonexperimental study that observes subjects on the response and explanatory variables. The study samples the population of interest, but merely observes rather than applies treatments to those subjects.

Since association does not imply causation, with observational studies we must be aware of potential **lurking variables** that influence the association. In an experiment, a researcher uses **randomization** in assigning experimental units (the subjects) to the treatments. This helps to balance the treatments on lurking variables. A randomized experiment can provide support for causation. To reduce bias, experiments should be **double-blind**, with neither the subject nor the data collector knowing to which treatment a subject was assigned.

A **sample survey** is a type of nonexperimental study that takes a sample from a population. Good methods for selecting samples incorporate random sampling.

- A **simple random sample** of n subjects from a population is one in which each possible sample of size n has the same chance of being selected.
- A **cluster random sample** takes a simple random sample of clusters (such as city blocks) and uses subjects in those clusters as the sample.
- A **stratified random sample** divides the population into separate groups, called **strata**, and then selects a simple random sample from each stratum.

Be cautious of results from studies that use a convenience sample, such as a **volunteer sample**, which Internet polls use. Even with random sampling, **biases** can occur due to **sample undercoverage, nonresponse** by many subjects in the sample, and **responses that are biased** because of question wording or subjects' lying.

Most **sample surveys** take a cross section of subjects at a particular time. A **census** is a complete enumeration of an entire population. **Prospective** studies follow subjects into the future, as is true with many experiments. Medical studies often use **retrospective** observational studies, which look at subjects' behavior in the past.

- A **case-control study** is an example of a retrospective study. Subjects who have a response outcome of interest, such as cancer, serve as cases. Other subjects not having that outcome serve as controls. The cases and controls are compared on an explanatory variable, such as whether they had been smokers.

A categorical explanatory variable in an experiment is also called a **factor**. **Multifactor experiments** have at least two explanatory variables. With a **completely randomized design**, subjects are assigned randomly to categories of each explanatory variable. Some designs instead use **blocks**—sets of experimental units that are matched. **Matched-pairs designs** have two observations in each block. Often this is the same subject observed for each of two treatments, such as in a crossover study.

Data are only as good as the method used to obtain them. Unless we use a good study design, conclusions may be faulty.

CHAPTER PROBLEMS

Practicing the Basics

4.56 Cell phones If you want to conduct a study with humans to see if cell phone use makes brain cancer more likely, explain why an observational study is more realistic than an experiment.

4.57 Observational versus experimental study Without using technical language, explain the difference between observational and experimental studies to someone who has not studied statistics. Illustrate with an example, using it also to explain the possible weaknesses of an observational study.

4.58 Unethical experimentation Give an example of a scientific question of interest for which it would be unethical to conduct an experiment. Explain how you could instead conduct an observational study.

4.59 Spinal fluid proteins and Alzheimer's A research study published in 2010 in the Archives of Neurology investigated the relationship between the results of a spinal fluid test and the presence of Alzheimer's disease. The study included 114 patients with normal memories, 200 with memory problems, and 102 with Alzheimer's disease. Each individual's spinal fluid was analyzed to detect the presence of two types of proteins. Almost everyone with Alzheimer's had the proteins in their spinal fluid. Nearly three quarters of the group with memory problems had the proteins, and each such member developed Alzheimer's within five years. About one third of those with normal memories had the proteins, and the researchers suspect that those individuals will develop memory problems and eventually Alzheimer's.

a. Identify the explanatory and response variable(s).

b. Was this an experimental or nonexperimental study? Why?

c. Would it be possible to design this study as an experiment? Explain why or why not.

4.60 Unemployment rate  The Gallup U.S. Employment poll reported on March 15, 2011, a national unemployment rate of 10.4%. (*Source:* Data from www.gallup.com/poll/125639/Gallup_Daily_Workforce.aspx.)

a. Gallup claims that the margin of error of the poll is +/− 0.7 percentage points. Estimate approximately on how many responses the poll was based?

b. In trying to identify the population under consideration, we must recognize that different organizations have different definitions of the term *unemployment*. Visit the Gallup Web site at www.gallup.com/poll/125639/Gallup-Daily-Workforce.aspx and report how they define unemployment.

4.61 Fear of asbestos Your friend reads about a study in an epidemiology journal that estimates that the chance is 15 out of 1 million that a teacher who works for 30 years in a school with typical asbestos levels gets cancer from asbestos. However, she also knows about a teacher who died recently who may have had asbestos exposure. In deciding whether to leave teaching, should she give more weight to the study or to the story she heard about the teacher who died? Why?

4.62 NCAA men's basketball poll The last four teams of the Southeast region of the 2011 NCAA Men's Basketball Tournament were Butler (located in Indiana), Brigham Young University (located in Utah), Florida, and Wisconsin. The sports Web site espn.com asked visitors of the site which team would win the Southeast region. Nationwide results are depicted on the map that follows. It was reported that 44% of the more than 3300 Indiana resident respondents believed Butler would win the regional, and 78% of the more than 5600 Wisconsin resident respondents believed Wisconsin would win.

a. What are the estimated margins of error associated with the Indiana and Wisconsin polls?

b. Explain why the percentages within Indiana and Wisconsin vary so drastically from the nationwide percentages displayed in the figure.

c. It was reported that between 42.3% and 45.7% of Indiana residents believed Butler was likely to win. What type of potential bias prevented these results from being representative of the entire population of Indiana residents?

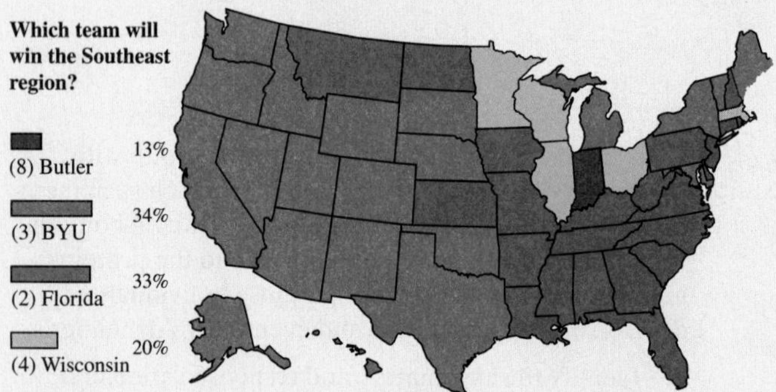

© ESPN.com (2012). Graphic feature used with permission.

4.63 Sampling your fellow students You are assigned to direct a study on your campus to discover factors that are associated with strong academic performance. You decide to identify 20 students who have perfect GPAs of 4.0, and then measure explanatory variables for them that you think may be important, such as high school GPA and average amount of time spent studying per day.

a. Explain what is wrong with this study design.

b. Describe a study design that would provide more useful information.

4.64 Beware of Internet polling An Internet survey at a newspaper Web site reports that only 14% of respondents believe in gun control. Mention a lurking variable that could bias the results of such an online survey, and explain how it could affect the results.

4.65 Comparing female and male students You plan to sample from the 3500 undergraduate students who are enrolled at the University of Rochester to compare the proportions of female and male students who would like to see the United States have a female president.

a. Describe the steps for how you would proceed, if you plan a simple random sample of 80 students. Illustrate, by picking the first three students for the sample.

b. Suppose that you use random numbers to select students, but stop selecting females as soon as you have 40, and you stop selecting males as soon as you have 40. Is the resulting sample a simple random sample? Why or why not?

c. What type of sample is the sample in part b? What advantage might it have over a simple random sample?

4.66 Football discipline A large southern university had problems with 17 football players being disciplined for team rule violations, arrest charges, and possible NCAA violations. The online *Atlanta Journal Constitution* ran a poll with the question, "Has the football coach lost control over his players?" having possible responses, "Yes, he's been too lenient," and "No, he can't control everything teenagers do."

a. Was there potential for bias in this study? If so, what types of bias?

b. The poll results after two days were

Yes	6012	93%
No	487	7%

Does this large sample size guarantee that the results are unbiased? Explain.

4.67 Obesity in metro areas A Gallup poll tracks obesity in the United States for the most and least obese metro areas in the United States. The poll, based on more than 200,000 responses between January and December of 2010, reported that certain chronic conditions are more prevalent in the most obese metro areas. The table that follows presents a summary of the findings.

	10 Most Obese Metro Areas	10 Least Obese Metro Areas
Diabetes	14.6%	8.5%
High blood pressure	35.8%	25.6%
High cholesterol	28.5%	24.1%
Heart attack	5.9%	3.4%

Source: www.gallup.com/poll/146669/Adults-Obese-Metro-Areas-Nationwide.aspx

a. Are we able to conclude from the Gallup poll that obesity causes a higher incidence of these conditions?

b. What is a possible variable other than obesity that may be associated with these chronic conditions?

4.68 What's your favorite poem? In fall 1995, the BBC in Britain requested viewers to call the network and indicate their favorite poem. Of 7500 callers, more than twice as many voted for Rudyard Kipling's *If* than for any other poem. The BBC then reported that this was the clear favorite.

a. Since any person could call, was this sample a simple random sample? Explain.

b. Was this a volunteer sample? Explain.

c. If the BBC truly wanted to determine Brits' favorite poem, how could they do so more reliably?

4.69 Video games mindless? "Playing video games not so mindless." This was the headline of a *CNN* news report[16] about a study that concluded that young adults who regularly play video games demonstrated better visual skills than young adults who do not play regularly. Sixteen young men volunteered to take a series of tests that measured their visual skills; those who who had played video games in the previous six months performed better on the tests than those who hadn't played.

a. What are the explanatory and response variables?

b. Was this an observational study or an experiment? Explain.

c. Specify a potential lurking variable. Explain your reasoning.

4.70 Physicians' health study Read about the first Physicians' Health Study at phs.bwh.harvard.edu.

a. Explain whether it was (i) an experiment or an observational study and (ii) a retrospective or prospective study.

b. Identify the response variable and the explanatory variable(s), and summarize results.

4.71 Aspirin prevents heart attacks? During the 1980s approximately 22,000 physicians over the age of 40 agreed to participate in a long-term study called the Physicians' Health Study. One question investigated was whether aspirin helps to lower the rate of heart attacks. The physicians were randomly assigned to take aspirin or take placebo.

a. Identify the response variable and the explanatory variable.

b. Explain why this is an experiment, and identify the treatments.

c. There are other explanatory variables, such as the amount of exercise a physician got, that we would expect to be associated with the response variable. Explain how such a variable is dealt with by the randomized nature of the experiment.

4.72 Exercise and heart attacks Refer to Exercise 4.71. One potential confounding variable was the amount of exercise the physicians got. The randomization should have balanced the treatment groups on exercise. The contingency table shows the relationship between whether the physician exercised vigorously and the treatments.

	Exercise Vigorously?		
Treatment	**Yes**	**No**	**Total**
Aspirin	7,910	2,997	**10,907**
Placebo	7,861	3,060	**10,921**

a. Find the conditional proportions (recall Section 3.1) in the categories of this potential confounder (amount of exercise) for each treatment group. Are they similar?

b. Do you think that the randomization process did a good job of achieving balanced treatment groups in terms of this potential confounder? Explain.

4.73 Smoking and heart attacks Repeat the previous exercise, considering another potential confounding variable—whether the physicians smoked. The contingency table cross-classifies treatment group by smoking status.

	Smoking Status			
Treatment	**Never**	**Past**	**Current**	**Total**
Aspirin	5,431	4,373	1,213	**11,017**
Placebo	5,488	4,301	1,225	**11,014**

4.74 Aspirin, beta-carotene, and heart attacks In the study discussed in the previous three exercises, this completely randomized study actually used two factors: whether received aspirin or placebo and whether received beta-carotene or placebo. Draw a table or a flow chart to portray the four treatments for this study.

4.75 Bupropion and nicotine patch study results The subjects for the study described in Example 12 were evaluated for abstinence from cigarette smoking at the end of 12 months. The table shows the percentage in each group that were abstaining.

Group	**Abstinence Percentage**	**Sample Size**
Nicotine patch only	16.4	244
Bupropion only	30.3	244
Nicotine patch with bupropion	35.5	245
Placebo only	15.6	160

a. Find the approximate margin of error for the abstinence percentage in each group. Explain what a margin of error means.

b. Based on the results in part a, does it seem as if the difference between the bupropion-only and placebo-only treatments is statistically significant? Explain.

c. Based on the results in part a, does it seem as if the difference between the bupropion-only and nicotine patch with bupropion treatments is statistically significant? Explain.

d. Based on the results in parts a, b, and c, how would you summarize the results of this experiment?

4.76 Prefer Coke or Pepsi? You want to conduct an experiment with your class to see if students prefer Coke or Pepsi.

a. Explain how you could do this, incorporating ideas of blinding and randomization, (i) with a completely randomized design and (ii) with a matched pairs design.

b. Which design would you prefer? Why?

4.77 Comparing gas brands The marketing department of a major oil company wants to investigate whether cars get better mileage using their gas (Brand A) than from an

[16]See www.cnn.com/2003/TECH/fun/games/05/28/action.video.ap/index.html.

independent one (Brand B) that has cheaper prices. The department has 20 cars available for the study.

a. Identify the response variable, the explanatory variable, and the treatments.

b. Explain how to use a completely randomized design to conduct the study.

c. Explain how to use a matched-pairs design to conduct the study. What are the blocks for the study?

d. Give an advantage of using a matched-pairs design.

4.78 Samples not equally likely in a cluster sample? In a cluster random sample with equal-sized clusters, every subject has the same chance of selection. However, the sample is not a simple random sample. Explain why not.

4.79 Nursing homes You plan to sample residents of registered nursing homes in your county. You obtain a list of all 97 nursing homes in the county, which you number from 01 to 97. Using random numbers, you choose five of the nursing homes. You obtain lists of residents from those five homes and interview all the residents in each home.

a. Are the nursing homes clusters or strata?

b. Explain why the sample chosen is not a simple random sample of the population of interest to you.

4.80 Multistage health survey A researcher wants to study regional differences in dental care. He takes a multistage sample by dividing the United States into four regions, taking a simple random sample of ten schools in each region, randomly sampling three classrooms in each school, and interviewing all students in those classrooms about whether they've been to a dentist in the previous year. Identify each stage of this sampling design, indicating whether it involves stratification or clustering.

4.81 Hazing Hazing within college fraternities is a continuing concern. Before a national meeting of college presidents, a polling organization is retained to conduct a survey among fraternities on college campuses, gathering information on hazing for the meeting. The investigators from the polling organization realize that it is not possible to find a reliable sampling frame of all fraternities. Using a list of all college institutions, they randomly sample 30 of them, and then interview the officers of each fraternity at each of these institutions that has fraternities. Would you describe this as a simple random sample, cluster random sample, or stratified random sample of the fraternities? Explain.

4.82 Marijuana and schizophrenia Studies show that marijuana is extremely popular among individuals who suffer from psychotic disorders such as schizophrenia. A recent study published in the *British Journal of Psychiatry* followed 80 marijuana smokers, 42 who had schizophrenia. Over the course of six days, the participants were asked to periodically record their moods. All participants reported feeling generally happier while using marijuana, but the increase was stronger in the group with schizophrenia.

a. Identify the explanatory and response variables.

b. Is this an example of an observational study or an experiment? If observational, is the study retrospective or prospective?

4.83 Marijuana and schizophrenia, continued Many research studies such as the one discussed in Exercise 4.82 focus on a link between marijuana use and psychotic disorders such as schizophrenia. Studies have found that people with schizophrenia are twice as likely to smoke marijuana as those without the disorder. Data also suggest that individuals who smoke marijuana are twice as likely to develop schizophrenia as those who do not use the drug. Contributing to the apparent relationship, a comprehensive review done in 2007 of the existing research reported that individuals who merely try marijuana increase their risk of developing schizophrenia by 40%. Meanwhile, the percentage of the population who has tried marijuana has increased dramatically in the United States over the past 50 years, whereas the percentage of the population affected by schizophrenia has remained constant at about 1%. What might explain this puzzling result?

4.84 Twins and breast cancer Excessive cumulative exposure to ovarian hormones is believed to cause breast cancer. A study (*New England J. Medic.* 2003; 348: 2313–2322) used information from 1811 pairs of female twins, one or both of whom had breast cancer. Paired twins were compared with respect to age at puberty, when breast cancer was first diagnosed, and other factors. Their survey did not show an association between hereditary breast cancer and hormone exposure.

a. What type of observational study was used by the researchers?

b. Why do you think the researchers used this design instead of a randomized experiment?

Concepts and Investigations

4.85 Cell phone use Using the Internet, find a study about cell phone use and its potential risk when used by drivers of automobiles.

a. Was the study an experiment or was it an observational study?

b. Identify the response and explanatory variables.

c. Describe any randomization or control conducted in the study as well as attempts to take into account lurking variables.

d. Summarize conclusions of the study. Do you see any limitations of the study?

4.86 Read a medical journal Go to a Web site for an online medical journal, such as *British Medical Journal* (www.bmj.com). Pick an article in a recent issue.

a. Was the study an experiment, or was it an observational study?

b. Identify the response variable and the primary explanatory variable(s).

c. Describe any randomization or control conducted in the study as well as attempts to take into account lurking variables.

d. Summarize conclusions of the study. Do you see any limitations of the study?

4.87 Internet poll Find an example of results of an Internet poll. Do you trust the results of the poll? If not, explain why not.

4.88 Search for an observational study Find an example of an observational study from a newspaper, journal, the Internet, or some other medium.

a. Identify the explanatory and response variables and the population of interest.

b. What type of observational study was conducted? Describe how the data were gathered.

c. Were lurking variables considered? If so, discuss how. If not, can you think of potential lurking variables? Explain how they could affect the association.

d. Can you identify any potential sources of bias? Explain.

4.89 Search for an experimental study Find an example of a randomized experiment from a newspaper, journal, the Internet, or some other media.

a. Identify the explanatory and response variables.

b. What were the treatments? What were the experimental units?

c. How were the experimental units assigned to the treatments?

d. Can you identify any potential sources of bias? Explain.

4.90 Judging sampling design In each of the following situations, summarize negative aspects of the sample design.

a. A newspaper asks readers to vote at its Internet site to determine if they believe government expenditures should be reduced by cutting social programs. Based on 1434 votes, the newspaper reports that 93% of the city's residents believe that social programs should be reduced.

b. A congresswoman reports that letters to her office are running 3 to 1 in opposition to the passage of stricter gun control laws. She concludes that approximately 75% of her constituents oppose stricter gun control laws.

c. An anthropology professor wants to compare attitudes toward premarital sex of physical science majors and social science majors. She administers a questionnaire to her Anthropology 437, Comparative Human Sexuality class. She finds no appreciable difference in attitudes between the two majors, so she concludes that the two student groups are about the same in their views about premarital sex.

d. A questionnaire is mailed to a simple random sample of 500 household addresses in a city. Ten are returned as bad addresses, 63 are returned completed, and the rest are not returned. The researcher analyzes the 63 cases and reports that they represent a "simple random sample of city households."

4.91 More poor sampling designs Repeat the previous exercise for the following scenarios:

a. A principal in a large high school wants to sample student attitudes toward a proposal that seniors must pass a general achievement test in order to graduate. She lists all of the first-period classes. Then, using a random number table, she chooses a class at random and interviews every student in that class about the proposed test.

b. A new restaurant opened in January. In June, after six months of operation, the owner applied for a loan to improve the building. The loan application asked for the annual gross income of the business. The owner's record book contains receipts for each day of operation since opening. She decides to calculate the average daily receipt based on a sample of the daily records and multiply that by the number of days of operation in a year. She samples every Friday's record. The average daily receipt for this sample was then used to estimate the yearly receipts.

4.92 Age for legal alcohol You want to investigate the opinions students at your school have about whether the age for legal drinking of alcohol should be 18.

a. Write a question to ask about this in a sample survey in such a way that results would be biased. Explain why it would be biased.

b. Now write an alternative question that should result in unbiased responses.

4.93 Quota sampling An interviewer stands at a street corner and conducts interviews until obtaining a quota in various groups representing the relative sizes of the groups in the population. For instance, the quota might be 50 factory workers, 100 housewives, 60 elderly people, 30 Hispanics, and so forth. This is called **quota sampling**. Is this a random sampling method? Explain, and discuss potential advantages or disadvantages of this method. (The Gallup organization used quota sampling until it predicted, incorrectly, that Dewey would easily defeat Truman in the 1948 presidential election.)

4.94 Smoking and heart attacks A Reuters story (April 2, 2003) reported that "The number of heart attack victims fell by almost 60% at one hospital six months after a smoke-free ordinance went into effect in the area (Helena, Montana), a study showed, reinforcing concerns about second-hand smoke." The number of hospital admissions for heart attack dropped from just under seven per month to four a month during the six months after the smoking ban.

a. Did this story describe an experiment or an observational study?

b. In the context of this study, describe how you could explain to someone who has never studied statistics that association does not imply causation. For instance, give a potential reason that could explain this association.

4.95 Issues in clinical trials A randomized clinical trial is planned for AIDS patients to investigate whether a new treatment provides improved survival over the current standard treatment. It is not known whether it will be better or worse.

a. Why do researchers use randomization in such experiments, rather than letting the subjects choose which treatment they will receive?

b. When patients enrolling in the study are told the purpose of the study, explain why they may be reluctant to be randomly assigned to one of the treatments.

c. If a researcher planning the study thinks the new treatment is likely to be better, explain why he or she may have an ethical dilemma in proceeding with the study.

4.96 Compare smokers with nonsmokers Example 9 and Table 4.3 described a case-control study on smoking and lung cancer. Explain carefully why it is not sensible to use the study's proportion of smokers who had lung cancer (that is, 688/(688 + 650)) and proportion of nonsmokers who had lung cancer (21/(21 + 59)) to estimate corresponding proportions of smokers and nonsmokers who have lung cancer in the overall population.

4.97 Is a vaccine effective? A vaccine is claimed to be effective in preventing a rare disease that occurs in about one of every 100,000 people. Explain why a randomized clinical trial comparing 100 people who get the vaccine to 100 people who do not get it is unlikely to be worth doing. Explain how you could use a case-control study to investigate the efficacy of the vaccine.

4.98 Distinguish helping and hindering among infants Exercise 1.10 discussed a study at Yale University's Infant Cognition Center. Researchers were interested in determining whether infants had the ability to distinguish between the actions of helping and hindering. Each infant in the study was shown two videos. One video included a figure performing a helping action and the other included a figure performing a hindering action. Infants were presented with two toys resembling the helping and hindering figures from the videos and allowed to choose one of the toys to play with. The researchers conjectured that, even at a very young age, the infants would tend to choose the helpful object.

a. How might the results of this study been biased had each infant been shown the video with the helpful figure before being shown the video with the hindering figure?

b. How could such a potential bias be eliminated?

4.99 Distinguish helping and hindering among infants, continued In the previous exercise, we considered how showing each baby in the study the two videos in the same order might create a bias. In fact, of the 16 babies in the study, half were shown the videos in one order while the other half was shown the videos in the opposite order. Explain how to use the table of random digits to randomly divide the 16 infants into two groups of 8.

4.100 Distinguish helping and hindering among infants, continued Fourteen of the 16 infants in the Yale study elected to play with a toy resembling the helpful figure as opposed to one resembling the hindering figure. Is this convincing evidence that infants tend to prefer the helpful figure? Use the Simulating the Probability of Head with a Fair Coin applet to investigate the approximate likelihood of the observed results of 14 out of 16 infants choosing the helpful figure, if in fact infants are indifferent between the two figures. To perform a simulation, set $n = 1$, push the flip button 16 times and observe how often you obtain a head out of 16 tosses. Repeat this simulation for a total of 10 simulations. Out of the 10 simulations, how often did you obtain 14 or more heads out of 16 tosses? Are your results convincing evidence that infants actually tend to exhibit a preference?

For Exercises 4.101–4.108, select the best response.

4.101 Multiple choice: What's a simple random sample? A simple random sample of size n is one in which

a. Every nth member is selected from the population.

b. Each possible sample of size n has the same chance of being selected.

c. There is *exactly* the same proportion of women in the sample as is in the population.

d. You keep sampling until you have a fixed number of people having various characteristics (e.g., males, females).

4.102 Multiple choice: Getting a random sample When we use random numbers to take a simple random sample of 50 students from the 20,000 students at a university,

a. It is impossible to get the random number 00000 or 99999, since they are not random sequences.

b. If we get 20001 for the first random number, for the second random number, that number is less likely to occur than the other possible five-digit random numbers.

c. The draw 12345 is no more or less likely than the draw 11111.

d. Since the sample is random, it is impossible that it will be nonrepresentative, such as having only females in the sample.

4.103 Multiple choice: Be skeptical of medical studies? An analysis of published medical studies about heart attacks (Crossen, 1994, p. 168) noted that in the studies having randomization and strong controls for bias, the new therapy provided improved treatment 9% of the time. In studies without randomization or other controls for bias, the new therapy provided improved treatment 58% of the time.

a. This result suggests it is better not to use randomization in medical studies, because it is harder to show that new ideas are beneficial.

b. Some newspaper articles that suggest a particular food, drug, or environmental agent is harmful or beneficial should be viewed skeptically, unless we learn more about the statistical design and analysis for the study.

c. This result shows the value of case-control studies over randomized studies.

d. The randomized studies were poorly conducted, or they would have found the new treatment to be better much more than 9% of the time.

4.104 Multiple choice: Opinion and question wording A recent General Social Survey asked subjects if they supported legalized abortion in each of seven different circumstances. The percentage who supported legalization varied between 42.4% (if the woman wants it for any reason) to 89.2% (if the woman's health is seriously endangered by the pregnancy). This indicates that

a. Responses can depend greatly on the question wording.

b. Nonexperimental studies can never be trusted.

c. The sample must not have been randomly selected.

d. The sample must have had problems with response bias.

4.105 Multiple choice: Campaign funding When the Yankelovich polling organization asked, "Should laws be passed to eliminate all possibilities of special interests giving huge sums of money to candidates?" 80% of the sample answered yes. When they posed the question, "Should laws be passed to prohibit interest groups from contributing to campaigns, or do groups have a right to contribute to the candidate they support?" only 40% said yes (*Source: A Mathematician Reads the Newspaper,* by J. A. Paulos, New York: Basic Books, 1995, p. 15). This example illustrates problems that can be caused by

a. Nonresponse

b. Bias in the way a question is worded

c. Sampling bias

d. Undercoverage

4.106 Multiple choice: Emotional health survey An Internet poll conducted in the United Kingdom by Netdoctor.co.uk asked individuals to respond to an "emotional health survey" (see www.hfienberg.com/clips/pollspiked.htm). There were 400 volunteer respondents. Based on the results, the British Broadcasting Corporation (BBC) reported that "Britons are miserable—it's official." This conclusion reflected the poll responses, of which one quarter feared a "hopeless future," one in three felt "downright miserable," and nearly one in ten thought "their death would make things better for others." Which of the following is *not* correct about why these results may be misleading?

a. Many people who access a medical Web site and are willing to take the time to answer this questionnaire may be having emotional health problems.

b. Some respondents may not have been truthful or may have been Internet surfers who take pleasure in filling out a questionnaire multiple times with extreme answers.

c. The sample is a volunteer sample rather than a random sample.

d. It's impossible to learn useful results about a population from a sample of only 400 people.

4.107 Multiple choice: Sexual harassment In 1995 in the United Kingdom, the Equality Code used by the legal profession added a section to make members more aware of sexual harassment. It states that "research for the Bar found that over 40 percent of female junior tenants said they had encountered sexual harassment during their time at the Bar." This was based on a study conducted at the University of Sheffield that sent a questionnaire to 334 junior tenants at the Bar, of whom 159 responded. Of the 159, 67 were female. Of those females, 3 said they had experienced sexual harassment as a major problem, and 24 had experienced it as a slight problem.

a. The quoted statement might be misleading because the nonresponse was large.

b. No one was forced to respond, so everyone had a chance to be in the sample, which implies it was a simple random sample.

c. This was an example of a completely randomized experiment, with whether a female junior tenant experienced sexual harassment as the response variable.

d. This was a retrospective case-control study, with those who received sexual harassment as the cases.

4.108 Multiple choice: Effect of response categories A study (N. Schwarz et al., *Public Opinion Quarterly*, vol. 49, 1985, p. 388) asked German adults how many hours a day they spend watching TV on a typical day. When the possible responses were the six categories (up to $\frac{1}{2}$ hour, $\frac{1}{2}$ to 1 hour, 1 to $1\frac{1}{2}$ hours,..., more than $2\frac{1}{2}$ hours), 16% of respondents said they watched more than $2\frac{1}{2}$ hours per day. When the six categories were (up to $2\frac{1}{2}$ hours, $2\frac{1}{2}$ to 3 hours,..., more than 4 hours), 38% said they watched more than $2\frac{1}{2}$ hours per day.

a. The samples could not have been random, or this would not have happened.

b. This shows the importance of question design, especially when people may be uncertain what the answer to the question really is.

c. This study was an experiment, not an observational study.

4.109 Sample size and margin of error ◆◆

a. Find the approximate margin of error when $n = 1$.

b. Show the two possible percentage outcomes you can get with a single observation. Explain why the result in part a means that with only a single observation, you have essentially no information about the population percentage.

c. How large a sample size is needed to have a margin of error of about 5% in estimating a population percentage? (*Hint*: Take the formula for the approximate margin of error, and solve for the sample size.)

4.110 Systematic sampling ◆◆ A researcher wants to select 1% of the 10,000 subjects from the sampling frame. She selects subjects by picking one of the first 100 on the list at random, and then skipping 100 names to get the next subject, skipping another 100 names to get the next subject, and so on. This is called a **systematic random sample**.

a. With simple random sampling, (i) every subject is equally likely to be chosen, and (ii) every possible sample of size n is equally likely. Indicate which, if any, of (i) and (ii) are true for systematic random samples. Explain.

b. An assembly-line process in a manufacturing company is checked by using systematic random sampling to inspect 2% of the items. Explain how this sampling process would be implemented.

4.111 Complex multistage GSS sample ◆◆ Go to the Web site for the GSS, www.norc.org/GSS+Website/, click on *Documentation*, and then click on *Sampling Design and Weighting*. There you will see described the complex multistage design of the GSS. Explain how the GSS uses (a) clustering, (b) stratification, and (c) simple random sampling.

4.112 Mean family size You'd like to estimate the mean size ◆◆ of families in your community. Explain why you'll tend to get a smaller sample mean if you sample n families than if you sample n individuals (asking them to report their family size). (*Hint*: When you sample individuals, explain why you are more likely to sample a large family than a small family. To think of this, it may help to consider the case $n = 1$ with a population of two families, one with 10 people and one with only 2 people.)

4.113 Capture–recapture Biologists and naturalists often use ◆◆ sampling to estimate sizes of populations, such as deer or fish, for which a census is impossible. Capture–recapture is one method for doing this. A biologist wants to count the deer population in a certain region. She captures 50 deer, tags each, and then releases them. Several weeks later, she captures 125 deer and finds that 12 of them were tagged. Let N = population size, M = size of first sample, n = size of second sample, R = number tagged in second sample. The table shows how results can be summarized.

		In First Sample?		
		Yes (tagged)	**No (not tagged)**	**Total**
In Second	**Yes**	R		n
Sample?	**No**			
	Total	M		N

a. Identify the values of M, n, and R for the biologist's experiment.

b. One way to estimate N lets the sample proportion of tagged deer equal the population proportion of tagged deer. Explain why this means that

$$\frac{R}{n} = \frac{M}{N},$$

and hence that the estimated population size is $N = (M \times n)/R$.

c. Estimate the number of deer in the deer population using the numbers given.

d. The U.S. Census Bureau uses capture–recapture to make adjustments to the census by estimating the undercount. The capture phase is the census itself (persons are "tagged" by having returned their census form and being recorded as counted) and the recapture phase (the second sample) is the postenumerative survey (PES) conducted after the census. Label the table in terms of the census application.

Student Activities

4.114 Munchie capture–recapture Your class can use the capture–recapture method described in the previous exercise to estimate the number of goldfish in a bag of Cheddar Goldfish. Pour the Cheddar Goldfish into a paper bag, which represents the pond. Sample 10 of them. For this initial sample, use Pretzel Goldfish to replace them, to represent the tagged fish. Then select a second sample and derive an estimate for N, the number of Cheddar Goldfish in the original bag. See how close your estimate comes to the actual number of fish in the bag. (Your teacher will count the population of Cheddar Goldfish in the bag before beginning the sampling.) If the estimate is not close, what could be responsible, and what would this reflect as difficulties in a real-life application such as sampling a wildlife population?

4.115 Margin of error Activity 2 in Chapter 1 used the TRY Sample from a Population applet to simulate randomness and variability. We'll use that applet again here, but with a much larger sample size, 1000.

a. For a population proportion of 0.50 for outcome 1, simulate a random sample of size 1000. What is the sample proportion of outcome 1? Do this 10 separate times, keeping track of the 10 sample proportions.

b. Find the approximate margin of error for a sample proportion based on 1000 observations.

c. Using the margin of error found in part b and the 10 sample proportions found in part a, form 10 intervals of believable values for the true proportion. How many of these intervals captured the actual population proportion, 0.50?

d. Collect the 10 intervals from each member of the class. (If there are 20 students, 200 intervals will be collected.) What percentage of these intervals captured the actual population proportion, 0.50?

4.116 Activity: Sampling the states This activity illustrates how sampling bias can result when you use a nonrandom sample, even if you attempt to make it representative: You are in a geography class, discussing center and variability for several characteristics of the states in the contiguous United States. A particular value of center is the mean area of the states. A map and a list of the states with their areas (in square miles) are shown in the figure and table that follow. Area for a state includes dry land and permanent inland water surface.

Although we could use these data to calculate the actual mean area, let's explore *how well sampling performs in estimating the mean area* by sampling five states and finding the sample mean.

a. The most convenient sampling design is to use our eyes to pick five states from the map that we think have areas representative of all the states. Do this, picking five states that you believe have areas representative of the actual mean area of the states. Compute their sample mean area.

b. Collect the sample means for all class members. Construct a dot plot of these means. Describe the distribution of sample means. Note the shape, center, and variability of the distribution.

c. Another possible sampling design is simple random sampling. Select five random numbers between 01 and 48. Go to the table and take a simple random sample of five states. Compute the sample mean area.

d. Collect the sample means from part c of all class members. Construct a dot plot of the sample means using the same horizontal scale as in part b. Describe this distribution of sample means. Note the shape, center, and variability of the distribution.

e. The true mean total land area for the 48 states can be calculated from the accompanying table by dividing the total at the bottom of the table by 48. Which sampling method, using your eyes or using random selection, tended to be better at estimating the true population mean? Which method seems to be less biased? Explain.

f. Write a short summary comparing the two distributions of sample means.

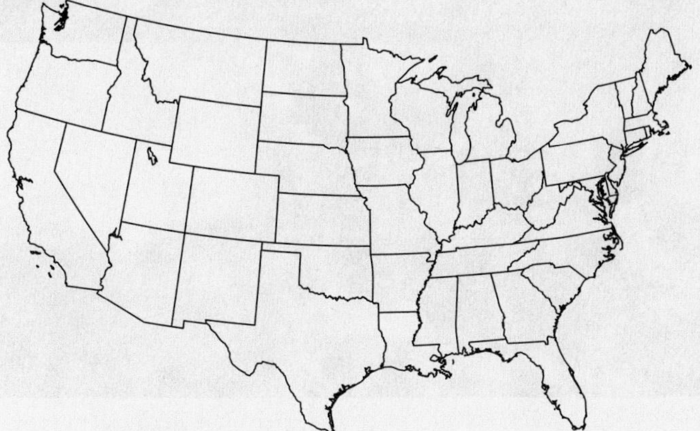

Map of the continental United States.

Areas of the 48 States in the Continental U.S.

State	Area (square miles)
Alabama	52,419
Arizona	113,998
Arkansas	53,179
California	163,696
Colorado	104,094
Connecticut	5,543
Delaware	2,489
Florida	65,755
Georgia	59,425
Idaho	83,570
Illinois	57,914
Indiana	36,418

(Continued)

State	Area (square miles)
Iowa	56,272
Kansas	82,277
Kentucky	40,409
Louisiana	51,840
Maine	35,385
Maryland	12,407
Massachusetts	10,555
Michigan	96,716
Minnesota	86,939
Mississippi	48,430
Missouri	69,704
Montana	147,042
Nebraska	77,354
Nevada	110,561
New Hampshire	9,350
New Jersey	8,721
New Mexico	121,589
New York	54,556
North Carolina	53,819
North Dakota	70,700
Ohio	44,825
Oklahoma	69,898
Oregon	98,381
Pennsylvania	46,055
Rhode Island	1,545
South Carolina	32,020
South Dakota	77,116
Tennessee	42,143
Texas	268,581
Utah	84,899
Vermont	9,614
Virginia	42,774
Washington	71,300
West Virginia	24,230
Wisconsin	65,498
Wyoming	97,814
U.S. TOTAL	**3,119,819**

BIBLIOGRAPHY

Burrill, G., C. Franklin, L. Godbold, and L. Young (2003). *Navigating through Data Analysis in Grades 9–12*, NCTM

Crossen, C. (1994). *Tainted Truth: The Manipulation of Fact in America*. New York: Simon & Schuster.

Hepworth, S. J., et al. (2006). "Mobile phone use and risk of glioma in adults: case control study." *BMJ* 332: 883–887

Leake, J. (2001) "Scientists Link Eye Cancer to Mobile Phones," *The London Times*, Jan. 14. www.emf-health.com /reports-eyecancer.htm.

Stang, A., et al. (2001). "The possible role of radio frequency radiation in the development of uveal melanoma." *Epidemiology* 12(1): 7–12.

Volkow, N. (2011). "Effect of Cell Phone Radiofrequency Signal Exposure on Brain Flucose Metabolism." *JAMA* 305(8): 808–813.

Part Review 1

Gathering and Exploring Data

In Chapters 1–4, we learned that *statistics* consists of methods for conducting research studies and for analyzing and interpreting the data produced by those studies. This review section is not comprehensive, but it gives examples of questions you should be able to answer about the main concepts in these chapters. After you read each question, think about how you would answer it. The questions are followed by brief summaries or hints as well as references to sections of the text where you can find more detail to help strengthen your understanding of these concepts.

Review Questions

- What is the difference between *descriptive statistics* and *inferential statistics*?

  Section 1.2 explained that we collect data for a **sample** of subjects. They are usually just a small part of the **population**, which is the set of *all* subjects in which we're interested. We use **descriptive statistics** to summarize the sample data with numbers and graphs. In future chapters, we'll use **inferential statistics** to make decisions and predictions about the entire population, based on the sample data.

- What is the difference between a *categorical* variable and a *quantitative* variable?

  From Section 2.1, a **categorical variable** has observations that fall into one of a set of categories, such as a preferred place to shop (with categories: local mall, downtown, Internet). A **quantitative variable** takes numerical values, such as grade point average.

- How can you describe a set of data *graphically*? What's the advantage/disadvantage of using a *histogram* compared to other graphs?

  Section 2.2 showed that for *categorical* variables, data are displayed using **pie charts** and **bar graphs**. For *quantitative* variables, a **histogram** is a graph of a frequency table, showing bars above intervals of values. The **stem-and-leaf plot** (a vertical line dividing the final digit, the leaf, from the stem) and **dot plot** (dots above the number line) show the individual observations. The histogram does not show the individual observations but more easily handles large amounts of data.

- How can you numerically summarize categorical data?

Section 2.1 showed that for categorical data, the data are summarized using frequencies (or counts) of categories, the proportions of observations in a category, or the percentage of observations in a category. We describe the category with the most observations as the modal category.

- How can you describe the *center* of a set of data numerically? Why is the *mean* sometimes less appropriate than the *median*?

Right-Skewed Distribution

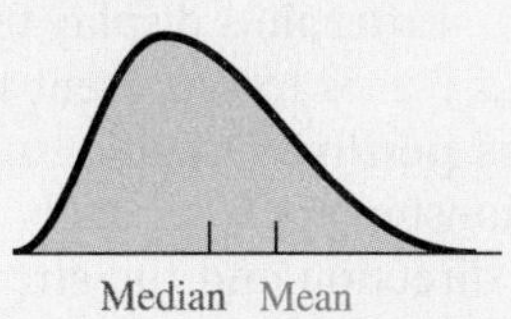

Chapter 2 showed that for quantitative variables, numerical summaries describe the **center** and **variability** of the data and relative **positions. Measures of the center** describe a "representative" or "typical" observation (Section 2.3). The **mean** is the balance point of a distribution and is calculated as the sum of the observations divided by the number of observations. The **median** divides the ordered data set into two parts of equal numbers of observations, half below and half above that point. For highly **skewed** distributions or distributions having extreme **outliers** in one direction, the mean is drawn in the direction of the longer tail of the distribution and may be a misleading summary.

- How can you describe the variability of a set of data numerically? How can you interpret the value of a *standard deviation*?

Section 2.4 introduced us to **measures** that describe the variability of the data. The **range** is the difference between the largest and smallest observations. The **deviation** for an observation is its distance from the mean. The square root of the average squared deviation, called the **standard deviation**, describes a typical distance from the mean. For a bell-shaped distribution, by the **empirical rule,** about (68%, 95%, all) the data fall within (1, 2, 3) standard deviations of the mean.

- What are measures of *position*, and how do a *boxplot* and *interquartile range* summarize positions?

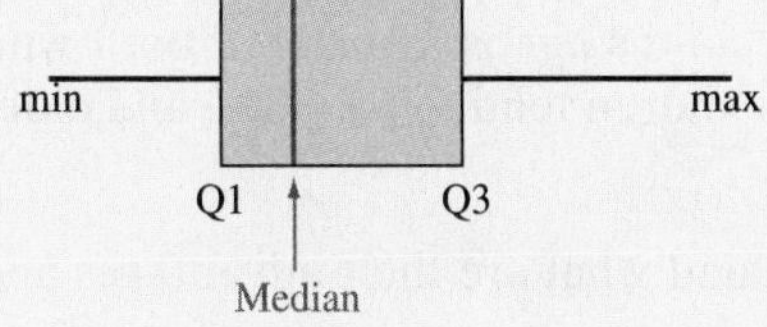

Section 2.5 showed that the **percentiles** are **measures of position** that tell us points above which or below which a certain percentage of the data fall. The lower quarter of the observations fall below the 25th percentile, called the **first quartile (Q1)**. The upper quarter fall above the 75th percentile, called the **third quartile (Q3)**. These two quartiles span the middle half of the data. The **five-number summary** consists of these quartiles, the median (which is the 50th percentile and second quartile), and the minimum and maximum values. The **box plot** portrays this five-number summary, using a box for the middle half of the data between Q1 and Q3, while highlighting potential outliers. The **interquartile range (IQR)** is a measure of variability that equals the distance between Q3 and Q1.

- What is a *z-score*? What values of z would be unusual if a distribution is bell shaped?

The ***z*-score** tells us the number of standard deviations that an observation falls from the mean. For example $z = -1.4$ means that the observation falls 1.4 standard deviations below the mean. For an approximately bell-shaped distribution, since all or nearly all the observations fall within 3 standard deviations of the mean, z-scores above 3 or below -3 would be unusual.

- What's the difference between a *response variable* and an *explanatory variable*, and why do we distinguish between them?

Section 3.1 explained that in practice, most studies have more than one variable and explore **associations** between the variables. For example, do students who spend more time studying tend to have higher GPAs? The statistical analysis studies how the value of a **response variable** (the outcome of interest, such as GPA) depends on the value of an **explanatory variable** (such as time spent studying).

■ What is a *contingency table*?

Belief in Life After Death by Gender		
	Yes	No
Female	90	10
Male	87	13

Section 3.1 showed that for *categorical variables*, **contingency tables** summarize the counts of observations at the various combinations of categories of the two variables. They can show the percentages in the various categories for the response variable (such as yes or no for whether one believes in life after death) separately for each category of the explanatory variable (such as gender or religious affiliation).

■ How can you describe the *association* between two *quantitative* variables graphically and numerically?

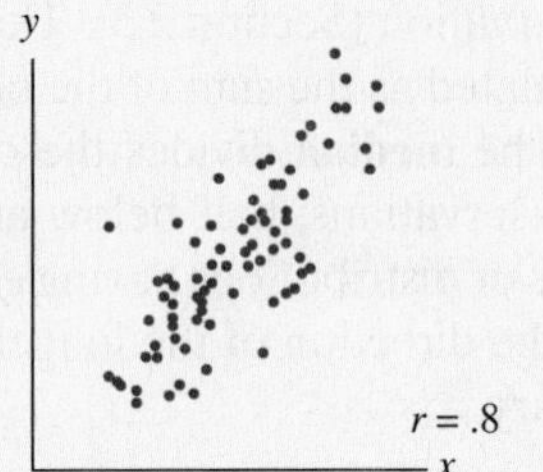

Section 3.2 showed that for *quantitative variables*, **scatterplots** display the data, with one point for each subject, using the *x*- and *y*-axes to represent the two variables. They show whether the association is **positive** (trending upward) or **negative** (trending downward). When a relationship approximately follows a straight line, the **correlation**, *r*, describes the direction and the strength of association. It satisfies $-1 \leq r \leq +1$, with values farther from 0 representing stronger straight-line associations.

■ How can you use an equation to describe the relationship between two *quantitative* variables?

Section 3.3 showed that a **regression line** $\hat{y} = a + bx$ describes how the predicted value $\hat{y}$ of the response variable relates to the explanatory variable *x* when the scatterplot indicates a linear relationship. The slope *b* of this line describes the effect on $\hat{y}$ of a 1-unit increase in *x*.

■ Why does *association not imply causation*?

Section 3.4 explained how an association may occur merely because of a **lurking variable** that is associated with both of the variables. For example, we might observe a positive association between a math achievement test score and height for children from various grade levels in a school. But this association may be explained by the way that a child's *age* is associated both with math achievement and with height: Older children tend to be taller, and older children tend to score better on the test.

■ What are the main ways of *gathering data*, and what are their advantages and disadvantages?

Chapter 4 explained that an **experiment** randomly assigns subjects to experimental conditions called **treatments** and then observes the outcome. Randomly assigning units to treatments in an experiment balances groups with respect to lurking variables and leaves two possible causes for a difference in the response to the treatments: either the treatments or random variation. Random assignment in experiments allows for cause-and-effect conclusions. An **observational study** merely observes an available sample of subjects without conducting an experiment. Another type of nonexperimental study uses a **sample survey** to take a sample of subjects from a population and observe them, usually by obtaining answers to questions on a questionnaire. With **random sampling**, each subject in the population has an opportunity of being in the sample. Random sampling enables us to make inferences from the sample to the population.

■ Explain the difference among *sampling bias, nonresponse bias*, and *response bias* in sample surveys.

Section 4.2 showed that **sampling bias** can occur from using nonrandom samples (such as volunteer samples) or having undercoverage (when some subjects have no chance of being sampled).

Severe **nonresponse bias** can occur when many sampled subjects refuse to participate, and **response bias** occurs when subjects respond incorrectly (perhaps lying) or a question asked is confusing or misleading.

Here's an example of the sort of exercise you should be able to answer at this stage of the course:

Example

Time Spent on the Internet

Picture the Scenario

When the General Social Survey (GSS) last asked about the number of hours a week spent on the Internet (variable denoted WWWHRS) software summarized the observations for the 2778 subjects in the sample, for each gender, by

Group	N	Mean	Median	StDev	Minimum	Maximum
Female	1574	5.07	1.0	9.42	0	60
Male	1204	6.62	2.0	11.3	0	60

Questions to Explore

a. Identify the response variable and explanatory variable. Indicate whether each variable is quantitative or categorical.

b. The GSS takes a multistage sample that incorporates randomness at each stage. Identify the population and the sample. Why doesn't the GSS instead use the simpler and cheaper approach of collecting data from people who visit the GSS Web site?

c. Do you think the distribution of time spent on the Internet was bell shaped? Why or why not? What would you predict for the shape of the distribution?

d. The GSS also observes family income. For this sample, time spent on the Internet had a correlation of 0.09 with family income. Interpret.

Think It Through

a. The response variable is the number of hours a week spent on the Internet. This variable takes numerical values, so it is quantitative. The explanatory variable is gender. Its values (female, male) are categories, so it is a categorical variable.

b. For the GSS, the population consists of all adult Americans. The sample is the 2778 adult Americans who gave responses in the GSS. The hypothetical sample the GSS could use of those visitors to the GSS Web site who agree to provide data would be a *volunteer sample.* Such a sample usually has *sampling bias*, not being a good representation of the population. For example, those who visit the GSS Web site might tend to spend more time on the Internet than the general population, perhaps because they have more free time or are younger or are more educated. Samples that instead incorporate randomness tend to be more representative of the population.

c. Time spent on the Internet can take only nonnegative values. Since the standard deviation is larger than the mean (for each gender), the lowest possible value of 0 is less than 1 standard deviation below the mean. If the distribution were bell shaped, observations could fall 2 or even 3 standard deviations from the mean. Also, the mean is quite a bit larger than the median. These factors suggest that this distribution is skewed to the right.

d. The correlation falls between −1 and +1. Values near 0 indicate weak associations, in terms of straight-line trends. As family income increases, there is only a very weak tendency for time spent on the Internet to increase.

Insight

The mean and the median both indicate that, in this sample, males tended to spend more time on the Internet than females. Can we conclude that this difference is also true in the entire population? In the next part of this book, we'll learn how to answer this question.

Try Exercise R1.2

Part 1 Review Exercises

Practicing the Basics

R1.1 Believe in astrology? A General Social Survey asked whether astrology—the study of star signs—has some scientific truth (GSS question SCITEST3). Of 1245 subjects in the sample, 651 responded definitely or probably true, and 594 responded definitely or probably not true.

a. Identify the sample and the population of interest.

b. Identify the variable observed. Is it quantitative, or categorical?

c. A parameter of interest is the population proportion who would respond definitely or probably true. This can be approximated by the sample proportion. Find this sample proportion.

R1.2 Time spent in housework When the General Social Survey last asked the number of hours the respondent spent a week on housework (variable RHHWORK), the responses were summarized for each gender by

TRY

Group	N	Mean	Median	StDev	Minimum	Maximum
Female	391	12.7	10.0	11.6	0	84
Male	292	8.4	5.0	9.5	0	60

a. Identify the response variable and explanatory variable. Indicate whether each is quantitative or categorical.

b. Identify the population and the sample.

c. Do you think the distributions were bell shaped? Why or why not? What would you predict for the shape of the distributions?

R1.3 Best long-term investment A public opinion poll conducted by *Gallup*[1] interviewed 1077 adults by random telephone dialing during April 7–11, 2011. These adults were asked, "Which of the following do you think is the best long-term investment?" The possible choices were real estate, savings accounts/CDs, stocks/mutual funds, and bonds.

a. Is this variable quantitative, or categorical? Why?

b. The percentages in the four categories were 33% (real estate), 24% (savings accounts/CDs), 24% (stocks/mutual funds), and 12% (bonds). The other 7% of the sample did not respond. Are the given percentages statistics or parameters? Why?

R1.4 Pay more to reduce global warming? In a recent survey of Europeans by Eurobarometer about energy issues and global warming,[2] one question asked, "Would you be willing to pay more for energy produced from renewable sources than for energy produced from other sources?" The percentage of yes responses varied among countries between 10% (in Bulgaria) to 60% (in Luxembourg). Of the 631 subjects interviewed in the UK, 45% said yes. For all 48 million adults in the UK, that percentage who would answer yes was predicted to fall between 41% and 49%. Identify in this discussion (a) a statistic, (b) a parameter, (c) a descriptive statistical analysis, and (d) an inferential statistical analysis.

R1.5 Religions According to www.adherents.com, in 2006 the number of followers of the world's five largest religions were 2.1 billion for Christianity, 1.3 billion for Islam, 0.9 billion for Hinduism, 0.4 billion for Confucianism, and 0.4 billion for Buddhism.

a. Construct a relative frequency distribution.

b. Sketch a bar graph.

c. Can you find a mean, median, or mode for these data? If so, do so and interpret. If not, explain why not.

R1.6 Highest degree The table summarizes the distribution of the highest degree completed in the U.S. population of age 25 years and over *(2005 American Community Survey)*.

a. Find and interpret the modal category.

b. Is the median appropriate for these ordered categories? If so, find it. If not, explain why not.

[1] www.gallup.com/poll/147206/Stock-Market-Investments-Lowest-1999.aspx.

[2] *Attitudes Towards Energy*, published January 2006 at ec.europa.eu/public_opinion.

(See Exercise 2.133 for a discussion about ordered categorical data.)

Highest Degree	Frequency (millions)	Percentage
Not a high school graduate	30	16.4%
High school only	56	30.6%
Some college, no degree	38	20.8%
Associate's degree	14	7.7%
Bachelor's degree	32	17.5%
Master's degree	13	7.1%

R1.7 Newspaper reading The 2008 General Social Survey asked respondents, "How often do you read the newspaper?" The possible responses were (every day, a few times a week, once a week, less than once a week, never). The counts in those categories were (431, 300, 207, 200, 191).

a. Identify the modal category and the median response for these ordered categories.

b. For the scores (7, 3, 1, 0.5, 0) for the categories, find the sample mean number of times per week that a newspaper is read. Interpret, and compare to the mean of 4.4 for the 1994 GSS.

R1.8 Earnings by gender According to the *2005 American Community Survey* taken by the U.S. Bureau of the Census, the median earnings in the past 12 months was $32,168 for females and $41,965 for males, whereas the mean was $39,890 for females and $56,724 for males.

a. Does this suggest that the distribution of income for each gender is symmetric, skewed to the right, or skewed to the left? Explain.

b. The results refer to 73.8 million females and 83.4 million males. Find the overall mean income.

R1.9 Females in labor force In *Human Development Report 2005*, the United Nations reported an index of female economic activity. This index specifies employment as a percentage of male employment. The value for the U.S. and Canada was 83, which indicates that the number of females in the work force was 83% of the number of males in the work force.

a. In Eastern Europe, the values were Czech Republic 83, Estonia 82, Hungary 72, Latvia 80, Lithuania 80, Poland 81, Slovakia 84, and Slovenia 81. Summarize female economic activity in Eastern Europe by the mean and median.

b. In South America, the values were Argentina 48, Bolivia 58, Brazil 52, Chile 50, Colombia 62, Ecuador 40, Guyana 51, Paraguay 44, Peru 45, Uruguay 68, and Venezuela 55. Does female economic activity tend to be lower in South America than in Eastern Europe? Explain.

c. Consider the variables, nation and female economic activity. Which is quantitative? Which is the response variable?

R1.10 Females working in Europe The United Nations reports that in Western Europe female employment as a percentage of male employment is Austria 66, Germany 71, Norway 86, Belgium 67, Greece 60, Portugal 72, Cyprus 63, Ireland 54, Spain 58, Denmark 85, Italy 60, Sweden 90, Finland 87, Luxembourg 58, U.K. 76, France 78, and Netherlands 68.

a. Construct a graph to compare these values to the values for South America in the previous exercise. Interpret, explaining how this plot shows which group of values tend to be larger.

b. For the Western Europe nations, these values have $\bar{x} = 70.5$ and $s = 11.4$. By contrast, for Eastern Europe, $\bar{x} = 80.4$ and $s = 3.7$. Explain how to compare the s values for Western Europe and Eastern Europe.

R1.11 Golf scoring An article[3] about factors that affect scoring in golf analyzed recent data for professional golfers on the PGA tour. The correlation with scoring average was 0.05 for driving distance, 0.15 for driving accuracy, 0.62 for percentage of greens reached in regulation, 0.31 for average number of putts per round, and 0.63 for average number of putts taken on holes for which the green was reached in regulation. (Note: For golf, lower scores mean better performance. Reaching a green "in regulation" means taking 1 shot on a par 3 hole, 2 shots on a par 4 hole, or 3 shots on a par 5 hole.)

a. Does driving distance seem to have a strong or a weak association with scoring average? Explain.

b. Explain what it means to have a negative correlation between scoring average and percentage of greens reached in regulation.

c. Of the factors reported, choose the two that seem to be most important in contributing to the scoring average.

R1.12 Holiday time Excluding the United States, the national mean number of holiday and vacation days in a year for the Organisation for Economic Co-operation and Development (OECD) (advanced industrialized) nations is approximately bell shaped with a mean of 35 days and standard deviation of 3 days.[4]

a. Using the empirical rule, report the range within which all or nearly all values fall of the annual number of holiday and vacation days for OECD nations.

b. The observation for the United States is 19. If this observation is included with the others, (i) will the mean increase or decrease? and (ii) will the standard deviation increase or decrease? Explain your reasoning.

c. Using the mean and standard deviation for the other countries, how many standard deviations is the U.S. observation from the mean? Interpret.

R1.13 Infant mortality According to the OECD Health Data for 2010[5] infant mortality rates (number of infant deaths, per 1000 live births) for OECD nations in 2008 had minimum = 1.8 (Luxembourg), lower quartile = 2.775,

[3]R. Quinn, *Teaching Statistics,* vol. 28, Spring 2006, pp. 10–13.

[4]Table 8.9 in www.stateofworkingamerica.org, from The Economic Policy Institute.

[5]www.oecd.org/document/16/0,3343,en_2649_34631_2085200_1_1_1_1,00.html.

median = 3.8, upper quartile = 5.025, and maximum = 17 (Turkey). Interpret this five-number summary.

R1.14 Murder rates The table shows part of a printout for analyzing murder rates (per 100,000) in the 50 U.S. states and the District of Columbia in 2005. The first column refers to the entire data set, and the second column deletes the observation for D.C.

a. Why are the means and standard deviations so different but the medians the same for the two sets of data?

b. Of the range and the interquartile range, which is most affected by outliers? Why? Illustrate with these data.

n	51	*n*	50
Mean	5.6	Mean	4.8
Std Dev	6.05	Std Dev	2.57
Maximum	44	Maximum	13
75% Q3	6	75% Q3	6
Median	5	Median	5
25% Q1	3	25% Q1	3
Minimum	1	Minimum	1

R1.15 Using water A report[6] indicated that annual water consumption for nations in the OECD was skewed to the right, with values (in cubic meters per capita) having a median of about 500 and ranging from about 200 in Denmark to 1700 in the United States Which is the most realistic value?

a. For the standard deviation: −10, 0, 10, 300, 1000. Why?

b. For the IQR: −10, 0, 10, 350, 1500. Why?

R1.16 Energy consumption The United Nations publication *Energy Statistics Yearbook* (unstats.un.org/unsd/energy) lists consumption of energy. For the 25 nations that made up the European Union (EU) in 2006, the energy values (in kilograms per capita) had a mean of 4998 and a standard deviation of 1786.

a. Italy had a value of 4222. How many standard deviations from the mean was it?

b. The value for the United States was 11,067. Relative to the distribution for the EU, find its z-score. Interpret.

c. If the distribution of EU energy values were bell shaped, would a value of 11,067 be unusually high? Why?

R1.17 Human contacts When the General Social Survey asked subjects of age 18–25 in 2004 how many people they were in contact with at least once a year (NUMCNTCT), the responses had mean = 20.2, mode = 10, standard deviation = 28.7, minimum = 0, lower quartile = 5, median = 10, upper quartile = 25, maximum = 300.

a. Based on these values, predict the shape of the distribution. Explain.

b. The five values above 50 were 80, 80, 100, 100, 300. Sketch a box plot of the data, identifying outliers. (Assume 50 is the largest observation before the values of 80, 100, and 300.)

[6]OECD Key Environmental Indicators 2005.

R1.18 Attacked in Iraq A study[7] of American armed forces who had served in Iraq or Afghanistan found that the event of being attacked or ambushed was reported by 1139 of 1961 Army members who had served in Afghanistan, 789 of 883 Army members who had served in Iraq, and 764 of 805 Marines who had served in Iraq.

a. Display these data in the form of a contingency table with three rows and two columns, identifying the response variable and the explanatory variable.

b. What proportion was attacked, given that one served in the (i) Army in Iraq and (ii) Marines in Iraq?

R1.19 Opinion about homosexuality The contingency table shows results from answers to two questions on the 2008 General Social Survey.

a. Of those who reported being fundamentalist in religion, what percent thought homosexual relations are always wrong?

b. For those who reported being liberal in religion, what percent thought homosexual relations are always wrong?

c. Based on parts a and b, does there seem to be an association between these two variables? Explain.

Opinion about Homosexual Relations

Religion	Always Wrong	Almost Always Wrong	Sometimes Wrong	Not Wrong at All	Total
Fundamentalist	282	7	19	60	368
Liberal	116	16	34	215	381

R1.20 Iris blossoms Exercise 2.26 discussed a data set containing various measurements on different varieties of iris blossoms. Two of those varieties are the *setosa* and *versicolor.* The following graph depicts the lengths and widths of the petals and sepals of 50 blossoms of each variety. Note how the graph incorporates information by using different colors and symbols to represent each measurement.

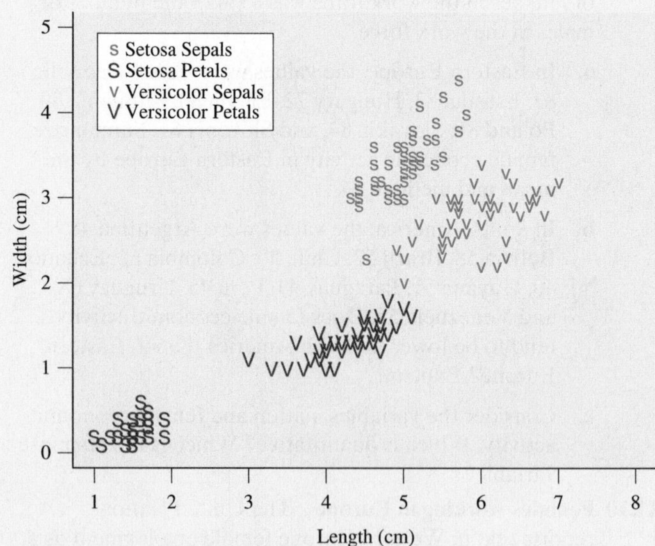

[7]C. Hoge et al. (2004), *New England J. Med.*, 351:13–21.

a. What are some comparisons that can be made based on the graph?

b. Which variety exhibits the most variation in petal length?

c. For which species is the correlation between petal length and petal width the strongest?

d. Consider an iris for which the petal length is 4 cm. Is the iris more likely to be a *setosa* or a *versicolor*? Consider another iris for which the sepal length is 5.5 cm. Is the iris more likely to be a *setosa* or a *versicolor*? Does knowing the length of the petal or sepal provide more information about which type of species a blossom might be?

R1.21 How much is a college degree worth? An Associated Press story (October 27, 2006) reported the U.S. Census Bureau as stating that college graduates made an average of $51,554 in 2004, compared with $28,645 for adults with a high school diploma. Suppose you performed a regression analysis with y = annual income and x = number of years of education.

a. Assuming four years of time in college, predict what you would get for the slope.

b. Suppose older people tend to have higher incomes but tend to be less likely to be college graduates. Could age be a lurking variable that is responsible for the association between x and y? Explain.

R1.22 Yours and mother's education For 1777 observations from the 2008 GSS on y = number of years of education (EDUC) and x = number of years of mother's educations (MAEDUC), $\hat{y} = 9.592 + 0.35x$.

a. Predict the number of years of education of a person whose mother had 10 years of education.

b. Is the correlation between these variables positive, or negative? Why?

R1.23 Child poverty For the 20 advanced, industrialized countries in the OECD, the prediction equation $\hat{y} = 22 - 1.3x$ recently related y = child poverty rate to x = social expenditure as a percent of gross domestic product.[8] The y-values ranged from 2.8% (Finland) to 21.9% (U.S). The x-values ranged from 2% (U.S.) to 16% (Denmark).

a. Interpret the y-intercept and the slope.

b. Find the predicted poverty rates for the United States and for Denmark.

c. The correlation is −0.79. Interpret.

R1.24 TV and GPA A survey[9] of 50 college students in an introductory psychology class observed self reports of y = high school GPA and x = weekly number of hours viewing television. The study reported $\hat{y} = 3.44 - 0.03x$.

a. Interpret the intercept and the slope.

b. Find the predicted GPA of a student who watched 20 hours a week of TV.

c. The study reported a correlation of −0.49. Interpret.

R1.25 U.S. child poverty Look at Figure 2 in the pdf file found at www.ajph.org/cgi/reprint/93/4/652?ck=nck, which is a scatterplot for U.S. states with correlation 0.53 between x = child poverty rate and y = child mortality rate. For the graph with Federally-Referenced Child Poverty Rate on the x-axis, approximate the y-intercept and slope of the prediction equation shown there.

R1.26 Ginger for pain relief It has long been thought that ginger can remedy a variety of conditions such as colds and upset stomachs. A study at the University of Georgia investigated ginger as a potential pain reliever following strenuous exercise activities.[10] As part of the study, volunteers were randomly assigned to take either ginger or placebo for 11 days. The study was double-blind in the sense that neither the researchers nor the volunteers knew which treatments were administered to which volunteers. On the eighth day, each volunteer was asked to engage in an exercise activity that places stress on the flexor muscles. At the end of the 11 days, researchers assessed the pain levels of the volunteers and found that ginger reduced pain by 25% compared to the placebo group.

a. Explain why randomization was used.

b. Explain why those not taking ginger were given a placebo. What is the purpose of blinding the volunteers to their treatments? The researchers?

R1.27 Fewer vacations and death A study using a 20-year follow-up of women participants in the Framingham Heart Study[11] found that less frequent vacationing was associated with greater frequency of deaths from heart attacks. A later study[12] questioned whether this could be explained by the effects of a lurking variable such as socioeconomic status (SES). Explain why if higher SES is responsible both for lower mortality and for more frequent vacations, this association could disappear.

R1.28 Education and a long life? An article[13] in *The New York Times* summarized research studies dealing with human longevity. It noted that consistently across studies, life length was positively associated with educational attainment. Many researchers believe education is the most important variable in explaining how long a person lives. Is having more education responsible for having a longer life?

a. Explain how the causation could potentially go in the other direction, for example if in some societies, many children die at an early age and never have a chance to attend school much, if at all.

b. Suppose education leads to greater wealth, which then (possibly for a variety of reasons) leads to living longer. Explain how this could be responsible for this association.

R1.29 Taxes and global warming When a *New York Times*/CBS News poll in 2006 asked whether the interviewee would be in favor of a new gasoline tax, only 12% said

[8]*Source:* Figure 8H in www.stateofworkingamerica.org, from The Economic Policy Institute.

[9]www.iusb.edu/~journal/2002/hershberger/hershberger.html.

[10]www.onlineathens.com/stories/052510/new_644007510.shtml.

[11]E.D. Eaker et al. (1992), *Amer. J. Epidemiology,* vol. 135, pp. 835–864.

[12]B. B. Gump and K.A. Matthews (2000), *Psychosomatic Medicine,* vol. 62, pp. 608–612.

[13]By G. Kolata, January 3, 2007.

yes. When the tax was presented as reducing U.S. dependence on foreign oil, 55% said yes, and when asked about a gas tax that would help reduce global warming, 59% said yes.[14] Explain what type of bias this example reflects.

Concepts and Investigations

R1.30 Executive pay An article[15] about how incomes had increased dramatically for the top executives of American companies in recent years reported, "In 2000–03 their mean annual pay was $8.5 million and the median $4.1 million. The median is a better measure than the mean because the mean is skewed by a few huge payments."

a. Explain this reasoning, and explain why the median is not influenced by the few huge payments.

b. Would the standard deviation or the IQR be influenced by the few huge payments? Explain.

R1.31 Fat, sugar, and health After pointing out that diets high in fats and sugars (bad for our health) are more affordable than diets high in fruit and vegetables (good for our health), a recent study[16] reported that every extra 100 g of fats and sweets eaten decreased diet costs by an average of 0.22 euros, whereas every extra 100 g of fruit and vegetables eaten increased diet costs by an average of 0.23 euros. Approximate the slope if we fit a regression line to y = cost and

a. x = fats and sweets eaten (in 100s of grams).

b. x = fruits and vegetables eaten (in 100s of grams).

c. In parts a and b, indicate whether the correlation would be positive or negative, and explain why.

R1.32 Effects of nuclear fallout An article appearing in the *St. Louis Post Dispatch* in October 2009 discusses a long-running investigation of the effect of nuclear fallout on children born in and around St. Louis in the 1960s. One phase of the study, referred to as the St. Louis Tooth Survey, collected baby teeth from more than 300,000 children in the St. Louis area. Analysis of those teeth concluded that children born in St. Louis in 1964 after the start of atomic testing in Nevada had 50 times more of the radioactive strontium-90 isotope in their baby teeth than children born in 1950, before the atomic testing began. In a more recent phase of the study, a New York-based research group determined that male tooth donors who ended up with cancer as adults had, on average, more than double the amount of strontium-90 of healthy donors.

a. This an example of an observational study. Explain obstacles scientists would encounter preventing them from carrying out a similar study in an experimental framework.

b. How might you try to convince a skeptic who says that the results of this study do not prove anything because everything is based on observational data?

R1.33 Sneezing at benefits of echinacea A study by Dr. R. Turner et al. published in *New England Journal of Medicine* in July 2005 claimed that taking echinacea to ward off a cold or reduce its symptoms is a waste of time and money. In the study, 339 healthy volunteers were split randomly into two groups. One was further divided into subgroups who received three different doses of the echinacea extract. The second group, which was also divided into subgroups, received an inactive placebo. Five days later, both groups were exposed to a rhinovirus through a nasal spray. Among those receiving the herb, 81 to 92% caught colds, compared with 85 to 92% of those given the placebo. An Associated Press story (July 28, 2005) about the study quoted Kevin Park, manager at Life Spring Health Foods and Juice Bar, as saying he would continue to sell echinacea, because "Customers keep coming back and telling me good things." Use this example to explain the difference between anecdotal evidence and randomized experiments, and explain why results of randomized experiments are much more trustworthy.

R1.34 Compulsive buying A study[17] of compulsive buying behavior conducted a national telephone survey in 2004 of adults ages 18 and over. The study found that lower income subjects were more likely to be compulsive buyers. The authors reported, "Compulsive buyers did not differ significantly from other respondents in mean total credit card balances, but the compulsive buyers' lower income was a confounding factor." Explain the meaning of this sentence.

R1.35 Internet time and age How does the amount of time spent on the Internet depend on age? At the GSS Web site sda.berkeley.edu/GSS, conduct a regression and correlation analysis for the variables WWWHR and AGE with the most recent GSS data. Prepare a short report summarizing your analysis.

---

[14]Column by T. Friedman, *The New York Times*, March 2, 2006.

[15]*The Economist*, January 20, 2007.

[16]E. Frazao and E. Golan (2005), *Evidence-Based Healthcare and Public Health*, vol. 9, pp. 104–107.

[17]Koran et al. (2006), *Amer. J. Psychiatry*, vol. 163, p. 1806.

Probability, Probability Distributions, and Sampling Distributions

Part

2

5 Probability in Our Daily Lives

Example 1

Failing a Test for Illegal Drug Use

Picture the Scenario

Many employers require potential employees to take a diagnostic test for illegal drug use. Such tests are also used to detect drugs used by athletes. For example, at the 2008 Summer Olympic Games in Beijing, 6 out of 4500 sampled specimens tested positive for a banned substance.

But diagnostic tests have a broader use. They can be used to detect certain medical conditions. For example, one test for detecting HIV is the ELISA screening test.

Questions to Explore

- Given that a person recently used drugs, how can we estimate the likelihood that a diagnostic test will correctly predict drug use?
- Suppose a diagnostic test says that a person has recently used drugs. How likely is it that the person truly did use drugs?

Thinking Ahead

We'll learn how to answer the preceding questions using probability methods; the answers might be quite different from what you might expect. In Example 8, we'll look at data on the performance of a diagnostic test for detecting Down syndrome in pregnant women. In Example 15, we'll analyze results of random drug tests for air traffic controllers.

Learning how to find probabilities and how to interpret them will help you, as a consumer of information, understand how to assess probabilities in many uncertain aspects of your life.

In everyday life, often you must make decisions when you are uncertain about the outcome. Should you invest money in the stock market? Should you get extra collision insurance on your car? Should you start a new business, such as opening a pizza place across from campus? Daily, you face even mundane decisions, such as whether or not to carry an umbrella with you in case it rains.

This chapter introduces **probability**—the way we quantify uncertainty. You'll learn how to measure the chances of the possible outcomes for **random phenomena**—everyday choices for which the outcome is uncertain. Using probability, for instance, you can find the chance of winning the lottery. You can find the likelihood that an employer's drug test correctly detects whether or not you've used drugs. You can measure the uncertainty that comes with randomized experiments and with random sampling in surveys. The ideas in this chapter set the foundation for how we'll use probability in the rest of the book to make inferences based on data.

5.1 How Probability Quantifies Randomness

As we discovered in Chapter 4, there's an essential component that statisticians rely on to avoid bias in gathering data. This is **randomness**—randomly assigning subjects to treatments or randomly selecting people for a sample. Randomness also applies to the outcomes of a response variable. The possible outcomes are known, but it's uncertain which outcome will occur for any given observation.

We've all employed randomization in games. Some popular randomizers are rolling dice, spinning a wheel, and flipping a coin. Randomization helps to make a game fair, each player having the same chances for the possible outcomes. Rolls of dice and flips of coins are simple ways to represent the randomness of randomized experiments and sample surveys. For instance, the head and tail outcomes of a coin flip can represent drug and placebo when a medical study randomly assigns a subject to receive one of two treatments.

With a *small* number of observations, outcomes of *random phenomena* may look quite different from what you expect. For instance, you may expect to see a random pattern with different outcomes; instead, exactly the same outcome may happen a few times in a row. That's not necessarily a surprise, as unpredictability for any given observation is the essence of randomness. We'll discover, however, that with a *large* number of observations, summary statistics "settle down" and get increasingly closer to particular numbers. For example, with 4 tosses of a coin, we wouldn't be surprised to find all 4 tosses resulting in heads. However, with 100 tosses, we would be surprised to see all 100 tosses resulting in heads. As we make more observations, the proportion of times that a particular outcome occurs gets closer and closer to a certain number we would expect. This long-run proportion provides the basis for the definition of *probability*.

In Words

Phenomena is any observable occurrence.

Example 2

Randomness

The Fairness of Rolling Dice

Picture the Scenario

The board game you've been playing has a die that determines the number of spaces moved on the board. After you've rolled the die 100 times, the number 6 has appeared 23 times, more frequently than each of the other 5 numbers, 1 through 5. At one point, it turns up three times in a row, resulting in you winning a game. Your opponent then complains that the die favors the number 6 and is not a fair die.

Questions to Explore

a. If a fair die is rolled 100 times, how many 6s do you expect?

b. Would it be unusual for a 6 to be rolled 23 times in 100 rolls? Would it be surprising to roll three 6s in a row at some point?

Think It Through

a. With many rolls of a fair die, each of the six numbers would appear about equally often. A 6 should occur about one sixth of the time. In 100 rolls, we expect a 6 to come up about $(1/6)100 = 16.7 \approx 17$ times.

b. How can we determine whether or not it is unusual for 6 to come up 23 times out of 100 rolls, or three times in a row at some point? One way uses simulation. We could roll the die 100 times and see what happens, roll it another 100 times and see what happens that time, and so on. Does a 6 appear 23 (or more) times in many of the simulations? Does a 6 occur three times in a row in many simulations?

This simulation using a die would be tedious. Fortunately, we can use an applet or other software to simulate rolling a fair die. Each simulated roll of a die is called a **trial**. After each trial, we record whether or not a 6 occurred. We will keep a running record of the proportion of times that a 6 has occurred. At each value for the number of trials, this is called a **cumulative proportion**. Table 5.1 shows partial results of one simulation of 100 rolls. To find the cumulative proportion after a certain number of trials, divide the number of 6s at that stage by the number of trials. For example, by the eighth roll (trial), there had been three 6s in eight trials, so the cumulative proportion is $3/8 = 0.375$. Figure 5.1 plots the cumulative proportions against the trial number.

Did You Know?

The singular of *dice* is *die*. For a proper die, numbers on opposite sides add to 7, and when 4 faces up, the die can be turned so that 2 faces the player, 1 is at the player's right, and 6 is at the player's left. (*Ainslie's Complete Hoyle,* New York: Simon & Schuster, 2003) ◀

Table 5.1 Simulation Results of Rolling a Fair Die 100 Times

Each trial is a simulated roll of the die, with chance 1/6 of a 6. At each trial, we record whether a 6 occurred as well as the cumulative proportion of 6s by that trial.

Trial	6 Occurs?	Cumulative Proportion of 6s	
1	yes	1/1	= 1.0
2	no	1/2	= 0.500
3	yes	2/3	= 0.667
4	no	2/4	= 0.500
5	yes	3/5	= 0.600
6	no	3/6	= 0.500
7	no	3/7	= 0.429
8	no	3/8	= 0.375
⋮	⋮	⋮	⋮
30	no	9/30	= 0.300
31	no	9/31	= 0.290
32	yes	10/32	= 0.313
33	yes	11/33	= 0.333
34	yes	12/34	= 0.353
35	no	12/35	= 0.343
⋮	⋮	⋮	⋮
99	no	22/99	= 0.220
100	yes	23/100	= 0.230

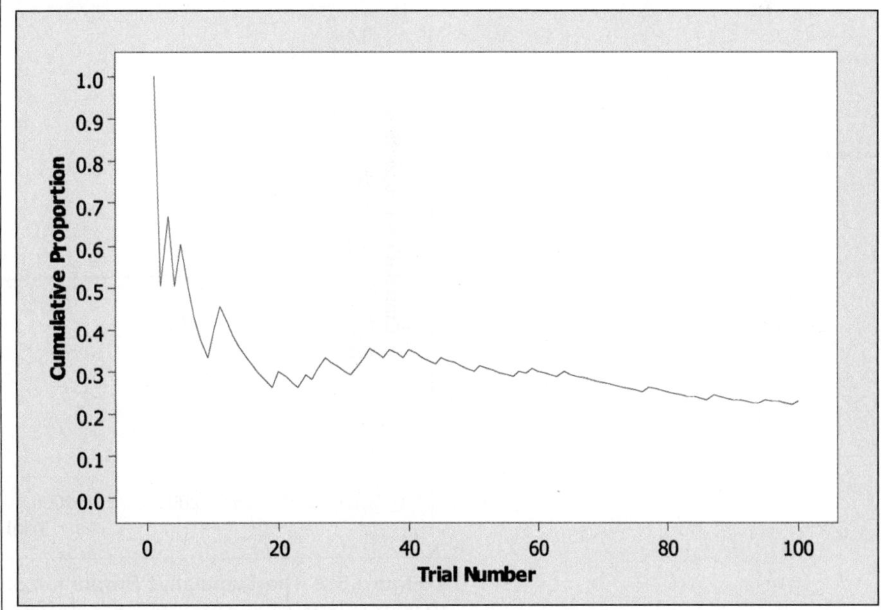

▲ **Figure 5.1 The Cumulative Proportion of Times a 6 Occurs, for a Simulation of 100 Rolls of a Fair Die.** The horizontal axis of this MINITAB figure reports the number of the trial, and the vertical axis reports the cumulative proportion of 6s observed by then. **Question** The first four rolls of the die were 6, 2, 6, and 5. How can you find the cumulative proportion of 6s after each of the first four trials?

In this simulation of 100 rolls of a die, a 6 occurred 23 times, different from the expected value of about 17. From Table 5.1, a 6 appeared three times in a row for trials 32 through 34.

Insight

One simulation does not prove anything. It suggests, however, that rolling three 6s in a row out of 100 rolls may not be highly unusual. It also shows that 23 rolls with a 6 out of 100 trials can occur. To find out whether 23 rolls with a 6 is unusual, we need to repeat this simulation many times. In Chapter 6, we will learn about the binomial distribution, which allows us to compute the likelihood for observing 23 (or more) 6s out of 100 trials.

Try Exercise 5.10

Activity

You can try this yourself using the Simulating the Probability of Rolling a 6 applet on the text CD. This is designed to generate "binary" data, which means that each trial has only *two* possible outcomes, such as "6" or "not 6." See Activity 1 on page 213.

Long-Run Behavior of Random Outcomes

The number of 6s in 100 rolls of a die can vary. One time we might get 19 rolls with 6s, another time we might get 22, another time 13, and so on. So what do we mean when we say that there's a one-in-six chance of a 6 on any given roll?

Let's continue the simulation from Example 2 and Table 5.1. In that example we stopped after only 100 rolls, but now let's continue simulating for a very large number of rolls. Figure 5.2 shows the cumulative proportion of 6s plotted against the trial number from trial 100, up to a total of 10,000 rolls. As the trial number increases, the cumulative proportion of 6s gradually settles down. After 10,000 simulated rolls, Figure 5.2 shows that the cumulative proportion is *very* close to the value of 1/6.

With a relatively short run, such as 10 rolls of a die, the cumulative proportion of 6s can fluctuate a lot. It need not be close to 1/6 after the final trial. However, as the number of trials keeps increasing, the proportion of times the number 6 occurs becomes more predictable and less random: It gets closer and closer to 1/6. *With random phenomena, the proportion of times that something happens is highly random and variable in the short run but very predictable in the long run.*

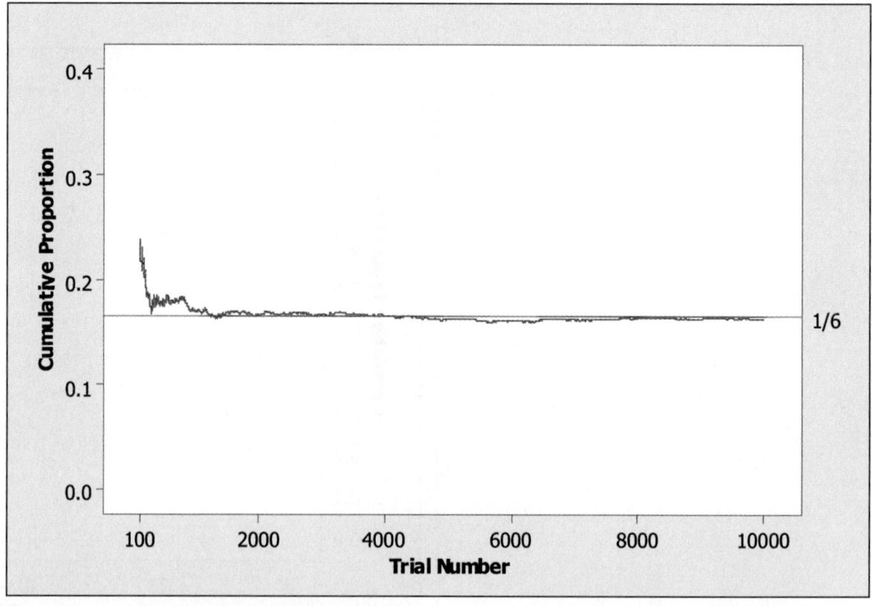

▲ **Figure 5.2 The Cumulative Proportion of Times that a 6 Occurs for a Simulation of 10,000 Rolls of a Fair Die.** Figure 5.1 showed the first 100 trials, and this MINITAB figure shows results of trials 100 through 10,000. As the trial number gets larger, the cumulative proportion gets closer to 1/6. The probability of a 6 on any single roll is defined to be this *long-run* value. **Question** What would you expect for the cumulative proportion of heads after you flipped a balanced coin 10,000 times?

Activity 1

Simulate Many Rolls of a Die

Since one simulation is insufficient to tell us what is typical, go to the Simulating the Probability of Rolling a 6 applet on the text CD, which can conduct simulations of this type in which each trial has two possible outcomes. Set the sample size equal to 100. How many 6s did you observe out of the 100 simulated rolls? After simulating 100 rolls, how close was the cumulative proportion of 6s to the expected value of 1/6?

Do the same simulation 25 times to get a feeling for how the sample cumulative proportion at 100 simulated rolls compares to the expected value of 1/6 (that is, 16.7%). You've probably seen that it's possible for 23 or more (that is, at least 23%) of the 100 rolls to result in 6s. If you keep doing this simulation over and over again, about 6.3% of the time you will get 23 or more 6s out of the 100 rolls. Also, about 30% of the time, you will see at least three 6s in a row somewhere out of the 100 rolls.

Now, change the sample size for each simulation to 1000 and simulate rolling the die 1000 times. Here's a prediction: The cumulative proportion of 6s at 1000 simulated rolls will tend to fall *closer* to its expected value of 1/6. We'll now be very surprised if at least 23% of the 1000 rolls result in 6s.

Probability Quantifies Long-Run Randomness

Did You Know?

The **law of large numbers** is what helps keep casinos in business. For a short time, a gambler may be lucky and make money, but the casino knows that, in the long run, it's going to come out ahead. ◀

In 1689, the Swiss mathematician Jacob Bernoulli proved that as the number of trials increases, the proportion of occurrences of any given outcome approaches a particular number (such as 1/6) in the long run. To show this, he assumed that the outcome of any one trial does not depend on the outcome of any other trial. Bernoulli's result is known as the **law of large numbers.**

We will interpret the **probability** of an outcome to represent long-run results. Imagine a randomized experiment or a random sampling of subjects that provides a very long sequence of observations. Each observation does or does not have that outcome. The probability of the outcome is the proportion of times that it occurs, in the long run.

Probability

With a randomized experiment or a random sample or other random phenomenon (such as a simulation), the **probability** of a particular outcome is the proportion of times that the outcome would occur in a long run of observations.

When we say that a roll of a die has outcome 6 with probability 1/6, this means that the proportion of times that a 6 would occur in a long run of observations is 1/6. The probability would also be 1/6 for each of the other possible outcomes: 1, 2, 3, 4, or 5.

A weather forecaster might say that the probability of rain today is 0.70. This means that in a large number of days with atmospheric conditions like those today, the proportion of days in which rain occurs is 0.70. Since a probability is a *proportion*, it takes a value between 0 and 1. Sometimes probabilities are expressed as percentages, such as when the weather forecaster reports the probability of rain as 70%. Probabilities then fall between 0 and 100, but we'll mainly use the proportion scale.

Why does probability refer to the *long run*? Because we can't accurately assess a probability with a small number of trials. If you sample 10 people and they are all right-handed, you can't conclude that the probability of being right-handed equals 1.0. As we've seen, there can be a lot of variability in the cumulative proportion for small samples. It takes a much larger sample of people to predict accurately the proportion of people in the population who are right-handed.

Independent Trials

With random phenomena, many believe that when some outcome has not happened in a while, it is *due* to happen: Its probability goes up until it happens. In many rolls of a fair die, if a particular value (say, 5) has not occurred in a long time, some think it's due and that the chance of a 5 on the next roll is greater than 1/6. If a family has four girls in a row and is expecting another child, are they due to get a boy? Does the next child have more than a 1/2 chance of being a boy?

Example 2 showed that over a short run, observations may deviate from what is expected (remember three 6s in a row?). But with many random phenomena, such as outcomes of rolling a die or having children, what happens on previous trials does not affect the trial that's about to occur. The trials are **independent** of each other.

> **Independent Trials**
>
> Different trials of a random phenomenon are **independent** if the outcome of any one trial is not affected by the outcome of any other trial.

With independent trials, whether you get a 5 on one roll of a fair die does not affect whether you get a 5 on the following roll. It doesn't matter if you had 20 rolls in a row that are not 5s, the next roll still has probability 1/6 of being a 5. The die has no memory. If you have lost many bets in a row, don't assume that you are due to win if you continue to gamble. The *law of large numbers*, which gamblers invoke as the *law of averages*, only guarantees *long-run* performance. Over the short amount of time in which a gambler's money can disappear, the variability may well exceed what you expect.

Finding Probabilities

In practice, we sometimes can find probabilities by making assumptions about the nature of the random phenomenon. For instance, by symmetry, it may be reasonable to assume that the possible outcomes are *equally likely.* In rolling a die, we might assume that based on the physical makeup of the die, each of the six sides has an equal chance. Then the probability of rolling any particular number equals 1/6. If we flip a coin and assume that the coin is balanced, then the probability of flipping a tail (or a head) equals 1/2. Notice that, like proportions, *the total of the probabilities for all the possible outcomes equals 1.*

> **In Practice** The Sample Proportion Estimates the Actual Probability
>
> In theory, we could observe several trials of a random phenomenon and use the sample proportion of times an outcome occurs as its probability. In practice, this is imperfect. The sample proportion merely *estimates* the actual probability, and only for a *very large* number of trials is it necessarily close. In Chapters 7 and 8, we'll see how the sample size determines just how good that estimate is.

Types of Probability: Relative Frequency and Subjective Probability

We've defined the probability of an outcome as a long-run proportion (relative frequency) of times that the outcome occurs in a very large number of trials. However, this definition is not always helpful. Before the launch of the first space shuttle, how could NASA scientists assess the probability of success? No data were available about long-run observations of past flights. If you decide to start a new type of business when you graduate, you won't have a long run of trials with which to estimate the probability that the business is successful.

In such situations, you must rely on **subjective** information rather than solely on **objective** information such as data. You assess the probability of an outcome by taking into account all the information you have. Such probabilities are not based on a long run of trials. In this **subjective definition of probability**, the probability of an outcome is defined to be a *personal probability*—your degree of belief that the outcome will occur, based on the available information. A branch of statistics uses subjective probability as its foundation. It is called **Bayesian statistics**, in honor of Thomas Bayes, a British clergyman who discovered a probability rule on which it is based. The subjective approach is less common than the approach we discuss in this text. We'll merely warn you to be wary of anyone who gives a subjective probability of 1 (certain occurrence) or of 0 (certain nonoccurrence) for some outcome. As Benjamin Franklin said, nothing is certain but death and taxes!

5.1 Practicing the Basics

5.1 Probability Explain what is meant by the long-run relative frequency definition of probability.

5.2 Testing a coin Your friend decides to flip a coin repeatedly to analyze whether the probability of a head on each flip is 1/2. He flips the coin 10 times and observes a head 7 times. He concludes that the probability of a head for this coin is $7/10 = 0.70$.

a. Your friend claims that the coin is not balanced, since the probability is not 0.50. What's wrong with your friend's claim?

b. If the probability of flipping a head is actually 1/2, what would you have to do to ensure that the cumulative proportion of heads falls very close to 1/2?

5.3 Vegetarianism You randomly sample 10 people in your school, and none of them is a vegetarian. Does this mean that the probability of being a vegetarian for students at your school equals 0? Explain.

5.4 Airline accident deaths For the 10-year period between 2000 and 2010, the average number of deaths due to accidents involving U.S. commercial airline carriers has been about 46 per year. Over that same period, the average number of passengers has been more than 600 million per year.

a. Can you consider this a long run or short run of trials? Explain.

b. Estimate the probability of dying on a particular flight. (By contrast, for a trip by auto in a Western country, the probability of death in a 1000-mile trip is about 1 in 42,000, or more than 50 times the flight's estimated probability.)

c. As of April 2011, the last fatal accident involving a U.S. carrier occurred in Buffalo, New York, in February 2009, a span of more than two years. Mary comments that she is currently afraid of flying because the airlines are "due for an accident." Comment on Mary's reasoning.

5.5 NBA Championship probability Each year, the Web site espn.go.com/nba/hollinger/playoffodds displays the probabilities of professional basketball teams achieving certain goals. For example, at the end of the 2009–2010 regular season, the site listed the following probabilities (expressed as percentages) of each of the 16 playoff teams winning the NBA Championship. (Note that the site uses the term *odds* to represent in this context a probability.)

Orlando	29.5	Denver	3.4
Cleveland	12.9	Oklahoma City	2.5
Phoenix	10.9	Boston	1.8
San Antonio	10.8	Miami	1.6
Utah	9.6	Dallas	1.4
Atlanta	6.5	Charlotte	0.7
LA Lakers	3.9	Milwaukee	0.6
Portland	3.7	Chicago	0.1

The Web site explains that a “computer plays out the remainder of the season 5000 times to see the potential range of projected outcomes.”

a. Note the sum of the probabilities for the 16 teams is 99.9. Why do you think the sum differs from 100?

b. Interpret Orlando’s probability of 29.5%, which was calculated from the 5000 simulations. Is it based on the relative frequency or the subjective interpretation of probability?

5.6 Random digits Consider a random number generator designed for equally likely outcomes. Which of the following is *not* correct, and why?

a. For each random digit generated, each integer between 0 and 9 has probability 0.10 of being selected.

b. If you generate 10 random digits, each integer between 0 and 9 must occur exactly once.

c. If you generated a very large number of random digits, then each integer between 0 and 9 would occur close to 10% of the time.

d. The cumulative proportion of times that a 0 is generated tends to get closer to 0.10 as the number of random digits generated gets larger and larger.

5.7 Polls and sample size A pollster wants to estimate the proportion of Canadian adults who support the prime minister’s performance on the job. He comments that by the law of large numbers, to ensure a sample survey’s accuracy, he does not need to worry about the method for selecting the sample, only that the sample has a very large sample size. Do you agree with the pollster’s comment? Explain.

5.8 Heart transplant Before the first human heart transplant, Dr. Christiaan Barnard of South Africa was asked to assess the probability that the operation would be successful. Did he need to rely on the relative frequency definition or the subjective definition of probability? Explain.

5.9 Life on other planets? Is there intelligent life on other planets in the universe? If you are asked to state the probability that there is, would you need to rely on the relative frequency or the subjective definition of probability? Explain.

5.10 Applet for coin flipping Use the Simulating the Probability of Head With a Fair Coin applet on the text CD or other software to illustrate the long-run definition of probability by simulating short-term and long-term results of flipping a balanced coin.

TRY

a. Set the sample size to $n = 10$. Run the applet 10 times, and record the cumulative proportion of heads for each of the 10 simulations.

b. Now set the sample size $n = 100$. Run the applet 10 times, and record the 10 cumulative proportions of heads for the separate sets. Do they vary much?

c. Now set $n = 1000$. Run the applet 10 times, and record the 10 cumulative proportions of heads. Do they vary more, or less, than the proportions in part b based on $n = 100$?

d. Summarize the effect of the number of trials n on the variability of the proportion. How does this reflect what’s implied by the law of large numbers?

5.11 Unannounced pop quiz A teacher announces a pop quiz for which the student is completely unprepared. The quiz consists of 100 true-false questions. The student has no choice but to guess the answer randomly for all 100 questions.

a. Simulate taking this quiz by random guessing. Number a sheet of paper 1 to 100 to represent the 100 questions. Write a T (true) or F (false) for each question, by predicting what you think would happen if you repeatedly flipped a coin and let a tail represent a T guess and a head represent an F guess. (Don’t actually flip a coin, but merely write down what you think a random series of guesses would look like.)

b. How many questions would you expect to answer correctly simply by guessing?

c. The table shows the 100 correct answers. The answers should be read across rows. How many questions did you answer correctly?

Pop Quiz Correct Answers

T	F	T	T	F	F	T	T	T	T	T	F	T	F	F	T	T	F	T	F
F	F	F	F	F	F	F	T	F	F	T	F	T	F	F	T	F	T	T	F
T	F	F	F	F	F	T	F	T	T	F	T	T	T	F	F	F	F	F	T
T	F	F	T	F	F	T	T	T	T	F	F	F	F	F	F	F	T	F	F
F	F	T	F	F	T	T	F	F	T	F	T	F	T	T	T	T	F	F	F

d. The above answers were actually randomly generated by the Simulating the Probability of Head With a Fair Coin applet on the text CD. What percentage were true, and what percentage would you expect? Why are they not necessarily identical?

e. Are there groups of answers within the sequence of 100 answers that appear nonrandom? For instance, what is the longest run of Ts or Fs? By comparison, which is the longest run of Ts or Fs within your sequence of 100 answers? (There is a tendency in guessing what randomness looks like to identify too few long runs in which the same outcome occurs several times in a row.)

5.12 Stock market randomness An interview in an investment magazine (*In the Vanguard*, Autumn 2003) asked mathematician John Allen Paulos, “What common errors do investors make?” He answered, “People tend not to believe that markets move in random ways. Randomness is difficult to recognize. If you have people write down 100 Hs and Ts to simulate 100 flips of a coin, you will always be able to tell a sequence generated by a human from one generated by real coin flips. When humans make up the sequence, they don’t put in enough consecutive Hs and Ts, and they don’t make the lengths of those runs long enough or frequent enough. And that is one of the reasons people look at patterns in the stock market and ascribe significance to them.” (© The Vanguard Group, Inc., used with permission.)

a. Suppose that on each of the next 100 business days the stock market has a 1/2 chance of going up and a 1/2 chance of going down, and its behavior one day is independent of its behavior on another day. Use the Simulating the Stock Market applet on the text CD or other software to simulate whether the market goes up or goes down for each of the next 100 days. What is the longest sequence of consecutive moves up or consecutive moves down that you observe?

b. Run the applet nine more times, with 100 observations for each run, and each time record the longest sequence of consecutive moves up or consecutive moves down that you observe. For the 10 runs, summarize the proportion of times that the longest sequence was 1, 2, 3, 4, 5, 6, 7, 8, or more. (Your class may combine results to estimate this more precisely.)

c. Based on your findings, explain why if you are a serious investor you should not get too excited if sometime in the next few months you see the stock market go up for five days in a row or go down for five days in a row.

5.2 Finding Probabilities

We've learned that probability enables us to quantify uncertainty and randomness. Now, let's explore some basic rules that help us find probabilities.

Sample Spaces

The first step is to list all the possible outcomes. The set of possible outcomes for a random phenomenon is called the **sample space**.

> Sample Space
>
> For a random phenomenon, the **sample space** is the set of all possible outcomes.

For example, when you roll a die once, the sample space consists of the six possible outcomes, {1,2,3,4,5,6}. When you flip a coin twice, the sample space consists of the four possible outcomes, {HH, HT, TH, TT}, where, for instance, TH represents a tail on the first flip and a head on the second flip.

Example 3

Sample space

Multiple-Choice Pop Quiz

Picture the Scenario

Your statistics instructor decides to give an unannounced pop quiz with three multiple-choice questions. Each question has five options, and the student's answer is either correct (C) or incorrect (I). If a student answered the first two questions correctly and the last question incorrectly, the student's outcome on the quiz can be symbolized by CCI.

Question to Explore

What is the sample space for the correctness of a student's answers on this pop quiz?

Think It Through

One technique for listing the outcomes in a sample space is to draw a **tree diagram**, with branches showing what can happen on different trials. For a student's performance on three questions, the tree has three sets of branches, as Figure 5.3 shows.

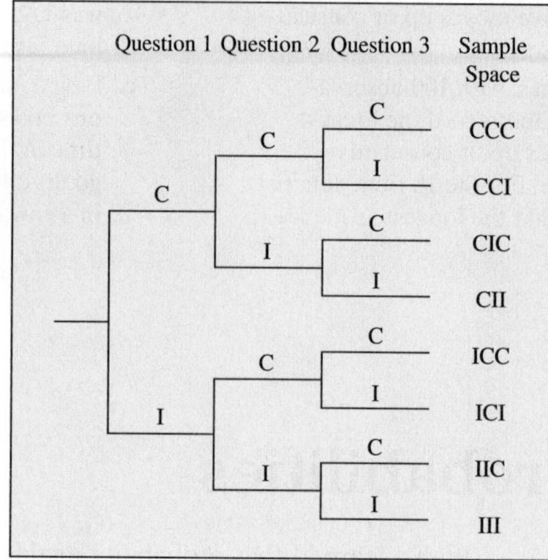

◀ **Figure 5.3 Tree Diagram for Student Performance on a Three-Question Pop Quiz.** Each path from the first set of two branches to the third set of eight branches determines an outcome in the sample space. **Question** How many possible outcomes would there be if the quiz had four questions?

From the tree diagram, a student's performance has eight possible outcomes:

{CCC, CCI, CIC, CII, ICC, ICI, IIC, III}.

Insight

The number of branches doubles at each stage. There are 2 branches for question 1, 2 × 2 = 4 branches at question 2, and 2 × 2 × 2 = 8 branches at question 3.

Try Exercise 5.13

How many possible outcomes are in a sample space when there are repeated trials? To determine this, multiply the number of possible outcomes for each trial. A pop quiz with three questions has 2 × 2 × 2 = 8 possible outcomes denoting whether each answer is correct or incorrect. With four questions, there are 2 × 2 × 2 × 2 = 16 possible outcomes.

What if we want to consider the actual responses made? With four multiple-choice questions and five possible responses on each, there are 5 × 5 × 5 × 5 = 625 possible response sequences. The tree diagram is ideal for visualizing a small number of outcomes. As the number of trials or the number of possible outcomes on each trial increases, it becomes impractical to construct a tree diagram, so we'll also learn about methods for finding probabilities without having to list entire sample spaces.

Events

We'll sometimes need to specify a particular group of the outcomes in a sample space, that is, a *subset* of the outcomes. An example is the subset of outcomes for which a student passes the pop quiz, by answering at least two of the three questions correctly. A subset of a sample space is called an **event**.

Event

An **event** is a subset of the sample space. An **event** corresponds to a particular outcome or a group of possible outcomes.

Events are usually denoted by letters from the beginning of the alphabet, such as A and B, or by a letter or string of letters that describes the event. For a student taking the three-question pop quiz, some possible events are

A = student answers all three questions correctly = {CCC}

B = student passes (at least two correct) = {CCI, CIC, ICC, CCC}.

Finding Probabilities of Events

Each outcome in a sample space has a probability. So does each event. To find such probabilities, we list the sample space and specify plausible assumptions about its outcomes. Sometimes, for instance, we can assume that the outcomes are equally likely. The probabilities for the outcomes in a sample space must follow two basic rules:

- The probability of each individual outcome is between 0 and 1.
- The total of all the individual probabilities equals 1.

Example 4

Probability for a sample space

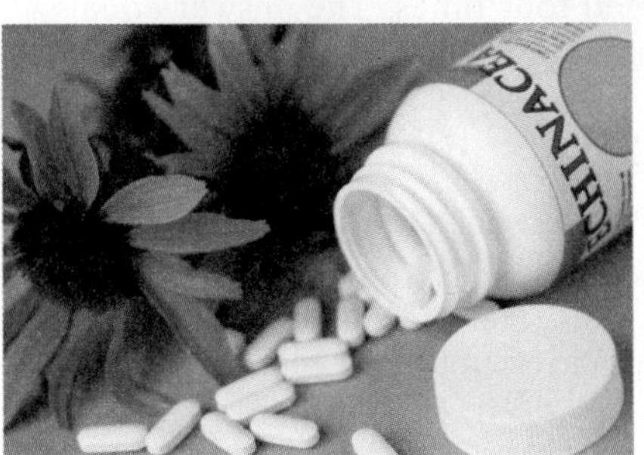

Treating Colds

Picture the Scenario

The University of Wisconsin is conducting a randomized experiment[1] to compare an herbal remedy (echinacea) to a placebo for treating the common cold. The response variables are the cold's severity and its duration. Suppose a particular clinic in Madison, Wisconsin, has four volunteers, of whom two are men (Jamal and Ken) and two are women (Linda and Mary). Two of these volunteers will be randomly chosen to receive the herbal remedy, and the other two will receive the placebo.

Questions to Explore

a. Identify the possible samples to receive the herbal remedy. For each possible sample, what is the probability that it is the one chosen?

b. What's the probability of the event that the sample chosen to receive the herbal remedy consists of one man and one woman?

Think It Through

a. The six possible samples to assign to the herbal remedy are {(Jamal, Ken), (Jamal, Linda), (Jamal, Mary), (Ken, Linda), (Ken, Mary), (Linda, Mary) }. This is the sample space for randomly choosing two of the four people. For a simple random sample, every sample is equally likely. Since there are six possible samples, each one has probability 1/6. These probabilities fall between 0 and 1, and their total equals 1, as is necessary for probabilities for a sample space.

b. The event in which the sample chosen has one man and one woman consists of the outcomes {(Jamal, Linda), (Jamal, Mary), (Ken, Linda), (Ken, Mary)}. These are the possible pairings of one man with one woman. Each outcome has probability 1/6, so the probability of this event is $4(1/6) = 4/6 = 2/3$.

Insight

When each outcome is equally likely, the probability of a single outcome is simply 1/(number of possible outcomes), such as 1/6 in part a of Think It Through. The probability of an event is then (number of outcomes in the event)/(number of possible outcomes), such as 4/6 in part b of Think It Through above.

Try Exercise 5.19

[1]See www.fammed.wisc.edu/research/news0503d.html.

This example shows that to find the probability for an event, we can (1) find the probability for each outcome in the sample space, and (2) add the probabilities of each outcome that the event contains.

SUMMARY: Probability of an Event

The probability of an event A, denoted by P(A), is obtained by adding the probabilities of the individual outcomes in the event.

- When all the possible outcomes are equally likely,

$$P(A) = \frac{\text{number of outcomes in event A}}{\text{number of outcomes in the sample space}}$$

In Example 4, to find the probability of choosing one man and one woman, we first determined the probability of each of the six possible outcomes. Because the probability is the same for each, 1/6, and because the event contains four of those outcomes, the probability is 1/6 added four times. The answer equals 4/6, the number of outcomes in the event divided by the number of outcomes in the sample space.

In Practice Equally Likely Outcomes Are Unusual

Except for simplistic situations such as random sampling or flipping balanced coins or rolling fair dice, different outcomes are not usually equally likely. Then, probabilities are often estimated, using sample proportions from simulations or from large samples of data.

Example 5

Probability for a sample space

Tax Audit

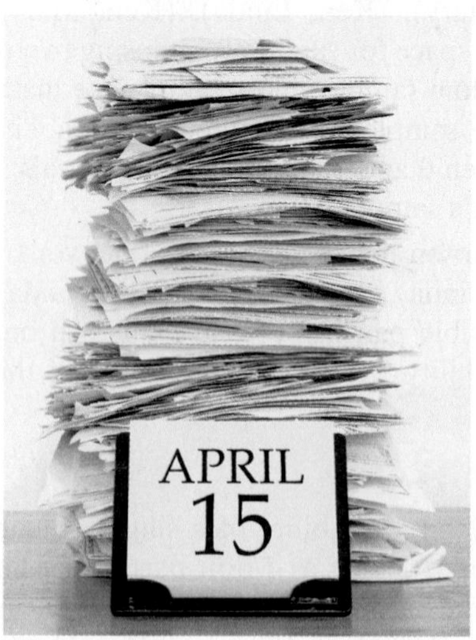

Picture the Scenario

April 15 is tax day in the United States—the deadline for filing federal income tax forms. The main factor in the amount of tax owed is a taxpayer's income level. Each year, the IRS audits a sample of tax forms to verify their accuracy. Table 5.2 is a contingency table that cross-tabulates the 138.2 million long-form federal returns received in 2008 by the taxpayer's income level and whether the tax form was audited.

Table 5.2 Contingency Table Cross-Tabulating Tax Forms by Income Level and Whether Audited or Not

There were 138.2 million returns filed. The frequencies in the table are reported in thousands. For example, 1260 represents 1,260,000 tax forms that reported income under $200,000 and were audited.

	Audited		
Income Level	**Yes**	**No**	**Total**
Under $200,000	1,260	132,147	**33,407**
$200,000–$1,000,000	131	4,311	**4,442**
More than $1,000,000	22	371	**393**
Total	**1,413**	**136,829**	**138,242**

Source: www.irs.gov/pub/irs-news/2008_enforcement.pdf

Recall

Section 3.1 introduced **contingency tables** to summarize the relationship between two categorical variables. ◀

Questions to Explore

a. What is the sample space?

b. For a randomly selected taxpayer in 2008, what is the probability of (i) an audit, (ii) an income of more than $1,000,000?

Think It Through

a. The sample space is the set of possible outcomes. These are the six cells for income level and audited combinations in this table, such as (Under $200,000, Yes), (Under $200,000, No), ($200,000–$1,000,000, Yes), and so forth.

b. If a taxpayer was randomly selected, from Table 5.2,

(i) The probability of an audit was 1413/138,242 = 0.0102.

(ii) The probability of an income of more than $1,000,000 was 393/138,242 = 0.0028.

Insight

Even though audits are not welcome news to a taxpayer, in 2008 only about 1% of taxpayers were audited. Is this typical? Inspecting similar data for earlier years shows that this percentage increased considerably in recent years. For instance, in 2002 the percentage audited was 0.4%, considerably less than the value in 2008.

Try Exercise 5.23, parts a and b

Basic Rules for Finding Probabilities About a Pair of Events

Some events are expressed as the outcomes that (a) are *not* in some other event, or (b) are in one event *and* in another event, or (c) are in one event *or* in another event. We'll next learn how to calculate probabilities for these three cases.

The Complement of an Event For an event A, the rest of the sample space that is *not* in that event is called the **complement** of A.

In Words

A^c reads as "**A-complement.**" The *c* in the superscript denotes the term complement. You can think of A^c as meaning "not A."

Complement of an Event

The **complement** of an event A consists of all outcomes in the sample space that are *not* in A. It is denoted by A^c. The probabilities of A and of A^c add to 1, so

$$P(A^c) = 1 - P(A).$$

In Example 5, for instance, the event of having an income *less than* $200,000 is the complement of the event of having an income of $200,000 or more. Because the probability that a randomly selected taxpayer had an income of $200,000 or more is 0.035, the probability of income less than $200,000 is 1 − 0.035 = 0.965.

Figure 5.4 illustrates the complement of an event. The box represents the entire sample space. The event A is the oval in the box. The complement of A, which is shaded, is everything else in the box that is not in A. Together, A and A^c cover the sample space. Because an event and its complement contain all possible outcomes, their total probability is 1, and the probability of either one of them is 1 minus the probability of the other. A diagram like Figure 5.4 that uses areas inside a box to represent events is called a **Venn diagram**.

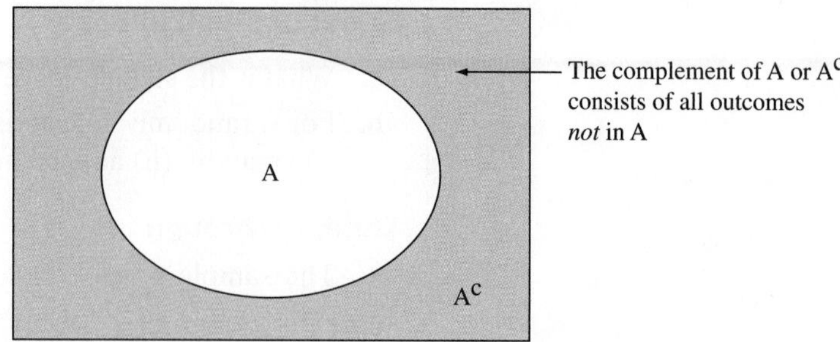

▲ **Figure 5.4 Venn Diagram Illustrating an Event A and Its Complement A^C.**
Question Can you sketch a Venn diagram of two events A and B such that they share some common outcomes, but some outcomes are only in A or only in B?

To find the probability of an event, it's sometimes easier to find the probability of its complement and then subtract that probability from 1. An example is when we need to find the probability that *at least one* of several events will occur. It's usually easier to find the probability of its complement, that *none* of these events will occur.

Example 6

Complement of an event

Women on a Jury

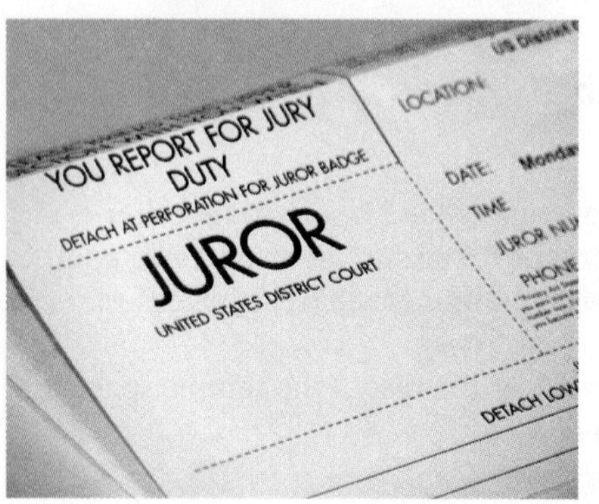

Picture the Scenario

A jury of 12 people is chosen for a trial. The defense attorney claims it must have been chosen in a biased manner because 50% of the city's adult residents are female, yet the jury contains no women.

Questions to Explore

If the jury were randomly chosen from the population, what is the probability that the jury would have (a) no females, (b) at least one female?

Think It Through

Let's use a symbol with 12 letters, with F for female and M for male, to represent a possible jury selection. For instance, MFMMMMMMMMMM denotes the jury in which only the second person selected is female. The number of possible outcomes is $2 \times 2 \times 2 \times \ldots \times 2$, that is, 2 multiplied 12 times, which is $2^{12} = 4096$. This is a case in which listing the entire sample space is not practical. Since the population is 50% male and 50% female, these 4096 possible outcomes are equally likely.

a. Only 1 of the 4096 possible outcomes corresponds to a no-female jury, namely, MMMMMMMMMMMM. So the probability of this outcome is 1/4096, or 0.00024. This is extremely unlikely, if a jury is truly chosen by random sampling.

b. As noted previously, it would be tedious to list all possible outcomes in which at least one female is on the jury. But this is not necessary. The event that the jury contains *at least one* female is the complement of the event that it contains *no* females. Thus,

$$P(\text{at least one female}) = 1 - P(\text{no females}) = 1 - 0.00024 = 0.99976.$$

Insight

You might instead let the sample space be the possible values for the *number* of females on the jury, namely 0,1,2, ... 12. But these outcomes are not equally likely. For instance, only one of the 4096 possible samples has 0 females, but 12 of them have 1 female: The female could be the first person chosen, or the second (as in MFMMMMMMMMMM), or the third, and so on. Chapter 6 will show a formula (binomial) that gives probabilities for this alternative sample space.

Try Exercise 5.16

Disjoint Events Events that do not share any outcomes in common are said to be **disjoint**.

Did You Know?

Disjoint events are also referred to as **mutually exclusive** events. We will use only the term "disjoint." ◄

Disjoint Events

Two events, A and B, are **disjoint** if they do not have any common outcomes.

Example 3 discussed a pop quiz with three questions. The event that the student answers exactly one question correctly is {CII, ICI, IIC}. The event that the student answers exactly two questions correctly is {CCI, CIC, ICC}. These two events have no outcomes in common, so they are disjoint. In a Venn diagram, they have no overlap. (Figure 5.5). By contrast, neither is disjoint from the event that the student answers the first question correctly, which is {CCC, CCI, CIC, CII}, because this event has outcomes in common with each of the other two events.

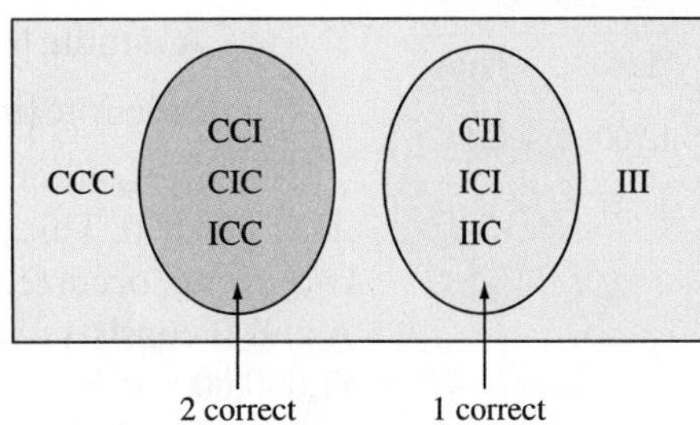

▲ **Figure 5.5 Venn Diagram Illustrating Disjoint Events.** The event of a student answering exactly one question correctly is disjoint from the event of answering exactly two questions correctly. **Question** Identify on this figure the event that the student answers the first question correctly. Is this event disjoint from either of the two events identified in the Venn diagram?

Consider an event A and its complement, A^c. They share no common outcomes, so they are disjoint events.

Intersection and Union of Events Some events are composed from other events. For instance, for two events A and B, the event that *both* occur is also an event. Called the **intersection** of A and B, it consists of the outcomes that are in both A and B. By contrast, the event that the outcome is in A *or* B or both is the **union** of A and B. It is a larger set, containing the intersection as well as outcomes that are in A but not in B and outcomes that are in B but not in A. Figure 5.6 shows Venn diagrams illustrating the intersection and union of two events.

In Words

Intersection means A *and* B (the "overlap" of the events). **Union** means A *or* B or both.

Intersection and Union of Two Events

The **intersection** of A and B consists of outcomes that are in both A *and* B.

The **union** of A and B consists of outcomes that are in A *or* B or both. In probability, "A *or* B" denotes that A occurs or B occurs or both occur.

Intersection

A B

A and B

(a)

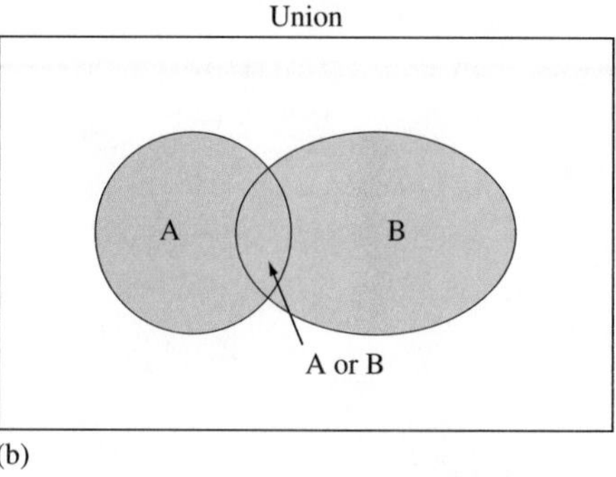

(b)

▲ **Figure 5.6 The Intersection and the Union of Two Events.** Intersection means A occurs *and* B occurs, denoted "A and B." The intersection consists of the shaded "overlap" part in Figure 5.6 (a). Union means A occurs *or* B occurs *or* both occur, denoted "A or B." It consists of all the shaded parts in Figure 5.6 (b). **Question** How could you find P(A or B) if you know P(A), P(B), and P(A and B)?

For instance, for the three-question pop quiz, consider the events:

A = student answers first question correctly = {CCC, CCI, CIC, CII}

B = student answers two questions correctly = {CCI, CIC, ICC}.

Then the intersection, A *and* B, is {CCI, CIC}, the two outcomes common to A and B. The union, A *or* B, is {CCC, CCI, CIC, CII, ICC}, the outcomes that are in A or in B or in both A and B.

How do we find probabilities of intersections and unions of events? Once we identify the possible outcomes, we can use their probabilities. For instance, for Table 5.2 (shown in the margin) for 138.2 million tax forms, let

	Audited	
Income	**Yes**	**No**
Under \$200,000	1,260	132,147
\$200,000–\$1,000,000	31	4,311
More than \$1,000,000	22	371

A denote {audited = yes}

B denote {income ≥ \$1,000,000}

The intersection A and B is the event that a taxpayer is audited *and* has income ≥ \$1,000,000. This probability is simply the proportion for the cell in which these two events occurred, namely P(A and B) = 22/138,242 = 0.0002. The union of A and B consists of all those who either were audited or had income greater than \$1,000,000 or both. From the table, these were (1260 + 31 + 22 + 371) people, so the probability is 1684/138,242 = 0.012. We can formalize rules for finding the union and intersection of two events, as we'll see in the next two subsections.

Addition Rule: Finding the Probability that Event A or Event B Occurs

Since the union A or B contains outcomes from A and from B, we can add P(A) to P(B). However, this sum counts the outcomes that are in *both* A and B (their intersection) twice (Figure 5.7). We need to subtract the probability of the intersection from P(A) + P(B) so that it is only counted once. If there is no overlap, that is, if the events are disjoint, no outcomes are common to A and B. Then we can simply add the probabilities.

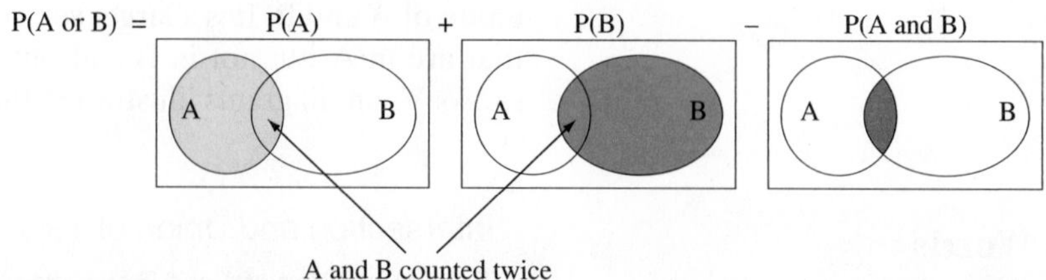

▲ **Figure 5.7 The Probability of the Union, Outcomes in A or B or Both.** Add P(A) to P(B) and subtract P(A and B) to adjust for outcomes counted twice. **Question** When does P(A or B) = P(A) + P(B)?

Addition Rule: Probability of the Union of Two Events

For the **union** of two events, P(A or B) = P(A) + P(B) − P(A and B).
If the events are **disjoint**, then P(A and B) = 0, so P(A or B) = P(A) + P(B).

For example, consider a family with two children. The sample space possibilities for the genders of the two children are {FF, FM, MF, MM}, where the first letter in a symbol is the first child's gender and the second letter is the second child's gender (F = female, M = male). Let A = {first child a girl} and B = {second child a girl}. Then, assuming the four possible outcomes in the sample space are equally likely, P(A) = P({FF, FM}) = 0.50, P(B) = P({FF, MF}) = 0.50, and P(A and B) = P({FF}) = 0.25. The event A or B is the event that the first child is a girl, or the second child is a girl, or both, that is, *at least* one child is a girl. Its probability is

$$P(\text{A or B}) = P(\text{A}) + P(\text{B}) - P(\text{A and B}) = 0.50 + 0.50 - 0.25 = 0.75.$$

Multiplication Rule: Finding the Probability that Events A and B Both Occur

The probability of the intersection of events A and B has a formula to be introduced in Section 5.3. In the special case discussed next, it equals P(A) × P(B).

Consider a basketball player who shoots two free throws. Let M1 denote making free throw 1, and let M2 denote making free throw 2. For any given free throw, suppose he has an 80% chance of making it, so P(M1) = P(M2) = 0.80. What is the probability of M1 and M2, making free throw 1 *and* free throw 2? In the long run of many pairs of free throws, suppose that for 80% of the cases in which he made the first free throw, he also made the second. Then the percentage of times he made both is the 80% of the 80% of times he made the first one, for a probability of 0.80 × 0.80 = 0.64 (that is, 64%).

This multiplication calculation is valid only under an assumption, **independent trials**: Whether a player makes the second free throw is independent of whether he makes the first. The chance of making the second is 80%, regardless of whether or not he made the first. For pro basketball players, independence is approximately true: Whether a player makes his first shot has almost no influence on whether he makes the second one. See Exercise 5.34 for data.

Recall

Independent trials means that what happens on one trial is not influenced by what happens on any other trial. ◀

To find the probability of the intersection of two events, we can multiply probabilities whenever the events are independent. We'll see a formal definition of independent events in the next section, but it essentially means that whether one event occurs does not affect the probability that the other event occurs.

Multiplication Rule: Probability of the Intersection of Independent Events

For the **intersection** of two **independent** events, A and B,

$$P(\text{A and B}) = P(\text{A}) \times P(\text{B}).$$

The paradigm for independent events is repeatedly flipping a coin or rolling a die, where what happens on one trial does not affect what happens on another. For instance, for two rolls of a die,

$$P(\text{6 on roll 1 } \textit{and} \text{ 6 on roll 2}) = P(\text{6 on roll 1}) \times P(\text{6 on roll 2}) = \frac{1}{6} \times \frac{1}{6} = \frac{1}{36}.$$

This multiplication rule extends to more than two independent events.

Multiplication rule

Example 7

Guessing yet Passing a Pop Quiz

Picture the Scenario

For a three-question multiple-choice pop quiz, a student is totally unprepared and randomly guesses the answer to each question. If each question has five options, then the probability of selecting the correct answer for any given question is 1/5, or 0.20. With guessing, the response on one question is not influenced by the response on another question. Thus, whether one question is answered correctly is independent of whether or not another question is answered correctly.

Questions to Explore

a. Find the probabilities of the possible student outcomes for the quiz, in terms of whether each response is correct (C) or incorrect (I).

b. Find the probability that the student passes, answering *at least two* questions correctly.

Think It Through

a. For each question P(C) = 0.20 and P(I) = 1 − 0.20 = 0.80. The probability that the student answers all three questions correctly is

$$P(CCC) = P(C) \times P(C) \times P(C) = 0.20 \times 0.20 \times 0.20 = 0.008.$$

This would be unusual. Similarly, the probability of answering the first two questions correctly and the third question incorrectly is

$$P(CCI) = P(C) \times P(C) \times P(I) = 0.20 \times 0.20 \times 0.80 = 0.032.$$

This is the same as P(CIC) and P(ICC), the other possible ways of getting two correct. Figure 5.8 is a tree diagram showing how to multiply probabilities to find the probabilities for all eight possible outcomes.

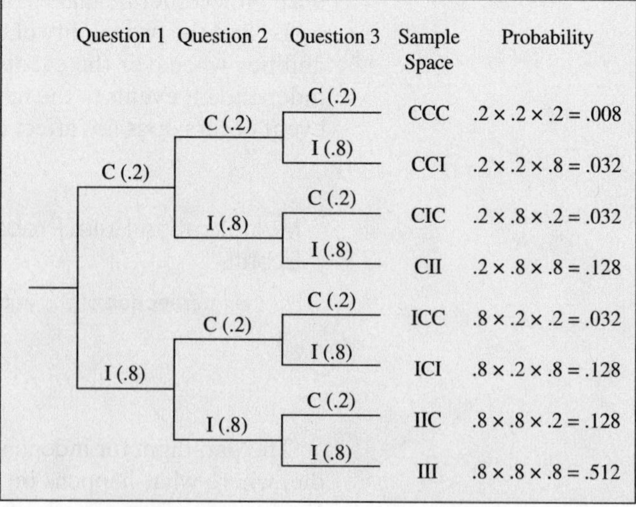

▲ **Figure 5.8 Tree Diagram for Guessing on a Three-Question Pop Quiz.** Each path from the first set of branches to the third set determines one sample space outcome. Multiplication of the probabilities along that path gives its probability, when trials are independent. **Question** Would you expect trials to be independent if a student is *not* merely guessing on every question? Why or why not?

b. The probability of *at least* two correct responses is

$$P(CCC) + P(CCI) + P(CIC) + P(ICC) = 0.008 + 3(0.032) = 0.104.$$

In summary, there is only about a 10% chance of passing when a student randomly guesses the answers.

Insight

As a check, you can see that the probabilities of the eight possible outcomes sum to 1.0. The probabilities indicate that it is in a student's best interests not to rely on random guessing.

Try Exercise 5.15

Events Often Are Not Independent

In practice, events need not be independent. For instance, on a quiz with only two questions, the instructor found the following proportions for the actual responses of her students (I = incorrect, C = correct):

Outcome:	II	IC	CI	CC
Probability:	0.26	0.11	0.05	0.58

Recall

From Section 3.1, the proportions expressed in contingency table form

	2nd Question	
1st Question	**C**	**I**
C	0.58	0.05
I	0.11	0.26

A and B

Let A denote {first question correct} and let B denote {second question correct}. Based on these probabilities,

$$P(A) = P(\{CI, CC\}) = 0.05 + 0.58 = 0.63$$

$$P(B) = P(\{IC, CC\}) = 0.11 + 0.58 = 0.69$$

and

$$P(A \text{ and } B) = P(\{CC\}) = 0.58.$$

If A and B were independent, then

$$P(A \text{ and } B) = P(A) \times P(B) = 0.63 \times 0.69 = 0.43.$$

Since P(A and B) actually equaled 0.58, A and B were not independent.

Responses to different questions on a quiz are typically not independent. Most students do not guess randomly. Students who get the first question correct may have studied more than students who do not get the first question correct, and thus they may also be more likely to get the second question correct.

In Practice Make Sure that Assuming Independence Is Realistic

Don't assume that events are independent unless you have given this assumption careful thought and it seems plausible. In Section 5.3, you will learn more about how to find probabilities when events are not independent.

Probability Rules

In this section, we have developed several rules for finding probabilities. Let's summarize them.

SUMMARY: Rules for Finding Probabilities

- The probability of each individual outcome is between 0 and 1, and the total of all the individual probabilities equals 1. The **probability of an event** is the sum of the probabilities of the individual outcomes in that event.

- For an event A and its **complement** A^c (not in A), $P(A^c) = 1 - P(A)$.
- The **union** of two events (that is, A occurs or B occurs or both) has

$$P(A \text{ or } B) = P(A) + P(B) - P(A \text{ and } B).$$

- When A and B are **independent**, the **intersection** of two events has

$$P(A \text{ and } B) = P(A) \times P(B).$$

- Two events A and B are **disjoint** when they have no common elements. Then

$$P(A \text{ and } B) = 0, \text{and thus } P(A \text{ or } B) = P(A) + P(B).$$

5.2 Practicing the Basics

5.13 Student union poll Part of a student opinion poll at a university asks students what they think of the quality of the existing student union building on the campus. The possible responses were great, good, fair, and poor. Another part of the poll asked students how they feel about a proposed fee increase to help fund the cost of building a new student union. The possible responses to this question were in favor, opposed, and no opinion. (TRY)

a. List all potential outcomes in the sample space for someone who is responding to both questions.

b. Show how a tree diagram can be used to display the outcomes listed in part a.

5.14 Random digit A single random digit is selected using software or a random number table.

a. State the sample space for the possible outcomes.

b. State the probability for each possible outcome, based on what you know about the way random numbers are generated.

c. Each outcome in a sample space must have probability between 0 and 1, and the total of the probabilities must equal 1. Show that your assignment of probabilities in part b satisfies this rule.

5.15 Pop quiz A teacher gives a four-question unannounced true-false pop quiz, with two possible answers to each question. (TRY)

a. Use a tree diagram to show the possible response patterns, in terms of whether any given response is correct or incorrect. How many outcomes are in the sample space?

b. An unprepared student guesses all the answers randomly. Find the probabilities of the possible outcomes on the tree diagram.

c. Refer to part b. Using the tree diagram, evaluate the probability of passing the quiz, which the teacher defines as answering *at least* three questions correctly.

5.16 More true-false questions Your teacher gives a true-false pop quiz with 10 questions. (TRY)

a. Show that the number of possible outcomes for the sample space of possible sequences of 10 answers is 1024.

b. What is the complement of the event of getting *at least* one of the questions wrong?

c. With random guessing, show that the probability of getting *at least* one question wrong is 0.999.

5.17 Horse racing bets Two friends decide to go to the track and place some bets. One friend remarks that in an upcoming race, the number 5 horse is paying 50 to 1. This means that anyone who bets on the 5 horse receives $50 for each $1 bet, if in fact the 5 horse wins the race. He goes on to mention that it is a great bet, because there are only eight horses running in the race, and therefore the probability of horse 5 winning must be 1/8. Is the last statement true or false? Explain.

5.18 Two girls A couple plans to have two children. Each child is equally likely to be a girl or boy, with gender independent of that of the other child.

a. Construct a sample space for the genders of the two children.

b. Find the probability that both children are girls.

c. Answer part b if in reality, for a given child, the chance of a girl is 0.49.

5.19 Three children A couple plans on having three children. Suppose that the probability of any given child being female is 0.5, and also suppose that the genders of each child are independent events. (TRY)

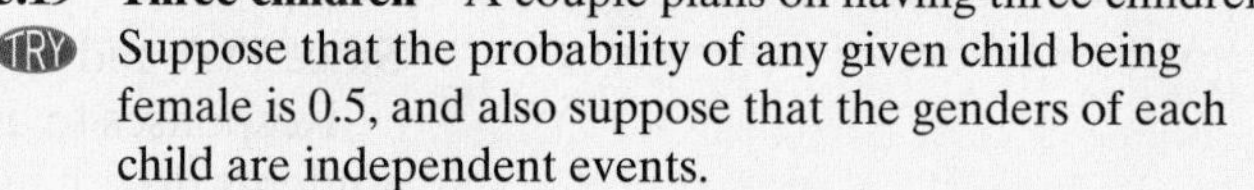

a. Write out all outcomes in the sample space for the genders of the three children.

b. What should be the probability associated with each outcome?

Using the sample space constructed in part a, find the probability that the couple will have

c. two girls and one boy.

d. at least one child of each gender.

5.20 Wrong sample space A couple plans on having four children. The father notes that the sample space for the number of girls the couple can have is 0, 1, 2, 3, and 4. He goes on to say that since there are five outcomes in the sample space, and since each child is equally likely to be a boy or girl, all five outcomes must be equally likely. Therefore, the probability of all four children being girls is 1/5. Explain the flaw in his reasoning.

5.21 Insurance Every year the insurance industry spends considerable resources assessing risk probabilities. To accumulate a risk of about one in a million of death, you can drive 100 miles, take a cross country plane flight, work as a police officer for 10 hours, work in a coal mine for 12 hours, smoke two cigarettes, be a nonsmoker but live with a smoker for two weeks, or drink 70 pints of beer in a year (Wilson and Crouch, 2001, pp. 208–209). Show that a risk of about one in a million of death is also approximately the probability of flipping 20 heads in a row with a balanced coin.

5.22 Washer and dryer purchase At a local appliance store, the probability that a customer will purchase a new washing machine is 0.031. The probability that a customer will purchase a new dryer is 0.029. Is this information sufficient to determine the probability a customer will purchase a new washing machine and a new dryer? If so, find the probability; if not, explain why not. (*Hint:* Considering the buying practices of consumers in this context is it reasonable to assume these events are independent?)

5.23 Seat belt use and auto accidents Based on records of automobile accidents in a recent year, the Department of Highway Safety and Motor Vehicles in Florida reported the counts who survived (S) and died (D), according to whether they wore a seat belt (Y = yes, N = no). The data are presented in the contingency table shown.

TRY

Outcome of auto accident by whether subject wore seat belt

Wore Seat Belt	Survived (S)	Died (D)	Total
Yes (Y)	412,368	510	**412,878**
No (N)	162,527	1,601	**164,128**
Total	**574,895**	**2,111**	**577,006**

a. What is the sample space of possible outcomes for a randomly selected individual involved in an auto accident? Use a tree diagram to illustrate the possible outcomes. (*Hint:* One possible outcome is YS.)

b. Using these data, estimate (i) P(D), (ii) P(N).

c. Estimate the probability that an individual did not wear a seat belt and died.

d. Based on part a, what would the answer to part c have been if the events N and D were independent? So, are N and D independent, and if not, what does that mean in the context of these data?

5.24 Protecting the environment When the General Social Survey most recently asked subjects whether they are a member of an environmental group (variable GRNGROUP) and whether they would be very willing to pay higher prices to protect the environment (variable GRNPRICE), the results were as shown in the table. For a randomly selected American adult:

a. Estimate the probability of being (i) a member of an environmental group and (ii) willing to pay higher prices to protect the environment.

b. Estimate the probability of being both a member of an environmental group *and* very willing to pay higher prices to protect the environment.

c. Given the probabilities in part a, show that the probability in part b is larger than it would be if the variables were independent. Interpret.

d. Estimate the probability that a person is a member of an environmental group *or* very willing to pay higher prices to protect the environment. Do this (i) directly using the counts in the table and (ii) by applying the appropriate probability rule to the estimated probabilities found in parts a and b.

		Pay Higher Prices (GRNPRICE)		
		Yes	No	**Total**
Environmental Group Member (GRNGROUP)	Yes	69	15	**84**
	No	435	276	**711**
	Total	**504**	**291**	**795**

5.25 Global warming and trees A survey asks subjects whether they believe that global warming is happening (yes or no) and how much fuel they plan to use annually for automobile driving in the future, compared to their past use (less, about the same, more).

a. Show the sample space of possible outcomes by drawing a tree diagram that first gives the response on global warming and then the response on fuel use.

b. Let A be the event of a "yes" response on global warming and let B be the event of a "less" response on future fuel use. Suppose P(A and B) > P(A)P(B). Indicate whether A and B are independent events, and explain what this means in nontechnical terms.

5.26 Catalog sales You are the marketing director for a museum that raises money by selling gift items from a mail-order catalog. For each catalog sent to a potential customer, the customer's entry in the data file is Y if they ordered something and N if they did not (Y = yes, N = no). After you have mailed the fall and the winter catalogs, you estimate the probabilities of the buying patterns based on those who received the catalog as follows:

Outcome (fall, winter):	YY	YN	NY	NN
Probability:	0.30	0.10	0.05	0.55

a. Display the outcomes and their probabilities in a contingency table, using the rows for the (Y, N) outcomes for the fall catalog and the columns for the (Y, N) outcomes for the winter catalog.

b. Let F denote buying from the fall catalog and W denote buying from the winter catalog. Find P(F) and P(W).

c. Explain what the event "F and W" means, and find P(F and W).

d. Are F and W independent events? Explain why you would not normally expect customer choices to be independent.

5.27 Arts and crafts sales A local downtown arts and crafts shop found from past observation that 20% of the people who enter the shop actually buy something. Three potential customers enter the shop.

a. How many outcomes are possible for whether the clerk makes a sale to each customer? Construct a tree diagram to show the possible outcomes. (Let Y = sale, N = no sale.)

b. Find the probability of at least one sale to the three customers.

c. What did your calculations assume in part b? Describe a situation in which that assumption would be unrealistic.

5.3 Conditional Probability: The Probability of A Given B

As Example 1 explained, many employers require potential employees to take a diagnostic test for drug use. The diagnostic test has two categorical variables of interest: (1) whether or not the person has recently used drugs (yes or no), and (2) whether or not the diagnostic test shows that the person has used them (yes or no). Suppose the diagnostic test predicts that the person has recently used drugs. What's the probability that the person truly did use drugs?

This section introduces **conditional probability**, which deals with finding the probability of an event when you know that the outcome was in some particular part of the sample space. Most commonly, it is used to find a probability about a category for one variable (for instance, a person being a drug user), when we know the outcome on another variable (for instance, a test result showing drug use).

Finding the Conditional Probability of an Event

	Audited		
Income	**Yes**	**No**	**Total**
Under $200,000	1260	132,147	**133,407**
$200,000–$1,000,000	131	4311	**4442**
More than $1,000,000	22	371	**393**
Total	**1413**	**136,829**	**138,242**

Example 5 showed a contingency table on income and whether a taxpayer is audited by the Internal Revenue Service. The table is shown again in the margin. We found that the probability that a randomly selected taxpayer was audited equaled $1413/138{,}242 = 0.0102$. Were the chances higher if a taxpayer was at the highest income level? From the margin table, the number having income $\geq \$1{,}000{,}000$ was 393. Of them, 22 were audited, for a probability of $22/393 = 0.056$. This is substantially higher than 0.0102, indicating that those earning the most are the most likely to be audited.

In practice, tables often provide probabilities rather than counts for the outcomes in the sample space. Table 5.3 shows probabilities for the cells, based on the cell frequencies in the contingency table, for the six possible outcomes. For example, the probability of having income $\geq \$1{,}000{,}000$ *and* being audited was $22/138{,}242 = 0.00016$. The probability of income $\geq \$1{,}000{,}000$ was $393/138{,}242 = 0.0029$. So, of those at the highest income category, the proportion $0.00016/0.0029 = 0.055$ were audited. This is the same answer we obtained earlier using the cell frequencies, apart from a small rounding error.

Table 5.3 Probabilities of Taxpayers at the Six Possible Combinations of Income Level and Audited

Each frequency in Table 5.2 was divided by 138,242 to obtain the cell probabilities shown here, such as $1260/138{,}242 = 0.0091$.

	Audited		
Income Level	**Yes**	**No**	**Total**
	These 6 probabilities sum to 1.0		
Under $200,000	0.0091	0.9559	**0.9650**
$200,000–$1,000,000	0.0009	0.0312	**0.0321**
More than $1,000,000	0.0002	0.0027	**0.0029**
Total	**0.0102**	**0.9898**	**1.0000**

Let event A denote {audited = yes} and let event B denote {income $\geq \$1{,}000{,}000$}. The people in the highest income group and who are also audited make up the intersection event A and B (that is, audited = yes *and* income ≥ \$1,000,000). So, given B, the probability of A is the proportion of the cases in the intersection of A and B out of the cases in B. This is P(A and B)/P(B). This ratio is the **conditional probability** of the event A given the event B.

Conditional Probability

For events A and B, the **conditional probability** of event A, given that event B has occurred, is

$$P(A \mid B) = \frac{P(A \text{ and } B)}{P(B)}.$$

P(A | B) is read as "the probability of event A given event B." The vertical slash represents the word "given." Of the times that B occurs, **P(A|B)** is the proportion of times that A also occurs.

From Table 5.3 with events A (audited = yes) and B (income ≥ \$1,000,000),

$$P(A|B) = \frac{P(A \text{ and } B)}{P(B)} = \frac{0.0002}{0.0029} = 0.069.$$

Given that a taxpayer has income ≥ \$1,000,000, the chances of being audited are 0.055.

In Section 3.1, in learning about contingency tables, we saw that we could find **conditional proportions** for a categorical variable at any particular category of a second categorical variable. These enable us to study how the outcome on a response variable depends on the outcome on an explanatory variable. The conditional probabilities just found are merely conditional proportions. They refer to the population of taxpayers, treating audit status as a response variable and income level as an explanatory variable.

	Audited	
Income	**Yes**	**No**
Under \$200,000	0.0091	0.9559
\$200,000–\$1,000,000	0.0009	0.0312
More than \$1,000,000	0.0002	0.0027

The total of these cell probabilities over the 6 cells is 1.0.

We could find similar conditional probabilities on audit status at each given level of income. We'd then get the results shown in Table 5.4. Using the cell probabilities in Table 5.3 (which refer to intersections of income events and audit events, as shown in the margin table), we get each conditional probability by dividing a cell probability for a particular audit status by the row total that is the probability of income at that level. In each row of Table 5.4, the conditional probabilities sum to 1.0.

Table 5.4 Conditional Probabilities on Audited Given the Income Level

Each cell probability in Table 5.3 was divided by the row marginal total probability to obtain the conditional probabilities shown here, such as 0.0091/(0.0091 + 0.9559) = 0.0091/0.9650 = 0.0094.

	Audited		
Income Level	**Yes**	**No**	**Total**
Under \$200,000	0.0094 ← 0.0094 = 0.0091/0.9650	0.9906	**1.000**
\$200,000–\$1,000,000	0.0280	0.9720	**1.000**
More than \$1,000,000	0.0690	0.9310	**1.000**

Figure 5.9 is a graphical illustration of the definition of conditional probability. "Given event B" means that we restrict our attention to the outcomes in that event. This is the set of outcomes in the denominator. The proportion of those cases in which A occurred are those outcomes that are in event A as well as B. So the intersection of A and B is in the numerator.

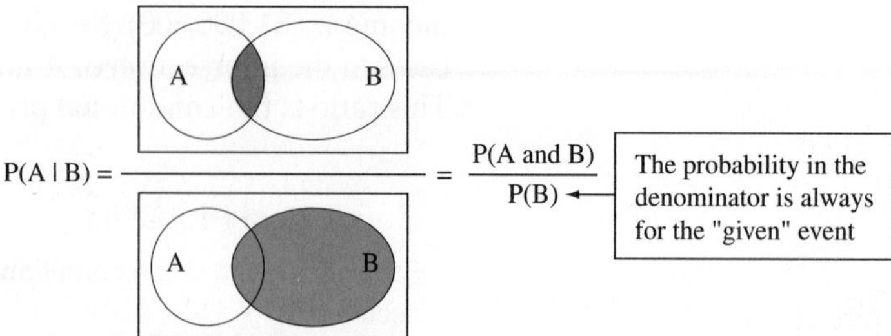

▲ **Figure 5.9** **Venn Diagram of Conditional Probability of Event A Given Event B.** Of the cases in which B occurred, P(A | B) is the proportion in which A also occurred. **Question** Sketch a representation of P(B | A). Is P(A | B) necessarily equal to P(B | A)?

Example 8

Conditional probability

The Triple Blood Test for Down Syndrome

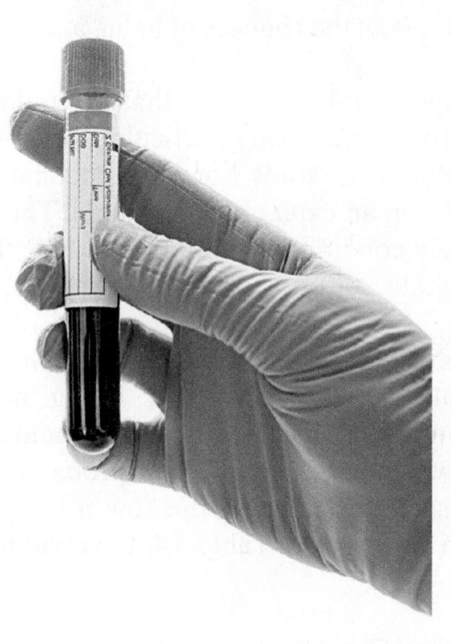

Picture the Scenario

A diagnostic test for a condition is said to be **positive** if it states that the condition is present and **negative** if it states that the condition is absent. How accurate are diagnostic tests? One way to assess accuracy is to measure the probabilities of the two types of possible error:

False positive: Test states the condition is present, but it is actually absent.

False negative: Test states the condition is absent, but it is actually present.

The Triple Blood Test screens a pregnant woman and provides as estimated risk of her baby being born with the genetic disorder Down syndrome. This syndrome, which occurs in about 1 in 800 live births, arises from an error in cell division that results in a fetus having an extra copy of chromosome 21. It is the most common genetic cause of mental impairment. The chance of having a baby with Down syndrome increases after a woman is 35 years old.

A study[2] of 5282 women aged 35 or over analyzed the Triple Blood Test to test its accuracy. It was reported that of the 5282 women, "48 of the 54 cases of Down syndrome would have been identified using the test and 25 percent of the unaffected pregnancies would have been identified as being at high risk for Down syndrome (these are false positives)."

Questions to Explore

a. Construct the contingency table that shows the counts for the possible outcomes of the blood test and whether the fetus has Down syndrome.

b. Assuming the sample is representative of the population, estimate the probability of a positive test for a randomly chosen pregnant woman 35 years or older.

c. Given that the diagnostic test result is positive, estimate the probability that Down syndrome truly is present.

[2]J. Haddow et al., *New England Journal of Medicine*, vol. 330, pp. 1114–1118, 1994.

Think It Through

a. We'll use the following notation for the possible outcomes of the two variables:

Down syndrome status: D = Down syndrome present, D^c = unaffected

Blood test result: POS = positive, NEG = negative.

Table 5.5 shows the four possible combinations of outcomes. From the article quote, there were 54 cases of Down syndrome. This is the first row total. Of them, 48 tested positive, so 54 − 48 = 6 tested negative. These are the counts in the first row. There were 54 Down cases out of $n = 5282$, so 5282 − 54 = 5228 cases were unaffected, event D^c. That's the second row total. Now, 25% of those 5228, or 0.25 × 5228 = 1307, would have a positive test. The remaining 5228 − 1307 = 3921 would have a negative test. These are the counts for the two cells in the second row.

Table 5.5 Contingency Table for Triple Blood Test of Down Syndrome

	Blood Test		
Down Syndrome Status	**POS**	**NEG**	**Total**
D (Down)	48	6	**54**
D^c (unaffected)	1307	3921	**5228**
Total	**1355**	**3927**	**5282**

b. From Table 5.5, the estimated probability of a positive test is P(POS) = 1355/5282 = 0.257.

c. The probability of Down syndrome, given that the test is positive, is the conditional probability, P(D | POS). Conditioning on a positive test means we consider only the cases in the first column of Table 5.5. Of the 1355 who tested positive, 48 cases actually had Down syndrome, so P(D | POS) = 48/1355 = 0.035. Let's see how to get this from the definition of conditional probability,

$$P(D \mid POS) = \frac{P(D \text{ and } POS)}{P(POS)}.$$

Since P(POS) = 0.257 from part b and P(D and POS) = 48/5282 = 0.0091, we estimate P(D | POS) = 0.0091/0.257 = 0.035. In summary, of the women who tested positive, fewer than 4% actually had fetuses with Down syndrome. This is somewhat comforting news to a woman who has a positive test result.

Insight

So why should a woman undergo this test, as most positives are false positives? From Table 5.5, P(D) = 54/5282 = 0.0102, so we estimate about a 1% chance of Down syndrome for women aged 35 or over. Also from Table 5.5, P(D | NEG) = 6/3927 = 0.0015, a bit more than 1 in 1000. A woman can have much less worry about Down syndrome if she has a negative test result because the chance of Down is then a bit more than 1 in 1000, compared to 1 in 100 overall.

In 2011, researchers announced a new and promising blood test for detecting Down's syndrome using DNA. Researchers noted, however, that more research is needed to improve the test's accuracy and that the small-scale study needed to be expanded to a larger-scale study of the population (www.technologyreview.com/biomedicine/18139/page1/).

Try Exercises 5.34 and 5.37

Caution

The P(D | NEG) is not the same as the false negative rate. We found in Example 8 that the P(D | NEG) = 0.0015. The false negative rate is found by evaluating P(NEG | D) = 6/54 = 0.11. Be careful to watch the event being conditioned upon. ◀

In Practice Conditional Probabilities in the Media

When you read or hear a news report that uses a probability statement, be careful to distinguish whether it is reporting a conditional probability. Most statements are conditional on some event and must be interpreted in that context. For instance, probabilities reported by opinion polls are often conditional on a person's gender, race, or age group.

Multiplication Rule for Finding P(A and B)

From Section 5.2, when A and B are independent events, P(A and B) = P(A) × P(B). The definition of conditional probability provides a more general formula for P(A and B) that holds regardless of whether A and B are independent. We can rewrite the definition P(A|B) = P(A and B)/P(B), multiplying both sides of the formula by P(B), to get P(B) × P(A|B) = P(B) × [P(A and B)/P(B)] = P(A and B), so that

$$P(\text{A and B}) = P(\text{B}) \times P(\text{A}|\text{B}).$$

Multiplication Rule for Evaluating P(A and B)

For events A and B, the probability that A and B both occur equals

$$P(\text{A and B}) = P(\text{B}) \times P(\text{A}|\text{B}).$$

Applying the conditional probability formula to P(B|A), we also see that

$$P(\text{A and B}) = P(\text{A}) \times P(\text{B}|\text{A}).$$

Example 9

Multiplication rule

Double Faults in Tennis

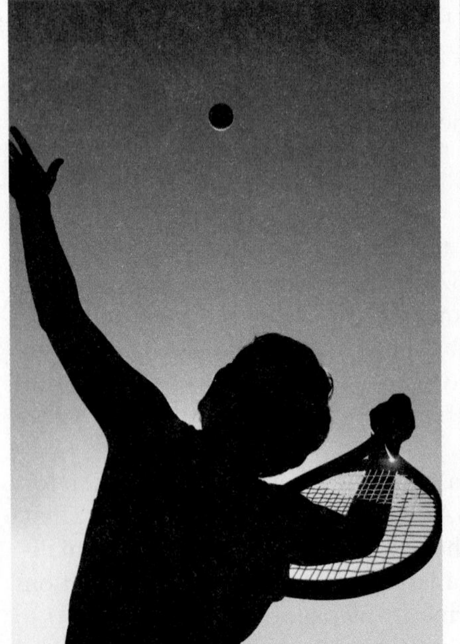

Picture the Scenario

In a tennis match, on a given point, the player who is serving has two chances to hit the ball in play. The ball must fall in the correct marked box area on the opposite side of the net. A serve that misses that box is called a *fault.* Most players hit the first serve very hard, resulting in a fair chance of making a fault. If they do make a fault, they hit the second serve less hard and with some spin, making it more likely to be successful. Otherwise, with two misses—a *double fault*—they lose the point.

Question to Explore

The 2010 men's champion in the Wimbledon tournament was Rafael Nadal of Spain. During the tournament, he made 68% of his first serves. He faulted on the first serve 32% of the time (100 − 68 = 32). Given that he made a fault with his first serve, he made a fault on his second serve only 9% of the time. Assuming these are typical of his serving performance, what is the probability that he makes a double fault when he serves?

Think It Through

Let F1 be the event that Nadal makes a fault with the first serve, and let F2 be the event that he makes a fault with the second serve. We know P(F1) = 0.32 and

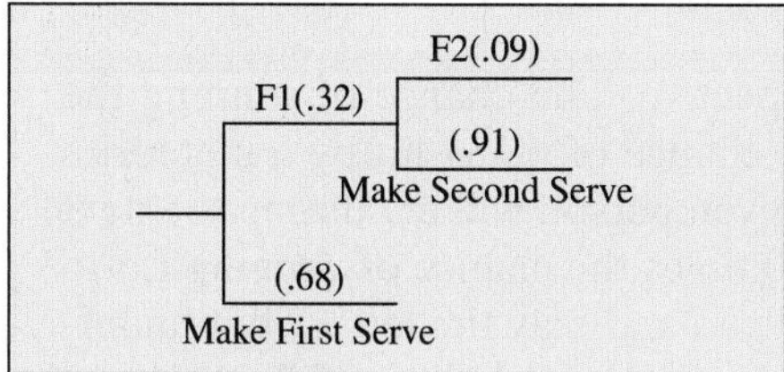

$P(F2 \mid F1) = 0.09$, as shown in the margin figure. The event that Nadal makes a double fault is "F1 and F2." From the multiplication rule, its probability is

$$P(\text{F1 and F2}) = P(F_1) \times P(F_2 \mid F_1) = 0.32 \times 0.09 = 0.029.$$

Insight

Nadal makes a fault on his first serve 32% of the time, and in 9% of those cases, he makes a fault on his second serve. He makes a double fault in 9% of the 32% of points in which he faults on the first serve, or in the proportion $0.09 \times 0.32 = 0.029$, just under 3% of his service points.

Try Exercise 5.40

Did You Know?

In tennis, you only serve a second time if you fault on the first serve. ◀

Sampling With or Without Replacement

In many sampling processes, once subjects are selected from a population, they are not eligible to be selected again. This is called *sampling without replacement.* At any stage of such a sampling process, probabilities of potential outcomes depend on the previous outcomes. Conditional probabilities are then used in finding probabilities of the possible samples.

Example 10

Conditional probability

Winning Lotto

Picture the Scenario

The biggest jackpot in state lotteries, typically millions of dollars, comes from the Lotto game. In Lotto South, available in Georgia, Kentucky, and Virginia, six numbers are randomly sampled without replacement from the integers 1 to 49. For example, a possible sample is (4, 9, 23, 26, 40, 46). Their order of selection is not important.

Question to Explore

You buy a Lotto South ticket. What is the probability that it is a winning ticket, having the six numbers chosen?

Think It Through

The probability of winning is the probability that the six numbers chosen are the six that you have on your ticket. Since your ticket has 6 of the 49 numbers that can be selected for the first number, $P(\text{you have 1st number}) = 6/49$. Given that you have the first number, for the second trial there are 5 numbers left that you hold out of 48 possible, so $P(\text{have 2nd number} \mid \text{have 1st number}) = 5/48$. Given that you have the first two, for the third trial there are 4 numbers left that you hold out of 47 possible, so $P(\text{have 3rd number} \mid \text{have 1st and 2nd numbers}) = 4/47$. Continuing with this logic, using an extension of the multiplication rule with conditional probabilities,

$$P(\text{have all 6 numbers}) = P(\text{have 1st and 2nd and 3rd and 4th and 5th and 6th})$$

$$= P(\text{have 1st})\, P(\text{have 2nd} \mid \text{have 1st})\, P(\text{have 3rd} \mid \text{have 1st and 2nd}) \ldots$$

$$\ldots P(\text{have 6th} \mid \text{have 1st and 2nd and 3rd and 4th and 5th})$$

$$= (6/49) \times (5/48) \times (4/47) \times (3/46) \times (2/45) \times (1/44)$$

$$= 720/10{,}068{,}347{,}520 = 0.00000007.$$

This is about 1 chance in 14 million.

Did You Know?

Comparable to the probability of 0.00000007 of winning Lotto South are the annual probabilities of about 0.0000004 of being hit by a meteorite, 0.0000002 of dying in a tornado, and 0.00000016 of dying by a lightning strike (Wilson and Crouch 2001, p. 200). It's also roughly the probability that a person of average mortality will die in the next three minutes. ◀

Insight

Let's give this small number some perspective. The chance of winning the jackpot in Lotto South is less than your chance of being hit by a meteorite in the next year. The probability that a given person will die during the next year in a car crash is about 0.00015, 2100 times the chance of winning Lotto South. If you have money to spare, go ahead and play the lottery but understand why many call it "sport for the mathematically challenged."

Try Exercise 5.44

Lotto South uses *sampling without replacement.* Once a number is chosen, it cannot be chosen again. If, by contrast, Lotto South allowed numbers to be picked more than once, the sampling scheme would be *sampling with replacement.*

After each observation without replacement, the population remaining is reduced by one and the conditional probability of a particular outcome changes. It can change considerably when the population size is small, as we saw in the Lotto example. With large population sizes, reducing the population by one does not much affect the probability from one trial to the next. In practice, when selecting random samples, we usually sample without replacement. With populations that are large compared to the sample size, the probability at any given observation depends little on the previous observations. Probabilities of possible samples are then quite similar for sampling without replacement and sampling with replacement.

Independent Events Defined Using Conditional Probability

Two events A and B are **independent** if the probability that one occurs is not affected by whether or not the other event occurs. This is expressed more formally using conditional probabilities.

Recall

Two events A and B are also independent if P(A and B) = P(A) × P(B). ◀

Independent Events, in Terms of Conditional Probabilities

Events A and B are **independent** if P(A | B) = P(A), or equivalently, if P(B | A) = P(B). If either holds, then the other does too.

For instance, let's consider the genders of two children in a family (F = female, M = male). The sample space is {FF, FM, MF, MM}. Suppose these four outcomes are equally likely, which is approximately true in practice. Let A denote {first child is female} and let B denote {second child is female}. Then P(A) = 1/2, since two of the four possible outcomes have a female for the first child. Likewise, P(B) = 1/2. Also, P(A and B) = 1/4, since this corresponds to the single outcome, FF. So, from the definition of conditional probability,

$$P(B \mid A) = P(A \text{ and } B)/P(A) = (1/4)/(1/2) = 1/2.$$

Thus, P(B | A) = 1/2 = P(B), so A and B are independent events. Given A (that the first child is female), the probability that the second child was female is 1/2 since one outcome (FF) has this out of the two possibilities (FF, FM). Intuitively, the gender of the second child does not depend on the gender of the first child.

In sampling without replacement, outcomes of different trials are dependent. For instance, in the Lotto game described in Example 10, let A denote {your first number is chosen} and B denote {your second number is chosen}. Then P(A) = 6/49, but P(A | B) = 5/48 (since there are 5 possibilities out of 48 numbers), which differs slightly from P(A).

Checking for independence

Example 11

Two Events from Diagnostic Testing

Picture the Scenario

Table 5.5 showed a contingency table relating the result of a diagnostic blood test (POS = positive, NEG = negative) to whether or not a woman's fetus has Down syndrome (D = Down syndrome, D^c = unaffected). The estimated cell probabilities based on the frequencies in that table are shown in the marginal table.

	Blood Test		
Status	**POS**	**NEG**	**Total**
D	0.009	0.001	**0.010**
D^c	0.247	0.742	**0.990**
Total	**0.257**	**0.743**	**1.00**

Questions to Explore

a. Are the events POS and D independent or dependent?

b. Are the events POS and D^c independent or dependent?

Think It Through

a. The probability of a positive test result is 0.257. However, the probability of a positive result, given Down syndrome, is

$$P(\text{POS} \mid \text{D}) = P(\text{POS and D})/P(\text{D}) = 0.009/0.010 = 0.90.$$

Since $P(\text{POS} \mid \text{D}) = 0.90$ differs from $P(\text{POS}) = 0.257$, the events POS and D are dependent.

b. Likewise,

$$P(\text{POS} \mid \text{D}^c) = P(\text{POS and D}^c)/P(\text{D}^c) = 0.247/0.990 = 0.250.$$

This differs slightly from $P(\text{POS}) = 0.257$, so POS and D^c are also dependent events.

Insight

As we'd expect, the probability of a positive result depends on whether the fetus has Down syndrome, and it's much higher if the fetus does. The diagnostic test would be worthless if the disease status and the test result were independent.

Generally, if A and B are dependent events, then so are A and B^c, and so are A^c and B, and so are A^c and B^c. For instance, if A depends on whether B occurs, then A also depends on whether B does not occur. So, once we find that POS and D are dependent events, we know that POS and D^c are dependent events also.

Try Exercise 5.45

We can now justify the formula given in Section 5.2 for the probability of the intersection of two independent events, namely $P(\text{A and B}) = P(\text{A}) \times P(\text{B})$. This is a special case of the multiplication rule for finding P(A and B),

$$P(\text{A and B}) = P(\text{A}) \times P(\text{B} \mid \text{A}).$$

If A and B are independent, then $P(\text{B} \mid \text{A}) = P(\text{B})$, so the multiplication rule simplifies to

$$P(\text{A and B}) = P(\text{A}) \times P(\text{B}).$$

SUMMARY: Checking for Independence

Here are three ways to determine if events A and B are independent:

- Is $P(A \mid B) = P(A)$?
- Is $P(B \mid A) = P(B)$?
- Is $P(A \text{ and } B) = P(A) \times P(B)$?

If any of these is true, then the others are also true and the events A and B are independent.

Students often struggle to distinguish between the concepts of disjoint events and independence events. This is because the words seem to have a similar connotation. In fact, their precise meanings when referring to events of a sample space are very different, as illustrated in the following example.

Example 12

Checking for independence

Distinguishing Between Disjoint and Independent Events

Picture the Scenario

Consider three events: walking, chewing gum, and tying one's shoe. Let W = walking, C = chewing gum, and T = tying one's shoe. Suppose that while Joe works at his job in the campus cafeteria, the probability that he is walking at any given time is 0.1, the probability he is chewing gum is 0.3, and the probability he is tying his shoe is 0.001. Suppose also that for Joe, events W and C are independent.

Question to Explore

a. What does it mean contextually for events W and C to be independent?

b. What is the probability of the intersection of W and C? Are W and C disjoint events?

c. Of the events W, C, and T, which pair of events is Joe least likely to be doing at the same time? What is the probability of the intersection of those two events? Are they disjoint events?

Think It Through

a. Independence of W and C means that whether or not Joe is chewing gum does not depend on whether or not he is walking. For example, Joe is just as likely to be chewing gum while walking as chewing gum while not walking.

b. Since W and C are independent, $P(W \text{ and } C) = P(W) \times P(C) = 0.1 \times 0.3 = 0.03$. Since $P(W \text{ and } C) \neq 0$, events W and C, although independent, are not disjoint.

c. We already know it is possible for Joe to walk and chew gum at the same time. It is also conceivable for him to be tying his shoe and chewing gum at the same time, so that P(T and C) is likely to be greater than zero. Meanwhile, it is simply not possible for him to walk and tie his shoe at the same time. (If you don't believe this, give it a try!) Therefore, $P(W \text{ and } T) = 0$. The events walking and tying one's shoe are disjoint.

Insight

It is quite possible that for a different individual, the events W and C are not independent. Suppose for example that Jennifer only chews gum while she is walking back and forth to class. Whether or not she is chewing gum depends on whether or not she is walking. In fact, if she is not walking, then she is not chewing gum. For Joe (or anyone else), whether or not he is walking strongly depends on whether or not he is tying his shoe. If he is tying his shoe, then we know he is not walking. Likewise, if he is walking, then we know he is not tying his shoe. These two disjoint events, or any other pair of disjoint events, cannot be independent.

Try Exercise 5.45

5.3 Practicing the Basics

5.28 Alcohol and college students A 2007 study by the National Center on Addiction and Substance Abuse at Columbia University reported that for college students, the estimated probability of being a binge drinker was 0.50 for males and 0.34 for females. Using notation, express each of these as a conditional probability.

5.29 Spam Because of the increasing nuisance of spam e-mail messages, many start-up companies have emerged to develop e-mail filters. One such filter was recently advertised as being 95% accurate. The way the advertisement is worded, 95% accurate could mean that (a) 95% of spam is blocked, (b) 95% of valid e-mail is allowed through, (c) 95% of the e-mail allowed through is valid, or (d) 95% of the blocked e-mail is spam. Let S denote {message is spam}, and let B denote {filter blocks message}. Using these events and their complements, identify each of these four possibilities as a conditional probability.

5.30 Audit and low income Table 5.3 on audit status and income follows. Show how to find the probability of:

a. Being audited, given that the taxpayer is in the lowest income category.

b. Being in the lowest income category, given that the taxpayer is audited.

	Audited	
Income	**Yes**	**No**
Under \$200,000	0.0091	0.9559
\$200,000 − \$1,000,000	0.0009	0.0312
More than \$1,000,000	0.0002	0.0027

5.31 Religious affiliation The 2011 Statistical Abstract of the United States[3] provides information on individuals' self-described religious affiliations. The information for 2008 is summarized in the following table (all numbers are in thousands).

Christian		
	Catholic	57,199
	Baptist	36,148
	Christian (no denomination specified)	16,834
	Methodist/Wesleyan	11,366
	Other Christian	51,855
Jewish		2,680
Muslim		1,349
Buddhist		1,189
Other non-Christian		3,578
No Religion		34,169
Refused to Answer		11,815
Total Adult Population in 2008		228,182

a. Find the probability that a randomly selected individual is identified as Christian.

b. Given that an individual identifies as Christian, find the probability that the person is Catholic.

c. Given that an individual answered, find the probability the individual is identified as following no religion.

5.32 Cancer deaths Current estimates are that about 25% of all deaths are due to cancer, and of the deaths that are due to cancer, 30% are attributed to tobacco, 40% to diet, and 30% to other causes.

a. Define events, and identify which of these four probabilities refer to conditional probabilities.

b. Find the probability that a death is due to cancer and tobacco.

5.33 Revisiting seat belts and auto accidents The following table is from Exercise 5.23 classifying auto accidents by survival status (S = survived, D = died) and seat belt status of the individual involved in the accident.

[3]*Source:* Data from www.census.gov/compendia/statab/2011/tables/11s0075.xls.

	Outcome		
Belt	**S**	**D**	**Total**
Yes	412,368	510	**412,878**
No	162,527	1,601	**164,128**
Total	**574,895**	**2,111**	**577,006**

a. Estimate the probability that the individual died (D) in the auto accident.

b. Estimate the probability that the individual died, given that the person (i) wore and (ii) did not wear a seat belt. Interpret results.

c. Are the events of dying and wearing a seat belt independent? Justify your answer.

5.34 Go Celtics! Larry Bird, who played pro basketball for the Boston Celtics, was known for being a good shooter. In games during 1980–1982, when he missed his first free throw, 48 out of 53 times he made the second one, and when he made his first free throw, 251 out of 285 times he made the second one.

TRY

a. Form a contingency table that cross tabulates the outcome of the first free throw (made or missed) in the rows and the outcome of the second free throw (made or missed) in the columns.

b. For a given pair of free throws, estimate the probability that Bird (i) made the first free throw and (ii) made the second free throw. (*Hint*: Use counts in the (i) row margin and (ii) column margin.)

c. Estimate the probability that Bird made the second free throw, given that he made the first one. Does it seem as if his success on the second shot depends strongly, or hardly at all, on whether he made the first?

5.35 Identifying spam An article[4] in www.networkworld.com about evaluating e-mail filters that are designed to detect spam described a test of MailFrontier's Anti-Spam Gateway (ASG). In the test, there were 7840 spam messages, of which ASG caught 7005. Of the 7053 messages that ASG identified as spam, they were correct in all but 48 cases.

a. Set up a contingency table that cross classifies the actual spam status (with the rows "spam" and "not spam") by the ASG filter prediction (with the columns "predict message is spam" and "predict message is not spam"). Using the information given, enter counts in three of the four cells.

b. For this test, given that a message is truly spam, estimate the probability that ASG correctly detects it.

c. Given that ASG identifies a message as spam, estimate the probability that the message truly was spam.

5.36 Homeland security According to an article in *The New Yorker* (March 12, 2007), the Department of Homeland Security in the United States is experimenting with installing devices for detecting radiation at bridges, tunnels, roadways, and waterways leading into Manhattan. The New York Police Department (NYPD) has expressed concerns that the system would generate too many false alarms.

a. Form a contingency table that cross classifies whether a vehicle entering Manhattan contains radioactive material and whether the device detects radiation. Identify the cell that corresponds to the false alarms the NYPD fears.

b. Let A be the event that a vehicle entering Manhattan contains radioactive material. Let B be the event that the device detects radiation. Sketch a Venn diagram for which each event has similar (not the same) probability but the probability of a false alarm equals 0.

c. For the diagram you sketched in part b, explain why $P(A|B) = 1$, but $P(B|A) < 1$.

5.37 Down syndrome again Example 8 discussed the Triple Blood Test for Down syndrome, using data summarized in a table shown again below.

TRY

	Blood Test		
Down	**POS**	**NEG**	**Total**
D	48	6	**54**
D^c	1307	3921	**5228**
Total	**1355**	**3927**	**5282**

a. Given that a test result is negative, show that the probability the fetus actually has Down syndrome is $P(D|NEG) = 0.0015$.

b. Is $P(D|NEG)$ equal to $P(NEG|D)$? If so, explain why. If not, find $P(NEG|D)$.

5.38 Is a bad job better than no job? Workers specified as actively disengaged are those who are emotionally disconnected from their work and workplace. A Gallup poll conducted in December 2010[5] surveyed individuals who were either unemployed or who were actively disengaged in their current position. Individuals were asked to classify themselves as thriving or struggling. The poll reported that 42% of the actively disengaged group claimed to be thriving, compared to 48% of the unemployed group.

a. Are these percentages (probabilities) ordinary or conditional? Explain, by specifying events to which the probabilities refer.

b. Of the individuals polled, 1266 were unemployed and 400 were actively disengaged. Create a contingency table showing counts for job status and self-classification.

c. Create a tree diagram such that the first branching represents job status and the second branching represent self-classification. Be sure to include the appropriate percentages on each branch.

5.39 Happiness in marriage Are people happy in their marriages? The table shows results from the 2008 General Social Survey for married adults classified by gender and level of happiness.

[4] www.networkworld.com/reviews/2003/0915spamstats.html.

[5] www.gallup.com/poll/146867/Workers-Bad-Jobs-Worse-Wellbeing-Jobless.aspx.

	Level of Happiness			
Gender	**Very Happy**	**Pretty Happy**	**Not too Happy**	**Total**
Male	183	243	43	**469**
Female	215	247	38	**500**
Total	**398**	**490**	**81**	**969**

a. Estimate the probability that a married adult is very happy.

b. Estimate the probability that a married adult is very happy, (i) given that their gender is male and (ii) given that their gender is female.

c. For these subjects, are the events being very happy and being a male independent? (Your answer will apply merely to this sample. Chapter 11 will show how to answer this for the population.)

5.40 Serena Williams serves Serena Williams won the 2010 TRY Wimbledon Ladies' Singles Championship. For the seven matches she played in the tournament, her total number of first serves was 379, total number of good first serves was 256, and total number of double faults was 15.

a. Find the probability that her first serve is good.

b. Find the conditional probability of double faulting, given that her first serve resulted in a fault.

c. On what percentage of her service points does she double fault?

5.41 Shooting free throws Pro basketball player Shaquille O'Neal is a poor free-throw shooter. Consider situations in which he shoots a pair of free throws. The probability that he makes the first free throw is 0.50. Given that he makes the first, suppose the probability that he makes the second is 0.60. Given that he misses the first, suppose the probability that he makes the second one is 0.40.

a. What is the probability that he makes both free throws?

b. Find the probability that he makes *one* of the two free throws (i) using the multiplicative rule with the two possible ways he can do this and (ii) by defining this as the complement of making neither or both of the free throws.

c. Are the results of the free throws independent? Explain.

5.42 Drawing cards A standard card deck has 52 cards consisting of 26 black and 26 red cards. Three cards are dealt from a shuffled deck, *without replacement.*

a. True or false: The probability of being dealt three black cards is $(1/2) \times (1/2) \times (1/2) = 1/8$. If true, explain why. If false, show how to get the correct probability.

b. Let A = first card red and B = second card red. Are A and B independent? Explain why or why not.

c. Answer parts a and b if each card is replaced in the deck after being dealt.

5.43 Drawing more cards A standard deck of poker playing cards contains four suits (clubs, diamonds, hearts, and spades) and 13 different cards of each suit. During a hand of poker, 5 of the 52 cards have been exposed. Of the exposed cards, 3 were diamonds. Tony will have the opportunity to draw two more cards, and he has surmised that in order to win the hand, each of those two cards will need to be diamonds. What is Tony's probability of winning the hand? (Assume the two unexposed cards are not diamonds.)

5.44 Big loser in Lotto Example 10 showed that the prob- TRY ability of having the winning ticket in Lotto South was 0.00000007. Find the probability of holding a ticket that has zero winning numbers out of the 6 numbers selected (without replacement) for the winning ticket out of the 49 possible numbers.

5.45 Family with two children For a family with two children, TRY let A denote {first child is female}, let B denote (at least one child is female}, and let C denote {both children are female}.

a. Show that $P(C \mid A) = 1/2$.

b. Are A and C independent events? Why or why not?

c. Find $P(C \mid B)$.

d. Describe what makes $P(C \mid A)$ different than $P(C \mid B)$.

5.46 Checking independence In three independent flips of a balanced coin, let A denote {first flip is a head}, B denote {second flip is a head}, C denote {first two flips are heads}, and D denote {three heads on the three flips}.

a. Find the probabilities of A, B, C, and D.

b. Which, if any, pairs of these events are independent? Explain.

5.4 Applying the Probability Rules

Probability relates to many aspects of your daily life—for instance, when you make decisions that affect your financial well being and when you evaluate risks due to lifestyle decisions. Objectively or subjectively, you need to consider questions such as: What's the chance that the business you're thinking of starting will succeed? What's the chance that the extra collision insurance you're thinking of getting for your car will be needed?

We'll now apply the basics of probability to coincidence in our lives.

Is a Coincidence Truly an Unusual Event?

Some events in our lives seem to be coincidental. One of the authors, who lives in Georgia, once spent a summer vacation in Newfoundland, Canada. One day during that trip, she made an unplanned stop at a rest area and observed a camper with a Georgia license tag from a neighboring county to her hometown. She discovered that the camper's owner was a patient of her physician husband. Was this meeting as coincidental as it seemed?

Events that seem coincidental are often not so unusual when viewed in the context of *all* the possible random occurrences at all times. Lots of topics can trigger an apparent coincidence—the person you meet having the same last name as yours, the same birth place, the same high school or college, the same profession, the same birthday, and so on. It's really not so surprising that in your travels you will sometime have a coincidence, such as seeing a friend or meeting someone who knows somebody you know.

With a large enough sample of people or times or topics, seemingly surprising things are actually quite sure to happen. Events that are rare per person occur rather commonly with large numbers of people. If a particular event happens to one person in a million each day, then in the United States, we expect about 300 such events a day and more than 100,000 every year. The one in a million chance regularly occurs, however surprised we may be if it should happen to us.

If you take a coin now and flip it 10 times, you would probably be surprised to get 10 heads. But if you flipped the coin for a long time, would you be surprised to get 10 heads in a row at some point? Perhaps you would, but you should not be. For instance, if you flip a balanced coin 2000 times, then you can expect the longest run of heads during those flips to be about 10. When a seemingly unusual event happens to you, think about whether it is like seeing 10 heads on the next 10 flips of a coin, or more like seeing 10 heads in a row sometime in a long series of flips. If the latter, it's really not such an unusual occurrence.

Did You Know?

If something has a very large number of opportunities to happen, occasionally it will happen, even if it's highly unlikely at any one observation. This is a consequence of the **law of large numbers**. ◀

Activity 2

Matching Birthdays

If your class is small to moderate in size (say, fewer than about 60 students), the instructor may ask you to state your birth dates. Does a pair of students in the class share the same birthday?

Coincidence and Seemingly Unusual Patterns

Once we have data, it's easy to find patterns: 10 heads in a row, 10 tails in a row, 5 tails followed by 5 heads, and so forth. Our minds are programmed to look for patterns. Out of the huge number of things that can happen to us and to others over time, it's not surprising to occasionally see patterns that seem unusual.

To illustrate that an event you may perceive as a coincidence is actually not that surprising, let's answer the question, "What is the chance that at least two people in your class have the same birthday?"

Coincidence

Example 13

Matching Birthdays

Picture the Scenario

Suppose a class has 25 students. Since there are 365 possible birth dates (without counting February 29), our intuition tells us that the probability is small that there will be any birthday matches. Assume that the birth date of a student is equally likely to be any one of the 365 days in a year and that students' birth dates are independent (for example, there are no twins in the class).

Question to Explore

What is the probability that *at least* two of the 25 students have the same birthday? Is our intuition correct that the probability of a match is small?

Think It Through

The event of *at least one* birthday match includes one match, two matches, or more. To find the probability of *at least one* match, it is simpler to find the complement probability of *no* matches. Then

$$\text{P(at least one match)} = 1 - \text{P(no matches)}.$$

To begin, suppose a class has only two students. The first student's birthday could be any of 365 days. Given that student's birthday, the chance that the second student's birthday is different is 364/365 because 364 of the possible 365 birthdays are different. The probability is $1 - 364/365 = 1/365$ that the two students share the same birthday.

Now suppose a class has three students. The probability that all three have different birthdays is

$$\text{P(no matches)} = \text{P(students 1 and 2 and 3 have different birthdays)} =$$
$$\text{P(students 1 and 2 different)} \times \text{P(student 3 different} \mid \text{students 1 and}$$
$$\text{2 different)} = (364/365) \times (363/365).$$

The second probability in this product equals 363/365 because there are 363 days left for the third student that differ from the different birthdays of the first two students.

For 25 students, similar logic applies. By the time we get to the 25th student, for that student's birthday to differ from the other 24, there are 341 choices left out of 365 possible birthdays. So

$$\text{P(no matches)} = \text{P(students 1 and 2 and 3} \ldots \text{ and 25 have different birthdays)}$$
$$= (364/365) \times (363/365) \times (362/365) \times \quad \ldots \quad \ldots \quad \ldots \quad \times (341/365).$$

↑	↑	↑	↑	↑
student 2, given 1	student 3, given 1,2	student 4, given 1,2,3	next 20 students	student 25, given 1, …, 24

This product equals 0.43. Using the probability for the complement of an event,

$$\text{P(at least one match)} = 1 - \text{P(no matches)} = 1 - 0.43 = 0.57.$$

The probability exceeds 1/2 of at least one birthday match in a class of 25 students.

Insight

Is this probability higher than you expected? It should not be once you realize that with 25 students, there are 300 *pairs* of students who can share the same birthday (see Exercise 5.49.) Remember that with lots of opportunities for something to happen, coincidences are really not so surprising.

Did you have a birthday match in your class? Was it coincidence? If the number of students in your class is at least 23, the probability of at least one match is greater than 1/2. For a class of 50 students, the probability of at least one match is 0.97. For 100 students, it is 0.9999997 (there are then 4950 different *pairs* of students). Here are a couple of other facts about matching birthdays that may surprise you:

- With 88 people, there's a 1/2 chance that at least three people have the same birthday.
- With 14 people, there's a 1/2 chance that at least two people have a birthday within a day of each other in the calendar year.

Try Exercise 5.47

Sometimes a cluster of occurrences of some disease, like cancer, in a neighborhood will cause worry in residents that there is some environmental cause. But *some* disease clusters will appear around a nation just by chance. If we look at a large number of places and times, we should expect some disease clusters. By themselves, they seem unusual, but viewed in a broader context, they may not be. Epidemiologists are statistically trained scientists who face the difficult task of determining which events can be explained by ordinary random variation and which cannot.

In Practice, Probability Models Approximate Reality

We've now found probabilities in many idealized situations. In practice, it's often not obvious when different outcomes are equally likely or different events are independent. When calculating probabilities, it is advisable to specify a **probability model** that spells out all the assumptions made.

Probability Model

A **probability model** specifies the possible outcomes for a sample space and provides assumptions on which the probability calculations for events composed of those outcomes are based.

The next example illustrates a probability model used together with the rules for finding probabilities.

Example 14

Probability model

Safety of the Space Shuttle

Picture the Scenario

Out of the first 113 space shuttle missions, there were two failures, the *Challenger* disaster on the 25th flight (January 28, 1986) and the *Columbia* disaster on the 113th flight (January 16, 2003). Since then, much attention has focused on estimating probabilities of success or failure (disaster) for a mission. But before the first flight, there were no trials to provide data for estimating probabilities.

Did You Know?

The space shuttle missions have served an important role in space exploration during the years 1981–2011 with 135 flights. ◀

Question to Explore

Based on all the information available, a scientist is willing to predict the probability of success for any particular mission. How can you use this to find the probability of *at least one* failure in a total of 100 missions?

Think It Through

Let S1 denote the event that the first mission is successful, S2 the event that the second mission is successful, and so on up to S100, the event that mission 100 is successful. If all these events occur, there is no failure in the 100 flights. The event of *at least one* failure in 100 flights is the complement of the event that they are all successful (0 failures). By the probability for complementary events,

$$\text{P(at least 1 failure)} = 1 - \text{P(0 failures)}$$
$$= 1 - \text{P(S1 and S2 and S3} \ldots \text{ and S100)}.$$

The intersection of events S1 through S100 is the event that *all* 100 missions are successful.

We now need a probability model to evaluate P(S1 and S2 and S3... and S100). If our probability model assumes that the 100 shuttle flights are *independent*, then the multiplication rule for independent events implies that

$$\text{P(S1 and S2 and S3} \ldots \text{ and S100)} =$$
$$\text{P(S1)} \times \text{P(S2)} \times \text{P(S3)} \times \ldots \times \text{P(S100)}.$$

If our probability model assumes that each flight has the *same probability* of success, say P(S), then this product equals $[P(S)]^{100}$. To proceed further, we need a value for P(S). According to a discussion of this issue in the PBS series *Against All Odds: Inside Statistics*, a risk assessment study by the Air Force used the estimate P(S) = 0.971. With it,

$$\text{P(at least 1 failure)} = 1 - [P(S)]^{100} = 1 - [0.971]^{100} = 1 - 0.053 = 0.947.$$

Different estimates of P(S) can result in very different answers. For instance, an *Against All Odds* episode mentioned that a NASA study estimated that P(S) = 0.9999833. Using this,

$$\text{P(at least 1 failure)} = 1 - [P(S)]^{100} = 1 - [0.9999833]^{100}$$
$$= 1 - 0.998 = 0.002.$$

This was undoubtedly overly optimistic. We see that different probability models can result in drastically different probability assessments.

Insight

The answer above depended strongly on the assumed probability of success for each mission. Since there were two failures in the first 113 flights, after that flight an estimate of P(S) was 111/113 = 0.982. But the assumptions of independence and of the same probability for each flight may also be suspect. For instance, other variables (e.g., temperature at launch, experience of crew, age of craft used, quality of O-ring seals) could affect that probability.

Try Exercise 5.56

In practice, probability models merely *approximate* reality. They are rarely *exactly* satisfied. For instance, in the matching birthday example, our probability model ignored February 29, assumed each of the other 365 birthdays are equally likely, and assumed that students' birth dates are independent. So, the answer found is only approximate. Whether probability calculations using a particular probability model are accurate depends on whether assumptions in that model are close to the truth or unrealistic.

Probabilities and Diagnostic Testing

We've seen the important role that probability plays in diagnostic testing, illustrated in Section 5.3 (Example 8) with the pregnancy test for Down syndrome. Table 5.6 summarizes conditional probabilities about the result of a diagnostic test, given whether some condition or state (such as Down syndrome) is present. We let S denote that the state is present and S^c denote that it is not present.

Recall

The vertical slash means "given." **P(POS | S)** denotes the conditional probability of a positive test result, given that the state is present. ◀

Table 5.6 Probabilities of Correct and Incorrect Results in Diagnostic Testing

The probabilities in the body of the table refer to the test result, conditional on whether or not the state (S) is truly present. The sensitivity and specificity are the probabilities of the two types of correct diagnoses.

	Diagnostic Test Result		
State Present?	**Positive (POS)**	**Negative (NEG)**	**Total Probability**
Yes (S)	Sensitivity P(POS | S)	False negative rate P(NEG | S)	**1.0**
No (S^c)	False positive rate P(POS | S^c)	Specificity P(NEG | S^c)	**1.0**

In Table 5.6, **sensitivity** and **specificity** refer to correct test results, given the actual state. For instance, given that the state tested for is present, sensitivity is the probability the test detects it by giving a positive result, that is, P(POS | S).

Medical journal articles that discuss diagnostic tests commonly report the sensitivity and specificity. However, what's more relevant to you once you take a diagnostic test are the conditional probabilities that condition on the test result. If a diagnostic test for Down syndrome is positive, you want to know the probability that Down syndrome is truly present. If you know the sensitivity and specificity and how often the state occurs, can you find P(S | POS) using the rules of probability?

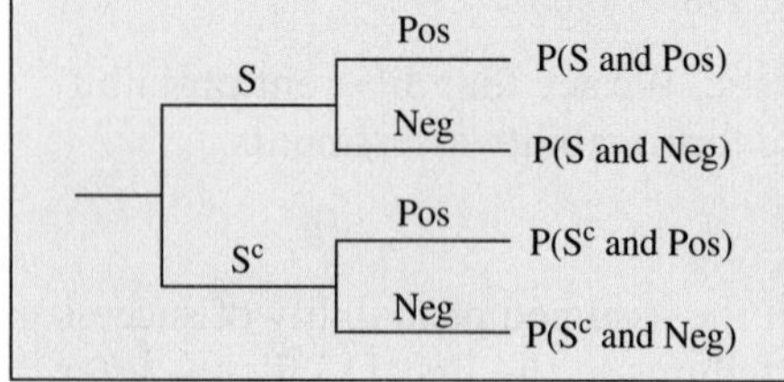

The easiest way to do this is with a tree diagram, as shown in the margin. The first branches show the probabilities of the two possible states, P(S) and P(S^c). The next set of branches show the known conditional probabilities, such as P(POS | S), for which we are given the true state. Then the products P(S)P(POS | S) and P(S^c)P(POS | S^c) give intersection probabilities P(S and POS) and P(S^c and POS), which can be used to get probabilities such as

$$P(POS) = P(S \text{ and } POS) + P(S^c \text{ and } POS).$$

Then you can find P(S | POS) = P(S and POS)/P(POS).

Example 15

Diagnostic tests

Random Drug Testing of Air Traffic Controllers

Picture the Scenario

Air traffic controllers monitor the flights of aircraft and help to ensure safe takeoffs and landings. In the United States, air traffic controllers are required to undergo periodic random drug testing. A urine test is used as an initial screening due to its low cost and ease of implementation. One such urine

test, the *Triage Panel for Drugs of Abuse plus TCA*,[6] detects the presence of drugs. Its sensitivity and specificity have been reported[7] as 0.96 and 0.93. Based on past drug testing of air traffic controllers, the FAA reports that the probability of drug use at a given time is approximately 0.007 (less than 1%). This is called the **prevalence** of drug use.

Questions to Explore

a. A positive test result puts the air traffic controller's job in jeopardy. What is the probability of a positive test result?

b. Find the probability an air traffic controller truly used drugs, given that the test is positive.

Think It Through

a. We've been given the probability of drug use, P(S) = 0.007, the sensitivity P(POS | S) = 0.96, and the specificity P(NEG | S^c) = 0.93. Figure 5.10 shows a tree diagram that is useful for visualizing these probabilities and for finding P(POS).

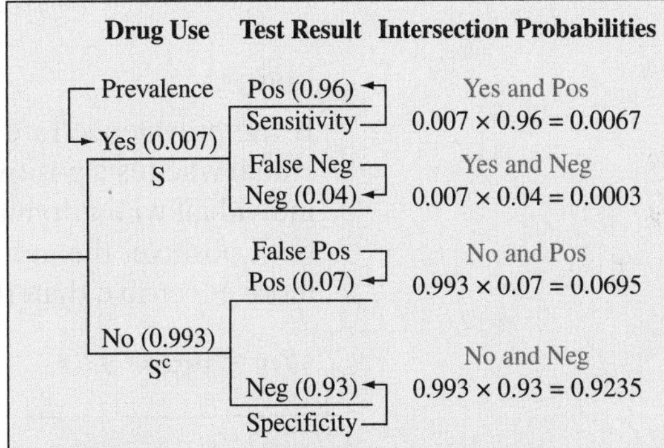

▲ **Figure 5.10** Tree Diagram for Random Drug Testing of Air Traffic Controllers. The first set of branches show the probabilities for drug use. The second set of branches shows the conditional probabilities for the test result, given whether the person used drugs or not. Multiplication of the probabilities along each path gives the probabilities of intersections of the events.

From the tree diagram and the multiplicative rule for intersection probabilities,

$$P(S \text{ and } POS) = P(S)P(POS \mid S) = 0.007 \times 0.96 = 0.0067.$$

The other path with a positive test result has probability

$$P(S^c \text{ and } POS) = P(S^c)P(POS \mid S^c) = 0.993 \times 0.07 = 0.0695.$$

To find P(POS), we add the probabilities of these two possible positive test paths. Thus,

$$P(POS) = P(S \text{ and } POS) + P(S^c \text{ and } POS) = 0.0067 + 0.0695 = 0.0762.$$

There's nearly an 8% chance of the test suggesting that the person used drugs.

b. The probability of drug use, given a positive test, is P(S | POS). From the definition of conditional probability,

$$P(S \mid POS) = P(S \text{ and } POS)/P(POS) = 0.0067/0.0762 = 0.09.$$

When the test is positive, only 9% of the time had the person actually used drugs.

---

[6]Screening assay from Biosite Diagnostics, San Diego, California.

[7]M. Peace et al., *Journal of Analytical Toxicology*, vol. 24 (2000).

If you're uncertain about how to answer part a and part b with a tree diagram, you can construct a contingency table, as shown in Table 5.7. The table shows the summary value of P(S) = 0.007 in the right margin, the other right-margin value P(S^c) = 1 − 0.007 = 0.993 determined by it, and the intersection probabilities P(S and POS) = 0.0067 and P(S^c and POS) = 0.0695 found in part a. From that table, of the proportion 0.0762 of positive cases, 0.0067 truly had used drugs, so the conditional probability is P(S|POS) = P(S and POS)P(POS) = 0.0067/0.0762 = 0.09

Table 5.7 Contingency Table-Cell Probabilities for Air Controller Drug Test

	Drug Test Result		
State Present?	**POS**	**NEG**	**Total**
Yes (S)	0.0067	0.0003	0.007
NO (S^c)	0.0695	0.9235	0.993
Total	**0.0762**	**0.9238**	**1.000**

Insight

If the prevalence rate is truly near 0.007, the chances are low that an individual who tests positive is actually a drug user. Does a positive test mean the individual will automatically lose his or her job? No, if the urine test comes back positive, the individual is given a second test that is more accurate but more expensive than the urine test.

Try Exercise 5.57

Are you surprised that P(S|POS) is so small (only 0.09) for the air controllers? To help understand the logic behind this, it's a good idea to show on a tree diagram what you'd expect to happen with a typical group of air controllers. Figure 5.11 shows a tree diagram for what we'd expect to happen for 1000 of them.

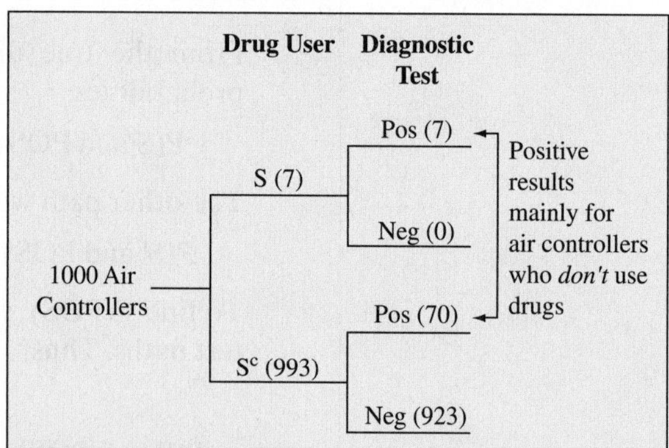

▲ **Figure 5.11 Expected Results of Drug Tests for 1000 Air Controllers.** This diagram shows typical results when the proportion 0.007 (7 in 1000) are drug users (event S), a positive result has probability 0.96 for those who use drugs, and a negative result has probability 0.93 for those who do not use drugs. Most positive results occur with individuals who are *not* using drugs because there are so many such individuals (more than 99% of the population). **Question** How would results change if a higher percentage of the population were using drugs?

Since the proportion 0.007 (which is 7 in 1000) of the target population uses drugs, the tree diagram shows 7 drug users and 993 nondrug users out of the 1000 air controllers. For the 7 drug users, there's a 0.96 chance the test detects the drug use. So we'd expect all 7 or perhaps 6 of the 7 to be detected with the test (Figure 5.11 shows 7). For the 993 nondrug users, there's a 0.93 chance the test is negative. So we'd expect about $0.93 \times 993 = 923$ individuals to have a negative result and the other $993 - 923 = 70$ nondrug users to have a positive result, as shown on Figure 5.11.

In summary, Figure 5.11 shows $7 + 70 = 77$ individuals having a positive test result, but only 7 were actually drug users. Of those with a positive test result, the proportion who truly were drug users is $7/77 = 0.09$. What's happening is that the 7% of errors for the large majority of individuals who *do not* use drugs is much larger than the 96% of correct decisions for the small number of individuals who *do* use drugs.

If the prevalence rate were $P(S) = 0.15$ (let's hope it's not really this high!) instead of 0.007, you can verify that $P(S \mid POS) = 0.71$, rather than 0.09. The probability that the person truly used drugs is then much higher, 0.71. In fact, the chance that a positive test result is truly correct depends very much on the prevalence rate. The lower the prevalence rate, the lower the chance. The more drug-free the population, the less likely an individual who tests positive for drugs truly used them.

In Example 15, we started with probabilities of the form $P(POS \mid S)$ and used them to find a conditional probability $P(S \mid POS)$ that reverses what is given and what is to be found. The method used to find this reverse conditional probability can be summarized in a single formula (see Exercise 5.112), known as **Bayes's rule**. We have not shown that formula here because it is easier to understand the logic behind evaluating this conditional probability using tree diagrams or contingency tables.

Probability Answers to Questions Can Be Surprising

The results in Example 15 on drug testing of air traffic controllers may seem surprising, but actually they are not uncommon. For instance, consider the mammogram for detecting breast cancer in women. One recent study[8] estimated sensitivity = 0.86 and specificity = 0.88. Of the women who receive a positive mammogram result, what proportion actually have breast cancer? In Exercise 5.57 you can work out that it may be only about 0.07.

These diagnostic testing examples point out that when probability is involved, answers to questions about our daily lives are sometimes not as obvious as you may think. Consider the question, "Should a woman have an annual mammogram?" Medical choices are rarely between certainty and risk, but rather between different risks. If a woman fails to have a mammogram, this is risky—she may truly have breast cancer. If she has a mammogram, there's the risk of a false positive or false negative, such as the risk of needless worry and additional invasive procedures (often biopsy and other treatments) due to a false positive.

Likewise, with the Triple Blood Test for Down syndrome, most positive results are false, yet the recommendation after receiving a positive test may be to follow up the test with an amniocentesis, a procedure that gives a more definitive diagnosis but has the risk of causing a miscarriage. With some diagnostic tests (such as the PSA test for prostate cancer), testing can detect the disease, but the evidence is unclear about whether or not early detection has much, if any, effect on life expectancy.

Simulation to Estimate a Probability

Some probabilities are very difficult to find with ordinary reasoning. In such cases, one way to approximate an answer is to simulate. We have carried out simulations in previous chapters, in chapter exercises, and in Activity 1 in this chapter. The steps for a simulation are as follows:

- Identify the random phenomenon to be simulated.
- Describe how to simulate observations of the random phenomenon.

[8] W. Barlow et al., *J. Natl. Cancer Inst.*, 2002, vol. 94, p. 1151.

- Carry out the simulation many times.
- Summarize results and state the conclusion.

Example 16

Using the table of random digits

Estimating Probabilities

Picture the Scenario

At carnivals and entertainment parks, vendors often try to get people to play games of chance. Suppose a vendor shows you a box that contains seven \$1 bills, one \$5, one \$10 bill, and one \$20 bill. If you decide to play the game, you reach into the box blindfolded and pull out two of the bills. You get to keep whichever bills you choose as a prize. The price to play the game is \$10.

At this moment, you have exactly \$10, but you want to purchase a \$20 T-shirt from a different vendor. Therefore, the only way you can buy the T-shirt is to play the game and win at least \$20. Before deciding whether to play, you would like to know the probability of winning at least \$20.

Question to Explore

How can you use the table of random digits to estimate your probability of winning at least \$20? What is the estimated probability?

Think It Through

We proceed according to the following four steps:

Step 1: *Identify the random phenomenon to be simulated.* We want to simulate playing the bill-grabbing game.

Step 2: *Describe how to simulate observations of the random phenomenon.* There are 10 different bills in the box. There are also 10 different digits between 0 and 9. We can assign each of the bills to a different digit between 0 and 9. One way to do so would be to let the digits 0 through 6 represent a \$1 bill, 7 represent the \$5 bill, 8 represent the \$10 bill, and 9 represent the \$20 bill.

Now we can simulate choosing two bills from the box by simply reading two numbers from the table of random digits and seeing which two bills the digits represent. Let's start at the beginning of the 8th row of the table (although in practice you should begin at a randomly selected starting point). An excerpt of the table follows.

The first two digits we encounter are 9 and 6. These correspond to a \$20 bill and a \$1 bill, respectively. Our total winnings are \$20 + \$1 = \$21, so in this instance of the simulation, the winnings are more than \$20.

Line/Col	1	2	3	4	5	6	7	8
8	96301	91977	05463	07972	18876	20922	94595	56869
9	89579	14342	63661	10281	17453	18103	57740	84378
10	85475	36857	53342	53988	53060	59533	38867	62300

Step 3: *Carry out the simulation many times.* In general, the more times we repeat the simulation, the more reliable our estimated result will be. In practice, we can program a computer to assist with the simulation and may want to simulate the random phenomenon a thousand, or even hundreds of thousands of times. While learning

the underlying ideas of simulation in conjunction with the random digit table, we will repeat the random phenomenon 20 times to obtain the estimated probability.

Picking up where we left off in Step 2, the next two digits in the table are 3 and 0, each of which corresponds to a \$1 bill. In this instance of the simulation, our winnings are \$1 + \$1 = \$2 and are less than \$20 (in fact, we lost \$8 total between the \$10 fee and the \$2 winnings).

The next two blocks of two digits are 1 and 9, which correspond in each case to total winnings of \$21.

Next we come to two consecutive 7s. It may be tempting to say that this corresponds to choosing two \$5 bills. However, the bills are chosen from the box *without replacement,* which means that it is only possible to choose a single \$5 bill. To resolve this matter, we use the first 7, which corresponds to choosing the \$5 bill. We simply skip over the second 7 and move to the next digit, a 0 (a \$1 bill). Our total winnings for this instance are \$5 + \$1 = \$6.

Results of the first 20 repetitions are summarized in the following table:

Repetition	1	2	3	4	5	6	7	8	9	10
Digits from Table	96	30	19	19	70	54	63	07	97	21
Winnings	21	2	21	21	6	2	2	6	25	2
At least \$20 won?	Yes	No	Yes	Yes	No	No	No	No	Yes	No

Repetition	11	12	13	14	15	16	17	18	19	20
Digits from Table	87	62	09	29	45	95	56	86	98	95
Winnings	15	2	21	21	2	21	2	11	30	21
At least \$20 won?	No	No	Yes	Yes	No	Yes	No	No	No	Yes

Note that we encountered double digits two more times throughout the simulation. We treat 88 in the same way as 77, since as with the \$5 bill, there is only one \$10 bill in the box. We also encountered a 22. We may be tempted to think that we can use both of the 2s since there are multiple \$1 bills in the box. But remember, each bill corresponds to a digit. Once we arrive at the first 2 and choose the \$1 bill associated with it, it is no longer possible to choose that same bill second. As was the case when we encountered 77 and 88, we skip the second 2.

Another point worth noting is that when we ran out of digits at the end of line 8, we moved to the beginning of line 9.

Step 4: *Summarize results and state conclusion.* In 8 of the 20 simulated outcomes, winnings were at least \$20. An estimate of the probability of winning at least \$20 is thus 8 / 20 = 0.40.

Insight

Simulation is a powerful tool in that we can estimate probabilities without having to use and understand more complex mathematical probability rules. We could actually determine the probability in question without having to rely on simulation by using more advanced probability rules. The mathematical probability is given to be 0.20.[9] Our simulated probability differs somewhat from the true probability, but this disparity can be fixed easily by using

[9]For students interested in pursuing further reading, we have used a concept from the field of probability known as DeMorgan's Law.

more than the relatively small number of 20 repetitions in the simulation. Exercise 5.64 will explore adding more repetitions.

This example demonstrated how to use simulation in estimating probabilities for assessing the risk of playing a game of chance in terms of gains versus loses in winnings. Simulation can also be used for evaluating the risk of making business decisions. Exercise 5.120 presents a possible scenario for assessing risk in terms of saving a business enterprise.

Try Exercises 5.63 and 5.64

Probability Is the Key to Statistical Inference

The concepts of probability hold the key to methods for conducting statistical inference–making conclusions about populations using sample data. To help preview this connection, let's consider an opinion poll. Suppose a poll indicates that 45% of those sampled favor legalized gambling. What's the probability that this sample percentage falls within a certain margin of error, say plus or minus 3%, of the true population percentage? The next two chapters will build on the probability foundation of this chapter and enable us to answer such a question. We will study how to evaluate the probabilities of all the possible outcomes in a survey sample or an experiment. This will then be the basis of the statistical inference methods we'll learn about in Chapters 8–10.

5.4 Practicing the Basics

5.47 Birthdays of presidents Of the first 44 presidents of the United States (George Washington through Barack Obama), two had the same birthday (Polk and Harding). Is this highly coincidental? Answer by finding the probability of at least one birthday match among 44 people.
TRY

5.48 Matching your birthday You consider your birth date to be special since it falls on January 1. Suppose your class has 25 students.

a. Is the probability of finding at least one student with a birthday that matches yours greater, the same, or less than the probability found in Example 13 of a match for at least two students? Explain.

b. Find that probability.

5.49 Lots of pairs Show that with 25 students, there are 300 *pairs* of students who can have the same birthday. So it's really not so surprising if at least two students have the same birthday. (*Hint*: You can pair 24 other students with each student, but how can you make sure you don't count each pair twice?)

5.50 Holes in one at Masters The Augusta National Golf Course in Augusta, Georgia hosts the Masters Tournament each April. The course consists of four par 3s, ten par 4s, and four par 5s. The par 4s and par 5s are long enough so that no golfer has a realistic chance of getting a hole in one, but the par 3s are each short enough so that the possibility of a hole in one does exist. Over the 75-year history of the tournament, golfers have teed off on par 3s approximately 70,000 times, and a total of 73 holes in one have been recorded. For a given golfer, suppose the probability of getting a hole in one on each of the par 3s at Augusta are as follows:

Hole Number	P(hole in one)
4	0.0005
6	0.0015
12	0.0005
16	0.0025

a. For a randomly selected golfer, find the probability of no holes in one during a round of golf. Assume independence from one hole to the next.

b. For a randomly selected golfer, find the probability of no holes in one during the next 20 rounds of golf. Assume independence from one round to the next.

c. Use your answer in part b to find the probability of making at least one hole in one during the next 20 rounds of golf.

5.51 Corporate bonds A simple way for a company to raise money to fund its operations is by selling corporate bonds. Suppose an investor buys a bond from a company for \$7500. As part of the terms of the bond, the company will repay the investor \$2000 at the end of each of the next five years. It seems like a good deal for the investor; the problem, however, lies in the fact that the company may not be able to afford to make the bond payments. In such a case, the company is said to default on the issue of the bond. Suppose that the probabilities of default in each of the next one-year periods are 0.05, 0.07, 0.07, 0.07, and 0.09, and also that defaulting is independent from one year to the next. What is the probability the company does not default during the five-year term of the bond?

5.52 Horrible 11 on 9/11 The digits in 9/11 add up to 11 $(9 + 1 + 1)$, American Airlines flight 11 was the first to hit the World Trade Towers (which took the form of the number 11), there were 92 people on board $(9 + 2 = 11)$, September 11 is the 254th day of the year $(2 + 5 + 4 = 11)$, and there are 11 letters in Afghanistan, New York City, the Pentagon, and George W. Bush (see article by L. Belkin, *New York Times*, August 11, 2002). How could you explain to someone who has not studied probability that, because of the way we look for patterns out of the huge number of things that happen, this is not necessarily an amazing coincidence?

5.53 Coincidence in your life State an event that has happened to you or to someone you know that seems highly coincidental (such as seeing a friend while on vacation). Explain why that event may not be especially surprising, once you think of all the similar types of events that could have happened to you or someone that you know, over the course of several years.

5.54 Monkeys typing Shakespeare Since events of low probability eventually happen if you observe enough trials, a monkey randomly pecking on a typewriter could eventually write a Shakespeare play just by chance. Let's see how hard it would be to type the title of *Macbeth* properly. Assume 50 keys for letters and numbers and punctuation. Find the probability that the first seven letters that a monkey types are macbeth. (Even if the monkey can type 60 strokes a minute and never sleeps, if we consider each sequence of seven keystrokes as a trial, we would wait on the average over 100,000 years before seeing this happen!)

5.55 A *true* coincidence at DisneyWorld Wisconsin has 5.4 million residents. On any given day, the probability is 1/5000 that a randomly selected Wisconsin resident decides to visit DisneyWorld in Florida.

a. Find the probability that they all will decide to go tomorrow, in which case DisneyWorld has more than 5.4 million people in line when it opens in the morning.

b. What assumptions did your solution in part a make? Are they realistic? Explain.

5.56 Rosencrantz and Guildenstern TRY In the opening scene of Tom Stoppard's play *Rosencrantz and Guildenstern Are Dead*, about two Elizabethan contemporaries of Hamlet, Guildenstern flips a coin 91 times and gets a head each time. Suppose the coin was balanced.

a. Specify the sample space for 91 coin flips, such that each outcome in the sample space is equally likely. How many outcomes are in the sample space?

b. Show Guildenstern's outcome for this sample space. Show the outcome in which only the second flip is a tail.

c. What's the probability of the event of getting a head 91 times in a row?

d. What's the probability of at least one tail, in the 91 flips?

e. State the probability model on which your solutions in parts c and d are based.

5.57 Mammogram diagnostics TRY Breast cancer is the most common form of cancer in women, affecting about 10% of women at some time in their lives. There is about a 1% chance of having breast cancer at a given time (that is, $P(S) = 0.01$ for the state of having breast cancer at a given time). The chance of breast cancer increases as a woman ages, and the American Cancer Society recommends an annual mammogram after age 40 to test for its presence. Of the women who undergo mammograms at any given time, about 1% are typically estimated to actually have breast cancer. The likelihood of a false test result varies according to the breast density and the radiologist's level of experience. For use of the mammogram to detect breast cancer, typical values reported are sensitivity $= 0.86$ and specificity $= 0.88$.

a. Construct a tree diagram in which the first set of branches shows whether a woman has breast cancer and the second set of branches shows the mammogram result. At the end of the final set of branches, show that $P(S \text{ and } POS) = 0.01 \times 0.86 = 0.0086$, and report the other intersection probabilities also.

b. Restricting your attention to the two paths that have a positive test result, show that $P(POS) = 0.1274$.

c. Of the women who receive a positive mammogram result, what proportion actually have breast cancer?

d. The following tree diagram illustrates how $P(S|POS)$ can be so small, using a typical group of 100 women who have a mammogram. Explain how to get the frequencies shown on the branches, and explain why this suggests that $P(S|POS)$ is only about 0.08.

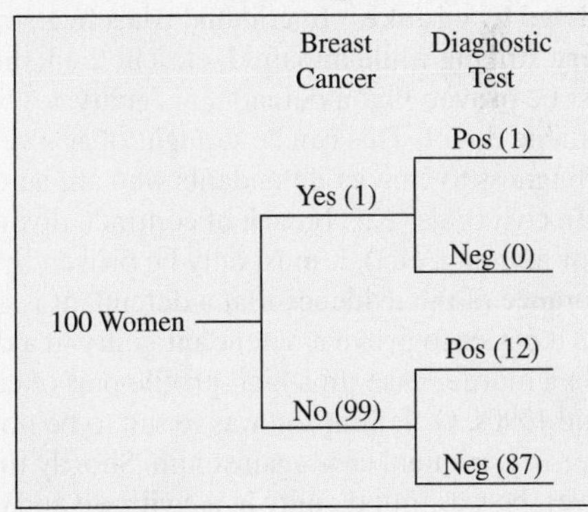

Typical results of mammograms for 100 women

5.58 More screening for breast cancer Refer to the previous exercise. For young women, the prevalence of breast cancer is lower. Suppose the sensitivity is 0.86 and the specificity is 0.88, but the prevalence is only 0.001.

a. Given that a test comes out positive, find the probability that the woman truly has breast cancer.

b. Show how to use a tree diagram with frequencies for a typical sample of 1000 women to explain to someone who has not studied statistics why the probability found in part a is so low.

c. Of the cases that are positive, explain why the proportion in error is likely to be larger for a young population than for an older population.

5.59 Was OJ actually guilty? Former pro football star O. J. Simpson was accused of murdering his wife. In the trial, a defense attorney pointed out that although Simpson had been guilty of earlier spousal abuse, annually only about 40 women are murdered per 100,000 incidents of partner abuse. This means that P(murdered by partner|partner abuse) = 40/100,000. More relevant, however, is P(murdered by partner| partner abuse and women murdered). Every year it is estimated that 5 of every 100,000 women in the United States who suffer partner abuse are killed by someone other than their partner (Gigerenzer, 2002, p. 144). Part of a tree diagram is shown starting with 100,000 women who suffer partner abuse.

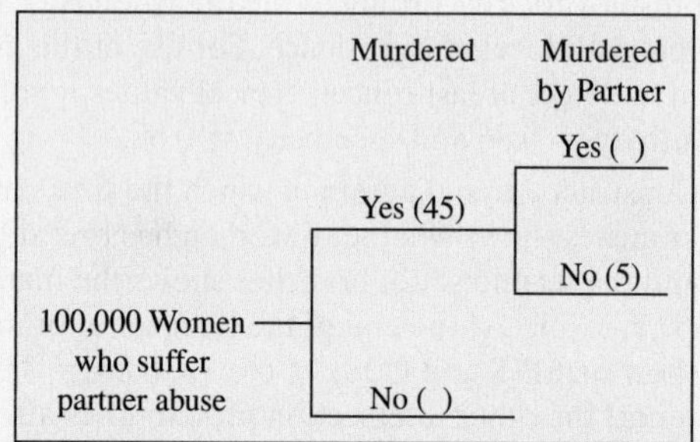

a. Based on the results stated, explain why the numbers 45 and 5 are entered as shown on two of the branches.

b. Fill in the two blanks shown in the tree diagram.

c. Conditional on partner abuse and the woman being murdered (by someone), explain why the probability the woman was murdered by her partner is 40/45. Why is this so dramatically different from P(murdered by partner|partner abuse) = 40/100,000?

5.60 Convicted by mistake In criminal trials (e.g., murder, robbery, driving while impaired, etc.) in the United States, it must be proven that a defendant is guilty beyond a reasonable doubt. This can be thought of as a very strong unwillingness to convict defendants who are actually innocent. In civil trials (e.g., breach of contract, divorce hearings for alimony, etc.), it must only be proven by a preponderance of the evidence that a defendant is guilty. This makes it easier to prove a defendant guilty in a civil case than in a murder case. In a high-profile pair of cases in the mid 1990s, O. J. Simpson was found to be not guilty of murder in a criminal case against him. Shortly thereafter, however, he was found guilty in a civil case and ordered to pay damages to the families of the victims.

a. In a criminal trial by jury, suppose the probability the defendant is convicted, given guilt, is 0.95, and the probability the defendant is acquitted, given innocence, is 0.95. Suppose that 90% of all defendants truly are guilty. Given that a defendant is convicted, find the probability he or she was actually innocent. Draw a tree diagram or construct a contingency table to help you answer.

b. Repeat part a, but under the assumption that 50% of all defendants truly are guilty.

c. In a civil trial, suppose the probability the defendant is convicted, given guilt is 0.99, and the probability the defendant is acquitted, given innocence, is 0.75. Suppose that 90% of all defendants truly are guilty. Given that a defendant is convicted, find the probability he or she was actually innocent. Draw a tree diagram or construct a contingency table to help you answer.

5.61 DNA evidence compelling? DNA evidence can be extracted from biological traces such as blood, hair, and saliva. "DNA fingerprinting" is increasingly used in the courtroom as well as in paternity testing. Given that a person is innocent, suppose that the probability of their DNA matching that found at the crime scene is only 0.000001, one in a million. Further, given that a person is guilty, suppose that the probability of their DNA matching that found at the crime scene is 0.99. Jane Doe's DNA matches that found at the crime scene.

a. Find the probability that Jane Doe is actually innocent, if absolutely her probability of innocence is 0.50. Interpret this probability. Show your solution by introducing notation for events, specifying probabilities that are given, and using a tree diagram to find your answer.

b. Repeat part a if the unconditional probability of innocence is 0.99. Compare results.

c. Explain why it is very important for a defense lawyer to explain the difference betweenP(DNA match|person innocent) and P(person innocent|DNA match).

5.62 Triple Blood Test Example 8 about the Triple Blood Test for Down syndrome found the results shown in the table below.

	Blood Test		
Down	**POS**	**NEG**	**Total**
Yes	48	6	**54**
No	1307	3921	**5228**
Total	**1355**	**3927**	**5282**

a. Estimate the probability that Down syndrome occurs (Down = Yes).

b. Find the estimated (i) sensitivity and (ii) specificity.

c. Find the estimated (i) P(Yes|POS) and (ii) P(No|NEG). (Note: These probabilities are the predictive values.)

d. Explain how the probabilities in parts b and c give four ways of describing the probability that a diagnostic test makes a correct decision.

5.63 Simulating donations to local blood bank The director of a local blood bank is in need of a donor with type AB blood. The distribution of blood types for Americans is estimated by the American Red Cross to be

TRY

Type	A	B	O	AB
Probability	40%	10%	45%	5%

The director decides that if more than 20 donors are required before the first donor with AB blood appears, she will need to issue a special appeal for AB blood.

a. Conduct a simulation 10 times, using the Random Numbers applet on the text CD or a calculator or software, to estimate the probability this will happen.

Show all steps of the simulation, including any assumptions that you make. Refer to Example 16 as a model for carrying out this simulation.

b. In practice, you would do at least 1000 simulations to estimate this probability well. You'd then find that the probability of exceeding 20 donors before finding a type AB donor is 0.36. Actually, simulation is not needed. Show how to find this probability using the methods of this chapter.

5.64 Probability of winning $20 In Example 16, we estimated the probability of winning at least $20 in the game was 0.40. Meanwhile, we concluded analytically that the actual probability was 0.20.

TRY

a. Explain what caused the fairly large disparity between our estimated result and the actual result.

b. The simulation in the example consisted of 20 repetitions. Pick up where we left off in the random number table in the example, and conduct another 80 repetitions, for a total of 100. What is the estimated probability based on these 100 repetitions? Is it closer to the actual probability?

c. What should tend to happen to the difference between the actual and the estimated probabilities as the number of repetitions in the simulation increases?

5.65 Probability of winning $20 In Example 16 and in the previous exercise, we used random numbers to estimate the probability of winning at least $20 in the game. Recall that to do so, for each repetition we simply recorded a Yes or No, pending whether or not we won at least $20 on the given repetition. At the end of the series of repetitions, we essentially determined the proportion of Yes answers in our sequence of repetitions. Simulation can actually be used to answer many more questions than just those about probability. For example, suppose that rather than estimating the probability of winning at least $20, we wanted to estimate the expected winnings from playing the game. We need to make two minor adjustments to the approach used in the example. Instead of recording Yes or No for each repetition, record the actual dollar amount won. At the end of the series of repetitions, simply calculate the average winnings across all repetitions. This provides an estimate for the amount we can expect to win by playing the game.

a. Beginning at line 5 of the table of random digits, carry out 20 repetitions and report the average winnings.

b. How does the estimated average winnings compare with the price to play the game?

Chapter Review

ANSWERS TO THE CHAPTER FIGURE QUESTIONS

Figure 5.1 The cumulative proportion after each roll is evaluated as the frequency of 6s rolled through that trial (roll) divided by the total number of trials (rolls). The cumulative proportions are 1/1 = 1 for roll 1, 1/2 = 0.50 for roll 2, 2/3 = 0.67 for roll 3, and 2/4 = 0.50 for roll 4.

Figure 5.2 It should be very, very close to 1/2.

Figure 5.3 $2 \times 2 \times 2 \times 2 = 16$ branches.

Figure 5.4 Venn diagrams will vary. One possible Venn diagram follows.

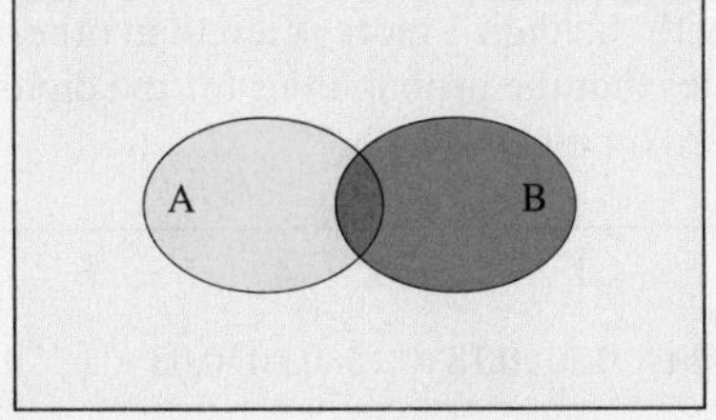

Figure 5.5 The favorable outcomes to the event that the student answers the first question correctly are {CCC,CCI,CIC,CII} This event is not disjoint from either of the two labeled events in the Venn diagram.

Figure 5.6 P(A or B) = P(A) + P(B) − P(A and B)

Figure 5.7 P(A or B) = P(A) + P(B) when the events A and B are disjoint.

Figure 5.8 We would not expect the trials to be independent if the student is not guessing. If the student is using prior knowledge, the probability of getting a correct answer on a question, such as question 3, may be more likely if the student answers questions 1 and 2 correctly.

Figure 5.9 P(A|B) is not necessarily equal to P(B|A). They have the same numerator, P(A and B), but different denominators, P(B) and P(A) respectively, which may not be equal.

Figure 5.11 As the prevalence rate increases, of the individuals that test positive, a higher percentage of positive test results will occur with individuals that actually use drugs. For example, if 5% of the population uses drugs, of the 115 individuals out of 1000 that would be expected to test positive, 48 would be actual drug users while 67 would be nondrug users.

CHAPTER SUMMARY

In this chapter, we've seen how to quantify uncertainty and randomness. With many independent trials of random phenomena, outcomes do show certain regularities. The proportion of times an outcome occurs, in the long run, is its **probability**.

The **sample space** is the set of all possible outcomes of a random phenomenon. An **event** is a subset of the sample space. Two events A and B are **disjoint** if they have no common elements.

We find probabilities using basic rules:

- The **probability** of each individual outcome falls between 0 and 1, and the total of all the individual probabilities equals 1.
- The **probability of an event** is the sum of the probabilities of individual outcomes in that event.
- For an event A and its **complement A^c** (the outcomes not in A),
$$P(A^c) = 1 - P(A).$$
- P(A or B) = P(A) + P(B) − P(A and B). This simplifies to P(A or B) = P(A) + P(B) when the events are disjoint.
- P(A and B) = P(A) × P(B | A), where P(B | A) denotes the **conditional probability** of event B, given that event A occurs. Equivalently, the conditional probability satisfies
$$P(B \mid A) = \frac{P(A \text{ and } B)}{P(A)}.$$
- Likewise, P(A and B) = P(B)P(A | B), and P(A | B) = P(A and B)/P(B).
- When A and B are **independent**, P(A | B) = P(A) and P(B | A) = P(B) Then, also P(A and B) = P(A) × P(B).

A **probability model** states certain assumptions and, based on them, finds the probabilities of events of interest.

Understanding probability helps us to make informed decisions in our lives. In the next four chapters, we'll see how an understanding of probability gives us the basis for making statistical inferences about a population.

SUMMARY OF NEW NOTATION IN CHAPTER 5

A, B, C Events

A^c Complement of event A (the outcomes *not* in A)

P(A) Probability of event A

P(A | B) Conditional probability of event A, given event B (| denotes "given")

CHAPTER PROBLEMS

Practicing the Basics

5.66 Peyton Manning completions As of the end of the 2010 NFL season, Indianapolis Colts quarterback Peyton Manning, throughout his 13-year career, completed 65% of all of his pass attempts. Suppose the probability each pass attempted in the next season has probability 0.65 of being completed.

a. Does this mean that if we watch Manning throw 100 times in the upcoming season, he would complete exactly 65 passes? Explain.

b. Explain what this probability means in terms of observing him over a longer period, say for 1000 passes over the course of the next two seasons assuming Manning is still at his typical playing level. Would it be surprising if his completion percentage over a large number of passes differed significantly from 65%?

5.67 Due for a boy? A couple has five children, all girls. They are expecting a sixth child. The father tells a friend that by the law of large numbers the chance of a boy is now much greater than 1/2. Comment on the father's statement.

5.68 P(life after death) Explain the difference between the relative frequency and subjective definitions of probability. Illustrate by explaining how it is possible to give a value for (a) the probability of life after death, (b) the probability that in the morning you remember at least one dream that you had in the previous night's sleep, based on what you observe every morning for the next year.

5.69 Choices for lunch For the set lunch at Amelia's Restaurant, customers can select one meat dish, one vegetable, one beverage, and one dessert. The menu offers two meats (beef and chicken), three vegetables (corn, green beans, or potatoes), three beverages (cola, ice tea, or coffee), and one dessert (Amelia's apple pie).

a. Use a tree diagram to list the possible meals and to determine how many there are.

b. In practice, would it be sensible to treat all the outcomes in the sample space as equally likely for the customer selections we'd observe? Why or why not?

5.70 Caught doctoring the books After the major accounting scandals with Enron, a large energy company, the question may be posed, "Was there any way to examine Enron's accounting books to determine if they had been 'doctored'?" One way uses Benford's law, which states that in a variety of circumstances, numbers as varied as populations of small towns, figures in a newspaper or magazine, and tax returns and other business records begin with the digit 1 more often than other digits. This law states that the probabilities for the digits 1 through 9 are approximately:

Digit	1	2	3	4	5	6	7	8	9
Probability	0.30	0.18	0.12	0.10	0.08	0.07	0.06	0.05	0.04

a. If we were to randomly pick one of the digits between 1 and 9 using a random number table or software, what is the probability for each digit?

b. When people attempt to fake numbers, there's a tendency to use 5 or 6 as the initial digit more often than predicted by Benford's law. What is the probability of 5 or 6 as the first digit by (i) Benford's law and (ii) random selection?

5.71 Life after death In a General Social Survey, in response to the question "Do you believe in life after death?" 1455 answered yes and 332 answered no.

a. Based on this survey, estimate the probability that a randomly selected adult in the United States believes in life after death.

b. A married couple is randomly selected. Estimate the probability that both subjects believe in life after death.

c. What assumption did you make in answering part b? Explain why that assumption is probably unrealistic, making this estimate unreliable.

5.72 Death penalty jury In arguing against the death penalty, Amnesty International has pointed out supposed inequities, such as the many times a black person has been given the death penalty by an all-white jury. If jurors are selected randomly from an adult population, find the probability that all 12 jurors are white when the population is (a) 90% white and (b) 50% white.

5.73 Driver's exam Three 15-year-old friends with no particular background in driver's education decide to take the written part of the Georgia Driver's Exam. Each exam was graded as a pass (P) or a failure (F).

a. How many outcomes are possible for the grades received by the three friends together? Using a tree diagram, list the sample space.

b. If the outcomes in the sample space in part a are equally likely, find the probability that all three pass the exam.

c. In practice, the outcomes in part a are not equally likely. Suppose that statewide 70% of 15-year-olds pass the exam. If these three friends were a random sample of their age group, find the probability that all three pass.

d. In practice, explain why probabilities that apply to a random sample are not likely to be valid for a sample of three friends.

5.74 Independent on coffee? Students in a geography class are asked whether they've visited Europe in the past 12 months and whether they've flown on a plane in the past 12 months.

a. For a randomly selected student, would you expect these events to be independent or dependent? Explain.

b. How would you explain to someone who has never studied statistics what it means for these events to be either independent or dependent?

c. Students in a different class were asked whether they've visited Italy in the past 12 months and whether they've visited France in the past 12 months. For a randomly selected student, would you expect these events to be independent or dependent? Explain.

d. Students in yet another class were asked whether they've been to a zoo in the past 12 months and whether they drink coffee. For a randomly selected student, would you expect these events to be independent or dependent? Explain.

e. If you had to rank the pairs of events in parts a, c, and d in terms of the strength of any dependence, which pair of events is most dependent? Least dependent?

5.75 Health insurance According to a 2006 Census Bureau report, 59% of Americans have private health insurance, 25% have government health insurance (Medicare or Medicaid or military health care), and 16% have no health insurance.

a. Estimate the probability that a patient has health insurance.

b. Given that a subject has health insurance, estimate the probability it is private.

5.76 Teens and drugs In August 2006 the Center on Addiction and Substance Abuse (CASA) at Columbia University reported results of a survey of 1297 teenagers about their views on the use of illegal substances. Twenty percent of the teens surveyed reported going to clubs for music or dancing at least once a month. Of them, 26% said drugs were usually available at these club events. Which of these percentages estimates a conditional probability? For each that does, identify the event conditioned on and the event to which the probability refers.

5.77 Teens and parents In the CASA teen survey described in the previous exercise, 33% of teens reported that parents are never present at parties they attend. Thirty-one percent of teens who say parents are never present during parties report that marijuana is available at the parties they attend, compared to only 1% of teens who say parents are present at parties reporting that marijuana is available at the parties they attend.

a. Which of these percentages estimates a conditional probability? For each that does, identify the event conditioned on and the event to which the probability refers.

b. For the variables parents present (yes or no) and marijuana available (yes or no), construct a contingency table showing counts for the 1297 teenagers surveyed.

c. Using the contingency table, given that marijuana is not available at the party, estimate the probability that parents are present.

5.78 Laundry detergent A manufacturer of laundry detergent has introduced a new product that it claims to be more environmentally sound. An extensive survey gives the percentages shown in the table.

Result of Advertising for New Laundry Product

	Tried the New Product	
Advertising Status	**Yes**	**No**
Seen the ad	10%	25%
Have not seen ad	5%	60%

a. Estimate the probability that a randomly chosen consumer would have seen advertising for the new product and tried the product.

b. Given that a randomly chosen consumer has seen the product advertised, estimate the probability that the person has tried the product.

c. Let A be the event that a consumer has tried the product. Let B be the event that a consumer has seen the product advertised. Express the probabilities found in part a and b in terms of A and B.

d. Are A and B independent? Justify your answer.

5.79 Board games and dice A board game requires players to roll a pair of balanced dice for each player's turn. Denote the outcomes by (die 1, die 2), such as (5, 3) for a 5 on die 1 and a 3 on die 2.

a. List the sample space of the 36 possible outcomes for the two dice.

b. Let A be the event that you roll doubles (that is, each die has the same outcome). List the outcomes in A, and find its probability.

c. Let B be the event that the sum on the pair of dice is 7. Find P(B).

d. Find the probability of (i) A and B, (ii) A or B, and (iii) B given A.

e. Are events A and B independent, disjoint, or neither? Explain.

5.80 Roll two more dice Refer to the previous exercise. Define event D as rolling an odd sum with two dice.

a. B denotes a sum of 7 on the two dice. Find P(B and D). When an event B is contained within an event D, as here, explain why P (B and D) = P (B).

b. Find P(B or D). When an event B is contained within an event D, explain why P (B or D) = P (D).

5.81 Conference dinner Of the participants at a conference, 50% attended breakfast, 90% attended dinner, and 40% attended both breakfast and dinner. Given that a participant attended breakfast, find the probability that she also attended dinner.

5.82 Waste dump sites A federal agency is deciding which of two waste dump projects to investigate. A top administrator estimates that the probability of federal law violations is 0.30 at the first project and 0.25 at the second project. Also, he believes the occurrences of violations in these two projects are disjoint.

a. What is the probability of federal law violations in the first project or in the second project?

b. Given that there is not a federal law violation in the first project, find the probability that there is a federal law violation in the second project.

c. In reality, the administrator confused disjoint and independent, and the events are actually independent. Answer parts a and b with this correct information.

5.83 A dice game Consider a game in which you roll two dice, and you win if the total is 7 or 11 and you lose if the total is 2, 3, or 12. You keep rolling until one of these totals occurs. Using conditional probability, find the probability that you win.

5.84 No coincidences Over time, you have many conversations with a friend about your favorite actress, favorite musician, favorite book, favorite TV show, and so forth for 100 separate topics. On any given topic, there's only a 0.02 probability that you agree. If you did agree on a topic, you would consider it to be coincidental.

a. If whether you agree on any two different topics are independent events, find the probability that you and your friend *never* have a coincidence on these 100 topics.

b. Find the probability that you have a coincidence on at least one topic.

5.85 Amazing roulette run? A roulette wheel in Monte Carlo has 18 even-numbered slots, 18 odd-numbered slots, a slot numbered zero, and a double zero slot. On August 18, 1913, it came up even 26 times in a row.[10] As more and more evens occurred, the proportion of people betting on an odd outcome increased, as they figured it was "due."

a. Comment on this strategy.

b. Find the probability of 26 evens in a row if each slot is equally likely.

c. Suppose that over the past 100 years there have been 1000 roulette wheels, each being used hundreds of times a day. Is it surprising if sometime in the previous 100 years one of these wheels had 26 evens in a row? Explain.

5.86 Death penalty and false positives For the decision about whether to convict someone charged with murder and give the death penalty, consider the variables reality (defendant innocent, defendant guilty) and decision (convict, acquit).

a. Explain what the two types of errors are in this context.

b. Jurors are asked to convict a defendant if they feel the defendant is guilty "beyond a reasonable doubt." Suppose this means that given the defendant is executed, the probability that he or she truly was guilty is 0.99. For the 1234 people put to death from the time the death penalty was reinstated in 1977 until December 2010, find the probability that (i) they were all truly guilty, and (ii) at least one of them was actually innocent.

c. How do the answers in part b change if the probability of true guilt is actually 0.95?

5.87 Screening smokers for lung cancer An article about using a diagnostic test (helical computed tomography) to screen adult smokers for lung cancer warned that a negative test may cause harm by providing smokers with false reassurance, and a false-positive test results in an unnecessary operation opening the smoker's chest (Mahadevia et al., *JAMA*, vol. 289, pp. 313–322, 2003). Explain what false negatives and false positives mean in the context of this diagnostic test.

5.88 Screening for heart attacks Biochemical markers are used by emergency room physicians to aid in diagnosing patients who have suffered acute myocardial infarction (AMI), or what's commonly referred to as a heart attack. One type of biochemical marker used is creatine kinase (CK). Based on a review of published studies on the effectiveness of these markers (by E. M. Balk et al., *Annals of Emergency Medicine*, vol. 37, pp. 478–494, 2001), CK had an estimated sensitivity of 37% and specificity of 87%. Consider a population having a prevalence rate of 25%.

a. Explain in context what is meant by the sensitivity equaling 37%.

[10] *What Are the Chances?* by B. K. Holland (Johns Hopkins University Press, 2002, p. 10).

b. Explain in context what is meant by the specificity equaling 87%.

c. Construct a tree diagram for this diagnostic test. Label the branches with the appropriate probabilities.

5.89 Screening for colorectal cancer Gigerenzer (2002, p. 105) reported that on the average, "Thirty out of every 10,000 people have colorectal cancer. Of these 30 people with colorectal cancer, 15 will have a positive hemoccult test. Of the remaining 9,970 people without colorectal cancer, 300 will still have a positive hemoccult test."

a. Sketch a tree diagram or construct a contingency table to display the counts.

b. Of the 315 people mentioned above who have a positive hemoccult test, what proportion actually have colorectal cancer? Interpret.

5.90 Color blindness For genetic reasons, color blindness is more common in men than women: 5 in 100 men and 25 in 10,000 women suffer from color blindness.

a. Define events, and identify in words these proportions as conditional probabilities.

b. If the population is half male and half female, what proportion of the population is color blind? Use a tree diagram or contingency table with frequencies to portray your solution, showing what you would expect to happen with 20,000 people.

c. Given that a randomly chosen person is color blind, what's the probability that person is female? Use the tree diagram or table from part b to find the answer.

5.91 HIV testing For a combined ELISA–Western blot blood test for HIV positive status, the sensitivity is about 0.999 and the specificity is about 0.9999 (Gigerenzer 2002, pp. 124, 126).

a. Consider a high-risk group in which 10% are truly HIV positive. Construct a tree diagram to summarize this diagnostic test.

b. Display the intersection probabilities from part a in a contingency table.

c. A person from this high-risk group has a positive test result. Using the tree diagram or the contingency table, find the probability that this person is truly HIV positive.

d. Explain why a positive test result is more likely to be in error when the prevalence is lower. Use tree diagrams or contingency tables with frequencies for 10,000 people with 10% and 1% prevalence rates to illustrate your arguments.

5.92 Prostate cancer A study of the PSA blood test for diagnosing prostate cancer in men (by R. M. Hoffman et al., *BMC Family Practice*, vol. 3, p. 19, 2002) used a sample of 2620 men who were 40 years and older. When a positive diagnostic test result was defined as a PSA reading of at least 4, the sensitivity was estimated to be 0.86 but the specificity only 0.33.

a. Suppose that 10% of those who took the PSA test truly had prostate cancer. Given that the PSA was positive, use a tree diagram and/or contingency table to estimate the probability that the man truly had prostate cancer.

b. Illustrate your answer in part a by using a tree diagram or contingency table with frequencies showing what you would expect for a typical sample of 1000 men.

c. Lowering the PSA boundary to 2 for a positive result changed the sensitivity to 0.95 and the specificity to 0.20. Explain why, intuitively, if the cases increase for which a test is positive, the sensitivity will go up but the specificity will go down.

5.93 U Win A fast food chain is running a promotion to try to increase sales. Each customer who purchases a meal combo receives a game piece that contains one of the letters U, W, I, or N. If a player collects one of all four letters, that player is the lucky winner of a free milkshake. Suppose that of all game pieces, 10% contain the letter U, 30% contain W, 30% contain I, and 30% contain N. Use the table of random digits to estimate the probability that you will collect enough game pieces with your next five combo meal purchases to receive the free shake. Clearly describe the steps of your simulation. Your simulation should consist of at least 20 repetitions.

5.94 Win again Exercise 5.65 discussed how to use simulation and the table of random digits to estimate an *expected* value. Referring to the previous exercise, conduct a simulation consisting of at least 20 repetitions to estimate the expected number of combo meals one would need to purchase in order to win the free shake. Clearly describe the steps of your simulation.

Concepts and Investigations

5.95 Simulate law of large numbers Using the Simulating the Probability of a Head With a Fair Coin applet on the text CD or other software, simulate the flipping of a balanced coin.

a. Report the cumulative proportion of heads after (i) 10 flips, (ii) 100 flips, (iii) 1000 flips, and (iv) 10,000 flips. Explain how the results illustrate the law of large numbers and the long-run relative frequency definition of probability.

b. Using the Simulating the Probability of Rolling a 3 or 4 applet or other software, simulate the roll of a die with a success defined as rolling a 3 or 4, using (i) 10 rolls, (ii) 100 rolls, (iii) 1000 rolls, and (iv) 10,000 rolls. Summarize results.

5.96 Illustrate probability terms with scenarios

a. What is a sample space? Give an example of a sample space for a scenario involving (i) a designed experiment and (ii) an observational study.

b. What are disjoint events? Give an example of two events that are disjoint.

c. What is a conditional probability? Give an example of two events in your everyday life that you would expect to be (i) independent and (ii) dependent.

5.97 Short term versus long run Short-term aberrations do not affect the long run. To illustrate, suppose that you flip a coin 10 times and you get 10 heads. Find the cumulative proportion of heads, including these first 10 flips, if (a) in the 100 flips after those 10 you get 50 heads (so there are now 60 heads in 110 flips), (b) in the 1000 flips after those 10 you get 500 heads, and (c) in the 10,000 flips after those 10 you get 5000 heads. What number is the cumulative proportion tending toward as n increases?

5.98 Risk of space shuttle After the Columbia space shuttle disaster, a former NASA official who faulted the way the agency dealt with safety risk warned (in an AP story, March 7, 2003) that NASA workers believed, "If I've flown 20 times, the risk is less than if I've flown just once."

a. Explain why it would be reasonable for someone to form this belief, from the way we use empirical evidence and margins of error to estimate an unknown probability.

b. Explain a criticism of this belief, using coin flipping as an analogy.

5.99 Mrs. Test Mrs. Test (see www.mrstest.com) sells diagnostic tests for various conditions. Their Web site gives only imprecise information about the accuracy of the tests. The test for pregnancy is said to be "over 99% accurate." Describe at least four different probabilities to which this could refer.

5.100 Marijuana leads to heroin? Nearly all heroin addicts have used marijuana sometime in their lives. So, some argue that marijuana should be illegal because marijuana users are likely to become heroin addicts. Use a Venn diagram to illustrate the fallacy of this argument, by sketching sets for M = marijuana use and H = heroin use such that $P(M \mid H)$ is close to 1 but $P(H \mid M)$ is close to 0.

5.101 Stay in school Suppose 80% of students finish high school. Of them, 50% finish college. Of them, 20% get a masters' degree. Of them, 30% get a Ph.D.

a. What percentage of students get a Ph.D.?

b. Explain how your reasoning in part a used a multiplication rule with conditional probabilities.

c. Given that a student finishes college, find the probability of getting a Ph.D.

5.102 How good is a probability estimate? In Example 8 about Down syndrome, we estimated the probability of a positive test result (predicting that Down syndrome is present) to be $P(POS) = 0.257$, based on observing 1355 positive results in 5282 observations. How good is such an estimate? From Section 4.2, $1/\sqrt{n}$ is an approximate margin of error in estimating a proportion with n observations.

a. Find the approximate margin of error to describe how well this proportion estimates the true probability, P(POS).

b. The *long run* in the definition of probability refers to letting n get very large. What happens to this margin of error formula as n keeps growing, eventually toward infinity? What's the implication of this?

5.103 Protective bomb Before the days of high security at airports, there was a legendary person who was afraid of traveling by plane because someone on the plane might have a bomb. He always brought a bomb himself on any plane flight he took, believing that the chance would be astronomically small that two people on the same flight would both have a bomb. Explain the fallacy in his logic, using ideas of independent events and conditional probability.

5.104 Streak shooter Sportscaster Maria Coselli claims that players on the New York Knicks professional basketball team are streak shooters. To make her case, she looks at the statistics for all the team's players over the past three games and points out that one of them (Joe Smith) made six shots in a row at one stage. Coselli argues, "Over a season, Smith makes only 45% of his shots. The probability that he would make six in a row if the shots were independent is $(0.45)^6 = 0.008$, less than one in a hundred. This would be such an unusual occurrence that we can conclude that Smith is a streak shooter."

a. Explain the flaws in Coselli's logic.

b. Use this example to explain how some things that look highly coincidental may not be so when viewed in a wider context.

Problems 5.105–5.109 may have more than one correct answer.

5.105 Multiple choice Choose ALL correct responses. For two events A and B, $P(A) = 0.5$ and $P(B) = 0.2$. Then P(A or B) equals

a. 0.10, if A and B are independent

b. 0.70, if A and B are independent

c. 0.60, if A and B are independent

d. 0.70, if A and B are disjoint

5.106 Multiple choice Which of the following is always true?

a. If A and B are independent, then they are also disjoint.

b. $P(A \mid B) + P(A \mid B^c) = 1$

c. If $P(A \mid B) = P(B \mid A)$, then A and B are independent.

d. If A and B are disjoint, then A and B cannot occur at the same time.

5.107 Multiple choice: Coin flip A balanced coin is flipped 100 times. By the law of large numbers:

a. There will almost certainly be *exactly* 50 heads and 50 tails.

b. If we got 100 heads in a row, almost certainly the next flip will be a tail.

c. For the 100 flips, the probability of getting 100 heads equals the probability of getting 50 heads.

d. It is absolutely impossible to get a head every time.

e. None of the above.

5.108 Multiple choice: Dream come true You have a dream in which you see your favorite movie star in person. The very next day, you are visiting Manhattan and you see her walking down Fifth Avenue.

a. This is such an incredibly unlikely coincidence that you should report it to your local newspaper.

b. This is somewhat unusual, but given the many dreams you could have in your lifetime about people you know or know of, it is not an incredibly unlikely event.

c. If you had not had the dream, you would definitely not have seen the film star the next day.

d. This proves the existence of ESP.

5.109 Multiple choice: Comparable risks Mammography is estimated to save about 1 life in every 1000 women. "Participating in annual mammography screening ... has roughly the same effect on life expectancy as reducing the distance one drives each year by 300 miles" (Gigerenzer 2002, pp. 60, 73). Which of the following do you think has the closest effect on life expectancy as that of smoking throughout your entire life?

a. Taking a commercial airline flight once a year

b. Driving about ten times as far every year as the average motorist

c. Drinking a cup of coffee every day

d. Eating a fast-food hamburger once a month

e. Never having a mammogram (if you are a woman)

5.110 True or false Answer true of false for each part.

a. When you flip a coin ten times, you are more likely to get *the* sequence HHHHHTTTTT than the sequence HHHHHHHHHH.

b. When you flip a coin ten times, you are more likely to get *a* sequence that contains five heads than a sequence that contains ten heads.

5.111 True or false When you flip a balanced coin twice, there are three things that can happen: 0 heads, 1 head, or 2 heads. Since there are three possible outcomes, they each have probability 1/3. (This was claimed by the French mathematician, Jean le Rond d' Alembert, in 1754. To reason this, write down the sample space for the possible sequence of results for the two flips.)

5.112 Driving versus flying In the United States in 2002, about 43,000 people died in auto crashes and 0 people died in commercial airline accidents. G. Gigerenzer (2002, p. 31) states, "The terrorist attack on September 11, 2001, cost the lives of some 3,000 people. The subsequent decision of millions to drive rather than fly may have cost the lives of many more." Explain the reasoning behind this statement.

5.113 Prosecutor's fallacy An eyewitness to the crime says that the person who committed it was male, between 15 and 20 years old, Hispanic, and drove a blue Honda. The prosecutor points out that a proportion of only 0.001 people living in that city match all those characteristics, and one of them is the defendant. Thus, the prosecutor argues that the probability that the defendant is not guilty is only 0.001. Explain what is wrong with this logic. (*Hint:* Is $P(\text{match}) = P(\text{not guilty} \mid \text{match})$?)

5.114 Generalizing the addition rule ◆◆ For events A, B, and C such that each pair of events is disjoint, use a Venn diagram to explain why

$$P(A \text{ or } B \text{ or } C) = P(A) + P(B) + P(C).$$

5.115 Generalizing the multiplication rule ◆◆ For events A, B, and C, explain why $P(A \text{ and } B \text{ and } C) =$

$$P(A) \times P(B \mid A) \times P(C \mid A \text{ and } B).$$

5.116 Bayes's rule ◆◆ Suppose we know $P(A), P(B \mid A)$, and $P(B^c \mid A^c)$, but we want to find $P(A \mid B)$.

a. Using the definition of conditional probability for $P(A \mid B)$ and for $P(B \mid A)$, explain why $P(A \mid B) = P(A \text{ and } B)/P(B) = [P(A)P(B \mid A)]/P(B)$.

b. Splitting the event that B occurs into two parts, according to whether or not A occurs, explain why

$$P(B) = P(B \text{ and } A) + P(B \text{ and } A^c).$$

c. Using part b and the definition of conditional probability, explain why

$$P(B) = P(A)P(B \mid A) + P(A^c)P(B \mid A^c).$$

d. Combining what you have shown in parts a–c, reason that

$$P(A \mid B) = \frac{P(A)P(B \mid A)}{P(A)P(B \mid A) + P(A^c)P(B \mid A^c)}.$$

This formula is called **Bayes's rule**. It is named after a British clergyman who discovered the formula in 1763.

Student Activities

5.117 Simulating matching birthdays Do you find it hard to believe that the probability of at least one birthday match in a class of 25 students is 0.57? Let's simulate the answer. Using the Random Numbers applet on the text CD, each student in the class should simulate 25 numbers between 1 and 365. Observe whether there is at least one match of two numbers out of the 25 simulated numbers.

a. Combine class results. What proportion of the simulations had at least one birthday match?

b. Repeat part a for a birthday match in a class of 50 students. Again, combine class results. What is the simulated probability of at least one match?

5.118 Simulate table tennis In table tennis, the first person to get at least 21 points while being ahead of the opponent by at least two points wins the game. In games between you and an opponent, suppose successive points are independent, and suppose the probability of your winning any given point is 0.40.

a. Do you have any reasonable chance to win a game? Showing all steps, simulate two games, using the Random Numbers applet on the text CD or other software. (*Hint:* Let the integers 0–3 represent winning a point and 4–9 represent losing a point.)

b. Combining results for all students in the class, estimate the probability of winning.

5.119 Which tennis strategy is better? A tennis match can consist of the best of three sets (that is, the winner is the first to win two sets) or the best of five sets (the winner is the first to win three sets). Which would you be better off playing if you are the weaker player and have probability 0.40 of winning any particular set?

a. Simulate 10 matches of each type, using the Random Numbers applet on the text CD or other software. Show the steps of your simulation, and specify any assumptions.

b. Combining results for all students in the class, estimate the probability of the weaker player winning the match under each type of match.

5.120 Saving a business The business you started last year has only $5000 left in capital. A week from now you need to repay a $10,000 loan or you will go bankrupt.

You see two possible ways to raise the money you need. You can ask a large company to invest the \$10,000, wooing them with the \$5000 you have left. You guess your probability of success is 0.20. Or you could ask, in sequence, several small companies to each invest \$2000, spending \$1000 on each to woo them. You guess your probability of success is 0.20 for each appeal. What's the probability you will raise enough money to repay the loan, with each strategy? Which is the better strategy, to be bold with one large bet or to be cautious and use several smaller bets? With the bold strategy, the probability of success is simply 0.20. With the cautious strategy, we need to use simulation to find the probability of success.

a. Simulate the cautious strategy 10 times using the Random Numbers applet on the text CD or other software. Show the steps of the simulations and specify any assumptions.

b. Combining simulation results for all students in the class, estimate the probability of the cautious strategy being successful.

c. Which strategy would you choose? The bolder strategy or the cautious strategy? Explain.

BIBLIOGRAPHY

Against All Odds: Inside Statistics (1989). Produced by Consortium for Mathematics and Its Applications, for PBS. See www.learner.org

Gigerenzer, G. (2002). *Calculated Risks.* New York: Simon & Schuster.

Wilson, R., and E. A. C. Crouch (2001). *Risk-Benefit Analysis.* Cambridge, MA: Harvard Univ. Press.

Probability Distributions

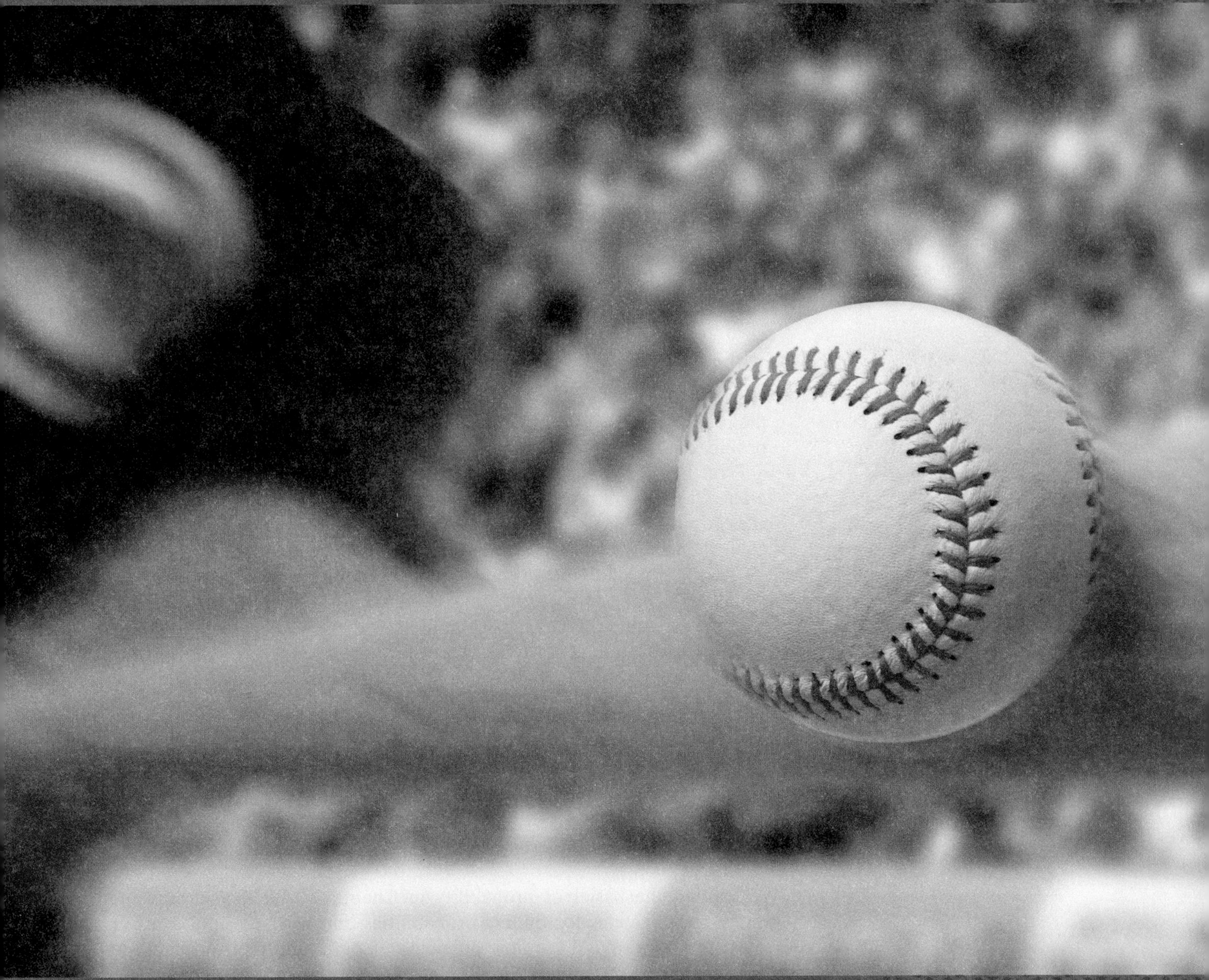

Example 1

Gender Discrimination

Picture the Scenario

A 2001 lawsuit by seven female employees claimed that Wal-Mart Stores, Inc. promoted female workers less frequently than male workers. This lawsuit grew into the largest sex-discrimination class-action lawsuit in U.S. history, representing more than 1.5 million current and former female Wal-Mart employees, before it was eventually decided, in 2011, in favor of Wal-Mart by the Supreme Court.[1,2]

Question to Explore

Suppose that 10 individuals are chosen from a large pool of qualifying employees for a promotion program, and the pool has an equal number of female and male employees. When all 10 individuals chosen for promotion are male, the female employees claim that the program is gender-biased. How can we investigate statistically the validity of the women's claim?

Thinking Ahead

Other factors being equal, the probability of selecting a female is 0.50 and the probability of selecting a male is 0.50 for each choice of an individual for promotion. If the employees are selected randomly in terms of gender, about half of the employees picked should be females and about half should be male. Due to ordinary sampling variation, however, it need not happen that exactly 50% of those selected are female. A simple analogy is in flipping a coin 10 times. We won't necessarily see exactly 5 heads.

Because none of the 10 employees chosen for promotion are female, we might be inclined to support the women's claim. The question we need to consider is, would these results be unlikely if there were no gender bias? An equivalent question is, if we flip a coin 10 times, would it be very surprising if we got 0 heads?

In this chapter, we'll apply the probability tools of the previous chapter to answer such questions. In Examples 12 and 13, we'll revisit this gender discrimination question. In Example 14, we'll use probability arguments to analyze whether racial profiling may have occurred when police officers make traffic stops.

We learned the basics of probability in Chapter 5. We'll next study tables, graphs, and formulas for finding probabilities that will be useful to us for the rest of the book. We'll see how possible outcomes and their probabilities are summarized in a **probability distribution.** We'll then study two commonly used probability distributions—the **normal** and the **binomial.** We'll see later that the normal distribution, which has a bell-shaped graph, plays a key role in statistical inference.

[1] *Source:* Background information taken from www.reuters.com/article/2010/12/06/us-walmart-lawsuit-discrimination-idUSTRE6B531W20101206.

[2] www.cnn.com/2011/US/06/20/scotus.wal.mart.discrimination/index.html?hpt=hp_t1.

6.1 Summarizing Possible Outcomes and Their Probabilities

Recall

The characteristics we measure are called **variables** because their values vary from subject to subject. If the possible outcomes are a set of separate numbers, the variable is **discrete** (for example, a variable expressed as "the number of..." with possible values 0, 1, 2,...). If the variable can take any value in an interval, such as the proportion of a tree leaf that shows a fungus (any number between 0 and 1), it is **continuous**.◀

With proper methods of gathering data, the numerical values that a variable assumes should be the result of some random phenomenon. For example, the randomness may involve selecting a random sample from a population or performing a randomized experiment. In such cases, we call the variable a **random variable**.

> Random Variable
>
> A **random variable** is a numerical measurement of the outcome of a random phenomenon. Often, the randomness results from the use of random sampling or a randomized experiment to gather the data.

We've used letters near the end of the alphabet, such as x, to symbolize variables. We'll also use letters such as x for the possible value of a random variable. When we refer to the random variable itself, rather than a particular value, we'll use a capital letter, such as X. For instance, $X =$ number of heads in three flips of a coin is a random variable, whereas $x = 2$ is one of its possible values.

Because a random variable refers to the outcome of a random phenomenon, each possible outcome has a specific probability of occurring. The **probability distribution** of a random variable specifies its possible values and their probabilities. An advantage of a variable being a random variable is that it's possible to specify such probabilities. Without randomness, we cannot predict the probabilities of the possible outcomes in the long run.

Probability Distributions of Discrete Random Variables

When a random variable has separate possible values, such as 0, 1, 2, 3 for the number of heads in three flips of a coin, it is called **discrete**. The **probability distribution** of a discrete random variable assigns a probability to each possible value. Each probability falls between 0 and 1, and the sum of the probabilities of all possible values equals 1. We let $P(x)$ denote the probability of a possible value x, such as P(2) for the probability that the random variable takes the value 2.

Recall

From Chapter 5, a **probability** is a long-run **proportion**. So, it can take any value in the interval from 0 up to 1. The probabilities for all the possible outcomes add up to 1.0, so that's the sum of the probabilities in a probability distribution. ◀

> Probability Distribution of a Discrete Random Variable
>
> A **discrete** random variable X takes a set of separate values (such as 0, 1, 2,...). Its **probability distribution** assigns a probability $P(x)$ to each possible value x.
>
> - For each x, the probability $P(x)$ falls between 0 and 1.
> - The sum of the probabilities for all the possible x values equals 1.

For instance, let X be a random digit selected using software as part of the process of identifying subjects to include in a random sample. The possible values for X are $x = 0, 1, 2, \ldots, 8, 9$. Each digit is equally likely, so the probability distribution is

$$P(0) = P(1) = P(2) = P(3) = P(4) =$$
$$P(5) = P(6) = P(7) = P(8) = P(9) = 0.10.$$

Each probability falls between 0 and 1, and the probabilities add up to 1.0. If X is the outcome of rolling a balanced die, then the possible values for X are $x = 1,2,3,4,5,6$, and $P(1) = P(2) = P(3) = P(4) = P(5) = P(6) = 1/6$.

Random variables can also be **continuous**, having possible values that are an interval rather than a set of separate numbers. We'll learn about continuous random variables and their probability distributions later in this section.

Example 2

Probability distributions

The Number of Home Runs in a Game

Picture the Scenario

Before moving to San Francisco after the 1957 season, the Giants played in New York and New Jersey. Between 1903 and 1954, the team won five World Series championships. Following the move to San Francisco, however, fans would have to wait a long time for a sixth championship. That championship finally came in 2010, the first for the city of San Francisco in 53 years of what many fans referred to as "torture!"

Many statistics are recorded for sporting events such as professional baseball games. One such statistic is the number of home runs per game. A reasonable question for a fan to ask before watching a game is how many home runs we might expect a team to hit. For a given San Francisco Giants game, let X represent the number of home runs the Giants hit. Table 6.1 shows the probability distribution of X for Giants games in the 2010 regular season.[3]

Table 6.1 Probability Distribution of Number of Home Runs in a Game for San Francisco Giants

Number of Home Runs (x)	Number of Occurrences	Probability
0	63 out of 162 games	63/162 = 0.3889
1	51	51/162 = 0.3148
2	36	36/162 = 0.2222
3	9	9/162 = 0.0556
4	3	3/162 = 0.0185
5 or more	0	0/162 = 0.0000

Questions to Explore

a. Show how Table 6.1 satisfies the two properties needed for a probability distribution.

b. What is the probability of at least three home runs in a game?

Think It Through

a. In Table 6.1 the two conditions in the definition of a probability distribution are satisfied:

(i) For each x, the probability $P(x)$ falls between 0 and 1.

(ii) The sum of the probabilities for all the possible x values equals 1.

---

[3] *Source:* Data from www.hittrackeronline.com/detail.php?id=2011_220&type=ballpark.

Recall

P(3) denotes the probability that the random variable takes the value 3, that is, the probability that the number of home runs in a game equals 3.◀

b. The probability of at least three home runs in a game is

$$P(3) + P(4) + P(5\text{ or more}) = 0.0556 + 0.0185 + 0 = 0.0741.$$

Insight

We can display the probability distribution of a discrete random variable with a *graph* and sometimes with a *formula.* Figure 6.1 displays a graph for the probability distribution in Table 6.1.

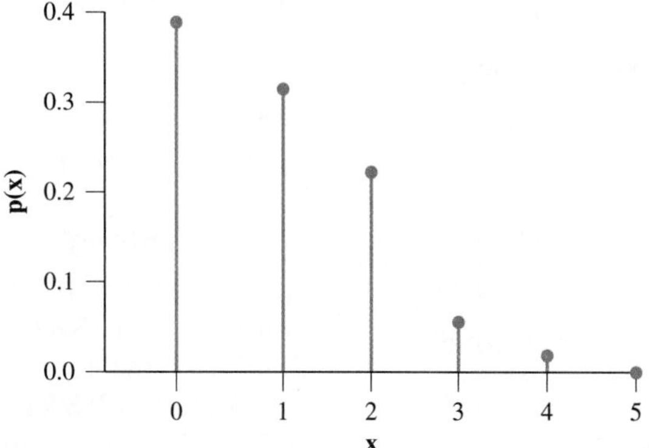

▲ **Figure 6.1** MINITAB Histogram for Probability Distribution of X = Number of Home Runs by the San Francisco Giants in a Game. The bar for each possible value of X has a height equal to its probability. **Question** How would you describe the shape of the distribution?

Try Exercises 6.1 and 6.2

As in describing sample data, a graphical display is useful for revealing the three key components of a distribution: center, variability, and shape. How do we find the corresponding summary measures of the center and variability of a probability distribution?

The Mean of a Probability Distribution

To describe characteristics of a probability distribution, we can use any of the numerical summaries defined in Chapter 2. These include mean, median, quartiles, and standard deviation. It is most common to use the *mean* to describe the center and the *standard deviation* to describe the variability.

Recall

Section 1.2 defined a **parameter** to be a numerical summary of the population, such as a population mean or a population proportion.◀

Recall that numerical summaries of populations are called **parameters**. You can think of a **population distribution** as merely being a type of probability distribution—one that applies for selecting a subject at random from a population. Like numerical summaries of populations, numerical summaries of probability distributions are referred to as **parameters**. Most parameters are denoted by Greek letters. The mean of a probability distribution is denoted by μ and the standard deviation is denoted by σ.

In Words

μ is the Greek letter mu, pronounced "mew." σ is the lowercase Greek letter sigma. The corresponding Roman letter *s* is used for the standard deviation of *sample* data. Recall that the sample mean is denoted by $\bar{x}$.

Suppose we repeatedly observe values of a random variable, such as repeatedly noting the outcome when we roll a die. The mean μ of the probability distribution for that random variable is the value we would get, in the long run, for the average of those values. This long-run interpretation parallels the interpretation (in Section 5.1) of probability itself as a summary of the long-run behavior of a random phenomenon.

Mean of a probability distribution

Recall

Table 6.1 shows

Number of Home Runs	Probability
0	0.3889
1	0.3148
2	0.2222
3	0.0556
4	0.0185
5 or more	0.0000

Example 3

The Expected Number of Home Runs in a Game

Picture the Scenario

Let's refer back to the probability distribution for x = number of home runs the San Francisco Giants hit in a game. The table is shown again in the margin.

Question to Explore

Find the mean of this probability distribution and interpret it.

Think It Through

Because P(0) = 0.3889, over the long run you expect $x = 0$ (that is, no home runs) in 38.89% of the games. Likewise, you expect $x = 1$ home run 31.48% of the time, and so forth. In 162 games, for example, you expect $x = 0$ about 0.3889(162) = 63 times, $x = 1$ about 0.3148(162) = 51 times, and so forth. Because the mean equals the total of the observations divided by the sample size, for 162 games we can calculate the mean as

$$\mu = \frac{\overbrace{(0+0+\ldots+0)}^{63\text{ terms}}+\overbrace{(1+1+\ldots+1)}^{51\text{ terms}}+\overbrace{(2+2+\ldots+2)}^{36\text{ terms}}+\overbrace{(3+3+\ldots+3)}^{9\text{ terms}}+(4+4+\ldots+4)}{162}$$

$$= \frac{63(0) + 51(1) + 36(2) + 9(3) + 3(4)}{162} = 1.$$

Rather than adding a number 51 times, you can multiply the number by 51, such as 51(1). If this probability distribution applies in a large number of games, you'd expect the mean number of home runs hit by the Giants to be 1, or one home run per game on average. Across all teams in major league baseball in 2010, the mean number of home runs per game was 0.9506, so the Giants hit slightly more home runs per game than the overall average.

Insight

Because 63/162 = 0.3889, 51/162 = 0.3148, and so forth, the calculation has the form

$$\mu = \frac{63(0) + 51(1) + 36(2) + 9(3) + 3(4)}{162}$$

$$= 0(0.3889) + 1(0.3148) + 2(0.2222) + 3(0.0556) + 4(0.0185)$$

$$= 0 \times P(0) + 1 \times P(1) + 2 \times P(2) + 3 \times P(3) + 4 \times P(4).$$

Each possible value x is multiplied by its probability. In fact, for any discrete random variable, the mean of its probability distribution results from multiplying each possible value x by its probability $P(x)$ and then adding.

Try Exercise 6.3

In Words

To get the **mean** of a probability distribution, multiply each possible value of the random variable by its probability, and then add all these products.

Mean of a Discrete Probability Distribution

The **mean of a probability distribution** for a discrete random variable is

$$\mu = \Sigma x P(x),$$

where the sum is taken over all possible values of x.

For the San Francisco Giants' home runs in baseball, as we saw in the example,

$$\mu = \Sigma x\mathrm{P}(x) = 0\mathrm{P}(0) + 1\mathrm{P}(1) + 2\mathrm{P}(2) + 3\mathrm{P}(3) + 4\mathrm{P}(4)$$
$$= 0(0.3889) + 1(0.3148) + 2(0.2222) + 3(0.5556) + 4(0.0185) = 1.$$

In Words

A **weighted average** is used when each x value is not equally likely. If a particular x value is more likely to occur, it has a larger influence on the mean, which is the balance point of the distribution.

The mean $\mu = \Sigma x\mathrm{P}(x)$ is called a **weighted average**: Values of x that are more likely receive greater weight $\mathrm{P}(x)$. It does not make sense to take a simple average of the possible values of x, $(0 + 1 + 2 + 3 + 4)/5 = 2$, because some outcomes are much more likely than others.

Consider the special case in which the outcomes are *equally likely*. Suppose there are n such possible outcomes, each with probability $1/n$, such as $n = 6$ outcomes for rolling a die, each with probability 1/6. Then

$$\mu = \Sigma x\mathrm{P}(x) = \Sigma x(1/n) = (\Sigma x)/n.$$

The formula $(\Sigma x)/n$ is the same as the ordinary one for a sample mean. So the formula $\mu = \Sigma x\mathrm{P}(x)$ generalizes the ordinary formula to allow different outcomes to not be equally likely and to apply to probability distributions as well as to sample data.

Recall

Section 2.3 showed how to find the **sample mean** to describe the center of quantitative data. ◀

The mean of the probability distribution of a random variable X is also called the **expected value of X**. The expected value reflects not what we'll observe in a *single* observation, but rather what we expect for the *average* in a long run of observations. In the preceding example, the expected value of the number of San Francisco Giants home runs in a game is $\mu = 1$. As with means of sample data, the mean of a probability distribution doesn't have to be one of the possible values for the random variable. We will not see exactly one home run in each game, but the long-run average of observing some games with 0 home runs, some with 1, some with 2, and so forth, is 1.

Expected gains/losses

Example 4

Responding to Risk

Picture the Scenario

Are you a risk-averse person who prefers the sure thing to a risky action that could give you a better or a worse outcome? Or are you a risk taker, willing to gamble in hopes of achieving the better outcome?

Questions to Explore

a. You are given \$1000 to invest. You must choose between (i) a sure gain of \$500, and (ii) a 0.50 chance of a gain of \$1000 and a 0.50 chance to gain nothing. What is the expected gain with each strategy? Which do you prefer?

b. You are given \$2000 to invest. You must choose between (i) a sure loss of \$500, and (ii) a 0.50 chance of losing \$1000 and a 0.50 chance to lose nothing. What is the expected loss with each strategy? Which do you prefer?

Risk-Taking Probability Distribution

Gain x	P(x)
0	0.50
1000	0.50

Think It Through

a. The expected gain is \$500 with the sure strategy (i), since that strategy has gain \$500 with probability 1.0. With the risk-taking strategy (ii), the probability distribution is shown in the margin. The expected gain is $\mu = \Sigma xP(x) = \$0(0.50) + \$1000(0.50) = \$500$.

b. The expected loss is \$500 with the sure strategy (i). With the risk-taking strategy (ii), the expected loss is $\mu = \Sigma xP(x) = \$0(0.50) + \$1000(0.50) = \$500$.

Insight

In each case, the expected values are the same with each strategy. Yet most people prefer the sure-gain strategy (i) in case a but the risk-taking strategy (ii) in case b. They are risk averse in case a but risk taking in case b. This preference was explored in research by Daniel Kahneman and Amos Tversky, for which Kahneman won the Nobel Prize in 2002.

Try Exercise 6.7

Summarizing the Variability of a Probability Distribution

As with distributions of sample data, it's useful to summarize both the *center* and the variability of a probability distribution. For instance, suppose two different investment strategies have the same expected payout. Does one strategy have more variability in its payoffs?

The **standard deviation** of a probability distribution, denoted by σ, measures the variability from the mean. Larger values for σ correspond to greater variability. Roughly, σ describes how far the random variable falls, on the average, from the mean of its distribution. We won't deal with the formula for calculating σ until we study particular types of probability distributions in Sections 6.2 and 6.3. (See Exercise 6.85.)

Example 5

Variability of a probability distribution

Risk Taking Entails More Variability

Picture the Scenario

Let's revisit the first scenario in Example 4. You are given \$1000 to invest and must choose between (i) a sure gain of \$500, and (ii) a 0.50 chance of a gain of \$1000 and a 0.50 chance to gain nothing. Table 6.2 shows the probability distribution of the gain X for the two strategies.

Table 6.2 Probability Distribution of X = Gain

Sure Strategy		Risk Taking	
x	P(x)	x	P(x)
500	1.0	0	0.50
		1000	0.50

Questions to Explore

Example 4 showed that both strategies have the same mean, namely $500. Which of the two probability distributions would have the larger standard deviation?

Think It Through

There is *no* variability for the sure strategy. The standard deviation of this probability distribution is 0, the smallest possible value. With the risk-taking strategy, no matter what the outcome (either $0 or $1000) the result will be $500 from the mean of $500. In fact, the standard deviation of the probability distribution for the risk-taking strategy is $500. With that strategy, there's much more variability in what can happen.

Insight

In practice, different investment strategies are often compared by their variability. Not only is an investor interested in the expected return of an investment but also the consistency of the yield in the investment from year to year.

Try Exercise 6.9

Recall

We saw in the Insight for Example 12 of Section 2.3 that the proportion is a special case of a mean calculated for observations that equal 1 for the outcome of interest and 0 otherwise.◀

Probability Distributions of Categorical Variables

In examples so far, variables have been quantitative rather than categorical. Partly this is because a random variable is defined to be a *numerical* measurement of the outcome of a random phenomenon. However, we'll see that for categorical variables having only two categories, it's useful to represent the two possible outcomes by the numerical values 0 and 1.

For example, suppose you've conducted a marketing survey of many potential customers to estimate the probability a customer would buy a new product you are developing and plan to sell. Your study estimates that the probability is 0.20. If you denote the possible outcomes (success, failure) for whether or not a customer buys the product by (1, 0), then the probability distribution of X for the outcome is

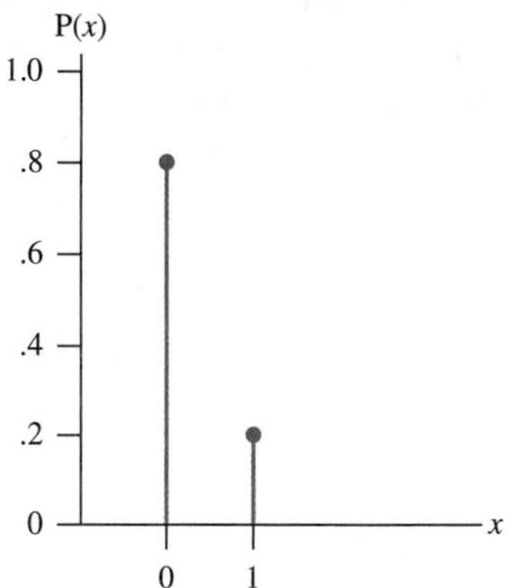

x	P(x)
0	0.80
1	0.20

The graph for this probability distribution is shown in the margin.

The mean of this probability distribution is

$$\mu = \Sigma x P(x) = 0(0.80) + 1(0.20) = 0.20.$$

The mean is the probability of success. *For random variables that have possible values 0 and 1, the mean is the probability of the outcome designated by 1.* We'll find this result to be quite useful in the next chapter.

Probability Distributions of Continuous Random Variables

A random variable is called **continuous** when its possible values form an interval. For instance, a recent study by the U.S. Census Bureau analyzed the time that people take to commute to work. Commuting time can be measured with real number values, such as between 0 and 150 minutes.

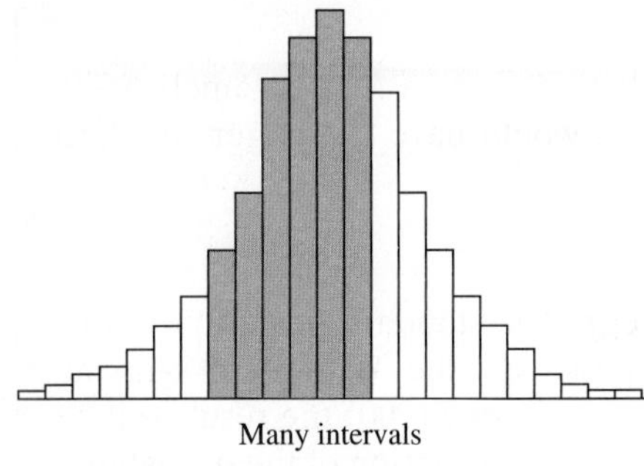

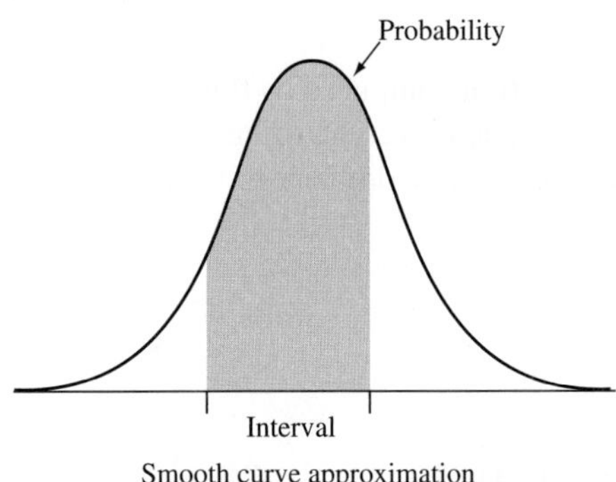

Probability distributions of continuous random variables assign probabilities to any interval of the possible values. For instance, a probability distribution for commuting time provides the probability that the travel time is less than 15 minutes or that the travel time is between 30 and 60 minutes. The probability that a random variable falls in any particular interval is between 0 and 1, and the probability of the interval that contains all the possible values equals 1.

When a random variable is continuous, the intervals of values for the bars of a histogram can be chosen as desired. For instance, one possibility for commuting time is {0 to 30, 30 to 60, 60 to 90, 90 to 120, 120 to 150}, quite wide intervals. By contrast, using {0 to 1,1 to 2,2 to 3, . . . ,149 to 150} gives lots of very narrow intervals. As the number of intervals increases, with their width narrowing, the shape of the histogram gradually approaches a smooth curve. We'll use such curves to portray probability distributions of continuous random variables. See the graphs in the margin.

Probability Distribution of a Continuous Random Variable

A **continuous** random variable has possible values that form an interval. Its **probability distribution** is specified by a curve that determines the probability that the random variable falls in any particular interval of values.

- Each interval has probability between 0 and 1. This is the area under the curve, above that interval.
- The interval containing all possible values has probability equal to 1, so the total area under the curve equals 1.

Figure 6.2 shows a graph for a probability distribution of X = commuting time for workers in the United States who commute to work. Historically, in surveys about travel to work, a commute of 45 minutes has been the maximum time that people would be willing to spend. However, the 2000 U.S. Census Bureau report[4] suggested that of the 97% of U.S. workers who do not work at home, 15% of the commuters have a commuting time above 45 minutes. The shaded area in Figure 6.2 refers to the probability of values higher than 45.0, for those who travel to work. This area equals 15% of the total area under the curve. So the probability is 0.15 of spending more than 45 minutes commuting to work. The report stated that the probability distribution is skewed to the right, with a mean

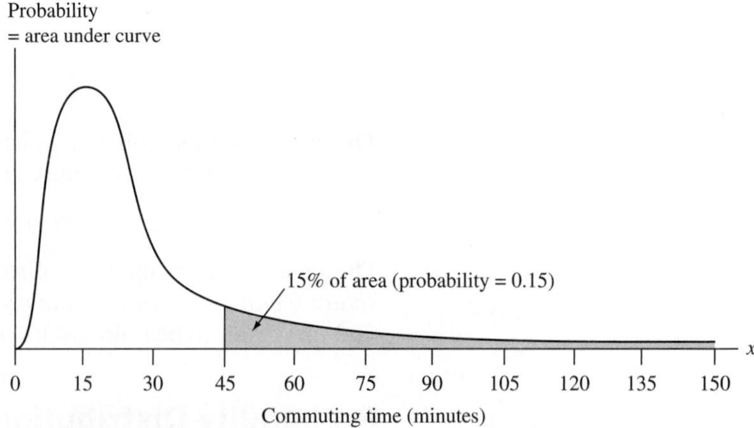

▲ **Figure 6.2 Probability Distribution of Commuting Time.** The area under the curve for values higher than 45 is 0.15. **Question** Identify the area under the curve represented by the probability that commuting time is less than 15 minutes, which equals 0.29.

---

[4]*Source:* Data from *Journey to Work: 2000*, issued March 2004 by U.S. Census Bureau.

of 25.5 minutes. Once the 2010 census data is released, a comparison can be made of the percentages for 2010 to the 2000 commuting times for possible changes in the distribution.

For continuous random variables, we need to round off our measurements. Probabilities are given for *intervals* of values rather than individual values. In measuring commuting time, the U.S. Census Bureau asked, "How many minutes did it usually take to get from home to work last week?" and gave as possible responses 0, 1, 2, 3, ... Then, for instance, a commuting time of 24 minutes actually means the interval of real numbers that round to the integer value 24. This is the area under the curve for the interval between 23.50 and 24.50. *In practice, the probability that the commuting time falls in some given interval, such as above 45 minutes, or between 30 and 60 minutes, is of greater interest than the probability it equals some particular single value.*

Did You Know?

For continuous random variables, the area above a single value equals 0. For example, the probability that the commuting time equals 23.7693611045. . . minutes is 0, but the probability that commuting time is between 23.5 and 24.5 minutes is positive.◀

In Practice Continuous Variables Are Measured in a Discrete Manner

In practice, **continuous** variables are measured in a **discrete** manner because of rounding. With rounding, a continuous random variable can take on a large number of separate values. A probability distribution for a continuous random variable is used to approximate the probability distribution for the possible rounded values.

Example 6

Probability distribution

Height

Picture the Scenario

Figure 6.3 shows a histogram of heights of females at the University of Georgia, with a smooth curve superimposed. (*Source:* Data from University of Georgia.)

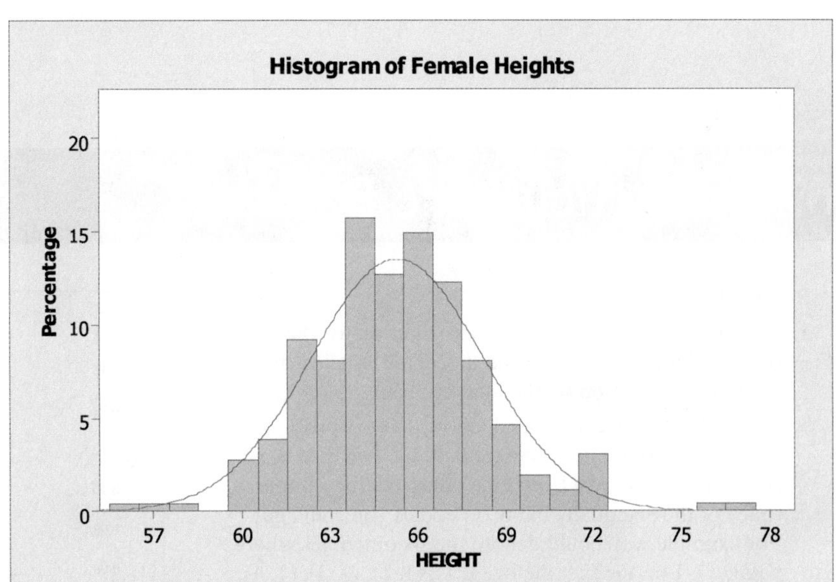

▲ **Figure 6.3 Histogram of Female Student Heights with Bell-Shaped Curve Superimposed.** Height is continuous but is measured as discrete by rounding to the nearest inch. The smooth curve approximates the probability distribution for height, treating it as a continuous random variable. **Question** How would you describe the shape, center, and variability of the distribution?

Question to Explore

What does the smooth curve represent?

Think It Through

In theory, height is a continuous random variable. In practice, it is measured here by rounding to the nearest inch. For instance, the bar of the histogram above 64 represents heights between 63.5 and 64.5 inches, which were rounded to 64 inches. The histogram gives the data distribution for the discrete way height is actually measured in a sample. The smooth curve uses the data distribution to approximate the probability distribution (the population distribution) that height would have if we could actually measure all female heights precisely as a continuous random variable. The area under this curve between two points approximates the probability that height falls between those points.

Insight

The histogram is a graphical representation of the sample. The smooth curve is a graphical representation of the population. We observe the smooth curve has a bell shape. In the next section we'll study a probability distribution that has this shape and we'll learn how to find probabilities for it.

Try Exercise 6.13

Sometimes we can use a sample space to help us find a probability distribution, as you'll see in Exercises 6.1 and 6.2 and examples later in the chapter. Sometimes the probability distribution is available with a *formula*, a *table*, or a *graph*, as we'll see in the next two sections. Sometimes, though, it's necessary to use simulation to approximate the distribution, as shown in Activity 1 at the end of the chapter.

6.1 Practicing the Basics

6.1 Rolling dice

a. State in a table the probability distribution for the outcome of rolling a balanced die. (This is called the **uniform distribution** on the integers 1,2, . . . ,6.)

b. Two balanced dice are rolled. Show that the probability distribution for X = total on the two dice is as shown in the figure. (*Hint*: First construct the sample space of the 36 equally likely outcomes you could get. For example, you could denote the six outcomes where you get a 1 on the first die by (1, 1), (1, 2), (1, 3), (1, 4), (1, 5), (1, 6), where the second number in a pair is the number you get on the second die.)

c. Show that the probabilities in part b satisfy the two conditions for a probability distribution.

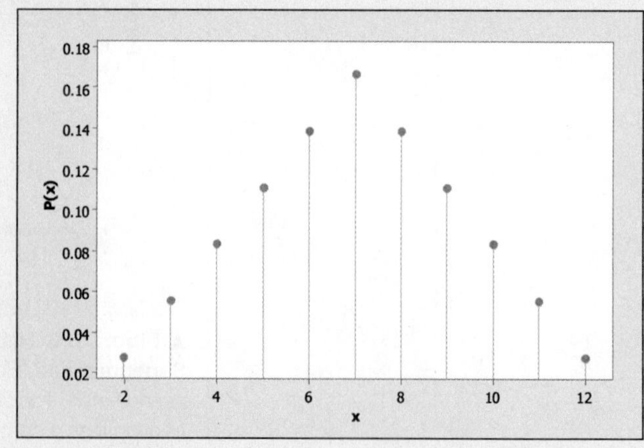

6.2 **Move first in Monopoly** In *Monopoly*, dice are used to determine which player gets to move first. Suppose there are two players in the game. Each player rolls a die and the player with the higher number gets to move first. If the numbers are the same, the players roll again.
TRY

a. Using the sample space {(1,1), (1,2), (1,3), (1,4), (1,5), (1,6), (2,1),... (6,5), (6,6)} of the 36 equally likely outcomes for the two dice, show that the probability distribution for the maximum of the two numbers is as shown in the table. *Hint:* For each outcome in the sample space, indicate the value of X assigned to that outcome.

Probability Distribution of X = Maximum Number for Rolls of Two Dice

x	P(x)
1	1/36
2	3/36
3	5/36
4	7/36
5	9/36
6	11/36

b. Show that the two conditions in the definition of a probability distribution are satisfied.

6.3 **San Francisco Giants hitting** The table shows the probability distribution of the number of bases for a randomly selected time at bat for a San Francisco Giants player in 2010 (excluding times when the player got on base because of a walk or being hit by a pitch). In 74.29% of the at-bats the player was out, 17.04% of the time the player got a single (one base), 5.17% of the time the player got a double (two bases), 0.55% of the time the player got a triple, and 2.95% of the time the player got a home run.
TRY

a. Verify that the probabilities give a legitimate probability distribution.

b. Find the mean of this probability distribution.

c. Interpret the mean, explaining why it does not have to be a whole number, even though each possible value for the number of bases is a whole number.

San Francisco Giants Hitting

Number of Bases	Probability
0	0.7429
1	0.1704
2	0.0517
3	0.0055
4	0.0295

6.4 **Grade distribution** An instructor always assigns final grades such that 20% are A, 40% are B, 30% are C, and 10% are D. The grade point scores are 4 for A, 3 for B, 2 for C, and 1 for D.

a. Specify the probability distribution for the grade point score of a randomly selected student of this instructor.

b. Find the mean of this probability distribution. Interpret it.

6.5 **Selling houses** Let X represent the number of homes a real estate agent sells during a given month. Based on previous sales records, she estimates that $P(0) = 0.68$, $P(1) = 0.19$, $P(2) = 0.09$, $P(3) = 0.03$, $P(4) = 0.01$, with negligible probability for higher values of x.

a. Explain why it does not make sense to compute the mean of this probability distribution as $(0 + 1 + 2 + 3 + 4)/5 = 2.0$.

b. Find the correct mean.

6.6 **Playing the lottery** The state of Ohio has several state-wide lottery options. One is the Pick 3 game in which you pick one of the 1000 three-digit numbers between 000 and 999. The lottery selects a three-digit number at random. With a bet of \$1, you win \$500 if your number is selected and nothing (\$0) otherwise. (Many states have a very similar type of lottery.) (*Source:* Background information from www.ohiolottery.com.)

a. With a single \$1 bet, what is the probability that you win \$500?

b. Let X denote your winnings for a \$1 bet, so $x = \$0$ or $x = \$500$. Construct the probability distribution for X.

c. Show that the mean of the distribution equals 0.50, corresponding to an expected return of 50 cents for the dollar paid to play. Interpret the mean.

d. In Ohio's Pick 4 lottery, you pick one of the 10,000 four-digit numbers between 0000 and 9999 and (with a \$1 bet) win \$5000 if you get it correct. In terms of your expected winnings, with which game are you better off—playing Pick 4, or playing Pick 3 in which you win \$500 for a correct choice of a three-digit number? Justify your answer.

6.7 **Which wager do you prefer?** A roulette wheel consists of 38 numbers, 0 through 36 and 00. Of these, 18 numbers are red, 18 are black, and 2 are green (0 and 00). You are given \$10 and told that you must pick one of two wagers, for an outcome based on a spin of the wheel: (1) Bet \$10 on number 23. If the spin results in 23, you win \$350 and also get back your \$10 bet. If any other number comes up, you lose your \$10, or (2) Bet \$10 on black. If the spin results in any one of the black numbers, you win \$10 and also get back your \$10 bet. If any other color comes up, you lose your \$10.
TRY

a. Without doing any calculation, which wager would you prefer? Explain why. (There is no correct answer. Peoples' choices are based on their individual preferences and risk tolerances.)

b. Find the expected outcome for each wager. Which wager is better in this sense?

6.8 **Gambling some more** Consider a game of poker being played with a standard 52-card deck (four suits, each of which has 13 different denominations of cards). At a certain point in the game, six cards have been exposed. Of the six, four are diamonds. Your opponent makes a bet of \$20, and you must decide whether to call the bet. If you do call the bet, you will receive one more card. If that final card turns out to be another diamond, you will win \$100. If not, you will lose the hand as well as the \$20 you called in order to receive the final card. On the other hand, if you do not call the bet, the hand ends immediately, your opponent wins, and you neither win nor lose any more money.

a. Specific the probability distribution for X = expected winnings.

b. Find the expected value of X. Based on the expected value, should you call the \$20 bet and receive one more card or not call the bet?

6.9 **Ideal number of children** Let X denote the response of a randomly selected person to the question, "What is the ideal number of children for a family to have?" The probability distribution of X in the United States is approximately as shown in the table, according to the gender of the person asked the question.

TRY

Probability Distribution of X = Ideal Number of Children

x	P(x) Females	P(x) Males
0	0.01	0.02
1	0.03	0.03
2	0.55	0.60
3	0.31	0.28
4	0.11	0.08

Note that the probabilities do not sum to exactly 1 due to rounding error.

a. Show that the means are similar, 2.50 for females and 2.39 for males.

b. The standard deviation for the females is 0.770 and 0.758 for the males. Explain why a practical implication of the values for the standard deviations is that males hold slightly more consistent views than females about the ideal family size.

6.10 **Profit and the weather** From past experience, a wheat farmer living in Manitoba, Canada finds that his annual profit (in Canadian dollars) is \$80,000 if the summer weather is typical, \$50,000 if the weather is unusually dry, and \$20,000 if there is a severe storm that destroys much of his crop. Weather bureau records indicate that the probability is 0.70 of typical weather, 0.20 of unusually dry weather, and 0.10 of a severe storm. In the next year, let X be the farmer's profit.

a. Construct a table with the probability distribution of X.

b. What is the probability that the profit is \$50,000 or less?

c. Find the mean of the probability distribution of X. Interpret.

d. Suppose the farmer buys insurance for \$3000 that pays him \$20,000 in the event of a severe storm that destroys much of the crop and pays nothing otherwise. Find the probability distribution of his profit. Find the mean, and summarize the effect of buying this insurance.

6.11 **Selling at the right price** Some companies, such as DemandTec, have developed software to help retail chains set prices that optimize their profits. An Associated Press story (April 28, 2007) about this software described a case in which a retail chain sold three similar power drills: one for \$90, a better one for \$120, and a top-tier one for \$130. Software predicted that by selling the middle-priced drill for only \$110, the cheaper drill would seem less a bargain and more people would buy the middle-price drill.

a. For the original pricing, suppose 50% of sales were for the \$90 drill, 20% for the \$120 drill, and 30% for the \$130 drill. Construct the probability distribution of X = selling price for the sale of a drill, and find its mean and interpret,

b. For the new pricing, suppose 30% of sales were for the \$90 drill, 40% for the \$110 drill, and 30% for the \$130 drill. Is the mean of the probability distribution of selling price higher with this new pricing strategy? Explain.

6.12 **Uniform distribution** A random number generator is used to generate a real number between 0 and 1, equally likely to fall anywhere in this interval of values. (For instance, 0.3794259832 . . . is a possible outcome.)

a. Sketch a curve of the probability distribution of this random variable, which is the continuous version of the **uniform distribution** (see Exercise 6.1).

b. What is the mean of this probability distribution?

c. Find the probability that this random variable falls between 0.25 and 0.75.

6.13 **TV watching** A social scientist uses the General Social Survey to study how much time per day people spend watching TV. The variable denoted by TVHOURS at the GSS Web site measures this using the values 0, 1, 2, . . . , 24.

TRY

a. Explain how, in theory, TV watching is a continuous random variable.

b. An article about the study shows two histograms, both skewed to the right, to summarize TV watching for females and males. Since TV watching is in theory continuous, why were histograms used instead of curves?

c. If the article instead showed two curves, explain what they would represent.

6.2 Probabilities for Bell-Shaped Distributions

Some probability distributions merit special attention because they are useful for many applications. They have formulas or tables that provide probabilities of the possible outcomes. We next learn about a probability distribution, called the **normal distribution**, that is commonly used for continuous random variables. It is characterized by a particular[5] symmetric, bell-shaped curve with two parameters—the mean μ and the standard deviation σ.

[5]A mathematical formula specifies *which* bell-shaped curve is the normal distribution, but it is complex and we'll not use it in this text. In Chapter 8, we'll learn about another bell-shaped distribution, one with "thicker tails" than the normal has.

Did You Know?

The color blue is used throughout the text for graphs of normal distributions. For more information, refer to "A Guide to Learning From the Art in This Text" located at the very front of the text. ◀

The Normal Distribution: A Probability Distribution with a Bell-Shaped Curve

Figure 6.3 at the end of the previous section showed that heights of female students at the University of Georgia have approximately a bell-shaped distribution. The approximating curve describes a probability distribution with a mean of 65.0 inches and a standard deviation of 3.5 inches. In fact, adult female heights in North America have approximately a normal distribution with $\mu = 65.0$ inches and $\sigma = 3.5$ inches. Adult male heights have approximately a normal distribution with $\mu = 70.0$ inches and $\sigma = 4.0$ inches. Adult males tend to be a bit taller (since $70 > 65$) with heights varying from the mean a bit more (since $4.0 > 3.5$). See Figure 6.4.

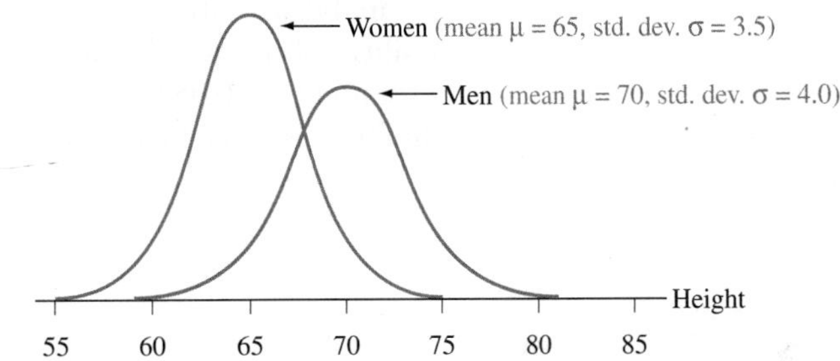

▲ **Figure 6.4 Normal Distributions for Women's Height and Men's Height.** For each different combination of μ and σ values, there is a normal distribution with mean μ and standard deviation σ. **Question** Given that $\mu = 70$ and $\sigma = 4$, within what interval do almost all of the men's heights fall?

For any real number for the mean μ and any positive number for the standard deviation σ, there is a normal distribution with that mean and standard deviation.

> **Normal Distribution**
>
> The **normal distribution** is symmetric, bell-shaped, and characterized by its mean μ and standard deviation σ. The probability within any particular number of standard deviations of μ is the same for all normal distributions. This probability equals 0.68 within 1 standard deviation, 0.95 within 2 standard deviations, and 0.997 within 3 standard deviations. See Figure 6.5.

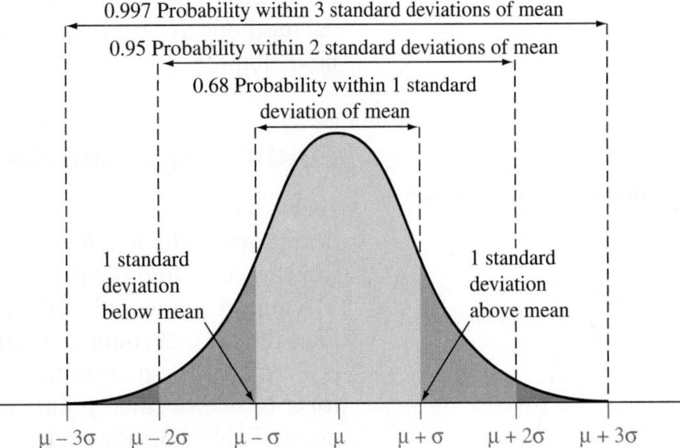

▲ **Figure 6.5 The Normal Distribution.** The probability equals 0.68 within 1 standard deviation of the mean, 0.95 within 2 standard deviations, and 0.997 within 3 standard deviations. **Question** How do these probabilities relate to the empirical rule?

To illustrate, for adult female heights, $\mu = 65.0$ and $\sigma = 3.5$ inches. Since

$$\mu - 2\sigma = 65.0 - 2(3.5) = 58.0 \text{ and } \mu + 2\sigma = 65.0 + 2(3.5) = 72.0,$$

about 95% of the female heights fall between 58 inches and 72 inches (6 feet). For adult male heights, $\mu = 70.0$ and $\sigma = 4.0$ inches. About 95% fall between $\mu - 2\sigma = 70.0 - 2(4.0) = 62$ inches and $\mu + 2\sigma = 70.0 + 2(4.0) = 78$ inches (6 1/2 feet).

Caution

From Section 2.5, the ***z*-score** for an observation is the number of standard deviations that it falls from the mean. The *z*-score can be used with any distribution for a quantitative variable. This includes both normal and nonnormal distributions. ◀

The property of the normal distribution in the definition tells us probabilities within 1, 2, and 3 standard deviations of the mean. The multiples 1, 2, and 3 of the number of standard deviations from the mean are denoted by the symbol z in general. For instance, $z = 2$ for 2 standard deviations. For each fixed number z, the probability within z standard deviations of the mean is the area under the normal curve between $\mu - z\sigma$ and $\mu + z\sigma$ as shown in Figure 6.6. For every normal distribution, this probability is 0.68 for $z = 1$, so 68% of the area (probability) of a normal distribution falls between $\mu - \sigma$ and $\mu + \sigma$. Similarly, this probability is 0.95 for $z = 2$, and nearly 1.0 for $z = 3$ (that is, between $\mu - 3\sigma$ and $\mu + 3\sigma$). The total probability for any normal distribution equals 1.0.

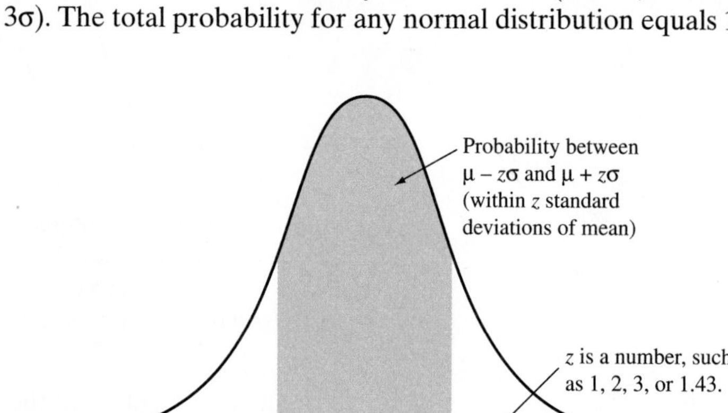

▲ **Figure 6.6 The Probability between $\mu - z\sigma$ and $\mu + z\sigma$.** This is the area highlighted under the curve. It is the same for every normal distribution and depends only on the value of z. Figure 6.5 showed this for $z = 1, 2$, and 3, but z does not have to be an integer—it can be any number.

The normal distribution is the most important distribution in statistics, partly because many variables have approximately normal distributions. The normal distribution is also important because it approximates many discrete distributions well when there are a large number of possible outcomes. The main reason for the prominence of the normal distribution is that many statistical methods use it even when the data are not bell shaped. We'll see why in the next chapter.

Recall

A cumulative proportion (probability) was defined in Section 5.1.

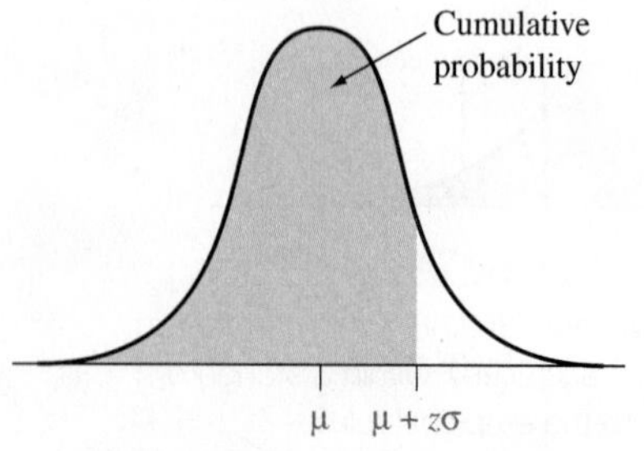

Finding Probabilities for the Normal Distribution

As we'll discuss, the probabilities 0.68, 0.95, and 0.997 within 1, 2, and 3 standard deviations of the mean are no surprise, because of the empirical rule. But what if we wanted to find the probability within, say, 1.43 standard deviations?

Table A at the end of the text enables us to find normal probabilities. It tabulates the normal **cumulative probability**, the probability of falling *below* the point $\mu + z\sigma$ (see the margin figure). The leftmost column of Table A lists the values for z to one decimal point, with the second decimal place listed above the columns. Table 6.3 shows a small excerpt from Table A. The tabulated probability for $z = 1.43$ falls in the row labeled 1.4 and in the column labeled 0.03. It equals 0.9236. For every normal distribution, the probability that falls below $\mu + 1.43\sigma$ equals 0.9236. Figure 6.7 illustrates.

Table 6.3 Part of Table A for Normal Cumulative (Left-Tail) Probabilities

The top of the table gives the second digit for z. The table entry is the probability falling below $\mu + z\sigma$, for instance, 0.9236 below $\mu + 1.43\sigma$ for $z = 1.43$.

Second Decimal Place of z										
z	.00	.01	.02	.03	.04	.05	.06	.07	.08	.09
0.0	.5000	.5040	.5080	.5120	.5160	.5199	.5239	.5279	.5319	.5359
...										
1.3	.9032	.9049	.9066	.9082	.9099	.9115	.9139	.9147	.9162	.9177
1.4	.9192	.9207	.9222	.9236	.9251	.9265	.9278	.9292	.9306	.9319
1.5	.9332	.9345	.9357	.9370	.9382	.9394	.9406	.9418	.9429	.9441

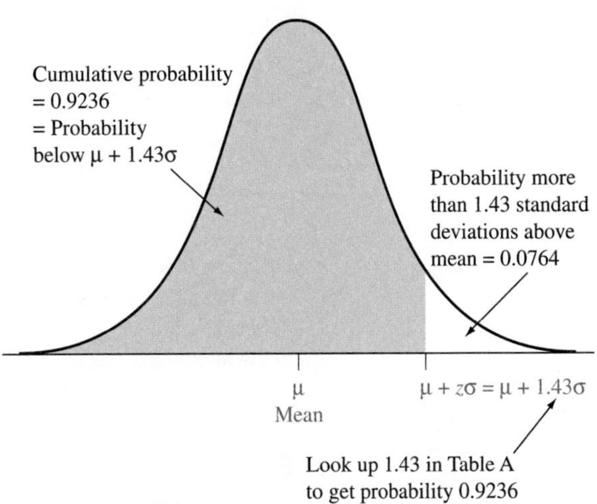

▲ **Figure 6.7 The Normal Cumulative Probability, Less than z Standard Deviations above the Mean.** Table A lists a cumulative probability of 0.9236 for $z = 1.43$, so 0.9236 is the probability less than 1.43 standard deviations above the mean of any normal distribution (that is, below $\mu + 1.43\sigma$). The complement probability of 0.0764 is the probability *above* $\mu + 1.43\sigma$ in the right tail.

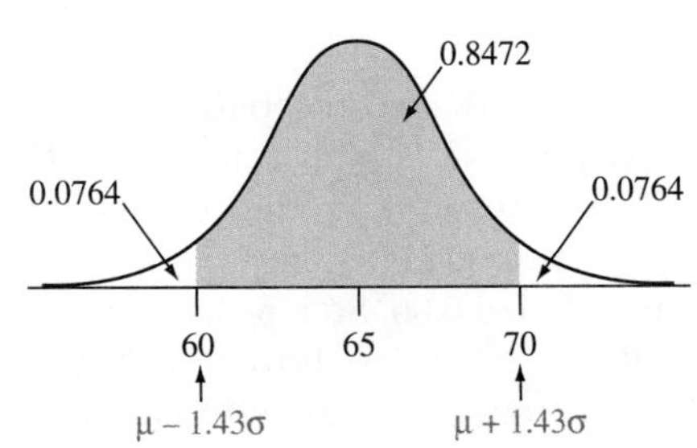

Since an entry in Table A is a probability *below* $\mu + z\sigma$, one minus that probability is the probability *above* $\mu + z\sigma$. For example, the right-tail probability above $\mu + 1.43\sigma$ equals $1 - 0.9236 = 0.0764$. By the symmetry of the normal curve, this probability also refers to the left tail below $\mu - 1.43\sigma$, which you'll find in Table A by looking up $z = -1.43$. The negative z-scores in the table refer to cumulative probabilities for random variable values *below* the mean.

Since the probability is 0.0764 in each tail, the total probability *more than* 1.43 standard deviations from the mean equals $2(0.0764) = 0.1528$. The total probability equals 1, so the probability falling *within* 1.43 standard deviations of the mean equals $1 - 0.1528 = 0.8472$, about 85%. For instance, 85% of women in North America have height between $\mu - 1.43\sigma = 65.0 - 1.43(3.5) = 60$ inches and $\mu + 1.43\sigma = 65 + 1.43(3.5) = 70$ inches (that is, between 5 feet and 5 feet, 10 inches).

Normal Probabilities and the Empirical Rule

The empirical rule states that for an approximately bell-shaped distribution, about 68% of observations fall within 1 standard deviation of the mean, 95%

within 2 standard deviations, and all or nearly all within 3. In fact, those percentages came from probabilities calculated for the normal distribution.

Recall

The empirical rule was introduced in Section 2.4.◀

For instance, a value that is 2 standard deviations below the mean has $z = -2.00$. The cumulative probability below $\mu - 2\sigma$ listed in Table A opposite $z = -2.00$. is 0.0228. The right-tail probability above $\mu + 2\sigma$ also equals 0.0228, by symmetry. See Figure 6.8. The probability falling more than 2 standard deviations from the mean in either tail is $2(0.0228) = 0.0456$. Thus, the probability that falls *within* 2 standard deviations of the mean equals $1 - 0.0456 = 0.9544$. When a variable has a normal distribution, 95.44% of the distribution (95%, rounded) falls within 2 standard deviations of the mean.

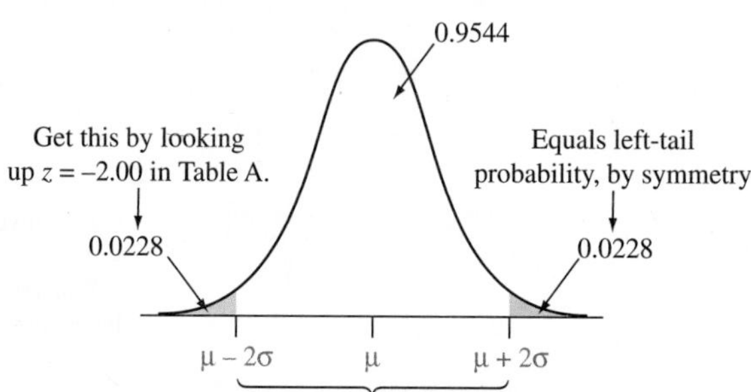

▲ **Figure 6.8 Normal Probability within 2 Standard Deviations of the Mean.** Probabilities in one tail determine probabilities in the other tail by symmetry. Subtracting the total two-tail probability from 1.0 gives probabilities within a certain distance of the mean. **Question** Can you do the analogous calculation for *3* standard deviations?

The approximate percentages that the empirical rule lists are the percentages for the normal distribution, rounded. For instance, you can verify that the probability within 1 standard deviation of the mean of a normal distribution equals 0.68. (*Hint:* Get the left-tail probability by looking up $z = -1.00$, double it for the two-tail probability, and then subtract from 1.) The probability within 3 standard deviations of the mean equals 0.997, or 1.00 rounded off. The empirical rule stated the probabilities as being *approximate* rather than *exact* because that rule referred to *all approximately* bell-shaped distributions, not just the normal.

How Can We Find the Value of *z* for a Certain Cumulative Probability?

In practice, we'll sometimes need to find the value of z that corresponds to a certain normal cumulative probability. How can we do this? To illustrate, let's find the value of z for a cumulative probability of 0.025. We look up the cumulative probability of 0.025 in the body of Table A. It corresponds to $z = -1.96$, since it is in the row labeled -1.9 and in the column labeled 0.06. So a probability of 0.025 lies below $\mu - 1.96\sigma$. Likewise, a probability of 0.025 lies above $\mu + 1.96\sigma$. A total probability of 0.050 lies more than 1.96σ from μ. Precisely 95.0% of a normal distribution falls within 1.96 standard deviations of the mean. (See figure in margin.) We've seen previously that 95.44% falls within 2.00 standard deviations, and we now see that precisely 95.0% falls within 1.96 standard deviations.

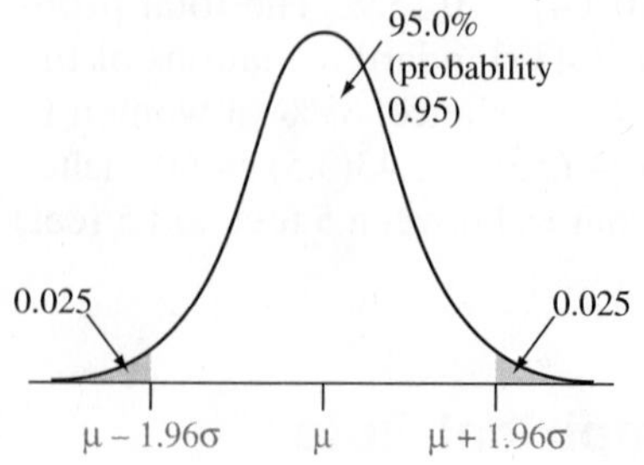

Cumulative probability and a percentile value

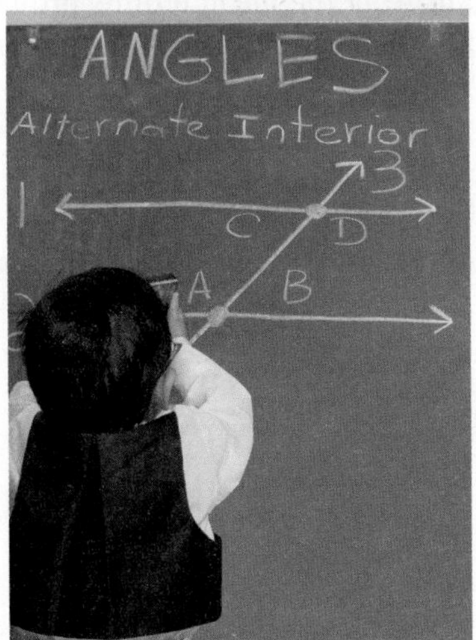

Recall

The **percentile** was defined in Section 2.5. For the 98th percentile, 98% of the distribution falls below that point. ◀

Example 7

Mensa IQ Scores

Picture the Scenario

Mensa[6] is a society of high-IQ people whose members have IQ scores at the 98th percentile or higher. The Stanford-Binet IQ scores that are used as the basis for admission into Mensa are approximately normally distributed with a mean of 100 and a standard deviation of 16.

Questions to Explore

a. How many standard deviations above the mean is the 98th percentile?

b. What is the IQ score for that percentile?

Think It Through

a. For a value to represent the 98th percentile, its cumulative probability must equal 0.98, by the definition of a percentile. See Figure 6.9.

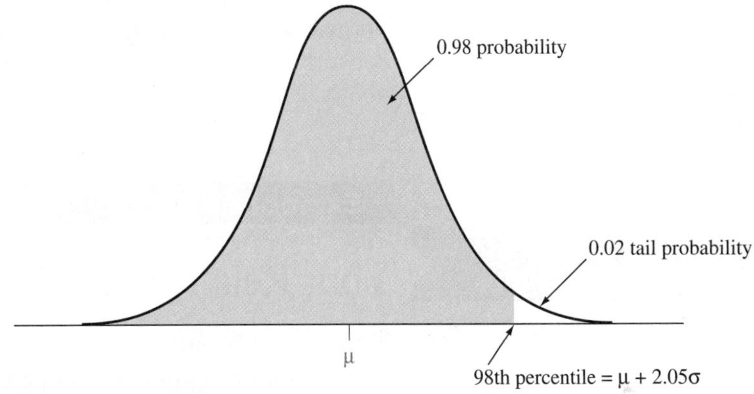

▲ **Figure 6.9 The 98th Percentile for a Normal Distribution.** This is the value such that 98% of the distribution falls below it and 2% falls above. **Question** Where is the second percentile located?

The cumulative probability of 0.980 in the body of Table A corresponds to $z = 2.05$. The 98th percentile is 2.05 standard deviations above the mean, at $\mu + 2.05\sigma$.

b. Since $\mu = 100$ and $\sigma = 16$, the 98th percentile of IQ scores equals

$$\mu + 2.05\sigma = 100 + 2.05(16) = 133.$$

In summary, 98% of the IQ scores fall below 133, and an IQ score of at least 133 is required to join Mensa.

Insight

About 2% of IQ scores are higher than 133. By symmetry, about 2% of IQ scores are lower than $\mu - 2.05\sigma = 100 - 2.05(16) = 67$. This is the second percentile. The remaining 96% of the IQ scores fall between 67 and 133, which is the region within 2.05 standard deviations of the mean.

It's also possible to use software to find normal probabilities or z-scores. The margin shows a screen shot using the TI-83+/84 to find the 98th percentile of IQ.

```
invNorm(.98,100,
16)
              132.86
█
```

TI-83+/84 output

Try Exercises 6.18 and 6.27, part b

[6]See www.mensa.org.

Using z = Number of Standard Deviations to Find Probabilities

We've used the symbol z to represent the *number of standard deviations* a value falls from the mean. If we have a value x of a random variable, how can we figure out the number of standard deviations it falls from the mean μ of its probability distribution? The difference between x and μ equals $x - \mu$. The **z-score** expresses this difference as a number of standard deviations, using $z = (x - \mu)/\sigma$.

Recall

Section 2.5 showed that for sample data, the number of standard deviations that a value x falls from the mean of the sample is

$$z = \frac{x - \text{mean}}{\text{standard deviation}}.$$

For example, if $z = 1.8$, then the value x falls 1.8 standard deviations *above* the mean, and if $z = -1.8$, then x falls 1.8 standard deviations *below* the mean. ◄

z-Score for a Value of a Random Variable

The **z-score** for a value x of a random variable is the number of standard deviations that x falls from the mean μ. It is calculated as

$$z = \frac{x - \mu}{\sigma}.$$

The formula for the z-score is useful when we are given the value of x for some normal random variable and need to find a probability relating to that value. We convert x to a z-score and then use a normal table to find the appropriate probability. The next two examples illustrate.

Example 8

Using z-scores to find probabilities

Your Relative Standing on the SAT

Picture the Scenario

The Scholastic Aptitude Test (SAT), a college entrance examination, has three components: critical reading, mathematics, and writing. The scores on each component are approximately normally distributed with mean $\mu = 500$ and standard deviation $\sigma = 100$. The scores range from 200 to 800 on each component.

Questions to Explore

a. If your SAT score from one of the three components was $x = 650$, how many standard deviations from the mean was it?

b. What percentage of SAT scores was higher than yours?

Think It Through

a. The SAT score of 650 has a z-score of $z = 1.50$ because 650 is 1.50 standard deviations above the mean. In other words, $x = 650 = \mu + z\sigma = 500 + z(100)$, where $z = 1.50$. We can find this directly using the formula

$$z = \frac{x - \mu}{\sigma} = \frac{650 - 500}{100} = 1.50.$$

b. The percentage of SAT scores higher than 650 is the right-tail probability above 650, for a normal random variable with mean $\mu = 500$ and standard deviation $\sigma = 100$. From Table A, the z-score of 1.50 has cumulative probability 0.9332. That's the probability *below* 650, so the right-tail probability above it is $1 - 0.9332 = 0.0668$. (See figure in margin.) Only about 7% of SAT test scores fall above 650. In summary,

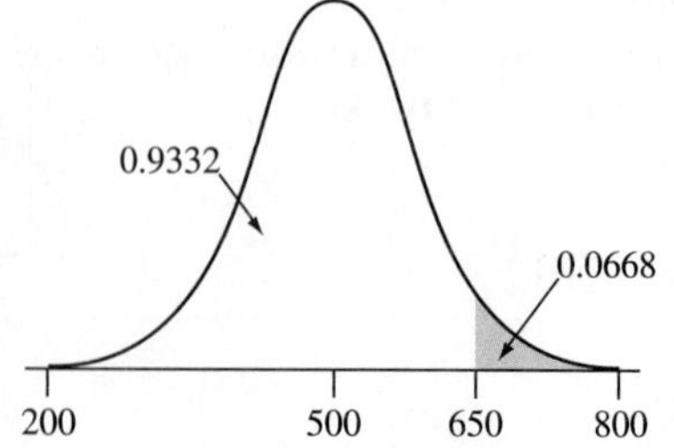

a score of 650 was well above average, in the sense that relatively few students scored higher.

Insight

Positive z-scores occur when the value x falls *above* the mean μ. Negative z-scores occur when x falls *below* the mean. For instance, an SAT score = 350 has a z-score of

$$z = \frac{x - \mu}{\sigma} = \frac{350 - 500}{100} = -1.50.$$

The SAT score of 350 is 1.50 standard deviations *below* the mean. The probability that an SAT score falls below 350 is also 0.0668. Figure 6.10 illustrates.

Try Exercise 6.23

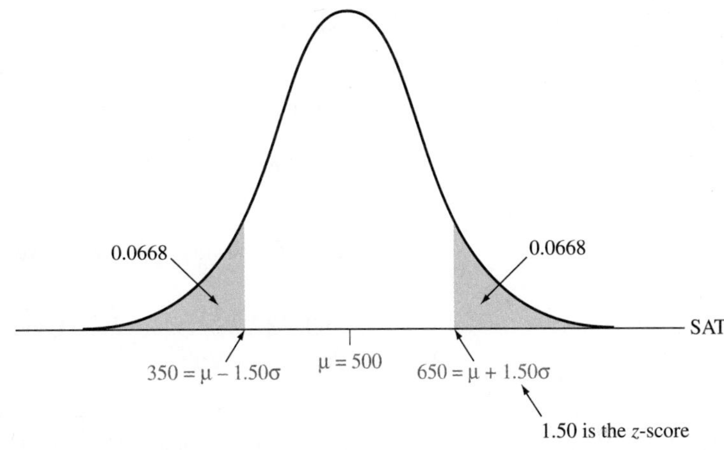

▲ **Figure 6.10 Normal Distribution for SAT.** The SAT scores of 650 and 350 have z-scores of 1.50 and −1.50 because they fall 1.50 standard deviations above and below the mean. **Question** Which SAT scores have $z = 3.0$ and $z = -3.0$?

Using z-scores to find probabilities

Example 9

The Proportion of Students Who Get a B

Picture the Scenario

On the midterm exam in introductory statistics, an instructor always gives a grade of B to students who score between 80 and 90.

Question to Explore

One year, the scores on the exam have approximately a normal distribution with mean 83 and standard deviation 5. About what proportion of students earn a B?

Think It Through

A midterm exam score of 90 has a z-score of

$$z = \frac{x - \mu}{\sigma} = \frac{90 - 83}{5} = 1.40.$$

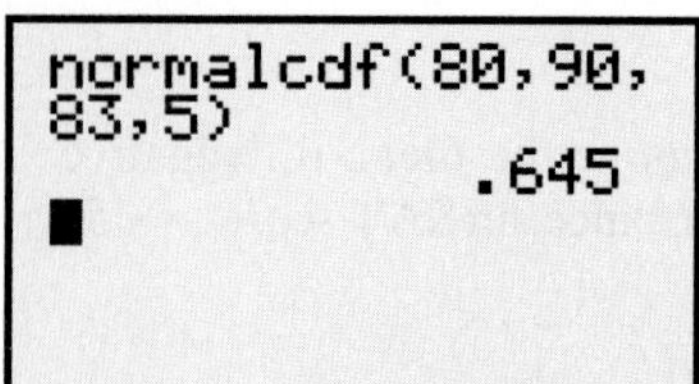

TI-83+/84 output

Its cumulative probability of 0.9192 (from Table A) means that about 92% of the exam scores were below 90. Similarly, an exam score of 80 has a z-score of

$$z = \frac{x - \mu}{\sigma} = \frac{80 - 83}{5} = -0.60.$$

Its cumulative probability of 0.2743 means that about 27% of the exam scores were below 80. See the normal curves that follow. Therefore about $0.9192 - 0.2743 = 0.6449$, or about 64%, of the exam scores were in the B range.

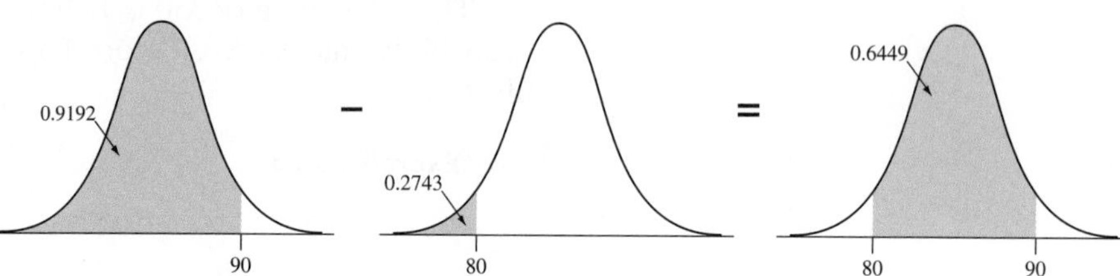

Insight

Here, we took the difference between two separate cumulative probabilities to find a probability between two points.

Try Exercise 6.30

Here's a summary of how we've used z-scores so far:

SUMMARY: Using z-Scores to Find Normal Probabilities or Random Variable x Values:

- If we're given a value x and need to find a probability, convert x to a z-score using $z = (x - \mu)/\sigma$, use a table of normal probabilities (or software, or a calculator) to get a cumulative probability and then convert it to the probability of interest.
- If we're given a probability and need to find the value of x, convert the probability to the related cumulative probability, find the z-score using a normal table (or software, or a calculator), and then evaluate $x = \mu + z\sigma$.

In Example 7, we used the equation $x = \mu + z\sigma$ to find a percentile score (namely, 98th percentile $= \mu + 2.05\sigma = 100 + 2.05(16) = 133$). In Examples 8 and 9, we used the equation $z = (x - \mu)/\sigma$ to determine how many standard deviations certain scores fell from the mean, which enabled us to find probabilities relating to those scores.

Another use of z-scores is for comparing observations from different normal distributions in terms of their relative distances from the mean.

Example 10

Comparing z-scores

Comparing Test Scores That Use Different Scales

Picture the Scenario

There are two primary standardized tests used by college admissions, the SAT and the ACT.[7]

[7]See www.sat.org and www.act.org.

Question to Explore

When you applied to college, you scored 650 on an SAT exam, which had mean $\mu = 500$ and standard deviation $\sigma = 100$. Your friend took the comparable ACT, scoring 30. For that year, the ACT had $\mu = 21.0$ and $\sigma = 4.7$. How can we compare these scores to tell who performed better?

Think It Through

The test scores of 650 and 30 are not directly comparable because the SAT and ACT have different means and different standard deviations. But we can convert them to z-scores and analyze how many standard deviations each falls from the mean.

With $\mu = 500$ and $\sigma = 100$, we saw in Example 8 that an SAT test score of $x = 650$ converts to a z-score of $z = (x - \mu)/\sigma = (650 - 500)/100 = 1.50$. With $\mu = 21.0$ and $\sigma = 4.7$, an ACT score of 30 converts to a z-score of

$$z = \frac{x - \mu}{\sigma} = \frac{30 - 21}{4.7} = 1.91.$$

The ACT score of 30 is a bit higher than the SAT score of 650, since ACT = 30 falls 1.91 standard deviations above its mean, whereas SAT = 650 falls 1.50 standard deviations above its mean. In this sense, the ACT score is better even though its numerical value is smaller.

Insight

The SAT and ACT tests both have approximately normal distributions. From Table A, $z = 1.91$ (for the ACT score of 30) has a cumulative probability of 0.97. Of all students who took the ACT, only about 3% scored above 30. From Table A, $z = 1.50$ (for the SAT score of 650) has a cumulative probability of 0.93. Of all students who took the SAT, about 7% scored above 650.

Try Exercise 6.31

The Standard Normal Distribution has Mean = 0 and Standard Deviation = 1

Many statistical methods refer to a particular normal distribution called the **standard normal distribution.**

Standard Normal Distribution

The **standard normal distribution** is the normal distribution with mean $\mu = 0$ and standard deviation $\sigma = 1$. It is the distribution of normal z-scores.

For the standard normal distribution, the number falling z standard deviations above the mean is $\mu + z\sigma = 0 + z(1) = z$, simply the z-score itself. For instance, the value of 2.0 is two standard deviations above the mean, and the value of -1.3 is 1.3 standard deviations below the mean. As Figure 6.11 shows, the original values are the same as the z-scores, since

$$z = \frac{x - \mu}{\sigma} = \frac{x - 0}{1} = x.$$

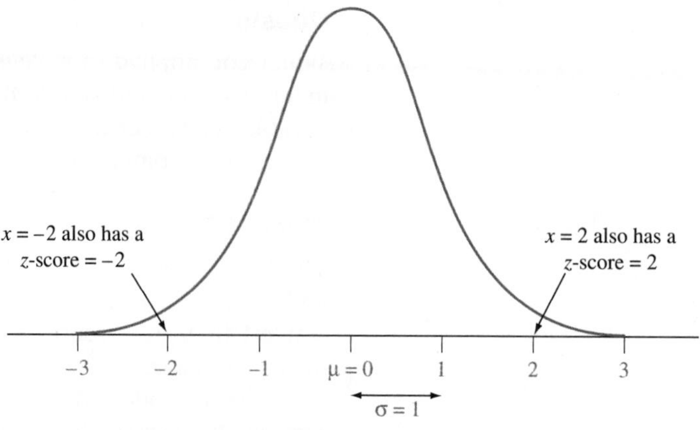

▲ **Figure 6.11 The Standard Normal Distribution.** This has mean = 0 and standard deviation = 1. The random variable value *x* is the same as its *z*-score. **Question** What are the limits within which almost all its values fall?

Examples 8 and 10 dealt with SAT scores, having $\mu = 500$ and $\sigma = 100$. Suppose we convert each SAT score x to a z-score by using $z = (x - \mu)/\sigma = (x - 500)/100$. Then $x = 650$ converts to $z = 1.50$, and $x = 350$ converts to $z = -1.50$. When the values for a normal distribution are converted to z-scores, those z-scores have a mean of 0 and have a standard deviation of 1. That is, the entire set of z-scores has the standard normal distribution.

z-Scores and the Standard Normal Distribution

When a random variable has a normal distribution and its values are converted to *z*-scores by subtracting the mean and dividing by the standard deviation, the *z*-scores have the **standard normal** distribution (mean = 0, standard deviation = 1).

This result will be useful for statistical inference in upcoming chapters.

6.2 Practicing the Basics

6.14 Probabilities in tails For a normal distribution, use Table A, software, or a calculator to find the probability that an observation is

a. at least 1 standard deviation above the mean.

b. at least 1 standard deviation below the mean.

c. In each case, sketch a curve and show the tail probability.

6.15 Tail probability in graph For the normal distribution shown below, use Table A, software, or a calculator to find the probability that an observation falls in the shaded region.

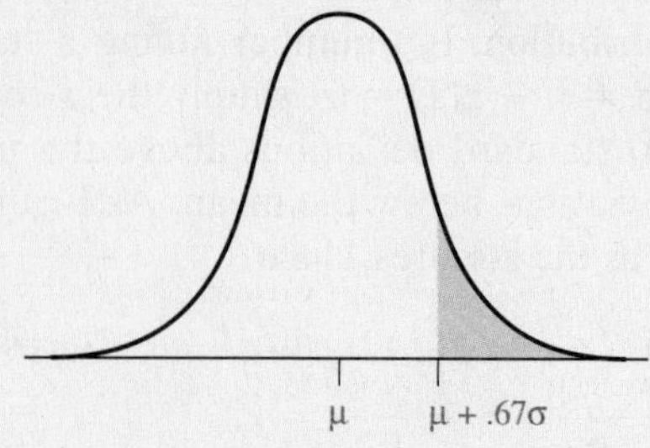

6.16 Empirical rule Verify the empirical rule by using Table A, software, or a calculator to show that for a normal distribution, the probability (rounded to two decimal places) within

a. 1 standard deviation of the mean equals 0.68.

b. 2 standard deviations of the mean equals 0.95.

c. 3 standard deviations of the mean is very close to 1.00.

In each case, sketch a normal distribution, identifying on the sketch the probabilities you used to show the result.

6.17 Central probabilities For a normal distribution, use Table A to verify that the probability (rounded to two decimal places) within

a. 1.64 standard deviations of the mean equals 0.90.

b. 2.58 standard deviations of the mean equals 0.99.

c. Find the probability that falls within 0.67 standard deviations of the mean.

d. Sketch these three cases on a single graph.

6.18 ***z*-score for given probability in tails** For a normal distribution,
TRY

a. Find the z-score for which a total probability of 0.02 falls more than z standard deviations (in either direction) from the mean, that is, below $\mu - z\sigma$ or above $\mu + z\sigma$.

b. For this z, explain why the probability more than z standard deviations above the mean equals 0.01.

c. Explain why $\mu + 2.33\sigma$ is the 99th percentile.

6.19 **Probability in tails for given *z*-score** For a normal distribution,

a. Show that a total probability of 0.01 falls more than $z = 2.58$ standard deviations from the mean.

b. Find the z-score for which the two-tail probability that falls more than that many standard deviations from the mean in either direction equals (a) 0.05, (b) 0.10. Sketch the two cases on a single graph.

6.20 ***z*-score for right-tail probability**

a. For the normal distribution shown below, find the z-score.

b. Find the value of z (rounding to two decimal places) for right-tail probabilities of (i) 0.05 and (ii) 0.005.

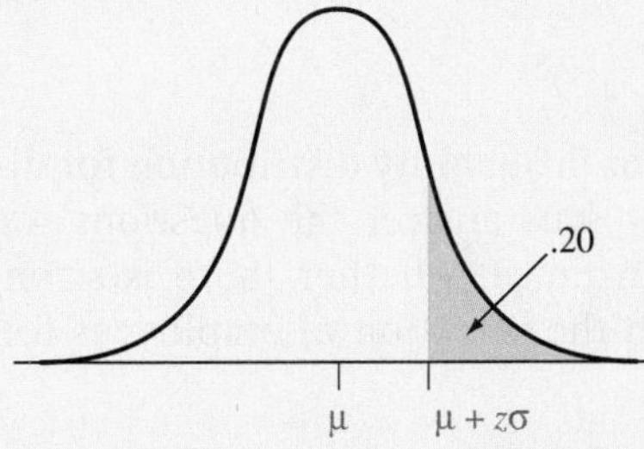

6.21 ***z*-score and central probability** Find the z-score such that the interval within z standard deviations of the mean (between $\mu - z\sigma$ and $\mu + z\sigma$) for a normal distribution contains

a. 50% of the probability.

b. 90% of the probability.

c. Sketch the two cases on a single graph.

6.22 **Female heights** The normal distribution for women's height in North America has $\mu = 65$ inches, $\sigma = 3.5$ inches. Most major airlines have height requirements for flight attendants (www.cabincrewjobs.com). Although exceptions are made, the minimum height requirement is 62 inches. What proportion of adult females in North America are not tall enough to be a flight attendant?

6.23 **Blood pressure** A World Health Organization study (the MONICA project) of health in various countries reported that in Canada, systolic blood pressure readings have a mean of 121 and a standard deviation of 16. A reading above 140 is considered to be high blood pressure.
TRY

a. What is the z-score for a blood pressure reading of 140?

b. If systolic blood pressure in Canada has a normal distribution, what proportion of Canadians suffers from high blood pressure?

c. What proportion of Canadians has systolic blood pressures in the range from 100 to 140?

d. Find the 90th percentile of blood pressure readings.

6.24 **Working hours** According to *Current Population Reports*, self-employed individuals in the United States work an average of 44.6 hours per week, with a standard deviation of 14.5. If this variable is approximately normally distributed, find the proportion of the self-employed who work more than 40 hours per week. Sketch a graph, and mark off on it 44.6, 40, and the region to which the answer refers.

6.25 **Energy use** An energy study in Gainesville, Florida, found that in March 2006, household use of electricity had a mean of 673 and a standard deviation of 556 kilowatt-hours. (*Source:* Data from Todd Kamhoot, Gainesville Regional Utilities.)

a. Suppose the distribution of energy use was normal. Using a table, calculator, or software that can give normal probabilities, find the proportion of households with electricity use greater than 1000 kilowatt-hours.

b. Based on the mean and standard deviation given, do you think that the distribution of energy use actually is normal? Why or why not?

6.26 **Apartment rentals** According to www.mynewplace.com, in January 2011 the average monthly rental rate for one-bedroom apartments in Ann Arbor, Michigan, was \$796. (Background information from www.mynewplace.com/city/ann_arbor_apartments_for_rent_michigan.) Suppose rental rates across all one-bedroom apartments in Ann Arbor follow approximately a normal distribution, with a standard deviation of \$150. Using a table, calculator, or software that can provide normal probabilities, find the approximate proportion of one-bedroom apartments for which the rental rate:

a. is at least \$1000 a month.

b. is less than \$500 a month.

c. is between \$500 and \$1000 a month.

6.27 **MDI** The Mental Development Index (MDI) of the Bayley Scales of Infant Development is a standardized measure used in observing infants over time. It is approximately normal with a mean of 100 and a standard deviation of 16.
TRY

a. What proportion of children has an MDI of (i) at least 120? (ii) at least 80?

b. Find the MDI score that is the 99th percentile.

c. Find the MDI score such that only 1% of the population has MDI below it.

6.28 **Quartiles and outliers** Refer to the previous exercise.

a. Find the z-score corresponding to the lower quartile (Q1) of a normal distribution.

b. Find and interpret the lower quartile and upper quartile of the MDI.

c. Find the interquartile range (IQR) of MDI scores.

d. Section 2.5 defined an observation to be a potential outlier if it is more than $1.5 \times$ IQR below Q1 or above Q3. Find the intervals of MDI scores that would be considered potential outliers.

6.29 **Murder rates** In 2008, the murder rates (per 100,000 residents) for the 50 states and the District of Columbia (D.C.) had a mean of 5.39 and a standard deviation of 4.434 (*Statistical Abstract of the United States*).[8]

---

[8]*Source:* Data from www.census.gov/compendia/statab/2011/tables/11s0304.pdf.

a. D.C. had a murder rate of 31.4. Find its z-score. If the distribution were roughly normal, would this be unusually high? Explain.

b. Based on the mean and standard deviation, do you think that the distribution of murder rates is approximately normal? Why or why not?

6.30 **Tall enough to ride?** A new roller coaster at an TRY amusement park requires individuals to be at least 4' 8" (56 inches) tall to ride. It is estimated that the heights of 10-year-old boys are normally distributed with $\mu = 54.5$ inches and $\sigma = 4.5$ inches.

a. What proportion of 10-year-old boys is tall enough to ride the coaster?

b. A smaller coaster has a height requirement of 50 inches to ride. What proportion of 10 year-old-boys is tall enough to ride this coaster?

c. What proportion of 10-year-old boys is tall enough to ride the coaster in part b but not tall enough to ride the coaster in part a?

6.31 **SAT versus ACT** SAT math scores follow a normal distri- TRY bution with an approximate $\mu = 500$ and $\sigma = 100$. Also ACT math scores follow a normal distrubution with an approximate $\mu = 21$ and $\sigma = 4.7$. You are an admissions officer at a university and have room to admit one more student for the upcoming year. Joe scored 600 on the SAT math exam, and Kate scored 25 on the ACT math exam. If you were going to base your decision solely on their performances on the exams, which student should you admit? Explain.

6.32 **Relative height:** Refer to the normal distributions for women's height ($\mu = 65$, $\sigma = 3.5$) and men's height ($\mu = 70$, $\sigma = 4.0$). A man's height of 75 inches and a woman's height of 70 inches are both 5 inches above their means. Which is relatively taller? Explain why.

6.3 Probabilities When Each Observation Has Two Possible Outcomes

We next study the most important probability distribution for discrete random variables. Learning about it helps us answer the questions we asked at the beginning of the chapter—for instance, whether there is strong evidence of discrimination against women in the selection of employees for management training.

The Binomial Distribution: Probabilities for Counts with Binary Data

In many applications, each observation is **binary:** It has one of two possible outcomes. For instance, a person may

- accept, or decline, an offer from a bank for a credit card,
- have, or not have, health insurance,
- vote yes or no in a referendum, such as whether to recall a governor from office.

With a sample, we summarize such variables by counting the *number* or the *proportion* of cases with an outcome of interest. For instance, with a sample of size $n = 5$, let the random variable X denote the number of people who vote yes about some issue in a referendum. The possible values for X are 0, 1, 2, 3, 4, and 5. Under certain conditions, a random variable X that counts the number of observations of a particular type has a probability distribution called the **binomial**.

Consider n cases, called **trials**, in which we observe a binary random variable. This is a *fixed number*, such as $n = 5$ for a sample of five voters. The number X (trials in which the outcome of interest occurs) can take any one of the integer values $0, 1, 2, \ldots, n$. The binomial distribution gives probabilities for these possible values of X when the following three conditions hold:

Conditions for Binomial Distribution

- Each of n trials has two possible outcomes. The outcome of interest is called a success and the other outcome is called a failure.
- Each trial has the same probability of a success. This is denoted by p, so the probability of a failure is denoted by $1 - p$.
- The n trials are independent. That is, the result for one trial does not depend on the results of other trials.

The **binomial random variable** X is the number of successes in the n trials.

Flipping a coin n times, where n is determined in advance, is a prototype for the binomial distribution:

- Each trial is a flip of the coin. There are two possible outcomes for each flip, head or tail. Let's identify (arbitrarily) head as success.
- The probability p of a head equals 0.50 for each flip if head and tail are equally likely.
- The flips are independent, since the result for any specific flip does not depend on the outcomes of previous flips.

The binomial random variable X counts the number of heads (the outcome of interest) in the n flips. With $n = 3$ coin flips, $X =$ number of heads could equal 0, 1, 2, or 3.

Example 11

Binomial probabilities

An ESP Experiment

Picture the Scenario

John Doe claims to possess extrasensory perception (ESP). An experiment is conducted in which a person in one room picks one of the integers 1, 2, 3, 4, 5 at random and concentrates on it for one minute. In another room, John Doe identifies the number he believes was picked. The experiment is done with three trials. After the third trial, the random numbers are compared with John Doe's predictions. Doe got the correct result twice.

Question to Explore

If John Doe does not actually have ESP and is merely guessing the number, what is the probability that he'd make a correct guess on two of the three trials?

Think It Through

Let $X =$ number of correct guesses in $n = 3$ trials. Then $X = 0, 1, 2,$ or 3. Let p denote the probability of a correct guess for a given trial. If Doe is guessing, $p = 0.2$ for Doe's prediction of one of the five possible integers. Then, $1 - p = 0.8$ is the probability of an incorrect prediction on a given trial. Denote the outcome on a given trial by S or F, representing success or failure for whether Doe's guess was correct or not. Table 6.4 shows the eight outcomes in the sample space for this experiment. For instance, FSS represents a correct guess on the second and third trials. It also shows their probabilities by using the multiplication rule for independent events.

Recall

From Section 5.2, for independent events, P(A and B) = P(A)P(B). Thus, P(FSS) = P(F)P(S)P(S) = 0.8 × 0.2 × 0.2.◀

Table 6.4 Sample Space and Probabilities for Three Guesses

The probability of a correct guess is 0.2 on each of the three trials, if John Doe does not have ESP.

Outcome	Probability	Outcome	Probability
SSS	$0.2 \times 0.2 \times 0.2 = (0.2)^3$	SFF	$0.2 \times 0.8 \times 0.8 = (0.2)^1(0.8)^2$
SSF	$0.2 \times 0.2 \times 0.8 = (0.2)^2(0.8)^1$	FSF	$0.8 \times 0.2 \times 0.8 = (0.2)^1(0.8)^2$
SFS	$0.2 \times 0.8 \times 0.2 = (0.2)^2(0.8)^1$	FFS	$0.8 \times 0.8 \times 0.2 = (0.2)^1(0.8)^2$
FSS	$0.8 \times 0.2 \times 0.2 = (0.2)^2(0.8)^1$	FFF	$0.8 \times 0.8 \times 0.8 = (0.8)^3$

The three ways John Doe could make two correct guesses in three trials are SSF, SFS, and FSS. Each of these has probability equal to $(0.2)^2(0.8) = 0.032$. The total probability of two correct guesses is

$$3(0.2)^2(0.8) = 3(0.032) = 0.096.$$

Insight

In terms of the probability $p = 0.2$ of a correct guess on a particular trial, the solution $3(0.2)^2(0.8)$ for $x = 2$ correct in $n = 3$ trials equals $3p^2(1-p)^1 = 3p^x(1-p)^{n-x}$. The multiple of 3 represents the number of ways that two successes can occur in three trials (SSF or SFS or FSS). You can use similar logic to evaluate the probability that $x = 0$, or 1, or 3. Try $x = 1$, for which you should get P(1) = 0.384.

Try Exercise 6.33, part a

The formula for binomial probabilities When the number of trials n is large, it's tedious to write out all the possible outcomes in the sample space. But there's a formula you can use to find binomial probabilities for *any n*.

Probabilities for a Binomial Distribution

Denote the probability of success on a trial by p. For n independent trials, the probability of x successes equals

$$P(x) = \frac{n!}{x!(n-x)!} p^x(1-p)^{n-x}, \; x = 0, 1, 2, \ldots, n.$$

The symbol $n!$ is called ***n* factorial**. It represents $n! = 1 \times 2 \times 3 \times \ldots \times n$, the product of all integers from 1 to n. That is, $1! = 1$, $2! = 1 \times 2 = 2$, $3! = 1 \times 2 \times 3 = 6$, $4! = 1 \times 2 \times 3 \times 4 = 24$, and so forth. Also, 0! is defined to be 1. For given values for p and n, you can find the probabilities of the possible outcomes by substituting values for x into the binomial formula.

Let's use this formula to find the answer for Example 11 about ESP:

- The random variable X represents the number of correct guesses (successes) in $n = 3$ trials of the ESP experiment.
- The probability of a correct guess in a particular trial is $p = 0.2$.
- The probability of exactly two correct guesses is the binomial probability with $n = 3$ trials, $x = 2$ correct guesses, and $p = 0.2$ probability of a correct guess for a given trial,

$$P(2) = \frac{n!}{x!(n-x)!} p^x(1-p)^{n-x} = \frac{3!}{2!1!}(0.2)^2(0.8)^1 = 3(0.04)(0.8) = 0.096.$$

Did You Know?

The term with factorials at the start of the binomial formula is

$$\binom{n}{x} = \frac{n!}{x!(n-x)!},$$

which is also called the **binomial coefficient**. It is the number of outcomes that have x successes in n trials, such as the

$$\binom{n}{x} = \binom{3}{2} = \frac{3!}{2!1!} = 3$$

outcomes (SSF, SFS, and FSS) that have $x = 2$ successes in $x = 3$ trials in Example 11.◀

What's the role of the different terms in this binomial formula?

- The factorial term tells us the number of possible outcomes that have $x = 2$ successes. Here, $[3!/(2!1!)] = (3 \times 2 \times 1)/(2 \times 1)(1) = 3$ tells us there were three possible outcomes with two successful guesses, namely SSF, SFS, and FSS.
- The term $(0.2)^2(0.8)^1$ with the exponents gives the probability for each such sequence. Here, the probability is $(0.2)^2(0.8) = 0.032$ for each of the three sequences having $x = 2$ successful guesses, for a total probability of $3(0.032) = 0.096$.

Try to calculate P(1) by letting $x = 1$ in the binomial formula with $n = 3$ and $p = 0.2$. You should get 0.384. You'll see that there are again three possible sequences, now each with probability $(0.2)^1(0.8)^2 = 0.128$. Table 6.5 summarizes the calculations for all four possible x values. You can also find binomial probabilities using statistical software, such as MINITAB or a calculator with statistical functions.

Table 6.5 The Binomial Distribution for $n = 3, p = 0.20$

When $n = 3$, the binomial random variable X can take any integer value between 0 and 3. The total probability equals 1.0.

x	$P(x) = [n!/(x!(n - x)!)]p^x(1 - p)^{n-x}$
0	$0.512 = [3!/(0!3!)](0.2)^0(0.8)^3$
1	$0.384 = [3!/(1!2!)](0.2)^1(0.8)^2$
2	$0.096 = [3!/(2!1!)](0.2)^2(0.8)^1$
3	$0.008 = [3!/(3!0!)](0.2)^3(0.8)^0$

Example 12

Binomial distribution

Testing for Gender Bias in Promotions

Picture the Scenario

While the binomial distribution is useful for finding probabilities for such trivial pursuits as flipping a coin or checking ESP claims, it's also an important tool in helping us understand more serious issues. For instance, Example 1 introduced a case involving possible discrimination against female employees. A group of women employees has claimed that female employees are less likely than male employees of similar qualifications to be promoted.

Question to Explore

Suppose the large employee pool that can be tapped for management training is half female and half male. In a group recently selected for promotion, none of the 10 individuals chosen were female. What would be the probability of 0 females in 10 selections, if there truly were no gender bias?

Think It Through

Other factors being equal, at each choice the probability of selecting a female equals 0.50 and the probability of selecting a male equals 0.50. Let X denote the number of females selected for promotion in a random sample of 10 employees. Then, the possible values for X are $0, 1, \ldots, 10$, and X has the binomial distribution with $n = 10$ and $p = 0.50$. For each x between 0 and

10, we can find the probability that x of the 10 people selected are female. We use the binomial formula for that x value with $n = 10$ and $p = 0.50$, namely

$$P(x) = \frac{n!}{x!(n-x)!}p^x(1-p)^{n-x} = \frac{10!}{x!(10-x)!}(0.50)^x(0.50)^{10-x},\ x = 0, 1, 2, \ldots, 10.$$

The probability that no females are chosen ($x = 0$) equals

$$P(0) = \frac{10!}{0!10!}(0.50)^0(0.50)^{10} = (0.50)^{10} = 0.001.$$

Any number raised to the power of 0 equals 1. Also, $0! = 1$, and the 10! terms in the numerator and denominator divide out, leaving $P(0) = (0.50)^{10}$. If the employees were chosen randomly, it is very unlikely (one chance in a thousand) that none of the 10 selected for promotion would have been female.

Insight

In summary, because this probability is so small, seeing no women chosen would make us highly skeptical that the choices were random with respect to gender.

Try Exercise 6.43

Binomial Probability Distribution for $n = 10$ and $p = 0.5$

x	P(x)	x	P(x)
0	0.001	6	0.205
1	0.010	7	0.117
2	0.044	8	0.044
3	0.117	9	0.010
4	0.205	10	0.001
5	0.246		

```
binompdf(10,.5,6
)
                .205
```

TI-83+/84 output

In this example, we found the probability that $x = 0$ female employees would be chosen for promotion. To get a more complete understanding of just which outcomes are likely, you can find the probabilities of *all* the possible x values. The table in the margin lists the entire binomial distribution for $n = 10$, $p = 0.50$. In the table, the least likely values for x are 0 and 10. If the employees were randomly selected, it is highly unlikely that 0 females or 10 females would be selected. If your statistical software provides binomial probabilities, see if you can verify that $P(0) = 0.001$ and $P(1) = 0.010$. The margin shows a screen shot of the TI-83+/84 calculator providing the binomial probability of $x = 6$.

The left graph in Figure 6.12 (next page) shows the binomial distribution with $n = 10$ and $p = 0.50$. It has a symmetric appearance around $x = 5$. For instance, $x = 10$ has the same probability as $x = 0$. The binomial distribution is perfectly symmetric only when $p = 0.50$. When $p \neq 0.50$, the binomial distribution has a skewed appearance. The degree of skew increases as p gets closer to 0 or 1. To illustrate, the graph on the right in Figure 6.12 shows the binomial distribution for $n = 10$ when $p = 0.9$. If 90% of the people who might be promoted were female, it would not be especially surprising to observe 10, 9, 8, or even 7 females in the sample, but the probabilities drop sharply for smaller x-values.

Check to See If Binomial Conditions Apply

Before you use the binomial distribution, check that its three conditions apply. These are (1) binary data (success or failure), (2) the same probability of success for each trial (denoted by p), and (3) a fixed number n of independent trials. To judge this, ask yourself whether the observations resemble coin flipping, the

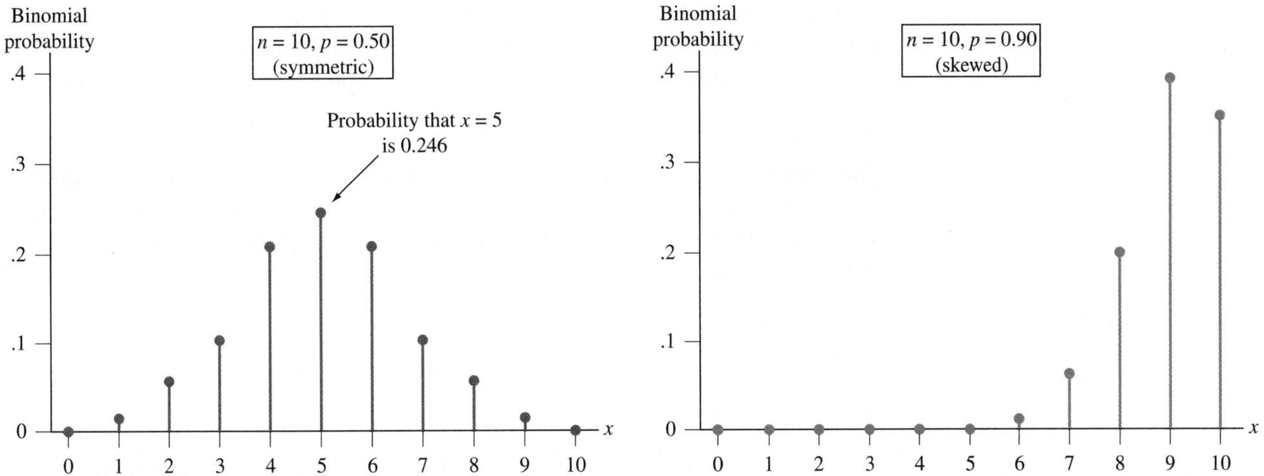

▲ **Figure 6.12 Binomial Distributions when $n = 10$ for $p = 0.5$ and for $p = 0.9$.** The binomial probability distribution is symmetric when $p = 0.5$, but it can be quite skewed for p near 0 or near 1. **Question** How do you think the distribution would look if $p = 0.1$?

simple prototype for the binomial. For instance, it seems plausible to use the binomial distribution for Example 12 on gender bias in selecting employees for promotion. In this instance,

- The data are binary because (female, male) plays the role of (head, tail).
- If employees are randomly selected, the probability p of selecting a female on any given trial is 0.50.
- With random sampling of 10 employees, the outcome for one trial does not depend on the outcome of another trial.

Binomial sampling

Example 13

Gender Bias in Promotions

Picture the Scenario

Consider the gender bias investigation in Example 12. Suppose the population of individuals to choose for promotion contained only four people, two men and two women (instead of the very large pool of employees), and the number chosen was $n = 2$.

Question to Explore

Do the binomial conditions apply for calculating the probability, under random sampling, of selecting 0 women in the two choices for promotion?

Think It Through

For the first person selected, the probability of a woman is 2/4 = 0.50. The usual sampling is "sampling without replacement," in which the first person selected is no longer in the pool for future selections. So, if the first person selected is a woman, the conditional probability the second person selected is male equals 2/3 since the pool of individuals now has one woman and two men. To find the probability of 0 women being selected, the probability of a male being selected first is 2/4. Given that the first person selected was male, the conditional probability that the second person selected is male equals 1/3, since

Recall

As Section 5.3 discussed, if once subjects are selected from a population they are not eligible to be selected again, this is called *sampling without replacement.* Example 10 in Section 5.3 showed the effect on sampling from a small population versus a large population. ◀

the pool of individuals now has one man and two women. So, the outcome of the second selection *depends* on that of the first. The trials are *not* independent, which the binomial requires. In summary, the binomial conditions do not apply.

Insight

This example suggests a caution with applying the binomial to a random sample from a population. For trials to be sufficiently "close" to independent with common probability p of success, the population size must be large relative to the sample size.

Try Exercise 6.46

In Practice Population and Sample Sizes to Use the Binomial

For sampling n separate subjects from a population (that is, sampling without replacement), the exact probability distribution of the number of successes is too complex to discuss in this text, but the binomial distribution approximates it well when n is less than 10% of the population size. In practice, sample sizes are usually small compared to population sizes, and this guideline is satisfied.

The margin shows a guideline about the relative sizes of the sample and population for which the binomial formula works well. For example, suppose your school has 4000 students. Then the binomial formula is adequate as long as the sample size is less than 10% of 4000, which is 400. Why? Because the probability of success for any one observation will be similar regardless of what happens on other observations. Likewise, Example 12 dealt with the selection of 10 employees for promotion when the employee pool for promotion was very large. Again, the sample size was less than 10% of the population size, so using the binomial is valid.

Mean and Standard Deviation of the Binomial Distribution

Example 12 applied the binomial distribution for the number of women selected for promotion when $n = 10$ and $p = 0.50$. If p truly equals 0.50, out of 10 selections, what do you expect for the number of women selected for promotion?

As with any discrete probability distribution, we can use the formula $\mu = \Sigma x P(x)$ to find the mean. However, finding the mean μ and standard deviation σ is actually simpler for the binomial distribution. There are special formulas based on the number of trials n and the probability p of success on each trial.

Binomial Mean and Standard Deviation

The binomial probability distribution for n trials with probability p of success on each trial has mean μ and standard deviation σ given by

$$\mu = np, \quad \sigma = \sqrt{np(1 - p)}.$$

The formula for the mean makes sense. If the probability of success is p for a given trial, then we expect about a proportion p of the n trials to be successes, or about np total. If we sample $n = 10$ people from a population in which half are female, then we expect that about $np = 10(0.50) = 5$ in the sample will be female.

When the number of trials n is large, it can be tedious to calculate binomial probabilities of all the possible outcomes. Often, it's adequate merely to use the mean and standard deviation to describe where most of the probability falls. The binomial distribution has a bell shape when n is large (as explained in a guideline at the end of this section), so in that case, we can use the normal distribution to approximate the binomial distribution and conclude that nearly all the probability falls between $\mu - 3\sigma$ and $\mu + 3\sigma$.

Binomial distributions

Example 14

Checking for Racial Profiling

Picture the Scenario

In 2006, the New York City Police Department (NYPD) confronted approximately 500,000 pedestrians for suspected criminal violations. Of those confronted, 88.9% were non-white.[9] Meanwhile, according to the 2006 American Community Survey conducted by the U.S. Census Bureau, of the more than 8 million individuals living in New York City, 44.6% were white.

Question to Explore

Are the data presented above evidence of racial profiling in police officers' decisions to confront particular individuals?

Think It Through

We'll treat the 500,000 confrontations as $n = 500{,}000$ trials. From the fact that 44.6% of the population was white, we can deduce that the other 55.4% was non-white. Then, if there is no racial profiling, the probability that any given confrontation should involve a non-white suspect is $p = 0.554$ (other things being equal, such as the rate of engaging in criminal activity). Suppose also that successive confrontations are independent. (They would not be, for example, if once an individual were stopped, the police followed that individual and repeatedly stopped her or him.) Under these assumptions, for the 500,000 police confrontations, the number of non-whites confronted has a binomial distribution with $n = 500{,}000$ and $p = 0.554$.

The binomial distribution with $n = 500{,}000$ and $p = 0.554$ has

$$\mu = np = 500{,}000(0.554) = 277{,}000,$$
$$\text{and } \sigma = \sqrt{np(1-p)} = \sqrt{500{,}000(0.554)(0.446)} = 351.$$

Recall

From Section 6.2, when a distribution has a normal distribution, nearly 100% of the observations fall within 3 standard deviations of the mean.◀

Next we see that this binomial distribution is approximated reasonably well by the normal distribution because the number of trials is so large. So, the probability within 3 standard deviations of the mean is close to 1.0. This is the interval between

$$\mu - 3\sigma = 277{,}000 - 3(351) = 275{,}947$$
$$\text{and } \mu + 3\sigma = 277{,}000 + 3(351) = 278{,}053.$$

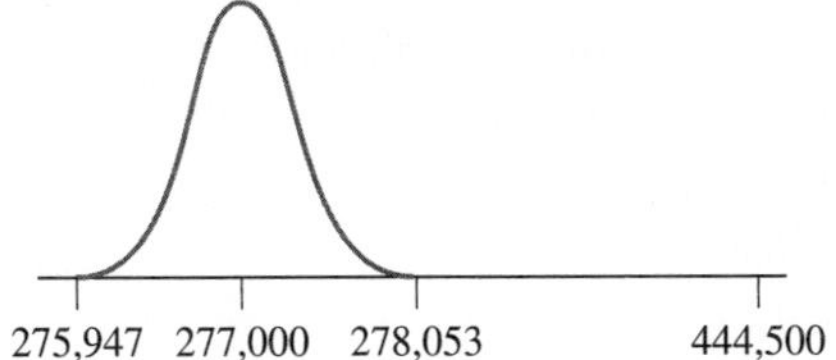

If no racial profiling is taking place, we would not be surprised if between about 275,947 and 278,053 of the 500,000 people stopped were non-white. See the smooth curve approximation for the binomial in the margin. However, 88.9% of all stops, or $500{,}000(0.889) = 444{,}500$ involved non-whites. This suggests that the number of non-whites stopped is much higher than we would expect if the probability of confronting a pedestrian were the same for each resident, regardless of their race.

Insight

By this approximate analysis, we would not expect to see so many non-whites stopped if there were truly a 0.554 chance that each confrontation involved

[9] *Source:* Background data from www.racialprofilinganalysis.neu.edu/background/.

a non-white. If we were to use software to do a more precise analysis by calculating the binomial probabilities of *all* possible values 0, 1, 2, … … 500,000 when $n = 500{,}000$ and $p = 0.544$, we'd find that the probability of getting 444,500 or a larger value out in the right tail of the distribution is 0 to the precision of many decimal places (that is, 0.00000000…).

The controversial subject of racial profiling has received nationwide attention. In April 2007, the Bureau of Justice Statistics released a report[10] based on interviews by the U.S. Census Bureau of nearly 64,000 people, which found that black, white, and Hispanic drivers are about equally likely to be pulled over by police. However, the study found that blacks and Hispanics were much more likely to be searched than whites, and blacks were more than twice as likely as whites to be arrested.

Try Exercise 6.42

In Practice When the Binomial Distribution Is Approximately Normal

The binomial distribution can be approximated well by the normal distribution when n is large enough that the expected number of successes, np, and the expected number of failures, $n(1 - p)$, are both at least 15.

The solution in Example 14 treated the binomial distribution as having approximately a normal distribution. This holds when n is sufficiently large. The margin shows a guideline.[11]

In Example 14, of those stopped, the expected number who were non-white was $np = 500{,}000(0.554) = 277{,}000$. The expected number who were white was $n(1 - p) = 500{,}000(0.446) = 223{,}000$. Both exceed 15, so this binomial distribution has approximately a normal distribution.

[10] *Source:* Data from www.ojp.usdoj.gov/bjs/pub/press/cpp05pr.htm.

[11] A lower bound of 15 is actually a bit higher than needed. The binomial is bell shaped even when both np and $n(1 - p)$ are about 10. We use 15 here because it ties in better with a guideline in coming chapters for using the normal distribution for inference about proportions.

6.3 Practicing the Basics

6.33 ESP Jane Doe claims to possess extrasensory perception (ESP). She says she can guess more often than not the outcome of a flip of a balanced coin in another room. In an experiment, a coin is flipped three times. If she does not actually have ESP, find the probability distribution of the number of her correct guesses.
TRY

a. Do this by constructing a sample space, finding the probability for each point, and using them to construct the probability distribution.

b. Do this using the formula for the binomial distribution.

6.34 More ESP In Example 11 on ESP, John Doe had to predict which of five numbers was chosen in each of three trials. Doe did not actually have ESP. Explain why this experiment satisfies the three conditions for the binomial distribution by answering parts a–c.

a. For the analogy with coin flipping, what plays the role of (head, tail)?

b. Explain why it is sensible to assume the same probability of a correct guess on each trial.

c. Explain why it is sensible to assume independent trials.

6.35 Symmetric binomial Construct a graph similar to that in Figure 6.1 for each of the following binomial distributions:

a. $n = 4$ and $p = 0.50$.

b. $n = 4$ and $p = 0.30$.

c. $n = 4$ and $p = 0.10$.

d. Which if any of the graphs in parts a–c are symmetric? Without actually constructing the graph, would the case $n = 10$ and $p = 0.50$ be symmetric or skewed?

e. Which of the graphs in parts a–c is the most heavily skewed? Without actually constructing the graph, would the case $n = 4$ and $p = 0.01$ exhibit more or less skewness than the graph in part c?

6.36 Number of girls in a family Each newborn baby has a probability of approximately 0.49 of being female and

0.51 of being male. For a family with four children, let X = number of children who are girls.

a. Explain why the three conditions are satisfied for X to have the binomial distribution.

b. Identify n and p for the binomial distribution.

c. Find the probability that the family has two girls and two boys.

6.37 It's Just Lunch The Internet site www.ItsJustLunch.com advertises itself as a dating service for busy professionals that has set up over two million first dates for lunch or drinks after work. An advertisement for this site stated that a survey of their users found that a woman has chance 1 in 8 of a second date if she has not heard from the man within 24 hours of their first date. On Saturday, Shawna had a luncheon date with Jack and a dinner date with Lawrence. By Sunday evening she had not heard from either of them. Based on the information claimed by www.ItsJustLunch.com, construct a table with the probability distribution of X = the number of these men (0, 1, or 2) with whom she has a second date. (*Source:* Background information from www.ItsJustLunch.com.)

6.38 Passing by guessing A quiz in a statistics course has four multiple-choice questions, each with five possible answers. A passing grade is three or more correct answers to the four questions. Allison has not studied for the quiz. She has no idea of the correct answer to any of the questions and decides to guess at random for each.

a. Find the probability she lucks out and answers all four questions correctly.

b. Find the probability that she passes the quiz.

6.39 NBA shooting In the National Basketball Association, the top free throw shooters usually have probability of about 0.90 of making any given free throw.

a. During a game, one such player (Dolph Schayes) shot 10 free throws. Let X = number of free throws made. What must you assume in order for X to have a binomial distribution? (Studies have shown that such assumptions are well satisfied for this sport.)

b. Specify the values of n and p for the binomial distribution of X in part a.

c. Find the probability that he made (i) all 10 free throws and (ii) 9 free throws.

6.40 Season performance Refer to the previous exercise. Over the course of a season, this player shoots 400 free throws.

a. Find the mean and standard deviation of the probability distribution of the number of free throws he makes.

b. By the normal distribution approximation, within what range would you expect the number made to almost certainly fall? Why?

c. Within what range would you expect the *proportion* made to fall?

6.41 Is the die balanced? A balanced die with six sides is rolled 60 times.

a. For the binomial distribution of X = number of 6s, what is n and what is p?

b. Find the mean and the standard deviation of the distribution of X. Interpret.

c. If you observe $x = 0$, would you be skeptical that the die is balanced? Explain why, based on the mean and standard deviation of X.

d. Show that the probability that $x = 0$ is 0.0000177.

6.42 Exit poll TRY An exit poll is taken of 3000 voters in a statewide election. Let X denote the number who voted in favor of a special proposition designed to lower property taxes and raise the sales tax. Suppose that in the population, exactly 50% voted for it.

a. Explain why this scenario would seem to satisfy the three conditions needed to use the binomial distribution. Identify n and p for the binomial.

b. Find the mean and standard deviation of the probability distribution of X.

c. Using the normal distribution approximation, give an interval in which you would expect X almost certainly to fall, if truly $p = 0.50$. (*Hint*: You can follow the reasoning of Example 14 on racial profiling.)

d. Now, suppose that the exit poll had $x = 1706$. What would this suggest to you about the actual value of p?

6.43 Jury duty TRY The juror pool for the upcoming murder trial of a celebrity actor contains the names of 100,000 individuals in the population who may be called for jury duty. The proportion of the available jurors on the population list who are Hispanic is 0.40. A jury of size 12 is selected at random from the population list of available jurors. Let X = the number of Hispanics selected to be jurors for this jury.

a. Is it reasonable to assume that X has a binomial distribution? If so, identify the values of n and p. If not, explain why not.

b. Find the probability that no Hispanic is selected.

c. If no Hispanic is selected out of a sample of size 12, does this cast doubt on whether the sampling was truly random? Explain.

6.44 Poor, poor, Pirates On September 7, 2008, the Pittsburgh Pirates lost their 82nd game of the 2008 season and tied the 1933–1948 Philadelphia Phillies major sport record (baseball, football, basketball, and hockey) for most consecutive losing seasons at 16. One year later on September 7, 2009, they lost their 82nd game of the 2009 season, and the record became theirs alone. The only way things could get much worse for the Pirates was to lose their 82nd game earlier in the season. Sure enough, on August 21, 2010, they lost their 82nd game of the 2010 season, extending their streak to 18 consecutive seasons. A major league baseball season consists of 162 games, so for the Pirates to end their streak, they will eventually need to win at least 81 games in a season.

a. Over the course of the streak, the Pirates have won approximately 42% of their games. For simplicity, assume the number of games they win in a given season follows a binomial distribution with $n = 162$ and $p = 0.42$. What is their expected number of wins in a season?

b. What is the probability that the Pirates will win at least 81 games in a given season? (You may use technology to find the exact binomial probability or use the normal distribution to approximate the probability by finding a z-score for 81 and then evaluating the appropriate area under the normal curve.)

c. Can you think of any factors that might make the binomial distribution an inappropriate model for the number of games won in a season?

6.45 Checking guidelines For Example 12 on the gender distribution of promotions, the population size was more than one thousand, half of whom were female. The sample size was 10.

a. Check whether the guideline was satisfied about the relative sizes of the population and the sample, thus allowing you to use the binomial for the probability distribution for the number of females selected.

b. Check whether the guideline was satisfied for this binomial distribution to be approximated well by a normal distribution.

6.46 Class sample Four of the 20 students (20%) in a class are
TRY fraternity or sorority members. Five students are picked at random. Does $X =$ the number of students in the sample who are fraternity or sorority members have the binomial distribution with $n = 5$ and $p = 0.20$? Explain why or why not.

6.47 Binomial needs fixed *n* For the binomial distribution, the number of trials n is a fixed number. Let X denote the number of girls in a randomly selected family in Canada that has three children. Let Y denote the number of girls in a randomly selected family in Canada (that is, the number of children could be any number). A binomial distribution approximates well the probability distribution for one of X and Y, but not for the other.

a. Explain why.

b. Identify the case for which the binomial applies, and identify n and p.

6.48 Binomial assumptions For the following random variables, explain why at least one condition needed to use the binomial distribution is unlikely to be satisfied.

a. $X =$ number of people in a family of size 4 who go to church on a given Sunday, when any one of them goes 50% of the time in the long run (binomial, $n = 4, p = 0.50$). (*Hint:* Is the independence assumption plausible?)

b. $X =$ number voting for the Democratic candidate out of the 100 votes in the first precinct that reports results, when 60% of the population voted for the Democrat in that state (binomial, $n = 100, p = 0.60$). (*Hint:* Is the probability of voting for the Democratic candidate the same in each precinct?)

c. $X =$ number of females in a random sample of four students from a class of size 20, when half the class is female (binomial, $n = 4, p = 0.50$).

Chapter Review

ANSWERS TO THE CHAPTER FIGURE QUESTIONS

Figure 6.1 The shape of the distribution is skewed to the right.

Figure 6.2 The area under the curve from 0 to 15 should be shaded.

Figure 6.3 The shape of the distribution is approximately symmetric and bell shaped. The center of the distribution is about 65 inches. The variability as described by the standard deviation is about 3.5 inches.

Figure 6.4 We evaluate $70 \pm 3(4)$, which gives an interval for the men's heights from 58 inches to 82 inches.

Figure 6.5 These probabilities are similar to the percentages stated for the empirical rule. The empirical rule states that for an approximately bell-shaped distribution, approximately 68% of the observations fall within 1 standard deviation of the mean, 95% within 2 standard deviations of the mean, and nearly all within 3 standard deviations of the mean.

Figure 6.8 For 3 standard deviations, the probability in one tail is 0.0013. By symmetry, this is also the probability in the other tail. The total tail area is 0.0026, which subtracted from 1 gives an answer of 0.997.

Figure 6.9 The second percentile is the value located on the left side of the curve such that 2% of the distribution falls below it and 98% falls above.

Figure 6.10 The SAT score with $z = -3$ is $500 - 3(100) = 200$. The SAT score with $z = 3$ is $500 + 3(100) = 800$.

Figure 6.11 The values of -3 and 3.

Figure 6.12 The distribution with $p = 0.1$ is skewed to the right.

CHAPTER SUMMARY

- A **random variable** is a numerical measurement of the outcome of a random phenomenon. As with ordinary variables, random variables can be **discrete** (taking separate values) or **continuous** (taking an interval of values).

- A **probability distribution** specifies probabilities for the possible values of a random variable. Probability distributions have summary measures of the center and the variability, such as the mean μ and standard deviation σ. The mean (also called **expected value**) for a discrete random variable is

$$\mu = \Sigma x P(x),$$

where $P(x)$ is the probability of the outcome x and the sum is taken over all possible outcomes.

- The **normal distribution** is the probability distribution of a continuous random variable that has a symmetric bell-shaped graph specified by the parameters mean (μ) and standard deviation (σ). For any z, the probability within z standard deviations of μ is the same for every normal distribution.

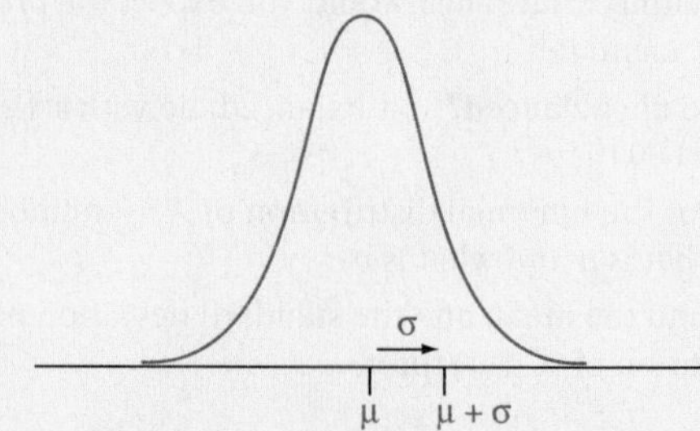

- The **z-score** for an observation x equals

$$z = (x - \mu)/\sigma.$$

It measures the number of standard deviations that x falls from the mean μ. For a normal distribution, the z-scores have the **standard normal distribution**, which has mean = 0 and standard deviation = 1.

- The **binomial distribution** is the probability distribution of the discrete random variable that measures the number of successes X in n independent trials, with probability p of a success on a given trial. It has

$$P(x) = \frac{n!}{x!(n-x)!}p^x(1-p)^{n-x}, x = 0, 1, 2, \ldots, n.$$

The mean is $\mu = np$ and standard deviation is

$$\sigma = \sqrt{np(1-p)}.$$

SUMMARY OF NEW NOTATION IN CHAPTER 6

$P(x)$	Probability that a random variable takes value x
σ	Standard deviation of the probability distribution or population distribution
μ	Mean of the probability distribution or population distribution
$p, 1-p$	Probabilities of the two possible outcomes of a binary variable

CHAPTER PROBLEMS

Practicing the Basics

6.49 Grandparents Let X = the number of living grandparents that a randomly selected adult American has. According to recent General Social Surveys, its probability distribution is approximately P(0) = 0.71, P(1) = 0.15, P(2) = 0.09, P(3) = 0.03, P(4) = 0.02.

a. Does this refer to a discrete or a continuous random variable? Why?

b. Show that the probabilities satisfy the two conditions for a probability distribution.

c. Find the mean of this probability distribution.

6.50 Straight or boxed? Consider a Pick-3 lottery such as the one described in Exercise 6.6. Suppose your birthday is May 14, and like many people, you decide to bet \$1 on your birthday number (i.e., 514) on your birthday. There are two options to play, straight or boxed. If you choose to play straight, you win \$500 if and only if the number chosen is 514. If you choose to play boxed, you win \$80 if the number chosen contains the digits 5, 1, and 4 in any order. Which option would you prefer to play and why?

6.51 NJ Lottery New Jersey has a six-way combination lottery game in which you pick three digits (each can be any of $0, 1, 2, \ldots, 9$) and you win if the digits the lottery picks are the same as yours in any of the six possible orders. For a dollar bet, a winner receives \$45.50. Two possible strategies are (a) to pick three different digits, or (b) to pick the same digit three times. Let X = winnings for a single bet.

a. Find the probability distribution of X and its mean for strategy a.

b. Find the probability distribution of X and its mean for strategy b.

c. Which is the better strategy? Why?

6.52 Are you risk averse? You need to choose between two alternative programs for dealing with the outbreak of a deadly disease. In program 1, 200 people are saved. In program 2, there is a 2/3 chance that no one is saved and a 1/3 chance that 600 people are saved.

a. Find the expected number of lives saved with each program.

b. Now you need to choose between program 3, in which 400 people will die, and program 4, in which there is a 1/3 chance that no one will die and a 2/3 chance that 600 people will die. Find the expected number of deaths with each program.

c. Explain why programs 1 and 3 are similar and why 2 and 4 are similar. (If you had to choose, would you be like most people and be risk averse in part a, choosing program 1, and risk taking in part b, choosing program 4?)

6.53 Flyers' insurance An insurance company sells a policy to airline passengers for \$1. If a flyer dies on a given flight (from a plane crash), the policy gives \$100,000 to the chosen beneficiary. Otherwise, there is no return. Records show that a passenger has about a one in a million chance of dying on any given flight. You buy a policy for your next flight.

a. Specify the probability distribution of the amount of money the beneficiary makes from your policy.

b. Find the mean of the probability distribution in part a. Interpret.

c. Explain why the company is very likely to make money in the long run.

6.54 Normal probabilities For a normal distribution, find the probability that an observation is

a. Within 1.96 standard deviations of the mean.

b. More than 2.33 standard deviations from the mean.

6.55 *z*-scores Find the z-score such that the interval within z standard deviations of the mean contains probability (a) 0.95 and (b) 0.99 for a normal distribution. Sketch the two cases on a single graph.

6.56 *z*-score and tail probability

a. Find the z-score for the number that is less than only 1% of the values of a normal distribution. Sketch a graph to show where this value is.

b. Find the z-scores corresponding to the (i) 90th and (ii) 99th percentiles of a normal distribution.

6.57 Quartiles If z is the positive number such that the interval within z standard deviations of the mean contains 50% of a normal distribution, then

a. Explain why this value of z is about 0.67.

b. Explain why for any normal distribution the first and third quartiles equal $\mu - 0.67\sigma$ and $\mu + 0.67\sigma$.

c. The interquartile range, IQR, relates to σ by $IQR = 2 \times 0.67\sigma$. Explain why.

6.58 Cholesterol The American Heart Association reports that a total cholesterol score of 240 or higher represents high risk of heart disease. A study of postmenopausal women reported a mean of 220 and standard deviation of 40. If the total cholesterol scores have a normal distribution, what proportion of the women fall in the high-risk category? (*Source:* Data from *Clin. Drug. Invest.*, 2000, vol. 20, pp. 207–214. The American Heart Association.)

6.59 Female heights Female heights in North America follow a normal distribution with $\mu = 65$ inches and $\sigma = 3.5$ inches. Find the proportion of females who are

a. under five feet.

b. over six feet.

c. between 60 and 70 inches.

d. Repeat parts a–c for North American males, the heights of whom are normally distributed with $\mu = 70$ inches and $\sigma = 4$ inches.

6.60 Cloning butterflies The wingspans of recently cloned monarch butterflies follow a normal distribution with mean 9 inches and standard deviation 0.75 inches. What proportion of the butterflies has a wingspan

a. less than 8 inches?

b. wider than 10 inches?

c. between 8 and 10 inches?

d. Ten percent of the butterflies have a wingspan wider than how many inches?

6.61 Gestation times For 5459 pregnant women using Aarhus University Hospital in Denmark in a two-year period who reported information on length of gestation until birth, the mean was 281.9 days, with standard deviation 11.4 days. A baby is classified as premature if the gestation time is 258 days or less. (Data from *British Medical Journal*, July 24, 1993, p. 234.) If gestation times are normally distributed, what's the proportion of babies born in that hospital prematurely?

6.62 Water consumption A study of water use in Gainesville, Florida, indicated that in 2006 residential water consumption had a mean of 78 and a standard deviation of 119, in thousands of gallons. (*Source:* Data from Todd Kamhoot, Gainesville Regional Utilities.)

a. If the distribution of water consumption were approximately normal, then what proportion of the residences used less than 100,000 gallons in 2006? (Assume negative infinity to 100 in calculating the proportion.)

b. In fact, the distribution of water use was not actually normal. What shape do you expect the distribution to have? Why?

6.63 Winter energy use Refer to the previous exercise. Water use is much less in Florida in the winter. To make water use values comparable from different seasons, in a given month each home's use is converted to a z-score. For each home with a z-score greater than 1.5 (water use more than 1.5 standard deviations above the mean), their bill contains a note suggesting a reduction in water use, to conserve resources. If the distribution of water use is normal, what proportion of households receives this note?

6.64 Global warming Suppose that weekly use of gasoline for motor vehicle travel by adults in North America has approximately a normal distribution with a mean of 20 gallons and a standard deviation of 6 gallons. Many people who worry about global warming believe that Americans should pay more attention to energy conservation. Assuming that the standard deviation and the normal shape are unchanged, to what level must the mean reduce so that 20 gallons per week is the third quartile rather than the mean?

6.65 Fast-food profits Mac's fast-food restaurant finds that its daily profits have a normal distribution with mean \$140 and standard deviation \$80.

a. Find the probability that the restaurant loses money on a given day (that is, daily profit less than 0).

b. Find the probability that the restaurant makes money for the next seven days in a row. What assumptions must you make for this calculation to be valid? (*Hint:* Use the binomial distribution.)

6.66 Metric height A Dutch researcher reads that male height in the Netherlands has a normal distribution with $\mu = 72.0$ inches and $\sigma = 4.0$ inches. She prefers to convert this to the metric scale (1 inch = 2.54 centimeters). The mean and standard deviation then have the same conversion factor.

a. In centimeters, would you expect the distribution still to be normal? Explain.

b. Find the mean and standard deviation in centimeters. (*Hint:* What does 72.0 inches equal in centimeters?)

c. Find the probability that height exceeds 200 centimeters.

6.67 Manufacturing tennis balls According to the rules of tennis, a tennis ball is supposed to weigh between 56.7 grams (2 ounces) and 58.5 grams (2 1/16 ounces). A machine for manufacturing tennis balls produces balls with a mean of 57.6 grams and a standard deviation of 0.3 grams, when it is operating correctly. Suppose that the distribution of the weights is normal.

a. If the machine is operating properly, find the probability that a ball manufactured with this machine satisfies the rules.

b. After the machine has been used for a year, the process still has a mean of 57.6, but because of wear on certain parts the standard deviation increases to 0.6 grams. Find the probability that a manufactured ball satisfies the rules.

---

[12] *Source:* Background information from abcnews.go.com/Health/change-bride-marriage-uproar-judgment/story?id=12860570&page=1.

6.68 Bride's choice of surname According to a study done by the *Lucy Stone League* and reported by *ABC News*[12] in February 2011, 90% of brides take the surname of their new husband. Ann notes that of her four best friends who recently married, none kept her own name. If they had been a random sample of brides, how likely would this have been to happen?

6.69 Yale babies In a study carried out at the Infant Cognition Center at Yale University, researchers showed 16 infants two videos: one featured a character that could be perceived as helpful, and the other featured a character that could be perceived as hindering. After the infants viewed the videos, the researchers presented the infants with two objects that resembled the figures from the videos and allowed the infants to choose one to play with. The researchers assumed that the infants would not exhibit a preference and would make their choices by randomly choosing one of the objects. Fourteen of the 16 infants chose the helpful object. If the assumption that infants choose objects randomly were true, what is the probability that 14 or more of the infants would have chosen the helpful object? Could this be considered evidence that the infants must actually be exhibiting a preference for the helpful object? (*Hint:* Use the binomial distribution.)

6.70 Weather A weather forecaster states, "The chance of rain is 50% on Saturday and 50% again on Sunday. So there's a 100% chance of rain sometime over the weekend." If whether or not it rains on Saturday is independent of whether or not it rains on Sunday, find the actual probability of rain *at least once* during the weekend as follows:

a. Answer using methods from Chapter 5, such as by listing equally likely sample points or using the formula for the probability of a union of two events.

b. Answer using the binomial distribution.

6.71 Dating success Based on past experience, Julio believes he has a 60% chance of success when he calls a woman and asks for a date.

a. State assumptions needed for the binomial distribution to apply to the number of times he is successful on his next five requests.

b. If he asks the same woman each of the five times, is it sensible to treat these requests as independent trials?

c. Under the binomial assumptions, state n and p and the mean of the distribution.

6.72 Canadian lottery In one Canadian lottery option, you bet on one of the million six-digit numbers between 000000 and 999999. For a \$1 bet, you win \$100,000 if you are correct. In playing n times, let X be the number of times you win.

a. Find the mean of the distribution of X, in terms of n.

b. How large an n is needed for you to expect to win once (that is, $np = 1$)?

c. If you play the number of times n that you determined in part b, show that your expected winnings is \$100,000 but your expected profit is − \$900,000.

6.73 Female driving deaths In a given year, the probability that an adult American female dies in a motor vehicle accident equals 0.0001. (*Source:* Data from *Statistical Abstract of the United States*, 2001.)

a. In a city having 1 million adult American females, state assumptions for a binomial distribution to apply to X = the number of them who die in the next year from motor vehicle accidents. Identify n and p for that distribution.

b. If the binomial applies, find the mean and standard deviation of X.

c. Based on the normal distribution, find the interval of possible outcomes within three standard deviations of the mean that is almost certain to occur.

d. Refer to the assumptions for the analysis in part a. Explain at least one way they may be violated.

6.74 Males drive more poorly? Refer to the previous exercise. The probability of a motor vehicle death for adult American males is 0.0002. Repeat part b and c for a city having 1 million of them, and compare results to those for females.

6.75 Which distribution for sales? A salesperson uses random digit dialing to call people and try to interest them in applying for a charge card for a large department store chain. From past experience, she is successful on 2% of her calls. In a typical working day, she makes 200 calls. Let X be the number of calls on which she is successful.

a. What type of distribution does X have: normal, binomial, discrete probability distribution but not binomial, or continuous probability distribution but not normal?

b. Find the mean and standard deviation of X. Interpret the mean.

c. Find the probability that on a given day she has 0 successful calls.

Concepts and Investigations

6.76 Family size in Gaza The Palestinian Central Bureau of Statistics (www.pcbs.gov.ps) asked mothers of age 20–24 about the ideal number of children. For those living on the Gaza Strip, the probability distribution is approximately P(1) = 0.01, P(2) = 0.10, P(3) = 0.09, P(4) = 0.31, P(5) = 0.19, and P(6 or more) = 0.29. Because the last category is open-ended, it is not possible to calculate the mean exactly. Explain why you can find the *median* of the distribution, and find it. (*Source:* Data from www.pcbs.gov.ps.)

6.77 Longest streak made In basketball, when the probability of making a free throw is 0.50 and successive shots are independent, the probability distribution of the longest streak of shots made has $\mu = 4$ for 25 shots, $\mu = 5$ for 50 shots, $\mu = 6$ for 100 shots, and $\mu = 7$ for 200 shots.

a. How does the mean change for each doubling of the number of shots taken? Interpret.

b. What would you expect for the longest number of consecutive shots made in a sequence of (i) 400 shots and (ii) 3200 shots?

c. For a long sequence of shots, the probability distribution of the longest streak is approximately bell shaped and σ equals approximately 1.9, no matter how long the sequence (Schilling, 1990). Explain why the longest number of consecutive shots made has more than a 95% chance of falling within about 4 of its mean, whether we consider 400 shots, 3200 shots, or 1 million shots.

6.78 Stock market randomness Based on the previous exercise and what you have learned in this and the previous chapter (for example, Exercise 5.12), if you are a serious investor, explain why you should not get too excited if sometime in the next year the stock market goes up for seven days in a row.

6.79 Airline overbooking For the Boston to Chicago route, an airline flies a Boeing 737–800 with 170 seats. Based on past experience, the airline finds that people who purchase a ticket for this flight have 0.80 probability of showing up for the flight. They routinely sell 190 tickets for the flight, claiming it is unlikely that more than 170 people show up to fly.

a. Provide statistical reasoning they could use for this decision.

b. Describe a situation in which the assumptions on which their reasoning is based may not be satisfied.

6.80 Babies in China The sex distribution of new babies is close to 50% each, with the percentage of males usually being just slightly higher. In China in recent years, the percentage of female births seems to have dropped, a combination of policy limiting family size, the desire of most families to have at least one male child, the possibility of determining sex well before birth, and the availability of abortion. Suppose that historically 49% of births in China were female but birth records in a particular town for the past year show 800 females born and 1200 males. Conduct an investigation to determine if the current probability of a female birth in this town is less than 0.49, by using the mean and standard deviation of the probability distribution of what you would observe with 2000 births if it were still 0.49.

6.81 Multiple choice: Guess answers A question has four possible answers, only one of which is correct. You randomly guess the correct response. With 20 such questions, the distribution of the number of incorrect answers

a. is binomial with $n = 20$ and $p = 0.25$.

b. is binomial with $n = 20$ and $p = 0.50$.

c. has mean equal to 10.

d. has probability $(.75)^{20}$ that all 20 guesses are incorrect.

6.82 Multiple choice: Terrorist coincidence? On 9/11/2002, the first anniversary of the terrorist destruction of the World Trade Center in New York City, the winning three-digit New York State Lottery number came up 9-1-1. The probability of this happening was

a. 1/1000.

b. $(1/1000)^2 = 0.000001$.

c. 1 in a billion.

d. 3/10.

6.83 ♦♦ **SAT and ethnic groups** Lake Wobegon Junior College admits students only if they score above 1200 on the sum of their critical reading, mathematics, and writing scores. Applicants from ethnic group A have a mean of 1500 and a standard deviation of 300 on this test, and applicants from ethnic group B have a mean of 1350 and a standard deviation of 200. Both distributions are approximately normal.

a. Find the proportion not admitted for each ethnic group.

b. Both ethnic groups have the same size. Of the students who are not admitted, what proportion is from group B?

c. A state legislator proposes that the college lower the cutoff point for admission to 600, thinking that of the students who are not admitted, the proportion from ethnic group B would decrease. If this policy is implemented, determine the effect on the answer to part b, and comment.

6.84 ♦♦ **College acceptance** The National Center for Educational Statistics reported that in 2009 the ACT college placement and admission examination had a mean of 21.1 and standard deviation of 5.1. (*Source:* Data from nces.ed.gov/programs/digest/d09/tables/dt09_147.asp.)

a. Which probability distribution would you expect to be most appropriate for describing the scores: the normal, or the binomial? Why?

b. A college requires applicants to have an ACT score in the top 20% of all scores. Using the distribution you chose in part a, find the lowest ACT score a student could get to meet this requirement.

c. Of five students picked at random from those taking the ACT, find the probability that *none* score high enough to satisfy the admission standard you found in part b.

6.85 ♦♦ **Standard deviation of a discrete probability distribution** The **variance** of a probability distribution of a random variable is a weighted average of its squared distances from the mean μ. For discrete random variables, it equals

$$\sigma^2 = \Sigma(x - \mu)^2 P(x).$$

Multiply each possible squared deviation $(x - \mu)^2$ by its probability $P(x)$, and then add. The **standard deviation** σ is the positive square root of the variance. Suppose $x = 1$ with probability p and $x = 0$ with probability $(1 - p)$, so that $\mu = p$. Since $(x - \mu)^2$ equals $(0 - p)^2 = p^2$ when $x = 0$ and $(1 - p)^2$ when $x = 1$, derive that $\sigma^2 = p(1 - p)$ and $\sigma = \sqrt{p(1 - p)}$, the special case of the binomial σ with $n = 1$.

6.86 ♦♦ **Binomial probabilities** Justify the $p^x(1 - p)^{n-x}$ part of the binomial formula for the probability $P(x)$ of a particular sequence with x successes, using what you learned in Section 5.2 about probabilities for intersections of independent events.

6.87 ♦♦ **Waiting time for doubles** Most discrete random variables can take on a finite number of values. Let X = the number of rolls of two dice necessary until doubles (the same number on each die) first appears. The possible values for this discrete random variable (called the **geometric**) are 1, 2, 3, 4, 5, 6, 7, and so on, still separate values (and discrete) but now an infinite number of them.

a. Using intersections of independent events, explain why $P(1) = 1/6$, $P(2) = (5/6)(1/6)$, and $P(3) = (5/6)^2(1/6)$.

b. Find $P(4)$, and explain how to find $P(x)$ for an arbitrary positive integer x.

6.88 Geometric mean Exercise 6.44 discussed the ongoing futility streak of the Pittsburgh Pirates. In particular, if the Pirates chance of winning any single game is 0.42, then at the beginning of a new season, the probability of them winning at least 81 games, and hence not adding another losing season to the streak, is about 0.024. Exercise 6.87 mentions a type of random variable called the *geometric random variable.* A simple formula exists for calculating the mean of a geometric variable; if the probability of success on any given trial is p, then the expected number of trials required for the first success to occur is $1/p$. If the first such trial is taken to be the 2011 Major League Baseball season, during what year would we expect the Pirates to break their streak?

Student Activities

6.89  **Best of seven games** In professional baseball, basketball, and hockey in North America, the final two teams in the playoffs play a "best of seven" series of games. The first team to win four games is the champion. Use simulation with the Random Numbers applet on the text CD to approximate the probability distribution of the number of games needed to declare a champion, when (a) the teams are evenly matched and (b) the better team has probability 0.90 of winning any particular game. In each case, conduct 10 simulations. Then combine results with other students in your class and estimate the mean number of games needed in each case. In which case does the series tend to be shorter?

Activity 1

What "Hot Streaks" Should We Expect in Basketball?

In basketball games, TV commentators and media reporters often describe a player as being "hot" if he or she makes several shots in a row. Yet statisticians have shown that for players at the professional level, the frequency and length of streaks of good (or poor) shooting are similar to what we'd expect if the success of the shots were random, with the outcome of a particular shot being independent of previous shots.[13]

Shaquille O'Neal was one of the top players in the National Basketball Association, but he was a poor free-throw shooter. He made about 50% of his free-throw attempts over the course of a season. Let's suppose that he had probability 0.50 of making any particular free throw that he took in a game. Suppose also whether or not he made any particular free throw was independent of his previous ones. He took 20 free throws during the game. Let X denote the longest streak he made in a row during the game. This is a discrete random variable, taking the possible values $x = 0, 1, 2, 3, \ldots, 20$. Here, $x = 0$ if he made none of the 20, $x = 1$ if he never made more than 1 in a row, $x = 2$ if his longest streak was 2 in a row, and so forth.

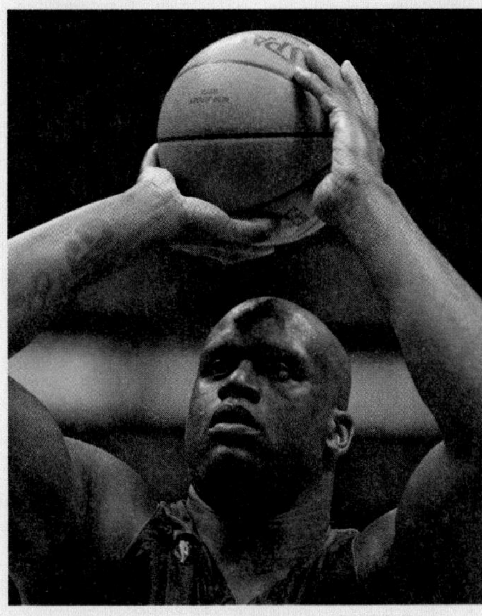

What is the probability distribution of X? This is difficult to find using probability rules, so let's approximate it by simulating 10 games with $n = 20$ free throws in each, using either coin flipping or the Random Numbers applet on the text CD. Representing each of O'Neal's shots by the flip of a coin, we treat a head as making a shot and a tail as missing it. We simulate O'Neal's 20 free throws by flipping a coin 20 times. We did this and got

TTHHHTHTTTHTHHTHTTHT

This corresponds to missing shots 1 and 2, then making shots 3, 4, and 5, missing shot 6, and so forth. The simulated value of $X =$ the longest streak of shots made is the longest sequence of heads in a row. In the sequence just shown, the longest streak made is $x = 3$, corresponding to the heads on flips 3, 4, and 5.

You would do the 20 coin flips 10 separate times to simulate what would happen in 10 games with 20 free throws in each. This would be a bit tedious. Instead, we suggest that you use your own judgment to write down quickly 10 sets of 20 H and T symbols (as shown above for one set) on a sheet of paper to reflect the sort of results you would expect for 10 games with 20 free throws in each. After doing this, find the 10 values of $X =$ longest streak of Hs, using each set of 20 symbols. Do you think your instructor would be able to look at your 200 Hs and Ts and figure out that you did not actually flip the coin?

A more valid way to do the simulation uses software, such as the Random Numbers applet on the text CD. For each digit, we could let 0–4 represent making a shot (H) and 5–9 as missing it (T). Use this to simulate 10 games with $n = 20$ free

[13]For instance, see articles by A. Tversky and T. Gilovich, *Chance*, vol. 2 (1989), pp. 16–21 and 31–34.

throws in each. When we did this, we got the following results for the first three games:

Coin flips	x
HTHHTHTTTHTHHHHTTHTT	4
HTTTHTHHTHHHTTHTHHTH	3
TTHTHHHTHTTHTTTTHTHH	3

In practice, to get accurate results you have to simulate a *huge* number of games (at least a thousand), for each set of 20 free throws observing the longest streak of successes in a row. Your results for the probability distribution would then approximate[14] those shown in the table.

Probability Distribution of X = Longest Streak of Successful Free Throws

The distribution refers to 20 free throws with a 0.50 chance of success for each. All potential x values higher than 9 had a probability of 0.00 to two decimal places and a total probability of only 0.006.

x	$P(x)$	x	$P(x)$
0	0.00	5	0.13
1	0.02	6	0.06
2	0.20	7	0.03
3	0.31	8	0.01
4	0.23	9	0.01

[14]*Source:* Background material from M. F. Schilling, "The longest run of heads," *The College Mathematics Journal,* vol. 21, 1990, pp. 196–207.

The probability that O'Neal never made more than four free throws in a row equals $P(0) + P(1) + P(2) + P(3) + P(4) = 0.76$. This would usually be the case. The mean of the probability distribution is $\mu = 3.7$.

The longest streak of successful shots tends to be longer however, with a larger number of total shots. With 200 shots, the distribution of the longest streak has a mean of $\mu = 7$. Although it would have been a bit unusual for O'Neal to make his next seven free throws in a row; it would not have been at all unusual if he made seven in a row sometime in his next 200 free throws. Making seven in a row was then not really a hot streak, but merely what we expect by random variation. Sports announcers often get excited by streaks that they think represents a "hot hand" but merely represent random variation.

With 200 shots, the probability is only 0.03 that the longest streak equals four or less. Look at the 10 sets of 20 H and T symbols that you wrote on a sheet of paper to reflect the results you expected for 10 games of 20 free throws each. Find the longest streak of Hs out of the string of 200 symbols. Was your longest streak four or less? If so, your instructor could predict that you faked the results rather than used random numbers, because there's only a 3% chance of this if they were truly generated randomly. Most students underestimate the likelihood of a relatively long streak.

This concept relates to the discussion in Section 5.4 about coincidences. By itself, an event may seem unusual. But when you think of all the possible coincidences and all the possible times they could happen, it is probably not so unusual.

Sampling Distributions

7

Example 1

Predicting Election Results Using Exit Polls

Picture the Scenario

An exit poll is an opinion poll in which voters are randomly sampled after leaving the voting booth. Using exit polls, polling organizations predict winners after learning how a small number of people voted, often only a few thousand out of possibly millions of voters. What amazes many people is that these predictions almost always turn out to be correct.

In California in November 2010, the gubernatorial race pitted the Republican candidate Meg Whitman against the Democratic candidate, Jerry Brown. The exit poll on which TV networks relied for their projections found that, after sampling 3889 voters, 53.1% said they voted for Brown, 42.4% for Whitman, and 4.5% for other/no answer (www.cnn.com/ELECTION/2010). At the time of the exit poll, the percentage of the entire voting population (nearly 9.5 million people) that voted for Brown was unknown. In determining if they could predict a winner, the TV networks had to decide whether or not the exit polls gave enough evidence to predict that the population percentage voting in favor of Brown was enough to win the election.

Questions to Explore

- How close can we expect a sample percentage to be to the population percentage? For instance, if 53.1% of 3889 sampled voters supported Brown, how close to 53.1% is the percentage of the entire population of 9.5 million voters who voted for him?
- How does the sample size influence our analyses? For instance, could we sample 100 voters instead of 3889 voters and make an accurate inference about the population percentage voting for Brown. On the other hand, is 3889 enough voters or do we need tens of thousands of voters in our sample?

Thinking Ahead

In this chapter, we'll apply the probability tools of the previous two chapters to analyze how likely it is that sample results will be close to population values. We'll see why the results for 3889 voters allow us to make a reasonable prediction about the outcome for the entire population of nearly 9.5 million voters. This prediction will be an example of the use of inferential statistics.

Recall

From Section 1.2, a **statistic** is a numerical summary of sample data, such as the proportion in an exit poll who voted for Brown. A **parameter** is a numerical summary of a population, such as the proportion of all California voters who voted for Brown. ◄

We learned about probability in Chapter 5 and about probability distributions such as the normal distribution in Chapter 6. We'll next study how probability, and in particular the normal probability distribution, provides the basis for making statistical inferences. As Chapter 1 explained, inferential methods use statistics from sample data to make decisions and predictions about a population. The population summaries are called parameters. Inferential methods are the main focus of the rest of this book.

7.1 How Sample Proportions Vary Around the Population Proportion

In Chapter 6, we used probability distributions with known parameter values to find probabilities of possible outcomes of a random variable. In practice, we seldom know the values of parameters. They are estimated using sample data from surveys, observational studies, or experiments. However, elections provide a context in which we eventually know the population parameter (after election day). Before election day, candidates are interested in gauging where they stand with voters and so they rely on surveys (polls) to help in predicting whether they will receive the necessary percentage to win. On election day, TV networks rely on exit polls to assist in making early evening predictions before all the votes are counted.

Let's consider the California gubernatorial election mentioned in Example 1. Before all the votes were counted, the proportion of the population of voters who voted for Jerry Brown was an unknown parameter. An exit poll of 3889 voters reported that the sample proportion who voted for him was 0.531. This statistic *estimates* the unknown parameter in this scenario, the population proportion that voted for Jerry Brown. How do we know if the sample proportion from the California exit poll is a good estimate, falling close to the population proportion? The total number of voters was over nine million, and the poll sampled a minuscule portion of them. This section introduces a type of probability distribution called the **sampling distribution** that helps us determine how close to the population parameter a sample statistic is likely to fall.

A Sampling Distribution Shows How Sample Statistics Such as a Sample Proportion Vary

Recall

Section 6.1 introduced the term **random variable.** An uppercase letter, such as X, is used to refer to the random variable. A lowercase letter, such as x, refers to particular values of the random variable. ◀

For an exit poll of randomly selected voters, a person's vote is a categorical random variable because the outcome (the vote) varies from voter to voter and falls into a category (the candidate). Let $X =$ vote outcome, with $x = 1$ for Jerry Brown and $x = 0$ for all other responses (Meg Whitman, other candidates, no response). In this case, we say that the outcome is **binary.** That is, each voter has one of two options: vote for Brown or do not vote for Brown. The sample proportion is a numerical summary of the binary outcome and is a statistic. It can also be regarded as a random variable because a sample proportion varies from sample to sample. For example, before the exit poll sample was selected, the value of the sample proportion who voted for Brown was unknown. It was a random variable. The proportion of surveyed voters who voted for him in one exit poll probably differed from the proportion for another sample of voters in another exit poll.

For the exit poll in Example 1, the possible values of the random variable X (0 and 1) and how often these values occurred (0.469 and 0.531) give the **data distribution** for this one sample. For each random sample of voters taken, a different data distribution will result.

For all the voters in this election, final results showed that Brown officially received 0.538 proportion of the vote, Whitman received 0.409 proportion, and others 0.053 proportion. These are the population proportions. So, for the population, the possible values of the random variable X (0 and 1) and how often these values occurred (0.462 and 0.538) give the **population distribution.** Note that $x = 0$ includes Whitman, all other candidates, and no response. We graphically represent the population distribution and the specific data distribution from this exit poll in as shown in Figure 7.1.

We have considered one exit poll that gave a sample proportion for Brown that was similar to the actual population proportion. Would we see sample

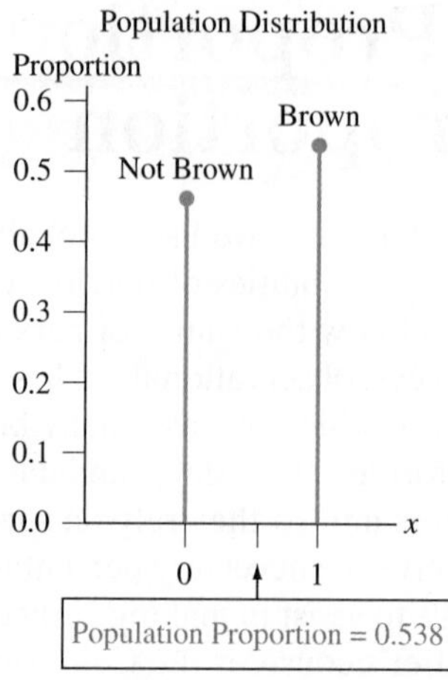

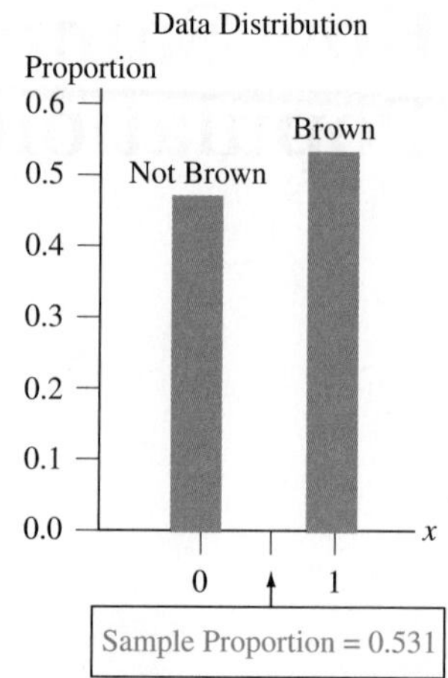

▲ **Figure 7.1** The Population (9.5 Million Voters) and Data ($n = 3889$) Distributions of Candidate Preference (0 = Not Brown, 1= Brown). **Question** Why do these look so similar?

proportions for Brown in different polls similar to the population proportion, such as maybe 0.54 in one poll, 0.57 in another, 0.51 in a third? Or might they tend to be quite different, such as 0.40 in one, 0.59 in another, and 0.65 in a third? To answer, we need to learn about a probability distribution that provides probabilities for the possible values of the sample proportion, represented by a random variable. A probability distribution for a statistic such as a sample proportion is called a **sampling distribution.**

In Words

Individual polls ask different people and therefore use distinct samples. It makes sense then that specific samples have their own sample proportion values. The **sampling distribution** shows all possible values of the sample proportion and how often these sample proportions are expected to occur in random sampling.

Sampling Distribution

The **sampling distribution** of a statistic is the probability distribution that specifies probabilities for the possible values the statistic can take.

For sampling 3889 voters in an exit poll, imagine *all* the distinct samples of 3889 voters you could possibly get. Each such sample has a value for the sample proportion voting for Brown. Suppose you could look at each possible sample, find the sample proportion for each one, and then construct the frequency distribution of those sample proportion values (graphically, construct a histogram since sample proportions are numerical values). This would be the sampling distribution of the sample proportion.

A sampling distribution is merely a type of probability distribution. Rather than giving probabilities for an observation for an individual subject (as in a population or data distribution), it gives probabilities for the value of a statistic for a sample of subjects. The next activity shows how to use simulation to figure out what a sampling distribution looks like.

Activity 1

Simulating a Sampling Distribution for a Sample Proportion

How much might a sample proportion vary from sample to sample? How would you describe the shape, center, and variability of the possible sample proportion values?

To answer these questions, we can simulate random samples of a given size. In this activity, we'll simulate samples, first using random numbers and then using the Sampling Distributions applet on the text CD. We'll carry out this simulation using the actual population proportion (rounded to two decimal places), $p = 0.54$, that voted for Jerry Brown in the 2010 California gubernatorial election.

In Practice Using Technology to Generate Random Numbers

Two-digit random numbers between 00 and 99 can be generated using calculators, for example with the randInt function on the TI-83+/84.

Simulating Using Random Numbers

We can simulate selecting an individual voter at random from a population by picking a random number. To simulate random sampling from a population in which exactly 54% of all voters voted for Brown, we identify the 54 two-digit numbers between 00 and 53 as voting for Brown and the 46 two-digit numbers between 54 and 99 as voting for another candidate. That is,

00 01 02 03 04 . . . 53	54 . . . 99
Vote for Brown	**Vote for another candidate**

Then Brown has a 54% chance of selection on each choice of a two-digit number.

Let's simulate taking a random sample of size $n = 6$. We will use a small sample size since sampling by hand is time-intensive. The first two digits of the first column of the random number table in Chapter 4 (Table 4.1, part of the first column of which is shown in the margin) provide the random numbers 10, 22, 24, 42, 37, 77. The first number, 10, falls between 00 and 53, so in this simulation, the first person sampled voted for Brown. Of the first six people selected, five voted for Brown (they have numbers between 00 and 53), for a sample proportion of 5/6.

Random Numbers
10480
22368
24130
42167
37570
77921

Let's take another random sample of size 6. Again using the last two digits of the random numbers shown in the margin, we get the random numbers 80, 68, 30, 67, 70, 21. Of these six people, two—numbers 30 and 21—voted for Brown, a sample proportion of 2/6. We could do this many more times to simulate how much sample proportions for sample of size $n = 6$ tend to vary from sample to sample. We already suspect that with a small sample size, there will be much variability in the sample proportions from one sample to the next. If possible, combine your simulation results with the results of classmates to better gauge the amount of variability occurring from sample to sample with a small sample size.

Typically exit polls interview more than six people. In the second part of the activity, we'll simulate polls closer to the sample size used in the exit poll of the California governor's race. Since the actual sample size was 3889, we will round to a sample size of $n = 4000$ people. Selecting 4000 two-digit random numbers simulates the votes of 4000 randomly sampled voters of the 9.5 million voters in the population.

Rather then laboriously using random numbers, it is easier to use the Sampling Distributions applet on the text CD to randomly generate a sample of voters. Access that applet now to follow our explanation here.

Simulating Using the Sampling Distributions Applet

In the applet,

- Select Binary for the parent population (since the outcome recorded for each voter is binary—the vote was for Brown or for someone else). Then select Binary Custom from the menu.
- Set the population proportion as $p = 0.54$.
- Select $n = 4000$ for the sample size.
- Select N = 1 to take one sample of this size.
- Click the *Sample* button once.

Figure 7.2 shows what part of the applet screen looks after you make these selections and take one sample. Note that your screen may look somewhat different because you will have a different random sample than shown here.

When we conducted this simulation, we had 2209 voting for Brown and 1791 voting for an alternative candidate. (*Note*: This is the outcome of a *binomial* random variable with $n = 4000$ and $p = 0.54$.) The sample proportion voting for Brown was 0.5523 This particular estimate seems pretty good, since the estimate of 0.5523 is near the population proportion of 0.54. Were we simply lucky? We repeated the process and used this applet to generate another sample with $n = 4000$ and $p = 0.54$. The second time the sample proportion voting for Brown was 0.5345, different but also quite good, near 0.54.

Now use the applet to take a random sample of size 4000. Did you get a sample proportion close to 0.54? Perform this simulation of a random sample of size 4000

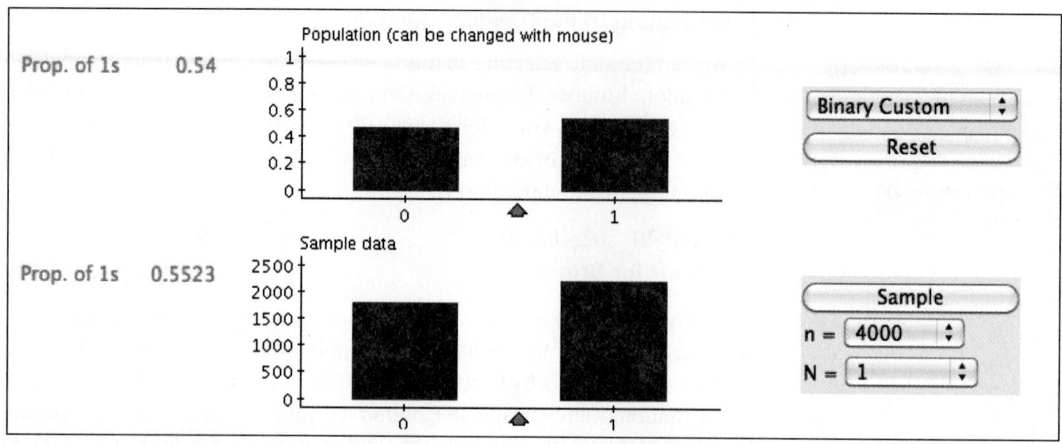

▲ **Figure 7.2 The Result of Taking One Sample of Size $n = 4000$ Using the Sampling Distributions Applet.** The population proportion has been set at $p = 0.54$ in the first menu. The sample size has been set at $n = 4000$ in the second menu. The symbol N refers to how many samples of size 4000 are taken, in this case only one. The first plot shows the population distribution. The second plot shows the data distribution for the one sample, which consist of 1791 zero values and 2209 one values.

Caution

When using the applet, N is the symbol representing the number of simulations being performed. The symbol n represents the sample size for each sample being simulated. ◀

ten times, each time observing (from the graphs) the counts and the corresponding sample proportion of yes votes. Record your 10 sample proportions.

We next ran the applet by setting N = 10,000, to simulate the process of picking 4000 people 10,000 separate times, so we could search for a pattern in the results. Figure 7.3 shows a histogram of the 10,000 outcomes for the sample proportion. The simulated sample proportions resulted in a bell-shape distribution around the true population value of $p = 0.54$. Nearly all the sample proportions fell between 0.515 and 0.565, that is, within 0.025 of the true population value of 0.54. Did the 10 sample proportion values in your simulation all fall between 0.515 and 0.565?

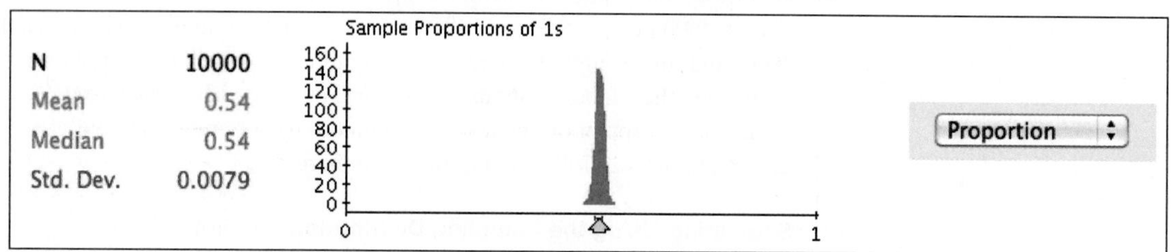

▲ **Figure 7.3 Results of Simulating the Sample Proportion Voting for Jerry Brown in the 2010 California Gubernatorial Election.** 10,000 random samples of 4000 subjects each were simulated from a population in which $p = 0.54$. You can do this simulation with the Sampling Distributions applet by setting Binary: $p = 0.54$. for the population graph and $n = 4000$ and N = 10,000 on the Sample Data menu. **Question** What does it mean that the sample proportion values for this distribution fall between about 0.515 and 0.565?

Return to the Sampling Distributions applet and perform 10,000 simulations yourself by clicking on the Reset button and setting $p = 0.54$, $n = 4000$, and N = 10,000 in the applet menu. Then click on *Sample.* (Be patient as it will take a minute or so to generate a response.) Compare the simulated sampling distribution of the 10,000 sample proportions in the applet graph that you ran to the simulated sampling distribution in Figure 7.3. Are the results similar, showing a bell shape centered at 0.54?

In summary, you are simulating how much the sample proportion can vary from sample to sample of size 4000, when the population proportion equals 0.54.

The graph of the population distribution in Figure 7.2 from using the applet should be identical to the graph of the population distribution in Figure 7.1. The graph of the data distribution from using the applet is not identical to the data distribution in Figure 7.1, but it is similar. The two data distribution graphs vary because they represent two different random samples from the population. The graph in Figure 7.3 represents a simulated sampling distribution; that is, the 10,000 sample proportions from the 10,000 random samples simulated from the population distribution.

Try Exercises 7.1 and 7.11

SUMMARY: Population Distribution, Data Distribution, Sampling Distribution

- **Population distribution:** This is the distribution from which we take the sample. Values of its parameters, such as the population proportion *p* for a categorical variable, are fixed but usually unknown; that is, the specific population parameter does not vary. The population parameter is what we'd like to learn about and to make a prediction about.
- **Data distribution:** This is the distribution of the sample data and is the distribution we actually see in practice. It's described by sample statistics such as a sample proportion. Since the statistic from one random sample to another random sample will vary, data distributions will also differ from one sample to another. With random sampling, the larger the sample size *n*, the more closely the data distribution resembles the population distribution.
- **Sampling distribution:** This is the distribution of a sample statistic such as a sample proportion. With random sampling, the sampling distribution provides probabilities (likelihood of occurrence) for all the possible values of the statistic. The sampling distribution provides the key for telling us how close a sample statistic, such as the sample proportion, falls to the corresponding unknown parameter such as the actual population proportion.

A Sampling Distribution Shows How a Statistic Varies with Many Similar Studies

Sampling distributions describe the variability that occurs from sample to sample (or study to study) using statistics to estimate population parameters. If different polling organizations each take a sample and estimate the population proportion that voted a certain way, they will get varying estimates because their samples have different people. Figure 7.3 portrayed the variability for different samples of size $n = 4000$.

Sampling distributions help us to predict how close a statistic falls to the parameter it estimates. For instance, for a random sample of size 4000, the simulation with the applet showed that the probability is high that a sample proportion falls within 0.025 of the population proportion. In practice, we won't have to perform simulations to figure out the sampling distribution of a statistic. We'll learn in the rest of this section and in the next section about results that tell us the expected shape of the sampling distribution and its mean and standard deviation. These results will also show us how much the precision of estimation improves with larger samples.

In Practice We Generally Use Only One Sample From a Population

Typically we would observe only *one* sample of the given size *n*, not many. But we'll learn about results that tell us how much a statistic *would* vary from sample to sample if we had many samples each of size *n*.

Every sample statistic is a numerical variable and has a sampling distribution. Besides the sample proportion, there is a sampling distribution of a sample mean, a sampling distribution of a sample median, a sampling distribution of a sample standard deviation, and so forth. We'll focus in this section on the sample proportion and in the next section on the sample mean.

Describing the Sampling Distribution of a Sample Proportion

Because a statistic is a numerical random variable, we can describe its sampling distribution by focusing on the key features of shape, center, and variability. We typically use the mean to describe center and the standard deviation to describe variability. For the sampling distribution of a sample proportion, the mean and standard deviation depend on the sample size *n* and the population proportion *p*.

Mean and Standard Deviation of the Sampling Distribution of a Proportion

For a random sample of size *n* from a population with proportion *p* of outcomes in a particular category, the sampling distribution of the sample *proportion* in that category has

$$\text{mean} = p, \text{ standard deviation} = \sqrt{\frac{p(1-p)}{n}}.$$

Recall

You can regard the *n* observations as *n* trials for a binomial distribution. The population proportion *p* is then the probability of success on any given trial. ◀

Mean and standard deviation

Example 2

Exit Poll of California Voters Revisited

Picture the Scenario

Example 1 discussed an exit poll of 3889 voters for the 2010 California gubernatorial election.

Question to Explore

Election results showed that 53.8% of the population of all voters voted for Brown. What was the mean and standard deviation of the sampling distribution of the sample proportion who voted for him? Interpret these two measures.

Think It Through

For the sample of 3889 voters, the sample proportion who voted for Brown could be any of the following values: 0, 1/3889, 2/3889, 3/3889..., 3888/3889, 1. But some of these values are more likely to occur than others, and the sampling distribution describes the frequency of these values. Now, if we know the population proportion, then we can describe the sampling distribution of the sample proportion more specifically by finding its mean and standard deviation. In this case, the population proportion is $p = 0.538$ so the sampling distribution of the sample proportion has mean $= p = 0.538$ and standard deviation =

$$\sqrt{\frac{(0.538)(0.462)}{3889}} = 0.008 \text{ or approximately } 0.01.$$

There is variability around the mean, and we would probably not observe exactly a sample proportion of 0.538 voting for Brown. For example, we might observe a sample proportion of 0.53 or 0.55, both values falling approximately a standard deviation from the mean. In many exit polls of 3889 voters each, the sample proportion voting for Brown would vary from poll to poll, and the standard deviation of those sample proportion values would be approximately 0.01.

Insight

Why would we care about finding a sampling distribution? Because it shows us how close to the population proportion we're really interested in a sample proportion is likely to fall. Here, the standard deviation of the sampling distribution is very small (0.008). This small value tells us that with $n = 3889$, the sample proportion will probably fall quite close to the population proportion.

Try Exercises 7.3 and 7.5

Now we know the mean and standard deviation of the sampling distribution and can describe its expected center (mean) and variability from that mean. What can we say about its shape? Our simulation in Activity 1 suggested that the possible sample proportions pile up in a bell shape around the population proportion. We observed in Section 6.3 that for a large sample size n, the binomial distribution for the number of n observations that fall in a particular category has approximately a normal distribution (See Recall in the margin). *When this happens, the sampling distribution of the sample proportion also has approximately a normal distribution.*

Recall

We learned in Section 6.3 that the binomial distribution is approximately normal when the expected numbers of successes and failures, np and $n(1 - p)$, are both at least 15. Here $np = 3889(0.538) = 2092$ and $n(1 - p) = 3889(0.462) = 1797$ are both much larger than 15. ◄

Example 3

Using the sample proportion to predict

Predicting the Election Outcome

Picture the Scenario

Let's now conduct an analysis that uses the actual exit poll of 3889 voters for the 2010 California gubernatorial election. In that exit poll, 53.1% of the 3889 voters sampled said they voted for Jerry Brown.

Questions to Explore

a. Given that the actual population proportion supporting Brown was 0.538, what are the values of the sample proportion we would expect to observe from random sampling?

b. Based on the results of the exit poll, would you have been willing to predict Brown as the winner on election night while the votes were still being counted?

Think It Through

a. We use the fact, as noted previously, that the sampling distribution of the sample proportion is approximately a normal distribution. We found in Example 2 that when $p = 0.538$, this distribution has a mean of $p = 0.538$ and standard deviation $= 0.008$.

Figure 7.4 shows the sampling distribution. Nearly the entire distribution falls within 3 standard deviations of the mean, or within about $3(0.008) = 0.024$ of the mean of 0.538.

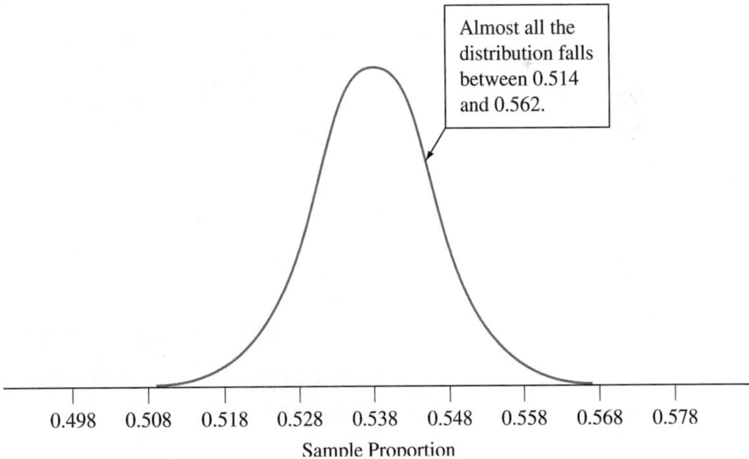

▲ **Figure 7.4 Sampling Distribution of Sample Proportion Voting for Brown in 2010 California Gubernatorial Election.** This graph shows where we expect sample proportions to fall for random samples of size $n = 3889$, when the population proportion $p = 0.538$ voted for Brown. Nearly all the distribution falls between 0.514 and 0.562. The observed sample proportion of 0.531 is within this expected range of values for the sample proportion given the actual population proportion of 0.538. We also observe that the expected range of values for the sample proportions gives sample proportions all greater than 0.50, a majority of the votes. **Question** What would the sampling distribution look like if instead $p = 0.60$? Within what range would the sample proportion then be likely to fall?

So, given $p = 0.538$, it is likely that the sample proportion from a random sample taken from this population will fall within 0.024

of $p = 0.538$, that is, between 0.514 and 0.562. We observe that the sample proportion, $\hat{p} = 0.531$ from the randomly selected exit poll resulted in a sample proportion that fell in this range.

b. On election night, the polling agency does not know the actual population proportion. However, we know that the expected variability in the sampling distribution of the sample proportion is given by $\sqrt{p(1 - p)/n}$. Our best estimate of the population proportion on election day is the sample proportion from the exit poll. Therefore, we could estimate the expected standard deviation in the sampling distribution by substituting $\hat{p}$ for p and evaluating $\sqrt{(0.531)(0.469)/3889}$, resulting in 0.008. We know that in a bell-shaped distribution, we expect to find nearly 100% of our values within 3 standard deviations of the mean (in this case, the population proportion). If we take the sample proportion 0.531 from the exit poll, then add and subtract 3 standard deviations, we can find a range of plausible values for the actual population proportion as $0.531 \pm 3(0.008)$, which gives 0.507 to 0.557. We observe that all the plausible values estimated for the population proportion of voters who will vote for Brown are above the value of 0.50 and give Brown a majority over any other candidate. Therefore, we would expect the polling agency to tell the TV network it can feel confident in predicting Brown as the winner.

Insight

In this election, when all 9.5 million votes were tallied, 53.8% voted for Brown. The exit poll prediction that he would win was correct. Jerry Brown was elected as California governor.

If the sample proportion favoring Brown had been closer to 0.50, we would have been unwilling to make a prediction. For instance, a sample proportion of 0.51 for a sample of size 3889 would not have been unlikely if $p = 0.50$ or slightly less than 0.50. Also, if the polling agency used a smaller sample size, let's say $n = 800$ instead of $n = 3889$, then a sample proportion of 0.531 would not have been unlikely if $p = 0.50$ or slightly less than 0.50. *A smaller sample size creates more variability in the sampling distribution and therefore creates less precision in making a prediction about the actual population proportion.*

Try Exercise 7.6

SUMMARY: Sampling Distribution of a Sample Proportion

For a random sample of size n from a population with proportion p, the sampling distribution of the sample proportion has

$$\text{mean} = p, \text{ standard deviation} = \sqrt{\frac{p(1-p)}{n}}.$$

If n is sufficiently large so that the expected numbers of outcomes of the two types, np in the category of interest and $n(1 - p)$ not in that category, are both at least 15, then the sampling distribution of a sample proportion is approximately normal.

When we simulated a sampling distribution in Activity 1 for a sample size of 4000, we found that almost certainly the sample proportion falls within 0.024 of the population proportion $p = 0.538$. We now can see why this happens. We know that the standard deviation of this simulated sampling distribution is

$$\text{Standard deviation} = \sqrt{p(1 - p)/n} = \sqrt{(0.538)(0.462)/3889} = 0.008.$$

In Practice **Exit Polls Use a Multistage Type of Random Sampling**

Exit polls use a multistage type of random sampling, not simple random sampling. The standard deviation formulas in this chapter technically apply only to simple random samples. However, they provide good approximations for the standard deviations with many multistage samples and are often used for such sampling schemes.

Because the simulated sampling distribution is approximately normal, the probability is very close to 1.0 that the sample proportions would fall within 3 standard deviations of 0.538, that is, within 3(0.008) = 0.024.

In practice, we seldom know the population parameter. An election is an exception in that we eventually know the population parameter. This allows the luxury of being able to compare the actual population proportion, for example, to the estimated proportions from surveys before the votes are counted. The previous examples showed through simulation how we can expect the sampling distribution of sample proportions to behave in the long run with repeated random samples. In particular, a sampling distribution describes the long-run variability of a statistic (such as a sample proportion) in comparison to the parameter of interest. In practice, typically only one sample is taken from the population, not repeated samples. However, now knowing the expected behavior of the sampling distribution allows us to use only one sample to make a prediction about the population parameter. We will explore how to utilize the sampling distribution for these predictions in Chapters 8 and 9. It is also important to remember that this expected behavior for the sampling distribution is only appropriate for a randomly selected sample.

7.1 Practicing the Basics

7.1 Simulating the exit poll Simulate an exit poll of 100 voters, using the Sampling Distributions applet on the text CD, assuming that the population proportion is 0.53. Refer to Activity 1 for guidance on using the applet.

a. What sample proportion did you get? Why do you not expect to get exactly 0.53?

b. Simulate this exit poll 10,000 times (set N = 10,000 on the applet menu). Keep the sample size at $n = 100$ and $p = 0.53$. Describe the graph of the 10,000 sample proportion values.

c. Use a formula from this section to predict the value of the standard deviation of the sample proportions that you generated in part b.

d. Now change the population proportion to 0.7, keeping the sample size $n = 100$. Simulate the exit poll 10,000 times. How would you say the results differ from those in part b?

7.2 Simulate condo solicitations A company that is selling condos in Florida plans to send out an advertisement for the condos to 4000 potential customers, in which they promise a free weekend at a resort on the Florida coast in exchange for agreeing to attend a four-hour sales presentation. The company would like to know how many people will accept this invitation. Its best guess is that there is a 10% chance that any particular customer will accept the offer. The company decides to simulate how small a proportion could actually accept the offer, if this is the case. Simulate this scenario for the company, using the Sampling Distributions applet on the text CD, assuming that the population proportion is 0.10. Refer to Activity 1 for guidance on using the applet.

a. Perform N = 1 simulation for a sample of size 4000. What sample proportion did you get? Why do you not expect to get exactly 0.10?

b. Simulate now N = 10,000 times. Keep the sample size at $n = 4000$ and $p = 0.10$. Describe the graph of the 10,000 sample proportion values. Does it seem likely that the sample proportion will fall very close to 0.10?

7.3 Exit poll sample distribution Consider the sampling distribution you were simulating in parts a and b of the previous exercise, assuming $p = 0.10$ with samples of size 4000 each. Using the appropriate formulas from this section, find the mean and the standard deviation of the sampling distribution of the sample proportion:

a. For a random sample of size $n = 4000$, as in that exercise. (*Hint*: n is the size of each sample, not the number of simulations.)

b. For a random sample of size $n = 1000$.

c. For a random sample of size $n = 250$. Summarize the effect of the sample size on the size of the standard deviation of the sampling distribution of the sample proportion.

7.4 iPhone apps For the population of individuals who own an iPhone, suppose $p = 0.25$ is the proportion that has a given app. For a particular iPhone owner, let $x = 1$ if they have the app and $x = 0$ otherwise. For a random sample of 50 people who have an iPhone:

a. State the population distribution (that is, the probability distribution of X for each observation).

b. Find the mean of the sampling distribution of the sample proportion who have the app among the 50 people.

c. Find the standard deviation of the sampling distribution of the sample proportion who have the app among the 50 people.

d. Explain what the standard deviation in part c describes.

7.5 Other scenario for exit poll Refer to Examples 1 and 2 about the exit poll, for which the sample size was 3889. In that election, 40.9% voted for Whitman.

a. Define a binary random variable X taking values 0 and 1 that represents the vote for a particular voter

(1 = vote for Whitman and 0 = another candidate). State its probability distribution, which is the same as the population distribution for X.

b. Find the mean and standard deviation of the sampling distribution of the proportion of the 3889 people in the sample who voted for Whitman.

7.6 Exit poll and *n* Refer to the previous exercise.

TRY **a.** In part b, if the sampling distribution of the sample proportion had mean 0.409 and the standard deviation 0.008, give an interval of values within which the sample proportion will almost certainly fall. (*Hint:* You can use the approximate normality of the sampling distribution.)

b. The sample proportion for Whitman from the exit poll was $\hat{p} = 0.424$. Using part a, was this one of the plausible values expected in an exit poll? Why?

7.7 Random variability in baseball A baseball player in the major leagues who plays regularly will have about 500 at-bats (that is, about 500 times he can be the hitter in a game) during a season. Suppose a player has a 0.300 probability of getting a hit in an at-bat. His batting average at the end of the season is the number of hits divided by the number of at-bats. This batting average is a sample proportion, so it has a sampling distribution describing where it is likely to fall.

a. Describe the shape, mean, and standard deviation of the sampling distribution of the player's batting average after a season of 500 at-bats.

b. Explain why a batting average of 0.320 or of 0.280 would not be especially unusual for this player's year-end batting average. (That is, you should not conclude that someone with a batting average of 0.320 one year is necessarily a better hitter than a player with a batting average of 0.280.)

7.8 Relative frequency of heads Construct the sampling distribution of the sample proportion of heads obtained in the experiment of flipping a balanced coin:

a. Once. (*Hint*: The possible sample proportion values are 0, if the flip is a tail, and 1, if the flip is a head. What are their probabilities?)

b. Twice. (*Hint*: The possible samples are {HH, HT, TH, TT}, so the possible sample proportion values are 0, 0.50, and 1.0, corresponding to 0, 1, or 2 heads. What are their probabilities?)

c. Three times. (*Hint*: There are 8 possible samples.)

d. Refer to parts a–c. Sketch the sampling distributions and describe how the shape is changing as the number of flips n increases.

7.9 Experimental medication As part of a drug research study, individuals suffering from arthritis take an experimental pain relief medication. Suppose that 25% of all individuals who take the new drug experience a certain side effect. For a given individual, let X be either 1 or 0, depending on whether s/he experienced the side effect or not, respectively.

a. If $n = 3$ people take the drug, find the probability distribution of the proportion who will experience the side effect.

b. Referring to part a, what are the mean and standard deviation of the sample proportion?

c. Repeat part b for a group of $n = 10$ individuals; $n = 100$. What happens to the mean and standard deviation of the sample proportion as n increases?

7.10 Effect of *n* on sample proportion The figure illustrates two sampling distributions for sample proportions when the population proportion $p = 0.50$.

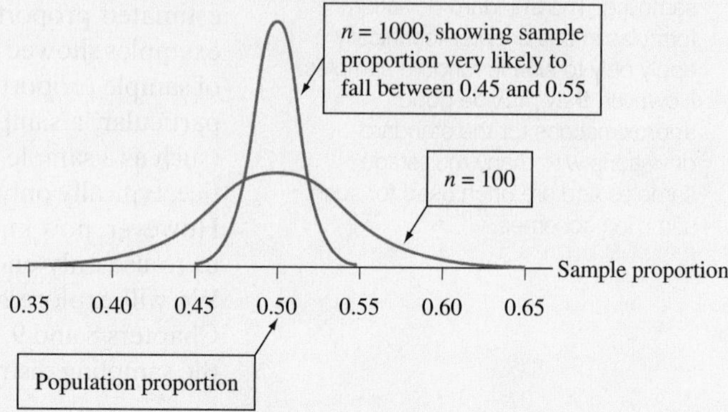

a. Find the standard deviation for the sampling distribution of the sample proportion with (i) $n = 100$ and (ii) $n = 1000$.

b. Explain why the sample proportion would be very likely (as the figure suggests) to fall (i) between 0.35 and 0.65 when $n = 100$, and (ii) between 0.45 and 0.55 when $n = 1000$. (*Hint:* Recall that for an approximately normal distribution, nearly the entire distribution is within 3 standard deviations of the mean.)

c. Explain how the results in part b indicate that the sample proportion tends to more precisely estimate the population proportion when the sample size is larger.

7.11 Syracuse full-time students You'd like to estimate the proportion of the 14,201 (www.syr.edu/about/facts.html) undergraduate students at Syracuse University who are full-time students. You poll a random sample of 100 students, of whom 94 are full-time. Unknown to you, the proportion of all undergraduate students who are full-time students is 0.951. Let X denote a random variable for which $x = 1$ denotes full-time student and for which $x = 0$ denotes part-time student.

TRY **a.** Describe the data distribution. Sketch a graph representing the data distribution.

b. Describe the population distribution. Sketch a graph representing the population distribution.

c. Find the mean and standard deviation of the sampling distribution of the sample proportion for a sample of size 100. Explain what this sampling distribution represents. Sketch a graph representing this sampling distribution.

7.12 Gender distributions At a university, 60% of the 7,400 students are female. The student newspaper reports results of a survey of a random sample of 50 students about various topics involving alcohol abuse, such as participation in binge drinking. They report that their sample contained 26 females.

a. Explain how you can set up a binary random variable X to represent gender.

b. Identify the population distribution of gender at this university. Sketch a graph.

c. Identify the data distribution of gender for this sample. Sketch a graph.

d. Identify the sampling distribution of the sample proportion of females in the sample. State its mean and standard deviation for a random sample of size 50. Sketch a graph.

7.13 Shapes of distributions

a. With random sampling, does the shape of the data distribution tend to resemble more closely the sampling distribution or the population distribution? Explain.

b. Explain carefully the difference between a data distribution of sample observations and the sampling distribution of the sample proportion for a binary variable that can take values only of 0 and 1. [*Hint*: Consider the graphs constructed in the two previous exercises.]

7.2 How Sample Means Vary Around the Population Mean

Recall

Population Distribution: The distribution of the random variable for the population from which we sample.
Data Distribution: The distribution of the sample data from one sample and the distribution we see in practice.
Sampling Distribution: The distribution of the sample statistic (sample proportion or sample mean) from repeated random samples.

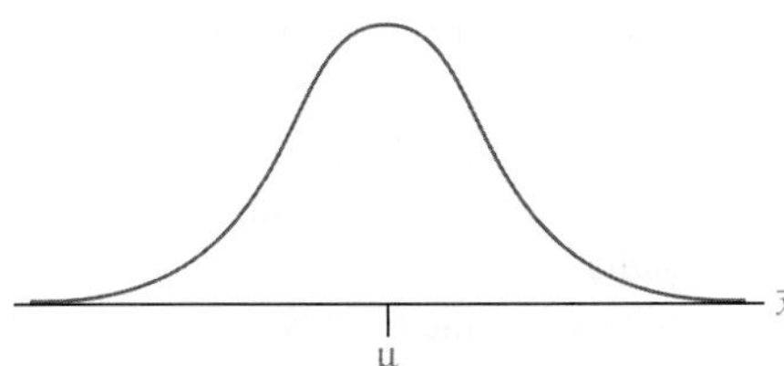

The sampling distribution of $\bar{x}$: The sample mean $\bar{x}$ fluctuates from sample to sample around the population mean μ.

The previous section discussed three types of distributions: the population distribution, the data distribution, and the sampling distribution (see Recall in margin). The sampling distribution is a probability distribution for the possible values of a statistic. We learned how much a sample proportion (a statistic summarizing sample categorical data) can vary among different random samples. Because the sample mean is so commonly used as a statistic to summarize sample numerical (quantitative) data, we'll now pay special attention to the sampling distribution of the sample mean. We'll discover results allowing us to predict how close a particular sample mean $\bar{x}$ falls to the population mean μ.

As with the sampling distribution of the sample proportion, there are two main results about the sampling distribution of the sample mean:

- One result provides formulas for its mean and standard deviation of the sampling distribution.
- The other indicates that its shape is often approximately a normal distribution, as we observed in the previous section for the sample proportion.

Let's return to the Sampling Distributions applet to investigate with simulation how the sampling distribution for the sample mean might look with repeated random sampling from a population represented by a quantitative random variable.

Simulating a Sampling Distribution for a Sample Mean

How much might a sample mean vary from sample to sample? How would you describe the shape, center, and variability of the possible sample mean values? To answer these questions, we can use the Sampling Distributions applet.

Activity 2

Simulating a Sampling Distribution for a Sample Mean From a Bell-Shaped Distribution

We use the Sampling Distribution applet to simulate selecting random samples of a given size from a population represented by a quantitative random variable with a bell-shaped distribution, population mean μ = 25, and population standard deviation σ = 8.22.

Suppose this random variable represents the distribution of scores on a personality test. We will first select the random samples using a smaller sample size of 10, then a larger sample size of 100.

Simulating Using the Sampling Distributions Applet

In the applet,

- Select Bell Shaped for the parent population (the applet by default sets the population mean at 25 and the population standard deviation at 8.22).
- Select n = 10 for the sample size.
- Select N = 1 to take one sample of this size.
- Click the *Sample* button once.

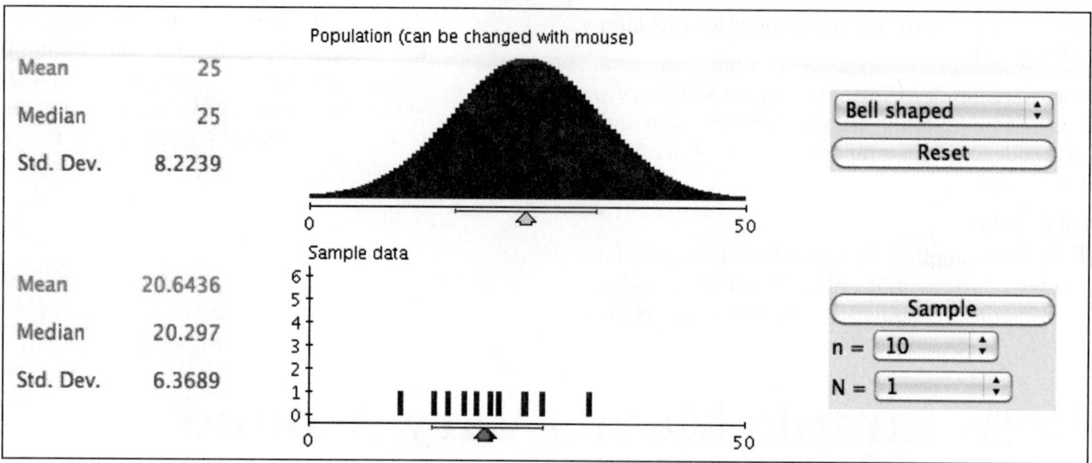

▲ **Figure 7.5 The Result of Taking One Sample of Size 10 Using the Sampling Distributions Applet.** The population has been set at bell shaped with a mean of 25 and a standard deviation of 8.22 in the first menu. The sample size has been set at $n = 10$ in the second menu. The symbol N refers to how many samples of size 10 are taken, in this case only 1. The first plot shows the **population distribution.** The second plot shows the **data distribution** for the one sample, which has a sample mean of 20.64 and a sample standard deviation of 6.37.

Figure 7.5 shows what part of the applet screen looks after you make these selections and take one sample.

After using the applet to take a random sample of size 10, did you get a sample mean close to 25? Perform this simulation of a random sample of size 10 ten times, each time observing (from the graphs of the data distribution) the sample means and the sample standard deviations. Record your 10 sample means and standard deviations.

We next ran the applet by setting N = 10,000, to simulate the process of picking 10 observations in 10,000 separate samples so that we could search for a pattern in the results. Figure 7.6 shows a histogram of the 10,000 outcomes for the sample means from the repeated sampling. The simulated sample means resulted in a bell-shaped distribution around the true population mean of $\mu = 25$. Nearly all the sample means fell between 18 and 33, that is, within approximately 8 of the true population mean of 25. Did the 10 sample mean values in your previous simulations all fall between 18 and 33?

Return to the Sampling Distributions applet and perform 10,000 simulations yourself by clicking on the Reset button and setting Bell Shaped, $n = 100$, and N = 10,000 in the applet menu. (*Note*: You are increasing the sample size from $n = 10$ to $n = 100$.) Click on *Sample.* (Be patient as it will take a minute or two to generate a response.) Compare the simulated sampling distribution of the 10,000 sample means in the applet graph that you created to the simulated sampling distribution in Figure 7.6. Are the results similar, showing a bell shape centered at 25? Is the variability the same or different in your graph compared to Figure 7.6? Explain clearly how increasing the sample size from 10 to 100 affected the simulated sampling distribution you created.

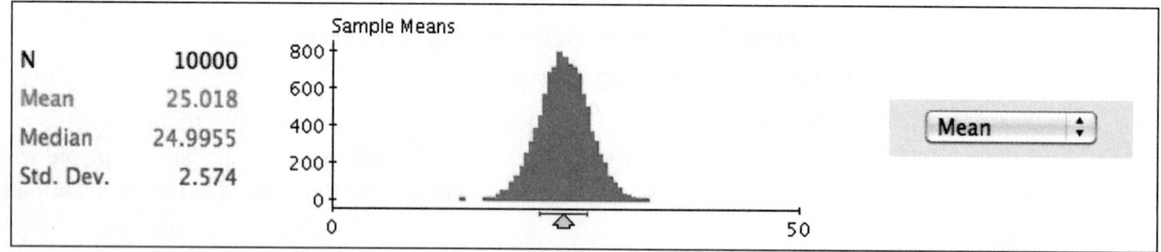

▲ **Figure 7.6 Results of Simulating Repeated Random Sampling From a Bell-Shaped Population Distribution.** 10,000 random samples of 10 scores each were simulated from a population in which $\mu = 25$. You can do this simulation with the Sampling Distributions applet by setting Bell Shaped for the population graph and $n = 10$ and $N = 10{,}000$ on the Sample Data menu.

Try Exercise 7.27

We observe from our simulation in Activity 2 that when we repeatedly sample from a population that is bell shaped, even with a small sample size of $n = 10$, the shape of the simulated sampling distribution is also bell shaped with a mean close to the population mean of 25 and a standard deviation of approximately 2.6, which mathematically is the population standard deviation of 8.22 divided by $\sqrt{n}$. We can state the following result in the summary box.

SUMMARY: Expected Behavior of the Sampling Distribution of the Sample Mean When the Population Distribution is Normally Distributed

For a random sample of size n from a normally distributed population having mean μ and standard deviation σ, then regardless of the sample size n, the sampling distribution of the sample mean $\bar{x}$ is also normally distributed with its center described by the **population mean** μ and the variability described by the standard deviation of the sampling distribution, which equals **the population standard deviation divided by the square root of the sample size, $\sigma/\sqrt{n}$.**

What if the population distribution is not bell shaped? Will the sampling distribution of the sample mean still have a bell shape?

Simulating a Sampling Distribution for a Sample Mean from a Non-Bell-Shaped Distribution

The simulation in Activity 2 for the sample mean with $n = 10$ showed a bell shape for the sampling distribution. This was not surprising since we were sampling from a population distribution with a bell shape. Is it typical for sampling distributions of a sample mean to have bell shapes even if the population distribution is not bell shaped?

Example 4

Population, data, and sampling distributions

Rolling Dice

Picture the Scenario

Let's consider a scenario in which the probability distribution of the random variable X is highly skewed. Exercise 6.2 in Chapter 6 derived the probability distribution of X = maximum value on two rolls of a die. The figure in the margin displays the probability distribution. This probability distribution is called the *population distribution* because in practice it describes a population from which we take a sample. It has a population mean of 4.5 and a population standard deviation of 1.4.

Suppose you roll a pair of dice and record the maximum of the two numbers shown. Then you roll this pair of dice 29 more times, each time recording the maximum. For the 30 rolls, you find the mean of the sample of 30 maximum values. This is one simulated statistic, the sample mean, from $n = 30$ observations for this skewed population distribution. For example, the first sample of size 30 might give us the results 5, 6, 6, 4, 5, 2, 3, 4, 1, 6, 3, 6, 5, 2, 4, 4, 6, 3, 5, 4, 5 , 6, 2, 6, 1, 5, 4, 5, 6, 4 that have a sample mean of 4.3 and a sample standard deviation of 1.55. The graph of this data distribution is in the margin on the next page.

You could then repeat this process for a very large number of simulated samples (such as N = 10,000) of 30 observations of the maximum. Figure 7.7 displays the histogram of the sample means for 10,000 simulated samples of size 30 from this skewed population distribution. In theory, the sampling distribution refers to an infinite number of samples each of size 30. Using the software MINITAB, we took 10,000 samples each of size 30, which gives similar results.

Questions to Explore

a. Identify the population distribution, and state its shape, mean, and standard deviation.

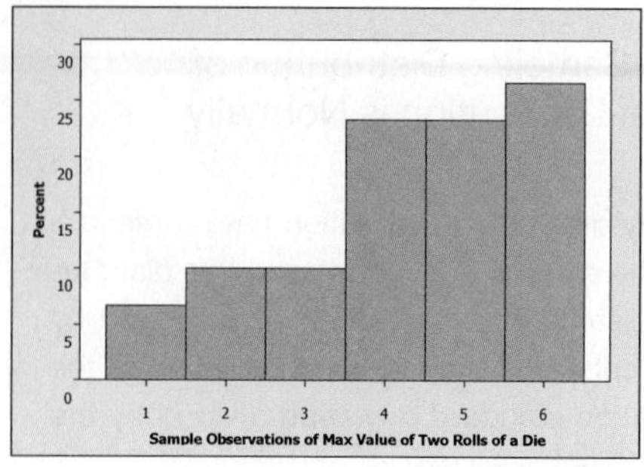

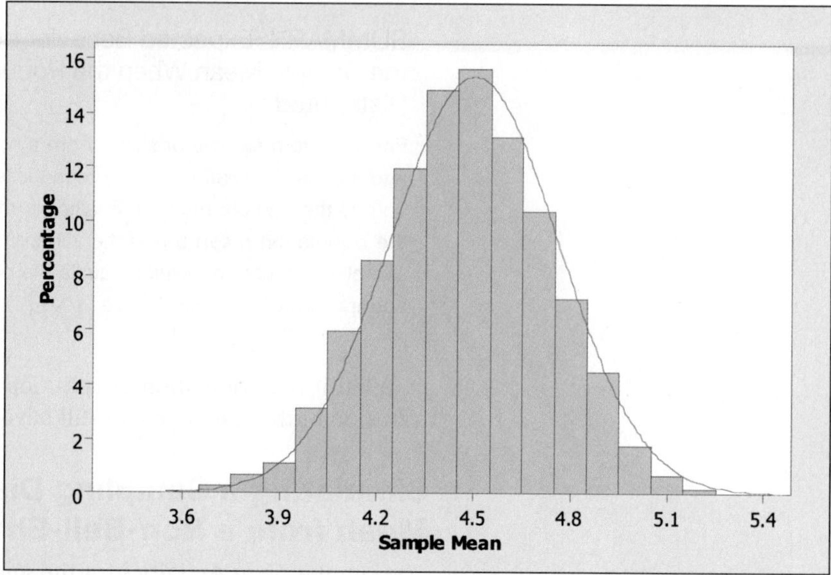

▲ **Figure 7.7** MINITAB Histogram of Simulated Sample Means With $n = 30$. Each sample was simulated from a discrete probability (population) distribution (shown previously in the margin) that is skewed to the left with population mean = 4.5 and population standard deviation = 1.4. The superimposed normal distribution has mean 4.5 and standard deviation 0.26 and approximates the simulated sampling distribution. **Question** How does this simulated sampling distribution relate with respect to shape, center, and variability to the population distribution from which the samples were taken?

b. Identify the data distribution, and state its shape, mean, and standard deviation.

c. Identify the sampling distribution of $\bar{x}$, and state its shape, mean, and standard deviation.

Think It Through

a. The *population distribution* is given by the probability distribution of the random variable X = maximum value on two rolls of a die (see figure in margin on previous page). It is skewed to the left and described by the population mean of 4.5 and the population standard deviation of 1.4.

b. The *data distribution* is the collection of 30 values in the sample. It is described by the sample mean of 4.3 and the sample standard deviation of 1.55. It looks similar to the population distribution, being skewed to the left (see figure in margin).

c. We will use the simulated *sampling distribution* in Figure 7.7 to predict the behavior of the sampling distribution. Unlike the population and data distributions that are skewed left, the simulated sampling distribution is bell shaped and displays much less variability. The histogram in Figure 7.7 represents a simulated sampling distribution of 10,000 random samples and has a bell-shaped distribution superimposed with the same mean as the expected theoretical sampling distribution (population mean = 4.5) and the same standard deviation as the expected standard deviation of the theoretical sampling distribution, which is $\frac{\sigma}{\sqrt{n}} = \frac{1.4}{\sqrt{30}} = 0.26$.

Try Exercise 7.23 and Conduct the Simulation in Exercise 7.26

Describing the Behavior of the Sampling Distribution for the Sample Mean for Any Population

Here we've seen a surprising result: Even when a population distribution is not bell shaped, the sampling distribution of the sample mean $\bar{x}$ can have a bell shape. We also observe that the mean of the sampling distribution of the sample mean appears to be the same as the population mean μ, and the standard deviation of the sampling distribution for the sample mean appears to be $\frac{\sigma}{\sqrt{n}}$. This bell shape is a consequence of the **central limit theorem (CLT).** It states that the sampling distribution of the sample mean $\bar{x}$ often has approximately a normal distribution. This result applies no matter what the shape of the population distribution from which the samples are taken. For relatively large sample sizes, the sampling distribution is bell shaped even if the population distribution is highly discrete or highly skewed. We observed this in Figure 7.7 with a skewed, highly discrete population, using $n = 30$, which is not all that large but is the size of sample we sometimes see in practice.

Mean and Standard Deviation of the Sampling Distribution of Sample Mean $\bar{x}$

For a random sample of size n from a population having mean μ and standard deviation σ, the sampling distribution of the sample mean $\bar{x}$ has its center described by the **population mean** μ and the variability described by the **standard deviation of the sampling distribution** which equals $\frac{\sigma}{\sqrt{n}}$.

The Central Limit Theorem (CLT): Describes the Expected Shape of the Sampling Distribution for Sample Mean $\bar{x}$

For a random sample of size n from a population having mean μ and standard deviation σ, then as the sample size n increases, the sampling distribution of the sample mean $\bar{x}$ approaches an approximately normal distribution.

In Words

As the sample size increases, the sampling distribution of the sample mean has a more bell-shaped appearance.

What is amazing about the central limit theorem is that no matter what the shape of the population distribution, the sampling distribution of the sample mean approaches an approximately normal distribution. In fact, for most population distributions, the bell shape is approached very quickly as the sample size n increases. Thus, if our sample size is sufficiently large, we don't need to worry or know about the shape of the population distribution in order to work with the sampling distribution of the sample mean.

Figure 7.8 displays sampling distributions of the sample mean $\bar{x}$ for four different shapes for the population distribution from which samples are taken. The population shapes are shown at the top of the figure; below them are portrayed the sampling distributions of $\bar{x}$ for random sampling of sizes $n = 2, 5,$ and 30. Even if the population distribution itself is uniform (column 1 of the figure) or U-shaped (column 2) or skewed (column 3), the sampling distribution of the sample mean has approximately a bell shape when n is at least 30 and sometimes for n as small as 5. In addition, the variability of the sampling distribution noticeably decreases as n increases, because the standard deviation of the sampling distribution decreases.

In Practice Expect Bell Shape When Sample Size at Least 30

The sampling distribution of $\bar{x}$ takes more of a bell shape as the random sample size n increases. The more skewed the population distribution, the larger n must be before the shape is close to normal (bell shape). In practice, the sampling distribution is usually close to bell shape when the sample size n is at least 30.

If the population distribution is normally distributed, then the sampling distribution is normally distributed for all sample sizes. The rightmost column of Figure 7.8 shows this case.

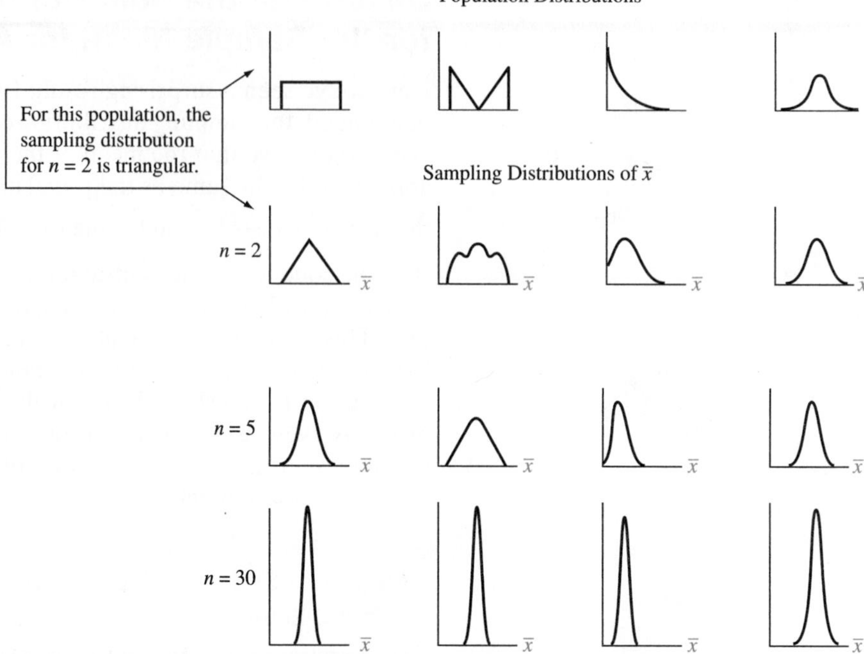

▲ **Figure 7.8 Four Population Distributions and the Corresponding Sampling Distributions of $\bar{x}$.** Regardless of the shape of the population distribution, the sampling distribution becomes more bell shaped as the random sample size n increases. You can use the Sampling Distributions applet on the text CD to simulate how the first population distribution shown has a sampling distribution for the sample mean that becomes more nearly normal as n increases. See Exercise 7.25.

Example 5

The variability of means

Weekly Mean Sales

Picture the Scenario

Aunt Erma's Restaurant in the North End of Boston specializes in pizza that is baked in a wood-burning oven. The sales of food and drink in this restaurant vary from day to day. Past records indicate that the daily sales follow a probability (population) distribution with a mean of $\mu = \$900$ and a standard deviation of $\sigma = \$300$.

Questions to Explore

The mean sales for the seven days in a week are computed each week. Each week is considered a sample of size $n = 7$ days. The weekly sample means are observed over time to check for substantial increases or decreases in sales assuming the population distribution described in the scenario.

a. What would we expect the weekly sample mean sales amounts to fluctuate around (in dollars)?

b. How much variability would you expect in the weekly sample mean sales figures? Find the standard deviation of the sampling distribution of the sample mean, and interpret this standard deviation.

Think It Through

a. We would expect the weekly sample means to fluctuate around the mean of the assumed population distribution. This is $\mu = \$900$.

b. The sampling distribution of the sample mean for $n = 7$ has mean \$900. Its standard deviation equals

$$\frac{\sigma}{\sqrt{n}} = \frac{300}{\sqrt{7}} = 113.4$$

If we were to observe the sample mean sales for several weeks, we would expect the sample means to vary around \$900, with variability from the mean described by the standard deviation of \$113.40.

Insight

Figure 7.9 portrays a possible population distribution for daily sales that is somewhat symmetric and unimodal. It also shows the sampling distribution of the mean sales $\bar{x}$ for $n = 7$. There is less variability from week to week in the mean sales than there is from day to day in the daily sales (as portrayed by the population distribution).

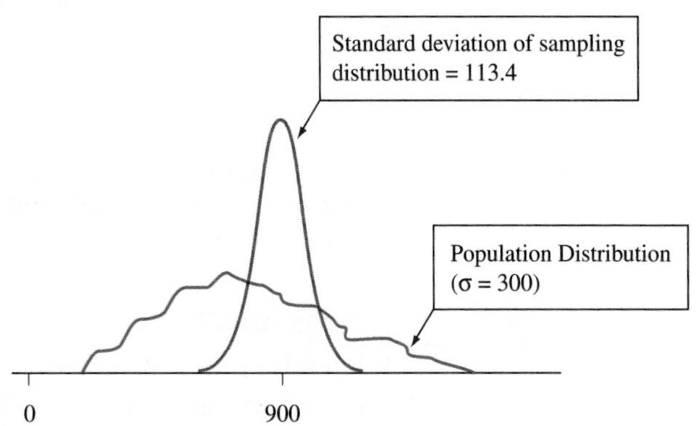

▲ **Figure 7.9 A Population Distribution for Daily Sales and the Sampling Distribution of Weekly Mean Sales $\bar{x}$ ($n = 7$).** There is more variability day to day in the daily sales than week to week in the weekly mean sales.

The assumptions for the probability (population) model in this example are a bit unrealistic. Some days of the week (such as weekends) may be busier than others, and certain periods of the year may be busier (such as summer). Also, in practice we would not know values of population parameters such as the population mean and standard deviation. However, this example shows that even under some simplifying assumptions, we'd still expect to see quite a bit of variability from week to week in the sample mean sales. We should not be surprised if the mean is \$800 this week, \$1000 next week, and so forth.

Knowing how to find a standard deviation for the sampling distribution gives us a mechanism for understanding how much variability to expect in sample statistics that we observe in our jobs or in our daily lives.

Try Exercises 7.14 and 7.15

Effect of *n* on the Standard Deviation of the Sampling Distribution

Let's consider again the formula $\frac{\sigma}{\sqrt{n}}$ for the standard deviation of the sample mean. Notice that as the sample size n increases, the denominator increases, so the standard deviation of the sample mean decreases. This relationship has an important practical implication: **With larger samples, the sample mean tends to fall closer to the population mean.**

Example 6

Expected winnings

Long-Run Consequence of Playing Roulette

Picture the Scenario

During an episode of the 2011 MTV series, *The Real World*, filmed in Las Vegas, one of the roommates (Adam) decided to play roulette. He had the option to bet smaller amounts of money on multiple spins but decided to bet his entire bankroll of \$400 on a single spin, placing the entire amount on red. As luck would have it, he won.

In Exercise 7 of Chapter 6, we learned that 18 of the 38 numbers on a roulette wheel are red. Let X denote the possible winnings associated with a single \$400 bet on red. If the ball lands on red, Adam wins \$400; otherwise, he loses \$400. Thus the possible values of x are \$400 and –\$400. This probability distribution has a mean of

$$\Sigma x P(x) = \frac{18}{38}(400) + \frac{20}{38}(-400) = -\$21.05.$$

So he can expect to lose an average of \$21.05 on a single spin. This is clearly unfavorable for Adam, and one might wonder why he would be willing to make such a bet.

Questions to Explore

Suppose rather than betting \$400 on a single spin, Adam decides to bet \$10 on red on each of the next 40 spins.

a. Find the mean and standard deviation of the sampling distribution of his sample mean winnings for 40 spins.

b. Find the mean and standard deviation of the sampling distribution of his sample mean winnings for 400 spins.

Think It Through

a. As before, let X denote the possible winnings associated with a single bet on red. The possible values of x are \$10 and –\$10, and the expected winnings of a single spin is

$$\mu = \Sigma x P(x) = \frac{18}{38}(10) + \frac{20}{38}(-10) = -\$0.53.$$

Additionally, the standard deviation, σ, of the winnings on a single \$10 bet equals approximately \$10.05.

The sampling distribution of the sample mean winnings, $\bar{X}$ has the same mean as the probability distribution of X for a single spin. When placing bets on 40 different spins, the standard deviation of the sampling distribution of $\bar{X}$ is $\sigma/\sqrt{n}$. Thus the mean and standard deviation of the sampling distribution of $\bar{X}$ for 40 spins are –\$0.53 and $\$10.05/\sqrt{40} = \$1.59$.

b. The mean of this sampling distribution is still –\$0.53. The standard deviation of this sampling distribution is $\frac{\sigma}{\sqrt{n}} = \frac{10.05}{\sqrt{400}} = 0.503$. (approximately 50 cents).

Recall

The sampling distribution of $\bar{x}$ has mean equal to the population mean μ and standard deviation equal to $\frac{\sigma}{\sqrt{n}}$, where σ is the population standard deviation. For large n by the central limit theorem, this sampling distribution is approximately a normal distribution. ◀

Insight

As n increases, the standard deviation of the sampling distribution of $\bar{X}$ becomes smaller. Thus, as the number of spins played increases, the winnings per spin tend to get closer and closer to the mean of the probability distribution for each spin, namely –$0.53. This phenomenon stems from the *law of large numbers* and is generally bad news for roulette players and good news for casino owners.

Try Exercise 7.16, part b

This example reinforces what we observed in the previous section about the effect of increased sample size on the precision of a sample proportion in estimating a population proportion. The larger the sample size, the smaller is the standard deviation of the sampling distribution.

Example 7

Probability

Coming Out Ahead When Playing Roulette

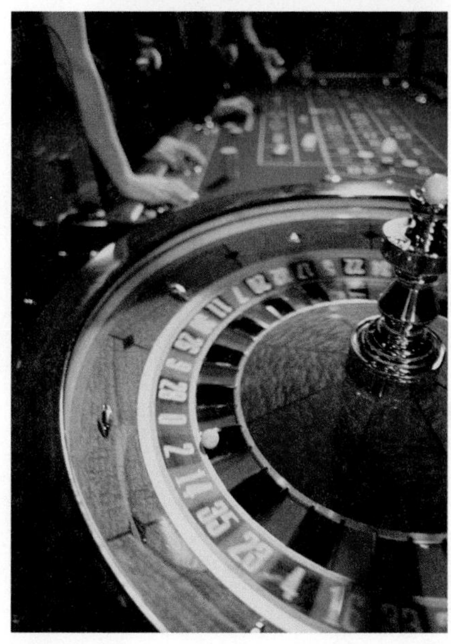

Picture the Scenario

In Example 6, we considered two options for playing roulette. You could place a $400 bet for the ball to land on red during a single spin. In this case, you either win $400 or lose $400, and the outcome is based on the outcome of a single spin. The second option was to bet $10 on red for each of 40 different spins. In this case, your overall winnings depend on the outcomes of each of the 40 spins.

Questions to Explore

Let's see how your probability of coming out ahead (winning more than you pay to play) depends on how often you play. What's the probability that you come out ahead if (a) you bet $400 on red on a single spin; (b) you bet $10 on red on each of 40 different spins.

Think It Through

You come out ahead if your mean winnings per spin $\bar{x}$, is greater than zero.

a. In playing $n = 1$ time, you come out ahead only if you win. Since 18 of the 38 numbers on the wheel are red, your probability of winning and thus coming out ahead is simply 18/38.

b. When playing $n = 40$ different spins, to determine the probability of coming out ahead, we must find the probability that $\bar{x}$ is greater than zero. To do so, we rely on the central limit theorem to determine the sampling distribution of $\bar{x}$. Since n is large, the sampling distribution of $\bar{x}$ is approximately normal. In Example 6, we found the mean and standard deviation of the distribution to be –$0.53 and $1.59, respectively. Figure 7.10 shows this distribution.

The probability that, on average, you come out ahead after 40 spins (the sample mean exceeds $0) is approximately the area (called probability) above 0 in the right tail of this figure. So we need to find the probability that

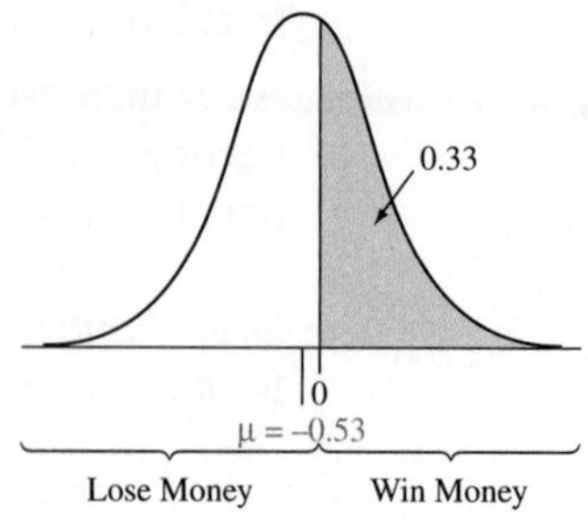

▲ **Figure 7.10 Approximate Sampling Distribution of the Sample Mean $\bar{x}$ when μ = −0.53 and Standard Deviation of Sampling Distribution = 1.59.** We use this distribution to approximate the probability that $\bar{x}$ falls above \$0 when you play 40 spins. **Question** Does this distribution have the same shape as the probability distribution for each game? Explain.

a normal random variable with mean –0.53 and standard deviation 1.59 takes a value above 0. The value 0 has a z-score of

$$z = (\text{value} - \text{mean})/(\text{standard deviation of sampling distribution}) = [0 - (-0.53)]/1.59 = 0.33.$$

From Table A, the cumulative probability, or the area under the standard normal curve to the left of $z = 0.33$ is 0.6293. Thus, the right-tail probability is $1 - 0.6293 = 0.3707$. The probability you will come out ahead after 40 spins is 0.37.

Insight

The probability ($18/38 = 0.474$) of coming out ahead in winnings is larger if we just make the single bet in part a. However, the probability is 0.526 of losing \$400. Meanwhile, the larger the value of n, the less likely you will come out ahead, but it also becomes less likely that you will lose a relatively large sum of money. We can see using the normal distribution that the probability is 0.95 that the average winnings per spin is within 2 standard deviations from the mean, or between $-0.53 - 2(1.59) = -\$3.71$ and $-0.53 + 2(1.59) = \$2.65$. Over the course of 40 spins, this translates to overall winnings between $-\$3.71(40) = -\$148.40$ and $\$2.65(40) = \$106.00$. Because the potential loss associated with part b will be smaller in magnitude than that of part a, we can think of this approach as more risk averse in terms of long-run winnings.

Try Exercises 7.16, part c, and 7.17

In Practice We Rarely Know the Exact Characteristics of the Population Distribution

Rarely does a researcher know the exact characteristics of the population distribution (shape, mean, and standard deviation). With random sampling, we hope that the data distribution will look similar to the population distribution. In the next chapter, we will show how in practice the statistician uses the data distribution to estimate the mean and standard deviation of the population distribution. In turn, we can use the sample standard deviation from the data distribution to estimate the standard deviation of the sampling distribution of the sample mean. This standard deviation helps us predict how close the sample mean falls to the population mean.

The Central Limit Theorem Helps Us Make Inferences

Sampling distributions are fundamental to statistical inference for making decisions and predictions about populations based on sample data. The central limit theorem and the formula for the standard deviation of the sample mean $\bar{x}$ have many implications. These include the following:

- When the sampling distribution of the sample mean $\bar{x}$ is approximately normal, $\bar{x}$ falls within 2 standard deviations of the population mean μ with probability close to 0.95, and $\bar{x}$ almost certainly falls within 3 standard deviations of μ. Results of this type are vital to inferential methods that predict how close sample statistics fall to unknown population parameters.

- For large n, the sampling distribution is approximately normal even if the population distribution is not. This enables us to make inferences about population means regardless of the shape of the population distribution. This is helpful in practice, because usually we don't know the shape of the population distribution. Often, it is quite skewed.

Using the fundamental concepts we have learned in this chapter for the sampling distribution of a sample proportion and for the sampling distribution of a sample mean will allow us to develop inference procedures for population proportions and population means in the upcoming chapters.

7.2 Practicing the Basics

7.14 Education of the self-employed According to a recent Current Population Reports, the population distribution of number of years of education for self-employed individuals in the United States has a mean of 13.6 and a standard deviation of 3.0.

TRY

a. Identify the random variable X whose population distribution is described here.

b. Find the mean and standard deviation of the sampling distribution of $\bar{x}$ for a random sample of size 100. Interpret the results.

c. Repeat part b for $n = 400$. Describe the effect of increasing n.

7.15 Rolling one die Let X denote the outcome of rolling a die.

TRY

a. Construct a graph of the (i) probability distribution of X and (ii) sampling distribution of the sample mean for $n = 2$. (You can think of (i) as the population distribution you would get if you could roll the dice an infinite number of times. The first column of Figure 7.8 portrays a case like this with the sampling distribution for $n = 2$. It shows also how the sampling distribution becomes more bell shaped as n increases to 5 and to 30.)

b. The probability distribution of X has mean 3.50 and standard deviation 1.71. Find the mean and standard deviation of the sampling distribution of the sample mean for (i) $n = 2$, (ii) $n = 30$. What is the effect of n on the sampling distribution?

7.16 Playing roulette A roulette wheel in Las Vegas has 38 slots. If you bet a dollar on a particular number, you'll win \$35 if the ball ends up in that slot and \$0 otherwise. Roulette wheels are calibrated so that each outcome is equally likely.

TRY

a. Let X denote your winnings when you play once. State the probability distribution of X. (This also represents the population distribution you would get if you could play roulette an infinite number of times.) It has mean 0.921 and standard deviation 5.603.

b. You decide to play once a minute for 12 hours a day for the next week, a total of 5040 times. Show that the sampling distribution of your sample mean winnings has mean = 0.921 and standard deviation = 0.079.

c. Refer to part b. Using the central limit theorem, find the probability that with this amount of roulette playing, your mean winnings is at least \$1, so that you have not lost money after this week of playing. (*Hint*: Find the probability that a normal random variable with mean 0.921 and standard deviation 0.079 exceeds 1.0.)

7.17 Canada lottery In one lottery option in Canada (*Source:* Lottery Canada), you bet on a six-digit number between 000000 and 999999. For a \$1 bet, you win \$100,000 if you are correct. The mean and standard deviation of the probability distribution for the lottery winnings are $\mu = 0.10$ (that is, 10 cents) and $\sigma = 100.00$. Joe figures that if he plays enough times every day, eventually he will strike it rich, by the law of large numbers. Over the course of several years, he plays 1 million times. Let $\bar{x}$ denote his average winnings.

TRY

a. Find the mean and standard deviation of the sampling distribution of $\bar{x}$.

b. About how likely is it that Joe's average winnings exceed \$1, the amount he paid to play each time? Use the central limit theorem to find an approximate answer.

7.18 Income of farm workers For the population of farm workers in New Zealand, suppose that weekly income has a distribution that is skewed to the right with a mean of $\mu = \$500$ (N.Z.) and a standard deviation of $\sigma = \$160$. A researcher, unaware of these values, plans to randomly sample 100 farm workers and use the sample mean annual income $\bar{x}$ to estimate μ.

a. Show that the standard deviation of $\bar{x}$ equals 16.0.

b. Explain why it is almost certain that the sample mean will fall within \$48 of \$500.

c. The sampling distribution of $\bar{x}$ provides the probability that $\bar{x}$ falls within a certain distance of μ, regardless of the value of μ. Show how to calculate the probability that $\bar{x}$ falls within \$20 of μ for all such workers. (*Hint*:

Using the standard deviation, convert the distance 20 to a z-score for the sampling distribution.)

7.19 Unusual sample mean income? The previous exercise reported that for the population, $\mu = \$500$ and $\sigma = \$160$, and that the sample mean income for a random sample of 100 farm workers would have a standard deviation of 16.0. Sketch the sampling distribution of the sample mean and find the probability that the sample mean falls above \$540.

7.20 Restaurant profit? Jan's All You Can Eat Restaurant charges \$8.95 per customer to eat at the restaurant. Restaurant management finds that its expense per customer, based on how much the customer eats and the expense of labor, has a distribution that is skewed to the right with a mean of \$8.20 and a standard deviation of \$3.

a. If the 100 customers on a particular day have the characteristics of a random sample from their customer base, find the mean and standard deviation of the sampling distribution of the restaurant's sample mean expense per customer.

b. Find the probability that the restaurant makes a profit that day, with the sample mean expense being less than \$8.95. (*Hint:* Apply the central limit theorem to the sampling distribution in part a.)

7.21 Survey accuracy A study investigating the relationship between age and annual medical expenses randomly samples 100 individuals in Davis, California. It is hoped that the sample will have a similar mean age as the entire population.

a. If the standard deviation of the ages of all individuals in Davis is $\sigma = 15$, find the probability that the mean age of the individuals sampled is within two years of the mean age for all individuals in Davis. (*Hint*: Find the sampling distribution of the sample mean age and use the central limit theorem. You don't have to know the population mean to answer this, but if it makes it easier, use a value such as $\mu = 30$.)

b. Would the probability be larger, or smaller, if $\sigma = 10$? Why?

7.22 Blood pressure Vincenzo Baranello was diagnosed with high blood pressure. He was able to keep his blood pressure in control for several months by taking blood pressure medicine (amlodipine besylate). Baranello's blood pressure is monitored by taking three readings a day, in early morning, at mid-day, and in the evening.

a. During this period, the probability distribution of his systolic blood pressure reading had a mean of 130 and a standard deviation of 6. If the successive observations behave like a random sample from this distribution, find the mean and standard deviation of the sampling distribution of the sample mean for the three observations each day.

b. Suppose that the probability distribution of his blood pressure reading is normal. What is the shape of the sampling distribution? Why?

c. Refer to part b. Find the probability that the sample mean exceeds 140, which is considered problematically high. (*Hint*: Use the sampling distribution, not the probability distribution for each observation.)

7.23 Household size TRY According to the 2010 U.S. Census Bureau Current Population Survey (www.census.gov/population/www/socdemo/hh-fam/cps2010.html), the average number of people in family households which contain both family and non-family members is 4.43 with a standard deviation of 2.02. This is based on census information for the population. Suppose the Census Bureau instead had estimated this mean using a random sample of 225 homes. Suppose the sample had a sample mean of 4.2 and standard deviation of 1.9.

a. Identify the random variable X. Indicate whether it is quantitative or categorical.

b. Describe the center and variability of the population distribution. What would you predict as the shape of the population distribution? Explain.

c. Describe the center and variability of the data distribution. What would you predict as the shape of the data distribution? Explain.

d. Describe the center and variability of the sampling distribution of the sample mean for 225 homes. What would you predict as the shape of the sampling distribution? Explain.

7.24 Average monthly sales A large corporation employs 27,251 individuals. The average income in 2008 for all employees was \$74,550 with a standard deviation of \$19,872. You are interested in comparing the incomes of today's employees with those of 2008. A random sample of 100 employees of the corporation yields $\bar{x} = \$75,207$ and $s = \$18,901$.

a. Describe the center and variability of the population distribution. What shape does it probably have? Explain.

b. Describe the center and variability of the data distribution. What shape does it probably have? Explain.

c. Describe the center and variability of the sampling distribution of the sample mean for $n = 100$. What shape does it have? Explain.

d. Explain why it would not be unusual to observe an individual who earns more than \$100,000, but it would be highly unusual to observe a sample mean income of more than \$100,000 for a random sample size of 100 people.

7.25 Central limit theorem for uniform population Let's use the Sampling Distributions applet on the text CD to show that the first population distribution shown in Figure 7.8 has a more nearly normal sampling distribution for the mean as n increases. Select uniform for the population distribution.

a. Use the applet to create a sampling distribution for the sample mean using sample sizes $n = 2$. Take $N = 10,000$ repeated samples of size 2, and observe the histogram of the sample means. What shape does this sampling distribution have?

b. Now take N = 10,000 repeated samples of size 5. Explain how the variability and the shape of the sampling distribution changes as n increases from 2 to 5.

c. Now take N = 10,000 repeated samples of size 30. Explain how the variability and the shape of the sampling distribution changes as n increases from 2 to 30. Compare results from parts a–c to the first column of Figure 7.7.

d. Explain how the central limit theorem describes what you have observed.

7.26 CLT for skewed population Use the Sampling Distributions applet by selecting the skewed population distribution, which looks like the third column of Figure 7.8. Repeat part a–c of the previous exercise, and explain how the variability and shape of the sampling distribution of the sample mean changes as n changes from 2 to 5 to 30. Explain how the central limit theorem describes what you have observed.

7.27 Sampling distribution for normal population Use the Sampling Distributions applet by selecting the bell-shaped population distribution, which looks like the fourth column of Figure 7.8. Take N = 10,000 repeated samples of size 2, and plot the sample means. Is the sampling distribution normal even for $n = 2$? What does this tell you?

7.3 The Binomial Distribution Is a Sampling Distribution (Optional)

The binomial probability distribution, introduced in Section 6.3, is an example of a sampling distribution. It is the sampling distribution for the number of successes or counts in n independent trials. It describes the possible values for the number of successes, out of all the possible samples we could observe in the n trials. In practice, studies usually report the sample proportion (or percentage) of successes. The proportion is simpler to interpret because the possible values fall between 0 and 1 regardless of the value of n. That's why we focused on the proportion in this chapter. However, there is a close connection between results for the number of successes and results for the proportion of successes.

Consider a binomial random variable X for $n = 3$ trials, such as the number of heads in three tosses of a coin. The number x can equal 0, 1, 2, or 3. To find the proportion of heads in three tosses, we take the number of successes x and divide by the sample size n; i.e, the proportion is equal x/n. These correspond to sample proportions of 0, 1/3, 2/3, 1. The proportion values vary less, by a factor of $1/n = 1/3$, going from 0 to 1 instead of from 0 to 3. This is because we find the proportion by dividing the count by 3. The smaller amount of variability for the proportion values is illustrated by the sampling distributions shown in Figure 7.11, which result when the probability of a success is $p = 0.50$.

Because the sample proportion is the binomial random variable divided by the sample size, the formulas for the mean and the standard deviation of the sampling distribution of the proportion of successes are the formulas for the mean and standard deviation of the number of successes divided by n.

Mean and Standard Deviation of Sampling Distribution of a Proportion

For a binomial random variable with n trials and probability p of success for each, the sampling distribution of the *proportion* of successes has

$$\text{mean} = p, \quad \text{standard deviation} = \sqrt{p(1-p)/n}$$

To obtain these values using the binomial distribution, take the binomial mean np and binomial standard deviation $\sqrt{np(1-p)}$ of the *number* of successes and divide by n.

Recall

For a binomial random variable X = number of successes in n trials, Section 6.3 gave the formulas mean = np and standard deviation = $\sqrt{np(1-p)}$. ◀

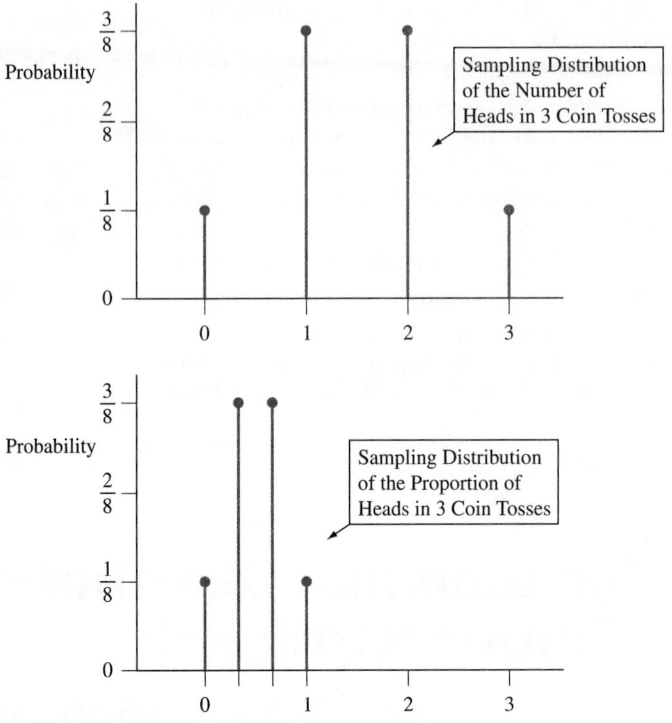

▲ **Figure 7.11** The Sampling Distribution of the Sample *Proportion* Displays Less Variability Than the Sampling Distribution of the *Number* of Successes.

Example 8

Proportions versus counts

2010 Election Exit Poll

Picture the Scenario

Examples 1–3 considered $n = 3889$ people and the number who said they voted for Jerry Brown was 2065. This corresponds to a sample proportion of 0.531. The actual proportion of the population who voted for Brown (determined after the election was over) was 0.538.

Questions to Explore

a. What would we expect for the behavior (shape, center, and variability) of the sampling distribution of counts (number who vote for Brown) in an exit poll of sample size $n = 3889$? Is the observed count from the exit poll a plausible value in this binomial sampling distribution?

b. How does the sampling distribution of counts compare to the sampling distribution of the sample proportion?

Think It Through

a. The number of people voting for Brown follows the binomial distribution. Given that the actual population proportion voting for Brown was $p = 0.538$, we expect the number voting for Brown to be a random variable with mean $= np = 3889(0.538) = 2092.3$ and standard deviation $= \sqrt{np(1 - p)} = \sqrt{3889(0.538)(0.462)} = 31.1$. Since $np \geq 15$ and $np(1 - p) \geq 15$ [2092 and 1797], this binomial distribution can be approximated by the normal distribution.

The observed count of 2065 is within 1 standard deviation of the expected mean 2092.3. The z-score is $z = (2065 - 2092.3)/31.1 = -0.88$. The observed number of successes for Brown in the exit poll is a very plausible value in the actual distribution of successes based on the election result.

b. For $n = 3889$ voters, we found in Section 7.1 (with the actual population proportion as $p = 0.538$) that the expected sampling distribution of the sample proportion has mean $= p = 0.538$, standard deviation $= \sqrt{p(1 - p)/n} = \sqrt{(0.538)(0.462)/3889} = 0.008$. e observe that the mean $= 0.538$ is the mean of the binomial divided by n; i.e., $2092.3/3889 = 0.538$. The standard deviation of the sample proportion is the standard deviation of the counts divided by n; i.e., $31.1/3889 = 0.008$. The shape of both sampling distributions is approximately normal.

Insight

When attempting to make predictions and inferences with categorical variables, we can make the same prediction whether we use the binomial sampling distribution for the number voting for Brown or the sampling distribution for the proportion voting for him. A critical role of the binomial distribution is how the sampling distribution of the sample proportion is derived from the sampling distribution of the count. In future chapters, we will mainly use the sampling distribution of the proportion.

Try Exercise 7.28

7.3 Practicing the Basics

7.28 Sampling distributions for the exit poll Refer to TRY Examples 1 and 2 about the California exit poll, for which the sample size was 3889 and 42.4% of the sample voted for Whitman. We found that after the election, 40.9% voted for Whitman in the population.

a. Identify n and p for the binomial distribution that is the sampling distribution of the number in the sample who voted for Whitman, and use n and p to find the mean and standard deviation of the sampling distribution of the number in the sample who voted for Whitman. (Use the value for p based on the population vote after the election was concluded.)

b. Find the mean and standard deviation of the sampling distribution of the proportion of the people in the sample who voted for Whitman.

c. Compare the mean and standard deviation in part b to the mean and standard deviation in part a.

d. What is the expected shape of both the sampling distribution in part a and in part b? Explain.

7.29 Simulating the sampling distribution of counts Use the Sampling Distributions applet. Set the population to Binary, $p = 0.6$. The graph of the population distribution for a categorical variable with p, the population proportion, equal to 0.60 should appear. Let's simulate a sample of size $n = 100$ from this population. Set $n = 100$, then click on *Sample.*

a. Using the second graph and the numerical summary to the side of the graph, how does the data distribution (the sample data) compare to the population distribution represented in the first graph?

b. Let's simulate 1000 samples of size $n = 100$ from this population. Set N = 1000. Then click on *Sample.* Let's first consider the counts of the successes (sum of the 1s) from each sample. Go to the third graph. This is a histogram of all the counts or number of successes in each simulated sample (there should be a total of 1001 samples) along with descriptive statistics in the box to the left of the third graph.

Describe the shape, center (mean), and variability (standard deviation) of this distribution. Note that this is a sampling distribution of counts with $n = 100$. What would you expect for the shape, mean and standard deviation of this sampling distribution of counts? How do these expected values compare to your simulated values?

c. Let's work with the proportions instead of counts. Go to the fourth graph. This is a histogram of all the sample proportions of 1s (successes) in each simulated sample along with descriptive statistics.

Describe the shape, center (mean), and variability (standard deviation) of the distribution. Note that this is a **sampling distribution of sample proportions with $n = 100$**. What would you expect for the shape, mean and standard deviation of this sampling distribution of the sample proportion? How do these expected values compare to your simulated values?

d. Compare the simulated sampling distribution of sample proportions to the simulated sampling distribution of counts with respect to shape, the means, and the standard deviations.

Chapter Review

ANSWERS TO FIGURE QUESTIONS

Figure 7.1 The larger the sample size, the more we expect the data distribution to resemble the population distribution.

Figure 7.3 For the 10,000 simulations of sample size 4000 each, none of the samples had sample proportions voting for Brown that were below 0.515 or above 0.565.

Figure 7.4 Bell shaped with a mean of 0.60 and a standard deviation of $\sqrt{(0.60)(0.40)/3889} = 0.008$. Nearly the entire distribution would fall within $3(0.008) = 0.024$ of 0.60, that is, between 0.576 and 0.624.

Figure 7.7 The probability distribution is highly skewed left. However, the sampling distribution of sample means from this probability distribution is bell shaped. The mean of the sampling distribution is the same as the mean of the probability distribution, 4.5. The standard deviation of the sampling distribution is smaller than the standard deviation of the probability distribution. The standard deviation of the sampling distribution equals the standard deviation of the probability distribution divided by the square root of the sample size n, that is, $\frac{\sigma}{\sqrt{n}} = \frac{1.4}{\sqrt{30}} = 0.26$.

Figure 7.10 The probability distribution and the sampling distribution do not have the same shape. The sampling distribution is bell shaped. The probability distribution is discrete, concentrated at two values, −\$10 with height of 20/38 and \$10 with height of 18/38.

CHAPTER SUMMARY

- The **sampling distribution** is the probability distribution of a sample statistic, such as a sample proportion or sample mean. With random sampling, it provides probabilities for all the possible values of the statistic.
- The **population distribution** is the probability distribution from which we take the sample. It is described by *parameters*, the values of which are usually unknown. The **data distribution** describes the sample data. It's the distribution we actually see in practice and is described by sample *statistics*, such as a sample proportion or a sample mean. The **sampling distribution** provides the key for telling us how close a sample statistic falls to the unknown parameter we'd like to make an inference about.

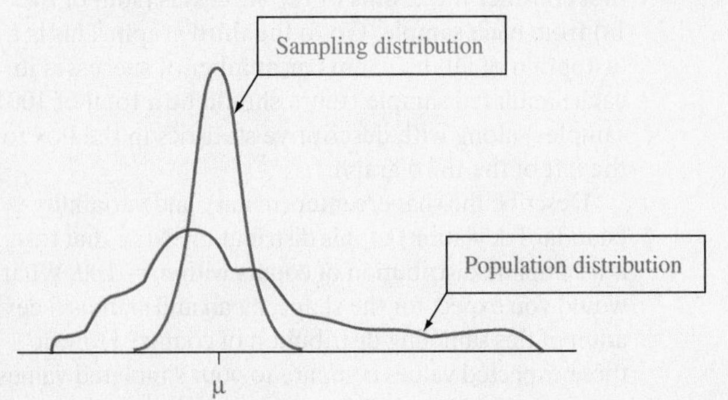

- The **sampling distribution of the sample mean** $\bar{x}$ is centered at the population mean μ. Its variability is described by $\sigma/\sqrt{n}$. For binary data (0 or 1), the sampling distribution of the sample proportion is centered at the population proportion p. The **standard deviation of the sample proportion** for an outcome simplifies to $\sqrt{p(1-p)/n}$.
- The **central limit theorem** states that for random samples of sufficiently large size (at least about 30 is usually enough), the **sampling distribution of the sample mean is approximately normal**. This theorem holds *no matter what the shape of the population distribution*. It applies to sample proportions as well, because the sample proportion is a sample mean when the possible values are 0 and 1. In that case, the sampling distribution of the sample proportion is approximately normal whenever n is large enough that both np and $n(1 - p)$ are at least 15. The bell-shaped appearance of the sampling distributions for most statistics is the main reason for the importance of the normal distribution.

SUMMARY OF NEW NOTATION IN CHAPTER 7

μ Mean of population distribution

σ Standard deviation of population distribution

p Population proportion

CHAPTER PROBLEMS

Practicing the Basics

7.30 Exam performance An exam consists of 50 multiple-choice questions. Based on how much you studied, for any given question you think you have a probability of $p = 0.70$ of getting the correct answer. Consider the sampling distribution of the sample proportion of the 50 questions on which you get the correct answer.

a. Find the mean and standard deviation of the sampling distribution of this proportion.

b. What do you expect for the shape of the sampling distribution? Why?

c. If truly $p = 0.70$, would it be very surprising if you got correct answers on only 60% of the questions? Justify your answer by using the normal distribution to approximate the probability of a sample proportion of 0.60 or less.

7.31 Blue eyes According to a *Boston Globe* story, only about 1 in 6 Americans have blue eyes, whereas in 1900 about half had blue eyes. (*Source:* Data from *The Boston Globe*, October 17, 2006.)

a. For a random sample of 100 living Americans, find the mean and standard deviation of the proportion that have blue eyes.

b. In a course you are taking with 100 students, half of the students have blue eyes. Would this have been a surprising result if the sample were a random sample of Americans? Answer by finding how many standard deviations that sample result falls from the mean of the sampling distribution of the proportion of 100 students who have blue eyes.

c. In part b, identify the population distribution, the data distribution, and the sampling distribution of the sample proportion.

7.32 Alzheimer's According to the Alzheimer's Association[1], as of 2011 Alzheimer's disease affects 1 in 8 Americans over the age of 65. A study is planned of health problems faced by the elderly. For a random sample of Americans over the age of 65, report the shape, mean, and standard deviation of the sampling distribution of the proportion who suffer from Alzheimer's disease, if the sample size is (a) 200 and (b) 800.

7.33 Basketball shooting In college basketball, a shot made from beyond a designated arc radiating about 20 feet from the basket is worth three points, instead of the usual two points given for shots made inside that arc. Over his career, University of Florida basketball player Lee Humphrey made 45% of his three-point attempts. In one game in his final season, he made only 3 of 12 three-point shots, leading a TV basketball analyst to announce that Humphrey was in a shooting slump.

a. Assuming Humphrey has a 45% chance of making any particular three-point shot, find the mean and standard deviation of the sampling distribution of the proportion of three-point shots he will make out of 12 shots.

b. How many standard deviations from the mean is this game's result of making 3 of 12 three-point shots?

c. If Humphrey was actually not in a slump but still had a 45% chance of making any particular three-point shot, explain why it would not be especially surprising for him to make only 3 of 12 shots. Thus, this is not really evidence of a shooting slump.

7.34 Baseball hitting Suppose a baseball player has a 0.200 probability of getting a hit in each time at-bat.

a. Describe the shape, mean, and standard deviation of the sampling distribution of the proportion of times the player gets a hit after 36 at-bats.

b. Explain why it would not be surprising if the player has a 0.300 batting average after 36 at-bats.

7.35 Exit poll CNN conducted an exit poll of 1751 voters in the 2010 Senatorial election in New York between Charles Schumer and Jay Townsend. It is possible that all 1751 voters sampled happened to be Charles Schumer supporters. Investigate how surprising this would be, if actually 65% of the population voted for Schumer, by

a. Finding the probability that all 1751 people voted for Schumer. (*Hint:* Use the binomial distribution.)

b. Finding the number of standard deviations that a sample proportion of 1.0 for 1751 voters falls from the population proportion of 0.65.

7.36 Aunt Erma's restaurant In Example 5 about Aunt Erma's Restaurant, the daily sales follow a probability distribution that has a mean of $\mu = \$900$ and a standard deviation of $\sigma = \$300$. This past week the daily sales for the seven days had a mean of \$980 and a standard deviation of \$276.

a. Identify the mean and standard deviation of the population distribution.

b. Identify the mean and standard deviation of the data distribution. What does the standard deviation describe?

[1]www.alz.org/documents_custom/2011_Facts_Figures_Fact_Sheet.pdf.

c. Identify the mean and the standard deviation of the sampling distribution of the sample mean for samples of seven daily sales. What does this standard deviation describe?

7.37 Home runs Based on data from the 2010 major league baseball season, X = number of home runs the San Francisco Giants hits in a game has a mean of 1.0 and a standard deviation of 1.0.

a. Do you think that X has a normal distribution? Why or why not?

b. Suppose that this year X has the same distribution. Report the shape, mean, and standard deviation of the sampling distribution of the mean number of home runs the team will hit in its 162 games.

c. Based on the answer to part b, find the probability that the mean number of home runs per game in this coming season will exceed 1.50.

7.38 Physicians' assistants The 2006 AAPA survey of the population of physicians' assistants who were working full time reported a mean annual income of \$84,396 and standard deviation of \$21,975. (*Source:* Data from 2006 AAPA survey [www.aapa.org].)

a. Suppose the AAPA had randomly sampled 100 physicians' assistants instead of collecting data for all of them. Describe the mean, standard deviation, and shape of the sampling distribution of the sample mean.

b. Using this sampling distribution, find the z-score for a sample mean of \$80,000.

c. Using parts a and b, find the probability that the sample mean would fall within approximately \$4000 of the population mean.

7.39 Bank machine withdrawals An executive in an Australian savings bank decides to estimate the mean amount of money withdrawn in bank machine transactions. From past experience, she believes that \$50 (Australian) is a reasonable guess for the standard deviation of the distribution of withdrawals. She would like her sample mean to be within \$10 of the population mean. Estimate the probability that this happens if she randomly samples 100 withdrawals. (*Hint:* Find the standard deviation of the sample mean. How many standard deviations does \$10 equal?)

7.40 PDI The scores on the Psychomotor Development Index (PDI), a scale of infant development, have a normal population distribution with mean 100 and standard deviation 15. An infant is selected at random.

a. Find the z-score for a PDI value of 90.

b. A study uses a random sample of 225 infants. Using the sampling distribution of the sample mean PDI, find the z-score corresponding to a sample mean of 90.

c. Explain why a PDI value of 90 is not surprising, but a sample mean PDI score of 90 for 225 infants would be surprising.

7.41 Number of sex partners According to recent General Social Surveys, in the United States the population distribution for adults of X = number of sex partners in the past 12 months has a mean of about 1.0 and a standard deviation of about 1.0.

a. Does X have a normal distribution? Explain.

b. For a random sample of 100 adults, describe the sampling distribution of $\bar{x}$ and give its mean and standard deviation. What is the effect of X not having a normal distribution?

7.42 ◆◆ **Using control charts to assess quality** In many industrial production processes, measurements are made periodically on critical characteristics to ensure that the process is operating properly. Observations vary from item to item being produced, perhaps reflecting variability in material used in the process and/or variability in the way a person operates machinery used in the process. There is usually a target mean for the observations, which represents the long-run mean of the observations when the process is operating properly. There is also a target standard deviation for how observations should vary around that mean if the process is operating properly. A **control chart** is a method for plotting data collected over time to monitor whether the process is operating within the limits of expected variation. A control chart that plots *sample means* over time is called an **$\bar{x}$-chart.** As shown in the following, the horizontal axis is the time scale and the vertical axis shows possible sample mean values. The horizontal line in the middle of the chart shows the target for the true mean. The upper and lower lines are called the **upper control limit** and **lower control limit,** denoted by **UCL** and **LCL.** These are usually drawn 3 standard deviations above and below the target value. The region between the LCL and UCL contains the values that theory predicts for the sample mean when the process is in control. When a sample mean falls above the UCL or below the LCL, it indicates that something may have gone wrong in the production process.

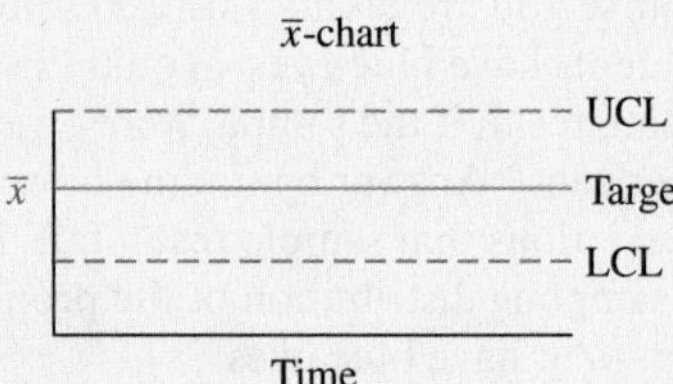

a. Walter Shewhart invented this method in 1924 at Bell Labs. He suggested using 3 standard deviations in setting the UCL and LCL to achieve a balance between having the chart fail to diagnose a problem and having it indicate a problem when none actually existed. If the process is working properly ("in statistical control") and if n is large enough that $\bar{x}$ has approximately a normal distribution, what is the probability that it indicates a problem when none exists? (That is, what's the probability a sample mean will be at least 3 standard deviations from the target, when that target is the true mean?)

b. What would the probability of falsely indicating a problem be if we used 2 standard deviations instead for the UCL and LCL?

c. When about nine sample means in a row fall on the same side of the target for the mean in a control chart, this is an indication of a potential problem, such as a shift up or a shift down in the true mean relative to the target value. If the process is actually in control and has a normal distribution around that mean, what is the probability that the next nine sample means in a row would (i) all fall above the mean and (ii) all fall above or all fall below the mean? (*Hint:* Use the binomial distribution, treating the successive observations as independent.)

7.43 Too little or too much cola? Refer to the previous exercise. When a machine for dispensing a cola drink into bottles is in statistical control, the amount dispensed has a mean of 500 ml (milliliters) and a standard deviation of 4 ml.

a. In constructing a control chart to monitor this process with periodic samples of size 4, how would you select the target line and the upper and lower control limits?

b. If the process actually deteriorates and operates with a mean of 491 ml and a standard deviation of 6 ml, what is the probability that the next value plotted on the control chart indicates a problem with the process, falling more than 3 standard deviations from the target? What do you assume in making this calculation?

Concepts and Investigations

7.44 CLT for custom population Use the Sampling Distributions applet by selecting the Custom population distribution. Using your mouse, create your own population distribution, making it far from bell shaped. Explain how the variability and shape of the sampling distribution of the sample mean changes as n changes from 2 to 5 to 30 to 100, taking N = 10,000 samples each time. Explain how the central limit theorem describes what you have observed.

7.45 What is a sampling distribution? How would you explain to someone who has never studied statistics what a sampling distribution is? Explain by using the example of polls of 1000 Canadians for estimating the proportion who think the prime minister is doing a good job.

7.46 What good is a standard deviation? Explain how the standard deviation of the sampling distribution of a sample proportion gives you useful information to help gauge how close a sample proportion falls to the unknown population proportion.

7.47 Purpose of sampling distribution You'd like to estimate the proportion of all students in your school who are fluent in more than one language. You poll a random sample of 50 students and get a sample proportion of 0.12. Explain why the standard deviation of the sampling distribution of the sample proportion gives you useful information to help gauge how close this sample proportion is to the unknown population proportion.

7.48 Comparing pizza brands The owners of Aunt Erma's Restaurant plan an advertising campaign with the claim that more people prefer the taste of their pizza (which we'll denote by A) than the current leading fast-food chain selling pizza (which we'll denote by D). To support their claim, they plan to randomly sample three people in Boston. Each person is asked to taste a slice of pizza A and a slice of pizza D. Subjects are blindfolded so they cannot see the pizza when they taste it, and the order of giving them the two slices is randomized. They are then asked which pizza tastes better. Use a symbol with three letters to represent the responses for each possible sample. For instance, ADD represents a sample in which the first subject sampled preferred pizza A and the second and third subjects preferred pizza D.

a. Identify the eight possible samples of size 3, and for each sample report the proportion that preferred pizza A.

b. In the entire Boston population, suppose that exactly half would prefer pizza A and half would prefer pizza D. Explain why the sampling distribution of the sample proportion who prefer Aunt Erma's pizza, when $n = 3$, is

Sample Proportion	Probability
0	1/8
1/3	3/8
2/3	3/8
1	1/8

c. In part b, we can also find the probabilities for each possible sample proportion value using the binomial distribution. Use the binomial with $n = 3$ and $p = 0.50$ to show that the probability of a sample proportion of 1/3 equals 3/8. (*Hint:* This equals the probability that $x = 1$ person out of $n = 3$ prefer pizza A. It's especially helpful to use the binomial formula when p differs from 0.50, since then the eight possible samples listed in part a would not be equally likely.)

7.49 Simulating pizza preference Use the Sampling Distributions applet on the text CD.

a. For the previous exercise, in which $n = 3$ people indicate their preferred pizza, simulate the sampling distribution of the *number* of people preferring pizza A when $p = 0.50$, by taking 10,000 samples of size 3 from a binary population. Sketch a plot of the sampling distribution that you generate.

b. Simulate the sampling distribution of sample *proportions* with $n = 3$ subjects, by taking 10,000 samples of size 3. Sketch a plot of the sampling distribution that you generate. Compare this generated sampling distribution to the sampling distribution in part a. Comment on the shape, means, and standard deviations. Does your sampling distribution for sample proportions look roughly like the one shown on the next page?

c. Repeat part b using a sample size $n = 100$ when $p = 0.50$. Compare the sampling distribution that you generated to the sampling distribution generated in part b. Comment on similarities and differences.

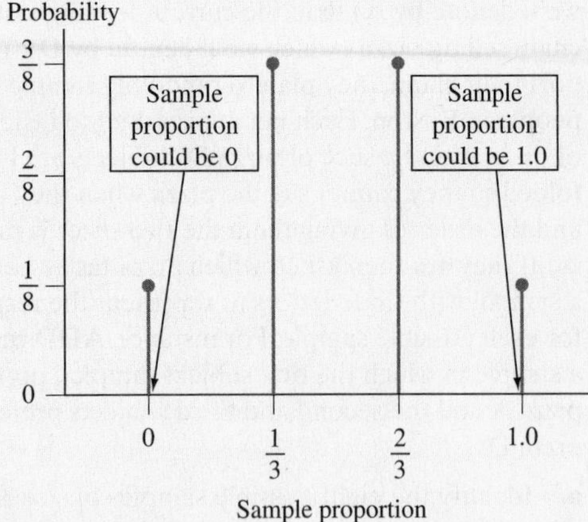

7.50 Winning at roulette Part b of Example 7 used the central limit theorem to approximate the probability of coming out ahead if you bet \$10 on red on each of 40 different roulette wheel spins. For each spin, the winnings are \$10 with probability 18/38 and −\$10 with probability 20/38. You are interested in the probability of winning at least \$100 over the course of the 40 spins.

a. The average winnings per spin need to be at least how much in order to win at least \$100?

b. Use the Central Limit Theorem to determine the approximate probability of winning at least \$100?

c. You must win at least how many of the spins in order to win at least \$100?

d. Using the binomial distribution, calculate the exact probability of winning at least \$100 in the 40 spins. How does your answer compare with that of part b?

7.51 True or false As the sample size increases, the standard deviation of the sampling distribution of $\bar{x}$ increases. Explain your answer.

7.52 Multiple choice: Standard deviation Which of the following is *not* correct? The standard deviation of a statistic describes

a. The standard deviation of the sampling distribution of that statistic.

b. The standard deviation of the sample data measurements.

c. How close that statistic falls to the parameter that it estimates.

d. The variability in the values of the statistic for repeated random samples of size *n*.

7.53 Multiple choice: CLT The central limit theorem implies

a. All variables have approximately bell-shaped data distributions if a random sample contains at least about 30 observations.

b. Population distributions are normal whenever the population size is large.

c. For sufficiently large random samples, the sampling distribution of $\bar{x}$ is approximately normal, regardless of the shape of the population distribution.

d. The sampling distribution of the sample mean looks more like the population distribution as the sample size increases.

7.54 Multiple choice: Sampling distribution The sampling distribution of a sample mean for a random sample size of 100 describes

a. How sample means tend to vary from random sample to random sample of size 100.

b. How observations tend to vary from person to person in a random sample of size 100.

c. How the data distribution looks like the population distribution when the sample size is larger than 30.

d. How the standard deviation varies among samples of size 100.

7.55 ◆◆ **Sample = population** Let X = GPA for students in your school.

a. What would the sampling distribution of the sample mean look like if you sampled *every* student in the school, so the sample size equals the population size? (*Hint:* The sample mean then equals the population mean.)

b. How does the sampling distribution compare to the population distribution if we take a sample of size $n = 1$?

7.56 ◆◆ **Standard deviation of a proportion** Suppose $x = 1$ with probability p, and $x = 0$ with probability $(1 - p)$. Then, x is the special case of a binomial random variable with $n = 1$, so that $\sigma = \sqrt{np(1 - p)} = \sqrt{p(1 - p)}$. With n trials, using the formula $\sigma/\sqrt{n}$ for a standard deviation of a sample mean, explain why the standard deviation of a sample proportion equals $\sqrt{p(1 - p)/n}$.

7.57 ◆◆ **Finite populations** The formula $\sigma/\sqrt{n}$ for the standard deviation of $\bar{x}$ actually is an approximation that treats the population size as *infinitely* large relative to the sample size n. The exact formula for a *finite* population size N is

$$\text{Standard deviation} = \sqrt{\frac{N-n}{N-1}}\frac{\sigma}{\sqrt{n}}.$$

The term $\sqrt{(N - n)/(N - 1)}$ is called the **finite population correction.**

a. When $n = 300$ students are selected from a college student body of size N = 30,000, show that the standard deviation equals $0.995\,\sigma/\sqrt{n}$. (When n is small compared to the population size N, the approximate formula works very well.)

b. If n = N (that is, we sample the entire population), show that the standard deviation equals 0. In other words, no sampling error occurs, since $\bar{x} = \mu$ in that case.

Student Activities

7.58 Simulate a sampling distribution The table provides the ages of all 50 heads of households in a small Nova Scotian fishing village. The distribution of these ages is characterized by $\mu = 47.18$ and $\sigma = 14.74$.

Name	Age	Name	Age	Name	Age	Name	Age
Alexander	50	Griffith	66	McTell	49	Shindell	33
Bell	45	Grosvenor	51	MacLeod	30	Staines	36
Black	23	Ian	57	Mayo	28	Stewart	25
Bok	28	Jansch	40	McNeil	31	Thames	29
Clancy	67	Kagan	36	Mitchell	45	Thomas	57
Cochran	62	Lavin	38	Morrison	43	Todd	39
Fairchild	41	Lunny	81	Muir	43	Travers	50
Finney	68	MacColl	27	Renbourn	54	Trickett	64
Fisher	37	Mallett	37	Rice	62	Tyson	76
Fraser	60	McCusker	56	Rogers, G.	67	Watson	63
Fricker	41	MacDonald	71	Rogers, S.	48	Young	29
Gaughan	70	McDonald	39	Rowan	32		
Graham	47	McTell	46	Rusby	42		

a. Each student or the class should construct a graphical display (stem-and-leaf plot, dotplot, or histogram) of the population distribution of the ages.

b. Each student should select nine random numbers between 01 and 50, with replacement. Using these numbers, each student should sample nine heads of households and find their sample mean age. Collect all sample mean ages. Using technology, construct a graph (using the same graphical display as in part a) of the simulated sampling distribution of the $\bar{x}$-values for all the student samples. Compare it to the distribution in part a.

c. Find the mean of the $\bar{x}$-values in part b. How does it compare to the value you would expect in a long run of repeated samples of size 9?

d. Find the standard deviation of the $\bar{x}$-values in part b. How does it compare to the value you would expect in a long run of repeated samples of size 9?

7.59 Coin-tossing distributions For a single toss of a balanced coin, let $x = 1$ for a head and $x = 0$ for a tail.

a. Construct the probability distribution for x, and calculate its mean. (You can think of this as the population distribution corresponding to a very long sequence of tosses.)

b. The coin is flipped 10 times, yielding 6 heads and 4 tails. Construct the data distribution.

c. Each student in the class should flip a coin 10 times and find the proportion of heads. Collect the sample proportion of heads from each student. Summarize the simulated sampling distribution by constructing a plot of all the proportions obtained by the students. Describe the shape and variability of the sampling distribution compared to the distributions in parts a and b.

d. If you performed the experiment in part c a huge number of times, what would you expect to get for the (i) mean and (ii) standard deviation of the sample proportions?

7.60 Sample versus sampling Each student should bring 10 coins to class. For each coin, observe its age, the difference between the current year and the year on the coin.

a. Using all the students' observations, the class should construct a histogram of the sample ages. What is its shape?

b. Now each student should find the mean for that student's 10 coins, and the class should plot the means of all the students. What type of distribution is this, and how does it compare to the one in part a? What concepts does this exercise illustrate?

Part Review

2

Probability, Probability Distributions, and Sampling Distributions

In Chapters 5, 6, and 7, we've seen how to quantify uncertainty and randomness. This review section gives examples of questions you should be able to answer about the main concepts in these chapters. The questions are followed by brief summaries or hints, as well as references to sections in the text where you can find more detail to help strengthen your understanding of these concepts.

Review Questions

- What is meant by the *probability* of an outcome?

  Section 5.1 explained that the **probability** of an outcome is the proportion of times it occurs, in the long run.

- What are the two properties that the probabilities of outcomes must satisfy?

  The probability of each outcome must fall between 0 and 1. The total of the probabilities of the possible outcomes equals 1.

Let's now review *probability rules*, from Sections 5.2 and 5.3:

- What is meant by the *complement* of an event?

  For an event A, its **complement** consists of all outcomes *not* in that event. For example, for the event that you have a job, the complement is that you do not have a job. The probability of the complement of A, denoted by A^c, relates to the probability of the event A by $P(A^c) = 1 - P(A)$. An event and its complement are **disjoint** events, because they have no common elements.

- For two *independent* events A and B, how do you find the probability that both A *and* B occur?

  When A and B are independent, $P(A \text{ and } B) = P(A) \times P(B)$. The events are independent if the probability that one occurs is not affected by the probability that the other occurs. For example, if A is the event that you get a head when you flip a coin and B is the event that you get a head when you flip it again, then $P(A) = P(B) = 0.5$ and $P(A \text{ and } B) = (0.5) \times (0.5) = 0.25$.

- For two events A and B, how do you find the probability that A *or* B occur?

$$P(\text{A or B}) = P(\text{A}) + P(\text{B}) - P(\text{A and B})$$

For example, the probability you get a head on the first flip of a coin or a head on the second flip equals $0.5 + 0.5 - 0.25 = 0.75$.

- What is meant by a *conditional probability*?

The **conditional probability** of the event B, given the event A, is the probability that B occurs if we know that A occurred. It is denoted by P(B|A). We can find it by

$$P(\text{B}|\text{A}) = \frac{P(\text{A and B})}{P(\text{A})}.$$

We often find a conditional probability using a contingency table.

- When events are *not independent*, how can we find the probability that A and B occur?

The multiplication rule tells us that $P(\text{A and B}) = P(\text{A}) \times P(\text{B}|\text{A})$. It is also the case that $P(\text{A and B}) = P(\text{B}) \times P(\text{A}|\text{B})$.

- How do we define *independent events* in terms of *conditional probabilities*?

When A and B are **independent,** $P(\text{A}|\text{B}) = P(\text{A})$ and $P(\text{B}|\text{A}) = P(\text{B})$. That is, whether A occurs does not depend on whether B occurs. In that case, also $P(\text{A and B}) = P(\text{A}) \times P(\text{B})$.

After Chapter 5 introduced the basic concepts of probability, Chapter 6 showed how to summarize and find probabilities using *probability distributions* for *random variables.*

- What is a *random variable*?

Section 6.1 explained that a **random variable** is a numerical measurement of the outcome of a random phenomenon. Random variables can be **discrete** (taking separate values) or **continuous** (taking an interval of values).

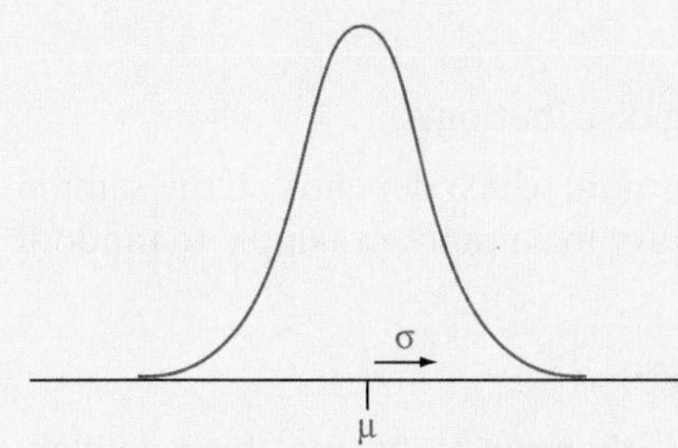

- What is meant by the term *probability distribution*?

A **probability distribution** specifies probabilities for all the possible values of a random variable. The distribution is summarized by a mean μ and a standard deviation σ to describe its center and its variability from the mean. (See figure in margin.)

- How would you describe the *normal distribution*?

Section 6.2 explained that the **normal distribution** is the probability distribution of a *continuous* random variable that has a certain symmetric bell-shaped graph. For any z, the probability of falling within z standard deviations of the mean is the same for every normal distribution. For example, the probability is 0.95 within $z = 1.96$ standard deviations of the mean.

Recall

We use z-scores to standardize the unit of measurement for a normal distribution. Once the units are standardized, we have a standard normal distribution with mean 0 and standard deviation of 1. ◀

- What is a ***z*-score**, and how can you find it for a particular observed value?

A z-score tells us the number of standard deviations and the direction that an observation falls from the mean. For a value x from a probability distribution with mean μ and standard deviation σ, the z-score is

$$z = \frac{x - \mu}{\sigma}.$$

Caution

Using z-scores doesn't imply we are working with a normal distribution. We can use z-scores to standardize unit of measurements in any distribution. ◀

For the normal distribution, for example, the cumulative probability below any given value depends on how many standard deviations that value falls from the mean, so the table of probabilities for the normal distribution uses z-scores.

- How would you describe the *binomial distribution*?

  From Section 6.3, the **binomial distribution** is the probability distribution of the *discrete* random variable that measures the number of successes X in n independent trials, with probability p of a success on a given trial. The probability of outcome x for X has the formula

$$P(x) = \frac{n!}{x!(n-x)!} p^x (1-p)^{n-x}, \quad x = 0, 1, 2, \ldots, n.$$

  The mean of the binomial distribution is $\mu = np$ and the standard deviation is $\sigma = \sqrt{np(1-p)}$. The sample proportion of successes is x/n.

In Chapter 7, we learned about a particular type of probability distribution, called a *sampling distribution*, that is the key to statistical inference. This chapter also explained why the normal distribution is so important for the statistical inference methods presented in the rest of this book.

- What is a *sampling distribution*?

  Section 7.1 explained that the **sampling distribution** is the probability distribution of a sample statistic, such as a sample proportion or a sample mean. It is the distribution we'd obtain if we repeatedly took random samples of a fixed size n, calculated the statistic each time, and formed a histogram of the values of the statistic. In practice, we won't need to do this, because there are results that tell us the shape, center, and variability of sampling distributions for various statistics.

- Why is a sampling distribution important in statistical inference?

  The sampling distribution tells us how close a sample statistic is likely to fall to the unknown value of the population parameter for which we want to make an inference. For instance, using a sampling distribution, we can find the margin of error within which a sample proportion in an exit poll is likely to fall from the proportion of the entire population that voted for a particular candidate.

- What is a standard deviation of a sampling distribution?

  A standard deviation of a sampling distribution, which depends on the sample size, tells us how much a statistic tends to vary from random sample to random sample of that size.

- What does the *central limit theorem* tell us?

  As Section 7.2 explained, the **central limit theorem** states that for relatively large random samples, the sampling distribution of the sample mean is approximately a normal distribution. The bell-shape for this sampling distribution happens *no matter what the shape of the population distribution.*

- Why is the *normal* distribution so important for *statistical inference*?

  As discussed previously, the sampling distribution is the key in statistical inference for determining how close sample statistics fall to population parameters. The bell-shaped appearance of the sampling distributions for statistics such as the sample mean and the sample proportion is the main reason that we'll use the normal distribution throughout the rest of the text for methods of statistical inference.

Here's an example of the type of exercise you should be able to answer at this stage of the course:

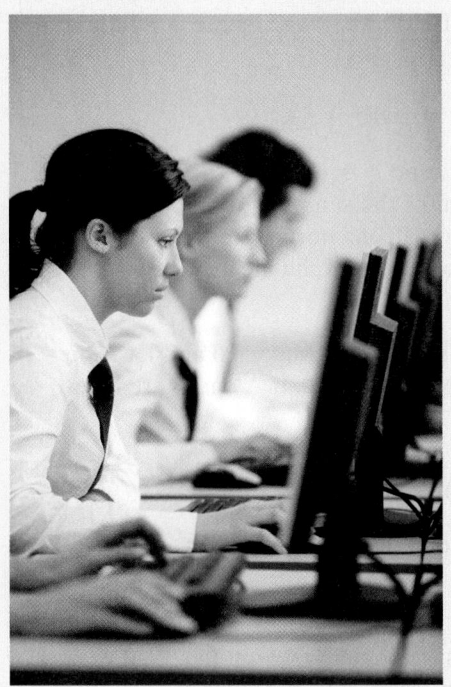

Example

Applying to Graduate School: GRE Performance

Picture the Scenario

The Graduate Record Examination (GRE) general test has components that measure verbal reasoning and quantitative reasoning. The verbal exam and the quantitative exam each have a minimum score of 200 and maximum score of 800. In recent years, the total score on the two exams has had approximately a normal distribution with a mean of about 1050 and standard deviation of about 200.

Questions to Explore

a. About what proportion of total scores fall (i) below 1200 and (ii) above 1200?

b. Of those GRE test-takers who scored above 1200, about what proportion of them scored above 1400?

c. A group of 25 students formed a study group to prepare for the GRE. For them, the mean of their 25 total scores is 1200. If they were a random sample of the students taking the exam, explain why this would have been a very unusual outcome.

d. In recent years, 40% of the verbal scores and 74% of the quantitative scores have been above 500. If performance on the two exams is independent, find the percentage of test-takers who would score above 500 both on the verbal exam and on the quantitative exam.

Think It Through

a. A total score of 1200 is $z = (1200 - 1050)/200 = 0.75$ standard deviation above the mean. From Table A or using a calculator or software, the cumulative probability below that point is 0.77. That is, about 77% of the total scores fall below 1200. The event that the score falls above 1200 is the complement of the event that it falls below 1200. So, its probability is about $1 - 0.77 = 0.23$, that is, about 23% of the total scores fall above 1200.

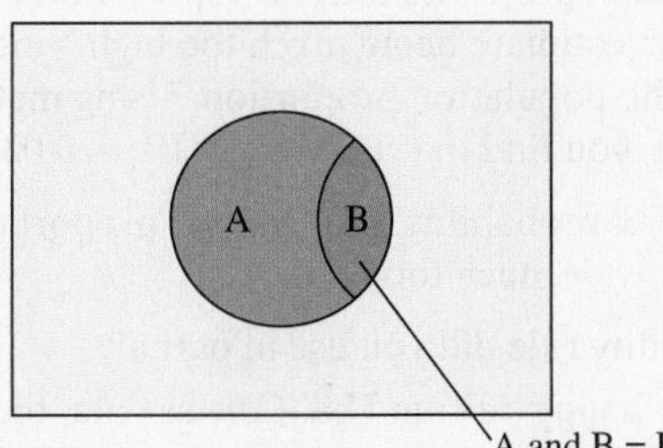

b. A score of 1400 is $z = (1400 - 1050)/200 = 1.75$ standard deviations above the mean. From Table A or software, it has a cumulative probability of 0.96, so the probability is about $1 - 0.96 = 0.04$ of scoring above 1400. Let A be the event of scoring above 1200 and let B be the event of scoring above 1400. Then, $P(B) = 0.04$, and from part a, $P(A) = 0.23$. We need to find the conditional probability $P(B|A)$, which equals P(A and B)/P(A). The event "A and B" of scoring above 1200 *and* above 1400 is just the event B of scoring above 1400 (See Venn Diagram in margin.) So,

$$P(B|A) = P(A \text{ and } B)/P(A) = 0.04/0.23 = 0.17.$$

In summary, of those who scored above 1200, about 17% scored above 1400.

c. Because the distribution of total GRE scores is approximately normal, the sampling distribution of the sample mean for a random sample of size $n = 25$ is also approximately normal. (If the distribution of GRE scores were *not* normal, the sampling distribution would

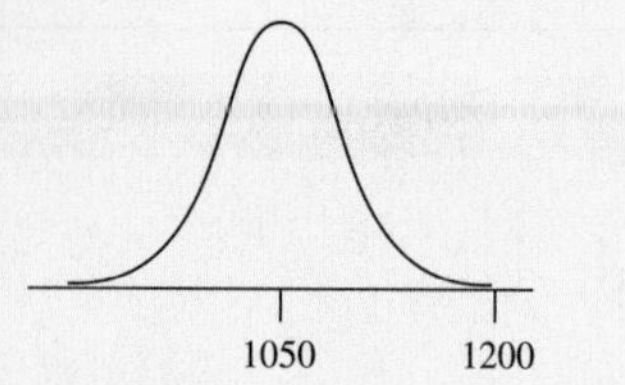

Sampling Distribution of a Sample Mean for GRE scores

still be approximately bell shaped, by the central limit theorem.) The sampling distribution has a mean of 1050, the same as the population mean, and a standard deviation of

$$\sigma/\sqrt{n} = 200/\sqrt{25} = 200/5 = 40.$$

A sample mean score of 1200 would be $z = (1200 - 1050)/40 = 3.75$ standard deviations above the mean of the sampling distribution. This would be very unusual, because nearly all of a normal sampling distribution falls within about 3 standard deviations of the mean. (See figure in margin.)

d. Let A be the event that the verbal score falls above 500 and let B be the event that the quantitative score is above 500. If A and B are independent events, then

$$P(\text{A and B}) = P(A) \times P(B) = 0.40 \times 0.74 = 0.296.$$

About 30% would score above 500 on both exams.

Insight

In reality, performance on the two exams is not independent. In recent years, the correlation between the verbal exam score and the quantitative exam score equals 0.23. So, there is a weak positive association between the two exam scores: Those who scored relatively high on one exam also had a slight tendency to score relatively high on the other exam.

Try Exercises R2.10 and R2.11

Part 2 Review Exercises

Practicing the Basics

R2.1 Vote for Jerry Brown? Let A represent the outcome that a randomly selected voter in the 2010 California gubernatorial election cast his or her vote for Jerry Brown.

a. If P(A) = 0.54, then find the probability of *not* voting for Brown

b. What probability rule did you use in part a?

R2.2 Correct inferences An inference method that will be presented in Chapter 8 (confidence intervals) often is used with the probability of a correct inference set at 0.95. Suppose the event A represents a prediction being correct that the interval (0.83, 0.89) contains the population proportion of men who believe in God. Let B represent a separate inference being correct that the interval (0.86, 0.91) contains the population proportion of women who believe in God.

a. Find the probability that *both* inferences are correct, assuming independence.

b. What probability rule did you use in part a?

R2.3 Embryonic stem cell research You take a survey to estimate the population proportion of people who believe that embryonic stem cell research should be banned by the federal government. Let A represent the sample proportion estimate being much too low—more than 0.10 *below* the population proportion. Let B represent the sample proportion estimate being much too high—more than 0.10 *above* the population proportion. Using methods from this text, you find that P(A) = P(B) = 0.03.

a. Find the overall probability your sample proportion is much too low *or* much too high.

b. What probability rule did you use in part a?

R2.4 Married and very happy From U.S. Census data, the probability that a randomly selected American adult is married equals 0.56. Of those who are married, General Social Surveys indicate that the probability a person reports being "very happy" when asked to choose among (very happy, pretty happy, not too happy) is 0.40.

a. Find the probability a person reports being both married *and* very happy.

b. What probability rule did you use in part a?

R2.5 Heaven and hell A General Social Survey estimates the probability that an American adult believes in heaven is 0.85.

a. Estimate the probability that an American adult does *not* believe in heaven.

b. Of those who believe in heaven, about 85% believe in hell. Estimate the probability a randomly chosen American adult believes in both heaven and hell.

R2.6 Environmentally green One year the GSS asked subjects whether they are a member of an environmental group (variable GRNGROUP) and whether they would be very willing to pay higher taxes to protect the environment (variable GRNTAXES). The table shows results.

		Pay Higher Taxes		
		Yes	**No**	**Total**
Member of	Yes	56	24	80
Environmental Group	No	307	438	745
	Total	363	462	825

a. Explain why 80/825 = 0.097 estimates the probability that a randomly selected American adult is a member of an environmental group.

b. Show that the estimated probability of being very willing to pay higher taxes to protect the environment is (i) 0.7, given that the person is a member of an environmental group and (ii) 0.412, given that a person is not a member of an environmental group.

c. Show that the estimated probability a person is both a member of an environmental group *and* very willing to pay higher taxes to protect the environment is 0.068 (i) directly using the counts in the table and (ii) using the probability estimates from parts a and b.

d. Show that the estimated probability a person answers yes to both questions or answers no to both questions is 0.599.

R2.7 UK lottery In the main game in the weekly UK national lottery, you select six different numbers from a list of 49 numbers. The six winning numbers are selected at random.

a. What is the probability that a given number you select is one of the six winning numbers?

b. The probability you get all six numbers correct is 1/13,983,816. If you play once a week for the next 50 years (i.e., 2600 times), what is the expected value for the number of times you would win the jackpot? (*Hint:* How do you find the mean of a binomial distribution?)

c. How many years would you need to play in order for the expected number of times you win to equal one? (But winning does not guarantee happiness. One recent winner in the UK bought himself an expensive sports car and died soon afterward when he crashed it.)

R2.8 SAT quartiles One section of the SAT college entrance exam has a normal distribution with a mean of 500 and a standard deviation of 100.

a. What proportion of scores fell above 650?

b. Find the lower quartile of the SAT scores.

R2.9 Fraternal bias? A fraternal organization admits 80% of all applicants who satisfy certain requirements. Of four members of a minority group who recently applied for admission, all met the requirements but none was accepted. Find the probability that none would be accepted if the same admissions standards were applied to the minority group, other things being equal.

R2.10 Verbal GRE scores In recent years, scores on the verbal portion of the GRE have had a mean of 467 and standard deviation of 118.

TRY

a. If the distribution is normal, about what proportion of verbal scores fall (i) below 500 and (ii) above 500?

b. If the distribution is normal, of those who scored above 500, about what proportion scored above 600?

c. A group of 30 students forms a study group to prepare for the exam. If they are a random sample of the students taking the exam, would it be surprising if the mean of their scores equals 600? Why?

d. In fact, the verbal GRE distribution is not normal but is slightly skewed right. Does this affect your answer in part c? Why or why not?

e. In recent years, 15% of the verbal scores and 53% of the quantitative scores have been above 600. If performance on the two exams is independent, find the percentage of test-takers who scored above 600 both on the verbal exam and on the quantitative exam.

R2.11 Quantitative GRE scores In recent years, scores on the quantitative portion of the GRE have had a mean of 591 and standard deviation of 148.

TRY

a. If the distribution is normal, about what proportion of quantitative scores fall (i) below 700 and (ii) above 700?

b. In fact, the distribution is not exactly normal. How can you tell this from the values of the mean and standard deviation, given that the scores fall between 200 and 800?

c. Describe the shape, mean, and standard deviation of the sampling distribution of the sample mean quantitative exam score for a random sample of 100 people who take this exam.

R2.12 Baseball hitting A particular major-league baseball player has a 0.280 probability of getting a hit each time he bats in a game.

a. After a season of 500 at-bats, describe the shape, mean, and standard deviation of the sampling distribution of the proportion of times the player gets a hit.

b. Would it be surprising if the player got a hit more than 30% of the time in this season? Explain your reasoning.

R2.13 Ending the war in Afghanistan In a Gallup poll taken during June 2011, 72% of a random sample of 1034 U.S. adults favored President Obama's plan to withdraw all American troops from Afghanistan by no later than 2014.[1]

a. Using the sample proportion to estimate the population proportion, p, find the standard deviation for the distribution of the sample proportion.

b. Interpret the standard deviation found in part a.

---

[1]www.gallup.com/poll/148313/Americans-Broadly-Favor-Obama-Afghanistan-Pullout-Plan.aspx.

R2.14 Estimating mean text time You plan to take a random sample of 36 students at your school to estimate the mean amount of time per week that students spend text messaging. The distribution of text time is not normal (for example, a certain percentage of students never text). Unknown to you, the mean weekly text time for the population of students is 100 minutes, with a standard deviation of 60 minutes.

a. What is the shape of the sampling distribution for the sample mean text time for the 36 students in your sample? How do you know this?

b. Identify the mean and standard deviation of this sampling distribution.

c. Find the probability that your sample mean is not a very good estimate of the population mean, falling more than 20 minutes above or 20 minutes below the population mean.

R2.15 Ice cream sales Joe DiMento's Gelateria in Irvine, California sells ice cream and related products. Past experience indicates that daily sales follow a probability distribution that has a mean of $\mu = \$1000$ and a standard deviation of $\sigma = \$300$. This past week the daily sales for the seven days had a mean of \$880 and a standard deviation of \$276.

a. Identify the mean and standard deviation of the population distribution.

b. Identify the mean and standard deviation of the data distribution.

c. Find the mean and the standard deviation of the sampling distribution of the sample mean for a random sample of seven daily sales. What does this standard deviation describe?

R2.16 Election poll For the fall 2010 senatorial election in California between Democratic incumbent Barbara Boxer and Republican challenger Carly Fiorina, the CNN exit poll of 3870 voters indicated that 2015 of those voters voted for Boxer and 1855 voted for Fiorina or another candidate.[2] Of the 9,534,523 voters in the actual election, 5,218,137 voted for Boxer and 4,316,386 did not vote for Boxer.

a. Identify the data distribution.

b. Identify the population distribution.

c. Identify the shape, mean, and standard deviation of the sampling distribution of the sample proportion for a random sample of 3870 voters.

R2.17 NY exit poll In an exit poll of 1751 voters in the 2010 senatorial election in New York State, 65% said they voted for Charles Schumer.[3] Based on this information, would you be willing to predict the winner of the election? Explain your reasoning.

R2.18 Rising materialism An Associated Press story (February 23, 2007) about UCLA's annual survey of college freshmen indicated that 73% of college freshmen in 2006 considered being financially well off to be very important, compared to 42% in 1966 (the first year the survey was done). It also reported that 81% of 18- to 25-year-olds in the United States see getting rich as a top goal in life. We need to know two things to be able to find standard deviations for these estimates. The first is whether the surveys took random samples. What is the second? (*Hint*: What else must you know to use the standard deviation formula for a sample proportion?)

Concepts and Investigations

R2.19 Breast cancer gene test On February 6, 2007, the MammaPrint test became the first genetic test for breast cancer to be formally approved by the Food and Drug Administration. It predicts whether women with early stage breast cancer have high risk of a relapse in 5–10 years. An Associated Press story about the test stated that when the MammaPrint predicts that a woman is at high risk of cancer returning, it is right only a quarter of the time, but when a woman is not predicted to be high risk, it is 95% accurate. Let H denote the event that MammaPrint predicts a woman has high risk of a relapse. Let B denote the event that the woman has a relapse of breast cancer within five years.

a. Express 0.25 as a conditional probability involving these events and/or their complements.

b. Express 0.95 as a conditional probability involving these events and/or their complements.

c. Suppose 20% of women are predicted to be at high risk. Construct a tree diagram showing what you would expect with a typical sample of 100 women. How many women total would you expect to relapse?

R2.20 Exit poll In 2010, voters in California voted on a ballot measure entitled Proposition 19. This special ballot measure was used to determine whether marijuana would be legalized in California. In a CNN exit poll of 3895 voters in the ballot measure, let X = the number in the exit poll who voted yes (legalize marijuana).[4]

a. Explain why this scenario would seem to satisfy the three conditions needed to use the binomial distribution.

b. If the population proportion voting yes for Proposition 19 had been 0.50, find the mean and standard deviation of the probability distribution of X.

c. Now, actually the exit poll had $x = 1809$. Do you have enough information to make a prediction about the result of the ballot measure? Why? (*Hint*: Using the normal distribution that approximates the binomial distribution, give an interval in which X would almost certainly fall if the population proportion were 0.50.)

R2.21 Sample means vary We'd like to collect data and find a sample mean to estimate a population mean. Explain what it means to say that "sample means vary from study to study." Explain how that variability is summarized by a distribution and by a numerical summary.

---

[2] www.cnn.com/ELECTION/2010/results/polls/#CAS01p1.

[3] www.cnn.com/ELECTION/2010/results/polls/#val=NYS01p1.

[4] www.cnn.com/ELECTION/2010/results/polls/#val=CAI01p1.

For questions R2.22–R2.25, assume a random sample is used to gather the data. Answer each question true or false where each question begins with: Then, as you collect more data,

R2.22 True or false: Data are normal? You expect a histogram of the data distribution to look more and more like a normal distribution.

R2.23 True or false: Data and population You expect the data distribution to resemble more closely the population distribution.

R2.24 True or false: Measuring variability By the standard deviation formula for the sampling distribution of a sample proportion, a sample proportion tends to get closer to the population proportion.

R2.25 True or false: CLT By the central limit theorem, the sampling distribution tends to take on more of a bell shape.

R2.26 Profit variability Mark Newman notices that his restaurant, The Bistro, averages about 100 customers a day during the week and about 200 customers a day on the weekend. For any day, the charge for food and drink varies among customers according to a probability distribution that is skewed to the right and has a mean of about \$12 and a standard deviation of about \$6. If the mean charge in a given day is less than \$10, the restaurant loses money, taking into account the cost of food and paying the staff. Select one of the following responses, and justify it.

a. The proportion of weekdays in which the mean charge is less than \$10 would be *larger* than the proportion of weekend days in which the mean charge is less than \$10.

b. The proportion of weekdays in which the mean charge is less than \$10 would be *smaller* than the proportion of weekend days in which the mean charge is less than \$10.

c. Not enough information is given to be able to predict whether days in which the restaurant loses money would be more common during the week or on weekends.

Part

3

Inferential Statistics

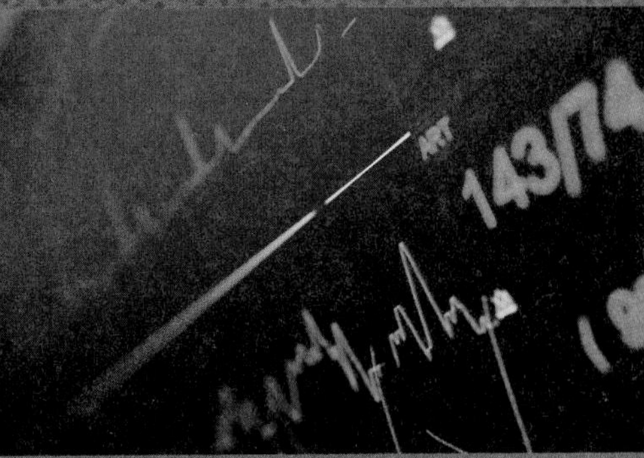

8 Statistical Inference: Confidence Intervals

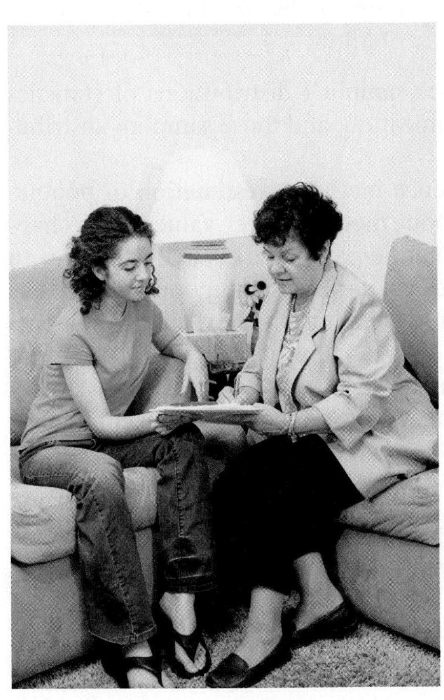

Example 1

Analyzing Data from the General Social Survey

Picture the Scenario

For more than 30 years, the National Opinion Research Center at the University of Chicago (www.norc.uchicago.edu) has conducted an opinion survey called the General Social Survey (GSS). The survey randomly samples about 2000 adult Americans. In a 90-minute in-person interview, the interviewer asks a long list of questions about opinions and behavior for a wide variety of issues. Other nations have similar surveys. For instance, every five years Statistics Canada conducts its own General Social Survey. *Eurobarometer* regularly samples about 1000 people in each country in the European Union.

Analyzing such data helps researchers learn about how people think and behave at a given time and track opinions over time. Activity 1 in Chapter 1 showed how to access the GSS data at sda.berkeley.edu/GSS.

Questions to Explore

Based on data from a recent GSS, how can you make an inference about

- The proportion of Americans who are willing to pay higher prices to protect the environment?
- The proportion of Americans who believe a wife should sacrifice her career for her husband's?
- The mean number of hours that Americans watch TV per day?

Thinking Ahead

We will analyze data from the GSS in examples and exercises throughout this chapter. For instance, in Example 3 we'll see how to estimate the proportion of Americans who are willing to pay higher prices to protect the environment.

We'll answer the other two questions above in Examples 2 and 6, and in exercises we'll explore opinions about issues such as whether it should or should not be the government's responsibility to reduce income differences between the rich and poor, whether a preschool child is likely to suffer if his or her mother works, and how politically conservative or liberal Americans are.

A sample of about 2000 people (as the GSS takes) is relatively small. For instance, in the United States, a survey of this size gathers data for less than 1 of every 100,000 people. How can we possibly make reliable predictions about the entire population with so few people?

We now have the tools to see how this is done. We're ready to learn about a powerful use of statistics: **statistical inference** about population parameters using sample data. Inference methods help us to predict how close a sample statistic falls to the population parameter. We can make decisions and predictions about populations even if we have data for relatively few subjects from that population. The previous chapter illustrated that it's often possible to predict the winner of an election in which millions of people voted, knowing only how a couple of thousand people voted.

Recall

A **statistic** describes a **sample**. Examples are the sample mean $\bar{x}$ and standard deviation s.

A **parameter** describes a **population**. Examples are the population mean μ and standard deviation σ. ◀

For statistical inference methods, you may wonder what's the relevance of learning about the role of randomization in gathering data (Chapter 4), concepts of probability (Chapter 5), and the normal distribution (Chapter 6) and its use as a sampling distribution (Chapter 7)? They're important for two primary reasons:

- Statistical inference methods use probability calculations that assume that the data were gathered with a random sample or a randomized experiment.

Recall

A **sampling distribution** specifies the possible values a statistic can take and their probabilities. ◄

Recall

We use the **proportion** to summarize the *relative frequency* of observations in a category for a categorical variable. The proportion equals the number in the category divided by the sample size. We use the **mean** as one way to summarize the *center* of the observations for a quantitative variable. ◄

- The probability calculations refer to a sampling distribution of a statistic, which is often approximately a normal distribution.

In other words, statistical inference uses sampling distributions of statistics calculated from data gathered using randomization, and those sampling distributions are often approximately normal.

There are two types of statistical inference methods—**estimation** of population parameters and **testing hypotheses** about the parameter values. This chapter discusses the first type, estimating population parameters. We'll learn how to estimate population proportions for categorical variables and population means for quantitative variables. For instance, a study dealing with how college students pay for their education might estimate the proportion of college students who work part time and the mean annual income for those who work. The most informative estimation method constructs an interval of numbers, called a **confidence interval,** within which the unknown parameter value is believed to fall.

8.1 Point and Interval Estimates of Population Parameters

Population parameters have two types of estimates, a point estimate and an interval estimate.

Point Estimate and Interval Estimate

A **point estimate** is a *single number* that is our best guess for the parameter.

An **interval estimate** is an *interval of numbers* within which the parameter value is believed to fall.

For example, one General Social Survey asked, "Do you believe in hell?" From the sample data, the point estimate for the proportion of adult Americans who would respond yes equals 0.73—more than 7 of 10. The adjective "point" in *point estimate* refers to using a single number or *point* as the parameter estimate.

An interval estimate, found with the method introduced in the next section, predicts that the proportion of *all* adult Americans who believe in hell falls between 0.71 and 0.75. That is, it predicts that the sample point estimate of 0.73 falls within a *margin of error* of 0.02 of the population proportion. Figure 8.1 illustrates this idea.

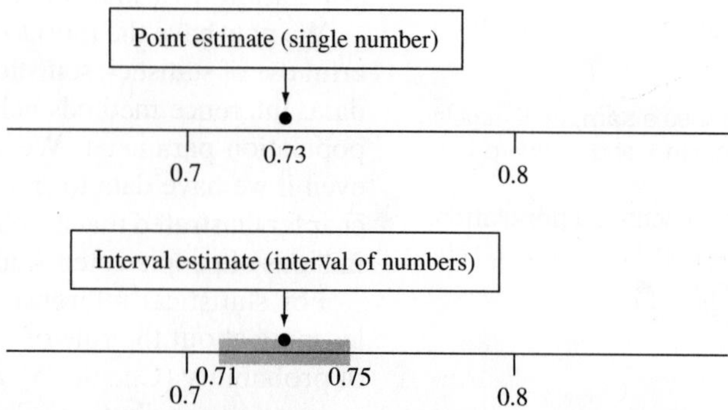

▲ **Figure 8.1 A Point Estimate Predicts a Parameter by a Single Number.** An **interval estimate** is an interval of numbers that are believable values for the parameter. **Question** Why is a point estimate alone not sufficiently informative?

A point estimate by itself is not sufficient because it doesn't tell us *how close* the estimate is likely to be to the parameter. An interval estimate is more useful. It incorporates a margin of error, so it helps us to gauge the accuracy of the point estimate.

Point Estimation: Making a Best Guess for a Population Parameter

Once we've collected the data, how do we find a point estimate, representing our best guess for a parameter value? The answer is straightforward—we can use an appropriate sample statistic. For example, for a population mean μ, the sample mean $\bar{x}$ is a point estimate of μ. For the population proportion, the sample proportion is a point estimate.

Point estimates are the most common form of inference reported by the mass media. For example, the Gallup organization conducts a monthly survey to estimate the U.S. president's popularity, and the mass media report the results. In early May 2011, this survey reported that 52% of the American public approved of President Obama's performance in office. This percentage was a *point estimate* rather than a parameter because Gallup used a sample of about 1550 people rather than the entire population. For simplicity, we'll usually use the term *estimate* in place of point estimate when there is no risk of confusing it with an interval estimate.

Properties of Point Estimators For any particular parameter, there are several possible point estimates. For a normal distribution, for instance, the center is the mean μ and the median since that distribution is symmetric. So, with sample data from a normal distribution, two possible estimates of that center value are the sample mean and the sample median. What makes a particular estimate better than others? *A good estimator* of a parameter has two desirable properties:

Property 1: A good estimator has a sampling distribution that is centered at the parameter. We define *center* in this case as the mean of that sampling distribution. An estimator with this property is said to be **unbiased.** From Section 7.2, we know that for random sampling the mean of the sampling distribution of the sample mean $\bar{x}$ equals the population mean μ. So, the sample mean $\bar{x}$ is an unbiased estimator of μ. Figure 8.2 recalls this result.

The sample proportion is an unbiased estimator of a population proportion. In polling about the president's popularity, we don't know whether any particular sample proportion supporting the president falls below or above the actual population proportion. We do know, however, that with random sampling the sample proportions tend to fall *around* the population proportion and that if

Recall

From Chapter 7, the standard deviation of the sampling distribution of the statistic describes the variability in the possible values of the statistic for the given sample size. It also tells us how much the statistic would vary from sample to sample of that size. ◀

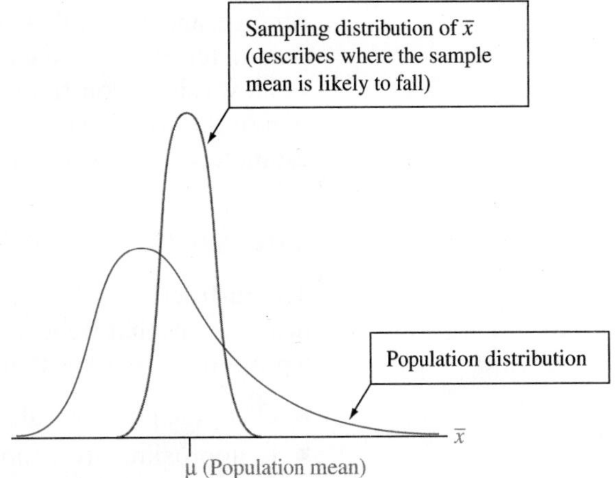

▲ **Figure 8.2 The Sample Mean $\bar{x}$ Is an Unbiased Estimator.** Its sampling distribution is centered at the parameter it estimates—the population mean μ. **Question** When is the sampling distribution bell shaped, as it is in this figure?

we took many samples, the mean of all these sample proportions would be very close to the population proportion.

Property 2: A good estimator has a *small standard deviation* compared to other estimators. This tells us the estimator tends to fall closer than other estimates to the parameter. For example, for estimating the center of a normal distribution, the sample mean has smaller standard deviation than the sample median (Exercise 8.123). The sample mean is a better estimator of this parameter. In this text, we'll use estimators that are unbiased (or nearly so, in practical terms) and that have relatively small standard deviation.

Interval Estimation: Constructing an Interval That Contains the Parameter (We Hope!)

A recent survey[1] of new college graduates estimated the mean salary for those who had taken a full-time job after graduation to equal \$50,500. Does \$50,500 seem plausible to you? Too high? Too low? Any individual point estimate may or may not be close to the parameter it estimates. For the estimate to be useful, we need to know how close it is likely to fall to the actual parameter value. Is the estimate of \$50,500 likely to be within \$1000 of the actual population mean? Within \$5000? Within \$10,000? Inference about a parameter should provide not only a point estimate but should also indicate its likely precision.

An **interval estimate** indicates precision by giving an interval of numbers around the point estimate. The interval is made up of numbers that are the most believable values for the unknown parameter, based on the data observed. For instance, perhaps a survey of new college graduates predicts that the mean salary of all the graduates working full-time falls somewhere between \$48,500 and \$52,500, that is, within a *margin of error* of \$2000 of the point estimate of \$50,500. An interval estimate is designed to contain the parameter with some chosen probability, such as 0.95. Because interval estimates contain the parameter with a certain degree of confidence, they are referred to as **confidence intervals.**

Confidence Interval

A **confidence interval** is an interval containing the most believable values for a parameter. The probability that this method produces an interval that contains the parameter is called the **confidence level.** This is a number chosen to be close to 1, most commonly 0.95.

The interval from \$48,500 to \$52,500 is an example of a confidence interval. It was constructed using a confidence level of 0.95. This is often expressed as a percentage, and we say that we have "95% confidence" that the interval contains the parameter. It is a **95% confidence interval.**

How can we construct a confidence interval? The key is the *sampling distribution* of the point estimate. This distribution tells us the probability that the point estimate will fall within any certain distance of the parameter.

The Logic behind Constructing a Confidence Interval

To construct a confidence interval, we'll put to work some results about sampling distributions that we learned in the previous chapter. Let's do this for estimating a proportion. We saw that the sampling distribution of a sample proportion:

- Gives the possible values for the sample proportion and their probabilities
- Is approximately a normal distribution, for large random samples, where $np \geq 15$ and $n(1 - p) \geq 15$

[1]By National Association of Colleges and Employers, www.naceweb.org.

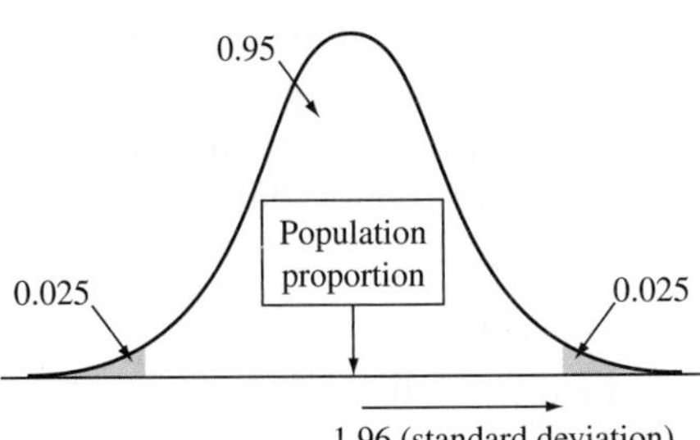

- Has mean equal to the population proportion, p
- Has standard deviation equal to $\sqrt{\frac{p(1-p)}{n}}$

Let's use these results to construct a 95% confidence interval for a population proportion. From Chapter 6, approximately 95% of a normal distribution falls within 2 standard deviations of the mean. More precisely, we saw in Section 6.2 that the mean plus and minus 1.96 standard deviations includes exactly 95% of a normal distribution. Since the sampling distribution of the sample proportion is approximately normal, with probability 0.95 the sample proportion falls within about 1.96 standard deviations of the population proportion (see the margin figure). The distance of 1.96 standard deviations is the **margin of error.** We've been using this term since Chapter 4. Let's take a closer look at how it's calculated.

Recall

Using normal cumulative probabilities (Table A or a calculator or software), $z = 1.96$ has cumulative probability 0.975, right-tail probability 0.025, two-tail probability 0.05, and central probability 0.95. ◄

Margin of Error

The **margin of error** measures how accurate the point estimate is likely to be in estimating a parameter. It is a multiple of the standard deviation of the sampling distribution of the estimate, such as 1.96 × (standard deviation) when the sampling distribution is a normal distribution.

Recall

Example 6 in Chapter 4 showed how the **margin of error** is reported in practice for a sample proportion. We approximated it there by $1/\sqrt{n}$. This is a rough approximation for 1.96 × (standard deviation), as shown following Example 9 in Section 8.4. ◄

Once the sample is selected, if the sample proportion *does* fall within 1.96 standard deviations of the population proportion, then the interval from

[sample proportion − 1.96(standard deviation)] to

[sample proportion + 1.96(standard deviation)]

contains the population proportion. In other words, with probability about 0.95, a sample proportion value occurs such that the interval

sample proportion ± 1.96(standard deviation)

contains the unknown population proportion. This interval of numbers is an approximate **95% confidence interval** for the population proportion.

In Words

"Sample proportion ± 1.96 (standard deviation)" represents taking the sample proportion and adding and subtracting 1.96 standard deviations.

Example 2

Margin of error

A Wife's Career

Picture the Scenario

One question on the General Social Survey asks whether you agree or disagree with the following statement: "It is more important for a wife to help her husband's career than to have one herself." The last time this question was asked, 19% of 1805 respondents agreed. So the sample proportion agreeing was 0.19. From a formula in the next section, we'll see that this point estimate has an approximate standard deviation = 0.01.

Questions to Explore

a. Find and interpret the margin of error for a 95% confidence interval for the population proportion who agreed with the statement about a woman's role in a marriage.

b. Construct the 95% confidence interval and interpret it in context.

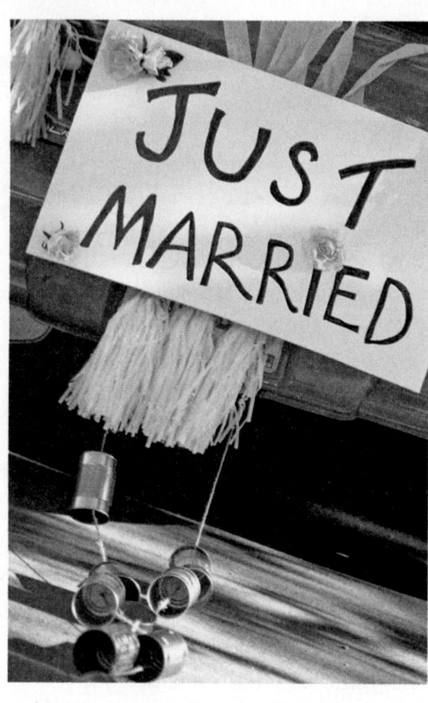

Think It Through

a. The margin of error for a 95% confidence interval for a population proportion equals 1.96 × (standard deviation), or 1.96(001), approximately 0.02. This means that with probability 0.95, the error in using the point estimate to predict the population proportion is no greater than 0.02.

b. The approximate 95% confidence interval is

$$\text{Sample proportion} \pm 1.96(\text{standard deviation}),$$

$$\text{which is } 0.19 \pm 1.96(0.01), \text{ or } 0.19 \pm 0.02.$$

This gives the interval of proportions from 0.17 to 0.21, denoted by (0.17, 0.21). In summary, using this 95% confidence interval, we predict that the population proportion who believed it is more important for a wife to help her husband's career than to have one herself was somewhere between 0.17 and 0.21.

Insight

In 1977, when this question was first asked on the GSS, the point estimate was 0.57 and the 95% confidence interval was (0.55, 0.59). The proportion of Americans who agree with this statement has decreased considerably since then.

Try Exercise 8.7

SUMMARY: A Confidence Interval

A confidence interval is constructed by taking a point estimate and adding and subtracting a margin of error. The margin of error is based on the standard deviation of the sampling distribution of that point estimate. When the sampling distribution is approximately normal, a 95% confidence interval has a margin of error equal to 1.96 standard deviations.

The sampling distribution of most point estimates is approximately normal when the random sample size is relatively large. Thus, this logic of taking the margin of error for a 95% confidence interval to be approximately 2 standard deviations applies with large random samples, such as those found in the General Social Survey. The next two sections show more precise details for estimating proportions and means.

Activity 1

Download Data from the General Social Survey

We saw in Chapter 1 that it's easy to download data from the GSS. Let's recall how.

- Go to the Web site sda.berkeley.edu/GSS/. Click on the link there for GSS—with No Weight as the default weight selection.
- The GSS name for the variable in Example 2 (whether a wife should sacrifice her career for her husband's) is FEHELP. Type FEHELP as the row variable name. Click on *Run the Table.* Now you'll see category counts and the percentages for all the years combined in which this question was asked.
- To download results only for the year 1998, go back to the previous menu and enter YEAR(1998) in the Selection Filter space. When you click again on *Run the Table*, you'll see that in 1998, 2.4% strongly agreed and 16.8% agreed, for a total of about 19% (a proportion of 0.19) in the two agree categories.

Now, create other results. If you open the Standard Codebook, you will see indexes for the variables. Look up a subject that interests you and find the GSS code name for a variable. For example, to find the percentage who believe in hell, enter HELL as the code name.

Try Exercise 8.8

8.1 Practicing the Basics

8.1 Health care A study dealing with health care issues plans to take a sample survey of 1500 Americans to estimate the proportion who have health insurance and the mean dollar amount that Americans spent on health care this past year.

a. Identify two population parameters that this study will estimate.

b. Identify two statistics that can be used to estimate these parameters.

8.2 Projecting winning candidate News coverage during a recent election projected that a certain candidate would receive 54.8% of all votes cast; the projection had a margin of error of $\pm 3\%$

a. Give a point estimate for the proportion of all votes the candidate will receive.

b. Give an interval estimate for the proportion of all votes the candidate will receive.

c. In your own words, state the difference between a point estimate and an interval estimate.

8.3 Believe in hell? Using the General Social Survey Web site sda.berkeley.edu/GSS with YEAR(2008) as the filter, find the point estimate of the population proportion of Americans who would have answered "yes, definitely" in 2008 when asked whether they believe in hell (variable HELL).

8.4 Help the poor? One question (called NATFAREY) on the General Social Survey for the year 2008 asks, "Are we spending too much, too little, or about the right amount on assistance to the poor?" Of the 998 people who responded in 2008, 695 said too little, 217 said about right, and 86 said too much.

a. Find the point estimate of the population proportion who would answer "about right."

b. The margin of error of this estimate is 0.05. Explain what this represents.

8.5 Watching TV In response to the GSS question in 2008 about the number of hours daily spent watching TV, the responses by the five subjects who identified themselves as Hindu were 3, 2, 1, 1, 1.

a. Find a point estimate of the population mean for Hindus.

b. The margin of error at the 95% confidence level for this point estimate is 0.7. Explain what this represents.

8.6 Nutrient effect on growth rate Researchers are interested in the effect of a certain nutrient on the growth rate of plant seedlings. Using a hydroponics grow procedure that utilized water containing the nutrient, they planted six tomato plants and recorded the heights of each plant 14 days after germination. Those heights, measured in millimeters, were as follows: 55.5, 60.3, 60.6, 62.1, 65.5, 69.2.

a. Find a point estimate of the population mean height of this variety of seedling 14 days after germination.

b. A method that we'll study in Section 8.3 provides a margin of error of 4.9 mm for a 95% confidence interval for the population mean height. Construct that interval.

c. Use this example to explain why a point estimate alone is usually insufficient for statistical inference.

8.7 Believe in heaven? TRY When a GSS asked 1326 subjects, "Do you believe in heaven?" (coded HEAVEN), the proportion who answered yes was 0.85. From results in the next section, the estimated standard deviation of this point estimate is 0.01.

a. Find and interpret the margin of error for a 95% confidence interval for the population proportion of Americans who believe in heaven.

b. Construct the 95% confidence interval. Interpret it in context.

8.8 Feel lonely often? TRY The GSS has asked "On how many days in the past seven days have you felt lonely?" At sda.berkeley.edu/GSS, enter LONELY as the variable, check Summary Statistics in the menu of table options, and click on *Run the Table* to see the responses.

a. Report the percentage making each response and the mean and standard deviation of the responses. Interpret.

b. The standard deviation of the sample mean can be estimated using this data as 0.06. Interpret the value of 0.06.

8.9 CI for loneliness Refer to the previous exercise. The margin of error for a 95% confidence interval for the population mean is 0.12. Construct that confidence interval, and interpret it.

8.10 Newspaper article Conduct a search to find a newspaper or magazine article that reported a margin of error. If you prefer, use the Internet for the search.

a. Specify the population parameter, the value of the sample statistic, the point estimate, and the size of the margin of error.

b. Explain how to interpret the margin of error.

8.2 Constructing a Confidence Interval to Estimate a Population Proportion

Let's now see how to construct a confidence interval for a population proportion. We'll apply the ideas discussed at the end of Section 8.1. In this case, the data are categorical, specifically *binary* (two categories), which means that each observation either falls or does not fall in the category of interest. We'll use the generic terminology "success" and "failure" for these two possible outcomes (as in the discussion of the binomial distribution in Section 6.3). We summarize the data by the sample proportion of successes and construct a confidence interval for the population proportion.

Did You Know?

If a categorical variable has more than two categories, it can still be considered binary by classifying one or more categories as a success and the remaining categories as a failure.

Finding the 95% Confidence Interval for a Population Proportion

The 2000 General Social Survey asked respondents if they would be willing to pay much higher prices to protect the environment. Of $n = 1154$ respondents, 518 said yes. The sample proportion of yes responses was $518/1154 = 0.45$, less than half. How can we construct a confidence interval for the population proportion that would respond yes?

We symbolize the population proportion by p. The point estimate of the population proportion is the *sample proportion*.

We symbolize the sample proportion by $\hat{p}$, called "p-hat."

In Words

It is traditional in statistics to use Greek letters for most parameters. Some books use the Greek letter π (pi) for a proportion. Since π already has its own meaning in math and statistics (3.14...), we use the symbol p. This is also the parameter for the probability of success in the binomial distribution (Section 6.3). The sample proportion $\hat{p}$ is read as "p-hat."

In statistics, the circumflex ("hat") symbol over a parameter symbol represents a point estimate of that parameter. Here, the sample proportion $\hat{p}$ is a point estimate, such as $\hat{p} = 0.45$ for the proportion willing to pay much higher prices to protect the environment.

For large random samples, the central limit theorem tells us that the sampling distribution of the sample proportion $\hat{p}$ is approximately normal. As discussed in the previous section, the z-score for a 95% confidence interval with the normal sampling distribution is 1.96. So there is about a 95% chance that $\hat{p}$ falls within 1.96 standard deviations of the population proportion p. A 95% confidence interval uses margin of error $=$ 1.96(standard deviation). The formula

[point estimate $\pm$ margin of error] becomes $\hat{p} \pm 1.96$(standard deviation).

The exact standard deviation of a sample proportion equals $\sqrt{p(1-p)/n}$. This formula depends on the unknown population proportion, p. In practice, we don't know p, and we need to estimate p to compute the standard deviation.

Did You Know?

The term *standard error* is often used for both the actual standard deviation of a statistic and the estimated standard deviation of a statistic. To help make the distinction of the actual versus the estimated, this textbook will use the term *standard error* for the estimated standard deviation of a statistic. ◀

In Practice The Standard Deviation of a Sampling Distribution for a Statistic Is Estimated

The exact value of the standard deviation of a sampling distribution for a statistic depends on the parameter value. In practice, the parameter value is unknown, so we find the standard deviation of the sampling distribution for a statistic by substituting an estimate of the parameter. The term *standard error* is commonly used for what is actually an "estimated standard deviation of a sampling distribution." Beginning here, we'll use **standard error** to refer to this estimated value because that's what we'll use in practice.

In Words

se = standard error

Standard Error

A standard error is an estimated standard deviation of a sampling distribution. We will use se as shorthand for standard error.

For example, for finding a confidence interval for a population proportion p, the standard error is

$$se = \sqrt{\hat{p}(1-\hat{p})/n}.$$

In Words

To find the **95% confidence interval,** you take the sample proportion and add and subtract 1.96 standard errors.

A 95% confidence interval for a population proportion p is

$$\hat{p} \pm 1.96(se), \text{ with } se = \sqrt{\hat{p}\frac{(1-\hat{p})}{n}},$$

where $\hat{p}$ denotes the sample proportion based on n observations.

Figure 8.3 shows the sampling distribution of $\hat{p}$ and how there's about a 95% chance that $\hat{p}$ falls within 1.96(*se*) of the population proportion *p*. This confidence interval is designed for large samples. We'll be more precise about what "large" means after the following example.

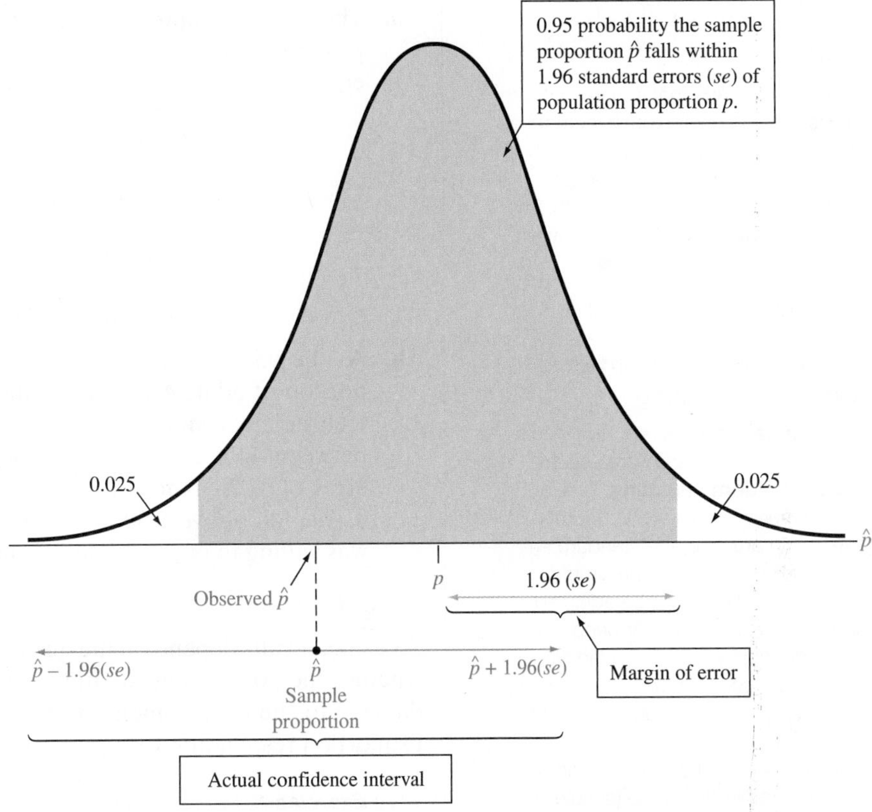

▲ **Figure 8.3 Sampling Distribution of Sample Proportion $\hat{p}$.** For large random samples, the sampling distribution is normal around the population proportion *p*, so $\hat{p}$ has probability 0.95 of falling within 1.96(*se*) of *p*. As a consequence, $\hat{p} \pm 1.96(se)$ is a 95% confidence interval for *p*. **Question** Why is the confidence interval $\hat{p} \pm 1.96(se)$ instead of $p \pm 1.96(se)$?

Example 3

Constructing a confidence interval

Paying Higher Prices to Protect the Environment

Picture the Scenario

Many people consider themselves "green," meaning that they support (in theory) environmental issues. But how do they act in practice? For instance, Americans' per capita use of energy is roughly double that of Western Europeans. If you live in North America, would you be willing to pay the same price for gas that Europeans do (often about $8 or more per gallon) if the government proposed a significant price hike as an incentive for conservation and for driving more fuel-efficient cars to reduce air pollution and its impact on global warming?

Questions to Explore

In 2010, the GSS asked subjects if they would be willing to pay much higher prices to protect the environment. Of $n = 1{,}361$ respondents, 637 indicated a willingness to do so.

Activity

To obtain these data for yourself, go to the GSS Web site sda.berkeley.edu/GSS and download data on the variable GRNPRICE for the 2010 survey.

Recall

$se = 0.0135$ means that if many random samples of 1361 people each were taken to gauge their opinion about this issue, the standard deviation of the sample proportions would be about 0.0135. The sample proportions would vary relatively little from sample to sample. ◀

In Practice Standard Errors and Type of Sampling

The **standard errors** reported in this book and by most software assume **simple random sampling.** The GSS uses a multistage cluster random sample. In practice, the standard error based on the formula for a simple random sample should be adjusted slightly, as explained in Appendix A of the codebook at the GSS Web site. It's beyond our scope to show the details, but when you request an analysis at the GSS Web site you can get the adjusted standard errors by checking Complex rather than SRS (simple random sample) for the sample design.

a. Find a 95% confidence interval for the population proportion of adult Americans willing to do so at the time of that survey.

b. Interpret that interval.

Think It Through

a. The sample proportion that estimates the population proportion p is $\hat{p} = 637/1361 = 0.468$. The standard error of the sample proportion $\hat{p}$ equals

$$se = \sqrt{\hat{p}(1 - \hat{p})/n} = \sqrt{(0.468)(0.532)/1361} = 0.0135.$$

Using this se, a 95% confidence interval for the population proportion is

$$\hat{p} \pm 1.96(se), \text{ which is } 0.468 \pm 1.96(0.0135)$$

$$= 0.468 \pm 0.026, \text{ or } (0.441, 0.494).$$

b. At the 95% confidence level, we estimate that the population proportion of adult Americans willing to pay much higher prices to protect the environment was at least 0.44 but no more than 0.49, that is, between 44% and 49%. The point estimate of 0.47 has a margin of error of 0.026. None of the numbers in the confidence interval (0.441, 0.494) fall above 0.50. So we infer that less than half the population was willing to pay much higher prices to protect the environment.

Insight

As usual, results depend on the question's wording. For instance, when asked whether the government should impose strict laws to make industry do less damage to the environment, a 95% confidence interval for the population proportion responding yes is (0.92, 0.95). (See Exercise 8.14.)

Try Exercise 8.13

Recall

From Section 6.3, the **binomial** random variable X counts the number of successes in n observations, and the sample proportion equals x/n, for instance, $637/1361 = 0.47$. ◀

Table 8.1 shows how MINITAB software reports the data summary and confidence interval. Here, X represents the number that *support* paying much higher prices. The notation reflects that this is the outcome of a random variable, specifically a binomial random variable. The heading "Sample p" stands for the sample proportion $\hat{p}$, "CI" stands for confidence interval, and "N" stands for the sample size (which we have denoted by n). In reporting results from such output, you should use only the first two or three significant digits. Report the confidence interval as (0.44, 0.49) or (0.441, 0.494) rather than (0.44153, 0.49455). Software's extra precision provides accurate calculations in finding se and the confidence interval. However, the extra digits are distracting when reported to others and do not tell them anything extra in a practical sense about the population proportion. *Likewise, if you do a calculation with a hand calculator, don't round off while doing the calculation or your answer may be affected, but do round off when you report the final answer.*

Table 8.1 MINITAB Output for 95% Confidence Interval for a Proportion for Example 3

(The italicized lines are added annotation to explain what's shown.)

X	N	Sample p	95.0% CI
637	1361	0.468038	(0.44153, 0.49455)
↑	↑	↑	↑
Category count	*Sample size*	*Sample proportion*	*Endpoints of confidence interval*

Recall

From Section 6.3, the binomial distribution is bell shaped when the expected counts np and $n(1 - p)$ of successes and failures are both at least 15. Here, we don't know p, and we use the guideline with the observed counts of successes and failures. ◀

Sample Size Needed for Validity of Confidence Interval for a Proportion

The confidence interval formula $\hat{p} \pm 1.96(se)$ applies with *large random samples.* This is because the sampling distribution of the sample proportion $\hat{p}$ is then approximately normal and the *se* estimate also tends to be good, allowing us to use the z-score of 1.96 from the normal distribution.

In practice, "large" means that *you should have at least 15 successes and at least 15 failures* for the binary outcome.[2] At the end of Section 8.4, we'll see how the large-sample method can fail when this guideline is not satisfied.

SUMMARY: Sample Size Needed for Large-Sample Confidence Interval for a Proportion

For the 95% confidence interval $\hat{p} \pm 1.96(se)$ for a proportion p to be valid, you should have at least 15 successes and 15 failures. This can also be expressed as

$$n\hat{p} \geq 15 \text{ and } n(1 - \hat{p}) \geq 15.$$

At the end of Section 8.4 we'll learn about a simple adjustment to this confidence interval formula that you should use when this guideline is violated.

This guideline was easily satisfied in Example 3. The binary outcomes had counts 637 willing to pay much higher prices and 724 (= 1361 − 637) unwilling, both much larger than 15.

Example 4

Constructing a confidence interval

The Proportion That *Won't* Pay Higher Prices to Protect the Environment

Picture the Scenario

Example 3 found a confidence interval for the population proportion p that *will* pay higher prices to support the environment. The population proportion that *won't* pay higher prices to support the environment is then $1 - p$.

Question to Explore

How can we construct a 95% confidence interval for the population proportion $1 - p$?

Think It Through

We can use the formula for a 95% confidence interval, because the GSS uses random sampling and because there's at least 15 successes and 15 failures (the counts are 637 and 724, with $n = 1361$). The estimate of $1 - p$ is $724/1361 = 0.532$, which is $1 - \hat{p} = 1 - 0.468$. When $n = 1361$ and $\hat{p} = 0.468$, we saw that the standard error of $\hat{p}$ is $se = \sqrt{\hat{p}(1 - \hat{p})/n} = 0.0135$. Similarly, the standard error for $1 - \hat{p} = 0.532$ is

$$\sqrt{\text{proportion}(1 - \text{proportion})/n} = \sqrt{(0.532)(0.468)/1361} = 0.0135.$$

[2]Many statistics texts use 5 or 10 as the minimum instead of 15, but recent research suggests that those sizes are too small (e.g., L. Brown et al., *Statistical Science*, vol. 16, pp. 101–133, 2001).

This is necessarily the same as the standard error of $\hat{p}$. A 95% confidence interval for the population proportion that won't pay higher prices is

$$\text{sample proportion} \pm 1.96(se), \text{ or } 0.532 \pm 1.96(0.0135),$$

$$\text{which is } 0.532 \pm 0.026, \text{ or } (0.51, 0.56).$$

In summary, we can be 95% confident that the population proportion that won't pay higher prices to support the environment falls between 0.51 and 0.56.

Insight

Now $0.51 = 1 - 0.49$ and $0.56 = 1 - 0.44$ where (0.44, 0.49) is the 95% confidence interval for the population proportion who *will* pay higher prices. For a binary variable, inferences about the second category (*won't* pay higher prices) follow directly from those for the first category (*will* pay higher prices) by subtracting each endpoint of the confidence interval from 1.0. We do not need to construct the confidence interval separately for both categories. The confidence interval for one determines the confidence interval for the other.

Try Exercise 8.16

Using a Confidence Level Other than 95%

Recall

99% confidence interval: From Section 6.2, for central probability 0.99, you look up the cumulative probability of 0.005 or $1 - 0.005 = 0.995$ in Table A, or use software or a calculator, to find $z = 2.58$. ◄

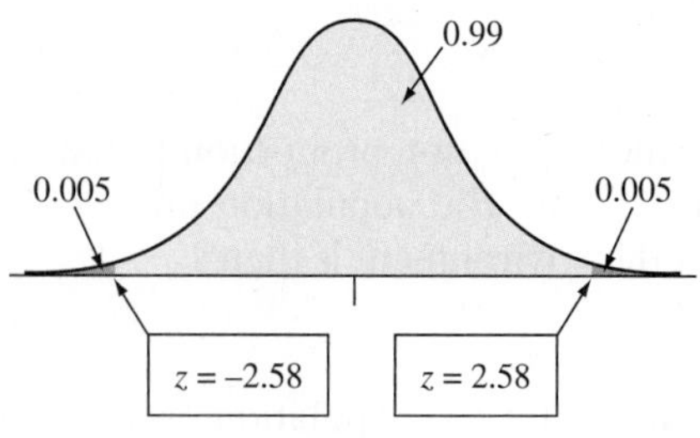

So far we've used a confidence level of 0.95, that is, "95% confidence." This means that there's a 95% chance that a sample proportion value $\hat{p}$ occurs such that the confidence interval $\hat{p} \pm 1.96(se)$ contains the unknown value of the population proportion p. With probability 0.05, however, the method produces a confidence interval that misses p. The population proportion then does *not* fall in the interval, and the inference is incorrect.

In practice, the confidence level 0.95 is the most common choice. But some applications require greater confidence. This is often true in medical research, for example. For estimating the probability p that a new treatment for a deadly disease works better than the treatment currently used, we would want to be extremely confident about any inference we make. To increase the chance of a correct inference (that is, having the interval contain the parameter value), we use a larger confidence level, such as 0.99.

Now, 99% of the normal sampling distribution for the sample proportion $\hat{p}$ occurs within 2.58 standard errors of the population proportion p. So, with probability 0.99, $\hat{p}$ falls within 2.58(se) of p. (See the margin figure.) A 99% confidence interval for p is $\hat{p} \pm 2.58(se)$.

Example 5

Confidence level

A Husband Choosing Not to Have Children

Picture the Scenario

A recent GSS asked "If the wife in a family wants children, but the husband decides that he does not want any children, is it all right for the husband to refuse to have children?" Of 598 respondents, 366 said yes and 232 said no.

Questions to Explore

a. Find a 99% confidence interval for the population proportion who would say yes.

b. How does it compare to the 95% confidence interval?

Think It Through

a. The assumptions for the method are satisfied in that the GSS sample was randomly selected, and there were at least 15 successes and 15 failures (366 yes and 232 no). The sample proportion who said yes is $\hat{p} = 366/598 = 0.61$. Its standard error is

$$se = \sqrt{\hat{p}(1 - \hat{p})/n} = \sqrt{(0.61)(0.39)/598} = 0.020.$$

The 99% confidence interval is

$$\hat{p} \pm 2.58(se), \text{ or } 0.61 \pm 2.58(0.020),$$

$$\text{which is } 0.61 \pm 0.05, \text{ or } (0.56, 0.66).$$

In summary, we can be 99% confident that between 56% and 66% of the U.S. adult population agree that it's all right for the husband to refuse to have children. Note that statistical software or a statistical calculator can calculate this confidence interval. Screen shots from the TI-83+/84 are shown in the margin.

```
1-PropZInt
 x:366
 n:598
 C-Level:.99
 Calculate
```

```
1-PropZInt
 (.56,.66)
 p̂=.61
 n=598.00
```

TI-83+/84 output

b. The 95% confidence interval is $0.61 \pm 1.96(0.020)$, which is 0.61 ± 0.04, or (0.57, 0.65). This is a bit narrower than the 99% confidence interval of (0.56, 0.66). Figure 8.4 illustrates.

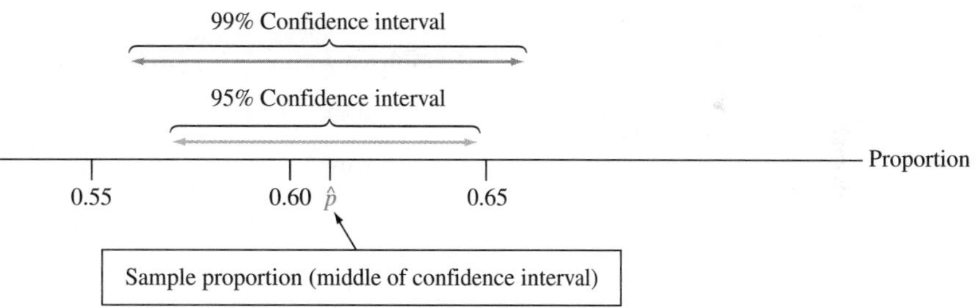

▲ **Figure 8.4** A 99% Confidence Interval Is Wider Than a 95% Confidence Interval. **Question** If you want greater confidence, why would you expect a wider interval?

Insight

The inference with a 99% confidence interval is less precise, with margin of error equal to 0.05 instead of 0.04. Having a greater margin of error is the sacrifice for gaining greater assurance (99% rather than 95%) of correctly inferring where p falls. A medical study that uses a higher confidence level will be less likely to make an incorrect inference, but it will not be able to narrow in as well on where the true parameter value falls.

Try Exercise 8.22

Why settle for anything less than 100% confidence? To be absolutely 100% certain of a correct inference (that is, of capturing the parameter value inside the confidence interval), the confidence interval must contain *all* possible values for the parameter. For example, a 100% confidence interval for the population proportion believing it is all right for a husband to refuse to have children goes from

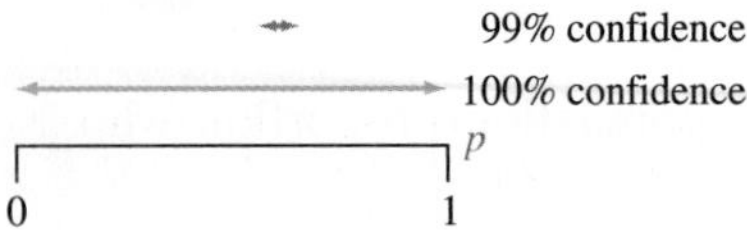

0.0 to 1.0. This inference would tell us only that some number between 0.0% and 100.0% of Americans feel this way. This obviously is not helpful. In practice, we settle for a little less than perfect confidence so we can estimate the parameter value more precisely (illustrated by the margin figure, showing 99% and 100% confidence intervals). It is far more informative to have 99% confidence that the population proportion is between 0.56 and 0.66 than to have 100% confidence that it is between 0.0 and 1.0.

In Practice Margin of Error Refers to 95% Confidence

When the news media report a **margin of error,** it is the margin of error for a 95% confidence interval.

In using confidence intervals, *we must compromise between the desired margin of error and the desired confidence of a correct inference.* As one gets better, the other gets worse. This is why you would probably not use a 99.9999% confidence interval. It would usually have too large a margin of error to tell you much about where the parameter falls (its z-score is 4.9). In practice, 95% confidence intervals are the most common.

Error Probability for the Confidence Interval Method

The general formula for the confidence interval for a population proportion is

sample proportion $\pm$ (z–score from normal table)(standard error), which in symbols is $\hat{p} \pm z(se)$.

The z-score depends on the confidence level. Table 8.2 shows the z-scores for the confidence levels usually used in practice. There is no need to memorize them. You can find them yourself, using Table A or a calculator or software.

Table 8.2 z-Scores for the Most Common Confidence Levels

The large-sample confidence interval for the population proportion is $\hat{p} \pm z(se)$.

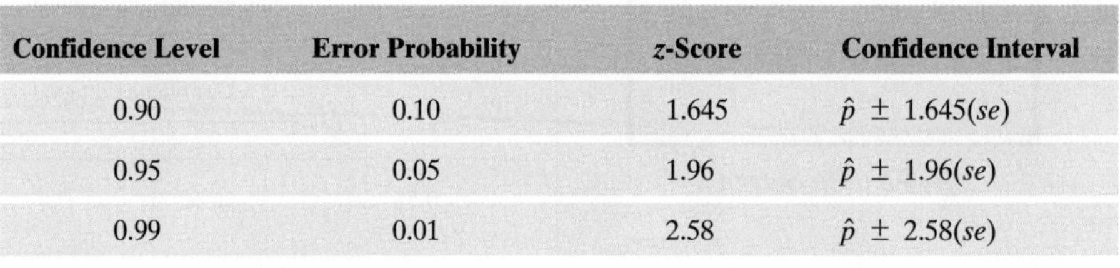

Confidence Level	Error Probability	z-Score	Confidence Interval
0.90	0.10	1.645	$\hat{p} \pm 1.645(se)$
0.95	0.05	1.96	$\hat{p} \pm 1.96(se)$
0.99	0.01	2.58	$\hat{p} \pm 2.58(se)$

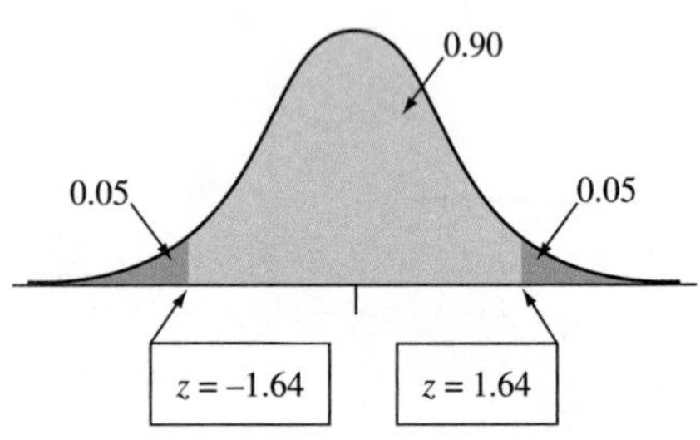

Let's review how by finding the z-score for a 90% confidence interval. When 0.90 probability falls within z standard errors of the mean, then 0.10 probability falls in the two tails and $0.10/2 = 0.05$ falls in each tail. Looking up 0.05 in the body of Table A, we find $z = -1.64$, or we find $z = 1.64$ if we look up the cumulative probability $1 - 0.05 = 0.95$ corresponding to tail probability 0.05 in the right tail. (See margin figure.) The 90% confidence interval equals $\hat{p} \pm 1.64(se)$. (More precise calculation using interpolation or software or a calculator gives z-score 1.645.) Try this again for 99% confidence, using either the table or a calculator or software (you should get $z = 2.58$).

Table 8.2 contains a column labeled "Error Probability." This is the probability that the method results in an incorrect inference, namely, that the data generates a confidence interval that does *not* contain the population proportion. The error probability equals 1 minus the confidence level. For example, when the confidence level equals 0.95, the error probability equals 0.05. The error probability is the two-tail probability under the normal curve for the given z-score. Half the error probability falls in each tail. For 95% confidence with its error probability of 0.05, the z-score of 1.96 is the one with probability $0.05/2 = 0.025$ in each tail.

SUMMARY: Confidence Interval for a Population Proportion p

A confidence interval for a population proportion p, using the sample proportion $\hat{p}$ and the standard error $se = \sqrt{\hat{p}(1 - \hat{p})/n}$ for sample size n, is

$$\hat{p} \pm z(se), \text{ which is } \hat{p} \pm z\sqrt{\hat{p}(1 - \hat{p})/n}.$$

For 90%, 95%, and 99% confidence intervals, z equals 1.645, 1.96, and 2.58. This method assumes

- Data obtained by randomization (such as a random sample or a randomized experiment)
- A large enough sample size n so that the number of successes and the number of failures, that is, $n\hat{p}$ and $n(1 - \hat{p})$, are both at least 15.

Effect of the Sample Size

We'd expect that estimation should be more precise with larger sample sizes. With more data, we know more about the population. The margin of error is $z(se) = z\sqrt{\hat{p}(1 - \hat{p})/n}$. This margin decreases as the sample size n increases, for a given value of $\hat{p}$. The larger the value of n, the narrower the interval.

To illustrate, in Example 5, suppose the survey result of $\hat{p} = 0.61$ for the proportion believing a husband can refuse to have children had resulted from a sample of size $n = 150$, only one-fourth the actual sample size of $n = 598$. Then the standard error would be

$$se = \sqrt{\hat{p}(1 - \hat{p})/n} = \sqrt{0.61(0.39)/150} = 0.040,$$

twice as large as the $se = 0.020$ we got for $n = 598$. The 99% confidence interval would be

$$\hat{p} \pm 2.58(se) = 0.61 \pm 2.58(0.040), \text{ which is } 0.61 \pm 0.10, \text{ or } (0.51, 0.71).$$

The margin of error of 0.10 is twice as large as the margin of error of 0.05 in the 99% confidence interval with the sample size $n = 598$ in Example 5. The confidence interval with larger n is narrower.

Because the standard error has the square root of n in the denominator, and because $\sqrt{4n} = 2\sqrt{n}$, *quadrupling* the sample size *halves* the standard error. That is, we must quadruple n, rather than double it, to halve the margin of error.

In summary, we've observed the following properties of a confidence interval:

SUMMARY: Effects of Confidence Level and Sample Size on Margin of Error

The **margin of error** for a confidence interval:

- Increases as the confidence level increases
- Decreases as the sample size increases.

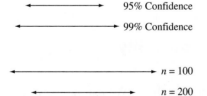

95% Confidence

99% Confidence

$n = 100$

$n = 200$

For instance, a 99% confidence interval is wider than a 95% confidence interval, and a confidence interval with 200 observations is narrower than one with 100 observations (see margin figure). These properties apply to *all* confidence intervals, not just the one for the population proportion.

Interpretation of the Confidence Level

In Example 3, the 95% confidence interval for the population proportion p willing to pay higher prices to protect the environment was (0.44, 0.49). The value of p is unknown to us, so we don't know that it actually falls in that interval. It may actually equal 0.42 or 0.51, for example.

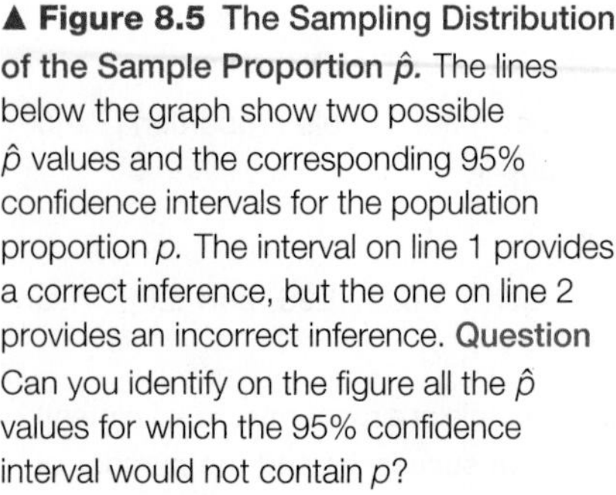

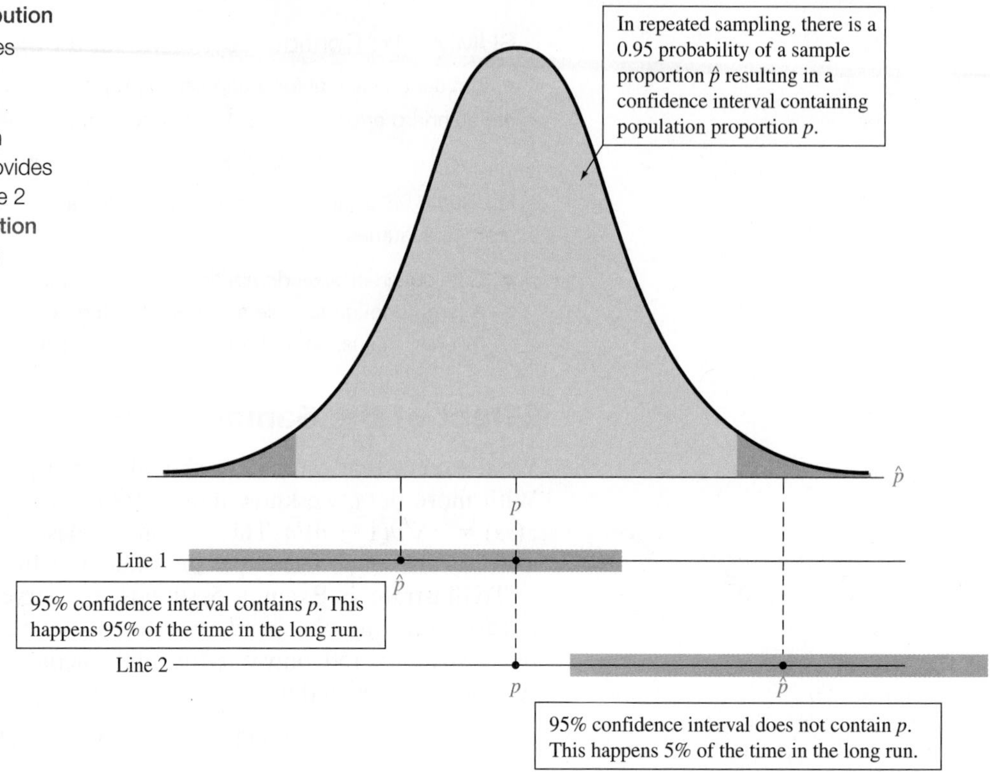
In repeated sampling, there is a 0.95 probability of a sample proportion $\hat{p}$ resulting in a confidence interval containing population proportion p.
$\hat{p}$
p
Line 1
$\hat{p}$
95% confidence interval contains p. This happens 95% of the time in the long run.
Line 2
p
$\hat{p}$
95% confidence interval does not contain p. This happens 5% of the time in the long run.

▲ **Figure 8.5 The Sampling Distribution of the Sample Proportion $\hat{p}$.** The lines below the graph show two possible $\hat{p}$ values and the corresponding 95% confidence intervals for the population proportion p. The interval on line 1 provides a correct inference, but the one on line 2 provides an incorrect inference. **Question** Can you identify on the figure all the $\hat{p}$ values for which the 95% confidence interval would not contain p?

So what does it mean to say that we have "95% confidence"? The meaning refers to a *long-run* interpretation—how the method performs when used over and over with many different random samples. If we used the 95% confidence interval method over time to estimate many population proportions, then *in the long run about 95% of those intervals would give correct results, containing the population proportion.* This happens because 95% of the sample proportions would fall within 1.96(se) of the population proportion. A graphical example is the $\hat{p}$ in line 1 of Figure 8.5.

Saying that a particular interval contains p with "95% confidence" signifies that in the long run, 95% of such intervals would provide a correct inference (containing the actual parameter value). On the other hand, in the long run, 5% of the time the sample proportion $\hat{p}$ *does not* fall within 1.96(se) of p. If that happens, then the confidence interval *does not* contain p, as is seen for $\hat{p}$ in line 2 of Figure 8.5.

By our choice of the confidence level, we can control the chance that we make a correct inference. If an error probability of 0.05 makes us too nervous, we can instead form a 99% confidence interval, which is in error only 1% of the time in the long run. But then we must settle for a wider confidence interval and less precision.

Long Run Versus Subjective Probability Interpretation of Confidence

You might be tempted to interpret a statement such as "We can be 95% confident that the population proportion p falls between 0.42 and 0.48" as meaning that the *probability* is 0.95 that p falls between 0.42 and 0.48. However, probabilities apply to statistics (such as in sampling distributions of the sample proportion), not to parameters. The estimate $\hat{p}$, not the parameter p, is the random variable having a sampling distribution and probabilities. The 95% confidence refers not to a probability for the population proportion p but rather to a probability that applies to the confidence interval *method* in its relative frequency sense: If we use it over and over for various samples, in the long run we make correct inferences 95% of the time.

Section 5.1 mentioned a *subjective* definition of probability that's an alternative to the relative frequency definition. This approach treats the *parameter* as a random variable. Statistical inferences based on the subjective definition of probability *do* make probability statements about parameters. For instance, with it you *can* say that the probability is 0.95 that p falls between 0.44 and 0.49. Statistical inference based on the subjective definition of probability is called *Bayesian statistics.*[3] It has gained in popularity in recent years, but it is beyond the scope of this text.

[3]The name refers to *Bayes's theorem,* which can generate probabilities of parameter values, given the data, from probabilities of the data, given the parameter values.

Activity 2

Let's Simulate the Performance of Confidence Intervals

Let's get a feel for how confidence intervals sometimes provide an incorrect inference. To do this, for a given population proportion value we will simulate taking many samples and forming a confidence interval for each sample. We can then check how often the intervals provide an incorrect inference. We can conduct the simulation using statistical software (such as MINITAB) or using an applet in which we can control the parameter value, the sample size, and the confidence level.

Try this by going to the Confidence Intervals for a Proportion applet on the text CD. We'll set the population $p = 0.50$, and see what happens when we take samples of size 50 and form 95% and 99% confidence intervals. At the menu, set the proportion value to $p = 0.5$ and set the sample size to $n = 50$. Click *Simulate* and 100 samples will be generated. The applet calculates the resulting 95% and 99% confidence intervals using the sample proportion from each sample. The applet summarizes the number of intervals that captured the true population proportion p. How many confidence intervals out of 100 intervals captured $p = 0.5$ at the 95% confidence level? At the 99% confidence level? How many would you expect at each confidence level to capture p, giving us a correct inference? How many intervals would you expect at each confidence level to provide an incorrect inference?

To get a feel for what happens in the long run, do this simulation 10,000 times by clicking the *Simulate* button multiple times until you have a cumulative total of 10,000 confidence intervals. You can add the counts for "Contained p" and "Did not contain p" to keep track of how many simulations you have performed. How many confidence intervals out of 10,000 intervals captured $p = 0.5$ at the 95% confidence level? At the 99% confidence level?

Try Exercises 8.25 and 8.26

8.2 Practicing the Basics

8.11 Unemployed Americans A Gallup poll taken during June 2011 estimated that 8.8% of U.S. adults were unemployed. The poll was based on the responses of 30,000 U.S. adults in the workforce. Gallup reported that the margin of error associated with the poll is ±0.3 percentage points. Explain how they got this result. (*Source:* www.gallup.com/poll/125639/Gallup-Daily-Workforce.aspx.)

8.12 Crime victims In 1994 (the most recent year asked), the General Social Survey asked, "During the last year, did anyone take something from you by using force—such as a stickup, mugging, or threat?" Of 1223 subjects, 31 answered yes and 1192 answered no.

a. Find the point estimate of the proportion of the population who were victims.

b. Find the standard error of this estimate.

c. Find the margin of error for a 95% confidence interval.

d. Construct the 95% confidence interval for the population proportion. Can you conclude that fewer than 10% of all adults in the United States were victims?

8.13 How green are you? TRY When the 2000 GSS asked subjects (variable GRNSOL) if they would be willing to accept cuts in their standard of living to protect the environment, 344 of 1170 subjects said yes.

a. Estimate the population proportion who would answer yes.

b. Find the margin of error for a 95% confidence interval for this estimate.

c. Find a 95% confidence interval for that proportion. What do the numbers in this interval represent?

d. State and check the assumptions needed for the interval in part c to be valid.

8.14 Make industry help environment? When the GSS recently asked subjects whether it should or should not be the government's responsibility to impose strict laws to make industry do less damage to the environment (variable GRNLAWS), 1403 of 1497 subjects said yes.

a. What assumptions are made to construct a 95% confidence interval for the population proportion who would say yes? Do they seem satisfied here?

b. Construct the 95% confidence interval. Interpret in context. Can you conclude whether or not a majority or minority of the population would answer yes?

8.15 Favor death penalty In the 2008 General Social Survey, respondents were asked if they favored or opposed the death penalty for people convicted of murder. Software shows results

```
Sample   X    N  Sample p        95% CI
     1 1263 1902 0.664038  (0.642811, 0.685265)
```

Here, X refers to the number of the respondents who were in favor.

a. Show how to obtain the value reported under "Sample p."

b. Interpret the confidence interval reported, in context.

c. Explain what the "95% confidence" refers to, by describing the long-run interpretation.

d. Can you conclude that more than half of all American adults were in favor? Why?

8.16 Oppose death penalty Refer to the previous exercise. TRY Show how you can get a 95% confidence interval for the proportion of American adults who were *opposed* to the death penalty from the confidence interval stated in the previous exercise for the proportion in favor. (*Hint:* The proportion opposed is 1 minus the proportion in favor.)

8.17 Stem cell research A Harris poll of a random sample of 2113 adults in the United States in October 2010 reported that 72% of those polled believe that stem cell research has merit. (*Source:* www.harrisinteractive.com/vault/Harris-Interactive-Poll-HealthDay-2010-10.pdf.) The results, presented using MINITAB software, are

```
   X      N     Sample p        95% CI
1521   2113      0.7198    (0.7007, 0.7390)
```

Here, X denotes the number who believed that stem cell research has merit.

a. Explain how to interpret "Sample p" and "95% CI" on this printout.

b. What is the 95% margin of error associated with the poll?

8.18 *z*-score and confidence level Which z-score is used in a (a) 90%, (b) 98%, and (c) 99.9% confidence interval for a population proportion?

8.19 Believe in ghosts A Harris poll of a random sample of 2303 adults in the United States in 2009 reported that 82% believe in God, 75% believe in heaven, 61% believe in hell, and 42% believe in ghosts.[4] (Interestingly, these numbers are all substantially smaller than the corresponding values in a similar poll taken by Harris in 2003.) The screen shot shows how the TI 83+/84 reports results of interval estimation at the 95% confidence level for the proportion who believe in ghosts. Explain how to interpret the confidence interval shown.

```
1-PropZInt
 (.400,.440)
 p̂=.420
 n=2303.000
```

8.20 Stem cell research and religion In Exercise 8.17, it was stated that 72% of the 2113 adults surveyed in a Harris poll believed that stem cell research has merit. In the same study, only 58% of those describing themselves as Republicans believed that it has merit. There were 1010 people who described themselves as Republican. The screen shot shows how the TI 83+/84 reports results of interval estimation at the 95% confidence level. Specify the population to which this inference applies and explain how to interpret the confidence interval.

```
1-PropZInt
 (.550,.611)
 p̂=.580
 n=1010.000
```

8.21 Fear of breast cancer A recent survey of 1000 American women between the ages of 45 and 64 asked them what medical condition they most feared. Of those sampled, 61% said breast cancer, 8% said heart disease, and the rest picked other conditions. By contrast, currently about 3% of female deaths are due to breast cancer, whereas 32% are due to heart disease.[5]

a. Construct a 90% confidence interval for the population proportion of women who most feared breast cancer. Interpret.

b. Indicate the assumptions you must make for the inference in part a to be valid.

8.22 Wife doesn't want kids The 1996 GSS asked, "If the hus- TRY band in a family wants children, but the wife decides that she does not want any children, is it all right for the wife to refuse to have children?" Of 699 respondents, 576 said yes.

a. Find a 99% confidence interval for the population proportion who would say yes. Can you conclude that the population proportion exceeds 75%? Why?

b. Without doing any calculation, explain whether the interval in part a would be wider or narrower than a

[4] www.harrisinteractive.com/vault/Harris_Poll_2009_12_15.pdf.

[5] See B. Lomborg, *The Skeptical Environmentalist*, Cambridge University Press, 2001, p. 222.

95% confidence interval for the population proportion who would say yes.

8.23 Exit poll predictions A national television network takes an exit poll of 1400 voters after each has cast a vote in a state gubernatorial election. Of them, 660 say they voted for the Democratic candidate and 740 say they voted for the Republican candidate.

a. Treating the sample as a random sample from the population of all voters, would you predict the winner? Base your decision on a 95% confidence interval.

b. Base your decision on a 99% confidence interval. Explain why you need stronger evidence to make a prediction when you want greater confidence.

8.24 Exit poll with smaller sample In the previous exercise, suppose the same proportions resulted from $n = 140$ (instead of 1400), with counts 66 and 74.

a. Now does a 95% confidence interval allow you to predict the winner? Explain.

b. Explain why the same proportions but with smaller samples provide less information. (*Hint:* What effect does n have on the standard error?)

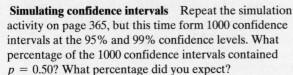

8.25 Simulating confidence intervals Repeat the simulation activity on page 365, but this time form 1000 confidence intervals at the 95% and 99% confidence levels. What percentage of the 1000 confidence intervals contained $p = 0.50$? What percentage did you expect?

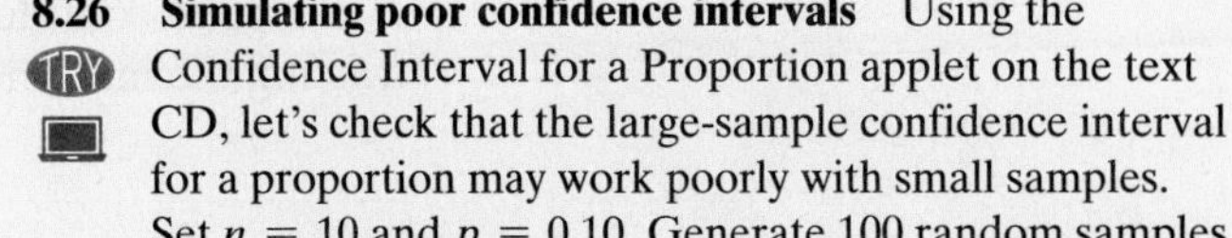

8.26 Simulating poor confidence intervals Using the Confidence Interval for a Proportion applet on the text CD, let's check that the large-sample confidence interval for a proportion may work poorly with small samples. Set $n = 10$ and $p = 0.10$. Generate 100 random samples, each of size 10, and for each one, form a 95% confidence interval for p.

a. How many of the intervals fail to contain the true value, $p = 0.10$?

b. How many would you expect not to contain the true value? What does this suggest?

c. To see that this is not a fluke, now take 1000 samples and see what percentage of 95% confidence intervals contain 0.10. (*Note:* For every interval formed, the number of successes is smaller than 15, so the large-sample formula is not adequate.)

d. Using the Sampling Distribution applet, generate 10,000 random samples of size 10 when $p = 0.10$. The applet will plot the empirical sampling distribution of the sample proportion values. Is it bell shaped and symmetric? Use this to help you explain why the large-sample confidence interval performs poorly in this case. (This exercise illustrates that assumptions for statistical methods are important, because the methods may perform poorly if we use them when the assumptions are violated.)

8.3 Constructing a Confidence Interval to Estimate a Population Mean

We've learned how to construct a confidence interval for a population proportion—a parameter that summarizes a categorical variable. Next we'll learn how to construct a confidence interval for a population mean—a summary parameter for a quantitative variable. We'll analyze GSS data to estimate the mean number of hours per day that Americans watch television. The method resembles that for a proportion. The confidence interval again has the form

$$\text{point estimate} \pm \text{margin of error}.$$

The margin of error again equals a multiple of a standard error. What do you think plays the role of the point estimate and the role of the standard error (se) in this formula?

How to Construct a Confidence Interval for a Population Mean

The sample mean $\bar{x}$ is the point estimate of the population mean μ. In Section 7.2, we learned that the standard deviation of the sample mean equals $\sigma/\sqrt{n}$, where σ is the population standard deviation. Like the standard deviation of the sample proportion, the standard deviation of the sample mean depends on a parameter whose value is unknown, in this case σ. In practice, we estimate σ by the sample standard deviation s. So, the estimated standard deviation used in confidence intervals is the standard error,

$$se = s/\sqrt{n}.$$

Recall

Section 7.2 introduced the standard deviation of the sampling distribution of $\bar{x}$, which describes how much the sample mean varies from sample to sample for a given size n. ◀

We get a confidence interval by taking the sample mean and adding and subtracting the margin of error. We'll see the details following the next example.

Constructing a confidence interval for a population mean

Activity

Try finding *n*, the mean, standard deviation, and *se* at the GSS Web site by entering the variable TVHOURS for the 2008 survey. (You should get exactly the same results if you enter year(2008) in the selection filter and pick *No Weight* in the weight menu.)

Example 6

Number of Hours Spent Watching Television

Picture the Scenario

How much of the typical person's day is spent in front of the TV? Excessive TV watching has been named as one factor, other than diet, for the increasing proportion of obese Americans. A recent General Social Survey asked respondents, "On the average day, about how many hours do you personally watch television?" A computer printout (from MINITAB) summarizes the results for the GSS variable, TV:

```
   N         Mean        StDev       SE Mean              95% Cl
1324       2.9800       2.6600        0.0731        (2.8366, 3.1234)
```

We see that the sample size was 1324, the sample mean was 2.98, the sample standard deviation was 2.66, the standard error of the sample mean was 0.0731, and the 95% confidence interval for the population mean time μ spent watching TV goes from 2.84 to 3.12 hours per day.

Questions to Explore

a. What do the sample mean and standard deviation suggest about the likely shape of the population distribution?

b. How did the software get the standard error? What does it mean?

c. Interpret the 95% confidence interval reported by software.

Think It Through

a. For the sample mean of $\bar{x} = 2.98$ and standard deviation of $s = 2.66$, the lowest possible value of 0 falls only a bit more than 1 standard deviation below the mean. This information suggests that the population distribution of TV watching may be skewed to the right. Figure 8.6 shows a histogram of the data, which suggests the same thing. The median was 2, the lower and upper quartiles were

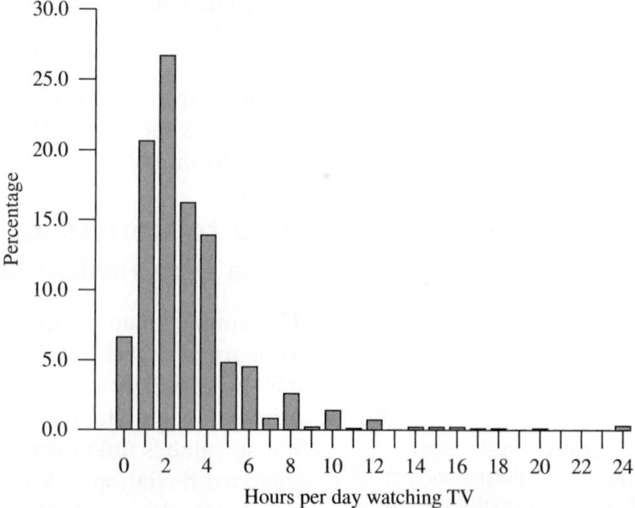

▲ **Figure 8.6 Histogram of Number of Hours a Day Watching Television.** Question Does the skew affect the validity of a confidence interval for the population mean?

1 and 4, the 95th percentile was 8, yet some subjects reported much higher values.

b. With sample standard deviation $s = 2.66$ and sample size $n = 1324$, the standard error of the sample mean is

$$se = s/\sqrt{n} = 2.66/\sqrt{1324} = 0.0731 \text{ hours.}$$

If many studies were conducted about TV watching, with $n = 1324$ for each, the sample mean would not vary much among those studies.

c. A 95% confidence interval for the population mean μ of TV watching in the United States is (2.84, 3.12) hours. We can be 95% confident that the mean amount of TV watched for Americans is between 2.84 to 3.12 hours of TV a day.

Insight

Because the sample size was relatively large, the estimation is precise and the confidence interval is quite narrow. The larger the sample size, the smaller the standard error and the subsequent margin of error.

Try Exercise 8.27

We have not yet seen how software found the margin of error for the confidence interval in Example 6. As with the proportion, the margin of error for a 95% confidence interval is roughly two standard errors. However, we need to introduce a new distribution similar to the normal distribution to give us a more precise margin of error. We'll find the margin of error by multiplying *se* by a score that is a bit larger than the *z*-score when *n* is small but very close to it when *n* is large.

The *t* Distribution and Its Properties

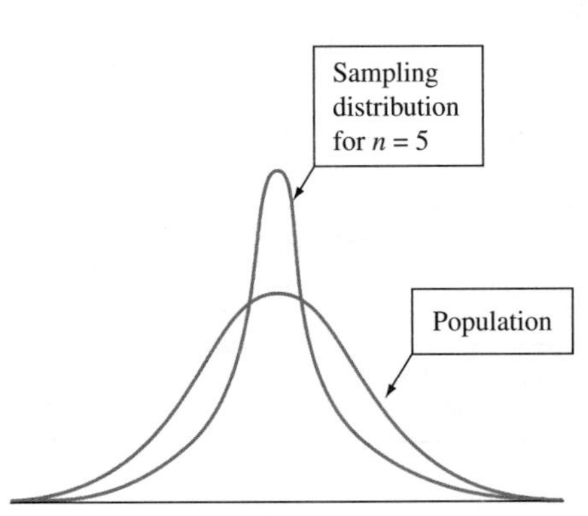

We'll now learn about a confidence interval that applies even for small sample sizes. A disadvantage is that it makes the assumption that the *population distribution of the variable is normal.* In that case, the sampling distribution of $\bar{x}$ is normal even for small sample sizes. (The right panel of Figure 7.8, which showed sampling distributions for various shapes of population distributions, illustrated this, as does the figure here in the margin.) When the population distribution is normal, the sampling distribution of $\bar{x}$ is normal for all *n*, not just large *n*.

Suppose we knew the standard deviation, $\sigma/\sqrt{n}$, of the sample mean. Then, with the additional assumption that the population is normal, with small *n* we could use the formula $\bar{x} \pm z(\sigma/\sqrt{n})$, for instance with $z = 1.96$ for 95% confidence. In practice, we don't know the population standard deviation σ. Substituting the sample standard deviation *s* for σ to get $se = s/\sqrt{n}$ then introduces extra error. This error can be sizeable when *n* is small. To account for this increased error, we must replace the *z*-score by a slightly larger score, called a ***t*-score**. The confidence interval is then a bit wider. *The t-score is like a z-score but it comes from a bell-shaped distribution that has slightly thicker tails than a normal distribution.* This distribution is called the ***t* distribution**.

In Practice The *t* Distribution Adjusts for Estimating σ

In practice, we estimate the standard deviation of the sample mean by $se = s/\sqrt{n}$. Then we multiply *se* by a *t*-score from the ***t* distribution** to get the margin of error for a confidence interval for the population mean.

Recall

From Section 6.2, the **standard normal** distribution has mean 0 and standard deviation 1. ◀

The *t* distribution resembles the *standard normal* distribution, being bell shaped around a mean of 0. Its standard deviation is a bit larger than 1, the precise value depending on what is called the **degrees of freedom**, denoted by ***df***. For inference about a population mean, the degrees of freedom equal $df = n - 1$, one less than the sample size. Before presenting this confidence interval for a mean, we list the major properties of the *t* distribution.

SUMMARY: Properties of the *t* Distribution

- The *t* distribution is bell shaped and symmetric about 0.
- The probabilities depend on the degrees of freedom, *df*. The *t* distribution has a slightly different shape for each distinct value of *df*, and different *t*-scores apply for each *df* value.
- The *t* distribution has thicker tails and has more variability than the standard normal distribution. The larger the *df* value, however, the closer it gets to the standard normal. Figure 8.7 illustrates this point. When *df* is about 30 or more, the two distributions are nearly identical.
- A *t*-score multiplied by the standard error gives the margin of error for a confidence interval for the mean.

Table B at the end of the text lists *t*-scores from the *t* distribution for the right-tail probabilities of 0.100, 0.050, 0.025, 0.010, 0.005, and 0.001. The table labels these by $t_{.100}, t_{.050}, t_{.025}, t_{.010}, t_{.005}$, and $t_{.001}$. For instance, $t_{.025}$ has probability 0.025 in the right tail, a two-tail probability of 0.05, and is used in 95% confidence intervals. Statistical software reports *t*-scores for any tail probability.

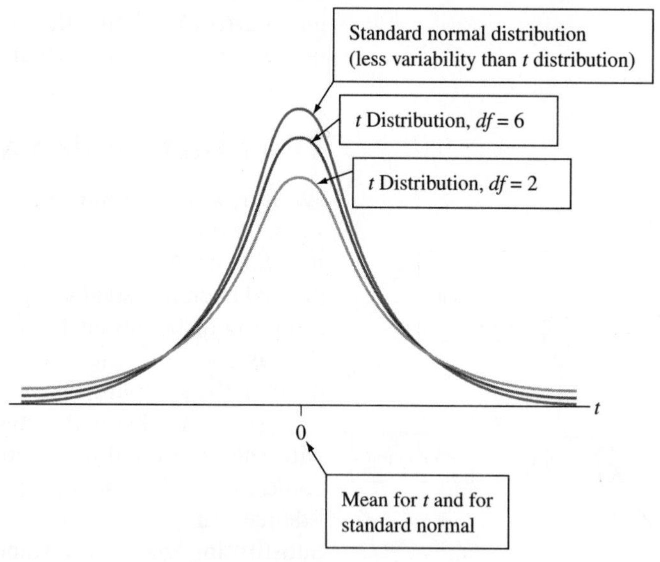

▲ **Figure 8.7 The *t* Distribution Relative to the Standard Normal Distribution.** The *t* distribution gets closer to the standard normal as the degrees of freedom (*df*) increase. The two are practically identical when $df \geq 30$. **Question** Can you find *z*-scores (such as 1.96) for a normal distribution on the *t* table (Table B)?

Table 8.3 is an excerpt from the *t* table (Table B). To illustrate its use, suppose the sample size is 7. Then the degrees of freedom $df = n - 1 = 6$. Row 6 of the *t* table shows the *t*-scores for $df = 6$. The column labeled $t_{.025}$ contains *t*-scores with right-tail probability equal to 0.025. With $df = 6$, this *t*-score is $t_{.025} = 2.447$. This means that 2.5% of the *t* distribution falls in the right tail above 2.447. By symmetry, 2.5% also falls in the left tail below $-t_{.025} = -2.447$. Figure 8.8 illustrates. When $df = 6$, the probability equals 0.95 between −2.447 and 2.447. This is the *t*-score for a 95% confidence interval when $n = 7$. The confidence interval is $\bar{x} \pm 2.447(se)$. You can also use some calculators and software to find *t*-scores.

Table 8.3 Part of Table B Displaying *t*-Scores

The scores have right-tail probabilities of 0.100, 0.050, 0.025, 0.010, 0.005, and 0.001. When $n = 7, df = 6$, and $t_{.025} = 2.447$ is the *t*-score with right-tail probability $= 0.025$ and two-tail probability $= 0.05$. It is used in a 95% confidence interval, $\bar{x} \pm 2.447(se)$.

	Confidence Level					
	80%	**90%**	**95%**	**98%**	**99%**	**99.8%**
df	$t_{.100}$	$t_{.050}$	$t_{.025}$	$t_{.010}$	$t_{.005}$	$t_{.001}$
1	3.078	6.314	12.706	31.821	63.657	318.3
...						
6	1.440	1.943	2.477	3.143	3.707	5.208
7	1.415	1.895	2.365	2.998	3.499	4.785

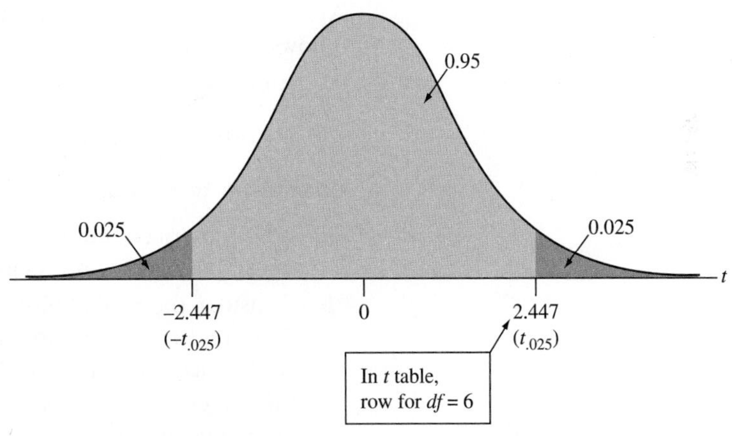

▲ **Figure 8.8** The *t* Distribution with $df = 6$. 95% of the distribution falls between −2.447 and 2.447. These *t*-scores are used with a 95% confidence interval when $n = 7$. **Question** Which *t*-scores with $df = 6$ contain the middle 99% of a *t* distribution (for a 99% confidence interval)?

Using the *t* Distribution to Construct a Confidence Interval for a Mean

The confidence interval for a mean has margin of error that equals a *t*-score times the standard error.

SUMMARY: 95% Confidence Interval for a Population Mean

A 95% confidence interval for the population mean μ is

$$\bar{x} \pm t_{.025}(se), \text{ where } se = s/\sqrt{n}.$$

Here, $df = n - 1$ for the *t*-score $t_{.025}$ that has right-tail probability 0.025 (total probability 0.05 in the two tails and 0.95 between $-t_{.025}$ and $t_{.025}$). To use this method, you need

- Data obtained by randomization (such as a random sample or a randomized experiment)
- An approximately normal population distribution.

Example 7

Using a *t* Distribution

Stock Market Activity on Different Days of the Week

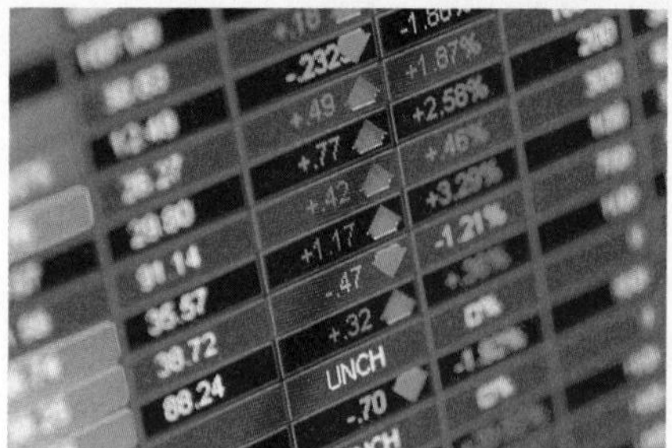

Picture the Scenario

Companies often finance their daily operations by selling shares of common stock. Each share of stock in a given company represents one share of ownership of that company. These shares of stock are bought and sold on various stock exchanges such as the New York Stock Exchange and the NASDAQ. One statistic monitored closely is the numbers of shares, or trading volume, that are bought and sold each day. In the absence of any events that trigger heavy trading, volume tends to be the highest early in the week and lightest late in the week. In this example, we compare the number of shares traded of General Electric stock on Mondays and Fridays during February through April of 2011. The trading volumes (rounded to the nearest million) are as follows:

Monday: 45, 43, 43, 66, 91, 53, 35, 45, 29, 64, 56

Fridays: 43, 41, 45, 46, 61, 56, 80, 40, 48, 49, 50, 41

Questions to Explore

a. Use numerical and graphical descriptive statistics to summarize trading activity for Mondays and Fridays.

b. Consider the probability distribution of trading volume on Mondays. State and check the assumptions for using these data to find a 95% confidence interval for the mean of that distribution.

c. Find the 95% confidence interval for Monday volume and interpret it. How does it compare to the 95% confidence interval for Friday volume, which you will find in Exercise 8.31 is (42.8, 57.2)?

Think It Through

a. When we use MINITAB and request descriptive statistics, we get the mean and standard deviation and the five-number summary using quartiles:

Day	N	Mean	StDev	Minimum	Q1	Median	Q3	Maximum
Monday	11	51.82	17.19	29.00	43.00	45.00	64.00	91.00
Friday	12	50.00	11.34	40.00	41.50	47.00	54.50	80.00

The centers are somewhat similar. The sample mean selling price is slightly higher for Monday than Friday, 51.82 million compared to 50.00 million. However, the sample median is slightly higher for Fridays, 47 million compared to 45 million. Because the mean is higher than the median for both distributions, this suggests that the distributions are right skewed. There is more variability in Monday trading volume, as we see by comparing the standard deviations.

Figure 8.9 shows a MINITAB dot plot for a day's trading volume. The plot shows the greater variability for the 11 Monday volumes as well as evidence of a slightly more skew to the right than Friday.

b. The confidence interval using the *t* distribution assumes a random sample from an approximately normal population distribution of selling prices. Unlike a survey such as the GSS, with the trading

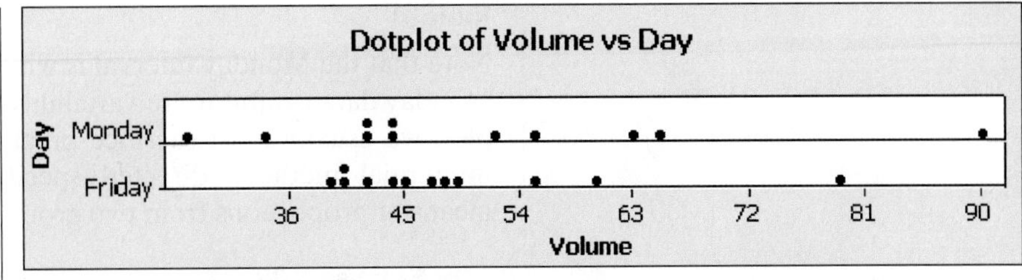

▲ **Figure 8.9 MINITAB Dot Plot for General Electric Trading Volume.** Data are shown for 11 Mondays and 12 Fridays during February, March, and April 2011. **Question** What does this plot tell you about how the trading volumes compare for these two days of the week?

volumes, the sample and population distinction is not as clear. We'll regard the observed volumes as independent observations from a probability distribution for all potential volumes over the given time period. That probability distribution reflects the variability in day-to-day trading volume. Inference treats that probability distribution like a population distribution and treats the observed volumes as a random sample from it.

We can get some information about the shape of the probability distribution of volume by looking at Figure 8.9. With only 11 observations for Monday, it's hard to tell much. Although we do observe one seemingly outlying observation, a bell-shaped, discrete population distribution could generate sample data such as shown in the first line of Figure 8.9. Later in this section, we'll discuss this assumption further.

c. Let μ denote the population mean for Monday volume. The estimate of μ is the sample mean of 45, 43, 43, 66, 91, 53, 35, 45, 29, 64, and 56, which is $x = 51.82$. This and the sample standard deviation of $s = 17.19$ are reported in part a. The standard error of the sample mean is

$$se = s/\sqrt{n} = 17.19/\sqrt{11} = 5.18.$$

Since $n = 11$, the degrees of freedom are $df = n - 1 = 10$. For a 95% confidence interval, from Table 8.3 we use $t_{.025} = 2.228$. The 95% confidence interval is

$$\bar{x} \pm t_{.025}se = 51.82 \pm 2.228(5.18),$$

which is 51.82 ± 11.541, or (40.27,63.37).

With 95% confidence, the range of believable values for the mean Monday volume is 40.27 million shares to 63.37 million shares. In the margin, a screen shot shows how the TI-83+/84 reports this confidence interval. Table 8.4 shows the way MINITAB reports it, with "SE Mean" being the standard error of the sample mean.

```
TInterval
 Inpt:Data Stats
 List:L1
 Freq:1
 C-Level:95
 Calculate
```

```
TInterval
 (40.268,63.368)
 x̄=51.8182
 Sx=17.1920
 n=11.0000
```

TI-83+/84 output

Table 8.4 MINITAB Output for 95% Confidence Interval for Mean

Variable	N	Mean	StDev	SE Mean	95% CI
Mondays	11	51.8182	17.1920	5.1836	(40.2685, 63.3679)

This confidence interval is not much different from the confidence interval (42.8, 57.2) for the mean of Friday volume. There's not enough information for us to conclude that one probability distribution has a higher mean than the other.

Insight

Note that the Monday interval is wider than the Friday interval because the Monday data exhibit more variability than the Friday data. With small samples, we usually must sacrifice precision. In Chapter 10, we'll learn about inferential methods directed specifically toward comparing population means or proportions from two groups.

Try Exercise 8.33

Finding a *t* Confidence Interval for Other Confidence Levels

The 95% confidence interval uses $t_{.025}$, the t-score for a right-tail probability of 0.025, since 95% of the probability falls between $-t_{.025}$ and $t_{.025}$. For 99% confidence, the error probability is 0.01, the probability is $0.01/2 = 0.005$ in each tail, and the appropriate t-score is $t_{.005}$. The top margins of Table B and Table 8.3 show both the t subscript notation and the confidence level.

For instance, the Monday volume sample in the preceding example has $n = 11$, so $df = 10$ and from Table 8.3, $t_{.005} = 3.169$. A 99% confidence interval for the mean Monday volume is

$$\bar{x} \pm t_{.005}(se) = 51.82 \pm 3.169(5.18),$$

which is 51.82 ± 16.42, or (35.40, 68.24).

As we saw with the confidence interval for a population proportion using a larger confidence level, this 99% confidence interval for a population mean is wider than the 95% confidence interval of (40.27, 63.37).

If the Population Is Not Normal, Is the Method Robust?

A basic assumption of the confidence interval using the t distribution is that the population distribution is normal. This is worrisome because many variables have distributions that are far from a bell shape. How problematic is it if we use the t confidence interval even if the population distribution is not normal? For large random samples, it's not problematic because of the central limit theorem. The sampling distribution is bell shaped even when the population distribution is not. But what about for small n?

For the confidence interval in Example 7 with $n = 11$ to be valid, we must assume that the probability distribution of Monday volume is normal. Does this assumption seem plausible? A dot plot, histogram, or stem-and-leaf plot gives us some information about the population distribution, but it is not precise when n is small and it tells us little when $n = 11$. Fortunately, the confidence interval using the t distribution is a **robust** method in terms of the normality assumption.

Robust Statistical Method

A statistical method is said to be **robust** with respect to a particular assumption if it performs adequately even when that assumption is modestly violated.

Even if the population distribution is not normal, confidence intervals using t-scores usually work quite well. The actual probability that the 95% confidence interval method provides a correct inference is close to 0.95 and gets closer as n increases.

Recall

Section 2.5 identified an observation as a potential **outlier** if it falls more than 1.5 × *IQR* below the first quartile or above the third quartile, or if it falls more than 3 standard deviations from the mean. ◀

The most important case when the t confidence interval method does *not* work well is *when the data contain extreme outliers.* Partly this is because of the effect on the method but also because the mean itself may not then be a representative summary of the center. In Example 7 with the 11 observations in the Monday volume sample, you can check that the only potentially outlying value is 91. However, 91 does not fall more than 1.5 × *IQR* above the third quartile. Another case that calls for caution is with binary data, in which case the mean is a proportion. Section 8.2 presented a separate method for binary data.

Caution

The t confidence interval method is not robust to violations of the random sampling assumption. The t method, like all inferential statistical methods, has questionable validity if the method for producing the data did not use randomization. ◀

In Practice Assumptions Are Rarely Perfectly Satisfied

Knowing that a statistical method is **robust** (that is, it still performs adequately) even when a particular assumption is violated is important because in practice assumptions are rarely perfectly satisfied. Confidence intervals for a mean using the ***t* distribution** are robust against most violations of the normal population assumption. However, you should check the data graphically to identify **outliers** that could affect the validity of the mean or its confidence interval. Also, unless the data production used **randomization,** statistical inference may be inappropriate.

The Standard Normal Distribution Is the *t* Distribution with *df* = ∞

Look at the table of t-scores (Table B in the Appendix), part of which is shown in Table 8.5. As df increases, you move down the table. The t-score decreases toward the z-score for a standard normal distribution. For instance, when df increases from 1 to 100 in Table B, the t-score $t_{.025}$ that has right-tail probability equal to 0.025 decreases from 12.706 to 1.984. This reflects the t distribution having less variability and becoming more similar in appearance to the standard normal distribution as df increases. The z-score with right-tail probability of 0.025 for the standard normal distribution is $z = 1.96$. When df is above about 30, the t-score is similar to this z-score. For instance, they both round to 2.0. The t-score gets closer and closer to the z-score as df keeps increasing. *You can think of the standard normal distribution as a t distribution with df = infinity.*

Did You Know?

z-score = *t-score* with *df* = ∞ (infinity). ◀

Table 8.5 Part of Table B Displaying *t*-Scores for Large *df* Values

The z-score of 1.96 is the t-score $t_{.025}$ with right-tail probability of 0.025 and $df = \infty$.

	Confidence Level					
	80%	90%	95%	98%	99%	99.8%
	Right-Tail Probability					
df	$t_{.100}$	$t_{.050}$	$t_{.025}$	$t_{.010}$	$t_{.005}$	$t_{.001}$
1	3.078	6.314	12.706	31.821	63.657	318.3
30	1.310	1.697	2.042	2.457	2.750	3.385
40	1.303	1.684	2.021	2.423	2.704	3.307
50	1.299	1.676	2.009	2.403	2.678	3.261
60	1.296	1.671	2.000	2.390	2.660	3.232
80	1.292	1.664	1.990	2.374	2.639	3.195
100	1.290	1.660	1.984	2.364	2.626	3.174
∞	1.282	1.645	1.960	2.326	2.576	3.090

Table 8.5 shows the first row and the last several rows of Table B. The last row lists the z-scores for various confidence levels, opposite $df = \infty$ (infinity). The t-scores are not printed for $df > 100$, but they are close to the z-scores. For instance, to get the confidence interval about TV watching in Example 6, for which the GSS sample had $n = 1324$, software uses the $t_{.025}$ score for $df = 1324 - 1 = 1323$, which is 1.96176. This is nearly identical to the z-score of 1.960 from the standard normal distribution.

Recall

The reason we use a t-score instead of a z-score in the confidence interval for a mean is that it accounts for the extra error due to estimating σ by s. ◀

You can get t-scores for *any df* value using software and many calculators, so you are not restricted to Table B. (For instance, MINITAB provides percentile scores for various distributions under the *CALC* menu.) If you don't have access to software, you won't be far off if you use a z-score instead of a t-score for *df* values larger than shown in Table B (above 100). For a 95% confidence interval you will then use

$$\bar{x} \pm 1.96(se) \text{ instead of } \bar{x} \pm t_{.025}(se).$$

You will not get *exactly* the same result that software would give, but it will be very, very close.

In Practice Use t for Inference about μ Whenever You Estimate

Statistical software and calculators use the t distribution for *all* cases when the sample standard deviation s is used to estimate the population standard deviation σ. The normal population assumption is mainly relevant for small n, but even then the t confidence interval is a robust method, working well unless there are extreme outliers or the data are binary.

On the Shoulders of...William S. Gosset

W.S. Gosset: Discovered the t distribution allowing statistical methods for working with small sample sizes.

How do you find the best way to brew beer if you have only small samples?

The statistician and chemist William S. Gosset was a brewer in charge of the experimental unit of Guinness Breweries in Dublin, Ireland. The search for a better stout in 1908 led him to the discovery of the t distribution. Only small samples were available from his experiments pertaining to the selection, cultivation, and treatment of barley and hops. The established statistical methods at that time relied on large samples and the normal distribution. Because of company policy forbidding the publication of company work in one's own name, Gosset used the pseudonym "Student" in articles he wrote about his discoveries. The t distribution became known as *Student's t* distribution, a name sometimes still used today.

8.3 Practicing the Basics

8.27 Females' ideal number of children The 2008 General Social Survey asked, "What do you think is the ideal number of children for a family to have?" The 678 females who responded had a median of 2, mean of 3.22, and standard deviation of 1.99.

TRY

a. What is the point estimate of the population mean?

b. Find the standard error of the sample mean.

c. The 95% confidence interval is (3.07, 3.37). Interpret.

d. Is it plausible that the population mean $\mu = 2$? Explain.

8.28 Males' ideal number of children Refer to the previous exercise. For the 604 males in the sample, the mean was 3.06 and the standard deviation was 1.92.

a. Find the point estimate of the population mean, and show that its standard error is 0.078.

b. The 95% confidence interval is 2.91 and 3.21. Explain what "95% confidence" means for this interval.

8.29 Using *t*-table Using Table B or software or a calculator, report the *t*-score which you multiply by the standard error to form the margin of error for a

a. 95% confidence interval for a mean with 5 observations.

b. 95% confidence interval for a mean with 15 observations.

c. 99% confidence interval for a mean with 15 observations.

8.30 Anorexia in teenage girls A study[6] compared various therapies for teenage girls suffering from anorexia, an eating disorder. For each girl, weight was measured before and after a fixed period of treatment. The variable measured was the change in weight, X = weight at the end of the study minus weight at the beginning of the study. The therapies were designed to aid weight gain, corresponding to positive values of X. For the sample of 17 girls receiving the family therapy, the changes in weight during the study were

$$11, 11, 6, 9, 14, -3, 0, 7, 22, -5, -4, 13, 13, 9, 4, 6, 11.$$

a. Plot these with a dot plot or box plot, and summarize.

b. Using a calculator or software, show that the weight changes have $\bar{x} = 7.29$ and $s = 7.18$ pounds.

c. Using a calculator or software, show that the standard error of the sample mean was $se = 1.74$.

d. To use the *t* distribution, explain why the 95% confidence interval uses the *t*-score equal to 2.120.

e. Let μ denote the population mean change in weight for this therapy. Using results from parts b, c, and d, show that the 95% confidence interval for μ is (3.6, 11.0). Explain why this suggests that the true mean change in weight is positive, but possibly quite small.

8.31 Stock market activity From Example 7, trading volumes for General Electric stock on Mondays and Fridays during February through April of 2011 were given as follows

Mondays: 45, 43, 43, 66, 91, 53, 35, 45, 29, 64, 56
Fridays: 43, 41, 45, 46, 61, 56, 80, 40, 48, 49, 50, 41

The Monday data have $\bar{x} = 51.82$, $s = 17.19$, Q1 = 43, Median = 45, Q3 = 64.

The Friday data have $\bar{x} = 50$, $s = 11.34$, Q1 = 41.5, Median = 47, Q3 = 54.5.

a. Construct a histogram or boxplot of the Friday data. What assumptions are needed to construct a 95% confidence interval for μ? Point out any assumptions that seem questionable.

b. Using the Friday data and appropriate given summary statistics, show that the 95% confidence interval is (42.795, 57.205). Interpret it in context.

c. Check whether this data set has any potential outliers according to the criterion of (i) 1.5 * IQR below Q1 or above Q3 and (ii) 3 standard deviations from the mean.

d. The value 80 is quite a bit larger than the others. Delete this observation, find the new mean and standard deviation, and use software to construct the 95% confidence interval for μ. How does it compare to the 95% confidence interval (42.796, 57.204) using all the data?

8.32 Heights of seedlings Exercise 8.6 reported heights (in mm) of 55.5, 60.3, 60.6, 62.1, 65.5, and 69.2 for six seedlings fourteen days after germination.

a. Using software or a calculator, verify that the 95% confidence interval for the population mean is (57.3, 67.1).

b. Name two things you could do to get a narrower interval than the one in part a.

c. Construct a 99% confidence interval. Why is it wider than the 95% interval?

d. On what assumptions is the interval in part a based? Explain how important each assumption is.

8.33 TV watching for Muslims TRY Having estimated the mean amount of time spent watching TV in Example 6, we might want to estimate the mean for various groups, such as different religious groups. Let's consider Muslims. A recent GSS had responses on TV watching from seven subjects who identified their religion as Muslim. Their responses on the number of hours of TV watching were 0, 1, 2, 2, 2, 2, 2, 3, 5 shown also in the accompanying MINITAB dot plot.

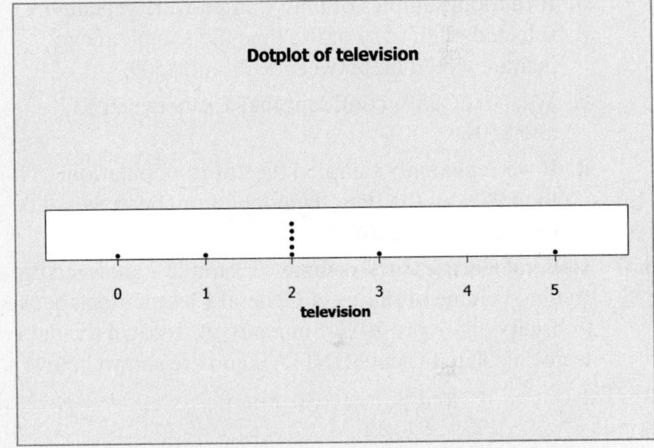

a. What must we assume to use these data to find a 95% confidence interval for the mean amount of TV watched by the population of American Muslims?

b. The table shows the way MINITAB reports results. Explain how to interpret the 95% confidence interval in context.

Variable	N	Mean	StDev	SE Mean	95% CI
television	9	2.111	1.364	0.455	(1.062, 3.160)

c. What is the main factor causing the confidence interval to be so wide?

8.34 Time spent on e-mail When the GSS asked in 2004, "About how many hours per week do you spend sending and answering e-mail?" the results were as shown in the TI-83+/84 screen shot. Explain how to interpret the information shown, and interpret the confidence interval at the 95% confidence level.

[6]Data courtesy of Prof. Brian Everitt, Institute of Psychiatry, London.

8.35 Grandmas using e-mail For the question about e-mail in the previous exercise, suppose seven females in the GSS sample of age at least 80 had the responses

$$0, 0, 1, 2, 5, 7, 14.$$

a. Using software or a calculator, find the sample mean and standard deviation and the standard error of the sample mean.

b. Find and interpret a 90% confidence interval for the population mean.

c. Explain why the population distribution may be skewed right. If this is the case, is the interval you obtained in part b useless, or is it still valid? Explain.

8.36 Wage discrimination? According to a union agreement, the mean income for all senior-level assembly-line workers in a large company equals \$500 per week. A representative of a women's group decides to analyze whether the mean income for female employees matches this norm. For a random sample of nine female employees, using software she obtains a 95% confidence interval of (371, 509). Explain what is wrong with each of the following interpretations of this interval.

a. We infer that 95% of the women in the population have income between \$371 and \$509 per week.

b. If random samples of nine women were repeatedly selected, then 95% of the time the sample mean income would be between \$371 and \$509.

c. We can be 95% confident that $\bar{x}$ is between \$371 and \$509.

d. If we repeatedly sampled the entire population, then 95% of the time the population mean would be between \$371 and \$509.

8.37 General electric stock volume Example 7 analyzed the trading volume of shares of General Electric stock between February and April 2011. Summary statistics of the data were calculated using MINITAB and are shown below:

Variable	N	Mean	SE Mean	StDev	Median
Monday	11	51.82	5.18	17.19	45.00
Friday	12	50.00	3.27	11.34	47.00

The 95% confidence intervals for the means are (40.2685, 63.3679) for Monday's volume and (42.7963, 57.2037) for Friday's. Interpret each of these intervals, and explain what you learn by comparing them.

8.38 How often read a newspaper? For the FL Student Survey data file on the text CD, software reports the results for responses on the number of times a week the subject reads a newspaper:

Variable	N	Mean	Std Dev	SE Mean	95.0% CI
News	60	4.1	3.0	0.387	(3.325, 4.875)

a. Is it plausible that $\mu = 7$, where μ is the population mean for all Florida students? Explain.

b. Suppose that the sample size had been 240, with $\bar{x} = 4.1$ and $s = 3.0$. Find a 95% confidence interval, and compare it to the one reported. Describe the effect of sample size on the margin of error.

c. Does it seem plausible that the population distribution of this variable is normal? Why?

d. Explain the implications of the term *robust* regarding the normality assumption made to conduct this analysis.

8.39 Political views The General Social Survey asks respondents to rate their political views on a seven-point scale, where 1 = extremely liberal, 4 = moderate, and 7 = extremely conservative. A researcher analyzing data from the 2008 GSS obtains MINITAB output:

Variable	N	Mean	StDev	SE Mean	95% CI
POLVIEWS	1933	4.1128	1.4327	0.0326	(4.0489, 4.1767)

a. Show how to construct the confidence interval from the other information provided.

b. Can you conclude that the population mean is higher than the moderate score of 4.0? Explain.

c. Would the confidence interval be wider, or narrower, (i) if you constructed a 99% confidence interval and (ii) if $n = 500$ instead of 1933?

8.40 Length of hospital stays A hospital administrator wants to estimate the mean length of stay for all inpatients using that hospital. Using a random sample of 100 records of patients for the previous year, she reports that "The sample mean was 5.3. In repeated random samples of this size, the sample mean could be expected to fall within 1.0 of the true mean about 95% of the time." Explain the meaning of this sentence from the report, showing what it suggests about the 95% confidence interval.

8.41 Effect of *n* Find the margin of error for a 95% confidence interval for estimating the population mean when the sample standard deviation equals 100, with a sample size of (i) 400 and (ii) 1600. What is the effect of the sample size?

8.42 Effect of confidence level Find the margin of error for estimating the population mean when the sample standard deviation equals 100 for a sample size of 400, using confidence level (i) 95% and (ii) 99%. What is the effect of the choice of confidence level?

8.43 Catalog mail-order sales A company that sells its products through mail-order catalogs wants information about the success of its most recent catalog. The company decides to estimate the mean dollar amount of items ordered from those who received the catalog. For a random sample of 100 customers from their files, only 5 made an order, so 95 of the response values were \$0. The overall mean of all 100 orders was \$10, with a standard deviation of \$10.

a. Is it plausible that the population distribution is normal? Explain, and discuss how much this affects the validity of a confidence interval for the mean.

b. Find a 95% confidence interval for the mean dollar order for the population of all customers who received this catalog. Normally, the mean of their sales per catalog is about \$15. Can we conclude that it declined with this catalog? Explain.

8.44 Number of children For the question, "How many children have you ever had?" use the GSS Web site sda.berkeley.edu/GSS with the variable CHILDS to find the sample mean and standard deviation for the 2008 survey.

a. Show how to obtain a standard error of 0.04 for a random sample of 2020 adults.

b. Construct a 95% confidence interval for the population mean. Can you conclude that the population mean is less than 2.0? Explain.

c. Discuss the assumptions for the analysis in part b and whether that inference seems to be justified.

8.45 Simulating the confidence interval Go to the Confidence Intervals for a Mean applet on the text CD. Choose the sample size of 50 and the skewed population distribution with $\mu = 100$. Generate 100 random samples, each of size 50, and for each one form a 95% confidence interval for the mean.

a. How many of the intervals fail to contain the true value?

b. How many would you expect not to contain the true value?

c. Now repeat the simulation using 10,000 random samples of size 50. Why do close to 95% of the intervals contain μ, even though the population distribution is quite skewed?

8.4 Choosing the Sample Size for a Study

Have you ever wondered how sample sizes are determined for polls? How does a polling organization know if it needs 10,000 people, 100 people, or some odd number such as 745 people? The simple answer is that this depends on how much precision is needed, as measured by the margin of error. The smaller the margin of error, the larger the sample size must be. We'll next learn how to determine which sample size has the desired margin of error. For instance, we'll find out how large an exit poll must be so that a 95% confidence interval for the population proportion voting for a candidate has a certain margin of error, such as 0.04.

The key results for finding the sample size for a random sample are as follows:

- The *margin of error* depends on the *standard error* of the sampling distribution of the point estimate.
- The *standard error* itself depends on the *sample size.*

So one of the main components of the margin of error is the sample size n. Once we specify a margin of error with a particular confidence level, we can determine the value of n that has a standard error giving that margin of error.

Choosing the Sample Size for Estimating a Population Proportion

How large should n be to estimate a population proportion? First we must decide on the desired *margin of error*—how close the sample proportion should be to the population proportion. Second, we must choose the *confidence level* for achieving that margin of error. In practice, 95% confidence intervals are most common. If we specify a margin of error of 0.04, this means that a 95% confidence interval should equal the sample proportion plus and minus 0.04.

Example 8

Choosing a sample size

Exit Poll

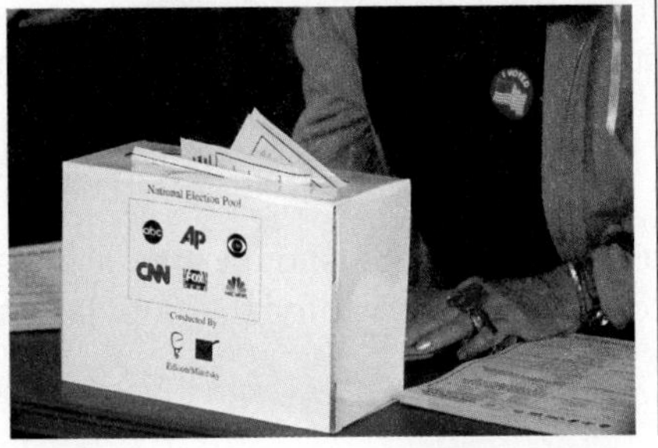

Picture the Scenario

A TV network plans to predict the outcome of an election between Levin and Sanchez using an exit poll that randomly samples voters on election day. They want a reasonably accurate estimate of the population proportion that voted for Levin. The final poll a week before election day estimated her to be well ahead, 58% to 42%, so they do not expect the outcome to be close. Since their finances for this project are limited, they don't want to collect a large sample if they don't need it. They decide to use a sample size for which the margin of error is 0.04, rather than their usual margin of error of 0.03.

Recall

From Section 8.2, a 95% confidence interval for a population proportion p is

$$\hat{p} \pm 1.96(se),$$

where $\hat{p}$ denotes the sample proportion and the standard error (se) is

$$se = \sqrt{\hat{p}(1 - \hat{p})/n}.$$ ◀

Question to Explore

What is the sample size for which a 95% confidence interval for the population proportion has a margin of error equal to 0.04?

Think It Through

The 95% confidence interval for a population proportion p is $\hat{p} \pm 1.96(se)$. So, if the sample size is such that $1.96(se) = 0.04$, then the margin of error will be 0.04. See Figure 8.10.

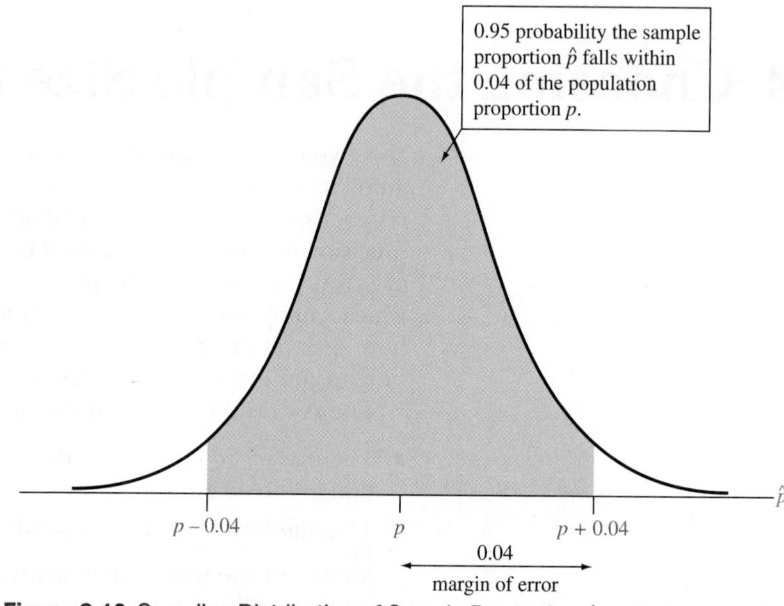

▲ **Figure 8.10 Sampling Distribution of Sample Proportion $\hat{p}$ such that a 95% Confidence Interval Has Margin of Error 0.04.** We need to find the value of n that has this margin of error. **Question** What must we assume for this distribution to be approximately normal?

Let's find the value of the sample size n for which $0.04 = 1.96(se)$. For a confidence interval for a proportion, the standard error is $\sqrt{\hat{p}(1 - \hat{p})/n}$. So the equation $0.04 = 1.96(se)$ becomes

$$0.04 = 1.96\sqrt{\hat{p}(1 - \hat{p})/n}.$$

To find the answer, we solve algebraically for n:

$$n = (1.96)^2\hat{p}(1 - \hat{p})/(0.04)^2.$$

(If your algebra is rusty, don't worry—we'll soon show a general formula.)

Now, we face a problem. We're doing this calculation *before* gathering the data, so we don't yet have a sample proportion $\hat{p}$. The formula for n depends on $\hat{p}$ because the standard error depends on it. Since $\hat{p}$ is unknown, we must substitute an educated guess for what we'll get once we gather the sample and analyze the data.

Since the latest poll *before* election day predicted that 58% of the voters preferred Levin, it is sensible to substitute 0.58 for $\hat{p}$ in this equation. Then we find

$$n = (1.96)^2\hat{p}(1 - \hat{p})/(0.04)^2 = (1.96)^2(0.58)(0.42)/(0.04)^2 = 584.9.$$

In summary, a random sample of size about $n = 585$ should give a margin of error of about 0.04 for a 95% confidence interval for the population proportion.

Insight

Sometimes, we may have no idea what to expect for $\hat{p}$. We may prefer not to guess the value it will take, as we did in this example. We'll next learn what we can do in those situations.

Try Exercise 8.47

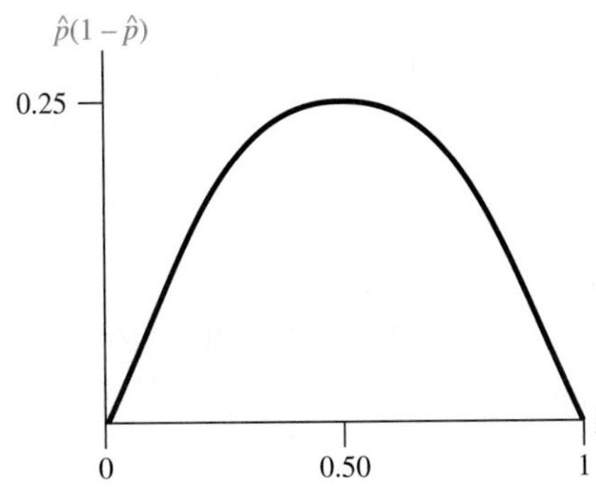

Selecting a Sample Size without Guessing a Value for $\hat{p}$

In the previous example, the solution for n was proportional to $\hat{p}(1 - \hat{p})$. The figure in the margin shows how that product depends on the value of $\hat{p}$. The largest possible value for $\hat{p}(1 - \hat{p})$ is 0.25, which occurs when $\hat{p} = 0.50$. You can check by plugging values of $\hat{p}$ into $\hat{p}(1 - \hat{p})$ that this product is near 0.25 unless $\hat{p}$ is quite far from 0.50. For example, $\hat{p}(1 - \hat{p}) = 0.24$ when $\hat{p} = 0.40$ or $\hat{p} = 0.60$.

In the formula for determining n, setting $\hat{p} = 0.50$ out of all the possible values to substitute for $\hat{p}$ gives the largest value for n. So doing this is the safe approach that guarantees we'll have enough data. In the election exit poll example, for a margin of error of 0.04, we then get

$$n = (1.96)^2\hat{p}(1 - \hat{p})/(0.04)^2$$
$$= (1.96)^2(0.50)(0.50)/(0.04)^2 = 600.25, \text{ rounded up to } 601.$$

This result compares to $n = 585$ from guessing that $\hat{p} = 0.58$. Using the slightly larger value of $n = 601$ ensures that the margin of error for a 95% confidence interval will not exceed 0.04, no matter what value $\hat{p}$ takes once we collect the data.

This safe approach is not always sensible, however. Substituting $\hat{p} = 0.50$ gives us an overly large solution for n if $\hat{p}$ actually falls far from 0.50. Suppose that based on other studies we expect $\hat{p}$ to be about 0.10. Then an adequate sample size to achieve a margin of error of 0.04 is

$$n = (1.96)^2\hat{p}(1 - \hat{p})/(0.04)^2$$
$$= (1.96)^2(0.10)(0.90)/(0.04)^2 = 216.09, \text{ rounded up to } 217.$$

A sample size of 601 would be much larger and more costly than needed.

General Sample Size Formula for Estimating a Population Proportion A general formula exists for determining the sample size, based on solving algebraically for n by setting the margin of error formula equal to the desired value. Let m denote the desired margin of error. This is $m = 0.04$ in the previous example. The general formula also uses the z-score for the confidence level.

SUMMARY: Sample Size for Estimating a Population Proportion

The random sample size n for which a confidence interval for a population proportion p has margin of error m (such as $m = 0.04$) is

$$n = \frac{\hat{p}(1 - \hat{p})z^2}{m^2}.$$

The z-score is based on the confidence level, such as $z = 1.96$ for 95% confidence. You either guess the value you'd get for the sample proportion $\hat{p}$ based on other information or take the safe approach of setting $\hat{p} = 0.5$.

Choosing sample size

Example 9

Using an Exit Poll When a Race Is Close

Picture the Scenario

An election is expected to be close. Pollsters planning an exit poll decide that a margin of error of 0.04 is too large.

Question to Explore

How large should the sample size be for the margin of error of a 95% confidence interval to equal 0.02?

Think It Through

Since the election is expected to be close, we expect $\hat{p}$ to be near 0.50. In the formula, we set $\hat{p} = 0.50$ to be safe. We also set the margin of error $m = 0.02$ and use $z = 1.96$ for a 95% confidence interval. The required sample size is

$$n = \frac{\hat{p}(1 - \hat{p})z^2}{m^2} = \frac{(0.50)(0.50)(1.96)^2}{(0.02)^2} = 2401.$$

The sample size of about 2400 is four times the sample size of 600 necessary to guarantee a margin of error of $m = 0.04$. *Reducing the margin of error by a factor of one-half requires quadrupling n.*

Insight

Example 1 in Chapter 7 described an exit poll for the California gubernatorial race in 2010 between Jerry Brown, Meg Whitman, and others. That exit poll used a sample size ($n = 3889$), approximately 1.5 times the value of 2401 just determined. Did they achieve a margin of error less than 0.02? From that example, the sample proportion who voted for Brown was 0.531. A 95% confidence interval for the population proportion p who voted for Brown is $\hat{p} \pm 1.96(se)$, with $se = \sqrt{\hat{p}(1 - \hat{p})/n} = \sqrt{(0.531)(0.469)/3889} = 0.008$. This gives the interval

$$0.531 \pm 1.96(0.008), \text{which is } 0.531 \pm 0.016, \text{or } (0.52, 0.55).$$

The margin of error was less than 0.02 providing even more precision. This was small enough to predict (correctly) that Brown would win, since the confidence interval values (0.52, 0.55) all fell above 0.50. Note that if the exit poll had used a sample size of $n = 2400$, the margin of error would have been 0.01997 which rounds to 0.02.

Try Exercise 8.46

Samples taken by polling organizations typically contain 1000–2000 subjects. This is large enough to estimate a population proportion with a margin of error of about 0.02 or 0.03. At first glance, it seems astonishing that a sample of this size from a population of perhaps many millions is adequate for predicting outcomes of elections, summarizing opinions on controversial issues, showing relative sizes of television audiences, and so forth. The basis for this inferential power lies in the standard error formulas for the point estimates, with random sampling. Good estimates result no matter how large the population size.[7]

Revisiting the Approximation $1/\sqrt{n}$ for the Margin of Error Chapter 4 introduced a margin of error approximation of $1/\sqrt{n}$ for estimating a population

[7]In fact, the mathematical derivations of these methods treat the population size as infinite. See Exercise 7.62.

proportion using random sampling. What's the connection between this approximation and the more exact margin of error we've used in this chapter? Let's take the margin of error $1.96\sqrt{\hat{p}(1-\hat{p})/n}$ for a 95% confidence interval, round the z-score to 2, and replace the sample proportion by the value of 0.50 that gives the maximum possible standard error. Then we get the margin of error

$$2\sqrt{0.50(0.50)/n} = 2(0.50)\sqrt{1/n} = 1/\sqrt{n}.$$

So for a 95% confidence interval, the margin of error is approximately $1/\sqrt{n}$ when $\hat{p}$ is near 0.50.

Choosing the Sample Size for Estimating a Population Mean

As with the population proportion, to derive the sample size for estimating a population mean, you set the margin of error equal to its desired value and solve for n. Recall that a 95% confidence interval for the population mean is

$$\bar{x} \pm t_{.025}(se),$$

where $se = s/\sqrt{n}$ and s is the sample standard deviation. If you don't know n, you also don't know the degrees of freedom and the t-score. However, we saw in Table B that when $df > 30$, the t-score is very similar to the z-score from a normal distribution, such as 1.96 for 95% confidence. Also, before collecting the data, we do not know the sample standard deviation s, which we use to estimate the population standard deviation σ. When we use a z-score (in place of the t-score), supply an educated guess for the standard deviation of the sample mean $\sigma/\sqrt{n}$, and then set $z(\sigma/\sqrt{n})$ equal to a desired margin of error m and solve for n, we get the following result:

In Words

The sample size n needed to estimate the population mean depends on how precisely you want to estimate it (by the choice of the margin of error m), how sure you want to be (by the confidence level, which determines the z-score), and how variable the data will be (by the guess for the population standard deviation σ).

SUMMARY: Sample Size for Estimating a Population Mean

The random sample size n for which a confidence interval for a population mean has margin of error approximately equal to m is

$$n = \frac{\sigma^2 z^2}{m^2}.$$

The z-score is based on the confidence level, such as $z = 1.96$ for 95% confidence. To use this formula, you guess the value for the population standard deviation σ.

In practice, you don't know the population standard deviation. You must substitute an educated guess for σ. Sometimes you can guess it using the sample standard deviation from a similar study already conducted. The next example shows another sort of reasoning to form an educated guess.

Example 10

Finding n

Estimating Mean Education in South Africa

Picture the Scenario

A social scientist studies adult South Africans living in townships on the outskirts of Cape Town, to investigate educational attainment in the black community. Educational attainment is the number of years of education completed. Many of the study's potential subjects were forced to leave Cape Town in 1966 when the government passed a law forbidding blacks to live in the inner cities. Under the apartheid system, black South African children were not required to attend school, so some residents had very little education.

Question to Explore

How large a sample size is needed so that a 95% confidence interval for the mean number of years of attained education has a margin of error equal to 1 year?

Recall

Section 2.4 noted that for an approximately symmetric, bell-shaped distribution, we can approximate the standard deviation by roughly a sixth of the range. ◀

Think It Through

No prior information is stated about the standard deviation of educational attainment for the township residents. As a crude approximation, we might guess that the education values will fall within a range of about 18 years, such as between 0 and 18 years. If the distribution is bell shaped, the range from $\mu - 3\sigma$ to $\mu + 3\sigma$ will contain all or nearly all the distribution. Since the distance from $\mu - 3\sigma$ to $\mu + 3\sigma$ equals 6σ, the range of 18 years would equal about 6σ. Then, solving $18 = 6\sigma$ for σ, $18/6 = 3$ is a crude guess for σ. So we'd expect a standard deviation value of about $\sigma = 3$.

The desired margin of error is $m = 1$ year. The required sample size is

$$n = \frac{\sigma^2 z^2}{m^2} = \frac{3^2(1.96)^2}{1^2} = 35.$$

We need to randomly sample about 35 subjects for a 95% confidence interval for mean educational attainment to have a margin of error of 1 year.

Insight

A more cautious approach would select the *largest* value for the standard deviation that is plausible. This will give the largest sensible guess for how large n needs to be. For example, we could reasonably predict that σ will be no greater than 4, since a range of 6 standard deviations then extends over 24 years. Then we get $n = (1.96)^2(4^2)/(1^2) = 62$. If we collect the data and the sample standard deviation is actually less than 4, we will have more data than we need. The margin of error will be even less than 1.0.

Try Exercise 8.52

Other Factors That Affect the Choice of the Sample Size

We've looked at two factors that play a role in determining a study's sample size.

- The first is the desired *precision*, as measured by the *margin of error m.*
- The second is the *confidence level*, which determines the z-score or t-score in the sample size formulas.

Other factors also play a role.

- A third factor is the *variability* in the data.

Let's look at the formula $n = \sigma^2 z^2/m^2$ for the sample size for estimating a mean. The greater the value expected for the standard deviation σ, the larger the sample size needed. If subjects have little variation (that is, σ is small), we need fewer data than if they have substantial variation. Suppose a study plans to estimate the mean level of education in several countries. Western European countries have relatively little variation, as students are required to attend school until the middle teen years. To estimate the mean to within a margin of error of $m = 1$, we need fewer observations than in South Africa.

- A fourth factor is *cost.*

Larger samples are more time consuming to collect. They may be more expensive than a study can afford. Cost is often a major constraint. You may need to ask, "Should we go ahead with the smaller sample that we can afford, even though the margin of error will be greater than we would like?"

Using a Small *n*

Sometimes, because of financial or ethical reasons, it's just not possible to take as large of a sample as we'd like. For example, each observation may result from an expensive experimental procedure. A consumer group that estimates the mean repair cost after a new-model automobile crashes into a concrete wall at 30 miles per hour would probably not want to crash a large sample of cars so they can get a narrow confidence interval!

If n must be small, how does that affect the validity of the confidence interval methods? The t methods for a mean are valid *for any n.* When n is small, though, you need to be extra cautious to look for extreme outliers or great departures from the normal population assumption (such as implied by highly skewed data). These can affect the results and the validity of using the mean as a summary of center.

For the confidence interval formula for a proportion, we've seen that we need at least 15 successes and at least 15 failures. Why? Otherwise, the central limit theorem no longer applies. The numbers of successes and failures must both be at least 15 for the normal distribution to approximate well the binomial distribution or the sampling distribution of a sample proportion (as discussed in Chapter 7). An even more serious difficulty is that the standard deviation of the sample proportion depends on the parameter we're trying to estimate. If the estimate $\hat{p}$ is far from p, as often happens for small n, then the estimate $se = \sqrt{\hat{p}(1 - \hat{p})/n}$ of the standard deviation of the sample proportion may also be far off. As a result, the confidence interval formula works poorly, as we'll see in the next example.

Example 11

Small sample CI

Proportion of University Students Who Own an iPod

Picture the Scenario

Apple, Inc.'s famous iPod revolutionized the way people listen to their favorite music. Since the launch of the first generation iPod in 2001, more than 300 million units have been sold worldwide. For a statistics class project, a student randomly selected and interviewed 20 students at his university to estimate the proportion of students who own an iPod. Each of the 20 students interviewed owned an iPod.

Question to Explore

What does a 95% confidence interval reveal about the proportion of students at the university who own an iPod?

Think It Through

Let p denote the proportion of students at the university who own an iPod. The sample proportion from the interviews was $\hat{p} = 20/20 = 1.0$. When $\hat{p} = 1.0$, then

$$se = \sqrt{\hat{p}(1 - \hat{p})/n} = \sqrt{1.0(0.0)/20} = 0.0.$$

The 95% confidence interval for the proportion of students at the university who own an iPod is

$$\hat{p} \pm z(se) = 1.0 \pm 1.96(0.0),$$

which is 1.0 ± 0.0, or (1.0, 1.0). This investigation told the student that he can be 95% confident that p falls between 1 and 1, that is, that $p = 1$. He was surprised by this result. It seemed unrealistic to conclude that *every* student at the university owns an iPod.

Insight

Do you trust this inference? Just because everyone in a small sample owns an iPod, would you conclude that everyone in the much larger population owns an iPod? We doubt it. Perhaps the population proportion p is close to 1, but it is almost surely not exactly equal to 1.

The confidence interval formula $\hat{p} \pm z\sqrt{\hat{p}(1 - \hat{p})/n}$ is valid only if the sample contains at least 15 individuals who own an iPod and 15 individuals who don't. The sample did not contain at least 15 individuals who don't, so we have to use a different method.

Try Exercise 8.55, parts a–c

Confidence Interval for a Proportion with Small Samples

Constructing a Small-Sample Confidence Interval for a Proportion p

Suppose a random sample does *not* have at least 15 successes and 15 failures. The confidence interval formula $\hat{p} \pm z\sqrt{\hat{p}(1 - \hat{p})/n}$ still is valid if we use it after adding 2 to the original number of successes and 2 to the original number of failures. This results in adding 4 to the sample size n.

The sample of size $n = 20$ in Example 11 contained 20 individuals who owned an iPod and 0 who did not. We can apply the confidence interval formula with $20 + 2 = 22$ individuals who own an iPod and $0 + 2 = 2$ who do not. The value of the sample size for the formula is then $n = 24$. *Now we can use the formula, even though we don't have at least 15 individuals who do not own an iPod.* We get

$$\hat{p} = 22/24 = 0.917,\ se = \sqrt{\hat{p}(1 - \hat{p})/n} = \sqrt{(0.917)(0.083)/24} = 0.056.$$

The resulting 95% confidence interval is

$$\hat{p} \pm 1.96(se), \text{ which is } 0.917 \pm 1.96(0.056), \text{ or } (0.807, 1.027).$$

A proportion cannot be greater than 1, so we report the interval as (0.807, 1.0). We can be 95% confident that the proportion of individuals at the university who own an iPod is at least 0.807.

This approach enables us to use a large-sample method even when the sample size is small. With it, the point estimate moves the sample proportion a bit toward 1/2 (e.g., from 1.0 to 0.917). This is particularly helpful when the ordinary sample proportion is 0 or 1, which we would not usually expect to be a believable estimate of a population proportion. Why do we add 2 to the counts of the two types? The reason is that the resulting confidence interval is then close to a confidence interval based on a more complex method (described in Exercise 8.121) that does not require estimating the exact standard deviation of a sample proportion.[8]

[8]See article by A. Agresti and B. Coull (who proposed this small-sample confidence interval), *The American Statistician*, vol. 52, pp. 119–126, 1998.

Finally, a word of caution: In the estimation of parameters, "margin of error" refers to the size of error resulting from having data from a random sample rather than the population—what's called *sampling error.* This is the error that the sampling distribution describes in showing how close the estimate is likely to fall to the parameter. But that's not the only source of potential error. Data may be missing for a lot of the target sample; some observations may be recorded incorrectly by the data collector; and some subjects may not tell the truth. When errors like these occur, the actual confidence level may be much lower than advertised. Be skeptical about a claimed margin of error and confidence level unless you know that the study was well conducted and these other sources of error are negligible.

8.4 Practicing the Basics

8.46 **South Africa study** The researcher planning the study in South Africa also will estimate the population proportion having at least a high school education. No information is available about its value. How large a sample size is needed to estimate it to within 0.07 with 95% confidence?
TRY

8.47 **Binge drinkers** A study at the Harvard School of Public Health found that 44% of 10,000 sampled college students were binge drinkers. A student at the University of Minnesota plans to estimate the proportion of college students at that school who are binge drinkers. How large a random sample would she need to estimate it to within 0.05 with 95% confidence, if before conducting the study she uses the Harvard study results as a guideline?
TRY

8.48 **Abstainers** The Harvard study mentioned in the previous exercise estimated that 19% of college students abstain from drinking alcohol. To estimate this proportion in your school, how large a random sample would you need to estimate it to within 0.05 with probability 0.95, if before conducting the study

a. You are unwilling to predict the proportion value at your school.

b. You use the Harvard study as a guideline.

c. Use the results from parts a and b to explain why strategy (a) is inefficient if you are quite sure you'll get a sample proportion that is far from 0.50.

8.49 **How many businesses fail?** A study is planned to estimate the proportion of businesses started in the year 2006 that had failed within five years of their start-up. How large a sample size is needed to guarantee estimating this proportion correct to within

a. 0.10 with probability 0.95?

b. 0.05 with probability 0.95?

c. 0.05 with probability 0.99?

d. Compare sample sizes for parts a and b, and b and c, and summarize the effects of decreasing the margin of error and increasing the confidence level.

8.50 **Canada and the death penalty** A poll in Canada in 1998 indicated that 48% of Canadians favor imposing the death penalty (Canada does not have it). A report by Amnesty International on this and related polls (www.amnesty.ca) did not report the sample size but stated, "Polls of this size are considered to be accurate within 2.5 percentage points 95% of the time." About how large was the sample size?

8.51 **Farm size** An estimate is needed of the mean acreage of farms in Ontario, Canada. A 95% confidence interval should have a margin of error of 25 acres. A study 10 years ago in this province had a sample standard deviation of 200 acres for farm size.

a. About how large a sample of farms is needed?

b. A sample is selected of the size found in part a. However, the sample has a standard deviation of 300 acres, rather than 200. What is the margin of error for a 95% confidence interval for the mean acreage of farms?

8.52 **Income of Native Americans** How large a sample size do we need to estimate the mean annual income of Native Americans in Onondaga County, New York, correct to within $1000 with probability 0.99? No information is available to us about the standard deviation of their annual income. We guess that nearly all of the incomes fall between $0 and $120,000 and that this distribution is approximately bell shaped.
TRY

8.53 **Population variability** Explain the reasoning behind the following statement: "In studies about a very diverse population, large samples are often necessary, whereas for more homogeneous populations smaller samples are often adequate." Illustrate for the problem of estimating mean income for all medical doctors in the United States compared to estimating mean income for all entry-level employees at McDonald's restaurants in the United States.

8.54 **Web survey to get large *n*** A newspaper wants to gauge public opinion about legalization of marijuana. The sample size formula indicates that they need a random sample of 875 people to get the desired margin of error. But surveys cost money, and they can only afford to randomly sample 100 people. Here's a tempting alternative: If they place a question about that issue on their Web site, they will get more than 1000 responses within a day at little cost. Are they better off with the random sample of 100 responses or the Web site volunteer sample of more than 1000 responses? (*Hint:* Think about the issues discussed in Section 4.2 about proper sampling of populations.)

8.55 Do you like tofu? You randomly sample five students at your school to estimate the proportion of students who like tofu. All five students say they like it.

TRY

a. Find the sample proportion who like it.

b. Find the standard error. Does its usual interpretation make sense?

c. Find a 95% confidence interval, using the large-sample formula. Is it sensible to conclude that *all* students at your school like tofu?

d. Why is it not appropriate to use the ordinary large-sample confidence interval in part c? Use a more appropriate approach, and interpret the result.

8.56 Alleviate PMS? A pharmaceutical company proposes a new drug treatment for alleviating symptoms of PMS (premenstrual syndrome). In the first stages of a clinical trial, it was successful for 7 out of 10 women.

a. Construct an appropriate 95% confidence interval for the population proportion.

b. Is it plausible that it's successful for only half the population? Explain.

8.57 Accept a credit card? A bank wants to estimate the proportion of people who would agree to take a credit card they offer if they send a particular mailing advertising it. For a trial mailing to a random sample of 100 potential customers, 0 people accept the offer. Can they conclude that fewer than 10% of their population of potential customers would take the credit card? Answer by finding an appropriate 95% confidence interval.

8.5 Using Computers to Make New Estimation Methods Possible

We've seen how to construct point and interval estimates of a population proportion and a population mean. Confidence intervals are relatively simple to construct for these parameters. For some parameters, it's not so easy because it's difficult to derive the sampling distribution or the standard error of a point estimate. We'll now introduce a relatively new simulation method for constructing a confidence interval that statisticians often use for such cases.

The Bootstrap: Using Simulation to Construct a Confidence Interval

When it is difficult to derive a standard error or a confidence interval formula that works well, you can "pull yourself up by your bootstraps" to attack the problem without using mathematical formulas. A recent computational invention, called the **bootstrap,** does just that.

The bootstrap is a simulation method that resamples from the observed data. It treats the data distribution as if it were the population distribution. You resample, *with replacement,* n observations from the data distribution. Each of the original n data points has probability $1/n$ of selection for each "new" observation. For this new sample of size n, you construct the point estimate of the parameter. You then resample another set of n observations from the original data distribution and construct another value of the point estimate. In the same way, you repeat this resampling process (using a computer) from the original data distribution a very large number of times, for instance, selecting 10,000 separate samples of size n and calculating 10,000 corresponding values of the point estimate.

This variability of resampled point estimates provides information about the accuracy of the original point estimate. For instance, a 95% confidence interval for the parameter is the 95% central set of the resampled point estimate values. These are the ones that fall between the 2.5th percentile and 97.5th percentile of those values.

Example 12

Bootstrap

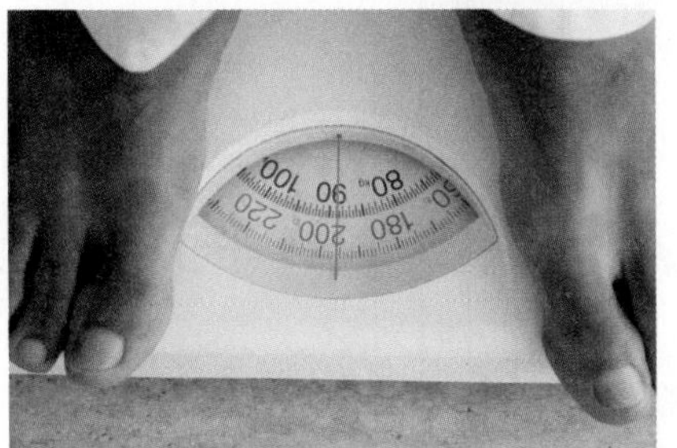

How Variable Are Your Weight Readings on a Scale?

Picture the Scenario

Instruments used to measure physical characteristics such as weight and blood pressure do not give the same value every time they're used in a given situation. The measurements vary. One of the authors recently bought a scale (called "Thinner") that is supposed to give precise weight readings. To investigate how much the weight readings tend to vary, he weighed himself ten times, taking a 30-second break after each trial to allow the scale to reset. He got the values (in pounds):

$$160.2, 160.8, 161.4, 162.0, 160.8$$
$$162.0, 162.0, 161.8, 161.6, 161.8.$$

These 10 trials have a mean of 161.44 and standard deviation of 0.63.

Because weight varies from trial to trial, it has a probability distribution. You can regard this distribution as describing long-run population values that you would get if you could conduct a huge number of weight trials. The sample mean and standard deviation estimate the center and variability of the distribution. Ideally, you would like the scale to be precise and give the same value every time. Then the population standard deviation would be 0.0, but it's not that precise in practice.

How could we get a confidence interval for the population standard deviation? The sample standard deviation s has an approximate normal sampling distribution for very large n. However, its standard error is highly sensitive to any assumption we make about the shape of the population distribution. It is safer to use the bootstrap method to construct a confidence interval.

Question to Explore

Using the bootstrap method, find a 95% confidence interval for the population standard deviation.

Think It Through

We sample from a distribution that has probability 1/10 at each of the values in the sample. For each new observation this corresponds to selecting a random digit and making the observation 160.2 if we get 0, 160.8 if we get 1,…, 161.8 if we get 9. This can be done with software. The bootstrap with 100,000 resamples of the data uses the following steps:

- Randomly sample 10 observations from this sample data distribution. We did this and got the 10 new observations

$$160.8, 161.4, 160.2, 161.8, 162.0, 161.8, 161.6, 161.8, 161.6, 160.2.$$

  For the 10 new observations, the sample standard deviation is $s = 0.67$.
- Repeat the preceding step, taking 100,000 separate resamples of size 10. This gives us 100,000 values of the sample standard deviation. Figure 8.11 shows a histogram of their values.
- Now identify the middle 95% of these 100,000 sample standard deviation values. For the 100,000 samples we took, the 2.5th percentile was 0.26 and the 97.5th percentile was 0.80. In other words, 95% of the resamples had sample standard deviation values between 0.26 and 0.80. This is our 95% bootstrap confidence interval for the population standard deviation.

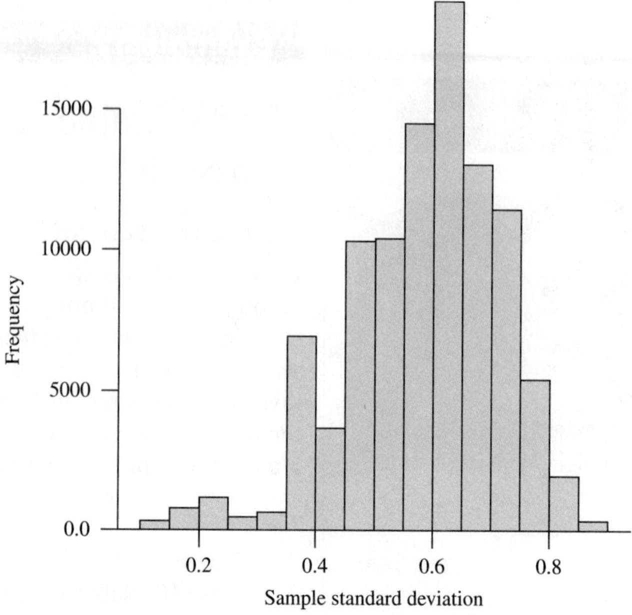

▲ **Figure 8.11 A Bootstrap Frequency Distribution of Standard Deviation Values.** These were obtained by taking 100,000 samples of size 10 each from the sample data distribution. **Question** What is the practical reason for using the bootstrap method?

In summary, the 95% confidence interval for σ is (0.26, 0.80). A typical deviation of a weight reading from the mean might be rather large, nearly a pound.

Insight

Figure 8.11 is skewed and is quite irregular. This appearance is because of the small sample size ($n = 10$). Such simulated distributions take a more regular shape when n is large, usually becoming symmetric and bell shaped when n is sufficiently large.

Try Exercise 8.59

The bootstrap method was invented by Brad Efron, a statistician at Stanford University, in 1979. It is now widely used, taking advantage of modern computing power. New statistical methods continue to be developed today. (The small-sample confidence interval for a proportion at the end of Section 8.4 was developed in 1998 by one of the authors with then-Ph.D. student Brent Coull.)

On the Shoulders of... Ronald A. Fisher

R. A. Fisher. Fisher was the statistician most responsible for the statistical methods used to analyze data today.

How do you conduct scientific inquiry, whether it be developing methods of experimental design, finding the best way to estimate a parameter, or answering specific questions such as which fertilizer works best?

Compared with other mathematical sciences, statistical science is a mere youth. The most important contributions were made between 1920 and 1940 by the British statistician Ronald A. Fisher (1890–1962). While working at an agricultural research station north of London, Fisher was the first statistician to show convincingly the importance of randomization in designing experiments. He also developed the theory behind point estimation and proposed several new statistical methods.

Fisher was involved in a wide variety of scientific questions, ranging from finding the best ways to plant crops to controversies (in his time) about whether smoking was harmful. He also did fundamental work in genetics and is regarded as a giant in that field as well.

Fisher had strong disagreements with others about the way statistical inference should be conducted. One of his main adversaries was Karl Pearson. Fisher corrected a major error Pearson made in proposing methods for contingency tables, and he also criticized Pearson's son's work on developing the theory of confidence intervals in the 1930s. Although Fisher often disparaged the ideas of other statisticians, he reacted strongly if anyone criticized him in return. Writing about Pearson, Fisher once said, "If peevish intolerance of free opinion in others is a sign of senility, it is one which he had developed at an early age."

8.5 Practicing the Basics

8.58 Why bootstrap? Explain the purpose of using the bootstrap method.

8.59 Estimating variability Refer to Example 12 about weight readings of a scale. For 10 successive trials on the next day, the weight values were

TRY

$$159.8, 159.8, 159.6, 159.0, 158.4$$
$$159.2, 158.8, 158.4, 158.8, 159.0.$$

Explain the steps of how you could use the bootstrap method to get a 95% confidence interval for a "long-run" standard deviation of such values.

8.60 Bootstrap the proportion We want a 95% confidence interval for the population proportion of students in a high school in Dallas, Texas, who can correctly find Iraq on an unlabeled globe. For a random sample of size 50, 10 get the correct answer.

a. Using software or the Sampling Distributions applet on the text CD, set the population menu to Binary and treat the sample proportion as the population proportion by setting the proportion parameter to 0.20 = 10/50 (Binary: $p = 0.2$). Take a random sample of size 50, and find the sample proportion of correct answers.

b. Take 100 resamples like the one in part a, each time calculating the sample proportion. Take one sample at a time, recording each sample proportion. Now, construct a 90% confidence interval by identifying the 5th and 95th percentiles of the sample proportions. This is the 90% bootstrap confidence interval.

c. Explain why the sample proportion does not fall exactly in the middle of the bootstrap confidence interval. (*Hint:* Is the sampling distribution symmetric or skewed?)

Chapter Review

ANSWERS TO THE CHAPTER FIGURE QUESTIONS

Figure 8.1 A point estimate alone will not tell us how close the estimate is likely to be to the parameter.

Figure 8.2 When the size of the random sample is relatively large, by the central limit theorem.

Figure 8.3 We don't know the value of p, the population proportion, to form the interval $p \pm 1.96(se)$. The population proportion p is what we're trying to estimate.

Figure 8.4 Having greater confidence means that we want to have greater assurance of a correct inference. Thus, it is natural that we would expect a wider interval of believable values for the population parameter.

Figure 8.5 The $\hat{p}$ values falling in the darker shaded left and right tails of the bell-shaped curve.

Figure 8.6 This skew should not affect the validity of the confidence interval for the mean because of the large sample size ($n = 899$).

Figure 8.7 Using Table B or Table 8.5, $t = 1.96$ when $df = \infty$ with right-tail probability = 0.025.

Figure 8.8 $t = -3.707$ and $t = 3.707$.

Figure 8.9 Monday has greater variability and slightly more skew than Friday. This is largely influenced by the value 90. The centers of the two distributions seem to be similar.

Figure 8.10 The sample size n is sufficiently large such that $np \geq 15$ and $n(1 - p) \geq 15$.

Figure 8.11 The bootstrap method is used when it is difficult to derive a standard deviation of a statistic or confidence interval formula by using mathematical techniques.

CHAPTER SUMMARY

We've now learned how to **estimate** the population proportion p for categorical variables and the population mean μ for quantitative variables.

- A **point estimate** (or **estimate**, for short) is our best guess for the unknown parameter value. An estimate of the population mean μ is the sample mean $\bar{x}$. An estimate of the population proportion p is the sample proportion $\hat{p}$.
- A **confidence interval** contains the most plausible values for a parameter. Confidence intervals for most parameters have the form

  Estimate $\pm$ margin of error,
  which is estimate $\pm$ (z- or t-score) $\times$ (se),

  where se is the standard error of the estimate. For the proportion, the score is a z-score from the normal distribution. For the mean, the score is a t-score from the ***t* distribution** with degrees of freedom $\boldsymbol{df = n - 1}$. The t-score is similar to a z-score when $df \geq 30$. Table 8.6 summarizes the point and interval estimation methods.
- The z- or t-score depends on the **confidence level**, the probability that the method produces a confidence interval that contains the population parameter value. For a proportion, for instance, since a probability of 0.95 falls within 1.96 standard errors of the center of the normal sampling distribution, we use $z = 1.96$ for 95% confidence. *To achieve greater confidence, we make the sacrifice of a larger margin of error and wider confidence interval.*
- For estimating a **mean**, the ***t* distribution** accounts for the extra variability due to using the sample standard deviation s to estimate the population standard deviation in finding a standard error. The t method assumes that the population distribution is normal. This ensures that the sampling distribution of $\bar{x}$ is bell shaped. This assumption is mainly important for small n, because when n is large the central limit theorem guarantees that the sampling distribution is bell shaped.
- For estimating a **proportion,** the formulas rely on the central limit theorem. For large random samples, this guarantees that the sample proportion has a normal sampling distribution. For estimating a proportion with small samples (fewer than 15 successes or fewer than 15 failures), the confidence interval formula $\hat{p} \pm z(se)$ still applies if we use it after adding 2 successes and 2 failures (and add 4 to n).
- Before conducting a study, we can **determine the sample size** $\boldsymbol{n}$ having a certain margin of error. Table 8.6 shows the sample size formulas. To use them, we must (1) select the margin of error m, (2) select the confidence level, which determines the z-score or t-score, and (3) guess the value the data will have for the sample standard deviation s (to estimate a population mean) or the sample proportion $\hat{p}$ (to estimate a population proportion). In the latter case, substituting $\hat{p} = 0.50$ guarantees that the sample size is large enough regardless of the value the sample has for $\hat{p}$.

Table 8.6 Estimation Methods for Means and Proportions

Parameter	Point Estimate	Standard Error	Confidence Interval	Sample Size for Margin of Error m
Proportion p	$\hat{p}$	$se = \sqrt{\hat{p}(1-\hat{p})/n}$	$\hat{p} \pm z(se)$	$n = [\hat{p}(1-\hat{p})z^2]/m^2$
Mean μ	$\bar{x}$	$se = s/\sqrt{n}$	$\bar{x} \pm t(se)$	$n = (\sigma^2 z^2)/m^2$

Note: The z- or t-score depends on the confidence level. The t-score has $df = n - 1$.

SUMMARY OF NOTATION

se = standard error

$\hat{p}$ = sample proportion

m = margin of error

$t_{.025}$ = t-score with right-tail probability 0.025

df = degrees of freedom

(= $n - 1$ for inference about a mean)

CHAPTER PROBLEMS

Practicing the Basics

8.61 Divorce and age of marriage A U.S. Census Bureau report[9] in 2009 estimated that for men between 20 and 24, 86.2% were never married. For women between 20 and 24, the corresponding value is 74.6%.

a. Are these point estimates or interval estimates?

b. Is the information given here sufficient to allow you to construct confidence intervals? Why or why not?

8.62 Approval rating for President Obama A July 2011 Gallup poll based on the responses of 1500 adults indicated that 46% of Americans approve of the job Barack Obama is doing as president. One way to summarize the findings of the poll is by saying, "It is estimated that 46% of Americans approve of the job Barack Obama is doing as president. This estimate has a margin of error of plus

[9] www.census.gov/hhes/socdemo/marriage/data/sipp/2009/tables.html.

or minus 3%." How could you explain the meaning of this to someone who has not taken a statistics course?

8.63 British monarchy In February 2002, the Associated Press quoted a survey of 3000 British residents conducted by YouGov.com. It stated, "Only 21% wanted to see the monarchy abolished, but 53% felt it should become more democratic and approachable. No margin of error was given." If the sample was random, find the 95% margin of error for each of these estimated proportions.

8.64 Born again A poll of a random sample of $n = 2000$ Americans by the Pew Research Center (www.people-press.org) indicated that 36% considered themselves "born-again" or evangelical Christians. How would you explain to someone who has not studied statistics:

a. What it means to call this a *point estimate*.

b. Why this does not mean that *exactly* 36% of *all* Americans consider themselves to be born-again or evangelical Christians.

8.65 Life after death The variable POSTLIFE in the 2008 General Social Survey asked, "Do you believe in life after death?" Of 1787 respondents, 1455 answered yes. A report based on these data stated that "81.4% of Americans believe in life after death. The margin of error for this result is plus or minus 1.85%." Explain how you could form a 95% confidence interval using this information, and interpret that confidence interval in context.

8.66 Female belief in life after death Refer to the previous exercise. The following printout shows results for the females in the sample, where $X =$ the number answering yes. Explain how to interpret each item, in context.

```
Sample    X    N    Sample p         95% CI
1       834  979    0.851890  (0.829639, 0.874140)
```

8.67 Vegetarianism *Time* magazine (July 15, 2002) quoted a poll of 10,000 Americans in which only 4% said they were vegetarians.

a. What has to be assumed about this sample to construct a confidence interval for the population proportion of vegetarians?

b. Construct a 99% confidence interval for the population proportion. Explain why the interval is so narrow, even though the confidence level is high.

c. In interpreting this confidence interval, can you conclude that fewer than 10% of Americans are vegetarians? Explain your reasoning.

8.68 Alternative therapies The Department of Public Health at the University of Western Australia conducted a survey in which they randomly sampled general practitioners in Australia.[10] One question asked whether the GP had ever studied alternative therapy, such as acupuncture, hypnosis, homeopathy, and yoga. Of 282 respondents, 132 said yes. Is the interpretation, "We are 95% confident that the percentage of all GPs in Australia who have ever studied alternative therapy equals 46.8%" correct or incorrect? Explain.

[10]This was reported at www.internethealthlibrary.com/Surveys/surveys-index-uk.htm.

8.69 Population data You would like to find the proportion of bills passed by Congress that were vetoed by the president in the last congressional session. After checking congressional records, you see that for the population of all 40 bills passed, 15 were vetoed. Does it make sense to construct a confidence interval using these data? Explain. (*Hint:* Identify the sample and population.)

8.70 Wife supporting husband Consider the statement that it is better for the man to work and the woman to tend the home, from the GSS (variable denoted FEFAM).

a. Go to the Web site, sda.berkeley.edu/GSS. Find the number who agreed or strongly agreed with that statement and the sample size for the year 2008.

b. Find the sample proportion and standard error.

c. Find a 99% confidence interval for the population proportion who would agree or strongly agree, and interpret it.

8.71 Legalize marijuana? The General Social Survey has asked respondents, "Do you think the use of marijuana should be made legal or not?" Go to the GSS Web site, sda.berkeley .edu/GSS. For the 2008 survey with variable GRASS:

a. Of the respondents, how many said "legal" and how many said "not legal"? Report the sample proportions.

b. Is there enough evidence to conclude whether a majority or a minority of the population support legalization? Explain your reasoning.

c. Now look at the data on this variable for all years by entering YEAR as the column variable. Describe any trend you see over time in the proportion favoring legalization.

8.72 Smoking A report in 2004 by the U.S. National Center for Health Statistics provided an estimate of 20.4% for the percentage of Americans over the age of 18 who were currently smokers. The sample size was 30,000. Assuming that this sample has the characteristics of a random sample, a 99.9% confidence interval for the proportion of the population who were smokers is (0.20, 0.21). When the sample size is extremely large, explain why even confidence intervals with large confidence levels are narrow.

8.73 Nondrinkers Refer to the previous exercise. The same study provided the following results for estimating the proportion of adult Americans who have been lifetime abstainers from drinking alcohol.

```
Sample      X      N     Sample p     95.0% CI
Nodrink   7380   30000    0.2460    (0.241, 0.251)
```

Explain how to interpret all results on this printout, in context.

8.74 U.S. popularity In 2007, a poll conducted for the BBC of 28,389 adults in 27 countries found that the United States had fallen sharply in world esteem since 2001 (www .globescan.com). The United States was rated third most negatively (after Israel and Iran), with 30% of those polled saying they had a positive image of the United States.

a. In Canada, for a random sample of 1008 adults, 56% said the United States is mainly a negative influence in the world. **True or false:** The 99% confidence interval of (0.52, 0.60) means that we can be 99% confident that between 52% and 60% of the population of all Canadian adults have a negative image of the United States.

b. In Australia, for a random sample of 1004 people, 66% said the United States is mainly a negative influence in the world. **True or false:** The 95% confidence interval of (0.63, 0.69) means that for a random sample of 100 people, we can be 95% confident that between 63 and 69 people in the sample have a negative image of the United States.

8.75 Grandpas using e-mail When the GSS asked in 2004, "About how many hours per week do you spend sending and answering e-mail?" the eight males in the sample of age at least 75 responded:

0, 1, 2, 2, 7, 10, 14, 15.

a. The TI-83+/84 screen shot shows results of a statistical analysis for finding a 90% confidence interval. Identify the results shown and explain how to interpret them.

b. Find and interpret a 90% confidence interval for the population mean.

c. Explain why the population distribution may be skewed right. If this is the case, is the interval you obtained in part b useless, or is it still valid? Explain.

```
TInterval
 (2.34,10.41)
 x̄=6.38
 Sx=6.02
 n=8.00
```

8.76 Travel to work As part of the 2000 census, the Census Bureau surveyed 700,000 households to study transportation to work. They reported that 76.3% drove alone to work, 11.2% carpooled, 5.1% took mass transit, 3.2% worked at home, 0.4% bicycled, and 3.8% took other means.

a. With such a large survey, explain why the margin of error for any of these values is extremely small.

b. The survey also reported that the mean travel time to work was 24.3 minutes, compared to 22.4 minutes in 1990. Explain why this is not sufficient information to construct a confidence interval for the population mean. What else would you need?

8.77 *t*-scores

a. Show how the *t*-score for a 95% confidence interval changes as the sample size increases from 10 to 20 to 30 to infinity.

b. What does the answer in part a suggest about how the *t* distribution compares to the standard normal distribution?

8.78 Buddhists watching TV In a recent GSS, the responses about the number of hours daily spent watching TV for the five subjects who identified themselves as Buddhists were

0, 5, 0, 1, 2.

a. Find the mean, standard deviation, and standard error.

b. Construct a 95% confidence interval for the population mean.

c. Specify the assumptions for the method. What can you say about their validity for these data?

8.79 Psychologists' income In 2003, the American Psychological Association conducted a survey (at research.apa.org) of a random sample of psychologists to estimate mean incomes for psychologists with various academic degrees and levels of experience. Of the 31 psychologists who received a masters degree in 2003, the mean income was $43,834 with a standard deviation of $16,870.

a. Construct a 95% confidence interval for the population mean. Interpret.

b. What assumption about the population distribution of psychologists' incomes does the confidence interval method make?

c. If the assumption about the shape of the population distribution is not valid, does this invalidate the results? Explain.

8.80 More psychologists Refer to the previous exercise. Interpret each item on the following printout that software reports for psychologists with a doctorate but with less than one year of experience.

```
Variable    N    Mean   StDev  SE Mean      95.0% CI
income     190  49411  15440   1120.1  (47204, 51618)
```

8.81 How long lived in town? The General Social Survey has asked subjects, "How long have you lived in the city, town, or community where you live now?" The responses of 1415 subjects in one survey had a mode of less than 1 year, a median of 16 years, a mean of 20.3 and a standard deviation of 18.2.

a. Do you think that the population distribution is normal? Why or why not?

b. Based on your answer in part a, can you construct a 95% confidence interval for the population mean? If not, explain why not. If so, do so and interpret.

8.82 How often do women feel sad? A recent GSS asked, "How many days in the past seven days have you felt sad?" The 816 women who responded had a median of 1, mean of 1.81, and standard deviation of 1.98. The 633 men who responded had a median of 1, mean of 1.42, and standard deviation of 1.83.

a. Find a 95% confidence interval for the population mean for women. Interpret.

b. Do you think that this variable has a normal distribution? Does this cause a problem with the confidence interval method in part a? Explain.

8.83 How often feel sad? Refer to the previous exercise. This question was asked of 10 students in a class at the University of Wisconsin recently. The responses were

0, 0, 1, 0, 7, 2, 1, 0, 0, 3.

Find and interpret a 90% confidence interval for the population mean, and indicate what you would have to assume for this inference to apply to the population of all University of Wisconsin students.

8.84 Happy often? The 1996 GSS asked, "How many days in the past seven days have you felt happy?" (This was the most recent year this question was posed.)

a. Using the GSS variable HAPFEEL, verify that the sample had a mean of 5.27 and a standard deviation of 2.05. What was the sample size?

b. Find the standard error for the sample mean.

c. Stating assumptions, construct and interpret a 95% confidence interval for the population mean. Can you conclude that the population mean is at least 5.0?

8.85 Revisiting mountain bikes Use the Mountain Bike data file from the text CD, shown also below.

a. Form a 95% confidence interval for the population mean price of all mountain bikes. Interpret.

b. What assumptions are made in forming the interval in part a? State at least one important assumption that does not seem to be satisfied, and indicate its impact on this inference.

Mountain Bikes

Brand and Model	Price($)
Trek VRX 200	1000
Cannondale SuperV400	1100
GT XCR-4000	940
Specialized FSR	1100
Trek 6500	700
Specialized Rockhop	600
Haro Escape A7.1	440
Giant Yukon SE	450
Mongoose SX 6.5	550
Diamondback Sorrento	340
Motiv Rockridge	180
Huffy Anorak 36789	140

8.86 eBay selling prices For eBay auctions of the iPad2 64GB 3G Wi-Fi units, a sample was taken in July 2011 where the Buy-it-Now prices were (in dollars):

1388, 1199, 1100, 1099, 1088, 1049, 1026, 999, 998, 978, 949, 930

a. Explain what a parameter might represent that you could estimate with these data.

b. Find the point estimate of μ.

c. Find the standard deviation of the data and the standard error of the sample mean. Interpret.

d. Find the 95% confidence interval for μ. Interpret the interval in context.

8.87 Income for families in public housing A survey is taken to estimate the mean annual family income for families living in public housing in Chicago. For a random sample of 29 families, the annual incomes (in hundreds of dollars) are as follows:

90 77 100 83 64 78 92 73 122 96 60 85 86 108 70
139 56 94 84 111 93 120 70 92 100 124 59 112 79

a. Construct a box plot of the incomes. What do you predict about the shape of the population distribution? Does this affect the possible inferences?

b. Using software, find point estimates of the mean and standard deviation of the family incomes of all families living in public housing in Chicago.

c. Obtain and interpret a 95% confidence interval for the population mean.

8.88 Females watching TV The GSS asked in 2008, "On the average day about how many hours do you personally watch television?" Software reports the results for females,

```
Variable   N    Mean  St Dev  SE Mean        95% CI
      TV  698  3.080   2.700    0.102  (2.879, 3.281)
```

a. Would you expect that TV watching has a normal distribution? Why or why not?

b. On what assumptions is the confidence interval shown based? Are any of them violated here? If so, is the reported confidence interval invalid? Explain.

c. What's wrong with the interpretation, "In the long run, 95% of the time females watched between 2.88 and 3.28 hours of TV a day."

8.89 Males watching TV Refer to the previous exercise. The 626 males had a mean of 2.87 and a standard deviation of 2.61. The 95% confidence interval for the population mean is (2.67, 3.08). Interpret in context.

8.90 Working mother In response to the statement on a recent General Social Survey, "A preschool child is likely to suffer if his or her mother works," suppose the response categories (strongly agree, agree, disagree, strongly disagree) had counts (104, 370, 665, 169). Scores $(2, 1, -1, -2)$ were assigned to the four categories, to treat the variable as quantitative. Software reported

```
Variable  N     Mean   St Dev SE Mean         95% CI
Response 1308 -0.1261 1.3105  0.0291 (-0.1832, -0.0689)
```

a. Explain what this choice of scoring assumes about relative distances between categories of the scale.

b. Based on this scoring, how would you interpret the sample mean of -0.1261?

c. Explain how you could also make an inference about proportions for these data.

8.91 Highest grade completed The 2008 GSS asked, "What is the highest grade that you finished and got credit for?" (variable EDUC). Of 2018 respondents, the mean was 13.4, the standard deviation was 3.1, and the proportion who gave responses below 12 (i.e., less than a high school education) was 0.166. Explain how you could analyze these data by making inferences about a population mean or about a population proportion, or both. Show how to implement one of these types of inference with these data.

8.92 Interpreting an interval for μ Refer to the previous exercise. For the 280 African Americans in the 2008 GSS, a 99% confidence interval for the mean of EDUC is (12.38, 13.16). Explain why the following interpretation is incorrect: 99% of all African Americans have completed grades between 12.38 and 13.16.

8.93 Sex partners in previous year The 2008 General Social Survey asked respondents how many sex partners they had in the previous 12 months (variable PARTNERS). Software summarizes the results of the responses by

```
Variable N    Mean   StDev  SE Mean      95% CI
partners 1766 1.1100 1.2200 0.0290 (1.0531, 1.1669)
```

a. Based on the reported sample size and standard deviation, verify the reported value for the standard error.

b. Based on these results, explain why the distribution was probably skewed to the right.

c. Explain why the skew need not cause a problem with constructing a confidence interval for the population mean, unless there are extreme outliers such as a reported value of 1000.

8.94 Men don't go to the doctor A survey of 1084 men age 18 and older in 1998 for the Commonwealth Fund (www.cmwf.org) indicated that more than half did not have a physical exam or a blood cholesterol test in the past year. A medical researcher plans to sample men in her community randomly to see if similar results occur. How large a random sample would she need to estimate this proportion to within 0.05 with probability 0.95?

8.95 Driving after drinking In December 2004, a report based on the National Survey on Drug Use and Health estimated that 20% of all Americans of ages 16 to 20 drove under the influence of drugs or alcohol in the previous year (AP, December 30, 2004). A public health unit in Wellington, New Zealand, plans a similar survey for young people of that age in New Zealand. They want a 95% confidence interval to have a margin of error of 0.04.

a. Find the necessary sample size if they expect results similar to those in the United States.

b. Suppose that in determining the sample size, they use the safe approach that sets $\hat{p} = 0.50$ in the formula for n. Then, how many records need to be sampled? Compare this to the answer in part a. Explain why it is better to make an educated guess about what to expect for $\hat{p}$, when possible.

8.96 Changing views of United States The June 2003 report on *Views of a Changing World*, conducted by the Pew Global Attitudes Project (www.people-press.org), discussed changes in views of the United States by other countries. In the largest Muslim nation, Indonesia, a poll conducted in May 2003 after the Iraq war began reported that 83% had an unfavorable view of America, compared to 36% a year earlier. The 2003 result was claimed to have a margin of error of 3 percentage points. How can you approximate the sample size the study was based on?

8.97 Mean property tax A tax assessor wants to estimate the mean property tax bill for all homeowners in Madison, Wisconsin. A survey 10 years ago got a sample mean and standard deviation of \$1400 and \$1000.

a. How many tax records should the tax assessor randomly sample for a 95% confidence interval for the mean to have a margin of error equal to \$100? What assumption does your solution make?

b. In reality, suppose that they'd now get a standard deviation equal to \$1500. Using the sample size you derived in part a, without doing any calculation, explain whether the margin of error for a 95% confidence interval would be less than \$100, equal to \$100, or more than \$100.

c. Refer to part b. Would the probability that the sample mean falls within \$100 of the population mean be less than 0.95, equal to 0.95, or greater than 0.95? Explain.

8.98 Kicking accuracy A football coach decides to estimate the kicking accuracy of a player who wants to join the team. Of 10 extra point attempts, the player makes all 10.

a. Find an appropriate 95% confidence interval for the probability that the player makes any given extra point attempt.

b. What's the lowest value that you think is plausible for that probability?

c. How would you interpret the random sample assumption in this context? Describe a scenario such that it would not be sensible to treat these 10 kicks as a random sample.

Concepts and Investigations

8.99 Religious beliefs A column by *New York Times* columnist Nicholas Kristof (August 15, 2003) discussed results of polls indicating that religious beliefs in the United States tend to be quite different from those in other Western nations. He quoted recent Gallup and Harris polls of random samples of about 1000 Americans estimating that 83% believe using the Virgin Birth of Jesus but only 28% believe in evolution. A friend of yours is skeptical, claiming that it's impossible to predict beliefs of over 200 million adult Americans by interviewing only 1000 of them. Write a one-page report using this context to show how you could explain about random sampling, the margin of error, and how a margin of error depends on the sample size.

8.100 TV watching and race For the number of hours of TV watching, the 2008 GSS reported a mean of 2.98 for the 1324 white subjects, with a standard deviation of 2.66. The mean was 4.38 for the 188 black subjects, with a standard deviation of 3.58. Analyze these data, preparing a short report in which you mention the methods used and the assumptions on which they are based, and summarize and interpret your findings.

8.101 Housework and gender Using data from the National Survey of Families and Households, a study (from S. South and G. Spitze, *American Sociological Review*, vol. 59, 1994, pp. 327–347) reported the descriptive statistics in the following table for the hours spent on housework. Analyze these data. Summarize results in a short report, including assumptions you made to perform the inferential analyses.

Gender	Sample Size	Mean	Std Dev
Men	4252	18.1	12.9
Women	6764	32.6	18.2

8.102 Women's role opinions When subjects in a recent GSS were asked whether they agreed with the following statements, the (yes, no) counts under various conditions were as follows:

- Women should take care of running their homes and leave running the country up to men: (275, 1556).
- It is better for everyone involved if the man is the achiever outside the home and the woman takes care of the home and the family: (627, 1208).
- A preschool child is likely to suffer if his or her mother works: (776, 1054).

Analyze these data. Prepare a one-page report stating assumptions, showing results of description and inference, and summarizing conclusions.

8.103 Types of estimates An interval estimate for a mean is more informative than a point estimate, because with an interval estimate you can figure out the point estimate, but with the point estimate alone you have no idea how wide the interval estimate is. Explain why this statement is correct, illustrating using the reported 95% confidence interval of (4.0, 5.6) for the mean number of dates in the previous month based on a sample of women at a particular college.

8.104 Width of a confidence interval Why are confidence intervals wider when we use larger confidence levels but narrower when we use larger sample sizes, other things being equal?

8.105 99.9999% confidence Explain why confidence levels are usually large, such as 0.95 or 0.99, but not extremely large, such as 0.999999. (*Hint:* What impact does the extremely high confidence level have on the margin of error?)

8.106 Need 15 successes and 15 failures To use the large-sample confidence interval for p, you need at least 15 successes and 15 failures. Show that the smallest value of n for which the method can be used is (a) 30 when $\hat{p} = 0.50$, (b) 50 when $\hat{p} = 0.30$, (c) 150 when $\hat{p} = 0.10$. That is, the overall n must increase as $\hat{p}$ moves toward 0 or 1. (When the true proportion is near 0 or 1, the sampling distribution can be highly skewed unless n is quite large.)

8.107 Outliers and CI For the observations 0, 1, 2, 2, 2, 2, 2, 3, 5 on TV watching for the Muslims in the sample considered in Exercise 8.33, a 95% confidence interval for the population mean for that group is (1.1, 3.2). Suppose the observation of 5 for the ninth subject was incorrectly recorded as 50. What would have been obtained for the 95% confidence interval? Compare to the interval (1.1, 3.2). How does this warn you about potential effects of outliers when you construct a confidence interval for a mean?

8.108 What affects n? Using the sample size formula $n = [\hat{p}(1 - \hat{p})z^2]/m^2$ for a proportion, explain the effect on n of (a) increasing the confidence level and (b) decreasing the margin of error.

8.109 Multiple choice: CI property Increasing the confidence level causes the margin of error of a confidence interval to **(a)** increase, **(b)** decrease, **(c)** stay the same.

8.110 Multiple choice: CI property 2 Other things being equal, increasing n causes the margin of error of a confidence interval to **(a)** increase, **(b)** decrease, **(c)** stay the same.

8.111 Multiple choice: Number of close friends Based on responses of 1467 subjects in a General Social Survey, a 95% confidence interval for the mean number of close friends equals (6.8, 8.0). Which *two* of the following interpretations are correct?

a. We can be 95% confident that $\bar{x}$ is between 6.8 and 8.0.

b. We can be 95% confident that μ is between 6.8 and 8.0.

c. Ninety-five percent of the values of X = number of close friends (for this sample) are between 6.8 and 8.0.

d. If random samples of size 1467 were repeatedly selected, then 95% of the time $\bar{x}$ would be between 6.8 and 8.0.

e. If random samples of size 1467 were repeatedly selected, then in the long run 95% of the confidence intervals formed would contain the true value of μ.

8.112 Multiple choice: Why z? The reason we use a z-score from a normal distribution in constructing a large-sample confidence interval for a proportion is that

a. For large random samples the sampling distribution of the sample proportion is approximately normal.

b. The population distribution is normal.

c. For large random samples the data distribution is approximately normal.

d. For any n we use the t distribution to get a confidence interval, and for large n the t distribution looks like the standard normal distribution.

8.113 Mean age at marriage A random sample of 50 records yields a 95% confidence interval of 21.5 to 23.0 years for the mean age at first marriage of women in a certain county. Explain what is wrong with each of the following interpretations of this interval.

a. If random samples of 50 records were repeatedly selected, then 95% of the time the sample mean age at first marriage for women would be between 21.5 and 23.0 years.

b. Ninety-five percent of the ages at first marriage for women in the county are between 21.5 and 23.0 years.

c. We can be 95% confident that $\bar{x}$ is between 21.5 and 23.0 years.

d. If we repeatedly sampled the entire population, then 95% of the time the population mean would be between 21.5 and 23.5 years.

8.114 Interpret CI For the previous exercise, provide the proper interpretation.

8.115 True or false Suppose a 95% confidence interval for the population proportion of students at your school who regularly drink alcohol is (0.61, 0.67). The inference is that you can be 95% confident that the sample proportion falls between 0.61 and 0.67.

8.116 True or false The confidence interval for a mean with a random sample of size n = 2000 is invalid if the population distribution is bimodal.

8.117 True or false If you have a volunteer sample instead of a random sample, then a confidence interval for a parameter is still completely reliable as long as the sample size is larger than about 30.

8.118 Women's satisfaction with appearance A special issue of *Newsweek* in March 1999 on women and their health reported results of a poll of 757 American women aged 18 or older. When asked, "How satisfied are you with your overall physical appearance?" 30% said very satisfied, 54% said somewhat satisfied, 13% said not too satisfied, and 3% said not at all satisfied. **True or false:** Since all these percentages are based on the same sample size, they all have the same margin of error.

8.119 ◆◆ **Opinions over time about the death penalty** For many years, the General Social Survey has asked respondents whether they favor the death penalty for persons convicted of murder. Support has been quite high in the United States, one of few Western nations that currently has the death penalty. The following figure uses the 20 General Social Surveys taken between 1975 and 2000 and plots the 95% confidence intervals for the population proportion in the United States who supported the death penalty in each of the 20 years of these surveys.

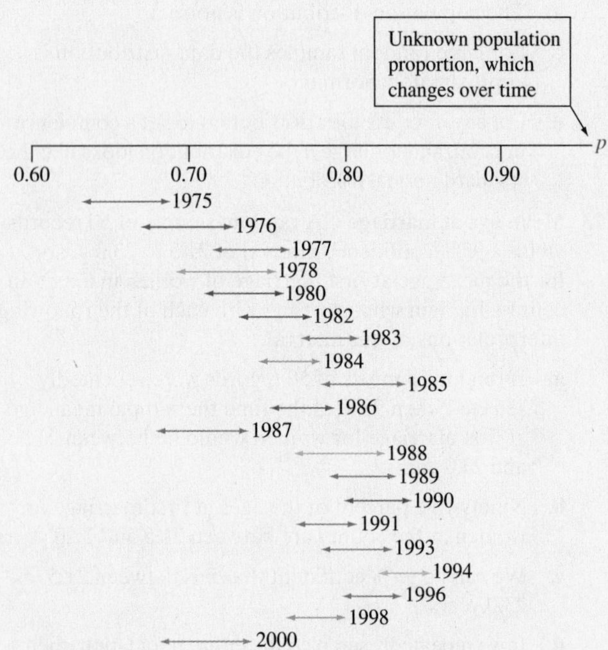

Twenty 95% confidence intervals for the population proportions supporting the death penalty.

a. When we say we have "95% confidence" in the interval for a particular year, what does this mean?

b. For 95% confidence intervals constructed using data for 20 years, let X = the number of the intervals that contain the true parameter values. Find the probability that $x = 20$, that is, all 20 inferences are correct. (*Hint:* You can use the binomial distribution to answer this.)

c. Refer to part b. Find the mean of the probability distribution of X.

d. What could you do differently so it is more likely that all 20 inferences are correct?

8.120 ◆◆ **Why called "degrees of freedom"?** You know the sample mean $\bar{x}$ of n observations. Once you know $(n - 1)$ of the observations, show that you can find the remaining one. In other words, for a given value of $\bar{x}$, the values of $(n - 1)$ observations determine the remaining one. In summarizing scores on a quantitative variable, having $(n - 1)$ *degrees of freedom* means that only that many observations are independent. (If you have trouble with this, try to show it for $n = 2$, for instance showing that if you know that $\bar{x} = 80$ and you know that one observation is 90, then you can figure out the other observation. The *df* value also refers to the divisor in $s^2 = \Sigma(x - \bar{x})^2/(n - 1)$.)

8.121 ◆◆ **Estimating *p* without estimating *se*** The large-sample confidence interval for a proportion substitutes $\hat{p}$ for the unknown value of p in the exact standard error of $\hat{p}$. A less approximate 95% confidence interval has endpoints determined by the p values that are 1.96 standard errors from the sample proportion, without estimating the standard error. To do this, you solve for p in the equation

$$|\hat{p} - p| = 1.96\sqrt{p(1 - p)/n}.$$

a. For Example 11 with no students without iPods in a sample of size 20, substitute $\hat{p}$ and n in this equation and show that the equation is satisfied at $p = 0.83337$ and at $p = 1$. So the confidence interval is (0.83887,1), compared to (1, 1) with $\hat{p} \pm 1.96(se)$.

b. Which confidence interval seems more believable? Why?

8.122 ◆◆ ***m* and *n*** Consider the sample size formula $n = [\hat{p}(1 - \hat{p})z^2]/m^2$ for estimating a proportion. When $\hat{p}$ is close to 0.50, for 95% confidence explain why this formula gives roughly $n = 1/m^2$.

8.123 ◆◆ **Median as point estimate** When the population distribution is normal, the population mean equals the population median. How good is the sample median as a point estimate of this common value? For a random sample, the estimated standard error of the sample median equals $1.25(s/\sqrt{n})$. If the population is normal, explain why the sample mean tends to be a better estimate than the sample median.

Student Activities

8.124 ◆◆ **Randomized response** To encourage subjects to make honest responses on sensitive questions, the method of *randomized response* is often used. Let's use your class to estimate the proportion who have had alcohol at a party. Before carrying out this method, the class should discuss what they would guess for the value of the proportion of students in the class who have had alcohol at a party. Come to a class consensus. Now each student should flip a coin, in secret. If it is a head, toss the coin once more and report the outcome, head or tails. If the first flip is a tail, report instead the response to whether you have had alcohol, reporting the response *head* if the true response is yes and reporting the response *tail* if the true response is no. Let p denote the true probability of the *yes* response on the sensitive question.

a. Explain why the numbers in the following table are the probabilities of the four possible outcomes.

b. Let $\hat{q}$ denote the sample proportion of subjects who report *head* for the second response. Explain why we can set $\hat{q} = 0.25 + p/2$ and hence use $\hat{p} = 2\hat{q} - 0.5$ to estimate p.

c. Using this approach with your class, estimate the probability of having had alcohol at a party. Is it close to the class guess?

Table for Randomized Response

	Second Response	
First Coin	**Head**	**Tail**
Head	0.25	0.25
Tail	$p/2$	$(1 - p)/2$

8.125 GSS project The instructor will assign the class a theme to study. Download recent results for variables relating to that theme from sda.berkeley.edu/GSS. Find and interpret confidence intervals for relevant parameters. Prepare a two-page report summarizing results.

9 Statistical Inference: Significance Tests About Hypotheses

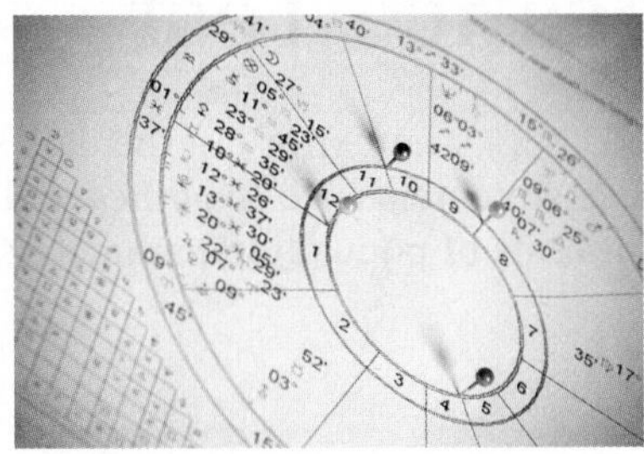

Example 1

Are Astrology Predictions Better than Guessing?

Picture the Scenario

Astrologers believe that the positions of the planets and the moon at the moment of your birth determine your personality traits. But have you ever seen any scientific evidence that astrology works? One scientific test of astrology used the following experiment: Each of 116 volunteers were asked to give their dates and times of birth. From this information, an astrologer prepared each subject's horoscope based on the positions of the planets and the moon at the moment of birth. Each volunteer also filled out a California Personality Index survey. Then the birth data and horoscope for one subject, together with the results of the personality survey for that individual and for two other participants randomly selected from the experimental group, were given to an astrologer. The astrologer was asked to predict which of the three personality charts matched the birth data and horoscope for the subject.[1]

Let p denote the probability of a correct prediction by an astrologer. Suppose an astrologer actually has no special predictive powers, as would be expected by those who view astrology as "quack science." The predictions then merely correspond to random guessing, that is, picking one of the three personality charts at random, so $p = 1/3$. However, the participating astrologers claimed that $p > 1/3$. They felt they could predict better than with random guessing.

Questions to Explore

- How can we use data from such an experiment to summarize the evidence about the claim by the astrologers?
- How can we decide, based on the data, whether or not the claims are believable?

Thinking Ahead

In this chapter, we'll learn how to use inferential statistics to answer such questions. We will use an inferential method called a **significance test** to analyze evidence in favor of the astrologers' claim. We'll analyze data from the astrology experiment in Examples 3, 5, and 13.

Therapeutic touch is another practice that some believe is "quack science." Its practitioners claim to be able to heal many medical conditions by using their hands to manipulate a human energy field above the patient's skin. We'll analyze data from an experiment to investigate these practitioners' claim in Examples 2, 6, and 14.

The significance test is the second major method for making statistical inference about a population. Like a confidence interval for estimating a parameter (the first major method), the significance test uses probability to provide a way to quantify how plausible a parameter is while controlling the chance of an incorrect inference. With significance tests, we'll be able to use data to answer questions such as:

- Does a proposed diet truly result in weight loss, on average?
- Is there evidence of discrimination against women in promotion decisions?
- Does one advertising method result in better sales, on average, than another advertising method?

[1] S. Carlson, *Nature,* vol. 318, pp. 419–425, 1985.

9.1 Steps for Performing a Significance Test

The main goal of many research studies is to check whether or not the data support certain statements or predictions. These statements are **hypotheses** about a population. They are usually expressed in terms of population parameters for variables measured in the study.

Hypothesis

In statistics, a **hypothesis** is a statement about a population, usually claiming that a parameter takes a particular numerical value or falls in a certain range of values.

For instance, the parameter might be a population proportion or a probability. Here's an example of a hypothesis for the astrology experiment in Example 1:

Hypothesis: Using a person's horoscope, the probability p that an astrologer can correctly predict which of three personality charts applies to that person equals 1/3. In other words, astrologers' predictions correspond to random guessing.

A *significance test* (or "test" for short) is a method for using data to summarize the evidence about a hypothesis. For instance, if a high proportion of the astrologers' predictions are correct, the data might provide strong evidence against the hypothesis that $p = 1/3$ in favor of an alternative hypothesis representing the astrologers' claim that $p > 1/3$.

Before conducting a significance test, we identify the variable measured and the population parameter of interest. For a categorical variable the parameter is a proportion, and for a quantitative variable the parameter is a mean. Section 9.2 shows the details for tests about proportions. Section 9.3 presents tests about means.

The Steps of a Significance Test

A significance test has five steps. In this section, we introduce the general ideas behind these steps.

Step 1: Assumptions

Each significance test makes certain assumptions or has certain conditions under which it applies. Foremost, a test assumes that the data production used randomization. Other assumptions may be about the sample size or about the shape of the population distribution.

Step 2: Hypotheses

Each significance test has two hypotheses about a population parameter: the null hypothesis and an alternative hypothesis.

In Words

H_0: null hypothesis (read as "H zero" or "H naught")
H_a: alternative hypothesis (read as "H a")
In everyday English, "null" is an adjective meaning "of no consequence or effect, amounting to nothing."

Null Hypothesis, Alternative Hypothesis

The **null hypothesis** is a statement that the parameter takes a particular value.

The **alternative hypothesis** states that the parameter falls in some alternative range of values.

The value in the null hypothesis usually represents *no effect.* The value in the alternative hypothesis then represents an effect of some type.

The symbol $\mathbf{H_0}$ denotes **null hypothesis** and the symbol $\mathbf{H_a}$ denotes **alternative hypothesis.**

For the experiment in Example 1, consider the hypothesis, "Based on any person's horoscope, the probability p that an astrologer can correctly predict which of three personality charts applies to that person equals 1/3." This hypothesis states that there is *no effect* in the sense that an astrologer's predictive power is no better than random guessing. This is a *null hypothesis.* It is symbolized by H_0: $p = 1/3$. If it is true, any difference that we observe between the sample proportion of correct guesses and 1/3 is due merely to ordinary sampling variability. The *alternative hypothesis* states that there *is* an effect—an astrologer's predictions are *better* than random guessing. It is symbolized by H_a: $p > 1/3$.

A null hypothesis has a *single* parameter value, such as H_0: $p = 1/3$. An alternative hypothesis has a *range* of values that are alternatives to the one in H_0, such as H_a: $p > 1/3$ or H_a: $p \neq 1/3$. You formulate the hypotheses for a significance test *before* viewing or analyzing the data.

Example 2

Null and alternative hypotheses

Therapeutic Touch Study

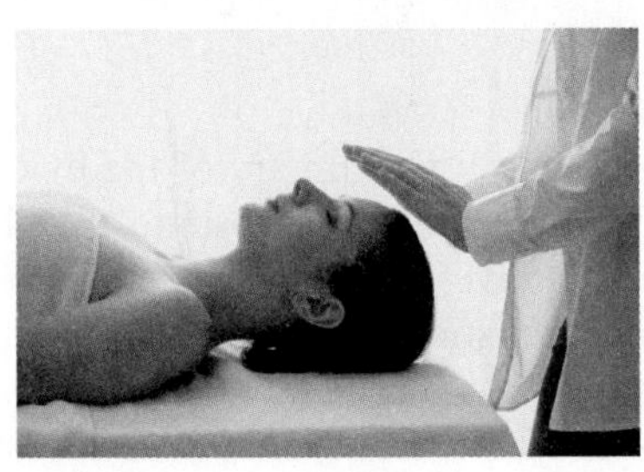

Picture the Scenario

Therapeutic touch (TT) practitioners claim to improve or heal many medical conditions by using their hands to manipulate a human energy field above the patient's skin. (The patient does not have to be touched.) A test investigating this claim used the following experiment: A TT practitioner was blindfolded. In each trial, the researcher placed her hand over either the right or left hand of the TT practitioner, the choice being determined by flipping a coin. The TT practitioner was asked to identify whether his or her right or left hand was closer to the hand of the researcher.[2] Let p denote the probability of a correct prediction by a TT practitioner. With random guessing, $p = 1/2$. However, the TT practitioners claimed that they could do better than random guessing. They claimed that $p > 1/2$.

Questions to Explore

Consider the hypothesis, "In any given trial, the probability p of guessing the correct hand is larger than 1/2." This is the claim of TT practitioners.

a. Is this a null or an alternative hypothesis?

b. How can we express the hypothesis that being a TT practitioner has no effect on the probability of a correct guess?

Think It Through

a. The hypothesis states that $p > 1/2$. It has a range of parameter values, so it is an alternative hypothesis, symbolized by H_a: $p > 1/2$.

b. The quoted hypothesis is an alternative to the null hypothesis that a TT practitioner's predictions are equivalent to random guessing. This no effect hypothesis is H_0: $p = 1/2$.

Try Exercises 9.1 and 9.2

In a significance test, the null hypothesis is presumed to be true unless the data give strong evidence against it. The burden of proof falls on the researcher who claims the alternative hypothesis is true. In the TT study, we assume that $p = 1/2$ unless the data provide strong evidence against it and in favor of the TT practitioners' claim that $p > 1/2$. An analogy may be found in a courtroom trial, in which a jury must decide the guilt or innocence of a defendant. The null hypothesis, corresponding to no effect, is that the defendant is innocent. The alternative hypothesis

[2]L. Rosa, et al., *JAMA,* vol. 279, pp. 1005–1010, 1998.

In Words

The courtroom provides an analogy for a significance test. H_0 is that the defendant is innocent. H_a is that the defendant is guilty. The jury presumes H_0 to be true unless the evidence (data) suggests otherwise.

is that the defendant is guilty. The jury presumes the defendant is innocent unless the prosecutor can provide strong evidence that the defendant is guilty "beyond a reasonable doubt." The burden of proof is on the prosecutor to convince the jury that the defendant is guilty.

Step 3: Test Statistic

The parameter to which the hypotheses refer has a point estimate. A **test statistic** describes how far that point estimate falls from the parameter value given in the null hypothesis. Usually this distance is measured by the number of standard errors between the point estimate and the parameter.

Recall

The **standard error** of a point estimate is what we use in practice to describe the variability of the sampling distribution of that point estimate (Section 8.2). Denoted by *se*, it is used both in confidence intervals and in significance tests. ◀

For instance, consider the null hypothesis H_0: $p = 1/3$ that, based on a person's horoscope, the probability p that an astrologer can correctly predict which of three personality charts applies to that person equals 1/3. For the experiment described in Example 1, 40 of 116 predictions were correct. The estimate of the probability p is the sample proportion, $\hat{p} = 40/116 = 0.345$. The test statistic compares this point estimate to the value in the null hypothesis ($p = 1/3$), using a z-score that measures the number of standard errors that the estimate falls from the null hypothesis value of 1/3.

In Words

The **P-value** is a tail probability, beyond the observed test statistic value if we presume $\mathbf{H_0}$ is true. Smaller P-values provide stronger evidence against the null hypothesis.

Step 4: P-Value

To interpret a test statistic value, we use a probability summary of the evidence against the null hypothesis, H_0. Here's how we get it: We presume that H_0 is true, since the burden of proof is on the alternative, H_a. Then we consider the sorts of values we'd expect to get for the test statistic, according to its sampling distribution presuming H_0 is true. If the sample test statistic falls well out in a tail of the sampling distribution, it is far from what H_0 predicts. If H_0 were true, such a value would be unusual. When there are a large number of possible outcomes, any single one may be unlikely, so we summarize how far out in the tail the test statistic falls by the tail probability of that value and values even more extreme (meaning, even farther from what H_0 predicts). See Figure 9.1. This probability is called a **P-value**. The smaller the P-value, the stronger the evidence is against H_0.

P-value

The **P-value** is the probability that the test statistic equals the observed value or a value even more extreme. It is calculated by presuming that the null hypothesis H_0 is true.

In the astrology study, suppose a P-value is small, such as 0.01. This means that if H_0 were true (that an astrologer's predictions correspond to random guessing), it would be unusual to get sample data such as we observed. Such a P-value provides strong evidence against the null hypothesis of random guessing and in support of the astrologers' claim. On the other hand, if the P-value is not near 0, the data are consistent with H_0. For instance, a P-value such as 0.26 or 0.63 indicates

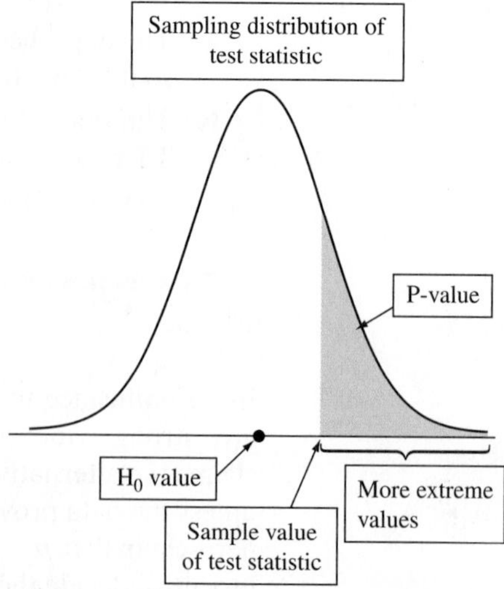

▲ **Figure 9.1 Suppose H_0 Were True. The P-value Is the Probability of a Test Statistic Value Like the Observed One or Even More Extreme.** This is the shaded area in the tail of the sampling distribution. **Question** Which gives stronger evidence against the null hypothesis, a P-value of 0.20 or of 0.01? Why?

that if the astrologer were actually randomly guessing, the observed data would not be unusual and could be attributed to random variation.

Step 5: Conclusion

The conclusion of a significance test reports the P-value and *interprets* what it says about the question that motivated the test. Sometimes this includes a decision about the validity of the null hypothesis H_0. For instance, based on the P-value, can we reject H_0 and conclude that astrologers' predictions are better than random guessing? As we'll discuss in the next section, we can reject H_0 in favor of H_a only when the P-value is very small, such as 0.05 or less.

SUMMARY: The Five Steps of a Significance Test

1. **Assumptions**
   First, specify the variable and parameter. The assumptions commonly pertain to the method of data production (randomization), the sample size, and the shape of the population distribution.
2. **Hypotheses**
   State the null hypothesis, H_0 (a single parameter value, usually no effect), and the alternative hypothesis, H_a (a set of alternative parameter values)
3. **Test statistic**
   The test statistic measures distance between the point estimate of the parameter and its null hypothesis value, usually by the number of standard errors between them.
4. **P-value**
   The P-value is the probability that the test statistic takes the observed value or a value more extreme if we presume H_0 is true. Smaller P-values represent stronger evidence against H_0.
5. **Conclusion**
   Report and interpret the P-value in the context of the study. Based on the P-value, make a decision about H_0 (either reject or do not reject H_0) if a decision is needed.

9.1 Practicing the Basics

9.1 **H_0 or H_a?** For parts a and b, is the statement a null hypothesis, or an alternative hypothesis?
TRY

a. In Canada, the proportion of adults who favor legalized gambling equals 0.50.

b. The proportion of all Canadian college students who are regular smokers is less than 0.24, the value it was 10 years ago.

c. Introducing notation for a parameter, state the hypotheses in parts a and b in terms of the parameter values.

9.2 **H_0 or H_a?** For each of the following, is the statement a null hypothesis or an alternative hypothesis? Why?
TRY

a. The mean IQ of all students at Lake Wobegon High School is larger than 100.

b. The probability of rolling a 6 with a particular die equals 1/6.

c. The proportion of all new business enterprises that remain in business for at least five years is less than 0.50.

9.3 **Burden of proof** For a new pesticide, should the Environmental Protection Agency (EPA) have the burden of proof to show that it is harmful to the environment, or should the producer of the pesticide have the burden of proof to show that it is not harmful to the environment? Give the analog of the null hypothesis and the alternative hypothesis if the burden of proof is on the EPA to show the new pesticide is harmful.

9.4 **Alabama GPA** Suppose the mean GPA of all students graduating from the University of Alabama in 1985 was 3.05. The registrar plans to look at records of students graduating in 2011 to see if mean GPA has changed. Define notation and state the null and alternative hypotheses for this investigation.

9.5 **Low-carbohydrate diet** A study plans to have a sample of obese adults follow a proposed low-carbohydrate diet for three months. The diet imposes limited eating of starches (such as bread and pasta) and sweets, but otherwise no limit on calorie intake. Consider the hypothesis,

The population mean of the values of weight change (= weight at start of study − weight at end of study) is a positive number.

a. Is this a null or an alternative hypothesis? Explain your reasoning.

b. Define a relevant parameter, and express the hypothesis that the diet has no effect in terms of that parameter. Is it a null or alternative hypothesis?

9.6 Examples of hypotheses Give an example of a null hypothesis and an alternative hypothesis about a (a) population proportion and (b) population mean.

9.7 *z* test statistic To test H_0: $p = 0.50$ that a population proportion equals 0.50, the test statistic is a *z*-score that measures the number of standard errors between the sample proportion and the H_0 value of 0.50. If $z = 3.6$, do the data support the null hypothesis, or do they give strong evidence against it? Explain.

9.8 P-value Indicate whether each of the following P-values gives strong evidence or not especially strong evidence against the null hypothesis.

a. 0.38

b. 0.001

9.2 Significance Tests About Proportions

For categorical variables, the parameters of interest are the population proportions in the categories. We'll use the astrology study to illustrate a significance test for population proportions.

Example 3

Hypotheses for a significance test

Are Astrologers' Predictions Better than Guessing?

Picture the Scenario

Many people take astrological predictions seriously, but there has never been any scientific evidence that astrology works. One scientific test of astrology used the experiment mentioned in Example 1: For each of 116 adult volunteers, an astrologer prepared a horoscope based on the positions of the planets and the moon at the moment of the person's birth. Each adult subject also filled out a California Personality Index (CPI) survey. For a given adult, his or her birth data and horoscope were shown to one of the participating astrologers in the experiment, together with the results of the personality survey for that adult and for two other adults randomly selected from the experimental group. The astrologer was asked which personality chart of the three subjects was the correct one for that adult, based on the horoscope.

The 28 participating astrologers were randomly chosen from a list prepared by the National Council for Geocosmic Research (NCGR), an organization dealing with astrology and respected by astrologers worldwide. The NCGR sampling frame consisted of astrologers with some background in psychology who were familiar with the CPI and who were held in high esteem by their astrologer peers. The experiment was double-blind: Each subject was identified by a random number, and neither the astrologers nor the experimenter knew which number corresponded to which subject. The chapter of the NCGR that recommended the participating astrologers claimed that the probability of a correct guess on any given trial in the experiment was larger than 1/3, the value for random guessing. (In fact, they felt that it would exceed 1/2.)

Question to Explore

Put this investigation in the context of a significance test by stating null and alternative hypotheses.

Think It Through

The variable specifying the outcome of any given trial in the experiment is categorical. The categories are correct prediction and incorrect prediction.

For each person, let p denote the probability of a correct prediction by the astrologer. We can regard this as the population proportion of correct guesses for the population of people and population of astrologers from which the study participants were sampled. Hypotheses refer to the probability p of a correct prediction. With random guessing, $p = 1/3$. If the astrologers can predict better than random guessing then $p > 1/3$. To test the hypothesis of random guessing against the astrologers' claim that $p > 1/3$, we would test H_0: $p = 1/3$ against H_a: $p > 1/3$.

Insight

In the experiment, the astrologers were correct with 40 of their 116 predictions. We'll see how to use these data to test these hypotheses as we work through the five steps of a significance test for a proportion in the next subsection.

Try Exercises 9.9 and 9.10

Steps of a Significance Test About a Population Proportion

This section presents a significance test about a population proportion that applies with relatively large samples. Here are the five steps of the test:

Step 1: Assumptions

- The variable is categorical.
- The data are obtained using randomization (such as a random sample or a randomized experiment).
- The sample size is sufficiently large that the sampling distribution of the sample proportion $\hat{p}$ is approximately normal. The approximate normality happens when the *expected numbers of successes and failures are both at least 15, using the null hypothesis value for p.*

Recall

Section 7.1 introduced the sampling distribution of a sample proportion. It has mean p and standard deviation

$$\sqrt{p(1-p)/n}$$

and is well approximated by the normal distribution when

$$np \geq 15 \text{ and } n(1-p) \geq 15. ◀$$

The sample size guideline is the one we used in Chapter 6 for judging when the normal distribution approximates the binomial distribution well. For the astrology experiment with $n = 116$ trials, when H_0 is true that $p = 1/3$, we expect $116(1/3) = 38.7 \approx 39$ correct guesses and $116(2/3) = 77.3 \approx 77$ incorrect guesses. Both of these are above 15, so the sample size guideline is satisfied.

As with other statistical inference methods, if randomization is not used, the validity of the results is questionable. A survey should use random sampling. An experiment should use principles of randomization and blinding with the study subjects, as was done in the astrology study. In that study, the astrologers were randomly selected, but the subjects evaluated were people (mainly students) who volunteered for the study. Consequently, any inference applies to the population of astrologers but only to the particular subjects in the study. If the study could have chosen the subjects randomly as well, then the inference would extend more broadly to *all* people.

Step 2: Hypotheses

The null hypothesis of a test about a proportion has the form

$$H_0: p = p_0,$$

In Words

p_0 is read as "p-zero," or "the null-hypothesized proportion value."

where p_0 represents a particular proportion value between 0 and 1. In Example 3, the null hypothesis of no effect states that the astrologers' predictions correspond to random guessing. This is H_0: $p = 1/3$. The null hypothesis value p_0 is 1/3.

The alternative hypothesis refers to alternative parameter values from the number in the null hypothesis. One possible alternative hypothesis has the form

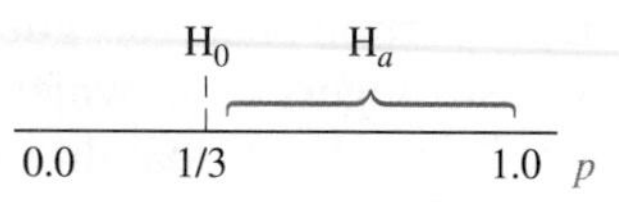

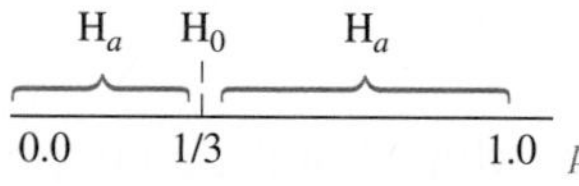

$$H_a: p > p_0.$$

This is used when a test is designed to detect whether p is *larger* than the number in the null hypothesis. In the astrology experiment, the astrologers claimed they could predict *better* than by random guessing. Their claim corresponds to $H_a: p > 1/3$. This is called a **one-sided** alternative hypothesis because it has values falling only on one side of the null hypothesis value. (See margin figure.) We'll use this H_a below. The other possible one-sided alternative hypothesis is $H_a: p < p_0$, such as $H_a: p < 1/3$.

A **two-sided** alternative hypothesis has the form

$$H_a: p \neq p_0.$$

It includes *all* the other possible values, both below and above the value p_0 in H_0. It states that the population proportion *differs* from the number in the null hypothesis. An example is $H_a: p \neq 1/3$, which states that the population proportion equals some number other than 1/3. (See margin figure.)

In summary, for Example 3 we'll use $H_0: p = 1/3$ and the one-sided $H_a: p > 1/3$.

Step 3: Test Statistic

The test statistic measures how far the sample proportion $\hat{p}$ falls from the null hypothesis value p_0, relative to what we'd expect if H_0 were true. The sampling distribution of the sample proportion has mean equal to the population proportion p and standard deviation equal to $\sqrt{p(1-p)/n}$. When H_0 is true, $p = p_0$, so the sampling distribution has mean p_0 and standard error $se_0 = \sqrt{p_0(1-p_0)/n}$. (We use the zero subscript here on se to reflect using the value p_0 rather than $\hat{p}$ for estimating p in the standard error, as we're presuming H_0 to be true in conducting the test.) The test statistic is

Recall

The standard error estimates the standard deviation of the sampling distribution of $\hat{p}$. Here in the context of a significance test, we are estimating how much $\hat{p}$ would tend to vary from sample to sample of size n if the null hypothesis were true. ◀

$$z = \frac{\hat{p} - p_0}{se_0} = \frac{\hat{p} - p_0}{\sqrt{\dfrac{p_0(1-p_0)}{n}}} = \frac{\text{Sample proportion} - \text{Null hypothesis proportion}}{\text{Standard error when null hypothesis is true}}.$$

This z-score measures the number of standard errors between the sample proportion $\hat{p}$ and the null hypothesis value p_0.

In testing $H_0: p = 1/3$ for the astrology experiment with $n = 116$ trials, the standard error is $\sqrt{p_0(1-p_0)/n} = \sqrt{[(1/3)(2/3)]/116} = 0.0438$. The astrologers were correct with 40 of their 116 predictions, a sample proportion of $\hat{p} = 0.345$. The test statistic is

Recall

The numbers shown in formulas are rounded, such as 0.345 for 40/116 = 0.34482758..., but calculations are done without rounding. ◀

$$z = \frac{\hat{p} - p_0}{se_0} = \frac{\hat{p} - p_0}{\sqrt{\dfrac{p_0(1-p_0)}{n}}} = \frac{0.345 - 1/3}{0.0438} = 0.26.$$

The sample proportion of 0.345 is only 0.26 standard errors above the null hypothesis value of 1/3.

Step 4: P-value

Does $z = 0.26$ give much evidence against $H_0: p = 1/3$ and in support of $H_a: p > 1/3$? The P-value summarizes the evidence. It describes how unusual the data would be if H_0 were true, that is, if the probability of a correct prediction were 1/3. The P-value is the probability that the test statistic takes a value like the observed test statistic or even more extreme, if actually $p = 1/3$.

Figure 9.2 shows the approximate sampling distribution of the z test statistic when H_0 is true. This is the **standard normal distribution**. For the astrology study,

Recall

The **standard normal** is the normal distribution with mean = 0 and standard deviation = 1. See the end of Section 6.2. ◀

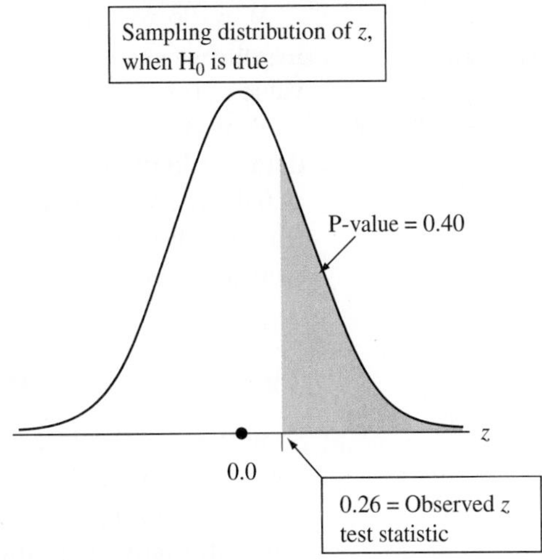

▲ **Figure 9.2 Calculation of P-value, When $z = 0.26$ for Testing H_0: $p = 1/3$ Against H_a: $p > 1/3$.** Presuming that H_0 is true, the P-value is the right-tail probability of a test statistic value even more extreme than observed. **Question** Logically, why are the *right-tail* z-scores considered to be the *more extreme* values for testing H_0 against H_a: $p > 1/3$?

$z = 0.26$. Values even farther out in the right tail, above 0.26, are even more extreme, in that they provide even stronger evidence against H_0.

When a random variable has a large number of possible values, the probability of any single value is usually very small. (For the normal distribution, the probability of a single value, such as $z = 0.26$, is *zero,* because the area under the curve above a single point is 0.) So the P-value is taken to be the probability of a *region* of values, specifically the more extreme values, $z > 0.26$. From Table A or using software, the right-tail probability above 0.26 is 0.40. This P-value of 0.40 tells us that if H_0 were true, the probability would be 0.40 that the test statistic would be more extreme than the observed value.

Step 5: Conclusion

We summarize the test by reporting and interpreting the P-value. The P-value of 0.40 is not especially small. It does *not* provide strong evidence against H_0: $p = 1/3$ and in favor of the astrologers' claim that $p > 1/3$. The sample data are the sort we'd expect to see if $p = 1/3$ (that is, if astrologers were randomly guessing the personality type). Thus, it is plausible that $p = 1/3$. We would not conclude that astrologers have special predictive powers.

Try Exercise 9.16

Interpreting the P-value

A significance test analyzes the strength of the evidence against the null hypothesis, H_0. We start by presuming that H_0 is true, putting the *burden of proof* on H_a. The approach taken is the indirect one of *proof by contradiction.* To convince ourselves that H_a is true, we must show the data contradict H_0, by showing they'd be unusual if H_0 were true. We analyze whether the data would be unusual if H_0 were true by finding the P-value. If the P-value is small, the data contradict H_0 and support H_a.

In the astrology study, the P-value for H_a: $p > 1/3$ is the right-tail probability of 0.40 from the sampling distribution of the z statistic. This P-value also approximates the probability that the sample proportion $\hat{p}$ takes a value that is as least as far above the null hypothesis value of 1/3 as the observed value of $\hat{p} = 0.345$ (See the margin figure). Since the P-value is not small, if truly $p = 1/3$, it would not be unusual to observe $\hat{p} = 0.345$. Based on the data, it is believable that the astrologers' predictions merely correspond to random guessing.

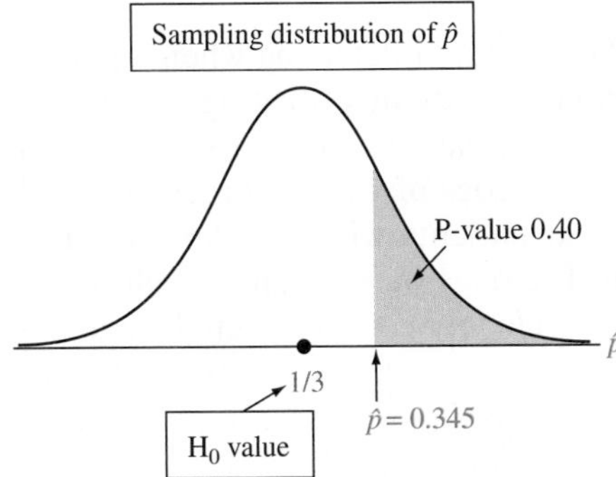

Caution

When interpreting the P-value, always include the conditional statement that presumes the hypothesized value in the null hypothesis is true. ◀

Why do we find the total probability in the *right tail*? Because the alternative hypothesis H_a: $p > 1/3$ has values *above* (that is, to the right of) the null hypothesis value of 1/3. It's the relatively *large* values of $\hat{p}$ that support this alternative hypothesis.

Why do smaller P-values indicate stronger evidence against H_0? Because the data would then be more unusual if H_0 were true. For instance, if we got a P-value of 0.01, then we might be more impressed by the astrologers' claims. When H_0 is true, the P-value is roughly equally likely to fall anywhere between 0 and 1. By contrast, when H_0 is false, the P-value is more likely to be near 0 than near 1.

Two-Sided Significance Tests

Sometimes we're interested in investigating whether a proportion falls above or below some point. For instance, can we conclude whether the population proportion who voted for a particular candidate is above 1/2, or below 1/2? We then use a *two-sided* alternative hypothesis. This has the form H_a: $p \neq p_0$, such as H_a: $p \neq 1/2$.

For two-sided tests, the values that are more extreme than the observed test statistic value are ones that fall farther in the tail in *either* direction. The P-value is the *two-tail* probability under the standard normal curve because these are the test statistic values that provide even stronger evidence in favor of H_a: $p \neq p_0$ than the observed value. We calculate this by finding the tail probability in a single tail and then doubling it, since the distribution is symmetric. See Figure 9.3.

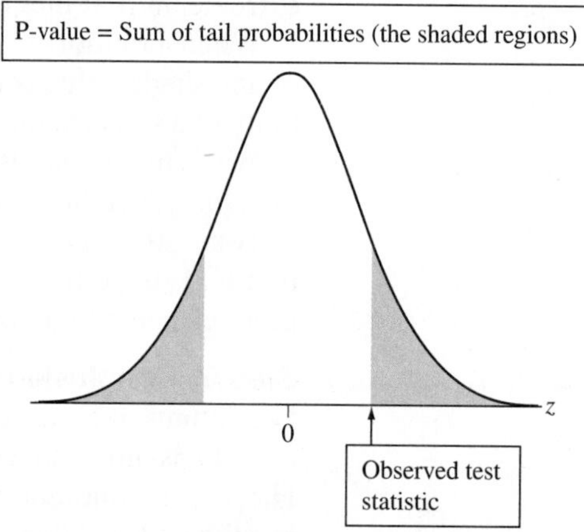

▲ **Figure 9.3 For the Two-Sided Alternative H_a: $p \neq p_0$, the P-value Is a Two-Tail Probability. Question** Logically, why are both the left-tail values and the right-tail values the more extreme values for H_a: $p \neq p_0$?

Example 4

Two-sided significance test

Dogs Detecting Cancer by Smell

Picture the Scenario

Recent research suggests that dogs may be helpful in detecting when a person has cancer. In this example, we describe a study investigating whether dogs can be trained to distinguish a patient with bladder cancer by smelling certain compounds in the patient's urine.[3] Six dogs of varying breeds were trained to discriminate between urine from patients with bladder cancer and urine from control patients without it. The dogs were taught to indicate which among several specimens was from the bladder cancer patient by lying beside it.

[3]Article by C. M. Willis et al., *British Medical Journal,* vol. 329, September 25, 2004.

An experiment was conducted to analyze how the dogs' ability to detect the correct urine specimen compared to what would be expected with random guessing. Each of the six dogs was tested with nine trials. In each trial, one urine sample from a bladder cancer patient was randomly placed among six control urine samples. In the total of 54 trials with the six dogs, the dogs made the correct selection 22 times.

Let p denote the probability that a dog makes the correct selection on a given trial. Since the urine from the bladder cancer patient was one of seven specimens, with random guessing we can write $p = 1/7$.

Question to Explore

Did this study provide strong evidence that the dogs' predictions were better or worse than with random guessing? Specifically, is there strong evidence that $p > 1/7$, with dogs able to select better than with random guessing, or that $p < 1/7$, with dogs' selections being poorer than random guessing?

Think It Through

The outcome of each trial is binary. The categories are correct selection and incorrect selection. Since we want to test whether the probability of a correct selection differs from random guessing, the hypotheses are

$$H_0: p = 1/7 \text{ and } H_a: p \neq 1/7.$$

The null hypothesis represents no effect, the selections being like random guessing. The alternative hypothesis says there is an effect, the selections differing from random guessing. You might instead use $H_a: p > 1/7$, if you expect the dogs' predictions to be better than random guessing. Most medical studies, however, use two-sided alternative hypotheses. This represents an open-minded research approach that recognizes that if an effect exists, it could be negative rather than positive.

The sample proportion of correct selections by the dogs was $\hat{p} = 22/54 = 0.407$, for the sample size $n = 54$. The null hypothesis value is $p_0 = 1/7$. When $H_0: p = 1/7$ is true, the expected counts are $np_0 = 54(1/7) = 7.7$ correct selections and $n(1-p_0) = 54(6/7) = 46.3$ incorrect selections. The first of these is not larger than 15, so according to the sample size guideline for step 1 of the test, n is not large enough to use the large-sample test. Later in this section, we'll see that the *two-sided* test is robust when this assumption is not satisfied. So we'll use the large-sample test, under the realization that the P-value is a good approximation for the P-value of a small-sample test mentioned later.

Recall

From Section 8.3, a method is **robust** with respect to a particular assumption if it works well even when that assumption is violated. ◀

The standard error is $se_0 = \sqrt{p_0(1 - p_0)/n} = \sqrt{(1/7)(6/7)/54} = 0.0476$. The test statistic for $H_0: p = 1/7$ (= 0.143) equals

$$z = \frac{\hat{p} - p_0}{se_0} = \frac{0.407 - (1/7)}{0.0476} = 5.6.$$

The sample proportion, 0.407, falls more than 5 standard errors above the null hypothesis value of 1/7.

Figure 9.4 shows the approximate sampling distribution of the z test statistic when H_0 is true. The test statistic value of 5.6 is well out in the right tail. Values farther out in the tail, above 5.6, are even more extreme. The P-value is the total two-tail probability of the more extreme outcomes, above 5.6 or below −5.6. From software, the cumulative probability in the right tail above $z = 5.6$ is 0.00000001, and the probability in the two tails equals 2(0.00000001) = 0.00000002. (Some software, such as MINITAB, rounds off and reports the P-value as 0.000.) This tiny P-value provides extremely strong evidence against $H_0: p = 1/7$, and we would conclude p is not equal to 1/7.

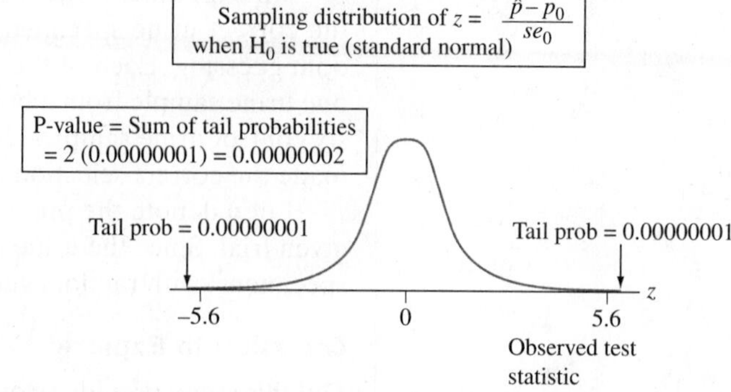

▲ **Figure 9.4 Calculation of P-value, When $z = 5.6$, for Testing H_0: $p = 1/7$ Against H_a: $p \neq 1/7$.** Presuming H_0 is true, the P-value is the two-tail probability of a test statistic value even more extreme than observed. **Question** Is the P-value of 0.00000002 strong evidence supporting H_0 or strong evidence against H_0?

When the P-value in a two-sided test is small, the point estimate tells us the direction in which the parameter appears to differ from the null hypothesis value. In summary, since the P-value is very small and $\hat{p} > 1/7$, the evidence strongly suggests that the dogs' selections are *better* than random guessing.

Recall

See Section 4.2 for potential difficulties with convenience samples, which use subjects who are conveniently available. ◀

Insight

Recall that one assumption for this significance test is randomization for obtaining the data. This study is like most medical studies in that its subjects were a *convenience sample* rather than a random sample from some population. It is not practical that a study can identify the population of all people who have bladder cancer and then randomly sample them for an experiment. Likewise, the dogs were not randomly sampled. As a result, any inferential predictions are tentative. They are valid only to the extent that the patients and the dogs in the experiment are representative of their populations. The predictions become more conclusive if similar results occur in other studies with other samples. In medical studies, even though the sample is not random, it is important to employ randomization in any experimentation, for instance in the placement of the bladder cancer patient's urine specimen among the six control urine specimens.

Try Exercises 9.14 and 9.18

Summary of How the Alternative Hypothesis Determines the P-value

The P-value is the probability of the values that are more extreme than the observed test statistic value. What is "more extreme" depends on the alternative hypothesis. For the two-sided alternative hypothesis, more extreme values fall in each direction, so the P-value is a two-tail probability. A one-sided alternative states that the parameter falls in a particular direction relative to the null hypothesis value. The P-value then uses only the tail in that direction. For instance, in the astrology study, the alternative H_a: $p > 1/3$ used only the *right* tail. For the alternative H_a: $p < 1/3$ in the other direction, we would have used the probability to the *left* of the observed test statistic value. See Figure 9.5, which illustrates with the one-sided P-values of 0.40 for H_a: $p > 1/3$ and 0.60 for H_a: $p < 1/3$. The P-values for the two one-sided alternative hypotheses add up to 1.0 for a given data set.

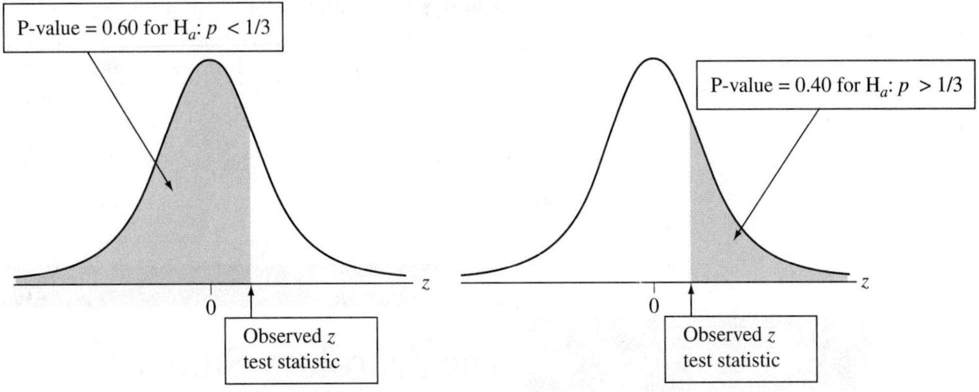

▲ **Figure 9.5 Calculation of P-value for One-Sided Alternative Hypotheses.** **Question** For a given one-sided H_a, how do we know which tail to use for finding the tail probability?

SUMMARY: P-values for Different Alternative Hypotheses

Alternative Hypothesis	P-value
$H_a: p > p_0$	Right-tail probability
$H_a: p < p_0$	Left-tail probability
$H_a: p \neq p_0$	Two-tail probability

The Significance Level Tells Us How Strong the Evidence Must Be

Sometimes we need to make a decision about whether or not the data provide sufficient evidence to reject H_0. Before seeing the data, we decide how small the P-value would need to be to reject H_0. For example, we might decide that we will reject H_0 if the P-value $\leq$ 0.05. The cutoff point of 0.05 is called the **significance level**. It is shown in Figure 9.6.

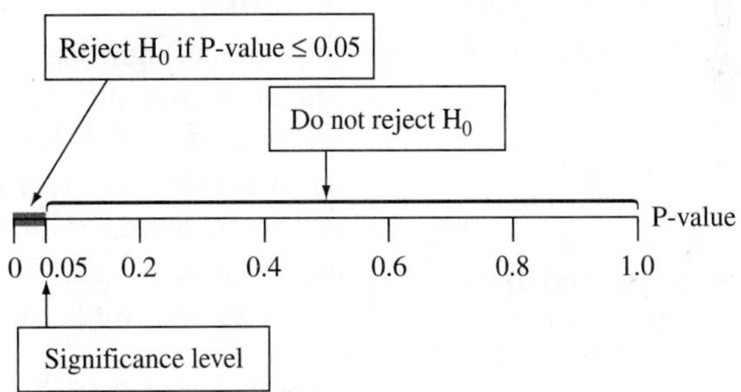

▲ **Figure 9.6 The Decision in a Significance Test.** Reject H_0 if the P-value is less than or equal to a chosen **significance level**, usually 0.05.

Significance Level

The **significance level** is a number such that we reject H_0 if the P-value is less than or equal to that number. In practice, the most common significance level is 0.05.

Table 9.1 summarizes the two possible outcomes for a test decision when the significance level is 0.05. We either reject H_0 or do not reject H_0. If the P-value if larger than 0.05, the data do not contradict H_0 sufficiently for us to reject it. Then, H_0 is still believable to us. If the P-value is $\leq$ 0.05, the data provide enough evidence to reject H_0. Recall that H_a had the burden of proof, and in this case we feel that the proof is sufficient. When we reject H_0, we say the results are **statistically significant**.

Table 9.1 Possible Decisions in a Test of Significance

P-value	Decision About H_0
≤ 0.05	Reject H_0
> 0.05	Do not reject H_0

Example 5

Significance levels

The Astrology Study

Picture the Scenario

Let's continue our analysis of the astrology study from Example 3. The parameter p is the probability that an astrologer picks the correct one of three personality charts, based on the horoscope and birth data provided. We tested H_0: $p = 1/3$ against H_a: $p > 1/3$ and got a P-value of 0.40.

Questions to Explore

What decision would we make for a significance level of (a) 0.05? (b) 0.50?

Think It Through

a. For a significance level of 0.05, the P-value of 0.40 is *not* less than 0.05. So we do not reject H_0. The evidence is not strong enough to conclude that the astrologers' predictions are better than random guessing.

b. For a significance level of 0.50, the P-value of 0.40 *is* less than 0.50. So we reject H_0 in favor of H_a. We conclude that this result provides sufficient evidence to support $p > 1/3$. The results are statistically significant at the 0.50 significance level.

Insight

In practice, significance levels are typically close to 0, such as 0.05 or 0.01. The reason is that the significance level is also a type of error probability. We'll see in Section 9.4 that when H_0 is true, the significance level is the probability of making an error by rejecting H_0. We used the significance level of 0.50 in part b above for illustrative purposes, but the value of 0.05 used in part a is much more typical. The astrologers' predictions are consistent with random guessing, and we would not reject H_0 for these data.

Try Exercise 9.19

In Practice Report the P-value

Report the P-value, rather than merely indicating whether the results are statistically significant. Learning the actual P-value is more informative than learning only whether the test is "statistically significant at the 0.05 level." The P-values of 0.01 and 0.049 are both statistically significant in this sense, but the first P-value provides much stronger evidence that the result is statistically significant than the second P-value.

Now that we've studied all five steps of a significance test about a proportion, let's use them in a new example, much like those you'll see in the exercises. In this example and in the exercises, it may help you to refer to the following summary box.

SUMMARY: Steps of a Significance Test for a Population Proportion p

1. Assumptions

- Categorical variable, with population proportion p defined in context.
- Randomization, such as a simple random sample or a randomized experiment, for gathering data.
- n large enough to expect at least 15 successes and 15 failures under H_0 (that is $np_0 \geq 15$ and $n(1 - p_0) \geq 15$). This is mainly important for one-sided tests.

2. Hypotheses

Null: H_0: $p = p_0$, where p_0 is the hypothesized value.

Alternative: H_a: $p \neq p_0$ (two-sided) or H_a: $p < p_0$ (one-sided) or H_a: $p > p_0$ (one-sided)

3. Test statistic

$$z = \frac{\hat{p} - p_0}{se_0} \text{ with } se_0 = \sqrt{p_0(1 - p_0)/n}$$

4. P-value

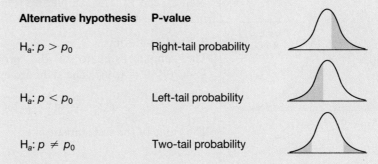

Alternative hypothesis	P-value
H_a: $p > p_0$	Right-tail probability
H_a: $p < p_0$	Left-tail probability
H_a: $p \neq p_0$	Two-tail probability

5. Conclusion

Smaller P-values give stronger evidence against H_0. If a decision is needed, reject H_0 if the P-value is less than or equal to the preselected significance level (such as 0.05). Relate the conclusion to the context of the study.

Example 6

Conducting a significance test

Therapeutic Touch Experiment

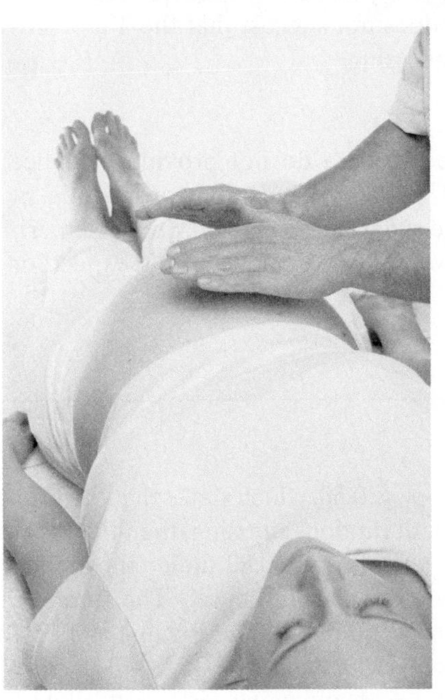

Picture the Scenario

Let's revisit the therapeutic touch (TT) experiment from Example 2. Each trial investigated whether a TT practitioner could correctly identify (while blindfolded) which of her hands was closer to the hand of a researcher. The researcher determined hand placement by flipping a coin. In a set of 150 trials with 15 TT practitioners (10 trials each), the TT practitioners were correct with 70 of their 150 predictions.

Questions to Explore

How strong is the evidence to support the TT practitioners' claim that they can predict the correct hand better than with random guessing? What decision would be made for a 0.05 significance level?

Think It Through

Let's follow the five steps of a significance test to organize our response to these questions:

1. Assumptions:

- The response is categorical, with outcomes correct and incorrect for a TT practitioner's prediction. Let p denote the probability of a correct prediction.
- We'll treat the 150 trials as a random sample of the possible trials that could occur in such an experiment with these TT practitioners. The inference will apply to the 15 particular TT practitioners used in this experiment. If the study randomly selected them from the population

Recall

H_0 contains a single number, whereas H_a has a range of values. ◄

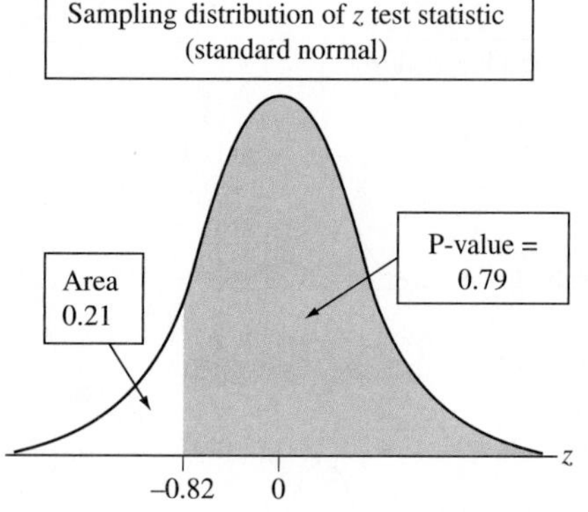

1-PropZTest
prop>.50
z=-.82
p=.79
p̂=.47
n=150.00

TI-83+/84 output

of TT practitioners, then the inference would apply to the entire population of TT practitioners.

- We'll discuss the sample size requirement in the next step.

2. Hypotheses:

If a TT practitioner merely guesses, then $p = 0.50$. However, the TT practitioners claimed that $p > 0.50$. To check their claim, we'll test $H_0: p = 0.50$ against $H_a: p > 0.50$. Supposing H_0 is true, of 150 trials we expect $150(0.50) = 75$ correct predictions and $150(0.50) = 75$ incorrect ones. These both exceed 15. The sample size is large enough to use the large-sample significance test.

3. Test statistic:

The sample estimate of the probability p of a correct decision is $\hat{p} = 70/150 = 0.467$. The standard error of $\hat{p}$ when $H_0: p = 0.50$ is true is

$$se_0 = \sqrt{p_0(1 - p_0)/n} = \sqrt{0.50(0.50)/150} = 0.0408.$$

The value of the test statistic is

$$z = \frac{\hat{p} - p_0}{se_0} = \frac{0.467 - 0.50}{0.0408} = -0.82.$$

The sample proportion is 0.82 standard error below the null hypothesis value.

4. P-value:

For testing $H_0: p = 0.50$ against $H_a: p > 0.50$, the P-value is the *right-tail* probability above $z = -0.82$ in the standard normal distribution (see the margin figure). From software or a normal table, this is $1 - 0.21 = 0.79$. TI-83+/84 output is also provided in the margin.

5. Conclusion:

The P-value of 0.79 is not small. If the null hypothesis $H_0: p = 0.50$ were true, the data we observed would not be unusual. With a 0.05 significance level, the evidence is not strong enough to reject H_0. It seems plausible that $p = 0.50$. In summary, this experiment does not suggest that the TT practitioners can predict better than random guessing.

Insight

The P-value is larger than 1/2 because the data do not provide evidence against $H_0: p = 0.50$ in favor of $H_a: p > 0.50$. The sample proportion of 0.467 actually falls in the direction of the *other* tail, $p < 0.50$. In this experiment, the TT practitioners' predictions were *worse* than what we'd expect for random guessing.

Try Exercise 9.17

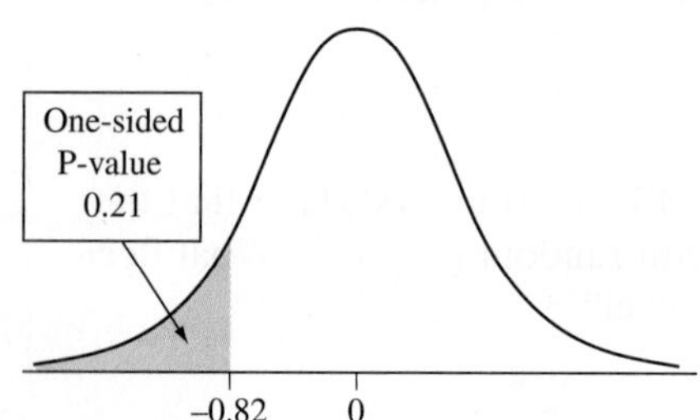

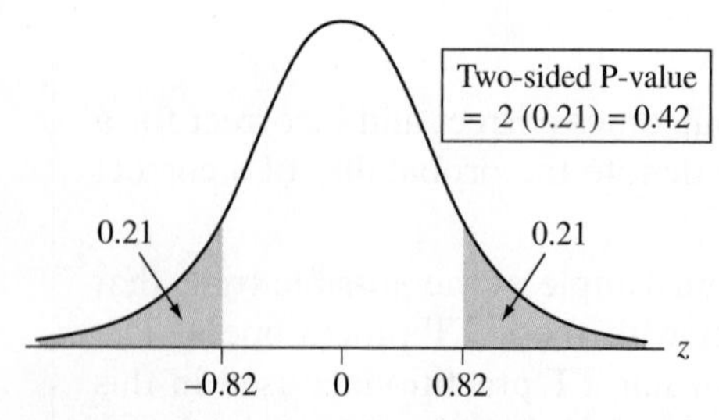

If we consider the one-sided alternative $H_a: p < 0.50$, which states that the probability of a correct prediction is *worse* than with random guessing, then the P-value is a *left-tail* probability, everything to the left of $z = -0.82$ under the standard normal curve. Now the P-value is 0.21. (See the margin figure.) The sum of the P-values for the one-sided alternatives always equals 1.0. The P-value $= 0.79$ for $H_a: p > 0.50$, the P-value $= 0.21$ for $H_a: p < 0.50$, and $0.79 + 0.21 = 1.0$.

In practice, many researchers would take an objective view and allow for predictions to possibly be better or possibly be worse than random guessing, by using the two-sided $H_a: p \neq 0.50$. The P-value is then the two-tail probability, $2(0.21) = 0.42$. (See the bottom margin figure.)

In Practice Picking H_a and Using Software for Significance Tests

In practice, you should pick H_a before seeing the data. You can use software to do the test. For the one-sided test of Example 6 with H_a: $p > 0.50$, MINITAB reports the results shown in Table 9.2. Unless requested otherwise, by default, software reports the two-sided P-value.

Some software reports the P-value to several decimals, such as 0.792892. We recommend rounding it to two or three decimal places, for instance, to 0.79, before reporting it. Reporting a P-value as 0.792892 suggests greater accuracy than actually exists, since the normal sampling distribution is only *approximate.*

Table 9.2 MINITAB Output for One-Sided Test in Example 6

Test of p = 0.5 vs p > 0.5

X	N	Sample p	Z-Value	P-Value
70	150	0.466667	− 0.82	0.793
↑	↑	↑	↑	
Category Count	*Sample Size*	*Sample Proportion*	*Test Statistic*	

"Do Not Reject H_0" Does Not Mean "Accept H_0"

Caution

When the P-value does not provide strong evidence to reject H_0, do not conclude that the hypothesized value is true. We can only conclude it is plausible. ◀

A small P-value means that the sample data would be unusual if H_0 were true. If the P-value is not small, such as 0.79 for the one-sided test in Example 6 on TT practitioners, the null hypothesis is plausible. In this case, the conclusion is reported as do not reject H_0 because the data do not contradict H_0.

"*Do not reject* H_0" *is not the same as saying "accept* H_0." The population proportion has many plausible values besides the number in the null hypothesis. For instance, consider Example 6, with p the probability of a correct prediction by a TT practitioner. We did not reject H_0: $p = 0.50$. Thus, p may equal 0.50, but other values are also believable. A 95% confidence interval for p is

$$\hat{p} \pm 1.96\sqrt{\hat{p}(1-\hat{p})/n}, \text{ or } 0.467 \pm 1.96\sqrt{(0.467)(0.533)/150},$$

which equals (0.39, 0.55). Even though insufficient evidence exists to reject H_0, it is improper to accept it and conclude that $p = 0.50$. It is plausible that p is as low as 0.39 or as high as 0.55.

An analogy here is again that of a courtroom trial. The null hypothesis is that the defendant is innocent. The alternative hypothesis is that the defendant is guilty. If the jury acquits the defendant, this does not mean that it *accepts* the defendant's claim of innocence. It merely means that innocence is plausible because guilt has not been established *beyond a reasonable doubt.*

Saying "Accept H_a" When It Contains All the Plausible Values The null hypothesis contains a single possible value for the parameter. Saying "Do not reject H_0" instead of "Accept H_0" emphasizes that that value is merely one of many plausible ones. Because of sampling error, there is a range of believable values besides the H_0 value. Saying accept H_a *is* permissible for the alternative hypothesis. When the P-value is sufficiently small (no greater than the significance level), the entire range of believable values falls within the range of numbers contained in H_a.

Use p_0 Instead of $\hat{p}$ in the Standard Error for Significance Tests In calculating the standard error in the test statistic, we substituted the null hypothesis value p_0 for the population proportion p in the formula $\sqrt{p(1-p)/n}$ for the actual standard deviation of the sample proportion. *The parameter values for sampling distributions in significance tests are those from* H_0, *since the* P*-value is calculated presuming that* H_0 *is true.* So we use p_0 in standard errors for tests. This practice differs from confidence intervals, in which the sample proportion $\hat{p}$ substitutes for p (which is unknown) in the actual standard deviation. When we estimate a confidence interval,[4] we do not have a hypothesized value for p, so that method substitutes the point estimate $\hat{p}$ for p. To differentiate between the two cases, we've denoted $se_0 = \sqrt{p_0(1-p_0)/n}$ and $se = \sqrt{\hat{p}(1-\hat{p})/n}$.

Deciding Between a One-Sided and a Two-Sided Test?

In practice, two-sided tests are more common than one-sided tests. Even if we think we know the direction of an effect, two-sided tests can also detect an effect that falls in the opposite direction. For example, in a medical study, even if we think a drug will perform better than a placebo, using a two-sided alternative allows us to detect if the drug is actually worse, perhaps because of bad side effects. However, as we saw with the astrology and TT examples, in some scenarios a one-sided test is natural.

Guidelines in Forming the Alternative Hypothesis

- In deciding between one-sided and two-sided alternative hypotheses in a particular exercise or in practice, *consider the context of the real problem.*

For instance, in the astrology experiment, to test whether someone can guess *better* than with random guessing, we used the values $p > 1/3$ in the alternative hypothesis corresponding to that possibility. An exercise that says "test whether the population proportion *differs* from 0.50" suggests a two-sided alternative, H_a: $p \neq 0.50$, to allow for p to be larger or smaller than 0.50. "Test whether the population proportion is *larger* than 0.50" suggests the one-sided alternative, H_a: $p > 0.50$.

- In most research articles, significance tests use two-sided P-values.

Partly this reflects an objective approach to research that recognizes that an effect could go in either direction. Using an alternative hypothesis in which an effect can go in either direction is regarded as the most even-handed way to perform the test. In using two-sided P-values, researchers avoid the suspicion that they chose H_a when they saw the direction in which the data occurred. That is not ethical and would be cheating, as we'll discuss in Section 9.5.

In Practice Tests Are Usually Two-Sided

For the reasons just discussed, two-sided tests are more common than one-sided tests. Most examples and exercises in this book reflect the way tests are most often used and employ two-sided alternatives. *In practice, you should use a two-sided test unless you have a well-justified reason for a one-sided test.*

- Confidence intervals are two-sided.

The practice of using a two-sided test coincides with the ordinary approach for confidence intervals, which are two-sided, obtained by adding and subtracting some quantity from the point estimate. There is a way to construct one-sided confidence intervals, for instance, concluding that a population proportion is *at least* equal to 0.70. In practice, though, two-sided confidence intervals are much more common.

[4] If we instead conduct the test by substituting the sample proportion in the standard deviation of the sample proportion, the standard normal approximation for the sampling distribution of z is much poorer.

SUMMARY: Three Basic Facts When Specifying Hypotheses

- The null hypothesis has an equal sign (such as H_0: $p = 0.5$), but the alternative hypothesis does not.
- You shouldn't pick H_a based on looking at the data.
- The hypotheses always refer to population parameters, not sample statistics.

So *never* express a hypothesis using sample statistic notation, such as H_0: $\hat{p} = 0.5$. There is no need to conduct inference about statistics such as the sample proportion $\hat{p}$, because you can find their values exactly from the data.

The Binomial Test for Small Samples

The test about a proportion applies normal sampling distributions for the sample proportion $\hat{p}$ and the z test statistic. Therefore, it is a *large-sample* test because the central limit theorem implies approximate normality of the sampling distribution for large random samples. The guideline is that the expected numbers of successes and failures should be at least 15, when H_0 is true; that is, $np_0 \geq 15$ and $n(1 - p_0) \geq 15$.

In practice, the large-sample z test performs well for *two-sided* alternatives even for small samples. When p_0 is below 0.3 or above 0.7 and n is small, the sampling distribution is quite skewed. However, a tail probability that is smaller than the normal probability in one tail is compensated by a tail probability that is larger than the normal probability in the other tail. Because of this, the P-value from the two-sided test using the normal table approximates well a P-value from a small-sample test.[5]

For one-sided tests, when p_0 differs from 0.5, the large-sample test does not work well when the sample size guideline ($np_0 \geq 15$ and $n(1 - p_0) \geq 15$) is violated. In that case, you should use a small-sample test. This test uses the binomial distribution with parameter value p_0 to find the exact probability of the observed value and all the more extreme values, according to the direction in H_a. Since one-sided tests with small n are not common in practice, we will not study the binomial test here. Exercises 9.25 and 9.26 show how to do it.

[5]With small samples, this test works much better than the confidence interval of Section 8.2. Having the actual, rather than an estimated standard deviation of the sample proportion makes a great difference.

9.2 Practicing the Basics

9.9 Psychic A person who claims to be psychic says that the probability p that he can correctly predict the outcome of the roll of a die in another room is greater than 1/6, the value that applies with random guessing. If we want to test this claim, we could use the data from an experiment in which he predicts the outcomes for n rolls of the die. State hypotheses for a significance test, letting the alternative hypothesis reflect the psychic's claim.
TRY

9.10 Believe in astrology? You plan to apply significance testing to your own experiment for testing astrology, in which astrologers have to guess which of four personality profiles is the correct one for someone who has a particular horoscope. Define notation and state hypotheses, letting one hypothesis reflect the possibility that the astrologers' predictions could be better than random guessing.
TRY

9.11 Get P-value from z For a test of H_0: $p = 0.50$, the z test statistic equals 1.04.

a. Find the P-value for H_a: $p > 0.50$.

b. Find the P-value for H_a: $p \neq 0.50$.

c. Find the P-value for H_a: $p < 0.50$. (*Hint:* The P-values for the two possible one-sided tests must sum to 1.)

d. Do any of the P-values in part a, part b, or part c give strong evidence against H_0? Explain.

9.12 Get more P-values from z Refer to the previous exercise. Suppose $z = 2.50$ instead of 1.04.

a. Find the P-value for (i) H_a: $p > 0.50$, (ii) H_a: $p \neq 0.50$, and (iii) H_a: $p < 0.50$.

b. Do any of the P-values in part a provide strong evidence against H_0? Explain.

9.13 Find test statistic and P-value For a test of H_0: $p = 0.50$, the sample proportion is 0.35 based on a sample size of 100.

a. Show that the test statistic is $z = -3.0$.

b. Find the P-value for H_a: $p < 0.50$.

c. Does the P-value in part b give much evidence against H_0? Explain.

9.14 Dogs and cancer A recent study[6] considered whether dogs could be trained to detect if a person has lung cancer or breast cancer by smelling the subject's breath. The researchers trained five ordinary household dogs to distinguish, by scent alone, exhaled breath samples of 55 lung and 31 breast cancer patients from those of 83 healthy controls. A dog gave a correct indication of a cancer sample by sitting in front of that sample when it was randomly placed among four control samples. Once trained, the dogs' ability to distinguish cancer patients from controls was tested using breath samples from subjects not previously encountered by the dogs. (The researchers blinded both dog handlers and experimental observers to the identity of breath samples.) Let p denote the probability a dog correctly detects a cancer sample placed among five samples, when the other four are controls.

TRY

a. Set up the null hypothesis that the dog's predictions correspond to random guessing.

b. Set up the alternative hypothesis to test whether the probability of a correct selection *differs* from random guessing.

c. Set up the alternative hypothesis to test whether the probability of a correct selection is *greater than* with random guessing.

d. In one test with 83 Stage I lung cancer samples, the dogs correctly identified the cancer sample 81 times. The test statistic for the alternative hypothesis in part c was $z = 17.7$. Report the P-value to three decimal places, and interpret. (The success of dogs in this study made researchers wonder whether dogs can detect cancer at an earlier stage than conventional methods such as MRI scans.)

9.15 Religion important in your life? Americans ages 18 to 29 are considered to be less religious than older Americans. According to recent studies by the Pew Forum on Religion & Public Life, fewer young adults are affiliated with a specific religion than older people today. And, compared with their elders, fewer young people say that religion is very important in their lives. Yet, many young people still believe in traditional religious concepts and practices. Pew Research Center surveys show, for example, that "young adults' beliefs about life after death and the existence of heaven, hell and miracles closely resemble the beliefs of older people today." According to GSS (General Social Survey) results from a random sample of 1,679 subjects, 45% in the 18–29 age group pray daily (an increase of 5% over the 1990s), while 55% pray less often.[7] The MINITAB output shows the results for a significance test for which the alternative hypothesis is that the percentage of 18–29-year-olds who pray daily differs from 50%. State and interpret the five steps of a significance test in this context, using information shown in the output to provide the particular values for the hypothesis, test statistic, and P-value.

Test and CI for One Proportion

```
Test of p = 0.5 vs p not = 0.5

                                                      Exact
Sample    X     N    Sample p      95% CI            P-Value
                                 (0.426277,
  1      756  1679   0.450268    0.474433)            0.000
```

9.16 Another test of astrology Examples 1, 3, and 5 referred to a study about astrology. Another part of the study used the following experiment: Professional astrologers prepared horoscopes for 83 adults. Each adult was shown three horoscopes, one of which was the one an astrologer prepared for them and the other two were randomly chosen from ones prepared for other subjects in the study. Each adult had to guess which of the three was theirs. Of the 83 subjects, 28 guessed correctly.

TRY

a. Defining notation, set up hypotheses to test that the probability of a correct prediction is 1/3 against the astrologers' claim that it exceeds 1/3.

b. Show that the sample proportion = 0.337, the standard error of the sample proportion for the test is 0.052, and the test statistic is $z = 0.08$.

c. Find the P-value. Would you conclude that people are more likely to select their horoscope than if they were randomly guessing, or are results consistent with random guessing?

9.17 Another test of therapeutic touch Examples 2 and 6 described a study about therapeutic touch (TT). A second run of the same experiment in the study used 13 TT practitioners who had to predict the correct hand in each of 10 trials.

TRY

a. Defining notation, set up hypotheses to test that the probability of a correct guess is 0.50 against the TT practitioners' claim that it exceeds 0.50.

b. The 130 trials had 53 correct guesses. Find and interpret the test statistic.

c. Report the P-value. (*Hint:* Since the sample proportion is in the opposite direction from the practitioners' claim, you should get a P-value > 0.50.) Indicate your decision, in the context of this experiment, using a 0.05 significance level.

d. Check whether the sample size was large enough to make the inference in part c. Indicate what assumptions you would need to make for your inferences to apply to all TT practitioners and to all subjects.

9.18 Testing a headache remedy Studies that compare treatments for chronic medical conditions such as headaches can use the same subjects for each treatment. This type of study is commonly referred to as a crossover design. With a crossover design, each person crosses over from using one treatment to another during the study. One such study

TRY

[6] M. McCulloch et al., *Integrative Cancer Therapies*, vol 5, p. 30, 2006.

[7] Accessed from pewforum.org/Age/Religion-Among-the-Millennials.aspx.

considered a drug (a pill called Sumatriptan) for treating migraine headaches in a convenience sample of children.[8] The study observed each of 30 children at two times when he or she had a migraine headache. The child received the drug at one time and a placebo at the other time. The order of treatment was randomized and the study was double-blind. For each child, the response was whether the drug or the placebo provided better pain relief. Let p denote the proportion of children having better pain relief with the drug, in the population of children who suffer periodically from migraine headaches. Can you conclude that $p > 0.50$, with more than half of the population getting better pain relief with the drug, or that $p < 0.50$, with less than half getting better pain relief with the drug (i.e., the placebo being better)? Of the 30 children, 22 had more pain relief with the drug and 8 had more pain relief with the placebo.

a. For testing H_0: $p = 0.50$ against H_a: $p \neq 0.50$, show that the test statistic $z = 2.56$.

b. Show that the P-value is 0.01. Interpret.

c. Check the assumptions needed for this test, and discuss the limitations due to using a convenience sample rather than a random sample.

9.19 Gender bias in selecting managers For a large supermarket chain in Florida, a women's group claimed that female employees were passed over for management training in favor of their male colleagues. The company denied this claim, saying they picked the employees from the eligible pool at random to receive this training. Statewide, the large pool of more than 1000 eligible employees who can be tapped for management training is 40% female and 60% male. Since this program began, 28 of the 40 employees chosen for management training were male and 12 were female.

TRY

a. In a significance test, the random sampling assumption is the claim of the company. In defining hypotheses, explain why this company's claim of a lack of gender bias is a no effect hypothesis. State the null and alternative hypotheses for a test to investigate the strength of evidence to support the women's claim.

b. The table shows results of using MINITAB to do a large-sample analysis. Explain why the large-sample analysis is justified, and show how software obtained the test statistic value.

```
Test of p = 0.60 vs. not = 0.60
X    N    Sample p    95.0% CI        Z-Value  P-Value
28   40   0.70000    (0.558,0.842)    1.29     0.1967
```

c. To what alternative hypothesis does the P-value in the table refer? Use it to find the P-value for the alternative hypothesis you specified in part a, and interpret it.

d. What decision would you make for a 0.05 significance level? Interpret.

9.20 Gender discrimination Refer to the previous exercise.

a. Explain why the alternative hypothesis of bias against males is H_a: $p < 0.60$.

b. Show that the P-value for testing H_0: $p = 0.60$ against H_a: $p < 0.60$ equals 0.90. Interpret. Why is it large?

[8]Data based on those in a study by M. L. Hamalainen et al., reported in *Neurology*, vol. 48, pp. 1100–1103, 1997.

9.21 Garlic to repel ticks A study (*J. Amer. Med. Assoc.*, vol. 284, p. 831, 2000) considered whether daily consumption of 1200 mg of garlic could reduce tick bites. The study used a crossover design with a sample of Swedish military conscripts, half of whom used placebo first and garlic second and half the reverse. The authors did not present the data, but the effect they described is consistent with garlic being more effective with 37 subjects and placebo being more effective with 29 subjects. Does this suggest a real difference between garlic and placebo, or are the results consistent with random variation? Answer by:

a. Identifying the relevant variable and parameter. (*Hint:* The variable is categorical with two categories. The parameter is a population proportion for one of the categories.)

b. Stating hypotheses for a large-sample two-sided test and checking that sample size guidelines are satisfied for that test.

c. Finding the test statistic value.

d. Finding and interpreting a P-value and stating the conclusion in context.

9.22 Exit-poll predictions According to an exit poll in the 2008 Vermont gubernatorial election, 54.5% of the sample size of 837 reported voting for the Republican candidate Douglas. Is this enough evidence to predict who won? Test that the population proportion who voted for Douglas was 0.50 against the alternative that it differed from 0.50. Answer by:

a. Identifying the variable and parameter, and defining notation.

b. Stating hypotheses and checking assumptions for a large-sample test.

c. Reporting the P-value and interpreting it. (The test statistic equals 2.662.)

d. Explaining how to make a decision for the significance level of 0.05.

9.23 Which cola? The 49 students in a class at the University of Florida made blinded evaluations of pairs of cola drinks. For the 49 comparisons of Coke and Pepsi, Coke was preferred 29 times. In the population that this sample represents, is this strong evidence that a majority prefers one of the drinks? Refer to the following MINITAB printout.

```
Test of p = 0.50 vs. not = 0.50
X    N    Sample p    95.0% CI         Z-Value  P-Value
29   49   0.5918     (0.454, 0.729)    1.286    0.1985
```

a. Explain how to get the test statistic value that MINITAB reports.

b. Explain how to get the P-value. Interpret it.

c. Based on the result in part b, does it make sense to accept H_0? Explain.

d. What does the 95% confidence interval tell you that the test does not?

9.24 How to sell a burger A fast-food chain wants to compare two ways of promoting a new burger (a turkey burger). One way uses a coupon available in the store. The other way uses a poster display outside the store. Before the promotion, their marketing research group matches 50 pairs of stores. Each pair has two stores with similar sales volume and customer demographics. The store in a pair that uses coupons is randomly chosen, and after a month-long promotion, the increases in sales of the turkey burger are compared for the two stores. The increase was higher for 28 stores using coupons and higher for 22 stores using the poster. Is this strong evidence to support the coupon approach, or could this outcome be explained by chance? Answer by performing all five steps of a two-sided significance test about the population proportion of times the sales would be higher with the coupon promotion.

9.25 ♦♦ **A binomial headache** A null hypothesis states that the population proportion p of headache sufferers who have more pain relief with aspirin than with another pain reliever equals 0.50. For a crossover study with 10 subjects, all 10 have more relief with aspirin. If the null hypothesis were true, by the binomial distribution the probability of this sample result equals $(0.50)^{10} = 0.001$. In fact, this is the small-sample P-value for testing H_0: $p = 0.50$ against H_a: $p > 0.50$. Does this P-value give (a) strong evidence in favor of H_0 or (b) strong evidence against H_0? Explain why.

9.26 ♦♦ **P-value for small samples** Example 4, on whether dogs can detect bladder cancer by selecting the correct urine specimen (out of seven), used the normal sampling distribution to find the P-value. The normal distribution P-value approximates a P-value using the binomial distribution. That binomial P-value is more appropriate when either expected count is less than 15. In Example 4, n was 54, and 22 of the 54 selections were correct.

a. If H_0: $p = 1/7$ is true, X = number of correct selections has the binomial distribution with $n = 54$ and $p = 1/7$. Why?

b. For H_a: $p > 1/7$, with $x = 22$, the P-value using the binomial is $P(22)+P(23)+ \cdots +P(54)$, where $P(x)$ denotes the binomial probability of outcome x with $p = 1/7$. (This equals 0.0000019.) Why would the P-value be this sum rather than just P(22)?

9.3 Significance Tests About Means

For quantitative variables, significance tests often refer to the population mean μ. We illustrate the significance test about means with the following example.

Significance test about a population mean

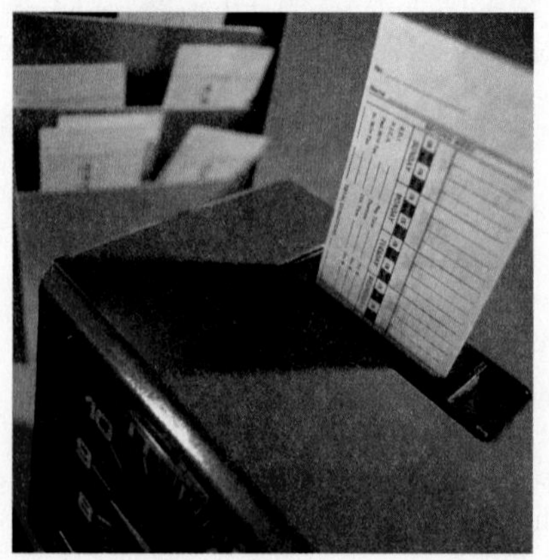

Example 7

The 40-Hour Work Week

Picture the Scenario

Since the Fair Labor Standards Act was passed in 1938, the standard work week in the United States has been 40 hours.

In recent years, the standard work week has fallen to less than 40 hours in most of Western Europe and in Australia. But many believe that the work-oriented culture in the United States has resulted in pressure among workers to put in longer hours than the 40-hour standard. In industries such as investment banking, a 40-hour work week is considered "slacker" behavior and may result in losing a job.

Question to Explore

How could we frame a study of working hours using a significance test that can detect whether the mean work week for the U.S. working population equals 40 hours or differs from 40 hours? State the null and alternative hypotheses for that test.

Think It Through

The response variable, length of work week, is quantitative. Hypotheses refer to the mean work week μ for the U.S. working population, whether it is 40 hours or differs from 40 hours. To use a significance test to analyze the strength of evidence about this question, we'll test H_0: $\mu = 40$ against H_a: $\mu \neq 40$.

Insight

For those who were working in 2008, the General Social Survey asked, "How many hours did you work last week?" For 636 men, the mean was 45.5 with a standard deviation of 15.16. For 567 women, the mean was 38.08, with a standard deviation of 12.57. We'll test H_0: $\mu = 40$ against H_a: $\mu \neq 40$ as we learn about the five steps of a significance test about means in the next subsection. We'll analyze whether the difference between the sample mean work week and the historical standard of 40 hours can be explained by random variability. We'll do the analysis for the sample of women, and you can practice a similar analysis using the data for men in Exercise 9.31.

Try Exercise 9.31, parts a and b

Steps of a Significance Test About a Population Mean

A significance test about a mean has the same five steps as a test about a proportion: Assumptions, hypotheses, test statistic, P-value, and conclusion. We will mention modifications as they relate to the test for a mean. However, the reasoning is the same as that for the significance test for a proportion.

Five Steps for a Significance Test About a Population Mean

Step 1: Assumptions

The three basic assumptions of a test about a mean are as follows:

- The variable is quantitative.
- The data production employed randomization.
- The *population distribution* is approximately normal. This assumption is most crucial when n is small and H_a is one-sided, as discussed later in the section.

For our study about the length of the work week, the variable is the number of hours worked in the past week, which is quantitative. The GSS used random sampling. The figure below is a histogram for the data collected from women. This graph

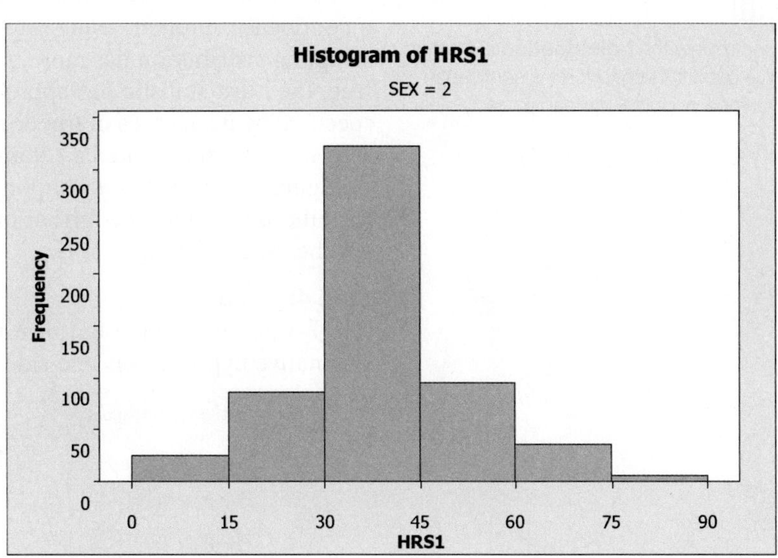

does not show any dramatic deviation from a normal shape. The sample size is large enough here ($n = 567$) that this assumption is less important than the others.

Step 2: Hypotheses

The null hypothesis in a test about a population mean has the form

$$H_0: \mu = \mu_0,$$

where μ_0 denotes a particular value for the population mean. The two-sided alternative hypothesis

$$H_a: \mu \neq \mu_0$$

includes values both below and above the number μ_0 listed in H_0. Also possible are the one-sided alternative hypotheses,

$$H_a: \mu > \mu_0 \text{ or } H_a: \mu < \mu_0.$$

For instance, let μ denote the mean work week for the U.S. female working population. If the mean equals the historical standard, then $\mu = 40$. If today's working pressures have forced this above 40, then $\mu > 40$. To test that the population mean equals the historical standard against the alternative that it is greater than that, we test $H_0: \mu = 40$ against $H_a: \mu > 40$. In practice, the two-sided alternative $H_a: \mu \neq 40$ is more common and lets us take an objective approach that can detect whether the mean is larger or smaller than the historical standard.

Recall

From Section 7.2, the standard deviation of the sample mean is $\sigma/\sqrt{n}$, where σ = population standard deviation. In practice, σ is unknown, so from Section 8.3 we estimate the standard deviation by the standard error

$$se = s/\sqrt{n}. ◀$$

Step 3: Test Statistic

The test statistic is the distance between the sample mean $\bar{x}$ and the null hypothesis value μ_0, as measured by the number of standard errors between them. This is measured by

$$\frac{(\bar{x} - \mu_0)}{se} = \frac{\text{sample mean} - \text{null hypothesis mean}}{\text{standard error of sample mean}}.$$

In practice, as in forming a confidence interval for a mean (Section 8.3), the standard error is $se = s/\sqrt{n}$. The test statistic is

$$t = \frac{(\bar{x} - \mu_0)}{se} = \frac{(\bar{x} - \mu_0)}{s/\sqrt{n}}.$$

In the length of work week study, for women the sample mean $\bar{x} = 38.08$ and the sample standard deviation $s = 12.57$. The standard error $se = s/\sqrt{n} = 12.57/\sqrt{567} = 0.528$. The test statistic equals

$$t = (\bar{x} - \mu_0)/se = (38.08 - 40)/0.528 = -3.64.$$

Recall

You can review the ***t* distribution** in Section 8.3. We used it there to form a confidence interval for a population mean μ. ◀

We use the symbol t rather than z for the test statistic because, as in forming a confidence interval, using s to estimate σ introduces additional error: The t sampling distribution has more variability than the standard normal. When H_0 is true, the t test statistic has approximately the *t distribution.* The t distribution is specified by its degrees of freedom, which equal $n-1$ for inference about a mean. This test statistic is called a ***t* statistic**.

Figure 9.7 shows the t sampling distribution. The farther $\bar{x}$ falls from the null hypothesis mean μ_0, the farther out in a tail the t test statistic falls, and the stronger the evidence is against H_0.

Step 4: P-Value

The P-value is a single tail or a two-tail probability depending on whether the alternative hypothesis is one-sided or two-sided.

Alternative Hypothesis	P-value
$H_a: \mu \neq \mu_0$	Two-tail probability from t distribution
$H_a: \mu > \mu_0$	Right-tail probability from t distribution
$H_a: \mu < \mu_0$	Left-tail probability from t distribution

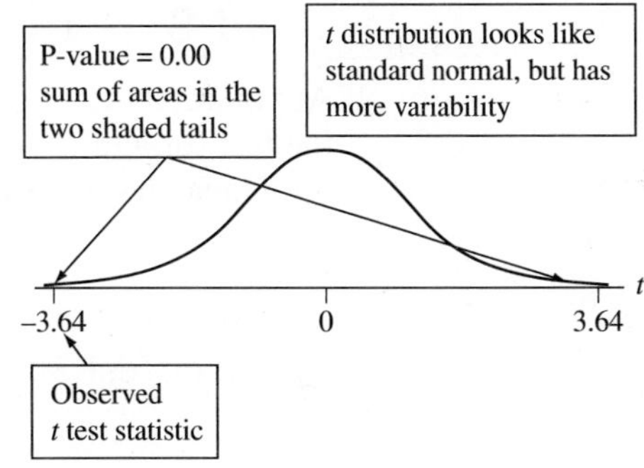

▲ **Figure 9.7 The t Distribution of the t Test Statistic.** There is stronger evidence against H_0 when the t test statistic falls farther out in a tail. The P-value for a two-sided H_a is a two-tail probability (shaded in figure). **Question** Why is it that t-scores farther out in the tails provide stronger evidence against H_0?

For the length of work week study for women with H_a: $\mu \neq 40$, the P-value is the two-tail probability of a test statistic value farther out in either tail than the observed value of −3.64. See Figure 9.7. This probability is double the single-tail probability. Table 9.3 shows the way MINITAB software reports results. Since $n = 567$, $df = n - 1 = 566$. The P-value is 0.00. This is the two-tail probability of the t test statistic values below −3.64 and above +3.64 when $df = 566$.

Table 9.3 MINITAB Output for Analyzing Data From Study of Work Week

```
Test of mu = 40 vs not = 40

Variable   N     Mean    StDev  SE Mean        95% CI            T       P

HRS1_2    567   38.078  12.574   0.528  (37.040, 39.115)   -3.64   0.000
```

Step 5: Conclusion

The conclusion of a significance test reports the P-value and *interprets* what it says about the question that motivated the test. Sometimes this includes a decision about the validity of H_0. We reject the null hypothesis when the P-value is less than or equal to the preselected significance level. In this study, the P-value of 0 provides strong evidence against the null hypothesis. If we had preselected a significance level of 0.05, this would be enough evidence to reject H_0: $\mu = 40$ in favor of H_a: $\mu \neq 40$.

Table 9.3 also shows a 95% confidence interval for μ of (37.040, 39.115). This shows the believable values for the population mean length of work week for working women. The interval does not contain 40. From this interval of values, we can infer that the population mean of 40 is not a plausible value for the mean number of hours in a work week for women.

Try Exercises 9.31 and 9.32

When research scientists conduct a significance test, their report would not show all results of a printout but would instead present a simple summary such as, "The evidence that the work week for women in recent time differs from 40 hours was statistically significant (sample mean = 38.08, standard deviation = 12.57, $n = 567$, $t = -3.64$, P-value = 0.00)." Often results are summarized even further, such as, "There was statistically significant evidence that the work week differs from 40 hours (P-value < 0.05)." This brief a summary is undesirable because it does not show the estimated mean or the actual P-value. Don't condense your conclusions this much.

The *t* Statistic and *z* Statistic Have the Same Form

As you read the examples in this section, notice the parallel between each step of the test for a mean and the test for a proportion. For instance, the *t* test statistic for a mean has the same form as the *z* test statistic for a proportion, namely,

Form of Test Statistic

$$\frac{\text{Estimate of parameter} - \text{H}_0 \text{ value of parameter}}{\text{Standard error of estimate}}$$

For the test about a mean, the estimate $\bar{x}$ of the population mean μ replaces the estimate $\hat{p}$ of the population proportion p, the H_0 mean μ_0 replaces the H_0 proportion p_0, and the standard error of the sample mean replaces the standard error of the sample proportion.

SUMMARY: Steps of a Significance Test for a Population Mean μ

1. **Assumptions**
   - Quantitative variable, with population mean μ defined in context
   - Data are obtained using randomization, such as a simple random sample or a randomized experiment
   - Population distribution is approximately normal (Mainly needed for one-sided tests with small *n*)
2. **Hypotheses**
   Null: H_0: $\mu = \mu_0$, where μ_0 is the hypothesized value (such as H_0: $\mu = 0$)
   Alternative: H_a: $\mu \neq \mu_0$ (two-sided) or H_a: $\mu < \mu_0$ (one-sided) or H_a: $\mu > \mu_0$ (one-sided)
3. **Test statistic**
   $$t = \frac{(\bar{x} - \mu_0)}{se}, \text{ where } se = s/\sqrt{n}$$
4. **P-value** Use *t* distribution with $df = n - 1$

Alternative Hypothesis	P-value
H_a: $\mu \neq \mu_0$	Two-tail probability
H_a: $\mu > \mu_0$	Right-tail probability
H_a: $\mu < \mu_0$	Left-tail probability

5. **Conclusion**
   Smaller P-values give stronger evidence against H_0 and supporting H_a. If using a significance level to make a decision, reject H_0 if P-value is less than or equal to the significance level (such as 0.05). Relate the conclusion to the context of the study.

Performing a One-Sided Test About a Population Mean

One-sided alternative hypotheses apply for a prediction that μ differs from the null hypothesis value in a certain direction. For example, H_a: $\mu > 0$ predicts that the true mean is *larger* than the null hypothesis value of 0. Its P-value is the probability of a *t* value *larger* than the observed value, that is, in the *right* tail. Likewise, for H_a: $\mu < 0$ the P-value is the *left*-tail probability. In each case, again, $df = n - 1$.

Example 8

One-sided significance test about a mean

Weight Change in Anorexic Girls

Picture the Scenario

A recent study compared different psychological therapies for teenage girls suffering from anorexia, an eating disorder that causes them to become dangerously underweight.[9] Each girl's weight was measured before and after a period of therapy. The variable of interest was the weight change, defined as weight at the end of the study minus weight at the beginning of the study. The weight change was positive if the girl gained weight and negative if she lost weight.

In this study, 29 girls received cognitive behavioral therapy. This form of psychotherapy stresses identifying the thinking that causes the undesirable behavior and replacing it with thoughts designed to help improve this behavior. Table 9.4 shows the data. The weight changes for the 29 girls had a sample mean of $\bar{x} = 3.00$ pounds and standard deviation of $s = 7.32$ pounds.

Table 9.4 Weights of Anorexic Girls (in Pounds) Before and After Treatment

This example uses the weight change as the variable of interest.

	Weight				Weight				Weight		
Girl	**Before**	**After**	**Change**	**Girl**	**Before**	**After**	**Change**	**Girl**	**Before**	**After**	**Change**
1	80.5	82.2	1.7	11	85.0	96.7	11.7	21	83.0	81.6	−1.4
2	84.9	85.6	0.7	12	89.2	95.3	6.1	22	76.5	75.7	−0.8
3	81.5	81.4	−0.1	13	81.3	82.4	1.1	23	80.2	82.6	2.4
4	82.6	81.9	−0.7	14	76.5	72.5	−4.0	24	87.8	100.4	12.6
5	79.9	76.4	−3.5	15	70.0	90.9	20.9	25	83.3	85.2	1.9
6	88.7	103.6	14.9	16	80.6	71.3	−9.3	26	79.7	83.6	3.9
7	94.9	98.4	3.5	17	83.3	85.4	2.1	27	84.5	84.6	0.1
8	76.3	93.4	17.1	18	87.7	89.1	1.4	28	80.8	96.2	15.4
9	81.0	73.4	−7.6	19	84.2	83.9	−0.3	29	87.4	86.7	−0.7
10	80.5	82.1	1.6	20	86.4	82.7	−3.7				

Question to Explore

Conduct a significance test for finding the strength of evidence supporting the effectiveness of the cognitive behavioral therapy, that is, to determine whether it results in a positive mean weight change.

Think It Through

The response variable, weight change, is quantitative. A significance test about the population mean weight change assumes that the population distribution of weight change is normal. We'll learn how to check this assumption later in the section. The test also assumes randomization for the data production. The anorexia study is like the dogs detecting cancer study in Example 4 in that its subjects were a convenience sample. As a result, inferences are highly tentative. They are more convincing if researchers can argue that the girls in the

[9]The data are courtesy of Prof. Brian Everitt, Institute of Psychiatry, London.

sample are representative of the population of girls who suffer from anorexia. The study did employ randomization in assigning girls to one of three therapies, only one of which (cognitive behavioral) is considered in this example.

Hypotheses refer to the population mean weight change μ, namely whether it is 0 (the no effect value) or is positive. To use a significance test to analyze the strength of evidence about the therapy's effect, we'll test H_0: $\mu = 0$ against H_a: $\mu > 0$.

Now let's find the test statistic. Using the information from the start of this example, the standard error $se = s/\sqrt{n} = 7.32/\sqrt{29} = 1.36$. The test statistic equals

$$t = (\bar{x} - \mu_0)/se = (3.00 - 0)/1.36 = 2.21,$$

with $df = n - 1 = 28$. Because H_a predicts that the mean is *above* 0, the P-value is the right-tail probability *above* the test statistic value of 2.21. This is 0.018, or 0.02 rounded to two decimal places. See Figure 9.8 and TI output in margin. The P-value of 0.02 means that if H_0: $\mu = 0$ were true, it would be unusual (about a 2% chance) to observe a t statistic of 2.21 or even larger in the positive direction. A P-value of 0.02 provides substantial evidence against H_0: $\mu = 0$ and in favor of H_a: $\mu > 0$. If we needed to make a decision and had preselected a significance level of 0.05, we would reject H_0: $\mu = 0$ in favor of H_a: $\mu > 0$.

```
T-Test
 μ>0.00
 t=2.21
 p=.02
 x̄=3.00
 Sx=7.32
 n=29.00
```

TI-83+/84 output

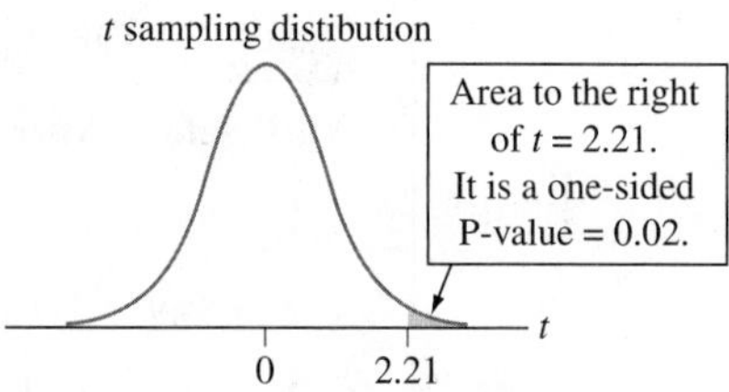

▲ **Figure 9.8 P-value for Testing H_0: $\mu = 0$ against H_0: $\mu > 0$. Question** Why are large positive, rather than large negative, t test statistic values the ones that support H_0: $\mu > 0$?

In summary, anorexic girls on this therapy appear to gain weight, on average. This conclusion is tentative, however, for two reasons. First, as with nearly all medical studies, this study used a convenience sample rather than a random sample. Second, we have not yet investigated the shape of the distribution of weight changes. We'll study this issue later in the section.

Insight

We used a one-sided alternative hypothesis in this test to detect a positive effect of the therapy. In practice, the two-sided alternative H_a: $\mu \neq 0$ is more common and lets us take an objective approach that can detect either a positive or a negative effect. For H_a: $\mu \neq 0$, the P-value is the two-tail probability, so $P = 2(0.02) = 0.04$. This two-sided test also provides relatively strong evidence against the no effect null hypothesis.

Try Exercise 9.30

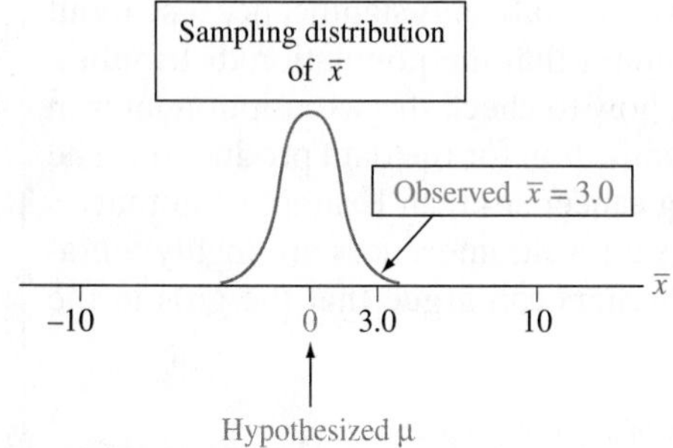

For the one-sided H_a: $\mu > 0$ in this example, the P-value is the probability of t values above the observed t test statistic value of 2.21. Equivalently, it is the probability of a sample mean weight change $\bar{x} \geq 3.0$ (the observed value), if H_0 were true that $\mu = 0$. (See the figure in the margin.)

We've seen that if $\mu = 0$, it would be unusual to get a sample mean of 3.0. If in fact μ were a number *less than* 0, it would be even more unusual. For example, a sample value of $\bar{x} = 3.0$ is even more unusual when $\mu = -5$ than when $\mu = 0$,

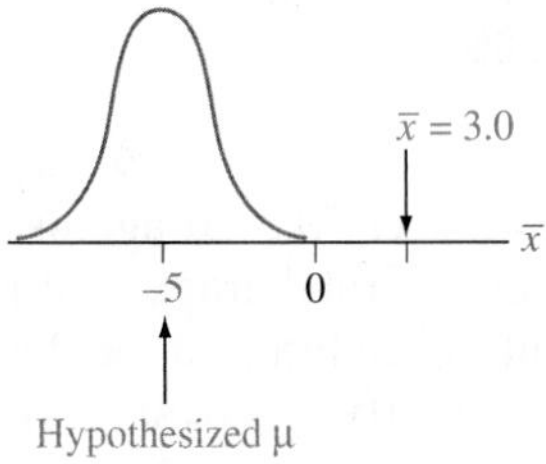

since 3.0 is farther out in the tail of the sampling distribution of $\bar{x}$ when $\mu = -5$ than when $\mu = 0$. (See the figure in the margin.)

Thus, when we reject $H_0: \mu = 0$ in favor of $H_a: \mu > 0$, we can also reject the broader null hypothesis of $H_0: \mu \leq 0$. In other words, we conclude that $\mu = 0$ is false and that $\mu < 0$ is false. However, statements of null hypotheses use a *single* number in the null hypothesis because a single number is entered in the test statistic to compare to the sample mean.

Using the *t* Table to Approximate a P-Value

Statistical software and many calculators can find the P-value for you, as illustrated in the margin TI-83+/84 screen shot for the anorexia study of Example 8. If you have the test statistic value but are not using software, you can use the *t* table (Table B in the Appendix). Table B is not detailed enough to provide an *exact* tail probability, but it provides enough information to determine whether a one-tail probability is greater than or less than 0.100, 0.050, 0.025, 0.010, 0.005, or 0.001.

Let's see how to do this for the *t* test statistic from the anorexia study, $t = 2.21$ based on $df = 28$. Look at the row of Table B for $df = 28$. You will see

df	$t_{.100}$	$t_{.050}$	$t_{.025}$	$t_{.010}$	$t_{.005}$	$t_{.001}$
28	1.313	1.701	2.048	2.467	2.763	3.408

Note that $t = 2.21$ falls between 2.048 and 2.467. Now, the value $2.048 = t_{.025}$ has a right-tail probability of 0.025 and the value $2.467 = t_{.010}$ has a right-tail probability of 0.010. So the right-tail probability for $t = 2.21$ falls between 0.010 and 0.025. Figure 9.9 illustrates. In fact, software tells us that the actual P-value is 0.018.

To get the P-value for the two-sided H_a, double the one-sided P-value, because you want a two-tail probability. For these data, we double 0.01 and 0.025 to report $0.02 < \text{P-value} < 0.05$. So, we can reject H_0 at the 0.05 significance level. In fact, software told us that the actual P-value = 0.036.

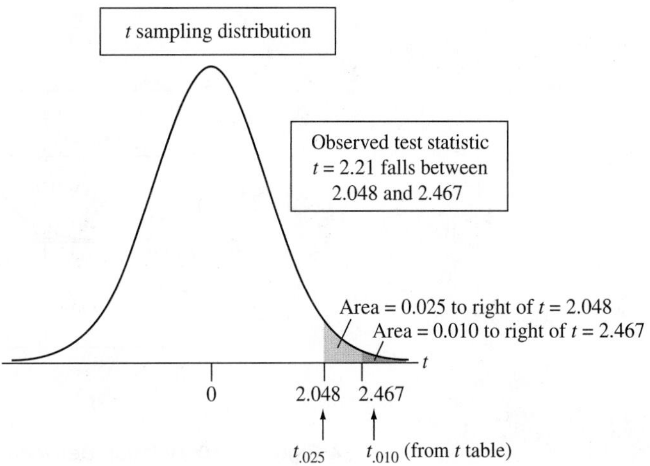

▲ **Figure 9.9 For *df* = 28, *t* = 2.21 has a Right-Tail Probability Between 0.010 and 0.025.** The two-sided P-value falls between 2(0.010) = 0.02 and 2(0.025) = 0.05. **Question** Using software or a calculator, can you show that the actual right-tail probability equals 0.018 and that the two-sided P-value equals 0.036?

Results of Two-Sided Tests and Results of Confidence Intervals Agree

For the anorexia study, we got a P-value of 0.04 for testing H_0: $\mu = 0$ against H_a: $\mu \neq 0$ for the mean weight change with the cognitive behavioral therapy. With the 0.05 significance level, we would reject H_0. A 95% confidence interval for the population mean weight change μ is $\bar{x} \pm t_{.025}(se)$, which is $3.0 \pm 2.048(1.36)$, or (0.2, 5.8) pounds. The confidence interval shows just how different from 0 the population mean weight change is likely to be. It is estimated to fall between 0.2 and 5.8 pounds. We infer that the population mean weight change μ is positive because all the numbers in this interval are greater than 0, but the effect of the therapy may be very small.

Both the significance test and the confidence interval suggested that μ differs from 0. In fact, conclusions about means using two-sided significance tests are consistent with conclusions using confidence intervals. If a two-sided test says you can reject the hypothesis that $\mu = 0$, then 0 is not in the corresponding confidence interval.

If* P-value $\leq$ 0.05 *in a Two-Sided Test, a 95% Confidence Interval Does Not Contain the* H_0 *Value In the anorexia study of Example 8, the two-sided test of H_0: $\mu = 0$ has P-value $= 0.04 \leq 0.05$. This small P-value says we can reject H_0 at the 0.05 significance level. The 95% confidence interval for μ is (0.2, 5.8). The interval does not contain 0, so this method also suggests that μ is not exactly equal to 0.

By contrast, suppose that the P-value > 0.05 in a two-sided test of H_0: $\mu = 0$, so we cannot reject H_0 at the 0.05 significance level. Then, a 95% confidence interval for μ will contain 0. Both methods will show that the value of 0 is a plausible one for μ.

Confidence Intervals and Two-Sided Tests About Means Are Consistent Figure 9.10 illustrates why decisions from two-sided tests about means are consistent with confidence intervals.

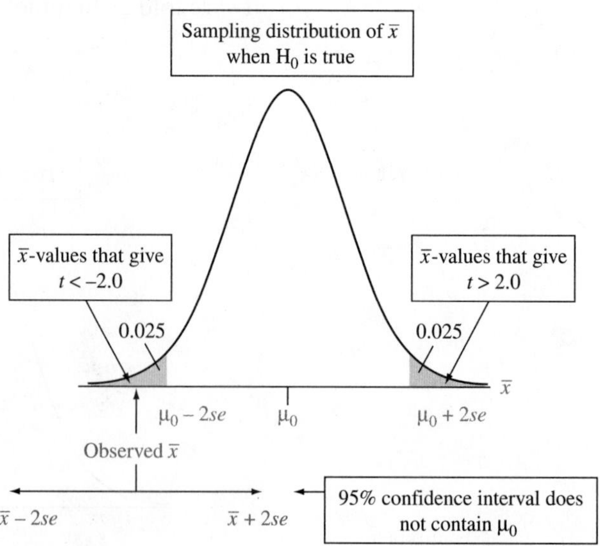

▲ **Figure 9.10 Relation Between Confidence Interval and Significance Test.** With large samples, if the sample mean falls more than about two standard errors from μ_0, then μ_0 does not fall in the 95% confidence interval and also μ_0 is rejected in a test at the 0.05 significance level. **Question** Inference about proportions does not have an *exact* equivalence between confidence intervals and tests. Why? (*Hint:* Are the same standard error values used in the two methods?)

With large samples, the t-score for a 95% confidence interval is approximately 2, so the confidence interval is roughly $\bar{x} \pm 2(se)$. If this interval does not contain a particular value μ_0, then the sample mean $\bar{x}$ falls more than 2 standard errors from μ_0, which means that the test statistic $t = (\bar{x} - \mu_0)/se$ is larger than 2 in absolute value. Consequently, the two-tail P-value is less than 0.05, so we would reject the hypothesis that $\mu = \mu_0$.

When the Population Does Not Satisfy the Normality Assumption

Recall

From Section 7.2, with random sampling the **sampling distribution of the sample mean is approximately normal** for large *n*, by the central limit theorem. ◀

For the test about a mean, the third assumption states that the population distribution should be approximately normal. This ensures that the sampling distribution of the sample mean $\bar{x}$ is normal and, after using s to estimate σ, the test statistic has the t distribution. For large samples (roughly about 30 or higher), this assumption is usually not important. Then, an approximate normal sampling distribution occurs for $\bar{x}$ regardless of the population distribution. (Remember the central limit theorem? See recall box in margin.) Chapter 15 presents a type of statistical method, called *nonparametric,* that does not require the normal assumption. However, the normal assumption made with small samples for the t test is not crucial when we use a two-sided H_a.

In Practice Two-Sided Inferences Are Robust

Two-sided inferences using the *t* distribution are *robust* against violations of the normal population assumption. They still usually work well if the actual population distribution is not normal. The test does not work well for a one-sided test with small *n* when the population distribution is highly skewed.

Checking for Normality in the Anorexia Study Figure 9.11 shows a histogram and a box plot of the data from the anorexia study. With small *n,* such plots are very rough estimates of the population distribution. It can be difficult to determine whether the population distribution is approximately normal. However, Figure 9.11 does suggest skew to the right, with a small proportion of girls having considerable weight gains.

A two-sided t test still works quite well even if the population distribution is skewed. So, we feel comfortable with the two-sided test summarized in the Insight step of Example 8. However, this plot makes us cautious about using a one-sided test for these data. The sample size is not large ($n = 29$), and the histogram in Figure 9.11 shows substantial skew, with the box plot highlighting six quite large weight change values.

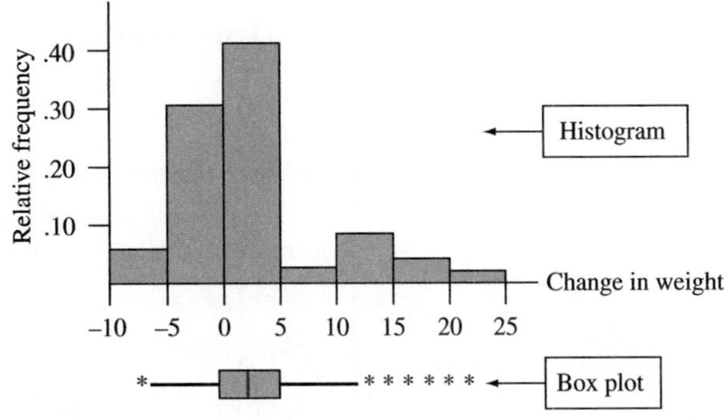

▲ **Figure 9.11** Histogram and Box Plot of Weight Change for Anorexia Sufferers.
Question What do these plots suggest about the shape of the population distribution of weight change?

Regardless of Robustness, Look at the Data Whether n is small or large, you should look at the data to check for severe skew or for outliers that occur primarily in one direction. They could cause the sample mean to be a misleading measure. For the anorexia data, the median weight change is only 1.4 pounds, less than the mean of 3.0 because of the skew to the right. Even though the significance test indicated that the population mean weight change was positive, the sample median is another indication that the size of the effect could be small. You also need to be cautious about any conclusion if it changes after removing an extreme outlier from the data set (see Exercise 9.40).

Effect of Sample Size on P-values

The result of a significance test can depend strongly on how much data you have. For a given sample effect, the test statistic tends to be larger as the sample size increases, as the next example shows.

Example 9

Sample size and P-values

Testing Software

Picture the Scenario

Random numbers are used for conducting simulations and for identifying subjects to be chosen for a random sample. A difficult programming challenge is writing a computer program that can generate random numbers properly. Each digit must be equally likely to be 0, 1, 2,..., 9, and the digits in the sequence must be independent: The probability of each digit is 0.10, regardless of which digits were generated in the past. One test of software for generating random numbers checks if the mean of the generating process differs from the mean of 4.50 that holds when each digit 0, 1,..., 9 truly has probability 0.10 of occurring each time.

Questions to Explore

Consider the test of H_0: $\mu = 4.50$ against H_a: $\mu \neq 4.50$ when a sequence of random digits has sample mean 4.40 and standard deviation 2.90, if (i) $n = 100$ and (ii) $n = 10{,}000$. Table 9.5 shows software output for the two cases.

a. Show how the *se* values and test statistic values were obtained in the two cases.

b. Explain the practical implications about the effect of n on results of a test.

Table 9.5 Effect on a Significance Test of Increasing the Sample Size *n*

For a given size of effect (such as $\bar{x} - \mu_0 = -0.10$), as n increases, the test statistic increases in absolute value and the P-value decreases.

Test of mu = 4.50 versus not = 4.50

	Variable	N	Mean	StDev	SE Mean	T	P
(i) $n = 100 \rightarrow$	Variable	N	Mean	StDev	SE Mean	T	P
	RanDigit	100	4.40	2.90	0.290	−0.345	0.731
(ii) $n = 10{,}000 \rightarrow$	Variable	N	Mean	StDev	SE Mean	T	P
	RanDigit	10,000	4.40	2.90	0.029	−3.45	0.0006

Think It Through

a. In both cases, $\bar{x} = 4.40$, $s = 2.90$, and the sample effect as measured by $\bar{x} - \mu_0 = 4.40 - 4.50 = -0.10$ is the same. However, the standard error is smaller with a larger sample size. The *se* values are

$$\text{(i) } se = s/\sqrt{n} = 2.90/\sqrt{100} = 0.29,$$

$$\text{(ii) } se = s/\sqrt{n} = 2.90/\sqrt{10{,}000} = 0.029.$$

The test statistics then are also quite different:

$$\text{(i) } t = \frac{(\bar{x} - \mu_0)}{se} = \frac{4.40 - 4.50}{0.29} = -0.345,$$

$$\text{(ii) } t = \frac{(\bar{x} - \mu_0)}{se} = \frac{4.40 - 4.50}{0.029} = -3.45.$$

From Table 9.5, the P-values are dramatically different, 0.73 compared to 0.0006. The same effect, $\bar{x} - \mu_0 = -0.10$, but based on a larger sample size, results in a much smaller P-value.

b. For a given sample effect, larger sample sizes produce larger test statistics (in absolute value). Why does this happen? As n increases, the standard error in the denominator of the t statistic decreases. So the t statistic itself increases. The two-sided P-value is then smaller. This makes sense: We can be more certain that a given sample effect reflects a true population effect if the sample size is large than if it is small.

Insight

An implication of this result is that, for large n, statistical significance may not imply an important result in practical terms. For instance, you can get a small P-value even if the sample mean falls quite near the null hypothesis value. We'll discuss this further in Section 9.5.

Try Exercise 9.29

9.3 Practicing the Basics

9.27 Which t has P-value = 0.05? A t test for a mean uses a sample of 15 observations. Find the t test statistic value that has a P-value of 0.05 when the alternative hypothesis is (a) $H_a: \mu \neq 0$, (b) $H_a: \mu > 0$, and (c) $H_a: \mu < 0$?

9.28 Practice mechanics of a t test A study has a random sample of 20 subjects. The test statistic for testing $H_0: \mu = 100$ is $t = 2.40$. Find the approximate P-value for the alternative, (a) $H_a: \mu \neq 100$, (b) $H_a: \mu > 100$, and (c) $H_a: \mu < 100$.

9.29 Effect of n TRY Refer to the previous exercise. If the same sample mean and standard deviation had been based on $n = 5$ instead of $n = 20$, the test statistic would have been $t = 1.20$.

a. Would the P-value for $H_a: \mu \neq 100$ be larger, or smaller, than when $t = 2.40$? Why?

b. Other things being equal, explain why larger sample sizes result in smaller P-values.

9.30 Low carbohydrate diet TRY In a recent study,[10] 272 moderately obese subjects were randomly assigned to one of three diets: low-fat, restricted-calorie; Mediterranean, restricted-calorie; or low-carbohydrate, non-restricted-calorie. The prediction was that subjects on a low-carbohydrate diet would lose weight, on the average. After two years, the mean weight loss was 5.5 kg for the 109 subjects in the low-carbohydrate group with a standard deviation of 7.0kg. The MINITAB output shows results of a significance test for testing $H_0: \mu = 0$ against $H_a: \mu \neq 0$, where μ is the population mean weight change. Note that weight change is determined by calculating after weight − before weight.

```
                     SE
  N    Mean  StDev  Mean        95% CI           T      P
109  -5.500  7.000  0.670  (-6.829,-4.171)  -8.20  0.000
```

[10] *Am. J. Med.*, vol. 359, 2008, pp. 229–44.

a. Identify the P-value for this test.

b. How is the P-value interpreted?

c. Would the P-value and 95% confidence interval lead to the same conclusion about H_0? Explain.

9.31 Men at work When the 636 male workers in the 2008 GSS were asked how many hours they worked in the previous week, the mean was 45.5 with a standard deviation of 15.16. Does this suggest that the population mean work week for men exceeds 40 hours? Answer by:

a. Identifying the relevant variable and parameter.

b. Stating null and alternative hypotheses.

c. Reporting and interpreting the P-value for the test statistic value of $t = 9.15$.

d. Explaining how to make a decision for the significance level of 0.01.

9.32 Young workers When the 127 workers aged 18–25 in the 2008 GSS were asked how many hours they worked in the previous week, the mean was 37.47 with a standard deviation of 13.63. Does this suggest that the population mean work week for this age group differs from 40 hours? Answer by:

a. Identifying the relevant variable and parameter.

b. Stating null and alternative hypotheses.

c. Finding and interpreting the test statistic value.

d. Reporting and interpreting the P-value and stating the conclusion in context.

9.33 Lake pollution An industrial plant claims to discharge no more than 1000 gallons of wastewater per hour, on the average, into a neighboring lake. An environmental action group decides to monitor the plant, in case this limit is being exceeded. Doing so is expensive, and only a small sample is possible. A random sample of four hours is selected over a period of a week. The observations (gallons of wastewater discharged per hour) are

$$2000, 1000, 3000, 2000.$$

a. Show that $\bar{x} = 2000$, $s = 816.5$, and standard error $= 408.25$.

b. To test H_0: $\mu = 1000$ vs. H_a: $\mu > 1000$, show that the test statistic equals 2.45.

c. Using Table B or software, show that the P-value is less than 0.05, so there is enough evidence to reject the null hypothesis at the 0.05 significance level.

d. Explain how your one-sided analysis in part b implicitly tests the broader null hypothesis that $\mu \leq 1000$.

9.34 Weight change for controls A disadvantage of the experimental design in Example 8 on weight change in anorexic girls is that girls could change weight merely from participating in a study. In fact, girls were randomly assigned to receive a therapy or to serve in a control group, so it was possible to compare weight change for the therapy group to the control group. For the 26 girls in the control group, the weight change had $\bar{x} = -0.5$ and $s = 8.0$. Repeat all five steps of the test of H_0: $\mu = 0$ against H_a: $\mu \neq 0$ for this group, and interpret the P-value.

9.35 Crossover study A crossover study of 13 children suffering from asthma (*Clinical and Experimental Allergy,* vol. 20, pp. 429–432, 1990) compared single inhaled doses of formoterol (F) and salbutamol (S). The outcome measured was the child's peak expiratory flow (PEF) 8 hours following treatment. The data on PEF follow:

Child	F	S	Child	F	S	Child	F	S	Child	F	S
1	310	270	5	410	380	9	330	365	13	220	90
2	385	370	6	370	300	10	250	210			
3	400	310	7	410	390	11	380	350			
4	310	260	8	320	290	12	340	260			

Let μ denote the population mean of the difference between the PEF values for the F and S treatments. Use a calculator or software for the following analyses:

a. Form the 13 difference scores, for instance $310 - 270 = 40$ for child 1 and $330 - 365 = -35$ for child 9, always taking F − S. Construct a dot plot or a box plot. Describe the sample data distribution.

b. Carry out the five steps of the significance test for a mean of the difference scores, using H_0: $\mu = 0$ and H_a: $\mu \neq 0$.

c. Discuss whether or not the assumptions seem valid for this example. What is the impact of using a convenience sample?

9.36 Too little or too much wine? Wine-pouring vending machines, previously available in Europe and international airports, have become popular in the last few years in the United States. They are even approved to dispense wine in some Walmart stores. The available pouring options are a 5-ounce glass, a 2.5-ounce half-glass, and a 1-ounce taste. When the machine is in statistical control (see Exercise 7.42), the amount dispensed for a full glass is 5.1 ounces. Four observations are taken each day, to plot a daily mean over time on a control chart to check for irregularities. The most recent day's observations were 5.05, 5.15, 4.95, and 5.11. Could the difference between the sample mean and the target value be due to random variation, or can you conclude that the true mean is now different from 5.1? Answer by showing the five steps of a significance test, making a decision using a 0.05 significance level.

9.37 Selling a burger In Exercise 9.24, a fast-food chain compared two ways of promoting a turkey burger. In a separate experiment with 10 pairs of stores, the difference in the month's increased sales between the store that used coupons and the store with the outside poster had a mean of \$3000. Does this indicate a true difference between mean sales for the two advertising approaches? Answer by using the output shown to test that the population mean difference is 0, carrying out the five steps of a significance test. Make a decision using a 0.05 significance level.

```
Test of mu = 0 vs. mu not = 0
Variable   N        Mean          StDev      SE Mean
Sales      10       3000          4000       1264.91
Variable          95.0% CI            T           P
Sales          (138.8, 5861.2)     2.372     0.04177
```

9.38 Assumptions important? Refer to the previous exercise.

a. Explain how the result of the 95% confidence interval shown in the table agrees with the test decision using the 0.05 significance level.

b. Suppose you instead wanted to perform a one-sided test, because the study predicted that the increase in sales would be higher with coupons. Explain why the normal population assumption may possibly be problematic.

9.39 Anorexia in teenage girls Example 8 described a study about various therapies for teenage girls suffering from anorexia. For each of 17 girls who received the family therapy, the changes in weight were

11, 11, 6, 9, 14, −3, 0, 7, 22, −5, −4, 13, 13, 9, 4, 6, 11.

a. Plot these data with a dot plot or box plot, and summarize.

b. Verify that the weight changes have $\bar{x} = 7.29$, $s = 7.18$, and $se = 1.74$ pounds.

c. Give all steps of a significance test about whether the population mean was 0, against an alternative designed to see if there is any effect.

9.40 Sensitivity study Ideally, results of a statistical analysis should not depend greatly on a single observation. To check this, it's a good idea to conduct a **sensitivity study.** This entails redoing the analysis after deleting an outlier from the data set or changing its value to a more typical value and checking whether results change much. If results change little, this gives us more faith in the conclusions that the statistical analysis reports. For the weight changes in Table 9.4 from the anorexia study (shown again here and also in the Anorexia data file on the text CD), the greatest reported value of 20.9 pounds was a severe outlier. Suppose this observation was actually 2.9 pounds but was incorrectly recorded. Redo the two-sided test of that example, and summarize how the results differ. Does the ultimate conclusion depend on that single observation?

Weight Changes in Anorexic Girls		
1.7	11.7	−1.4
0.7	6.1	−0.8
−0.1	1.1	2.4
−0.7	−4.0	12.6
−3.5	20.9	1.9
14.9	−9.3	3.9
3.5	2.1	0.1
17.1	1.4	15.4
−7.6	−0.3	−0.7
1.6	−3.7	

9.41 Test and CI Results of 99% confidence intervals are consistent with results of two-sided tests with which significance level? Explain the connection.

9.4 Decisions and Types of Errors in Significance Tests

In significance tests, the P-value summarizes the evidence about H_0. A P-value such as 0.001 casts strong doubt on H_0 being true because if it were true the observed data would be very unusual.

When we need to decide if the evidence is strong enough to reject H_0, we've seen that the key is whether or not the P-value falls below a prespecified **significance level**. The significance level is usually denoted by the Greek letter α (alpha). In practice, $\alpha = 0.05$ is most common: We reject H_0 if the P-value ≤ 0.05. We do not reject H_0 if the P-value > 0.05. The smaller α is, the stronger the evidence must be to reject H_0. To avoid bias, we select α *before* looking at the data.

Two Potential Types of Errors in Test Decisions

Because of sampling variability, decisions in significance tests always have some uncertainty. A decision can be in error. For instance, in the anorexia study of Example 8, we got a P-value of 0.04 for testing H_0: $\mu = 0$ of no weight change, on average, against H_a: $\mu \neq 0$. With significance level $\alpha = 0.05$ we rejected H_0 and concluded the mean was not equal to 0. The data indicated that anorexic girls have a positive mean weight change when undergoing the therapy. If the therapy truly has a positive effect, this is a correct decision. In reality, though, perhaps the therapy has no effect, and the population mean weight change (unknown to us) is actually 0.

Tests have two types of potential errors called **Type I** and **Type II errors.**

Type I and Type II Errors

When H_0 is true, a **Type I error** occurs when H_0 is rejected.
When H_0 is false, a **Type II error** occurs when H_0 is not rejected.

If the anorexia therapy actually has no effect (that is, if $\mu = 0$), we've made a Type I error. H_0 was actually true, but we rejected it. A consequence of committing this Type I error would be implementing a therapy that actually has no effect on helping with weight gain and thus gives patients false hope.

Consider the experiment about astrology in Example 5. In that study H_0: $p = 1/3$ corresponded to random guessing by the astrologers. We got a P-value of 0.40. With significance level = 0.05, we do not reject H_0. If truly $p = 1/3$, this is a correct decision. However, if astrologers actually can predict better than random guessing (so that $p > 1/3$), we've made a Type II error, failing to reject H_0 when it is false. A consequence of committing this Type II error would be calling these astrologers fakes when actually they have predictive powers.

When we make a decision, there are four possible results. These refer to the two possible decisions combined with the two possibilities for whether H_0 is true. Table 9.6 summarizes these four possibilities. In practice, when we plan to make a decision in a significance test, it's important to know the probability of an incorrect decision.

Table 9.6 The Four Possible Results of a Decision in a Significance Test

Type I and Type II errors are the two possible incorrect decisions. We make a correct decision if we do not reject H_0 when it is true or if we reject it when it is false.

	Decision	
Reality About H_0	**Do not reject H_0**	**Reject H_0**
H_0 true	Correct decision	Type I error
H_0 false	Type II error	Correct decision

Type I error occurs if we reject H_0 when it is actually true.

Type II error occurs if we do not reject H_0 when it is actually false.

An Analogy: Decision Errors in a Legal Trial The two types of errors can occur with any decision having two options, one of which is incorrect. For instance, consider a decision in a legal trial. The null hypothesis tested is the defendant's claim of innocence. The alternative hypothesis is that the defendant is guilty. The jury rejects H_0 if it decides that the evidence is sufficient to convict. The defendant is then judged guilty. A Type I error, rejecting a true null hypothesis, occurs in convicting a defendant who is actually innocent. Not rejecting H_0 means the defendant is acquitted (judged not guilty). A Type II error, not rejecting H_0 even though it is false, occurs in acquitting a defendant who is actually guilty. See Table 9.7. A potential consequence of a Type I error is sending an innocent person to jail. A potential consequence of a Type II error is setting free a guilty person.

Table 9.7 Possible Results of a Legal Trial

	Legal Decision	
Defendant	**Acquit**	**Convict**
Innocent (H_0)	Correct decision	Type I error
Guilty (H_a)	Type II error	Correct decision

The Significance Level Is the Probability of a Type I Error

When H_0 is actually true, let's see how to find the probability of a Type I error. This is the probability of rejecting H_0, even though it is actually true. We'll find this for the two-sided test about a proportion.

With the $\alpha = 0.05$ significance level, we reject H_0 if the P-value ≤ 0.05. For two-sided tests about a proportion, the two-tail probability that forms the P-value is ≤ 0.05 whenever the test statistic z satisfies $|z| \geq 1.96$. The collection of test statistic values for which a test rejects H_0 is called the **rejection region**. These are the z test statistic values that occur when the sample proportion falls at least 1.96 standard errors from the null hypothesis value. They are the values we'd least expect to observe if H_0 were true. Figure 9.12 illustrates.

In Words

The $|z| \geq 1.96$ is the same as expressing $z < -1.96$ or $z > 1.96$.

Now, if H_0 is actually true, the sampling distribution of the z test statistic is the standard normal. Therefore, the probability of rejecting H_0, which is the probability that $|z| \geq 1.96$, is exactly 0.05. But this is precisely the significance level.

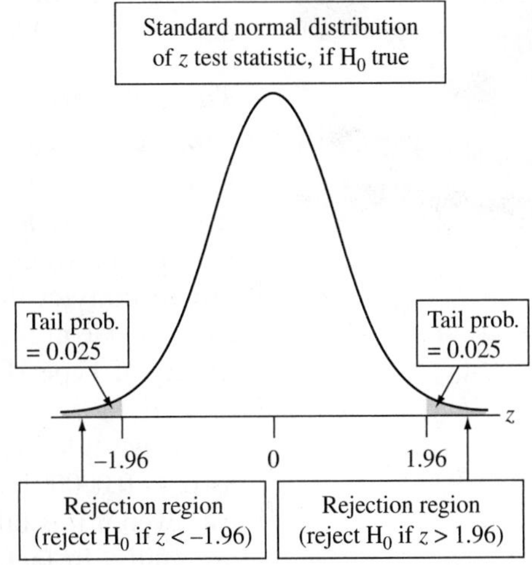

▲ **Figure 9.12 The Rejection Region Is the Set of Test Statistic Values That Reject H_0.** For a two-sided test about a proportion with significance level 0.05, these are the z test statistic values that exceed 1.96 in absolute value. **Question** If instead we use a one-sided H_a: $p > p_0$, what is the rejection region?

P(Type I error) = Significance level α

Suppose H_0 is true. The probability of rejecting H_0, thereby making a Type I error, equals the significance level for the test.

In Practice We Don't Know If a Decision Is Correct

In practice, we don't know whether or not a decision in a significance test is correct, just as we don't know whether or not a particular confidence interval truly contains an unknown parameter value. However, we can control the *probability* of an incorrect decision for either type of inference.

We can control the probability of a Type I error by our choice of the significance level. The more serious the consequences of a Type I error, the smaller α should be. In practice, $\alpha = 0.05$ is most common, just as a probability of error of 0.05 in interval estimation is most common (that is, 95% confidence intervals). However, this number may be too high when a decision has serious implications. For example, suppose a convicted defendant gets the death penalty. Then, if a defendant is actually innocent, we would hope that the probability of conviction is smaller than 0.05.

Although we don't know if the decision in a particular test is correct, we justify the method in terms of the long-run proportions of Type I and Type II errors. We'll learn how to calculate P(Type II error) later in the chapter.

Probability of a Type I error

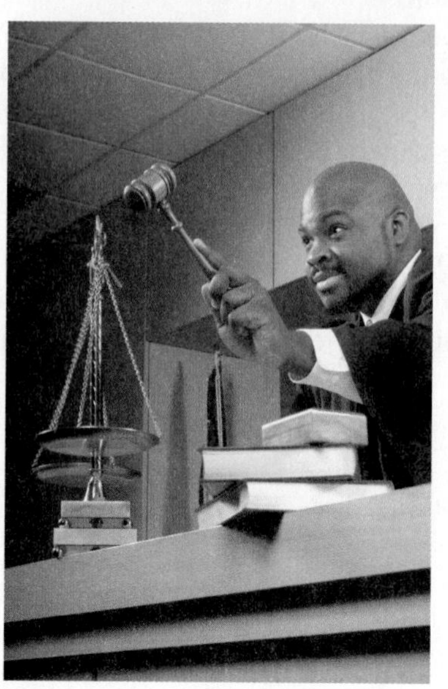

Example 10

Type I Errors in Legal Verdicts

Picture the Scenario

In an ideal world, Type I or Type II errors would not occur. In practice, however, whether in significance tests or in applications such as courtroom trials or medical diagnoses, errors do happen. It can be surprising and disappointing how often they occur, as we saw in the diagnostic testing examples of Chapter 5. Likewise, we've all read about defendants who were given the death penalty but were later determined to be innocent, but we don't have reliable information about how often this occurs.

Question to Explore

When we make a decision, why don't we use an extremely small probability of Type I error such as $\alpha = 0.000001$? For instance, why don't we make it almost impossible to convict someone who is really innocent?

Think It Through

When we make α smaller in a significance test, we need a smaller P-value to reject H_0. It then becomes harder to reject H_0. But this means that it will also be harder even if H_0 is false. The stronger the evidence that is required to convict someone, the more likely it becomes that we will fail to convict defendants who are actually guilty. In other words, the smaller we make the probability of Type I error, the *larger* the probability of Type II error becomes (that is, failing to reject H_0 even though it is false).

If we tolerate only an extremely small chance of a Type I error (such as $\alpha = 0.000001$), then the test may be unlikely to reject the null hypothesis even if it is false—for instance, unlikely to convict someone even if they are guilty. In fact, some of our laws are set up to make Type I errors very unlikely, and as a consequence some truly guilty individuals are not punished for their crimes.

Insight

This reasoning reflects a fundamental relation between the probabilities of the two types of errors, for a given sample size *n*:

As P(Type I Error) Goes Down, P(Type II Error) Goes Up

The two probabilities are inversely related.

Try Exercise 9.49, part c

Except in the final section of this chapter, we will not calculate the probability of a Type II error. This calculation can be complex. In practice, to make a decision in a test, we only need to set the probability of Type I error, which is the significance level α.

These days, most research articles merely report the P-value rather than a decision about whether to reject H_0. From the P-value, readers can see the strength of evidence against H_0 and make their own decisions.

Activity 1

Why Is 0.05 Commonly Used as a Significance Level?

A crossover study compares a drug for children who suffer from migraine headaches with a placebo. The study observed each child at two times when he or she had a migraine headache. The child received the drug at one time and a placebo at the other time. Let p denote the probability that the pain relief is better with the drug. You will decide whether or not you can reject H_0: $p = 0.50$ in favor of H_a: $p \neq 0.50$. Ahead of time, you have no idea whether the drug will be better, or worse, than the placebo.

Consider the following:

- The first child does better with the placebo. Would you reject H_0?
- The second child also does better with the placebo. Would you now reject H_0?
- The third child also does better with the placebo. Would you now reject H_0?
- The fourth child also does better with the placebo. Would you now reject H_0?
- The fifth child also does better with the placebo. Are you ready yet to reject H_0?

If you are like many people, by the time you see the fifth straight success for the placebo over the drug, you are willing to predict that the placebo is better. If the null hypothesis that $p = 0.50$ is actually true, then by the binomial distribution the probability this happens is $(0.50)^5 = 1/32 = 0.03$. For a two-sided test, this result gives a P-value $= 2(0.03) = 0.06$, close to 0.05. So, for many people, it takes a P-value near 0.05 before they feel there is enough evidence to reject a null hypothesis. This may be one reason the significance level of 0.05 has become common over the years in a wide variety of disciplines that use significance tests.

Try Exercise 9.125

9.4 Practicing the Basics

9.42 Dr. Dog In the experiment in Example 4, we got a P-value = 0.000 for testing H_0: $p = 1/7$ about dogs' ability to diagnose urine from bladder cancer patients.

a. For the significance level 0.05, what decision would you make?

b. If you made an error in part a, what type of error was it? Explain what the error means in context of the Dr. Dog experiment.

9.43 Error probability A significance test about a proportion is conducted using a significance level of 0.05. The test statistic equals 2.58. The P-value is 0.01.

a. If H_0 were true, for what probability of a Type I error was the test designed?

b. If this test resulted in a decision error, what type of error was it?

9.44 Astrology errors Example 3, in testing H_0: $p = 1/3$ against H_a: $p > 1/3$, analyzed whether astrologers could predict the correct personality chart (out of three possible ones) for a given horoscope better than by random guessing. In the words of that example, what would be (a) a Type I error and (b) a Type II error?

9.45 Anorexia errors Example 8 tested a therapy for anorexia, using hypotheses H_0: $\mu = 0$ and H_a: $\mu \neq 0$ about the population mean weight change μ. In the words of that example, what would be (a) a Type I error and (b) a Type II error?

9.46 Anorexia decision Refer to the previous exercise. When we test H_0: $\mu = 0$ against H_a: $\mu > 0$, we get a P-value of 0.02.

a. What would the decision be for a significance level of 0.05? Interpret in context.

b. If the decision in part a is in error, what type of error is it?

c. Suppose the significance level were instead 0.01. What decision would you make, and if it is in error, what type of error is it?

9.47 Errors in the courtroom Consider the test of H_0: The defendant is not guilty against H_a: The defendant is guilty.

a. Explain in context the conclusion of the test if H_0 is rejected.

b. Describe the consequence of a Type I error.

c. Explain in context the conclusion of the test if you fail to reject H_0.

d. Describe the consequence of a Type II error.

9.48 Errors in medicine Consider the test of H_0: The new drug is safe against H_a: the new drug is not safe.

a. Explain in context the conclusion of the test if H_0 is rejected.

b. Describe the consequence of a Type I error.

c. Explain in context the conclusion of the test if you fail to reject H_0.

d. Describe the consequence of a Type II error.

9.49 Decision errors in medical diagnostic testing Consider
TRY medical diagnostic testing, such as using a mammogram to detect if a woman may have breast cancer. Define the null hypothesis of no effect as the patient does not have the disease. Define rejecting H_0 as concluding that the

patient has the disease. See the table for a summary of the possible outcomes:

Medical Diagnostic Testing

	Medical Diagnosis	
Disease	**Negative**	**Positive**
No (H_0)	Correct	Type I error
Yes (H_a)	Type II error	Correct

a. When a radiologist interprets a mammogram, explain why a Type I error is a false positive, predicting that a woman has breast cancer when actually she does not.

b. A Type II error is a false negative. What does this mean, and what is the consequence of such an error to the woman?

c. A radiologist wants to decrease the chance of telling a woman that she may have breast cancer when actually she does not. Consequently, a positive test result will be reported only when there is *extremely* strong evidence that breast cancer is present. What is the disadvantage of this approach?

9.50 Detecting prostate cancer Refer to the previous exercise about medical diagnoses. A *New York Times* article (February 17, 1999) about the PSA blood test for detecting prostate cancer stated: "The test fails to detect prostate cancer in 1 in 4 men who have the disease."

a. For the PSA test, explain what a Type I error is, and explain the consequence to a man of this type of error.

b. For the PSA test, what is a Type II error? What is the consequence to a man of this type of error?

c. To which type of error does the probability of 1 in 4 refer?

d. The article also stated that if you receive a positive result, the probability that you do not actually have prostate cancer is 2/3. Explain the difference between this and the conditional probability of a Type I error, given that you do not actually have prostate cancer.

9.51 Which error is worse? Which error, Type I or Type II, would usually be considered more serious for decisions in the following tests? Explain why.

a. A trial to test a murder defendant's claimed innocence, when conviction results in the death penalty.

b. A medical diagnostic procedure, such as a mammogram.

9.5 Limitations of Significance Tests

Chapters 8 and 9 have presented the two primary methods of statistical inference—confidence intervals and significance testing. We will use both of these methods throughout the rest of the book. Of the two methods, confidence intervals can be more useful, for reasons we'll discuss in this section. Significance tests have more potential for misuse. We'll now summarize the major limitations of significance tests.

Statistical Significance Does Not Mean Practical Significance

When we conduct a significance test, its main relevance is studying whether the true parameter value is

- above, or below, the value in H_0, and
- sufficiently different from the value in H_0 to be of practical importance.

A significance test gives us information about whether or not the parameter differs from the H_0 value and its direction from that value, but we'll see now that it does not tell us about the practical significance (or importance) of the finding.

There is an important distinction between *statistical significance* and *practical significance.* A small P-value, such as 0.001, is highly statistically significant, giving strong evidence against H_0. It does not, however, imply an *important* finding in any practical sense. The small P-value means that if H_0 were true, the observed data would be unusual. It does not mean that the true value of the parameter is far from the null hypothesis value in practical terms. In particular, whenever the sample size is large, small P-values can occur when the point estimate is near the parameter value in H_0, as the following example shows.

Statistical significance and practical significance

Example 11

Political Conservatism and Liberalism in America

Picture the Scenario

Where do Americans say they fall on the conservative–liberal political spectrum? The General Social Survey asks, "I'm going to show you a seven-point scale on which the political views that people might hold are arranged from extremely liberal, point 1, to extremely conservative, point 7. Where would you place yourself on this scale?" Table 9.8 shows the scale and the distribution of 1933 responses for the GSS in 2008.

Table 9.8 Responses of 1933 Subjects on a Seven-Point Scale of Political Views

Category	Count
1. Extremely liberal	69
2. Liberal	240
3. Slightly liberal	221
4. Moderate, middle of road	740
5. Slightly conservative	268
6. Conservative	327
7. Extremely conservative	68

This categorical variable has seven categories that are ordered in terms of degree of liberalism or conservatism. Categorical variables that have *ordered* categories are called **ordinal variables**. Sometimes we treat an ordinal variable in a quantitative manner by assigning scores to the categories and summarizing the data by the mean. This summarizes whether observations gravitate toward the conservative or the liberal end of the scale. For the category scores of 1 to 7, as in Table 9.8, a mean of 4.0 corresponds to the moderate outcome. A mean below 4 shows a propensity toward liberalism, and a mean above 4 shows a propensity toward conservatism. The 1933 observations in Table 9.8 have a mean of 4.11 and a standard deviation of 1.43.

In Words

An ordinal variable is a categorical variable for which the categories are ordered from low to high in some sense. Table 9.8 lists categories of political views from low to high in terms of conservatism.

Questions to Explore

a. Do these data indicate that the population has a propensity toward liberalism or toward conservatism? Answer by conducting a significance test that compares the population mean to the moderate value of 4.0, by testing H_0: $\mu = 4.0$ against H_a: $\mu \neq 4.0$.

b. Does this test show (i) statistical significance? (ii) practical significance?

Think It Through

a. For the sample data, the standard error $se = s/\sqrt{n} = 1.43/\sqrt{1933} = 0.0325$. The test statistic for H_0: $\mu = 4.0$ equals

$$t = (\bar{x} - \mu_0)/se = (4.11 - 4.00)/0.0325 = 3.38.$$

Its two-sided P-value is 0.001. There is strong evidence that the population mean differs from 4.0. The data indicates that true mean exceeds 4.0, with the true mean falling on the conservative side of moderate.

b. Although the P-value is small, on a scale of 1 to 7, the sample mean of 4.11 is close to the value of 4.0 in H_0. It is only 11% of the distance from the moderate score of 4.0 to the slightly conservative score of 5.0. Although the difference of 0.11 between the sample mean of 4.11 and the null hypothesis mean of 4.0 is highly significant statistically, this difference is small in practical terms. We'd regard a mean of 4.11 as moderate on this 1 to 7 scale. In summary, there's statistical significance but not practical significance.

Insight

As we also saw in Example 9, with large samples P-values can be small even when the sample estimate falls near the parameter value in H_0. The P-value measures the extent of evidence about H_0, not how far the true parameter value happens to be from H_0. Always inspect the difference between the sample estimate and the null hypothesis value to gauge the practical implications of a test result.

Try Exercise 9.53

Significance Tests Are Less Useful Than Confidence Intervals

Although significance tests are useful, most statisticians believe that this method has been overemphasized in research.

- A significance test merely indicates whether the particular parameter value in H_0 (such as $\mu = 0$) is plausible.

When a P-value is small, the significance test indicates that the hypothesized value is not plausible, but it tells us little about which potential parameter values *are* plausible.

- A confidence interval is more informative because it displays the entire set of believable values.

A confidence interval shows if H_0 may be badly false by showing if the values in the interval are far from the H_0 value. It helps us to determine whether or not the difference between the true value and the H_0 value has practical importance.

Let's illustrate with Example 11 and the 1 to 7 scale for political beliefs. A 95% confidence interval for μ is $\bar{x} \pm 1.96(se) = 4.11 \pm 1.96(0.0325)$, or (4.05, 4.17). We can conclude that the difference between the population mean and the moderate score of 4.0 is small. Figure 9.13 illustrates. Although the P-value of 0.001 provided strong evidence against H_0: $\mu = 4.0$, in practical terms the confidence interval shows that H_0 is not wrong by much.

In contrast, if $\bar{x}$ had been 6.11 (instead of 4.11), the 95% confidence interval would equal (6.05, 6.17). This confidence interval indicates a substantial difference from 4.0, the mean response being near the conservative score rather than near the moderate score.

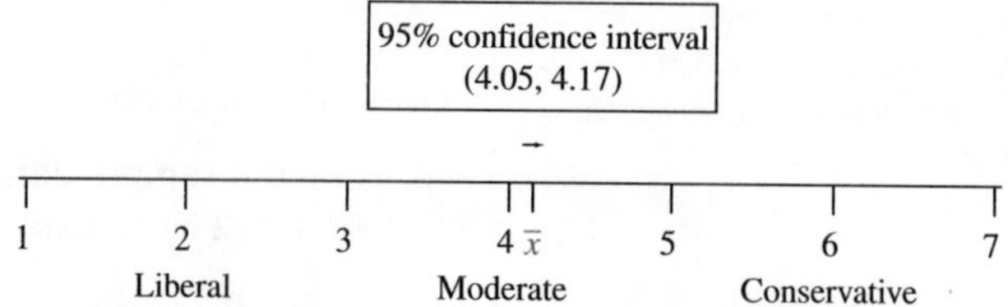

▲ **Figure 9.13 Statistical Significance but not Practical Significance.** In testing H_0: $\mu = 4.0$, the P-value = 0.001, but the confidence interval of (4.05, 4.17) shows that μ is very close to the H_0 value of 4.0. **Question** For H_0: $\mu = 4.0$, does a sample mean of 6.11 and confidence interval of (6.05, 6.17) indicate (a) statistical significance? (b) practical significance?

Misinterpretations of Results of Significance Tests

Unfortunately, results of significance tests are often misinterpreted. Here are some important comments about significance tests, some of which we've already discussed:

- **"Do not reject H_0" does not mean "Accept H_0."** If you get a P-value above 0.05 when the significance level is 0.05, you cannot conclude that H_0 is correct. We can never accept a single value, which H_0 contains, such as $p = 0.50$ or $\mu = 0$. A test merely indicates whether or not a particular parameter value is plausible. A confidence interval shows that there is a *range* of plausible values, not just a single one.
- **Statistical significance does not mean practical significance.** A small P-value does not tell us if the parameter value differs by much in practical terms from the value in H_0.
- **The P-value cannot be interpreted as the probability that H_0 is true.** The P-value is

  P(test statistic takes observed value or beyond in tails | H_0 true),
  NOT P(H_0 true | observed test statistic value).

  We've been calculating probabilities about test statistic values, not about parameters. It makes sense to find the probability that a test statistic takes a value in the tail, but the probability that a population mean = 0 does not make sense because probabilities do not apply to parameters (since parameters are fixed numbers and not random variables). The null hypothesis H_0 is either true or not true, and we simply do not know which.[11]

Recall

P(A|B) denotes the conditional probability of event A, given event B. To find a P-value, we condition on H_0 being true (that is, we presume it is true), rather than find the probability it is true. ◀

- **It is misleading to report results only if they are "statistically significant."** Some research journals have the policy of publishing results of a study only if the P-value ≤ 0.05. Here's a danger of this policy: If there truly is no effect, but 20 researchers independently conduct studies about it, we would expect about $20(0.05) = 1$ of them to obtain significance at the 0.05 level merely by chance. (When H_0 is true, about 5% of the time we get a P-value below 0.05 anyway.) If that researcher then submits results to a journal but the other 19 researchers do not, the article published will be a Type I error—reporting an effect when there really is not one. The popular media may then also report on this study, causing the general public to hear about an effect that does not actually exist.
- **Some tests may be statistically significant just by chance.** You should never scan pages and pages of computer output for results that are statistically significant and report only those. If you run 100 tests, even if all the null hypotheses are correct, you would expect to get P-values of 0.05 or less about $100(0.05) = 5$ times. Keep this in mind and be skeptical of reports of significance that might merely reflect ordinary random variability. For instance, suppose an article reports an unusually high rate of a rare type of cancer in your town. It could be due to some cause such as air or water pollution. However, if researchers found this by looking at data for *all* towns and cities nationwide, it could also be due to random variability. Determining which is true may not be easy.
- **True effects may not be as large as initial estimates reported by the media.** Even if a statistically significant result is a true effect, the true effect may be smaller than suggested in the first article about it. For instance, often several researchers perform similar studies, but the results that get attention are the most extreme ones. This sensationalism may come about because the researcher who is the first to publicize the result is the one who got the most impressive sample result, perhaps way out in the tail of the sampling

---

[11] It is possible to find probabilities about parameter values using an approach called *Bayesian statistics*, but this requires extra assumptions and is beyond the scope of this text.

No effect | True effect | Reported effect

distribution of all the possible results. Then the study's estimate of the effect may be greater than later research shows it to be. See the margin figure and the next example.

Type I errors

Example 12

Medical "Discoveries"

Picture the Scenario

What can be done with heart attack victims to increase their chance of survival? In the 1990s, trials of a clot-busting drug called anistreplase suggested that it doubled the chance of survival. Likewise, another study estimated that injections of magnesium could double the chance of survival. However, a much larger study of heart attack survival rates among 58,000 patients indicated that this optimism was premature. The actual effectiveness of anistreplase seemed to be barely half that estimated by the original trial, and magnesium injections seemed to have no effect at all.[12] The anistreplase finding is apparently an example of a true effect not being as large as the initial estimate, and the report from the original magnesium study may well have been a Type I error.

Question to Explore

In medical studies, suppose that a true effect exists only 10% of the time. Suppose also that when an effect truly exists, there's a 50% chance of making a Type II error and failing to detect it. These were the hypothetical percentages used in an article in a medical journal.[13] The authors noted that many medical studies have a high Type II error rate because they are not able to use a large sample size. Assuming these rates, could a substantial percentage of medical "discoveries" actually be Type I errors? Given that H_0 is rejected, approximate the P(Type I error) by considering a tree diagram of what you would expect to happen with 1000 medical studies that test various hypotheses.

Think It Through

Figure 9.14 is a tree diagram of what's expected. A true effect exists 10% of the time, or in 100 of the 1000 studies. We do not get a small enough P-value to detect this true effect 50% of the time, that is, in 50 of these 100 studies. No effect will be reported for the other 50 of the 100 that do truly have an

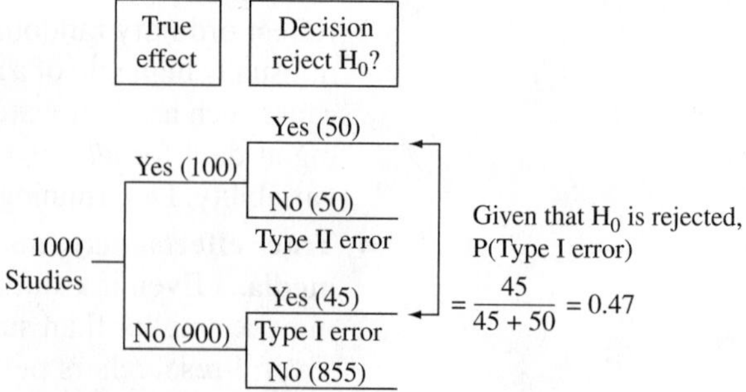

▲ **Figure 9.14 Tree Diagram of Results of 1000 Hypothetical Medical Studies.** This assumes that a true effect exists 10% of the time and that there's a 50% chance of a Type II error when an effect truly does exist.

[12] "The Great Health Hoax," by R. Matthews, in *The Sunday Telegraph,* September 13, 1998.

[13] By J. Sterne, G. Smith, and D.R. Cox, *BMJ,* vol. 322, pp. 226–231 (2001).

effect. Now, for the 900 cases in which there truly is no effect, with the usual significance level of 0.05, we expect 5% of the 900 studies (that is, 45 studies) to incorrectly reject the null hypothesis and predict that there is an effect. So, of the original 1000 studies, we expect 50 to report an effect that is truly there, but we also expect 45 to report an effect that does not really exist. If the assumptions are reasonable, then a proportion of $45/(45+50) = 0.47$ of medical studies that report effects (that is, reject H_0) are actually reporting Type I errors. In summary, nearly half the time when an effect is reported, there actually is no effect in the population!

Insight

Be skeptical when you hear reports of new medical advances. The true effect may be weaker than reported, or there may actually be no effect at all.

Try Exercise 9.57

Ethics in Data Analysis

Statistics, like any field, has standards for ethical behavior. The comments just made have implications about proper and improper use of significance tests. For example, when you conduct a study, it's not ethical to perform lots and lots of significance tests but only report results when the P-value is small. Even when *all* null hypotheses tested are true, just by chance small P-values occasionally occur. You should report *all* the analyses you have conducted. In fact, before you collect the data, you should formulate a research plan that sets out exactly what analyses you will conduct.

The discussion in Section 9.2 about how to decide between a one-sided and a two-sided test mentioned that it is not ethical to choose a one-sided H_a after seeing the direction in which the data fall. This is because the sampling distribution used in a significance test presumes only that H_0 is true, and it allows sample data in each direction. If we first looked at the data, the valid sampling distribution would be one that is conditioned on the extra information about the direction of the sample data. For example, if the test statistic is positive, the proper sampling distribution would not have possible negative values. Again, before you collect the data, your research plan should set out the details of the hypotheses you plan to test.

Finally, remember that tests and confidence intervals use *sample* data to make inferences about *populations.* If you have data for an entire population, statistical inference is not necessary. For instance, if your college reports that the class of entering freshmen had a mean high school GPA of 3.60, there is no need to perform a test or construct a confidence interval. You already know the mean for that population.

On the Shoulders of... Jerzy Neyman and the Pearsons

Jerzy Neyman and Egon Pearson. Neyman and Pearson developed statistical theory for making decisions about hypotheses.

How can you build a framework for making decisions?

The methods of confidence intervals and hypothesis testing were introduced in a series of articles beginning in 1928 by Jerzy Neyman (1894–1981) and Egon Pearson (1895–1980). Neyman emigrated from Poland to England and then to the United States, where he established a top-notch statistics department at the University of California at Berkeley. He helped develop the theory of statistical inference, and he applied the theory to scientific questions in a variety of areas, such as agriculture, astronomy, biology, medicine, and weather modification. For instance, late in his career, Neyman's analysis of data from several randomized experiments showed that cloud-seeding can have a considerable effect on rainfall.

Much of Neyman's theoretical research was done with Egon Pearson, a professor at University College, London.

Pearson's father, Karl Pearson, had developed one of the first statistical tests in 1900 to study various hypotheses, including whether the outcomes on a roulette wheel were equally likely. (We'll study his test, the *chi-squared test*, in Chapter 11.) Neyman and the younger Pearson developed the decision-making framework that introduced the two types of errors and the most powerful significance tests for various hypotheses.

9.5 Practicing the Basics

9.52 Misleading summaries? Two researchers conduct separate studies to test H_0: $p = 0.50$ against H_a: $p \neq 0.50$, each with $n = 400$.

a. Researcher A gets 220 observations in the category of interest, and $\hat{p} = 220/400 = 0.550$ and test statistic $z = 2.00$. Show that the P-value $= 0.046$ for Researcher A's analysis.

b. Researcher B gets 219 in the category of interest, and $\hat{p} = 219/400 = 0.5475$ and test statistic $z = 1.90$. Show that the P-value $= 0.057$ for Researcher B's analysis.

c. Using $\alpha = 0.05$, indicate in each case from part a and part b whether the result is "statistically significant." Interpret.

d. From part a, part b, and part c, explain why important information is lost by reporting the result of a test as "P-value ≤ 0.05" versus "P-value > 0.05," or as "reject H_0" versus "do not reject H_0," instead of reporting the actual P-value.

e. Show that the 95% confidence interval for p is (0.501, 0.599) for Researcher A and (0.499, 0.596) for Researcher B. Explain how this method shows that, in practical terms, the two studies had very similar results.

9.53 Practical significance TRY A study considers if the mean score on a college entrance exam for students in 2010 is any different from the mean score of 500 for students who took the same exam in 1985. Let μ represent the mean score for all students who took the exam in 2010. For a random sample of 25,000 students who took the exam in 2010, $\bar{x} = 498$ and $s = 100$.

a. Show that the test statistic is $t = -3.16$.

b. Find the P-value for testing H_0: $\mu = 500$ against H_a: $\mu \neq 500$.

c. Explain why the test result is statistically significant but not practically significant.

9.54 Effect of *n* Example 11 analyzed political conservatism and liberalism in the United States. Suppose that the sample mean of 4.11 and sample standard deviation of 1.43 were from a sample size of only 25, rather than 1933.

a. Find the test statistic.

b. Find the P-value for testing H_0: $\mu = 4.0$ against H_a: $\mu \neq 4.0$. Interpret.

c. Show that a 95% confidence interval for μ is (3.5, 4.7).

d. Together with the results of Example 11, explain what this illustrates about the effect of sample size on (i) the size of the P-value (for a given mean and standard deviation) and (ii) the width of the confidence interval.

9.55 Fishing for significance A marketing study conducts 60 significance tests about means and proportions for several groups. Of them, 3 tests are statistically significant at the 0.05 level. The study's final report stresses only the tests with significant results, not mentioning the other 57 tests. What is misleading about this?

9.56 Selective reporting In 2004, New York Attorney General Eliot Spitzer filed a lawsuit against GlaxoSmithKline pharmaceutical company, claiming that the company failed to publish results of one of their studies that showed that an antidepressant drug (Paxil) may make adolescents more likely to commit suicide. Partly as a consequence, editors of 11 medical journals agreed to a new policy to make researchers and companies register all clinical trials when they begin, so that negative results cannot later be covered up. The *International Journal of Medical Journal Editors* wrote, "Unfortunately, selective reporting of trials does occur, and it distorts the body of evidence available for clinical decision-making." Explain why this controversy relates to the argument that it is misleading to report results only if they are "statistically significant." (*Hint:* See the subsection of this chapter on misinterpretations of significance tests.)

9.57 How many medical discoveries are Type I errors? TRY Refer to Example 12. Using a tree diagram, given that H_0 is rejected, approximate P(Type I error) under the assumption that a true effect exists 20% of the time and that there's a 30% chance of a Type II error.

9.58 Interpret medical research studies

a. An advertisement by Schering Corp. in 1999 for the allergy drug Claritin mentioned that in a clinical trial, the proportion who showed symptoms of nervousness was not significantly greater for patients taking Claritin than for patients taking placebo. Does this mean that the population proportion having nervous symptoms is exactly the same using Claritin and using placebo? How would you explain this to someone who has not studied statistics?

b. An article in the medical journal *BMJ* (by M. Petticrew et al., published November 2002) found no evidence to back the commonly held belief that a positive attitude can lengthen the lives of cancer patients. The authors noted that the studies that had indicated a benefit from some coping strategies tended to be smaller studies with weaker designs. Using this example and the text discussion, explain why you need to have some skepticism when you hear that new research suggests that some therapy or drug has an impact in treating a disease.

9.6 The Likelihood of a Type II Error (Not Rejecting H_0, Even Though It's False)

The probability of a Type I error is the significance level α of the test. Given that H_0 is true, when $\alpha = 0.05$, the probability of rejecting H_0 equals 0.05.

Given that H_0 is false, a Type II error results from *not* rejecting H_0. This probability has more than one value because H_a contains a range of possible values for the parameter. Each value in H_a has its own probability of a Type II error. Let's see how to find the probability of a Type II error at a particular value.

Example 13

Finding P (Type II error)

Part of a Study Design

Picture the Scenario

Examples 1, 3, and 5 discussed an experiment to test astrologers' predictions. For each person's horoscope, an astrologer must predict which of three personality charts is the actual one. Let p denote the probability of a correct prediction. Consider the test of H_0: $p = 1/3$ (astrologers' predictions are like random guessing) against H_a: $p > 1/3$ (better than random guessing), using the 0.05 significance level. Suppose an experiment plans to use $n = 116$ people, as this experiment did.

Questions to Explore

a. For what values of the sample proportion can we reject H_0?

b. The National Council for Geocosmic Research claimed that p would be 0.50 or higher. If truly $p = 0.50$, for what values of the sample proportion would we make a Type II error, failing to reject H_0 even though it's false?

c. If truly $p = 0.50$, what is the probability that a significance test based on this experiment will make a Type II error?

Think It Through

a. For testing H_0: $p = 1/3$ with $n = 116$, the standard error for the test statistic is

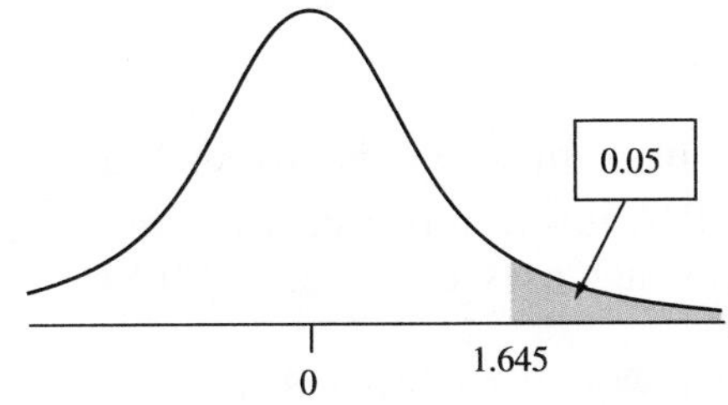

$$se_0 = \sqrt{p_0(1-p_0)/n} = \sqrt{[(1/3)(2/3)]/116} = 0.0438.$$

For H_a: $p > 1/3$, a test statistic of $z = 1.645$ has a P-value (right-tail probability) of 0.05. If $z \geq 1.645$, the P-value is ≤ 0.05 and we can reject H_0. That is, we reject H_0 when $\hat{p}$ falls at least 1.645 standard errors above $p_0 = 1/3$,

$$\hat{p} \geq 1/3 + 1.645(se_0) = 1/3 + 1.645(0.0438) = 0.405.$$

Figure 9.15 (on following page) shows the sampling distribution of $\hat{p}$ and this rejection region. The figure is centered at 1/3, because the test statistic is calculated and the rejection region is formed presuming that H_0 is correct.

b. When H_0 is false, a Type II error occurs when we fail to reject H_0. From part a and Figure 9.15, we do not reject H_0 if $\hat{p} < 0.405$.

c. If the true value of p is 0.50, then the true sampling distribution of $\hat{p}$ is centered at 0.50, as Figure 9.16 shows. The probability of a Type II error is the probability that $\hat{p} < 0.405$ when $p = 0.50$.

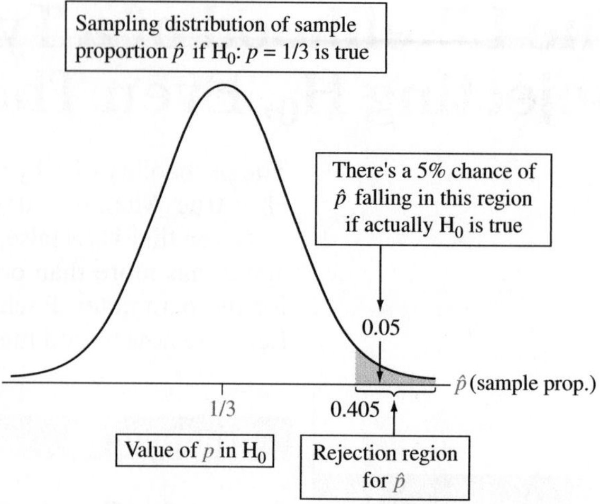

▲ **Figure 9.15 For Sample Proportion $\hat{p}$ Above 0.405, Reject H_0: $p = 1/3$ Against H_a: $p > 1/3$ at the 0.05 Significance Level.** When the true $p > 1/3$, a Type II error occurs if $\hat{p} < 0.405$, since then the P-value > 0.05 and we do not reject H_0. **Question** Why does each possible value of p from H_a have a separate probability of a Type II error?

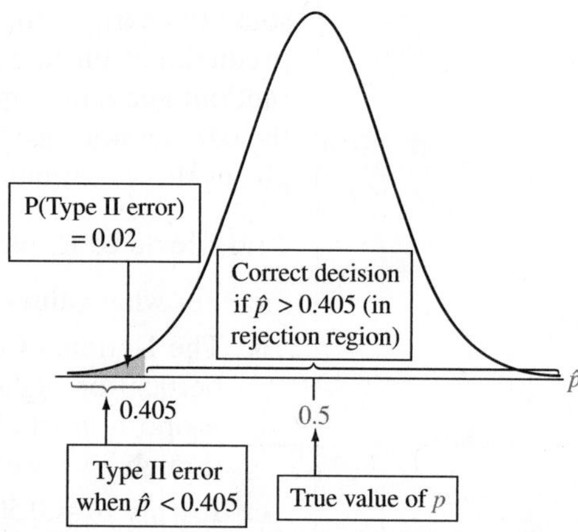

▲ **Figure 9.16 Calculation of P(Type II Error) when True Parameter $p = 0.50$.** A Type II error occurs if the sample proportion $\hat{p} < 0.405$. **Question** If the value of p decreases to 0.40, will the probability of Type II error decrease or increase?

When $p = 0.50$, the standard error of $\hat{p}$ for a sample size of 116 is $\sqrt{0.50(0.50)/116} = 0.0464$. (Note that this differs a bit from the standard error for the test statistic, which uses 1/3 instead of 0.50 for p.)

For a normal sampling distribution with mean 0.50 and standard error 0.0464, the $\hat{p}$ value of 0.405 has a z-score of

$$z = (0.405 - 0.50)/0.0464 = -2.04.$$

Using Table A or software, the left-tail probability below -2.04 for the standard normal distribution equals 0.02. In summary, when $p = 0.50$ the probability of making a Type II error and failing to reject H_0: $p = 1/3$ is only 0.02.

Figure 9.17 shows the two figures together that we used in our reasoning. The normal distribution with mean 1/3 was used to find the rejection region, based on what we expect for $\hat{p}$ when H_0: $p = 1/3$ is true. The normal distribution with mean 0.50 was used to find the probability that $\hat{p}$ fails to fall in the rejection region even though $p = 0.50$ (that is, a Type II error occurs).

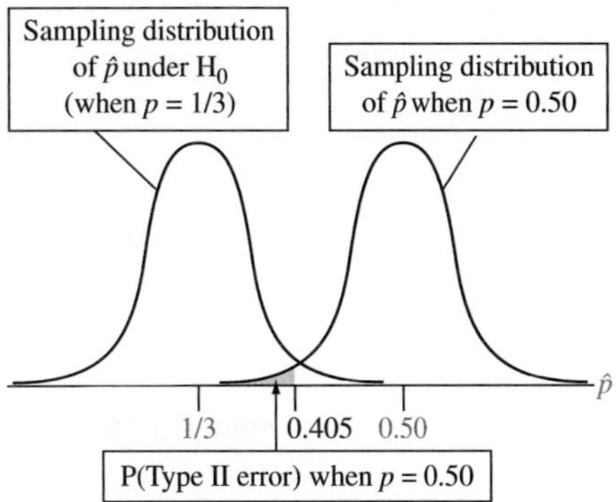

▲ **Figure 9.17** Sampling Distributions of $\hat{p}$ When H_0: $p = 1/3$ Is True and When $p = 0.50$. **Question** Why does the shaded area under the left tail of the curve for the case $p = 0.50$ represent P(Type II error) for that value of p?

Insight

If astrologers truly had the predictive power claimed by the National Council for Geocosmic Research, the experiment would have been very likely to detect it (probability $1 - 0.02 = 0.98$).

Try Exercise 9.60

The probability of a Type II error increases when the true parameter value moves closer to H_0. To verify this, try to find the probability of a Type II error when $p = 0.40$ instead of 0.50. You should get P(Type II error) = 0.54. (Remember to recompute the exact standard error, now using 0.40 for p.) If an astrologer can predict better than random guessing but not *much* better, we may not detect it with this experiment. Figure 9.18 plots P(Type II error) for various values of p above 1/3. The farther the parameter value falls from the number in H_0, the less likely a Type II error.

SUMMARY: For a fixed significance level α, P(Type II error) decreases

- as the parameter value moves farther into the H_a values and away from the H_0 value
- as the sample size increases

Also, recall that P(Type II error) increases as α decreases. One reason that extremely small values, such as $\alpha = 0.001$, are not common is that P(Type II error) is then too high. We may be unlikely to reject H_0, even though it is false.

Before conducting a study, researchers should find P(Type II error) for the size of effect they want to be able to detect. If P(Type II error) is high, it may not be worth conducting the study unless they can use a larger sample size and lower the probability. They won't know the value of the parameter, so they won't know the actual P(Type II error). It may be large if n is small and if the true parameter value is not far from the value in H_0. This may be the reason a particular significance test does not obtain a small P-value and reject H_0.

The Power of a Test

When H_0 is false, you want the probability of rejecting it to be high. The probability of rejecting H_0 when it is false is called the **power** of the test. For a particular value of the parameter from the range of alternative hypothesis values,

$$\text{Power} = 1 - \text{P(Type II error)}$$

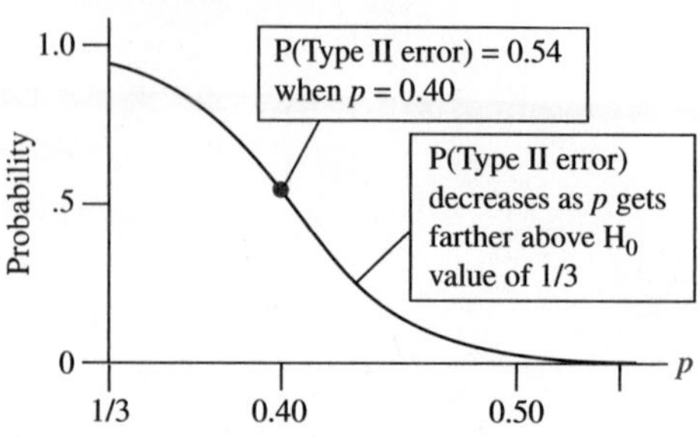

▲ **Figure 9.18 Probability of a Type II Error for Testing H_0: $p = 1/3$ Against H_a: $p > 1/3$.** This is plotted for the values of p in H_a when the significance level $= 0.05$.
Question If the sample size n increases, how do you think this curve would change?

In Example 13, for instance, P(Type II error) $= 0.02$ when $p = 0.50$. Therefore, the power of the test at $p = 0.50$ is $1 - 0.02 = 0.98$. The higher the power, the better, so a test power of 0.98 is quite good. Figure 9.18 illustrates power values related to Example 13.

In practice, it is ideal for studies to have high power while using a relatively small significance level, such as 0.05 or 0.01. Before granting research support, many agencies (such as the National Institutes of Health) expect research scientists to show that for the planned study reasonable power (usually, at least 0.80) exists at values of the parameter that are considered practically significant. For a fixed α, the power increases in the same cases that P(Type II error) decreases, namely as the sample size increases and as the parameter value moves farther into the H_a values away from the H_0 value.

Example 14

Power of the test

The TT Experiment

Picture the Scenario

In Example 6 on the therapeutic touch (TT) experiment, the data did not support the TT practitioners' claim to be able to detect a human energy field. The P-value was not small for testing H_0: $p = 0.50$ against H_a: $p > 0.50$, where p is the probability of a correct prediction about which hand was near the researcher's hand. The medical journal article about the study stated, "The statistical power of this experiment was sufficient to conclude that if TT practitioners could reliably detect a human energy field, the study would have demonstrated this." For the test of H_0: $p = 0.50$ with one of the sets of trials, the power was reported as 0.96 if actually $p = 2/3$.

Questions to Explore

a. How should the probability of 0.96 be interpreted?
b. In context, what is a Type II error for this experiment?
c. If $p = 2/3$, what is the probability of committing a Type II error?

Think It Through

a. The power of 0.96 is the probability of correctly rejecting H_0 when it is false. If the actual probability of correct predictions by TT practitioners was 2/3, there was a 96% chance of data such that the significance test performed would reject H_0.

b. A Type II error occurs if we do not reject H_0: $p = 0.50$, when actually TT practitioners *can* predict correctly more than half the time. The consequence would be to question the truthfulness of what they have been practicing for over 30 years, when they actually do have some ability.

c. If $p = 2/3$, the value of P(Type II error) is $1 -$ (Power at $p = 2/3$). This is $1 - 0.96 = 0.04$. A Type II error was unlikely if truly $p = 2/3$.

Insight

Based on the data, the researchers remained skeptical that TT practitioners could detect a human energy field. Because of the strong power at a value of p that TT practitioners claim is plausible, the researchers felt justified in concluding that "TT claims are groundless and further use of TT by health professionals is unjustified."

Try Exercise 9.64

Activity 2

Let's Simulate the Performance of Significance Tests

Let's get a feel for the two possible errors in significance tests. For a given population proportion value you can simulate many samples and perform a significance test for each. You can then check how often the tests make an incorrect inference. You can conduct the simulation using statistical software (such as MINITAB) or using a statistical applet that allows you to control the null hypothesis value, the true parameter value, the sample size, and the significance level.

Try this by going to the Hypothesis Tests for a Proportion applet on the text CD. Set the null hypothesis as $p = 1/3 = 0.33$, for a one-sided test (H_a: > 0.33) with sample size $n = 116$, a case Example 3 on the astrology experiment considered. At the menu, set the true proportion value to $p = 0.33$. Click *Simulate* and 100 samples will be generated. The applet calculates the resulting z statistic and P-value for the sample proportion from each sample. The applet summarizes the number of significance tests that reject the null hypothesis at the 0.05 and 0.01 significance levels. How many significance tests out of 100 tests rejected the null hypothesis at the 0.05 significance level? At the 0.01 significance level? How many would you expect at each significance level to reject the null, giving us an incorrect inference, a Type I error?

To get a feel for what happens "in the long run," do this simulation 10,000 times by clicking the *Simulate* button multiple times until you have a cumulative total of 10,000 significance tests. How many significance tests out of 10,000 tests incorrectly rejected the null at the 0.05 significance level? At the 0.01 significance level?

Next, change the value of p to 0.50, but keep the null hypothesis as $p = 0.33$, so H_0 is actually false. Perform a single significance test. Now if you get a P-value above 0.05 or 0.01 and fail to reject H_0, you made a Type II error, since H_0 is actually false. Perform the significance test 10,000 times. What percentage of times did you make a Type II error using a significance level of 0.05? By Example 13, this should happen only about 2% of the time.

Try Exercise 9.65

9.6 Practicing the Basics

9.59 Two sampling distributions A study is designed to test H_0: $p = 0.50$ against H_a: $p > 0.50$, taking a random sample of size $n = 100$, using significance level 0.05.

a. Show that the rejection region consists of values of $\hat{p} > 0.582$.

b. Sketch a single picture that shows (i) the sampling distribution of $\hat{p}$ when H_0 is true and (ii) the sampling distribution of $\hat{p}$ when $p = 0.60$. Label each sampling distribution with its mean and standard error, and highlight the rejection region.

c. Find P(Type II error) when $p = 0.60$.

9.60 Gender bias in selecting managers Exercise 9.19 tested TRY the claim that female employees were passed over for management training in favor of their male colleagues. Statewide, the large pool of more than 1000 eligible employees who can be tapped for management training is 40% female and 60% male. Let p be the probability of selecting a female for any given selection. For testing H_0: $p = 0.40$ against H_a: $p < 0.40$ based on a random sample of 50 selections, using the 0.05 significance level, verify that:

a. A Type II error occurs if the sample proportion falls less than 1.645 standard errors below the null hypothesis value, which means that $\hat{p} > 0.286$.

b. When $p = 0.20$, a Type II error has probability 0.06.

9.61 Balancing Type I and Type II errors Recall that for the same sample size the smaller the probability of a Type I error, α, the larger the P(Type II error). Let's check this for Example 13. There we found P(Type II error) for testing H_0: $p = 1/3$ (astrologers randomly guessing) against H_a: $p > 1/3$ when actually $p = 0.50$, with $n = 116$. If we use $\alpha = 0.01$, verify that:

a. A Type II error occurs if the sample proportion falls less than 2.326 standard errors above the null hypothesis value, which means $\hat{p} < 0.435$.

b. When $p = 0.50$, a Type II error has probability 0.08. (By comparison, Example 13 found P(Type II error) = 0.02 when $\alpha = 0.05$, so we see that P(Type II error) increased when P(Type I error) decreased.)

9.62 P(Type II error) large when p close to H_0 For testing H_0: $p = 1/3$ (astrologers randomly guessing) against H_a: $p > 1/3$ with $n = 116$, Example 13 showed that P(Type II error) = 0.02 when $p = 0.50$. Now suppose that $p = 0.35$. Recall that P(Type I error) = 0.05.

a. Show that P(Type II error) = 0.89.

b. Explain intuitively why P(Type II error) is large when the parameter value is close to the value in H_0 and decreases as it moves farther from that value.

9.63 Type II error with two-sided H_a In Example 13 for testing H_0: $p = 1/3$ (astrologers randomly guessing) with $n = 116$ when actually $p = 0.50$, suppose we used H_a: $p \neq 1/3$. Then show that:

a. A Type II error occurs if $0.248 < \hat{p} < 0.419$.

b. The probability is 0.00 that $\hat{p} < 0.248$ and 0.96 that $\hat{p} > 0.419$.

c. P(Type II error) = 0.04.

9.64 Power of TT Consider Example 14 about the power of the test used in the TT experiment for testing H_0: $p = 0.50$ against H_a: $p > 0.50$, where p is the probability of a correct prediction about which hand was nearer the researcher's hand. In a significance test planned for a second set of trials, the power was reported as 0.98 at $p = 2/3$.

TRY

a. How should the probability of 0.98 be interpreted?

b. In context, what is a Type II error for this experiment?

c. What is the probability of committing a Type II error if $p = 2/3$?

9.65 Simulating Type II errors Refer to the simulation in Activity 2 at the end of the section.

TRY

a. Repeat the simulation, now assuming that actually $p = 0.45$. In a large number of simulations, what proportion of the time did you make a Type II error at the significance level of 0.05? What does theory predict for this proportion?

b. Do you think that the proportion of Type II errors will increase, or decrease, if actually $p = 0.50$? Check your intuition by conducting another simulation.

c. Refer to part a. Do you think that the proportion of Type II errors will increase, or decrease, if the sample size in each simulation is 50 instead of 116? Check your intuition by conducting another simulation.

Chapter Review

ANSWERS TO THE CHAPTER FIGURE QUESTIONS

Figure 9.1 A P-value of 0.01 gives stronger evidence against the null hypothesis. This smaller P-value (0.01 compared to 0.20) indicates it would be more unusual to get the observed sample data or a more extreme value if the null hypothesis is true.

Figure 9.2 It is the relatively large values of $\hat{p}$ (with their corresponding right-tail z-scores) that support the alternative hypothesis of $p > 1/3$.

Figure 9.3 Values more extreme than the observed test statistic value can be either relatively smaller or larger to support the alternative hypothesis $p \neq p_0$. These smaller or larger values are ones that fall farther in either the left or right tail.

Figure 9.4 A P-value = 0.00000002 is strong evidence against H_0.

Figure 9.5 The tail probability is found based on the direction stated in H_a. If H_a is $p > p_0$, the right tail is used. If H_a is $p < p_0$, the left tail is used.

Figure 9.7 The t-score indicates the number of standard errors $\bar{x}$ falls from the null hypothesis mean. The farther out a t-score falls in the tails, the farther the sample mean falls from the null hypothesis mean, providing stronger evidence against H_0.

Figure 9.8 For H_a: $\mu > 0$, the t test statistics providing stronger evidence against H_0 will be t statistics corresponding to the relatively large $\bar{x}$ values falling to the right of the hypothesized mean of 0. The t test statistics for these values of $\bar{x}$ are positive.

Figure 9.9 Answers will vary.

Figure 9.10 The standard error for the significance test is computed using p_0, the hypothesized proportion. The standard error for the confidence interval is computed using $\hat{p}$, the sample proportion. The values p_0 and $\hat{p}$ are not necessarily the same; therefore, there is not an exact equivalence.

Figure 9.11 The plots suggest that the population distribution of weight change is skewed to the right.

Figure 9.12 Using a significance level of 0.05, the rejection region consists of values of $z \geq 1.645$.

Figure 9.13 A sample mean of 6.11 indicates both statistical and practical significance.

Figure 9.15 A Type II error results from not rejecting H_0 even though H_0 is false. The probability of a Type II error has more than one value because H_a contains a range of possible values for the parameter p.

Figure 9.16 The probability of a Type II error will increase.

Figure 9.17 A correct decision for the case $p = 0.50$ is to reject H_0: $p = 1/3$. The shaded area under the left tail of the curve for the case $p = 0.50$ represents the probability we would not reject H_0: $p = 1/3$, which is the probability of an incorrect decision for this situation. This shaded area is the probability of committing a Type II error.

Figure 9.18 By increasing the sample size n, the probability of a Type II error decreases. The curve would more quickly approach the horizontal axis, indicating that the Type II error probability more rapidly approaches a probability of 0 as p increases.

CHAPTER SUMMARY

A significance test helps us to judge whether or not a particular value for a parameter is believable. Each significance test has five steps:

1. **Assumptions** The most important assumption for any significance test is that the data result from a *random sample* or *randomized experiment.* Specific significance tests make other assumptions, such as those summarized below for significance tests about means and proportions.
2. **Null and alternative hypotheses about the parameter** Null hypotheses have the form H_0: $p = p_0$ for a proportion and H_0: $\mu = \mu_0$ for a mean, where p_0 and μ_0 denote particular values, such as $p_0 = 0.4$ and $\mu_0 = 0$. The most common alternative hypothesis is **two-sided**, such as H_a: $p \neq 0.4$. **One-sided** hypotheses such as H_a: $p > 0.4$ and H_a: $p < 0.4$ are also possible.
3. **Test statistic** This measures how far the sample estimate of the parameter falls from the null hypothesis value. The z statistic for proportions and the t statistic for means have the form

$$\text{Test statistic} = \frac{\text{Parameter estimate} - \text{Null hypothesis value}}{\text{Standard error}}$$

This measures the number of standard errors that the parameter estimate ($\hat{p}$ or $\bar{x}$) falls from the null hypothesis value (p_0 or μ_0).

4. **P-value** This is a probability summary of the evidence that the data provide about the null hypothesis. It equals the probability that the test statistic takes a value like the observed one or even more extreme if the hypothesized value in H_0 is true.
   - It is calculated by presuming that H_0 is true.
   - The test statistic values that are "more extreme" depend on the alternative hypothesis. When H_a is two-sided, the P-value is a two-tail probability. When H_a is one-sided, the P-value is a one-tail probability.
   - When the P-value is small, the observed data would be unusual if H_0 were true. The smaller the P-value, the stronger the evidence against H_0.
5. **Conclusion** A test concludes by interpreting the P-value in the context of the study. Sometimes a decision is needed, using a fixed **significance level** α, usually $\alpha = 0.05$. Then we reject H_0 if the P-value $\leq \alpha$. Two types of error can occur:
   - A **Type I error** results from rejecting H_0 when it is true. When H_0 is true, the significance level = P(Type I error).
   - When H_0 is false, a **Type II error** results from failing to reject H_0.

SUMMARY: Significance Tests for Population Proportions and Means

	Parameter	
	Proportion	**Mean**
1. Assumptions	Categorical variable Randomization Expected numbers of successes and failures ≥ 15	Quantitative variable Randomization Approximately normal population
2. Hypotheses	H_0: $p = p_0$ H_a: $p \neq p_0$ (two-sided) H_a: $p > p_0$ (one-sided) H_a: $p < p_0$ (one-sided)	H_0: $\mu = \mu_0$ H_a: $\mu \neq \mu_0$ (two-sided) H_a: $\mu > \mu_0$ (one-sided) H_a: $\mu < \mu_0$ (one-sided)
3. Test statistic	$z = \frac{\hat{p} - p_0}{se_0}$ $(se_0 = \sqrt{p_0(1 - p_0)/n})$	$t = \frac{\bar{x} - \mu_0}{se}$ $(se = s/\sqrt{n})$
4. P-value	Two-tail (H_a: $p \neq p_0$) or right tail (H_a: $p > p_0$) or left tail (H_a: $p < p_0$) probability from standard normal distribution	Two-tail (H_a: $\mu \neq \mu_0$) or right tail (H_a: $\mu > \mu_0$) or left tail (H_a: $\mu < \mu_0$) probability from t distribution ($df = n - 1$)
5. Conclusion	Interpret P-value in context Reject H_0 if P-value $\leq \alpha$	Interpret P-value in context Reject H_0 if P-value $\leq \alpha$

Where We're Going

To introduce statistical inference, Chapters 8 and 9 presented methods for a single proportion or mean. In practice, inference is used more commonly to compare parameters for different groups (for example, females and males). The next chapter shows how to compare proportions and how to compare means for two groups.

SUMMARY OF NOTATION

H_0 = null hypothesis, H_a = alternative hypothesis,

p_0 = null hypothesis value of proportion,

μ_0 = null hypothesis value of mean

α = significance level

= probability of Type I error

(usually 0.05; the P-value must be $\leq \alpha$ to reject H_0)

CHAPTER PROBLEMS

Practicing the Basics

9.66 **H_0 or H_a?** For each of the following hypotheses, explain whether it is a null hypothesis or an alternative hypothesis:

a. For females, the population mean on the political ideology scale is equal to 4.0.

b. For males, the population proportion who support the death penalty is larger than 0.50.

c. The diet has an effect, the population mean change in weight being less than 0.

d. For all Subway (submarine sandwich) stores worldwide, the difference between sales this month and in the corresponding month a year ago has a mean of 0.

9.67 **ESP** A person who claims to possess extrasensory perception (ESP) says she can guess more often than not the outcome of a flip of a balanced coin. Out of 20 flips, she guesses correctly 12 times. Would you conclude that she truly has ESP? Answer by reporting all five steps of a significance test of the hypothesis that each of her guesses has probability 0.50 of being correct against the alternative that corresponds to her having ESP.

9.68 **Free-throw accuracy** Consider all cases in which a pro basketball player shoots two free throws and makes one and misses one. Which do you think is more common: making the first and missing the second, or missing the first and making the second? One of the best shooters was Larry Bird of the Boston Celtics. During 1980–1982 he made only the first free throw 34 times and made only the second 48 times (A. Tversky and T. Gilovich, *Chance,* vol. 2, pp. 16–21, 1989). Does this suggest that one sequence was truly more likely than the other sequence for Larry Bird? Answer by conducting a significance test, (a) defining notation and specifying assumptions and hypotheses, (b) finding the test statistic, and (c) finding the P-value and interpreting in this context.

9.69 **Brown or Whitman?** California's governor election in 2010 had two major candidates, Brown and Whitman.

a. For a random sample of 650 voters in one exit poll, 360 voted for Brown and 240 for Whitman. Conduct all five steps of a test of H_0: $p = 0.50$ against H_a: $p \neq 0.50$, where p denotes the probability that a randomly selected voter prefers Brown. Would you be willing to predict the outcome of the election? Explain how to make a decision, using a significance level of 0.05.

b. Suppose the sample size had been 50 voters, of whom 28 voted for Brown. Show that the sample proportion is similar to the sample proportion in part a, but the test statistic and P-value are very different. Are you now willing to predict the outcome of the election?

c. Using part a and part b, explain how results of a significance test can depend on the sample size.

9.70 **Protecting the environment?** When the 2000 General Social Survey asked, "Would you be willing to pay much higher taxes in order to protect the environment?" (variable GRNTAXES), 369 people answered yes and 483 answered no. (We exclude those who made other responses.) Let p denote the population proportion who would answer yes. MINITAB shows the following results to analyze whether a majority or minority of Americans would answer yes:

```
Test of p = 0.5 vs p not = 0.5
  X    N  Sample p        95% CI        Z-Value P-Value
369  852  0.433099 (0.3998,0.4663)  −3.91    0.000
```

a. What are the assumptions for the significance test? Do they seem to be satisfied for this application? Explain.

b. For this printout, specify the null and alternative hypotheses that are tested, and report the point estimate of p and the value of the test statistic.

c. Report and interpret the P-value in context.

d. According to the P-value, is it plausible that $p = 0.50$? Explain.

e. Explain an advantage of the confidence interval shown over the significance test.

9.71 **Frustration and the federal government** A Pew Research Center poll in 2010 asked a random sample of 2505 adults about their attitudes and opinions concerning the U.S. government (www.people-press.org). When asked whether they felt content, frustrated or angry, 56% said that they were frustrated. Let p denote the population proportion who would say they are frustrated. Conduct all five steps of a test of H_0: $p = 0.50$ against H_a: $p \neq 0.50$. Can you reject H_0 using a significance level of 0.05?

9.72 Plant inheritance In an experiment on chlorophyll inheritance in maize (corn), of the 1103 seedlings of self-fertilized green plants, 854 seedlings were green and 249 were yellow. Theory predicts the ratio of green to yellow is 3 to 1. Show all five steps of a test of the hypothesis that 3 to 1 is the true ratio. Interpret the P-value in context.

9.73 Ellsberg paradox You are told that a ball will be randomly drawn from one of two boxes (A and B), both of which contain black balls and red balls, and if a red ball is chosen, you will win $100. You are also told that Box A contains half black balls and half red balls, but you are not told the proportions in Box B. Which box would you pick?

a. Set up notation and specify hypotheses to test whether the population proportion who would pick Box A is 0.50.

b. For a random sample of 40 people, 36 pick Box A. Can you make a conclusion about whether the proportion for one option is higher in the population? Explain all steps of your reasoning. (Logically those who picked Box A would seem to think that Box B has greater chance of a black ball. However, a paradox first discussed by Daniel Ellsberg predicts that if they were now told that they would instead receive $100 if a black ball is chosen, they would overwhelmingly pick Box A again, because they prefer definite information over ambiguity.)

9.74 Start a hockey team A fraternity at a university lobbies the administration to start a hockey team. To bolster its case, it reports that of a simple random sample of 100 students, 83% support starting the team. Upon further investigation, their sample has 80 males and 20 females. Should you be skeptical of whether the sample was random, if you know that 55% of the student body population was male? Answer this by performing all steps of a two-sided significance test.

9.75 Interest charges on credit card A bank wants to evaluate which credit card would be more attractive to its customers: One with a high interest rate for unpaid balances but no annual cost, or one with a low interest rate for unpaid balances but an annual cost of $40. For a random sample of 100 of its 52,000 customers, 40 say they prefer the one that has an annual cost. Software reports the following:

```
Test of p = 0.50 vs p not = 0.50
 X    N  Sample p     95.0% CI       Z-Value  P-Value
40  100  0.40000  (0.304,0.496)      -2.00   0.04550
```

Explain how to interpret all results on the printout. What would you tell the company about what the majority of its customers prefer?

9.76 Jurors and gender A jury list contains the names of all individuals who may be called for jury duty. The proportion of the available jurors on the list who are women is 0.53. If 40 people are selected to serve as candidates for being picked on the jury, show all steps of a significance test of the hypothesis that the selections are random with respect to gender.

a. Set up notation and hypotheses, and specify assumptions.

b. 5 of the 40 selected were women. Find the test statistic.

c. Report the P-value, and interpret.

d. Explain how to make a decision using a significance level of 0.01.

9.77 Type I and Type II errors Refer to the previous exercise.

a. Explain what Type I and Type II errors mean in the context of that exercise.

b. If you made an error with the decision in part d, is it a Type I or a Type II error?

9.78 Levine = author? The authorship of an old document is in doubt. A historian hypothesizes that the author was a journalist named Jacalyn Levine. Upon a thorough investigation of Levine's known works, it is observed that one unusual feature of her writing was that she consistently began 10% of her sentences with the word "whereas." To test the historian's hypothesis, it is decided to count the number of sentences in the disputed document that begin with the word *whereas.* Out of the 300 sentences in the document, none begin with that word. Let p denote the probability that any one sentence written by the unknown author of the document begins with the word "whereas."

a. Conduct a test of the hypothesis H_0: $p = 0.10$ against H_a: $p \neq 0.10$. What conclusion can you make, using a significance level of 0.05?

b. What assumptions are needed for that conclusion to be valid? (F. Mosteller and D. Wallace conducted an investigation similar to this to determine whether Alexander Hamilton or James Madison was the author of 12 of the *Federalist Papers.* See *Inference and Disputed Authorship: The Federalist,* Addison-Wesley, 1964.)

9.79 Practice steps of test for mean For a quantitative variable, you want to test H_0: $\mu = 0$ against H_a: $\mu \neq 0$. The 10 observations are

$$3, 7, 3, 3, 0, 8, 1, 12, 5, 8.$$

a. Show that (i) $\bar{x} = 5.0$, (ii) $s = 3.71$, (iii) standard error $= 1.17$, (iv) test statistic $= 4.26$, and (v) $df = 9$.

b. The P-value is 0.002. Make a decision using a significance level of 0.05. Interpret.

c. If you had instead used H_a: $\mu > 0$, what would the P-value be? Interpret it.

d. If you had instead used H_a: $\mu < 0$, what would the P-value be? Interpret it. (*Hint:* Recall that the two one-sided P-values should sum to 1.)

9.80 Two ideal children? Is the ideal number of children equal to 2, or higher or lower than that? For testing that the mean response from a recent GSS equals 2.0 for the question, "What do you think is the ideal number of children to have?" software shows results:

```
Test of mu = 2.0 vs not = 2.0
Variable     N    Mean  StDev  SE Mean      T       P
Children  1131   2.490   0.88   0.0262  18.73  0.0000
```

a. Report the test statistic value, and show how it was obtained from other values reported in the table.

b. Explain what the P-value represents, and interpret its value.

9.81 Hours at work When all subjects in the 2008 GSS who were working full- or part-time were asked how many hours they worked in the previous week at all jobs (variable HRS1), software produced the following analyses:

```
Test of mu = 40 vs not = 40
   N     Mean    StDev  SE Mean    95% CI        T       P
                                  (41.171,
1202  41.990  14.480   0.418    42.809)   4.76   0.000
```

For this printout,

a. State the hypotheses.

b. Explain how to interpret the values of (i) SE Mean, (ii) T, and (iii) P.

c. Show the correspondence between a decision in the test using a significance level of 0.05 and whether 40 falls in the confidence interval reported.

9.82 Females liberal or conservative? Example 11 compared mean political beliefs (on a 1 to 7 point scale) to the moderate value of 4.0, using GSS data. Test whether the population mean equals 4.00 for females, for whom the sample mean was 3.98 and standard deviation was 1.45 for a sample of size 819. Carry out the five steps of a significance test, reporting and interpreting the P-value in context.

9.83 Blood pressure When Vincenzo Baranello's blood pressure is in control, his systolic blood pressure reading has a mean of 130. For the last six times he has monitored his blood pressure, he has obtained the values

$$140, 150, 155, 155, 160, 140.$$

a. Does this provide strong evidence that his true mean has changed? Carry out the five steps of the significance test, interpreting the P-value.

b. Review the assumptions that this method makes. For each assumption, discuss it in context.

9.84 Increasing blood pressure In the previous exercise, suppose you had predicted that if the mean changed, it would have increased above the control value. State the alternative hypothesis for this prediction, and report and interpret the P-value.

9.85 Tennis balls in control? When it is operating correctly, a machine for manufacturing tennis balls produces balls with a mean weight of 57.6 grams. The last eight balls manufactured had weights

$$57.3, 57.4, 57.2, 57.5, 57.4, 57.1, 57.3, 57.0$$

a. Using a calculator or software, find the test statistic and P-value for a test of whether the process is in control against the alternative that the true mean of the process now differs from 57.6.

b. For a significance level of 0.05, explain what you would conclude. Express your conclusion so it would be understood by someone who never studied statistics.

c. If your decision in part b is in error, what type of error have you made?

9.86 Catalog sales A company that sells products through mail-order catalogs wants to evaluate whether the mean sales for their most recent catalog were different from the mean of \$15 from past catalogs. For a random sample of 100 customers, the mean sales were \$10, with a standard deviation of \$10. Find a P-value to provide the extent of evidence that the mean differed with this catalog. Interpret.

9.87 Wage claim false? Management claims that the mean income for all senior-level assembly-line workers in a large company equals \$500 per week. An employee decides to test this claim, believing that it is actually less than \$500. For a random sample of nine employees, the incomes are

$$430, 450, 450, 440, 460, 420, 430, 450, 440.$$

Conduct a significance test of whether the population mean income equals \$500 per week against the alternative that it is less. Include all assumptions, the hypotheses, test statistic, P-value, and interpret the result in context.

9.88 CI and test Refer to the previous exercise.

a. For which significance levels can you reject H_0? (i) 0.10, (ii) 0.05, or (iii) 0.01.

b. Based on the answers in part a, for which confidence levels would the confidence interval contain 500? (i) 0.90, (ii) 0.95, or (iii) 0.99.

c. Use part a and part b to illustrate the correspondence between results of significance tests and results of confidence intervals.

d. In the context of this study, what is (i) a Type I error and (ii) a Type II error?

9.89 CI and test connection The P-value for testing $H_0: \mu = 100$ against $H_a: \mu \neq 100$ is 0.043.

a. What decision is made using a 0.05 significance level?

b. If the decision in part a is in error, what type of error is it?

c. Does a 95% confidence interval for μ contain 100? Explain.

9.90 Religious beliefs A journal article that deals with changes in religious beliefs over time states, "For these subjects, the difference in their responses on the scale of religiosity between age 16 and the current survey was statistically significant (P-value < 0.05)."

a. How would you explain to someone who has never taken a statistics course what it means for the result to be "statistically significant" with a P-value < 0.05?

b. Explain why it would have been more informative if the authors provided the actual P-value rather than merely indicating that it is below 0.05.

c. Can you conclude that a practically *important* change in religiosity has occurred between age 16 and the time of the current survey? Why or why not?

9.91 How to reduce chance of error? In making a decision in a significance test, a researcher worries about rejecting H_0 when it may actually be true.

a. Explain how the researcher can control the probability of this type of error.

b. Why should the researcher probably not set this probability equal to 0.00001?

9.92 Legal trial errors Consider the analogy discussed in Section 9.4 between making a decision about a null hypothesis in a significance test and making a decision about the innocence or guilt of a defendant in a criminal trial.

a. Explain the difference between Type I and Type II errors in the trial setting.

b. In this context, explain intuitively why decreasing the chance of Type I error increases the chance of Type II error.

9.93 P(Type II error) with smaller *n* Consider Example 13 about testing H_0: $p = 1/3$ against H_a: $p > 1/3$ for the astrology study, with $n = 116$. Find P(Type II error) for testing H_0: $p = 1/3$ against H_a: $p > 1/3$ when actually $p = 0.50$, if the sample size is 60 instead of 116. Do this by showing that

a. The standard error is 0.061 when H_0 is true.

b. The rejection region consists of $\hat{p}$ values above 0.433.

c. When $p = 0.50$, the probability that $\hat{p}$ falls below 0.433 is the left-tail probability below -1.03 under a standard normal curve. What is the answer? Why would you expect P(Type II error) to be larger when n is smaller?

Concepts and Investigations

9.94 Student data Refer to the FL Student Survey data file on the text CD. Test whether the (a) population mean political ideology (on a scale of 1 to 7, where 4 = moderate) equals or differs from 4.0 and (b) population proportion favoring affirmative action equals or differs from 0.50. For each part, write a one-page report showing all five steps of the test, including what you must assume for each inference to be valid.

9.95 Class data Refer to the data file your class created in Activity 3 at the end of Chapter 1. For a variable chosen by your instructor, conduct inferential statistical analyses. Prepare a report, summarizing and interpreting your findings. In this report, also use graphical and numerical methods presented earlier in this text to describe the data.

9.96 Gender of best friend A GSS question asked the gender of your best friend. Of 1381 people interviewed, 147 said their best friend had the opposite gender, and 1234 said their best friend had the same gender. Prepare a short report in which you analyze these data using a confidence interval and a significance test. Which do you think is more informative? Why?

9.97 Baseball home team advantage In major league baseball's 2010 season, the home team won 1359 games and the away team won 1071 (www.mlb.com).

a. Although these games are not a random sample, explain how you could think of these data as giving you information about some long-run phenomenon.

b. Analyze these data using (i) a significance test and (ii) a confidence interval. Which method is more informative? Why?

9.98 Statistics and scientific objectivity The president of the American Statistical Association stated, "Statistics has become the modern-day enforcer of scientific objectivity. Terms like randomization, blinding, and 0.05 significance wield a no-doubt effective objectivity nightstick." He also discussed how learning what effects *aren't* in the data is as important as learning what effects *are* significant. In this vein, explain how statistics provides an objective framework for testing the claims of what many believe to be quack science, such as astrology and therapeutic touch. (*Source:* Bradley Efron, *Amstat News,* July 2004, p. 3.)

9.99 Two-sided or one-sided? A medical researcher gets a P-value of 0.056 for testing H_0: $\mu = 0$ against H_a: $\mu \neq 0$. Since he believes that the true mean is positive and is worried that his favorite journal will not publish the results because they are not "significant at the 0.05 level," he instead reports in his article the P-value of 0.028 for H_a: $\mu > 0$. Explain what is wrong with

a. Reporting the one-sided P-value after seeing the data.

b. The journal's policy to publish results only if they are statistically significant.

9.100 Interpret P-value An article in a marketing journal states that "no statistically significant difference was found between men and women in the proportion who said they would consider buying a hybrid car (P-value = 0.63)." In practical terms, how would you explain to someone who has not studied statistics what this means?

9.101 Subgroup lack of significance A crossover study on comparing a magnetic device to placebo for reducing pain in 54 people suffering from low back or knee pain (neuromagnetics.mc.vanderbilt.edu/publications) reported a significant result overall, the magnetic device being preferred to placebo. However, when the analysis was done separately for the 27 people with shorter illness duration and the 27 people with longer illness duration, results were not significant. Explain how it might be possible that an analysis could give a P-value below 0.05 using an entire sample but not with subgroups, even if the subgroups have the same effects. (*Hint:* What is the impact of the smaller sample size for the subgroups?)

9.102 Vitamin E and prostate cancer A study in Finland in 1998 suggested that vitamin E pills reduced the risk of prostate cancer in a group of male smokers. This was only one effect in a series of reported successes for vitamin E and other health conditions. A more recent study reported that selenium or vitamin E, used alone or in combination, did not prevent prostate cancer in a population of relatively healthy men (*JAMA*, April 2009). Discuss the factors that can cause different medical studies to come to different conclusions.

9.103 Overestimated effect When medical stories in the mass media report dangers of certain agents (e.g., coffee drinking), later research often suggests that the effects may not exist or are weaker than first believed. Explain how this could happen if some journals tend to publish only statistically significant results.

9.104 Choosing α An alternative hypothesis states that a newly developed drug is better than the one currently used to treat a serious illness. If we reject H_0, the new drug will be prescribed instead of the current one.

a. Why might we prefer to use a smaller significance level than 0.05, such as 0.01?

b. What is a disadvantage of using $\alpha = 0.01$ instead of 0.05?

9.105 Why not accept H_0? Explain why the terminology "do not reject H_0" is preferable to "accept H_0."

9.106 Report P-value It is more informative and potentially less misleading if you conclude a test by reporting and interpreting the P-value rather than by merely indicating whether or not you reject H_0 at the 0.05 significance level. One reason is that a reader can then tell whether the result is significant *at any significance level.* Give another reason.

9.107 Significance Explain the difference between *statistical significance* and *practical significance.* Make up an example to illustrate your reasoning.

9.108 More doctors recommend An advertisement by Company A says that three of every four doctors recommend pain reliever A over all other brands combined.

a. If the company based this claim on interviewing a random sample of doctors, explain how they could use a significance test to back up the claim.

b. Explain why this claim would be more impressive if it is based on a (i) random sample of 40 doctors than if it is based on a random sample of 4 doctors and (ii) random sample of 40 doctors nationwide than the sample of all 40 doctors who work in a particular hospital.

9.109 Medical diagnosis error Consider the medical diagnosis of breast cancer with mammograms. An AP story (September 19, 2002) said that a woman has about a 50% chance of having a false-positive diagnosis over the course of 10 annual mammography tests. Relate this result to the chance of eventually making a Type I error if you do lots of significance tests.

9.110 Bad P-value interpretations A random sample of size 1000 has $\bar{x} = 104$. The significance level α is set at 0.05. The P-value for testing H_0: $\mu = 100$ against H_a: $\mu \neq 100$ is 0.057. Explain what is incorrect about each of the following interpretations of this P-value, and provide a proper interpretation.

a. The probability that the null hypothesis is correct equals 0.057.

b. The probability that $\bar{x} = 104$ if H_0 is true equals 0.057.

c. If in fact $\mu \neq 100$ so H_0 is false, the probability equals 0.057 that the data would show at least as much evidence against H_0 as the observed data.

d. The probability of a Type I error equals 0.057.

e. We can accept H_0 at the $\alpha = 0.05$ level.

f. We can reject H_0 at the $\alpha = 0.05$ level.

9.111 Interpret P-value One interpretation for the P-value is that it is the smallest value for the significance level α for which we can reject H_0. Illustrate using the P-value of 0.057 from the previous exercise.

9.112 Incorrectly posed hypotheses What is wrong with expressing hypotheses about proportions and means in a form such as H_0: $\hat{p} = 0.50$ and H_0: $\bar{x} = 0$?

9.113 Multiple choice: Small P-value The P-value for testing H_0: $\mu = 100$ against H_a: $\mu \neq 100$ is 0.001. This indicates that

a. There is strong evidence that $\mu = 100$.

b. There is strong evidence that $\mu \neq 100$, since if μ were equal to 100, it would be unusual to obtain data such as those observed.

c. The probability that $\mu = 100$ is 0.001.

d. The probability that $\mu = 100$ is the significance level, usually taken to be 0.05.

9.114 Multiple choice: Probability of P-value When H_0 is true in a *t* test with significance level 0.05, the probability that the P-value falls ≤ 0.05

a. equals 0.05.

b. equals 0.95.

c. equals 0.05 for a one-sided test and 0.10 for a two-sided test.

d. can't be specified, because it depends also on P(Type II error).

9.115 Multiple choice: Pollution Exercise 9.33 concerned an industrial plant that may be exceeding pollution limits. An environmental action group took four readings to analyze whether the true mean discharge of wastewater per hour exceeded the company claim of 1000 gallons. When we make a decision in the one-sided test using $\alpha = 0.05$:

a. If the plant is not exceeding the limit, but actually $\mu = 1000$, there is only a 5% chance that we will conclude that they are exceeding the limit.

b. If the plant is exceeding the limit, there is only a 5% chance that we will conclude that they are not exceeding the limit.

c. The probability that the sample mean equals exactly the observed value would equal 0.05 if H_0 were true.

d. If we reject H_0, the probability that it is actually true is 0.05.

e. All of the above.

9.116 Multiple choice: Interpret P(Type II error) For a test of H_0: $\mu = 0$ against H_a: $\mu > 0$ based on $n = 30$ observations and using $\alpha = 0.05$ significance level, P(Type II error) $= 0.36$ at $\mu = 4$. Identify the response that is *incorrect.*

a. At $\mu = 5$, P(Type II error) < 0.36.

b. If $\alpha = 0.01$, then at $\mu = 4$, P(Type II error) > 0.36.

c. If $n = 50$, then at $\mu = 4$, P(Type II error) > 0.36.

d. The power of the test is 0.64 at $\mu = 4$.

9.117 True or false It is always the case that P(Type II error) = 1 − P(Type I error).

9.118 True or false If we reject H_0: $\mu = 0$ in a study about change in weight on a new diet using $\alpha = 0.01$, then we also reject it using $\alpha = 0.05$.

9.119 True or false A study about the change in weight on a new diet reports P-value = 0.043 for testing H_0: $\mu = 0$ against H_a: $\mu \neq 0$. If the authors had instead reported a 95% confidence interval for μ, then the interval would have contained 0.

9.120 True or false A 95% confidence interval for μ = population mean IQ is (96, 110). So, in the test of H_0: $\mu = 100$ against H_a: $\mu \neq 100$, the P-value > 0.05.

9.121 True or false For a fixed significance level α, the probability of a Type II error increases when the sample size increases.

9.122 True or false The P-value is the probability that H_0 is true.

9.123 ◆◆ **Standard error formulas** Suppose you wanted to test H_0: $p = 0.50$, but you had 0 successes in n trials. If you had found the test statistic using the $se = \sqrt{\hat{p}(1 - \hat{p})/n}$ designed for confidence intervals, show what happens to the test statistic. Explain why $se_0 = \sqrt{p_0(1 - p_0)/n}$ is a more appropriate se for tests.

9.124 ◆◆ **Rejecting true H_0?** A medical researcher conducts a significance test whenever she analyzes a new data set. Over time, she conducts 100 independent tests.

a. Suppose the null hypothesis is true in every case. What is the distribution of the number of times she rejects the null hypothesis at the 0.05 level?

b. Suppose she rejects the null hypothesis in five of the tests. Is it plausible that the null hypothesis is correct in every case? Explain.

Student Activities

9.125 TRY Refer to Activity 1 at the end of Section 9.4. Each student should indicate how many successive "wins" by the (a) placebo over the drug would be necessary before he or she would feel comfortable rejecting H_0: $p = 0.5$ in favor of H_a: $p \neq 0.5$ and concluding that $p < 0.5$ and (b) drug over the placebo would be necessary before he or she would feel comfortable rejecting H_0: $p = 0.5$ in favor of H_a: $p \neq 0.5$ and concluding that $p > 0.5$. The instructor will compile a "distribution of significance levels" for the two cases. Are they the same? In principle, should they be?

9.126 Refer to Exercise 8.124, "Randomized response," in Chapter 8. Before carrying out the method described there, the class was asked to hypothesize or predict what they believe is the value for the population proportion of students that have had alcohol at a party. Use the class estimate for $\hat{p}$ to carry out the significance test for testing this hypothesized value. Discuss whether to use a one-sided or two-sided alternative hypothesis. Describe how the confidence interval formed in Exercise 8.124 relates to the significance test results.

10 Comparing Two Groups

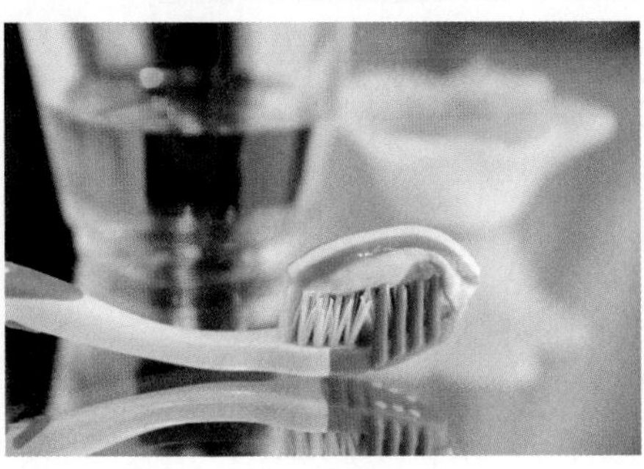

Example 1

Making Sense of Studies Comparing Two Groups

Picture the Scenario

A smile is a universal greeting, a way to communicate with others even if we don't speak their language. Making our smiles as bright as possible with teeth as white as possible has become very desirable. We want our teeth and our smiles to be as perfect looking as those we see on TV and in magazines. Today, teeth-whitening products can be obtained at the dentist or at the drugstore. Numerous claims have been made about the ability of the products to whiten teeth. As with any other claims, some of which follow pseudoscience methods, it is difficult to sort out which products work best or if they work at all.

Studies that investigate claims like weight loss, teeth whiteners, or binge drinkers involve a comparison of two groups or two treatments (such as before and after weight, before and after teeth whiteness, or comparing males and females who binge drink).

Questions to Explore

- How can we use data from an experiment to summarize the evidence about the claims by tooth whitener manufacturers?
- How can we decide, based on the data, whether or not the claims are believable?

Thinking Ahead

This chapter shows how to compare two groups on a categorical or quantitative outcome. To do this, we'll use the inferential statistical methods that the previous two chapters introduced—confidence intervals and significance tests.

For categorical variables, the inferences compare proportions. In Examples 2–4 we'll look at aspirin and placebo treatments, studying their effects on proportions getting cancer. Exercise 10.40 examines a study investigating the "test of whiteness" between whitening gel and toothpaste.

For quantitative variables, the inferences compare means. In Examples 6–8 we'll examine the extent of nicotine addiction for female and male teenage smokers and we'll compare nicotine addiction for teenagers who smoke with teenagers who have smoked in the past but no longer do so.

In reading this book, you are becoming an educated consumer of information based on statistics. By now, you know to be skeptical of studies that do not or could not use randomization in the sampling procedure or the experimental design. By the end of this chapter, you'll know about other potential pitfalls. You'll be better able to judge how much credence to give to claims made in newspaper stories. Such stories nearly always report only "statistically significant" results. Occasionally such a report may be a Type I error, claiming an effect that actually does not exist in the population. Some may predict that effects are larger than they truly are in that population. Does drinking tea really help heart attack victims as much as reported? Or, as Figure 10.1 suggests, is this merely today's random medical news?

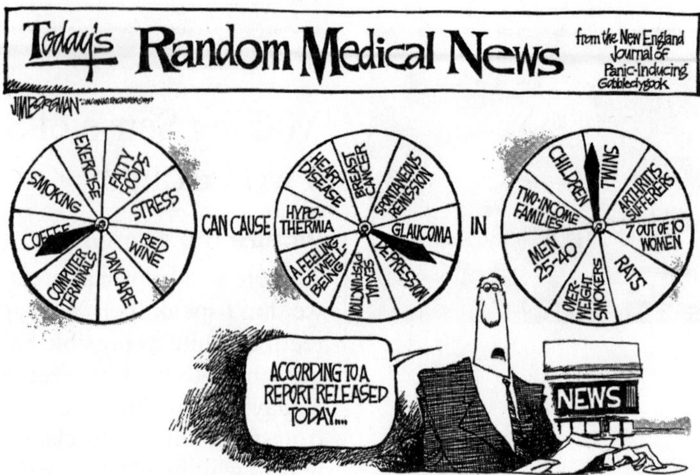

▲ **Figure 10.1** Today's Random Medical News

Bivariate Analyses: A Response Variable and a Binary Explanatory Variable

Consider a study that compares female and male college students on the proportion who say they have participated in binge drinking. The two groups being compared, females and males, are the categories of a *binary* variable—gender. The general category of statistical methods used when we have two variables is called **bivariate** methods. Special cases of these methods are used to compare two groups, where one of the two variables is the outcome variable, and the other is a binary variable that specifies the categories.

The outcome variable on which comparisons are made is called the **response variable**. The binary variable that specifies the groups is the **explanatory variable**. The beginning of Chapter 3 provides a review of how to distinguish between these two types of variables. Recall that a **binary variable** has two possible outcomes. In the previous example, with gender, the two outcomes would be male or female. Statistical methods analyze how the outcome on the response variable *depends on* or is *explained by* the value of the explanatory variable. In our example, participation in binge drinking (yes or no) is the response variable and gender is the explanatory variable. Our interest is in studying how binge drinking depends on gender, not how gender depends on binge drinking.

Dependent and Independent Samples

Most comparisons of groups use **independent samples** from the groups. The observations in one sample are *independent* of those in the other sample. For instance, randomized experiments that randomly allocate subjects to two treatments have independent samples. An example is the study mentioned in Example 1 comparing cancer rates for aspirin and placebo groups: The observation for any subject taking aspirin is independent of the observation for any subject taking placebo. Another way independent samples occur is when an observational study separates subjects into groups according to their value for an explanatory variable, such as smoking status. If the overall sample was randomly selected, then the groups (for instance, smokers and nonsmokers) can be treated as independent random samples.

When the two samples have the same subjects, they are **dependent**. Dependent samples also result when the data are **matched pairs**—each subject in one sample is matched with a subject in the other sample. An example is a set of married couples, the men being in one sample and the women in the other. Also, dependent samples occur when each subject is observed at two times, so the two samples have the same people. An example is a diet study in which subjects' weights are measured before and after the diet. A particular subject's weight before and after the diet, such as (144 lb, 127 lb), form a matched pair. Data from dependent samples need different statistical methods than data from independent samples. We'll study them in Section 10.4. Sections 10.1–10.3 show how to analyze independent samples, first for a categorical response variable and then for a quantitative response variable.

10.1 Categorical Response: Comparing Two Proportions

For a *categorical response variable*, inferences compare groups in terms of their population proportions in a particular category. Let p_1 represent the population proportion for the first group and p_2 the population proportion for the second group. We can compare the groups by their difference, $(p_1 - p_2)$. This is estimated by the difference of the sample proportions, $(\hat{p}_1 - \hat{p}_2)$. Let n_1 and n_2 denote the sample sizes for the two groups.

Example 2

Compare two proportions

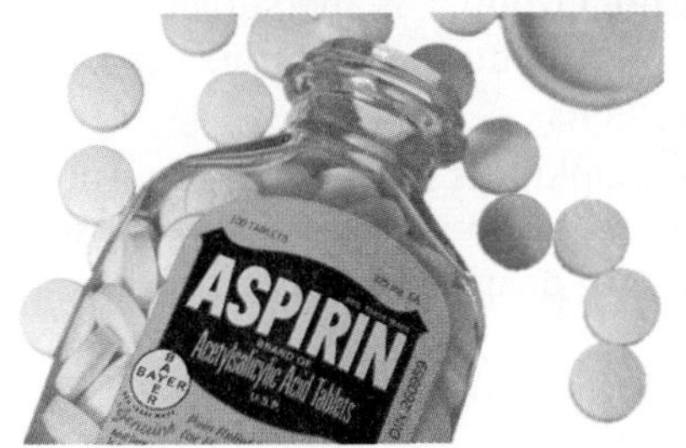

Aspirin, the Wonder Drug

Picture the Scenario

Here are two recent titles of newspaper articles about beneficial effects of aspirin:

"Small doses of aspirin can lower the risk of heart attacks"
"Aspirin could lower risk of colon cancer"

Most of us think of aspirin as a simple pill that helps relieve pain. In recent years, though, researchers have been on the lookout for new ways that aspirin may be helpful. Studies have shown that taking aspirin regularly may possibly forestall Alzheimer's disease and may increase the chance of survival for a person who has suffered a heart attack. Other studies have suggested that aspirin may protect against cancers of the pancreas, colon, and prostate. In the past decade, a growing number of studies have addressed the use of aspirin-like drugs to prevent cancer.[1]

Increasing attention has focused on aspirin since a landmark five-year study (Physicians Health Study Research Group, Harvard Medical School) about whether regular aspirin intake reduces deaths from heart disease. Studies have shown that treatment with daily aspirin for five years or longer reduces risk of colorectal cancer. These studies suggest that aspirin might reduce the risk of other cancers as well. Results of a recent meta-analysis combined the results of eight related studies with a minimum duration of treatment of four years or longer to determine the effects of aspirin on the risk of cancer death. A **meta-analysis** combines the results of several studies that address a set of related statistical questions. After analyzing the individual studies, the researchers assumed the different studies were measuring the same effect and pooled the results of the different

[1] *The Lancet*, vol. 377, January 2011, pp. 31–41.

Recall

Section 4.3 discussed the importance of **randomization** in experimental design and introduced **double blinding.** Exercises 4.71–4.73 in that chapter show other data from this study. ◀

studies. All experimental trials used were randomized and double-blind. The combined results provided evidence that daily aspirin reduced deaths due to several common cancers during and after the trials. We will explore some of these results.

Table 10.1 shows the study results. This is a **contingency table**, a data summary for categorical variables introduced in Section 3.1. Of the 25,570 individuals studied, 347 of those in the control group died of cancer, while 327 in the aspirin treatment died of cancer within 20 years following the study.

Table 10.1 Whether or Not Subject Died of Cancer, for Placebo and Aspirin Treatment Groups

	Death from Cancer		
Group	Yes	No	Total
Placebo	347	11,188	11,535
Aspirin	327	13,708	14,035

Questions to Explore

a. What is the response variable, and what are the groups to compare?

b. What are the two population parameters to compare? Estimate the difference between them using the data in Table 10.1.

Think It Through

a. In Table 10.1, the response variable is whether or not the subjects died of cancer, with categories yes and no. Group 1 is the subjects who took placebo and Group 2 is the subjects who took aspirin. They are the categories of the explanatory variable.

b. For the population from which this sample was taken, the proportion who died of cancer is represented by p_1 for taking placebo and p_2 for taking aspirin. The sample proportions of death from cancer were

$$\hat{p}_1 = 347/11535 = 0.030$$

for the $n_1 = 11,535$ in the placebo group and

$$\hat{p}_2 = 327/14035 = 0.023$$

for the $n_2 = 14,035$ in the aspirin group. Since $(\hat{p}_1 - \hat{p}_2) = 0.030 - 0.023 = 0.007$, the proportion of those who died of cancer was 0.007 higher for those who took placebo. In percentage terms, the difference was $3.0\% - 2.3\% = 0.7\%$, less than 1 percent.

Insight

The sample proportion of subjects who died of cancer was smaller for the aspirin group. But we really want to know if this result is true also for the population. To make an inference about the difference of population proportions, $(p_1 - p_2)$, we need to learn how much the difference $(\hat{p}_1 - \hat{p}_2)$ between the sample proportions would tend to vary from study to study. This is described by the standard error of the sampling distribution for the difference between the sample proportions.

Try Exercises 10.2 and 10.3, part a

The Standard Error for Comparing Two Proportions

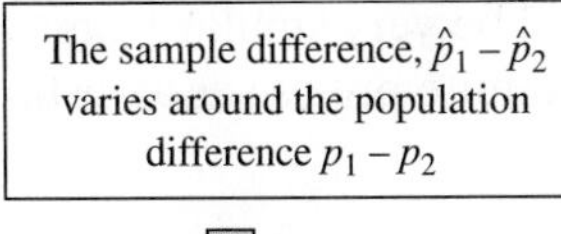

The sample difference, $\hat{p}_1 - \hat{p}_2$
varies around the population
difference $p_1 - p_2$

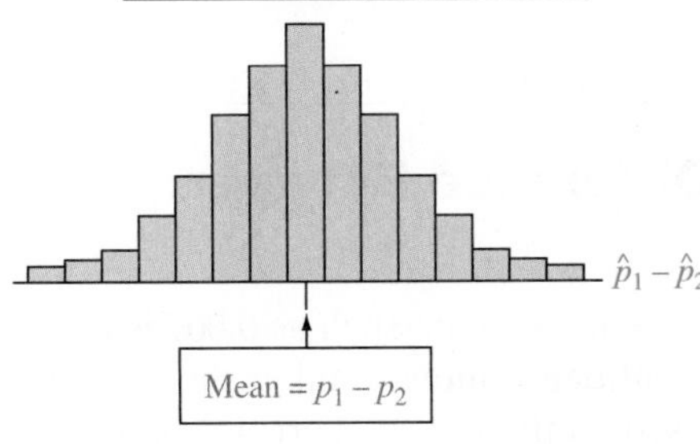

Just as a sample proportion has a standard error that describes how well it estimates a population proportion, so does the difference $(\hat{p}_1 - \hat{p}_2)$ between two sample proportions. This estimate would vary from study to study. The standard error describes the variability around the mean of the sampling distribution of the estimate. (See margin figure.) It is interpreted as the standard deviation of the estimates $(\hat{p}_1 - \hat{p}_2)$ from different randomized experiments of a particular sample size.

The formula for the standard error of $(\hat{p}_1 - \hat{p}_2)$ is

$$se = \sqrt{\frac{\hat{p}_1(1 - \hat{p}_1)}{n_1} + \frac{\hat{p}_2(1 - \hat{p}_2)}{n_2}}.$$

We'll see where this formula comes from at the end of the section. For now, notice that if you ignore one of the two samples (and half of this formula), you get the usual standard error for a proportion, as shown in the marginal Recall.

Recall

From Section 8.2, for using a confidence interval to estimate a proportion p, the **standard error** of a sample **proportion** $\hat{p}$ is

$$\sqrt{\frac{\hat{p}(1 - \hat{p})}{n}}. \blacktriangleleft$$

Standard error

Example 3

Cancer Death Rates for Aspirin and Placebo

Picture the Scenario

In Example 2, the sample proportions that died of cancer were $\hat{p}_1 = 347/11535 = 0.030$ for placebo (Group 1) and $\hat{p}_2 = 327/14035 = 0.023$ for aspirin (Group 2). The estimated difference was $\hat{p}_1 - \hat{p}_2 = 0.030 - 0.023 = 0.007$.

Questions to Explore

a. What is the standard error of this estimate?

b. How should we interpret this standard error?

Think It Through

a. Using the standard error formula given above with the values just recalled,

$$se = \sqrt{\frac{\hat{p}_1(1 - \hat{p}_1)}{n_1} + \frac{\hat{p}_2(1 - \hat{p}_2)}{n_2}}.$$

$$\sqrt{\frac{0.030(1 - 0.030)}{11535} + \frac{0.023(1 - 0.023)}{14035}} = 0.002.$$

b. Consider all the possible random samples of size about 11,535 for the placebo group and 14,035 for the aspirin group that could participate in this four-year study with follow-up. The difference $(\hat{p}_1 - \hat{p}_2)$ between the sample proportions of cancer deaths would not always equal 0.007 but would vary from sample to sample. The standard deviation of the $\hat{p}_1 - \hat{p}_2$ values for the different possible samples would equal about 0.002.

Insight

From the *se* formula, we see that *se* decreases as n_1 and n_2 increase. The standard error is very small for these data because the sample sizes were so large. This means that the $(\hat{p}_1 - \hat{p}_2)$ values would be very similar from study to study. It also implies that $(\hat{p}_1 - \hat{p}_2) = 0.007$ is quite precise as an estimate of the difference of population proportions.

Try Exercise 10.3, part b

Confidence Interval for the Difference Between Two Population Proportions

Recall

Review Section 8.2 for a discussion of the **confidence interval** for a single **proportion**. This has form

$$\hat{p} \pm z(se),$$

with $z = 1.96$ for 95% confidence. ◀

The standard error helps us predict how close an estimate such as 0.007 is likely to be to the population value $(p_1 - p_2)$. A confidence interval takes the estimate and adds and subtracts a margin of error based on the standard error. For a large random sample, recall that the sampling distribution of a sample proportion is approximately normal. This is due to the central limit theorem. The sampling distribution of $(\hat{p}_1 - \hat{p}_2)$ is also approximately normal when each sample size is large. See Figure 10.2.

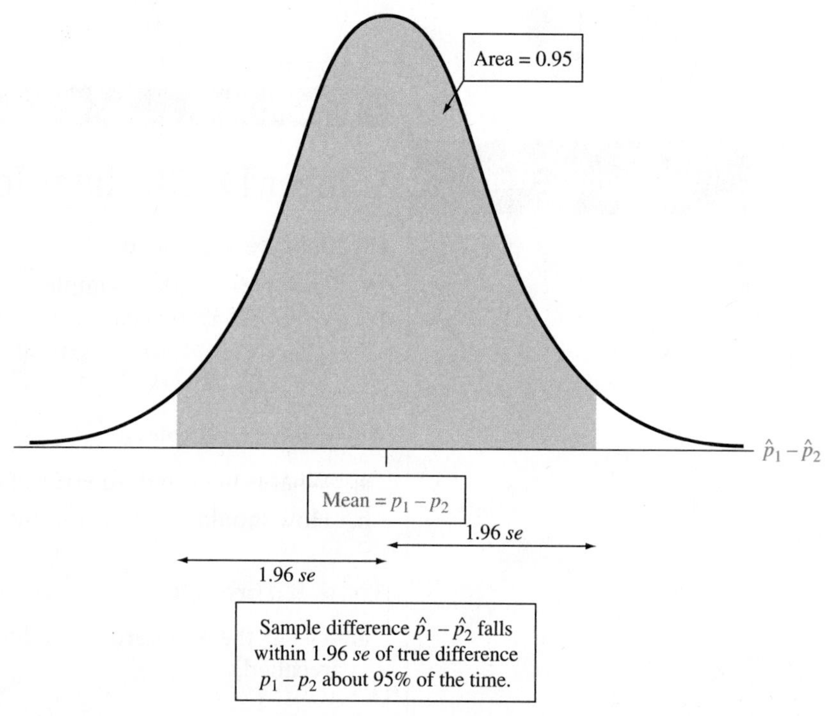

▲ **Figure 10.2 Sampling Distribution of the Estimate $(\hat{p} - \hat{p}_2)$ of the Difference Between Two Population Proportions.** With large random samples, this is approximately normal about the difference in the populations, $(p_1 - p_2)$. **Questions** For the aspirin and cancer death study (Examples 2 and 3), to what do $(\hat{p}_1 - \hat{p}_2)$ and $(p_1 - p_2)$ refer? What is the mean of this sampling distribution if cancer death rates are identical for aspirin and placebo?

As in the single proportion case, to get the margin of error, we multiply the standard error by a z-score from the normal distribution table. The confidence interval has the form

$$(\hat{p}_1 - \hat{p}_2) \pm z(se).$$

SUMMARY: Confidence Interval for the Difference Between Two Population Proportions

A confidence interval for the difference $(p_1 - p_2)$ between two population proportions is

$$(\hat{p}_1 - \hat{p}_2) \pm z(se), \text{ where } se = \sqrt{\frac{\hat{p}_1(1 - \hat{p}_1)}{n_1} + \frac{\hat{p}_2(1 - \hat{p}_2)}{n_2}}.$$

The z-score depends on the confidence level, such as $z = 1.96$ for 95% confidence. To use this method, you need

- A categorical response variable for two groups
- Independent random samples for the two groups, either from random sampling or a randomized experiment
- Large enough sample sizes n_1 and n_2 so that, in each sample, there are at least 10 successes and at least 10 failures. The confidence interval for a single proportion required at least 15 successes and 15 failures. Here, the method works well with slightly smaller samples, at least 10 of each type in each group.

Example 4

Confidence interval

Comparing Cancer Death Rates for Aspirin and Placebo

Picture the Scenario

In the aspirin and cancer study, the estimated difference between placebo and aspirin in the proportions dying of cancer was $\hat{p}_1 - \hat{p}_2 = 0.003 - 0.023 = 0.007$. In Example 3 we found that this estimate has a standard error of 0.002.

	Death from Cancer	
Group	**Yes**	**No**
Placebo	347	11,188
Aspirin	327	13,708

Question to Explore

What can we say about the difference of population proportions of cancer deaths for those taking placebo versus those taking aspirin? Construct a 95% confidence interval for $(p_1 - p_2)$, and interpret.

Think It Through

From Table 10.1, shown again in the margin, the four outcome counts (347, 327, 11188, and 13708) were at least 10 for each group, so the large-samples confidence interval method is valid. A 95% confidence interval for $(p_1 - p_2)$ is

$$(\hat{p}_1 - \hat{p}_2) \pm 1.96(se), \text{ or } 0.007 \pm 1.96(0.002),$$
$$\text{which is } 0.007 \pm 0.004, \text{ or } (0.003, 0.011).$$

Suppose this experiment could be conducted with the entire population. The inference at the 95% confidence level that $(p_1 - p_2)$ is between 0.003 and 0.011 means that the population proportion p_1 of cancer deaths for those taking placebo would be between 0.003 higher and 0.011 higher than the population proportion p_2 of cancer deaths for those taking aspirin. Since both endpoints of the confidence interval (0.003, 0.011) for $(p_1 - p_2)$ are positive, we infer that $(p_1 - p_2)$ is positive. This means that p_1 is larger than p_2: The population proportion of cancer deaths is larger when subjects take the placebo than when they take aspirin. Table 10.2 shows how MINITAB reports this result and the margin shows screen shots from the TI-83+/84.

In Words

The interval (0.003,0.011) for $(p_1 - p_2)$ predicts that the population proportion for the first group is between 0.003 and 0.011 larger than the population proportion for the second group.

```
2-PropZInt
 x1:347
 n1:11535
 x2:327
 n2:14035
 C-Level:95
 Calculate
```

```
2-PropZInt
 (.00279,.01078)
 p̂1=.030082358
 p̂2=.023298896
 n1=11535.00000
 n2=14035.00000
```

TI-83+/84 output

Table 10.2 MINITAB Output for Confidence Interval Comparing Proportions

```
Number of Cancer Deaths          Observed p̂
        ↓                            ↓
Sample    X      N          Sample p
  1      347    11535       0.030082
  2      327    14035       0.023299
Difference = p(1) - p(2)
Estimate for difference: 0.00678346  ← This is (p̂1 - p̂2)
95% CI for difference: (0.00279030, 0.0107766)
```

Insight

All the numbers in the confidence interval fall near 0. This suggests that the population difference $(p_1 - p_2)$ is small. However, this difference, small as it is, may be important in public health terms. For instance, projected over a population of 200 million adults (as in the United States or in Western Europe), a decrease over a twenty-year period of 0.01 in the proportion of people dying from cancer would mean two million fewer people dying from cancer.

This cancer study provided some of the first evidence that aspirin reduces deaths from several common cancers. Benefit was consistent across the different trial populations from the different randomized studies, suggesting that the findings have broader scope of generalization. However, it is important to replicate the study to see if results are similar or different for populations used in this study and other populations. Also, because there were fewer women than men in the study, findings about the effect of aspirin use and cancers related to women (such as breast cancer) were limited.

This example shows how the use of statistics can result in conclusions that benefit public health. An article[2] about proper and improper scientific methodology stated, "The most important discovery of modern medicine is not vaccines or antibiotics, it is the randomized double-blind study, by means of which we know what works and what doesn't."

Try Exercise 10.6

Recall

For **99% confidence** we use $z = 2.58$, because 99% of the standard normal distribution falls between -2.58 and 2.58. ◀

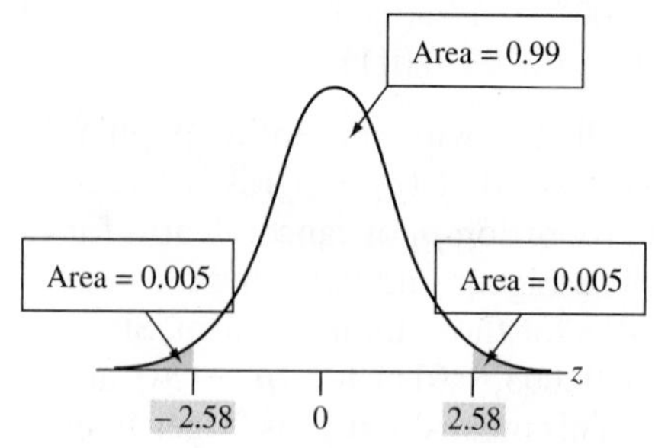

As in the one-sample case, we can use a higher confidence level, but the interval is then wider. For instance, a 99% confidence interval uses $z = 2.58$ (see margin figure) and equals

$$(\hat{p}_1 - \hat{p}_2) \pm 2.58(se), \text{ or } 0.007 \pm 2.58(0.002),$$

$$\text{which is } 0.007 \pm 0.005, \text{ or } (0.002, 0.012).$$

We can be 99% confident that the population proportion of cancer deaths is between 0.002 higher and 0.012 higher for the placebo treatment than for the aspirin treatment. For this example, the confidence interval is slightly wider than the 95% confidence interval of (0.003, 0.011). With very large sample sizes, it is often worth the slight widening, to obtain greater confidence.

Interpreting a Confidence Interval That Compares Proportions

Example 4 illustrated how to interpret a confidence interval for $(p_1 - p_2)$ in context. Whether a particular group is called Group 1 or Group 2 is completely arbitrary. If we reverse the labels, each endpoint of the confidence interval reverses sign. For instance, if we instead call aspirin Group 1 and placebo Group 2, then $(\hat{p}_1 - \hat{p}_2) = 0.0023 - 0.0030 = -0.007$. The 95% confidence interval is then $(-0.011, -0.003)$ instead of (0.003, 0.011). Since this confidence interval contains entirely negative numbers, we infer that $(p_1 - p_2)$ is negative; that is, p_1 is smaller than p_2. Since Group 1 is now aspirin, the interval $(-0.011, -0.003)$ predicts that the population proportion of cancer deaths is between 0.011 less and 0.003 less for aspirin than for placebo. (The negative signs translate to the first proportion being *less* than the second.)

In addition, if the confidence interval contains 0, then it is plausible that $(p_1 - p_2) = 0$, that is, $p_1 = p_2$. The population proportions might be equal. In such a case insufficient evidence exists to infer which of p_1 or p_2 is larger. See Figure 10.3.

[2]By R. L. Park, *The Chronicle of Higher Education*, January 31, 2003.

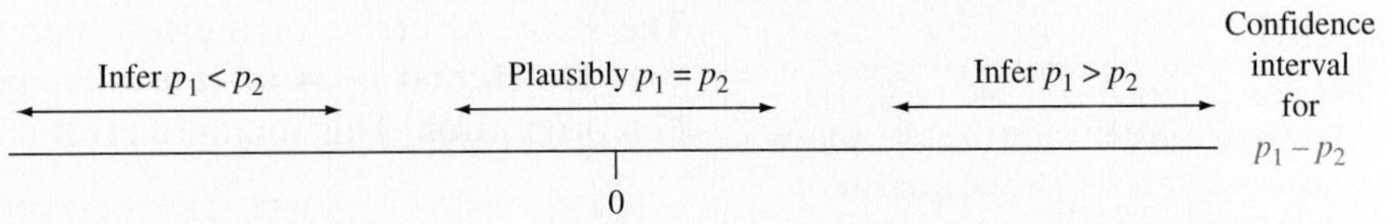

Infer $p_1 < p_2$ | Plausibly $p_1 = p_2$ | Infer $p_1 > p_2$ | Confidence interval for $p_1 - p_2$

0

▲ **Figure 10.3 Three Confidence Intervals for Difference Between Proportions, $p_1 - p_2$.** When the confidence interval for $p_1 - p_2$ contains 0 (the middle interval above), the population proportions may be equal. When it does not contain 0, we can infer which population proportion is larger. **Question** For the two confidence intervals that do not contain 0, how can we tell which population proportion is predicted to be larger?

For instance, Exercise 10.6 summarizes a study comparing population proportions for a placebo group and a treatment group. The study had a much smaller sample size than the aspirin and cancer study, so its standard error was larger. Its 95% confidence interval for $(p_1 - p_2)$ is $(-0.005, 0.034)$. Since the interval contains 0, the population proportions might well be equal. This interval infers that the population proportion p_1 for the placebo group is as much as 0.005 lower or as much as 0.034 higher than the population proportion p_2 for the treatment group. A *negative* value for $(p_1 - p_2)$, such as -0.005, means that p_1 could be *below* p_2, whereas a *positive* value for $(p_1 - p_2)$, such as $+0.034$, means that p_1 could be *above* p_2.

SUMMARY: Interpreting a Confidence Interval for a Difference of Proportions

- Check whether 0 falls in the confidence interval. If so, it is plausible (but not necessary) that the population proportions are equal.
- If all values in the confidence interval for $(p_1 - p_2)$ are positive, you can infer that $(p_1 - p_2) > 0$, or $p_1 > p_2$. The interval shows just how much larger p_1 might be. If all values in the confidence interval are negative, you can infer that $(p_1 - p_2) < 0$, or $p_1 < p_2$.
- The magnitude of values in the confidence interval tells you how *large* any true difference is. If all values in the confidence interval are near 0, the true difference may be relatively small in practical terms.

Significance Tests Comparing Population Proportions

Caution

Although technology will report the endpoints of a confidence interval for a difference of proportions to several decimal places, it is simpler to report and interpret these confidence intervals using at most 3 decimal places for the endpoints. ◄

Another way to compare two population proportions p_1 and p_2 is with a significance test. The null hypothesis is H_0: $p_1 = p_2$, the population proportion taking the same value for each group. In terms of the difference of proportions, this is H_0: $(p_1 - p_2) = 0$, *no difference*, or *no effect.*

Under the presumption for H_0 that $p_1 = p_2$, we estimate the common value of p_1 and p_2 by the proportion of the *total* sample in the category of interest. We denote this by $\hat{p}$. For example, if Group 1 had 7 successes in $n_1 = 20$ observations and Group 2 had 5 successes in $n_2 = 10$ observations, then $\hat{p}_1 = 7/20 = 0.35, \hat{p}_2 = 5/10 = 0.50$, and $\hat{p} = (7 + 5)/(20 + 10) = 12/30 = 0.40$. The proportion $\hat{p}$ is called a **pooled estimate** because it pools the total number of successes and total number of observations from the two samples. Whenever the sample sizes n_1 and n_2 are roughly equal, it falls about half way between $\hat{p}_1$ and $\hat{p}_2$. Otherwise, it falls closer to the sample proportion that has the larger sample size.

Recall

Section 9.2 presented the **significance test** for a single **proportion**. The *z* test statistic measures the number of standard errors that the sample proportion falls from the value in the null hypothesis. ◄

The test statistic measures the number of standard errors that the sample estimate $(\hat{p}_1 - \hat{p}_2)$ of $(p_1 - p_2)$ falls from its null hypothesis value of 0:

$$z = \frac{\text{Estimate} - \text{Null hypothesis value}}{\text{Standard error}} = \frac{(\hat{p}_1 - \hat{p}_2) - 0}{se_0}.$$

Recall

A **P-value** (introduced in Section 9.1) is the probability that the test statistic equals the observed value or a value even more extreme (in one tail for a one-sided H_a or both tails for a two-sided H_a) presuming that H_0 is true. Smaller P-values provide stronger evidence against H_0. ◀

The standard error for the test, denoted by se_0, is based on the presumption stated in H_0 that $p_1 = p_2$. It uses the pooled estimate $\hat{p}$ to estimate each population proportion. This standard error is

$$se_0 = \sqrt{\frac{\hat{p}(1-\hat{p})}{n_1} + \frac{\hat{p}(1-\hat{p})}{n_2}} = \sqrt{\hat{p}(1-\hat{p})\left(\frac{1}{n_1} + \frac{1}{n_2}\right)}.$$

In practice, when the sample proportions are close, se_0 is very close to se used in a confidence interval, which does not presume equal proportions.

As usual, the P-value for H_0: $p_1 = p_2$ depends on whether the alternative hypothesis is two-sided, H_a: $p_1 \neq p_2$, or one-sided, H_a: $p_1 > p_2$ or H_a: $p_1 < p_2$. When it is two-sided, the P-value is the two-tail probability beyond the observed z test statistic value from the standard normal distribution. This is the probability, presuming H_0 to be true, of obtaining results more extreme than observed in either direction.

In Practice Sample Size Guidelines for Significance Tests

Significance tests comparing proportions use the **sample size** guideline from confidence intervals: Each sample should have at least about 10 successes and 10 failures. Note that *two-sided tests* are robust against violations of this condition. In that case you can use the test with smaller samples. In practice, the two-sided test works well if there are at least five successes and five failures in each sample.

SUMMARY: Two-Sided Significance Test for Comparing Two Population Proportions

1. **Assumptions**
   - A categorical response variable for two groups
   - Independent random samples, either from random sampling or a randomized experiment
   - n_1 and n_2 are large enough that there are at least five successes and five failures in each group if using a two-sided alternative
2. **Hypotheses**
   Null H_0: $p_1 = p_2$ (that is, $p_1 - p_2 = 0$)
   Alternative H_a: $p_1 \neq p_2$ (one-sided H_a also possible; see after Example 5)
3. **Test Statistic**

$$z = \frac{(\hat{p}_1 - \hat{p}_2) - 0}{se_0} \text{ with } se_0 = \sqrt{\hat{p}(1-\hat{p})\left(\frac{1}{n_1} + \frac{1}{n_2}\right)},$$

   where $\hat{p}$ is the pooled estimate.
4. **P-value**
   P-value = Two-tail probability from standard normal distribution (Table A) of values even more extreme than observed z test statistic presuming the null hypothesis is true
5. **Conclusion**
   Smaller P-values give stronger evidence against H_0 and supporting H_a. Interpret the P-value in context. If a decision is needed, reject H_0 if P-value $\leq$ significance level (such as 0.05).

Example 5

Compare two population proportions

TV Watching and Aggressive Behavior

Picture the Scenario

A study[3] considered whether greater levels of television watching by teenagers were associated with a greater likelihood of aggressive behavior. The researchers randomly sampled 707 families in two counties in northern New York state and made follow-up observations over 17 years. Table 10.3 shows results about whether a sampled teenager later conducted any aggressive act against another person, according to a self-report by that person or by his or her parent.

[3]By J.G. Johnson et al., *Science*, vol. 295, March 29, 2002, pp. 2468–2471.

Table 10.3 TV Watching by Teenagers and Later Aggressive Acts

	Aggressive Act		
TV Watching	**Yes**	**No**	**Total**
Less than 1 hour per day	5	83	88
At least 1 hour per day	154	465	619

We'll identify Group 1 as those who watched less than 1 hour of TV per day, on average, as teenagers. Group 2 consists of those who averaged at least 1 hour of TV per day, as teenagers. Denote the population proportion committing aggressive acts by p_1 for the lower level of TV watching and by p_2 for the higher level of TV watching.

Questions to Explore

a. Find and interpret the P-value for testing $H_0: p_1 = p_2$ against $H_a: p_1 \neq p_2$.

b. Make a decision about H_0 using the significance level of 0.05.

Recall

It was noted in the Chapter 10 introduction that if a sample is randomly selected and the selected subjects are separated into groups according to their value for an explanatory variable, then the groups can be treated as independent random samples. ◀

Think It Through

a. The study used random sampling and then classified teenagers by the level of TV watching, so the samples were independent random samples. Each count in Table 10.3 is at least five, so we can use a large-sample test. The sample proportions of aggressive acts were $\hat{p}_1 = 5/88 = 0.057$ for the lower level of TV watching and $\hat{p}_2 = 154/619 = 0.249$ for the higher level. Under the null hypothesis presumption that $p_1 = p_2$, the pooled estimate of the common value p is $\hat{p} = (5 + 154)/(88 + 619) = 159/707 = 0.225$.

The standard error for the test is

$$se_0 = \sqrt{\hat{p}(1-\hat{p})\left(\frac{1}{n_1}+\frac{1}{n_2}\right)} = \sqrt{(0.225)(0.775)\left(\frac{1}{88}+\frac{1}{619}\right)} = 0.0476.$$

The test statistic for $H_0: p_1 = p_2$ is

$$z = \frac{(\hat{p}_1 - \hat{p}_2) - 0}{se_0} = \frac{(0.057 - 0.249) - 0}{0.0476} = \frac{-0.192}{0.0476} = -4.04.$$

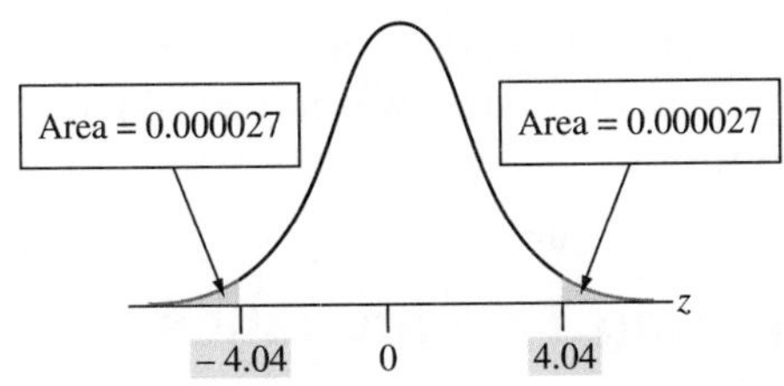

For the two-sided alternative hypothesis, the P-value is the two-tail probability from the standard normal distribution. A z-score of -4.04 is far out in the left tail. See the margin figure. From tables (such as Table A) or software, it has a P-value $= 2(0.000027) = 0.000054$, or 0.0001 rounded to four decimal places. Extremely strong evidence exists against the null hypothesis that the population proportions committing aggressive acts are the same for the two levels of TV watching. The study provides strong evidence in support of H_a.

b. Since the P-value is less than 0.05, we can reject H_0. We support $H_a: p_1 \neq p_2$ and conclude that the population proportions of aggressive acts differ for the two groups. The sample values suggest that the population proportion is higher for the higher level of TV watching. The final row of Table 10.4 shows how MINITAB reports this result and the margin of the next page shows screen shots from the TI-83+/84.

```
2-PropZTest
 x1:5
 n1:88
 x2:154
 n2:619
 p1:≠p2 <p2 >p2
 Calculate Draw
```

```
2-PropZTest
 p1≠p2
 z=-4.036
 p=5.442E-5
 p̂1=.057
 p̂2=.249
↓p̂=.225
```

TI-83+/84 output

Table 10.4 MINITAB Output for Example 5 on TV Watching and Aggression

Sample	X	N	Sample p
1	5	88	0.056818
2	154	619	0.248788

```
Difference = p(1) - p(2)
Estimate for difference: - 0.191970
95% CI for difference: (-0.251124, -0.132816)
Test for difference = 0(vs not = 0):z = - 4.04 P - Value = 0.000
```

Insight

This was an observational study. In practice, it would be impossible to conduct an experimental study by randomly assigning teenagers to watch little TV or watch much TV over several years. Also, just because a person watches more TV does not imply they watch more violence. We must be cautious of effects of lurking variables when we make conclusions. It is not proper to conclude that greater levels of TV watching *cause* later aggressive behavior. For instance, perhaps those who watched more TV had lower education levels, and perhaps lower education levels are associated with a greater likelihood of aggressive acts.

Try Exercise 10.8

Recall

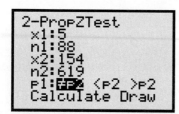

An **observational** study merely observes the explanatory variable, such as TV watching. An **experimental** study randomly allocates subjects to its levels. Sections 4.1 and 4.4 discussed limitations of observational studies, such as effects that lurking variables may have on the association. ◄

Caution

Considering a one-sided alternative would be questionable for this data since one of the counts is less than 10. ◄

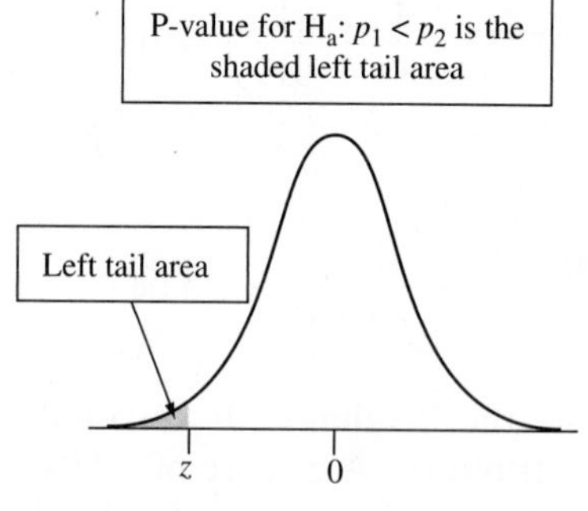

P-value for H_a: $p_1 > p_2$ is the shaded right tail area

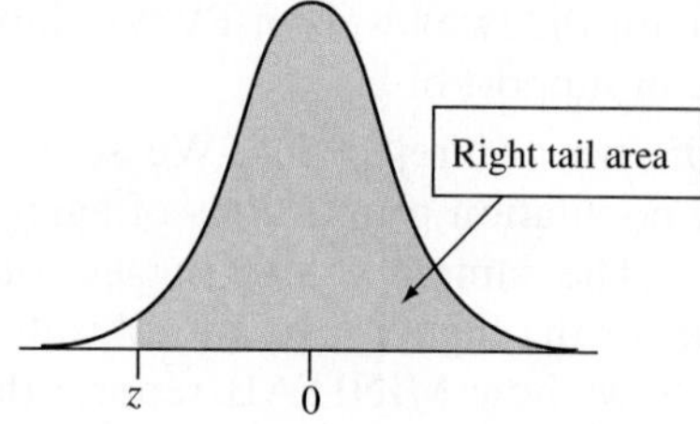

A One-Sided Alternative Hypothesis Example 5 requested a test for the two-sided alternative, H_a: $p_1 \neq p_2$. Two-sided P-values are typically reported in journals. However, suppose the researchers specifically predicted that greater levels of TV watching by teenagers were associated with greater likelihood of committing aggressive acts years later. Then they may have preferred to use the one-sided alternative, H_a: $p_1 < p_2$. This states that the population proportion of aggressive acts is higher at the higher level of TV watching. Equivalently, it is H_a: $(p_1 - p_2) < 0$.

This one-sided H_a predicts $(\hat{p}_1 - \hat{p}_2) < 0$ and the test statistic $z < 0$. The P-value is then a left-tail probability below the observed test statistic, $z = -4.04$. See the margin figure. From Table A or software, this is about 0.00003. The conclusion is that there was significantly greater probability of aggression for those who had watched more TV.

The one-sided H_a: $p_1 > p_2$ predicts $(\hat{p}_1 - \hat{p}_2) > 0$ and the test statistic $z > 0$. The P-value is then a right-tail probability. This alternative was not considered relevant for this study. See the margin figure.

The Standard Error for Comparing Two Statistics

Now that we've seen how to compare proportions inferentially, let's learn where the *se* formulas come from. Whenever we estimate a difference between two population parameters, a general rule specifies the standard error.

Standard Error of the Difference Between Two Estimates

For two estimates from independent samples, the standard error is

$$se(\text{estimate 1} - \text{estimate 2}) = \sqrt{[se(\text{estimate 1})]^2 + [se(\text{estimate 2})]^2}.$$

Each estimate has sampling error, as measured by the the standard error. The standard error of the difference of the estimates is determined as the square root of the sum of the squared standard errors of the two estimates.

Notice that we *add* squared standard errors under the square root sign, rather than subtract. For example, if we're comparing two proportions, then the standard error of the difference is *larger* than the standard error for either sample proportion alone. Why is this true? In practical terms, $(\hat{p}_1 - \hat{p}_2)$ is often farther from $(p_1 - p_2)$ than $\hat{p}_1$ is from p_1 or $\hat{p}_2$ is from p_2. For instance, in the aspirin and cancer study, *suppose that*

$$p_1 = p_2 = 0.0265 \text{ (unknown to us)},$$

but the sample proportions were

$$\hat{p}_1 = 0.030 \text{ and } \hat{p}_2 = 0.023,$$

as was actually observed for the sample data. Then the errors of estimation were

$$\hat{p}_1 - p_1 = 0.03 - 0.0265 = 0.0035 \text{ and } \hat{p}_2 - p_2 = 0.023 - 0.0265 = -0.0035,$$

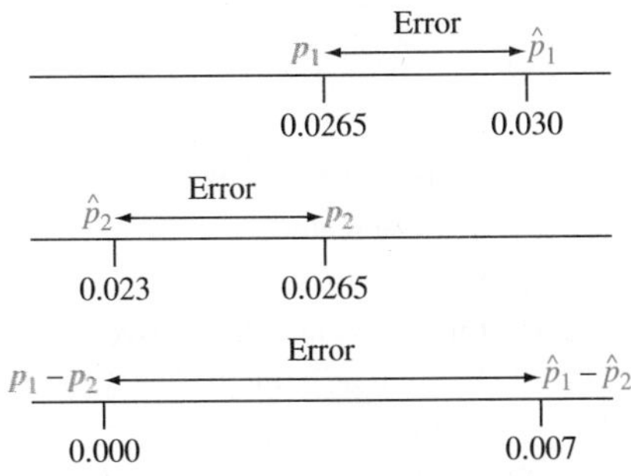

each estimate being off by a distance of 0.0035. But then $(\hat{p}_1 - \hat{p}_2) = 0.030 - 0.023 = 0.007$. That is, the estimate is 0.007 from $(p_1 - p_2) = 0$, larger than the error for either proportion individually. See the margin figure.

Let's apply the formula for the standard error of the difference between two estimates to the comparison of two proportions. The standard error of a single sample proportion $\hat{p}$ is $se = \sqrt{\hat{p}(1 - \hat{p})/n}$. Then, for two samples, the *se* of $(\hat{p}_1 - \hat{p}_2)$ is

$$se = \sqrt{[se(\text{estimate 1})]^2 + [se(\text{estimate 2})]^2} = \sqrt{\frac{\hat{p}_1(1 - \hat{p}_1)}{n_1} + \frac{\hat{p}_2(1 - \hat{p}_2)}{n_2}}.$$

We use this *se* formula to construct a confidence interval for the difference between two population proportions.

Small-Sample Inference for Comparing Proportions

The confidence interval for a difference of proportions specifies that each sample should have at least 10 outcomes of each type. For smaller sample sizes, the method may not work well: The sampling distribution of $(\hat{p}_1 - \hat{p}_2)$ may not be close to normal, and the estimate of the standard error may be poor. We won't cover this case here, but it is briefly discussed in Exercise 10.127. There are also small-sample significance tests for comparing proportions. We will discuss one, called *Fisher's exact test*, in the next chapter. Such tests have their own disadvantages, however, and for two-sided alternatives, the large-sample test usually performs quite well for small samples also.

10.1 Practicing the Basics

10.1 Unemployment rate According to the Bureau of Labor Statistics, the official unemployment rate was 15.5% among Blacks and 7.9% among Whites as of March 2011. During the recession of 2009–2011, the Black levels of unemployment have been similar or in many locations higher than those during the Great Depression era (www.bls.gov/news.release/empsit.t02.htm).

a. Identify the response variable and the explanatory variable.

b. Identify the two groups that are the categories of the explanatory variable.

c. The unemployment statistics are based on a sample of individuals. Were the samples of white individuals and black individuals independent samples, or dependent samples? Explain.

10.2 Sampling sleep The 2009 Sleep in America poll of a random sample of 1000 adults reported that respondents slept an average of 6.7 hours on weekdays and 7.1 hours on

TRY

weekends, and that 28% of respondents got eight or more hours of sleep on weekdays whereas 44% got eight or more hours of sleep on weekends (www.sleepfoundation.org).

a. To compare the means or the percentages using inferential methods, should you treat the samples on weekdays and weekends as independent samples, or as dependent samples? Explain.

b. To compare these results to polls of other people taken in previous years, should you treat the samples in the two years as independent samples, or as dependent samples? Explain.

10.3 Binge drinking The PACE project (pace.uhs.wisc.edu) TRY at the University of Wisconsin in Madison deals with problems associated with high-risk drinking on college campuses. Based on random samples, the study states that the percentage of UW students who reported bingeing at least three times within the past two weeks was 42.2% in 1999 ($n = 334$) and 21.2% in 2009 ($n = 843$).

a. Estimate the difference between the proportions in 1999 and 2009, and interpret.

b. Find the standard error for this difference. Interpret it.

c. Construct and interpret a 95% confidence interval to estimate the true change, explaining how your interpretation reflects whether the interval contains 0.

d. State and check the assumptions for the confidence interval in part c to be valid.

10.4 Less smoking now? The National Health Interview Survey conducted of 27,603 adults by the U.S. National Center for Health Statistics in 2009 indicated that 20.6% of adults were current smokers. A similar study conducted in 1991 of 42,000 adults indicated that 25.6% were current smokers.

a. Find and interpret a point estimate of the difference between the proportion of current smokers in 1991 and the proportion of current smokers in 2009.

b. A 99% confidence interval for the true difference is (0.042, 0.058). Interpret.

c. What assumptions must you make for the interval in part b to be valid?

10.5 Do you believe in miracles? Let p_1 and p_2 denote the population proportions of males and females in the United States who answer, yes, definitely when asked whether they believe in miracles. Estimate these by going to the General Social Survey Web site (sda.berkeley.edu/GSS), selecting GSS with *No Weight* as the default weight selection, and then entering the variable MIRACLES for the rows, SEX for the columns, and YEAR (2008) as the selection filter.

a. Report point estimates of p_1 and p_2.

b. Construct a 95% confidence interval for $(p_1 - p_2)$, specifying the assumptions you make to use this method. Interpret.

c. Based on the interval in part b, explain why the proportion believing in miracles may have been quite a bit larger for females, or it might have been only moderately larger.

10.6 Aspirin and heart attacks in Sweden A Swedish study TRY used 1360 patients who had suffered a stroke. The study randomly assigned each subject to an aspirin treatment or a placebo treatment.[4] The table shows MINITAB output, where X is the number of deaths due to heart attack during a follow-up period of about 3 years. Sample 1 received the placebo and sample 2 received aspirin.

a. Explain how to obtain the values labeled "Sample p."

b. Explain how to interpret the value given for "estimate for difference."

c. Explain how to interpret the confidence interval, indicating the relevance of 0 falling in the interval.

d. If we instead let sample 1 refer to the aspirin treatment and sample 2 the placebo treatment, explain how the estimate of the difference and the 95% confidence interval would change. Explain how then to interpret the confidence interval. (Note that the output below would change for the analysis of this difference.)

Deaths due to heart attacks in Swedish study

```
Sample      X        N        Sample p
1           28       684      0.040936
2           18       676      0.026627
Difference = p(1) - p(2)
Estimate for difference: 0.0143085
95% CI for difference: (-0.00486898,
0.0334859)
Test for difference = 0 (vs not = 0):
Z = 1.46 P - Value = 0.144
```

10.7 Swedish study test Refer to the previous exercise.

a. State the hypotheses that were tested.

b. Explain how to interpret the P-value for the test.

c. Report the P-value for the one-sided alternative hypothesis that the chance of death due to heart attack is lower for the aspirin group.

d. Even though the difference between the sample proportions was larger than in the Physicians Health Study (Examples 2–4), the results are less statistically significant. Explain how this could be. (*Hint:* How do the sample sizes compare for the two studies? How does this affect the standard error and thus the test statistic and P-value?)

10.8 Significance test for aspirin and cancer deaths study In TRY the study for cancer death rates, consider the null hypothesis that the population proportion of cancer deaths p_1 for placebo is the same as the population proportion p_2 for aspirin. The sample proportions were $\hat{p}_1 = 347/11{,}535 = 0.030$ and $\hat{p}_2 = 327/14{,}035 = 0.023$.

a. For testing H_0: $p_1 = p_2$ against H_a: $p_1 \neq p_2$, show that the pooled estimate of the common value p under H_0 is $\hat{p} = 0.027$ and the standard error is 0.002.

b. Show that the test statistic is $z = 3.5$.

c. Find and interpret the P-value in context.

10.9 Drinking and unplanned sex In the study of binge drinking mentioned in Exercise 10.3, the percent who said they had engaged in unplanned sexual activities because of drinking alcohol was 30.7% in 1999 and 23.0% in 2009. Is this change statistically significant at the 0.05 significance level?

a. Specify assumptions, notation, and hypotheses for a two-sided test.

b. Show how to find the pooled estimate of the proportion to use in a test. (*Hint:* You need the count in each year rather than the proportion. In 1999, note this is $(0.307)334 = 103$.) Interpret this estimate.

c. Find the standard error for the difference used in the test. Interpret it.

---

[4]Based on results described in *Lancet*, vol. 338, pp. 1345–1349 (1991).

d. Find the test statistic and P-value. Make a decision using a significance level of 0.05.

10.10 Comparing marketing commercials Two TV commercials are developed for marketing a new product. A volunteer test sample of 200 people is randomly split into two groups of 100 each. In a controlled setting, Group A watches commercial A and Group B watches commercial B. In Group A, 25 say they would buy the product. In group B, 20 say they would buy the product. The marketing manager who devised this experiment concludes that commercial A is better. Is this conclusion justified? Analyze the data. If you prefer, use software (such as MINITAB) for which you can enter summary counts.

a. Show all steps of your analysis, including assumptions.

b. Comment on the manager's conclusion, and indicate limitations of the experiment.

10.11 Hormone therapy for menopause The Women's Health Initiative conducted a randomized experiment to see if hormone therapy was helpful for postmenopausal women. The women were randomly assigned to receive the estrogen plus progestin hormone therapy or a placebo. After five years, 107 of the 8506 on the hormone therapy developed cancer and 88 of the 8102 in the placebo group developed cancer. Is this a significant difference?

a. Set up notation, and state assumptions and hypotheses.

b. Find the test statistic and P-value, and interpret. (If you prefer, use software, such as MINITAB, for which you can conduct the analysis by entering summary counts.)

c. What is your conclusion, for a significance level of 0.05? (The study was planned to be eight years long but was stopped after five years because of increased heart and cancer problems for the therapy group. This shows a benefit of doing two-sided tests, as results sometimes have the opposite direction from the expected one.)

10.12 TV watching A researcher predicts that the percentage of people who do not watch TV is higher now than before the advent of the Internet. Let p_1 denote the population proportion of American adults in 1975 who reported watching no TV. Let p_2 denote the corresponding population proportion in 2008.

a. Set up null and alternative hypotheses to test the researcher's prediction.

b. According to General Social Surveys, 57 of the 1483 subjects sampled in 1975 and 87 of the 1324 subjects sampled in 2008 reported watching no TV. Find the sample estimates of p_1 and p_2.

c. Show steps of a significance test. Explain whether the results support the reseacher's claim.

10.13 Living poorly A survey of residents of several Western countries estimated the percentage in each country who reported going without food or without health care or without adequate clothing sometime in the past year. For example, the estimated percentage was 33% in the United States (the highest in the survey) and 9% in Japan. Are the samples from the United States and Japan independent samples or dependent samples? Why?

10.2 Quantitative Response: Comparing Two Means

We can compare two groups on a *quantitative response variable* by comparing their means. What does the difference between the sample means tell us about the difference between the population means? We'll find out in this section.

Compare two population means

Example 6

Teenagers on Nicotine

Picture the Scenario

A recent study evaluated how addicted teenagers become to nicotine once they start smoking.[5] The 30-month study involved a random sample of 679 seventh-graders from two Massachusetts cities. Of this sample, 332 students who had ever smoked were the subjects. The response variable was constructed from a questionnaire called the Hooked on Nicotine Checklist (HONC). This is a list of ten questions such as "Have you ever tried to quit but couldn't?" and "Is it hard to keep from smoking in places where it is banned, like school?" The HONC score is the total number of questions to which a student answered yes during the study. Each student's HONC score falls between 0 and 10. The higher the score, the more hooked that student is on nicotine.

[5] J. DiFranza et al., *Archives of Pediatric and Adolescent Medicine*, vol. 156, 2002 pp. 397–403.

The study considered explanatory variables, such as gender, that might be associated with the HONC score. Figure 10.4, taken from the journal article about this study, shows the sample data distributions of the HONC scores for females and males who had ever smoked. Table 10.5 reports some descriptive statistics.

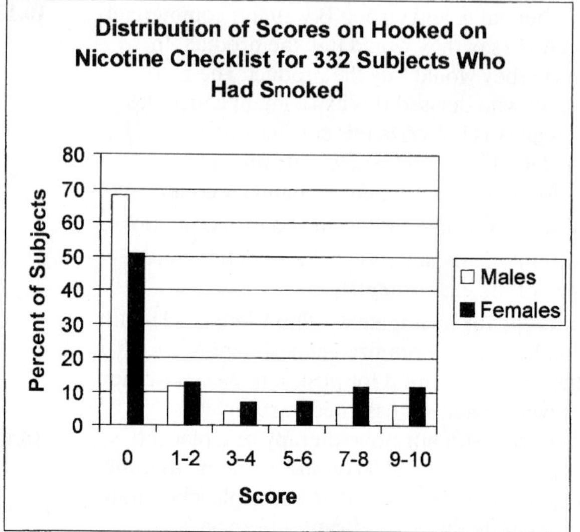

▲ **Figure 10.4** Sample Data Distribution of Hooked on Nicotine Checklist (HONC) Scores for Teenagers Who Have Smoked. **Question** Which group seems to have greater nicotine dependence, females or males? Explain your choice using the graph.

Table 10.5 Summary of Hooked on Nicotine Checklist (HONC) Scores, by Gender

		HONC Score	
Group	**Sample Size**	**Mean**	**Standard Deviation**
Females	150	2.8	3.6
Males	182	1.6	2.9

Question to Explore

How can we compare the sample HONC scores for females and males?

Think It Through

From Figure 10.4, a substantial proportion of teenagers who had ever smoked show no nicotine dependence. For both females and males the sample data distribution is highly skewed to the right. Because of the skew, we could use the median to summarize the sample HONC scores, but that has limited usefulness for such highly discrete data: The median is 0 both for boys and girls. We will estimate the population means and the difference between them. For Table 10.5, let's identify females as Group 1 and males as Group 2. Then $\bar{x}_1 = 2.8$ and $\bar{x}_2 = 1.6$. We estimate $(\mu_1 - \mu_2)$ by $(\bar{x}_1 - \bar{x}_2) = 2.8 - 1.6 = 1.2$. On average, females answered yes to about one more question on the HONC scale than males did.

Insight

This analysis uses descriptive statistics only. We'll next see how to make inferences about the difference between population means. You can conduct an inferential comparison of HONC scores by gender in Exercise 10.25. How do you think the nonnormality of HONC distributions affects inference about the population means?

Try Exercise 10.23, parts a and b

Recall

From the formula box at the end of Section 10.1, for two estimates from independent samples,

$se(\text{estimate 1} - \text{estimate 2}) = \sqrt{[se(\text{est. 1})]^2 + [se(\text{est. 2})]^2}$,

the square root of the sum of squared standard errors of the two estimates. ◀

Standard Error for Comparing Two Means

How well does the difference $(\bar{x}_1 - \bar{x}_2)$ between two sample means estimate the difference between two population means? This is described by the standard error of the sampling distribution of $(\bar{x}_1 - \bar{x}_2)$.

We can find the standard error of $(\bar{x}_1 - \bar{x}_2)$ using the general rule in the formula in the margin. The standard error of a single sample mean is $se = s/\sqrt{n}$, where s is the sample standard deviation. Let s_1 and s_2 denote the sample standard deviations for the first and second samples. Then, for independent samples, the standard error of $(\bar{x}_1 - \bar{x}_2)$ is

$$se = \sqrt{[se(\bar{x}_1)]^2 + [se(\bar{x}_2)]^2} = \sqrt{\frac{s_1^2}{n_1} + \frac{s_2^2}{n_2}}.$$

Compare means and find standard error

Example 7

Nicotine Dependence

Picture the Scenario

Another explanatory variable in the teenage smoking study was whether a subject was still a smoker when the study ended. The study had 75 smokers and 257 ex-smokers at the end of the study. The HONC means describing nicotine addiction were 5.9 ($s = 3.3$) for the smokers and 1.0 ($s = 2.3$) for the ex-smokers.

Questions to Explore

a. Compare smokers and ex-smokers on their mean HONC scores.

b. What is the standard error of the difference in sample mean HONC scores? How do you interpret that *se*?

Think It Through

a. Let's regard the smokers as Group 1 and ex-smokers as Group 2. Then, $\bar{x}_1 = 5.9$ and $\bar{x}_2 = 1.0$. Since $(\bar{x}_1 - \bar{x}_2) = 5.9 - 1.0 = 4.9$, on average, smokers answered yes to nearly five more questions than ex-smokers did on the 10-question HONC scale. That's a large sample difference.

b. Applying the formula for the *se* of $(\bar{x}_1 - \bar{x}_2)$ to these data,

$$se = \sqrt{\frac{s_1^2}{n_1} + \frac{s_2^2}{n_2}} = \sqrt{\frac{(3.3)^2}{75} + \frac{(2.3)^2}{257}} = 0.41.$$

This describes the variability of the sampling distribution of $(\bar{x}_1 - \bar{x}_2)$, as shown in the margin figure. If other random samples of this size were taken from the study population, the difference between the sample means would vary from study to study. The standard deviation of the values for $(\bar{x}_1 - \bar{x}_2)$ would equal about 0.41.

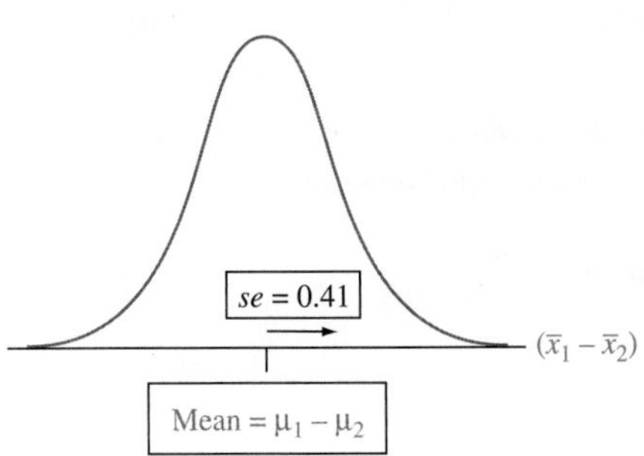

Insight

This standard error will be used in confidence intervals and in significance tests for comparing means.

Try Exercise 10.16, part b

Confidence Interval for the Difference Between Two Population Means

Recall

See Section 8.3 to review the **confidence interval** for a single **mean**, the ***t* distribution** and *t*-scores from it. The 95% confidence interval is

$$\bar{x} \pm t_{.025}(se),$$

where $se = s/\sqrt{n}$. Recall that $df = n - 1$ for inference about a single mean. ◀

As usual, a confidence interval takes the estimate and adds and subtracts a margin of error. For large random samples:

- The sampling distribution of a sample mean is approximately normal, by the central limit theorem.
- Likewise, $(\bar{x}_1 - \bar{x}_2)$ has a sampling distribution that is approximately normal.
- For 95% confidence, the margin of error is about two standard errors, so the confidence interval for the difference $(\mu_1 - \mu_2)$ between the population means is approximately

$$(\bar{x}_1 - \bar{x}_2) \pm 2(se).$$

More precisely, the multiple of *se* is a *t*-score from a table of *t* distribution values because of the extra variability that results from estimating parameters in finding the *se*.

The degrees of freedom for the *t*-score multiple of *se* depends on the sample standard deviations and the sample sizes. The formula is messy and does not give insight into the method, so we leave it as a footnote.[6] If $s_1 = s_2$ and $n_1 = n_2$, it simplifies to $df = (n_1 + n_2 - 2)$. This is the sum of the *df* values for single-sample inference about each group, or $df = (n_1 - 1) + (n_2 - 1) = n_1 + n_2 - 2$. *Generally, df falls somewhere between* $n_1 + n_2 - 2$ *and the minimum of* $(n_1 - 1)$ *and* $(n_2 - 1)$.

In Practice Software Finds *df* for Comparing Means

Software calculates the *df* value for comparing means for you. When s_1 and s_2 are similar and n_1 and n_2 are close, *df* is close to $n_1 + n_2 - 2$. Without software, you can take *df* to be the smaller of $(n_1 - 1)$ and $(n_2 - 1)$ and this will be safe, as the *t*-score will be larger than you actually need.

Recall

From Section 8.3, a method is **robust** with respect to a particular assumption if it works well even when that assumption is violated. ◀

When either sample size is small (roughly $n_1 < 30$ or $n_2 < 30$), we cannot rely on the central limit theorem. In that instance, the method makes the assumption that the population distributions are normal, so $(\bar{x}_1 - \bar{x}_2)$ has a bell-shaped sampling distribution. In practice, the method is *robust*, and it works quite well even if the distributions are not normal. This is subject to the usual caveat: We need to be on the lookout for outliers that might affect the means or their usefulness as a summary measure.

SUMMARY: Confidence Interval for Difference Between Population Means

For two samples with sizes n_1 and n_2 and standard deviations s_1 and s_2, a 95% confidence interval for the difference $(\mu_1 - \mu_2)$ between the population means is

$$(\bar{x}_1 - \bar{x}_2) \pm t_{.025}(se), \text{ with } se = \sqrt{\frac{s_1^2}{n_1} + \frac{s_2^2}{n_2}}.$$

---

[6] It is $df = \dfrac{\left(\frac{s_1^2}{n_1} + \frac{s_2^2}{n_2}\right)^2}{\frac{1}{n_1 - 1}\left(\frac{s_1^2}{n_1}\right)^2 + \frac{1}{n_2 - 1}\left(\frac{s_2^2}{n_2}\right)^2}$, called the Welch-Satterthwaite formula.

Software provides $t_{.025}$, the t-score with right-tail probability 0.025 (total probability = 0.95 between $-t_{.025}$ and $t_{.025}$).

This method assumes:

- Independent random samples from the two groups, either from random sampling or a randomized experiment.
- An approximately normal population distribution for each group. (This is mainly important for small sample sizes, and even then the method is robust to violations of this assumption.)

Example 8

Confidence interval

Nicotine Addiction

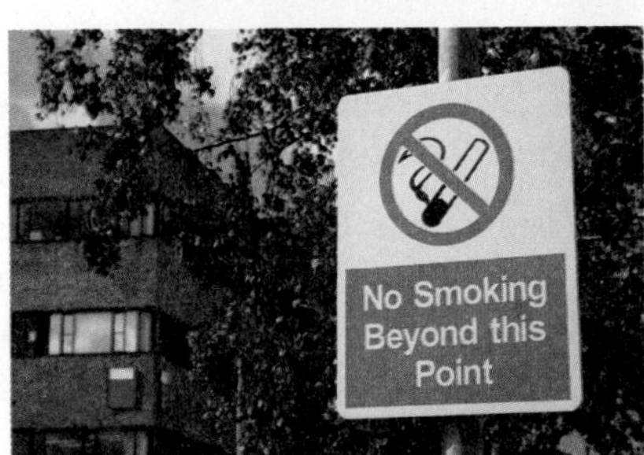

Picture the Scenario

For the teenage smoking study, let's make an inference comparing the population mean nicotine addiction (as summarized by the HONC score) for smokers and ex-smokers. From Example 7, $\bar{x}_1 = 5.9$ with $s_1 = 3.3$ for $n_1 = 75$ smokers, and $\bar{x}_2 = 1.0$ with $s_2 = 2.3$ for $n_2 = 257$ ex-smokers.

Questions to Explore

a. Were the HONC sample data distributions for smokers and ex-smokers approximately normal? How does this affect inference?

b. Software reports a 95% confidence interval of (4.1, 5.7) for the difference between the population mean HONC score for smokers and ex-smokers. Show how it obtained this confidence interval and interpret it.

Think It Through

a. For the ex-smokers (Group 2), the values $\bar{x}_2 = 1.0$ and $s_2 = 2.3$ suggest that the HONC distribution is far from bell-shaped because the lowest possible HONC score of 0 is less than 1 standard deviation below the mean. This is not problematic. With large samples, the confidence interval method does not require normal population distributions because the central limit theorem implies that the sampling distribution is approximately normal. (Recall that even for small samples, the method is robust when population distributions are not normal.)

b. For these sample sizes and s_1 and s_2 values, software reports that $df = 95$. The t-score for a 95% confidence interval is $t_{.025} = 1.985$. (As an approximation, Table B reports the value $t_{.025} = 1.984$ for $df = 100$. Also, the smaller of $(n_1 - 1)$ and $(n_2 - 1)$ is 74, so if you did not have software to find df, you could use the t-score with $df = 74$.)

Denote the population mean HONC score by μ_1 for smokers and μ_2 for ex-smokers. From Example 7, $(\bar{x}_1 - \bar{x}_2) = 4.9$ has a standard error of $se = 0.41$. The 95% confidence interval for $(\mu_1 - \mu_2)$ is

$$(\bar{x}_1 - \bar{x}_2) \pm 1.985(se), \text{ or } 4.9 \pm 1.985(0.41),$$

$$\text{which equals } 4.9 \pm 0.8, \text{ or } (4.1, 5.7).$$

We can be 95% confident that plausible values for $(\mu_1 - \mu_2)$, the difference between the population mean HONC scores for smokers and for

TI-83+/84 output

ex-smokers, falls between 4.1 and 5.7. We can infer that the population mean for the smokers is between 4.1 higher and 5.7 higher than for the ex-smokers. The margin shows screen shots from the TI-83+/84.

Insight

We are 95% confident that the smokers answer yes to about 4 to 6 more of the 10 questions, on average, than the ex-smokers. For the 10-point HONC scale, this is quite a substantial difference.

Try Exercise 10.23

Interpret a Confidence Interval for a Difference of Means

Besides interpreting the confidence interval comparing two means in context by stating plausible values for the difference in the two means, you can judge the implications using the same criteria as in comparing two proportions:

- *Check whether or not 0 falls in the interval.* When it does, 0 is a plausible value for $(\mu_1 - \mu_2)$, meaning that possibly $\mu_1 = \mu_2$. Figure 10.5 illustrates. For example, suppose the confidence interval for $(\mu_1 - \mu_2)$, equals $(-4.1, 5.7)$. Then, the population mean for smokers may be as much as 4.1 lower or as much as 5.7 higher than the population mean for ex-smokers.

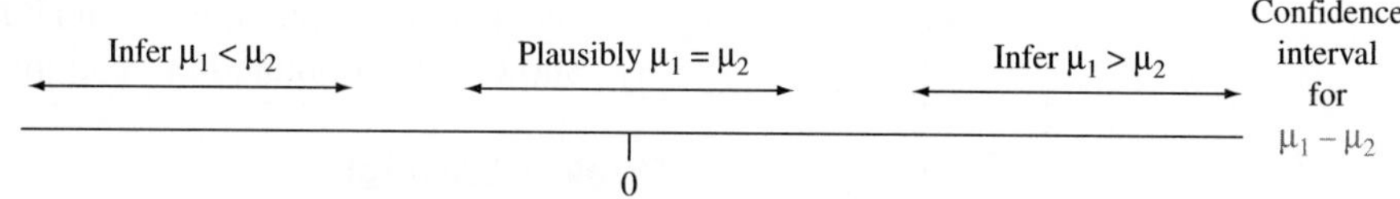

▲ **Figure 10.5 Three Confidence Intervals for the Difference Between Two Means.** When the interval contains 0, it is plausible that the population means may be equal. Otherwise, we can predict which is larger. **Question** Why do positive numbers in the confidence interval for $(\mu_1 - \mu_2)$ suggest that $\mu_1 > \mu_2$?

- *A confidence interval for* $(\mu_1 - \mu_2)$ *that contains only positive numbers suggests that* $(\mu_1 - \mu_2)$ *is positive.* We then infer that μ_1 is larger than μ_2. The 95% confidence interval in Example 8 is (4.1, 5.7). Since $(\mu_1 - \mu_2)$ is the difference between the mean for smokers and the mean for ex-smokers, we infer that the population mean is higher for smokers.
- *A confidence interval for* $(\mu_1 - \mu_2)$ *that contains only negative numbers suggests that* $(\mu_1 - \mu_2)$ *is negative.* We then infer that μ_1 is smaller than μ_2.
- *Which group is labeled 1 and which is labeled 2 is arbitrary.* If you change this, the confidence interval has the same endpoints but with different sign. For instance, (4.1, 5.7) becomes $(-5.7, -4.1)$. The margin shows screen shots from the TI-83+/84 for the data from the previous example with the group numbers switched.

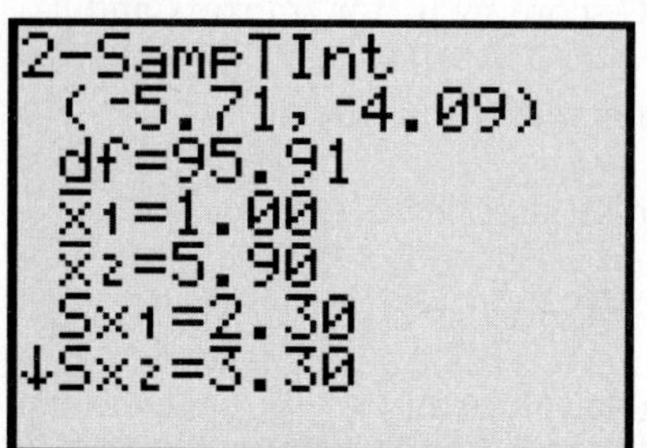

TI-83+/84 output

Significance Tests Comparing Population Means

Another way to compare two population means is with a significance test of the null hypothesis H_0: $\mu_1 = \mu_2$ of equal means. The assumptions of the test are the same as for a confidence interval—independent random samples and approximately normal population distributions for each group. When n_1 and n_2 are at least about 30 each, the normality assumption is not important because of the central limit theorem. For two-sided alternatives, the test is robust against violations of the normal assumption even when sample sizes are small. However, one-sided t tests are not trustworthy if a sample size is below 30 and the population distribution is highly skewed.

The test uses the usual form for a test statistic,

$$\frac{\text{Estimate of parameter} - \text{Null hypothesis value of parameter}}{\text{Standard error of estimate}}.$$

Recall

Section 9.3 presented the **significance test** for a single **mean**. The t test statistic measures the number of standard errors that the sample mean falls from the value in the null hypothesis. ◀

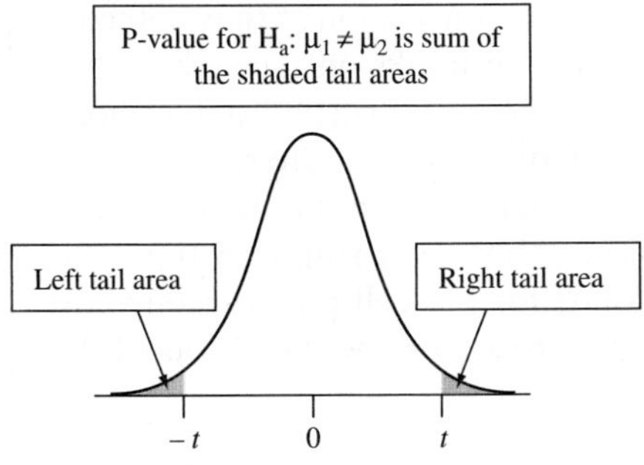

Treating the difference $(\mu_1 - \mu_2)$, as the parameter, we test that $(\mu_1 - \mu_2) = 0$, or, the null hypothesis value of the parameter $(\mu_1 - \mu_2)$, is 0. The estimate of the parameter $(\mu_1 - \mu_2)$, is $(\bar{x}_1 - \bar{x}_2)$. The standard error of $(\bar{x}_1 - \bar{x}_2)$ is the same for a significance test as for a confidence interval. The test statistic is

$$t = \frac{(\bar{x}_1 - \bar{x}_2) - 0}{se}, \text{ where } se = \sqrt{\frac{s_1^2}{n_1} + \frac{s_2^2}{n_2}}.$$

When H_0 is true, this statistic has approximately a t sampling distribution. Software can determine the *df* value.

The P-value for the test depends on whether the alternative hypothesis is two-sided, H_a: $\mu_1 \neq \mu_2$, or one-sided, H_a: $\mu_1 > \mu_2$ or H_a: $\mu_1 < \mu_2$. For the two-sided alternative, the P-value is the two-tail probability beyond the observed t value. That is, we find the probability of results more extreme in either direction under the presumption that H_0 is true. See the margin figure.

SUMMARY: Two-Sided Significance Test for Comparing Two Population Means

1. **Assumptions**
   - A quantitative response variable for two groups
   - Independent random samples, either from random sampling or a randomized experiment
   - Approximately normal population distribution for each group. (This is mainly important for small sample sizes, and even then the two-sided test is robust to violations of this assumption.)
2. **Hypotheses**

   H_0: $\mu_1 = \mu_2$

   H_a: $\mu_1 \neq \mu_2$ (one-sided H_a: $\mu_1 > \mu_2$ or H_a: $\mu_1 < \mu_2$ also possible)
3. **Test Statistic**

   $$t = \frac{(\bar{x}_1 - \bar{x}_2) - 0}{se} \text{ where } se = \sqrt{\frac{s_1^2}{n_1} + \frac{s_2^2}{n_2}}$$
4. **P-value**

   P-value = Two-tail probability from t distribution of values even more extreme than observed t test statistic, presuming the null hypothesis is true with *df* given by software.
5. **Conclusion**

   Smaller P-values give stronger evidence against H_0 and supporting H_a. Interpret the P-value in context, and if a decision is needed, reject H_0 if P-value $\leq$ significance level (such as 0.05).

Compare population means using significance test

Example 9

Cell Phone Use While Driving Reaction Times

Picture the Scenario

An experiment[7] investigated whether cell phone use impairs drivers' reaction times, using a sample of 64 students from the University of Utah. Students were randomly assigned to a cell phone group or to a control group, 32 to each. On a simulation of driving situations, a target flashed red or green at irregular periods. Participants pressed a brake button as soon as they detected a red light. The control group listened to radio or books-on-tape while they performed the simulated driving. The cell phone group carried out a phone conversation about a political issue with someone in a separate room.

The experiment measured each subject's mean response time over many trials. Averaged over all trials and subjects, the mean response time was 585.2 milliseconds (a bit over half a second) for the cell phone group and 533.7 milliseconds for the control group. Figure 10.6 shows box plots of the responses for the two groups.

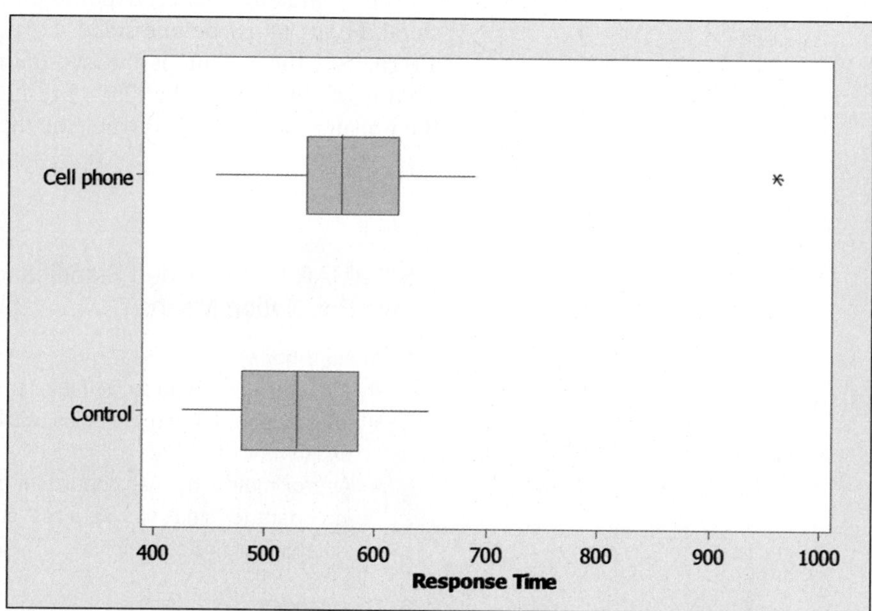

▲ **Figure 10.6** MINITAB Box Plots of Response Times for Cell Phone Study.
Question Does either box plot show any irregularities that could affect the analysis?

Denote the population mean response time by μ_1 for the cell phone group and by μ_2 for the control group. Table 10.6 shows how MINITAB reports inferential comparisons of those two means. The margin on the next page contains screen shots from the TI-83+/84.

Questions to Explore

a. Show how MINITAB got the test statistic for testing H_0: $\mu_1 = \mu_2$ against H_a: $\mu_1 \neq \mu_2$.

b. Report and interpret the P-value, and state the decision you would make about the population using a 0.05 significance level.

[7]Data courtesy of David Strayer, University of Utah. Example based on Experiment 1a in article by D. Strayer and W. Johnston, *Psych. Science*, vol. 21, 2001, pp. 462–466.

```
2-SampTTest
 Inpt:Data Stats
 x̄1:585.2
 Sx1:89.6
 n1:32
 x̄2:533.7
 Sx2:65.3
↓n2:32
```

```
2-SampTTest
↑n1:32
 x̄2:533.7
 Sx2:65.3
 n2:32
 μ1:≠μ2 <μ2 >μ2
 Pooled:No Yes
 Calculate Draw
```

```
2-SampTTest
 μ1≠μ2
 t=2.628
 p=.011
 df=56.685
 x̄1=585.200
↓x̄2=533.700
```

TI-83+/84 output

Table 10.6 MINITAB Output Comparing Mean Response Times for Cell Phone and Control Groups

```
               N          Mean          StDev
Cell
phone         32         585.2          89.6

Control       32         533.7          65.3

Difference = mu (Cell phone) − mu (Control)

Estimate for difference: 51.5172

95% CI for difference: (12.2393, 90.7951)

T-Test of difference = 0 (vs not = ):

T-Value = 2.63 P-Value = 0.011 DF = 56
```

c. What do the box plots tell us about the suitability of these analyses? What effect does the outlier for the cell phone group have on the analysis?

Think It Through

a. We estimate the difference $\mu_1 - \mu_2$ by $(\bar{x}_1 - \bar{x}_2) = 585.2 - 533.7 = 51.5$, shown in Table 10.6. The standard error of this estimate is

$$se = \sqrt{\frac{s_1^2}{n_1} + \frac{s_2^2}{n_2}} = \sqrt{\frac{(89.6)^2}{32} + \frac{(65.3)^2}{32}} = 19.6.$$

The test statistic for H_0: $\mu_1 = \mu_2$ equals

$$t = \frac{(\bar{x}_1 - \bar{x}_2)}{se} = \frac{51.5}{19.6} = 2.63.$$

b. The P-value is the two-tail probability from a t distribution. Table 10.6 reports $df = 56$ and a P-value $= 0.01$. If H_0 were true, the probability would be 0.01 of getting a t test statistic this large or even larger in either tail. The P-value is less than 0.05, so we can reject H_0. Suppose the entire population used a cell phone in this experiment, or the entire population did not use a cell phone. A population mean applies in each case. Then we have enough evidence to conclude that the population mean response times would differ between the cell phone and control groups. The sample means suggest that the population mean is higher for the cell phone group.

c. The t inferences assume normal population distributions. The box plots do not show any substantial skew, but there is an extreme outlier for the cell phone group. One subject in that group had a very slow mean reaction time. Because this observation is so far from the others in that group, it's a good idea to make sure the results of the analysis aren't affected too strongly by that single observation.

If we delete the extreme outlier for the cell phone group, software reports

```
                 N          Mean          StDev
Cell phone      31         573.1          58.9

Control         32         533.7          65.3

Estimate for difference: 39.4265

95% CI for difference: (8.1007, 70.7522)

T-Test of difference = 0 (vs not = ):

T-Value = 2.52 P-Value = 0.015 DF = 60
```

The mean and standard deviation for the cell phone group now decrease substantially. However, the t test statistic is not much different, and the P-value is still small, 0.015, leading to the same conclusion.

Insight

Even though the difference between the sample means decreased from 51.5 to 39.4 when we deleted the outlier, the standard error also got smaller (you can check that it equals 15.7) because of the smaller standard deviation for the cell phone group after removing the outlier. That's why the t test statistic did not change much. In practice, you should not delete outliers from a data set without sufficient cause (for example, if it seems the observation was incorrectly recorded). However, it's a good idea to check for *sensitivity* of an analysis to an outlier, as we did here, by repeating the analysis without it. If the results change much, it means that the inference including the outlier is on shaky ground.

Try Exercise 10.25

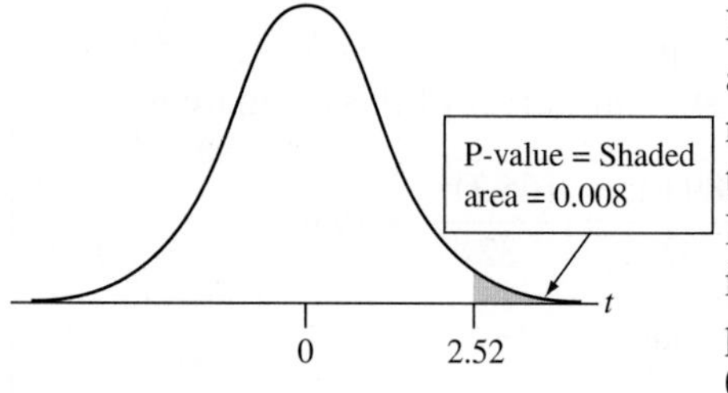

Example 9 used a two-sided alternative, which is the way that research results are usually reported in journal articles. But the researchers thought that the mean response time would be *greater* for the cell phone group than for the control. So, for their own purposes, they could find the P-value for the one-sided H_a: $\mu_1 > \mu_2$ (that is, $\mu_1 - \mu_2 > 0$), which predicts a higher mean response time for the cell phone group. For the analysis without the outlier, the P-value is the probability to the *right* of $t = 2.52$. This is half the two-sided P-value, namely $0.015/2 = 0.008$. See margin figure. There is very strong evidence in favor of this alternative.

Connection Between Confidence Intervals and Tests

We learn even more by constructing a confidence interval for $(\mu_1 - \mu_2)$. Let's do this for the data set without the outlier, for which $df = 60$ and the difference of sample means of 39.4 had a standard error of 15.7. The t-score with $df = 60$ for a 95% confidence interval is $t_{.025} = 2.000$. The confidence interval is

$$(\bar{x}_1 - \bar{x}_2) \pm t_{.025}(se), \text{ which is } 39.4 \pm 2.000(15.7), \text{ or } (8.1, 70.8).$$

This result is more informative than the test, because it shows the range of realistic values for the difference between the population means. The interval (8.1, 70.8) for $\mu_1 - \mu_2$ is quite wide. It tells us that the population means could be similar, or the mean response may be as much as about 71 milliseconds higher for the cell phone group. This is nearly a tenth of a second, which could be crucial in a practical driving situation.

The confidence interval (8.1, 70.8) does not contain 0. This inference agrees with the significance test that the population mean response times differ. Recall that Section 9.3 showed that the result of a two-sided significance test about a mean is consistent with a confidence interval for that mean. The same is true in comparing two means. If a test rejects H_0: $\mu_1 = \mu_2$, then the confidence interval for $(\mu_1 - \mu_2)$ having the same error probability does not contain 0.

In Example 9 without the outlier, the P-value = 0.015, so we rejected H_0: $\mu_1 = \mu_2$ at the 0.05 significance level. Likewise, the 95% confidence interval for $(\mu_1 - \mu_2)$ of (8.1, 70.8) does not contain 0, the null hypothesis value.

By contrast, when a 95% confidence interval for $(\mu_1 - \mu_2)$ *contains* 0, then 0 is a plausible value for $(\mu_1 - \mu_2)$ In a test of H_0: $\mu_1 = \mu_2$ against H_a: $\mu_1 \neq \mu_2$, the P-value would be > 0.05. The test would not reject H_0 at the 0.05 significance level and would conclude that the population means may be equal.

10.2 Practicing the Basics

10.14 Laptops lower GPA? A February 2007 story on www.redandblack.com stated, "Students who use laptops in class have lower GPAs, according to a study by Cornell University."

a. Suppose this conclusion was based on a significance test comparing means. Defining notation in context, identify the groups and the population means and state the null hypothesis for the test.

b. Suppose the study conclusion was based on a P-value of 0.01 obtained for the significance test mentioned in part a. Explain what you could learn from a confidence interval comparing the means that you are not able to learn from this P-value.

10.15 Address global warming You would like to determine what students at your school would be willing to do to help address global warming and the development of alternatively fueled vehicles. To do this, you take a random sample of 100 students. One question you ask them is, "How high of a tax would you be willing to add to gasoline (per gallon) in order to encourage drivers to drive less or to drive more fuel-efficient cars?" You also ask, "Do you believe (yes or no) that global warming is a serious issue that requires immediate action such as the development of alternatively fueled vehicles?" In your statistical analysis, use inferential methods to compare the mean response on gasoline taxes (the first question) for those who answer yes and for those who answer no to the second question. For this analysis,

a. Identify the response variable and the explanatory variable.

b. Are the two groups being compared independent samples or dependent samples? Why?

c. Identify a confidence interval you could form to compare the groups, specifying the parameters used in the comparison.

10.16 Housework for women and men Do women tend to spend more time on housework than men? If so, how much more? Based on data from the National Survey of Families and Households, one study reported the results in the table for the number of hours spent in housework per week. (*Source*: Data from A. Lincoln, *Journal of Marriage and Family*, vol. 70, 2008, pp. 806–814.)

TRY

Housework Hours

Gender	Sample Size	Mean	Standard Deviation
Women	476	33.0	21.9
Men	496	19.9	14.6

a. Based on this study, calculate how many more hours, on the average, women spend on housework than men.

b. Find the standard error for comparing the means. What factor causes the standard error to be small compared to the sample standard deviations for the two groups?

c. Calculate the 95% confidence interval comparing the population means for women and men. Interpret the result including the relevance of 0 being within the interval or not.

d. State the assumptions upon which the interval in part c is based.

10.17 More confident about housework Refer to part c in the previous exercise.

a. Show that a 99% confidence interval is (10.0, 16.2). (*Hint:* For such large sample sizes, the t-score is practically identical to a z-score.)

b. Explain why this interval is wider than the 95% confidence interval.

10.18 Employment by gender The study described in Exercise 10.16 also evaluated the weekly time spent in employment. This sample comprises men and women with a high level of labor force attachment. Software shows the results.

```
Gender    N      Mean    StDev   SE Mean
Men       496    47.54   9.92    0.45
Women     476    42.01   6.53    0.30
Difference = mu (Men) - mu (Women)
95% CI for difference: (4.477, 6.583)
T-Test of difference = 0 (vs not = ):
T-Value = 10.30 P-Value = 0.000
```

a. Does it seem plausible that employment has a normal distribution for each gender? Explain.

b. What effect does the answer to part a have on inference comparing population means? What assumptions are made for the inferences in this table?

c. Explain how to interpret the confidence interval.

d. Refer to part c. Do you think that the population means are equal? Explain.

10.19 Ideal number of children The Web site sda.berkeley.edu/GSS for the General Social Survey enables you to compare means for groups. Compare female and male responses on the question, "What is the ideal number of children for a family to have?" in 2008. Identify CHLDIDEL as the dependent variable, SEX as the row variable, use YEAR(2008) as the selection variable, and check the Summary statistics box.

a. The 677 females who responded had a mean of 3.22 and standard deviation of 1.99. What were the results for males?

b. Report the 95% confidence interval for the difference between the population means for females and males. Interpret.

10.20 Pay by gender The study described in the Exercise 10.18 also evaluated personal income by gender. The average earnings per year was \$35,800 for men with a standard deviation of \$14,600. The average earnings for women was

$23,900 with a standard deviation of $21,900. Software shows the following results:

Two-Sample T-Test and CI

```
Sample    N    Mean   StDev  SE Mean
1       496   35800   14600     656
2       476   23900   21900    1004
Difference = mu (1) - mu (2)
Estimate for difference:  11900
95% CI for difference:  (9547, 14253)
T-Test of difference = 0 (vs not =): T-Value = 9.93
P-Value = 0.000 DF = 822
```

a. Does it seem plausible that employment has a normal distribution for each gender? Explain.

b. What effect does the answer to part a have on inference comparing population means? What assumptions are made for the inferences in this table?

c. A 95% confidence interval for the difference in the population means for women and men is ($9,547, $14,253). Interpret, indicating the relevance of $0 not falling in the interval.

10.21 Bulimia CI A study of bulimia among college women (J. Kern and T. Hastings, *Journal of Clinical Psychology*, vol. 51, 1995, p. 499) studied the connection between childhood sexual abuse and a measure of family cohesion (the higher the score, the greater the cohesion). The sample mean on the family cohesion scale was 2.0 for 13 sexually abused students ($s = 2.1$) and 4.8 for 17 nonabused students ($s = 3.2$).

a. Find the standard error for comparing the means.

b. Construct a 95% confidence interval for the difference between the mean family cohesion for sexually abused students and non-abused students. Interpret.

10.22 Chelation useless? Chelation is an alternative therapy for heart disease that uses repeated intravenous administration of a human-made amino acid in combination with oral vitamins and minerals. Practitioners believe it removes calcium deposits from buildup in arteries. However, the evidence for a positive effect is anecdotal or comes from nonrandomized, uncontrolled studies. A double-blind randomized clinical trial comparing chelation to placebo used a treadmill test in which the response was the length of time until a subject experienced ischemia (lack of blood flow and oxygen to the heart muscle).

a. After 27 weeks of treatment, the sample mean time for the chelation group was 652 seconds. A 95% confidence interval for the population mean for chelation minus the population mean for placebo was −53 to 36 seconds. Explain how to interpret the confidence interval.

b. A significance test comparing the means had P-value = 0.69. Specify the hypotheses for this test, which was two-sided.

c. The authors concluded from the test, "There is no evidence to support a beneficial effect of chelation therapy" (M. Knudtson et al., JAMA, vol. 287, p. 481, 2002). Explain how this conclusion agrees with inference based on the values in the confidence interval.

10.23 Some smoked but didn't inhale Refer to Examples 6–8 on nicotine dependence for teenage smokers. Another explanatory variable was whether a subject reported inhaling when smoking. The table reports descriptive statistics.

	HONC Score		
Group	**Sample Size**	**Mean**	**Standard Deviation**
Inhalers	237	2.9	3.6
Noninhalers	95	0.1	0.5

a. Explain why (i) the overwhelming majority of noninhalers must have had HONC scores of 0 and (ii) on average, those who reported inhaling answered yes to nearly three more questions than those who denied inhaling.

b. Might the HONC scores have been approximately normal for each group? Why or why not?

c. Find the standard error for the estimate $(\bar{x}_1 - \bar{x}_2) = 2.8$. Interpret.

d. The 95% confidence interval for $(\mu_1 - \mu_2)$ is (2.3, 3.3). What can you conclude about the population means for inhalers and noninhalers?

10.24 Inhaling affect HONC? Refer to the previous exercise.

a. Show that the test statistic for H_0: $\mu_1 = \mu_2$ equals $t = 11.7$. If the population means were equal, explain why it would be nearly impossible by random variation to observe this large a test statistic.

b. What decision would you make about H_0, at common significance levels? Can you conclude which group had higher mean nicotine dependence? How?

c. State the assumptions for the inference in this exercise.

10.25 Females or males more nicotine dependent? Refer to Example 6, "Teenagers on Nicotine." Of those who had tried tobacco, the mean HONC score was 2.8 ($s = 3.6$) for the 150 females and 1.6 ($s = 2.9$) for the 182 males.

a. Find a standard error for comparing the sample means. Interpret.

b. Find the test statistic and P-value for H_0: $\mu_1 = \mu_2$ against H_a: $\mu_1 \neq \mu_2$. Interpret, and explain what (if any) effect gender has on the mean HONC score.

c. Do you think that the HONC scores were approximately normal for each gender? Why or why not? How does this affect the validity of the analysis in part b?

10.26 Female and male monthly smokers Refer to the previous exercise. A subject was called a monthly smoker if he or she had smoked cigarettes over an extended period of time. The 74 female monthly smokers had a mean HONC score of 5.4 ($s = 3.5$), and the 71 male monthly smokers had a mean HONC score of 3.9 ($s = 3.6$). Using software (such as MINITAB) that can conduct analyses using summary statistics, repeat parts b and c of the previous exercise.

10.27 TV watching and gender For the 2008 General Social Survey, a comparison of females and males on the number of hours a day that the subject watched TV gave

Group	N	Mean	StDev	SE Mean	95% CI
Females	698	3.08	2.70	0.102	(2.88, 3.28)
Males	626	2.87	2.61	0.104	(2.67, 3.08)

a. Set up the hypotheses of a significance test to analyze whether the population means differ for females and males.

b. Using software (such as MINITAB) in which you can use summarized data, conduct all parts of the significance test. Interpret the P-value, and report the conclusion for a significance level of 0.05.

c. If you were to construct a 95% confidence interval comparing the means, would it contain 0? Answer based on the result of part b, without actually finding the interval.

d. Do you think that the distribution of TV watching is approximately normal? Why or why not? Does this affect the validity of your inferences? What assumptions do the methods make?

10.28 Student survey Refer to the FL Student Survey data file on the text CD. Use the number of times reading a newspaper as the response variable and gender as the explanatory variable. The observations are as follows:

Females: 5, 3, 6, 3, 7, 1, 1, 3, 0, 4, 7, 2, 2, 7, 3, 0, 5, 0, 4, 4, 5, 14, 3, 1, 2, 1, 7, 2, 5, 3, 7

Males: 0, 3, 7, 4, 3, 2, 1, 12, 1, 6, 2, 2, 7, 7, 5, 3, 14, 3, 7, 6, 5, 5, 2, 3, 5, 5, 2, 3, 3

Using software,

a. Construct and interpret a plot comparing responses by females and males.

b. Construct and interpret a 95% confidence interval comparing population means for females and males.

c. Show all five steps of a significance test comparing the population means.

d. State and check the assumptions for part b and part c.

10.29 Study time A graduate teaching assistant for Introduction to Statistics (STA 2023) at the University of Florida collected data from students in one of her classes in spring 2007 to investigate whether study time per week (average number of hours) differed between students in the class who planned to go to graduate school and those who did not. The data were as follows:

Graduate school: 15, 7, 15, 10, 5, 5, 2, 3, 12, 16, 15, 37, 8, 14, 10, 18, 3, 25, 15, 5, 5

No graduate school: 6, 8, 15, 6, 5, 14, 10, 10, 12, 5

Using software or a calculator,

a. Find the sample mean and standard deviation for each group. Interpret.

b. Find the standard error for the difference between the sample means. Interpret.

c. Find a 95% confidence interval comparing the population means. Interpret.

10.30 More on study time Refer to the data in the previous exercise.

a. Show all steps of a two-sided significance test of the null hypothesis that the population mean is equal for the two groups. Interpret results in context.

b. Based on the explanation given in the previous exercise, do you think that the sample was a random sample or a convenience sample? Why? (In practice, you should not place much faith in inferences using such samples!)

10.31 Time spent on social networks As part of a class exercise, an instructor at a major university asks her students how many hours per week they spend on social networks. She wants to investigate if time spent on social networks differs for male and female students at this university. The results for those age 21 or under were:

Males: 5, 7, 9, 10, 12, 12, 12, 13, 13, 15, 15, 20

Females: 5, 7, 7, 8, 10, 10, 11, 12, 12, 14, 14, 14, 16, 18, 20, 20, 20, 22, 23, 25, 40

a. Using software or a calculator, find the sample mean and standard deviation for each group. Interpret.

b. Find the standard error for the difference between the sample means.

c. Find and interpret a 90% confidence interval comparing the population means.

10.32 More time on social networks In the previous exercise, plot the data. Do you see any outliers that could influence the results? Remove the most extreme observation from each group and redo the analyses. Compare results and summarize the influence of the extreme observations.

10.33 Normal assumption The methods of this section make the assumption of a normal population distribution. Why do you think this is more relevant for small samples than for large samples?

10.3 Other Ways of Comparing Means and Comparing Proportions

We've now learned about the primary methods for comparing two proportions or two means. We'll next study an alternative method for comparing means that is useful when we can assume that two groups have similar variability on the response variable. This method, focuses on inference about the *difference* between the parameters. Later we'll also illustrate the use of a ratio of parameters, as an alternative way of comparing two proportions or two means.

Comparing Means, Assuming Equal Population Standard Deviations

An alternative t method to the one described in Section 10.2 is sometimes used when, under the null hypothesis, it is reasonable to expect the *variability* as well as the mean to be the same. For example, consider a study comparing a drug to a placebo. If the drug has no effect, then we expect the entire distributions of the response variable to be identical for the two groups, not just the mean. This method requires an extra assumption in addition to the usual ones of independent random samples and approximately normal population distributions:

The population standard deviations are equal, that is, $\sigma_1 = \sigma_2$ (see Figure 10.7).

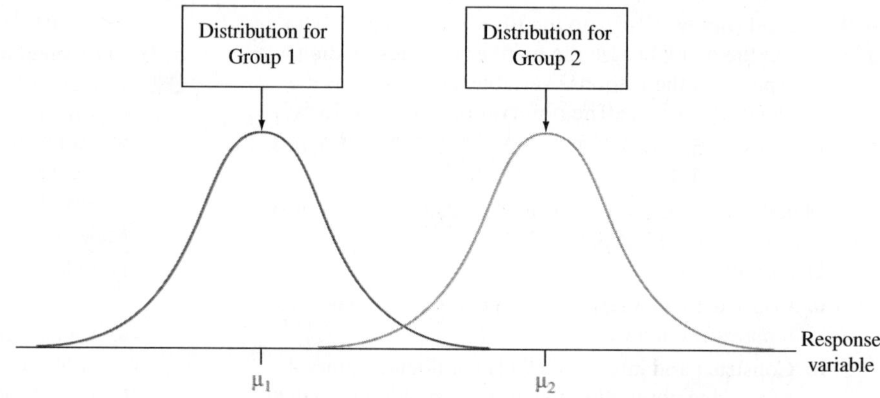

▲ **Figure 10.7** **Two Groups With Equal Population Standard Deviations.** Question What would a graph of the sample data look like to make you doubt this assumption?

This alternative method estimates the common value σ of σ_1 and σ_2 by

$$s = \sqrt{\frac{(n_1 - 1)s_1^2 + (n_2 - 1)s_2^2}{n_1 + n_2 - 2}}.$$

The estimate s, called the **pooled standard deviation**, combines information from the two samples to provide a single estimate of variability. The term inside the square root is a weighted average of the squares of the two sample standard deviations. When $n_1 = n_2$, it's an average, the sum of the squared standard deviations divided by 2. The estimate s falls between s_1 and s_2. The degrees of freedom for this method are $df = n_1 + n_2 - 2$, which appears in the denominator of the formula for s.

In Practice Robustness of Two-Sided Inferences

Confidence intervals and two-sided tests using this alternative method are robust. They work well even when the population distributions are not normal and when the population standard deviations are not exactly equal. This is particularly true when the sample sizes are similar and not extremely small. In practice, however, these alternative methods are not usually used if one sample standard deviation is more than double the other one.

SUMMARY: Comparing Population Means, Assuming Equal Population Standard Deviations

Using the pooled standard deviation estimate s of $\sigma = \sigma_1 = \sigma_2$, the standard error of $(\bar{x}_1 - \bar{x}_2)$ simplifies to

$$se = \sqrt{\frac{s^2}{n_1} + \frac{s^2}{n_2}} = s\sqrt{\frac{1}{n_1} + \frac{1}{n_2}}.$$

Otherwise, inference formulas look the same as those that do not assume $\sigma_1 = \sigma_2$:

- A 95% confidence interval for $(\mu_1 - \mu_2)$ is $(\bar{x}_1 - \bar{x}_2) \pm t_{.025}(se)$.

- The test statistic for H_0: $\mu_1 = \mu_2$ is $t = (\bar{x}_1 - \bar{x}_2)/se$.

  These methods have $df = n_1 + n_2 - 2$. To use them, you assume

- Independent random samples from the two groups, either from random sampling or a randomized experiment
- An approximately normal population distribution for each group (This is mainly important for small sample sizes, and even then the confidence interval and two-sided test are usually robust to violations of this assumption.)
- $\sigma_1 = \sigma_2$ (In practice, this type of inference is not usually relied on if one sample standard deviation is more than double the other one.)

Example 10

Assume equal population standard deviations

Arthroscopic Surgery

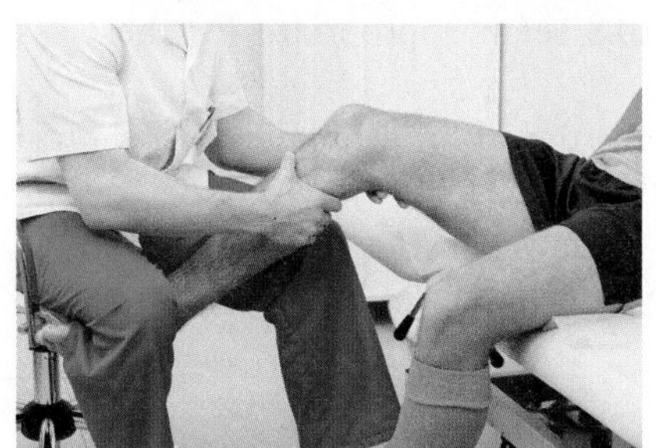

Picture the Scenario

A random trial study assessed the usefulness of arthroscopic surgery.[8] Over a three-year period, patients suffering from osteoarthritis who had at least moderate knee pain were recruited from a medical center in Houston. Patients were randomly assigned to one of three groups, to receive one of two types of arthroscopic surgeries or a surgery that was actually a placebo procedure. In the arthroscopic surgeries, the lavage group had the joint flushed with fluid but instruments were not used to remove tissue, whereas the debridement group had tissue removal as well. In the placebo procedure, the same incisions were made in the knee as with surgery and the surgeon manipulated the knee as if surgery was being performed, but none was actually done. The study was double-blind.

A knee-specific pain scale was created for the study. Administered two years after the surgery, it ranged from 0 to 100, with higher scores indicating more severe pain. Table 10.7 shows summary statistics. The pain scale was the response variable, and the treatment group was the explanatory variable.

Table 10.7 Summary of Knee Pain Scores

The descriptive statistics compare lavage and debridement arthroscopic surgery to a placebo (fake surgery) treatment.

	Knee Pain Score		
Group	**Sample Size**	**Mean**	**Standard Deviation**
1. Placebo	60	51.6	23.7
2. Arthroscopic—lavage	61	53.7	23.7
3. Arthroscopic—debridement	59	51.4	23.2

Denote the population mean of the pain scores by μ_1 for the placebo group and by μ_2 for the lavage arthroscopic group. Most software gives you the option of assuming equal population standard deviations. Table 10.8 is the MINITAB output for the two-sample t inferences. (We'll consider the debridement arthroscopic group in an exercise.)

[8] By J. B. Moseley et al., *New England Journal of Medicine*, vol. 347, 2002, pp. 81–88.

Table 10.8 MINITAB Output for Comparing the Mean Knee Pain for Placebo and Arthroscopic Surgery Groups

This analysis assumes equal population standard deviations.

```
Sample    N    Mean    StDev    SE Mean
  1      60    51.6    23.7       3.1
  2      61    53.7    23.7       3.0

Difference = mu(1) − mu(2)
Estimate for difference:  − 2.10000
95% CI for difference: (−10.63272,6.43272)
T-Test of difference = 0 (vs not = ):
T-Value =  − 0.49 P-Value = 0.627 DF = 119
Both use Pooled StDev = 23.7000
```

Questions to Explore

a. Does the t test inference in Table 10.8 seem appropriate for these data?

b. Show how to find the pooled standard deviation estimate of σ, the standard error, the test statistic, and its *df*.

c. Identify the P-value for testing H_0: $\mu_1 = \mu_2$ against H_a: $\mu_1 \neq \mu_2$. With the 0.05 significance level, can you reject H_0? Interpret.

Think It Through

a. If the arthroscopic surgery has no real effect, its population distribution of pain score should be the same as for the placebo. Not only will the population means be equal, but so will the population standard deviations. So, in testing whether this surgery has no effect, we will use a method that assumes equal population standard deviations. Table 10.8 shows that the sample standard deviations are identical. This is mere coincidence, and we would not typically expect this even if the population standard deviations were identical. But it suggests that we can use the method assuming equal population standard deviations.

b. The pooled standard deviation estimate of the common value σ of $\sigma_1 = \sigma_2$ is

$$s = \sqrt{\frac{(n_1 - 1)s_1^2 + (n_2 - 1)s_2^2}{n_1 + n_2 - 2}} =$$

$$\sqrt{\frac{(60 - 1)23.7^2 + (61 - 1)23.7^2}{60 + 61 - 2}} = 23.7,$$

shown at the bottom of Table 10.8. We estimate the difference $\mu_1 - \mu_2$ by $(\bar{x}_1 - \bar{x}_2) = 51.6 - 53.7 = -2.1$. The standard error of $(\bar{x}_1 - \bar{x}_2)$ equals

$$se = s\sqrt{\frac{1}{n_1} + \frac{1}{n_2}} = 23.7\sqrt{\frac{1}{60} + \frac{1}{61}} = 4.31.$$

The test statistic for H_0: $\mu_1 = \mu_2$ equals

$$t = \frac{(\bar{x}_1 - \bar{x}_2)}{se} = \frac{51.6 - 53.7}{4.31} = -0.49.$$

Its $df = (n_1 + n_2 - 2) = (60 + 61 - 2) = 119$. When H_0 is true, the test statistic has the t distribution with $df = 119$.

c. The two-sided P-value equals 0.63. So, if the true population means were equal, by random variation, it would not be surprising to observe a test statistic of this size. The P-value is larger than 0.05, so we cannot reject H_0. Consider the population of people who suffer from osteoarthritis and could conceivably receive one of these treatments. In summary, we don't have enough evidence to conclude that the mean pain level differs for the placebo treatment and the arthroscopic surgery treatment.

Insight

Table 10.8 reports a 95% confidence interval for $(\mu_1 - \mu_2)$ of $(-10.6, 6.4)$. Because the interval contains 0, it is plausible that there is no difference between the population means. This is the same conclusion as that for the significance test. We infer that the population mean for the knee pain score could be as much as 10.6 lower or as much as 6.4 higher for the placebo treatment than for the arthroscopic surgery. On the pain scale with range 100, this is a relatively small difference in practical terms.

Try Exercise 10.36

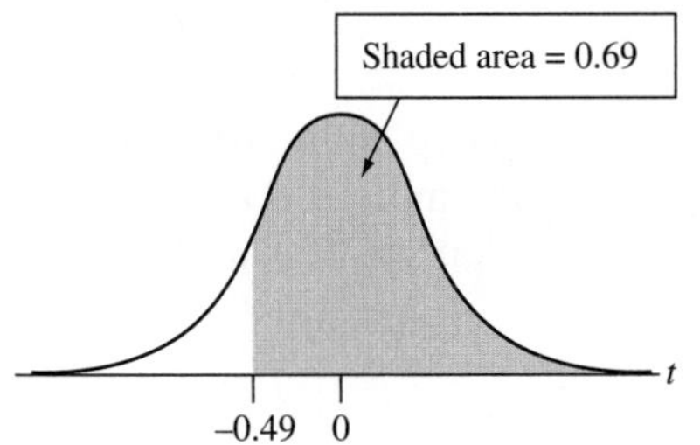

Suppose the researchers specifically predicted that the lavage surgery would give *better* pain relief than the placebo. Then they would find the P-value for the one-sided H_a: $\mu_1 > \mu_2$ (that is, $\mu_1 - \mu_2 > 0$), which predicts higher mean pain with the placebo. The P-value is the probability to the *right* of $t = -0.49$. Software reports this as 0.69. See margin figure. There is no evidence in favor of this alternative. The P-value exceeds 0.50 because the sample mean pain was actually *lower* for the placebo, not higher.

Comparing Population Standard Deviations

Many texts and software present a statistic denoted by F for testing the hypothesis that the population standard deviations are equal. This test assumes that the population distributions are normal. Unfortunately, this F-test is *not* a robust method.

In Practice *F* Test for Comparing Standard Deviations Is Not Robust

The ***F* test for comparing standard deviations** of two populations performs poorly if the populations are not close to normal. Consequently, statisticians do not recommend it for general use. If the data show evidence of a potentially large difference in standard deviations, with one of the sample standard deviations being at least double the other, it is better to use the two-sample *t*-inferences (of Section 10.2) that do not have this extra assumption.

The Ratio of Proportions: The Relative Risk

Group	Cancer Death Yes	Cancer Death No	Prop. Yes
Placebo	347	11,188	0.030
Aspirin	327	13,708	0.023

Examples 2–4 discussed data from the study comparing the proportions of cancer deaths between a group who took placebo and a group that took aspirin. See the margin table. The difference $\hat{p}_1 - \hat{p}_2 = 0.030 - 0.023 = 0.007$ was small. However, this inequality seems more substantial when viewed by their *ratio*, which is a statistic commonly reported in medical journals.

Ratio of Proportions (Relative Risk)

The **ratio of proportions** for two groups is $\hat{p}_1/\hat{p}_2$. In medical applications for which the proportion refers to a category that is an undesirable outcome, such as death or having a heart attack, this ratio is called the **relative risk**.

The ratio of proportions describes the sizes of the proportions *relative* to each other. For the cancer study, $\hat{p}_1 = 0.030$ for placebo, $\hat{p}_2 = 0.023$ for aspirin, and

$$\text{sample relative risk} = \hat{p}_1/\hat{p}_2 = 0.030/0.023 = 1.30.$$

This means that the proportion of the placebo group who had a cancer death was 1.30 times the proportion of the aspirin group who had a cancer death.

The population relative risk is p_1/p_2. When $p_1 = p_2$, the relative risk equals 1. The sample relative risk of 1.30 is above 1, indicating an effect.

Recall that the labeling of the groups is arbitrary. If we had labeled Group 1 as those taking aspirin, then the relative risk is 0.023/0.30 = 0.77. The proportion of the aspirin group who had a cancer death was 0.77 times the proportion of the placebo group who died from cancer. But 0.77 = 1/1.30. *Whether the relative risk equals a particular number or its reciprocal merely depends on the group labeling.*

Medical journal articles often report relative risk, especially when both proportions are close to 0. Software can form a confidence interval[9] for a population relative risk.

Example 11

Relative risk

Alcohol Consumption and Risk of Stroke

Picture the Scenario

A recent article in a medical journal[10] stated, "Compared with participants who had less than one drink per week, those who drank more had a reduced overall risk of stroke (relative risk, 0.79; 95% confidence interval, 0.66 to 0.94)."

Question to Explore

How do you interpret the relative risk value and the reported confidence interval?

Think It Through

Let $\hat{p}_1$ = sample proportion of the regular alcohol drinkers who had strokes and $\hat{p}_2$ = sample proportion of the light or nondrinkers who had strokes. Then the sample relative risk of 0.79 means that $\hat{p}_1/\hat{p}_2 = 0.79$. So $\hat{p}_1 = 0.79\hat{p}_2$. The proportion of the regular alcohol drinkers who had strokes was 0.79 times the proportion of the light or nondrinkers who had strokes. Since the relative risk was less than 1.0, the proportion of strokes was smaller for the regular alcohol drinkers. (How can you determine which is Group 1 and which is Group 2 for the ratio $\hat{p}_1/\hat{p}_2$? The quoted sentence says that those who drank more had a *reduced* risk of stroke. Since the relative risk of 0.79 is less then 1.0, this means that the group that drank more is in the numerator for this calculation, so it is Group 1.)

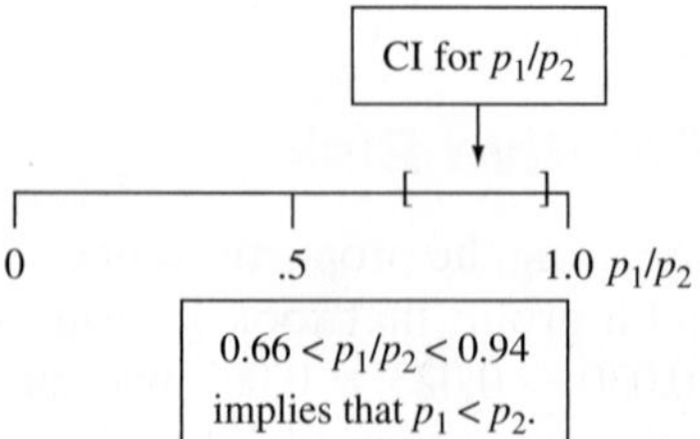

The 95% confidence interval for the population relative risk of stroke was (0.66, 0.94). Since all numbers in the interval are less than 1.0, we can infer that $p_1/p_2 < 1.0$ (see the margin figure). That is, it seems that $p_1 < p_2$. We can infer that the population proportion of strokes is smaller for those who drink alcohol regularly.

Insight

Regular alcohol drinking seems to correspond to a reduction in the incidence of stroke. Since the upper endpoint of the confidence interval is 0.94,

[9]The formula for the confidence interval is complex, and we do not cover it in this text.

[10]*New England Journal of Medicine*, vol. 341, 1999, pp. 1557–1564.

however, the population relative risk could be close to 1.0. So the effect of drinking might be weak. Another reason to be cautious is that this was an observational study, rather than a randomized clinical trial like in the study comparing aspirin and placebo. In any case, the conclusion does not imply that it's good to drink a lot of alcohol every day.

Try Exercise 10.41

The Ratio of Means

You can also use the ratio to compare means. For example, the National Center for Health Statistics recently reported that the mean weight for adult American women was 140 pounds in 1962 and 164 pounds in 2002. Since 164/140 = 1.17, the mean in 2002 was 1.17 times the mean in 1962. This can also be expressed by saying that the mean increased by 17% over this 40-year period.

10.3 Practicing the Basics

10.34 Body dissatisfaction Female college student participation in athletics has increased dramatically over the past few decades. Sports medicine providers are aware of some unique health concerns of athletic women, including disordered eating. A study (M. Reinking and L. Alexander, *Journal of Athletic Training,* vol. 40, 2005, p. 47–51) compared disordered-eating symptoms and their causes for collegiate female athletes (in lean and nonlean sports) and nonathletes. The sample mean of the body dissatisfaction assessment score was 13.2 ($s = 8.0$) for 16 lean sport athletes (those sports that place value on leanness, including distance running, swimming, and gymnastics) and 7.3 ($s = 6.0$) for the 68 nonlean sport athletes. Assuming equal population standard deviations,

a. Find the standard error for comparing the means.

b. Construct a 95% confidence interval for the difference between the mean body dissatisfaction for lean sport athletes and nonlean sport athletes. Interpret.

10.35 Body dissatisfaction test Refer to the previous exercise.

a. Find the P-value for testing whether the population means are equal. Use a two-sided alternative.

b. Summarize assumptions for the analysis in part a. Do you think the normality assumption is justified? If not, what is the consequence of violating it?

10.36 Surgery verses placebo for knee pain Refer to Example 10, "Arthroscopic Surgery." Here we show MINITAB output comparing mean knee pain scores for the placebo (Group 1) to debridement arthroscopic surgery (Group 2).

TRY

a. State and interpret the result of the confidence interval.

b. State all steps and interpret the result of the significance test.

c. Based on the confidence interval and test, would you conclude that the arthroscopic surgery works better than placebo? Explain.

```
Sample    N      Mean    StDev    SEMean
1         60     51.6    23.7     3.1
2         59     51.4    23.2     3.0
Difference = mu(1) - mu(2)
Estimate for difference: 0.200000
95% CI for difference: (-8.316130,8.716130)
T-Test of difference = 0 (vs not = ):
T-Value = 0.05 P-Value = 0.963 DF = 117
Both use Pooled StDev = 23.4535
```

10.37 Comparing clinical therapies A clinical psychologist wants to choose between two therapies for treating severe cases of mental depression. She selects six patients who are similar in their depressive symptoms and in their overall quality of health. She randomly selects three of the patients to receive Therapy 1, and the other three receive Therapy 2. She selects small samples for ethical reasons—if her experiment indicates that one therapy is superior, she will use that therapy on all her other depression patients. After one month of treatment, the improvement in each patient is measured by the change in a score for measuring severity of mental depression. The higher the score, the better. The improvement scores are

Therapy 1: 30, 45, 45
Therapy 2: 10, 20, 30

Analyze these data (you can use software, if you wish), assuming equal population standard deviations.

a. Show that $\bar{x}_1 = 40$, $\bar{x}_2 = 20$, $s = 9.35$, $se = 7.64$, $df = 4$, and a 95% confidence interval comparing the means is (−1.2, 41.2).

b. Explain how to interpret what the confidence interval tells you about the therapies. Why do you think that it is so wide?

c. When the sample sizes are very small, it may be worth sacrificing some confidence to achieve more precision. Show that a 90% confidence interval is (3.7, 36.3). At this confidence level, can you conclude that Therapy 1 is better?

10.38 Clinical therapies 2 Refer to the previous exercise.

a. For the null hypothesis, H_0: $\mu_1 = \mu_2$, show that $t = 2.62$ and the two-sided P-value = 0.059. Interpret.

b. What decision would you make in the test, using a (i) 0.05 and (ii) 0.10 significance level? Explain what this means in the context of the study.

c. Suppose the researcher had predicted ahead of time that Therapy 1 would be better. To which H_a does this correspond? Report the P-value for it, and make a decision with significance level 0.05.

10.39 Vegetarians more liberal? When a sample of social science graduate students at the University of Florida gave their responses on political ideology (ranging from 1 = very liberal to 7 = very conservative), the mean was 3.18 ($s = 1.72$) for the 51 nonvegetarian students and 2.22 ($s = 0.67$) for the 9 vegetarian students. Software for comparing the means provides the printout, which shows results first for inferences that assume equal population standard deviations and then for inferences that allow them to be unequal.

```
Sample    N     Mean    StDev   SEMean
  1      51     3.18    1.72     0.24
  2       9     2.22    0.67     0.22
Difference = mu (1) − mu (2)
95% CI for difference: (−0.209716, 2.129716)
T-Test of difference = 0 (vs not = ):
T-Value = 1.64  P-Value = 0.106  DF = 58
Both use Pooled StDev = 1.6162
95% CI for difference: (0.289196, 1.630804)
T-Test of difference = 0 (vs not = ):
T-Value = 2.92 P-Value = 0.007 DF = 30
```

a. Explain why the results of the two approaches differ so much. Which do you think is more reliable?

b. State your conclusion about whether the true means are plausibly equal.

10.40 Teeth whitening results One scientific "test of whiteness" study mentioned in Example 1 tested the effect of a self applied tooth-whitening peroxide gel system in a randomized, controlled clinical trial.[11] The 58 adults assigned to the gel whitening group applied the gel after normal brushing according to the manufacturer's instructions. The fluoride toothpaste group was instructed to brush twice a day. The procedure was repeated for both groups twice a day for 14 days. An experienced examiner determined the tooth shades comparing each tooth to the shade tabs from an accepted shade scale (Vita shade guide) at the start of the experiment to create a baseline and then after one and two weeks of product application. Changes between the baseline score and the one- and two-week assessments were expressed as a difference of the respective Vita score, with a positive difference indicating an improvement in tooth whiteness. The results of the study are shown in the table in the next column.

[11] *Source*: Data from L. Z. Collins et. al., *Journal of Dentistry*, vol. 32, 2004, pp. 13–17.

Mean Vita Shade Score Recorded at Two Weeks and Change From Baseline

Group	n	Two Weeks Mean Vita Shade (s.d.)	Change From Baseline (s.d.)	Treatment Difference
Xtra White whitening gel	58	6.80 (2.48)	1.02 (1.32)	0.67 ($p < 0.05$)
Toothpaste only	59	7.01 (2.19)	0.35 (1.29)	

a. State the hypotheses that were tested for the change from baseline means.

b. The P-value is reported as < 0.05 for the test comparing the means. Explain how to interpret this value.

c. Calculate the pooled standard error, the t statistic, and the resulting P-value.

d. The ratio of the change from baseline sample means was 2.91. Interpret this ratio.

10.41 Fish and heart disease A study in Rotterdam (European *Journal of Heart Failure*, vol. 11, 2009, pp. 922–928) followed the health of 7983 subjects over 20 years to determine whether intake of fish could be associated with a decreased risk of heart failure in a general population of men and women aged 55 years and older. Results showed that the dietary intake of fish was not significantly related to heart failure incidence. Even for a high daily fish consumption of more than 20 grams a day there appeared no added protection against heart failure. Incidence rates were similar in those who consumed no fish (incidence rate of 11 per 1000), moderate fish (median 9 g per day, 12.3 per 1000) or high fish (9.9 per 1000). The relative risk of heart failure in the high intake groups (medium and high fish) was 0.96 when compared with no intake with a 95% confidence interval of (0.78, 1.18). Explain how to interpret the (a) reported relative risk and (b) confidence interval for the relative risk.

TRY

10.42 Aspirin and heart attacks For the Swedish study described in Exercise 10.6, during the follow-up period 28 of 684 taking placebo died from a heart attack and 18 of 676 taking aspirin died from a heart attack.

a. Show that the relative risk for those taking placebo compared to those taking aspirin is 1.54. Interpret.

b. Software reports a 95% confidence interval of (0.86, 2.75) for the population relative risk. Interpret the lower endpoint of 0.86.

c. Explain the relevance of the confidence interval containing 1.0.

10.43 Obesity and cancer Medscape Medical News (July 2, 2010) reported that persons living in the Asia- Pacific region who are obese have a significantly increased risk for mortality from cancer compared with the risk for individuals of normal weight, according to a new report published online June 30 in *Lancet Oncology.* Among individuals with a body mass index (BMI) [body mass index = BMI (weight in pounds)/(height in inches)] higher than 18.5 kg/m^2, the authors found that there was "a positive and continuous association between BMI and all-cancer mortality." Compared with persons with a normal weight, the relative risk for cancer-related mortality was 1.06 (95% confidence interval [CI], 1.00 – 1.12) for those who were overweight and 1.21 (95% CI, 1.09 – 1.36) among obese persons. Fill in the blanks in the interpretation: "In the population of those who are obese, we are 95% confident that the increase in cancer-related mortality was between ___% and ___%.

10.44 Reducing risk summaries and interpretations

a. A *New York Times* health article (August 13, 2009) about a new study reported in the *Journal of American Medical Association* stated that, "patients with colorectal cancer who were regular aspirin users had a much better chance of surviving than nonusers, and were almost one-third less likely to die of the disease, while those who began using aspirin for the first time after the diagnosis cut their risk of dying by almost half." Which descriptive measure does "one-third" refer to? Interpret.

b. An AP story (September 10, 2003) stated, "Brisk walking for just an hour or two weekly can help older women reduce their risk of breast cancer by 18 percent," according to an analysis of 74,171 women in the Women's Health Initiative study. Explain how to form a sample relative risk measure that equals 0.82, for comparing those who briskly walk for an hour or two weekly to those who do not take such walks. *Note:* In this part and the next one, it may help you to use the result that Percent reduction in risk $= (1 - \text{relative risk}) * 100\%$.

c. A *Science Daily* article (Nov. 17, 2006) reported that taking low-dose aspirin daily reduces the risk of heart attack and stroke, as well as the risk of dying, among patients who previously have had a heart attack or stroke, according to analysis by Duke University Medical Center cardiologists. The researchers found that patients who took low-dose aspirin had a 26% reduction in the risk of a nonfatal heart attack and a 25% reduction in the risk of stroke compared with similar heart patients who did not take aspirin. State this result by defining and interpreting a relative risk for comparing those taking an aspirin a day to those who do not take an aspirin a day for the risk of nonfatal heart attacks.

d. An AP story (February 9, 2011) reported on a preliminary study presented at the International Stroke Conference in California stating that "daily diet soda drinkers (there were 116 in the study) had a 48 percent higher risk of stroke or heart attack than people who drank no soda of any kind (901 people, or 35% of total participants). That's after taking into account rates of smoking, diabetes, waistline size, and other differences among the groups. No significant differences in risk were seen among people who drank a mix of diet and regular soda." Report the relative risk value for stroke or heart attack in this statement, and interpret.

10.45 Obesity in children Childhood obesity continues to be a leading public health concern that disproportionately affects low-income and minority children. According to the National Center for Health Statistics, obesity prevalence among low-income, preschool-aged children increased steadily from 12.4% in 1998 to 14.5% in 2003, but subsequently remained essentially the same, with a 14.6% prevalence in 2008. Compare the percentages for 1998 and 2008 using a ratio, and interpret.

10.46 Comparing median income According to the U.S. Census, the median individual yearly income for whites in the United States was \$33,808 in 1990, almost three times the median individual yearly income for Hispanics, which was \$12,028 for that same year. In 2008, the median income for whites increased to \$35,120 and for Hispanics to \$16,417.

a. Show how the researcher got the value to be approximately 3, and explain what summary measure is estimated by this value.

b. Calculate the same value as part a for the 2008 numbers.

c. Why do you think the Census Bureau used the median instead of the mean for this comparison?[12]

[12]Based on results described in *Lancet*, vol. 338, 1991, pp. 1345–1349.

10.4 Analyzing Dependent Samples

With **dependent samples**, each observation in one sample has a matched observation in the other sample. The observations are called **matched pairs**. We've already seen examples of matched-pairs data. In Example 8 of Chapter 9, each of a sample of anorexic girls had her weight measured before and after a treatment for anorexia. The weights before the treatment form one group of observations. The weights after the treatment form the other group. The same girls were in each sample, so the samples were dependent.

By contrast, in Examples 2, 5, and 6 of this chapter, the samples were *independent*. The two groups consisted of different people, and there was no matching of an observation in one sample with an observation in the other sample.

Example 12

Dependent samples

Cell Phones and Driving

Picture the Scenario

In this chapter, Example 9 analyzed whether the use of cell phones impairs reaction times in a driving skills test. The analysis used independent samples—one group used cell phones and a separate control group did not use them. An alternative design uses the same subjects for both groups. Reaction times are measured when subjects performed the driving task without using cell phones and then again while the same subjects used cell phones.

Table 10.9 shows the mean of the reaction times (in milliseconds) for each subject under each condition. Figure 10.8 shows box plots of the data for the two conditions.

Table 10.9 Reaction Times on Driving Skills Before and While Using Cell Phone

The difference score is the reaction time using the cell phone minus the reaction time not using it, such as 636 − 604 = 32 milliseconds.

	Using Cell Phone?				Using Cell Phone?		
Student	No	Yes	Difference	Student	No	Yes	Difference
1	604	636	32	17	525	626	101
2	556	623	67	18	508	501	−7
3	540	615	75	19	529	574	45
4	522	672	150	20	470	468	−2
5	459	601	142	21	512	578	66
6	544	600	56	22	487	560	73
7	513	542	29	23	515	525	10
8	470	554	84	24	499	647	148
9	556	543	−13	25	448	456	8
10	531	520	−11	26	558	688	130
11	599	609	10	27	589	679	90
12	537	559	22	28	814	960	146
13	619	595	−24	29	519	558	39
14	536	565	29	30	462	482	20
15	554	573	19	31	521	527	6
16	467	554	87	32	543	536	−7

Questions to Explore

a. Summarize the sample data distributions for the two conditions.

b. To compare the mean response times using statistical inference, should we treat the samples as independent or dependent?

Think It Through

a. The box plots show that the reaction times tend to be larger for the cell phone condition. But each sample data distribution has one extremely large outlier.

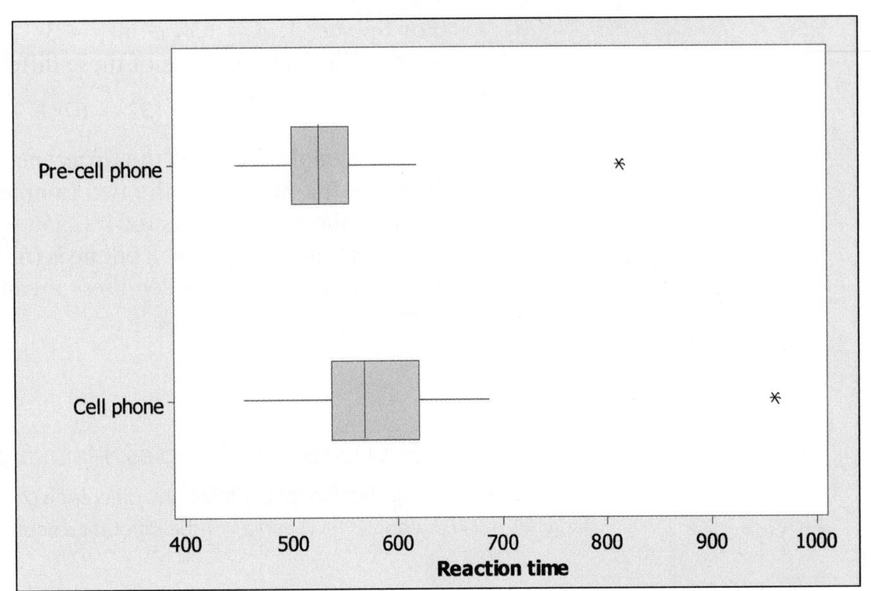

▲ **Figure 10.8** MINITAB Box Plots of Observations on Reaction Times. Question From the data file or Table 10.9, what do the outliers for the two distributions have in common?

b. The pre–cell phone (no cell phone) responses in Table 10.9 were made by the same subjects as the responses using cell phones. These are matched-pairs data because each control observation (Sample 1) pairs with a cell phone observation (Sample 2). Because this part of the study used the same subjects for each sample, the samples are dependent.

Insight

The data file and Table 10.9 show that subject number 28 had a large reaction time in each case. This one subject had a very slow reaction time, regardless of the condition.

Try Exercise 10.47, part a

Why would we use dependent instead of independent samples? A major benefit is that sources of potential bias are controlled so we can make a more accurate comparison. Using matched pairs keeps many other factors fixed that could affect the analysis. Often this results in the benefit of smaller standard errors.

From Figure 10.8, for instance, the sample mean was higher when subjects used cell phones. This did not happen because the subjects using cell phones tended to be older than the subjects not using them. Age was not a lurking variable, because each sample had the same subjects. When we used independent samples (different subjects) for the two conditions in Example 9, the two samples could differ somewhat on characteristics that might affect the results, such as physical fitness or gender or age. With independent samples, studies ideally use randomization to assign subjects to the two groups attempting to minimize the extent to which this happens. With observational studies, however, this is not an option.

Compare Means with Matched Pairs: *Use Paired Differences*

For each matched pair in Table 10.9, we construct a new variable consisting of a difference score (d for difference),

$$d = \text{reaction time using cell phone} - \text{reaction time without cell phone}.$$

For Subject 1, $d = 636 - 604 = 32$. Table 10.9 also shows these 32 difference scores. The sample mean of these difference scores, denoted by $\bar{x}_d$, is

$$\bar{x}_d = (32 + 67 + 75 + \cdots - 7)/32 = 50.6.$$

The sample mean of the difference scores necessarily equals the difference between the means for the two samples. In Table 10.9, the mean reaction time without the cell phone is $(604 + 556 + 540 + \cdots + 543)/32 = 534.6$, the mean reaction time using the cell phone is $(636 + 623 + 615 + \cdots + 536)/32 = 585.2$, and the difference between these means is 50.6. This is also the mean of the differences, $\bar{x}_d$.

For Dependent Samples, Mean of Differences = Difference of Means

For dependent samples, the difference $(\bar{x}_1 - \bar{x}_2)$ between the means of the two samples equals the mean $\bar{x}_d$ of the difference scores for the matched pairs.

In Words

For **dependent** samples, we calculate the **difference** scores and then use the one-sample methods of Chapters 8 and 9.

Likewise, the difference $(\mu_1 - \mu_2)$ between the population means is identical to the parameter μ_d that is the population mean of the difference scores. So the sample mean $\bar{x}_d$ of the differences not only estimates μ_d, the population mean difference, but also the difference $(\mu_1 - \mu_2)$. We can base inference about $(\mu_1 - \mu_2)$ on inference about the population mean of the difference scores. *This simplifies the analysis since it reduces a two-sample problem to a one-sample analysis using the difference scores.*

Let n denote the number of observations in each sample. This equals the number of difference scores. The 95% confidence interval for the population mean difference is

$$\bar{x}_d \pm t_{.025}(se) \text{ with } se = s_d/\sqrt{n},$$

Recall

You can review **confidence intervals for a mean** in Section 8.3 and **significance tests for a mean** in Section 9.3. ◀

where $\bar{x}_d$ is the sample mean of the difference scores and s_d is their standard deviation. The t-score comes from the t table with $df = n - 1$.

Likewise, to test the hypothesis H_0: $\mu_1 = \mu_2$ of equal means, we can conduct the single-sample test of H_0: $\mu_d = 0$ with the difference scores. The test statistic is

$$t = \frac{\bar{x}_d - 0}{se} \text{ with } se = s_d/\sqrt{n}.$$

This compares the sample mean of the differences to the null hypothesis value of 0. The standard error is the same as for the confidence interval. The $df = n - 1$. Since this test uses the difference scores for the pairs of observations, it is called a **paired-difference *t* test**. Software can do the computations for us.

Because these paired-difference inferences are special cases of single-sample inferences about a population mean, they make the same assumptions:

- The sample of difference scores is a random sample from a population of such difference scores.
- The difference scores have a population distribution that is approximately normal. This is mainly important for small samples (less than about 30) and for one-sided inferences.

Confidence intervals and two-sided tests are **robust**: They work quite well even if the normality assumption is violated. One-sided tests do not work well when the sample size is small and the distribution of differences is highly skewed.

Example 13

Matched-pairs analysis

Cell Phones and Driver Reaction Time

Picture the Scenario

The matched-pairs data in Table 10.9 showed the reaction times for the sampled subjects using and not using cell phones. Figure 10.9 is a box plot of the $n = 32$ difference scores. Table 10.10 shows MINITAB output for these data. The first part of Table 10.10 shows the mean, standard deviation, and standard error for each sample and for the differences. The second part shows inference about the mean of the differences.

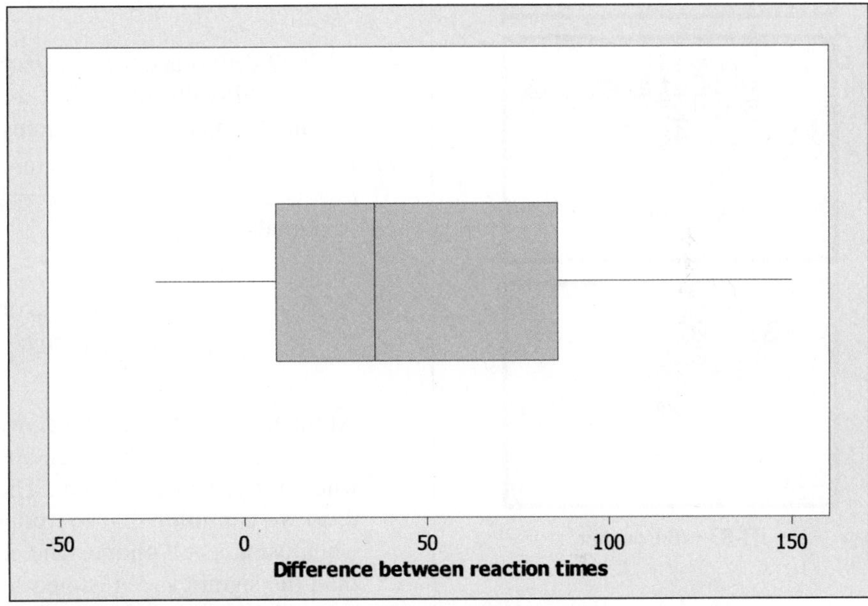

▲ **Figure 10.9** MINITAB Box Plot of Difference Scores from Table 10.9. Question How is it that some of the scores plotted here are negative?

Table 10.10 Software Output for Matched-Pairs Analysis With Table 10.9

The next page shows screen shots from the TI-83+/84.

```
Paired T for Cell phone - Pre-cell phone

                    N     Mean     StDev    SE Mean
Cell phone         32   585.188   89.646    15.847
Pre-cell phone     32   534.563   66.448    11.746
Difference         32   50.6250   52.4858   9.2783

95% CI for mean difference: (31.7019, 69.5481)
T-Test of mean difference = 0 (vs not = 0):
T-Value = 5.46 P-Value = 0.000
```

Questions to Explore

a. How can you conduct and interpret the significance test reported in Table 10.10?

b. How can you construct and interpret the confidence interval reported in Table 10.10?

```
T-Test
μ≠0.000
t=5.456
p=5.803E-6
x̄=50.625
Sx=52.486
n=32.000
```

```
TInterval
Inpt:Data Stats
x̄:50.625
Sx:52.4858
n:32
C-Level:.95
Calculate
```

```
TInterval
(31.702,69.548)
x̄=50.625
Sx=52.486
n=32.000
```

TI-83+/84 output

Think It Through

a. The box plot shows skew to the right for the difference scores. Two-sided inference is robust to violations of the assumption of a normal population distribution. The box plot does not show any severe outliers that would raise questions about the validity of using the mean to summarize the difference scores.

The sample mean difference is $\bar{x}_d = 50.6$, and the standard deviation of the difference scores is $s_d = 52.5$. The standard error is $se = s_d/\sqrt{n} = 52.5/\sqrt{32} = 9.28$. The t test statistic for the significance test of H_0: $\mu_d = 0$ (and hence equal population means for the two conditions) against H_a: $\mu_d \neq 0$ is

$$t = \bar{x}_d/se = 50.6/9.28 = 5.46.$$

With 32 difference scores, $df = n - 1 = 31$. Table 10.10 reports the two-sided P-value of 0.000. There is extremely strong evidence that the population mean reaction times are different.

b. For a 95% confidence interval for $\mu_d = \mu_1 - \mu_2$, with $df = 31$, $t_{.025} = 2.040$. We can use $se = 9.28$ from part a. The confidence interval equals

$$\bar{x}_d \pm t_{.025}(se), \text{ or } 50.6 \pm 2.040(9.28),$$
$$\text{which equals } 50.6 \pm 18.9, \text{ or } (31.7, 69.5).$$

At the 95% confidence level, we infer that the population mean when using cell phones is between about 32 and 70 milliseconds higher than when not using cell phones. The confidence interval does not contain 0, so we can infer that the population mean reaction time is greater when using a cell phone. The confidence interval *is* more informative than the significance test because it predicts just how large the difference might be.

The article about the study did not indicate whether the subjects were randomly selected, so inferential conclusions are tentative.

Insight

The study also showed that reaction times were similar with hands-free versus hand-held cell phones. It also showed that the probability of missing a simulated traffic signal doubled when students used cell phones.

A later related study[13] showed that the mean reaction time of students using cell phones was similar to that of elderly drivers not using cell phones. The study concluded, "The net effect of having younger drivers converse on a cell phone was to make their average reactions equivalent to those of older drivers who were not using a cell phone."

Try Exercises 10.47 and 10.48

In summary, comparing means with matched-pairs data is easy. We merely use methods we've already learned about in Chapters 8 and 9 for single samples.

---

[13]D. L. Strayer and F. A. Drews, *Human Factors*, vol. 46, 2004, pp. 640–649.

SUMMARY: Comparing Means of Dependent Samples

To compare means with dependent samples, construct confidence intervals and significance tests using the single sample of difference scores,

$$d = \text{observation in Sample 1} - \text{observation in Sample 2}.$$

The 95% confidence interval $\bar{x}_d \pm t_{.025}(se)$ and the test statistic $t = (\bar{x}_d - 0)/se$ are the same as for a single sample. The assumptions are also the same: A random sample or a randomized experiment and a normal population distribution of difference scores.

Comparing Proportions with Dependent Samples

We next present methods for comparing proportions with dependent samples. We can do this by conducting single-sample inference with difference scores.

Example 14

Compare proportions with dependent samples

Beliefs in Heaven and Hell

Picture the Scenario

A recent General Social Survey asked subjects whether they believed in heaven and whether they believed in hell. For the 1314 subjects who responded, Table 10.11 shows the data in contingency table form. The rows of Table 10.11 are the response categories for belief in heaven. The columns are the same categories for belief in hell. In the U.S. adult population, let p_1 denote the proportion who believe in heaven and let p_2 denote the proportion who believe in hell.

Table 10.11 Beliefs in Heaven and Hell

	Belief in Hell		
Belief in Heaven	**Yes**	**No**	**Total**
Yes	955	162	**1117**
No	9	188	**197**
Total	**964**	**350**	**1314**

Questions to Explore

a. How can we estimate p_1, p_2, and their difference?

b. Are the samples used to estimate p_1 and p_2 independent or dependent, samples?

c. How can we use a sample mean of difference scores to estimate $(p_1 - p_2)$?

Think It Through

a. The counts in the two margins of Table 10.11 summarize the responses. Of the 1314 subjects, 1117 said they believed in heaven, so $\hat{p}_1 = 1117/1314 = 0.85$. Of the same 1314 subjects, 964 said they believed in hell, so $\hat{p}_2 = 964/1314 = 0.73$. Since $(\hat{p}_1 - \hat{p}_2) = .85 - .73 = 0.12$, we estimate that 12% more people believe in heaven than in hell.

Recall

Section 2.3 and Section 7.1 showed that a **proportion** is a **mean** when we code the two possible outcomes by 1 and 0. ◀

b. Each sample of 1314 responses refers to the same 1314 subjects. Any given subject's response on belief in heaven can be matched with that subject's response on belief in hell. So the samples for these two proportions are dependent.

c. Recall that a proportion is a mean when we code the responses by 1 and 0. For belief in heaven or belief in hell, let 1 = yes and 0 = no. Then, for belief in heaven, 1117 responses were 1 and 197 responses were 0. The mean of the 1314 observations was $[1117(1) + 197(0)]/1314 = 1117/1314 = 0.85$, which is $\hat{p}_1$. For belief in hell, 964 responses were 1 and 350 responses were 0, and the mean was $\hat{p}_2 = 964/1314 = 0.73$.

Table 10.12 shows the possible paired binary responses, for the 0 and 1 coding. The difference scores for the heaven and hell responses are $d = 1, 0,$ or -1, as shown. The sample mean of the 1314 difference scores equals

$$[0(955) + 1(162) - 1(9) + 0(188)]/1314 = 153/1314 = 0.12.$$

This equals the difference of proportions, $\hat{p}_1 - \hat{p}_2 = 0.85 - 0.73 = 0.12$. In summary, the mean of the difference scores equals the difference between the sample proportions.

Table 10.12 Using 0 and 1 Responses for Binary Matched-Pairs Data

Heaven	Hell	Interpretation	Difference d	Frequency
1	1	Believe in heaven and in hell	$1 - 1 = 0$	955
1	0	Believe in heaven but not in hell	$1 - 0 = 1$	162
0	1	Believe in hell but not in heaven	$0 - 1 = -1$	9
0	0	Do not believe in heaven or hell	$0 - 0 = 0$	188

Insight

We've converted two samples with 1314 binary observations each to a single sample of 1314 difference scores. We can use single-sample methods with the differences, as we just did for matched-pairs analysis of means.

Try Exercise 10.58

Confidence Interval Comparing Proportions With Matched-Pairs Data

To conduct inference about the difference $(p_1 - p_2)$ between the population proportions, we can use the fact that the sample difference $(\hat{p}_1 - \hat{p}_2)$ is the mean of difference scores when we code the responses by 1 and 0. We can find a confidence interval for $(p_1 - p_2)$ by finding a confidence interval for the population mean of difference scores.

Example 15

Confidence interval

Beliefs in Heaven and Hell

Picture the Scenario

We continue our analysis of the data on belief in heaven and/or hell. When software analyzes the 1314 paired difference scores, we get the results in Table 10.13:

Table 10.13 Software Output for Analyzing Difference Scores from Table 10.12 to Compare Beliefs in Heaven and Hell

	N	Mean	StDev	SE Mean
Difference	1314	0.1218	0.5706	0.0157

```
95% CI for mean difference: (0.091, 0.153)
```

Questions to Explore

a. Explain how software got the confidence interval from the other results shown.

b. Interpret the reported 95% confidence interval.

Think It Through

a. We've already seen that the sample mean of the 1314 difference scores is $(\hat{p}_1 - \hat{p}_2) = 0.12$. Table 10.13 reports that the standard error of the sample mean difference is $se = 0.0157$. For $n = 1314$, $df = 1313$, and $t_{.025} = 1.962$. The population mean difference is $(p_1 - p_2)$, the difference between the population proportion p_1 who believe in heaven and the population proportion p_2 who believe in hell. A 95% confidence interval for $(p_1 - p_2)$ equals

$$\text{Sample mean difference} \pm t_{.025}(se), \text{ which is}$$
$$0.12 \pm 1.962(0.0157), \text{ or } (0.09, 0.15).$$

b. We can be 95% confident that the population proportion p_1 believing in heaven is between 0.09 higher and 0.15 higher than the population proportion p_2 believing in hell. Since the interval contains only positive values, we infer that $p_1 > p_2$.

Recall

With such a large *df*, you could use the *z*-score of 1.96 if your software or calculator cannot find the *t*-score for you. For such a very large *n*, the nonnormality of the data is not problematic. ◀

Insight

As in the matched-pairs inferences comparing two means, we conducted the inference comparing two proportions by using inference for a single parameter—namely, the population mean of the differences.

Try Exercise 10.59

$(\hat{p}_1 - \hat{p}_2) = 0$			
Heaven	**Hell**	***d***	**Freq.**
1	1	0	955
1	0	1	85
0	1	−1	85
0	0	0	188

Belief in Heaven	Belief in Hell	
	Yes	No
Yes	955	162
No	9	188

$(p_1 - p_2) = 0$ if cells with counts 162 and 9 have equal population counts.

McNemar Test Comparing Proportions With Matched-Pairs Data

For the difference scores on the binary response, $(\hat{p}_1 - \hat{p}_2) = 0$ if the number of subjects having $d = 1$ equals the number of subjects having $d = -1$. Then the sample mean of the differences equals 0. See the margin table for an example. Equivalently, the number believing in heaven but *not* in hell equals the number believing in hell but *not* in heaven. Similarly, in the population, if H_0: $p_1 = p_2$ is true, then the population mean of the difference scores is 0. So we can test H_0: $(p_1 - p_2) = 0$ by testing whether the population mean of difference scores is 0.

For testing H_0: $p_1 = p_2$ with two dependent samples, there's actually a simple way of calculating a test statistic. It uses the two counts in the contingency table that have the differences of 1 and −1, that is, the frequencies of yes on one response and no on the other. This test statistic equals their difference divided by the square root of their sum. For Table 10.11,

$$z = \frac{162 - 9}{\sqrt{162 + 9}} = 11.7.$$

We use the symbol z, because it has an approximate standard normal distribution when H_0: $p_1 = p_2$ is true.

For $z = 11.7$, the two-sided P-value equals 0.000000.... The sample has *extremely* strong evidence that the proportion p_1 believing in heaven is higher than the proportion p_2 believing in hell. The inference that $p_1 > p_2$ agrees with the conclusion in Example 15 based on the confidence interval for $(p_1 - p_2)$. The advantage of the confidence interval is that it indicates *how different* the proportions are likely to be. In percentage terms, we infer that the belief in heaven is about 9% to 15% higher.

The z test comparing proportions with dependent samples is often called **McNemar's test**, in honor of a psychologist (Quinn McNemar) who proposed it in 1947. It is a large-sample test. It applies when the sum of the two counts used in the test is at least 30 (the sum is 162 + 9 = 171 above). For two-sided tests, however, the test is robust and works well even for small samples.

SUMMARY: McNemar Test Comparing Proportions From Dependent Samples

Hypotheses: H_0: $p_1 = p_2$, H_a can be two-sided or one-sided.

Test Statistic: For the two counts for the frequency of yes on one response and no on the other, the z test statistic equals their difference divided by the square root of their sum. The sum of the counts should be at least 30, but in practice the two-sided test works well even if this is not true.

P-value: For H_a: $p_1 \neq p_2$, two-tail probability of z test statistic values more extreme than observed z, using standard normal distribution.

Example 16

McNemar's test

Speech Recognition Systems

Picture the Scenario

Research in comparing the quality of different speech recognition systems uses a series of isolated words as a benchmark test, finding for each system the proportion of words for which an error of recognition occurs. Table 10.14 shows data from one of the first articles[14] that showed how to conduct such a test. The article compared speech recognition systems called generalized minimal distortion segmentation (GMDS) and continuous density hidden Markov model (CDHMM). Table 10.14 shows the counts of the four possible sequences to test outcomes for the two systems with a given word, for a test using 2000 words.

[14]From S. Chen and W. Chen, *IEEE Transactions on Speech and Audio Processing*, vol. 3, 1995, pp. 141–145.

Table 10.14 Results of Test Using 2000 Words to Compare Two Speech Recognition Systems (GMDS and CDHMM)

	CDHMM		
GMDS	**Correct**	**Incorrect**	**Total**
Correct	1921	58	**1979**
Incorrect	16	5	**21**
Total	**1937**	**63**	**2000**

Question to Explore

Conduct McNemar's test of the null hypothesis that the probability of a correct outcome is the same for each system.

Think It Through

The article about this test did not indicate how the words were chosen. Inferences are valid if the 2000 words were a random sample of the possible words on which the systems could have been tested. Let p_1 denote the population proportion of correct results for GMDS and let p_2 denote the population proportion correct for CDHMM. The test statistic for McNemar's test of H_0: $p_1 = p_2$ against H_a: $p_1 \neq p_2$ is

$$z = \frac{58 - 16}{\sqrt{58 + 16}} = 4.88.$$

The two-sided P-value is 0.000001. There is extremely strong evidence against the null hypothesis that the correct detection rates are the same for the two systems.

Insight

We learn more by estimating parameters. A confidence interval for $p_1 - p_2$ indicates that the GMDS system is better, but the difference in correct detection rates is small. See Exercise 10.56.

Try Exercise 10.57

10.4 Practicing the Basics

10.47 Does exercise help blood pressure? Several recent TRY studies have suggested that people who suffer from abnormally high blood pressure can benefit from regular exercise. A medical researcher decides to test her belief that walking briskly for at least half an hour a day has the effect of lowering blood pressure. She conducts a small pilot study. If results from it are supportive, she will apply for funding for a larger study. She randomly samples three of her patients who have high blood pressure. She measures their systolic blood pressure initially and then again a month later after they participate in her exercise program. The table shows the results.

Subject	Before	After
1	150	130
2	165	140
3	135	120

a. Explain why the three before observations and the three after observations are dependent samples.

b. Find the sample mean of the before scores, the sample mean of the after scores, and the sample mean of d = before − after. How are they related?

c. Find a 95% confidence interval for the difference between the population means of subjects before and after going through such a study. Interpret.

10.48 Test for blood pressure Refer to the previous exercise. TRY The output shows some results of using software to analyze the data with a significance test.

```
Paired T for Before—After

            N     Mean    StDev    SE Mean
Before      3    150.0     15.0      8.660
After       3    130.0     10.0      5.774
Difference  3     20.0      5.0      2.887
T-Test of mean difference = 0 (vs not = 0):
T-Value = 6.93 P-Value = 0.020
```

a. State the hypotheses to which the reported P-value refers.

b. Explain how to interpret the P-value. Does the exercise program seem beneficial to lowering blood pressure?

c. What are the assumptions on which this analysis is based?

10.49 Social activities for students As part of her class project, a student at the University of Florida randomly sampled 10 fellow students to investigate their most common social activities. As part of the study, she asked the students to state how many times they had done each of the following activities during the previous year: going to a movie, going to a sporting event, or going to a party. The table shows the data.

Frequency of Attending Movies, Sports Events, and Parties

	Activity		
Student	**Movies**	**Sports**	**Parties**
1	10	5	25
2	4	0	10
3	12	20	6
4	2	6	52
5	12	2	12
6	7	8	30
7	45	12	52
8	1	25	2
9	25	0	25
10	12	12	4

a. To compare the mean movie attendance and mean sports attendance using statistical inference, should we treat the samples as independent or dependent? Why?

b. The figure is a dot plot of the $n = 10$ difference scores for movies and sports. Does this show any irregularities that would make statistical inference unreliable?

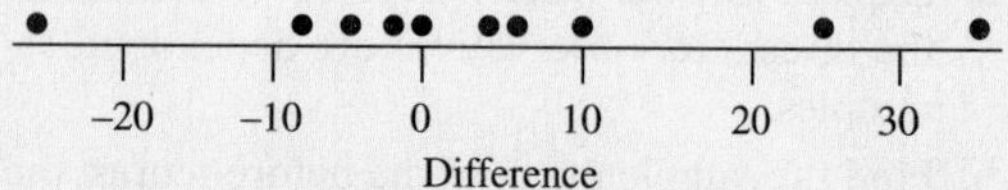

Dot plot of difference scores for Exercise 10.49.

c. Using the MINITAB output shown for these data, show how the 95% confidence interval was obtained from the other information given in the printout. Interpret the interval.

d. Show how the test statistic shown on the printout was obtained from the other information given. Report the P-value, and interpret in context.

MINITAB Output for Inferential Analyses:

```
Paired T for movies - sports

            N      Mean      StDev     SEMean
movies     10   13.0000    13.1740     4.1660
sports     10    9.0000     8.3799     2.6499
Difference 10   4.00000   16.16581    5.11208
95% CI for mean difference: (-7.56432,15.56432)
T-Test of mean difference = 0 (vs not = 0):
T-Value = 0.78 P-Value = 0.454
```

10.50 More social activities Refer to the previous exercise. The output shows the result of comparing the mean responses on parties and sports.

```
Paired T for parties - sports

            N      Mean     StDev    SE Mean
parties    10   21.8000   18.5760     5.8742
sports     10    9.0000    8.3799     2.6499
Difference 10   12.8000   22.5477     7.1302
95% CI for mean difference: (-3.3297, 28.9297)
T-Test of mean difference = 0 (vs not = 0):
T-Value = 1.80 P-Value = 0.106
```

a. Explain how to interpret the reported 95% confidence interval.

b. State the hypotheses to which the P-value refers, and interpret its value.

c. Explain the connection between the results of the test and the confidence interval.

d. What assumptions are necessary for these inferences to be appropriate?

10.51 Movies versus parties Refer to the previous two exercises. Using software, compare the responses on movies and parties using (a) all steps of a significance test and (b) a 95% confidence interval. Interpret results in context.

10.52 Freshman 15 a myth? The freshman 15 is the name of a common belief that college students, particularly women, gain an average of 15 pounds during their first year of college. A recent study (*Journal of American Health*, vol. 58, 2009, pp. 223–231) found that female students weighed an average of 133.0 pounds at the start of the school year and an average of 135.1 pounds at the end of the school year. The standard deviation of weights was about 30 pounds at each time. Other reports have been consistent with this finding.

a. Estimate the change in the mean weight

b. Is this sufficient information to find a confidence interval or conduct a test about the change in the mean? If not, what else do you need to know?

10.53 Checking for freshman 15 Refer to the previous exercise. Suppose that the change in weight scores for the 132 freshmen had a standard deviation of 2.0 pounds.

a. Explain how this standard deviation could be so much less than the standard deviation of 30 for the weight scores at each time.

b. Is 15 pounds a plausible mean weight change in the population of freshman women? Answer by constructing a 95% confidence interval for the population mean change in weight or conducting a significance test of the hypothesis that the mean weight change in the population equals 15. Interpret.

c. What assumptions are necessary for the inference in part b?

10.54 Internet book prices Anna's project for her introductory statistics course was to compare the selling prices of textbooks at two Internet bookstores. She first took a random sample of 10 textbooks used that term in courses at her college, based on the list of texts compiled by the college bookstore. The prices of those textbooks at the two Internet sites were

Site A: \$115, \$79, \$43, \$140, \$99, \$30, \$80, \$99, \$119, \$69
Site B: \$110, \$79, \$40, \$129, \$99, \$30, \$69, \$99, \$109, \$66

a. Are these independent samples or dependent samples? Justify your answer.

b. Find the mean for each sample. Find the mean of the difference scores. Compare, and interpret.

c. Using software or a calculator, construct a 90% confidence interval comparing the population mean prices of all textbooks used that term at her college. Interpret.

10.55 Comparing book prices 2 For the data in the previous exercise, use software or a calculator to perform a significance test comparing the population mean prices. Show all steps of the test, and indicate whether you would conclude that the mean price is lower at one of the two Internet bookstores.

10.56 Comparing speech recognition systems Table 10.14 in Example 16, repeated here, showed results of an experiment comparing the results of two speech recognition systems, GMDS and CDHMM.

	CDHMM	
GMDS	**Correct**	**Incorrect**
Correct	1921	58
Incorrect	16	5

a. Estimate the population proportion p_1 of correct results for GMDS and p_2 of correct results for CDHMM.

b. Software reports a 95% interval for $p_1 - p_2$ of (0.013, 0.029). Interpret.

10.57 Treat juveniles as adults? The table that follows refers to a sample of juveniles convicted of a felony in Florida. Matched pairs were formed using criteria such as age and the number of prior offenses. For each pair, one subject was handled in the juvenile court and the other was transferred to the adult court. The response of interest was whether the juvenile was rearrested within a year.

a. Are the outcomes for the courts independent samples or dependent samples? Explain.

b. Estimate the population proportions of rearrest for the adult and juvenile courts.

c. Test the hypothesis that the population proportions rearrested were identical for the adult and juvenile court assignments. Use a two-sided alternative, and interpret the P-value.

	Juvenile Court	
Adult Court	**Rearrest**	**No Rearrest**
Rearrest	158	515
No Rearrest	290	1134

Source: Data provided by Larry Winner.

10.58 Obesity now and in 20 years Many medical studies have used a large sample of subjects from Framingham, Massachusetts who have been followed since 1948. A recent study (*Annals of Internal Medicine*, vol. 138, pp. 24–32, 2003) gave the contingency table shown for weight at a baseline time and then 20 years later.

	20 Years after Baseline	
Baseline	**Normal**	**Overweight**
Normal	695	368
Overweight	87	827

a. Find the sample proportion with normal weight at (i) baseline and (ii) 20 years later.

b. Identify the two samples and whether they are independent or dependent. Explain.

10.59 Change coffee brand? A study was conducted to see if an advertisement campaign would increase market share for Sanka instant decaffeinated coffee (R. Grover and V. Srinivasan, *J. Marketing Research*, vol. 24, 1987, pp. 139–153). Subjects who use instant decaffeinated coffee were asked which brand they bought last. They were asked this before the campaign and after it. The results are shown in the table.

a. Estimate the population proportion choosing Sanka for the (i) first purchase and (ii) second purchase.

b. Explain how each proportion in part a can be found as a sample mean and how the estimated difference of population proportions is a difference of sample means.

c. The table also shows results of a confidence interval for the difference between the population proportions. Explain how to interpret it.

Two Purchases of Coffee

	Second Purchase	
First Purchase	**Sanka**	**Other Brand**
Sanka	155	49
Other brand	76	261

```
                N    Mean    StDev    SE Mean
Difference    541   0.050    0.478      0.021
95% CI for mean difference: (0.010, 0.090)
```

10.60 President's popularity Last month a random sample of 1000 subjects was interviewed and asked whether they thought the president was doing a good job. This month the same subjects were asked this again. The results are: 450 said yes each time, 450 said no each time, 60 said yes on the first survey and no on the second survey, and 40 said no on the first survey and yes on the second survey.

a. Form a contingency table showing these results.

b. Estimate the proportion giving a favorable rating (i) last month and (ii) this month.

c. Show how each of the proportions in part b can be obtained as a sample mean.

d. Find the test statistic and P-value for applying McNemar's test that the population proportion was the same each month. Interpret.

10.61 Heaven and hell Results of polls about belief in heaven and hell depend strongly on question wording. For instance, a survey conducted by the International Social Science Program at National Opinion Research Center (NORC) asked Americans whether they "definitely believe in heaven" or "definitely believe in hell" (see religioustolerance.org). The percentage who answered yes was 63.1% for heaven and 49.6% for hell.

a. Estimate the difference between the population proportions believing in heaven and believing in hell.

b. In this survey, suppose that 65 subjects believed in heaven but not in hell, whereas 35 believed in hell but not in heaven. Report the (i) assumptions, (ii) hypotheses, (iii) test statistic, (iv) P-value, and (v) conclusion for testing that the probability of belief is the same for heaven and for hell.

10.62 Heaven and hell around the world Refer to the previous exercise. Results in this poll also depended strongly on the country in which the question was asked. For instance, the percentages believing in (heaven, hell) were (52%, 26%) in Ireland, (32%, 19%) in New Zealand, (28%, 22%) in Italy, (25%, 13%) in Great Britain, (21%, 11%) in the Netherlands, and (18%, 9%) in Germany. If for a given country you know the sample size was 1000, does this give you enough information to compare the two proportions inferentially for that country? Explain.

10.5 Adjusting for the Effects of Other Variables

When a practically significant difference exists between two groups, can we identify a reason for the difference? As Chapters 3 and 4 explained, an association may be due to a *lurking variable* not measured in the study. Example 5 in Section 10.1 indicated that teenagers who watch more TV have a tendency later in life to commit more aggressive acts. Why? Might education be a lurking variable? Perhaps teenagers who watch more TV tend to attain lower educational levels, and perhaps lower education tends to be associated with higher levels of aggression.

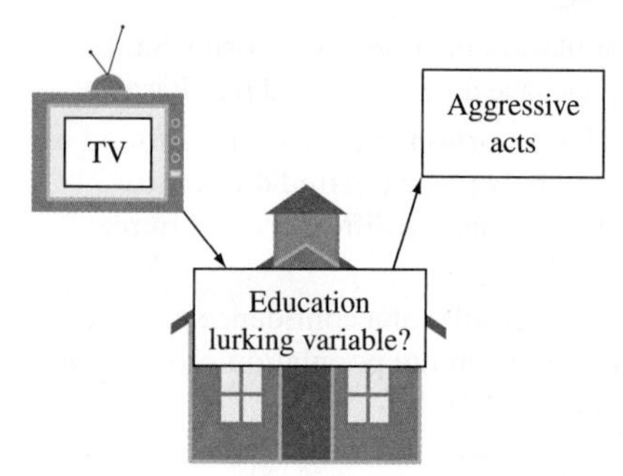

To investigate why differences occur between groups, we must measure potential lurking variables and use them in the statistical analysis. For instance, suppose we planned to study TV watching and aggression, and we thought that educational level was a potential lurking variable. Then, in the study we would want to measure educational level.

Let's examine the TV study again. Suppose we did so using categories (did not complete high school, did not attend college, attended college). We could then evaluate TV watching and aggressive behavior separately for subjects in these three educational levels and obtain the results in Table 10.15.

This analysis uses three variables—level of TV watching, committing of aggressive acts, and educational level. This takes us from a **bivariate** analysis (two variables) to a **multivariate** analysis (more than two variables): Whether the subject has committed aggressive acts is the response variable, the level of TV watching is the explanatory variable, and the educational level is called a **control variable**.

Recall

You can review Section 3.4 to see how **lurking variables** can influence an association. ◀

Table 10.15 Percentage Committing Aggressive Acts, According to Level of Teenage TV Watching, Controlling for Educational Level

For example, for those who attended college who watched TV less than 1 hour per day, 2% committed an aggressive act.

	Educational Level		
TV Watching	**Less than High School**	**High School**	**College**
Less than 1 hour per day	8%	4%	2%
At least 1 hour per day	30%	20%	10%

> **Control variable**
>
> A **control variable** is a variable that is held constant in a multivariate analysis.

Table 10.15 treats educational level as a control variable. Within each of the three education groups, educational level is approximately constant.

Statistical Control

To analyze whether an association can be explained by a third variable, we treat that third variable as a control variable. We conduct the ordinary bivariate analysis, such as comparing two proportions, while holding that control variable constant at fixed values. Then, whatever association occurs cannot be due to effects of the control variable, because in each part of the analysis it is not allowed to vary.

In Table 10.15, at each educational level, the percentage committing an aggressive act is higher for those who watched more TV. If these were the actual results, then the association between TV watching and aggressive acts was not because of education. In each part of Table 10.15, educational level is approximately constant, so changes in aggression as TV watching varies are not due to changes in educational level. To conduct inference, we could use the methods of this chapter to compare the population proportions of aggression at each education level.

Conducting a statistical analysis while holding a control variable constant occurs in scientific laboratory experiments, too. Such experiments hold certain variables constant so they do not affect results. For instance, a chemistry experiment might control temperature by holding it constant throughout the experiment. Otherwise, a sudden increase or decrease in the temperature might cause a reaction that would not happen normally. Unlike experiments, observational studies cannot assign subjects to particular values of variables that we want to control, such as educational level. But we can approximate an experimental control by grouping together observations with similar values on the control variable, as just described. This is a **statistical control**, rather than an experimental control.

The next example shows that if we control for a variable, the results can look quite different than if we do not take that control variable into account.

Example 17

Control variables

Death Penalty and Race

Picture the Scenario

The United States is one of only a few Western nations that still imposes the death penalty. Are those convicted of murder more likely to get the death penalty if they are black than if they are white?

Table 10.16 comes from one of the first studies on racial inequities of the death penalty.[15] The 326 subjects were defendants convicted of homicide in Florida murder trials. The variables are the defendant's race and whether the defendant received the death penalty. The contingency table shows that about 12% of white defendants and about 10% of black defendants received the death penalty.

[15]M. Radelet, *American Sociological Review*, vol. 46, 1981, pp. 918–927.

Table 10.16 Defendant's Race and Death Penalty Verdict for Homicide Cases in Florida

	Death Penalty			
Defendant's Race	Yes	No	Total	Percentage Yes
White	19	141	160	11.9
Black	17	149	166	10.2

In this study, the difference between the percentages of white defendants and black defendants who received the death penalty is small (1.7%), but the percentage was lower for black defendants.

Is there some explanation for these results? Does a control variable lurk that explains why relatively fewer black defendants got the death penalty in Florida? Researchers who study the death penalty stress that the *victim's* race is often an important factor. So, let's control for victim's race. We'll construct a table like Table 10.16 *separately* for cases in which the victim was white and for cases in which the victim was black. Table 10.17 shows this three-variable table.

Table 10.17 Defendant's Race and Death Penalty Verdict, Controlling for Victim's Race

		Death Penalty			
Victim's Race	Defendant's Race	Yes	No	Total	Percentage Yes
White	White	19	132	151	12.6
	Black	11	52	63	17.5
Black	White	0	9	9	0.0
	Black	6	97	103	5.8

Table 10.17 shows the effect of a defendant's race on the death penalty verdict while controlling for victim's race, or when victim's race is kept constant at white or kept constant at black. Table 10.17 decomposes Table 10.16 into two separate contingency tables, one for each victim's race. You can recover Table 10.16 by summing corresponding entries, $19 + 0 = 19$, $132 + 9 = 141$, $11 + 6 = 17$, and $52 + 97 = 149$.

Questions to Explore

a. Summarize the association between a defendant's race and the death penalty verdict in Florida, after controlling for victim's race.

b. Describe the difference in the results between ignoring and controlling victim's race.

c. Why does the control for victim's race make such a difference?

Think It Through

a. The top part of Table 10.17 lists cases in which the victim's race was white. Relatively more black defendants than white defendants got the death penalty in Florida. The difference between the percentages is $17.5 - 12.6 = 4.9$. The bottom part of the table lists cases in which the victim's race was black. Again, relatively more black defendants got the death penalty. The difference between the percentages is 5.8 − 0.0 = 5.8. In summary, controlling for victim's race, more black defendants than white defendants received the death penalty in Florida.

b. Table 10.16 showed that a *larger* percentage of white defendants than black defendants got the death penalty. By contrast, controlling for victim's race, Table 10.17 showed that a *smaller* percentage of white defendants than black defendants got the death penalty. This was true for each victim's race category.

c. In Table 10.17, look at the percentages who got the death penalty. When the victim was white, they are quite a bit larger (12.6 and 17.5) than when the victim was black (0.0 and 5.8). That is, defendants who killed a white person were more likely to get the death penalty. Now look at the totals for the four combinations of victim's race and defendant's race. The most common cases are white defendants having white victims (151 times) and black defendants having black victims (103 times). In summary, white defendants usually had white victims and black defendants usually had black victims. Killing a white person was more likely to result in the death penalty than killing a black person. These two factors operating together produce an overall association that shows (in Table 10.16) that a higher percentage of white defendants than black defendants got the death penalty.

Insight

Overall, relatively more white defendants in Florida got the death penalty than black defendants. Controlling for victim's race, however, relatively more black defendants got the death penalty. The effect changes direction after we control for victim's race. This shows that the association at each level of a control variable can have a different direction than overall when that third variable is ignored instead of controlled. This phenomenon is called **Simpson's paradox**.

In Table 10.17, the death penalty was imposed most often when a black defendant had a white victim. By contrast, it was never imposed when a white defendant had a black victim. Similar results have occurred in other studies of the death penalty. See Exercises 10.64 and 3.58.

Recall

Section 3.4 first introduced **Simpson's paradox.** It is named after a British statistician who wrote an article in 1951 about mathematical conditions under which the association can change direction when you control for a variable. ◀

Try Exercise 10.64

With statistical control imposed for a variable, the results tend to change considerably when that control variable has a strong association both with the response variable and the explanatory variable. In Example 17, for instance, victim's race was the control variable. Victim's race had a noticeable association both with the death penalty response (the death penalty was given more often when the victim was white) and with defendant's race (defendants and victims usually had the same race).

Recall

From the end of Section 3.4, we say that defendant's race and victim's race are confounded in their effects on the death penalty verdict. ◀

In Practice Control for Variables Associated Both With the Response and Explanatory Variables

In determining which variables to control in observational studies, *researchers choose variables that they expect to have a practically significant association with both the response variable and the explanatory variable.* A statistical analysis that controls such variables can have quite different results than when those variables are ignored.

Statistical control is also relevant when the response variable is quantitative. The difference between means for two groups can change substantially after controlling for a third variable. See Exercise 10.66.

Activity 1

Interpreting Newspaper Articles

This chapter began by stating that many newspaper articles describe research studies that use the statistical methods of this chapter. At this stage of your statistics course, you should be able to read and understand those articles much better than you could before you knew the basis of such terms as margin of error, statistical significance, and randomized clinical trial.

For the next three days, read a daily newspaper such as *The New York Times*. Make a copy of any article you see that refers to a study that used statistics to help make its conclusions. For one article, prepare a one-page report that answers the following questions:

- What was the purpose of the study?
- Was the study experimental or observational? Can you provide any details about the study design and sampling method?
- Identify explanatory and response variables.
- Can you tell whether the statistical analysis used (1) independent samples or dependent samples and (2) a comparison of proportions or a comparison of means?
- Can you think of any lurking variables that could have affected the results?
- Are there any limitations that would make you skeptical to put much faith in the conclusions of the study?

You might find additional information to answer these questions by browsing the Web or going to the research journal that published the results.

10.5 Practicing the Basics

10.63 Benefits of drinking A *USA Today* story (May 22, 2010) about the medical benefits of moderate drinking of alcohol stated that a major French study links those who drink moderately to a lower risk for cardiovascular disease but challenges the idea that moderate drinking is the cause. "Instead, the researchers say, people who drink moderately tend to have a higher social status, exercise more, suffer less depression and enjoy superior health overall compared to heavy drinkers and lifetime abstainers. A causal relationship between cardiovascular risk and moderate drinking is not at all established." The study looked at the health status and drinking habits of 149,773 French adults.

a. Explain how this story refers to an analysis of three types of variables. Identify those variables.

b. Suppose socioeconomic status is treated as a control variable when we compare moderate drinkers to abstainers in their heart attack rates. Explain how this analysis shows that an effect of an explanatory variable on a response variable can change at different values of a control variable.

10.64 Death penalty in Kentucky TRY A study of the death penalty in Kentucky reported the results shown in the table. (*Source*: Data from T. Keil and G. Vito, *Amer. J. Criminal Justice*, vol. 20, 1995, pp. 17–36.)

a. Find and compare the percentage of white defendants with the percentage of black defendants who received the death penalty, when the victim was (i) white and (ii) black.

b. In the analysis in part a, identify the response variable, explanatory variable, and control variable.

c. Construct the summary 2 × 2 table that ignores, rather than controls, victim's race. Compare the overall percentages of white defendants and black defendants who got the death penalty (ignoring, rather than controlling, victim's race). Compare to part a.

d. Do these data satisfy Simpson's paradox? If not, explain why not. If so, explain what is responsible for Simpson's paradox occurring.

e. Explain, without doing the calculations, how you could inferentially compare the proportions of white and black defendants who get the death penalty (i) ignoring victim's race and (ii) controlling for victim's race.

Victim's	Defendant's	Death Penalty		
Race	**Race**	**Yes**	**No**	**Total**
White	White	31	360	391
	Black	7	50	57
Black	White	0	18	18
	Black	2	106	108

10.65 Basketball paradox The following list summarizes shooting percentage in the 2001–2002 season in the National Basketball Association by Brent Barry and Shaquille O'Neal.

2-point shots

- Brent Barry: 58.8% (237/403)
- Shaquille O'Neal: 58.0% (712/1228)

3-point shots

- Brent Barry: 42.4% (164/387)
- Shaquille O'Neal: 0% (0/1)

Overall

- Brent Barry: 50.8% (401/790)
- Shaquille O'Neal: 57.9% (712/1229)

(Data from www.nba.com and article by T. P. Ryan, *Chance,* vol. 16, 2003, p. 16.)

a. Treating the type of shot as a control variable, whether a shot is made as the response variable, and the player as the explanatory variable, explain how these results illustrate Simpson's paradox.

b. Explain how O'Neal could have the higher overall percentage, even though he made a lower percentage of each type of shot.

10.66 Teacher salary, gender, and academic rank The American Association of University Professors (AAUP) reports yearly on faculty salaries for all types of higher education institutions across the United States. The following table lists the mean salary, in thousands of dollars, of full-time instructional faculty on nine-month contracts at four-year public institutions of higher education in 2010, by gender and academic rank. Regard salary as the response variable, gender as the explanatory variable, and academic rank as the control variable.

Mean Salary (Thousands of Dollars) for Men and Women Faculty Members

	Academic Rank				
Gender	**Professor**	**Associate**	**Assistant**	**Instructor**	**Overall**
Men	109.2	77.8	66.1	46.0	84.4
Women	96.2	72.7	61.8	46.9	68.8

a. Find the difference between men and women faculty members on their mean salary (i) overall and (ii) after controlling for academic rank.

b. The overall difference between the mean salary of men and women faculty members was larger than the difference at each academic rank. What could be a reason for this? (Simpson's paradox does not hold here, but the gender effect does weaken when we control for academic rank.)

10.67 Family size in Canada The table shows the mean number of children in Canadian families, classified by whether the family was English speaking or French speaking and by whether the family lived in Quebec or in another province.

Mean Number of Children in Canada

Province	English Speaking	French Speaking
Quebec	1.64	1.80
Other	1.97	2.14
Overall	1.95	1.85

a. Overall, compare the mean number of children for English-speaking and French-speaking families.

b. Compare the means, controlling for province (Quebec, Others).

c. How is it possible that for each level of province the mean is higher for French-speaking families, yet overall the mean is higher for English-speaking families? Which paradox does this illustrate?

10.68 Heart disease and age In the United States, the median age of residents is lowest in Utah. At each age level, the death rate from heart disease is higher in Utah than in Colorado. Overall, the death rate from heart disease is lower in Utah than Colorado. Are there any contradictions here, or is this possible? Explain.

10.69 Breast cancer over time The percentage of women who get breast cancer sometime during their lifetime is higher now than in 1900. Suppose that breast cancer incidence tends to increase with age, and suppose that women tend to live longer now than in 1900. Explain why a comparison of breast cancer rates now with the rate in 1900 could show different results if we control for the age of the woman.

Chapter Review

ANSWERS TO THE CHAPTER FIGURE QUESTIONS

Figure 10.2 $\hat{p}_1 - \hat{p}_2$ represents the difference between the sample proportions suffering a cancer death with placebo and with aspirin. $p_1 - p_2$ represents the difference between the population proportions. The mean of this sampling distribution is zero if cancer death rates are identical for aspirin and placebo.

Figure 10.3 If the interval contains only negative values, $p_1 - p_2$ is predicted to be negative, so p_1 is predicted to be smaller than p_2. If the interval contains only positive values, p_1 is predicted to be larger than p_2.

Figure 10.4 Females seem to have the greater nicotine dependence. The female percentages are higher for scores 1 through 10 (a higher score indicates higher nicotine dependence) and the male percentage is higher for score 0.

Figure 10.5 Positive numbers in the confidence interval for $(\mu_1 - \mu_2)$ suggest that $\mu_1 - \mu_2 > 0$, and thus that $\mu_1 > \mu_2$.

Figure 10.6 The box plot for the cell phone group indicates an extreme outlier. It's important to check that the results of the analysis aren't affected too strongly by that single observation.

Figure 10.7 One sample distribution will have more variability around the mean than the other.

Figure 10.8 The outliers for the two distributions are both from subject number 28.

Figure 10.9 This box plot represents a student's difference score in reaction time using the cell phone minus the reaction time not using it. Some students had a slower reaction time when not using the cell phone, resulting in a negative difference.

CHAPTER SUMMARY

This chapter introduced inferential methods for comparing two groups.

- For categorical data, we compare the *proportions* of individuals in a particular category. Confidence intervals and significance tests apply to the difference between the population proportions, $(p_1 - p_2)$, for the two groups. The test of H_0: $p_1 = p_2$ analyzes whether the population proportions are equal. If the test has a small P-value, or if the confidence interval for $(p_1 - p_2)$ does not contain 0, we conclude that they differ.

- An alternative way of comparing proportions uses the **relative risk,** which is the *ratio* of proportions. The population relative risk p_1/p_2 is 1.0 when the population proportions are equal.
- For quantitative data, we compare the *means* for the two groups. Confidence intervals and significance tests apply to $(\mu_1 - \mu_2)$.

Table 10.18 summarizes two-sided estimation and testing methods for large, **independent random samples.** This is the most common case in practice.

Table 10.18 Comparing Two Groups, for Large, Independent Random Samples

	Type of Response Variable	
	Categorical	**Quantitative**
Estimation		
1. Parameter	$p_1 - p_2$	$\mu_1 - \mu_2$
2. Point estimate	$(\hat{p}_1 - \hat{p}_2)$	$(\bar{x}_1 - \bar{x}_2)$
3. Standard error (*se*)	$\sqrt{\frac{\hat{p}_1(1-\hat{p}_1)}{n_1} + \frac{\hat{p}_2(1-\hat{p}_2)}{n_2}}$	$\sqrt{\frac{s_1^2}{n_1} + \frac{s_2^2}{n_2}}$
4. 95% confidence int.	$(\hat{p}_1 - \hat{p}_2) \pm 1.96(se)$	$(\bar{x}_1 - \bar{x}_2) \pm t_{.025}(se)$
Significance Test		
1. Assumptions	Randomization with at least five of each type for each group if using a 2-sided alternative	Randomization from Normal population (test robust)
2. Hypotheses	$H_0: p_1 = p_2\ (p_1 - p_2 = 0)$ $H_a: p_1 \neq p_2$	$H_0: \mu_1 = \mu_2\ (\mu_1 - \mu_2 = 0)$ $H_a: \mu_1 \neq \mu_2$
3. Test statistic	$z = (\hat{p}_1 - \hat{p}_2)/se_0$ $\left[se_0 = \sqrt{\hat{p}(1-\hat{p})\left(\frac{1}{n_1} + \frac{1}{n_2}\right)}\right.$ with $\hat{p}$ = pooled proportion]	$t = (\bar{x}_1 - \bar{x}_2)/se$ [can also use se $= s\sqrt{\frac{1}{n_1} + \frac{1}{n_2}}$ with s = pooled standard deviation]
4. P-value	Two-tail probability from standard normal distribution	Two-tail probability from *t* distribution
5. Conclusion	Interpret P-value in context. Reject H_0 if P-value $\leq \alpha$	Interpret P-value in context. Reject H_0 if P-value $\leq \alpha$

Both for differences of proportions and differences of means, confidence intervals have the form

Estimate ± (*z*- or *t*-score)(standard error),

such as for 95% confidence intervals,

$$(\hat{p}_1 - \hat{p}_2) \pm 1.96(se), \text{ and } (\bar{x}_1 - \bar{x}_2) \pm t_{.025}(se).$$

For significance tests, the test statistic equals the estimated difference divided by the standard error, $z = (\hat{p}_1 - \hat{p}_2)/se_0$ and $t = (\bar{x}_1 - \bar{x}_2)/se$.

With **dependent samples,** each observation in one sample is matched with an observation in the other sample. We compare means or proportions by analyzing the mean of the differences between the paired observations. The confidence interval and test procedures apply one-sample methods to the difference scores.

At this stage, you may feel confused about which method to use for any given exercise. It may help if you use the following checklist. Ask yourself, do you have

- Means or proportions (quantitative or categorical variable)?
- Independent samples or dependent samples?
- Confidence interval or significance test?
- Large n_1 and n_2 or not?

In practice, most applications have large, independent samples, and confidence intervals are more useful than tests. So the following cases are especially important: confidence intervals comparing proportions with large independent samples, and confidence intervals comparing means with large independent samples.

SUMMARY OF NOTATION

A subscript index identifies the group number. For instance, n_1 = sample size for Group 1; n_2 = sample size for Group 2. Likewise for population means μ_1, μ_2, sample means $\bar{x}_1, \bar{x}_2$, sample standard deviations s_1, s_2, population proportions p_1, p_2, sample proportions $\hat{p}_1, \hat{p}_2$. For dependent samples, $\bar{x}_d$ and μ_d are sample mean and population mean of difference scores, and s_d is standard deviation of difference scores.

CHAPTER PROBLEMS

Practicing the Basics

10.70 Pick the method Steve Solomon, the owner of Leonardo's Italian restaurant, wonders whether a redesigned menu will increase, on the average, the amount that customers spend in the restaurant. For the following scenarios, pick a statistical method from this chapter that would be appropriate for analyzing the data, indicating whether the samples are independent or dependent, which parameter is relevant, and what inference method you would use:

a. Solomon records the mean sales the week before the change and the week after the change and then wonders whether the difference is "statistically significant."

b. Solomon randomly samples 100 people and shows them each both menus, asking them to give a rating between 0 and 10 for each menu.

c. Solomon randomly samples 100 people and shows them each both menus, asking them to give an overall rating of positive or negative to each menu.

d. Solomon randomly samples 100 people and randomly separates them into two groups of 50 each. He asks those in Group 1 to give a rating to the old menu and those in Group 2 to give a rating to the new menu, using a 0 to 10 rating scale.

10.71 Opinion about America A report (December 4, 2002) by the Pew Research Center on *What the World Thinks in 2002* reported that, "The American public is strikingly at odds with publics around the world in its views about the U.S. role in the world and the global impact of American actions." Conclusions were based on polls in several countries in 1999/2000 and in 2002. One of the largest changes reported was in Pakistan, where the percentage of interviewed subjects who had a favorable view of the United States was 23% in 1999/2000 and 10% in 2002. The sample sizes were apparently about 1000 for the first survey and 2032 for the second.

a. Identify the response variable and the explanatory variable.

b. If the two surveys used separate samples of subjects, should we treat these samples as dependent or as independent in order to conduct inference? Explain.

c. The same report indicated that in the 1999/2000 survey the percentage in Pakistan who thought the spread of American ideas and customs was good was 2% (n = 2032). To compare this to the percentage who had a favorable view of the United States in that survey, should you treat the samples as dependent, or as independent? Why?

10.72 More overweight over time The Centers for Disease Control (www.cdc.gov) periodically takes large randomized surveys to track health of Americans. In a survey of 11,207 adults in 1976–1980, 47% were overweight (body mass index BMI $\geq$ 25). In a survey of 4431 adults in 2003–2004, 66% were overweight.

a. Estimate the change in the population proportion who are overweight, and interpret.

b. The standard error for estimating this difference equals 0.009. What is the main factor that causes *se* to be so small?

c. The 95% confidence interval comparing the population proportions is (0.17, 0.21). Interpret, taking into account whether or not 0 is in this interval.

10.73 Marijuana and gender In a survey conducted by Wright State University, senior high school students were asked if they had ever used marijuana. The table shows results of one analysis, where X is the count who said yes. Assuming these observations can be treated as a random sample from a population of interest,

a. Interpret the reported estimate and the reported confidence interval. Explain how to interpret the fact that 0 is not in the confidence interval.

b. Explain how the confidence interval would change if males were Group 1 and females were Group 2. Express an interpretation for the interval in that case.

```
Sample        X      N       Sample p
1. Female    445    1120     0.3973
2. Male      515    1156     0.4455
estimate for p(1) - p(2):    -0.0482
95% CI for p(1) - p(2):(-0.0887, -0.0077)
```

10.74 Gender and belief in afterlife The table shows results from the 2008 General Social Survey on gender and whether or not one believes in an afterlife.

	Belief in Afterlife		
Gender	**Yes**	**No**	**Total**
Female	599	111	710
Male	425	168	593

a. Denote the population proportion who believe in an afterlife by p_1 for females and by p_2 for males. Estimate p_1, p_2, and $(p_1 - p_2)$.

b. Find the standard error for the estimate of $(p_1 - p_2)$. Interpret.

c. Construct a 95% confidence interval for $(p_1 - p_2)$. Can you conclude which of p_1 and p_2 is larger? Explain.

d. Suppose that, unknown to us, $p_1 = 0.81$ and $p_2 = 0.72$. Does the confidence interval in part c contain the parameter it is designed to estimate? Explain.

10.75 Belief depend on gender? Refer to the previous exercise.

a. Find the standard error of $(\hat{p}_1 - \hat{p}_2)$ for a test of H_0: $p_1 = p_2$.

b. For a two-sided test, find the test statistic and P-value, and make a decision using significance level 0.05. Interpret.

c. Suppose that actually $p_1 = 0.81$ and $p_2 = 0.72$. Was the decision in part b in error?

d. State the assumptions on which the methods in this exercise are based.

10.76 Females or males have more close friends? A recent GSS reported that the 486 surveyed females had a mean of 8.3 close friends ($s = 15.6$) and the 354 surveyed males had a mean of 8.9 close friends ($s = 15.5$).

a. Estimate the difference between the population means for males and females.

b. The 95% confidence interval for the difference between the population means is 0.6 ± 2.1. Interpret.

c. For each gender, does it seem like the distribution of number of close friends is normal? Why? How does this affect the validity of the confidence interval in part b?

10.77 Heavier horseshoe crabs more likely to mate? A study of a sample of horseshoe crabs on a Florida island (J. Brockmann, *Ethology*, vol. 102, 1996, pp. 1–21) investigated the factors that were associated with whether or not female crabs had a male crab mate. Basic statistics, including the five-number summary on weight (kg) for the 111 female crabs who had a male crab nearby and for the 62 female crabs who did not have a male crab nearby, are given in the table. Assume that these horseshoe crabs have the properties of a random sample of all such crabs.

Summary Statistics for Weights of Horseshoe Crabs

	n	Mean	Std. Dev.	Min	Q1	Med	Q3	Max
Mate	111	2.6	0.6	1.5	2.2	2.6	3.0	5.2
No Mate	62	2.1	0.4	1.2	1.8	2.1	2.4	3.2

a. Sketch box plots for the weight distributions of the two groups. Interpret by comparing the groups with respect to shape, center, and variability.

b. Estimate the difference between the mean weights of female crabs who have mates and female crabs who do not have mates.

c. Find the standard error for the estimate in part b.

d. Construct a 90% confidence interval for the difference between the population mean weights, and interpret.

10.78 TV watching and race The 2008 GSS asked about the number of hours you watch TV per day. An analysis that evaluates this by race shows the results (note the codes: 1 = Black and 2 = White):

```
Sample     N      Mean     StDev     SE Mean
1        188      4.38      3.58        0.26
2       1014      2.76      2.39       0.075
Difference = mu (1) - mu (2)
Estimate for difference: 1.620
95% CI for difference: (1.085, 2.155)
T-Test of difference = 0 (vs not =):
T-Value = 5.96 P-Value = 0.000
```

a. Do you believe that TV watching has a normal distribution for each race? Why or why not? What effect does this have on inference comparing population means?

b. Explain how to interpret the reported confidence interval. Can you conclude that one population mean is higher? If so, which one? Explain.

c. On what assumptions is this inference based?

10.79 Test TV watching by race Refer to the previous exercise.

a. Specify the hypotheses that are tested in the output shown.

b. Report the value of the test statistic and the P-value. Interpret.

c. Make a decision, using the 0.05 significance level.

d. Explain the connection between the result of this significance test in part c and the result of the confidence interval in the previous exercise.

10.80 Time spent in housework When the General Social Survey last asked the number of hours the respondent spent a week on housework (variable RHHWORK), the responses were summarized for females and males by

Group	N	Mean	StDev
Female	391	12.7	11.6
Male	292	8.4	9.5

a. Identify the response variable and explanatory variable. Indicate whether each is quantitative or categorical.

b. A 95% confidence interval for the difference between the population means for females and males is (2.7, 5.9). Interpret.

c. The significance test comparing the two population means against the one-sided alternative that the population mean is higher for women has P-value = 0.000 rounded to three decimal places. Interpret.

10.81 Time spent on Internet In 2006, the General Social Survey asked about the number of hours a week spent on the World Wide Web (variable denoted WWWTIME). Some results are as follows:

Group	N	Mean	StDev	SE Mean
Male	1196	6.2	9.9	0.285
Female	1569	4.9	8.6	0.216

Difference = mu(Male) – mu (Female)
Estimate for difference: 1.3
99% CI for difference: (0.4, 2.2)
T-Test of difference = 0 (vs not =):
T-value = 3.62 P-value = 0.000

a. Identify the response variable and explanatory variable. Indicate whether each variable is quantitative or categorical.

b. Report and interpret the 99% confidence interval shown for the difference between the population means for males and females.

c. Report and interpret the steps of the significance test comparing the two population means. Make a decision, using significance level 0.01.

10.82 Test–CI connection In the previous exercise, explain how the result of the 99% confidence interval in part b corresponds to the result of the decision using significance level 0.01 in part c.

10.83 Sex roles A study of the effect of the gender of the tester on sex-role differentiation scores[16] in Manhattan gave a random sample of preschool children the Occupational Preference Test. Children were asked to give three choices of what they wanted to be when they grew up. Each occupation was rated on a scale from 1 (traditionally feminine) to 5 (traditionally masculine), and a child's score was the mean of the three selections. When the tester was male, the 50 girls had $\bar{x} = 2.9$ and $s = 1.4$, whereas when the tester was female, the 90 girls had $\bar{x} = 3.2$ and $s = 1.2$. Show all steps of a test of the hypothesis that the population mean is the same for female and male testers, against the alternative that they differ. Report the P-value and interpret.

10.84 How often do you feel sad? A recent General Social Survey asked, "How many days in the past seven days have you felt sad?" Software comparing results for men and women who responded showed the following results.

```
Gender   N     Mean    StDev   SE Mean
Female  816    1.81    1.98    .06931
Male    633    1.42    1.83    .07274
Difference = mu (Female) - mu (Male)
Estimate for difference: 0.39
95% CI for difference: (0.232,0.548)
T-Test of difference = 0 (vs not =):
T-Value = 4.84 P-Value = 0.000
```

a. Explain how to interpret the P-value. Do you think that the population means may be equal? Why?

b. Explain how to interpret the confidence interval for the difference between the population means. What do you learn from the confidence interval that you cannot learn from the test?

[16]Obtained from Bonnie Seegmiller, a psychologist at Hunter College.

c. What assumptions are made for these inferences? Can you tell from the summary statistics shown whether any of the assumptions is seriously violated? What's the effect?

10.85 Parental support and household type A recent study interviewed youths with a battery of questions that provides a summary measure of perceived parental support. This measure had sample means of 46 ($s = 9$) for the single-mother households and 42 ($s = 10$) for the households with both biological parents. One conclusion in the study stated, "The mean parental support was 4 units higher for the single-mother households. If the true means were equal, a difference of at least this size could be expected only 2% of the time. For samples of this size, 95% of the time one would expect this difference to be within 3.4 of the true value." Explain how this conclusion refers to the results of (a) a confidence interval and (b) a significance test.

10.86 Car bumper damage An automobile company compares two different types of front bumpers for their new model by driving sample cars into a concrete wall at 20 miles per hour. The response is the amount of damage to the car, as measured by the repair costs, in hundreds of dollars. Due to the costs, the study uses only six cars, obtaining results for three bumpers of each type. The results are in the table. Conduct statistical inference (95% confidence interval or significance test with significance level 0.05) about the difference between the population means, using software if you wish. Can you conclude that the mean is higher for one bumper type?

Bumper A	Bumper B
11	1
15	3
13	4

10.87 Teenage anorexia Example 8 in Section 9.3 described a study that used a cognitive behavioral therapy to treat a sample of teenage girls who suffered from anorexia. The study observed the mean weight change after a period of treatment. Studies of that type also usually have a control group that receives no treatment or a standard treatment. Then researchers can analyze how the change in weight compares for the treatment group to the control group. In fact, the anorexia study had a control group that received a standard treatment. Teenage girls in the study were randomly assigned to the cognitive behavioral treatment (Group 1) or to the control group (Group 2). The figure shows box plots of the weight changes for the two groups (displayed vertically). The output shows how MINITAB reports inferential comparisons of those two means.

MINITAB Output Comparing Mean Weight Changes

```
group        N     Mean    StDev    SE Mean
cog_behav   29     3.01    7.31       1.4
control     26    -0.45    7.99       1.6
Difference = mu (cog_behav) - mu (control)
Estimate for difference: 3.45690
95% CI for difference:(-0.70643, 7.62022)
T-Test of difference = 0 (vs not =):
T-Value = 1.67 P-Value = 0.102 DF = 50
```

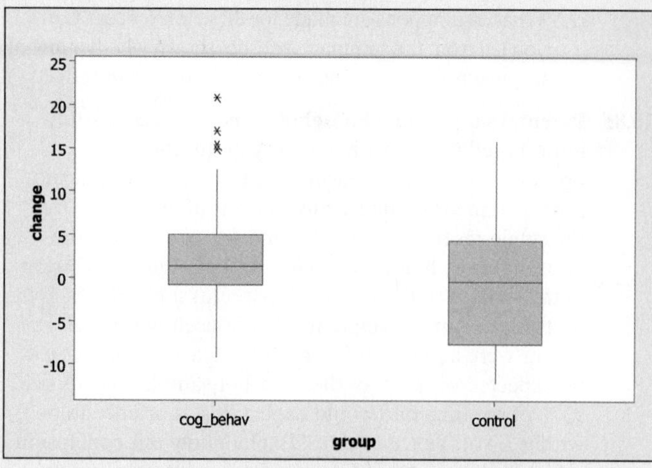

Box plots of weight change for anorexia study.

a. Report and interpret the P-value for testing $H_0: \mu_1 = \mu_2$ against $H_a: \mu_1 \neq \mu_2$.

b. Summarize the assumptions needed for the analysis in part a. Based on the box plots, would you be nervous if you had to perform a one-sided test instead? Why?

c. The reported 95% confidence interval tells us that if the population mean weight change is less for the cognitive behavioral group than for the control group, it is just barely less (less than 1 pound), but if the population mean change is greater, it could be nearly 8 pounds greater. Explain how to get this interpretation from the interval reported.

d. Explain the correspondence between the confidence interval and the decision in the significance test for a 0.05 significance level.

10.88 Anorexia paradox? Refer to the previous exercise. For the control group, the 26 changes in weights have a mean of −0.45. From Example 8 in Chapter 9, for the therapy group the sample mean change in weight of 3.0 pounds was significantly different from 0. However, the previous exercise showed it is not significantly different from the mean change for the control group, even though that group had a *negative* sample mean change. How do you explain this paradox? (*Hint:* From Section 10.2, how does the *se* value for estimating a *difference* between two means compare to the *se* value for estimating a single mean?)

10.89 Surgery versus placebo for knee pain Refer to Example 10 on whether arthroscopic surgery is better than placebo. The following table shows the pain scores one year after surgery. Using software (such as MINITAB) that can conduct analyses using summary statistics, compare the placebo to the debridement group using a 95% confidence interval. Use the method that assumes equal population standard deviations. Explain how to interpret the interval found using software.

		Knee Pain Score	
Group	**Sample Size**	**Mean**	**Standard Deviation**
Placebo	60	48.9	21.9
Arthroscopic—lavage	61	54.8	19.8
Arthroscopic—debridement	59	51.7	22.4

10.90 More knee pain Refer to the previous exercise. Compare the placebo to the debridement group using a significance test. State the assumptions, and explain how to interpret the P-value.

10.91 Anorexia again Refer to Exercise 10.87, comparing mean weight changes in anorexic girls for cognitive behavioral therapy and a control group. The MINITAB output shows results of doing analyses while assuming equal population standard deviations.

MINITAB output for comparing mean weight changes

```
group        N     Mean     StDev    SE Mean
cog_behav   29     3.01      7.31        1.4
control     26    -0.45      7.99        1.6
Difference = mu (cog_behav) - mu (control)
Estimate for difference: 3.45690
95% CI for difference:(-0.68014, 7.59393)
T-Test of difference = 0(vs not =):
T-Value = 1.68 P-Value = 0.100 DF = 53
Both use Pooled StDev = 7.6369
```

a. Interpret the reported confidence interval.

b. Interpret the reported P-value.

c. What would be the P-value for $H_a: \mu_1 > \mu_2$? Interpret it.

d. What assumptions do these inferences make?

10.92 Breast-feeding helps IQ? A Danish study of individuals born at a Copenhagen hospital between 1959 and 1961 reported higher mean IQ scores for adults who were breast-fed for longer lengths of time as babies (E. Mortensen et al., *JAMA*, vol. 287, 2002, pp. 2365–2371). The mean IQ score was 98.1 ($s = 15.9$) for the 272 subjects who had been breast-fed for no longer than a month and 108.2 ($s = 13.1$) for the 104 subjects who had been breast-fed for seven to nine months.

a. With software that can analyze summarized data, use an inferential method to analyze whether the corresponding population means differ, assuming the population standard deviations are equal. Interpret.

b. Was this an experimental or an observational study? Can you think of a potential lurking variable?

10.93 Tomatoes and prostate cancer A study published by the National Cancer Institute (V. Kirsh et al., 2005) reported on the effect of lycopene and tomato product intake on the risk of prostate cancer. The study followed the health history and dietary habits of 29,361 men during an average of 4.2 years of follow-up over which time 1338 of the men developed prostate cancer. Results indicated that there was no evidence that lycopene consumption protects from prostate cancer except for those with a family history of prostate cancer. In this case, risks tended to be lower with greater consumption of spaghetti/tomato sauce (relative risk for at least two servings per week versus less than one serving per month = 0.68, 95% CI = 0.31 to 1.51, P-value = 0.12). Explain how to interpret the (a) relative risk value and (b) confidence interval.

10.94 Drink tea A study followed patients for four years after they had a heart attack (K. Mukamal et al., *Circulation*, vol. 105, 2002, pp. 2476–2481). Patients were interviewed about their weekly caffeinated tea consumption during the year before the heart attack. The relative risk of another heart attack during the study period for those who reported drinking at least 14 cups a week, compared to those who reported not drinking tea, was 0.56, with a 95% confidence interval of 0.37 to 0.84.

a. Interpret the confidence interval for the population relative risk.

b. According to these results, can we conclude that in a corresponding population, tea drinking reduces the chance of another heart attack? Explain.

10.95 Prison rates The U.S. Department of Justice (www.ojp.usdoj.gov/bjs) gives incarceration rates in the nation's prisons for various groups.

a. In 2006, the incarceration rate was 1 per 109 male residents and 1 per 1563 female residents. Find the ratio of incarceration rates, and interpret.

b. In 2006, the incarceration rate was 1694 per 100,000 black residents, and 252 per 100,000 white residents. Find the ratio of incarceration rates, and interpret.

10.96 Australian cell phone use In Western Australia, handheld cell phone use while driving has been banned since 2001, but hands-free devices are legal. A study (published in the *British Medical Journal* in 2005) of 456 drivers in Perth who had been in a crash that put them in a hospital emergency room observed if they were using a cell phone before the crash and if they were using a cell phone during an earlier period when no accident occurred. Thus, each driver served as his or her own control group in the study.

a. In comparing rates of cell phone use for those in accidents and those not in accidents, should we use methods for independent samples or for dependent samples? Explain.

b. The study found that using a hands-free cell phone quadruples the risk of getting into a crash with serious injuries. Identify the statistic used to compare the groups, and interpret it.

10.97 Improving employee evaluations Each of a random sample of 10 customer service representatives from a large department store chain answers a questionnaire about how they respond to various customer complaints. Based on the responses, a summary score measures how positively the employees react to complaints. This is measured both before and after employees undergo an intensive training course designed to improve such scores. A report about the study states, "The mean was significantly higher after taking the training course [$t = 3.40$ ($df = 9$), P-value < 0.05]." Explain how to interpret this to someone who has never studied statistics. What else should have been reported, to make this more informative?

10.98 Which tire is better? A tire manufacturer believes that a new tire it is introducing (Brand A) will have longer wear than the comparable tire (Brand B) sold by its main competitor. To get evidence to back up its claim in planned advertising, the manufacturer conducts a study. On each of four cars it uses a tire of Brand A on the left front and a tire of Brand B on the right front. The response is the number of thousands of miles until a tire wears out, according to a tread marking on the tire. The sample mean response is 50 for Brand A and 40 for Brand B.

a. Show a pattern of four pairs of observations with these means for which you think Brand A would be judged better according to statistical inference. You do not need to actually conduct the inference. (*Hint:* What affects the value of the test statistic, other than the sample means and n?)

b. How could the design of this study be improved?

10.99 Effect of alcoholic parents A study[17] compared personality characteristics between 49 children of alcoholics and a control group of 49 children of nonalcoholics who were matched on age and gender. On a measure of well-being, the 49 children of alcoholics had a mean of 26.1 ($s = 7.2$) and the 49 subjects in the control group had a mean of 28.8 ($s = 6.4$). The difference scores between the matched subjects from the two groups had a mean of 2.7 ($s = 9.7$).

a. Are the groups to be compared independent samples or dependent samples? Why?

b. Show all steps of a test of equality of the two population means for a two-sided alternative hypothesis. Report the P-value and interpret.

c. What assumptions must you make for the inference in part b to be valid?

10.100 CI versus test Consider the results from the previous exercise.

a. Construct a 95% confidence interval to compare the population means.

b. Explain what you learn from the confidence interval that you do not learn from the significance test.

10.101 Breast augmentation and self-esteem A researcher in the College of Nursing, University of Florida, hypothesized that women who undergo breast augmentation surgery would gain an increase in self-esteem. The article about the study[18] indicated that for the 84 subjects who volunteered for the study, the scores on the Rosenberg Self-Esteem Scale were 20.7 before the surgery (std. dev. = 6.3) and 24.9 after the surgery (std. dev = 4.6). The author reported that a paired difference significance test had $t = 9.8$ and a P-value below 0.0001.

a. Were the samples compared dependent samples, or independent samples? Explain.

b. Can you obtain the stated t statistic from the values reported for the means, standard deviation, and sample size? Why or why not?

10.102 Internet use As part of her class project in a statistics course, a student decided to study ways in which her fellow students use the Internet. She randomly sampled 5 of the 165 students in her course and asked them, "In the

[17] D. Baker and L. Stephenson, *Journal of Clinical Psychology*, vol. 51, 1995.
[18] By C. Figueroa-Haas, *Plastic Surgical Nursing*, vol. 27, 2007, p. 16.

past week, how many days did you use the Internet to (a) read news stories, (b) communicate with friends using e-mail or text messaging?" The table shows the results. Using software, construct a confidence interval or conduct a significance test to analyze these data. Interpret, indicating the population to which the inferences extend.

Pair	News	Communicating
1	2	5
2	3	7
3	0	6
4	5	5
5	1	4

10.103 TV or rock music a worse influence? In a recent General Social Survey, subjects were asked to respond to the following: "Children are exposed to many influences in their daily lives. What kind of influence does each of the following have on children? 1. Programs on network television, 2. rock music." The possible responses were (very negative, negative, neutral, positive, very positive). The responses for 12 of the sampled subjects, using scores (−2, −1, 0, 1, 2) for the possible responses are given in the table:

Subject	TV	Rock	Subject	TV	Rock	Subject	TV	Rock
1	0	−1	5	−1	−1	9	−1	−1
2	0	0	6	−2	−2	10	0	1
3	1	−2	7	−1	0	11	1	−1
4	0	1	8	1	−1	12	−1	−2

a. When you compare responses for TV and Rock, are the samples independent or dependent? Why?

b. Use software to construct a 95% confidence interval for the population difference in means. Interpret in the context of the variables studied.

c. Use software to conduct a significance test of equality of the population means for TV and Rock. Interpret the P-value.

10.104 Influence of TV and movies Refer to the previous exercise. The GSS also asked about the influence of movies. The responses for these 12 subjects were −1, 1, 0, 2, 0,−2, −1, 0,−1, 1, 1, −1. The results of using MINITAB to compare the influence of movies and TV are shown below. Explain how to interpret (a) the confidence interval and (b) the significance test results.

```
Paired T for movies - tv

            N      Mean      StDev     SE Mean
movies     12  -0.083333  1.164500  0.336162
tv         12  -0.250000  0.965307  0.278660
Difference 12   0.166667  0.937437  0.270615
95% CI for mean difference:
(-0.428952, 0.762286)
T-Test of mean difference = 0 (vs not = 0):
T-Value = 0.62 P-Value = 0.551
```

10.105 Crossover study The table summarizes results of a crossover study to compare results of low-dose and high-dose analgesics for relief of menstrual bleeding (B. Jones and M. Kenward, *Statistics in Medicine*, vol. 6, 1987, pp. 555–564).

a. Find the sample proportion of successes for each dose of the analgesic.

b. Find the P-value for testing that the probability of success is the same with each dose, for a two-sided alternative. Interpret. What assumptions does this inference make?

	High Dose	
Low Dose	**Success**	**Failure**
Success	53	8
Failure	16	9

10.106 Belief in ghosts and in astrology A poll by Louis Harris and Associates of 1249 Americans indicated that 36% believe in ghosts and 37% believe in astrology.

a. Is it valid to compare the proportions using inferential methods for independent samples? Explain.

b. Do you have enough information to compare the proportions using methods for dependent samples? If yes, do so. If not, explain what else you would need to know.

10.107 Fast food, TV, and obesity A March 2003 article in the *Gainesville Sun* about a study at Boston's Children's Hospital had the headline, "New study: Lots of fast food and TV triple risk of obesity." The response variable was binary, namely whether a child was obese. Define, in terms of this study, the sample statistic to which the "triple risk" refers.

10.108 Seat belts help? For automobile accidents in a recent year in Maine, injuries occurred to 3865 of 30,902 subjects not wearing seat belts, and to 2409 of 37,792 subjects wearing seat belts. Find and interpret the relative risk for these data.

10.109 Death penalty paradox Exercise 3.58 showed results of another study about the death penalty and race. The data are repeated here.

a. Treating victim's race as the control variable, show that Simpson's paradox occurs.

b. Explain what causes the paradox to happen.

		Death Penalty	
Vic Race	**Def Race**	**Yes**	**No**
W	W	53	414
	B	11	37
B	W	0	16
	B	4	139

10.110 Death rate paradoxes The crude death rate is the number of deaths in a year, per size of the population, multiplied by 1000.

a. According to the U.S. Bureau of the Census, in 1995 Mexico had a crude death rate of 4.6 (i.e., 4.6 deaths per 1000 population) while the United States had a crude death rate of 8.4. Explain how this overall death rate could be higher in the United States even if the United States had a lower death rate than Mexico for people of each specific age.

b. For each age level, the death rate is higher in South Carolina than in Maine. Overall, the death rate is higher in Maine (H. Wainer, *Chance*, vol. 12, 1999, p. 44). Explain how this could be possible.

10.111 Income and gender For a particular Big Ten university, the mean income for male faculty is \$8000 higher than the mean income for female faculty. Explain how this difference could disappear:

a. Controlling for number of years since received highest degree. (*Hint:* What if relatively few female professors were hired until recent years?)

b. Controlling for college of employment.

Concepts and Investigations

10.112 Student survey Refer to the FL Student Survey data file on the text CD. Using software, prepare a short report summarizing the use of confidence intervals and significance tests (including checking assumptions) to compare males and females in terms of opinions about whether there is life after death.

10.113 Review the medical literature Your instructor will pick a medical topic of interest to the class. Find a recent article of a medical journal that reports results of a research study on that topic. Describe the statistical analyses that were used in that article. Did the article use (a) descriptive statistics, (b) confidence intervals, and (c) significance tests? If so, explain how these methods were used. Prepare a one-page summary of your findings that you will present to your class.

10.114 Attractiveness and getting dates The results in the table are from a study of physical attractiveness and subjective well-being (E. Diener et al., *Journal of Personality and Social Psychology*, vol. 69, 1995, pp. 120–129). As part of the study, college students in a sample were rated by a panel on their physical attractiveness. The table presents the number of dates in the past three months for students rated in the top or bottom quartile of attractiveness. Analyze these data. Write a short report that summarizes the analyses and makes interpretations.

	No. Dates, Men			No. Dates, Women		
Attractiveness	Mean	StdDev	*n*	Mean	StdDev	*n*
More	9.7	10.0	35	17.8	14.2	33
Less	9.9	12.6	36	10.4	16.6	27

10.115 Pay discrimination against women? A *Time Magazine* article titled "Wal-Mart's Gender Gap" (July 5, 2004) stated that in 2001 women managers at Wal-Mart earned \$14,500 less than their male counterparts.

a. If these data are based on a random sample of managers at Wal-Mart, what more would you need to know about the sample to determine whether this is a "statistically significant" difference?

b. If these data referred to *all* the managers at Wal-Mart and if you can get the information specified in part a, is it relevant to conduct a significance test? Explain.

10.116 Treating math anxiety Two new programs were recently proposed at the University of Florida for treating students who suffer from math anxiety. Program A provides counseling sessions, one session a week for six weeks. Program B supplements the counseling sessions with short quizzes that are designed to improve student confidence. For ten students suffering from math anxiety, five were randomly assigned to each program. Before and after the program, math anxiety was measured by a questionnaire with 20 questions relating to different aspects of taking a math course that could cause anxiety. The study measured, for each student, the drop in the number of items that caused anxiety. The sample values were

Program A: 0, 4, 4, 6, 6
Program B: 6, 12, 12, 14, 16

Using software, analyze these data. Write a report, summarizing the analyses and interpretations.

10.117 Obesity and earnings An AP story (April 9, 2005) with headline Study: Attractive People Make More stated that "A study concerning weight showed that women who were obese earned 17 percent lower wages than women of average weight."

a. Identify the two variables stated to have an association.

b. Identify a control variable that might explain part or all of this association. If you had the original data including data on that control variable, how could you check whether the control variable does explain the association?

10.118 Multiple choice: Alcoholism and gender Suppose that a 99% confidence interval for the difference $p_1 - p_2$ between the proportions of men and women in California who are alcoholics equals (0.02, 0.09). Choose the best correct choice.

a. We are 99% confident that the proportion of alcoholics is between 0.02 and 0.09.

b. We are 99% confident that the proportion of men in California who are alcoholics is between 0.02 and 0.09 larger than the proportion of women in California who are.

c. We can conclude that the population proportions may be equal.

d. We are 99% confident that a minority of California residents are alcoholics.

e. Since the confidence interval does not contain 0, it is impossible that $p_1 = p_2$.

10.119 Multiple choice: Comparing mean incomes A study compares the population mean annual incomes for Hispanics (μ_1) and for whites (μ_2) having jobs in construction, using a 95% confidence interval for $\mu_1 - \mu_2$. Choose the best correct choice.

a. If the confidence interval is (−6000, −3000), then at this confidence level we conclude that the mean income for whites is less than for Hispanics.

b. If the confidence interval is (−3000, 1000), then the test of H_0: $\mu_1 = \mu_2$ against H_a: $\mu_1 \neq \mu_2$ with significance level 0.05 rejects H_0.

c. If the confidence interval is (−3000, 1000), then we can conclude that $\mu_1 = \mu_2$.

d. If the confidence interval is (−3000, 1000) then we are 95% confident that the population mean annual income for Hispanics is between 3000 less and 1000 more than the population mean annual income for whites.

10.120 Multiple choice: Sample size and significance If the sample proportions in Example 4 comparing cancer death rates for aspirin and placebo had sample sizes of only 1000 each, rather than about 11,000 each, then the 95% confidence interval for $(p_1 - p_2)$ would be (−0.007, 0.021) rather than (0.003, 0.011). This reflects that

a. When an effect is small, it may take very large samples to have much power for establishing statistical significance.

b. Smaller sample sizes are preferable, because there is more of a chance of capturing 0 in a confidence interval.

c. Confidence intervals get wider when sample sizes get larger.

d. The confidence interval based on small sample sizes must be in error, because it is impossible for the parameter to take a negative value.

10.121 True or false: Positive values in CI If a 95% confi-dence interval for $(\mu_1 - \mu_2)$ contains only positive numbers, then we can conclude that both μ_1 and μ_2 are positive.

10.122 True or false: Afford food? A 2003 survey by the Pew Research Center asked whether there have been times in the past year the respondent has been unable to afford food. Of advanced economies, the country with the highest response was the United States, 15%. Worldwide, the highest response was in Angola, 86%. Because the same question was asked in both countries, the samples are dependent.

10.123 True or false: Control for clinic Suppose there is a higher percentage of successes with Treatment A than with Treatment B at a clinic in Rochester, and there is a higher percentage of successes with Treatment A than with Treatment B at a clinic in Syracuse. For the overall sample (combining results for the two cities), there must be a higher percentage of successes with Treatment A than with Treatment B.

10.124 ◆◆ **Guessing on a test** A test consists of 100 true-false questions. Joe did not study, and on each question he randomly guesses the correct response. Jane studied a little and has a 0.60 chance of a correct response for each question.

a. Approximate the probability that Jane's score is nonetheless lower that Joe's. (*Hint:* Use the sampling distribution of the difference of sample proportions.)

b. Intuitively, do you think that the probability answer to part a would decrease or would increase if the test had only 50 questions? Explain.

10.125 ◆◆ **Standard error of difference** From the box formula for the standard error at the end of Section 10.1,

$$se(\text{estimate 1} - \text{estimate 2}) = \sqrt{[se(\text{estimate 1})]^2 + [se(\text{estimate 2})]^2},$$

if you know the *se* for each of two independent estimates, you can find the *se* of their difference. This is useful, because often articles report a *se* for each sample mean or proportion, but not the *se* or a confidence interval for their difference. Many medical studies have used a large sample of subjects from Framingham, Massachusetts, who have been followed since 1948. A study (*Annual of Internal Medicine*, vol. 138, 2003, pp. 24–32) estimated the number of years of life lost by being obese and a smoker. For females of age 40, adjusting for other factors, the number of years of life left were estimated to have a mean of 46.3 ($se = 0.6$) for nonsmokers of normal weight and a mean of 33.0 ($se = 1.8$) for smokers who were obese. Construct a 95% confidence interval for the population mean number of years lost. Interpret.

10.126 ◆◆ **Lots of variation in the reported importance of religion**

a. For comparisons of groups in which $n_1 = n_2$, with common value denoted by n, use the fact that the largest possible value of $\hat{p}(1 - \hat{p})$ occurs at $\hat{p} = 0.5$ to show that the margin of error for a large-sample 95% confidence interval for $(p_1 - p_2)$ can be no greater than $2\sqrt{0.5/n} = \sqrt{2/n}$.

b. A 2003 report by the Pew Research Center reported that the percent of people who find religion very important in their life is 59% in the United States, 30% in Canada, 33% in Great Britain, 27% in Italy, 21% in Germany, and 11% in France. Assuming that each country had a random sample of 1000 people, use the $\sqrt{2/n}$ bound to identify any pairs of countries for which the true percentages might not be different.

10.127 ◆◆ **Small-sample CI** The small-sample confidence interval for comparing two proportions is a simple adjustment of the large-sample one. Recall that for a small-sample confidence interval for a single proportion, we used the ordinary formula after adding four observations, two of each type (see the end of Section 8.4). In the two-sample case we also add four observations, two of each type, and then use the ordinary formula, by adding one observation of each type to each sample.[19] When the proportions are near 0 or near 1, results are more sensible than with the ordinary formula. Suppose there are no successes in either group, with $n_1 = n_2 = 10$.

a. With the ordinary formula, show that (i) $\hat{p}_1 = \hat{p}_2 = 0$, (ii) $se = 0$, and (iii) the 95% confidence interval for $p_1 - p_2$ is 0 ± 0, or (0, 0). Obviously, it is too optimistic to predict that the true difference is exactly equal to 0.

b. Find the 95% confidence interval using the small-sample method described here. Are the results more plausible?

10.128 ◆◆ **Graphing Simpson's paradox** The figure illustrates Simpson's paradox for Example 17 on the death penalty. For each defendant's race, the figure plots the percentage receiving the death penalty. Each percentage is labeled by a letter symbol giving the category of the victim's

[19] This is a new method. It was first proposed by A. Agresti and B. Caffo in the journal, *The American Statistician*, vol. 54, 2000, pp. 280–288.

race. Surrounding each observation is a circle of an area representing the number of observations on which that percentage is based. For instance, the W in the largest circle represents a percentage of 12.6 receiving the death penalty for cases with white defendants and white victims. That circle is largest because the number of cases at that combination (151) is largest. When we add results across victim's race to get a summary result ignoring that variable, the largest circles, having the greater number of cases, have greater influence. Thus, the summary percentages for each defendant's race, marked on the figure by an ×, fall closer to the center of the larger circles than to the center of the smaller circles.

a. Explain why a comparison of a pair of circles having the same letter in the middle indicates that the death penalty was more likely for black defendants, when we control for victim's race.

b. Explain why a comparison of the × marks shows that overall the death penalty was more likely for white defendants than black defendants.

c. For white defendants, why is the overall percentage who got the death penalty so close to the percentage for the case with white victims?

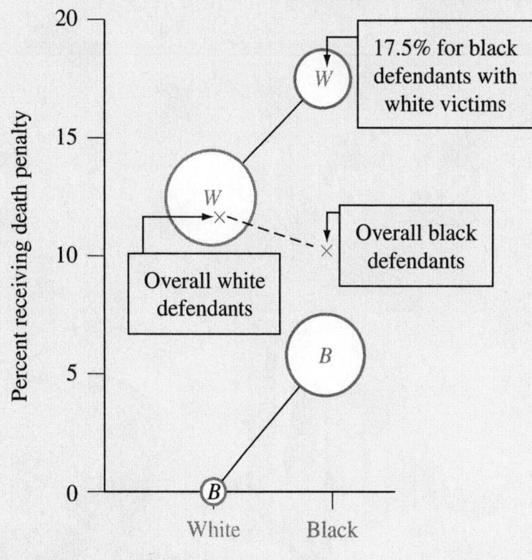

Percentage receiving death penalty by defendant's race, controlling victim's race (labels in center of circles) and ignoring victim's race (overall values marked by × for white defendants and black defendants).

Student Activities

10.129 Reading the medical literature Refer to Activity 2, which follows, about reading an article in a medical journal. Your instructor will pick a recent article at the Web site for the *British Medical Journal.* Prepare a short report in which you summarize the main conclusions of the article and explain how those conclusions depended on statistical analyses. After you have done this, your entire class will discuss this article and the statistical analyses.

Activity 2

Interpreting Statistics in a Medical Journal

Medical research makes frequent use of statistical methods shown in this chapter. At *British Medical Journal (BMJ)* bmj .com on the Internet, pull up the April 19, 2003, issue. Browse through the article on "Behavioural counseling to increase consumption of fruit and vegetables in low income adults: randomized trial" by A. Steptoe et al. You will see a variety of statistical analyses used. See if you can answer these questions:

- What was the objective of the study?
- Was the study experimental or observational? Summarize the design and the subjects used in the study.
- Identify response and explanatory variables.
- According to Table 2 in the article, the 135 subjects in the nutrition counseling group had a baseline mean of 3.67 and standard deviation of 2.00 for the number of portions per day they ate of fruits and vegetables. Over the 12 months of the study, the mean increased by 0.99. Report the 95% confidence interval for the change in the mean, and interpret. Report the corresponding results for the behavioral counseling group. Were the groups in this analysis independent or dependent samples?
- Adjusted for confounding variables, the difference between the change in the mean for the behavioral counseling group and the nutrition counseling group was 0.62. Report and interpret the 95% confidence interval for the true difference, and report the P-value for testing that the difference was 0. Were the groups in this analysis independent or dependent samples?
- According to Table 2, at the baseline for the nutrition counseling group, the proportion of subjects who ate at least five portions a day of fruits and vegetables was 0.267, and this increased by 0.287 over the 12 months. Report and interpret the 95% confidence interval for the change in the true proportion. Was this analysis for independent or dependent samples? Was this change in the proportion significantly larger, or smaller, than the change for the behavioral counseling group?
- What were the primary conclusions of the study?
- Describe two limitations of the study, as explained in the sections on "Representativeness of the sample" and "Limitations of the study."

Part Review

3 Statistical Inference Methods

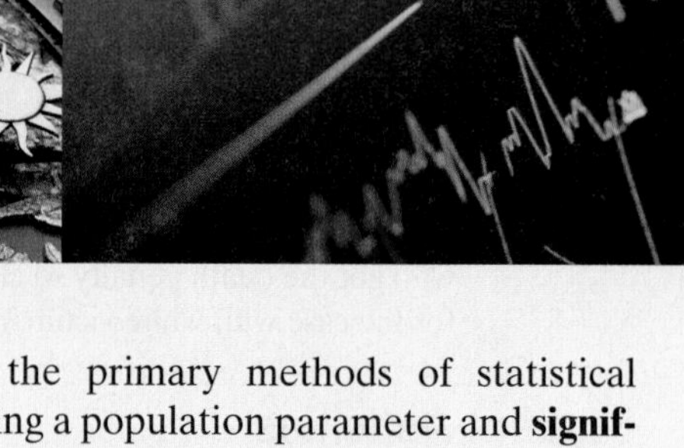

In Chapters 8–10 we've learned about the primary methods of statistical inference—**confidence intervals** for estimating a population parameter and **significance tests** for judging whether or not a particular value for a population parameter is believable. We've learned how to apply these methods to make inferences about population means and population proportions. We've also used these methods to compare population means or population proportions for two groups.

This review section gives examples of questions you should be able to answer about the main concepts in Chapters 8–10. The questions are followed by brief summaries or hints as well as references to sections in the text where you can find more detail to help you strengthen your understanding of these concepts.

Review Questions

- What is the difference between a *point estimate* and an *interval estimate*?

  Section 8.1 explained that a **point estimate** is a single number that is our best guess for the parameter. An **interval estimate** is an interval of numbers within which the parameter value is believed to fall.

- What are the point estimates of the population mean, population proportion, the difference between two population means, and the difference between two population proportions?

Parameter	**Point Estimate**
Population mean μ	Sample mean $\bar{x}$
Difference between two means $(\mu_1 - \mu_2)$	Difference between two means $(\bar{x}_1 - \bar{x}_2)$
Population proportion p	Sample proportion $\hat{p}$
Difference between two proportions $(p_1 - p_2)$	Difference between two proportions $(\hat{p}_1 - \hat{p}_2)$

- What is a *confidence interval?* What is its *confidence level*?

  Sections 8.2 and 8.3 showed that a **confidence interval** contains the most plausible values for a parameter. Confidence intervals for proportions and for means have the form:

$$\text{Point estimate} \pm \text{margin of error.}$$

This interval contains the parameter with a certain degree of confidence, called the **confidence level.** Most common are 95% confidence intervals. In the long run, when used in many studies, about 95% of such intervals contain the population parameter.

- What is the *margin of error* for a confidence interval?

  The margin of error is a multiple of the standard error *(se)* that equals about 2*(se)* with 95% confidence intervals. More precisely,

$$\text{margin of error} = (z\text{- or } t\text{-score}) \times (se).$$

  For the proportion (Section 8.2), the score is a z-score from the normal distribution. For the mean (Section 8.3), the score is a t-score from the ***t* distribution** with degrees of freedom $df = n - 1$. The z- or t-score depends on the confidence level. Each confidence interval has certain **assumptions** or conditions to be valid, such as randomization for gathering the data.

- What is a *significance test*, and what are the parts of a test?

  Section 9.1 explained that a **significance test** helps us to judge whether or not a particular value for a parameter is believable.

  Each test has certain **assumptions** or conditions to be satisfied, such as randomization for gathering the data.

  Each test has a **null hypothesis** that specifies a particular value for the parameter, such as H_o: $p = 0.50$, or H_o: $\mu_1 - \mu_2 = 0$. This usually corresponds to "no effect," in some sense. The alternative hypothesis contains alternative values for the parameter, such as the two-sided H_a: $p \neq 0.50$ or the one-sided H_a: $p > 0.50$ or the one-sided H_a: $p < 0.50$.

  Each test has a **test statistic** that measures how far the sample estimate of the parameter falls from the null hypothesis value. The z statistic for proportions and the t statistic for means have the form

$$\text{Test statistic} = \frac{\text{parameter estimate} - \text{null hypothesis value}}{\text{standard error}}.$$

  The test statistic measures the number of standard errors that the parameter estimate (such as $\hat{p}$) falls from the null hypothesis value (such as $p_0 = 0.50$).

  The **P-value** equals the probability that the test statistic takes a value like the observed one or even more extreme if we presume H_o is true. More extreme means "in the two tails" for a two-sided H_a such as H_a: $p \neq 0.50$ and "in one tail" for a one-sided H_a. The smaller the P-value, the stronger the evidence is against H_o.

  When we need to make a decision about H_o, we reject H_o if the P-value $\leq \alpha$ for a fixed **significance level** such as $\alpha = 0.05$. When H_o is true, the significance level is the probability that we make an error by rejecting H_o when we should NOT reject H_o. As Section 9.4 explained, such an error (rejecting H_o when H_o is actually true) is called a **Type I error**. When H_o is false, a **Type II error** results from failing to reject H_o when we should reject H_o.

- How are statistical inference methods used to *compare groups*?

  From Section 10.1, for a categorical response variable, we can compare groups by making an inference about the **difference between population proportions,** $(p_1 - p_2)$, for the two groups. From Section 10.2, for a quantitative response variable, we can compare groups by making an inference about the **difference between the population means**, $(\mu_1 - \mu_2)$, for the two groups.

- In comparing two groups, what is the difference between *independent* samples and *dependent* samples? Why do we distinguish between them?

  With **dependent samples**, each observation in one sample is matched with an observation in the other sample. An example is any study that observes the same people at two points in time (see Section 10.4). With **independent samples**, the observations in one sample are independent of those in the other sample. An example is a study that surveys people and classifies them into two groups according to some variable, such as sex or race. We distinguish between independent and dependent samples because we use different standard errors in the two cases. With dependent samples, for example, we calculate difference scores for the paired observations and then use methods for a single sample to find the standard error for the sample mean difference.

- What *assumptions* do we make about the method of gathering the data and the shape of the population distributions, for confidence intervals and significance tests about a parameter or comparing two parameters?

  All statistical inferences assume **randomization** for gathering the data. Inferences about means using the *t* distribution assume normal population distributions. This assumption is important mainly for small samples with one-sided tests. For large samples, sampling distributions are approximately normal for sample means, sample proportions, differences of sample means, and differences of sample proportions, by the central limit theorem, so the normality assumption is then relatively unimportant. For small samples, confidence intervals and *two-sided* tests are **robust** to violations of the normal population assumption. They work quite well even when that assumption is violated. However, with quantitative variables it's important to check for extreme outliers or skew because then the mean may be a misleading measure of the center.

- Why are *sampling distributions* important for methods of statistical inference?

  The **sampling distribution** is the key to how methods of statistical inference work. Chapter 7 explained that the sampling distribution describes how statistics such as sample proportions and sample means vary from sample to sample. Because of this, the sampling distribution determines the margin of error in using sample statistics to estimate population parameters. A z or t multiple of the **standard error of the sampling distribution** gives the **margin of error** for a **confidence interval** for a population proportion or population mean. Also, **test statistics** that summarize the evidence about a **null hypothesis** tell us how many standard errors the sample estimate of the parameter falls from its null hypothesis value.

  The following table summarizes inference methods for proportions and means:

Inference Methods for Proportions and Means

Parameter	Standard Error (*se*)	95% Confidence Interval	Significance Test: Test Statistic
Proportion p	$\sqrt{\hat{p}(1-\hat{p})/n}$	$\hat{p} \pm 1.96(se)$	$z = \dfrac{\hat{p} - p_0}{se_0}$
			$(se_0 = \sqrt{p_0(1-p_0)/n}\,)$
$p_1 - p_2$	$\sqrt{\dfrac{\hat{p}_1(1-\hat{p}_1)}{n_1} + \dfrac{\hat{p}_2(1-\hat{p}_2)}{n_2}}$	$(\hat{p}_1 - \hat{p}_2) \pm 1.96(se)$	$z = (\hat{p}_1 - \hat{p}_2)/se_0$
	(In test, $se_0 = \sqrt{\hat{p}(1-\hat{p})\left(\dfrac{1}{n_1} + \dfrac{1}{n_2}\right)}$ for pooled sample proportion $\hat{p}$)		
	Assumptions: Categorical variable, randomization, expected numbers of successes and failures each ≥ 15 for one sample and ≥ 10 for comparing two groups (two-sided tests are robust and work well with at least 5)		
Mean μ	$s/\sqrt{n}$	$\bar{x} \pm t_{.025}\,(se)$	$t = \dfrac{\bar{x} - \mu_0}{se}$
$\mu_1 - \mu_2$	$\sqrt{\dfrac{s_1^2}{n_1} + \dfrac{s_2^2}{n_2}}$	$(\bar{x}_1 - \bar{x}_2) \pm t_{.025}(se)$	$t = (\bar{x}_1 - \bar{x}_2)/se$
	Assumptions: Quantitative variable, randomization, approximately normal population distribution(s) (mainly needed for small samples with one-sided tests, for which the method is not robust)		

Note: For confidence levels other than 95%, for proportions replace 1.96 by the appropriate z-score and for means replace $t_{.025}$ by the appropriate t-score (which has $df = n - 1$ for a single mean).

Here's an example of an exercise you should be able to answer at this stage of the course, using ideas from previous chapters:

Example

Does Prayer Help Coronary Surgery Patients?

Picture the Scenario

A study about the effectiveness of prayer used as subjects a sample of patients scheduled to receive coronary artery bypass surgery at one of six U.S. hospitals.[1] The patients were randomly assigned to two groups. For one group, Christian volunteers were instructed to pray for a successful surgery with a quick, healthy recovery and no complications. The praying started the night before surgery and continued for two weeks. The response was whether or not medical complications occurred within 30 days of the surgery. The table summarizes results.

Whether or Not Complications Occurred for Heart Surgery Patients Who Did or Did Not Have Group Prayer

	Complications		
Prayer	**Yes**	**No**	**Total**
Yes	315	289	604
No	304	293	597
Total	**619**	**582**	**1201**

Questions to Explore

a. Suppose the probability of suffering medical complications is the same for each group. How can you use a (i) point estimate and (ii) 95% confidence interval, to estimate this probability?

b. Is there a difference in complication rates for the populations represented by these two groups? Answer using a significance test, and interpret.

c. In part b, explain what the sampling distribution for the difference between the complication rates represents. Explain how to interpret the standard error.

Think It Through

a. Denote by p the probability that a person having the coronary bypass surgery suffers complications within 30 days. From the table, a point estimate for p is $\hat{p} = 619/1201 = 0.515$. The standard error of this estimate is

$$\sqrt{\hat{p}(1-\hat{p})/n} = \sqrt{0.515(0.485)/1201} = 0.0144.$$

Because there are at least 15 observations in each category, we can find a 95% confidence interval for p using

$$\hat{p} \pm 1.96(se), \text{ which is } 0.515 \pm 1.96(0.0144).$$

[1] *Source*: Data from H. Benson et al., *American Heart Journal*, vol. 151, 2006, pp. 934–952.

The margin of error is 1.96(0.0144) = 0.028 and the confidence interval is (0.49, 0.54). We can be 95% confident that at least 49% but fewer than 54% of the corresponding population of individuals having coronary bypass surgery would suffer complications.

b. Let p_1 denote the probability that a person in the prayer group suffers complications. Let p_2 denote the probability that a person in the nonprayer group suffers complications. From the table, their point estimates are

$$\hat{p}_1 = 315/604 = 0.522, \hat{p}_2 = 304/597 = 0.509.$$

To compare population proportions p_1 and p_2, we conduct a significance test of H_0: $p_1 = p_2$ against H_a: $p_1 \neq p_2$. Under the presumption for H_0 that $p_1 = p_2$, the estimated common value of p_1 and p_2 is the pooled sample proportion for the entire sample, which part a showed is $\hat{p} = 0.515$.

The standard error for the test is

$$se_0 = \sqrt{\hat{p}(1-\hat{p})\left(\frac{1}{n_1} + \frac{1}{n_2}\right)} =$$

$$se_0 = \sqrt{0.515(0.485)\left(\frac{1}{604} + \frac{1}{597}\right)} = 0.0288.$$

The test statistic measures the number of standard errors that the estimate $\hat{p}_1 - \hat{p}_2$ falls from the null hypothesis value of 0 for $p_1 - p_2$. It equals

$$z = (\hat{p}_1 - \hat{p}_2)/se_0 = (0.522 - 0.509)/0.0288 = 0.45.$$

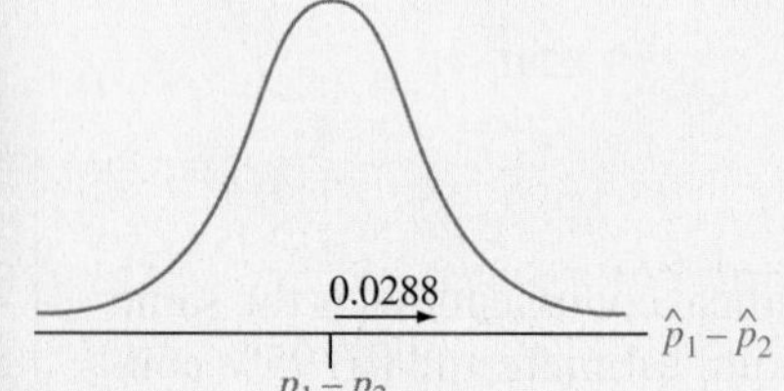

For H_a: $p_1 \neq p_2$, the P-value is the two-tail probability from the standard normal distribution beyond the observed test statistic value if we presume H_0 is true. A z-score of 0.45 has a two-sided P-value equal to 0.65. There is not much evidence against H_0. If prayer has no effect (that is, if H_0 is true), this sample outcome would not be unusual.

c. The sampling distribution represents how $(\hat{p}_1 - \hat{p}_2)$ would vary if this experiment were repeated a large number of times. The values of $(\hat{p}_1 - \hat{p}_2)$ would have approximately a normal distribution. The standard error of 0.0288 is the standard deviation of that distribution. So the value of 0.0288 describes how much the sample difference $(\hat{p}_1 - \hat{p}_2)$ tends to vary around the population difference $(p_1 - p_2)$.

Insight

Consider the population of people who undergo coronary bypass surgery and could receive one of these treatments. In summary, it is plausible that the probability of complications is the same for the prayer and nonprayer conditions. However, this study does not disprove the power of prayer. Recall that we cannot accept a null hypothesis. A confidence interval for $(p_1 - p_2)$ contains values other than 0. Can you think of lurking variables you would need to be concerned about in conducting such an experiment?

Try Exercise R3.3

Part 3 Review Exercises

Practicing the Basics

R3.1 Reincarnation In 2009, the Harris Poll reported results of a survey about religious beliefs. Of 2303 American adults surveyed, 20% believed in reincarnation.

a. Find the standard error for this estimate.

b. Without doing any calculation, explain how the standard error would change if the sample size had been only a fourth as large, $n = 575$. Explain the implication regarding how n must increase in order to increase precision of estimates.

R3.2 Environmental regulations In 2006, the Florida Poll, conducted by Florida International University (www.fiu.edu/orgs/ipor/ffp), asked whether current environmental regulations were too strict or not too strict. Of 1200 respondents, 229 said they were too strict. Find and interpret a 95% confidence interval for a relevant parameter at the time of that survey, indicating all assumptions on which your inference is based.

R3.3 Homosexual relations Since 1988, each year the Florida Poll has asked a random sample of 1200 Floridians whether sexual relations between two adults of the same sex is wrong. The percentage who respond that this is always wrong has decreased from 74% in 1988 to 54% in 2006.

TRY

a. Use a (i) point estimate and (ii) 95% confidence interval, to estimate the proportion of Floridians in 2006 who say this is always wrong

b. Can we conclude that the proportion of Floridians who say this is always wrong has changed from 1988 to 2006? Answer using a significance test. Make a decision, using significance level 0.05, and put in context.

R3.4 Random variability in baseball A professional baseball team wins 92 of its 162 games this year. Can we conclude that this team is a better than average team, having greater than a 0.50 chance of winning a game? Answer by putting this in the context of a one-sided significance test, stating all assumptions.

R3.5 Reduce services or raise taxes? These days, at the local, state, and national level, government often faces the problem of not having enough money to pay for the various services that it provides. One way to deal with this problem is to raise taxes. Another way is to reduce services. Which would you prefer? When the Florida Poll asked a random sample of 1200 Floridians in 2006, 52% said raise taxes and 48% said reduce services.

a. Can you conclude whether a majority or minority of Floridians preferred raising taxes? Explain your reasoning, including all assumptions for the method you use.

b. Suppose you wanted to estimate the population proportion of California adults who would prefer to raise taxes rather than reduce services. About how large a random sample would you need so that a 95% confidence interval estimating this proportion has a margin of error of 0.03?

R3.6 Florida poll The 2006 Florida Poll asked, "In general, do you think it is appropriate for state government to make laws restricting access to abortion?" Of 1200 randomly chosen adult Floridians, 396 said yes and 804 said no. Can you conclude whether a majority or minority of adult Floridians would say yes? Show all steps of a statistical inference to support your answer, including assumptions.

R3.7 "Don't ask, don't tell" opinions The military "don't ask, don't tell" policy was repealed in December 2010. Was this decision supported by a majority or a minority of the U.S. public? The results of a Gallup poll conducted in early May 2010 reported in the *Gallup News* (May 10, 2010) demonstrated widespread support for gays serving in the military. For the random sample of 1029 adults, the overall percentage of support was 70% (compared to a similar survey in 2004 that showed 64% support). Let p denote the population proportion of U.S. adults who believe that the policy should be repealed. For testing whether or not p equals 0.50,

a. State the hypotheses and assumptions for the method.

b. Find the test statistic, and interpret.

c. Find the P-value, and interpret.

d. With significance level = 0.05, state the decision in context.

R3.8 Compulsive buying Compulsive buying behavior is defined as an uncontrolled urge to buy. A recent study (J. Joireman et al., *The Journal of Consumer Affairs*, vol. 44, no. 1, 2010) investigated the relationship between compulsive buying tendencies and reported credit card debt. Participants were undergraduate business students from two colleges whose mean age was 21 years (range from 18 to 35). Participants completed a diagnostic screener for compulsive buying tendencies and also reported their credit card debt. A relationship was found between the compulsive buying tendencies and increased credit card debt. Of those students who had debt ($n = 74$), the credit card balance had a mean of \$1,333.61 and a standard deviation of \$1,706.22.

a. Find and interpret the standard error of the point estimate of the population mean credit card balance for undergraduate college students with debt.

b. Find and interpret a 90% confidence interval for the population mean.

c. Show all steps of a significance test of whether the population mean differs from \$1000. Interpret.

R3.9 Sex partners Some General Social Surveys ask respondents how many sex partners they have had since their eighteenth birthday. In 2008, when asked how many male sex partners they have had, the 166 females in the sample between the ages of 20 and 29 reported a mean of 4.99.

Software summarizes the results for this GSS variable (denoted by NUMMEN):

```
  N    Mean   StDev  SE Mean      95% Cl
166  4.990  8.370    0.650  (3.707, 6.273)
```

a. Show how software got the standard error from the other information reported.

b. Interpret the confidence interval shown.

c. Do you think this variable had roughly a normal distribution? Why or why not? Based on your answer, is the confidence interval valid?

d. Indicate an alternative measure of center that might be more suitable for these data, and indicate why it might be more suitable.

R3.10 Gas tax and global warming As part of her class project, a student at the University of Toronto randomly sampled 20 of her fellow students and asked them how much (in Canadian dollars) they would be willing to pay per gallon of gas in a special tax to encourage people to drive more fuel-efficient autos. The responses for the 20 students were as follows:

1.00, 0.50, 3.00, 0.00, 1.00, 0.25, 2.00, 0.50, 1.00, 1.50
0.00, 0.50, 0.00, 2.00, 1.00, 0.00, 1.00, 0.00, 0.25, 0.50

Using software or a calculator,

a. Find the sample mean and standard deviation. Interpret.

b. Find the standard error for the sample mean. Interpret.

c. Construct a 95% confidence interval for the population mean. Interpret.

d. Indicate what assumptions you made for the method in part c. Discuss whether the assumptions are fulfilled, and if not, explain the implication of the term "robustness" regarding whether the inference in part c is valid.

R3.11 Legal marijuana? The General Social Survey has asked respondents, "Do you think the use of marijuana should be made legal or not?" View results for all years at sda.berkeley.edu/GSS by entering the variables GRASS and YEAR.

a. Are the samples from the various years independent samples or dependent samples? Explain.

b. Describe any trend you see over time in the proportion favoring legalization.

R3.12 Renewable energy When a recent Eurobarometer survey[2] asked subjects in each European Union country whether they would be willing to pay more for energy produced from renewable sources than for energy produced from other sources, the percentage answering yes varied from a high of 52% in Denmark ($n = 1008$) to a low of 14% in Lithuania ($n = 1002$). For this survey,

a. Identify the response variable and the explanatory variable.

b. If you compare results inferentially for different countries, would you use a method for independent samples, or a method for dependent samples? Why?

R3.13 Laughter and blood flow A *Washington Post* article (March 15, 2005) summarized results of a study at the University of Maryland that suggested "a good laugh may help fend off heart attacks and strokes." A sample of 20 healthy people watched the violent opening scene of the movie *Saving Private Ryan* and the comedy movie *Kingpin.* Blood flow was reduced in 14 of the 20 people after watching the stressful film but in only 1 of the 20 after watching the comedy. If you compare the proportions having reduced blood flow for the two movies, are the samples independent samples or dependent samples? Explain.

R3.14 European views about Obama *Transatlantic Trends* is an annual survey of American and European public opinion,[3] with a random sample of about 1000 adults from each of 13 European countries each year. In the 2010 summary, the Obama overall approval rate in the European Union (EU) countries surveyed was 78%. In 2002, the EU approval rate for George W. Bush was 38%.

a. Explain what it would mean for these results to be based on (i) *independent* samples and (ii) *dependent* samples.

b. Which inferential statistical method could you use to analyze whether there had been a change in opinion about the U.S. president's approval rating for the population of all Europeans?

R3.15 Listening to rap music An AP story (February 1, 2007) about a University of Chicago survey of 1600 people aged 15 to 25 in several Midwest U.S. cities indicated that 58% of black youth, 45% of Hispanic youth, and 23% of white youth reported listening to rap music every day.

a. Suppose that a 95% confidence interval comparing the population proportions for black and Hispanic youths is (0.09, 0.17). Explain how to interpret this interval.

b. What value would the confidence interval in part b need to contain for it to be plausible that the population proportions are identical for black and Hispanic youths?

R3.16 Offensive portrayal of women The study mentioned in the previous exercise reported that 66% of black females and 57% of black males agreed that rap music videos portray black women in bad and offensive ways. True or false: Because both these groups had the same race, inferential methods comparing them must assume dependent rather than independent samples.

R3.17 Evolution The 2000 GSS asked whether human beings evolved from earlier species of animals (variable SCITEST4), with possible responses (definitely true, probably true, probably not true, definitely not true). Those who answered definitely not true were 190 of the 323 subjects who classified themselves fundamentalist in religious beliefs and 60 of the 309 subjects who classified themselves as liberal in religious beliefs. Is the difference between fundamentalists and liberals in the proportions who answer definitely not true statistically significant? Phrase your answer in the context of a significance test.

a. State the hypotheses.

b. Find the test statistic value.

c. Find and interpret the P-value.

d. Make a decision in context using significance level 0.05.

[2] ec.europa.eu/public_opinion/.

[3] See www.transatlantictrends.org.

R3.18 LAPD searches A 1999 Gallup poll found that 42% of African Americans felt they had been stopped by police because of their race, whereas only 6% of whites felt the same way. Los Angeles has been collecting data on motor vehicle stops and searches by its police department (LAPD). In the first half of 2005, 12,016 of 61,188 stops of African American drivers resulted in a search, whereas 5312 of 107,892 stops of white drivers resulted in a search.[4] Analyze these data **(a)** descriptively, **(b)** inferentially. In part b, be sure to specify assumptions, and interpret results.

R3.19 No time cooking For the British Time Use Survey in 2005, of those working full time, 45% of 1219 men and 26% of 733 women reported spending *no* time on cooking and washing up during a typical day. Find and interpret a 95% confidence interval for the difference between the population proportion of working men and women spending no time on cooking and washing.

R3.20 Degrading sexual song lyrics An AP story (August 7, 2006) about a research study regarding the impact on teens of sexual lyrics in songs reported, "Teens who said they listened to lots of music with degrading sexual messages were almost twice as likely to start having intercourse . . . within the following two years as were teens who listened to little or no sexually degrading music." The reported percentages were 51% and 29%.

a. What further information would you need to conduct a statistical inference to determine whether the difference reported in this study reflected a true difference in corresponding populations?

b. Suppose a 95% confidence interval for the difference between corresponding population proportions was (0.18, 0.26). Explain how to interpret it.

c. Suppose the P-value is 0.001 for testing the null hypothesis that the corresponding population proportions are equal. Interpret.

R3.21 Compulsive buying A study[5] of compulsive buying behavior (uncontrolled urges to buy) conducted a national telephone survey in 2004 of adults ages 18 and over. Of 1501 women, 90 were judged to be compulsive buyers according to the Compulsive Buying Scale. Of 800 men, 44 were judged to be compulsive buyers. Using results shown in the table,

a. Conduct all steps of a significance test to analyze whether one gender is more likely than the other to be compulsive buyers, including stating assumptions needed for this inference to be valid.

b. Report and interpret the confidence interval comparing the proportions of compulsive buyers for women and men.

```
Sample   X     N     Sample p
  1     90   1501    0.0600
  2     44    800    0.0550
Difference = p(1) - p(2)
Estimate for difference: 0.0050
95% CI for difference: (-0.015, 0.025)
Test for difference = 0(vs not = 0: z = 0.48 P-Value = 0.629
```

[4] L. Khadjavi, *Chance*, vol. 19, 2006, p. 45.

[5] Koran et al., *American Journal of Psychiatry*, vol. 163, 2006, p. 1806.

R3.22 Credit card balances Refer to the previous exercise on compulsive buying behavior. The total credit card balance had a mean of $3399 and standard deviation of $5595 for 100 compulsive buyers and a mean of $2837 and standard deviation of $6335 for 1682 other respondents.

a. Estimate the difference between the means for compulsive buyers and other respondents.

b. Find the standard error for the estimate in part a.

c. Compare the population means using a two-sided significance test. Interpret.

R3.23 Men and women's expectations on chores A recent study (S. Askari et al., *Psychology of Women Quarterly*, vol. 34, 2010, pp. 243–252) explored whether there was a discrepancy between young married adults' ideal and expected participation in household and child care chores. All participants in the study were unmarried, heterosexual, and with no children. Results showed that although men desired and expected a fairly equal division of labor, women projected that they would actually engage in a disproportionate amount of the household labor and child care. Additionally, women, but not men, expected to do significantly more chores than they ideally wanted. The results for expected percentage for household chores participation follow.

```
Two-Sample t-Test and CI
Sample      N      Mean     StDev    SE Mean
Women      218     69.4     14.3      0.97
Men        140     45.5     12.9      1.1
Difference = mu (women) - mu (men)
Estimate for difference: 23.92
95% CI for difference: (21.05, 26.79)
T-Test of difference = 0 (vs not = ):
T-Value = 16.41 P-Value = 0.000 DF = 316
```

a. Report and interpret the 95% confidence interval comparing the population means for women and men. Indicate the relevance of 0 not falling in the interval.

b. State the assumptions upon which the interval in part a is based.

R3.24 More expectations on chores Refer to the table in the previous exercise.

a. State the hypotheses tested.

b. Report and interpret the P-value.

c. For a 0.01 significance level, give your decision, in context.

d. Explain what a Type I error would represent, in this context.

R3.25 Loneliness At sda.berkeley.edu/GSS, consider responses to the question, "On how many days in the past seven days have you felt lonely?" (coded LONELY) for all the surveys in which this was asked. Use the range of years 1972–2010.

a. Find point estimates of the population means for males and for females (the categories of SEX).

b. Construct the 95% confidence interval for the population mean number of days in the past seven days that females have felt lonely, and interpret.

c. Construct the 95% confidence interval comparing the means for males and females, and interpret.

R3.26 Gas tax revisited Exercise R3.10 showed responses of a random sample of 20 students at the University of Toronto regarding how high a tax they would be willing to pay per gallon of gas to encourage people to drive more fuel-efficient autos. The responses, classified by sex of student, were as follows:

Female: 1.00, 0.50, 3.00, 0.00, 1.00, 0.25, 2.00, 0.50, 1.00, 1.50

Male: 0.00, 0.50, 0.00, 2.00, 1.00, 0.00, 1.00, 0.00, 0.25, 0.50

Using software or a calculator, conduct a significance test to analyze whether the population means differ for females and males. Interpret the P-value, and explain in context the decision you make using a 0.05 significance level.

R3.27 Sex partners and gender The table shows an analysis of GSS data about the number of sex partners reported in the past 12 months, by gender. (Note that male is coded as 1 and female is coded as 2.)

a. Explain how to interpret the P-value for the test.

b. Explain how to interpret the confidence interval. What do you learn from it that you cannot learn from the test?

c. What assumptions are made for these inferences?

```
Sample   N      Mean    StDev   SE Mean
1        276    1.19    1.35    0.081
2        365    0.850   0.710   0.037
Difference = mu (1) - mu (2)
Estimate for difference: 0.3400
95% CI for difference: (0.1643, 0.5157)

T-Test of difference = 0 (vs not = ):
T-Value = 3.81 P-Value = 0.000 DF = 389
```

R3.28 Are larger female crabs more attractive? A study[6] of the mating habits of horseshoe crabs investigated whether female crabs who had a male crab nearby (a potential mate) tended to be larger than the female crabs that did not have a male nearby. Size was measured by the width of their carapace shell. Of the 111 female crabs that had a male crab nearby, shell width had a mean of 26.9 cm ($s = 2.1$). Of the 62 female crabs that did not have a male crab nearby, shell width had a mean of 25.2 cm ($s = 1.7$). Assume that these horseshoe crabs have the properties of a random sample of all such crabs.

a. Estimate the difference between the mean shell widths for those with and without mates.

b. Construct a 90% confidence interval for the difference in mean widths in the population. Can you conclude that one mean is higher than the other? Explain.

c. What assumptions are necessary for this inference to be valid?

R3.29 Binge eating An AP story (February 1, 2007) about a Harvard University study estimated that 1% of women suffer from anorexia and 3.5% of women suffer from binge eating, defined as bouts of uncontrolled eating, well past the point of being full, that occur at least twice a week. Compare rates of binge eating and anorexia using the **(a)** difference of proportions and **(b)** relative risk. Interpret.

R3.30 Motor vehicle fatalities and race The National Center for Injury Prevention and Control, a division of the CDC, reported (*Morbidity and Mortality Weekly Report*, January 2011) that during 2007, approximately 44,000 persons were killed in motor vehicle crashes. The overall motor vehicle related age-adjusted death rate was 14.5 deaths per 100,000 population. For all racial/ethnic groups, males had death rates that were two to three times higher than the rates for females. The death rate (per 100,000 population) for white males was 21.5 while the death rate for white females was 8.8.

a. Summarize by the relative risk, and interpret.

b. The site reports 95% confidence intervals for the death rate (per 100,000 population) as (26.9, 31.2) for all American Indian/Alaskan Native and (14.8, 15.1) for all Whites and, and it estimates a 94% higher risk for the American Indian/Alaskan Native population. From the information given, explain why the estimated relative risk for comparing death rates for the American Indian/American Native population and the White population is 1.94.

R3.31 Improving math scores Each of a random sample of 10 college freshmen takes a mathematics aptitude test both before and after undergoing an intensive training course designed to improve such test scores. Then the scores for each student are paired, as shown in the table.

Student	Before	After
1	60	70
2	73	80
3	42	40
4	88	94
5	66	79
6	77	86
7	90	93
8	63	71
9	55	70
10	96	97

a. Compare the mean scores after and before the training course by (i) finding the difference of the sample means and (ii) finding the mean of the difference scores. Compare.

b. Calculate and interpret the P-value for testing whether the mean change equals 0.

c. Compare the mean scores after and before the training course by constructing and interpreting a 90% confidence interval for the population mean difference.

d. Explain the correspondence between the result of the significance test and the result of the confidence interval. What assumptions does each inference make?

R3.32 The McNemar test What is the purpose of the McNemar test? That is, what parameters does it compare, and what does it assume about the samples compared?

---

[6]By Jane Brockmann, Zoology Department, University of Florida.

Concepts and Investigations

R3.33 Student survey For the FL Student Survey data file on the text CD, we identify the weekly number of times reading a newspaper as the response variable and gender as the explanatory variable. The observations are as follows:

Females: 5, 3, 6, 3, 7, 1, 1, 3, 0, 4, 7, 2, 2, 7, 3, 0, 5, 0, 4, 4, 5, 14, 3, 1, 2, 1, 7, 2, 5, 3, 7

Males: 0, 3, 7, 4, 3, 2, 1, 12, 1, 6, 2, 2, 7, 7, 5, 3, 14, 3, 7, 6, 5, 5, 2, 3, 5, 5, 2, 3, 3

Using software, analyze these data. Write a one-page report summarizing your analyses and conclusions.

R3.34 Time Spent on WWW The GSS has asked about the number of hours a week spent on the World Wide Web, excluding e-mail (variable denoted WWWHR). State a research question you could address with these data. Go to sda.berkeley.edu/GSS and obtain and analyze the data you find. Prepare a short report summarizing your analysis and answering the question you posed.

R3.35 More people becoming isolated? When asked by the GSS about the number of people with whom the subject had discussed matters of importance over the past six months (variable called NUMGIVEN), the response of 0 was given by 8.9% of 1531 respondents in 1985 and by 25% of 1426 respondents in 2004. Analyze these data descriptively and inferentially, and interpret.

R3.36 Parental support and single mothers A study compared youths in single-mother households to youths in households having both biological parents. Answers to one set of questions were used to form a measure of perceived parental support. This measure had sample means of 46 ($s = 9$) for the single-mother households and 42 ($s = 10$) for the households with both biological parents. Consider the conclusion, "The mean parental support was 4 units higher for the single-mother households. If the true means were equal, a difference of at least this size could be expected only 2% of the time. For samples of this size, 95% of the time one would expect this difference to be within 3.4 of the true value." Explain how results would differ from those quoted here if you **(a)** formed a 99% confidence interval comparing means, **(b)** conducted a *one-sided* test that predicted a higher mean for single-mother households.

R3.37 Variability and inference Explain the reasoning behind the following statement: In studies about a very diverse population, large samples are often necessary, whereas for more homogeneous populations smaller samples are often adequate. Illustrate for the problem of estimating mean income for all lawyers in the United States compared to estimating mean income for all entry-level employees at Burger King restaurants in the United States.

R3.38 Survey about alcohol Your friend is interested in surveying students in your college to study whether or not a majority feels that the legal age for drinking alcohol should be reduced. He has never studied statistics. How would you explain to him the concepts of **(a)** null and alternative hypotheses, **(b)** P-value, **(c)** Type I error, and **(d)** Type II error?

R3.39 Freshman weight gain An AP story (October 23, 2006) indicated that a study at Brown University Medical School estimated an average weight gain during the freshman year of 5.6 pounds for males and 3.6 pounds for females. The story reported a statistical analysis in the study revealed that "males piled on significantly more pounds than females." How would you explain this sentence to someone who has never studied statistical significance testing?

R3.40 Practical significance? A report released on September 25, 2006, by the Collaborative on Academic Careers in Higher Education indicated that there is a notable gap between female and male academics in their confidence that tenure rules are clear, with men feeling more confident. The 4500 faculty members in the survey were asked to evaluate policies on a scale of 1 to 5 (very unclear to very clear). The mean response about the criteria for tenure was 3.51 for females and 3.55 for males, which was found to be statistically significant. Use this study to explain the distinction between *statistical significance* and *practical significance.*

R3.41 Overweight teenagers An AP story (February 6, 2007) quoted the National Center for Health Statistics as reporting that the percentage of teenagers who are overweight increased from about 6% in 1974 to about 18% in 2004 and that this was a statistically significant increase. Would you characterize this as practically significant also? Explain the distinction between *statistical significance* and *practical significance.*

R3.42 True or false Statistical inference methods using the t distribution are robust to violations of the random sampling assumption.

R3.43 Comparing literacy ◆◆ The International Adult Literacy Survey[7] was a 22-country study in which nationally representative samples of adults were interviewed and tested at home, using the same literacy test having scores that could range from 0 to 500. For those of age 16–25, some of the mean prose literacy scores were UK 273.5, New Zealand 276.8, Ireland 277.7, United States 277.9, Denmark 283.4, Australia 283.6, Canada 286.9, Netherlands 293.5, Norway 300.4, Sweden 312.1. The Web site does not provide sample sizes or standard deviations. Suppose it had reported a *standard error* of 2 for the mean for the United States and 3 for the mean for Canada. Is this sufficient information to construct a 95% confidence interval comparing the difference in population means for Canada and the United States? If so, construct the interval and interpret; if not, explain why not.

R3.44 Effect size ◆◆ In Exercise 10.87 in Chapter 10 on an anorexia study, the estimated difference between the mean weight gains between the therapy and the control group was $3.01 - (-0.45) = 3.46$ pounds. Was this large or small in practical terms? Keep in mind that the value we get for the estimated difference depends on the units of measurement. If converted to kilograms, the estimated difference would be 1.57. If converted to ounces, it would be 55.4. A standardized way to describe the difference divides the difference between the means by the estimated standard deviation for each group. This is called the **effect size**.

[7] www.nifl.gov/nifl/facts/IALS.html.

a. Find the effect size for sample means of 3.01 and −0.45 pounds when the estimated common standard deviation is $s = 7.64$ pounds.

b. Explain why the effect size in part a means that the difference between the sample means is less than half a standard deviation, a relatively small difference.

R3.45 Margins of error The 2006 publication *Attitudes towards European Union Enlargement* from Eurobarometer states, "The readers are reminded that survey results are *estimations*, the accuracy of which, everything being equal, rests upon the sample size and upon the observed percentage. With samples of about 1000 interviews, the real percentages vary within the following confidence limits":
◆◆

Observed	10%, 90%	20%, 80%	30%, 70%	40%, 60%	50%
Limits	1.9	2.5	2.7	3.0	3.1

a. Explain how the researchers got 3.0 points for 40% or 60%

b. Explain why the margin of error differs for different observed percentages.

c. Explain why the accuracy is the same for a particular percentage and for 100 minus that value (for example, both 40% and 60%).

R3.46 Prayer study The example in this review section discussed a study that found that prayer did not reduce the incidence of complications for coronary surgery patients.
◆◆

a. Just as association does not imply causality, so does a lack of association not imply a lack of causality, because there may be an alternative explanation. Illustrate this using this study.

b. A summary of the prayer study in *Time* magazine (December 4, 2006, p. 87) stated that "the prayers said by strangers were provided by the clergy and were all identical. Maybe that prevented them from being truly heartfelt. In short, the possible confounding factors in this study made it extraordinarily limited." Explain what the "possible confounding" means, in the context of this study.

Analyzing Association and Extended Statistical Methods

Part 4

11 Analyzing the Association Between Categorical Variables

Example 1

Happiness

Picture the Scenario

What contributes to your overall happiness? Is it love? Your health? Your friendships? The amount of money you make?

To investigate which variables are associated with happiness, we can use data from the General Social Survey (GSS). In each survey, the GSS asks, "Taken all together, would you say that you are very happy, pretty happy, or not too happy?" Table 11.1 uses the 2008 survey to cross-tabulate happiness with family income, here measured as the response to the question, "Compared with American families in general, would you say that your family income is below average, average, or above average?"

Table 11.1 Happiness and Family Income, from 2008 General Social Survey

	Happiness			
Income	**Not Too Happy**	**Pretty Happy**	**Very Happy**	**Total**
Above average	26	233	164	423
Average	117	473	293	883
Below average	172	383	132	687

Questions to Explore

- How can you determine if there is an association between happiness and family income in the population of all adult Americans?
- If there is an association, what is its nature? For example, do people with above-average family income tend to be happier than people with below-average family income?
- Can you think of a variable that might have a stronger association with happiness than family income?

Thinking Ahead

Both variables in Table 11.1 are categorical. The focus of this chapter is learning how to describe associations between categorical variables. To do this, we'll continue to apply the basic tools of statistical inference. These inferential methods help us answer questions such as, "Do people with higher family incomes tend to be happier?" and "Are married people happier than unmarried people?" We'll analyze Table 11.1 in Examples 3 and 4 and other data on happiness in Example 8 and in the exercises.

Recall

In the Chapter 3 introduction, we stated that two variables have an **association** if a particular value for one variable is more likely to occur with certain values of the other variable—for example, if being very happy is more likely to happen if a person has an above average income. Section 3.1 introduced the analysis of association for categorical variables. ◀

Association Between Variables

Let's recap where we are at this stage of our study. Chapter 7 introduced the fundamental concept of a *sampling distribution.* Chapter 8 showed how the sampling distribution is the basis of estimation, using *confidence intervals,* and Chapter 9 showed how it is the basis of *significance testing.* Those chapters focused on inference for a single proportion or a single mean.

Chapter 10 introduced methods for two variables—a response variable and a binary explanatory variable for which the two categories define two groups to compare (such as females and males). For a binary response variable, we compared two proportions, and for a quantitative response variable, we compared two means. Other methods apply when (a) both variables are quantitative or (b) both variables are categorical with possibly several categories. We'll study these cases in this chapter and in Chapter 12.

This chapter presents methods for investigating associations between two categorical variables. We'll refine what we mean by an association or lack of association, in the context of contingency tables such as Table 11.1 in Example 1.

11.1 Independence and Dependence (Association)

First, let's identify the response variable and the explanatory variable. In Table 11.1 it's more natural to study how happiness depends on income than how income depends on happiness. We'll treat happiness as the response variable and income as the explanatory variable.

Comparing Percentages

Table 11.1 is easier to digest if we convert the frequencies to percentages for the response categories. Within each category of income, Table 11.2 shows the percentages for the three categories of happiness. For example, of the 423 subjects who reported their family income as above average, 164 identified themselves as very happy. This is a proportion of $164/423 = 0.388$ or 39%. By contrast, 132 out of the 687 below-average income subjects said they were very happy, a percentage of 19%. The row totals in Table 11.1 are the basis of these percentage calculations. Within each row, the percentages sum to 100%.

Table 11.2 Conditional Percentages for Happiness Categories, Given Each Category of Family Income

The percentages are computed by dividing the counts in Table 11.1 by their row totals, and then multiplying by 100.

	Happiness			
Income	**Not Too Happy**	**Pretty Happy**	**Very Happy**	**Total**
Above average	6%	55%	39%	423 (100%)
Average	13%	54%	33%	883 (100%)
Below average	25%	56%	19%	687 (100%)

Recall

Section 3.1 introduced **conditional proportions,** which are proportions for the categories of a categorical response variable that are calculated **conditional** upon (i.e., "given") the value of another variable. ◄

The three percentages in a particular row are called **conditional percentages.** They refer to a sample data distribution of happiness, *conditional* on the category for family income. They form the **conditional distribution** for happiness, given a particular income level. For those who reported above-average family income, the conditional distribution of happiness is the set of percentages (6%, 55%, 39%) for the responses (not too happy, pretty happy, very happy). The proportions (0.06, 0.55, 0.39) are the estimated **conditional probabilities** of happiness, given above-average family income. For instance, given that a subject reported above-average family income, the estimated probability of being very happy is 0.39.

Let's compare the happiness percentages for the subjects who reported above-average income to those who reported below-average income. From Table 11.2, the higher-income subjects were more likely to be very happy (39% versus 19%) and less likely to be not too happy (6% versus 25%). Figure 11.1 uses a bar graph to portray the three conditional distributions.

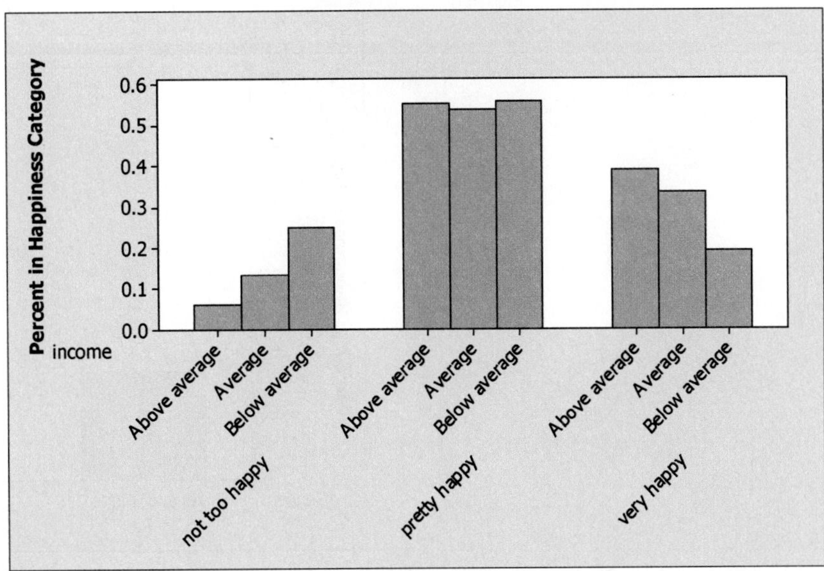

▲ **Figure 11.1 MINITAB Bar Graph of Conditional Distributions of Happiness, Given Income, from Table 11.2.** For each happiness category, you can compare the three categories of income on the percent in that happiness category. **Question** Based on this graph, how would you say that happiness depends on income?

We'll use the following guidelines when constructing tables with conditional distributions:

- Make the response variable the column variable. The response categories are shown across the top of the table.
- Compute conditional proportions for the response variable within each row by dividing each cell frequency by the row total. Those proportions multiplied by 100 are the conditional percentages.
- Include the total sample sizes on which the percentages are based. That way, readers who want to can reconstruct the cell counts.

Independence Versus Dependence (Association)

Table 11.2 shows conditional distributions for a sample. We'll use sample conditional distributions to make inferences about the corresponding population conditional distributions.

For instance, to investigate how happiness compares for females and males, we'd make inferences about the population conditional distributions of happiness for females and for males. Suppose that those population distributions are as shown in Table 11.3. (In fact, the percentages in Table 11.3 are consistent with data from the GSS. See Exercise 11.4.) We see that the percentage classified as very happy is the same for females and males, 30%. Similarly, the percentage classified as pretty happy (54%) and the percentage classified as not too happy (16%) is the same for females and for males. The probability that a person is classified in a particular happiness category is the same for females and males. The categorical variables happiness and gender are then said to be **independent.**

In Words

Happiness and gender are **independent** if (as in Table 11.3) the probability of a particular response on happiness is the same for females and males.

Table 11.3 Population Conditional Distributions Showing Independence

The conditional distribution of happiness is the same for each gender, namely (16%, 54%, 30%).

	Happiness			
Gender	Not Too Happy	Pretty Happy	Very Happy	Total
Female	16%	54%	30%	100%
Male	16%	54%	30%	100%
Overall	16%	54%	30%	100%

Source: Data from CSM, UC Berkeley.

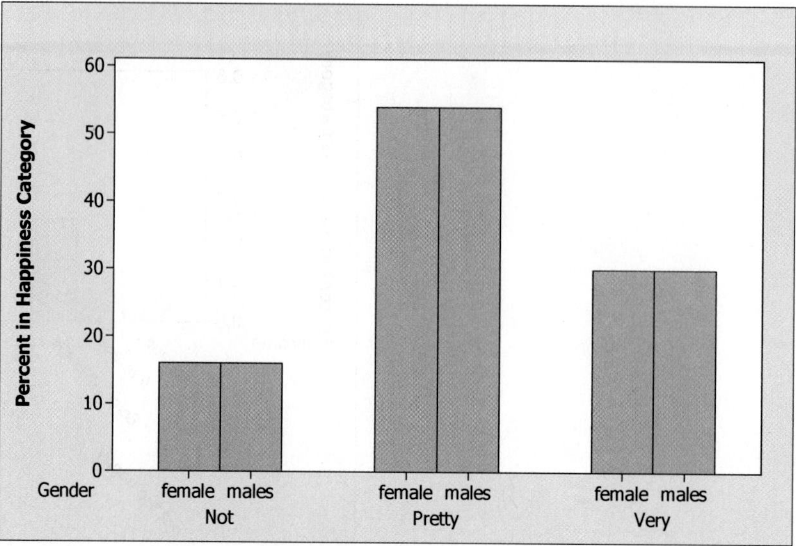

▲ **Figure 11.2** MINITAB Output of Population Conditional Distributions Showing Independence. **Question** What feature of the bar graph reflects the independence?

Independence and Dependence (Association)

Two categorical variables are **independent** if the population conditional distributions for one of them are identical at each category of the other. The variables are **dependent** (or **associated**) if the conditional distributions are not identical.

Recall

Section 5.3 in Chapter 5 defined A and B to be **independent events** if

$$P(A \mid B) = P(A).$$

Chapter 10 used **independent samples** to mean that observations in one sample are independent of those in the other sample. ◀

For two variables to be independent, the population percentage in any category of one variable is the same for all categories of the other variable, as in Table 11.3. Figure 11.2 portrays this for Table 11.3.

This definition extends independence of **events** to independence of **variables.** Recall that two *events* A and B are independent if P(A) is the same as the conditional probability of A, given B, denoted by $P(A \mid B)$. When two *variables* are independent, any event about one variable is independent of any event about the other variable. In Table 11.3, $P(A \mid B) = P(A)$ if A is an event about happiness and B is an event about gender. For example if A = very happy and B = female, then $P(A \mid B) = 0.30$. Notice in Table 11.3, this also equals P(A).

Example 2

Independence and dependence

Belief in Life After Death and Race

Picture the Scenario

Table 11.4 cross-tabulates belief in life after death with race, using data from the 2008 GSS. The table also shows sample conditional distributions for life after death, given race, and the overall percentages for the life after death categories.

Table 11.4 Sample Conditional Distributions for Belief in Life After Death, Given Race

	Postlife		
Race	**Yes**	**No**	**Total**
White	1132 (82.1%)	247 (17.9%)	1379 (100%)
Black	203 (82.2%)	44 (17.8%)	247 (100%)
Other	120 (74.5%)	41 (25.5%)	161 (100%)
Overall	1455 (81.4%)	332 (18.6%)	1787 (100%)

Source: Data from CSM, UC Berkeley.

Question to Explore

Are race and belief in life after death independent or dependent?

Think It Through

The conditional distributions in Table 11.4 are similar but not exactly identical. So it is tempting to conclude that the variables are dependent. For example, if A = yes for belief in life after death, we estimate $P(A \mid \text{white}) = 1132/1379 = 0.821$, $P(A \mid \text{black}) = 0.822$, $P(A \mid \text{other}) = 0.745$, whereas we estimate that $P(A) = 1455/1787 = 0.814$. However, the definition of independence between variables refers to the *population*. Since Table 11.4 refers to a *sample* rather than a population, it provides evidence but does not definitively tell us whether these variables are independent or dependent.

Even if the variables *were* independent, we would not expect the *sample* conditional distributions to be identical. Because of sampling variability, each sample percentage typically differs somewhat from the true population percentage. We would expect to observe *some* differences between sample conditional distributions such as we see in Table 11.4 even if *no* differences exist in the population.

Insight

If the observations in Table 11.4 were the entire population, the variables would be dependent. But the association would not necessarily be practically important, because the percentages are similar from one category to another of race. The next section presents a significance test of the hypothesis that the variables are independent.

Try Exercise 11.2

11.1 Practicing the Basics

11.1 Gender gap in politics? In the United States, is there a gender gap in political beliefs? That is, do women and men tend to differ in their political thinking and voting behavior? The table taken from the 2008 GSS relates gender and political party identification. Subjects indicated whether they identified more strongly with the Democratic or Republican party or as Independents.

	Political Party Identification			
Gender	**Democrat**	**Independent**	**Republican**	**Total**
Female	422	381	273	1076
Male	299	365	232	896

Source: Data from CSM, UC Berkeley.

a. Identify the response variable and the explanatory variable.
b. Construct a table that shows the conditional distributions of the response variable. Interpret.
c. Give a hypothetical example of population conditional distributions for which these variables would be independent.
d. Sketch bar graphs to portray the distributions in part b and in part c.

11.2 Beliefs of college freshmen Every year, a large-scale poll TRY of college freshmen conducted by the Higher Education Research Institute at UCLA asks their opinions about a variety of issues. In 2004, although women were more likely to rate their time management skills as "above average," they were also twice as likely as men to indicate that they frequently feel overwhelmed by all they have to do (36.4% versus 16.3%).

a. If results for the population of college freshmen were similar to these, would gender and feelings of being overwhelmed be independent or dependent?
b. Give an example of hypothetical population percentages for which these variables would be independent.

11.3 UGA enrollment statistics Data posted at the University of Georgia Web site indicated that of all female students in 2011, 78% were undergraduates, and of male students in 2011, 16% were graduate students. Let x denote gender of student and y denote type of student.

a. Which conditional distributions do these statistics refer to, those of y at given categories of x, or those of x at given categories of y? Set up a table with type of student as columns and gender of student as rows, showing the two conditional distributions.
b. Are x and y independent or dependent? Explain. (*Hint:* These results refer to the population.)

11.4 Happiness and gender The contingency table shown relates happiness and gender for the 2008 GSS.

Gender	Happiness			
	Not Too Happy	Pretty Happy	Very Happy	Total
Female	174	587	328	1089
Male	142	513	271	926

Source: Data from CSM, UC Berkeley.

a. Identify the response variable and the explanatory variable.

b. Construct a table or graph showing the conditional distributions. Interpret.

c. Give an example of population conditional distributions that would seem to be consistent with this sample and for which happiness and gender would be independent.

11.5 Marital happiness and income In the GSS, subjects who were married were asked about the happiness of their marriage, the variable coded as HAPMAR.

a. Go to the GSS Web site sda.berkeley.edu/GSS/ and construct a contingency table for 2008 relating HAPMAR (the column variable) to family income as measured in this section, by entering FINRELA (r: 1-2; 3; 4-5) as the row variable and YEAR(2008) in the selection filter, selecting No Weight from the weight menu.

b. Construct a table or graph that shows the conditional distributions of marital happiness, given family income. How would you describe the association?

c. Compare the conditional distributions to those in Table 11.2. For a given family income, what tends to be higher, general happiness or marital happiness for those who are married? Explain.

11.6 What is independent of happiness? Which one of the following variables would you think most likely to be independent of happiness: belief in an afterlife, family income, quality of health, region of the country in which you live, satisfaction with job? Explain the basis of your reasoning.

11.7 Sample evidence about independence Refer to the previous exercise. Go to the GSS Web site and construct a table relating happiness (HAPPY) to the variable you chose (AFTERLIF, FINRELA, HEALTH, REGION, or JOBSAT). Inspect the conditional distributions and indicate whether independence seems plausible, with the sample conditional distributions all being quite similar.

11.2 Testing Categorical Variables for Independence

How can we judge whether two categorical variables are independent or dependent? The definition of independence refers to the *population*. Could the observed *sample* association be due to sampling variation? Or would the observed results be unusual if the variables were truly independent?

A significance test can answer these questions. The hypotheses for the test are

H_0: The two variables are independent.

H_a: The two variables are dependent (associated).

The test assumes randomization and a large sample size. The test compares the cell counts in the contingency table with counts we would expect to see if the null hypothesis of independence were true.

Expected Cell Counts If the Variables Are Independent

The count in any particular cell is a random variable: Different samples have different values for the count. The mean of its distribution is called an **expected cell count.** This is found under the presumption that H_0 is true. That is, the expected cell counts are values that satisfy the null hypothesis of independence.

Table 11.5 shows the observed cell counts from Table 11.1, with the expected cell counts below them. How do we find the expected cell counts? Recall that when two variables are independent, any event about one variable is independent of any event about the other variable. Consider the events A = not too happy and B = above-average income. From the Total column and Total row (table margins) of Table 11.5 we estimate P(A) by 315/1993 and P(B) by 423/1993. The event

A and B (not too happy *and* above-average income) corresponds to the first cell in the table, with cell count of 26. Under the presumption of independence,

$$P(A \text{ and } B) = P(A) \times P(B),$$

Recall

From Section 5.2, two events A and B can also be defined as **independent** if

$P(A \text{ and } B) = P(A) \times P(B)$. ◀

which we estimate by (315/1993)(423/1993). For the $n = 1993$ subjects, the *number* of them we expect in this cell is 1993 times the cell's probability. So we estimate the **expected cell count** (presuming H_0 to be true) by

$$1993\left(\frac{315}{1993}\right)\left(\frac{423}{1993}\right) = \frac{315 \times 423}{1993} = 66.86.$$

If happiness were independent of income, we would expect about 67 subjects in that first cell. Since the observed cell count is 26, fewer people had above-average income and were not too happy than we would expect if the variables were independent.

Table 11.5 Happiness by Family Income, Showing Observed and Expected Cell Counts

We use the highlighted totals to get the expected count of 66.86 = (315 × 423)/1993 in the first cell.

	Happiness			
Income	**Not Too Happy**	**Pretty Happy**	**Very Happy**	**Total**
Above average	26	233	164	423
	66.86	231.13	125.01	
Average	117	473	293	883
	139.56	482.48	260.96	
Below average	172	383	132	687
	108.58	375.39	203.03	
Total	315 (15.8%)	1089 (54.6%)	589 (29.6%)	1993 (100%)

This discussion illustrates the general rule for calculating expected cell counts.

Expected Cell Count

For a particular cell, the **expected cell count** equals

$$\text{Expected cell count} = \frac{(\text{Row total}) \times (\text{Column total})}{\text{Total sample size}}.$$

For instance, the first cell in Table 11.5 has expected cell count = (315 × 423)/1993 = 66.86, the product of the row total (423) for that cell by the column total (315) for that cell, divided by the overall sample size (1993).

Let's see more about why this rule makes sense. For A = not too happy we estimated P(A) by 315/1993 = 0.158. If the variables were independent, we would expect this probability to be the same if we condition on any event about income. For instance, for B = above-average income, we expect $P(A \mid B) = P(A)$. Thus, 15.8% of the 423 who report above average income should be classified in the not too happy category. The expected cell count is then 0.158(423) = (315/1993)(423) = 0.6686. But this is the column total of 315 times the row total of 423 divided by the overall sample size of 1993.

Table 11.5 shows that for the overall sample, the percentages in the three happiness categories are (15.8%, 54.6%, 29.6%). You can check that the expected cell counts have these percentages in each of the three rows. For instance, for the first cell in row 1, 66.86/423 = 0.158. In fact, the expected frequencies are values that have the same row and column totals as the observed counts, but for which the conditional distributions are identical.

Chi-Squared Test Statistic

The test statistic for the test of independence summarizes how close the observed cell counts fall to the expected cell counts. Symbolized by X^2, it is called the **chi-squared statistic,** taking the name of its (approximate) sampling distribution. It is the oldest test statistic in use today. Introduced by the British statistician Karl Pearson in 1900, it is sometimes also called Pearson's chi-squared statistic.

In Words

The Greek letter chi is written as χ and pronounced "ki" (k + eye). Later in this section, we'll see that chi-squared (χ^2) is the name of the approximate sampling distribution of the statistic denoted here by X^2. The Roman letter denotes the statistic, and the Greek letter denotes the sampling distribution.

Chi-Squared Statistic

The **chi-squared statistic** summarizes how far the observed cell counts in a contingency table fall from the expected cell counts for a null hypothesis. Its formula is

$$X^2 = \Sigma \frac{(\text{observed count} - \text{expected count})^2}{\text{expected count}}.$$

For each cell, square the difference between the observed count and expected count, and then divide that square by the expected count. After calculating this term for every cell, sum the terms to find X^2.

When H_0: independence is true, the observed and expected cell counts tend to be close for each cell. Then X^2 has a relatively small value. If H_0 is false, at least some observed counts and expected counts tend to be far apart. For a cell where this happens, its value of [(observed count − expected count)^2/expected count] tends to be large. Then, X^2 has a relatively large value.

Example 3

Chi-squared statistic

Happiness and Family Income

Picture the Scenario

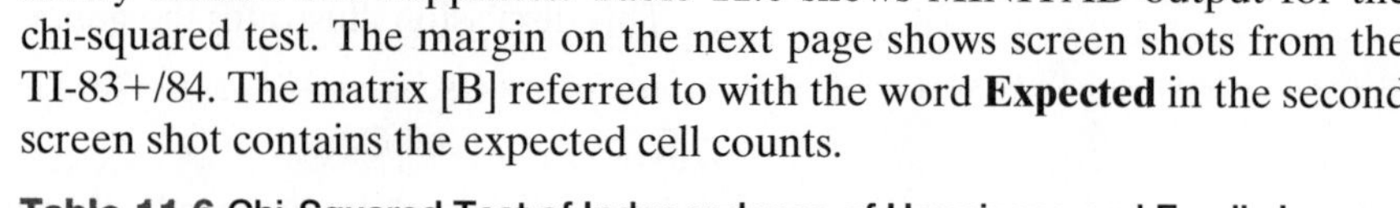

Table 11.5 showed the observed and expected cell counts for the data on family income and happiness. Table 11.6 shows MINITAB output for the chi-squared test. The margin on the next page shows screen shots from the TI-83+/84. The matrix [B] referred to with the word **Expected** in the second screen shot contains the expected cell counts.

Table 11.6 Chi-Squared Test of Independence of Happiness and Family Income

This table shows how MINITAB reports, for each cell, the observed and expected cell counts and the contribution [(observed − expected)^2/expected] to X^2, as well as the overall X^2 value and its P-value for testing H_0: independence.

```
Rows: income          Columns: happy

             not     pretty       very
above         26        233        164     423   <-- Observed cell count

           66.86     231.13     125.01           <-- Expected cell count

          24.968      0.015     12.160           <-- Contribution to X^2
average      117        473        293     883
          139.56     482.48     260.96
           3.647      0.186      3.935
below        172        383        132     687
          108.58     375.39     203.03
          37.039      0.154     24.851
Total        315       1089        589    1993
Cell contents:         Count
                       Expected count
                       Contribution to Chi-square
Pearson Chi-Square = 106.955, DF = 4, P-Value = 0.000
```

```
MATRIX[A] 3 x3
[ 26.000  233.00  164.00 ]
[ 117.00  473.00  293.00 ]
[ 172.00  383.00  132.00 ]
```

```
X²-Test
 Observed: [A]
 Expected: [B]
 Calculate Draw
```

```
X²-Test
 X²=106.9553
 p=3.2447E-22
 df=4.0000
```

```
X²-Test
 X²=106.9553
 p=3.2447E-22
 df=4.0000
```

TI-83+/84 output

Activity 1

When using software, try to re-create a table like Table 11.6. First create a data file with nine rows, such as shown below, and then choose an option to create a contingency table and find appropriate statistics.

Income	Happy	Count
above	not	26
above	pretty	233
above	very	164
average	not	117
average	pretty	473
average	very	293
below	not	172
below	pretty	383
below	very	132

Questions to Explore

a. State the null and alternative hypotheses for this test.

b. Report the X^2 statistic and explain how it was calculated.

Think It Through

a. The hypotheses for the chi-squared test are

H_0: Happiness and family income are independent.

H_a: Happiness and family income are dependent (associated).

The alternative hypothesis is that there's an association between happiness and family income.

b. Finding X^2 involves a fair amount of computation, and it's best to let software do the work for us. Beneath the observed and expected cell counts in a cell, MINITAB reports the contribution of that cell to the X^2 statistic. For the first cell (above average income and not too happy), for instance,

$$(\text{observed count} - \text{expected count})^2/\text{expected count} = (26 - 66.86)^2/66.86 = 25.0.$$

For all nine cells,

$$X^2 = \frac{(26-66.86)^2}{66.86} + \frac{(233-231.13)^2}{231.13} + \frac{(164-125.01)^2}{125.01} + \frac{(117-139.56)^2}{139.56} + \frac{(473-482.48)^2}{482.48} + \frac{(293-260.96)^2}{260.96} + \frac{(172-108.56)^2}{108.56} + \frac{(383-375.39)^2}{375.39} + \frac{(132-203.03)^2}{203.03}$$

$$= 24.968 + 0.015 + 12.160 + 3.647 + 0.186 + 3.935 + 37.039 + 0.154 + 24.851 = 106.955.$$

Insight

Table 11.6 has some large differences between observed and expected cell counts, so X^2 is large. The larger the X^2 value, the greater the evidence against H_0: independence and in support of the alternative hypothesis that happiness and income are associated. We next learn how to interpret the magnitude of the X^2 test statistic, so we know what is small and what is large and how to find the P-value.

Try Exercise 11.11

Chi-Squared Distribution

To find the P-value for the X^2 test statistic, we use the sampling distribution of the X^2 statistic. For large sample sizes, this sampling distribution is well approximated by the **chi-squared probability distribution.** Figure 11.3 shows several chi-squared distributions.

The main properties of the chi-squared distribution are as follows:

- **Always positive.** The chi-squared distribution falls on the positive part of the real number line. The X^2 test statistic cannot be negative since it sums squared

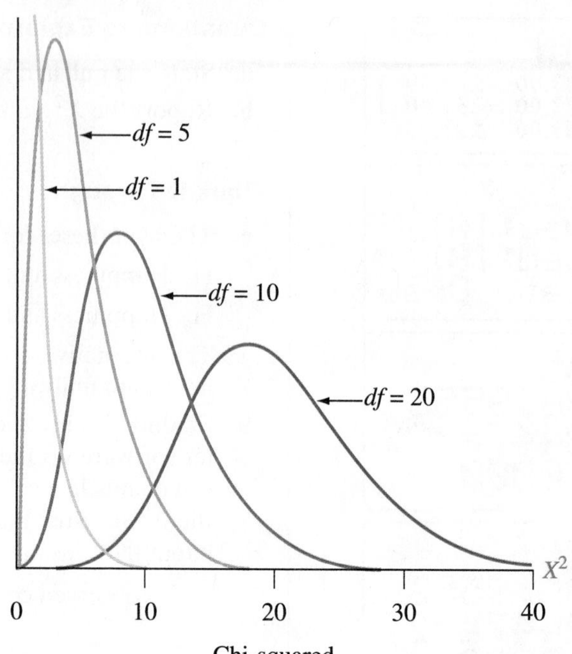

▲ **Figure 11.3 The Chi-Squared Distribution.** The curve has larger mean and standard deviation as the degrees of freedom increase. **Question** Why can't the chi-squared statistic be negative?

differences divided by positive expected frequencies. The minimum possible value, $X^2 = 0$, would occur if observed count = expected count in each cell.

- **Degrees of freedom from row and column.** The precise shape of the distribution depends on the **degrees of freedom** (*df*). For testing independence in a table with *r* rows and *c* columns (called an $r \times c$ table), the formula for the degrees of freedom is

$$df = (r - 1) \times (c - 1).$$

For example, Table 11.5 has $r = 3$ and $c = 3$, so $df = (3 - 1) \times (3 - 1) = 2 \times 2 = 4$. We'll see the reason behind this formula later in the section.

- **Mean equals *df*.** The mean of the distribution equals the *df* value. Since $df = (r - 1) \times (c - 1)$, larger numbers of rows and columns produce larger *df* values. Since larger tables have more terms in the summation for the X^2 statistic, the X^2 values also tend to be larger.
- **As *df* increases distribution goes to bell shaped.** The chi-squared distribution is skewed to the right. As *df* increases, the skew lessens and the chi-squared curve becomes more bell-shaped.
- **Large chi-square evidence against independence.** The larger the X^2 value, the greater the evidence against H_0: independence. Thus, the P-value equals the right-tail probability. It measures the probability that the X^2 statistic for a random sample would be larger than the observed value, if the variables are truly independent. Figure 11.4 depicts the P-value.

Table C at the back of the text lists values from the chi-squared distribution for various right-tail probabilities. These are X^2 test statistic values that have P-values equal to those probabilities. Table 11.7 shows an excerpt from Table C for small *df* values. For example, a 3×3 table has $df = 4$, for which Table 11.7 reports that an X^2 value of 9.49 has P-value = 0.05. In practice, software provides the P-value.

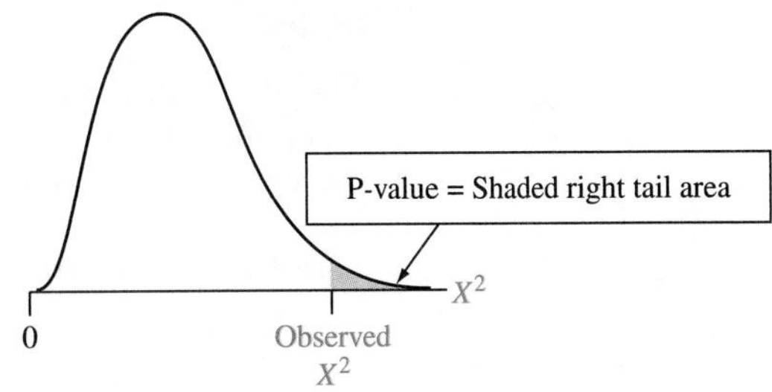

▲ **Figure 11.4 The P-value for the Chi-Squared Test of Independence.** This is the right-tail probability, above the observed value of the X^2 test statistic. **Question** Why do we not also use the left tail in finding the P-value?

Table 11.7 Rows of Table C Displaying Chi-Squared Values

The values have right-tail probabilities between 0.250 and 0.001. For a table with $r = 3$ rows and $c = 3$ columns, $df = (r - 1) \times (c - 1) = 4$, and 9.49 is the chi-squared value with a right-tail probability of 0.05.

	Right-Tail Probability						
df	.250	.100	.050	.025	.010	.005	.001
1	1.32	2.71	3.84	5.02	6.63	7.88	10.83
2	2.77	4.61	5.99	7.38	9.21	10.60	13.82
3	4.11	6.25	7.81	9.35	11.34	12.84	16.27
4	5.39	7.78	9.49	11.14	13.28	14.86	18.47

Example 4

Chi-squared distribution

Happiness and Income

Picture the Scenario

For testing the null hypothesis of independence of happiness and family income, Table 11.6 in Example 3 reported a test statistic value of $X^2 = 106.955$.

Questions to Explore

a. What is the P-value for the chi-squared test of independence for these data?

b. State your decision for a significance level of 0.05 and interpret in context.

Think It Through

a. Since the table has $r = 3$ rows and $c = 3$ columns, $df = (r - 1) \times (c - 1) = 4$. In Table 11.7, for $df = 4$ the largest chi-squared value shown is 18.47. It has tail probability = 0.001 (see the margin figure). Since $X^2 = 106.955$ falls well above this, it has a smaller right-tail probability. Thus, the P-value is < 0.001. The actual P-value would be 0 to many decimal places. MINITAB reports P-value = 0.000, as Table 11.6 showed.

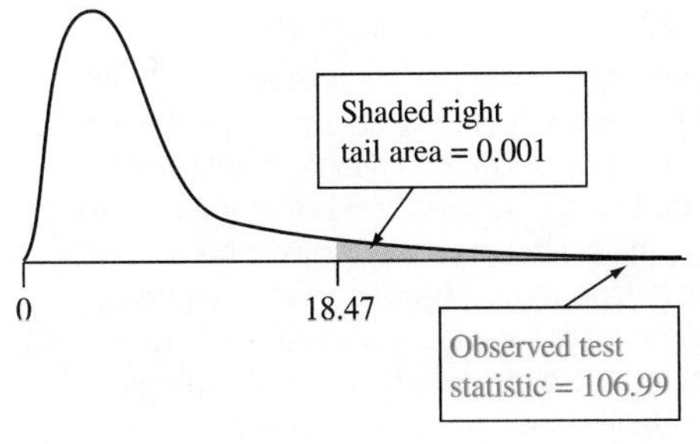

b. Since the P-value is below 0.05, we can reject H_0. Based on this sample, we have evidence to support that an association exists between happiness and family income in the population.

Insight

The extremely small P-value provides very strong evidence against H_0: independence. If the variables were independent, it would be highly unusual for a random sample to have this large a chi-squared statistic.

Try Exercise 11.10

Sample Size and the Chi-Squared Test

Presuming H_0 is true, the sampling distribution of the X^2 test statistic gets closer to the chi-squared distribution as the sample size increases. The approximation is good when each expected cell count exceeds about 5. Section 11.5 discusses the sample size issue further and presents a small-sample test.

The box summarizes the steps of the chi-squared test of independence.

SUMMARY: The Five Steps of the Chi-Squared Test of Independence

1. **Assumptions:** Two categorical variables
   Randomization, such as random sampling or a randomized experiment
   Expected count ≥ 5 in all cells (otherwise, use small-sample test in Section 11.5)
2. **Hypotheses:**
   H_0: The two variables are independent.
   H_a: The two variables are dependent (associated).
3. **Test statistic:**

$$X^2 = \sum \frac{(\text{observed count} - \text{expected count})^2}{\text{expected count}},$$

   where expected count = (row total $\times$ column total)/total sample size
4. **P-value:** Right-tail probability above observed X^2 value, for the chi-squared distribution with ***df*** $= (r - 1) \times (c - 1)$
5. **Conclusion:** Report P-value and interpret in context. If a decision is needed, reject H_0 when P-value $\leq$ significance level (such as 0.05).

As with other inferential methods, the chi-squared test assumes randomization, such as the simple random sampling that many surveys use. As explained in Section 4.4, the General Social Survey uses a rather complex multistage random sampling design. Its characteristics are similar, however, to those of a simple random sample. Inferences that assume simple random sampling, such as the chi-squared test, are routinely used with GSS data. The chi-squared test is also valid with randomized experiments. In that case, subjects are randomly assigned to different treatments, and the contingency table summarizes the subjects' responses on a categorical response variable. If the table compares groups, the random samples from those groups must be independent random samples. The test works best with large sample sizes.

In Practice Chi-Squared Is Also Used as a "Test of Homogeneity"

The **chi-squared test** does not depend on which is the response variable and which is the explanatory variable (if either). The steps of the test and the results are identical either way. When a response variable *is* identified and the population conditional distributions are identical, they are said to be **homogeneous.** The chi-squared test is then referred to as a **test of homogeneity.** The examples using the chi-squared test statistic considered thus far in the chapter have tested the notion of independence between two categorical variables. A test of independence is based on data on two categorical variables from a single random sample. In a test of homogeneity, data are observed on a single categorical variable from random samples selected independently from two or more populations or from an experiment where subjects are assigned at random to different treatment groups with a categorical response. The chi-squared statistic for a test of homogeneity has the same form as the chi-squared statistic for a test of independence. The differences in the two tests are in the nature of the hypotheses being tested and the study design employed for collecting data. Computationally, the chi-square test statistic and P-value are identical for the two tests. For simplicity, we will refer to the test statistic as a chi-squared test of independence for both study designs.

Chi-Squared and the Test Comparing Proportions in 2 × 2 Tables

In practice, contingency tables of size 2 × 2 are very common. These occur in summarizing responses of two groups on a binary response variable, such as belief in life after death (yes, no). For convenience, here, as in the discussion of the binomial distribution in Section 6.3, we label the two possible outcomes by the generic labels *success* and *failure.*

Denote the population proportion of success by p_1 in group 1 and p_2 in group 2. Then $(1 - p_1)$ and $(1 - p_2)$ are the population proportions of failures. Table 11.8 displays the notation. The rows are the groups to be compared and the columns are the response categories.

Table 11.8 A 2 × 2 Table Compares Two Groups on a Binary Response Variable

The proportions in a row form a population conditional distribution.

	Population Proportion		
Group	**Success**	**Failure**	**Total**
1	p_1	$1 - p_1$	1.0
2	p_2	$1 - p_2$	1.0

If the response variable is independent of the group, then $p_1 = p_2$, so the conditional distributions are identical. H_0: independence is, equivalently, the *homogeneity* hypothesis H_0: $p_1 = p_2$ of equality of population proportions.

Section 10.1 presented a z test of H_0: $p_1 = p_2$. The test statistic is the difference between the sample proportions divided by its standard error, that is,

$$z = (\hat{p}_1 - \hat{p}_2)/se_0,$$

where se_0 is the standard error estimated under the presumption that H_0 is true. In fact, using algebra, it can be shown that the chi-squared statistic for 2 × 2 tables is related to this z statistic by

$$X^2 = z^2.$$

In other words, square the z statistic, and you get the X^2 statistic. For H_a: $p_1 \neq p_2$, the z test has exactly the same P-value as the chi-squared test.

Example 5

Chi-squared test and 2 × 2 contingency tables

Aspirin and Cancer Death Rates Revisited

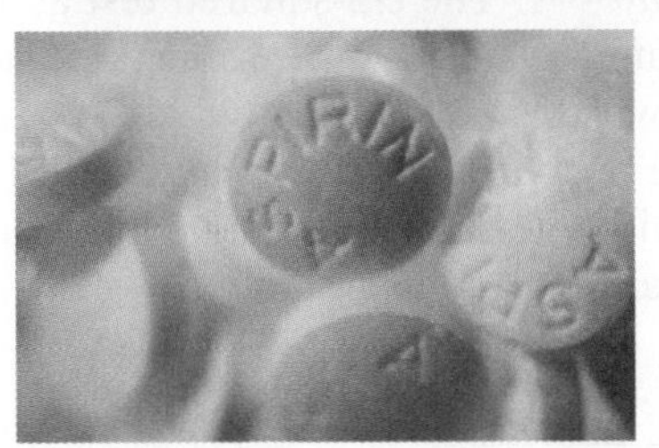

Picture the Scenario

Examples 2–4 in Chapter 10 discussed a meta-study on the effects of aspirin on cancer death rates, in which subjects were randomly assigned to take aspirin or placebo regularly. That example analyzed a 2 × 2 contingency table that compared the proportions of cancer deaths during a five-year period for those who took placebo and for those who took aspirin. Table 11.9 shows the data again with the expected cell counts, as part of MINITAB output for a chi-squared test of independence (or homogeneity).

Table 11.9 Annotated MINITAB Output for Chi-Squared Test of Independence of Group (Placebo, Aspirin) and Whether or Not Subject Died of Cancer

The same P-value results as with a two-sided Z test comparing the two population proportions.

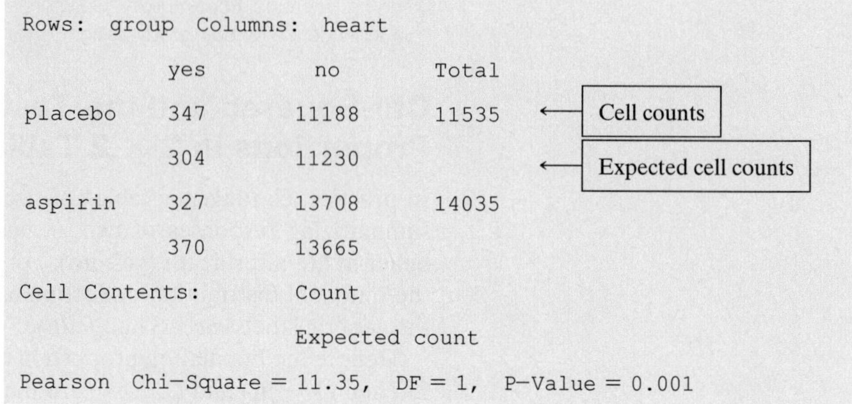

```
Rows:  group  Columns:  heart

              yes         no         Total

placebo       347        11188       11535   <- Cell counts

              304        11230               <- Expected cell counts

aspirin       327        13708       14035

              370        13665

Cell Contents:           Count

                         Expected count

Pearson  Chi-Square = 11.35,  DF = 1,  P-Value = 0.001
```

Denote the population proportion of cancer deaths by p_1 for the placebo treatment and by p_2 for the aspirin treatment. For H_0: $p_1 = p_2$, the z test of Section 10.1 has a test statistic value of $z \approx 3.45$ (Exercise 10.8). Its P-value is 0.001 for H_a: $p_1 \neq p_2$.

Questions to Explore

a. What are the hypotheses for the chi-squared test for these data?

b. Report the test statistic and P-value for the chi-squared test. How do these relate to results from the z test comparing the proportions?

Think It Through

a. The null hypothesis is that whether or not a someone dies of certain cancers is not associated with whether he or she takes placebo or aspirin. This is equivalent to H_0: $p_1 = p_2$, the population proportion of cancer deaths being the same for each group. The alternative hypothesis is that there's an association. This is equivalent to H_a: $p_1 \neq p_2$.

b. From Table 11.9, more subjects taking placebo died of cancer than we would expect if the variables were not associated (347 observed versus 304 expected). Fewer subjects taking aspirin died of cancer than we would expect if the variables were not associated (327 observed versus 370 expected). Table 11.9 reports $X^2 = 11.35$. This approximates the square of the z test statistic, $X^2 = z^2 = (3.45)^2 = 11.9$, which is close to the 11.35 value. (Note that the numerical values are not exact due to rounding error.) Table 11.9 reports a chi-squared P-value of 0.001. This is very strong evidence that the population proportions of cancer deaths differed for those taking aspirin and for those taking placebo. The sample proportions suggest that the aspirin group had a lower rate of cancer deaths than the placebo group.

Insight

For 2×2 tables, $df = (r - 1) \times (c - 1) = (2 - 1) \times (2 - 1) = 1$. Whenever a statistic has a standard normal distribution, the square of that statistic has a chi-squared distribution with $df = 1$. The chi-squared test and the two-sided z test necessarily have the same P-value and the same conclusion. Here, both the z and X^2 statistics show extremely strong evidence against the null hypothesis of equal population proportions. An advantage of the z test over X^2 is that it also can be used with one-sided alternative hypotheses. The direction of the effect is lost in squaring z and using X^2.

Try Exercise 11.18

For 2 × 2 tables, we don't really need the chi-squared test because we can use the z test. Why can't we use a z statistic to test independence for larger tables, for which $df > 1$? The reason is that a z statistic can be used only to compare a *single* estimate to a *single* null hypothesis value. Examples are a z statistic for comparing a sample proportion to a null hypothesis proportion, or a difference of sample proportions to a null hypothesis value of 0 for $p_1 - p_2$.

When a table is larger than 2 × 2 and thus $df > 1$, we need more than one difference parameter to describe the association. For instance, suppose Table 11.9 had three groups: placebo, regular-dose aspirin, and low-dose (children's) aspirin. Then the null hypothesis of no association corresponds to $p_1 = p_2 = p_3$, where p_3 is the population proportion of cancer death rates for the low-dose aspirin group. (See margin for 3 × 2 table.) The comparison parameters are $(p_1 - p_2)$, $(p_1 - p_3)$, and $(p_2 - p_3)$. We could use a z statistic for each comparison, but not a single z statistic for the overall test of independence (or homogeneity).

	Cancer Death	
Group	**Yes**	**No**
Placebo	p_1	$1 - p_1$
Regular dose	p_2	$1 - p_2$
Low dose	p_3	$1 - p_3$

Interpretation of Degrees of Freedom

The df value in a chi-squared test indicates how many parameters are needed to determine all the comparisons for describing the contingency table. For instance, we've just seen that a 3 × 2 table for comparing three treatments (placebo, regular aspirin, low-dose aspirin) on whether a subject died of cancer would have three parameters for making comparisons: $(p_1 - p_2)$, $(p_1 - p_3)$, and $(p_2 - p_3)$. We need to know only two of these to figure out the third. For instance, if we know $(p_1 - p_2)$ and $(p_2 - p_3)$, then

$$(p_1 - p_3) = (p_1 - p_2) + (p_2 - p_3).$$

A 3 × 2 table can use two parameters to determine all the comparisons, so its $df = 2$.

The df term also has the following interpretation: Given the row and column marginal totals in an $r \times c$ contingency table, the cell counts in a rectangular block of size $(r - 1) \times (c - 1)$ determine all the other cell counts. We illustrate using Table 11.10 from the GSS, which cross-tabulates political views by whether the subject would ever vote for a female for president. The table shows results for those subjects who called themselves extremely liberal, moderate, or extremely conservative.[1] For this 3 × 2 table, suppose we know the two counts in the upper left-hand $(3 - 1) \times (2 - 1) = 2 \times 1$ part of the table. Table 11.10 shows the cell counts in this block. Given these and the row and column totals, we can determine all the other cell counts.

Table 11.10 Illustration of Degrees of Freedom

A block of $(r - 1) \times (c - 1)$ cell counts determine the others, given the marginal totals. Here, a $(3 - 1) \times (2 - 1) = 2 \times 1$ block of cell counts in the first two rows and first column determines the others.

	Vote for Female President		
Political Views	**Yes**	**No**	**Total**
Extremely liberal	39	—	41
Moderate	435	—	469
Extremely conservative	—	—	45
Total	516	39	555

Source: Data from CSM, UC Berkeley.

[1] The 1998 GSS was the most recent one to ask the question about a female president.

	Vote Female?		
Political	Yes	No	Total
Liberal	39	2	41
Moderate	435	34	469
Conserv.	42	3	45
Total	516	39	555

For instance, the marginal total in row 1 is 41 and the first cell count is 39, so the second cell has $41 - 39 = 2$ subjects. Likewise, there are

$$469 - 435 = 34 \text{ subjects in the second cell in the second row, and}$$

$$516 - (39 + 435) = 42 \text{ subjects in the third cell in the first column.}$$

From these counts and the fact that the third row has 45 observations, there must be $45 - 42 = 3$ in the final cell in the third row and second column. To check that you understand, try reconstructing the table if you know only the cell counts in the two cells that are in the second and third rows and second column.

Once the marginal totals are fixed in a contingency table, a block of only $(r - 1) \times (c - 1)$ cell counts is free to vary, since these cell counts determine the remaining ones. The degrees of freedom value equals the number of cells in this block, or $df = (r - 1) \times (c - 1)$. (See margin for full table.)

Limitations of the Chi-Squared Test

Recall

From Section 9.5, a small P-value does not imply an important result in practical terms. Statistical significance is not the same as *practical* significance. ◀

The chi-squared test of independence, like other significance tests, provides limited information. If the P-value is very small, strong evidence exists against the null hypothesis of independence. We can infer that the variables are associated. The chi-squared statistic and the P-value tell us nothing, however, about the nature or the strength of the association. We know there is statistical significance, but the test alone does not indicate whether there is practical significance as well. A large X^2 means strong evidence of association, but not necessarily a strong association. We'll see how to investigate the strength of association in the next section.

SUMMARY: Misuses of the Chi-Squared Test

The chi-squared test is often misused. Some common misuses are applying it

- When some of the expected frequencies are too small.
- When separate rows or columns are dependent samples,[2] such as when each row of the table has the same subjects.
- To data that do not result from a random sample or randomized experiment.
- To data by classifying quantitative variables into categories. This results in a loss of information. It is usually more appropriate to analyze the data with methods for quantitative variables, like those the next chapter presents.

"Goodness of Fit" Chi-Squared Tests

So far we've used the chi-squared statistic to test the null hypothesis that two categorical variables are independent. In practice, the chi-squared statistic is used for a variety of hypotheses about categorical variables. For example, for three variables, the chi-squared statistic can test the hypothesis that two variables are independent at each fixed category of the third variable, such as happiness and income being independent both for women and for men (the two categories of the categorical variable, sex). What all of the various chi-squared tests have in common is that they all have a formula for finding the expected frequencies that satisfy that hypothesis and a formula for the *df* value for the test statistic.

The chi-squared statistic can be used even for a hypothesis involving a *single* categorical variable. For example, in one of his experiments, the geneticist Gregor Mendel crossed pea plants of pure yellow strain with plants of pure green strain. He predicted that a second-generation hybrid seed would have probability 0.75

[2]With dependent samples, McNemar's test (Section 10.4) is the appropriate test with binary variables, and extensions of it handle categorical variables with more than two categories.

of being yellow and 0.25 of being green (yellow being the dominant strain). The experiment produced 8023 seeds. If Mendel's theory is correct, we would expect to see 75% yellow seeds and 25% green seeds, that is, 8023(0.75) = 6017.25 yellow seeds and 8023(0.25) = 2005.75 green seeds. In the actual experiment, of the 8023 hybrid seeds, 6022 were yellow and 2001 were green. Are these results far enough from Mendel's predictions to give evidence against his theory? Applying the chi-squared statistics to test Mendel's hypothesis, we obtain

$$X^2 = \sum \frac{(\text{observed count} - \text{expected count})^2}{\text{expected count}} =$$

$$\frac{(6022 - 6017.25)^2}{6017.25} + \frac{(2001 - 2005.75)^2}{2005.75} = 0.015$$

When a hypothesis predicts a population proportion value for each category of a variable that has c categories, the chi-squared statistic has $df = c - 1$. For this example with $c = 2$ categories (green and yellow for the color of a hybrid seed), $df = 2 - 1 = 1$. The chi-squared statistic value of 0.015 has a P-value = 0.90. The data in Mendel's experiment did not contradict his hypothesis.

When testing particular proportion values for a categorical variable, the chi-squared statistic is referred to as a **goodness-of-fit statistic.** The statistic summarizes how well the hypothesized values predict what happens with the observed data. With $c = 2$ categories, the chi-squared statistic equals the square of the z statistic used in Chapter 9 to test a hypothesis about the proportion value for one of the two categories. For example, to test Mendel's hypothesis H_0: $p = 0.75$ that the probability is 0.75 that a second-generation hybrid seed is yellow, we find $\hat{p} = 6022/8023 = 0.7506$, $se_0 = \sqrt{p_0(1 - p_0)/n} = \sqrt{0.75(0.25)/8023} = 0.00483$, and

$$z = \frac{\hat{p} - p_0}{se_0} = \frac{0.7506 - 0.75}{0.00483} = 0.124. \text{ Note that } z^2 = 0.015.$$

The P-value using the chi-squared distribution with test statistic X^2 is identical to the two-sided P-value using the standard normal distribution with test statistic z. An advantage of the X^2 test is that it applies even when there are more than two categories. See Exercise 11.22 for an example.

A limitation of the goodness-of-fit test, like other significance tests, is that it does not tell you the ranges of plausible values for the proportion parameters. To do this, you can find confidence intervals for the individual population proportions (Section 8.2). For Mendel's experiment, a 95% confidence interval for the probability of a yellow hybrid seed is

$$0.7506 \pm 1.96(0.0048), \text{ which is } (0.74, 0.76).$$

Activity 2

Analyzing Contingency Tables of GSS Data

It's easy to analyze associations between categorical variables measured by the General Social Survey. Let's see how.

- Go to the GSS Web site, sda.berkeley.edu/GSS/. Click on *GSS*, with *No Weight* as the default weight selection.
- The GSS names for Table 11.1 are HAPPY (for general happiness) and FINRELA (for family income in relative terms). Enter HAPPY in the column space and FINRELA in the row space to construct a table with these as the column and row variables. Entering YEAR (2004) in the selection filter restricts the search to GSS data from the survey in 2004. Select *No Weight* from the Weight menu. You can rerun the analysis with the YEAR (2010) and compare results with the year 2004.
- Put a check in the row box . . . for the Percentaging option and in the Summary Statistics box for Table Options, and click on *Run the Table.*

You'll now get a contingency table relating happiness to family income for the 2003 survey. The table shows the cell counts and sample conditional distributions. It treats as the response variable the one you check for the Percentaging option. The output also shows some statistics (such as Chisq-P) for the Pearson chi-squared statistic) that we're using in this chapter to analyze contingency tables.

The family income variable has five categories (far below average, below average, average, above average, far above average). Table 11.1 combined categories 1 and 2 into a single category (for below average) and categories 4 and 5 into a single category (for above average). To create this collapsed table, type the family income variable as FINRELA(r:1-2;3;4-5) instead of FINRELA.

Now, create a table relating another variable to happiness. At the page on which you enter the variable names, click on *Standard Codebook* under the codebooks menu and you will see indexes for the variables. Look up a subject that interests you and find the GSS code name for a variable. For the table you create,

- Find the chi-squared statistic and *df* value.
- Report the P-value and interpret the results in context.

11.2 Practicing the Basics

11.8 What gives P-value = 0.05? How large a X^2 test statistic value provides a P-value of 0.05 for testing independence for the following table dimensions?

a. 2 × 2

b. 2 × 3

c. 2 × 5

d. 5 × 5

e. 3 × 9

11.9 Happiness and gender For the 2 × 3 table on gender and happiness in Exercise 11.4 (shown again following), software tells us that $X^2 = 0.46$ and the P-value = 0.79.

a. State the null and alternative hypothesis, in context, to which these results apply.

b. Interpret the P-value.

	Happiness		
Gender	**Not**	**Pretty**	**Very**
Female	174	587	328
Male	142	513	271

11.10 Marital happiness and income TRY In Exercise 11.5 when you used the GSS to download a 3 × 3 table for marital happiness and family income in 2008, you should have obtained results similar to the table shown below.

	Marital Happiness		
Income	**Not**	**Pretty**	**Very**
Above	123	105	7
Average	291	151	17
Below	172	83	6

a. State the null and alternative hypotheses for the test.

b. How large a X^2 value would give a P-value of exactly 0.05?

c. The chi-squared statistic for the table equals 12.837. Find and interpret the P-value.

d. State your decision for a significance level of 0.05, and interpret in context.

11.11 Life after death and gender In the 2008 GSS, 620 of 809 TRY males and 835 of 978 females indicated a belief in life after death. (*Source*: Data from CSM, UC Berkeley.)

a. Construct a 2 × 2 contingency table relating gender of respondent (SEX, categories male and female) as the rows to belief about life after death (POSTLIFE, categories yes and no) as the columns.

b. Find the four expected cell counts for the chi-squared test. Compare them to the observed cell counts, identifying cells having more observations than expected.

c. The data have $X^2 = 22.36$. Set up its calculation by showing how to substitute the observed and expected cell counts you found into its formula.

11.12 Basketball shots independent? In pro basketball games during 1980–1982, when Larry Bird of the Boston Celtics missed his first free throw, 48 out of 53 times he made the second one, and when he made his first free throw, 251 out of 285 times he made the second one.

a. Form a 2 × 2 contingency table that cross-tabulates the outcome of the first free throw (with categories made and missed) and the outcome of the second free throw (made and missed).

b. When we use MINITAB to analyze the contingency table, we get the result

```
Pearson Chi-Square = 0.273, DF = 1,
        P-Value = 0.602
```

Does it seem as if his success on the second shot depends on whether he made the first? Explain how to interpret the result of the chi-squared test.

11.13 Cigarettes and marijuana The table on the following page refers to a survey[3] in which senior high school students in Dayton, Ohio, were randomly sampled. It cross-tabulates whether a student had ever smoked cigarettes and whether a student had ever used marijuana. Analyze these data by

[3] *Source*: Data from personal communication from Harry Khamis, Wright State University.

(a) finding and interpreting conditional distributions with marijuana use as the response variable and (b) reporting all five steps of the chi-squared test of independence.

	Marijuana	
Cigarettes	**Yes**	**No**
Yes	914	581
No	46	735

11.14 Smoking and alcohol Refer to the previous exercise. A similar table relates cigarette use to alcohol use. The MINITAB output for the chi-squared test follows.

a. True or false: If we use cigarette use as the column variable instead of alcohol use, then we will get different values for the chi-squared statistic and the P-value shown in the table.

b. Explain what value you would get for the z statistic and P-value if you conducted a significance test of $H_0: p_1 = p_2$ against $H_a: p_1 \neq p_2$, where p_1 is the population proportion of non cigarette users who have drunk alcohol and p_2 is the population proportion of cigarette users who have drunk alcohol.

Dayton student survey

```
Row: cigarette Columns: alcohol
          no      yes
no       281      500
yes       46     1449
Pearson Chi-Square = 451.404,
DF = 1,  P-Value = 0.000
```

11.15 Help the environment In 2000 the GSS asked whether a subject is willing to accept cuts in the standard of living to help the environment (GRNSOL), with categories (vw = very willing, fw = fairly willing, nwu = neither willing nor unwilling, nvw = not very willing, nw = not at all willing). When this was cross-tabulated with gender (SEX), as shown below, $X^2 = 8.0$.

a. What are the hypotheses for the test to which X^2 refers?

b. Report r and c and the df value on which X^2 is based.

c. Is the P-value less than (i) 0.05? (ii) 0.10? Explain.

d. What conclusion would you make, using a significance level of (i) 0.05 and (ii) 0.10? State your conclusion in the context of this study.

	Help the Environment				
Sex	**vw**	**fw**	**nwu**	**nvw**	**nw**
F	34	149	160	142	168
M	30	131	152	98	106

11.16 What predicts being green? Refer to the previous exercise. What is associated with environmental opinion? Go to the GSS Web site sda.berkeley.edu/GSS/.

a. Find a variable that is significantly associated with GRNSOL, using the 0.05 significance level. (*Hint:* You could first try some demographic variables, such as gender (variable SEX) or race (RACE).)

b. For the variable you found, in which cells are there considerably (i) more and (ii) fewer, subjects than you would expect if the variables were independent?

11.17 Aspirin and heart attacks A Swedish study used 1360 patients who had suffered a stroke. The study randomly assigned each subject to an aspirin treatment or a placebo treatment. In this study heart attacks were suffered by 28 of the 684 subjects taking placebo and 18 of the 676 subjects taking aspirin.

a. Report the data in the form of a 2 × 2 contingency table.

b. Show how to carry out all five steps of the null hypothesis that having a heart attack is not associated with whether one takes placebo or aspirin. (You should get a chi-squared statistic equal to 2.1.) Interpret.

11.18 z test for heart attack study Refer to the previous exercise. The printout from MINITAB reports

TRY

```
Test for difference = 0 (vs not = 0):
          Z = 1.46 P-Value = 0.144
```

a. Define population proportions p_1 and p_2 and state the hypotheses for that test.

b. Explain how the result of the chi-squared test in (b) in the previous exercise corresponds to this z test result.

11.19 Happiness and life after death Are people who believe in life after death happier?

a. Go to the GSS Web site sda.berkeley.edu/GSS and download the contingency table for the 2008 survey relating happiness and whether you believe in life after death (variables denoted HAPPY and POSTLIFE). Construct the conditional distributions, using happiness as the response variable, and interpret.

b. Check the Summary Statistics box in GSS to show that $X^2 = 7.7$. What is the P-value? Interpret.

c. Based on the test in part b, what can you conclude?

11.20 What is independent of happiness? Refer to Exercises 11.6 and 11.7. For the variables that you thought might be independent,

a. At the GSS Web site, conduct all five steps of the chi-squared test.

b. Based on part a, which inference is most appropriate? (i) We accept the hypothesis that the variables are independent; (ii) the variables may be independent; (iii) the variables are associated.

11.21 Testing a genetic theory In an experiment on chlorophyll inheritance in corn, for 1103 seedlings of self-fertilized heterozygous green plants, 854 seedlings were green and 249 were yellow. Theory predicts that 75% of the seedlings would be green.

a. Specify a null hypothesis for testing the theory.

b. Find the value of the chi-squared goodness-of-fit statistic and report its *df*.

c. Report the P-value, and interpret.

11.22 Checking a roulette wheel Karl Pearson devised the chi-squared goodness-of-fit test partly to analyze data from an experiment to analyze whether a particular roulette wheel in Monte Carlo was fair, in the sense that each outcome was equally likely in a spin of the wheel. For a given European roulette wheel with 37 pockets (with numbers 0, 1, 2, ..., 36), consider the null hypothesis that the wheel is fair.

a. For the null hypothesis, what is the probability for each pocket?

b. For an experiment with 3700 spins of the roulette wheel, find the expected number of times each pocket is selected.

c. In the experiment, the 0 pocket occurred 110 times. Show the contribution to the X^2 statistic of the results for this pocket.

d. Comparing the observed and expected counts for all 37 pockets, we get $X^2 = 34.4$ Specify the *df* value, and indicate whether there is strong evidence that the roulette wheel is not balanced. (*Hint:* Recall that the *df* value is the mean of the distribution.)

11.3 Determining the Strength of the Association

Three questions are normally addressed in analyzing contingency tables:

- **Is there an association?** The chi-squared test of independence addresses this. When the P-value is small, we infer that the variables are associated.
- **How do the cell counts differ from what independence predicts?** To answer this question, we compare each observed cell count to the corresponding expected cell count. We'll say more about this in Section 11.4.
- **How strong is the association?** Analyzing the *strength* of the association reveals whether the association is an important one, or if it is statistically significant but weak and unimportant in practical terms. We now focus on this third question.

Measures of Association

For recent GSS data, Table 11.11 shows associations between opinion about the death penalty and two explanatory variables—gender and race. Both sides of Table 11.11 have large chi-squared statistics and small P-values for the test of independence. We can conclude that both gender and race are associated with the death penalty opinion.

Table 11.11 GSS Data Showing Race and Gender as Explanatory Variables for Opinion About the Death Penalty

	Opinion				Opinion		
Race	Favor	Oppose	n	Gender	Favor	Oppose	n
White	71%	29%	1473	Male	71%	29%	885
Black	46%	54%	259	Female	62%	38%	1017
Chi-squared = 65.55				Chi-squared = 17.78			
$df = 1$, P-value = 0.00				$df = 1$, P-value = 0.00			

Source: Data from CSM, UC Berkeley.

The X^2 test statistic is larger for the first table. Does this mean that opinion is more strongly associated with race than with gender? How can we quantify the strength of the association in a contingency table?

Measure of Association

A **measure of association** is a statistic or a parameter that summarizes the strength of the dependence between two variables.

A measure of association takes a range of values from one extreme to another as data range from the weakest to the strongest association. It is useful for comparing associations to determine which is stronger. Later in this section we'll see that although the chi-squared test tells us how strong the evidence is that an association truly exists in the population, it does not describe the strength of that association.

Difference of Proportions

An easily interpretable measure of association is the difference between the proportions making a particular response. The population difference of proportions is 0 whenever the conditional distributions are identical, that is, when the variables are independent. The difference of proportions falls between -1 and $+1$, where 1 and -1 represent the strongest possible association. Whether you get a negative or a positive value merely reflects how the rows are ordered. Section 10.1 showed how to use a sample difference of proportions value to conduct inference (such as a confidence interval) about a population value.

Table 11.12 shows two hypothetical contingency tables relating a subject's income and whether the subject responds to a promotional mailing from a bank offering a special credit card. Case A exhibits the weakest possible association—no association. Both the high-income and the low-income subjects have 60% rejecting the credit card. The difference of proportions is $0.60 - 0.60 = 0$. By contrast, case B exhibits the strongest possible association. All those with high income accept the card, whereas all those with low income decline it. The difference of proportions is $0 - 1.0 = -1$.

Table 11.12 Decision About Accepting a Credit Card Cross-Tabulated by Income

In Case A, the conditional distributions are the same in each row, so the difference of proportions is 0. In Case B the acceptance decision is completely dependent on income level, and the difference of proportions is -1.

	Case A Accept Credit Card			Case B Accept Credit Card		
Income	**No**	**Yes**	**Total**	**No**	**Yes**	**Total**
High	240 (60%)	160 (40%)	400 (100%)	0 (0%)	400 (100%)	400 (100%)
Low	360 (60%)	240 (40%)	600 (100%)	600 (100%)	0 (0%)	600 (100%)

	Opinion	
Race	**Favor**	**Oppose**
White	71%	29%
Black	46%	54%
Gender		
Male	71%	29%
Female	62%	38%

In practice, we don't expect data to follow either of these extremes, but the stronger the association, the larger the absolute value of the difference of proportions. For Table 11.11, shown again in the margin, the difference of proportions is $0.71 - 0.46 = 0.25$ for whites and blacks, and the difference of proportions is $0.71 - 0.62 = 0.09$ for males and females. The difference of 0.25 for whites and blacks is much larger than the difference of 0.09 for males and females.

Example 6

Strength of an association

Student Stress, Depression, and Gender

Picture the Scenario

Every year, the Higher Education Research Institute at UCLA conducts a large-scale survey of college freshmen on a variety of issues. From the survey of 283,000 freshmen in 2002, Table 11.13 compares females and males on the percent who reported feeling frequently stressed (overwhelmed by all they have to do) and the percent who reported feeling frequently depressed during the past year. P-values for chi-squared tests of independence were 0.000 (rounded) for each data set.

Table 11.13 Conditional Distributions of Stress and Depression, by Gender

	Stress				Depression		
Gender	Yes	No	Total	Gender	Yes	No	Total
Female	35%	65%	100%	Female	8%	92%	100%
Male	16%	84%	100%	Male	6%	94%	100%

Question to Explore

Which response variable, stress or depression, has the stronger sample association with gender?

Think It Through

The difference of proportions between females and males was $0.35 - 0.16 = 0.19$ for feeling stressed. It was $0.08 - 0.06 = 0.02$ for feeling depressed. Since 0.19 is much larger than 0.02, there is evidence of a greater difference between reports of females and males on their feeling stress than on depression. In the sample, stress has the stronger association with gender.

Insight

Although the difference of proportions can be as large as 1 or -1, in practice in comparisons of groups, it is rare to find differences near these limits.

Try Exercise 11.24

Recall

Section 10.3 introduced **relative risk.** It showed that software can report a confidence interval for a population relative risk. ◀

The Ratio of Proportions: Relative Risk

Another measure of association, introduced in Section 10.3, is the ratio of two proportions, p_1/p_2. In medical applications in which the proportion refers to an adverse outcome, it is called the **relative risk.**

Example 7

Relative risk

Seat Belt Use and Outcome of Auto Accidents

Picture the Scenario

Based on records of automobile accidents in a recent year, the Department of Highway Safety and Motor Vehicles in Florida reported Table 11.14 on the counts of nonfatalities and fatalities. The data have $X^2 = 2338$, with P-value $= 0.00000. \ldots$

Table 11.14 Outcome of Auto Accident by Whether or Not Subject Wore Seat Belt

	Outcome		
Wore Seat Belt	Survived	Died	Total
Yes	412,368	510	412,878
No	162,527	1601	164,128

Source: Department of Highway Safety and Motor Vehicles, Florida.

Question to Explore

Treating the auto accident outcome as the response variable, find and interpret the relative risk.

Think It Through

In Table 11.14, the adverse outcome is death, rather than survival. The relative risk is formed for that outcome. For those who wore a seat belt, the proportion who died equaled 510/412,878 = 0.00124. For those who did not wear a seat belt, the proportion who died was 1601/164,128 = 0.00975. The relative risk is the ratio, 0.00124/0.00975 = 0.127. The proportion of subjects wearing a seat belt who died was 0.127 times the proportion of subjects not wearing a seat belt who died.

Equivalently, since 0.00975/0.00124 = 1/0.127 = 7.9, the proportion of subjects *not wearing* a seat belt who died was nearly eight times the proportion of subjects *wearing* seat belts who died. This reciprocal value is the relative risk when the rows are interchanged, with no for seat-belt use in row 1.

Insight

The ordering of rows is arbitrary. Many people find it easier to interpret the relative risk for the ordering for which its value is *above* 1.0. A relative risk of 7.9 represents a strong association. This is far from the value of 1.0 that would occur if the proportion of deaths were the same for each group. Wearing a seat belt has a practically significant effect in enhancing the chance of surviving an auto accident.

Try Exercises 11.27, part c, and 11.30

In Example 7 the proportion of interest was close to 0 for each group, but the relative risk was far from 1. When this happens, an association may be practically important even though the difference of proportions is near 0 (it was 0.0085 in Example 7). If the chance of death is only a bit higher when one does not wear a seat belt, but that chance is nearly eight times the chance of death when wearing a seat belt, the association is practically significant. The relative risk is often more informative than the difference of proportions for comparing proportions that are both close to 0.

Properties of the Relative Risk

- The relative risk can equal any nonnegative number.
- When $p_1 = p_2$, the variables are independent and relative risk = 1.0.
- Values farther from 1.0 (in either direction) represent stronger associations. Two values for the relative risk represent the same strength of association, but in opposite directions, when one value is the reciprocal of the other. See Figure 11.5.

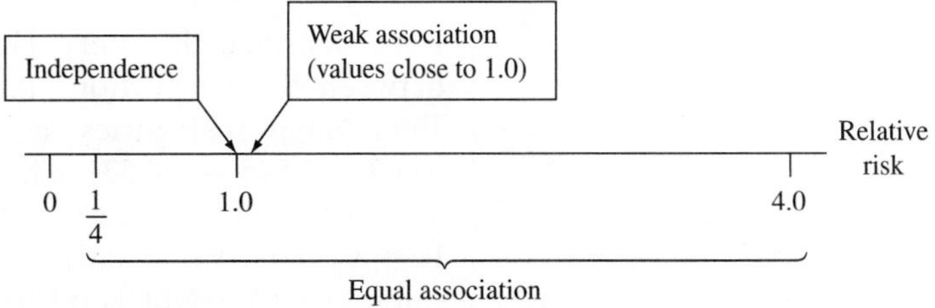

▲ **Figure 11.5 Values of the Relative Risk Farther from 1.0 Represent Stronger Association. Question** Why do relative risks that take reciprocal values, such as 4 and $\frac{1}{4}$, represent the same strength of association?

A relative risk of 8 is farther from independence and represents a stronger association than a relative risk of 4, and a relative risk of $\frac{1}{8}$ is farther from independence than a relative risk of $\frac{1}{4}$. But relative risk = 4.0 and relative risk = 1/4.0 = 0.25

represent the same strength of association. When relative risk $p_1/p_2 = 0.25$, then p_1 is 0.25 times p_2. Equivalently, p_2 is $1/0.25 = 4.0$ times p_1. Whether we get 4.0 or 0.25 for the relative risk depends merely on which group we call group 1 and which we call group 2.

Association in $r \times c$ Tables

The difference of proportions and the relative risk are defined for 2×2 tables. For $r \times c$ (r = # of rows, c = # of columns) tables, they can compare proportions in a particular response category for various pairs of rows.

Example 8

Association in $r \times c$ tables

General Happiness and Marital Happiness

Picture the Scenario

Earlier in the chapter we studied the association between happiness and family income. For married subjects, a possible predictor of happiness is their reported happiness with their marriage. Table 11.15 cross-tabulates these variables, using data from the 2008 GSS. It also reports conditional distributions with happiness as the response variable. The data have chi-squared statistic $X^2 = 263.9$, with $df = 4$ and a P-value of 0.0000....

Table 11.15 Happiness and Marital Happiness

The percentages are the conditional distributions of happiness in each row. When percentages are compared in rows 1 and 3, the highlighted cells reveal a strong association.

	Happiness			
Marital Happiness	Not Too Happy	Pretty Happy	Very Happy	Total
Not Too Happy	13 (43.3%)	15 (50.0%)	2 (6.7%)	30 (100%)
Pretty Happy	44 (12.8%)	259 (75.5%)	40 (11.7%)	343 (100%)
Very Happy	24 (4.0%)	215 (36.1%)	356 (59.8%)	595 (100%)

Source: Data from CSM, UC Berkeley.

Question to Explore

Describe the association between marital happiness and general happiness, using the not too happy and very happy categories.

Think It Through

Let's consider the Very Happy category of happiness. The difference between the proportions in this category for the Not Too Happy and Very Happy categories of marital happiness (that is, rows 1 and 3) is $0.067 - 0.598 = -0.531$. Those who are not too happy with their marriage are much less likely to be very happy overall. Consider next the Not Too Happy category of happiness. The difference between the proportions in this category for the Not Too Happy and Very Happy categories of marital happiness is $0.433 - 0.04 = 0.393$. This is also a substantial difference.

Likewise, we could use the *ratio* of proportions. For instance, consider the Not Too Happy category of happiness as an undesirable outcome. The relative risk of this outcome, comparing the Not Too Happy and Very Happy categories of marital happiness, is $0.433/0.04 = 10.825$. The estimated proportion who are not too happy is nearly 11 times larger for those who have not too happy marriages than for those who have very happy marriages.

This ratio of 10.825 is very far from 1.0, indicating that this part of the table reveals an extremely strong association.

Insight

It is unusual to find such strong associations. For instance, family income also has a substantial association with happiness, but it is not as strong. Using Table 11.1 (partly shown in the margin), you can check that the relative risk of being not too happy equals 4.07 when you compare the below-average and above-average levels of income. This is much less than the relative risk of 10.825 and is a weaker association.

Recall

From Table 11.1,

Income	Not Too Happy	Total
Below average	172	687
Above average	26	423

◀

Try Exercises 11.26 and 11.27

Chi-Squared and Association

You may be tempted to regard a large value of the X^2 statistic in the test of independence as indicating a strong association. This is a misuse of the statistic.

In Practice Large X^2 Does Not Mean There's a Strong Association

A large value for X^2 provides strong evidence that the variables are associated. It does not imply, however, that the variables have a strong association. This statistic merely indicates (through its P-value) how certain we can be that the variables are associated, not how strong that association is. For a given strength of association, larger X^2 values occur for larger sample sizes. Large X^2 values can occur with weak associations, if the sample size is sufficiently large.

Caution

If the sample size is sufficiently large, a X^2 value large enough to be considered statistically significant can occur even if the association between the two categorical variables is weak. ◀

For example, Cases A, B, and C in Table 11.16 are hypothetical tables relating gender and whether or not one attends religious services weekly. The association in each case is weak—the conditional distribution for females (51% yes, 49% no) is nearly identical to the conditional distribution for males (49% yes, 51% no). All three cases show the same weak strength of association: The difference between the proportions of females and males who say yes is $0.51 - 0.49 = 0.02$ in each.

Table 11.16 Response on Attending Religious Services Weekly by Gender, Showing Weak but Identical Associations

In each case the difference between proportions is small, only 0.02.

	Case A			Case B			Case C		
	Yes	No	n	Yes	No	n	Yes	No	n
Female	51%	49%	100	51%	49%	200	51%	49%	10,000
Male	49%	51%	100	49%	51%	200	49%	51%	10,000
Chi-squared = 0.08				Chi-squared = 0.16			Chi-squared = 8.0		
P-value = 0.78				P-value = 0.69			P-value = 0.005		

For the sample of size 200 in Case A, $X^2 = 0.08$, which has a P-value of 0.78. Case B has twice as large a sample size, with the cell counts doubling. For its sample of size 400, $X^2 = 0.16$, for which the P-value = 0.69. So, when the cell counts double, X^2 doubles. Similarly, for the sample size of 20,000 (100 times as large as $n = 200$) in Case C, $X^2 = 8.0$ (100 times as large as $X^2 = 0.08$). Then the P-value = 0.005.

In summary, for fixed conditional distributions, the value of X^2 is directly proportional to the sample size—larger values occur with larger sample sizes. Like other test statistics, the larger the X^2 statistic, the smaller the P-value and the stronger the evidence against the null hypothesis. However, a small P-value results from even a weak association when the sample size is large, as Case C shows.

11.3 Practicing the Basics

11.23 Party ID and race The table shows 2008 GSS data on race and political party identification.

	Party Identification			
Race	**Democrat**	**Independent**	**Republican**	***n***
Black	192	75	8	275
White	459	586	471	1516

Source: Data from CSM, UC Berkeley.

a. The chi-squared test of independence has $X^2 = 177.312$. Carry out the five steps of the test, and interpret the result in the context of these variables.

b. Estimate the difference between black subjects and white subjects in the proportion who identify themselves as Democrats. Interpret. Does this seem to be a weak (practically insignificant), or a strong (practically significant), association? Explain.

11.24 Party ID, race and gender Refer to the previous exercise and to the data in the table below from Exercise 11.1. Which has a stronger association with whether one identifies as a Democrat—race or gender? Justify your answer by comparing the difference between blacks and whites in the proportion identifying as Democrat to the difference between females and males in the proportion identifying as Democrat.

TRY

Data from Exercise 11.1

Gender	Democrat	*n*
Female	422	1076
Male	299	896

11.25 Happiness and highest degree The table shows 2008 GSS data on happiness and the highest degree attained.

	Happiness		
Highest Degree	**Not Too Happy**	**Pretty Happy**	**Very Happy**
< High school	87	137	72
HS, junior college	181	654	337
College, grad school	48	308	190

Source: Data from CSM, UC Berkeley.

a. The chi-squared test of independence has $X^2 = 64.41$. What conclusion would you make using a significance level of 0.05? Interpret.

b. Does this large chi-squared value mean there is a strong association between happiness and highest degree? Explain.

c. Estimate the difference between the lowest and highest education groups in the proportion who report being not too happy. Interpret.

d. Find and interpret the relative risk of being not too happy, comparing the lowest and highest education groups. Interpret.

11.26 Income and highest degree The table shows 2008 GSS data on family income and the subject's highest degree attained, for which $X^2 = 315.98$ (P-value $= 0.000$).

TRY

	Family Income		
Highest Degree	**Below Average**	**Average**	**Above Average**
< High school	176	98	18
HS, junior college	429	573	165
College, grad school	85	426	242

a. Estimate the difference between the lowest and highest education groups in the proportion who report (i) below average income and (ii) above average income. Interpret. Does this seem to be a practically significant, or practically insignificant, association?

b. Find and interpret the relative risk of a below-average income, comparing the lowest and highest education groups. Interpret.

11.27 Smoking and alcohol The table refers to a survey[4] of senior high school students in Dayton, Ohio. It cross-tabulates whether a student had ever smoked cigarettes and whether a student had ever drunk alcohol. The table also shows MINITAB output of conditional distributions.

TRY

a. Describe the strength of association using the difference between users and nonusers of cigarettes in the proportions who have used alcohol. Interpret.

b. Describe the strength of association using the difference between users and nonusers of alcohol in the proportions who have used cigarettes. Interpret.

c. Describe the strength of association using the relative risk of using cigarettes, comparing those who have used or not used alcohol. interpret.

Data from Dayton Student Survey

Rows: **cigarette**	**Columns:** **alcohol**		
	no	yes	
no	281	500	
	35.98	64.02	← For row conditional distribution
	85.93	25.65	← For column conditional distribution
yes	46	1449	
	3.08	96.92	
	14.07	74.35	
Cell Contents:		Count	
		% of Row	
		% of Column	

[4]*Source*: Data from survey from Harry Khamis, Wright State University.

11.28 Sex of victim and offender For murders in the United States in 2009, the table cross-tabulates the sex of the victim by the sex of the offender. Find and interpret a measure of association, treating sex of victim as the response variable.

	Sex of Victim	
Sex of Offender	**Female**	**Male**
Female	182	484
Male	1719	4078

Source: Data from www.fbi.gov.

11.29 Prison and gender According to the U.S. Department of Justice, in 2009 the incarceration rate in the nation's prisons was 949 per 100,000 male residents, and 67 per 100,000 female residents.

a. Find the relative risk of being incarcerated, comparing males to females. Interpret.

b. Find the difference of proportions of being incarcerated. Interpret.

c. Which measure do you think is more appropriate for these data? Why?

11.30 Risk of dying for teenagers According to summarized TRY data from 1999 to 2006 accessed from the Centers of Disease Control and Prevention, the annual probability that a male teenager at age 19 is likely to die is about 0.00135 and 0.00046 for females age 19. (www.cdc.gov)

a. Compare these rates using the difference of proportions, and interpret.

b. Compare these rates using the relative risk, and interpret.

c. Which of the two measures seems more useful when both proportions are very close to 0? Explain.

11.31 Death penalty associations Table 11.11, summarized again here, showed the associations between death penalty opinion and gender and race.

	Opinion	
Race	**Favor**	**Oppose**
White	71%	29%
Black	46%	54%
Gender		
Male	71%	29%
Female	62%	38%

a. True or false: The table with the larger X^2 statistic necessarily has the stronger association. Explain.

b. To make an inference about the strength of association in the population, you can construct confidence intervals around the sample differences of proportions. The 95% confidence intervals are (0.19, 0.32) comparing whites and blacks and (0.05, 0.13) comparing males and females in the proportions favoring the death penalty. In the *population*, can you make a conclusion about which variable is more strongly associated with the death penalty opinion? Explain.

11.32 Chi-squared versus measuring association For Table 11.15 on general happiness and marital happiness, the chi-squared statistic equals 263.9 ($df = 4$, P-value $= 0.000$). Explain the difference between the purpose of the chi-squared test and the descriptive analysis in Example 8 comparing conditional distributions using measures of association. (*Hint:* Is a chi-squared test a descriptive or inferential analysis?)

11.4 Using Residuals to Reveal the Pattern of Association

The chi-squared test and measures of association such as $p_1 - p_2$ and p_1/p_2 are fundamental methods for analyzing contingency tables. The P-value for X^2 summarizes the strength of evidence against H_0: independence. If the P-value is small, then we conclude that somewhere in the contingency table the population cell proportions differ from independence. The chi-squared test does not indicate, however, whether all cells deviate greatly from independence or perhaps only some of them do so.

Recall

Section 3.3 introduced a **residual** for quantitative variables as the difference between an observed response and a predicted response for a regression equation. For categorical variables, a residual compares an observed count to an expected count that represents what H_0 predicts. ◀

Residual Analysis

A cell-by-cell comparison of the observed counts with the counts that are expected when H_0 is true reveals the nature of the evidence. The difference between an observed and expected count in a particular cell is called a **residual.**

For example, Table 11.5 reported observed and expected cell counts for happiness and family income. The first column of this table is summarized in the margin on the next page, with the expected cell counts in parentheses. The first cell reports 26 subjects with above-average income who are not too happy.

Income	Not Too Happy
Above average	26 (66.86)
Average	117 (139.56)
Below average	172 (108.58)

The expected count is 66.86. The residual is 26 − 66.86 = −40.86. The residual is negative when, as in this cell, fewer subjects are in the cell than expected under H_0. The residual is positive when more subjects are in the cell than expected, as in the cell having observed count = 172 (below-average income subjects who are not too happy) and residual 172 − 108.58 = 63.42.

How do we know whether a residual is large enough to indicate strong evidence of a deviation from independence in that cell? To answer this question, we use an adjusted form of the residual that behaves like a z-score. It is called the **standardized residual.**

Standardized Residual

The **standardized residual** for a cell equals

(observed count − expected count)/*se*.

Here, *se* denotes a standard error for the sampling distribution of (observed count − expected count), when the variables are independent.

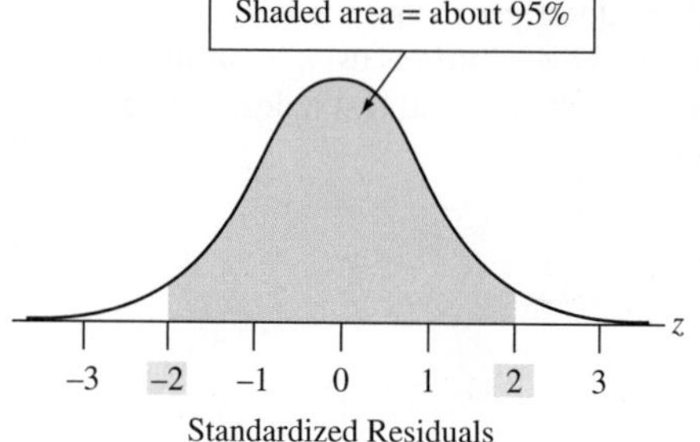

A standardized residual reports the number of standard errors that an observed count falls from its expected count. The *se* describes how much the (observed − expected) difference would tend to vary in repeated random sampling if the variables were independent. Its formula is complex, and we'll not give it here. Software can easily find standardized residuals for us.

When H_0: independence is true, the standardized residuals have approximately a standard normal distribution: They fluctuate around a mean of 0, with a standard deviation of 1. There is about a 5% chance that any particular standardized residual exceeds 2 in absolute value, and less than a 1% chance that it exceeds 3 in absolute value. See the margin figure. A large standardized residual provides evidence against independence in that cell. A value exceeding about 3 in absolute value provides strong evidence against independence. A value above +3 suggests that the population proportion for that cell is higher than what independence predicts. A value below −3 indicates that the population proportion for that cell is lower than what independence predicts.

When we inspect many cells in a table, some standardized residuals could be large just by random variation. Values below −3 or above +3, however, are quite convincing evidence of a true effect in that cell.

Standardized residuals

Example 9

Religiosity and Gender

Picture the Scenario

Table 11.16 shows some artificial tables comparing females and males on their religious attendance. Let's look at some actual data on religiosity. Table 11.17 displays observed and expected cell counts and the standardized residuals for the association between gender and response to the question, "To what extent do you consider yourself a religious person?" The possible responses were (very religious, moderately religious, slightly religious, not religious at all). The data are from the 2008 GSS. The table is in the form provided by MINITAB (which actually calls the standardized residuals "adjusted residuals").

Table 11.17 Religiosity by Gender, With Expected Counts and Standardized Residuals

Large positive standardized residuals (in green) indicate strong evidence of a higher population cell proportion than expected under independence. Large negative standardized residuals (in red) indicate strong evidence of a lower population proportion than expected.

Rows: gender Columns: religiosity

	Very	Moderately	Slightly	Not	All	
Female	241	499	208	131	1079	← Counts
	201.8	454.8	251.4	171.0	1079.0	← Expected counts
	4.513	4.016	−4.606	−4.916		← Standardized residuals
Male	133	344	258	186	921	
	172.2	388.2	214.6	146.0	921.0	
	−4.513	−4.016	4.606	4.916		
All	374	843	466	317	2000	

Cell Contents: Count
Expected count
Adjusted residual ← This is the name MINITAB uses for standardized residuals.

Source: Data from CSM, UC Berkeley.

Questions to Explore

a. How would you interpret the standardized residual of 4.513 in the first cell?

b. Interpret the standardized residuals in the entire table.

Think It Through

a. The cell for females who are very religious has observed count = 241, expected count = 201.8, and the standardized residual = 4.513. The difference between the observed and expected counts is more than 3 standard errors. In fact it exceeds 4 in absolute value, which tells us this cell shows a much greater discrepancy than we'd expect if the variables were truly independent. There's strong evidence that the population proportion for that cell (female and very religious) is higher than independence predicts.

b. Table 11.17 exhibits large positive residuals above 4 in four out of eight cells. These are the cells in which the observed count is much larger than the expected count. Those cells had strong evidence of more frequent occurrence than if religiosity and gender were independent. The table exhibits large negative residuals in the other four cells. In these cells, the observed count is much smaller than the expected count. There's strong evidence that the population has fewer subjects in these cells than if the variables were independent.

Insight

The standardized residuals help to describe the pattern of this association: Compared to what we'd expect if religiosity and gender were independent it appears based on this sample, there are more females who are very religious or moderately religious and more males who are slightly religious or not at all religious.

Try Exercise 11.33

In Practice Software Can Find Standardized Residuals

Most software has the option of reporting **standardized residuals.** For instance, they are available in MINITAB (where they are called *adjusted residuals*), in SPSS (where they are called *adjusted standardized residuals*), and at the GSS Web site (by checking Show *Z*-statistic).

11.4 Practicing the Basics

11.33 Standardized residuals for happiness and income TRY The table displays the observed and expected cell counts and the standardized residuals for testing independence of happiness and family income, for GSS data.

```
Rows: income     Columns:      happiness
            not    pretty       very      All
above        26       233        164      423
           66.9     231.1      125.0    423.0
         -6.136     0.206      4.681
average     117       473        293      883
          139.6     482.5      261.0    883.0
         -2.789    -0.859      3.167
below       172       383        132      687
          108.6     375.4      203.0    687.0
          8.193     0.721     -7.337
All         315      1089        589     1993
```

a. How would you interpret the standardized residual of 4.681?

b. Interpret the standardized residuals highlighted in green.

c. Interpret the standardized residuals highlighted in red.

11.34 Happiness and religious attendance The table shows MINITAB output for data from the 2008 GSS on happiness and frequency of attending religious services (1 = at most several times a year, 2 = once a month to nearly every week, 3 = every week to several times a week).

```
Rows: religion    Columns:      happiness
Not too happy   Pretty happy    Very happy
1   201             609            268
  4.057           1.731         -5.107
2    46             224            132
 -2.566           0.456          1.541
3    66             265            196
 -2.263          -2.376          4.385
Cell Contents:  Count
                Adjusted residual
Pearson Chi-Square = 36.445, DF = 4,
P-Value = 0.000
```

a. Based on the chi-squared statistic and P-value, give a conclusion about the association between the variables.

b. The numbers below the counts in the table are standardized residuals. Which cells have strong evidence that in the population there are more subjects than if the variables were independent?

c. Which cells have strong evidence that in the population there are fewer subjects than if the variables were independent?

11.35 Marital happiness and general happiness Table 11.15 showed the association between general happiness and marital happiness. The table shown here gives the standardized residuals for those data, in parentheses.

a. Explain what a relatively small standardized residual such as −0.1 in the second cell represents.

b. Identify the cells in which you'd infer that the population has more cases than would occur if happiness and marital happiness were independent.

Marital Happiness	Happiness: Not Too Happy	Pretty Happy	Very Happy
Not too happy	13 (7.0)	15 (−0.1)	2 (−3.9)
Pretty happy	44 (3.7)	259 (11.5)	40 (−13.8)
Very happy	24 (−6.2)	215 (−11.3)	356 (14.9)

11.36 Happiness and marital status For the 2008 GSS, the table shows cell counts and standardized residuals (in parentheses) for happiness and marital status. Summarize the association by indicating which marital statuses have strong evidence of (i) more and (ii) fewer people in the population in the Very Happy category than if the variables were independent.

Marital Status	Happiness: Very Happy	Pretty Happy	Not Too Happy
Married	398 (10.8)	490 (−3.6)	81 (−8.7)
Widowed	31 (−3.1)	95 (1.0)	37 (2.6)
Divorced	56 (−3.8)	169 (2.1)	55 (2.0)
Separated	11 (−2.6)	41 (0.7)	18 (2.4)
Never married	101 (−6.2)	304 (1.6)	123 (5.7)

Source: Data from CSM, UC Berkeley.

11.37 Gender gap? The table in Exercise 11.1 on gender and party identification is shown again. The largest standardized residuals in absolute value were 2.69 for females who identified as Democrats and −2.69 for males who identified as Democrats. Interpret.

Sex	Political Party		
	Dem	Indep	Repub
F	422	381	273
M	299	365	232

Source: Data from CSM, UC Berkeley.

11.38 Ideology and political party Go to the GSS Web site sda.berkeley.edu/GSS/. For the 2008 survey, construct the 7 × 8 contingency table relating political ideology (POLVIEWS) and party identification (PARTYID).

a. Summarize the results of carrying out the chi-squared test.

b. Find the standardized residuals (by checking Show *z* Statistic). What do you learn from these residuals that you did not learn from the chi-squared test?

11.5 Small Sample Sizes: Fisher's Exact Test

The chi-squared test of independence, like one- and two-sample *z* tests for proportions, is a large-sample test. When the expected frequencies are small, any of them being less than about 5, small-sample tests are more appropriate than the chi-squared test. For 2 × 2 contingency tables, **Fisher's exact test** is a small-sample test of independence.

Fisher's exact test of the null hypothesis that two variables are independent was proposed by the British statistician and geneticist, Sir Ronald Fisher. The test is called *exact* because it uses a sampling distribution that gives exact probability calculations rather than the approximate ones that use the chi-squared distribution.

The calculations for Fisher's exact test are complex and beyond the scope of this text. The principle behind the test is straightforward (see Exercise 11.75), however, and statistical software provides its P-value. The smaller the P-value, the stronger is the evidence that the variables are associated.

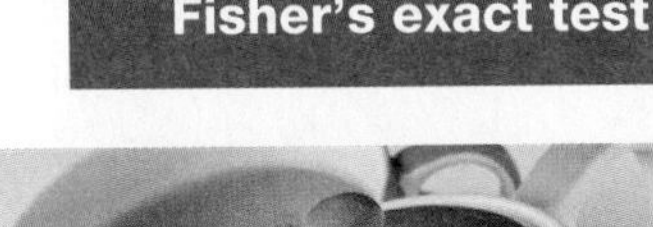

Example 10

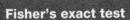

A Tea-Tasting Experiment

Picture the Scenario

In introducing his small-sample test of independence in 1935, Fisher described the following experiment: Fisher enjoyed taking a break in the afternoon for tea. One day, his colleague Dr. Muriel Bristol claimed that when drinking tea she could tell whether the milk or the tea had been added to the cup first (she preferred milk first). To test her claim, Fisher asked her to taste eight cups of tea, four of which had the milk added first and four of which had the tea added first. She was told there were four cups of each type and was asked to indicate which four had the milk added first. The order of presenting the cups to her was randomized.

Table 11.18 shows the result of the experiment. Dr. Bristol predicted three of the four cups correctly that had milk poured first.

Table 11.18 Result of Tea-Tasting Experiment

The table cross-tabulates what was actually poured first (milk or tea) by what Dr. Bristol predicted was poured first. She had to indicate which four of the eight cups had the milk poured first.

Actual	Prediction		Total
	Milk	Tea	
Milk	3	1	4
Tea	1	3	4
Total	4	4	8

For those cases where milk was poured first (row 1), denote p_1 as the probability of guessing that milk was added first. For those cases where tea was poured first, denote p_2 as the probability of guessing that milk was added first. If she really could predict when milk was poured first better than with random guessing, then $p_1 > p_2$. So we shall test $H_0: p_1 = p_2$ (prediction independent of actual pouring order) against $H_a: p_1 > p_2$.

Questions to Explore

a. Show the sample space of the five possible contingency table outcomes that could have occurred for this experiment.

b. Table 11.19 shows the result of using SPSS software to conduct the chi-squared test of the null hypothesis that her predictions were independent of the actual order of pouring. Is this test and its P-value appropriate for these data? Why or why not?

In Words

In statistics, **asymptotic** means "large sample." Mathematically, the theory behind large-sample statistical methods applies as the sample size grows toward infinity "asymptotically."

Table 11.19 Result of Fisher's Exact Test for Tea-Tasting Experiment

The chi-squared P-value is listed under Asymp. Sig. and the Fisher's exact test P-values are listed under Exact Sig. "Sig" is short for *significance* and "asymp." is short for *asymptotic.*

	Value	df	Asymp. Sig. (2-sided)	Exact Sig. (2-sided)	Exact Sig. (1-sided)
Pearson Chi-Square	2.000	1	.157		
Fisher's Exact Test				.486	.243

4 cells (100.0%) have expected count less than 5.

c. Table 11.19 also shows the result for Fisher's exact test of the same null hypothesis. The one-sided version of the test pertains to the alternative that her predictions are better than random guessing. How does the one-sided P-value relate to the probabilities for the five tables in part a? Does this P-value suggest that she had the ability to predict better than random guessing?

Think It Through

a. The possible sample tables have four observations in each row, because there were four cups with milk poured first and four cups with tea poured first. The tables also have four observations in each column, because Dr. Bristol was told there were four cups of each type and was asked to predict which four had the milk added first. Of the four cups where milk was poured first (row 1 of the table), the number she could predict correctly is 4, 3, 2, 1, or 0. These outcomes correspond to the sample tables:

	Guess Poured First									
Poured First	**Milk**	**Tea**	**Milk**	**Tea**	**Milk**	**Tea**	**Milk**	**Tea**	**Milk**	**Tea**
Milk	4	0	3	1	2	2	1	3	0	4
Tea	0	4	1	3	2	2	3	1	4	0

These are the five possible tables with totals of 4 in each row and 4 in each column.

b. The cell counts in Table 11.18 are small. Each cell has expected count $(4 \times 4)/8 = 2$. Since they are less than 5, the chi-squared test is not appropriate. In fact, Table 11.19 warns us that all four cells have expected counts less than 5. The table reports $X^2 = 2.0$ with

Recall

Expected cell count = (row total × column total) divided by total sample size. ◀

P-value = 0.157, but software provides results whether or not they are valid, and it's up to us to know when we can use a method.

c. The one-sided P-value reported by SPSS is for the alternative H_a: $p_1 > p_2$. It equals 0.243. This is the probability, presuming H_0 to be true, of the observed table (Table 11.18) and the more extreme table giving even more evidence in favor of $p_1 > p_2$. That more extreme table is the one with counts (4, 0) in the first row and (0, 4) in the second row, corresponding to guessing all four cups correctly. The P-value of 0.243 does not give much evidence against the null hypothesis.

Insight

This experiment did not establish an association between the actual order of pouring and Dr. Bristol's predictions. If she had predicted all four cups correctly, the one-sided P-value would have been 0.014. We might then believe her claim. With a larger sample, we would not need such extreme results to get a small P-value.

Try Exercise 11.41

	Guess		
Actual	**Milk**	**Tea**	**Total**
Milk	3	—	4
Tea	—	—	4
Total	4	4	8

In part a of Example 10, notice that for the given row and column marginal totals of four each, the count in the first cell determines the other three cell counts. For example, for the actual data (Table 11.18), knowing she guessed three of the four cups correctly that had milk poured first determines the other three cells (see the margin). Because the count in the first cell determines the others, that cell count is regarded as the test statistic for Fisher's exact test. The P-value finds the probability of that count and of the other possible counts for that cell that provide even more evidence in support of H_a. Table 11.19 indicates that the P-value is 0.486 for the two-alternative, H_a: $p_1 \neq p_2$. This alternative hypothesis investigates whether her predictions are better or worse than with random guessing. The P-value of 0.486 also does not give much evidence against the null hypothesis that her predictions are independent of the actual pouring order.

SUMMARY: Fisher's Exact Test of Independence for 2 × 2 Tables

1. **Assumptions:**
   Two binary categorical variables
   Randomization, such as random sampling or a randomized experiment
2. **Hypotheses:**
   H_0: The two variables are independent (H_0: $p_1 = p_2$)
   H_a: The two variables are associated
   (Choose H_a: $p_1 \neq p_2$ or H_a: $p_1 > p_2$ or H_a: $p_1 < p_2$).
3. **Test statistic:** First cell count (this determines the others, given the margin totals).
4. **P-value:** Probability that the first cell count equals the observed value or a value even more extreme than observed in the direction predicted by H_a.
5. **Conclusion:** Report P-value and interpret in context. If a decision is needed, reject H_0 when P-value ≤ significance level (such as 0.05).

The main use of Fisher's exact test is for small sample sizes, when the chi-squared test is not valid. However, Fisher's exact test can be used with *any* sample sizes. It also extends to tables of arbitrary size $r \times c$. The computations are complex, and unfortunately most software currently can perform it only for 2 × 2 tables.

Using the Ordering of Categories

Some categorical variables have a natural ordering of the categories. There is a low end and a high end of the categorical scale. Examples are education attained (less than high school, high school, some college, college degree, advanced degree), appraisal of a company's inventory level (too low, about right, too high), and the happiness and income variables used in this chapter. Such variables are called **ordinal** variables. The chi-squared test of independence treats the categories as unordered. It takes the same value regardless of the order the categories are listed in the table. This is because with any reordering of the categories, the set of row and column totals does not change, and so the expected frequency that goes with a particular observed cell count does not change.

When there is only a weak association, tests that use the ordering information are often more powerful than the chi-squared, giving smaller P-values and stronger evidence against H_0. One way to do this treats the two categorical variables as quantitative: Assign scores to the rows and to the columns, and then use the correlation to describe the strength of the association. A significance test can be based on the size of the correlation. This is beyond our scope here, but the next chapter presents inference methods for two quantitative variables.

11.5 Practicing the Basics

11.39 Keeping old dogs mentally sharp In an experiment with beagles ages 7–11, the dogs attempted to learn how to find a treat under a certain black-colored block and then relearn that task with a white-colored block. The control group of dogs received standard care and diet. The diet and exercise group were given dog food fortified with vegetables and citrus pulp and vitamin E and C supplements plus extra exercise and social play. All 12 dogs in the diet and exercise group were able to solve the entire task, but only 2 of the 8 dogs in the control group could do so. (Background material from N. W. Milgram et al., *Neurobiology of Aging,* vol. 26, 2005, pp. 77–90.)

a. Show how to summarize the results in a contingency table.

b. Conduct all steps of Fisher's exact test of the hypothesis that whether a dog can solve the task is independent of the treatment group. Use the two-sided alternative hypothesis. Interpret.

c. Why is it improper to conduct the chi-squared test for these data?

11.40 Tea-tasting results Consider the tea-tasting experiment of Example 10 and Table 11.18. Consider the possible sample table in which all four of her predictions about the cups that had milk poured first are correct. Using software, find the P-value for the one-sided alternative. Interpret the P-value.

11.41 Claritin and nervousness An advertisement by Schering Corporation for the allergy drug Claritin mentioned that in a pediatric randomized clinical trial, symptoms of nervousness were shown by 4 of 188 patients on Claritin and 2 of 262 patients taking placebo. Denote the population proportion who would show such symptoms by p_1 for Claritin and by p_2 for placebo. The computer printout shows results of significance tests for H_0: $p_1 = p_2$.

a. Report the P-value for the small-sample test, with H_a: $p_1 \neq p_2$. Interpret in the context of this study.

b. Is it appropriate to conduct the chi-squared test for these data? Why or why not?

Analyses of Claritin data

```
Rows: treatment Columns: nervousness

            yes      no

Claritin     4      184
Placebo      2      260

Statistic                        P-Value
Fisher's Exact Test (2-Tail)      0.24
Chi-squared = 1.55                0.21
```

11.42 AIDS and condom use Chatterjee et al. (1995, p. 132) described a study about the effect of condoms in reducing the spread of AIDS. This two-year Italian study followed heterosexual couples where one partner was infected with the HIV virus. Of 171 couples who always used condoms, 3 partners became infected with HIV, while of 55 couples who did not always use condoms, 8 partners became infected. Test whether the rates are significantly different.

a. Define p_1 and p_2 in this context, and specify the null and two-sided alternative hypotheses.

b. For these data, software reports

```
Pearson Chi-Square = 14.704, DF = 1,
P-Value = 0.0001
*Note* 1 cell with expected count less than 5
Fisher's exact test: P-Value = 0.0007
```

Report the result of the test that you feel is most appropriate for these data. For the test you chose, report the P-value and interpret in the context of this study.

Chapter Review

ANSWERS TO THE CHAPTER FIGURE QUESTIONS

Figure 11.1 The higher-income subjects were more likely to be very happy and less likely to be not too happy.

Figure 11.2 The height of the bars is the same for males and females at each level of happiness.

Figure 11.3 The chi-squared statistic can't be negative because it sums squared differences divided by positive expected frequencies.

Figure 11.4 We use only the right tail because larger chi-squared values provide greater evidence against H_0. Chi-squared values in the left tail represent small differences between observed and expected frequencies and do not provide evidence against H_0.

Figure 11.5 Whether we get $\frac{1}{4}$ or 4 for the relative risk depends merely on which group we call group 1 and which we call group 2.

CHAPTER SUMMARY

This chapter showed how to analyze the association between two categorical variables:

- By *describing the counts* in contingency tables using the **conditional distributions** across the categories of the response variable. If the population conditional distributions are identical, the two variables are **independent.**
- By using the **chi-squared** statistic to **test the null hypothesis of independence.**
- By *describing the strength of association with a* **measure of association** such as the **difference of proportions** and the **ratio of proportions** (the **relative risk**). When there is independence, a population difference of proportions equals 0 and a population relative risk equals 1. The stronger the association, the farther the measures fall from these baseline values.
- By *describing the pattern of association* by comparing observed and expected cell counts using **standardized residuals.** A standardized residual reports the number of standard errors that an observed count falls from an expected count. A value larger than about 3 in absolute value indicates that the cell provides strong evidence of association.

The **expected cell counts** are values with the same margins as the observed cell counts but that satisfy the null hypothesis of independence. The **chi-squared test statistic** compares the observed cell counts to the expected cell counts, using

$$X^2 = \sum \frac{(\text{observed count} - \text{expected count})^2}{\text{expected count}}.$$

Under the null hypothesis, the X^2 test statistic has a large-sample **chi-squared distribution.** The **degrees of freedom** depend on the number of rows r and the number of columns c through $df = (r - 1) \times (c - 1)$. The P-value is the right-tail probability above the observed value of X^2.

The expected cell counts for the chi-squared test should be at least 5. **Fisher's exact test** does not have a sample size restriction. It is used to test independence with samples that are too small for the chi-squared test.

The chi-squared statistic can also be used for a hypothesis involving a single categorical variable. For testing a hypothesis that predicts particular population proportion values for each category of the variable, the chi-squared statistic is referred to as a **goodness-of-fit** statistic.

The next chapter introduces methods for describing and making inferences about the association between two *quantitative* variables. We'll learn more about the correlation and regression methods that Chapter 3 introduced.

SUMMARY OF NOTATION

X^2 = chi-squared statistic for testing independence of two categorical variables

CHAPTER PROBLEMS

Practicing the Basics

11.43 Female for president? When recent General Social Surveys have asked, "If your party nominated a woman for president, would you vote for her if she were qualified for the job?" about 94% of females and 94% of males answered yes, the rest answered no. (*Source*: Data from CSM, UC Berkeley.)

a. For males and for females, report the conditional distributions on this response variable in a 2 × 2 table, using outcome categories (yes, no).

b. If results for the entire population are similar to these, does it seem possible that gender and opinion about having a woman president are independent? Explain.

11.44 Down syndrome diagnostic test The table shown, from Example 8 in Chapter 5, cross-tabulates whether a fetus has Down syndrome by whether or not the triple blood diagnostic test for Down syndrome is positive (that is, indicates that the fetus has Down syndrome).

a. Tabulate the conditional distributions for the blood test result, given the true Down syndrome status.

b. For the Down cases, what percentage was diagnosed as positive by the diagnostic test? For the unaffected cases, what percentage got a negative test result? Does the diagnostic test appear to be a good one?

c. Construct the conditional distribution on Down syndrome status, for those who have a positive test result. (*Hint:* You condition on the first column total and find proportions in that column.) Of those cases, what percentage truly have Down syndrome? Is the result surprising? Explain why this probability is small.

	Blood Test Result		
Down Syndrome Status	**Positive**	**Negative**	**Total**
D (Down)	48	6	54
D^c (unaffected)	1307	3921	5228
Total	1355	3927	5282

11.45 Down and chi-squared For the data in the previous exercise, $X^2 = 114.4$. Show all steps of the chi-squared test of independence.

11.46 What is *df*? The contingency table that follows has $df = 4$. Show this, by filling in the missing cell counts.

	A	B	C	D	E	Total
	24	21	12	10	—	100
	—	—	—	—	—	100
Total	40	40	40	40	40	200

11.47 Herbs and the common cold A recent randomized experiment of a multiherbal formula (Immumax) containing echinacea, garlic, ginseng, zinc and vitamin C, was found to improve cold symptoms in adults over a placebo group. "At the end of the study, eight (39%) of the placebo recipients and 18 (60%) of the Immumax recipients reported that the study medication had helped improve their cold symptoms (chi-squared P-value = 0.01)." (M. Yakoot et al., *International Journal of General Medicine*, vol. 4, 2011, pp. 45–51).

a. Identify the response variable and the explanatory variable and their categories for the 2 × 2 contingency table that provided this particular analysis.

b. How would you explain to someone who has never studied statistics how to interpret the parenthetical part of the quoted sentence?

11.48 Happiness and number of friends The table shows GSS data on happiness and the number of close friends, with expected cell counts given underneath the observed counts.

	Happiness			
Number of Close Friends	**Not Too Happy**	**Pretty Happy**	**Very Happy**	**Total**
0–1	34	92	43	169
	19.2	95.1	54.7	
2–5	83	418	200	701
	79.5	394.6	226.9	
6 or more	47	304	225	576
	65.3	324.3	186.4	
Total	164	814	468	1446

a. Suppose the variables were independent. Explain what this means in this context.

b. Explain what is meant by an expected cell count. Show how to get the expected cell count for the first cell, for which the observed count is 34.

c. Compare the expected cell frequencies to the observed counts. Based on this, what is a profile of subjects who tend to be (i) more happy than independence predicts and (ii) less happy than independence predicts.

11.49 Gender gap? Exercise 11.1 showed a 2 × 3 table relating gender and political party identification, shown again here. The chi-squared statistic for these data equals 8.294. Conduct all five steps of the chi-squared test.

	Political Party		
Sex	**Dem**	**Indep**	**Repub**
F	422	381	273
M	299	365	232

11.50 Job satisfaction and income A recent GSS was used to cross-tabulate income (<$15 thousand, $15–25 thousand, $25–40 thousand, >$40 thousand) in dollars with job satisfaction (very dissatisfied, little dissatisfied, moderately satisfied, very satisfied) for 96 subjects.

a. For these data, $X^2 = 6.0$. What is its *df* value, and what is its approximate sampling distribution, if H_0 is true?

b. For this test, the P-value is 0.74. Interpret in the context of these variables.

c. What decision would you make with a 0.05 significance level? Can you accept H_0 and conclude that job satisfaction is independent of income?

11.51 Aspirin and heart attacks for women A study in the *New England Journal of Medicine* compared cardiovascular events for treatments of low-dose aspirin or placebo among 39,876 healthy female health care providers for an average duration of about 10 years. Results indicated that women receiving aspirin and those receiving placebo did not differ for rates of a first major cardiovascular event, death from cardiovascular causes, or fatal or non-fatal heart attacks. However, women receiving aspirin had lower rates of stroke than those receiving placebo (Data from *N. Engl. J. Med.*, vol. 352, 2005, pp. 1293–1304).

Women's Aspirin Study Data

	Cardiovascular Events		
Group	**Mini-Stroke**	**Stroke**	**No Strokes**
Placebo	240	259	19443
Aspirin	185	219	19530

a. Use software to test independence. Show (i) assumptions, (ii) hypotheses, (iii) test statistic, (iv) P-value, (v) conclusion in the context of this study.

b. Describe the association by finding and interpreting the relative risk for the stroke category.

11.52 Women's role A recent GSS presented the statement, "Women should take care of running their homes and leave running the country up to men," and 14.8% of the male respondents agreed. Of the female respondents, 15.9% agreed. Of respondents having less than a high school education, 39.0% agreed. Of respondents having at least a high school education, 11.7% agreed.

a. Report the difference between the proportion of males and the proportion of females who agree.

b. Report the difference between the proportion at the low education level and the proportion at the high education level who agree.

c. Which variable, gender or educational level, seems to have the stronger association with opinion? Explain your reasoning.

11.53 Seat belt helps? The table refers to passengers in autos and light trucks involved in accidents in the state of Maine in a recent year.

a. Use the difference of proportions to describe the strength of association. Interpret.

b. Use the relative risk to describe the strength of association. Interpret.

Maine Accident Data

	Injury	
Seat Belt	**Yes**	**No**
No	3865	27,037
Yes	2409	35,383

Source: Dr. Cristanna Cook, Medical Care Development, Augusta, Maine.

11.54 Race and party ID The table shows data from an SPSS printout for some analyses of 2008 GSS data on race and party ID.

a. Interpret the expected count for the first cell.

b. Interpret the standardized residual of 12.5 for the first cell (SPSS calls it an adjusted residual).

c. How would you summarize to someone who has never studied statistics what you learn from the standardized residuals given in the four corner cells on this printout?

race* party_ID Crosstabulation

			party ID			
			democrat	independent	republican	Total
race	black	Count	192	75	8	275
		Expected Count	100.0	101.5	73.5	275.0
		Adjusted Residual	12.5	−3.6	−9.7	
	white	Count	459	586	471	1516
		Expected Count	551.0	559.5	405.5	1516.0
		Adjusted Residual	−12.5	3.6	9.7	
Total		Count	651	661	479	1791
		Expected Count	651.0	661.0	479.0	1791.0

Source: Data from CSM, UC Berkeley.

11.55 Happiness and sex A contingency table from the 2008 GSS relating happiness to number of sex partners in the previous year (0, 1, at least 2) had standardized residuals as shown in the table. Interpret the highlighted standardized residuals.

Results on Happiness and Sex

```
Rows:   partners  Columns: happy
           not        pretty       very
0          84          235          95
          (3.1)       (0.7)       (-3.3)
1          130         578          381
         (-5.2)      (-2.3)       (6.6)
2          58          160          41
          (3.4)       (2.3)       (-5.2)
```

Source: Data from CSM, UC Berkeley.

11.56 Education and religious beliefs When data from a recent GSS were used to form a 3 × 3 table that cross-tabulates highest degree (1 = less than high school, 2 = high school or junior college, 3 = bachelor or graduate) with religious beliefs (F = fundamentalist, M = moderate, L = liberal), the three largest standardized residuals were: 6.8 for the cell (3, F), 6.3 for the cell (3, L), and 4.5 for the cell (1, F). Summarize how to interpret these three standardized residuals.

11.57 TV and aggression From a study described in Example 5 in the previous chapter, the proportion of males committing aggressive acts was 4 out of 45 for those who watched less than 1 hour of TV a day, and 117 out of 315 for those who watched at least 1 hour per day. MINITAB reports the printout shown for conducting Fisher's exact test (two-sided) for these data.

a. State hypotheses for the test, defining the notation.

b. Report the P-value that software reports, and interpret in the context of this study.

Data on TV watching and aggression

```
Rows: TV  Columns: aggression
            no        yes
high        198       117
low         41        4
Fisher's exact test:
P-Value = 0.0000751
```

11.58 Botox side effects An advertisement for Botox Cosmetic by Allergan, Inc. for treating wrinkles appeared in several magazines. The back page of the ad showed the results of a randomized clinical trial to compare 405 people receiving Botox injections to 130 people receiving placebo in terms of the frequency of various side effects. The side effect of pain in the face was reported by 9 people receiving Botox and 0 people receiving placebo. Use software to conduct a two-sided test of the hypothesis that the probability of this side effect is the same for each group. Report all steps of the test, and interpret results.

Concepts and Applications

11.59 Student data Refer to the FL Student Survey data file on the text CD. Using software, create and analyze descriptively and inferentially the contingency table relating religiosity and belief in life after death. Summarize your analyses in a short report.

11.60 Marital happiness decreasing? At sda.berkeley.edu/GSS, cross-tabulate HAPMAR with YEAR, so you can see how conditional distributions on marital happiness in the GSS have changed since 1973. Using conditional distributions and standardized residuals, explain how these results show a very slight trend over time for fewer people to report being very happy. (You can get the standardized residuals by checking Show z-statistic at the Web site.)

11.61 Another predictor of happiness? Go to sda.berkeley.edu/GSS and find a variable that is associated with happiness, other than variables used in this chapter. Use methods from this chapter to describe and make inferences about the association, in a one-page report.

11.62 Pregnancy associated with contraceptive use? Whether or not a young married woman becomes pregnant in the next year is a categorical variable with categories (yes, no). Another categorical variable to consider is whether she and her partner use contraceptives with categories (yes, no). Would you expect these variables to be independent, or associated? Explain.

11.63 Babies and gray hair A young child wonders what causes women to have babies. For each woman who lives on her block, she observes whether her hair is gray and whether she has young children, with the results shown in the table that follows.

a. Construct the 2×2 contingency table that cross-tabulates gray hair (yes, no) with has young children (yes, no) for these nine women.

b. Treating has young children as the response variable, obtain the conditional distributions for those women who have gray hair and for those who do not. Does there seem to be an association?

c. Noticing this association, the child concludes that not having gray hair is what causes women to have children. Use this example to explain why association does not necessarily imply causation.

Woman	Gray Hair	Young Children
Andrea	No	Yes
Mary	Yes	No
Linda	No	Yes
Jane	No	Yes
Maureen	Yes	No
Judy	Yes	No
Margo	No	Yes
Carol	Yes	No
Donna	No	Yes

11.64 When is chi-squared not valid? Give an example of a contingency table for which the chi-squared test of independence should not be used.

11.65 Gun homicide in United States and Britain According to recent United Nations figures, the annual gun homicide rate is 62.4 per one million residents in the United States and 1.3 per one million residents in Britain.

a. Show how to compare the proportion of residents of the two countries killed annually by guns using the difference of proportions. Show how the results differ according to whether the United States or Britain is identified as Group 1.

b. Show how to compare the proportion of residents of the two countries using the relative risk. Show how the results differ according to whether the United States or Britain is identified as Group 1.

c. When both proportions are very close to 0, as in this example, which measure do you think is more useful for describing the strength of association? Why?

11.66 Colon cancer and race The State Center for Health Statistics for the North Carolina Division of Public Health released a report in 2010 that indicates that there are racial disparities in colorectal cancer incidence and mortality rates. The report states, "African Americans are less likely to receive appropriate screenings that reduce the risk of developing or dying from colorectal cancer." During 2002–2006, the rate of incidence for African Americans was 57.3 per 100,000 versus 46.5 per 100,000 for White residents (www.schs.state.nc.us/SCHS). African Americans were 19% more likely to have been diagnosed with colon cancer. Explain how to get this estimate.

11.67 True or false: $X^2 = 0$ The null hypothesis for the test of independence between two categorical variables is H_0: $X^2 = 0$, for the sample chi-squared statistic X^2. (*Hint:* Do hypotheses refer to a sample or the population?)

11.68 True or false: Group 1 becomes Group 2 Interchanging two rows or interchanging two columns in a contingency table has no effect on the value of the X^2 statistic.

11.69 True or false: Statistical but not practical significance Even when the sample conditional distributions in a contingency table are only slightly different, when the sample size is very large it is possible to have a large X^2 statistic and a very small P-value for testing H_0: independence.

11.70 Statistical versus practical significance In any significance test, when the sample size is very large, we have not necessarily established an important result when we obtain statistical significance. Explain what this means in the context of analyzing contingency tables with a chi-squared test.

11.71 Normal and chi-squared with $df = 1$ When $df = 1$, the P-value from the chi-squared test of independence is the same as the P-value for the two-sided test comparing two proportions with the z test statistic. This is because of a direct connection between the standard normal distribution and the chi-squared distribution with $df = 1$: Squaring a z-score yields a chi-squared value with $df = 1$ having chi-squared right-tail probability equal to the two-tail normal probability for the z-score.

a. Illustrate this with $z = 1.96$, the z-score with a two-tail probability of 0.05. Using the chi-squared table or software, show that the square of 1.96 is the chi-squared score for $df = 1$ with a P-value of 0.05.

b. Show the connection between the normal and chi-squared values with P-value = 0.01.

11.72 Multiple response variables Each subject in a sample of 100 men and 100 women is asked to indicate which of the following factors (one or more) are responsible for increases in crime committed by teenagers: A—the increasing gap in income between the rich and poor, B—the increase in the percentage of single-parent families, C—insufficient time that parents spend with their children. To analyze whether responses differ by gender of respondent, we cross-classify the responses by gender, as the table shows.

a. Is it valid to apply the chi-squared test of independence to these data? Explain.

b. Explain how this table actually provides information needed to cross-classify gender with each of three variables. Construct the contingency table relating gender to opinion about whether factor A is responsible for increases in teenage crime.

Three Factors for Explaining Teenage Crime

Gender	A	B	C
Men	60	81	75
Women	75	87	86

11.73 Standardized residuals for 2 × 2 tables The table that follows shows the standardized residuals in parentheses for GSS data about the statement, "Women should take care of running their homes and leave running the country up to men." The absolute value of the standardized residual is 13.2 in every cell. For chi-squared tests with 2 × 2 tables, since $df = 1$, only one nonredundant piece of information exists about whether an association exists. If observed count > expected count in one cell, observed count < expected count in the other cell in that row or column. Explain why this is true, using the fact that observed and expected counts have the same row and column totals. (In fact, in 2 × 2 tables, all four standardized residuals have absolute value equal to the square root of the X^2 test statistic.)

Year	Agree	Disagree
1974	509 (13.2)	924 (−13.2)
1998	280 (−13.2)	1534 (13.2)

11.74 Variability of chi-squared For the chi-squared distribution, the mean equals *df* and the standard deviation equals $\sqrt{2(df)}$.

a. Explain why, as a rough approximation, for a large *df* value, 95% of the chi-squared distribution falls within $df \pm 2\sqrt{2(df)}$.

b. With $df = 8$, show that $df \pm 2\sqrt{2(df)}$ gives the interval (0, 16) for *approximately* containing 95% of the distribution. Using the chi-squared table, show that *exactly* 95% of the distribution actually falls between 0 and 15.5.

11.75 Explaining Fisher's exact test A pool of six candidates for three managerial positions includes three females and three males. Denote the three females by F1, F2, F3 and the three males by M1, M2, M3. The result of choosing three individuals for the managerial positions is (F2, M1, M3).

a. Identify the 20 possible samples that could have been selected. Explain why the contingency table relating gender to whether chosen for the observed sample is

	Chosen For Position	
Gender	Yes	No
Male	2	1
Female	1	2

b. Let $\hat{p}_1$ denote the sample proportion of males selected and $\hat{p}_2$ the sample proportion of females. For the observed table, $\hat{p}_1 - \hat{p}_2 = (2/3) - (1/3) = 1/3$. Of the 20 possible samples, show that 10 have $\hat{p}_1 - \hat{p}_2 \geq 1/3$. (Note that, if the three managers were randomly selected, the probability would equal $10/20 = 0.50$ of obtaining $\hat{p}_1 - \hat{p}_2 \geq 1/3$. This is the reasoning that provides the one-sided P-value for Fisher's exact test with $H_a: p_1 > p_2$.)

11.76 Likelihood-ratio chi-squared For testing independence, most software also reports another chi-squared statistic, called **likelihood-ratio chi-squared.** It equals

$$G^2 = 2\sum\left[\text{observed count} \times \log\left(\frac{\text{observed count}}{\text{expected count}}\right)\right].$$

It has similar properties as the X^2 statistic, such as $df = (r - 1) \times (c - 1)$.

a. Show that $G^2 = X^2 = 0$ when each observed count = expected count.

b. Explain why in practice you would not expect to get *exactly* $G^2 = X^2 = 0$, even if the variables are truly independent.

Student Activities

11.77 Conduct a research study using the GSS Go to the GSS codebook at sda.berkeley.edu/GSS. Your instructor will assign a categorical response variable. Conduct a research study in which you find at least two other categorical variables that have both a statistically significant and a practically significant association with that response variable, using data for the most recent year in which the response variable was surveyed. Analyze each association using the methods of this chapter. Write a two-page report, describing your analyses and summarizing the results of your research study. Be prepared to discuss your results in a class discussion, because the class will discuss which variables seemed to be especially relevant.

BIBLIOGRAPHY

Alan Agresti (2007). *An Introduction to Categorical Data Analysis,* 2nd edition. Hoboken, New Jersey: Wiley.

Samprit Chatterjee, Mark S. Handcock, and Jeffrey S. Simonoff (1995). A Case Book for *A First Course in Statistics and Data Analysis.* Hoboken, New Jersey: Wiley.

12 Analyzing the Association Between Quantitative Variables: Regression Analysis

Example 1

Estimate a Person's Strength

Picture the Scenario

How can you measure a person's strength? One way is to find the *maximum* number of pounds that the individual can bench press. However, this technique can be risky for people who are unfamiliar with proper lifting techniques or who are inexperienced in using a bench press. Is there a variable that is easier to measure yet is a good predictor of the maximum bench press?

There have been many studies about strength using males but relatively few using females. One exception was a recent study of 57 female athletes in a Georgia high school. Several variables were measured, including ones that are easier and safer to assess than maximum bench press but are thought to correlate highly with it. One such variable is the number of times that a girl can lift a bench press set at only 60 pounds (a relatively low weight) before she becomes too fatigued to lift it again. The data are in the High School Female Athletes data file on the text CD. Let x = number of 60-pound bench presses performed (before fatigue) and let y = maximum bench press.

Questions to Explore

- How well can we predict an athlete's maximum bench press from knowing the number of 60-pound bench presses that she can perform?
- What can we say about the association between these variables in the population?

Thinking Ahead

The bench press variables are quantitative. This chapter presents methods for analyzing the association between two quantitative variables. Like the methods presented in the previous chapter for categorical variables, these methods enable us to answer questions such as:

- Could the variables x and y realistically be independent (in the population), or can we conclude that there is an association between them?
- If the variables are associated, how strong is the association?
- How does the outcome for the response variable depend on the value of the explanatory variable, and which observations are unusual?

The analyses that address these questions are collectively called a **regression analysis**. We'll conduct regression analyses of the data from the female athlete strength study using different variables in several examples in this chapter (Examples 2, 3, 6, 9, 11, 12, 15, and 16).

Chapter 3 introduced regression analysis. Here we'll review and learn more about using a **regression line** to predict the response variable y and the **correlation** to describe the strength of association. We'll then learn how to make inferences about the regression line for a population and how the variability of data points around the regression line helps us to predict how far from the line a value of y is likely to fall. Finally, we'll study an alternative regression analysis that is useful when two quantitative variables have a curved rather than a straight-line relationship.

Recall

From Section 3.3, a **regression line** is a straight line that predicts the value of a response variable y from the value of an explanatory variable x. From Section 3.2, the **correlation** is a summary measure of association that falls between −1 and +1. ◀

12.1 Model How Two Variables Are Related

In Words

y = response variable
Graph on vertical (y) axis.

x = explanatory variable
Graph on horizontal (x) axis.
Regression uses x to predict y.

As in most statistical analyses, the first step of a regression analysis is to identify the response and explanatory variables. In Example 1 we want to know how well we can predict an athlete's maximum bench press, so that's the response variable. The predictor, the number of 60-pound bench presses that an athlete can manage, is the explanatory variable. As in Chapter 3, we use y to denote the response variable and x to denote the explanatory variable.

The Scatterplot: Evaluating if a Straight-Line Trend Exists

Is there an association between the number of 60-pound bench presses and the maximum bench press? The first step in answering this question is to look at the data. Recall that a **scatterplot** shows, for each observation, a point representing its value on the two variables. The points are shown relative to the x (horizontal) and y (vertical) axes. The scatterplot shows if there is roughly a straight-line trend, in which case the relationship between x and y is said to be approximately **linear**.

Example 2

Scatterplot

The Strength Study

Picture the Scenario

Another variable in the strength study discussed in Example 1 measured the maximum bench press by giving a girl a 60-pound weight to lift and then increasing the weight in 5-pound increments until the girl could no longer lift it. The response outcome is the maximum weight that the girl bench pressed. Let's look at the High School Female Athletes data file on the text CD. The first four entries of the 57 lines for x = the number of 60-pound bench presses and y = maximum bench press show the values:

x	y
10	80
12	85
20	85
5	65

For the 57 girls in this study, these variables are summarized by

x: mean = 11.0, standard deviation = 7.1

y: mean = 79.9 lbs., standard deviation = 13.3 lbs.

Question to Explore

Using the entire data file for the 57 athletes, construct a scatterplot, and interpret it.

Think It Through

With any statistical software, you can construct a scatterplot after identifying in the data file the variables that play the role of x and y. Figure 12.1 shows the scatterplot that MINITAB produces. It shows that female athletes with higher numbers of 60-pound bench presses also tended to have higher values for the maximum bench press. The data points follow roughly an increasing linear trend. The margin shows a screen shot of the scatterplot from the TI-83+/84.

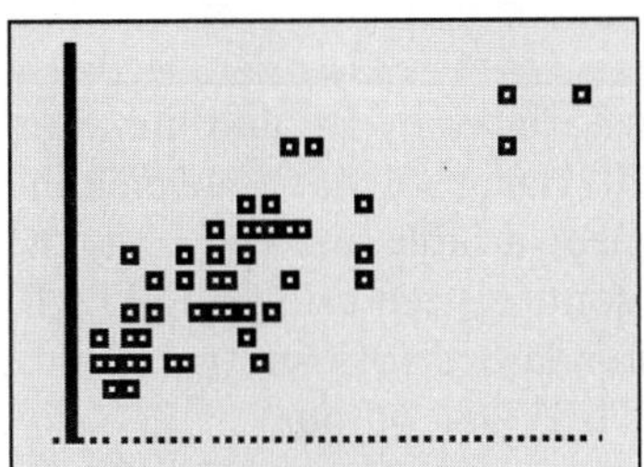

TI-83+/84 output

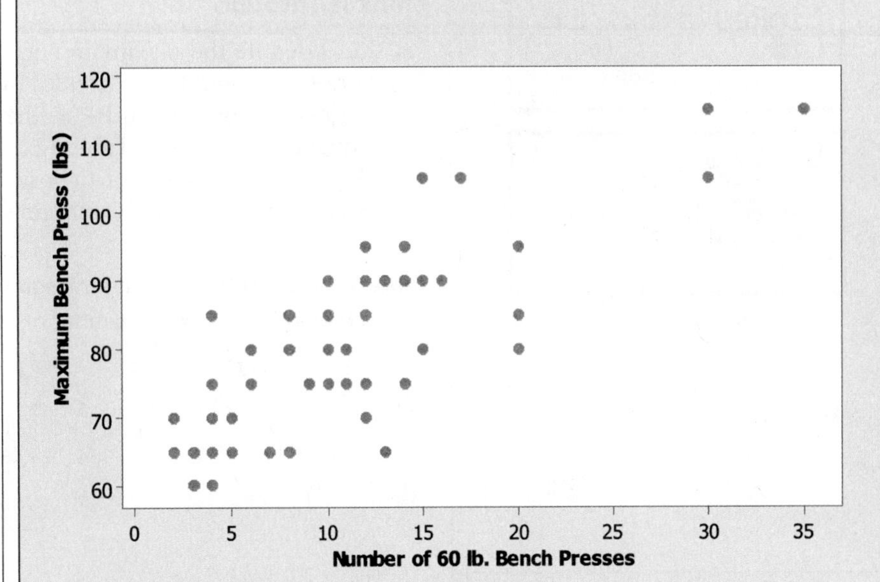

▲ **Figure 12.1** Scatterplot for y = Maximum Bench Press and x = Number of 60-Pound Bench Presses. **Question** How can you tell from this plot that (a) two athletes had y = maximum bench press of only 60 pounds, (b) 5-pound increments were used in determining maximum bench press—60, 65, 70, and so on?

Insight

The three data points at the upper right represent subjects who could do a large number of 60-pound bench presses. Their maximum bench presses were also high.

Try Exercise 12.9, part a

The Regression Line Equation Uses *x* to Predict *y*

In Words

$\hat{y}$ ("y-hat") denotes the predicted value you get for the response variable y by plugging a value into the equation for the explanatory variable x.

When the scatterplot shows a linear trend, a straight line fitted through the data points describes that trend. As in Chapter 3, we use the notation

$$\hat{y} = a + bx$$

for this line, called the **regression line**. The symbol $\hat{y}$ represents the *predicted value* of the response variable y. In the formula, a is the **y-intercept** and b is the **slope**.

Example 3

Regression line

Regression Line Predicting Maximum Bench Press

Picture the Scenario

Let's continue our analysis of the High School Female Athletes data.

Questions to Explore

a. Using software, find the regression line for y = maximum bench press and x = number of 60-pound bench presses.

b. Interpret the slope by comparing the predicted maximum bench press for subjects at the highest and lowest levels of x in the sample (35 and 2).

```
LinReg
 y=a+bx
 a=63.537
 b=1.491
```

TI-83+/84 output

Think It Through

a. Let's denote the maximum bench press variable by BP and the number of 60-pound bench presses by BP_60. Using software, we pick the regression option with BP as the response variable and BP_60 as the explanatory variable. Table 12.1 shows some output using MINITAB. We'll interpret some of this, such as standard errors and t statistics, later in the chapter. The margin shows TI-83+/84 output.

Table 12.1 MINITAB Printout for Regression Analysis of y = Maximum Bench Press (BP) and x = Number of 60-Pound Bench Presses (BP_60)

The regression equation is BP = 63.5 + 1.49 BP_60

Predictor	Coef	SE Coef	T	P
Constant	63.537	1.956	32.48	0.000
BP_60	1.4911	0.1497	9.96	0.000

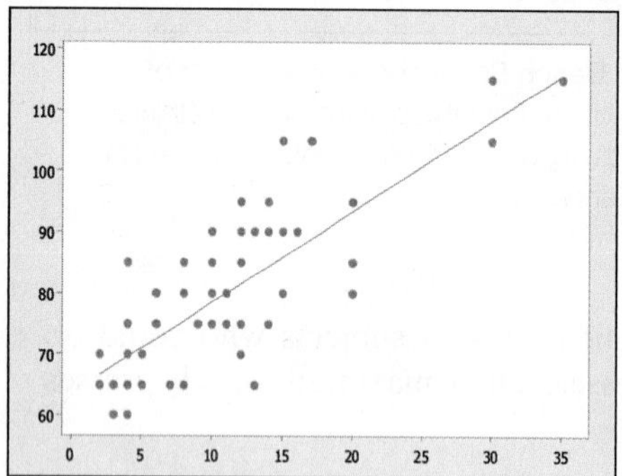

The output tells us that $\hat{y}$ = predicted maximum bench press (BP) relates to x = number of 60-pound bench presses (BP_60) by

$$\text{BP} = 63.5 + 1.49\ \text{BP\_60}, \text{ that is, } \hat{y} = 63.5 + 1.49x.$$

The y-intercept is 63.5 and the slope is 1.49. These are also shown in the column labeled "Coef," an abbreviation for *coefficient.* The slope appears opposite the variable name for which it is the coefficient, "BP_60." The predictor "Constant" refers to the y-intercept. The margin figure (Figure 12.1 reproduced) plots the regression line through the scatterplot.

b. The slope of 1.49 tells us that the predicted maximum bench press $\hat{y}$ increases by an average of 1 1/2 pounds for every additional 60-pound bench press an athlete can do. The impact on $\hat{y}$ of a 33-unit change in x, from the sample minimum of $x = 2$ to the maximum of $x = 35$, is $33(1.49) = 49.2$ pounds. An athlete who can do thirty-five 60-pound bench presses has a predicted maximum bench press nearly 50 pounds higher than an athlete who can do only two 60-pound bench presses. Those predicted values are $\hat{y} = 63.5 + 1.49(2) = 66.5$ pounds at $x = 2$ and $\hat{y} = 63.5 + 1.49(35) = 115.7$ pounds at $x = 35$.

Insight

The slope of 1.49 is positive: As x increases, the predicted value $\hat{y}$ increases. The association is *positive.* When the association is *negative*, the predicted value $\hat{y}$ *decreases* as x increases. When the slope = 0, the regression line is horizontal.

Try Exercise 12.1

Recall

The direction of the association (positive or negative) refers to the sign of the slope and whether the line slopes upward or downward. ◀

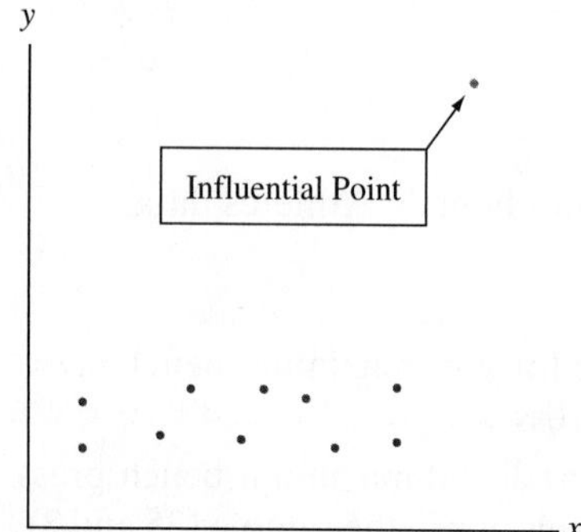

One reason for plotting the data before you do a regression analysis is to check for outliers. The regression line can be pulled toward an outlier and away from the general trend of points. We observed in Section 3.4 that an observation can be **influential** in affecting the regression line when two things happen (see margin figure):

- Its x value is low or high compared to the rest of the data.
- It does not fall in the straight-line pattern that the rest of the data have.

In Figure 12.1, three subjects had quite large values of x. However, their values of y fit in with the trend exhibited by the other data, so those points are not influential with respect to changing the equation of the regression line. However, as we will discuss later, the correlation may strengthen.

Residuals Are Prediction Errors for the Least Squares Line

The regression equation is often called a **prediction equation**, because substituting a particular value of x into the equation provides a prediction for y at that value of x. For instance, the third athlete in the data file could do x = twenty 60-pound bench presses. Her predicted maximum bench press is

$$\hat{y} = 63.5 + 1.49x = 63.5 + 1.49(20) = 93.3 \text{ pounds.}$$

The difference $y - \hat{y}$ between an observed outcome y and its predicted value $\hat{y}$ is the *prediction error*, called a **residual**. For instance, the third athlete in the data file had a maximum bench press of $y = 85$ pounds. Since her $\hat{y} = 93.3$ pounds, the prediction error is $y - \hat{y} = 85 - 93.3 = -8.3$.

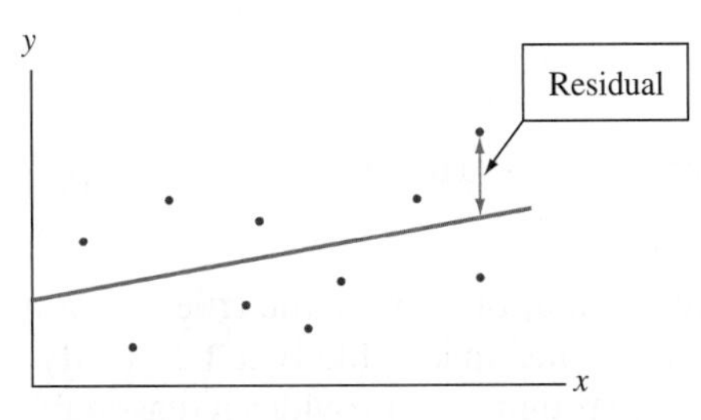

SUMMARY: Review of Residuals from Chapter 3

- Each observation has a residual. Some are positive, some are negative, some may be zero, and their average equals 0.
- In the scatterplot, a residual is the vertical distance between the data point and the regression line. The smaller the distance, the better the prediction. (See margin figure.)
- We can summarize how near the regression line the data points fall by

$$\text{sum of squared residuals} = \Sigma\,(\text{residual})^2 = \Sigma(y - \hat{y})^2.$$

- The regression line has a smaller sum of squared residuals than any other line. It is called the **least squares** line, because of this property.

Regression Model: A Line Describes How the Mean of *y* Depends on *x*

At a given value of x, the equation $\hat{y} = a + bx$ predicts a single value $\hat{y}$ of the response variable. However, we should not expect all subjects at that value of x to have the same value of y. Variability occurs in their y values.

For example, let x = number of years of education and y = annual income in dollars for the adult residents in the workforce of your hometown. For a random sample, suppose you find $\hat{y} = -20{,}000 + 4000x$. Those workers with $x = 12$ years of education have predicted annual income

$$\hat{y} = -20{,}000 + 4000(12) = 28{,}000.$$

It's not the case that every worker with 12 years of education would have annual income \$28,000. Income is not completely dependent upon education. One person may have income \$35,000, another \$15,000, and so forth. Instead, you can think of $\hat{y} = 28{,}000$ as estimating the mean annual income for all workers with $x = 12$.

Likewise, there's a mean of the annual income values at each separate education value. For those residents with 13 years of education, the estimated mean annual income is $-20{,}000 + 4000(13) = 32{,}000$. The slope is 4000, so the estimated mean goes up by \$4000 (from \$28,000 to \$32,000) for this one-year increase in education (from 12 to 13).

SUMMARY: The Regression Line Connects the Estimated Means of y at the Various x Values

$\hat{y} = a + bx$ describes the relationship between x and the estimated means of y at the various values of x.

Recall

Parameters describe the population. They are often denoted by Greek symbols. Here, α is *alpha*, β is *beta*, and μ_y is *mu-sub-y*. ◀

A similar equation describes the relationship in the population between x and the means of y. This **population regression equation** is denoted by

$$\mu_y = \alpha + \beta x.$$

Here, α is a population y-intercept and β is a population slope. These are parameters, so in practice their values are unknown. The parameter μ_y denotes the population mean of y for all the subjects at a particular value of x. It takes a different value at each separate x value. In practice, we estimate the population regression equation using the prediction equation for the sample data. We'll see how to conduct inference about the unknown parameters in Sections 12.3 and 12.4.

A straight line is the simplest way to describe the relationship between two quantitative variables. In practice, most relationships are not *exactly* linear. The equation $\mu_y = \alpha + \beta x$ merely *approximates* the actual relationship between x and the population means of y. It is a **model**.

Model

A **model** is a simple approximation for how variables relate in a population.

Figure 12.2 shows how a regression model can approximate the true relationship between x and the mean of y. The true relationship is unlikely to be exactly a straight line. That's not a problem, as long as a straight line provides a reasonably good approximation, as we expect of a model.

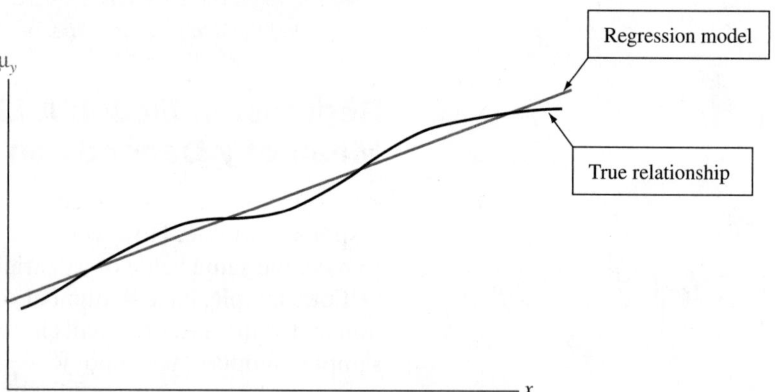

▲ **Figure 12.2 The Regression Model $\mu_y - \alpha + \beta x$ for the Means of y Is a Simple Approximation for the True Relationship. Question** Can you sketch a true relationship for which this model is a very *poor* approximation?

If the true relationship is far from a straight line, this regression model may be a poor one. Figure 12.3 illustrates. In that case, the figure shows a U-shaped relationship. It requires a parabolic curve rather than a straight line to describe it well. To check the straight-line assumption, you should always construct a scatterplot.

How could you get into trouble by using a straight-line regression model even when the true relationship is U-shaped? First, predictions about y would be poor. Second, inference about the association could be misleading. The variables are

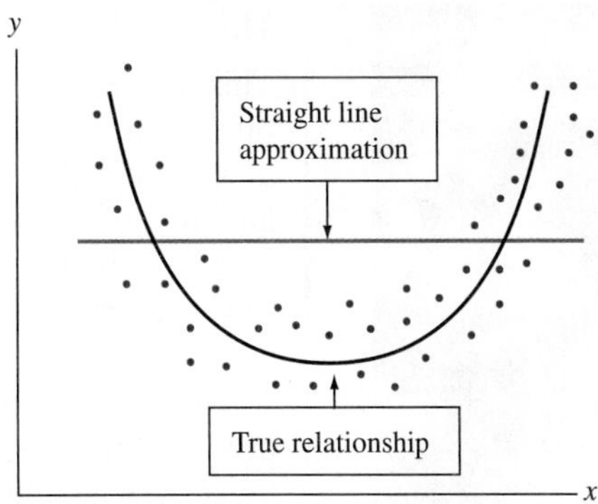

▲ **Figure 12.3 The Straight-Line Regression Model Provides a Poor Approximation When the Actual Relationship Is Highly Nonlinear.** Question What type of mathematical function might you consider using for a regression model in this case?

associated, since the mean of y changes as x does. The slope (and correlation) might not detect this, however, since the prediction line would be close to horizontal even though a substantial association exists.

The Regression Model Also Allows Variability About the Line

Recall

Chapter 11 used a **conditional distribution** to describe percentages in categories of a categorical response variable, at each value of an explanatory variable. Each conditional distribution refers to possible values of the response variable at a fixed value of the explanatory variable. ◀

At each fixed value of x, variability occurs in the y values around their mean μ_y. For instance, at education = 12 years, annual income varies around the mean annual income for all workers with 12 years of education. The probability distribution of y values at a fixed value of x is a **conditional distribution**. At each value of x, there is a conditional distribution of y values. A regression model also describes these distributions. An additional parameter σ describes the standard deviation of each conditional distribution.

In Words

A **regression model** uses a formula (usually a straight line) to approximate how the expected value (the mean) for y changes at different values of x. It also describes the variability of observations on y around the line.

SUMMARY: Regression Model

A **regression model** describes how the population mean μ_y of each conditional distribution for the response variable depends on the value x of the explanatory variable. A **straight-line regression model** uses the line $\mu_y = \alpha + \beta x$ to connect the means. The model also has a parameter σ that describes variability of observations around the mean of y at each x value.

Example 4

Regression model

Income and Education

Picture the Scenario

As described previously, suppose the regression line $\mu_y = -20{,}000 + 4000x$ models the relationship for the population of working adults in your hometown between x = number of years of education and the mean of y = annual income. This model tells us that income goes up as education does, on average, but how much variability is there? Suppose also that the conditional distribution of annual income at each value of x is modeled by a normal distribution, with standard deviation $\sigma = 13{,}000$.

Question to Explore

Use this regression model to describe the mean and variability around the mean for the conditional distributions of income at education values 12 and 16 years.

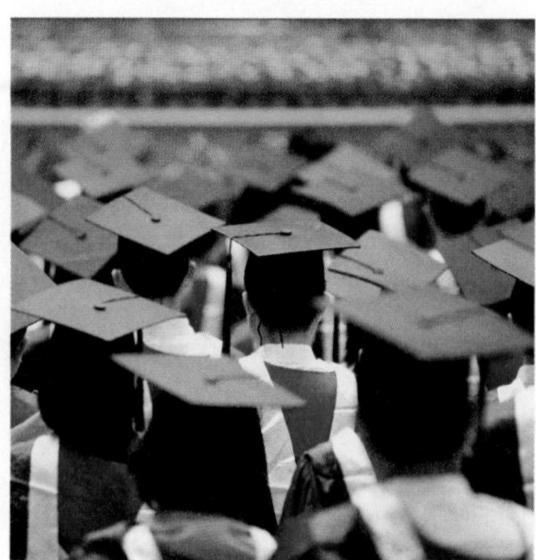

Think It Through

This regression model states that for workers with x years of education, their annual incomes have a normal distribution with a mean of $\mu_y = -20{,}000 + 4000x$ and a standard deviation of $\sigma = 13{,}000$. For those having a high school education ($x = 12$), the mean annual income is $\mu_y = -20{,}000 + 4000(12) = 28{,}000$ and the standard deviation is 13,000. Those with a college education ($x = 16$) have a mean annual income of $\mu_y = -20{,}000 + 4000(16) = 44{,}000$ and a standard deviation of 13,000.

Since the conditional distributions are modeled as normal, the model predicts that nearly all of the y values fall within 3 standard deviations of the mean. For instance, for those with 16 years of education, $\mu_y = 44{,}000$ and $\sigma = 13{,}000$. Since

$$44{,}000 - 3(13{,}000) = 5000 \text{ and } 44{,}000 + 3(13{,}000) = 83{,}000$$

the model predicts that nearly all of their annual incomes fall between \$5000 and \$83,000.

Insight

Figure 12.4 portrays this regression model. It plots the regression equation and the conditional distributions of y = income at $x = 12$ years and at $x = 16$ years of education. We'll see how to use data to estimate the variability, as described by the standard deviation σ, later in the chapter.

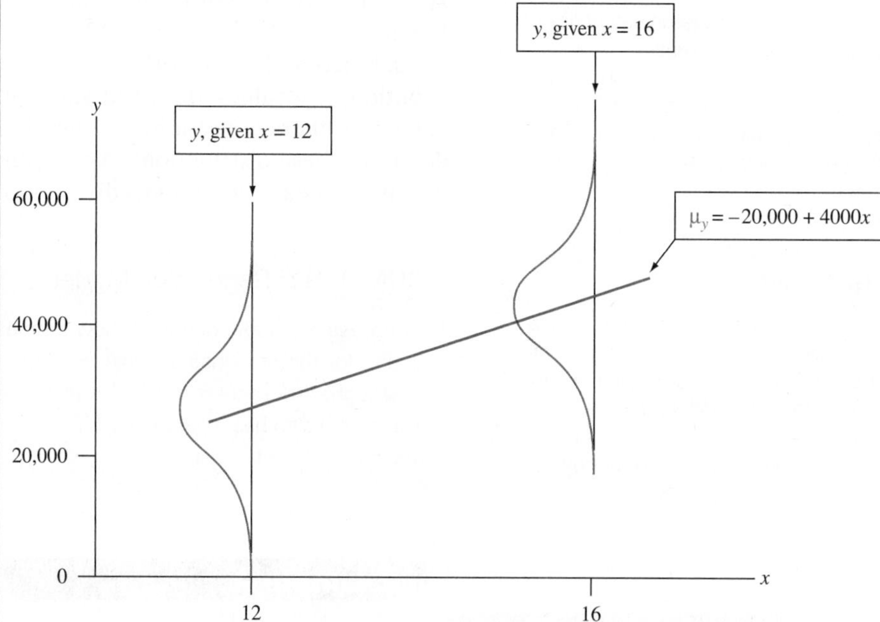

▲ **Figure 12.4 The Regression Model $\mu_y = -20{,}000 + 4000x$, with $\sigma = 13{,}000$, Relating the Means of y = Annual Income to x = Years of Education.** Question What do the bell-shaped curves around the line at $x = 12$ and at $x = 16$ represent?

Try Exercise 12.4

In Figure 12.4, each conditional distribution is normal, and each has the same standard deviation, $\sigma = 13{,}000$. But a model merely approximates reality. In practice, the means of the conditional distributions would not perfectly follow a straight line. Those conditional distributions would not be exactly normal. The standard deviation would not be exactly the same for each conditional distribution. For instance, the conditional distributions might be somewhat skewed to the right, and incomes might vary more for college graduates than for high school graduates.

In Practice A Model Is a Simple *Approximation* of Reality

A statistical **model** never holds exactly in practice. It is merely a simple approximation of reality. But even though it does not describe reality exactly, a model is useful if the true relationship is close to what the model predicts.

12.1 Practicing the Basics

12.1 **Car mileage and weight** The Car Weight and Mileage data file on the text CD shows the weight (in pounds) and mileage (miles per gallon) of 25 different model autos.

TRY

a. Identify the natural response variable and explanatory variable.

b. The regression of mileage on weight has MINITAB regression output

```
Predictor        Coef     SE Coef      T      P
Constant       45.645       2.603  17.54  0.000
weight     -0.0052220   0.0006271  -8.33  0.000
```

State the prediction equation and report the y-intercept and slope.

c. Interpret the slope in terms of a 1000-pound increase in the vehicle weight.

d. Does the y-intercept have any contextual meaning for these data? (*Hint:* The weight values range between 2460 and 6400 pounds.)

12.2 **Predicting car mileage** Refer to the previous exercise.

a. Find the predicted mileage for the Toyota Corolla, which weighs 2590 pounds.

b. Find the residual for the Toyota Corolla, which has observed mileage of 38.

c. Sketch a graphical representation of the residual in part b.

12.3 **Predicting maximum bench strength in males** For the Male Athlete Strength data file on the text CD, the prediction equation relating y = 1 repetition maximum bench press (1RMBP) in kilograms to x = repetitions to fatigue bench press (RTFBP) is $\hat{y} = 117.5 + 5.86x$.

a. Find the predicted 1RMBP for a male athlete with a RTFBP of 35, which was one of the highest RTFBP values.

b. Find the predicted 1RMBP for a male athlete with an RTFBP of 0, which was the lowest RTFBP value.

c. Interpret the y-intercept. Use the slope to describe how predicted 1RMBP changes as RTFBP increases from 0 to 35.

12.4 **Income higher with age** Suppose the regression line $\mu_y = -10{,}000 + 1000x$ models the relationship for the population of working adults in Canada between x = age and the mean of y = annual income (in Canadian dollars). The conditional distribution of y at each value of x is modeled as normal, with $\sigma = 5000$. Use this regression model to describe the mean and the variability around the mean for the conditional distribution at age (a) 20 years and (b) 50 years.

TRY

12.5 **Mu, not *y*** For a population regression equation, why is it more sensible to write $\mu_y = \alpha + \beta x$ instead of $y = \alpha + \beta x$? Explain with reference to the variables x = height and y = weight for the population of girls in elementary schools in your hometown.

12.6 **Parties and dating** Let y = number of parties attended in the past month and x = number of dates in the past month, measured for all single students at your school. Explain the mean and variability aspects of the regression model $\mu_y = \alpha + \beta x$, in the context of these variables. In your answer, explain why (a) it is more sensible to use a straight line to model the *means* of the conditional distributions rather than individual observations and (b) the model needs to allow variation around the mean.

12.7 **Study time and college GPA** Exercise 3.39 in Chapter 3 showed data collected at the end of an introductory statistics course to investigate the relationship between x = study time per week (average number of hours) and y = college GPA. The table here shows the data for the eight males in the class on these variables and on the number of class lectures for the course that the student reported skipping during the term.

a. Create a data file and use it to construct a scatterplot between x and y. Interpret.

b. Find the prediction equation and interpret the slope.

c. Find the predicted GPA for a student who studies 25 hours per week.

d. Find and interpret the residual for Student 2, who reported $x = 25$.

Student	Study Time	GPA	Skipped
1	14	2.8	9
2	25	3.6	0
3	15	3.4	2
4	5	3.0	5
5	10	3.1	3
6	12	3.3	2
7	5	2.7	12
8	21	3.8	1

12.8 **GPA and skipping class** Refer to the previous exercise. Now let x = number of classes skipped and y = college GPA.

a. Construct a scatterplot. Does the association seem to be positive or negative?

b. Find the prediction equation and interpret the y-intercept and slope.

c. Find the predicted GPA and residual for Student 1.

12.9 Predicting college GPA For the Georgia Student Survey data file on the text CD, look at college GPA and high school GPA.

TRY

a. Identify the response and explanatory variables and construct a scatterplot. What is the effect on this plot of several students having high school GPAs of exactly 4.0?

b. Find the sample prediction equation. Predict the college GPA for students having high school GPA of (i) 3.0 and (ii) 4.0. Explain how the slope relates to the answers.

12.10 Exercise and watching TV For the Georgia Student Survey file on the text CD, let y = exercise and x = watch TV (minutes per day).

a. Construct a scatterplot. Identify an outlier that could have an impact on the fit of the regression model. What would you expect its effect to be on the slope?

b. Fit the model with and without that observation. Summarize its impact.

12.2 Describe Strength of Association

In the straight-line regression model, the slope indicates whether the association is positive or negative and describes the trend. The slope does not, however, describe the *strength* of the association. Chapter 3 showed that the **correlation** does this.

Recall

You can review the **correlation** r and its properties in Section 3.2. Figure 3.7 displayed scatterplots having a variety of values for r. ◀

The Correlation Measures the Strength of Linear Association

The correlation, denoted by r, describes *linear association* between two variables.

SUMMARY: Properties of the Correlation, r

- The correlation r has the same sign as the slope b. Thus, $r > 0$ when the points in the scatterplot have an upward trend and $r < 0$ when the points have a downward trend.
- The correlation r always falls between -1 and $+1$, that is, $-1 \leq r \leq +1$.
- The larger the absolute value of r, the stronger the linear association, with $r = \pm 1$ when the data points all fall exactly on the regression line.

When you have several quantitative variables, software can provide a **correlation matrix**. This is a square table, listing the variables in the rows and again in the columns and showing the correlation between each pair. This table does not differentiate between response and explanatory variables, because *the correlation when we use y to predict x is the same as the correlation when we use x to predict y.*

Example 5

Strength of an association

Worldwide Internet Use

Picture the Scenario

The Twelve Countries data set on the text CD shows recent data for 12 countries on several variables. (*Source:* Data from www.internetworldstats.com and www.checkfacebook.com.) One of the variables in the file is Internet use, the percentage of adult residents who use the Internet. Which variables are strongly associated with Internet use? Let's consider three possibilities.

Unemployment rate: Total percentage of labor force unemployed

GDP: Gross domestic product, per capita, in thousands of U.S. dollars (a measure of a nation's economic development)

CO_2: Carbon dioxide emissions, per capita (a measure of air pollution)

Table 12.2 displays a correlation matrix for these four variables.

Table 12.2 Correlation Matrix for Internet Use, Unemployment Rate (%), Gross Domestic Product (GDP), and Carbon Dioxide (CO_2) Emissions

	Internet users (per 100 people)	Unemployment rate (%)	GDP per capita
Unemployment rate	0.238		
GDP per capita	0.938	0.163	
Carbon dioxide emissions	0.569	−0.107	0.740

Cell Contents: Pearson correlation

Source: Data from www.internetworldstats.com and www.checkfacebook.com.

Questions to Explore

a. Which variable has the strongest linear association with Internet use?

b. Which variable has the next strongest linear association with Internet use? Interpret.

Think It Through

a. The correlations with Internet use appear in the first column. The variable most strongly linearly associated with Internet use is the one having the largest correlation, in absolute value. This is GDP. Its correlation with Internet use is $r = 0.938$, a very strong positive association.

b. The variable Carbon Dioxide (CO_2) Emissions has the next strongest linear association with Internet use with correlation, $r = 0.569$. As CO_2 increases, there appears to be a moderate tendency for Internet use to increase.

Insight

We observe that the correlation ($r = 0.74$) between GDP and Carbon Dioxide Emissions indicates a strong positive association. This should not be surprising. Nations that are more economically advanced tend to have both higher GDP and CO_2. We have already observed that these two variables (GDP and CO_2) have a moderate to strong positive association with Internet use.

Try Exercise 12.12

Finding the Correlation and the Prediction Equation

Software uses the sample data to find the prediction equation $\hat{y} = a + bx$ and the correlation r. Sections 3.2 and 3.3 in Chapter 3 gave the formulas. We saw there that $r > 0$ and $b > 0$ when most observations are in the quadrants where x and y are both above their means or both below their means. See Figure 12.5. By contrast, $r < 0$ and $b < 0$ when subjects who are above the mean on one variable tend to be below the mean on the other variable.

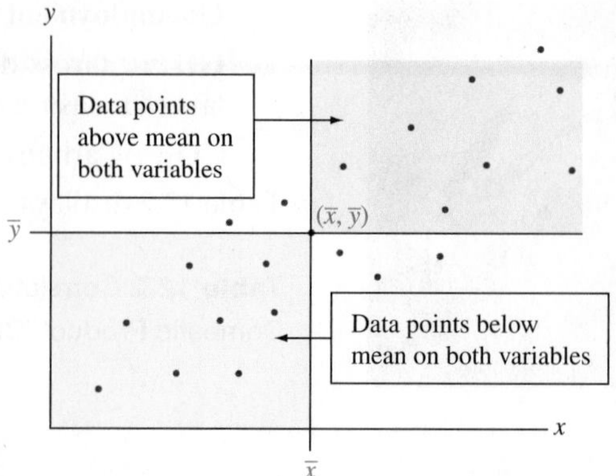

▲ **Figure 12.5 The Correlation *r* and the Slope *b* Are Positive When Most Data Points Are Above the Mean on Both Variables or Below the Mean on Both.** These are the areas shaded in this figure. **Question** Where do most data points usually fall when the slope and correlation are negative?

Recall

From Section 3.3,

$$b = r\left(\frac{s_y}{s_x}\right),$$

$$a = \bar{y} - b\bar{x}.$$ ◀

The formulas for the slope b and the y-intercept a (shown again in the margin) use the correlation r and the sample means $\bar{x}$ and $\bar{y}$ of the x values and the y values and the sample standard deviations s_x and s_y. The slope b is proportional to the correlation r, and we'll discuss their connection below. The formula $a = \bar{y} - b\bar{x}$ for the y-intercept can be rewritten as $\bar{y} = a + b\bar{x}$. This shows that if we substitute $x = \bar{x}$ into the regression equation $\hat{y} = a + bx$, then the predicted outcome is $\hat{y} = a + b\bar{x} = \bar{y}$. In words, *a subject who is average on x is predicted to be average on y.*

In Practice Check the Model Graphically; Use Software for Computations

Check whether it is sensible to do a regression analysis by looking at a scatterplot. If you see an approximately linear relationship, you can let technology do the computational work of finding the correlation, slope, and y-intercept.

The Correlation Is a Standardized Slope

Why do we need the correlation? Why can't we use the slope to describe the strength of the association? The reason is that the slope's numerical value depends on the units of measurement. Example 3 found a slope of 1.49 pounds in predicting y = maximum bench press using x = number of 60-pound bench presses. Suppose we instead measure y in kilograms (kg). Since 1 kg = 2.2 pounds, a slope of 1.49 pounds is equivalent to a slope of 1.49/2.2 = 0.68 kilograms. In grams, the slope would be 680. Whether a slope is a small number or a large number merely depends on the units of measurement.

The correlation is a *standardized* version of the slope. Unlike the slope b, the correlation does not depend on units of measurement. It takes the same value regardless of whether maximum bench press is measured in pounds, kilograms, or grams. The standardization adjusts the slope b for the way it depends on the standard deviations of x and y. Since the correlation r and slope b are related by $b = r(s_y/s_x)$, equivalently,

$$r = b\left(\frac{s_x}{s_y}\right).$$

SUMMARY: Relationship of Correlation and Slope

If the data have the same amount of variability for each variable, with $s_x = s_y$, then, $r = b$: The correlation and the slope are the same. (See margin figure.)

- The correlation r does not depend on the units of measurement.
- The correlation represents the value that the slope equals if the two variables have the same standard deviation.

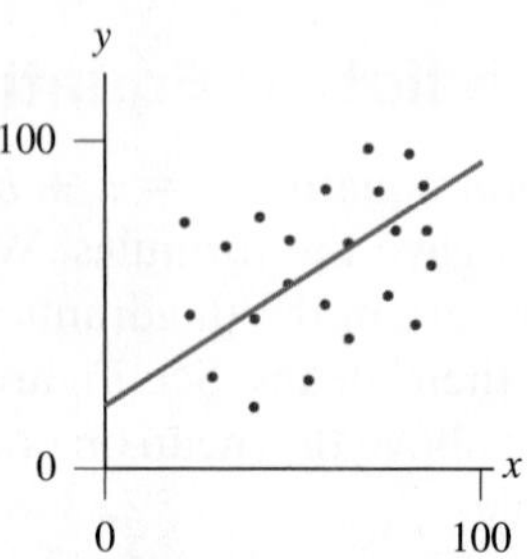

For instance, suppose the standard deviation equals 10 both for x = midterm exam score and also for y = final exam score of your statistics course. Then, if $\hat{y} = 24.0 + 0.70x$ (that is, slope = 0.70), the correlation is also 0.70 between the two exam scores.

In practice, the standard deviations are not usually identical. However, this tells us that the correlation represents what we would get for the slope of the regression line if the two variables did have the same standard deviations.

Example 6

Correlation

Predicting Strength

Picture the Scenario

For the female athlete strength study (Examples 1–3), x = number of 60-pound bench presses and y = maximum bench press had:

x: mean = 11.0, standard deviation = 7.1

y: mean = 79.9, standard deviation = 13.3 (both in pounds)

regression equation: $\hat{y} = 63.5 + 1.49x$.

Questions to Explore

a. Find the correlation r between these two variables.

b. Show that r does not change value if you measure y in kilograms.

Think It Through

a. The slope of the regression equation is $b = 1.49$. Since $s_x = 7.1$ and $s_y = 13.3$,

$$r = b\left(\frac{s_x}{s_y}\right) = 1.49\left(\frac{7.1}{13.3}\right) = 0.80.$$

The variables have a strong, positive association.

b. If y had been measured in kilograms, the y values would have been divided by 2.2, since 1 kg = 2.2 pounds. For instance, Subject 1 had $y = 80$ pounds, which is 80/2.2 = 36.4 kg. Likewise, the standard deviation s_y of 13.3 in pounds would have been divided by 2.2 to get 13.3/2.2 = 6.05 in kg. The slope of 1.49 would have been divided by 2.2, giving 0.68, since 1.49 pounds = 0.68 kg. Then

$$r = b(s_x/s_y) = 0.68(7.1/6.05) = 0.80.$$

The correlation is the same (0.80) if we measure y in pounds or in kilograms.

Insight

Now if we change units from kilograms to grams, s_y changes from 6.05 to 6050, b changes from 0.68 to 680, but again $r = 0.80$ because r does not depend on the units.

Try Exercise 12.15

Regression Toward the Mean

An important property of the correlation is that at any particular x value, the predicted value of y is relatively closer to its mean than x is to its mean.

If an x value is a certain number of standard deviations from its mean, then the predicted y is r times that many standard deviations from its mean.

Let's illustrate for $r = 0.50$ and a subject who is 1 standard deviation above the mean of x, that is with x value, $x = \bar{x} + s_x$. The predicted value of y is 0.50 standard deviation above the mean, $\bar{y}$. That is, at $x = \bar{x} + s_x$, the predicted outcome is $\hat{y} = \bar{y} + 0.50s_y$. The predicted y is relatively closer to its mean. See Figure 12.6, which also recalls that for subjects at the mean $\bar{x}$ of x, their predicted value of y is $\bar{y}$.

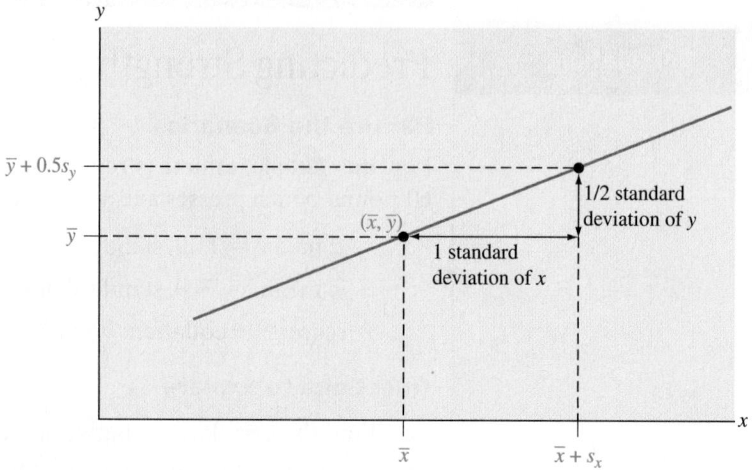

▲ Figure 12.6 The Predicted Value of y Is Closer to Its Mean $\bar{y}$ Than x Is to Its Mean $\bar{x}$ (in Number of Standard Deviations). When x is 1 standard deviaton above $\bar{x}$, the predicted y value is r standard deviations above $\bar{y}$ (shown in the figure for $r = 1/2$). **Question** When x is 2 standard deviations above $\bar{x}$, how many standard deviations does the predicted y value fall above $\bar{y}$ (if $r = 1/2$)?

Example 7

Regression toward the mean

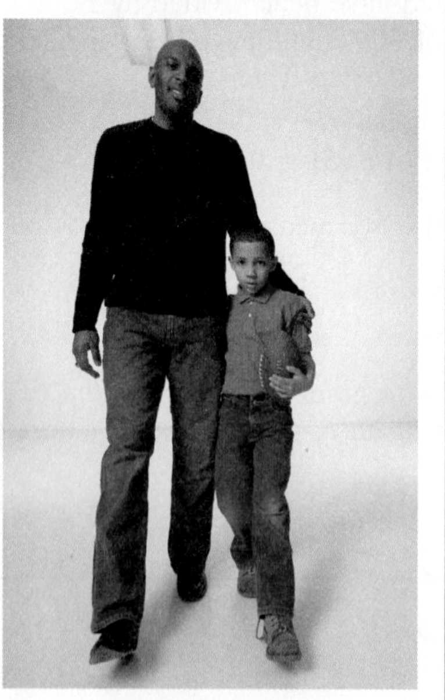

Tall Parents and Tall Children

Picture the Scenario

British scientist Francis Galton discovered the basic ideas of regression and correlation in the 1880s. He observed that very tall parents tended to have tall children, but on average not quite so tall. For instance, for all fathers with height 7 feet, their sons averaged 6 feet 5 inches when fully grown—taller than average, but not extremely tall. Likewise, for fathers with height 5 feet, perhaps their sons averaged 5 feet 5 inches—shorter than average, but not extremely short.

In his research, Galton accounted for gender height differences by multiplying each female height by 1.08, so heights of mothers and daughters had about the same mean as heights of fathers and sons. Then, for each son or daughter, he summarized their father's and mother's heights by parents' height = (father's height + mother's height)/2, the mean of their heights.

Question to Explore

Galton found a correlation of 0.5 between x = parents' height and y = child's height. How does his observation about very tall or very short parents with children who are not so very tall or so very short relate to the property about the correlation that a predicted value of y is relatively closer to its mean than x is to its mean?

Think It Through

From the property of r with $r = 0.5$, when x = parents' height is a certain number of standard deviations from its mean, then y = child's predicted height is *half* as many standard deviations from its mean. For example, if the parents' height is 2 standard deviations above the mean, the child is predicted to be 1 standard deviation above the mean (half as far, in relative terms, when $r = 0.5$). At $x = \bar{x} + 3s_x$, we predict $\hat{y} = \bar{y} + 1.5s_y$. In each case, on average a child's height is above the mean, but only half as far above the mean as their parent's height is above the parent's mean.

Insight

The correlation r is no greater than 1, in absolute value. So, a y value is predicted to be fewer standard deviations from its mean than x is from its mean.

Try Exercise 12.26

In Words

In English, "regression" means going back, or returning. Here, the predicted value of *y* is going back toward the mean.

In summary, the predicted y is relatively closer to its mean than x is to its mean. Because of this, there is said to be **regression toward the mean**. This is the origin of the name that Francis Galton chose for regression analysis.

For all cases in which parents' height is *extremely* tall, say 3 standard deviations above the mean (essentially at the upper limit of observed heights), it's no surprise that their children would tend to be shorter. For all cases in which parents' height is extremely short, say 3 standard deviations below the mean, it's no surprise that their children would tend to be taller. In both these cases, we'd expect regression toward the mean. What's interesting and perhaps surprising is that regression of y toward the mean happens not only at the very extreme values of x.

Example 8

Regression toward the mean

The Placebo Effect

Picture the Scenario

A clinical trial admits subjects suffering from high blood cholesterol (over 225 mg/dl). The subjects are randomly assigned to take either a placebo or a drug being tested for reducing cholesterol levels. After the three-month study, the mean cholesterol level for subjects taking the drug drops from 270 to 230. However, the researchers are surprised to see that the mean cholesterol level for the placebo group also drops, from 270 to 250.

Question to Explore

Explain how this placebo effect could merely reflect regression toward the mean.

Think It Through

For a group of people, a person's cholesterol reading at one time would likely be positively correlated with their reading three months later. So, for all people who are not taking the drug, a subject with relatively high cholesterol at one time would also tend to have relatively high cholesterol three months later. By regression toward the mean, however, subjects who are relatively high at one time will, on average, be lower at a later time. So, if a study gives placebo to people with relatively high cholesterol (that is, in the right-hand tail of the blood cholesterol distribution), on average we expect their values three months later to be lower.

Insight

Regression toward the mean is pervasive. In sports, excellent performance tends to be followed by good, but less outstanding, performance.

A football team that wins all its games in the first half of its schedule will probably not win all its games in the second half. A baseball player who hits 0.400 in the first month will probably not be hitting that high at the end of the season.

By contrast, the good news about regression toward the mean is that very poor performance tends to be followed by improved performance. If you got the worst score in your statistics class on the first exam, you probably did not do so poorly on the second exam (but you were probably still below the mean).

Try Exercise 12.23

The Squared Correlation (r^2) Describes Predictive Power

If you know how many times a person can bench press 60 pounds, can you predict well their maximum bench press? Another way to describe the strength of association refers to how close predictions for y tend to be to observed y values. The variables are strongly associated if you can predict y much better by substituting x values into the prediction equation $\hat{y} = a + bx$ than by merely using the sample mean $\bar{y}$ and ignoring x.

For a given person, the prediction error is the difference between the observed and predicted values of y.

- The error using the regression line to make a prediction is $y - \hat{y}$.
- The error using $\bar{y}$ to make a prediction is $y - \bar{y}$.

For each potential predictor ($\hat{y}$ and $\bar{y}$), some errors are positive, some errors are negative, some errors may be zero, and the sum of the errors for the sample equals 0. You can summarize the sizes of the errors by the sum of their squared values,

$$\text{Error summary} = \Sigma\,(\text{observed } y \text{ value} - \text{predicted } y \text{ value})^2.$$

When we predict y using $\bar{y}$ (that is, ignoring x), the error summary equals

$$\Sigma\,(y - \bar{y})^2.$$

This is called the **total sum of squares**. When we predict y using x with the regression equation, the error summary equals

$$\Sigma\,(y - \hat{y})^2.$$

Recall

$\Sigma\,(y - \bar{y})^2$ is the numerator of the variance of the y values. $\Sigma\,(y - \hat{y})^2$ is what's minimized in finding the least squares estimates for the regression equation. ◀

This is called the **residual sum of squares** because it sums the squared residuals.

When a strong linear association exists, the regression equation predictions ($\hat{y}$) tend to be much better than $\bar{y}$. Then, $\Sigma\,(y - \hat{y})^2$ is much less than $\Sigma\,(y - \bar{y})^2$. The difference between the two error summaries depends on the units of measure: It's different, for instance, with weight measured in kilograms than in pounds. We can eliminate this dependence on units by converting the difference to a proportion, by dividing by $\Sigma\,(y - \bar{y})^2$. This gives the summary measure of association,

$$r^2 = \frac{\Sigma\,(y - \bar{y})^2 - \Sigma\,(y - \hat{y})^2}{\Sigma\,(y - \bar{y})^2}.$$

This measure falls on the same scale as a proportion, 0 to 1.

The measure r^2 is interpreted as the **proportional reduction in error**. If $r^2 = 0.40$, for instance, the error using $\hat{y}$ to predict y is 40% smaller than the error using $\bar{y}$ to predict y. We use the notation r^2 for this measure because, in fact, it can be shown that this measure equals the square of the correlation r.

In Practice Get r^2 by Squaring the Correlation

If you know the correlation r, it is simple to calculate r^2, by squaring the correlation. The formula shown previously for r^2 is useful for interpretation of r^2, but it's not needed for calculation.

Example 9

The squared correlation

The Strength Study

Picture the Scenario

For the female athlete strength study, Example 6 showed that x = number of 60-pound bench presses and y = maximum bench press had a correlation of 0.80.

Question to Explore

Find and interpret r^2.

Think It Through

Since the correlation $r = 0.80$, $r^2 = (0.80)^2 = 0.64$. For predicting maximum bench press, the regression equation has 64% less error than $\bar{y}$ has. "Error" here refers to the summary given by the sum of squared prediction errors.

Insight

Since $r^2 = 0.64$ is quite far from 0, we can predict y much better using the regression equation than using $\bar{y}$. In this sense, the association is quite strong.

Try Exercise 12.16, part c

It's also possible to calculate r^2 directly from the definition. Software for regression routinely provides tables of sums of squares. Table 12.3 (showing partial output from MINITAB) is an example. The heading SS stands for "sum of squares." From Table 12.3, the residual sum of squared errors is $\Sigma(y - \hat{y})^2 = 3522.8$, and the total sum of squares is $\Sigma(y - \bar{y})^2 = 9874.6$. Thus,

$$r^2 = \frac{\Sigma(y - \bar{y})^2 - \Sigma(y - \hat{y})^2}{\Sigma(y - \bar{y})^2} = \frac{9874.6 - 3522.8}{9874.6} = 0.643.$$

Table 12.3 Annotated MINITAB Table Showing Sums of Squares and r^2 for Strength Study

Under the heading for Source, "Total" refers to $\Sigma(y - \bar{y})^2$ and "Residual Error" refers to $\Sigma(y - \hat{y})^2$. The "Regression" sum of squares equals their difference, which is the numerator of r^2.

Source	SS	
Regression	6351.8	← This is 9874.6 − 3522.8, the numerator of r^2
Residual Error	3522.8	← $\Sigma(y - \hat{y})^2$
Total	9874.6	← $\Sigma(y - \bar{y})^2$
R-Sq = 64.3%		← This is r^2, in percentage terms

(Annotation on SS column: ← SS stands for "sum of squares")

Normally, it is unnecessary to perform this computation, since most software reports r or r^2 or both.

A table that reports the sums of squares used in regression analysis is called an **analysis of variance table**, or **ANOVA table** for short. We'll discuss the ANOVA table further in Section 12.4.

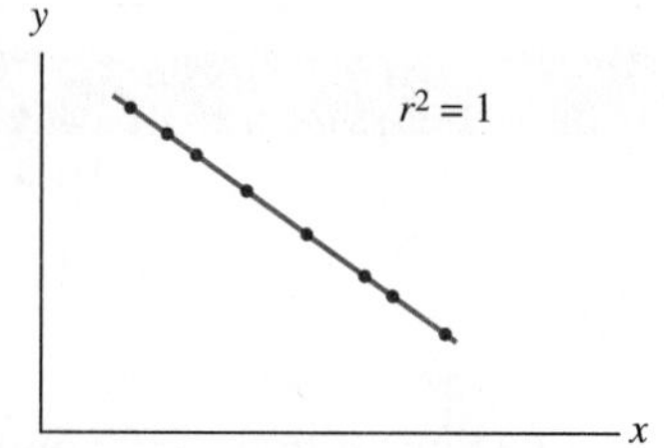

Properties of r^2

The r^2 measure, like the correlation r, measures the strength of *linear* association. We emphasize *linear* because r^2 compares predictive power of the *straight-line* regression equation to $\bar{y}$.

SUMMARY: Properties of r^2

- Since $-1 \leq r \leq 1$, r^2 falls between 0 and 1.
- $r^2 = 1$ when $\Sigma(y - \hat{y})^2 = 0$, which happens only when all the data points fall exactly on the regression line. There is then no prediction error using x to predict y (that is, $y = \hat{y}$ for each observation). This corresponds to $r = \pm 1$. (See margin figure.)
- $r^2 = 0$ when $\Sigma(y - \hat{y})^2 = \Sigma(y - \bar{y})^2$. This happens when the slope $b = 0$, in which case each $\hat{y} = \bar{y}$. The regression line and $\bar{y}$ then give the same predictions.
- The closer r^2 is to 1, the stronger the linear association: The more effective the regression equation $\hat{y} = a + bx$ then is compared to $\bar{y}$ in predicting y.

Caution

If finding the correlation r by evaluating $\sqrt{r^2}$, the sign (positive or negative) of the correlation r must be found by either looking at the scatterplot of the two variables or by knowing the slope of the corresponding least squares regression line. ◀

Because the calculation of prediction error in r^2 refers to squared distances, it relates to comparing the overall variance of the y values around $\bar{y}$ to the variance around the regression line. For example, $r^2 = 0.64$ means that the estimated variance around the regression line is 64% less than the overall variance of the y values. This is often phrased in the style, "64% of the variability in y is explained by x." This interpretation has the weakness, however, that variability in this case is summarized by the variance. Some statisticians find r^2 to be less useful than r, because (being based on sums of squares) it uses the square of the original scale of measurement. It's easier to interpret the original scale than a squared scale. This is a strong advantage of the standard deviation over the variance (i.e., r over r^2). Another disadvantage of r^2 is that the direction of the relationship is lost.

SUMMARY: Correlation r and Its Square r^2

Both the correlation r and its square r^2 describe the strength of association. They have different interpretations. The correlation falls between -1 and $+1$. It represents the slope of the regression line when x and y have equal standard deviations. It governs the extent of "regression toward the mean." The r^2 measure falls between 0 and 1 (or 0% and 100% when reported by software in percentage terms). It summarizes the reduction in sum of squared errors in predicting y using the regression line instead of using the mean of y.

Factors Affecting the Correlation

Section 3.4 showed that certain regression outliers can unduly influence the slope and the correlation. A single observation can have a large influence if its x value is unusually large or unusually small and if it falls quite far from the trend of the rest of the data.

Besides being influenced by outliers, the size of the correlation (and r^2) depends strongly on two other factors.

- First, if the subjects are grouped for the observations, such as when the data refer to county summaries instead of individual people, the correlation tends to increase in magnitude.

Suppose you want to find the correlation between number of years of education completed and annual income. You could measure these variables for a sample of individuals. Or you could use summary education and income measures such as means or medians for counties (or states or provinces). The scatterplot in Figure 12.7 shows that at the individual level, the correlation could be much weaker. Lots of variability in income exists for individuals, but not much variability in income (mean or median) exists for counties. The summary values for counties fall closer to a straight line.

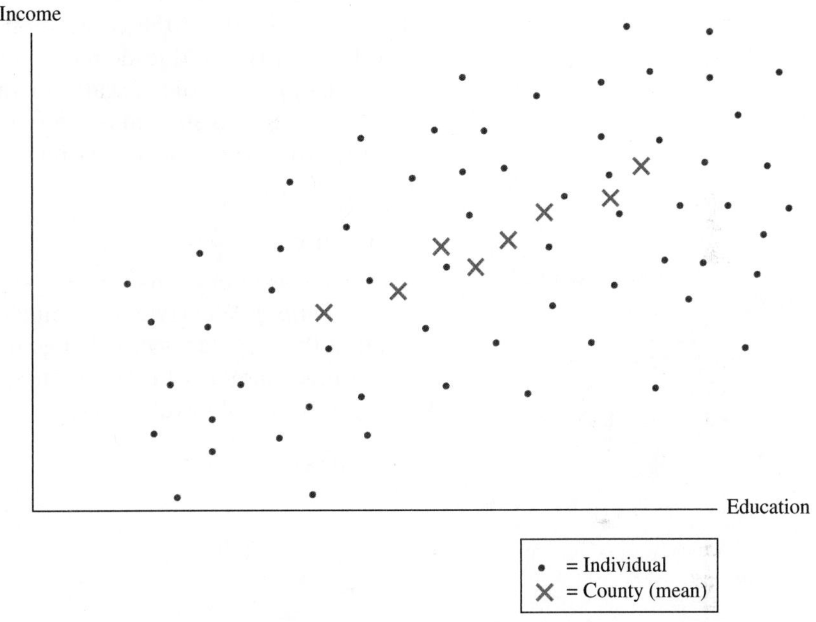

▲ **Figure 12.7 The Effect of Grouping of Subjects on the Correlation.** There is a stronger linear trend for the countywide data than for the data on individuals. **Question** Why would you expect much more variability for income of individuals than for mean income of counties?

Did You Know?

The **ecological fallacy** is a danger even with single variables. If the mean income in your town is \$50,000, you'd be wrong to assume that each worker makes \$50,000. Perhaps half are professionals making \$90,000 and half are service workers making \$10,000. ◀

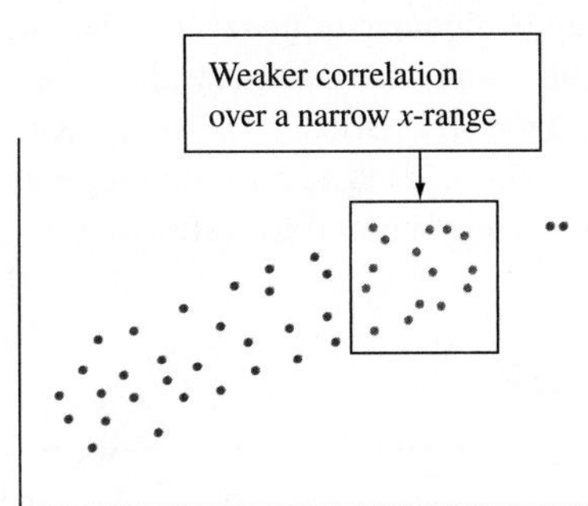

Because correlations can change dramatically when data are grouped, it is misleading to take results for groupings such as counties and extend them to individuals. If income and education are strongly correlated when we measure the summary means for each county, this does not imply they will be strongly correlated for measurements on individuals. Making predictions about individuals based on the summary results for groups is known as the **ecological fallacy** and should be avoided.

- Second, the size of the correlation depends on the range of x values sampled: The correlation tends to be smaller when we sample only a restricted range of x values than when we use the entire range. (See margin figure.)

By contrast, the prediction equation remains valid when we observe a restricted range of x values. The slope is usually not much affected by the range of x values, and we simply limit our predictions to that range. The correlation makes most sense, however, when x takes its full range of values. Otherwise, it can be misleading.

Example 10

Factors affecting the correlation

High School GPA Predicting College GPA

Picture the Scenario

Consider the correlation between high school GPA and later performance in college measured by college GPA.

Question to Explore

For which group would these variables have a stronger correlation: All students who graduate from Harvard University this year, or all students who graduate from college somewhere in the United States this year?

Think It Through

College GPA

Harvard sample

High school GPA

The magnitude of the correlation depends on the variability in high school GPA. For Harvard students, the high school GPAs will concentrate narrowly at the upper end of the scale. So, the correlation would probably be weak. By contrast, for all students who finish college, high school GPA values would range from very low to very high. We would likely see a stronger correlation for them.

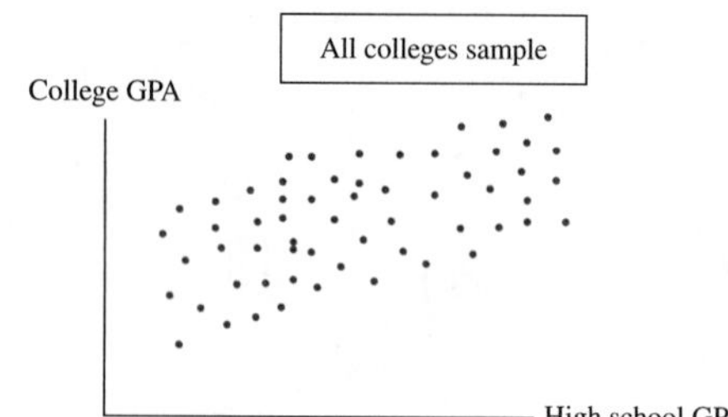

Insight

What reason explains this property of the correlation? Recall the formula, $r = b(s_x/s_y)$. When we use a much wider range of x values, s_x increases a lot. So, if the slope does not change much and if the variability of the y values is not much larger with the expanded sample, r will tend to increase because it is proportional in value to s_x.

Try Exercise 12.29

On the Shoulders of... Francis Galton

Francis Galton

Heights and peas—what do they have in common?

—Francis Galton (1822–1911)

Francis Galton, like his cousin Charles Darwin, was interested in genetics. In his work on inheritance, he collected data on parents' and children's heights. He found that, on average, the heights of children were not as extreme as those of the parents and tended toward the mean height. There was regression toward the mean. This refers to this average, not to every observation. It does not imply that over the course of many generations, heights will all vary less from the mean height.

Galton also conducted experiments examining the weight of sweet peas. The mean weight of offspring peas was closer than the weight of the parent group to the mean for all peas. Again, Galton observed regression toward the mean.

Galton discovered a way to summarize linear associations numerically. He named this numerical summary the "coefficient of correlation." In 1896, the British statistician Karl Pearson (who proposed the chi-squared test for contingency tables in 1900) derived the current method for estimating the correlation using sample data.

12.2 Practicing the Basics

12.11 Dollars and thousands of dollars If a slope is 1.50 when x = income in thousands of dollars, then what is the slope when x = income in dollars? (*Hint:* A \$1 change has only 1/1000 of the impact of a \$1000 change?)

12.12 Comparing slopes and correlations Suppose we want to TRY describe whether Internet use is more strongly associated with GDP or with unemployment rate.

a. Can we compare the slopes when GDP and unemployment rate each predict Internet use in separate regression equations? Why or why not?

b. According to the correlation matrix in Table 12.2, part of which is shown below, which is more strongly associated with Internet use? Why?

	INTERNET
GDP	0.938
Unemployment rate	0.238
CO_2	0.569

12.13 When can you compare slopes? Although the slope does not measure association, it *is* useful for comparing effects for two variables that have the *same* units. For the Internet Use data file of 33 nations on the text CD, let x = GDP (thousands of dollars per capita). For predicting y = percentage Internet penetration (the percentage of adult users), the prediction equation is $\hat{y} = 0.1239 + 0.0157x$. For predicting y = percentage Facebook penetration, the prediction equation is $\hat{y} = 0.081 + 0.0075x$.

a. Explain how to interpret the two slopes.

b. For these nations, explain why a one-unit increase in GDP has a slightly greater impact on the percentage using Facebook than on the percentage using the Internet.

12.14 Sketch scatterplot Sketch a scatterplot, identifying quadrants relative to the sample means as in Figure 12.5, for which (a) the slope and correlation would be negative and (b) the slope and correlation would be approximately zero.

12.15 Sit-ups and the 40-yard dash Is there a relation-TRY ship between x = how many sit-ups you can do and y = how fast you can run 40 yards (in seconds)? The MINITAB output of a regression analysis for the female athlete strength study is shown here.

```
Predictor     Coef      SE Coef    T     P
Constant      6.7065    0.1779 37.70 0.000
SIT-UPS     -0.024346 0.006349 -3.83 0.000
R-Sq = 21.1%
```

a. Find the predicted time in the 40-yard dash for a subject who can do (i) 10 sit-ups and (ii) 40 sit-ups. (The minimum and maximum in the study were 10 and 39.) Relate the difference in predicted times to the slope.

b. The number of sit-ups had mean 27.175 and standard deviation 6.887. The time in the 40-yard dash had mean 6.045 and standard deviation 0.365. Show how the correlation relates to the slope, and find its value.

12.16 Weight, height, and fat For the high school female TRY athlete strength study, the output shows a correlation matrix for height, weight, body mass index (BMI), and percentage of body fat (BF%).

	HT (IN)	WT (lbs)	BF%
WT (lbs)	0.553		
BF%	0.216	0.871	
BMI	0.153	0.898	0.945

a. Which pair of variables has the (i) strongest association and (ii) weakest association.

b. Interpret the sign and the strength of the correlation between height and weight.

c. Find and interpret r^2 for height and weight.

d. If height and weight were measured instead with metric units, would any results differ in parts a, b, and c? Explain.

12.17 Male and female strength For the Male Athlete Strength data file on the text CD, the output shows correlations for height, weight, and percentage body fat (BF%).

	Ht. (in.)	Wt. (lbs)
Wt. (lbs)	0.457	
BF %	0.232	0.883

a. Compare the correlations for females (Exercise 12.16) to males for weight and height, weight and body fat percentage, and height and body fat percentage.

b. Which pair of variables for men has the (i) strongest association and (ii) weakest association.

12.18 Verbal and math SAT All students who attend Lake Wobegon College must take the math and verbal SAT exams. Both exams have a mean of 500 and a standard deviation of 100. The regression equation relating y = math SAT score and x = verbal SAT score is $\hat{y} = 250 + 0.5x$.

a. Find the predicted math SAT score for a student who has the mean verbal SAT score of 500. (*Note:* At the x value equal to $\bar{x}$, the predicted value of y equals $\bar{y}$.)

b. Show how to find the correlation. Interpret its value as a standardized slope. (*Hint:* Both standard deviations are equal.)

c. Find r^2 and interpret its value.

12.19 SAT regression toward mean Refer to the previous exercise.

a. Predict the math SAT score for a student who has a verbal SAT = 800.

b. The correlation is 0.5. Interpret the prediction in part a in terms of regression toward the mean.

12.20 GPAs and TV watching For the Georgia Student Survey data file on the text CD, the table shows the correlation matrix for college GPA, high school GPA, and daily time spent watching TV.

```
         HSGPA   CGPA
CGPA     0.505
WatchTV  -0.333  -0.353
```

a. Interpret r and r^2 between time spent watching TV and college GPA.

b. One student is 2 standard deviations above the mean on high school GPA. How many standard deviations would you expect that student to be above the mean on college GPA? Use your answer to explain "regression toward the mean."

12.21 GPA and study time Refer to the association you investigated in Exercise 12.7 between study time and college GPA. Using software or a calculator with the data file you constructed for that exercise,

a. Find and interpret the correlation.

b. Find and interpret r^2.

12.22 GPA and skipping class Refer to the association you investigated in Exercise 12.8 between skipping class and college GPA. Using software or a calculator with the data file you constructed for that exercise,

a. Find the mean and standard deviation of each variable.

b. Report the slope of the prediction equation and the correlation.

c. Show how to find the correlation from the slope and the standard deviations.

12.23 Placebo helps cholesterol? TRY A clinical trial admits subjects suffering from high cholesterol, who are then randomly assigned to take a drug or a placebo for a 12-week study. For the population, without taking any drug, the correlation between the cholesterol readings at times 12 weeks apart is 0.70. The mean cholesterol reading at any given time is 200, with the same standard deviation at each time. Consider all the subjects with a cholesterol level of 300 at the start of the study, who take placebo.

a. What would you predict for their mean cholesterol level at the end of the study?

b. Does this suggest that placebo is effective for treating high cholesterol? Explain.

12.24 Does tutoring help? For a class of 100 students, the teacher takes the 10 students who performed poorest on the midterm exam and enrolls them in a special tutoring program. Both the midterm and final have a class mean of 70 with standard deviation 10, and the correlation is 0.50 between the two exam scores. The mean for the specially tutored students increases from 50 on the midterm to 60 on the final. Can we conclude that the tutoring program was successful? Explain, identifying the response and explanatory variables and the role of regression toward the mean.

12.25 What's wrong with your stock fund? Last year you looked at all the financial firms that had stock growth funds. You picked the growth fund that had the best performance last year (ranking at the 99th percentile on performance) and invested all your money in it this year. This year, with their new investments, they ranked only at the 65th percentile on performance. Your friend suggests that their stock picker became complacent or was burned out. Can you give another explanation?

12.26 Golf regression TRY In the first round of a golf tournament, five players tied for the lowest round, at 65. The mean score of all players was 75. If the mean score of all players is also 75 in the second round, what does regression toward the mean suggest about how well we can expect the five leaders to do, on the average, in the second round? (*Hint:* Suppose the standard deviation is also the same in each round.)

12.27 Car weight and mileage The Car Weight and Mileage data file on the text CD shows the weight and the mileage per gallon of gas of 25 cars of various models. The regression of mileage on weight has $r^2 = 0.75$. Explain how to interpret this in terms of how well you can predict a car's mileage if you know its weight.

12.28 Food and drink sales The owner of Bertha's Restaurant is interested in whether an association exists between the amount spent on food and the amount spent on drinks for the restaurant's customers. She decides to measure each variable for every customer in the next month. Each day she also summarizes the mean amount spent on food and the mean amount spent on drinks. Which correlation between amounts spent on food and drink do you think would be higher, the one computed for the 2500 customers in the next month, or the one computed using the means for the 30 days of the month? Why? Sketch a sample scatterplot showing what you expect for each case as part of your answer.

12.29 Yale and UConn TRY For which student body do you think the correlation between high school GPA and college GPA would be higher: Yale University or the University of Connecticut? Explain why.

12.30 Violent crime and single-parent families Use software to analyze the U.S. Statewide Crime data file on the text CD on y = violent crime rate and x = percentage of single parent families.

a. Construct a scatterplot. What does it show?

b. One point is quite far removed from the others, having a much higher value on both variables than the rest of the sample, but it fits in well with the linear trend exhibited by the rest of the points. Show that the correlation changes from 0.77 to 0.59 when you delete this observation. Why does it drop so dramatically?

12.31 Correlations for the strong and for the weak Refer to the High School Female Athlete and Male Athlete Strength data files on the text CD.

a. Find the correlation between number of 60-pound bench presses before fatigue and bench press maximum for females and between bench presses before fatigue and bench press maximum for males. Interpret.

b. Find the correlation using only the x values (i) below the median of 10 for females and below the median of 17 for males and (ii) above the median of 10 for females and above the median of 17 for males. Compare to the correlation in part a. Why are they so different?

12.3 Make Inferences About the Association

Section 12.1 showed how a regression line models a relationship between two quantitative variables, when the means of y at various values of x follow approximately a straight line. Section 12.2 showed how the correlation r and its square, r^2, describe the strength of the association.

The sample regression equation, r, and r^2 are descriptive parts of a regression analysis. The inferential parts of regression use the tools of confidence intervals and significance tests. They provide inference about the regression equation, the correlation, and r^2 in the population of interest. For instance, in studying ways to measure strength, we're interested not just in the 57 female athletes in the Georgia study described in Example 1 but in a population of female athletes.

Assumptions for Regression Analysis

Some assumptions are needed to use regression in a descriptive manner. Extra assumptions are needed to make inferences. Let's start with the descriptive part.

SUMMARY: Basic Assumption for Using Regression Line for Description

The population means of y at different values of x have a straight-line relationship with x, that is, $\mu_y = \alpha + \beta x$.

This assumption states that a straight-line regression model is valid. You can check it by constructing a scatterplot. Statistical inferences in regression analysis make two additional assumptions:

SUMMARY: Extra Assumptions for Using Regression to Make Statistical Inference

- The data were gathered using randomization, such as random sampling or a randomized experiment.
- The population values of y at each value of x follow a normal distribution, with the same standard deviation at each x value.

Recall

By definition, **models** merely approximate the true relationship. A relationship will not be exactly linear, with exactly normal distributions for y at each x and with exactly the same standard deviation of y values at each x value. A model is useful as long as the approximation is reasonably good. ◀

The randomness assumption applies for any statistical inference. For inferences to be valid, the sample must be representative of the population.

As in the case of inference about a mean, the assumption about normality is what leads to t test statistics having t distributions. This assumption is never exactly satisfied in practice. However, the closer reality resembles this ideal, the more appropriate are the confidence interval and test procedures for the regression model. This assumption is less important than the other two, especially when the sample size is large. In that case, an extended central limit theorem implies that sample slopes have bell-shaped sampling distributions, no matter what shape the population distribution of y has at each value of x.

Testing Independence Between Quantitative Variables

Under the assumptions for inference, suppose that the slope β of the regression line $\mu_y = \alpha + \beta x$, equals 0, as Figure 12.8 shows. Then, the mean of y is identical at each x value. In fact, the two quantitative variables are statistically independent: The outcome for y does not depend on the value of x. It does not help us to know the value of x if we want to predict the value of y.

Recall

Section 12.1 showed that a regression model allows **variability** around a line $\mu_y = \alpha + \beta x$, that describes how the mean of y changes as x changes.◀

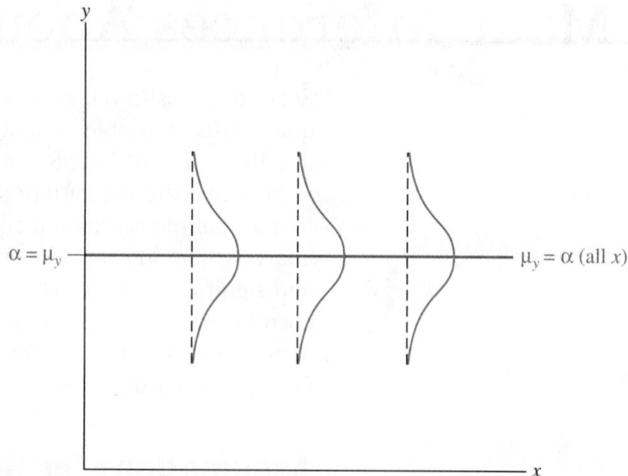

▲ **Figure 12.8 Quantitative Variables x and y Are Statistically Independent When the True Slope β = 0.** Each normal curve shown here represents the variability in y values at a particular value of x. When $\beta = 0$, the normal distribution of y is the same at each value of x.
Question How can you express the null hypothesis of independence between x and y in terms of a parameter from the regression model?

The null hypothesis that x and y are statistically independent is H_0: $\beta = 0$. Its significance test has the same purpose as the chi-squared test of independence for categorical variables (Chapter 11). It investigates how much evidence there is that the variables truly are significantly linearly associated. The smaller the P-value, the greater is the evidence.

Usually, the alternative hypothesis for the test of independence is two-sided, H_a: $\beta \neq 0$. Also possible is a one-sided alternative, H_a: $\beta > 0$ or H_a: $\beta < 0$, when you predict the direction of the association. The test statistic in each case equals

$$t = \frac{b - 0}{se},$$

where b is the sample slope and se denotes its standard error. Software supplies the slope b, the se, the t test statistic, and its P-value. The formula for se is rather complex, and we won't need it here (it's in Exercise 12.104). The form of the test statistic is the usual one for a t or z test. We take the estimate (b) of the parameter (β), subtract the null hypothesis value ($\beta = 0$), and divide by the standard error. The se describes the variability of the sampling distribution that measures how that estimate varies from sample to sample of size n (see the margin figure).

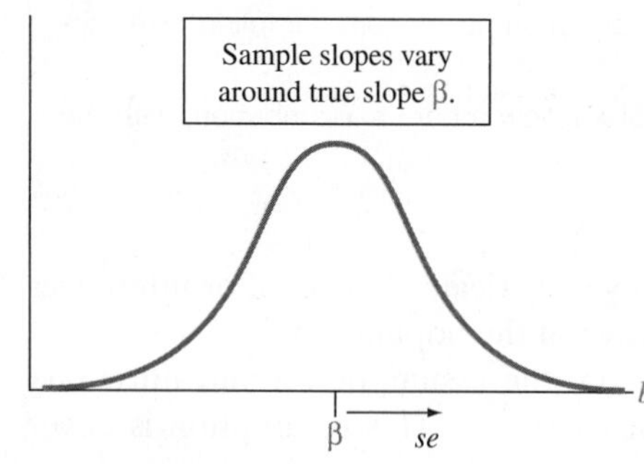

The symbol t is used for this test statistic because, under the assumptions stated, it has a t sampling distribution. The P-value for H_a: $\beta \neq 0$ is the two-tail probability from the t distribution (Table B). The degrees of freedom are $df = n - 2$. Software provides the P-value for any value of df.

Recall

Sections 8.3 and 9.3 used $df = n - 1$ for inference using the t distribution, so why is $df = n - 2$ here? We'll see that in regression, $df = n -$ number of parameters in equation for mean, μ_y.

The equation $\mu_y = \alpha + \beta x$ has two parameters (α and β) so $df = n - 2$. The inferences in Chapters 8 and 9 had one parameter (μ), so $df = n - 1$.◀

SUMMARY: Steps of Two-Sided Significance Test About a Population Slope β

1. **Assumptions:** (1) Population satisfies regression line $\mu_y = \alpha + \beta x$, (2) data gathered using randomization, (3) population y values at each x value have normal distribution, with same standard deviation at each x value.
2. **Hypotheses:** H_0: $\beta = 0$, H_a: $\beta \neq 0$
3. **Test statistic:** $t = (b - 0)/se$, where software supplies sample slope b and its se.
4. **P-value:** Two-tail probability of t test statistic value more extreme than observed, using t distribution with $df = n - 2$.
5. **Conclusions:** Interpret P-value in context. If a decision is needed, reject H_0 if P-value $\leq$ significance level (such as 0.05).

Example 11

Analyzing associations

60-Pound Strength and Bench Presses

Picture the Scenario

One purpose of the strength study introduced in Example 1 was to analyze whether x = number of times an athlete can lift a 60-pound bench press helps us predict y = maximum number of pounds the athlete can bench press. Table 12.4 shows the regression analysis for the 57 female athletes, with x denoted by BP_60 and y denoted by BP. The margin shows screen shots from the TI-83+/84.

Table 12.4 MINITAB Printout for Regression Analysis of y = Maximum Bench Press (BP) and x = Number of 60-Pound Bench Presses (BP_60)

The regression equation is BP = 63.5 + 1.49 BP_60

Predictor	Coef	SE Coef	T	P
Constant	63.537	1.956	32.48	0.000
BP_60	1.4911	0.150	9.96	0.000

R-Sq = 64.3%

```
LinRegTTest
 Xlist:L1
 Ylist:L2
 Freq:1
 β & ρ:≠0 <0 >0
 RegEQ:
 Calculate
```

```
LinRegTTest
 y=a+bx
 β≠0 and ρ≠0
 t=9.958
 p=6.481E-14
 df=55.000
↓a=63.537
```

```
LinRegTTest
 y=a+bx
 β≠0 and ρ≠0
↑b=1.491
 s=8.003
 r²=.643
 r=.802
```

TI-83+/84 output

Questions to Explore

a. Conduct a two-sided significance test of the null hypothesis of independence.

b. Report the P-value for the alternative hypothesis of a positive association, which is a sensible one for these variables. Interpret the results in context.

Think It Through

a. From Table 12.4, the prediction equation is $\hat{y} = 63.5 + 1.49x$. Here are the steps of a significance test of the null hypothesis of independence.

1. **Assumptions:** The scatterplot in Figure 12.1, shown again in the margin, revealed a linear trend, with scatter of points having similar variability (or spread) at different x values. The straight-line regression model $\mu_y = \alpha + \beta x$ seems appropriate. The 57 female athletes were *all* the female athletes at a particular high school. This was a convenience sample rather than a random sample of the population of female high school athletes. Although the goal was to make inferences about that population, inferences are tentative because the sample was not random.

2. **Hypotheses:** The null hypothesis that the variables are independent is H_0: $\beta = 0$. The two-sided alternative hypothesis of dependence is H_a: $\beta \neq 0$.

3. **Test statistic:** For the sample slope $b = 1.49$, Table 12.4 reports standard error, $se = 0.150$. This is listed under "SE Coef," in the row of the table for the predictor (BP_60). For testing H_0: $\beta = 0$, the t test statistic is

$$t = (b - 0)/se = 1.49/0.150 = 9.96.$$

This is extremely large. The sample slope falls nearly 10 standard errors above the null hypothesis value. The t test statistic appears

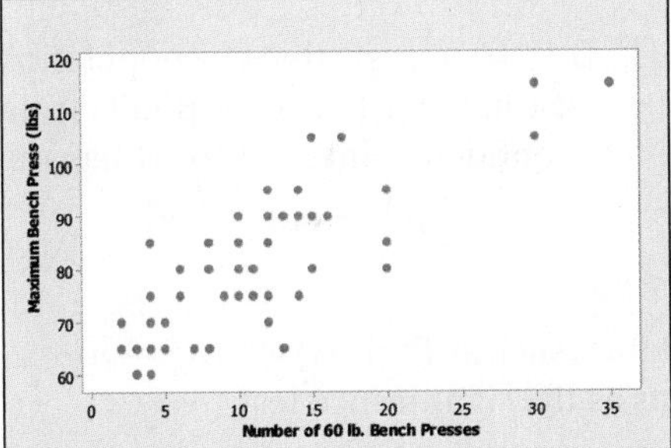

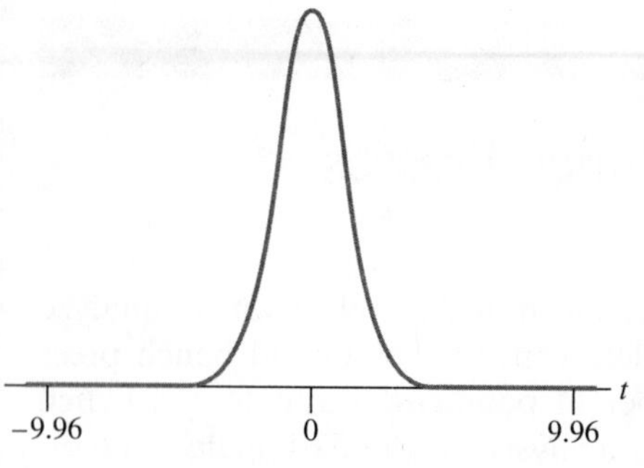

in Table 12.4 under the column labeled "T." The sample size equals $n = 57$, so $df = n - 2 = 55$.

4. **P-value:** The P-value, listed in Table 12.4 under the heading "P," is 0.000 rounded to three decimal places. Software reports the P-value for the two-sided alternative, H_a: $\beta \neq 0$. It is the two-tailed probability of "more extreme values" above 9.96 and below −9.96. See the margin figure.

5. **Conclusion:** If H_0 that the population slope β equals 0 were true, it would be extremely unusual to get a sample slope as far from 0 as $b = 1.49$. The P-value gives very strong evidence against H_0. We conclude that an association exists between the number of 60-pound bench presses and maximum bench press.

b. The one-sided alternative H_a: $\beta > 0$ predicts a positive association. For it, the P-value is halved, because it is then the right-tail probability of $t > 9.96$. This also equals 0.000, to three decimal places. On average, we infer that maximum bench press increases as the number of 60-pound bench presses increases.

Insight

In practice, studies must often rely on convenience samples. Results may be biased if the study subjects differ in an important way from those in the population of interest. Here, inference is reliable only to the extent that the sample is representative of the population of female high school athletes. This is a common problem with studies of this type, in which it would be difficult to arrange for a random sample of all subjects but a sample is conveniently available locally. We can place more faith in the inference if similar results occur in other studies.

The printout in Table 12.4 also contains a standard error and t statistic for testing that the population y-intercept α equals 0. This information is usually not of interest. Rarely is there any reason to test the hypothesis that a y-intercept equals 0.

Try Exercise 12.33, part a

You probably did not expect maximum bench press and number of 60-pound bench presses to be independent. The test confirms that they are positively associated. As usual, a confidence interval is more informative. It will help us learn about the actual size of the slope.

A Confidence Interval Tells Us How Closely We Can Estimate the Slope

A small P-value in the significance test of H_0: $\beta = 0$ suggests that the population regression line has a nonzero slope. To learn just how far the slope β falls from 0, we construct a confidence interval. A 95% confidence interval for β has the formula

$$b \pm t_{.025}(se).$$

The t-score is found by software or in a t table (such as Table B) with $df = n - 2$. This inference makes the same assumptions as the significance test.

Recall

For **95% confidence**, the error probability is 0.05, and the t-score with half this probability in the right tail is denoted $t_{.025}$ in the t table. ◀

Example 12

A 95% confidence interval for the slope

Estimating the Slope for Predicting Maximum Bench Press

Picture the Scenario

For the female athlete strength study, the sample regression equation is $\hat{y} = 63.5 + 1.49x$. From Table 12.4, the sample slope $b = 1.49$ has standard error $se = 0.150$.

Questions to Explore

a. Construct a 95% confidence interval for the population slope β.

b. What are the plausible values for the increase in maximum bench press, on average, for each additional 60-pound bench press that a female athlete can do?

Think It Through

a. For a 95% confidence interval, the $t_{.025}$ value for $df = n - 2 = 55$ is $t_{.025} = 2.00$. (If your software does not supply t-scores, you can find or approximate $t_{.025}$ from Table B.) The confidence interval is

$$b \pm t_{.025}(se) = 1.49 \pm 2.00(0.150),$$
which is 1.49 ± 0.30
or (1.2, 1.8).

b. We can be 95% confident that the population slope β falls between 1.2 and 1.8. On average, the maximum bench press increases by between 1.2 and 1.8 pounds for each additional 60-pound bench press that an athlete can do.

Insight

Confidence intervals and two-sided significance tests about slopes are consistent: When a two-sided test has P-value below 0.05, casting doubt on $\beta = 0$, the 95% confidence interval for β does not contain 0.

Try Exercise 12.33, part b

A confidence interval for β may not have a useful interpretation if a one-unit increase in x is relatively small (or large) in practical terms. In that case, we can estimate the effect for an increase in x that is a more relevant portion of the actual range of x values. For instance, an increase of 10 units in x has change 10β in the mean of y. To find the confidence interval for 10β, multiply the endpoints of the confidence interval for β by 10.

In the strength study, x = number of 60-pound bench presses varied between 2 and 35, with a standard deviation of 7.1. A change of 1 is very small. Let's estimate the effect of a 10-unit increase in x. Since the 95% confidence interval for β is (1.2, 1.8), the 95% confidence interval for 10β has endpoints $10(1.2) = 12$ and $10(1.8) = 18$. On average, we infer that the maximum bench press increases by at least 12 pounds and at most 18 pounds, for an increase of 10 in the number of 60-pound bench presses. For instance, the interval from 12 to 18 is also a 95% confidence interval for the difference between the mean maximum bench press for athletes who can do $x = 25$ bench presses and athletes who can do $x = 15$ (since $25 - 15 = 10$).

Inferences Also Apply to the Correlation

Since $r = b(s_x/s_y)$, the sample correlation $r = 0$ whenever the sample slope $b = 0$. Likewise, the population correlation equals 0 precisely when the population slope $\beta = 0$. We don't need to test whether or not the population correlation equals 0, because its t statistic is identical to the one for testing H_0: $\beta = 0$ using the sample slope b.

Software that reports the correlation matrix gives the option of getting the two-sided P-value for testing the significance of the correlation. Table 12.5 illustrates the way MINITAB reports the correlation matrix for the variables in the strength study (x = BP_60 and y = BP), including also two other variables that measure the maximum leg press (LP) and the number of 200-pound leg presses that the athlete can do (LP_200). For instance, the correlation between maximum bench press (BP) and the number of 60-pound bench presses (BP_60) equals 0.802. Its P-value for testing that the population correlation equals zero against the two-sided alternative is 0.000. This must be the same as the P-value (in Table 12.4 on p. 612) for the test for the slope. Table 12.5 lists the P-value underneath the correlation value. In fact, the P-value is 0.000 for the correlation for each of the six pairs of variables in this table.

Table 12.5 MINITAB Printout of Correlation Matrix for Strength Study

The value below a correlation is the P-value for testing that the population correlation equals 0.

	BP_60	BP	LP_200	
BP	0.802			
	0.000			
LP_200	0.611	0.578		
	0.000	0.000		
LP	0.669	0.501	0.793	← correlation
	0.000	0.000	0.000	← P-value

12.3 Practicing the Basics

12.32 *t*-score? A regression analysis is conducted with 25 observations.

a. What is the *df* value for inference about the slope β?

b. Which two t test statistic values would give a P-value of 0.05 for testing H_0: $\beta = 0$ against H_a: $\beta \neq 0$?

c. Which t-score would you multiply the standard error by in order to find the margin of error for a 95% confidence interval for β?

12.33 Predicting house prices For the House Selling Prices FL data file on the text CD, MINITAB results of a regression analysis are shown for 100 homes relating y = selling price (in dollars) to x = the size of the house (in square feet).

TRY

a. Show all steps of a two-sided significance test of independence. Could the sample association between these two variables be explained by random variation?

b. Show that a 95% confidence interval for the population slope is (64, 90).

c. A builder had claimed that the selling price increases $100, on average, for every extra square foot. Based on part b, what would you conclude about this claim?

House selling prices and size of home

Predictor	Coef	SE Coef	T	P
Constant	9161	10760	0.85	0.397
size	77.008	6.626	11.62	0.000

12.34 House prices in bad part of town Refer to the previous exercise. Of the 100 homes, 25 were in a part of town considered less desirable. For a regression analysis using y = selling price and x = size of house for these 25 homes,

a. You plan to test H_0: $\beta = 0$ against H_a: $\beta > 0$. Explain what H_0 means, and explain why a data analyst might choose a one-sided H_a for this test.

b. For this one-sided alternative hypothesis, how large would the t test statistic need to be in order to get a P-value equal to (i) 0.05 and (ii) 0.01?

12.35 Strength as leg press The high school female athlete strength study also considered prediction of y = maximum leg press (LP) using x = number of 200-pound leg presses (LP_200). MINITAB results of a regression analysis are shown.

```
The regression equation is LP = 234 + 5.27LP_200
Predictor    Coef      SE Coef      T       P
Constant     233.89    13.06       17.90   0.000
LP_200       5.2710    0.5469       9.64   0.000
R-Sq = 62.8%
```

a. Show all steps of a two-sided significance test of the hypothesis of independence.

b. Find a 95% confidence interval for the true slope. What do you learn from this that you cannot learn from the significance test in part a?

12.36 More boys are bad? A study of 375 women who lived in pre-industrial Finland (by S. Helle et al., *Science*, vol. 296, p. 1085, 2002) using Finnish church records from 1640 to 1870 found that there was roughly a linear relationship between y = lifelength (in years) and x = number of sons the woman had, with a slope estimate of -0.65 ($se = 0.29$).

a. Interpret the sign of the slope. Is the effect of having more boys good, or bad?

b. Show all steps of the test of the hypothesis that life length is independent of number of sons, for the two-sided alternative hypothesis. Interpret the P-value.

c. Construct a 95% confidence interval for the true slope. Interpret. Is it plausible that the effect is relatively weak, with true slope near 0?

12.37 More girls are good? Repeat the previous exercise using x = number of daughters the woman had, for which the slope estimate was 0.44 ($se = 0.29$).

12.38 CI and two-sided tests correspond Refer to the previous two exercises. Using significance level 0.05, what decision would you make? Explain how that decision is in agreement with whether 0 falls in the confidence interval. Do this for the data for both the boys and the girls.

12.39 Advertising and sales Each month, the owner of Café Gardens restaurant records y = monthly total sales receipts and x = amount spent that month on advertising, both in thousands of dollars. For the first four months of operation, the observations are as shown in the table. The correlation equals 0.857.

Advertising	Sales
0	4
1	7
2	8
5	9

a. Find the mean and standard deviation for each variable.

b. Using the formulas for the slope and the y-intercept or software, find the regression line.

c. The *se* of the slope estimate is 0.364. Test the null hypothesis that these variables are independent, using a significance level of 0.05.

12.40 GPA and study time—revisited Refer to the association you investigated in Exercises 12.7 and 12.21 between study time and college GPA. Using software with the data file you constructed, conduct a significance test of the hypothesis of independence, for the one-sided alternative of a positive population slope. Report the hypotheses, appropriate assumptions, sample slope, its standard error, the test statistic, the P-value, and interpret.

12.41 GPA and skipping class—revisited Refer to the association you investigated in Exercises 12.8 and 12.22 between skipping class and college GPA. Using software with the data file you constructed, construct a 90% confidence interval for the slope in the population. Interpret.

12.42 Student GPAs Refer to the Georgia Student Survey data file on the text CD. Treat college GPA as the response variable and high school GPA as the explanatory variable, and suppose these students are a random sample of all University of Georgia students.

a. Can you conclude that these variables are associated in that population? Show all steps of the relevant significance test with significance level 0.05, and interpret.

b. Find a 95% confidence interval for the population slope. Interpret the endpoints, and explain the correspondence with the result of the significance test in part a.

12.4 How the Data Vary Around the Regression Line

We've used regression to describe and to make inferences about the relationship between two quantitative variables. Now, we'll see what we can learn from the variability of the data around the regression line. We'll see that a type of residual helps us detect unusual observations. We'll also see that the sizes of the residuals affect how well we can predict y or the mean of y at any given value of x.

Recall

A **residual** is a prediction error—the difference $y - \hat{y}$ between an observed outcome y and its predicted value $\hat{y}$. For contingency tables, Section 11.4 constructed a **standardized residual** by dividing the difference between a cell count and its expected count by a standard error. In regression, an observation y takes the place of the cell count and the predicted value $\hat{y}$ takes the place of the expected cell count. ◀

Standardized Residuals

The magnitude of the residuals depends on the units of measurement for y. If we measure the bench press in kilograms instead of pounds, we'll get different residuals. A *standardized* version of the residual does not depend on the units. It equals the residual divided by a standard error that describes the sampling variability of the residuals. This ratio is called a **standardized residual**,

$$\text{Standardized residual} = \frac{y - \hat{y}}{se(y - \hat{y})}.$$

The *se* formula is complex, but we can rely on software to find this.

A standardized residual behaves like a z-score. It indicates how many standard errors a residual falls from 0. If the relationship is truly linear and if the standardized residuals have approximately a bell-shaped distribution, absolute values larger than about 3 should be quite rare. Often, observations with standardized residuals larger than 3 in absolute value represent outliers—observations that are far from what the model predicts.

Example 13

Standardized residual

Detecting an Underachieving College Student

Picture the Scenario

Two of the variables in the Georgia Student Survey data file on the text CD are college GPA and high school GPA (variables CGPA and HSGPA). These were measured for a sample of 59 students at the University of Georgia. Identifying y = CGPA and x = HSGPA, we find $\hat{y} = 1.19 + 0.64x$.

Question to Explore

MINITAB highlights observations that have standardized residuals with absolute value larger than 2 in a table of "unusual observations." Table 12.6 shows this data. Interpret the results for observation 59.

Table 12.6 Observations with Large Standardized Residuals in Student GPA Regression Analysis, as Reported by MINITAB

Obs	HSGPA	CGPA	Fit	Residual	St Resid ← standardized residuals
14	3.30	2.60	3.29	−0.69	−2.26R
28	3.80	2.98	3.61	−0.63	−2.01R
59	3.60	2.50	3.48	−0.98	− 3.14R

R denotes an observation with a large standardized residual.

Think It Through

Observation 59 is a student who had high school GPA $x = 3.60$, college GPA $y = 2.50$, predicted college GPA $\hat{y} = 3.48$ (the "fit"), and residual $= y - \hat{y} = -0.98$. The reported standardized residual of −3.14 indicates that the residual is 3.14 standard errors below 0. This student's actual college GPA is quite far below what the regression line predicts.

Insight

Based on his or her high school GPA and predicted college GPA, this student with an actual college GPA of 2.50 seems to be an underachiever in college.

Try Exercise 12.43

When a standardized residual is larger than about 3 in absolute value, check out if the observation is unusual in some way. Why does it fall away from the linear trend that the other points follow? Does it have too much influence on the results? As we've observed, a severe regression outlier can substantially affect the fit, especially when the value of the explanatory variable is relatively high or low and when the overall sample size is not very large.

Keep in mind that some large standardized residuals may occur just because of ordinary random variability. Table 12.6 reports three observations with standardized residuals having absolute value above 2 (labeled by R in the table). Even if the model is perfect we'd expect about 5% of the standardized residuals to have absolute value above 2 just by chance. So it's not at all surprising to find 3 such values out of 59 observations. This does not suggest that the model fits poorly.

Recall

About 5% of a normal distribution is more than 2 standard deviations from the mean. ◀

Checking the Response Distribution With a Histogram of Residuals

To detect unusual observations, it's helpful to construct a histogram of the residuals. This also helps us to check the inference assumption that the conditional distribution of *y* is normal. If this assumption is true, the residuals should have approximately a bell-shaped histogram. To check this, software can construct a histogram of the residuals or the standardized residuals.

Example 14

Using residuals to check our model assumptions

College GPA

Picture the Scenario

For the regression model of Example 13 predicting college GPA from high school GPA, Figure 12.9 is a MINITAB histogram of the standardized residuals. The margin shows a MINITAB boxplot of the standardized residuals.

Question to Explore

What does Figure 12.9 tell you?

Think It Through

The standardized residuals show some skew to the left. It may be that for a fixed value of high school GPA, college GPA tends to be skewed to the left, with some students doing much more poorly than the regression model would predict. The large negative standardized residual of −3.14 in Example 13, summarized by the left-most bar in Figure 12.9, may merely reflect skew in the distribution of college GPA. However, each of the three bars farthest to the left represents only a single observation, so this conclusion is tentative and requires a larger sample to check more thoroughly.

Insight

The sample size was 59. When n is not especially large, a graph like Figure 12.9 is an imprecise estimate of a corresponding graph for the population.

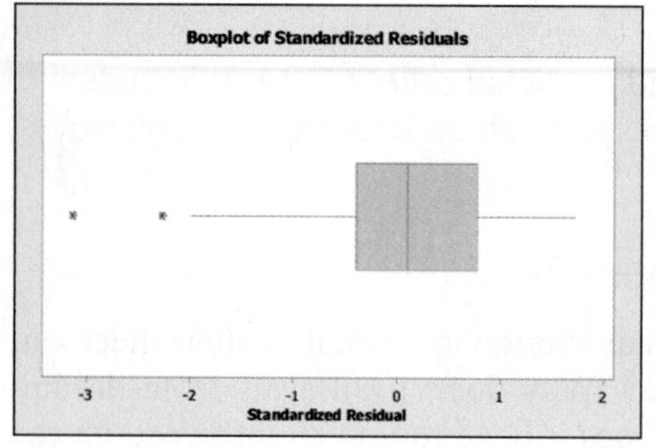

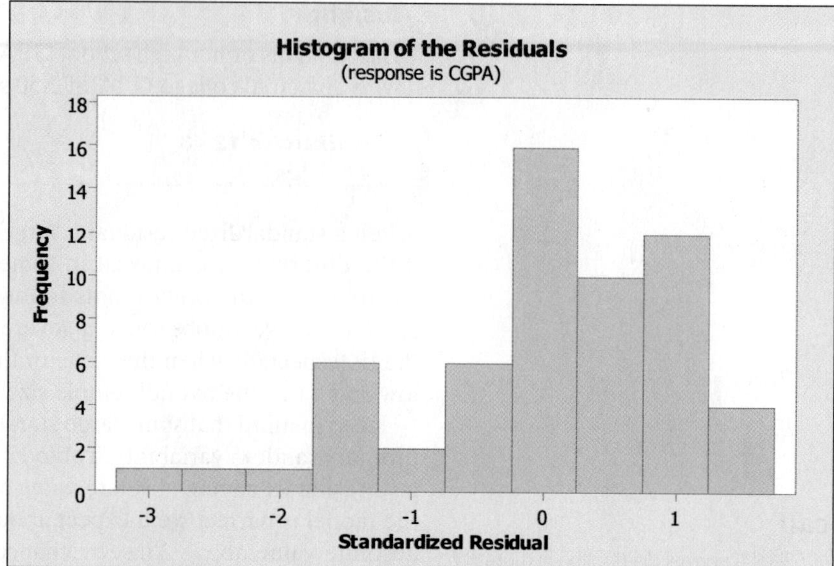

▲ **Figure 12.9** Histogram of Standardized Residuals for Regression Model Predicting College GPA. **Question** How many observations do the three left-most bars represent?

Although this graph shows some evidence of skew, much of this evidence is based on only three observations. This is not strong evidence that the conditional distribution is highly non-normal. In viewing such graphs, we need to be careful not to let a few observations influence our judgment too much. We're mainly looking for dramatic departures from the assumptions.

Try Exercise 12.45

If the distribution of the residuals is not normal, two-sided inferences about the slope parameter still work quite well. The t inferences are *robust.* The normality assumption is not as important as the assumption that the regression equation approximates well the true relationship between the predictor and the mean of y. If the model assumes a linear effect but the effect is actually U-shaped, for instance, descriptive and inferential statistics will be seriously faulty.

We'll study residuals in more detail in the next chapter. We'll learn there how to plot them to check other model assumptions.

Recall

From Section 12.1, σ **describes the variability** of y values from the mean of y at each fixed x value, that is, the variability of a **conditional distribution** of y values as shown in figure. ◀

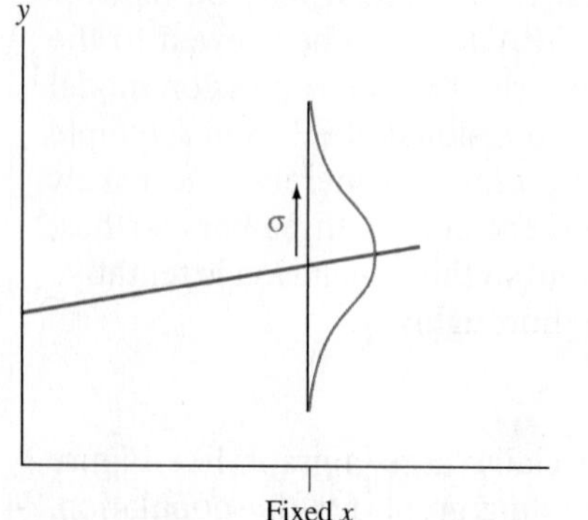

The Residual Standard Deviation and Mean Square Error (MSE)

Recall that the sample prediction equation $\hat{y} = a + bx$ estimates a population regression equation, $\mu_y = \alpha + \beta x$. For statistical inference, the regression model also assumes that the conditional distribution of y at a fixed value of x is normal, with the same standard deviation at each x. This standard deviation, denoted by σ, refers to the variability of y values for all subjects with the same value of x. This is a parameter that also can be estimated from the data.

The estimate of σ uses $\Sigma(y - \hat{y})^2$, the *residual sum of squares*, which summarizes sample variability about the regression line. The estimate, called the **residual standard deviation**, is

$$s = \sqrt{\frac{\Sigma(y - \hat{y})^2}{n - 2}}.$$

It describes the typical size of the residuals. The $n - 2$ term in the denominator is the df value of the t distribution used for inference about β in the previous section.

Example 15

Residual standard deviation

Variability of the Athletes' Strengths

Picture the Scenario

Let's return to the analysis of y = maximum bench press and x = number of 60-pound bench presses, for 57 female high school athletes. The prediction equation is $\hat{y} = 63.5 + 1.49x$. We'll see later in Table 12.8 that the residual sum of squares equals 3522.8.

Questions to Explore

a. Find the residual standard deviation of y.

b. Interpret the value you obtain in context at the sample mean value for x of 11.

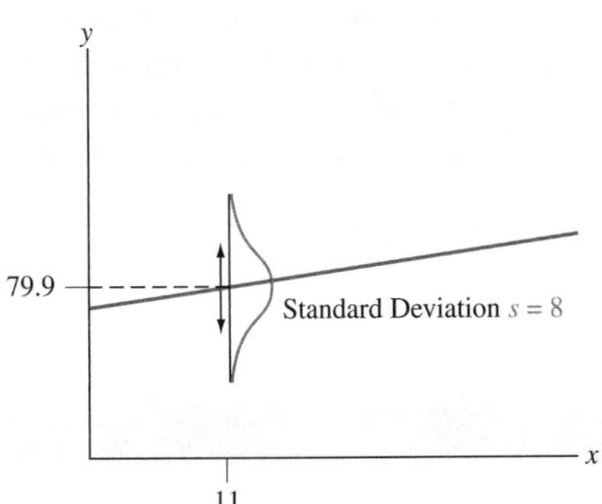

Think It Through

a. Since $n = 57$, $df = n - 2 = 55$. The residual standard deviation of the y values is

$$s = \sqrt{\frac{\Sigma(y - \hat{y})^2}{n - 2}} = \sqrt{\frac{3522.8}{55}} = \sqrt{64.1} = 8.0.$$

b. At any fixed value x of number of 60-pound bench presses, the model estimates that the maximum bench press values vary around a mean of $\hat{y} = 63.5 + 1.49x$ with a standard deviation of 8.0. See the figure in the margin.

At $x = 11$ (also the sample mean of the x values), the predicted maximum bench press is $\hat{y} = 63.5 + 1.49(11) = 79.9$. For female high school athletes who can do eleven 60-pound bench presses, we estimate that the maximum bench press values have a mean of about 80 pounds and a standard deviation of 8.0 pounds.

Recall

$$s_y = \sqrt{\frac{\Sigma(y - \bar{y})^2}{n - 1}}$$

Insight

Why does the residual standard deviation ($s = 8.0$) differ from the standard deviation ($s_y = 13.3$) of the 57 y values in the sample? The reason is that s_y refers to the variability of *all* the y values around their mean, not just those at a fixed x value. That is, $s_y = 13.3$ describes variability about the overall mean of $\bar{y} = 80$ for *all* 57 high school female athletes, whereas $s = 8.0$ describes variability at a fixed x value such as $x = 11$. See the margin figure.

When the correlation is strong, at a fixed value of x we see less variability than the overall sample has. For instance, at the fixed value $x = 11$, we describe the variability in maximum bench press values by $s = 8.0$, whereas overall we describe variability in maximum bench press values by $s_y = 13.3$.

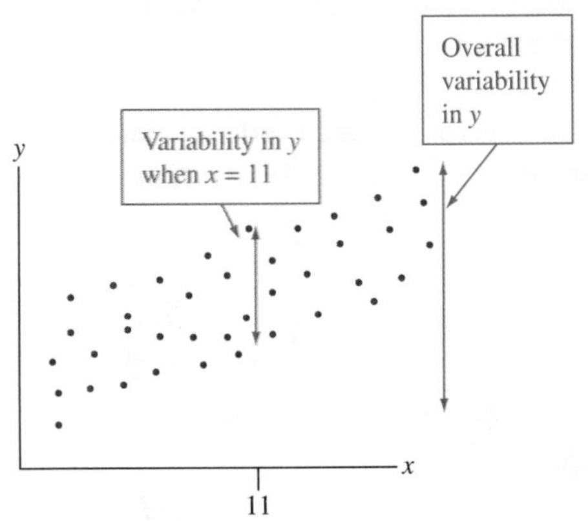

Try Exercise 12.49, part a

Using Intervals to Predict *y* Values and Their Mean at a Given Value of *x*

Consider all female high school athletes who can manage x = eleven 60-pound bench presses before fatigue. Based on the results for the sample, which do you think you can estimate more precisely: y = maximum bench press for a randomly chosen one of them, or μ_y = the mean of the y values for all of them? We'll see that the estimate is the same in each case, but the margins of error are quite different.

For the straight-line regression model, we estimate μ_y, the population mean of y at a given value of x, by $\hat{y} = a + bx$. How good is this estimate? We can use its *se* to construct a 95% **confidence interval for μ_y**. This interval is

$$\hat{y} \pm t_{.025}(se \text{ of } \hat{y}).$$

Again, the t-score has $df = n - 2$. The standard error formula is complex (we'll show a simple approximation later), and in practice we rely on software.

The estimate $\hat{y} = a + bx$ for the mean of y at a fixed value of x is also a prediction for the outcome on y for a particular subject at that value. With most regression software you can form an interval within which an outcome y itself is likely to fall. This interval is called a **prediction interval for y.**

What's the difference between the prediction interval for y and the confidence interval for μ_y? The prediction interval for y is an inference about where individual observations fall, whereas the confidence interval for μ_y is an inference about where a population mean falls. Use a prediction interval for y if you want to predict where a single observation on y will fall. Use a confidence interval for μ_y if you want to estimate the mean of y for everyone having a particular x value. These inferences make the same assumptions as the regression inferences of the previous section.

Example 16

Confidence and prediction intervals

Maximum Bench Press and Its Mean

Picture the Scenario

We've seen that the equation $\hat{y} = 63.5 + 1.49x$ predicts y = maximum bench press using x = number of 60-pound bench presses. For $x = 11$, Table 12.7 shows how MINITAB reports a confidence interval (CI) for the population mean of y and a prediction interval (PI) for a single y value. The predicted value, $\hat{y}$, is reported under the heading "Fit."

Table 12.7 MINITAB Output for Confidence Interval (CI) and Prediction Interval (PI) on Maximum Bench Press for Athletes Who Do Eleven 60-Pound Bench Presses before Fatigue

```
Predicted Values for New Observations

New Obs      Fit      SE Fit         95% CI               95% PI

      1     79.94     1.06      (77.81, 82.06)      (63.76, 96.12)

Values of Predictors for New Observations

New Obs     BP_60

      1     11.0
```

Questions to Explore

a. Using $\hat{y}$ and its *se* reported in Table 12.7, find and interpret a 95% *confidence interval* for the population mean of the maximum bench press values for all female high school athletes who can do $x = 11$ sixty-pound bench presses.

b. Report and interpret a 95% *prediction interval* for a single new observation on maximum bench press, for a randomly chosen female high school athlete with $x = 11$.

Think It Through

a. At $x = 11$, the predicted maximum bench press is $\hat{y} = 63.5 + 1.49(11) = 79.94$ pounds, the "fit." Table 12.7 reports a standard error for this estimate, $se = 1.06$, under the heading "SE Fit." With $n = 57$ we've seen that $df = n - 2 = 55$ and the t-score is $t_{.025} = 2.00$. So, the 95% confidence interval for the population mean of maximum bench press values at $x = 11$ is

$$\hat{y} \pm t_{.025}(se \text{ of } \hat{y}),$$
$$\text{which is } 79.94 \pm 2.00(1.06),$$
$$\text{or } 79.94 \pm 2.12,$$
$$\text{that is } (77.8, 82.1).$$

This is labeled as "95% CI" on the MINITAB printout. For all female high school athletes who can do eleven 60-pound bench presses, we are 95% confident that the mean of their maximum bench press values falls between about 78 and 82 pounds.

b. MINITAB reports the 95% prediction interval (63.8, 96.1) under the heading "95% PI." This predicts where maximum bench press y will fall for a randomly chosen female high school athlete having $x = 11$. Equivalently, this refers to where 95% of the corresponding population values fall. For all female high school athletes who can do eleven 60-pound bench presses, we predict that 95% of them have maximum bench press between about 64 and 96 pounds. Look at Figure 12.1, reproduced in the margin. Locate $x = 11$. Of all possible data points at that x value, we predict that 95% of them would fall between about 64 and 96.

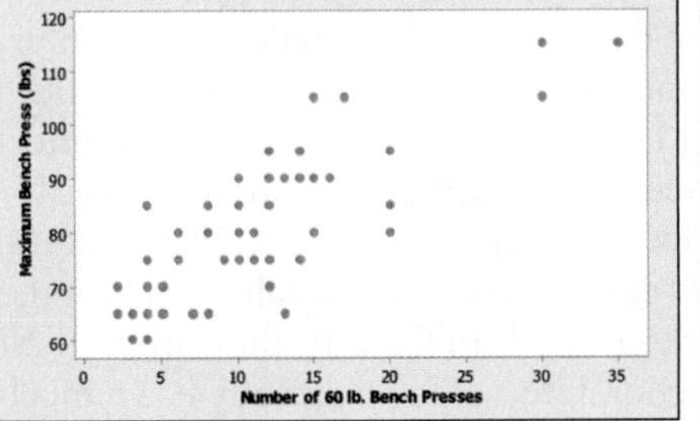

Insight

The 95% prediction interval (63.8, 96.1) predicts the maximum bench press y for a randomly chosen female high school athlete having $x = 11$. The 95% confidence interval (77.8, 82.1) estimates the mean of such y values for all female high school athletes having $x = 11$. The prediction interval for a single observation y is much wider than the confidence interval for the mean of y. In other words, you can estimate a population mean more precisely than you can predict a single observation.

Try Exercise 12.47

The margins of error for these intervals use the residual standard deviation s. For an approximately normal conditional distribution for y, about 95% of the observations fall within about 2 standard deviations of the true mean μ_y at a particular value of x. For large n, $\hat{y}$ is close to μ_y, especially near the mean of the x values. Also, for large n the residual standard deviation s estimates the standard deviation σ well. So we could predict that about 95% of the y values would fall within $\hat{y} \pm 2s$. In fact, this is roughly what software does to form a prediction interval, when n is large and x is at or near the mean. The margin of error is approximately a t-score times s for predicting an individual observation and a t-score times $s/\sqrt{n}$ for estimating a mean.

SUMMARY: Prediction Interval for y and Confidence Interval for μ_y at Fixed Value of x

For large samples with an x value equal to or close to the mean of x,

- The 95% **prediction interval** for y is approximately $\hat{y} \pm 2s$.
- The 95% **confidence interval** for μ_y is approximately

$$\hat{y} \pm 2(s/\sqrt{n}),$$

where s is the residual standard deviation. Software uses *exact* formulas. We show these *approximate* formulas here merely to give a sense of what these intervals do. *In practice, use software.*

For instance, for female student athletes, about 95% of the observations fall within about $2s = 2(8.0) = 16$ of the true mean at a particular value of x. At $x = 11$, which is the mean of x for the sample, $\hat{y} = 79.94 \approx 80$. So we predict that about 95% of the maximum bench press values would fall within $\hat{y} \pm 2s$, which is 80 ± 16, or between 64 and 96. This is the 95% PI result in Table 12.7. This formula is only an approximate one (because of using $\hat{y}$ as an approximation for μ_y and 2 as an approximation for the t-score), and you should rely on software for more precise results.

Use these confidence intervals and prediction intervals with caution. For these intervals to be valid, the true relationship must be close to linear, with about the same variability of y values at each fixed x value. You can get a rough visual check from the scatterplot. The variability of points around the regression line should be similar for the various possible x values. If it isn't, don't use these confidence and prediction intervals.

Because the model assumptions never hold exactly in practice, these inferences are sometimes inaccurate. For instance, suppose the variability in y values is smaller at small values of x and larger at large values of x. The prediction interval will have similar widths in each case. See Figure 12.10, which shows curves for the lower and upper endpoints of prediction intervals at all the possible x values. At a fixed x value, there should be a 95% chance a data point falls between the curves. In Case A, prediction intervals are justified. In Case B, they are not. At a large value of x in Case B, a 95% prediction interval for y actually has a much smaller than 95% chance (perhaps even less than 50%) of containing a value of y.

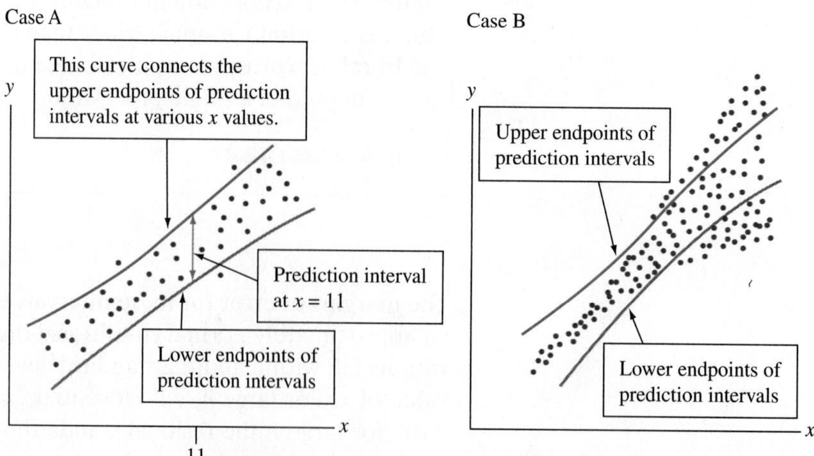

▲ **Figure 12.10 When a Regression Model Assumption Is Badly Violated, Prediction Intervals May Perform Poorly.** The curves plot how the endpoints of a 95% prediction interval change as x changes. For Case A, inferences are valid. For Case B, a 95% prediction interval has actual probability of containing y above 0.95 for small x and below 0.95 for large x.
Question How can you tell from the figure that Case B, will lead to inaccurate inferences?

The Analysis of Variance (ANOVA) Table Summarizes Variability

Software reports the sums of squares (SS) used in estimating standard deviations and variances (their squares) in an **analysis of variance (ANOVA) table**. Table 12.8 shows the ANOVA table that MINITAB software reports for the strength study.

Table 12.8 Analysis of Variance (ANOVA) Table for the Strength Study

Analysis of Variance

Source	DF	SS	MS	F	P
Regression	1	6351.8	6351.8	99.17	0.000
Residual Error	55	3522.8	64.1		
Total	56	9874.6			

In the ANOVA table, the *total sum of squares* is $\Sigma(y - \bar{y})^2$. It summarizes the total variability of the y values. The ANOVA table breaks the total SS into two parts. One part is the *residual sum of squares*, $\Sigma(y - \hat{y})^2$. It represents the error in using the regression line to predict y. The other part is called the *regression sum of squares*. Its formula is

$$\text{regression SS} = \Sigma(\hat{y} - \bar{y})^2, \text{and it equals } \Sigma(y - \bar{y})^2 - \Sigma(y - \hat{y})^2.$$

It summarizes how much less error there is in predicting y using the regression line compared to using $\bar{y}$. Recall that this difference is the numerator of r^2. The sum

$$\text{Regression SS} + \text{Residual SS} = \text{Total SS}.$$

In Table 12.8, regression SS + residual SS = 6351.8 + 3522.8 = 9874.6, the total SS. Each sum of squares in the ANOVA table has an associated degrees of freedom value. For instance, *df* for the residual sum of squares equals $n - 2 = 57 - 2 = 55$. The ratio of a sum of squares to its *df* value is called a **mean square**. It is listed in the ANOVA table under the heading MS.

In Table 12.8 the ratio of $\Sigma(y - \hat{y})^2 = 3522.8$ to its $df = 55$ gives 64.1. This ratio is the **mean square error**, often abbreviated as **MS error** or as **MSE**. It is roughly a mean of the squared errors in using the regression line to predict y. Equaling $s^2 = \Sigma(y - \hat{y})^2/(n - 2)$, it estimates the variance of the conditional distributions. In practice, it's easier to interpret its square root, s, the residual standard deviation of y.

Mean Square Error

The **mean square error** is the residual sum of squares divided by its *df* value (sample size − 2). Its square root s is a typical size of a residual (that is, a prediction error).

The ANOVA *F* Statistic Also Tests for an Association

The ANOVA table contains values labeled *F* and P. The *F* value is the ratio of the mean squares,

$$F = \frac{\text{Mean square for regression}}{\text{Mean square error}}.$$

This is an alternative test statistic for testing H_0: $\beta = 0$ against H_a: $\beta \neq 0$. In fact, it can be shown that *F is the square of the t statistic* for this null hypothesis. Using the sampling distribution of *F* values, which we'll discuss in the next chapter, we get exactly the same P-value as with the *t* test for the two-sided alternative hypothesis.

In Table 12.8, the *F* statistic equals the ratio of MS values, 6351.8/64.1 = 99.17. Table 12.4 listing the parameter estimates and their standard errors showed the *t* test:

Predictor	Coef	SE Coef	T	P
BP_60	1.4911	0.150	9.96	0.000

The F statistic of 99.17 here is the square of the t statistic value $t = 9.96$. The P-value, labeled as P on Table 12.8, is 0.000 to three decimal places, necessarily the same as the P-value for the two-sided t test.

Why do we need the F statistic if it does the same thing as a t statistic? In the next chapter, which generalizes regression to handle multiple predictors, we'll see that the main use of an F statistic is for testing effects of several predictors at once. We can use the t distribution to test the effect of only a single predictor in a given test.

12.4 Practicing the Basics

12.43 Poor predicted strengths The MINITAB output shows TRY the large standardized residuals for the female athlete strength study.

Large standardized residuals for strength study:

Obs	BP_60	BP	Fit	Residual	St Resid
10	15.0	105.00	85.90	19.10	2.41R
25	17.0	105.00	88.88	16.12	2.04R
37	13.0	65.00	82.92	−17.92	−2.26R

a. Explain how to interpret all the entries in the row of the output for athlete 10.

b. Out of 57 observations, is it surprising that 3 observations would have standardized residuals with absolute value above 2.0? Explain.

12.44 Loves TV and exercise For the Georgia Student Survey file on the text CD, let y = exercise and x = watch TV. One student reported watching TV an average of 180 minutes a day and exercising 60 minutes a day. This person's residual was 48.8 and standardized residual was 6.41.

a. Interpret the residual, and use it to find the predicted value of exercise.

b. Interpret the standardized residual.

12.45 Bench press residuals The figure is a histogram of the TRY standardized residuals for the regression of maximum bench press on number of 60-pound bench presses, for the high school female athletes.

a. Which distribution does this figure provide information about?

b. What would you conclude based on this figure?

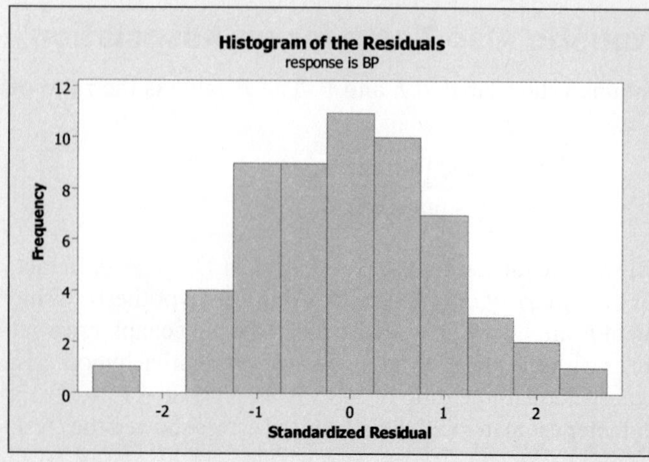

12.46 Predicting house prices The House Selling Prices FL data file on the text CD has several predictors of house selling prices. The table here shows the ANOVA table for a regression analysis of y = the selling price (in thousands of dollars) and x = the size of house (in thousands of square feet). The prediction equation is $\hat{y} = 9.2 + 77x$.

ANOVA table for selling price and size of house:

Source	DF	SS	MS	F	P
Regression	1	182220	182220	135.07	0.000
Residual Error	98	132213	1349		
Total	99	314433			

a. What was the sample size? (*Hint:* You can figure it out from the residual *df.*)

b. The sample mean house size was 1.53 thousand square feet. What was the sample mean selling price? (*Hint:* What does $\hat{y}$ equal when $x = \bar{x}$?)

c. Estimate the standard deviation of the selling prices for homes that have $x = 1.53$. Interpret.

d. Report an approximate prediction interval within which you would expect about 95% of the selling prices to fall, for homes of size $x = 1.53$.

12.47 Predicting clothes purchases For a random sample TRY of children from a school district in South Carolina, a regression analysis is conducted of y = amount spent on clothes in the past year (dollars) and x = year in school. MINITAB reports the tabulated results for observations at $x = 12$.

Predicted Values for New Observations

NewObs	Fit	SEFit	95% CI	95% PI
1	448.0	10.6	(427,469)	(101,795)

a. Interpret the value listed under "Fit."

b. Interpret the interval listed under "95% CI."

c. Interpret the interval listed under "95% PI."

12.48 CI versus PI Using the context of the previous exercise, explain the difference between the purpose of a 95% prediction interval (PI) for an observation and a 95% confidence interval (CI) for the mean of y at a given value of x. Why would you expect the PI to be wider than the CI?

12.49 ANOVA table for leg press Exercise 12.35 referred to TRY an analysis of leg strength for 57 female athletes, with y = maximum leg press and x = number of 200-pound leg presses until fatigue, for which $\hat{y} = 233.89 + 5.27x$.

The table shows ANOVA results from SPSS for the regression analysis.

ANOVA[b]

Model		Sum of Squares	df	Mean Square	F	Sig.
1	Regression	121082.4	1	121082.400	92.875	.000[a]
	Residual	71704.442	55	1303.717		
	Total	192786.8	56			

a. Show that the residual standard deviation is 36.1. Interpret it.

b. For this sample, $\bar{x} = 22.2$. For female athletes with $x = 22$, what would you estimate the variability to be of their maximum leg press values? If the y values are approximately normal, find an interval within which about 95% of them would fall.

12.50 Predicting leg press Refer to the previous exercise. MINITAB reports the tabulated results for observations at $x = 25$.

```
Predicted Values for New Observations
New Obs    Fit    SE Fit         95% CI              95% PI
   1     365.66    5.02   (355.61,375.72) (292.61,438.72)
```

a. Show how MINITAB got the "Fit" of 365.66.

b. Using the predicted value and *se* value, explain how MINITAB got the interval listed under "95% CI." Interpret this interval.

c. Interpret the interval listed under "95% PI."

12.51 Variability and F Refer to the previous two exercises.

a. In the ANOVA table, show how the Total SS breaks into two parts, and explain what each part represents.

b. From the ANOVA table, explain why the overall sample standard deviation of y values is $s_y = \sqrt{192787/56} = 58.7$. Explain the difference between the interpretation of this standard deviation and the residual s of 36.1.

c. Exercise 12.35 reported a t statistic of 9.64 for testing independence of these variables. Report the F test statistic from the table here, and explain how it relates to that t statistic.

12.52 Assumption violated For prediction intervals, an important inference assumption is a constant residual standard deviation of y values at different x values. In practice, the residual standard deviation often tends to be larger when μ_y is larger.

a. Sketch a hypothetical scatterplot for which this happens, using observations for the previous year on x = family income and y = amount donated to charity.

b. Explain why a 95% prediction interval would not work well at very small or at very large x values.

12.53 Understanding an ANOVA table For a random sample of U.S. counties, the ANOVA table shown refers to hypothetical data on x = percentage of the population aged over 50 and y = per capita expenditure (dollars) on education.

a. Fill in the blanks in the table.

b. For what hypotheses can the F test statistic be used?

Source	DF	SS	MS	F
Model	1	200000	______	______
Error	31	700000	______	
Total	32	900000		

12.54 Predicting GPA Refer to the Georgia Student Survey data file on the text CD. Regress y = college GPA on x = high school GPA.

a. Stating the necessary assumptions, find a 95% confidence interval for the mean college GPA for all University of Georgia students who have high school GPA = 3.6.

b. Find a 95% prediction interval for college GPA of a randomly chosen student having high school GPA = 3.6. Interpret.

c. Explain the difference between the purposes of the intervals in part a and part b.

12.55 GPA ANOVA Report the ANOVA table for the previous exercise.

a. Show how the Total SS breaks into two parts, and explain what each part represents.

b. Find the estimated residual standard deviation of y. Interpret it.

c. Find the sample standard deviation s_y of y values. Explain the difference between the interpretation of this standard deviation and the residual standard deviation in part b.

12.5 Exponential Regression: A Model for Nonlinearity

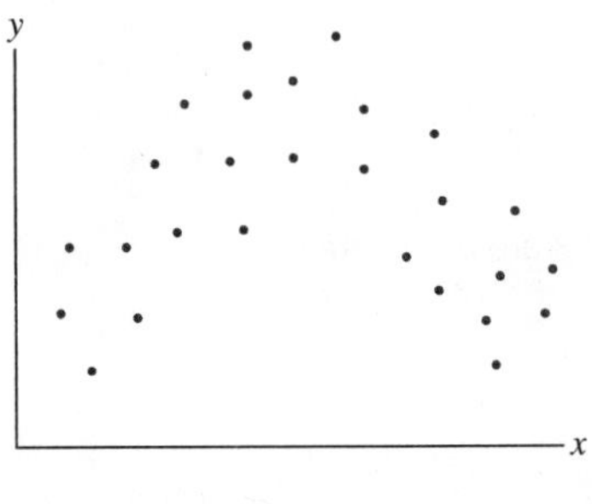

The straight line is by far the most common equation for a regression model. Sometimes, however, a scatterplot indicates substantial curvature in a relationship. In such cases, equations that provide curvature are more appropriate.

Occasionally a scatterplot has a parabolic appearance: As x increases, y tends to go up and then it goes back down (as shown in the margin figure), or the reverse. Then the regression model can use an equation giving the shape of a parabola. More often, y tends to continually increase or to continually decrease, but the trend shows curvature. Let's see a mechanism for how this can happen.

Exponential regression

Example 17

Growth in Population Size

Picture the Scenario

The population size of the United States has been growing rapidly in recent years, much of it due to immigration. According to the 2010 census, the population size was about 309 million on April 1, 2010.

Questions to Explore

a. Suppose that the rate of growth after 2010 is 2% a year. That is, the population is 2% larger at the end of each year than it was at the beginning of the year. Find the population size after (i) 1 year, (ii) 2 years, and (iii) 10 years.

b. Give a formula for the population size in terms of x = number of years since 2010.

Think It Through

a. With a 2% growth rate, the population size one year after 2010 is $309 \times 1.02 = 315.2$ million. That is, increasing the population size by 2% corresponds to multiplying it by 1.02. The population size after 2 years is 2% higher than this, or

$$315.2 \times 1.02 = (309 \times 1.02) \times 1.02 = 309 \times (1.02)^2 = 321.5 \text{ million.}$$

After 10 years, the population size is $309 \times 1.02 \times 1.02 \times 1.02 \times \ldots \times 1.02 = 309 \times (1.02)^{10} = 376.7$ million.

b. Can you see the pattern? For each additional year, we mulitply by another factor of 1.02. After x years, the population size is 309×1.02^x million.

Population Size

2000 2100

Insight

The population size formula, 309×1.02^x, is called **exponential growth**. Plotted, the response goes up faster than a straight line. See the margin figure. The amount of change in y per unit change in x increases as x increases. After 100 years (that is, in the year 2110), taking $x = 100$, population size $= 309 \times (1.02)^{100} = 2238.6$ million, more than 2.2 billion people!

Try Exercise 12.56

A statistical analysis that uses a regression function with x in the *exponent* is called **exponential regression**.

SUMMARY: Exponential Regression Model

An **exponential regression model** has the formula

$$\mu_y = \alpha\beta^x$$

for the mean μ_y of y at a given value of x, where α and β are parameters.

The formula 309×1.02^x for U.S. population size in Example 17 has the form $\alpha\beta^x$, with $\alpha = 309$, $\beta = 1.02$, and x = number of years since 2010. Exponential regression can model quantities that tend to increase by increasingly large amounts over time.

In the exponential regression equation, the explanatory variable x appears as the exponent of a parameter. Unlike with straight-line regression, the mean μ_y and the effect parameter β can take only positive values. As x increases, the mean μ_y continually increases when $\beta > 1$. It continually decreases when $0 < \beta < 1$. Figure 12.11 shows the shape for the two cases. We provide interpretations for the model parameters later in this section.

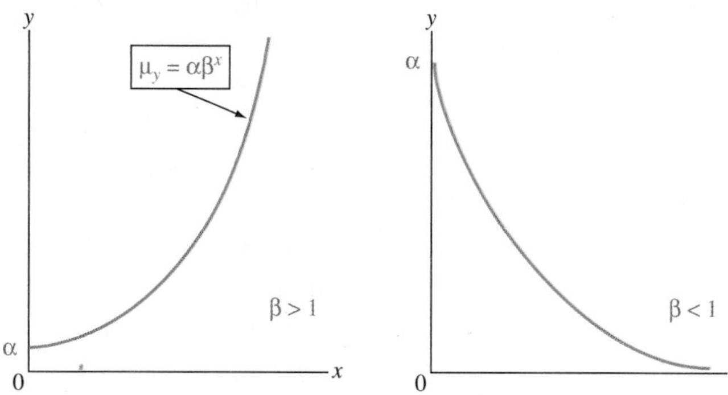

▲ **Figure 12.11** The Exponential Regression Curve for $\mu_y = \alpha\beta^x$. **Question** Why does μ_y decrease if $\beta = 0.5$, even though $\beta > 0$?

For exponential regression, the logarithm of the mean is a linear function of x. When the exponential regression model holds, a plot of the log of the y values versus x should show an approximate straight-line relation with x. Don't worry if you have forgotten logarithms. You will not need to use logarithms to understand how to fit the model. Software can do it. It's more important to know when it is appropriate and to be able to interpret the model fit.

Example 18

Exponential regression

Explosion in Number of Facebook Users

Picture the Scenario

Table 12.9 shows the number of people (in millions) worldwide using Facebook between 2004 and 2011. Figure 12.12 plots these values. They increase over time, and the amount of increase from one year to the next seems to itself increase over time.

Table 12.9 also shows the logarithm of the Facebook-user counts, using base-10 logs. Figure 12.13 plots these log values over time. They appear to grow approximately linearly. In fact, the correlation between the log of the population size and the date is 0.985, a very strong linear association. This suggests that growth in Facebook users over this time period was approximately exponential.

Question to Explore

Let x denote the number of days since December 1, 2004. That is, December 1, 2004 is $x = 0$, December 1, 2005 is $x = 365$, and so forth up to June 1, 2011, which is $x = 2373$. Software provides the exponential regression model fitted to y = number of Facebook users and x gives

$$\hat{y} = 1.9559 \times 1.00275^x.$$

Table 12.9 Number of Facebook Users Worldwide (in Millions)

Date	Number of Days Since December 1, 2004 x	Number of Users (in millions) y	Log Number Users Log(y)	Predicted Number $\hat{y}$
12/01/04	0	1	0	1.9559
12/01/05	365	5.5	0.740363	5.329316
12/01/06	730	12	1.079181	14.52099
04/01/07	851	20	1.30103	20.24461
10/01/07	1034	50	1.69897	33.46325
08/01/08	1339	100	2	77.32727
01/01/09	1492	150	2.176091	117.7094
02/01/09	1523	175	2.243038	128.1693
04/01/09	1582	200	2.30103	150.7133
07/01/09	1673	250	2.39794	193.5016
09/01/09	1735	300	2.477121	229.4194
12/01/09	1826	350	2.544068	294.5529
02/01/10	1888	450	2.653213	349.2278
07/01/10	2038	500	2.69897	527.2413
09/01/10	2100	550	2.740363	625.1079
01/01/11	2222	600	2.778151	873.8982
02/01/11	2253	650	2.812913	951.5544
06/01/11	2373	750	2.875061	1322.983

Source: Data from Facebook User Growth Chart—2004–2011 by Ben Foster (www.benphoster.com/facebook_user_growth_chart_2004_2010/).

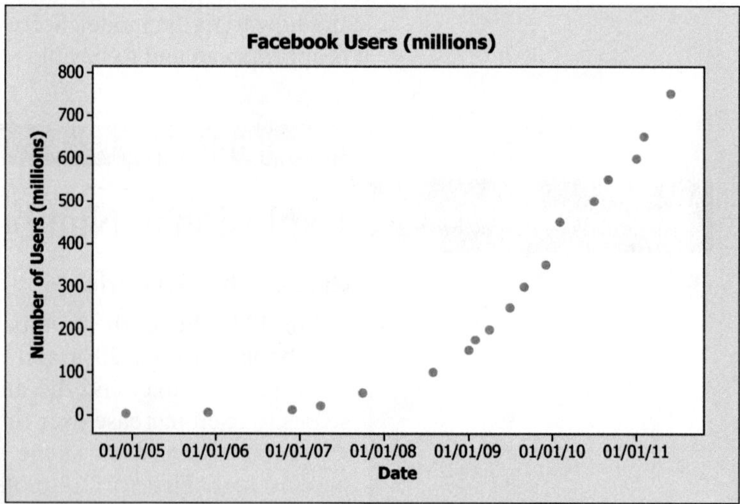

▲ **Figure 12.12** Plot of Number of Facebook Users (millions) from December 2004 to June 2011. *Source:* Graph by Ben Foster (twitter.com/benphoster) and updated at benphoster.com/facebookgrowth.

What does this equation predict for the number of Facebook users on Dec. 1, 2004, Dec. 1, 2007 (1095 days after Dec. 1, 2004), and on Dec. 1, 2015 (4017 days after Dec. 1, 2004)?

Think It Through

On December 1, 2004, $x = 0$, so the predicted number of Facebook users is $\hat{y} = 1.9559 \times 1.00275^0 = 1.9559$ million users.

For the day December 1, 2007, $x = 1095$ days, and the predicted number is $\hat{y} = 1.9559 \times 1.00275^{1095} = 39.566$ million users.

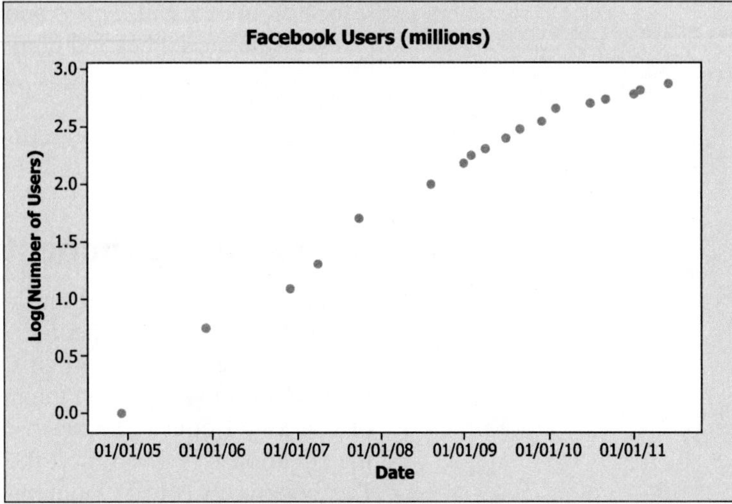

▲ **Figure 12.13** Plot of Log of Number of Facebook Users Between 2004 and 2011. When the log of the response has an approximate straight-line relationship with the explanatory variable, the exponential regression model is appropriate.

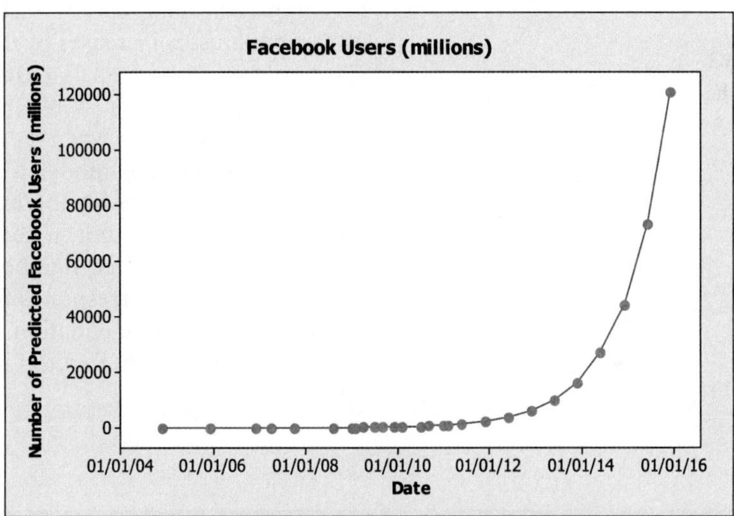

▲ **Figure 12.14** Plot of Predicted Number of Facebook Users Between 2004 and 2011. The values after June 2011 are extrapolations beyond scope of data.

December 1, 2015 is 4017 days after December 1, 2004, and the prediction is $\hat{y} = 1.9559 \times 1.00275^{4017} = 120{,}866.1428$ million users, that is, almost 120 billion people. You should be skeptical of this prediction because the world population size at the end of 2010 is about 6.8 billion people and projected to be between 7 and 8 billion people by 2015. It is impossible to have more Facebook users than people on the planet.

Insight

Table 12.9 shows the predicted value for each date recorded between 2004 and 2011. Comparing the observed values in Table 12.9 to the predicted values for this model, we see that the model fits the data pretty well. However, there is some indication that the actual growth was slowing a bit in 2010, as the observation for September 1, 2010 (550 million users), was quite a bit less than the predicted value (625.1079 million users). Figure 12.14 extends the graph to show also data for the years between 2011 and 2020. The prediction model extrapolates the data after June 2011 using the exponential model. It's extremely unlikely that the world population will be as high as the prediction for Facebook users. This is another example of the danger of extrapolating a regression model beyond predictor values for which we have data.

Try Exercise 12.58

Interpreting Exponential Regression Models

From what we've just seen in the previous example, here's how to interpret the parameters in the exponential regression model, $\mu_y = \alpha\beta^x$. The parameter α represents the mean of y when $x = 0$, since $\beta^0 = 1$. The parameter β represents the *multiplicative* effect on the mean of y for a one-unit increase in x. The mean of y at $x = 1$ equals the mean of y at $x = 0$ multiplied by β. For instance, for the equation $\hat{y} = 1.9559 \times 1.00275^x$ with $\beta = 1.00275$, the predicted number of Facebook users is 1.9559 million on December 1, 2004 (for which $x = 0$), and on December 2, 2004 (for which $x = 1$) it equals 1.9559 million times 1.00275.

By contrast, the parameter β in the straight-line model $\mu_y = \alpha + \beta x$ represents an *additive* effect on the mean of y for a one-unit increase in x. The mean of y at $x = 1$ equals the mean of y at $x = 0$ plus β. The straight-line model fitted to the Facebook users data was $\hat{y} = -211 + 0.330x$. This model predicts that the number of Facebook users increases by 0.330 million people every day. The model is not appropriate for these data. For example, it gives an inaccurate prediction for January 1, 2009: Plugging in $x = 1492$, it predicts $\hat{y} = -211 + 0.330(1492) = 281.36$ million Facebook users when the actual value was 150 million.

For the straight-line model, μ_y changes by the same quantity for each one-unit increase in x, whereas for the exponential model, μ_y changes by the same percentage for each one-unit increase. For the exponential regression model with Table 12.9, the predicted number of Facebook users multiplies by 1.00275 each day. This equation corresponds to a predicted 0.275% growth per day. With this exponential regression equation, over time the actual quantity of growth tends to get larger every day, as Figure 12.12 showed.

12.5 Practicing the Basics

12.56 Savings grow exponentially You invest $100 in a savings account with interest compounded annually at 10%. (TRY)

a. How much money does the account have after one year?

b. How much money does the account have after five years?

c. How much money does the account have after x years?

d. How many years does it take until your savings more than double in size?

12.57 Growth by year versus decade You want your savings to double in a decade.

a. Explain why 7.2% interest a year would do this. (*Hint:* What does $(1.072)^{10}$ equal?)

b. You might think that 10% interest a year would give 100% interest (that is, double your savings) over a decade. Explain why interest of 10% a year would actually cause your savings to multiply by 2.59 over a decade.

12.58 U.S. population growth The table shows the approximate U.S. population size (in millions) at 10-year intervals beginning in 1900. Let x denote the number of decades since 1900. That is, 1900 is $x = 0$, 1910 is $x = 1$, and so forth. The exponential regression model fitted to y = population size and x gives $\hat{y} = 81.14 \times 1.1339^x$. (TRY)

U.S. population sizes (in millions) from 1900 to 2010

Year	Population Size	Year	Population Size	Year	Population Size
1900	76.2	1940	132.1	1980	226.5
1910	92.2	1950	150.7	1990	248.7
1920	106.0	1960	179.3	2000	281.4
1930	122.8	1970	203.3	2010	308.7

Source: U.S. Census Bureau.

a. Show that the predicted population sizes are 81.14 million in 1900 and 323.3 million in 2010.

b. Explain how to interpret the value 1.1339 in the prediction equation.

c. The correlation equals 0.98 between the log of the population size and the year number. What does this suggest about whether or not the exponential regression model is appropriate for these data?

12.59 Future shock Refer to the previous exercise, for which predicted population growth was 14.18% per decade. Suppose the growth rate is now 15% per decade. Explain why the population size will (a) double after five decades, (b) quadruple after 100 years (10 decades), and (c) be 16 times its original size after 200 years. (The exponentially increasing function has the property that its doubling time is constant. This is a fast increase, even though the annual rate of growth seems small.)

12.60 Age and death rate Let x denote a person's age and let y be the death rate, measured as the number of deaths per thousand individuals of a fixed age within a period of a year. For men in the United States, these variables follow approximately the equation $\hat{y} = 0.32(1.078)^x$.

a. Interpret 0.32 and 1.078 in this equation.

b. Find the predicted death rate when age is (i) 20, (ii) 50, and (iii) 80.

c. In every how many years does the death rate double? (*Hint:* What is x such that $(1.078)^x = 2$?)

12.61 Leaf litter decay Ecologists believe that organic material decays over time according to an *exponential decay* model. This is the case $0 < \beta < 1$ in the exponential regression model, for which μ_y decreases over time. The rate of decay is determined by a number of factors, including composition of material, temperature, and humidity. In an experiment carried out by researchers at the University of Georgia Ecology Institute, leaf litter was allowed to sit for a 20-week period in a bag in a moderately forested area. Initially, the total weight of the organic mass in the bag was 75.0 kg. Each week, the remaining amount (y) was measured. The table shows the weight y by x = number of weeks of time that have passed.

x	y	x	y	x	y	x	y	x	y	x	y
0	75.0	1	60.9	2	51.8	3	45.2	4	34.7	5	34.6
6	26.2	7	20.4	8	14.0	9	12.3	11	8.2	15	3.1
20	1.4										

a. Construct a scatterplot. Why is a straight-line model inappropriate?

b. Show that the ordinary regression model gives the fit, $\hat{y} = 54.98 - 3.59x$. Find the predicted weight after $x = 20$ weeks. Does this prediction make sense? Explain.

c. Plot the log of y against x. Does a straight-line model now seem appropriate?

d. The exponential regression model has prediction equation $\hat{y} = 80.6(0.813)^x$. Find the predicted weight (i) initially and (ii) after 20 weeks.

e. Interpret the coefficient 0.813 in the prediction equation.

12.62 More leaf litter Refer to the previous exercise.

a. The correlation equals −0.890 between x and y, and −0.997 between x and $\log(y)$. What does this tell you about which model is more appropriate?

b. The half-life is the time for the weight remaining to be one-half of the original weight. Use the equation $\hat{y} = 80.6(0.813)^x$ to predict the half-life of this organic material. (*Hint:* By trial and error, find the value of x for which $(0.813)^x$ is about 1/2.)

Chapter Review

ANSWERS TO THE CHAPTER FIGURE QUESTIONS

Figure 12.1 (a) There are two dots at the value of 60 pounds on the vertical axis (representing maximum bench press) (b) All dots in the scatterplot are in rows corresponding to the values 60, 65, 70, etc., on the vertical axis.

Figure 12.2 Answers will vary. One possible sketch is Figure 12.3.

Figure 12.3 A mathematical function that has a parabolic shape.

Figure 12.4 The bell-shaped curves represent the conditional distributions of income at $x = 12$ years and $x = 16$ years. Each conditional distribution is assumed to be normal and to have the same standard deviation, $\sigma = \$13{,}000$.

Figure 12.5 This happens when most observations are in the quadrants where data points are above the mean on one variable and below the mean on the other variable, which are the upper-left and lower-right quadrants.

Figure 12.6 1 standard deviation.

Figure 12.7 The variability of the sampling distribution of a sample mean is much less than the variability of the distribution of individual observations.

Figure 12.8 H_0: $\beta = 0$.

Figure 12.9 Three observations, one for each bar.

Figure 12.10 The curves that represent the lower and upper endpoints of the prediction interval are too wide for the x values on the lower end (the 95% prediction interval for y has a greater than 95% chance of containing a value of y) and too narrow for the x values on the upper end (a 95% prediction interval for y has a much smaller than 95% chance of containing a value of y).

Figure 12.11 When x increases by 1, the mean μ_y is multiplied by β. When $0 < \beta < 1$, multiplying by β causes the mean to decrease.

CHAPTER SUMMARY

In Chapters 10–12 we've learned how to detect and describe *association between two variables.* Chapter 10 showed how to compare means or proportions for two groups. Chapter 11 dealt with *association between two categorical variables.* This chapter showed how to analyze linear *association between two quantitative variables.*

A regression analysis investigates the relationship between a quantitative explanatory variable x and a quantitative response variable y.

- At each value of x, there is a conditional distribution of y values that summarizes how y varies at that value. We denote the population mean of the conditional distribution of y by μ_y. The **regression model** $\mu_y = \alpha + \beta x$, with y-intercept α and slope β, uses a straight line to approximate the relationship between x and the mean μ_y of the conditional distribution of y at the different possible values of x. The *sample* **prediction equation** $\hat{y} = a + bx$ predicts y and estimates the mean of y at the fixed value of x.

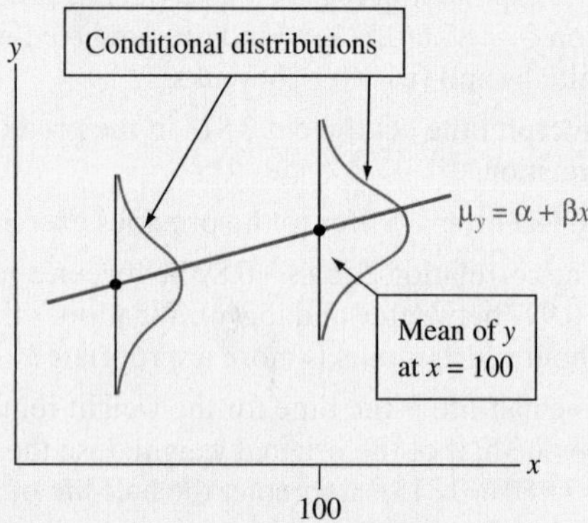

- The **correlation** r describes the strength of linear association. It has the same sign as the slope b but falls between -1 and $+1$. The weaker the correlation, the greater the *regression toward the mean*, with the y values tending to fall closer to their mean (in terms of the number of standard deviations) than the x values fall to their mean. The squared correlation, r^2, describes the proportional reduction in the sum of squared errors using the prediction equation $\hat{y} = a + bx$ to predict y compared to using the sample mean $\bar{y}$ to predict y.

- **Inference:** A significance test of $\mathbf{H_0: \beta = 0}$ for the population slope β tests **statistical independence** of x and y. It has test statistic

$$t = (b - 0)/se,$$

for the sample slope b and its standard error. Interval estimation is also useful for:

- A confidence interval for β.
- A confidence interval for μ_y, the mean of y at a given value of x.
- A **prediction interval** for a value of y at a given value of x.

These inferences all use the t distribution with $\mathbf{df = n - 2}$. Their basic assumptions are:

1. The population mean of y has a straight-line relationship with x.
2. The data were gathered using randomization.
3. The distribution of y at each value of x is normal, with the same standard deviation at each x value.

- An **ANOVA table** displays sums of squares, their *df* values, and an ***F* statistic** (which equals the square of the t statistic) that also tests H_0: $\beta = 0$. In this table, the **residual sum of squares** $\Sigma(y - \hat{y})^2$ takes each **residual** (prediction error) $y - \hat{y}$ and then squares and adds them. The residual SS divided by its *df* value of $n - 2$ is the **mean square error.** Its square root is the **residual standard deviation** estimate s of the variability as measured by σ of the conditional distribution of y at each fixed x. A residual divided by its *se* is a **standardized residual**, which measures the number of standard errors that a residual falls from 0. It helps us identify unusual observations.

- The **exponential regression** model $\mu_y = \alpha\beta^x$ has an increasing or a decreasing curved shape. For it, a one-unit increase in x has a *multiplicative* effect of β on the mean, rather than an *additive* effect as when $\mu_y = \alpha + \beta x$.

SUMMARY OF NOTATION

μ_y = population mean of conditional distribution of y values at fixed value of x

$\mu_y = \alpha + \beta x$ = population straight-line regression equation

$\hat{y} = a + bx$ = sample prediction equation that estimates the population regression equation

r^2 = proportional reduction in prediction error = square of correlation r

$\Sigma(y - \hat{y})^2$ = residual sum of squares, which summarizes how well $\hat{y}$ predicts y. This is the numerator of the mean squared error, and its square root is the numerator of the residual standard deviation s, and it is used in finding r^2.

$\Sigma(y - \bar{y})^2$ = total sum of squares, which summarizes how well $\bar{y}$ predicts y. This is used in finding r^2.

$\mu_y = \alpha\beta^x$ = exponential regression model: mean = α when $x = 0$ and the mean multiplies by β for each one-unit increase in x.

CHAPTER PROBLEMS

Practicing the Basics

12.63 Parties and sports Let y = number of parties attended in the past month and x = number of sports events watched in the past month, measured for all students at your school. Explain the mean and variability about the mean aspects of the regression model $\mu_y = \alpha + \beta x$, in the context of these variables. In your answer, explain why (a) it is more sensible to let $\alpha + \beta x$ represent the *means* of the conditional distributions rather than individual observations, (b) the model allows variation around the mean with its σ parameter.

12.64 Verbal–math correlation A report summarizing scores for students on a verbal aptitude test x and a mathematics aptitude test y states that $\bar{x} = 480$, $\bar{y} = 500$, $s_x = 80$, $s_y = 120$, and $r = 0.60$.

a. Find the slope of the regression line, based on its connection with the correlation.

b. Find the y-intercept of the regression line (*Hint:* Use its formula in Section 12.2), and state the prediction equation.

c. Find the prediction equation for predicting verbal test result using math test result.

12.65 Short people Do very short parents tend to have children who are even shorter, or short but not as short as they are? Explain, identifying the response and explanatory variables and the role of regression toward the mean.

12.66 Income and education in Florida The FL Crime data file on the text CD contains data for all counties in Florida on y = median annual income (thousand of dollars) for residents of the county and x = percent of residents with at least a high school education. The table shows some summary statistics and results of a regression analysis.

Income and Education for Florida Counties

Variable	Mean	Std Dev	Predictor	Parameter Estimate
INCOME	24.51	4.69	INTERCEPT	−4.60
EDUC	69.49	8.86	EDUC	0.42

a. Find the correlation r. Interpret (i) the sign and (ii) the magnitude.

b. Find the predicted median income for a county that is 1 standard deviation above the mean on x, and use it to explain the concept of regression toward the mean.

12.67 Bedroom residuals For the House Selling Prices FL data set on the text CD, when we regress y = selling price (in dollars) on x = number of bedrooms, we get the results shown in the printout.

a. One home with three bedrooms sold for \$338,000. Find the residual, and interpret.

b. The home in part a had a standardized residual of 4.02. Interpret.

House selling prices and number of bedrooms

```
Variable     Mean   StDev  Minimum   Q1    Median   Q3    Maximum
price      126698   56357    21000  86625 123750 155625   338000
Bed-        2.990   0.659    1.000  3.000  3.000  3.000    5.000
rooms
Predictor           Coef  SE Coef    T      P
Constant           33778    24637  1.37  0.173
Bedrooms           31077     8049  3.86  0.000
S = 52771.5       R-Sq = 13.2%
```

12.68 Bedrooms affect price? Refer to the previous exercise.

a. Explain what the regression parameter β means in this context.

b. Construct and interpret a 95% confidence interval for β.

c. Use the result of part b to form a 95% confidence interval for the difference in the mean selling prices for homes with $x = 4$ bedrooms and with $x = 2$ bedrooms. (*Hint:* How does this difference in means relate to the slope?)

12.69 Types of variability Refer to the previous two exercises.

a. Explain the difference between the residual standard deviation of 52,771.5 and the standard deviation of 56,357 reported for the selling prices.

b. Since they're not much different, explain why this means that number of bedrooms is not strongly associated with selling price. Support this by reporting and interpreting the value of r^2.

12.70 Exercise and college GPA For the Georgia Student Survey file on the text CD, let y = exercise and x = college GPA.

a. Construct a scatterplot. Identify an outlier that could influence the regression line. What would you expect its effect to be on the slope and the correlation?

b. Fit the model. Find the standardized residual for that observation. Interpret.

c. Fit the model without the outlying observation. Summarize its impact.

12.71 Bench press predicting leg press For the study of high school female athletes, when we use x = maximum bench press (BP) to predict y = maximum leg press (LP), we get the results that follow. The sample mean of BP was 80.

```
Predictor    Coef    SE Coef    T      P
Constant    173.98    41.75   4.17   0.000
BP          2.2158   0.5155   4.30   0.000
Predicted Values for New Observations
New Obs      Fit     SE Fit        95% CI              95% PI
   1       351.25     6.78   (337.65,364.84)   (247.70,454.80)
Values of Predictors for New Observations
  New Obs     BP
     1       80.0
```

a. Interpret the confidence interval listed under "95% CI."

b. Interpret the interval listed under "95% PI." What's the difference between the purpose of the 95% PI and the 95% CI?

12.72 Leg press ANOVA The analysis in the previous exercise has the ANOVA table shown.

```
Analysis of Variance
Source          DF      SS      MS       F       P
Regression       1   48483   48483   18.48   0.000
Residual Error  55  144304    2624
Total           56  192787
```

a. For those female athletes who had maximum bench press equal to the sample mean of 80 pounds, what is the estimated standard deviation of their maximum leg press values?

b. Assuming that maximum leg press has a normal distribution, show how to find an approximate 95% prediction interval for the y values at $x = 80$.

12.73 Savings grow You invest $1000 in an account having interest such that your principal doubles every 10 years.

a. How much money would you have after 50 years?

b. If you were still alive in 100 years, show that you'd be a millionaire.

c. Give the equation that relates y = principal to x = number of decades for which your money has been invested.

12.74 Florida population The population size of Florida (in thousands) since 1830 has followed approximately the exponential regression $\hat{y} = 46(1.036)^x$. Here, x = year − 1830 (so, $x = 0$ for 1830 and $x = 170$ for the year 2000).

a. What has been the approximate rate of growth per year?

b. Find the predicted population size in (i) 1830 and (ii) 2000.

c. Find the predicted population size in 2100. Do you think that this same formula will continue to hold between 2000 and 2100? Why?

12.75 World population growth The table shows the world population size (in billions) since 1900.

World Population Sizes (in billions)

Year	Population	Year	Population
1900	1.65	1975	4.07
1910	1.75	1980	4.43
1920	1.86	1985	4.83
1930	2.07	1990	5.26
1940	2.30	1995	5.67
1950	2.52	2000	6.07
1960	3.02	2005	6.47
1970	3.69	2010	6.85

Source: U.S. Census Bureau.

a. Let x denote the number of years since 1900. The exponential regression model fitted to y = population size and x gives $\hat{y} = 1.424 \times 1.014^x$. Show that the predicted population sizes are 1.42 billion in 1900 and 6.57 billion in 2010.

b. Explain why the fit of the model corresponds to a rate of growth rate of 1.4% per year.

c. For this model fit, explain why the predicted population size (i) doubles after 50 years and (ii) quadruples after 100 years.

d. The correlation equals 0.961 between the population size and the year number and it equals 0.991 between the log of the population size and the year number. What does this suggest about whether the straight-line regression model or the exponential regression model is more appropriate for these data?

12.76 Match the scatterplot Match each of the following scatterplots to the description of its regression and correlation. The plots are the same except for a single point. Justify your answer for each scatterplot. (*Hint:* Think about the possible effect of an outlier in the x-direction and an outlier in the y-direction relative to the regression line for the rest of the data.)

a. $r = -0.46$ $\hat{y} = 142 - 15.6x$

b. $r = -0.86$ $\hat{y} = 182 - 25.8x$

c. $r = -0.74$ $\hat{y} = 165 - 21.8x$

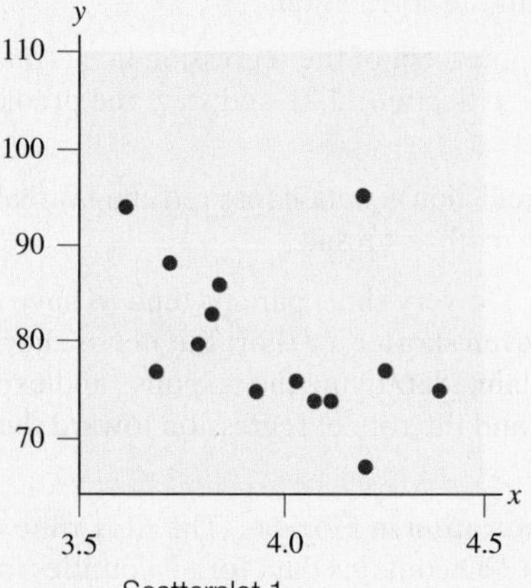

Scatterplot 1

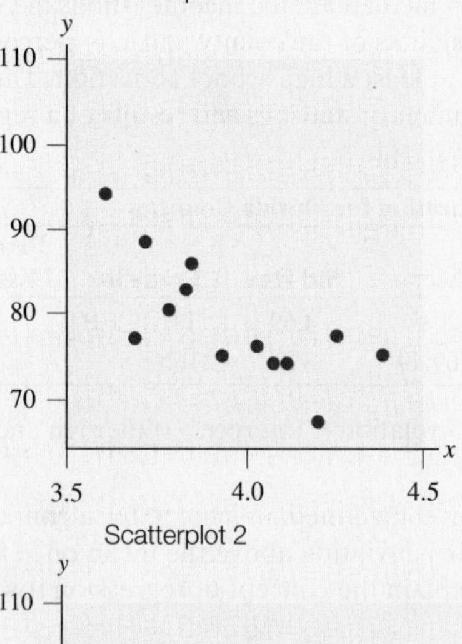

Scatterplot 2

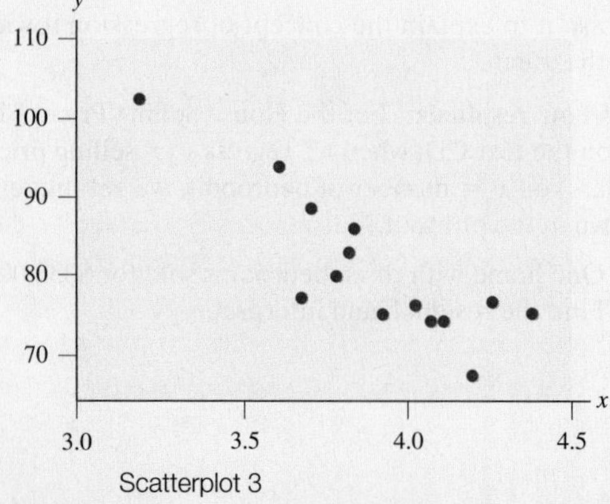

Scatterplot 3

Concepts and Investigations

12.77 Softball data The Softball data file on the text CD contains the records of a University of Georgia coed intramural softball team for 277 games over a 20-year period. (The players changed, but the team continued.) The variables include, for each game, the team's number of runs scored (RUNS), number of hits (HIT), number of errors (ERR), and the difference (DIFF) between the number of runs scored by the team and by the other team. Let DIFF be the response variable. Note that DIFF > 0 means the team won and DIFF < 0 means the team lost.

a. Construct a modified boxplot of DIFF. What do the three outlying observations represent?

b. Find the prediction equation relating DIFF to RUNS. Show that the team is predicted to win when RUNS = 8 or more.

c. Construct the correlation matrix for RUNS, HIT, ERR, and DIFF, and interpret.

d. Conduct statistical inference about the slope of the relationship between DIFF and RUNS.

12.78 Runs and hits Refer to the previous exercise. Conduct a regression analysis of y = RUNS and x = HIT. Does a straight-line regression model seem appropriate? Prepare a report

a. Using graphical ways of portraying the individual variables and their relationship.

b. Interpreting descriptive statistics for the individual variables and their relationship.

c. Viewing standardized residuals to find games with unusual results.

d. Conducting statistical inference about the slope of the relationship.

12.79 GPA and TV watching Using software with the FL Student Survey data file on the text CD, conduct regression analyses relating y = high school GPA and x = hours of TV watching. Prepare a two-page report, showing descriptive and inferential methods for analyzing the relationship.

12.80 Female athletes' speed For the High School Female Athletes data set on the text CD, conduct a regression analysis using the time for the 40-yard dash as the response variable and weight as the explanatory variable. Prepare a two-page report, indicating why you conducted each analysis and interpreting the results.

12.81 Football point spreads For a football game in the National Football League, let y = difference between number of points scored by the home team and the away team (so, $y > 0$ if the home team wins). Let x be the predicted difference according to the Las Vegas betting spread. For the 768 NFL games played between 2003 and 2006, MINITAB results of a regression analysis follow.[1]

a. Explain why you would expect the true y-intercept to be 0 and the true slope to be 1 if there is no bias in the Las Vegas predictions.

b. Based on the results shown in the table, is there much evidence that the sample fit differs from the model $\mu_y = \alpha + \beta x$ with $\alpha = 0$ and $\beta = 1$? Explain.

```
Predictor  Coef      SE Coef    T      P
Constant   -0.4022   0.5233   -0.77   0.442
LasVegas    1.0251   0.0824   12.44   0.000
R - Sq = 16.8%
```

12.82 Iraq war and reading newspapers A study by the Readership Institute[2] at Northwestern University used survey data to analyze how newspaper reader behavior was influenced by the Iraq war. The response variable was a Reader Behavior Score (RBS), a combined measure summarizing newspaper use frequency, time spent with the newspaper, and how much was read. Comparing RBS scores pre-war and during the war, the study noted that there was a significant increase in reading by light readers (mean RBS changing from 2.05 to 2.32) but a significant decrease in reading by heavy readers (mean RBS changing from 5.87 to 5.66). Identifying x = pre-war RBS and y = during-war RBS, explain how this finding could merely reflect regression toward the mean.

12.83 Sports and regression One of your relatives is a big sports fan but has never taken a statistics course. Explain how you could describe the concept of regression toward the mean in terms of a sports application, without using technical jargon.

12.84 Regression toward the mean paradox Does regression toward the mean imply that, over many generations, there are fewer and fewer very short people and very tall people? Explain your reasoning. (*Hint:* What happens if you look backward in time in doing the regressions?)

12.85 Height and weight Suppose the correlation between height and weight is 0.50 for a sample of males in elementary school, and also 0.50 for a sample of males in middle school. If we combine the samples, explain why the correlation will probably be larger than 0.50.

12.86 Income and education Explain why the correlation between x = number of years of education and y = annual income is likely to be smaller if we use a random sample of adults who have a Ph.D. than if we use a random sample of all adults.

12.87 Dollars and pounds Annual income, in dollars, was the response variable in a regression analysis. For a British version of a written report about the analysis, all responses were converted to British pounds sterling (£1 equaled \$2.00, when this was done).

a. How, if at all, does the slope of the prediction equation change?

b. How, if at all, does the correlation change?

c. How, if at all, does the t statistic change for testing the effect of a predictor?

[1] *Source:* Data from P. Everson, *Chance*, vol. 20, 2007, pp. 49–56.

[2] www.readership.org/consumers/data/FINAL_war_study.pdf.

12.88 All models are wrong The statistician George Box, who had an illustrious academic career at the University of Wisconsin, is often quoted as saying, "All models are wrong, but some models are useful." Why do you think that, in practice,

a. All models are wrong?

b. Some models are *not* useful?

12.89 *df* for *t* tests in regression In regression modeling, for *t* tests about regression parameters, $df = n -$ number of parameters in equation for the mean.

a. Explain why $df = n - 2$ for the model $\mu_y = \alpha + \beta x$.

b. Chapter 8 discussed how to estimate a single mean μ. Treating this as the parameter in a simpler regression model, $\mu_y = \mu$, with a single parameter, explain why $df = n - 1$ for inference about a single mean.

12.90 Assumptions What assumptions are needed to use the regression equation $\mu_y = \alpha + \beta x$, (a) to *describe* the relationship between two variables and (b) to make *inferences* about the relationship. In Case B, which assumption is least critical?

12.91 Assumptions fail? Refer to the previous exercise. In view of these assumptions, indicate why such a model would or would not be good in the following situations:

a. $x =$ year (from 1900 to 2005), $y =$ percentage unemployed workers in the United States. (*Hint:* Does *y* continually tend to increase or decrease?)

b. $x =$ age of subject, $y =$ subject's annual medical expenses. (*Hint:* Suppose expenses tend to be relatively high for the newborn and for the elderly.)

c. $x =$ per capita income, $y =$ life expectancy, for nations. (*Hint:* The increasing trend eventually levels off.)

12.92 Lots of standard deviations Explain carefully the interpretations of the standard deviations (a) s_y, (b) s_x, (c) residual standard deviation *s*, and (d) *se* of slope estimate *b*.

12.93 Decrease in home values A Freddie Mac quarterly statement (May 2010) reported that U.S. home sales for one of the central regions (including Illinois, Indiana, Ohio, and Wisconsin) have shown that home values decreased by 3.4% in the last previous year. What if someone interprets this information by saying, "The decrease in home sales is of concern. The decreased rate of 3.4% amounts to a decrease of 17% over five years and 34% over ten years."

a. Explain what is incorrect about this statement.

b. If, in fact, the current median house price for this region is $175,000 and in each of the next 10 years the house values decrease in price by 3.4% relative to the previous year, then what is the estimated house value after a decade? What percentage decrease occurs for the decade?

12.94 Population growth Exercise 12.58 about U.S. population growth showed a predicted growth rate of 13% per decade.

a. Show that this is equivalent to a 1.26% predicted growth *per year.*

b. Explain why the predicted U.S. population size (in millions) *x* years after 1900 is $81.137(1.0126)^x$

12.95 Multiple choice: Interpret *r* One can interpret $r = 0.30$ or the corresponding $r^2 = 0.09$ as follows:

a. A 30% reduction in error occurs in using *x* to predict *y*.

b. A 9% reduction in error occurs in using *x* to predict *y* compared to using $\bar{y}$ to predict *y*.

c. 9% of the time $\hat{y} = y$.

d. *y* changes 0.3 unit for every one-unit increase in *x*.

e. *x* changes 0.3 standard deviations when *y* changes 1 standard deviation.

12.96 Multiple choice: Correlation invalid The correlation is appropriate for describing association between two quantitative variables

a. Even when different people measure the variables using different units (e.g., kilograms and pounds).

b. When the relationship is highly nonlinear.

c. When the slope of the regression equation is 0 using nearly all the data, but a couple of outliers are extremely high on *y* at the high end of the *x*-scale.

d. When the sample has a much narrower range of *x* values than does the population.

e. When the response variable and explanatory variable are both categorical.

12.97 Multiple choice: Slope and correlation The slope of the least squares regression equation and the correlation are similar in the sense that

a. They both must fall between −1 and +1.

b. They both describe the strength of association.

c. Their squares both have proportional reduction in error interpretations.

d. They have the same *t* statistic value for testing H_0: Independence.

e. They both are unaffected by severe outliers.

12.98 Multiple choice: Regress *x* on *y* The regression of *y* on *x* has a prediction equation of $\hat{y} = -2.0 + 5.0x$ and a correlation of 0.3. Then, the regression of *x* on *y*

a. also has a correlation of 0.3.

b. could have a negative slope.

c. has $r^2 = \sqrt{0.3}$.

d. $= 1/(-2.0 + 5.0y)$.

12.99 Income and height University of Rochester economist Steven Landsburg surveyed economic studies in England and the United States that showed a positive correlation between height and income. The article stated that in the United States each one-inch increase in height was worth about $1500 extra earnings a year, on the average (*Toronto Globe and Mail,* 4/1/2002). The regression equation that links $y =$ annual earnings (in thousands of dollars) to $x =$ height (in inches)

a. has *y*-intercept = $1500.

b. has slope 1.5.

c. has slope 1/1500.

d. has slope 1500.

e. has correlation 0.150.

12.100 True or false The variables y = annual income (thousands of dollars), x_1 = number of years of education, and x_2 = number of years experience in job are measured for all the employees having city-funded jobs, in Knoxville, Tennessee. Suppose that the following regression equations and correlations apply:

i) $\hat{y} = 10 + 1.0x_1$, $r = 0.30$.

ii) $\hat{y} = 14 + 0.4x_2$, $r = 0.60$.

The correlation is -0.40 between x_1 and x_2. Which of the following statements are true and which are false?

a. The strongest sample association is between y and x_2.

b. A standard deviation increase in education corresponds to a predicted increase of 0.3 standard deviations in income.

c. There is a 30% reduction in error in using education, instead of $\bar{y}$, to predict y.

d. When x_1 is the predictor of y, the sum of squared residuals is larger than when x_2 is the predictor of y.

e. If $s = 8$ for the model using x_1 to predict y, then it is not unusual to observe an income of $100,000 for an employee who has 10 years of education.

12.101 Golf club velocity and distance A study about the ◆◆ effect of the swing on putting in golf (by C. M. Craig et al., *Nature*, vol. 405, 2000, pp. 295–296) showed a very strong linear relationship between y = putting distance and the square of x = club's impact velocity (r^2 is in the range 0.985 to 0.999).

a. For the model $\mu_y = \alpha + \beta x^2$, explain why it is sensible to set $\alpha = 0$.

b. If the appropriate model is $\mu_y = \beta x^2$, explain why if you double the velocity of the swing, you can expect to quadruple the putting distance.

12.102 Why is there regression toward the mean? Refer to the ◆◆ relationship $r = (s_x/s_y)b$ between the slope and correlation, which is equivalently $s_x b = rs_y$.

a. Explain why an increase in x of s_x units relates to a change in the predicted value of y of $s_x b$ units. (For instance, if $s_x = 10$, it corresponds to a change in $\hat{y}$ of $10b$.)

b. Based on part a, explain why an increase of one standard deviation in x corresponds to a change of only r standard deviations in the predicted y-variable.

12.103 r^2 and variances Suppose $r^2 = 0.30$. Since $\Sigma(y - \bar{y})^2$ is ◆◆ used in estimating the overall variability of the y values and $\Sigma(y - \hat{y})^2$ is used in estimating the residual variability at any fixed value of x, explain why approximately the estimated variance of the conditional distribution of y for a given x is 30% smaller than the estimated variance of the marginal distribution of y.

12.104 Standard error of slope The formula for the standard ◆◆ error of the sample slope b is $se = s/\Sigma(x - \bar{x})^2$, where s is the residual standard deviation of y.

a. Show that the smaller the value of s, the more precisely b estimates β.

b. Explain why a small s occurs when the data points show little variability about the prediction equation.

c. Explain why the standard error of b decreases as the sample size increases and when the x values display more variability.

12.105 Regression with an error term[3] An alternative to the ◆◆ regression formula $\mu_y = \alpha + \beta x$ expresses each y value, rather than the mean of the y values, in terms of x. This approach models an observation on y as

$$y = \text{mean} + \text{error} = \alpha + \beta x + \epsilon,$$

where the mean $\mu_y = \alpha + \beta x$ and the error $= \epsilon$. The **error term** denoted by ϵ (the Greek letter epsilon) represents the deviation of the observation from the mean, that is, ϵ = error = y − mean.

a. If an observation has $\epsilon > 0$, explain why the observation falls above the mean.

b. What does ϵ equal when the observation falls exactly at the mean? The ϵ term represents the error that results from using the mean value $(\alpha + \beta x)$ of y at a certain value of x for the prediction of the individual observation on y.

c. For the sample data and their prediction equation $\hat{y} = a + bx$, explain why an analogous equation to the population equation $y = \alpha + \beta x + \epsilon$ is $y = a + bx + e$, where e is the residual, $e = y - \hat{y}$. (The residual e estimates ϵ. We can interpret e as a sample residual and ϵ as a **population residual.**)

d. Explain why it does not make sense to use the simpler model, $y = \alpha + \beta x$, which does not have the error term.

12.106 Rule of 72 You invest $1000 at 6% compound interest ◆◆ a year. How long does it take until your investment is worth $2000?

a. Based on what you know about exponential regression, explain why the answer is the value of x for which $1000(1.06)^x = 2000$.

b. Using the property of logarithms that $\log(a^x) = x\log(a)$, show that the answer x satisfies $x[\log(1.06)] = \log(2)$, or $x = \log(2)/\log(1.06) = 12$.

c. The rule of 72 says that if you divide 72 by the interest rate, you will find approximately how long it takes your money to double. According to this rule, about how long (in years) does it take your money to double at an interest rate of (i) 1% and (ii) 18%?

Student Activities

12.107 Analyze your data Refer to the data file you created in Activity 3 at the end of Chapter 1. For variables chosen by your instructor, conduct a regression and correlation analysis. Report both descriptive and inferential statistical analyses, interpreting and summarizing your findings, and prepare to discuss the results in class.

[3]This formula is less useful than the one for the mean because it does not apply to regression models when y is not assumed to be normal, such as logistic regression for binary data.

Activity 1

SAT French Exam and Years of Study

In 2003, high school students taking the SAT French II language exam after two to four years of study scored, on average, 35 points higher for each additional year of study (*Source*: www.collegeboard.com/). The scores of students with two years of study averaged 505. For each number of years of study, the standard deviation was approximately 100. Let y = French II exam score and x = number of years of study beyond two years (so $x = 0, 1,$ or 2 for 2, 3, and 4 years of study).

1. Explain why the regression equation $\mu_y = \alpha + \beta x$ is $\mu_y = 505 + 35x$.
2. Suppose y has a normal conditional distribution. Using software, randomly select a student who has two years of French study, a student who has three years, and a student who has four years, by simulating from the appropriate normal distributions. Record each exam score.
3. Perform Step 2 at least 100 times. Store your results in a data file. You now have 100 values of y at each value of x.
4. Construct parallel dot plots and/or box plots for the sample conditional distributions. Compare the mean of the simulated distribution at each x to the actual mean, based on $\mu_y = 505 + 35x$. Is the variability similar for the three sample conditional distributions?
5. Fit a regression line, $\hat{y} = a + bx$, to the simulated values. How does this line compare to the actual regression line, $\mu_y = 505 + 35x$?

The simulated results in Step 4 and Step 5 are not *exactly* equal to those for the model because they use a sample, whereas $\mu_y = 505 + 35x$ refers to a population.

13 Multiple Regression

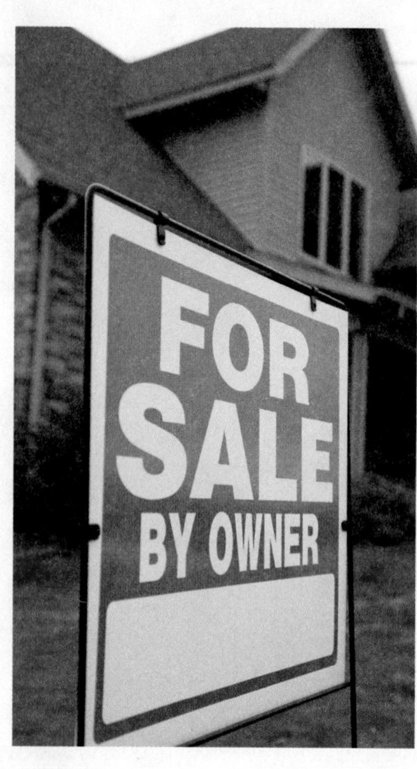

Example 1

Predicting the Selling Price of a House

Picture the Scenario

You are saving to buy a home, and you wonder how much it will cost. The House Selling Prices OR data file on the text CD has observations on 200 recent home sales in Corvallis, Oregon. Table 13.1 shows data for two homes.

Table 13.1 Selling Prices and Related Factors for a Sample of Home Sales

House	Selling Price	House Size (sq. ft)	Number of Bedrooms	Number of Bathrooms	Lot Size (sq. ft)	Year Built	Garage (Y/N)
1	$232,500	1679	3	1.5	10,019	1976	Y
2	$158,000	1292	1	1	217,800	1958	N

Variables listed are the selling price (in dollars), house size (in square feet), number of bedrooms, number of bathrooms, the lot size (in square feet), and whether or not the house has a garage. Table 13.2 reports the mean and standard deviation of these variables for all 200 home sales.

Table 13.2 Descriptive Statistics for Sales of 200 Homes

	Selling Price	House Size (sq. ft)	Number of Bedrooms	Number of Bathrooms	Lot Size (sq. ft)	Age
Mean	$267,466	2551	3.08	2.03	23,217	34.75
Standard deviation	$115,808	1238	1.10	0.77	47,637	24.44

Questions to Explore

In your community, if you know the values of such variables,

- How can you predict a home's selling price?
- How can you describe the association between selling price and the other variables?
- How can you make inferences about the population based on the sample data?

Thinking Ahead

You can find a regression equation to predict selling price by treating one of the other variables as the explanatory variable. However, since there are *several* explanatory variables, you might make better predictions by using *all* of them at once. That's the idea behind **multiple regression**. It uses *more than one* explanatory variable to predict a response variable. We'll learn how to apply multiple regression to a variety of analyses in Examples 2, 3, 8, 9, and 10 of this chapter.

Recall

Review the concept of *lurking variables* in Section 3.4. ◀

Besides helping you to better predict a response variable, multiple regression can help you analyze the association between two variables while controlling (keeping fixed) values of other variables. Such adjustment is important because the effect of an explanatory variable can change considerably after you account for potential lurking variables. A multiple regression analysis provides information not available from simple regression analyses involving only two variables at a time.

Section 13.1 presents the basics of multiple regression. Section 13.2 extends the concepts of correlation and r^2 for describing strength of association to multiple predictors (explanatory variables). Section 13.3 presents inference methods, and Section 13.4 shows how to check the model. Section 13.5 extends the model further to incorporate categorical explanatory variables. The final section presents a regression model for a categorical response variable.

Recall

This chapter builds strongly on the content in Chapter 12 on **regression** and **correlation** methods. You may want to review Sections 12.1 and 12.2 as you read Sections 13.1 and 13.2. ◀

13.1 Using Several Variables to Predict a Response

Chapter 12 modeled the mean μ_y of a quantitative response variable y and a quantitative explanatory variable x by the straight-line equation $\mu_y = \alpha + \beta x$. We refer to this model with a *single* predictor as **bivariate** (two-variable) **regression**, since it contains only two variables (x and y).

Now, suppose there are two predictors, denoted by x_1 and x_2. The bivariate regression equation generalizes to the **multiple regression** equation,

$$\mu_y = \alpha + \beta_1 x_1 + \beta_2 x_2.$$

In this equation, α, β_1, and β_2 are parameters. When we substitute values for x_1 and x_2, the equation specifies the population mean of y for all subjects with those values of x_1 and x_2. When there are additional predictors, each has a βx term.

Multiple Regression Model

The **multiple regression model** relates the mean μ_y of a quantitative response variable y to a set of explanatory variables $x_1, x_2, \ldots$. For three explanatory variables, for example, the multiple regression equation is

$$\mu_y = \alpha + \beta_1 x_1 + \beta_2 x_2 + \beta_3 x_3,$$

and the sample prediction equation is

$$\hat{y} = a + b_1 x_1 + b_2 x_2 + b_3 x_3.$$

With sample data, software estimates the multiple regression equation. It uses the method of least squares to find the best prediction equation (the one with the smallest possible sum of squared residuals).

Example 2

Multiple regression

Predicting Selling Price Using House Size and Number of Bedrooms

Picture the Scenario

For the house selling price data described in Example 1, MINITAB reports the results in Table 13.3 for a multiple regression analysis with selling price as the response variable and with house size and number of bedrooms as explanatory variables.

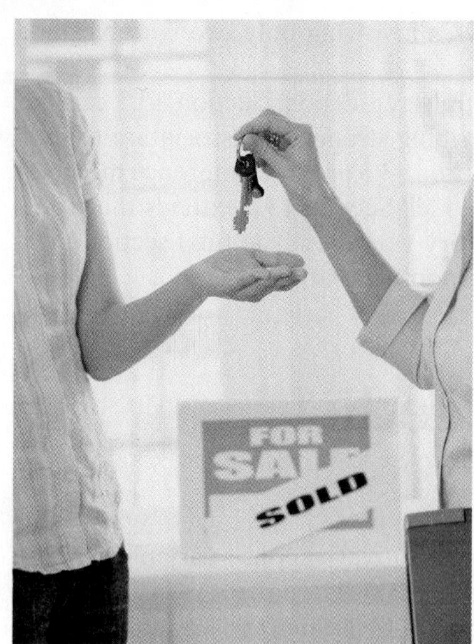

Table 13.3 Regression of Selling Price on House Size and Bedrooms

The regression equation is price = 60,102 + 63.0 house_size + 15,170 bedrooms

Predictor	Coef	SE Coef	T	P
Constant	60102	18623	3.23	0.001
House_size	62.983	4.753	13.25	0.000
Bedrooms	15170	5330	2.85	0.005

Questions to Explore

a. State the prediction equation.

b. The first home listed in Table 13.1 has house size = 1679 square feet, three bedrooms, and selling price $232,500. Find its predicted selling price and the residual (prediction error). Interpret the residual.

Think It Through

a. The response variable is y = selling price. Let x_1 = house size and x_2 = the number of bedrooms. From Table 13.3, the prediction equation is

$$\hat{y} = 60{,}102 + 63.0x_1 + 15{,}170x_2.$$

b. For $x_1 = 1679$ and $x_2 = 3$, the predicted selling price is

$$\hat{y} = 60{,}102 + 63.0(1679) + 15{,}170(3) = 211{,}389, \text{ that is, } \$211{,}389.$$

The residual is the prediction error,

$$y - \hat{y} = \$232{,}500 - \$211{,}389 = \$21{,}111.$$

This result tells us that the actual selling price was $21,111 higher than predicted.

Insight

The coefficients of house size and number of bedrooms are positive. As these variables increase, the predicted selling price increases, as we would expect.

Try Exercise 13.1

In Words

As in **bivariate regression**, we use the Greek letter β (beta) for parameters describing effects of explanatory variables, with β_1 and β_2 read as "beta one" and "beta two."

In Practice **The Number of Explanatory Variables You Can Use Depends on the Amount of Data**

In practice, you should not use many explanatory variables in a multiple regression model unless you have lots of data. A rough guideline is that *the sample size n should be at least 10 times the number of explanatory variables.* For example, to use two explanatory variables, you should have at least $n = 20$ observations.

Plotting Relationships

Always look at the data before doing a multiple regression analysis. Most software has the option of constructing scatterplots on a single graph for each pair of variables. This type of plot is called a **scatterplot matrix**.

Figure 13.1 shows a MINITAB scatterplot matrix for selling price, house size, and number of bedrooms. It shows each pair of variables twice. For a given pair, in one plot a variable is on the y-axis and in another it is on the x-axis. For instance, selling price of house is on the y-axis for the plots in the first row, whereas it is on the x-axis for the plots in the first column. Since selling price is the response variable for this example, the plots in the first row (where selling price is on the y-axis) are the ones of primary interest. These graphs show strong positive linear relationships between selling price and both house size and number of bedrooms. Because a scatterplot matrix shows each pair of variables twice, you only need to look at the plots in the upper-right triangle.

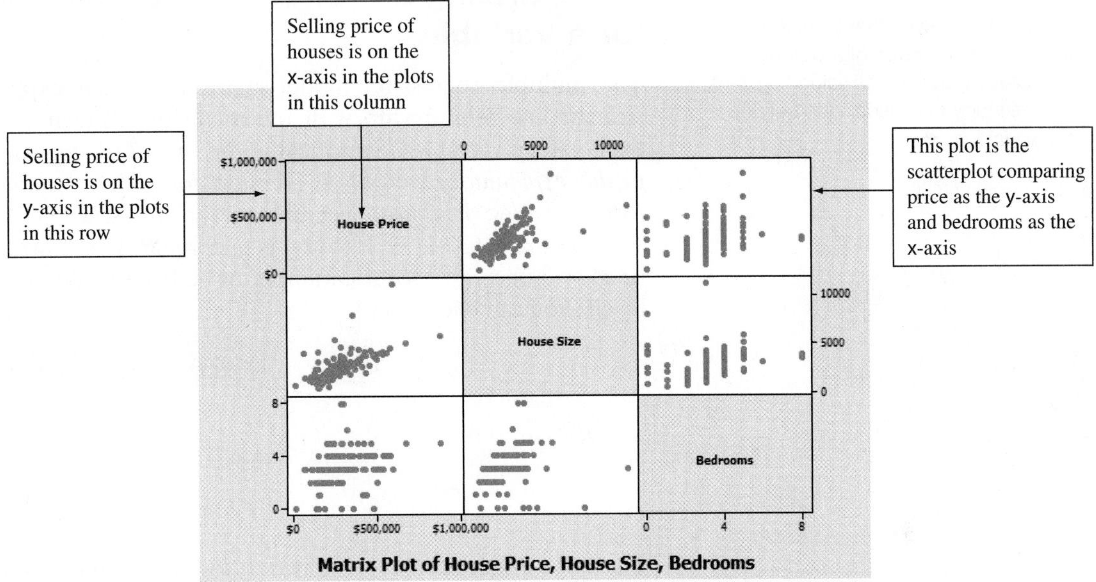

▲ **Figure 13.1** Scatterplot Matrix for Selling Price, House Size, and Number of Bedrooms. The middle plot in the top row has house size on the x-axis and selling price on the y-axis. The first plot in the second row reverses this, with selling price on the x-axis and house size on the y-axis. **Question** Why are the plots of main interest the ones in the top row?

Each scatterplot portrays only *two* variables. It's a two-dimensional picture. A multiple regression equation, which has *several* variables, is more difficult to portray graphically. Note also that the scatterplot involving the variable number of bedrooms as an explanatory variable has an appearance of columns of points. This is due to the highly discrete nature of this quantitative variable.

Interpretation of Multiple Regression Coefficients

The simplest way to interpret a multiple regression equation is to look at it in *two dimensions* as a function of a *single* explanatory variable. We can do this by fixing values for the other explanatory variable(s). For instance, let's fix x_1 = house size at the value 2000 square feet. Then the prediction equation simplifies to one with x_2 = number of bedrooms alone as the predictor,

$$\hat{y} = 60{,}102 + 63.0(2000) + 15{,}170x_2 = 186{,}102 + 15{,}170x_2.$$

For 2000-square-foot houses, the predicted selling price relates to number of bedrooms by $\hat{y} = 186{,}102 + 15{,}170x_2$. Since the slope coefficient of x_2 is 15,170, the predicted selling price increases by \$15,170 for every bedroom added.

Likewise, we could fix the number of bedrooms, and then describe how the predicted selling price depends on the house size. Let's consider houses with number of bedrooms x_2 = three bedrooms. The prediction equation becomes

$$\hat{y} = 60{,}102 + 63.0x_1 + 15{,}170(3) = 105{,}612 + 63.0x_1.$$

For houses with this number of bedrooms, the predicted selling price increases by \$63 for every additional square foot in house size or \$6300 for every 100 square foot in size increase.

Can we say an increase of one bedroom has a larger impact on the selling price (\$15,170) than an increase of a square foot in house size (\$63 per square foot)?

No, we cannot compare these slopes for these explanatory variables because their units of measurement are not the same. Slopes can't be compared when the units differ. We could compare house size and lot size directly if they had the same units of square feet.

Recall

Section 10.5 showed that in multivariate analyses, we can **control** a variable statistically by keeping its value constant while we study the association between other variables. ◀

Summarizing the Effect While Controlling for a Variable

The multiple regression model states that *each* explanatory variable has a straight line relationship with the mean of y, given fixed values of the other explanatory variables. Specifically, *the model assumes that the slope for a particular explanatory variable is identical for all fixed values of the other explanatory variables.* For instance, the coefficient of x_1 in the prediction equation $\hat{y} = 60{,}102 + 63.0x_1 + 15{,}170x_2$ is 63.0 regardless of whether we plug in $x_2 = 1$ or $x_2 = 2$ or $x_2 = 3$ for the number of bedrooms. When you fix x_2 at these three levels, you can check that:

x_2	$\hat{y} = 60{,}102 + 63.0x_1 + 15{,}170x_2$
1	$\hat{y} = 75{,}272 + 63.0x_1$
2	$\hat{y} = 90{,}442 + 63.0x_1$
3	$\hat{y} = 105{,}612 + 63.0x_1$

The slope effect of house size is 63.0 for each equation. Setting x_2 at a variety of values yields a collection of parallel lines, each having slope 63.0. See Figure 13.2.

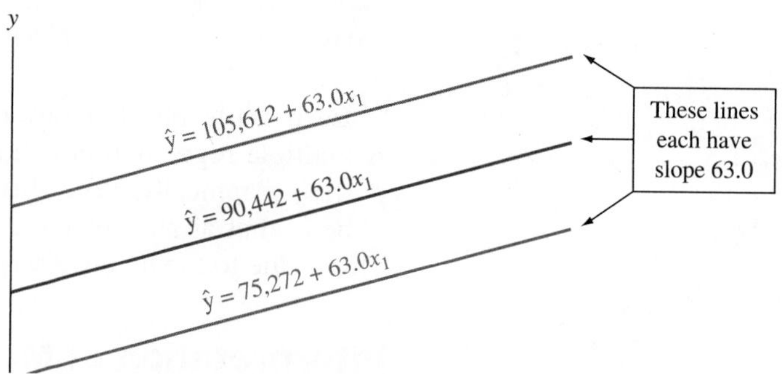

▲ **Figure 13.2 The Relationship Between $\hat{y}$ and x_1 for the Multiple Regression Equation $\hat{y} = 60{,}102 + 63.0x_1 + 15{,}170x_2$.** This shows how the equation simplifies when number of bedrooms $x_2 = 1$, or $x_2 = 2$, or $x_2 = 3$. **Question** The lines move upward (to higher $\hat{y}$-values) as x_2 increases. How would you interpret this fact?

When we fix the value of x_2 we are holding it constant: We are *controlling* for x_2. That's the basis of the major difference between the interpretation of slopes in multiple regression and in bivariate regression:

- In multiple regression, a slope describes the effect of an explanatory variable while *controlling* effects of the other explanatory variables in the model.
- Bivariate regression has only a single explanatory variable. So a slope in bivariate regression describes the effect of that variable while *ignoring* all other possible explanatory variables.

For example, the bivariate regression between y = selling price and x_1 = house size is $\hat{y} = 97{,}997 + 66.4x_1$. In this equation, x_2 = number of bedrooms and other possible predictors are ignored, not controlled. The equation describes the relationship for *all* the house sales in the data set. On the other hand, the equation $\hat{y} = 90{,}442 + 63.0x_1$ we obtained above by substituting $x_2 = 2$ into the multiple regression equation applies only for houses that have that number of bedrooms. In this case, number of bedrooms is controlled. This is why the slopes are different, 66.4 for bivariate regression and 63.0 for multiple regression.

One of the main uses of multiple regression is to identify potential lurking variables and control for them by including them as explanatory variables in the model. Doing so can have a major impact on a variable's effect. When we control a variable, we keep that variable from influencing the associations among the other variables in the study. As we've seen before, the direction of the effect can change after we control for a variable. Exercise 13.5 illustrates this for multiple regression modeling.

13.1 Practicing the Basics

13.1 Predicting weight For a study of University of Georgia female athletes, the prediction equation relating y = total body weight (in pounds) to x_1 = height (in inches) and x_2 = percent body fat is $\hat{y} = -121 + 3.50x_1 + 1.35x_2$.
TRY

a. Find the predicted total body weight for a female athlete at the mean values of 66 and 18 for x_1 and x_2.

b. An athlete with $x_1 = 66$ and $x_2 = 18$ has actual weight $y = 115$ pounds. Find the residual, and interpret it.

13.2 Does study help GPA? For the Georgia Student Survey file on the text CD, the prediction equation relating y = college GPA to x_1 = high school GPA and x_2 = study time (hours per day), is $\hat{y} = 1.13 + 0.643x_1 + 0.0078x_2$.

a. Find the predicted college GPA of a student who has a high school GPA of 3.5 and who studies three hours a day.

b. For students with fixed study time, what is the change in predicted college GPA when high school GPA increases from 3.0 to 4.0?

13.3 Predicting college GPA For all students at Walden University, the prediction equation for y = college GPA (range 0–4.0) and x_1 = high school GPA (range 0–4.0) and x_2 = college board score (range 200–800) is $\hat{y} = 0.20 + 0.50x_1 + 0.002x_2$.

a. Find the predicted college GPA for students having (i) high school GPA = 4.0 and college board score = 800 and (ii) $x_1 = 2.0$ and $x_2 = 200$.

b. For those students with $x_2 = 500$, show that $\hat{y} = 1.20 + 0.50x_1$.

c. For those students with $x_2 = 600$, show that $\hat{y} = 1.40 + 0.50x_1$. Thus, compared to part b, the slope for x_1 is still 0.50, and increasing x_2 by 100 (from 500 to 600) shifts the intercept upward by 100 × (slope for x_2) = 100(0.002) = 0.20 units.

13.4 Interpreting slopes on GPA Refer to the previous exercise.

a. Explain why setting x_2 at a variety of values yields a collection of parallel lines relating $\hat{y}$ to x_1. What is the value of the slope for those parallel lines?

b. Since the slope 0.50 for x_1 is larger than the slope 0.002 for x_2, does this imply that x_1 has a larger effect than x_2 on y in this sample? Explain.

13.5 Does more education cause more crime? The FL Crime data file on the text CD has data for the 67 counties in Florida on

y = crime rate: Annual number of crimes in county per 1000 population

x_1 = education: Percentage of adults in county with at least a high school education

x_2 = urbanization: Percentage in county living in an urban environment.

The figure shows a scatterplot matrix. The correlations are 0.47 between crime rate and education, 0.68 between crime rate and urbanization, and 0.79 between education and urbanization. MINITAB multiple regression results are also displayed.

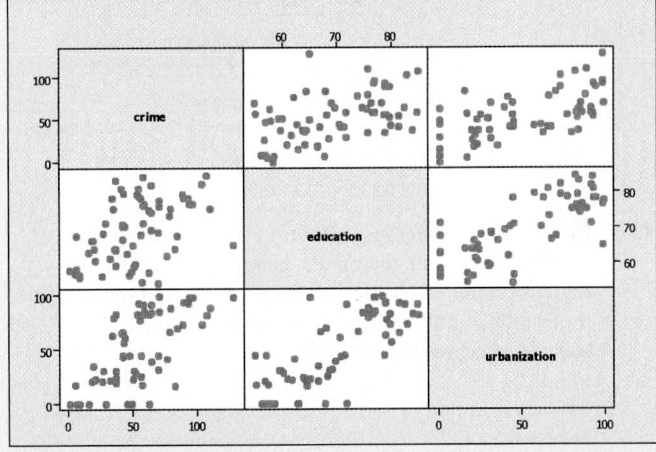

Scatterplot matrix for crime rate, education, and urbanization.

Multiple regression for y = crime rate, x_1 = education, and x_2 = urbanization.

Predictor	Coef	SE Coef	T	P
Constant	59.12	28.37	2.08	0.041
education	−0.5834	0.4725	−1.23	0.221
urbanization	0.6825	0.1232	5.54	0.000

a. Find the predicted crime rate for a county that has 0% in an urban environment and (i) 70% high school graduation rate and (ii) 80% high school graduation rate.

b. Use results from part a to explain how education affects the crime rate, controlling for urbanization, interpreting the slope coefficient −0.58 of education.

c. Using the prediction equation, show that the equation relating crime rate and education when urbanization is fixed at (i) 0, (ii) 50, and (iii) 100, is as follows:

x_2	$\hat{y} = 59.1 - 0.58x_1 + 0.68x_2$
0	$\hat{y} = 59.1 - 0.58x_1$
50	$\hat{y} = 93.2 - 0.58x_1$
100	$\hat{y} = 127.4 - 0.58x_1$

Sketch a plot with these lines and use it to interpret the effect of education on crime rate, controlling for urbanization.

d. The scatterplot matrix shows that education has a *positive* association with crime rate, but the multiple regression equation shows that the association is *negative* when we keep x_2 = urbanization fixed. The reversal in the association is an example of **Simpson's paradox** (See Example 16 in Sec. 3.4 and Example 17 in Sec. 10.5). Consider the hypothetical figure that follows. Sketch lines that represent (i) the bivariate relationship, ignoring the information on urbanization and (ii) the relationship for counties having urbanization = 50. Use this figure and the correlations provided to explain how Simpson's paradox can happen.

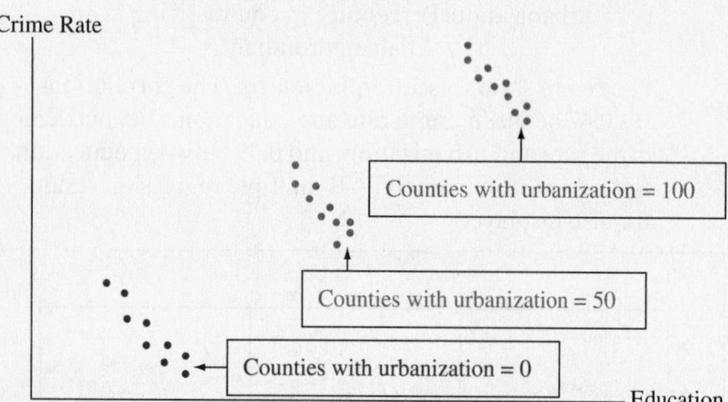

Hypothetical scatterplot for crime rate and education, labeling by urbanization.

13.6 Crime rate and income Refer to the previous exercise. MINITAB reports the results below for the multiple regression of y = crime rate on x_1 = median income (in thousands of dollars) and x_2 = urbanization.

Results of regression analysis

```
Predictor        Coef    SE Coef      T       P
Constant        39.97     16.35     2.44   0.017
income        -0.7906    0.8049    -0.98   0.330
urbanization   0.6418    0.1110     5.78   0.000
```

Correlations: crime, income, urbanization

```
               crime   urbanization
urbanization   0.677
income         0.434      0.731
```

a. Report the prediction equations relating crime rate to income at urbanization levels of (i) 0 and (ii) 100. Interpret.

b. For the bivariate model relating y = crime rate to x = income, MINITAB reports

```
crime = -11.6 + 2.61 income
```

Interpret the effect of income, according to the sign of its slope. How does this effect differ from the effect of income in the multiple regression equation?

c. The correlation matrix for these three variables is shown in the table. Use these correlations to explain why the income effect seems so different in the models in part a and part b.

d. Do these variables satisfy Simpson's paradox? Explain.

13.7 The economics of golf The earnings of a PGA Tour golfer are determined by performance in tournaments. A study analyzed tour data to determine the financial return for certain skills of professional golfers. The sample consisted of 393 golfers competing in one or both of the 2002 and 2008 seasons. The most significant factors that contribute to earnings were the percent of attempts a player was able to hit the green in regulation (GIR), the number of times that a golfer made par or better after hitting a bunker divided by the number of bunkers hit (SS), the average of putts after reaching the green (AvePutt), and the number of PGA events entered (Events). The resulting coefficients from multiple regression to predict 2008 earnings are:

Predictor	Coefficient
Constant	\$26,417,000
GIR	\$168,300
SS	\$33,859
AvePutt	−\$19,784,000
Events	−\$44,725

Source: Some data from K. Rinehart, *Major Themes in Economics*, 2009.

a. State the regression formula for a PGA Tour golfer's earnings for 2008.

b. Explain how to interpret the coefficients for AvePutt and Events.

c. Find the predicted total score for a golfer who had a GIR score of 60, SS score of 50, AvePutt is 1.5, for 20 Events.

13.8 Comparable number of bedrooms and house size effects In Example 2, the prediction equation between y = selling price and x_1 = house size and x_2 = number of bedrooms was $\hat{y} = 60{,}102 + 63.0x_1 + 15{,}170x_2$.

a. For fixed number of bedrooms, how much is the house selling price predicted to increase for each square foot increase in house size? Why?

b. For a fixed house size of 2000 square feet, how does the predicted selling price change for two, three, and four bedrooms?

13.9 Controlling can have no effect Suppose that the correlation between x_1 and x_2 equals 0. Then, for multiple regression with those predictors, it can be shown that the slope for x_1 is the same as in bivariate regression when x_1 is the only predictor. Explain why you would expect this to be true. (*Hint:* If you don't control x_2, would you expect it to have an impact on how x_1 affects y, if x_1 and x_2 have correlation of 0?)

13.10 House selling prices Using software with the House Selling Prices OR data file on the text CD, analyze y = selling price, x_1 = house size, and x_2 = lot size.

a. Construct box plots for each variable and a scatterplot matrix or scatter plots between y and each of x_1 and x_2. Interpret.

b. Find the multiple regression prediction equation.

c. If house size remains constant, what, if any, is the effect of an increase in lot size? Why do you think this is?

13.2 Extending the Correlation and R^2 for Multiple Regression

The correlation r and its square r^2 describe strength of association in a straight-line regression analysis. These measures can also describe association between y and a set of explanatory variables that predict y in a multiple regression model.

Multiple Correlation

To summarize how well a multiple regression model predicts y, we analyze how well the observed y values correlate with the predicted $\hat{y}$ values. As a set, the explanatory variables are strongly associated with y if the correlation between the y and $\hat{y}$ values is strong. Treating the $\hat{y}$ as a variable gives us a way of summarizing several explanatory variables by *one* variable, for which we can use its ordinary correlation with the y values. The correlation between the observed y values and the predicted $\hat{y}$ values from the multiple regression model is called the **multiple correlation**.

In Words

From Sections 3.2 and 12.2, the **correlation** describes the association between the response variable y and a single explanatory variable x. It is denoted by r (lowercase). The **multiple correlation** describes the association between y and a set of explanatory variables in a multiple regression model. It is denoted by R (uppercase).

Multiple Correlation, R

For a multiple regression model, the **multiple correlation** is the correlation between the observed y values and the predicted $\hat{y}$ values. It is denoted by R.

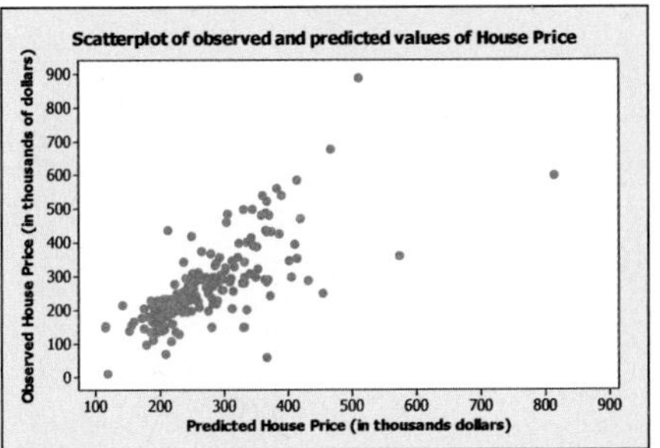

For each subject, the regression equation provides a predicted value $\hat{y}$. So, each subject has a y value and a $\hat{y}$ value. For the two houses listed in Table 13.1, Table 13.4 shows the actual selling price y and the predicted selling price $\hat{y}$ from the equation $\hat{y} = 60{,}102 + 63.0x_1 + 15{,}170x_2$ with house size and number of bedrooms as predictors. The correlation computed between all 200 pairs of y and $\hat{y}$ values is the multiple correlation, R. Software tells us that this equals 0.72. The scatterplot in the margin displays these pairs of y and $\hat{y}$ values.

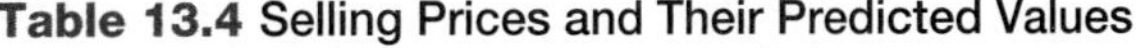

Table 13.4 Selling Prices and Their Predicted Values

These values refer to the two home sales listed in Table 13.1. The predictors are x_1 = house size and x_2 = number of bedrooms.

Home	Selling Price	Predicted Selling Price
1	232,500	$\hat{y} = 60{,}102 + 63.0(1679) + 15{,}170(3) = 211{,}389$
2	158,000	$\hat{y} = 60{,}102 + 63.0(1292) + 15{,}170(1) = 156{,}668$

Recall

For the first home sale listed in Table 13.1, $y = 232{,}500$, $x_1 = 1679$ and $x_2 = 3$. ◀

The larger the multiple correlation, the better are the predictions of y by the set of explanatory variables. For the housing data, $R = 0.72$ indicates a moderately strong association.

The predicted values $\hat{y}$ cannot correlate negatively with y. Otherwise, the predictions would be worse than merely using $\bar{y}$ to predict y. Therefore, *R falls between 0 and 1.* In this way, the multiple correlation R differs from the bivariate correlation r between y and a single variable x, which falls between -1 and $+1$.

R^2

Recall

Section 12.2 introduced r^2 **(r-squared)** as measuring the proportional reduction in prediction error from using $\hat{y}$ to predict y compared to using $\bar{y}$ to predict y. It falls between 0 and 1, with larger values representing stronger association. ◀

For predicting y, the statistic of R^2 describes the relative improvement from using the prediction equation instead of using the sample mean $\bar{y}$. The error in using the prediction equation to predict y is summarized by the *residual sum of squares*, $\Sigma(y - \hat{y})^2$. The error in using $\bar{y}$ to predict y is summarized by the *total sum of squares*, $\Sigma(y - \bar{y})^2$. The *proportional reduction in error* is

$$R^2 = \frac{\Sigma(y - \bar{y})^2 - \Sigma(y - \hat{y})^2}{\Sigma(y - \bar{y})^2}.$$

The better the predictions are using the regression equation, the larger R^2 is. The bivariate measure r^2 is the special case of R^2 applied to regression with one explanatory variable. As the notation suggests, for multiple regression R^2 is the square of the multiple correlation. Regression software reports R^2, and you can take the positive square root of it to get the multiple correlation, R.

Example 3

Multiple correlation and R^2

Predicting House Selling Prices

Picture the Scenario

For the 200 observations on y = selling price in thousands of dollars, using, x_1 = house size in thousands of square feet and x_2 = number of bedrooms, Table 13.5 shows the ANOVA (analysis of variance) table that MINITAB reports for the multiple regression model.

Table 13.5 ANOVA Table and R^2 for Predicting House Selling Price (in thousands of dollars) Using House Size (in thousands of square feet) and Number of Bedrooms

```
R-Sq = 52.4%

Analysis of Variance

Source               DF        SS
Regression            2   1399524
Residual Error      197   1269345
Total               199   2668870
```

Questions to Explore

a. Show how to use the sums of squares in the ANOVA table to find R^2 for this multiple regression model. Interpret.

b. Find and interpret the multiple correlation.

Think It Through

a. From the sum of squares (SS) column, the total sum of squares is $\Sigma(y - \bar{y})^2 = 2{,}668{,}870$. The residual sum of squares from using the multiple regression equation to predict y is $\Sigma(y - \hat{y})^2 = 1{,}269{,}345$. The value of R^2 is

$$R^2 = \frac{\Sigma(y - \bar{y})^2 - \Sigma(y - \hat{y})^2}{\Sigma(y - \bar{y})^2} = \frac{2{,}668{,}870 - 1{,}269{,}345}{2{,}668{,}870} = 0.524.$$

Using house size and number of bedrooms together to predict selling price reduces the prediction error by 52%, relative to using $\bar{y}$ alone to predict selling price. The R^2 statistic appears (in percentage form) in Table 13.5 under the heading "R-sq."

b. The multiple correlation between selling price and the two explanatory variables is $R = \sqrt{R^2} = \sqrt{0.524} = 0.72$. This equals the correlation for the 200 homes between the observed selling prices and the predicted selling prices from multiple regression. There's a moderately strong association between the observed and the predicted selling prices. In summary, house size and number of bedrooms are very helpful in predicting selling prices.

Insight

The bivariate r^2 value for predicting selling price is 0.51 with house size as the predictor. The multiple regression model has $R^2 = 0.52$, only a marginally larger value to using only house size as a predictor. We appear to be just as well off with the bivariate model using house size as the predictor as compared to using the multiple model. An advantage of using only the bivariate model is easier interpretation of the coefficients.

Try Exercise 13.11

Properties of R^2

The example showed that R^2 for the multiple regression model was larger than r^2 for a bivariate model using only one of the explanatory variables. In fact, *a key property of R^2 is that it cannot decrease when predictors are added to a model.*

The difference $\Sigma(y - \bar{y})^2 - \Sigma(y - \hat{y})^2$ that forms the numerator of R^2 also appears in the ANOVA table. It is called the *regression sum of squares.* A simpler formula for it is

$$\text{Regression SS} = \Sigma(\hat{y} - \bar{y})^2.$$

If each $\hat{y} = \bar{y}$, then the regression equation predicts no better than $\bar{y}$. Then regression SS $= 0$, and $R^2 = 0$.

The numerical values of the sums of squares reported in Table 13.5 result when y is measured in *thousands* of dollars (for instance, the first house y is 232.5 rather than 232,500, and $\hat{y}$ is 211.4 rather than 211,389). We truncated the numbers for convenience, because the sums of squares would otherwise be enormous with each one having six more zeros on the end! It makes no difference to the result. Another property of R and R^2 is that, like the correlation, *their values don't depend on the units of measurement.*

In summary, the properties of R^2 are similar to those of r^2 for bivariate models. Here's a list of the main properties:

SUMMARY: Properties of R^2

- R^2 falls between 0 and 1. The larger the value, the better the explanatory variables collectively predict y.

- $R^2 = 1$ only when all residuals are 0, that is, when all regression predictions are perfect (each $y = \hat{y}$), so residual SS $= \Sigma(y - \hat{y})^2 = 0$.
- $R^2 = 0$ when each $\hat{y} = \bar{y}$. In that case, the estimated slopes all equal 0, and the correlation between y and each explanatory variable equals 0.
- R^2 gets larger, or at worst stays the same, whenever an explanatory variable is added to the multiple regression model.
- The value of R^2 does not depend on the units of measurement.

Explanatory variable	r^2
House Size	0.505
Bedrooms	0.100
Lot Size	0.053
Baths	0.331
Garage	0.009
Age	0.029

Table 13.6 shows how R^2 increases as we add explanatory variables to a multiple regression model to predict y = house selling price. The single predictor in the data set that is most strongly associated with y is the house size ($r^2 = 0.505$). See the margin table. When we add number of bedrooms as a second predictor, R^2 goes up from 0.505 to 0.524. As other predictors are added, R^2 continues to go up, but not by much. Predictive power is not much worse with only house size as a predictor than with all six predictors in the regression model.

Table 13.6 R^2 Value for Multiple Regression Models for y = House Selling Price

Explanatory Variables in Model	R^2
House size	0.505
House size, Number of bedrooms	0.524
House size, Number of bedrooms, Lot size	0.524
House size, Number of bedrooms, Lot size, Number of bathrooms	0.604
House size, Number of bedrooms, Lot size, Number of bathrooms, Garage	0.608
House size, Number of bedrooms, Lot size, Number of bathrooms, Garage, Age	0.612

Although R^2 goes up by only small amounts when we add other variables after house size is already in the model, this does not mean that the other predictors are only weakly correlated with selling price. Because the predictors are themselves highly correlated, once one or two of them are in the model, the remaining ones don't help much in adding to the predictive power. For instance, lot size is highly positively correlated with number of bedrooms and with size of house. So, once number of bedrooms and size of house are included as predictors in the model, there's not much benefit to including lot size as an additional predictor.

In Practice R^2 Often Does Not Increase Much After a Few Predictors Are in the Model

When there are many explanatory variables but the correlations among them are strong, once you have included a few of them in the model, R^2 usually doesn't increase much more when you add additional ones. This does not mean that the additional variables are uncorrelated with the response variable but merely that they don't add much new power for predicting y, given the values of the predictors already in the model.

As in the bivariate case, a disadvantage of R^2 (compared to the multiple correlation R) is that its units are the *square* of the units of measurement. The R^2 of 0.524 in Example 3 implies that the estimated *variance* of y at fixed values of the predictors is 52.4% less than the overall variance of y.

13.2 Practicing the Basics

13.11 Predicting sports attendance Keeneland Racetrack in Lexington, Kentucky, has been a social gathering place since 1935. Every spring and fall thousands of people come to the racetrack to socialize, gamble, and enjoy the horse races that have become so popular in Kentucky. A study investigated the different factors that affect the attendance at Keeneland. To many people, the social aspect of Keeneland is equally as important as watching the races, if not more. The study investigates which factors significantly contribute to attendance by Keeneland visitors. (*Source*: Data from Gatton School of Business and Economics publication, Gatton College, 2009.)

TRY

ANOVA table for y = attendance at Keeneland

Source	DF	SS	MS	F	Significance F
Regression	6.00	2360879925	393479988	21.71	0.00
Residual	110.00	1993805006	18125500		
Total	116.00	4354684931			

a. Show how R^2 is calculated from the SS values, and report its value.

b. Interpret the R^2 value. Does the multiple regression equation help us predict the attendance much better than we could without knowing that equation?

c. Find the multiple correlation. Interpret.

13.12 Predicting weight Let's use multiple regression to predict total body weight (in pounds) using data from a study of University of Georgia female athletes. Possible predictors are HGT = height (in inches), %BF = percent body fat, and age. The display shows correlations among these explanatory variables.

	TBW	HGT	%BF
HGT	0.745		
%BF	0.390	0.096	
AGE	−0.187	−0.120	0.024

a. Which explanatory variable gives by itself the best predictions of weight? Explain.

b. With height as the sole predictor, $\hat{y} = -106 + 3.65$ (HGT) and $r^2 = 0.55$. If you add %BF as a predictor, you know that R^2 will be at least 0.55. Explain why.

c. When you add % body fat to the model, $\hat{y} = -121 + 3.50(\text{HGT}) + 1.35(\%\text{BF})$ and $R^2 = 0.66$. When you add age to the model, $\hat{y} = -97.7 + 3.43(\text{HGT}) + 1.36(\%\text{BF}) - 0.960(\text{AGE})$ and $R^2 = 0.67$. Once you know height and % body fat, does age seem to help you in predicting weight? Explain, based on comparing the R^2 values.

13.13 When does controlling have little effect? Refer to the previous exercise. Height has a similar estimated slope for each of the three models. Why do you think that controlling for % body fat and then age does not change the effect of height much? (*Hint:* How strongly is height correlated with the other two variables?)

13.14 Internet use For countries listed in the Twelve Countries data file on the text CD, y = Internet use (percent) is predicted by x_1 = per capita GDP (gross domestic product, in thousands of dollars) with $r^2 = 0.88$. Adding x_2 = carbon dioxide emissions per capita to the model yields the results in the following display.

Regression of Internet use on GDP and carbon dioxide emissions per capita

```
Predictor   Coef        SE Coef       T      P
Constant    2.843       5.473         0.52   0.616
GDP         0.0021875   0.0002779     7.87   0.000
CDE         -1.3849     0.7270       -1.90   0.089
S = 9.23490   R-Sq = 91.4%
Analysis of Variance
Source           DF        SS
Regression        2    8176.7
Residual Error    9     767.6
Total            11    8944.2
```

a. Report R^2 and show how it is determined by SS values in the ANOVA table.

b. Interpret its value as a proportional reduction in prediction error.

13.15 More Internet use In the previous exercise, $r^2 = 0.88$ when x_1 is the predictor and $R^2 = 0.914$ when both x_1 and x_2 are predictors. Why do you think that the predictions of y don't improve much when x_2 is added to the model? (The association of x_2 with y is $r = 0.5692$.)

13.16 Softball data For the Softball data set on the text CD, for each game the variables are a team's number of runs scored (RUNS), number of hits (HIT), number of errors (ERR), and the difference (DIFF) between the number of runs scored by that team and by the other team, which is the response variable. MINITAB reports

```
Difference = - 4.03 + 0.0260 Hits
+ 1.04 Run - 1.22 Errors
```

a. If you know the team's number of runs and number of errors in a game, explain why it does not help much to know how many hits the team has.

b. Explain why the result in part a is also suggested by knowing that $R^2 = 0.7594$ for this model, whereas $R^2 = 0.7593$ when only runs and errors are the explanatory variables in the model.

13.17 Slopes, correlations, and units In Example 2 on y = house selling price, x_1 = house size, and x_2 = number of bedrooms, $\hat{y} = 60{,}102 + 63.0x_1 + 15{,}170x_2$, and $R = 0.72$.

a. Interpret the value of the multiple correlation.

b. Suppose house selling prices are changed from dollars to *thousands* of dollars. Explain why if each house price in Table 13.1 is divided by 1000, then the prediction equation changes to $\hat{y} = 60.102 + 0.063x_1 + 15.170x_2$.

c. In part b, does the multiple correlation change to 0.00072? Justify your answer.

13.18 Predicting college GPA Using software with the Georgia Student Survey data file from the text CD, find and interpret the multiple correlation and R^2 for the relationship between y = college GPA, x_1 = high school GPA, and x_2 = study time.

13.3 Using Multiple Regression to Make Inferences

We've seen that multiple regression uses more than one explanatory variable to predict a response variable *y*. With it, we can study an explanatory variable's effect on *y* while controlling other variables that could affect the results. Now, let's turn our attention to using multiple regression to make inferences about the population.

Inferences require the same assumptions as in bivariate regression:

- The regression equation truly holds for the population means.
- The data were gathered using randomization.
- The response variable *y* has a normal distribution at each combination of values of the explanatory variables, with the same standard deviation.

The first assumption implies that there is a straight-line relationship between the mean of *y* and each explanatory variable, with the same slope at each value of the other predictors. We will see how to check this assumption and the third assumption in Section 13.4.

Estimating Variability Around the Regression Equation

A check for normality is needed to validate the third assumption; this is discussed in the next section. In addition to normality, a constant standard deviation for each combination of explanatory variables is assumed. This standard deviation is similar to the standard error for the explanatory variable in bivariate data. (Recall Section 12.4.) As in bivariate regression, a standard deviation parameter σ describes variability of the observations around the regression equation. Its sample estimate is

$$s = \sqrt{\frac{\text{Residual SS}}{df}} = \sqrt{\frac{\Sigma(y - \hat{y})^2}{n - (\text{number of parameters in regression equation})}}.$$

This *residual standard deviation* describes the typical size of the residuals. Its degrees of freedom are df = sample size n - number of parameters in the regression equation. Software reports s and its square, the **mean square error**.

Example 4

Estimating residual standard deviation

Female Athletes' Weight

Picture the Scenario

The College Athletes data set on the text CD comes from a study of 64 University of Georgia female athletes who participated in Division I sports. The study measured several physical characteristics, including total body

weight in pounds (TBW), height in inches (HGT), the percent of body fat (%BF), and age. Table 13.7 shows the ANOVA table for the regression of weight on height, % body fat, and age.

Table 13.7 ANOVA Table for Multiple Regression Analysis of Athlete Weights

Analysis of Variance

Source	DF	SS	MS	F	P
Regression	3	12407.9	4136.0	40.48	0.000
Residual Error	60	6131.0	102.2		
Total	63	18539.0			

S = 10.1086   R-Sq = 66.9%

Question to Explore

For female athletes at particular values of height, percent of body fat, and age, estimate the standard deviation of their weights.

Recall

Recall that the prediction equation was $\hat{y} = -97.7 + 3.43x_1 + 1.36x_2 - 0.96x_3$, so there were four parameters in the regression model. ◀

Think It Through

The SS column tells us that the residual sum of squares is 6131.0. There were $n = 64$ observations and 4 parameters in the regression model, so the DF column reports $df = n - 4 = 60$ opposite the residual SS. The mean square error is

$$s^2 = (\text{residual SS})/df = 6131.0/60 = 102.2.$$

It appears in the mean square (MS) column, in the row labeled "Residual Error." The residual standard deviation is $s = \sqrt{102.2} = 10.1$, identified as S. For athletes with certain fixed values of height, percent body fat, and age, the weights vary with a standard deviation of about 10 pounds.

Recall

For bivariate regression, Example 16 in Section 12.4 discussed confidence intervals for the mean and prediction intervals for y at a particular setting of x. These intervals can be formed with multiple regression as well. ◀

Insight

If the conditional distributions of weight are approximately bell shaped, about 95% of the weight values fall within about $2s = 20$ pounds of the true regression equation. More precisely, software can report *prediction intervals* within which a response outcome has a certain chance of falling. For instance, at $x_1 = 66, x_2 = 18$, and $x_3 = 20$, which are close to the mean predictor values, software reports $\hat{y} = 133.9$ and a 95% prediction interval of 133.9 ± 20.4.

Try Exercise 13.21

In the sample, weight has standard deviation = 17 pounds, describing variability around mean weight = 133 pounds (see Table 13.9). How is it that the *residual* standard deviation could be only 10 pounds? The residual standard deviation is smaller because it refers to variability at *fixed* values of the predictors. Weight varies less at given values of the predictors than it does overall. If we could predict weight perfectly knowing height, percent body fat, and age, the residual standard deviation would be 0.

The ANOVA table also reports another mean square, called a *mean square for regression*, or *regression mean square* for short. We'll next see how to use the mean squares to conduct a significance test about all the slope parameters together.

The Collective Effect of Explanatory Variables

Do the explanatory variables collectively have a statistically significant effect on the response variable y? With three predictors in a model, we can check this by testing

$$H_0: \beta_1 = \beta_2 = \beta_3 = 0.$$

This hypothesis states that the mean of y does not depend on *any* of the predictors in the model. That is, *y is statistically independent of all the explanatory variables.* The alternative hypothesis is

$$H_a\text{: At least one } \beta \text{ parameter is not equal to 0.}$$

This states that *at least one* explanatory variable is associated with y.

The null hypothesis that all the slope parameters equal 0 is equivalent to the hypothesis that the population values of the multiple correlation and R^2 equal 0. The equivalence occurs because the population values of R and R^2 equal 0 only in those situations in which all the β parameters equal 0.

The test statistic for H_0 is denoted by F. It equals the ratio of the mean squares from the ANOVA table,

$$F = \frac{\text{Mean square for regression}}{\text{Mean square error}}.$$

We won't need the formulas for the mean squares here, but the value of F is proportional to $R^2/(1 - R^2)$. As R^2 increases, the F test statistic increases.

When H_0 is true, the expected value of the F test statistic is approximately 1. When H_0 is false, F tends to be larger than 1. The larger the F test statistic, the stronger the evidence against H_0. The P-value is the right-tail probability from the sampling distribution of the F test statistic.

ANOVA Table from Table 13.7

Source	DF	SS	MS	F	P
Reg.	3	12407.9	4136.0	40.5	.000
Error	60	6131.0	102.2		
Total	63	18539.0			

From the ANOVA table in Table 13.7 (shown again in the margin) for the regression model predicting weight, the mean square for regression = 4136.0 and the mean square error = 102.2. Then the test statistic

$$F = 4136.0/102.2 = 40.5.$$

Before conclusions can be made about this test, we must first understand the F distribution better.

The *F* Distribution and Its Properties

The sampling distribution of the F test statistic is called the ***F* distribution**. The symbol for the F statistic and its distribution honors the eminent British statistician R. A. Fisher, who discovered the F distribution in 1922. Like the chi-squared distribution, the F distribution can assume only nonnegative values and is skewed to the right. See Figure 13.3. The mean is approximately 1.

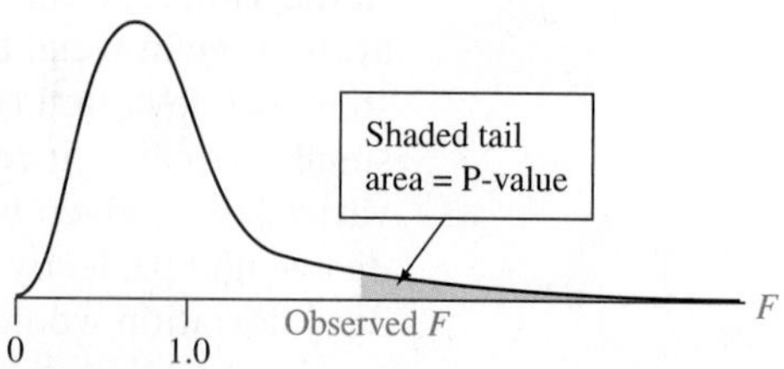

▲ **Figure 13.3 The *F* Distribution and the P-Value for *F* Tests.** **Question** Why does the P-value use the right tail but not the left tail?

The precise shape of the F distribution is determined by two degrees of freedom terms, denoted by df_1 and df_2. These are the df values in the ANOVA table for the two mean squares whose ratio equals F. The first one is

$$df_1 = \text{number of explanatory variables in the model.}$$

The second,

$$df_2 = n - \text{number of parameters in regression equation,}$$

is the df for the t tests about the individual regression parameters. Sometimes df_1 is called the **numerator *df***, because it is listed in the ANOVA table next to the mean square for regression, which goes in the numerator of the F test statistic. Likewise, df_2 is called the **denominator *df***, because it is listed in the ANOVA table next to the mean square error, which goes in the denominator of the F test statistic.

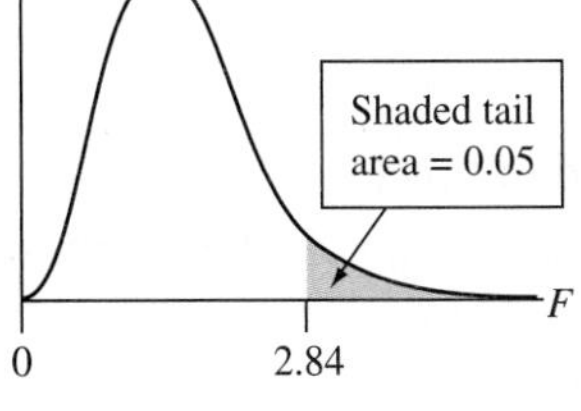

Table D at the end of the text lists the F values that have P-value = 0.05, for various df_1 and df_2 values. Table 13.8 shows a small excerpt. For instance, when $df_1 = 3$ and $df_2 = 40$, $F = 2.84$ has P-value = 0.05. When $F > 2.84$, the test statistic is farther out in the tail and the P-value < 0.05. (See margin figure.) Regression software reports the actual P-value. The F value for our example is extremely large ($F = 40.5$). See Table 13.7. With this large value of F the ANOVA table reports P-value = 0.000.

Table 13.8 An Excerpt of Table D Displaying F Values

These are the values that have right-tail probability equal to 0.05.

	df_1				
df_2	1	2	3	4	5
30	4.17	3.32	2.92	2.69	2.69
40	4.10	3.23	2.84	2.61	2.45
60	4.00	3.15	2.76	2.52	2.37

SUMMARY: F Test That All the Multiple Regression β Parameters = 0

1. **Assumptions:**
   - Multiple regression equation holds
   - Data gathered using randomization
   - Normal distribution for y with same standard deviation at each combination of predictors.
2. **Hypotheses:**

   H_0: $\beta_1 = \beta_2 = \ldots = 0$ (all the beta parameters in the model = 0)

   H_a: At least one β parameter differs from 0.
3. **Test statistic:** F = (mean square for regression)/(mean square error)
4. **P-value:** Right-tail probability above observed F test statistic value from F distribution with

   df_1 = number of explanatory variables,

   $df_2 = n -$ (number of parameters in regression equation).
5. **Conclusion:** The smaller the P-value, the stronger the evidence that at least one explanatory variable has an effect on y. If a decision is needed, reject H_0 if P-value ≤ significance level, such as 0.05. Interpret in context.

Example 5

F test for predictors

Athletes' Weight

Picture the Scenario

For the 64 female college athletes, the ANOVA table for the multiple regression predicting y = weight using x_1 = height, x_2 = percent body fat, and x_3 = age shows:

Source	DF	SS	MS	F	P
Regression	3	12407.9	4136.0	40.48	0.000
Residual Error	60	6131.0	102.2		

Questions to Explore

a. State and interpret the null hypothesis tested in this table.

b. From the *F* table, which *F* value would have a P-value of 0.05 for these data?

c. Report the observed test statistic and P-value. Interpret the P-value, and make a decision for a 0.05 significance level.

Think It Through

a. Since there are three explanatory variables, the null hypothesis is H_0: $\beta_1 = \beta_2 = \beta_3 = 0$. It states that weight is independent of height, percent body fat, and age.

b. In the DF column, the ANOVA table shows $df_1 = 3$ and $df_2 = 60$. The *F* table or software indicates that the *F* value with right-tail probability of 0.05 is 2.76. (See also Table 13.8.)

c. From the ANOVA table, the observed *F* test statistic value is 40.5. Since this is well above 2.76, the P-value is less than 0.05. The ANOVA table reports P-value = 0.000. If H_0 were true, it would be extremely unusual to get such a large *F* test statistic. We can reject H_0 at the 0.05 significance level. In summary, we conclude that at least one predictor has an effect on weight.

Insight

The *F* test tells us that *at least one* explanatory variable has an effect. The following section discusses how to follow up from the *F* test to investigate which explanatory variables have a statistically significant effect on predicting *y*.

Try Exercise 13.25

Inferences About Individual Regression Parameters

For the bivariate model, $\mu_y = \alpha + \beta x$, there's a *t* test for the null hypothesis H_0: $\beta = 0$ that *x* and *y* are statistically independent. Likewise, a *t* test applies to any slope parameter in multiple regression. Let's consider a particular parameter, say β_1.

If $\beta_1 = 0$, the mean of *y* is identical for all values of x_1, at fixed values for the other explanatory variables. So, H_0: $\beta_1 = 0$ states that *y and x_1 are statistically independent, controlling for the other variables.* This means that once the other

Recall

Section 12.3 showed that the ***t* test** statistic for H_0: $\beta = 0$ in the bivariate model is

$$t = \frac{(\text{sample slope} - 0)}{\text{std. error of sample slope}} = (b - 0)/se,$$

with $df = n - 2$. ◄

explanatory variables are in the model, it doesn't help to have x_1 in the model. The alternative hypothesis usually is two sided, H_a: $\beta_1 \neq 0$, but one-sided alternative hypotheses are also possible.

The test statistic for H_0: $\beta_1 = 0$ is

$$t = (b_1 - 0)/se,$$

where *se* is the standard error of the slope estimate b_1 of β_1. Software provides the *se* value, the *t* test statistic, and the P-value. If H_0 is true and the inference assumptions hold, the *t* test statistic has the *t* distribution. The degrees of freedom are

$$df = \text{sample size } n - \text{number of parameters in the regression equation.}$$

The degrees of freedom are also equal to those used to calculate residual standard deviation. The bivariate model $\mu_y = \alpha + \beta x$ has two parameters (α and β), so $df = n - 2$, as Section 12.3 used. The model with two predictors, $\mu_y = \alpha + \beta_1 x_1 + \beta_2 x_2$, has three parameters, so $df = n - 3$. Likewise, with three predictors, $df = n - 4$, and so on.

SUMMARY: Significance Test About a Multiple Regression Parameter (such as β_1)

1. **Assumptions:**
   - Each explanatory variable has a straight-line relation with μ_y, with the same slope for all combinations of values of other predictors in model
   - Data gathered with randomization (such as a random sample or a randomized experiment)
   - Normal distribution for *y* with same standard deviation at each combination of values of other predictors in model
2. **Hypotheses:**

$$H_0: \beta_1 = 0$$
$$H_a: \beta_1 \neq 0$$

When H_0 is true, *y* is independent of x_1, controlling for the other predictors.

3. **Test statistic:** $t = (b_1 - 0)/se$. Software supplies the slope estimate b_1, its *se*, and the value of *t*.
4. **P-value:** Two-tail probability from *t* distribution of values larger than observed *t* test statistic (in absolute value). The *t* distribution has

$$df = n - \text{number of parameters in regression equation}$$

(such as $df = n - 3$ when $\mu_y = \alpha + \beta_1 x_1 + \beta_2 x_2$, which has three parameters).

5. **Conclusion:** Interpret P-value in context; compare to significance level if decision needed.

Hypothesis test for multiple regression parameter β

Example 6

What Helps Predict a Female Athlete's Weight?

Picture the Scenario

The College Athletes data set (Examples 4 and 5) measured several physical characteristics, including total body weight in pounds (TBW), height in inches (HGT), the percent of body fat (%BF), and age. Table 13.9 shows summary statistics for these variables.

Table 13.9 Summary Statistics for Study of Female College Athletes

Variables are TBW = total body weight, HGT = height, %BF = percent body fat, and AGE.

Variable	Mean	StDev	Minimum	Q1	Median	Q3	Maximum
TBW	133.0	17.2	96.0	119.2	131.5	143.8	185.0
HGT	65.5	3.5	56.0	63.0	65.0	68.2	75.0
%BF	18.4	4.1	11.2	15.2	18.5	21.5	27.6
AGE	20.0	1.98	17.0	18.0	20.0	22.0	23.0

Table 13.10 shows results of fitting a multiple regression model for predicting weight using the other variables. The predictive power is good, with $R^2 = 0.669$.

Table 13.10 Multiple Regression Analysis for Predicting Weight

Predictors are HGT = height, %BF = body fat, and age of subject.

Predictor	Coef	SE Coef	T	P
Constant	−97.69	28.79	−3.39	0.001
HGT	3.4285	0.3679	9.32	0.000
%BF	1.3643	0.3126	4.36	0.000
AGE	−0.9601	0.6483	−1.48	0.144
R-Sq = 66.9%				

Questions to Explore

a. Interpret the effect of age on weight in the multiple regression equation.

b. In the population, does age help you to predict weight if you already know height and percent body fat? Show all steps of a significance test, and interpret.

Think It Through

a. Let $\hat{y}$ = predicted weight, x_1 = height, x_2 = %body fat, and x_3 = age. Then

$$\hat{y} = -97.7 + 3.43x_1 + 1.36x_2 - 0.96x_3.$$

The slope coefficient of age is −0.96. The sample effect of age on weight is negative, which may seem surprising, but in practical terms it is small: For athletes having fixed values of x_1 and x_2, the predicted weight decreases by only 0.96 pounds for a one-year increase in age, and the ages vary only from 17 to 23.

b. If $\beta_3 = 0$, then x_3 = age has *no* effect on weight in the population, controlling for height and body fat. The hypothesis that age does *not* help us better predict weight, if we already know height and body fat, is H_0: $\beta_3 = 0$. Here are the steps:

1. **Assumptions:** The 64 female athletes were a convenience sample, not a random sample. Although the goal was to make inferences about all female college athletes, inferences are tentative. We'll discuss the other assumptions and learn how to check them in Section 13.4.
2. **Hypotheses:** The null hypothesis is H_0: $\beta_3 = 0$. Since there's no prior prediction about whether the effect of age is positive or negative (for fixed values of x_1 and x_2), we use the two-sided H_a: $\beta_3 \neq 0$.

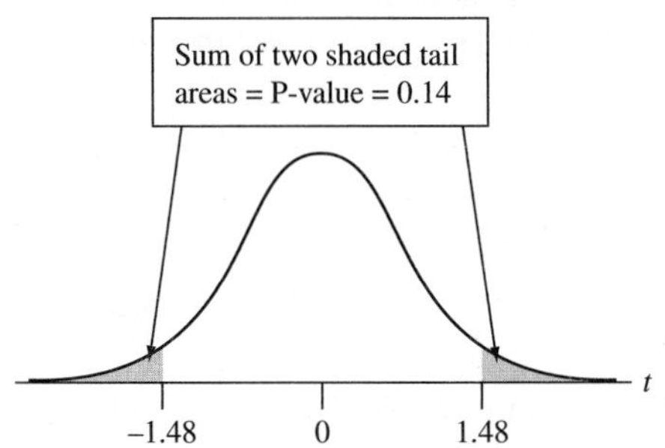

3. **Test statistic:** Table 13.10 reports a slope estimate of −0.960 for age and a standard error of $se = 0.648$. It also reports the t test statistic of

$$t = (b_3 - 0)/se = -0.960/0.648 = -1.48.$$

Since the sample size equals $n = 64$ and the regression equation has four parameters, the degrees of freedom are $df = n - 4 = 60$.

4. **P-value:** Table 13.10 reports P-value = 0.14. This is the two-tailed probability of a t statistic below −1.48 or above 1.48, if H_0 were true.

5. **Conclusion:** The P-value of 0.14 does not give much evidence against the null hypothesis that $\beta_3 = 0$. At common significance levels, such as 0.05, we cannot reject H_0. Age does not significantly predict weight if we already know height and percentage of body fat. These conclusions are tentative because the sample of 64 female athletes was selected using a convenience sample rather than a random sample.

Insight

By contrast, Table 13.10 shows that $t = 9.3$ for testing the effect of height (H_0: $\beta_1 = 0$) and $t = 4.4$ for testing the effect of %BF (H_0: $\beta_2 = 0$). Both P-values are 0.000. It *does* help to have each of these variables in the model, given the other two.

Try Exercise 13.19

As usual, a test merely tells us whether the null hypothesis is plausible. In Example 6 we saw that β_3 may equal 0, but what are its other plausible values? A confidence interval answers this question.

Confidence Interval for a Multiple Regression β Parameter

A 95% confidence interval for a β slope parameter in multiple regression equals

$$\text{Estimated slope} \pm t_{.025}(se).$$

The t-score has $df = n -$ number of parameters in regression equation, as in the t test. The assumptions are also the same as for the t test.

Confidence interval for β

Example 7

What's Plausible for the Effect of Age on Weight?

Picture the Scenario

For the college athletes data, consider the multiple regression analysis of y = weight and predictors x_1 = height, x_2 = %body fat, and x_3 = age.

Question to Explore

Find and interpret a 95% confidence interval for β_3, the effect of age while controlling for height and percent of body fat.

Think It Through

From the previous example, $df = 60$. For $df = 60$, $t_{.025} = 2.00$. From Table 13.10 (shown partly in the margin), the estimate of β_3 is −0.96, with $se = 0.648$. The 95% confidence interval equals

$$b_3 \pm t_{.025}(se) = -0.96 \pm 2.00(0.648),$$

Recall

From Table 13.10,

Predictor	Coef	SE
Constant	−97.7	28.8
HGT	3.43	0.368
%BF	1.36	0.313
AGE	−0.96	0.648 ◀

or -0.96 ± 1.30, roughly $(-2.3, 0.3)$.

At fixed values of x_1 and x_2, we infer that the population mean of weight changes very little (and may not change at all) for a one-year increase in age.

Insight

The confidence interval contains 0. Age may have *no* effect on weight, once we control for height and percent body fat. This is in agreement with not rejecting H_0: $\beta_3 = 0$ in favor of H_a: $\beta_3 \neq 0$ at the $\alpha = 0.05$ level in the significance test.

Try Exercise 13.20

Model Building

Unless the sample size is small and the correlation is weak between y and each explanatory variable, the F test usually has a small P-value. If the explanatory variables are chosen sensibly, at least one should have *some* predictive power. When the P-value is small, we can conclude merely that *at least* one explanatory variable affects y. The more narrowly focused t inferences about individual slopes judge *which* effects are nonzero and estimate their sizes.

In Practice The Overall F Test is Done Before the Individual t Inferences

As illustrated in the previous examples, the F test is typically performed *first* before looking at the individual t inferences. The F test result tells us if there is sufficient evidence to make it worthwhile to consider the individual effects. When there are many explanatory variables, doing the F test first provides protection from doing lots of t tests and having one of them be significant merely by random variation when, in fact, there truly are no effects in the population.

After finding the highly significant F test result of Example 5, we would study the individual effects, as we did in Examples 6 and 7. When we look at an individual t inference, suppose we find that the plausible values for a parameter are all relatively near 0. This was the case for the effect of age on weight (controlling for the other variables) as shown by the confidence interval in Example 7. Then, to simplify the model, you can remove that predictor, refitting the model with the other predictors. When we do this for y = weight, x_1 = height, and x_2 = percent body fat, we get

$$\hat{y} = -121.0 + 3.50x_1 + 1.35x_2.$$

Recall

The multiple regression equation using all three predictors was

$\hat{y} = -97.7 + 3.43x_1 + 1.36x_2 - 0.96x_3$.

The effects of height and percent body fat are similar in the equation without x_3 = age. ◀

For the simpler model, $R^2 = 0.66$. This is nearly as large as the value of $R^2 = 0.67$ with age also in the model. The predictions of weight are essentially as good without age in the model.

When you have several potential explanatory variables for a multiple regression model, how do you decide which ones to include? As just mentioned, the lack of statistical and practical significance is one criterion for deleting a term from the model. Many other possible criteria can also be considered. Most regression software has automatic procedures that successively add or delete predictor variables from models according to criteria such as statistical significance. Three of the most common procedures are **backward elimination, forward selection,** and **stepwise regression.** It is possible that all three procedures result in identical models; however, this is not always the case. In a situation where you must choose between models it is best to look at R^2 values and sensibility of model. These procedures must be used with great caution. There is no guarantee that the final model chosen is sensible. Model selection requires quite a bit of statistical sophistication, and for applications with many potential explanatory variables we recommend that you seek guidance from a statistician.

Caution

Procedures for regression model selection must be used with care. Consultation with a trained statistician is recommended. ◀

13.3 Practicing the Basics

13.19 Predicting GPA For the 59 observations in the Georgia Student Survey data file on the text CD, the result of regressing college GPA on high school GPA and study time follows.

TRY

College GPA, high school GPA, and study time

```
The regression equation is
CGPA = 1.13 + 0.643 HSGPA + 0.0078 Studytime
Predictor     Coef     SE Coef     T      P
Constant     1.1262    0.5690    1.98   0.053
HSGPA        0.6434    0.1458    4.41   0.000
Studytime    0.00776   0.01614   0.48   0.633
S = 0.3188   R-Sq = 25.8%
```

a. Explain in nontechnical terms what it means if the population slope coefficient for high school GPA equals 0.

b. Show all steps for testing the hypothesis that this slope equals 0.

13.20 Study time help GPA? Refer to the previous exercise.

TRY

a. Report and interpret the P-value for testing the hypothesis that the population slope coefficient for study time equals 0.

b. Find a 95% confidence interval for the true slope for study time. Explain how the result is in accord with the result of the test in part a.

c. Does the result in part a imply that in the corresponding population, study time has no association with college GPA? Explain. (*Hint:* What is the impact of also having HSGPA in the model?)

13.21 Variability in college GPA Refer to the previous two exercises.

TRY

a. Report the residual standard deviation. What does this describe?

b. Interpret the residual standard deviation by predicting where approximately 95% of the Georgia college GPAs fall when high school GPA = 3.80 and study time = 5.0 hours per day, which are the sample means.

13.22 Does leg press help predict body strength? Chapter 12 analyzed strength data for 57 female high school athletes. Upper body strength was summarized by the maximum number of pounds the athlete could bench press (denoted BP below, 1RM Bench in file). This was predicted well by the number of times she could do a 60-pound bench press (denoted BP_60 in output, BRTF(60) in file). Can we predict BP even better if we also know how many times an athlete can perform a 200-pound leg press? The table shows results after adding this second predictor (denoted LP_200 in output, LP RTF(200) in file) to the model.

Multiple regression analysis of strength data

```
Predictor    Coef    SE Coef     T       P
Constant    60.596   2.872     21.10   0.000
BP_60       1.3318   0.1876     7.10   0.000
LP_200      0.2110   0.1519     1.39   0.171

Analysis of Variance
Source       DF     SS       MS       F      P
Regression    2   6473.3   3236.6   51.39  0.000
Res. Error   54   3401.3     63.0
Total        56   9874.6
S = 7.93640   R-Sq = 65.6%
```

a. Does LP_200 have a significant effect on BP, if BP_60 is also in the model? Show all steps of a significance test to answer this.

b. Show that the 95% confidence interval for the slope for LP_200 equals 0.21 ± 0.30, roughly (−0.1, 0.5). Based on this interval, does LP_200 seem to have a strong impact, or a weak impact, on predicting BP, if BP_60 is also in the model?

c. Given that LP_200 is in the model, provide evidence from a significance test that shows why it *does* help to add BP_60 to the model.

13.23 Leg press uncorrelated with strength? The P-value of 0.17 in part a of the previous exercise suggests that LP_200 plausibly had no effect on BP, once BP_60 is in the model. Yet when LP_200 is the sole predictor of BP, the correlation is 0.58 and the significance test for its effect has a P-value of 0.000, suggesting very strong evidence of an effect. Explain why this is not a contradiction.

13.24 Interpret strength variability Refer to the previous two exercises. The sample standard deviation of BP was 13.3. The residual standard deviation of BP when BP_60 and LP_200 are predictors in a multiple regression model is 7.9.

a. Explain the difference between the interpretations of these two standard deviations.

b. If the conditional distributions of BP are approximately bell shaped, explain why most maximum bench press values fall within about 16 pounds of the regression equation, when the predictors BP_60 and LP_200 are near their sample mean values.

c. At $x_1 = 11$ and $x_2 = 22$, which are close to the sample mean values, software reports $\hat{y} = 80$ and a 95% prediction interval of 80 ± 16, or (64, 96). Is this interval an inference about where the population BP values fall or where the population *mean* of the BP values fall (for subjects having $x_1 = 11$ and $x_2 = 22$)? Explain.

d. Refer to part c. Would it be unusual for a female athlete with these predictor values to be able to bench press more than 100 pounds? Why?

13.25 Any predictive power? Refer to the previous three exercises.

TRY

a. State and interpret the null hypothesis tested with the *F* statistic in the ANOVA table given in Exercise 13.22.

b. From the *F* table (Table D), which *F* statistic value would have a P-value of 0.05 for these data?

c. Report the observed *F* test statistic and its P-value. Interpret the P-value, and make a decision for a 0.05 significance level. Explain in nontechnical terms what the result of the test means.

13.26 Predicting pizza revenue Aunt Erma's Pizza restaurant keeps monthly records of total revenue, amount spent on TV advertising, and amount spent on newspaper advertising.

a. Specify notation and formulate a multiple regression equation for predicting the monthly revenue. Explain how to interpret the parameters in the equation.

b. State the null hypothesis that you would test if you want to analyze whether TV advertising is helpful, for a given amount of newspaper advertising.

c. State the null hypothesis that you would test if you want to analyze whether *at least one* of the sources of advertising has some effect on monthly revenue.

13.27 Regression for mental health A study in Alachua County, Florida, investigated an index of mental health impairment, which had $\bar{y} = 27.3$ and $s = 5.5$. Two explanatory variables were x_1 = life events score (mean = 44.4, $s = 22.6$) and x_2 = SES (socioeconomic status, mean = 56.6, $s = 25.3$). Life events is a composite measure of the number and severity of major life events, such as death in the family, that the subject experienced within the past three years. SES is a composite index based on occupation, income, and education. The table shows data for 6 subjects. The complete data[1] for a random sample of 40 adults in the county is in the Mental Health data file on the text CD. Some regression results are also shown.

a. Find the 95% confidence interval for β_1.

b. Explain why the interval in part a means that an increase of 100 units in life events corresponds to anywhere from a 4- to 17-unit increase in mean mental impairment, controlling for SES. (This lack of precision reflects the small sample size.)

y = mental impairment, x_1 = life events index, and x_2 = socioeconomic status

y	x_1	x_2	y	x_1	x_2	y	x_1	x_2
17	46	84	26	50	40	30	44	53
19	39	97	26	48	52	31	35	38

```
Predictor      Coef     SE Coef        T        P
Constant     28.230      2.174     12.98    0.000
life          0.10326    0.03250     3.18    0.003
ses          -0.09748    0.02908    -3.35    0.002
S = 4.55644  R-Sq = 33.9%
```

[1]Based on a larger survey reported in a Ph.D. thesis by C. E. Holzer III at University of Florida.

```
Analysis of Variance
Source       DF       SS       MS      F      P
Regression    2    394.24   197.12   9.49   0.000
Res. Error   37    768.16    20.76
Total        39   1162.40
```

13.28 Mental health again Refer to the previous exercise.

a. Report the test statistic and P-value for testing H_0: $\beta_1 = \beta_2 = 0$.

b. State the alternative hypothesis that is supported by the result in part a.

c. Does the result in part a imply that necessarily *both* life events and SES are needed in the model? Explain.

13.29 More predictors for house price The MINITAB results are shown for predicting selling price using x_1 = size of home, x_2 = number of bedrooms, and x_3 = age.

Regression of selling price on house size, number of bedrooms, and age

```
The regression equation is
HP in thousands = 80.5 + 0.0626 House Size + 13.5 Bedrooms
- 0.418 Age
Predictor       Coef      SE Coef        T       P
Constant       80.49       21.81       3.69   0.000
House Size    0.062646    0.004732    13.24   0.000
Bedrooms      13.543       5.380       2.52   0.013
Age           -0.4177      0.2359     -1.77   0.078
```

a. State the null hypothesis for an F test, in the context of these variables.

b. The F statistic equals 74.23, with P-value = 0.000. Interpret.

c. Explain in nontechnical terms what you learn from the results of the t tests reported in the table for the three explanatory variables.

13.30 House prices Use software to do further analyses with the multiple regression model of y = selling price of home in thousands, x_1 = size of home, and x_2 = number of bedrooms, considered in Section 13.1. The data file House Selling Prices OR is on the text CD.

a. Report the F statistic and state the hypotheses to which it refers. Report its P-value, and interpret. Why is it not surprising to get a small P-value for this test?

b. Report and interpret the t statistic and P-value for testing H_0: $\beta_2 = 0$ against H_a: $\beta_2 > 0$.

c. Construct a 95% confidence interval for β_2, and interpret. This inference is more informative than the test in part b. Explain why.

13.4 Checking a Regression Model Using Residual Plots

In capsule form, recall that the three assumptions for inference with a multiple regression model are that (1) the regression equation approximates the true relationship between the predictors and the mean of y (that is, straight-line relationships between the mean of y and each explanatory variable, at fixed values of the other explanatory variables), (2) the data were gathered randomly, and (3) y has a normal distribution with the same standard deviation at each combination of predictors. Now, let's see how to check the assumptions about the regression equation and the distribution of y.

For bivariate regression, the scatterplot provides a simple visual check of whether the straight-line model is appropriate. For multiple regression, a plot of all the variables at once would require many dimensions. So instead, we study how the predicted values $\hat{y}$ compare to the observed values y. This is done using the residuals, $y - \hat{y}$.

Checking Shape and Detecting Unusual Observations

Recall

As discussed in Section 12.4, software can plot a histogram of the residuals or the *standardized residuals*, which are the residuals divided by their standard errors. ◄

Consider first the assumption that the conditional distribution of y is normal, at any fixed values of the explanatory variables. If this is true, the residuals should have approximately a bell-shaped histogram. Nearly all the standardized residuals should fall between about -3 and $+3$. A standardized residual below -3 or above $+3$ indicates a potential regression outlier.

When some observations have large standardized residuals, you should think about whether or not there is an explanation. Often this merely reflects skew in the conditional distribution of y, with a long tail in one direction. Other times the observations differ from the others on a variable that was not included in the model. Once that variable is added, those observations cease to be so unusual. For instance, suppose the data file from the weight study of Examples 4–7 also contained observations for a few males. Then we might have observed a few y values considerably above the others and with very large positive standardized residuals. These residuals would probably diminish considerably once the model also included gender as a predictor.

Example 8

Residuals

House Selling Price

Picture the Scenario

For the House Selling Price OR data set (Examples 1–3), Figure 13.4 is a MINITAB histogram of the standardized residuals for the multiple regression model predicting selling price by the house size and the number of bedrooms.

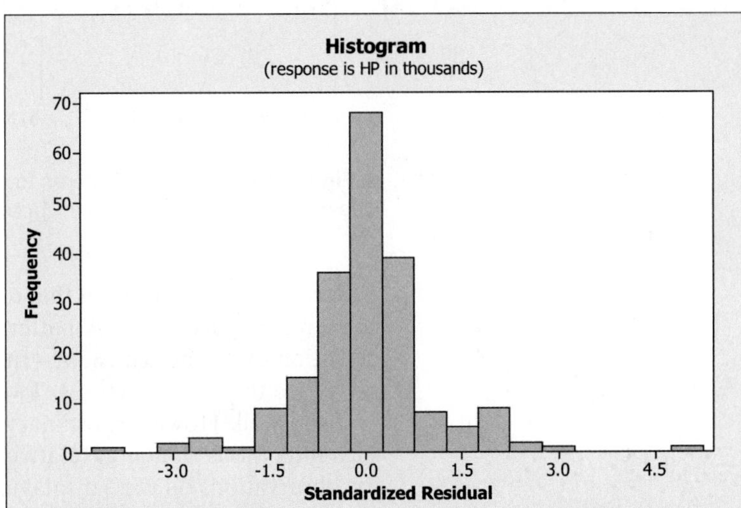

▲ **Figure 13.4** Histogram of Standardized Residuals for Multiple Regression Model Predicting Selling Price. **Question** Give an example of a shape for this histogram that would indicate that a few observations are highly unusual.

Question to Explore

What does Figure 13.4 tell you? Interpret.

Think It Through

The residuals are roughly bell shaped about 0. They fall mostly between about −3 and +3. The shape suggests that the conditional distribution of the response variable may have a bit of skew, but no severe nonnormality is indicated.

Insight

When n is small, don't make the mistake of reading too much into such plots. We're mainly looking for dramatic departures from the assumptions and highly unusual observations that stand apart from the others.

Try Exercise 13.31

Two-sided inferences about slope parameters are *robust*. The normality assumption is not as important as the assumption that the regression equation approximates the true relationship between the predictors and the mean of y. We consider that assumption next.

Plotting Residuals Against Each Explanatory Variable

Plots of the residuals against each explanatory variable help us check for potential problems with the regression model. We discuss this in terms of x_1, but you should view a plot for each predictor. Ideally, the residuals should fluctuate randomly about the horizontal line at 0, as in Figure 13.5a. There should be no obvious change in trend or change in variation as the values of x_1 increase.

By contrast, a scattering of the residuals as in Figure 13.5b suggests that y is actually *nonlinearly* related to x_1. It suggests that y tends to be *below* $\hat{y}$ for very small and very large x_1 values (giving negative residuals) and *above* $\hat{y}$ for medium-sized x_1 values (giving positive residuals).

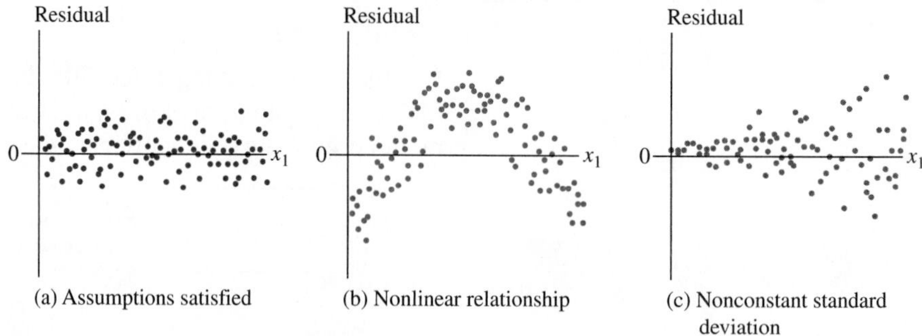

▲ **Figure 13.5 Possible Patterns for Residuals, Plotted Against an Explanatory Variable.** Question Why does the pattern in (b) suggest that the effect of x_1 is not linear?

A common occurrence is that residuals have increased variation as x_1 increases, as seen in Figure 13.5c. A pattern like this indicates that the residual standard deviation of y is not constant—the data display more variability at larger values of x_1 (see the margin figure). Two-sided inferences about slope parameters still perform well. However, ordinary prediction intervals are invalid. The width of such intervals is similar at relatively high and relatively low x_1 values, but in reality, observations on y are displaying more variability at high x_1 values than at low x_1 values. Figure 13.5a suggests that y is linearly related to x_1. It also suggests there is a constant variation as x_1 increases. Therefore, Assumption 1 and part of Assumption 3 are met.

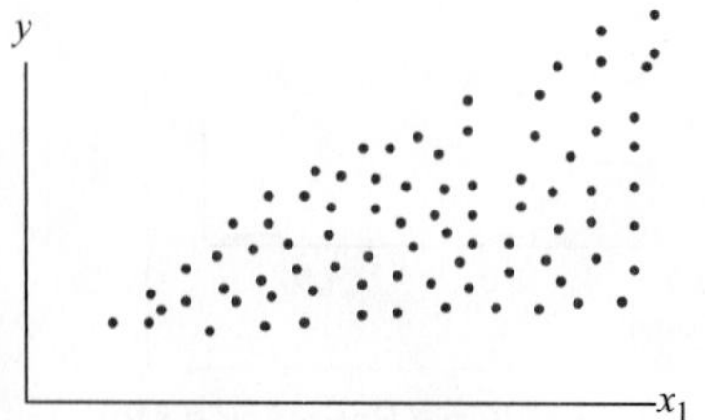

Scatterplot of data with nonconstant variance

In Practice Use Caution in Interpreting Residual Patterns

Residual patterns are often not as neat as the ones in Figure 13.5. Be careful not to let a single outlier or ordinary sampling variability overly influence your reading of a pattern from the plot.

Example 9

Plotting residuals

House Selling Price

Picture the Scenario

For the House Selling Price OR data set, Figure 13.6 is a residual plot for the multiple regression model relating selling price to house size and to number of bedrooms. It plots the standardized residuals against house size.

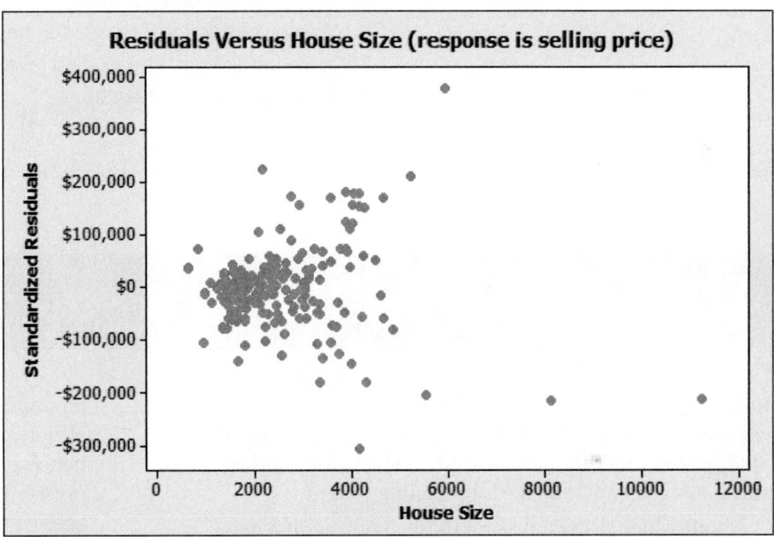

▲ **Figure 13.6** **Standardized Residuals of Selling Price Plotted Against House Size, for Model With House Size and Number of Bedrooms as Predictors.** Questions How does this plot suggest that selling price has more variability at higher house size values, for given number of bedrooms? You don't see number of bedrooms on the plot, so how do its values affect the analysis?

Question to Explore

Does this plot suggest any irregularities with the model?

Think It Through

As house size increases, the variability of the standardized residuals seems to increase. This suggests more variability in selling prices when house size is larger, for a given number of bedrooms. We must be cautious, though, because the few points with large negative residuals for the largest houses and the one point with a large positive residual may catch our eyes more than the others. Generally, it's not a good idea to allow a few points to overly influence your judgment about the shape of a residual pattern. However, there is definite evidence that the variability increases as house size increases. A larger data set would provide more evidence about this.

Insight

Nonconstant variability does not invalidate the use of multiple regression. It would, however, make us cautious about using prediction intervals. We would expect predictions about selling price to have smaller prediction errors when house size is small than when house size is large.

Try Exercise 13.32

We have seen several examples illustrating the components of multiple regression analysis. A summary of the entire process of multiple regression follows.

SUMMARY: The Process of Multiple Regression

Steps should include:

1. Identify response and potential explanatory variables
2. Create a multiple regression model; perform appropriate tests (F and t) to see if and which explanatory variables have a statistically significant effect in predicting y
3. Plot y versus $\hat{y}$ for resulting models and find R and R^2 values
4. Check assumptions (residual plot, randomization, residuals histogram)
5. Choose appropriate model
6. Create confidence intervals for slope
7. Make predictions at specified levels of explanatory variables
8. Create prediction intervals

13.4 Practicing the Basics

13.31 Body weight residuals Examples 4–7 used multiple regression to predict total body weight of college athletes. The figure shows the standardized residuals for another multiple regression model for predicting weight.

a. About which distribution do these give you information—the overall distribution of weight or the conditional distribution of weight at fixed values of the predictors?

b. What does the histogram suggest about the likely shape of this distribution? Why?

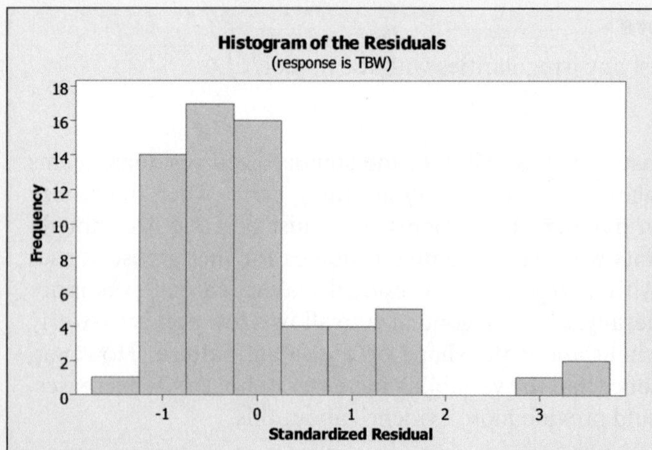

13.32 Strength residuals In Chapter 12, we analyzed strength data for a sample of female high school athletes. The following figure is a residual plot for the multiple regression model relating the maximum number of pounds the athlete could bench press (BP) to the number of 60-pound bench presses (BP_60) and the number of 200-pound leg presses (LP_200). It plots the standardized residuals against the values of LP_200.

a. You don't see BP_60 on the plot, so how do its values affect the analysis?

b. Explain how the plot might suggest less variability at the lower values of LP_200.

c. Suppose you remove the three points with standardized residuals around −2. Then is the evidence about variability in part b so clear? What does this suggest about cautions in looking at residual plots?

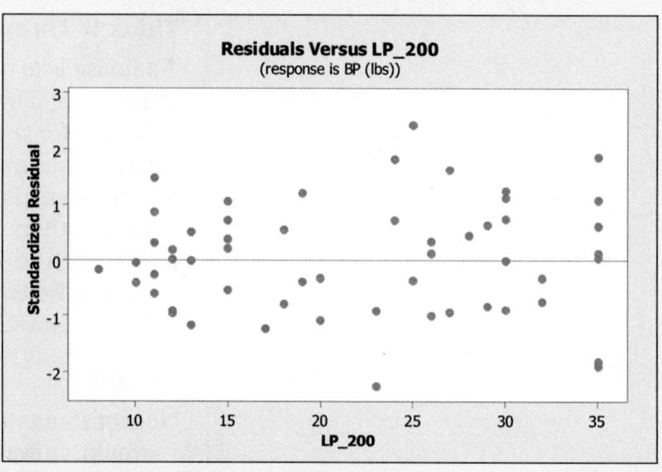

13.33 More residuals for strength Refer to the previous exercise. The following figure is a residual plot for the model relating maximum bench press to LP_200 and BP_60. It plots the standardized residuals against the values of BP_60. Does this plot suggest any irregularities with the model? Explain.

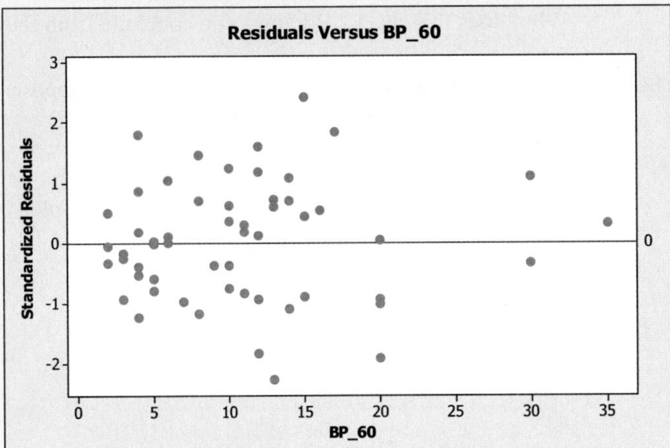

13.34 Nonlinear effects of age Suppose you fit a straight-line regression model to y = amount of time sleeping per day and x = age of subject. Values of y in the sample tend to be quite large for young children and for elderly people, and they tend to be lower for other people. Sketch what you would expect to observe for (a) the scatterplot of x and y and (b) a plot of the residuals against the values of age.

13.35 Driving accidents Suppose you fit a straight-line regression model to x = age of subjects and y = driving accident rate. Sketch what you would expect to observe for (a) the scatterplot of x and y and (b) a plot of the residuals against the values of age.

13.36 Why inspect residuals? When we use multiple regression, what's the purpose of doing a residual analysis? Why can't we just construct a single plot of the data for all the variables at once in order to tell whether the model is reasonable?

13.37 College athletes The College Athletes data set on the text CD comes from a study of University of Georgia female athletes. The response variable BP = maximum bench press (1RM in data set) has explanatory variables LBM = lean body mass (which is weight times 1 minus the proportion of body fat) and REP_BP = number of repetitions before fatigue with a 70-pound bench press (REPS70 in data set). Let's look at all the steps of a regression analysis for these data.

a. The first figure shows a scatterplot matrix. Which two plots in the figure describe the associations with BP as a response variable? Describe those associations.

b. Results of a multiple regression analysis are shown in the next column. Write down the prediction equation, and interpret the coefficient of REP_BP.

c. Report R^2, and interpret its value in the context of these variables.

d. Based on the value of R^2, report and interpret the multiple correlation.

e. Interpret results of the F test that BP is independent of these two predictors. Show how to obtain the F statistic from the mean squares in the ANOVA table.

f. Once REP_BP is in the model, does it help to have LBM as a second predictor? Answer by showing all steps of a significance test for a regression parameter.

g. Examine the histogram shown of the residuals for the multiple regression model. What does this describe, and what does it suggest?

h. Examine the plot shown of the residuals plotted against values of REP_BP. What does this describe, and what does it suggest?

i. From the plot in part h, can you identify a subject whose BP value was considerably lower than expected based on the predictor values? Identify by indicating the approximate values of REP_BP and the standardized residual for that subject.

Regression of maximum bench press on LBM and REP_BP

Predictor	Coef	SE Coef	T	P
Constant	55.012	7.740	7.11	0.000
LBM	0.16676	0.07522	2.22	0.030
REP_BP	1.6575	0.1090	15.21	0.000

S = 7.11957 R-Sq = 83.2%

Analysis of Variance

Source	DF	SS	MS	F	P
Regression	2	15283.0	7641.5	150.75	0.000
Res.	61	3092.0	50.7		
Total	63	18375.0			

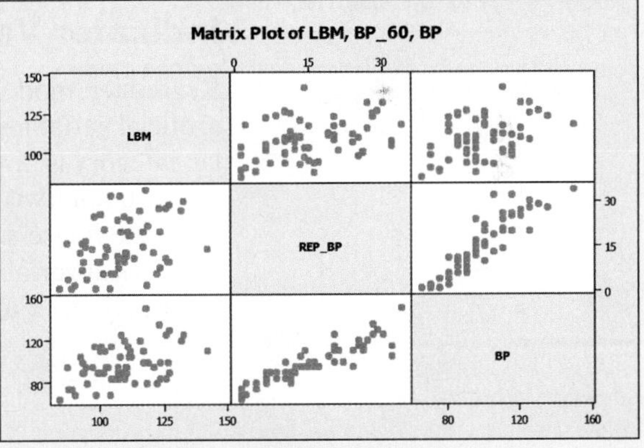

Scatterplot matrix for Exercise 13.37.

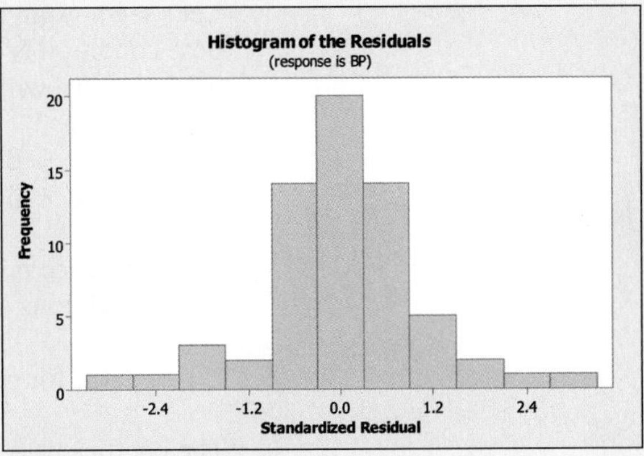

Residual plot for Exercise 13.37, part g.

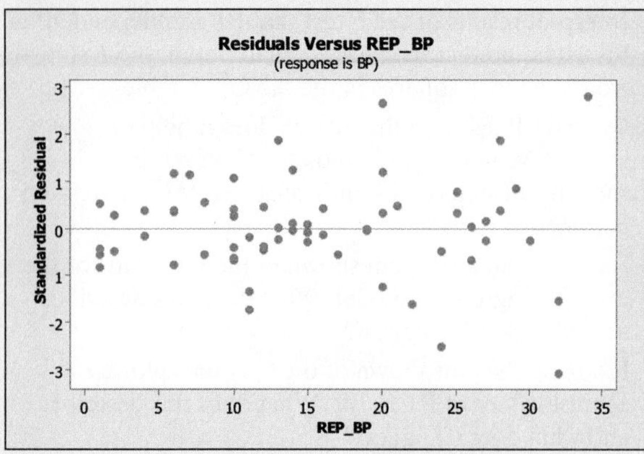

Residual plot for Exercise 13.37, part h.

13.38 House prices Use software with the House Selling Prices OR data file on the text CD to do residual analyses with the multiple regression model for y = house selling price (in thousands), x_1 = lot size, and x_2 = number of bathrooms.

a. Find a histogram of the standardized residuals. What assumption does this check? What do you conclude from the plot?

b. Plot the standardized residuals against the lot size. What does this check? What do you conclude from the plot?

13.39 Selling prices level off In the previous exercise, suppose house selling price tends to increase with a straight-line trend for small to medium size lots, but then levels off as lot size gets large, for a fixed value of number of bathrooms. Sketch the pattern you'd expect to get if you plotted the residuals against lot size.

13.5 Regression and Categorical Predictors

So far, we've studied regression models for quantitative variables. Next, we'll learn how to include a *categorical explanatory* variable. The final section shows how to perform regression for a *categorical response* variable.

Indicator Variables

Regression models specify categories of a categorical explanatory variable using artificial variables, called **indicator variables**. The indicator variable for a particular category is binary. It equals 1 if the observation falls into that category and it equals 0 otherwise.

In the house selling prices (Oregon) data set, the condition of the house is a categorical variable. It was measured with categories (good, not good), The indicator variable x for condition is

$$x = 1 \text{ if house in good condition}$$
$$x = 0 \text{ otherwise.}$$

This indicator variable indicates whether or not a home is in good condition. Let's see how an indicator variable works in a regression equation. To start, suppose condition is the only predictor of selling price. The regression model is then $\mu_y = \alpha + \beta x$, with x as just defined. Substituting the possible values 1 and 0 for x,

$$\mu_y = \alpha + \beta(1) = \alpha + \beta, \text{ if house is in good condition (so } x = 1)$$
$$\mu_y = \alpha + \beta(0) = \alpha, \quad \text{if house is not in good condition (so } x = 0).$$

The difference between the mean selling price for houses in good condition and other conditions is

$$(\mu_y \text{ for good condition}) - (\mu_y \text{ for other}) = (\alpha + \beta) - \alpha = \beta.$$

The coefficient β of the indicator variable x is the difference between the mean selling prices for homes in good condition and for homes not in good condition.

Indicator variables

Example 10

Including Condition in Regression for House Selling Price

Picture the Scenario

Let's now fit a regression model for y = selling price of home using x_1 = house size and x_2 = condition of the house. Table 13.11 shows MINITAB output.

Table 13.11 Regression Analysis of y = Selling Price Using x_1 = House Size and x_2 = Indicator Variable for Condition (Good, Not Good)

```
The regression equation is
House Price = 96271 + 66.5 House Size + 12927 Condition
Predictor       Coef      SE Coef       T        P
Constant       96271        13465     7.15    0.000
House Size    66.463        4.682    14.20    0.000
Condition      12927        17197     0.75    0.453
S = 81787.4 R-Sq = 50.6%
```

Questions to Explore

a. Find and plot the lines showing how predicted selling price varies as a function of house size, for homes in good condition or not in good condition.

b. Interpret the coefficient of the indicator variable for condition.

Think It Through

a. Table 13.11 reports the prediction equation,

$$\hat{y} = 96{,}271 + 66.5x_1 + 12{,}927x_2.$$

For homes not in good condition, $x_2 = 0$. The prediction equation for y = selling price using x_1 = house size then simplifies to

$$\hat{y} = 96{,}271 + 66.5x_1 + 12{,}927(0) = 96{,}271 + 66.5x_1.$$

For homes in good condition, $x_2 = 1$. The prediction equation then simplifies to

$$\hat{y} = 96{,}271 + 66.5x_1 + 12{,}927(1) = 109{,}198 + 66.5x_1.$$

Both lines have the same slope, 66.5. For homes in good condition or not in good condition, the predicted selling price increases by \$66.5 for each square-foot increase in house size. Figure 13.7 plots the two prediction equations. The quantitative explanatory variable, house size, is on the x-axis. The figure portrays a separate line for each category of condition (good, other).

b. At a fixed value of x_1 = house size, the difference between the predicted selling prices for homes in good (1) versus not good (0) condition is

$$(109{,}198 + 66.5x_1) - (96{,}271 + 66.5x_1) = 12{,}927.$$

This is precisely the coefficient of the indicator variable, x_2. For any fixed value of house size, we predict that the selling price is \$12,927 higher for homes that are good versus not in good condition.

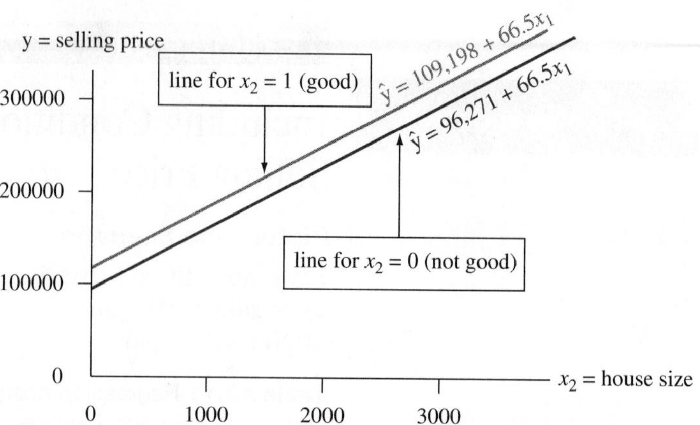

▲ **Figure 13.7 Plot of Equation Relating $\hat{y}$ = Predicted Selling Price to x_1 = House Size, According to x_2 = Condition (1 = Good, 0 = Not Good). Question** Why are the lines parallel?

Insight

Since the two lines have the same slope, they are parallel. The line for homes in good condition is above the other line because its *y*-intercept is larger. This means that for any fixed value of house size, the predicted selling price is higher for homes in good condition. The P-value of 0.453 for the test for the coefficient of the indicator variable suggests that this difference is not statistically significant.

Try Exercise 13.40

A categorical variable having more than two categories uses an additional indicator variable for each category. For instance, if we want to use all three categories good, average, fair in the model, we could use two indicator variables:

$$x_1 = 1 \text{ for houses in good condition, and } x_1 = 0 \text{ otherwise,}$$
$$x_2 = 1 \text{ for houses in average condition, and } x_2 = 0 \text{ otherwise,}$$

If $x_1 = x_2 = 0$, the house is in fair condition. We don't need a separate indicator variable for fair, as it would be redundant. We can tell whether a house is in the fair condition merely from seeing the values of x_1 and x_2. Generally, *a categorical explanatory variable in a regression model uses one fewer indicator variable than the number of categories.* For instance, with the two categories (good condition, other), we needed only a single indicator variable.

Why can't we specify the three categories merely by setting up a variable x that equals 1 for homes in good condition, 0 for homes in average condition, and -1 for fair condition? Because this would treat condition as *quantitative* rather than *categorical.* It would treat condition as if different categories corresponded to different *amounts* of the variable. But the variable measures *which* condition, not *how much* condition. Treating it as quantitative is inappropriate.

Determining if Interaction Exists

In Example 10, the regression equation simplified to two straight lines:

$$\hat{y} = 109{,}198 + 66.5x_1 \text{ for homes in good condition,}$$
$$\hat{y} = 96{,}271 + 66.5x_1 \text{ for homes in other conditions.}$$

Both equations have the same slope. The model forces the effect of x_1 = house size on selling price to be the same for both conditions.

Likewise, in a multiple regression model, *the slope of the relationship between the population mean of y and each explanatory variable is identical for all values of the other explanatory variables.* Such models are sometimes too simple. The effect of an explanatory variable may change considerably as the value of another explanatory variable in the model changes. The multiple regression model we've studied assumes this does not happen. When it does happen, there is **interaction.**

Interaction

For two explanatory variables, **interaction** exists between them if their effects on the response variable when the slope of the relationship between μ_y and one of them changes as the value of the other changes.

Suppose the *actual* population relationship between x_1 = house size and the mean selling price is

$$\mu_y = 100{,}000 + 50x_1 \text{ for homes in good condition,}$$

and

$$\mu_y = 80{,}000 + 35x_1 \text{ for homes in other conditions.}$$

The slope for the effect of x_1 differs (50 versus 35) for the two conditions. There is then interaction between house size and condition in their effects on selling price. See Figure 13.8.

How can you allow for interaction when you do a regression analysis? With two explanatory variables, one quantitative and one categorical (as in Example 10), you can fit a separate line with a different slope between the two quantitative variables for each category of the categorical variable.

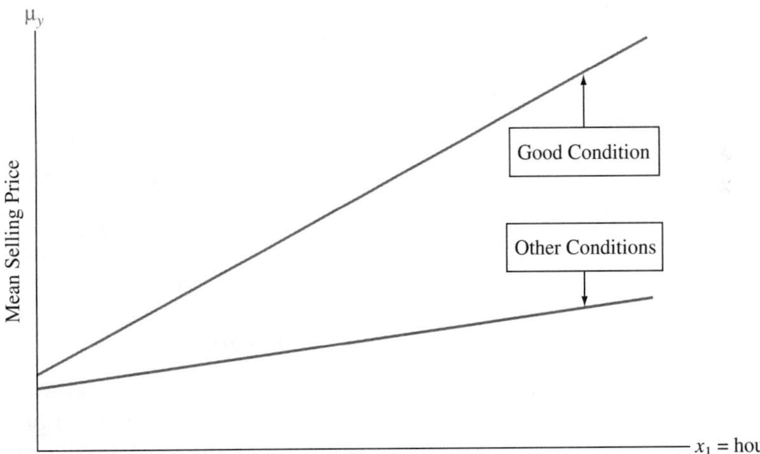

▲ **Figure 13.8 An Example of Interaction.** There's a larger slope between selling price and house size for homes in good condition than in other conditions.

Example 11

Interaction

Comparing Winning High Jumps for Men and Women

Picture the Scenario

Men have competed in the high jump in the Olympics since 1896 and women since 1928. Figure 13.9 shows how the winning high jump in the Olympics has changed over time for men and women. The High Jump data file on

the text CD contains the winning heights for each year. A multiple regression analysis of y = winning height (in meters) as a function of x_1 = number of years since 1928 (when women first participated in the high jump) and x_2 = gender (1 = male, 0 = female) gives $\hat{y} = 1.63 + 0.0057x_1 + 0.35x_2$.

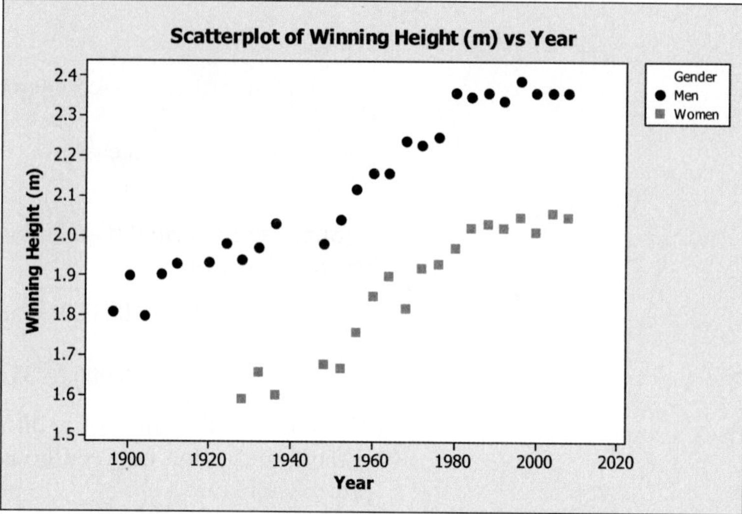

▲ **Figure 13.9** Scatterplot for the Winning High Jumps (in Meters) in the Olympic Games from 1896–2008.

Recall

These data were first analyzed in Example 11 in Chapter 3. ◀

Questions to Explore

a. Interpret the coefficient of year and the coefficient of gender in the equation.

b. To allow interaction, we can fit equations separately to the data for males and the data for females. We then get $\hat{y} = 1.98 + 0.0055x_1$ for males and $\hat{y} = 1.60 + 0.0065x_1$ for females. Describe the interaction by comparing slopes.

c. Describe the interaction allowed in part b by comparing predicted winning high jumps for males and females in 1928 and in 2008.

Think It Through

a. For the prediction equation $\hat{y} = 1.63 + 0.0057x_1 + 0.35x_2$, the coefficient of year is 0.0057. For each gender, the predicted winning high jump increases by 0.0057 meters per year. This seems small, but over a hundred years it projects to an increase of $100(0.0057) = 0.57$ meters, about 27 inches. The model does not allow interaction, as it assumes that the slope of 0.0057 is the same for each gender. The coefficient of gender is 0.35. In a given year, the predicted winning high jump for men is 0.35 meters higher than for women. Because this model does not allow interaction, the predicted difference between men and women is the same each year.

b. The slope of 0.0065 for females is higher than the slope of 0.0055 for males. So the predicted winning high jump increases a bit more for females than for males over this time period.

c. In 1928, $x_1 = 0$, and the predicted winning high jump was $\hat{y} = 1.98 + 0.0055(0) = 1.98$ for males and $\hat{y} = 1.60 + 0.0065(0) = 1.60$ for females, a difference of 0.38 meters. In 2008, $x_1 = 2008 - 1928 = 80$ and the predicted winning high jump was $1.98 +$

0.0055(80) = 2.42 for males and 1.60 + 0.0065(80) = 2.12 for females, a difference of 0.30 meters. The predicted difference between the winning high jumps of males and females decreased a bit between 1928 and 2008.

Insight

When we allow interaction, the estimated slope is a bit higher for females than for males. This is what caused the difference in predicted winning high jumps to be less in 2008 than in 1928. However, the slopes were not dramatically different, as Figure 13.9 shows that the points go up at similar rates for the two genders. The sample degree of interaction was not strong.

Try Exercises 13.46 and 13.47

How do we know whether the interaction shown by the sample is sufficiently large to indicate that there is interaction in the population? There is a significance test for checking this, but it is beyond the scope of this book. In practice, it's usually adequate to investigate interaction informally using graphics. For example, suppose there are two predictors, one quantitative and one categorical. Then you can plot y against the quantitative predictor, identifying the data points by the category of the categorical variable, as we did in Example 11. Do the points seem to go up or go down at quite different rates, taking into account sampling error? If so, it's safer to fit a separate regression line for each category, which then allows different slopes.

13.5 Practicing the Basics

13.40 Winning high jump Refer to Example 11 on winning Olympic high jumps. The prediction equation relating y = winning height (in meters) as a function of x_1 = number of years since 1928 and x_2 = gender (1 = male, 0 = female) is $\hat{y} = 1.63 + 0.0057x_1 + 0.348x_2$.

TRY

a. Using this equation, find the prediction equations relating winning height to year, separately for males and for females.

b. Find the predicted winning height in 2012 for (i) females, (ii) males, and show how the difference between them relates to a parameter estimate for the model.

13.41 Mountain bike prices The Mountain Bike data file on the text CD shows selling prices for mountains bikes. When y = mountain bike price ($) is regressed on x_1 = weight of bike (lbs) and x_2 = the type of suspension (0 = full, 1 = front end), $\hat{y} = 2741.62 - 53.752x_1 - 643.595x_2$.

a. Interpret the estimated effect of the weight of the bike.

b. Interpret the estimated effect of the type of suspension on the mountain bike.

13.42 Predict using house size and condition For the House Selling Prices OR data set, when we regress y = selling price (in thousands) on x_1 = house size and x_2 = condition (1 = Good, 0 = Not Good), we get the results shown.

Regression of selling price of house in thousands versus house size and condition

Predictor	Coef	SE Coef	T	P
Constant	96.27	13.46	7.15	0.000
House Size	0.066463	0.004682	14.20	0.000
Condition	12.93	17.20	0.75	0.453

S = 81.7874 R-Sq = 50.6% R-Sq(adj) = 50.1%

a. Report the regression equation. Find and interpret the separate lines relating predicted selling price to house size for good condition homes and for homes in not good condition.

b. Sketch how selling price varies as a function of house size, for homes in good condition and for homes in not good condition.

c. Estimate the difference between the mean selling price of homes in good and in not good condition, controlling for house size.

13.43 Quality and productivity The table shows data from 27 automotive plants on y = number of assembly defects per 100 cars and x = time (in hours) to assemble each vehicle. The data are in the Quality and Productivity file on the text CD.

Number of defects in assembling 100 cars and time to assemble each vehicle

Plant	Defects	Time	Plant	Defects	Time	Plant	Defects	Time
1	39	27	10	89	17	19	69	54
2	38	23	11	48	20	20	79	18
3	42	15	12	38	26	21	29	31
4	50	17	13	68	21	22	84	28
5	55	12	14	67	26	23	87	44
6	56	16	15	69	30	24	98	23
7	57	18	16	69	32	25	100	25
8	56	26	17	70	31	26	140	21
9	61	20	18	68	37	27	170	28

Source: Data from S. Chatterjee, M. Handcock, and J. Simonoff, *A Casebook for a First Course in Statistics and Data Analysis* (Wiley, 1995); based on graph in *The Machine that Changed the World*, by J. Womack, D. Jones, and D. Roos (Macmillan, 1990).

a. The prediction equation is $\hat{y} = 61.3 + 0.35x$. Find the predicted number of defects for a car having assembly time (i) 12 hours (the minimum) and (ii) 54 hours (the maximum).

b. The first 11 plants were Japanese facilities and the rest were not. Let x_1 = time to assemble vehicle and x_2 = whether facility is Japanese (1 = yes, 0 = no). The fit of the multiple regression model is $\hat{y} = 105.0 - 0.78x_1 - 36.0x_2$. Interpret the coefficients that estimate the effect of x_1 and the effect of x_2.

c. Explain why part a and part b indicate that Simpson's paradox has occurred.

d. Explain *how* Simpson's paradox occurred. To do this, construct a scatterplot between y and x_1 in which points are identified by whether the facility is Japanese. Note that the Japanese facilities tended to have low values for both x_1 and x_2.

13.44 Predicting hamburger sales A chain restaurant that specializes in selling hamburgers wants to analyze how y = sales for a customer (the total amount spent by a customer on food and drinks, in dollars) depends on the location of the restaurant, which is classified as inner city, suburbia, or at an interstate exit.

a. Construct indicator variables x_1 for inner city and x_2 for suburbia so you can include location in a regression equation for predicting the sales.

b. For part a, suppose $\hat{y} = 5.8 - 0.7x_1 + 1.2x_2$. Find the difference between the estimated mean sales in suburbia and at interstate exits.

13.45 Houses, size, and garage Use the House Selling Prices OR data file on the text CD to regress selling price in thousands on house size and whether the house has a garage.

a. Report the prediction equation. Find and interpret the equations predicting selling price using house size, for homes with and without a garage.

b. How do you interpret the coefficient of the indicator variable for whether the home has a garage?

13.46 House size and garage interact? Refer to the previous TRY exercise.

a. Explain what the no interaction assumption means for this model.

b. Sketch a hypothetical scatter diagram, showing points identified by garage or no garage, suggesting that there is actually a substantial degree of interaction.

13.47 Equal high jump for men and women Refer to TRY Example 11 and Exercise 13.40, with $\hat{y}$ = predicted winning high jump and x_1 = number of years since 1928. When equations are fitted *separately* for males and for females, we get $\hat{y} = 1.98 + 0.0055x_1$ for males and $\hat{y} = 1.60 + 0.0065x_1$ for females.

a. In allowing the lines to have different slopes, the overall model allows for ________________ between gender and year in their effects on the winning high jump. (Fill in the correct word.)

b. Show that both equations yield the same predicted winning high jump when $x_1 = 380$ (that is, 380 years after 1928, or in 2308).

c. Is it sensible to use this model to predict that in the year 2308 men and women will have about the same winning high jump? Why or why not?

13.48 Comparing sales You own a gift shop that has a campus location and a shopping mall location. You want to compare the regressions of y = daily total sales on x = number of people who enter the shop, for total sales listed by day at the campus location and at the mall location. Explain how you can do this using regression modeling

a. With a single model, having an indicator variable for location, that assumes the slopes are the same for each location.

b. With separate models for each location, permitting the slopes to be different.

13.6 Modeling a Categorical Response

The regression models studied so far are designed for a quantitative response variable y. When y is categorical, a different regression model applies, called **logistic regression**. Logistic regression can model

- A voter's choice in an election (Democrat or Republican), with explanatory variables of annual income, political ideology, religious affiliation, and race.

- Whether a credit card holder pays his or her bill on time (yes or no), with explanatory variables of family income and the number of months in the past year that the customer paid the bill on time.

We'll study logistic regression in this section for the special case of **binary** y.

The Logistic Regression Model

Recall

Section 2.3 (Example 12) showed that **a proportion for a category is a mean calculated for a binary variable** that has score 1 for that category and score 0 otherwise. ◀

Denote the possible outcomes for y by 0 and 1. We'll use the generic terms *failure* and *success* for these outcomes. The population mean of the 0 and 1 scores equals the population *proportion* of 1 outcomes (successes) for the response variable. That is, $\mu_y = p$, where p denotes the population proportion of successes. This proportion also represents the *probability* that a randomly selected subject has a success outcome. The model describes how p depends on the values of the explanatory variables.

For a single explanatory variable, x, the straight-line regression model is

$$p = \alpha + \beta x.$$

As Figure 13.10 shows, this model implies that p falls below 0 or above 1 for sufficiently small or large x-values. However, a proportion must fall between 0 and 1. Although the straight-line model may be valid over a restricted range of x-values, it is usually inadequate when there are multiple explanatory variables.

Figure 13.10 also shows a more realistic model. It has a curved, S-shape instead of a straight-line trend. The regression equation that best models this S-shaped curve is known as the **logistic regression equation.**

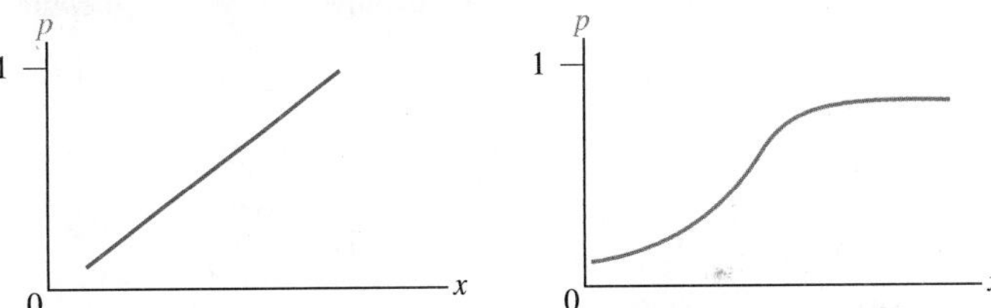

▲ **Figure 13.10 Two Possible Regressions for a Probability p of a Binary Response Variable.** A straight line is usually less appropriate than an S-shaped curve. **Question** Why is the straight-line regression model for a binary response variable often poor?

> **Logistic Regression**
>
> A regression equation for an S-shaped curve for the probability of success p is
>
> $$p = \frac{e^{\alpha+\beta x}}{1 + e^{\alpha+\beta x}}.$$
>
> This equation for p is called the **logistic regression** equation. Logistic regression is used when the response variable has only two possible outcomes (it's binary.)

Here, e raised to a power represents the *exponential function* evaluated at that number. Most calculators have an e^x key that provides values of e raised to a power. The model has two parameters, α and β. Since the numerator of the formula for p is smaller than the denominator, the model forces p to fall between 0 and 1. With a regression model having this S-shape, the probability p falls between 0 and 1 for all possible x values.

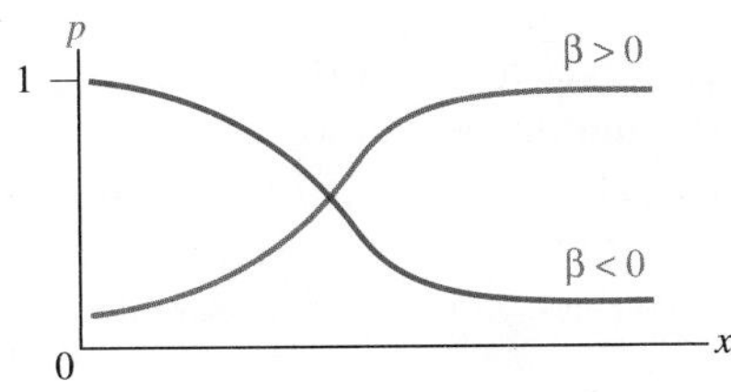

Like the slope of a straight line, the parameter β in this model refers to whether the mean of y increases or decreases as x increases. When $\beta > 0$, the probability p increases as x increases. When $\beta < 0$, the probability p decreases as x increases. See the margin figure. If $\beta = 0$, p does not change as x changes, so the curve flattens to a horizontal straight line. The steepness of the curve increases as the absolute value of β increases. However, unlike in the straight-line model, β is *not* a slope—the change in the mean per one-unit change in x. For this S-shaped curve, the rate at which the curve climbs or descends *changes* according to the value of x.

Software can estimate the parameters α and β in the logistic regression model. Calculators or software can find estimated probabilities $\hat{p}$ based on the model fit.

Example 12

Logistic regression model

Travel Credit Cards

Picture the Scenario

An Italian study with 100 randomly selected Italian adults considered factors associated with whether a person has at least one travel credit card. Table 13.12 shows results for the first 15 people on this response variable and on the person's annual income, in thousands of euros.[2] The complete data set is in the Credit Card and Income data file on the text CD. Let x = annual income and let y = whether the person has a travel credit card (1 = yes, 0 = no).

Table 13.12 Annual Income (in thousands of euros) and Whether Person Has a Travel Credit Card

The response y equals 1 if a person has a travel credit card and equals 0 otherwise. The complete data set is on the text CD.

Income	y	Income	y	Income	y
12	0	14	1	15	0
13	0	14	0	15	1
14	1	14	0	15	0
14	0	14	0	15	0
14	0	14	0	15	0

Source: Data from R. Piccarreta, Bocconi University, Milan (personal communication).

Questions to Explore

Table 13.13 shows what software provides for conducting a logistic regression analysis.

Table 13.13 Results of Logistic Regression for Italian Credit Card Data

Predictor	Coef	SE Coef	Z	P
Constant	−3.5180	0.71034	−4.95	0.000
income	0.1054	0.02616	4.03	0.000

a. State the prediction equation for the probability of owning a travel credit card, and explain whether annual income has a positive or a negative effect.

b. Find the estimated probability of having a travel credit card at the lowest and highest annual income levels in the sample, which were $x = 12$ and $x = 65$.

[2]The data were originally recorded in Italian lira but have been changed to euros and adjusted for inflation.

Think It Through

a. Substituting the α and β estimates from Table 13.13 into the logistic regression model formula, $p = e^{\alpha+\beta x}/(1 + e^{\alpha+\beta x})$, we get the equation for the estimated probability $\hat{p}$ of having a travel credit card,

$$\hat{p} = \frac{e^{-3.52+0.105x}}{1 + e^{-3.52+0.105x}}.$$

Because the estimate 0.105 of β (the coefficient of x) is positive, this sample suggests that annual income has a positive effect: The estimated probability of having a travel credit card is higher at higher levels of annual income.

b. For subjects with income $x = 12$ thousand euros, the estimated probability of having a travel credit card equals

$$\hat{p} = \frac{e^{-3.52+0.105(12)}}{1 + e^{-3.52+0.105(12)}} = \frac{e^{-2.26}}{1 + e^{-2.26}} = \frac{0.104}{1.104} = 0.09.$$

For $x = 65$, the highest income level in this sample, you can check that the estimated probability equals 0.97.

Insight

There is a strong effect. The estimated probability of having a travel credit card changes from 0.09 to 0.97 (nearly 1.0) as annual income changes over its range.

Using software, we could also fit the straight-line regression model. Its prediction equation is $\hat{p} = -0.159 + 0.0188x$. However, its $\hat{p}$ predictions are quite different at the low end and at the high end of the annual income scale. At $x = 65$, for instance, it provides the prediction $\hat{p} = 1.06$. This is a poor prediction because we know that a proportion must fall between 0 and 1.

Try Exercise 13.50

Ways to Interpret the Logistic Regression Model

The sign of β in the logistic regression model tells us whether the probability p increases or decreases as x increases. How else can we interpret the model parameters and their estimates? Here, we'll describe three ways.

1. To describe the effect of x, you can compare estimates of p at two different values of x. One possibility is to compare $\hat{p}$ found at the minimum and maximum values of x, as we did in Example 12. There we saw that the estimated probability changed from 0.09 to 0.97, a considerable change. An alternative is to instead use values of x to evaluate this probability that are not as affected by outliers, such as the first and third quartiles.
2. It's good to know the value of x at which $p = 0.50$, that is, that value of x for which each outcome is equally likely. This depends on the logistic regression parameters α and β. It can be shown that $x = -\alpha/\beta$ when $p = 0.50$. (Exercise 13.92).
3. The simplest way to use the logistic regression parameter β to interpret the *steepness* of the curve uses a straight-line approximation. Because the logistic regression formula is a curve rather than a straight line, β itself is no longer the ordinary slope. At the x-value where $p = 0.50$, the line drawn tangent to the logistic regression curve has slope $\beta/4$. See Figure 13.11. The value $\beta/4$ represents the approximate change in the probability p for a one-unit increase in x, when x is close to the value at which $p = 0.50$. Tangent lines at other points have weaker slopes (Exercise 13.91), close to 0 when p is near 0 or 1.

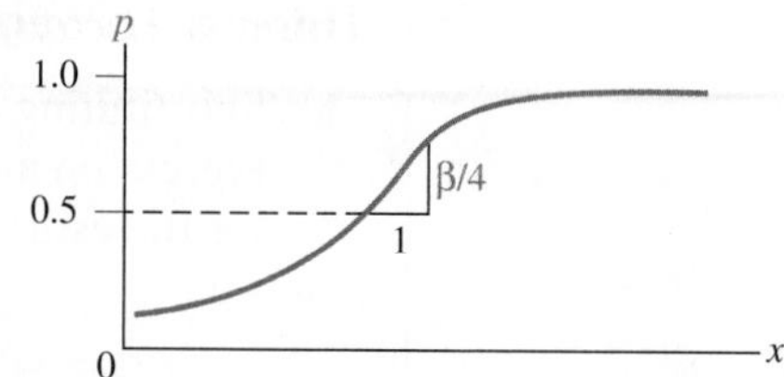

▲ **Figure 13.11** When the Probability $p = 0.50$, for a One-unit Change in x, p Changes by About $\beta/4$. Here, β is the coefficient of x from the logistic regression model.

Example 13

Interpreting a logistic regression model

Effect of Income on Credit Card Use

Picture the Scenario

For the Italian travel credit card study, Example 12 found that the equation

$$\hat{p} = \frac{e^{-3.52+0.105x}}{1 + e^{-3.52+0.105x}}$$

estimates the probability p of having such a credit card as a function of annual income x.

Questions to Explore

Interpret this equation by (a) finding the income value at which the estimated probability of having a credit card equals 0.50 and (b) finding the approximate rate of change in the probability at that income value.

Think It Through

a. The estimate of α is -3.52 and the estimate of β is 0.105. Substituting the estimates into the expression $-\alpha/\beta$ for the value of x at $p = 0.50$, we get $x = (3.52)/(0.105) = 33.5$. The estimated probability of having a travel credit card equals 0.50 when annual income equals €33,500.

b. A line drawn tangent to the logistic regression curve at the point where $p = 0.50$ has slope equal to $\beta/4$. The estimate of this slope is $0.105/4 = 0.026$. For each increase of €1000 in annual income near the income value of €33,500, the estimated probability of having a travel credit card increases by approximately 0.026.

Insight

Figure 13.12 shows the estimated logistic regression curve, highlighting what we've learned in Examples 12 and 13: The estimated probability increases from 0.09 to 0.97 between the minimum and maximum income values, and it equals 0.50 at an annual income of €33,500.

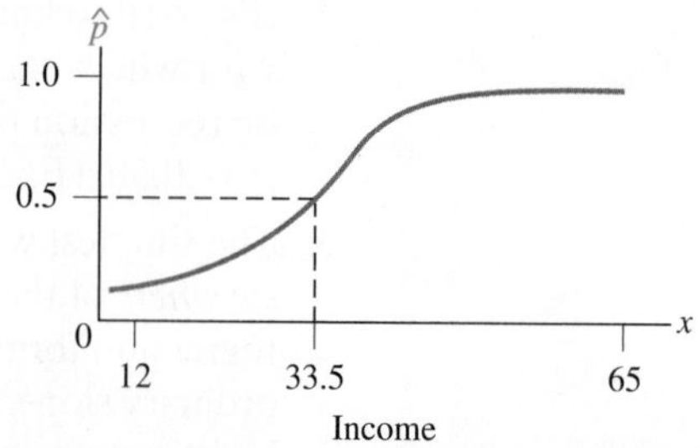

▲ **Figure 13.12** Logistic Regression Curve Relating Estimated Probability $\hat{p}$ of Having a Travel Credit Card to Annual Income x.

Try Exercise 13.54

Inference for Logistic Regression

Table 13.13 was

Predictor	Coef	SE	Z	P
Constant	−3.518	0.710	−4.95	0.000
income	0.105	0.0262	4.03	0.000

Software also reports a z test statistic for the hypothesis H_0: $\beta = 0$. When $\beta = 0$, the probability of possessing a travel credit card is the same at all income levels. The test statistic equals the ratio of the estimate b of β divided by its standard error. Sections 8.2 and 9.2 showed that **inference about proportions uses z test statistics** rather than t test statistics. From Table 13.13 (shown again in the margin), the z test statistic equals

$$z = (b - 0)/se = (0.105 - 0)/0.0262 = 4.0.$$

This has a P-value of 0.000 for H_a: $\beta \neq 0$. Since the sample slope is positive, for the population of adult Italians, there is strong evidence of a positive association between annual income and having a travel credit card.

The result of this test is no surprise. We would expect people with higher annual incomes to be more likely to have travel credit cards. Some software can construct a confidence interval for the probability p at various x levels. A 95% confidence interval for the probability of having a credit card equals (0.04, 0.19) at the lowest sample income level of €12,000, and (0.78, 0.996) at the highest sample income level of €65,000.

Multiple Logistic Regression

Just as ordinary regression extends to handle several explanatory variables, so does logistic regression. Also, logistic regression can include categorical explanatory variables using indicator variables.

Example 14

Multiple logistic regression

Estimating Proportion of Students Who've Used Marijuana

Picture the Scenario

Table 13.14 is a three-variable contingency table from a Wright State University survey asking senior high-school students near Dayton, Ohio, whether they had ever used alcohol, cigarettes, or marijuana. We'll treat marijuana use as the response variable and cigarette use and alcohol use as explanatory variables.

Table 13.14 Alcohol, Cigarette, and Marijuana Use for High School Seniors

		Marijuana Use	
Alcohol Use	**Cigarette Use**	**Yes**	**No**
Yes	Yes	911	538
	No	44	456
No	Yes	3	43
	No	2	279

Source: Data from Professor Harry Khamis, Wright State University (personal communication).

Questions to Explore

Let y indicate marijuana use, coded 1 = yes, 0 = no. Let x_1 be an indicator variable for alcohol use (1 = yes, 0 = no), and let x_2 be an indicator variable for cigarette use (1 = yes, 0 = no). Table 13.15 shows MINITAB output for a logistic regression model.

Table 13.15 MINITAB Output for Estimating the Probability of Marijuana Use Based on Alcohol Use and Cigarette Use

Predictor	Coef	SE Coef	Z	P
Constant	−5.30904	0.475190	−11.17	0.000
alcohol	2.98601	0.464671	6.43	0.000
cigarettes	2.84789	0.163839	17.38	0.000

a. Report the prediction equation, and interpret.

b. Find the estimated probability $\hat{p}$ of having used marijuana (i) for students who have not used alcohol or cigarettes and (ii) for students who have used both alcohol and cigarettes.

Think It Through

a. From Table 13.15, the logistic regression prediction equation is

$$\hat{p} = \frac{e^{-5.31+2.99x_1+2.85x_2}}{1 + e^{-5.31+2.99x_1+2.85x_2}}.$$

The coefficient of alcohol use (x_1) is positive (2.99). The indicator variable x_1 equals 1 for those who've used alcohol. Thus, the alcohol users have a higher estimated probability of using marijuana, controlling for whether they used cigarettes. Likewise, the coefficient of cigarette use (x_2) is positive (2.85), so the cigarette users have a higher estimated probability of using marijuana, controlling for whether they used alcohol. Table 13.15 tells us that for each predictor, the test statistic is large. In other words, the estimated effect is a large number of standard errors from 0. The P-values are both 0.000, so there is strong evidence that the corresponding population effects are positive also.

b. For those who have not used alcohol or cigarettes, $x_1 = x_2 = 0$. For them, the estimated probability of marijuana use is

$$\hat{p} = \frac{e^{-5.31+2.99(0)+2.85(0)}}{1 + e^{-5.31+2.99(0)+2.85(0)}} = \frac{e^{-5.31}}{1 + e^{-5.31}} = \frac{0.0049}{1.0049} = 0.005.$$

For those who have used alcohol and cigarettes, $x_1 = x_2 = 1$. For them, the estimated probability of marijuana use is

$$\hat{p} = \frac{e^{-5.31+2.99(1)+2.85(1)}}{1 + e^{-5.31+2.99(1)+2.85(1)}} = 0.629.$$

In summary, the probability that students have tried marijuana seems highly related on whether they've used alcohol and cigarettes.

Insight

Likewise, you can find the estimated probability of using marijuana for those who have used alcohol but not cigarettes (let $x_1 = 1$ and $x_2 = 0$) and for those who have not used alcohol but have used cigarettes. Table 13.16 summarizes results. We see that marijuana use is unlikely unless a student has used both alcohol and cigarettes.

Table 13.16 Estimated Probability of Marijuana Use, by Alcohol Use and Cigarette Use, Based on Logistic Regression Model

The sample proportions of marijuana use are shown in parentheses for the four cases.

	Cigarette Use	
Alcohol Use	**Yes**	**No**
Yes	0.629 (0.629)	0.089 (0.088)
No	0.079 (0.065)	0.005 (0.007)

Try Exercises 13.56 and 13.57

Checking the Logistic Regression Model

		Marijuana	
Alcohol	**Cigarette**	**Yes**	**No**
Yes	Yes	911	538
	No	44	456
No	Yes	3	43
	No	2	279

How can you check whether a logistic regression model fits the data well? When the explanatory variables are categorical, you can find the sample proportions for the outcome of interest. Table 13.16 also shows these for the marijuana use example. For instance, for those who had used alcohol and cigarettes, from Table 13.14 (shown again in the margin) the sample proportion who had used marijuana is $911/(911 + 538) = 0.629$. Table 13.16 shows that the sample proportions are close to the estimated proportions generated by the model. The model seems to fit well.

Table 13.16 may suggest the question, why bother to fit the model? Why not merely inspect a table of sample proportions? One important reason is that the model enables us to easily conduct inferences about the effects of explanatory variables, while controlling for other variables. For instance, if we want to test the effect of alcohol use on marijuana use, controlling for cigarette use, we can test H_0: $\beta_1 = 0$ in the model with alcohol use and cigarette use as predictors.

13.6 Practicing the Basics

13.49 Income and credit cards Example 12 used logistic regression to estimate the probability of having a travel credit card when x = annual income (in thousands of euros). At the mean income of €25,000, show that the estimated probability of having a travel credit card equals 0.29.

13.50 Hall of Fame induction Baseball's highest honor is election to the Hall of Fame. The history of the election process, however, has been filled with controversy and accusations of favoritism. Most recently, there is also the discussion about players who used performance enhancement drugs. The Hall of Fame has failed to define what the criteria for entry should be. Several statistical models have attempted to describe the probability of a player being offered entry into the Hall of Fame. How does hitting 400 or 500 home runs affect a player's chances of being enshrined? What about having a .300 average or 1500 RBI? One factor, the number of home runs, is examined by using logistic regression as the probability of being elected:

TRY

$$P(\text{HOF}) = \frac{e^{-6.7+0.0175\text{HR}}}{1 + e^{-6.7+0.0175\text{HR}}}.$$

a. Compare the probability of election for two players who are 10 home runs apart—say, 369 home runs versus 359 home runs.

b. Compare the probability of election for a player with 475 home runs versus the probability for a player with 465 home runs. (These happen to be the figures for Willie Stargell and Dave Winfield.)

13.51 Horseshoe crabs A study of horseshoe crabs by zoologist Dr. Jane Brockmann at the University of Florida used logistic regression to predict the probability that a female crab had a male partner nesting nearby. One explanatory variable was x = weight of the female crab (in kilograms). The results were

```
Predictor     Coef
Constant    -3.695
Weight       1.815
```

The quartiles for weight were Q1 = 2.00, Q2 = 2.35, and Q3 = 2.85.

a. Find the estimated probability of a male partner at Q1 and at Q3.

b. Interpret the effect of weight by estimating how much the probability increases over the middle half of the sampled weights, between Q1 and Q3.

13.52 More crabs Refer to the previous exercise. For what weight values do you estimate that a female crab has probability (a) 0.50, (b) greater than 0.50, and (c) less than 0.50, of having a male partner nesting nearby?

13.53 Voting and income A logistic regression model describes how the probability of voting for the Republican candidate in a presidential election depends on x, the voter's total family income (in thousands of dollars) in the previous year. The prediction equation for a particular sample is

$$\hat{p} = \frac{e^{-1.00+0.02x}}{1 + e^{-1.00+0.02x}}.$$

Find the estimated probability of voting for the Republican candidate when (a) income = \$10,000, (b) income = \$100,000. Describe how the probability seems to depend on income.

13.54 Equally popular candidates Refer to the previous exercise.

TRY

a. At which income level is the estimated probability of voting for the Republican candidate equal to 0.50?

b. Over what region of income values is the estimated probability of voting for the Republican candidate (i) greater than 0.50 and (ii) less than 0.50?

c. At the income level for which $\hat{p} = 0.50$, give a linear approximation for the change in the probability for each \$1000 increase in income.

13.55 Many predictors of voting Refer to the previous two exercises. When the explanatory variables are x_1 = family income, x_2 = number of years of education, and x_3 = gender (1 = male, 0 = female), suppose a logistic regression reports

```
Predictor      Coef      SE Coef
Constant      -2.40      0.12
income         0.02      0.01
education      0.08      0.05
gender         0.20      0.06
```

For this sample, x_1 ranges from 6 to 157 with a standard deviation of 25, and x_2 ranges from 7 to 20 with a standard deviation of 3.

a. Interpret the effects using the sign of the coefficient for each predictor.

b. Illustrate the gender effect by finding and comparing the estimated probability of voting Republican for (i) a man with 16 years of education and income \$40,000 and (ii) a woman with 16 years of education and income \$40,000.

13.56 Graduation, gender, and race The U.S. Census Bureau

TRY

lists college graduation numbers by race and gender. The table shows the data for graduating 25-year-olds.

College graduation

Group	Sample Size	Graduates
White females	31,249	10,781
White males	39,583	10,727
Black females	13,194	2,309
Black males	17,707	2,054

Source: J. J. McArdle and F. Hamagami, *J. Amer. Statist. Assoc.*, vol. 89 (1994), pp. 1107–1123. Data from U.S. Census Bureau, American Community Survey 2005–2007.

a. Identify the response variable.

b. Express the data in the form of a three-variable contingency table that cross-classifies whether graduated (yes, no), race, and gender.

c. When we use indicator variables for race (1 = white, 0 = black) and for gender (1 = female, 0 = male), the coefficients of those predictors in the logistic regression model are 0.975 for race and 0.375 for gender. Based on these estimates, which race and gender combination has the highest estimated probability of graduation? Why?

13.57 Death penalty and race The three-dimensional contin-

TRY

gency table shown is from a study of the effects of racial characteristics on whether or not individuals convicted of homicide receive the death penalty. The subjects classified were defendants in indictments involving cases with multiple murders in Florida between 1976 and 1987.

Death penalty verdict by defendant's race and victims' race

Defendant's Race	Victims' Race	Death Penalty Yes	Death Penalty No	Percent Yes
White	White	53	414	11.3
	Black	0	16	0.0
Black	White	11	37	22.9
	Black	4	139	2.8

Source: Data from M. L. Radelet and G. L. Pierce, *Florida Law Rev.*, vol. 43, 1991, pp. 1–34.

a. Based on the percentages shown, controlling for victims' race, for which defendant's race was the death penalty more likely?

b. Let y = death penalty verdict (1 = yes, 0 = no), let d be an indicator variable for defendant's race (1 = white, 0 = black), and let v be an indicator variable for victims' race (1 = white, 0 = black). The logistic regression prediction equation is

$$\hat{p} = \frac{e^{-3.596-0.868d+2.404v}}{1 + e^{-3.596-0.868d+2.404v}}.$$

According to this equation, for which of the four groups is the death penalty most likely? Explain your answer.

13.58 Death penalty probabilities Refer to the previous exercise.

a. Based on the prediction equation, when the defendant is black and the victims were white, show that the estimated death penalty probability is 0.233.

b. The model-estimated probabilities are 0.011 when the defendant is white and victims were black, 0.113 when the defendant and the victims were white, and 0.027 when the defendant and the victims were black. Construct a table cross-classifying defendant's race by victims' race and show the estimated probability of the death penalty for each cell. Use this to interpret the effect of defendant's race.

c. Collapse the contingency table over victims' race, and show that (ignoring victims' race) white defendants were more likely than black defendants to get the death penalty. Comparing this with what happens when you control for victims' race, explain how Simpson's paradox occurs.

Chapter Review

ANSWERS TO THE CHAPTER FIGURE QUESTIONS

Figure 13.1 Selling price is the response variable, and the graphs that show it as the response variable are in the first row.

Figure 13.2 The coefficient (15,170) of x_2 = number of bedrooms is positive, so increasing it has the effect of increasing the predicted selling price (for fixed house size).

Figure 13.3 We use only the right tail because larger F values provide greater evidence against H_0.

Figure 13.4 A histogram that shows a few observations that have extremely large or extremely small standardized residuals, well removed from the others.

Figure 13.5 The pattern suggests that y tends to be below $\hat{y}$ for very small and large x_1-values and above $\hat{y}$ for medium-sized x_1-values. The effect of x_1 appears to be better modeled by a mathematical function that has a parabolic shape.

Figure 13.6 As house size increases, the variability of the standardized residuals appears to increase, suggesting more variability in selling prices when house size is larger, for a given number of bedrooms. The number of bedrooms affect the analysis because they have an effect on the predicted values for the model, on which the residuals are based.

Figure 13.7 Each line has the same slope.

Figure 13.10 A straight line implies that p falls below 0 or above 1 for sufficiently small or large x-values. This is a problem, because a proportion must fall between 0 and 1.

CHAPTER SUMMARY

This chapter generalized regression to include more than one explanatory variable in the model. The **multiple regression model** relates the mean μ_y of a response variable y to several explanatory variables, for instance,

$$\mu_y = \alpha + \beta_1 x_1 + \beta_2 x_2$$

for two predictors. Advantages over bivariate regression include better predictions of y and being able to study the effect of an explanatory variable on y while *controlling* (keeping fixed) values of other explanatory variables in the model.

- The **regression parameters** (the betas) are slopes that represent effects of the explanatory variables while controlling for the other variables in the model. For instance, β_1 represents the change in the mean of y for a one-unit increase in x_1, at fixed values of the other explanatory variables.
- The **multiple correlation** R and its square (R^2) describe the predictability of the response variable y by the set of explanatory variables. The multiple correlation R equals the correlation between the observed and predicted y values. Its square, R^2, is the proportional reduction in error from predicting y using the prediction equation instead of using $\bar{y}$ (and ignoring x). Both R and R^2 fall between 0 and 1, with larger values representing stronger association.
- An ***F* statistic** tests H_0: $\beta_1 = \beta_2 = \ldots = 0$, which states that y is independent of all the explanatory variables in the model. The F test statistic equals a ratio of mean squares. A small P-value suggests that at least one explanatory variable affects the response.
- Individual ***t*** **tests** and **confidence intervals** for each β parameter analyze separate population effects of each explanatory variable, controlling for the other variables in the model.
- Categorical explanatory variables can be included in a regression model using **indicator variables.** With two categories, the indicator variable equals 1 when the observation is in the first category and 0 when it is in the second.
- For binary response variables, the **logistic regression model** describes how the probability of a particular category depends on the values of explanatory variables. An S-shaped curve describes how the probability changes as the value of a quantitative explanatory variable increases.

Table 13.17 summarizes the basic properties and inference methods for multiple regression and those that Chapter 12 introduced for bivariate regression, with quantitative variables.

Table 13.17 Summary of Bivariate and Multiple Regression

	Bivariate Regression	**Multiple Regression**	
Model formula	$\mu_y = \alpha + \beta x$	$\mu_y = \alpha + \beta_1 x_1 + \beta_2 x_2 + \ldots$	
		Simultaneous Effect	**Separate Effects**
Properties of measures	β = slope r = sample correlation $-1 \le r \le 1$	R = multiple correlation $0 \le R \le 1$	$\beta_1, \beta_2, \ldots$ are slopes for each predictor at fixed values of others
	r^2 is proportional reduction in error (PRE)	R^2 is PRE	
Hypotheses of no effect	H_0: $\beta = 0$	H_0: $\beta_1 = \beta_2 = \ldots = 0$	H_0:$\beta_1 = 0$, H_0:$\beta_2 = 0, \ldots$
Test statistic	$t = (b - 0)/se$ $df = n - 2$	F = (MS for regression)/MSE df_1 = no. predictors $df_2 = n -$ no. regression para.	$t = (b_1 - 0)/se, \ldots$ $df = n -$ no. regression para.

SUMMARY OF NOTATION

$x_1, x_2, \ldots$ = explanatory variables in multiple regression model
R = multiple correlation = correlation between observed y and predicted y values

R^2 = proportional reduction in prediction error, for multiple regression

F = test statistic for testing that all β parameters = 0 in regression model

$e^{\alpha+\beta x}$ = exponential function used in numerator and denominator of logistic regression equation, which models the probability p of a binary outcome by

$$p = e^{\alpha+\beta x}/(1 + e^{\alpha+\beta x}).$$

CHAPTER PROBLEMS

Practicing the Basics

13.59 House prices This chapter has considered many aspects of regression analysis. Let's consider several of them at once by using software with the House Selling Prices OR data file on the text CD to conduct a multiple regression analysis of y = selling price of home, x_1 = size of home, x_2 = number of bedrooms, x_3 = number of bathrooms.

a. Construct a scatterplot matrix. Identify the plots that pertain to selling price as a response variable. Interpret, and explain how the highly discrete nature of x_2 and x_3 affects the plots.

b. Fit the model. Write down the prediction equation, and interpret the coefficient of size of home by its effect when x_2 and x_3 are fixed.

c. Show how to calculate R^2 from SS values in the ANOVA table. Interpret its value in the context of these variables.

d. Find and interpret the multiple correlation.

e. Show all steps of the F test that selling price is independent of these predictors. Explain how to obtain the F statistic from the mean squares in the ANOVA table.

f. Report the t statistic for testing H_0: $\beta_2 = 0$. Report the P-value for H_a: $\beta_2 < 0$, and interpret. Why do you think this effect is not significant? Does this imply that the number of bedrooms is not associated with selling price?

g. Construct and examine the histogram of the residuals for the multiple regression model. What does this describe, and what does it suggest?

h. Construct and examine the plot of the residuals plotted against size of home. What does this describe, and what does it suggest?

13.60 Predicting body strength In Chapter 12, we analyzed strength data for a sample of female high school athletes. When we predict the maximum number of pounds the athlete can bench press using the number of times she can do a 60-pound bench press (BP_60), we get $r^2 = 0.643$. When we add the number of times an athlete can perform a 200-pound leg press (LP_200) to the model, we get $\hat{y} = 60.6 + 1.33(\text{BP\_60}) + 0.21(\text{LP\_200})$ and $R^2 = 0.656$.

a. Find the predicted value and residual for an athlete who has BP = 85, $BP_{60} = 10$, and $LP_{200} = 20$.

b. Find the prediction equation for athletes who have $LP_{200} = 20$, and explain how to interpret the slope for BP_60.

c. Note that $R^2 = 0.656$ for the multiple regression model is not much larger than $r^2 = 0.643$ for the bivariate model with LP_{200} as the only explanatory variable. What does this suggest?

13.61 Softball data Refer to the Softball data set on the text CD. Regress the difference (DIFF) between the number of runs scored by that team and by the other team on the number of hits (HIT) and the number of errors (ERR).

a. Report the prediction equation, and interpret the slopes.

b. From part a, approximately how many hits does the team need so that the predicted value of DIFF is positive (corresponding to a predicted win), if they can play error-free ball (ERR = 0)?

13.62 Violent crime A MINITAB printout is provided from fitting the multiple regression model to U.S. crime data for the 50 states (excluding Washington, D.C.), on y = violent crime rate, x_1 = poverty rate, and x_2 = percent living in urban areas.

a. Predict the violent crime rate for Massachusetts, which has violent crime rate = 476, poverty rate = 10.2%, and urbanization = 92.1%. Find the residual, and interpret.

b. Interpret the effects of the predictors by showing the prediction equation relating y and x_1 for states with (i) $x_2 = 0$ and (ii) $x_2 = 100$. Interpret.

Regression of violent crime rate on poverty rate and urbanization

Predictor	Coef	SE Coef	T	P
Constant	−270.7	121.1	−2.23	0.030
poverty	28.334	7.249	3.91	0.000
urbanization	5.416	1.035	5.23	0.000

13.63 Effect of poverty on crime Refer to the previous exercise. Now we add x_3 = percentage of single-parent families to the model. The SPSS table on the next page shows results. Without x_3 in the model, poverty has slope 28.33, and when x_3 is added, poverty has slope 14.95. Explain the differences in the interpretations of these two slopes.

Violent crime predicted by poverty rate, urbanization, and single-parent family

		Coefficients[a]			
		Unstandardized Coefficients			
Model		B	Std. Error	t	Sig.
1	(Constant)	−631.700	149.604	−4.222	.000
	poverty	14.953	7.540	1.983	.053
	urbanization	4.406	.973	4.528	.000
	singleparent	25.362	7.220	3.513	.001

[a] Dependent Variable: violent crime rate.

13.64 Modeling fertility For the World Data for Fertility and Literacy data file on the text CD, a MINITAB printout follows that shows fitting a multiple regression model for y = fertility, x_1 = adult literacy rate (both sexes), x_2 = combined educational enrollment (both sexes). Report the value of each of the following:

a. r between y and x_1

b. R^2

c. Total sum of squares

d. Residual sum of squares

e. Standard deviation of y

f. Residual standard deviation of y

g. Test statistic value for H_0: $\beta_1 = 0$

h. P-value for H_0: $\beta_1 = \beta_2 = 0$

```
Analysis of fertility, literacy, and combined
educational enrollment:
Variable                   N    Mean   SE Mean  StDev  Median
Adolescent fertility     142   60.42     3.73   44.50   49.55
Adult literacy rate
(both sexes)             142   80.80     1.64   19.56   88.40
Combined educ enrol
(both sexes)             142   68.59     1.32   15.79   71.20

Predictor                    Coef    SE Coef      T      P
Constant                   187.95     12.92   14.55  0.000
Adult literacy rate
(both sexes)              -1.2590    0.2416   -5.21  0.000
Combined educ
enrollment(both sexes)    -0.3762    0.2994   -1.26  0.211

S = 33.4554 R-Sq = 44.3%
Analysis of Variance
Source           DF      SS      MS      F      P
Regression        2  123582   61791  55.21  0.000
Residual Error  139  155577    1119
Total           141  279160
Correlations:
                      Fertility  Adult literacy
Adult literacy          -0.661
Combined educ enroll    -0.578       0.803
```

13.65 Significant fertility prediction? Refer to the previous exercise.

a. Show how to construct the F statistic for testing H_0: $\beta_1 = \beta_2 = 0$ from the reported mean squares, report its P-value, and interpret.

b. If these are the only nations of interest to us for this study, rather than a random sample of such nations, is this significance test relevant? Explain.

13.66 Motivation to study medicine What motivates someone to pursue the study of medicine? A student's interest and motivation to study medicine can depend on the strength of motivation and career-related values and approaches to learning. Validated and reliable questionnaires were used to obtain data from 116 first-year medical students. This study found no differences in strength of motivation based on sex, nationality, or age. It did find that the motivation to enter medical school was based on interpersonal factors such as wanting to help people, being respected and successful, and fulfilling a sense of achievement. Students motivated to perform better were driven either by positive self-esteem or by perceiving medicine as a means of enhancing their social status. Students who set their own goals and work toward these goals are more likely to succeed. A regression analysis for predicting motivation score is shown below.

	Coefficients	Std Error	P-value
Constant	44.677	6.749	0.000
Strategic approach to learning	0.252	0.065	0.000
Avoidance of role strain	−0.340	0.135	0.013
Prestige	0.451	0.117	0.000
Lack of purpose	−0.755	0.230	0.001
Income	−0.243	0.114	0.035

a. Find the equation relating predicted motivation score to the explanatory factors listed.

b. Interpret the signs of the coefficients in the equation for each of the explanatory variables as the explanatory variable relates to motivation.

13.67 Education and gender in modeling income Consider the relationship between $\hat{y}$ = annual income (in thousands of dollars) and x_1 = number of years of education, by x_2 = gender. Many studies in the United States have found that the slope for a regression equation relating y to x_1 is larger for men than for women. Suppose that in the population, the regression equations are $\mu_y = -10 + 4x_1$ for men and $\mu_y = -5 + 2x_1$ for women. Explain why these equations imply that there is *interaction* between education and gender in their effects on income.

13.68 Horseshoe crabs and width A study of horseshoe crabs found a logistic regression equation for predicting the probability that a female crab had a male partner nesting nearby using x = width of the carapace shell of the female crab (in centimeters). The results were

```
Predictor  Coef
Constant   -12.351
Width      0.497
```

a. For width, Q1 = 24.9 and Q3 = 27.7. Find the estimated probability of a male partner at Q1 and at Q3. Interpret the effect of width by estimating the increase in the probability over the middle half of the sampled widths.

b. At which carapace shell width level is the estimated probability of a male partner (i) equal to 0.50, (ii) greater than 0.50, and (iii) less than 0.50?

13.69 AIDS and AZT In a study (reported in *New York Times*, February 15, 1991) on the effects of AZT in slowing the development of AIDS symptoms, 338 veterans whose immune systems were beginning to falter after infection with the AIDS virus were randomly assigned either to receive AZT immediately or to wait until their T cells showed severe immune weakness. The study classified the veterans' race, whether they received AZT immediately, and whether they developed AIDS symptoms during the three-year study. Let x_1 denote whether used AZT (1 = yes, 0 = no) and let x_2 denote race (1 = white, 0 = black). A logistic regression analysis for the probability of developing AIDS symptoms gave the prediction equation

$$\hat{p} = \frac{e^{-1.074-0.720x_1+0.056x_2}}{1 + e^{-1.074-0.720x_1+0.056x_2}}.$$

a. Interpret the sign of the effect of AZT.

b. Show how to interpret the AZT effect for a particular race by comparing the estimated probability of AIDS symptoms for black veterans with and without immediate AZT use.

c. The *se* value was 0.279 for the AZT use effect. Does AZT use have a significant effect, at the 0.05 significance level? Show all steps of a test to justify your answer.

13.70 Factors affecting first home purchase The table summarizes results of a logistic regression model for predictions about first home purchase by young married households. The response variable is whether the subject owns a home (1 = yes, 0 = no). The explanatory variables are husband's income, wife's income (each in ten-thousands of dollars), the number of years the respondent has been married, the number of children aged 0–17 in the household, and an indicator variable that equals 1 if the subject's parents owned a home in the last year the subject lived in the parental home.

a. Explain why, other things being fixed, the probability of home ownership increases with husband's earnings, wife's earnings, the number of children, and parents' home ownership.

b. From the table, explain why the number of years married seems to show little evidence of an effect, given the other variables in the model.

Results of logistic regression for probability of home ownership

Variable	Estimate	Std. Error
Husband earnings	0.569	0.088
Wife earnings	0.306	0.140
No. years married	−0.039	0.042
No. children	0.220	0.101
Parents' home ownership	0.387	0.176

Source: Data from J. Henretta, "Family Transitions, Housing Market Context, and First Home Purchase," *Social Forces*, vol. 66, 1987, pp. 520–536.

Concepts and Investigations

13.71 Student data Refer to the FL Student Survey data file on the text CD. Using software, conduct a regression analysis using y = college GPA and predictors high school GPA and sports (number of weekly hours of physical exercise). Prepare a report, summarizing your graphical analyses, bivariate models and interpretations, multiple regression models and interpretations, inferences, checks of effects of outliers, and overall summary of the relationships.

13.72 Why regression? In 100–200 words, explain to someone who has never studied statistics the purpose of multiple regression and when you would use it to analyze a data set or investigate an issue. Give an example of at least one application of multiple regression. Describe how multiple regression can be useful in analyzing complex relationships.

13.73 Modeling salaries The table shows results of fitting a regression model to data on Oklahoma State University salaries (in dollars) of 675 full-time college professors of different disciplines with at least two years of instructional employment. All of the predictors are categorical (binary), except for years as professor, merit ranking, and market influence. The market factor represents the ratio of the average salary at comparable institutions for the corresponding academic field and rank to the actual salary at OSU. Prepare a summary of the results in a couple of paragraphs, interpreting the effects of the predictors. The levels of ranking for professors are assistant, associate, and full professor from low to high. An instructor ranking is nontenure track. Gender and race predictors were not significant in this study.

Modeling professor salaries

Variable	Estimate	Std. Error
Intercept	17.870	0.272
Nontenure track	−0.010	0.011
Instructor	−0.284	0.018
Associate professor	0.170	0.013
Full professor	0.407	0.018
Years as professor	0.004	0.001
Average merit rating	0.044	0.005
Business	0.395	0.015
Education	0.053	0.015
Engineering	0.241	0.014
Fine arts	0.000	0.018
Social science	0.077	0.013
Market influence	−7.046	0.268

Note: Dependent variable is the logarithm of annual salary. Model summary: Adjusted R_2 = 0.94; *F*-ration = 411.76; N = 675; P-value = 0.001.

Source: Some data from *New Directions for Institutional Research*, no. 140, Winter 2008 (www.interscience.wiley.com).

13.74 Multiple choice: Interpret parameter If $\hat{y} = 2 + 3x_1 + 5x_2 - 8x_3$, then controlling for x_2 and x_3, the change in the estimated mean of y when x_1 is increased from 10 to 20

a. equals 30.

b. equals 0.3.

c. Cannot be given—depends on specific values of x_2 and x_3.

d. Must be the same as when we ignore x_2 and x_3.

13.75 Multiple choice: Interpret indicator In the model $\mu_y = \alpha + \beta_1 x_1 + \beta_2 x_2$, suppose that x_2 is an indicator variable for gender, equaling 1 for females and 0 for males.

a. We set $x_2 = 0$ if we want a predicted mean without knowing gender.

b. The slope effect of x_1 is β_1 for males and β_2 for females.

c. β_2 is the difference between the population mean of y for females and for males.

d. β_2 is the difference between the population mean of y for females and males, for all those subjects having x_1 fixed, such as $x_1 = 10$.

13.76 Multiple choice: Regression effects Multiple regression is used to model y = annual income using x_1 = number of years of education and x_2 = number of years employed in current job.

a. It is possible that the coefficient of x_2 is positive in a bivariate regression but negative in multiple regression.

b. It is possible that the correlation between y and x_1 is 0.30 and the multiple correlation between y and x_1 and x_2 is 0.26.

c. If the F statistic for H_0: $\beta_1 = \beta_2 = 0$ has a P-value = 0.001, then we can conclude that *both* predictors have an effect on annual income.

d. If $\beta_2 = 0$, then annual income is independent of x_2 in bivariate regression.

13.77 True or false: R and R^2 For each of the following statements, indicate whether it is true or false. If false, explain why it is false.

a. The multiple correlation is always the same as the ordinary correlation computed between the values of the response variable and the values $\hat{y}$ predicted by the regression model.

b. The multiple correlation is like the ordinary correlation in that it falls between -1 and 1.

c. R^2 describes how well you can predict y using x_1 when you control for the other variables in the multiple regression model.

d. It's impossible for R^2 to go down when you add explanatory variables to a regression model.

13.78 True or false: Regression For each of the following statements, indicate whether it is true or false. If false, explain why it is false. In regression analysis:

a. The estimated coefficient of x_1 can be positive in the bivariate model but negative in a multiple regression model.

b. When a model is refitted after y = income is changed from dollars to euros, R^2, the correlation between y and x_1, the F statistics and t statistics will not change.

c. If $r^2 = 0.6$ between y and x_1 and if $r^2 = 0.6$ between y and x_2, then for the multiple regression model with both predictors $R^2 = 1.2$.

d. The multiple correlation between y and $\hat{y}$ can equal -0.40.

13.79 True or false: Slopes For data on y = college GPA, x_1 = high school GPA, and x_2 = average of mathematics and verbal entrance exam score, we get $\hat{y} = 2.70 + 0.45x_1$ for bivariate regression and $\hat{y} = 0.3 + 0.40x_1 + 0.003x_2$ for multiple regression. For each of the following statements, indicate whether it is true or false. Give a reason for your answer.

a. The correlation between y and x_1 is positive.

b. A one-unit increase in x_1 corresponds to a change of 0.45 in the predicted value of y, controlling for x_2.

c. Controlling for x_1, a 100-unit increase in x_2 corresponds to a predicted increase of 0.30 in college GPA.

13.80 Scores for religion You want to include religious affiliation as a predictor in a regression model, using the categories Protestant, Catholic, Jewish, Other. You set up a variable x_1 that equals 1 for Protestants, 2 for Catholics, 3 for Jewish, and 4 for Other, using the model $\mu_y = \alpha + \beta x_1$. Explain why this is inappropriate.

13.81 Lurking variable Give an example of three variables for which you expect $\beta \neq 0$ in the model $\mu_y = \alpha + \beta x_1$ but $\beta_1 = 0$ in the model $\mu_y = \alpha + \beta_1 x_1 + \beta_2 x_2$. (*Hint*: The bivariate effect of x_1 could be completely due to a lurking variable, x_2.)

13.82 Properties of R^2 Using its definition in terms of SS values, explain why $R^2 = 1$ only when all the residuals are 0, and $R^2 = 0$ when each $\hat{y} = \bar{y}$. Explain what this means in practical terms.

13.83 Why an F test? When a model has a very large number of predictors, even when none of them truly have an effect in the population, one or two may look significant in t tests merely by random variation. Explain why performing the F test first can safeguard against getting such false information from t tests.

13.84 Multicollinearity For the high school female athletes data file, regress the maximum bench press on weight and percent body fat.

a. Show that the F test is statistically significant at the 0.05 significance level.

b. Show that the P-values are both larger than 0.35 for testing the individual effects with t tests. (It seems like a contradiction when the F test tells us that at least one predictor has an effect but the t tests indicate that neither predictor has a significant effect. This can happen when the predictor variables are highly correlated, so a predictor has little impact when the other predictors are in the model. Such a condition is referred to as **multicollinearity.** In this example, the correlation is 0.871 between weight and percent body fat.)

13.85 Logistic versus linear For binary response variables, one reason that logistic regression is usually preferred over straight-line regression is that a fixed change in x often has a smaller impact on a probability p when p is near 0 or near 1 than when p is near the middle of its range. Let y refer to the decision to rent or to buy a home, with p = the probability of buying, and let x = weekly family income. In which case do you think an increase of \$100 in x has greater effect: when x = 50,000 (for which p is near 1), when x = 0 (for which p is near 0), or when x = 500? Explain how your answer relates to the choice of a linear versus logistic regression model.

13.86 **Adjusted R^2** When we use R^2 for a random sample to
◆◆ estimate a population R^2, it's a bit biased. It tends to be a bit too large, especially when n is small. Some software also reports

$$\text{Adjusted } R^2 = R^2 - \{p/[n - (p + 1)]\}(1 - R^2),$$

where p = number of predictor variables in the model. This is slightly smaller than R^2 and is less biased. Suppose $R^2 = 0.500$ for a model with $p = 2$ predictors. Calculate adjusted R^2 for the following sample sizes: 10, 100, 1000. Show that the difference between adjusted R^2 and R^2 diminishes as n increases.

13.87 ***R* can't go down** The least squares prediction equation
◆◆ provides predicted values $\hat{y}$ with the strongest possible correlation with y, out of all possible prediction equations of that form. Based on this property, explain why the multiple correlation R cannot decrease when you add a variable to a multiple regression model. (*Hint*: The prediction equation for the simpler model is a special case of a prediction equation for the full model that has coefficient 0 for the added variable.)

13.88 **Indicator for comparing two groups** Chapter 10 pre-
◆◆ sented methods for comparing means for two groups. Explain how it's possible to perform a significance test of equality of two population means as a special case of a regression analysis. (*Hint*: The regression model then has a single explanatory variable—an indicator variable for the two groups being compared. What does $\mu_1 = \mu_2$ correspond to in terms of a value of a parameter in this model?)

13.89 **Simpson's paradox** Let y = death rate and
◆◆ x = average age of residents, measured for each county in Louisiana and in Florida. Draw a hypothetical scatterplot, identifying points for each state, such that the mean death rate is higher in Florida than in Louisiana when x is ignored, but lower when it is controlled. (*Hint*: When you fit a line for each state, the line should be higher for Louisiana, but the y-values for Florida should have an overall higher mean.)

13.90 **Parabolic regression** A regression formula that gives a
◆◆ parabolic shape instead of a straight line for the relationship between two variables is

$$\mu_y = \alpha + \beta_1 x + \beta_2 x^2.$$

a. Explain why this is a multiple regression model, with x playing the role of x_1 and x^2 (the square of x) playing the role of x_2.

b. For x between 0 and 5, sketch the prediction equation (i) $\hat{y} = 10 + 2x + 0.5x^2$ and (ii) $\hat{y} = 10 + 2x - 0.5x^2$. This shows how the parabola is bowl-shaped or mound-shaped, depending on whether the coefficient x^2 is positive or negative.

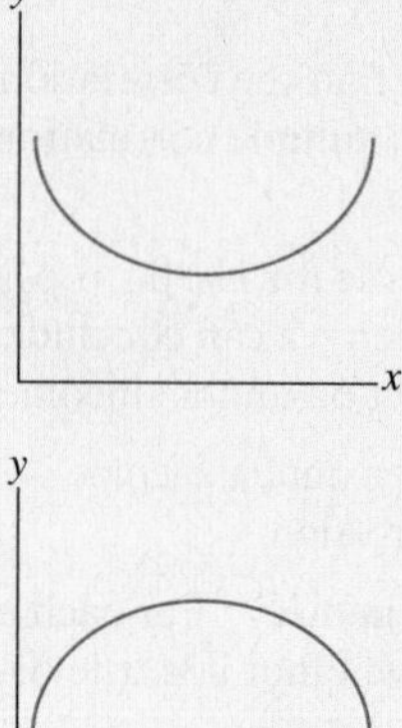

13.91 **Logistic slope** At the x value where the probability of
◆◆ success is some value p, the line drawn tangent to the logistic regression curve has slope $\beta p(1 - p)$.

a. Explain why the slope is $\beta/4$ when $p = 0.5$.

b. Show that the slope is weaker at other p values by evaluating this at $p = 0.1, 0.3, 0.7$, and 0.9. What does the slope approach as p gets closer and closer to 0 or 1? Sketch a curve to illustrate.

13.92 **When is $p = 0.50$?** When $\alpha + \beta x = 0$, so that
◆◆ $x = -\alpha/\beta$, show that the logistic regression equation $p = e^{\alpha+\beta x}/(1 + e^{\alpha+\beta x})$ gives $p = 0.50$.

Student Activities

13.93 **Class data** Refer to the data file your class created in Activity 3 in Chapter 1. For variables chosen by your instructor, fit a multiple regression model and conduct descriptive and inferential statistical analyses. Interpret and summarize your findings, and prepare to discuss these in class.

Comparing Groups: Analysis of Variance Methods

14

Example 1

Investigating Customer Satisfaction

Picture the Scenario

In recent years, many companies have increased the attention paid to measuring and analyzing customer satisfaction. Here are examples of two recent studies of customer satisfaction:

- A company that makes personal computers has provided a toll-free telephone number for owners of their PCs to call and seek technical support. For years the company had two service centers for these calls: San Jose, California, and Toronto, Canada. Recently the company outsourced many of the customer service jobs to a new center in Bangalore, India, because employee salaries are much lower there. The company wanted to compare customer satisfaction at the three centers.
- An airline has a toll-free telephone number that potential customers can call to make flight reservations. Usually the call volume is heavy and callers are placed on hold until an agent is free to answer. Researchers working for the airline recently conducted a randomized experiment to analyze whether callers would remain on hold longer if they heard (a) an advertisement about the airline and its current promotions, (b) recorded Muzak ("elevator music"), or (c) recorded classical music. Currently, messages are five minutes long and then repeated; the researchers also wanted to find out if it would make a difference if messages were instead 10 minutes long before repeating.

Questions to Explore

In the second study, the company's CEO had some familiarity with statistical methods, based on a course he took in college. He asked the researchers:

- In this experiment, are the sample mean times that callers stayed on hold before hanging up significantly different for the three recorded message types?
- What conclusions can you make if you take into account both the type of recorded message and whether it was repeated every five minutes or every ten minutes?

Thinking Ahead

Chapter 10 showed how to compare two means. In practice, there may be *several* means to compare, such as in the second example. This chapter shows how to use statistical inference to compare several means. We'll see how to determine whether a set of sample means is significantly different and how to estimate the differences among corresponding population means. To illustrate, we'll analyze data from the second study in Examples 2–4 and 7.

The methods introduced in this chapter apply when a quantitative response variable has a categorical explanatory variable. The categories of the explanatory variable identify the groups to be compared in terms of their means on the response variable. For example, the first study in Example 1 compared mean customer satisfaction for three groups—customers who call the service centers at the three locations. The response variable is customer satisfaction (on a scale of 0 to 10), and the explanatory variable is the service center location.

The inferential method for comparing means of several groups is called **analysis of variance**, denoted **ANOVA**. Section 14.1 shows that the name "analysis of variance" refers to the significance test's focus on two types of variability in the data. Section 14.2 shows how to construct confidence intervals comparing the means. It also shows that ANOVA methods are special cases of a multiple regression analysis.

Categorical explanatory variables in multiple regression and in ANOVA are often referred to as **factors**. ANOVA with a single factor, such as service center location, is called **one-way ANOVA**. Section 14.3 introduces ANOVA for two factors, called **two-way ANOVA**. The second study in Example 1 requires the use of two-way ANOVA to analyze how the mean telephone holding time varies across categories of recorded message type and categories defined by the length of time before repeating the message.

14.1 One-Way ANOVA: Comparing Several Means

The analysis of variance method compares means of *several* groups. Let g denote the number of groups. Each group has a corresponding population of subjects. The means of the response variable for the g populations are denoted by $\mu_1, \mu_2, \ldots, \mu_g$.

Hypotheses and Assumptions for the ANOVA Test Comparing Means

The analysis of variance is a significance test of the null hypothesis of equal population means,

$$H_0: \mu_1 = \mu_2 = \cdots = \mu_g.$$

An example is $H_0: \mu_1 = \mu_2 = \mu_3$ for testing population mean satisfaction at $g = 3$ service center locations. The alternative hypothesis is

$$H_a: \text{at least two of the population means are unequal.}$$

If H_0 is false, perhaps all the population means differ, but perhaps merely one mean differs from the others. The test analyzes whether the differences observed among the *sample* means could have reasonably occurred by chance, if the null hypothesis of equal *population* means were true.

The assumptions for the ANOVA test comparing population means are as follows:

- The population distributions of the response variable for the g groups are normal, with the same standard deviation for each group.
- Randomization (depends on data collection method): In a survey sample, independent random samples are selected from each of the g populations. For an experiment, subjects are randomly assigned separately to the g groups.

Under the first assumption, when the population means are equal, the population distribution of the response variable is the same for each group. The population distribution does not depend on the group to which a subject belongs. The ANOVA test is a test of independence between the quantitative response variable and the group factor.

Example 2

ANOVA hypotheses

Tolerance of Being on Hold?

Picture the Scenario

Let's refer back to the second scenario in Example 1. An airline has a toll-free telephone number for reservations. Often the call volume is heavy, and callers are placed on hold until a reservation agent is free to answer. The airline hopes a caller remains on hold until the call is answered, so as not to lose a potential customer.

The airline recently conducted a randomized experiment to analyze whether callers would remain on hold longer, on average, if they heard (a) an advertisement about the airline and its current promotions, (b) Muzak, or (c) classical music (Vivaldi's *Four Seasons*). The company randomly selected one out of every 1000 calls in a particular week. For each call, they randomly selected one of the three recorded messages to play and then measured the number of minutes that the caller remained on hold before hanging up (these calls were purposely not answered). The total sample size was 15. The company kept the study small, hoping it could make conclusions without alienating too many potential customers! Table 14.1 shows the data. It also shows the mean and standard deviation for each recorded message type.

Table 14.1 Telephone Holding Times by Type of Recorded Message

Each observation is the number of minutes a caller remained on hold before hanging up, rounded to the nearest minute.

Recorded Message	Holding Time Observations	Sample Size	Mean	Standard Deviation
Advertisement	5, 1, 11, 2, 8	5	5.4	4.2
Muzak	0, 1, 4, 6, 3	5	2.8	2.4
Classical	13, 9, 8, 15, 7	5	10.4	3.4

Questions to Explore

a. What are the hypotheses for the ANOVA test?

b. Figure 14.1 displays the sample means. Since these means are quite different, is there sufficient evidence to conclude that the population means differ?

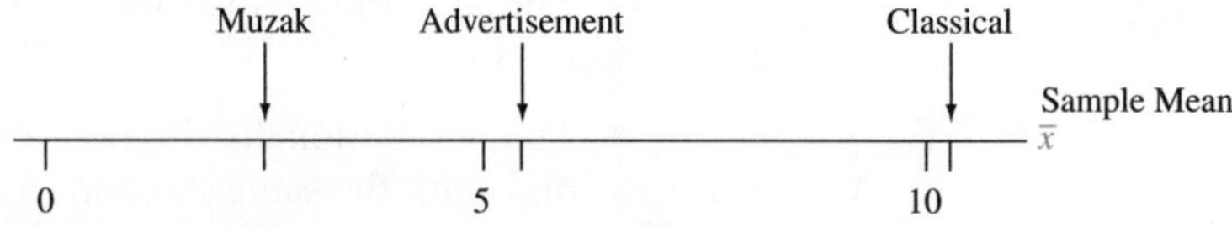

▲ **Figure 14.1 Sample Means of Telephone Holding Times for Callers Who Hear One of Three Recorded Messages.** Question Since the sample means are quite different, can we conclude that the population means differ?

Think It Through

a. Denote the holding time means for the population that these three random samples represent by μ_1 for the advertisement, μ_2 for Muzak, and μ_3 for classical music. ANOVA tests whether these are equal. The null hypothesis is H_0: $\mu_1 = \mu_2 = \mu_3$. The alternative hypothesis is that at least two of the population means are different.

b. The sample means are quite different. But even if the population means are equal, we expect the sample means to differ because of sampling variability. So these differences alone are not sufficient evidence to enable us to reject H_0.

Insight

The strength of evidence against H_0 will also depend on the sample sizes and the variability of the data. We'll next study how to test these hypotheses.

Try Exercise 14.1, parts a and b

Variability Between Groups and Within Groups Is the Key to Significance

The ANOVA method is used to compare population means. So, why is it called analysis of *variance*? The reason is that the test statistic uses evidence about two types of variability. Rather than presenting a formula now for this test statistic, which is rather complex, we'll discuss the reasoning behind it, which is quite simple.

Table 14.1 on the previous page listed data for three groups. Figure 14.2a shows the same data with dot plots. Suppose the sample data were different, as shown in Figure 14.2b. The data in Figure 14.2b have the same means as the data in Figure 14.2a but have smaller standard deviations within each group. Which case do you think gives stronger evidence against H_0: $\mu_1 = \mu_2 = \mu_3$?

What's the difference between the data in these two cases? The variability *between* pairs of sample means is the same in each case because the sample means are the same. However, the variability *within* each sample is much smaller in Figure 14.2b than in Figure 14.2a. The sample standard deviation is about 1.0 for each sample in Figure 14.2b whereas it is between 2.4 and 4.2 for the samples in Figure 14.2a. We'll see that the evidence against H_0: $\mu_1 = \mu_2 = \mu_3$ is stronger when the variability *within* each sample is smaller. Figure 14.2b has less variability within each sample than Figure 14.2a. Therefore, it gives stronger evidence against H_0. The evidence against H_0 is also stronger when the variability *between* sample

(a)

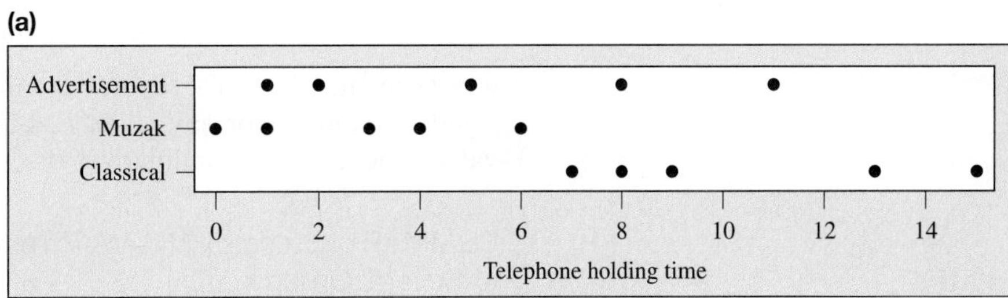

(b)

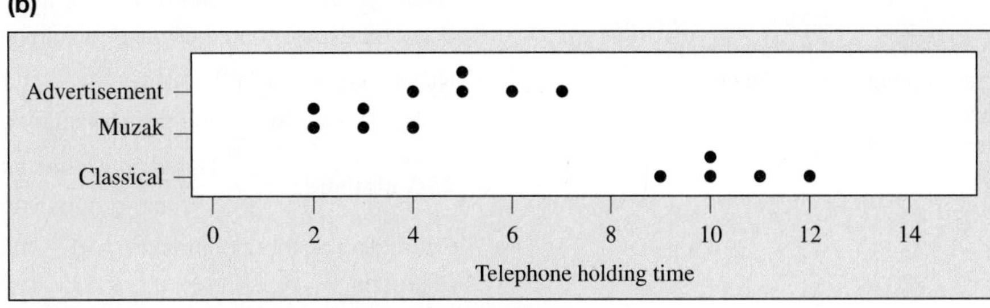

▲ **Figure 14.2 Data from Table 14.1 in Figure 14.2a and Hypothetical Data in Figure 14.2b That Have the Same Means but Less Variability Within Groups. Question** Do the data in Figure 14.2b give stronger or weaker evidence against H_0: $\mu_1 = \mu_2 = \mu_3$ than the data in Figure 14.2a. Why?

means increases (that is, when the sample means are farther apart) and as the sample sizes increase.

ANOVA F Test Statistic

The **analysis of variance (ANOVA) F test statistic** summarizes

$$F = \frac{\text{Between-groups variability}}{\text{Within-groups variability}}.$$

The larger the variability *between* groups relative to the variability *within* groups, the larger the F test statistic tends to be. For instance, $F = 6.4$ for the data in Figure 14.2a, whereas $F = 67.8$ for the data in Figure 14.2b. The between-groups variability is the same in each figure, but Figure 14.2b has much less within-groups variability and thus a larger test statistic value. Later in the section, we'll see the two types of variability described in the F test statistic are measured by estimates of *variances.*

The Test Statistic for Comparing Means Has the *F* Distribution

In Words

The P-value for an ANOVA F test statistic is the right-tail probability from the F distribution.

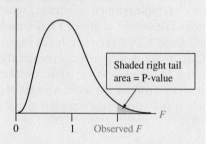

Shaded right tail area = P-value

0 1 Observed F F

When H_0 is true, the F test statistic has the F sampling distribution. The formula for the F test statistic is such that when H_0 is true, the F distribution has a mean of approximately 1. When H_0 is false, the F test statistic tends to be larger than 1, more so as the sample sizes increase. The larger the F test statistic value, the stronger the evidence against H_0.

Recall that we used the F distribution in the F test that the slope parameters of a multiple regression model are all zero (Section 13.3). As in that test, the P-value here is the probability (presuming that H_0 is true) that the F test statistic is larger than the observed F value. That is, it is the right-hand tail probability, as shown in the margin figure, representing results even more extreme than observed. The larger the F test statistic, the smaller the P-value.

In Section 13.3, we learned that the F distribution has two df values. For ANOVA with g groups and total sample size for all groups combined of $N = n_1 + n_2 + \cdots + n_g$,

$$df_1 = g - 1 \text{ and } df_2 = N - g.$$

Table D in the appendix reports F values having P-value of 0.05, for various df_1 and df_2 values. For any given F value and df values, software provides the P-value. The following summarizes the steps of an ANOVA F test.

Recall

The ***F* distribution** was introduced in Section 13.3. It's used for tests about several parameters (rather than a single parameter or a difference between two parameters, for which we can use a t test). ◀

SUMMARY: Steps of ANOVA F Test for Comparing Population Means of Several Groups

1. **Assumptions:** Independent random samples (either from random sampling or a randomized experiment), normal population distributions with equal standard deviations
2. **Hypotheses:** H_0: $\mu_1 = \mu_2 = \cdots = \mu_g$ (Equal population means for g groups), H_a: at least two of the population means are unequal.
3. **Test statistic:** $F = \frac{\text{Between-groups variability}}{\text{Within-groups variability}}.$

   F sampling distribution has $df_1 = g - 1$, $df_2 = N - g$ = total sample size − number of groups
4. **P-value:** Right-tail probability of above observed F value
5. **Conclusion:** Interpret in context. If decision needed, reject H_0 if P-value $\leq$ significance level (such as 0.05).

Example 3

One-way ANOVA

Telephone Holding Times

Picture the Scenario

Examples 1 and 2 discussed a study of the length of time that 15 callers to an airline's toll-free telephone number remain on hold before hanging up. The study compared three recorded messages: an advertisement about the airline, Muzak, and classical music. Let μ_1, μ_2, and μ_3 denote the population mean telephone holding times for the three messages.

F values with

df_2	df_1 1	2	3
6	5.99	5.14	4.76
12	4.75	3.88	3.49
18	4.41	3.55	3.16
24	4.26	3.40	3.01
30	4.17	3.32	2.92
120	3.92	3.07	2.68

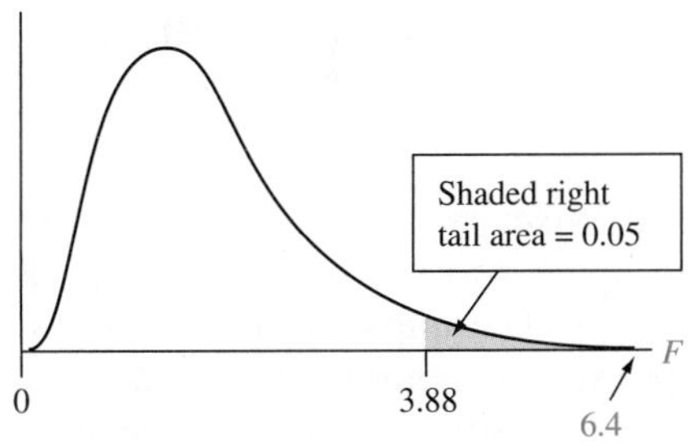

Questions to Explore

a. For testing H_0: $\mu_1 = \mu_2 = \mu_3$ based on this experiment, what value of the F test statistic would have a P-value of 0.05?

b. For the data in Table 14.1 on page 682, we'll see that software reports $F = 6.4$. Based on the answer to part a, will the P-value be larger, or smaller, than 0.05?

c. Can you reject H_0, using a significance level of 0.05? What can you conclude from this?

Think It Through

a. With $g = 3$ groups and a total sample size of $N = 15$ (5 in each group), the test statistic has

$$df_1 = g - 1 = 2 \text{ and } df_2 = N - g = 15 - 3 = 12.$$

From Table D (see the excerpt in the margin), with these df values an F test statistic value of 3.88 has a P-value of 0.05.

b. Since the F test statistic of 6.4 is farther out in the tail than 3.88 (see figure in margin), the right-tail probability above 6.4 is less than 0.05. So, the P-value is less than 0.05.

c. Since P-value < 0.05, there is sufficient evidence to reject H_0: $\mu_1 = \mu_2 = \mu_3$. We conclude that a difference exists among the three types of messages in the population mean amount of time that customers are willing to remain on hold.

Insight

We'll see that software reports a P-value $= 0.013$. This is quite strong evidence against H_0. If H_0 were true, there'd be only about a 1% chance of getting an F test statistic value larger than the observed F value of 6.4.

Try Exercise 14.2, parts a–c

The Variance Estimates and the ANOVA Table

Now let's take a closer look at the F test statistic. Denote the group sample means by $\bar{y}_1, \bar{y}_2, \ldots, \bar{y}_g$. (We use y rather than x because the quantitative variable is the response variable when used in a corresponding regression analysis.) We'll see that the F test statistic depends on these sample means, the sample standard deviations—$s_1, s_2, \ldots, s_g$—for the g groups, and the sample sizes.

One assumption for the ANOVA F test is that each population has the same standard deviation. Let σ denote the standard deviation for each of the g population distributions. The F test statistic for H_0: $\mu_1 = \mu_2 = \cdots = \mu_g$ is the ratio of two estimates of σ^2, the population *variance* for each group. Since we won't usually do computations by hand, we'll show formulas merely to give a better sense of what the F test statistic represents. The formulas are simplest when the group sample sizes are equal (as in Example 3), the case we'll show.

The estimate of σ^2 in the *denominator* of the F test statistic uses the variability *within* each group. The sample standard deviations $s_1, s_2, \ldots, s_g$ summarize the variation of the observations within the groups around their means.

- With equal sample sizes, the **within-groups estimate** of the variance σ^2 is the mean of the g sample variances for the g groups,

$$\text{Within-groups variance estimate } s^2 = \frac{s_1^2 + s_2^2 + \cdots + s_g^2}{g}.$$

When the sample sizes are not equal, the within-groups estimate is a *weighted average* of the sample variances, with greater weight given to samples with larger sample sizes. In either case, this estimate is unbiased: Its sampling distribution has σ^2 as its mean, regardless of whether H_0 is true.

The estimate of σ^2 in the *numerator* of the F test statistic uses the variability *between* each sample mean and the overall sample mean $\bar{y}$ for all the data.

- With equal sample sizes, n in each group, the **between-groups estimate** of the variance σ^2 is

$$\begin{array}{c}\text{Between-groups}\\ \text{variance estimate}\end{array} = \frac{n[(\bar{y}_1 - \bar{y})^2 + (\bar{y}_2 - \bar{y})^2 + \cdots + (\bar{y}_g - \bar{y})^2]}{g - 1}.$$

Did You Know?

Without n in the formula, the between-groups estimate is the sample variance of the g sample means. That sample variance of $\{\bar{y}_1, \bar{y}_2, \ldots, \bar{y}_g\}$ estimates the variance of the sampling distribution of each sample mean, which is σ^2/n. Multiplying by n then gives an estimate of σ^2 itself. See Exercise 14.66. ◄

If H_0 is true, this estimate is also unbiased. We expect this estimate to take a similar value as the within-groups estimate, apart from sampling error. If H_0 is false, however, the population means differ and the sample means tend to differ more greatly. Then, the between-groups estimate tends to overestimate σ^2.

The F test statistic is the ratio of these two estimates of the population variance,

$$F = \frac{\text{Between-groups estimate of } \sigma^2}{\text{Within-groups estimate of } \sigma^2}.$$

Computer software displays the two estimates in an ANOVA table similar to tables displayed in regression. Table 14.2 shows the basic format, illustrating the test in Example 3:

- The MS column contains the two estimates, which are called *mean squares*.
- The ratio of the two mean squares is the F test statistic, $F = 74.6/11.6 = 6.4$.

This F test statistic has P-value $= 0.013$, also shown in Table 14.2.

Recall

In Sections 12.4 and 13.3, we saw that a **mean square** (MS) is a ratio of a sum of squares (SS) to its *df* value. ◄

Table 14.2 ANOVA Table for *F* Test Using Data From Table 14.1

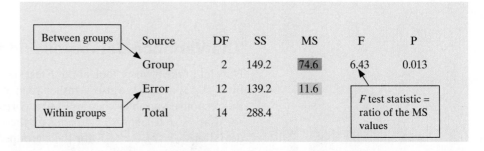

Source	DF	SS	MS	F	P
Group	2	149.2	74.6	6.43	0.013
Error	12	139.2	11.6		
Total	14	288.4			

Between groups (→ Group)

Within groups (→ Error)

F test statistic = ratio of the MS values

The mean square in the group row of the ANOVA table is based on variability *between* the groups. It is the *between-groups* estimate of the population variance σ^2. This is 74.6 in Table 14.2, listed under MS. The mean square error is the *within-groups* estimate s^2 of σ^2. This is 11.6 in Table 14.2. The "Error" label for this MS refers to the fact that it summarizes the error from not being able to predict subjects' responses exactly if we know only the group to which they belong.

Each mean square equals a sum of squares (in the SS column) divided by a degrees of freedom value (in the DF column). The *df* values for Group and Error in Table 14.2 are the *df* values for the *F* distribution, $df_1 = g - 1 = 3 - 1 = 2$ and $df_2 = N - g = 15 - 3 = 12$.

The sum of the between-groups sum of squares and the within-groups (Error) sum of squares is the **total sum of squares**. This is the sum of squares of the combined sample of *N* observations around the overall sample mean. The analysis of variance partitions the total sum of squares into two independent parts, the between-groups SS and the within-groups SS. It can be shown that the total sum of squares equals

$$\text{Total SS} = \Sigma(y - \bar{y})^2 = \text{between-groups SS} + \text{within-groups SS}.$$

In Table 14.2, for example,

$$\text{Total SS} = 288.4 = 149.2 + 139.2.$$

The total SS divided by $N - 1$ (the *df* for the total SS) is the sample variance when the data for the *g* groups are combined and treated as a single sample. The margin shows the TI-83+/84 output for this one-way ANOVA.

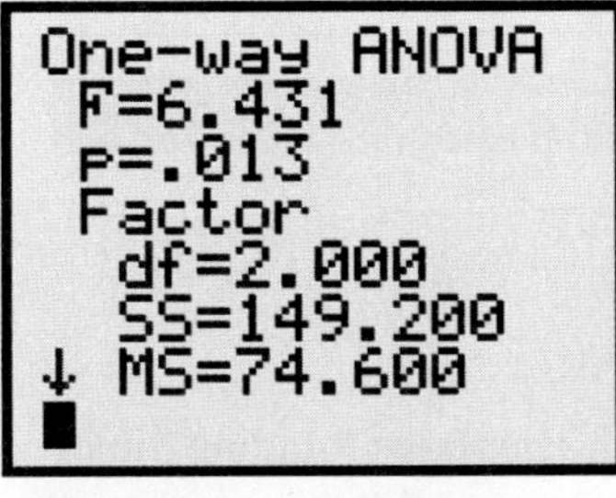

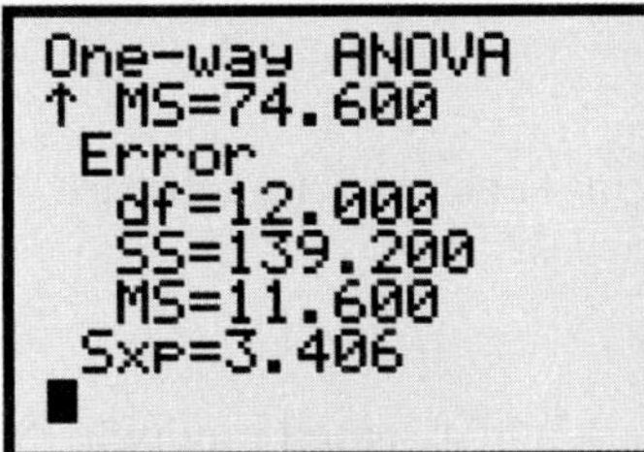

TI-83+/84 output

Assumptions and the Effects of Violating Them

The assumptions for the ANOVA *F* test comparing population means are:

(1) The population distributions of the response variable for the *g* groups are normal,

(2) Those distributions have the same standard deviation σ, and

(3) The data resulted from randomization.

Figure 14.3 portrays the population distribution assumptions.

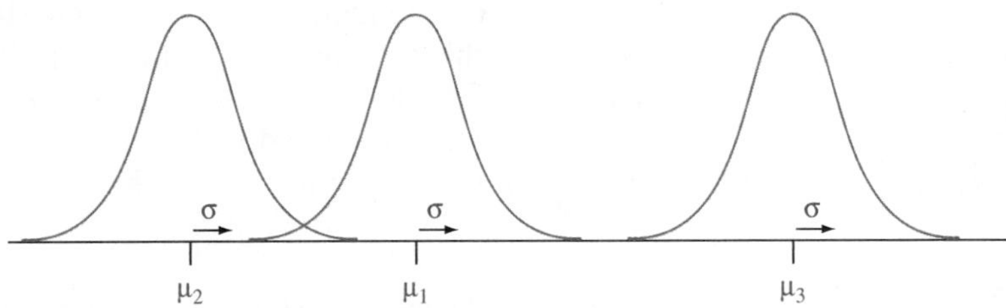

▲ **Figure 14.3 The Assumptions About the Population Distributions: Normal With Equal Standard Deviations.** Question In practice, what types of evidence would suggest that these assumptions are badly violated?

The assumptions that the population distributions are normal with identical standard deviations seem stringent. They are never satisfied exactly in practice. Moderate violations of the normality assumption are not serious. The *F* sampling distribution still provides a reasonably good approximation to the actual sampling distribution of the *F* test statistic. This becomes even more the case as the sample sizes increase, because the sampling distribution then has weaker dependence on the form of the population distributions.

Moderate violations of the equal population standard deviation assumption are also not serious. When the sample sizes are identical for the groups, the *F* test still works well even with severe violations of this assumption. When the sample sizes are not equal, the F test works quite well as long as the largest group standard deviation is no more than about twice the smallest group standard deviation.

You can construct box plots or dot plots for the sample data distributions to check for extreme violations of normality. Misleading results may occur with the F test if the distributions are highly skewed and the sample size N is small, or if there are relatively large differences among the standard deviations (the largest sample standard deviation being more than double the smallest one) and the sample sizes are unequal. When the distributions are highly skewed, the mean may not even be a relevant summary measure.[1]

In Practice Robustness of ANOVA F test

Since the ANOVA F test is robust to moderate breakdowns in the population normality and equal standard deviation assumptions, in practice it is used unless (1) graphical methods show extreme skew for the response variable or (2) the largest group standard deviation is more than about double the smallest group standard deviation and the sample sizes are unequal.

ANOVA assumptions

Example 4

Telephone Holding Time Study

Picture the Scenario

Let's check the assumptions for the F test on telephone holding times (Example 3).

Question to Explore

Is it appropriate to apply ANOVA to the data in Table 14.1 to compare mean telephone holding times for three message types?

Think It Through

Subjects were selected randomly for the experiment and assigned randomly to the three recorded messages. From Table 14.1 (summarized in the margin), the largest sample standard deviation of 4.2 is less than twice the smallest standard deviation of 2.4 (In any case, the sample sizes are equal, so this assumption is not crucial). The sample sizes in Table 14.1 are small, so it is difficult to make judgments about shapes of population distributions. However, the dot plots in Figure 14.2a on page 683 did not show evidence of severe nonnormality. Thus, ANOVA is suitable for these data.

Insight

As in other statistical inferences, the method used to gather the data is the most crucial factor. Inferences have greater validity when the data are from an experimental study that randomly assigned subjects to groups or from a sample survey that used random sampling.

Try Exercise 14.11, part c

Recall

Table 14.1 showed the following:

Recorded Message	Sample Size	Mean	Std. Dev.
Advert.	5	5.4	4.2
Muzak	5	2.8	2.4
Classical	5	10.4	3.4

The ANOVA methods in this chapter are designed for *independent* samples. Recall that for independent samples, the subjects in one sample are distinct from those in other samples. Separate methods, beyond the scope of this text, handle *dependent* samples.

Recall

See the beginning of Chapter 10 to review the distinction between **independent samples** and **dependent samples.**

[1]Chapter 15 discusses methods that apply when these assumptions are badly violated.

Recall

To compare two means while assuming equal population standard deviations, Section 10.3 showed the test statistic is

$$t = \frac{\bar{y}_1 - \bar{y}_2}{s\sqrt{\frac{1}{n_1} + \frac{1}{n_2}}},$$

where the standard deviation s pools information from within both samples. It has $df = n_1 + n_2 - 2 = N - g$ for $N = n_1 + n_2$ and $g = 2$. ◀

Using One *F* Test or Several *t* Tests to Compare the Means

With two groups, Section 10.3 showed how a t test can compare the means under the assumption of equal population standard deviations. That test also uses between-groups and within-groups variation. The t test statistic has the difference *between* the two group means in the numerator and a denominator based on pooling variability *within* the two groups. See the formula in the margin. In fact, if we apply the ANOVA F test to data from $g = 2$ groups, it can be shown that the F test statistic equals the square of this t test statistic. The P-value for the F test is exactly the same as the two-sided P-value for the t test. We can use either test to conduct the analysis.

When there are several groups, instead of using the F test why not use a t test to compare each pair of means? One reason for doing the F test is that using a *single* test rather than multiple tests enables us to control the probability of a Type I error. With a significance level of 0.05 in the F test, for instance, the probability of incorrectly rejecting a true H_0 is fixed at 0.05. When we do a separate t test for each pair of means, by contrast, a Type I error probability applies for *each* comparison. In that case, we are not controlling the *overall* Type I error rate for all the comparisons.

But the F test has its own disadvantages. With a small P-value, we can conclude that the population means are not identical. However, the result of the F test does not tell us *which* groups are different or *how* different they are. In Example 3, we have not concluded whether one recorded message works significantly better than the other two at keeping potential customers on the phone. We can address these issues using confidence intervals, as the next section shows.

14.1 Practicing the Basics

14.1 Hotel satisfaction The CEO of a company that owns five resort hotels wants to evaluate and compare satisfaction with the five hotels. The company's research department randomly sampled 125 people who had stayed at any of the hotels during the past month and asked them to rate their expectations of the hotel before their stay and to rate the quality of the actual stay at the hotel. Both observations used a rating scale of 0–10, with 0 = very poor and 10 = excellent. The researchers compared the hotels on the gap between prior expectation and actual quality, using the difference score, y = performance gap = (prior expectation score − actual quality score).

TRY

a. Identify the response variable, the factor, and the categories that form the groups.

b. State the null and alternative hypotheses for conducting an ANOVA.

c. Explain why the df values for this ANOVA are $df_1 = 4$ and $df_2 = 120$.

d. How large an F test statistic is needed to get a P-value = 0.05 in this ANOVA?

14.2 Satisfaction with banking A bank conducts a survey in which it randomly samples 400 of its customers. The survey asks the customers which way they use the bank the most: (1) interacting with a teller at the bank, (2) using ATMs, or (3) using the bank's Internet banking service. It also asks their level of satisfaction with the service they most often use (on a scale of 0 to 10 with 0 = very poor and 10 = excellent). Does mean satisfaction differ according to how they most use the bank?

TRY

a. Identifying notation, state the null and alternative hypotheses for conducting an ANOVA with data from the survey.

b. Report the df values for this ANOVA. Above what F test statistic values give a P-value below 0.05?

c. For the data, $F = 0.46$ and the P-value equals 0.63. What can you conclude?

d. What were the assumptions on which the ANOVA was based? Which assumption is the most important?

14.3 What's the best way to learn French? The following table shows scores on the first quiz (maximum score 10 points) for eighth-grade students in an introductory level French course. The instructor grouped the students in the course as follows:

Group 1: Never studied foreign language before but have good English skills

Group 2: Never studied foreign language before and have poor English skills

Group 3: Studied at least one other foreign language

French scores on the quiz

Group 1	Group 2	Group 3
4	1	9
6	5	10
8		5

	Group 1	Group 2	Group 3
Mean	6.0	3.0	8.0
Std. Dev.	2.000	2.828	2.646

Source	DF	SS	MS	F	P
Group	2	30.00	15.00	2.50	0.177
Error	5	30.00	6.00		
Total	7	60.00			

a. Defining notation and using results obtained with software, also shown in the table, report the five steps of the ANOVA test.

b. The sample means are quite different, but the P-value is not small. Name one important reason for this. (*Hint*: For given sample means, how do the results of the test depend on the sample sizes?)

c. Was this an experimental study, or an observational study? Explain how a lurking variable could be responsible for Group 3 having a larger mean than the others. (Thus, even if the P-value were small, it is inappropriate to assume that having studied at least one foreign language causes one to perform better on this quiz.)

14.4 What affects the *F* value? Refer to the previous exercise.

a. Suppose that the first observation in the second group was actually 9, not 1. Then the standard deviations are the same as reported in the table, but the sample means are 6, 7, and 8 rather than 6, 3, and 8. Do you think the *F* test statistic would be larger, the same, or smaller? Explain your reasoning, without doing any calculations.

b. Suppose you had the same means as shown in the table but the sample standard deviations were 1.0, 1.8, and 1.6, instead of 2.0, 2.8, and 2.6. Do you think the *F* test statistic would be larger, the same, or smaller? Explain your reasoning.

c. Suppose you had the same means and standard deviations as shown in the table but the sample sizes were 30, 20, and 30, instead of 3, 2, and 3. Do you think the *F* test statistic would be larger, the same, or smaller? Explain your reasoning.

d. In parts a, b, and c, would the P-value be larger, the same, or smaller? Why?

14.5 Outsourcing Example 1 at the beginning of this chapter mentioned a study to compare customer satisfaction at service centers in San Jose, California; Toronto, Canada; and Bangalore, India. Each center randomly sampled 100 people who called during a two-week period. Callers rated their satisfaction on a scale of 0 to 10, with higher scores representing greater satisfaction. The sample means were 7.6 for San Jose, 7.8 for Toronto, and 7.1 for Bangalore. The table shows the results of conducting an ANOVA.

a. Define notation and specify the null hypothesis tested in this table.

b. Explain how to obtain the *F* test statistic value reported in the table from the MS values shown, and report the values of df_1 and df_2 for the *F* distribution.

c. Interpret the P-value reported for this test. What conclusion would you make using a 0.05 significance level?

Customer satisfaction with outsourcing

Source	DF	SS	MS	F	P
Group	2	26.00	13.00	27.6	0.000
Error	297	140.00	0.47		
Total	299	60.00			

14.6 Good friends and astrological sign A recent General Social Survey asked subjects how many good friends they have. Is number of friends associated with the respondent's astrological sign (the 12 symbols of the Zodiac)? The ANOVA table for the GSS data reports $F = 0.61$ based on $df_1 = 11$, $df_2 = 813$.

a. Introduce notation, and specify the null hypothesis and alternative hypothesis for the ANOVA.

b. Based on what you know about the *F* distribution, would you guess that the test statistic value of 0.61 provides strong evidence against the null hypothesis? Explain.

c. Software reports a P-value of 0.82. Explain how to interpret this.

14.7 How many kids to have? A recent General Social Survey asked subjects, "What is the ideal number of kids for a family?" Do responses tend to depend on the subjects' religious affiliation? Results of an ANOVA are shown in the printout, for religion categories (Protestant, Catholic, Jewish, or Other).

a. Define notation and specify the null hypothesis tested in this printout.

b. Summarize the assumptions made to conduct this test.

c. Report the *F* test statistic value and the P-value for this test. Interpret the P-value.

d. Based on part c, can you conclude that *every* pair of religious affiliations has different population means for ideal family size? Explain.

Ideal number of kids in a family by religious affiliation

Source	DF	SS	MS	F	P
Religion	3	11.72	3.91	5.48	0.001
Error	1295	922.82	0.71		
Total	1298	934.54			

14.8 Smoking and personality A study about smoking and personality (by A. Terracciano and P. Costa, *Addiction*, vol. 99, 2004, pp. 472–481) used a sample of 1638 adults in the Baltimore Longitudinal Study on Aging. The subjects formed three groups according to smoking status (never, former, current). Each subject completed a personality questionnaire that provided scores on various personality scales designed to have overall means of about 50 and standard deviations of about 10. The table shows some results for three traits, giving the means with standard deviations in parentheses.

	Never smokers ($n = 828$)	Former smokers ($n = 694$)	Current smokers ($n = 116$)	*F*
Neuroticism	46.7 (9.6)	48.5 (9.2)	51.9 (9.9)	17.77
Extraversion	50.4 (10.3)	50.2 (10.0)	50.9 (9.4)	0.24
Conscientiousness	51.8 (10.1)	48.9 (9.7)	45.6 (10.3)	29.42

a. For the *F* test for the extraversion scale, using the 0.05 significance level, what conclusion would you make?

b. Refer to part a. Does this mean that the population means are necessarily equal?

14.9 Florida student data Refer to the Florida Student Survey data file on the text CD. For the response variable, use the number of weekly hours engaged in sports and other physical exercise. For the explanatory variable, use gender.

a. Using software, for each gender construct box plots and find the mean and standard deviation of the response variable.

b. Conduct an ANOVA. Report the hypotheses, *F* test statistic value, P-value, and interpret results.

c. To compare the means, suppose you instead used the two-sided *t* test from Section 10.3, which assumes equal population standard deviations. How would the *t* test statistic and P-value for that test relate to the *F* test statistic and P-value for ANOVA?

14.10 Software and French ANOVA Refer to Exercise 14.3. Using software,

a. Create the data file and find the sample means and standard deviations.

b. Find and report the ANOVA table. Interpret the P-value.

c. Change an observation in Group 2 so that the P-value will be smaller. Specify the value you changed, and report the resulting *F* test statistic and the P-value. Explain why the value you changed would have this effect.

14.11 Comparing therapies for anorexia The Anorexia data file on the text CD shows weight change for 72 anorexic teenage girls who were randomly assigned to one of three psychological treatments. Use software to analyze these data. (The change scores are given in the file for the control and cognitive therapy groups. You can create the change scores for the family therapy group.)

TRY

a. Construct box plots for the three groups. Use these and sample summary means and standard deviations to describe the three samples.

b. For the one-way ANOVA comparing the three mean weight changes, report the test statistic and P-value. Explain how to interpret.

c. State and check the assumptions for the test in part b.

14.2 Estimating Differences in Groups for a Single Factor

When an analysis of variance *F* test has a small P-value, the test does not specify *which* means are different or *how* different they are. In practice, we can estimate differences between population means with confidence intervals.

Confidence Intervals Comparing Pairs of Means

Recall

s^2 is an unbiased estimator of σ^2, regardless of whether H_0 is true in the ANOVA *F* test. ◀

Let *s* denote the estimate of the residual standard deviation. This is the square root of the within-groups variance estimate s^2 of σ^2 that is the denominator of the *F* test statistic in ANOVA. It is also the square root of the mean square error that software reports (the MS in the row for "Error" of any ANOVA table).

SUMMARY: Confidence Interval Comparing Means

For two groups *i* and *j*, with sample means $\bar{y}_i$ and $\bar{y}_j$ having sample sizes n_i and n_j, the 95% confidence interval for $\mu_i - \mu_j$ is

$$\bar{y}_i - \bar{y}_j \pm t_{.025}\, s\sqrt{\frac{1}{n_i} + \frac{1}{n_j}}.$$

The *t*-score from the *t* table has $df = N - g$ = total sample size − # groups.

The *df* value of $N - g$ for the *t*-score is also df_2 for the *F* test. This is the *df* for the MS error. For $g = 2$ groups, $N = n_1 + n_2$ and $df = N - g = (n_1 + n_2 - 2)$. This confidence interval is then identical to the one introduced in Section 10.3 for $(\mu_1 - \mu_2)$ based on a pooled standard deviation. In the context of follow-up analyses after the ANOVA *F* test where we form this confidence interval to compare a pair of means, some software (such as MINITAB) refers to this method of comparing means as the **Fisher method**.

When the confidence interval does not contain 0, we can infer that the population means are different. The interval shows just how different they may be.

Example 5

Fisher method

Number of Good Friends and Happiness

Picture the Scenario

Chapter 11 investigated the association between happiness and several categorical variables, using data from the General Social Survey. The respondents indicated whether they were very happy, pretty happy, or not too happy. Is happiness associated with having lots of friends? A recent GSS asked, "About how many good friends do you have?" Here, we could treat either happiness (variable HAPPY in GSS) or number of good friends (NUMFREND in GSS) as the response variable. If we choose number of good friends, then we are in the ANOVA setting, having a quantitative response variable and a categorical explanatory variable (happiness).

For each happiness category, Table 14.3 shows the sample mean, standard deviation, and sample size for the number of good friends. It also shows the ANOVA table for the F test comparing the population means. The small P-value of 0.03 suggests that at least two of the three population means are different.

Table 14.3 Summary of ANOVA for Comparing Mean Number of Good Friends for Three Happiness Categories

The analysis is based on GSS data.

	Very happy	Pretty happy	Not too happy
Mean	10.4	7.4	8.3
Standard deviation	17.8	13.6	15.6
Sample size	276	468	87

Source	DF	SS	MS	F	P
Group	2	1626.8	813.4	3.47	0.032
Error	828	193900.9	234.2		
Total	830	195527.7			

Question to Explore

Use 95% confidence intervals to compare the population mean number of good friends for the three pairs of happiness categories—very happy with pretty happy, very happy with not too happy, and pretty happy with not too happy.

Think It Through

From Table 14.3, the MS error is 234.2. The residual standard deviation is $s = \sqrt{234.2} = 15.3$, with $df = 828$ (listed in the row for the MS error). For a 95% confidence interval with $df = 828$, software reports $t_{.025} = 1.963$. For comparing the very happy and pretty happy categories, the confidence interval for $\mu_1 - \mu_2$ is

$$(\bar{y}_1 - \bar{y}_2) \pm t_{.025}\, s\sqrt{\frac{1}{n_1} + \frac{1}{n_2}} = (10.4 - 7.4) \pm 1.963(15.3)\sqrt{\frac{1}{276} + \frac{1}{468}}$$

which is 3.0 ± 2.3, or (0.7, 5.3).

We infer that the population mean number of good friends is between about 1 and 5 higher for those who are very happy than for those who are pretty happy. Since the confidence interval contains only positive numbers, this suggests that $\mu_1 - \mu_2 > 0$; that is, μ_1 exceeds μ_2. On the average, people who are very happy have more good friends than people who are pretty happy.

For the other comparisons, you can find

Very happy, not too happy: 95% CI for $\mu_1 - \mu_3$ is
$(10.4 - 8.3) \pm 3.7$, or $(-1.6, 5.8)$.
Pretty happy, not too happy: 95% CI for $\mu_2 - \mu_3$ is
$(7.4 - 8.3) \pm 3.5$, or $(-4.4, 2.6)$.

These two confidence intervals contain 0. So there's not enough evidence to conclude that a difference exists between μ_1 and μ_3 or between μ_2 and μ_3.

Insight

The confidence intervals are quite wide, even though the sample sizes are fairly large. This is because the sample standard deviations (and hence s) are large. Table 14.3 reports that the sample standard deviations are larger than the sample means, suggesting that the three distributions are skewed to the right. The margin figure shows a box plot of the overall sample data distribution of number of good friends, except for many large outliers. It would also be sensible to compare the median responses, but these are not available at the GSS website. Do non-normal population distributions invalidate this inferential analysis? We'll discuss this next.

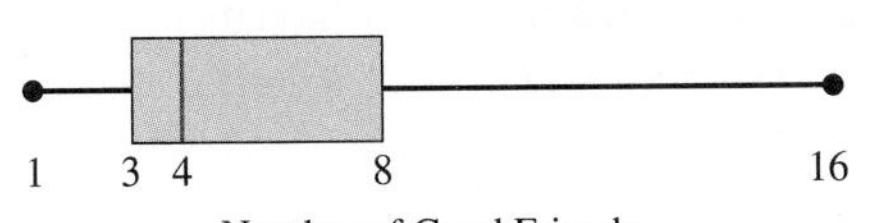

Try Exercise 14.12

The Effects of Violating Assumptions

Recall

From Section 10.2, a 95% confidence interval for $\mu_2 - \mu_1$ using separate standard deviations s_1 and s_2 is

$$\bar{y}_1 - \bar{y}_2 \pm t_{.025}\sqrt{\frac{s_1^2}{n_1} + \frac{s_2^2}{n_2}}.$$

Software supplies the *df* value and the confidence interval. ◀

The t confidence intervals have the same assumptions as the ANOVA F test: (1) normal population distributions, (2) identical standard deviations, and (3) data that resulted from randomization. These inferences also are not highly dependent on the normality assumption, especially when the sample sizes are large, such as in Example 5. When the standard deviations are quite different, with the ratio of the largest to smallest exceeding about 2, it is preferable to use the confidence interval formula from Section 10.2 (see the margin Recall) that uses *separate* standard deviations for the groups rather than a single pooled value s. That approach does not assume equal standard deviations.

In Example 5, the sample standard deviations are not very different (ranging from 13.6 to 17.8), and the GSS is a multistage random sample with properties similar to a simple random sample. Thus, Assumptions 2 and 3 are reasonably well satisfied. The sample sizes are fairly large, so Assumption 1 of normality is not crucial. It's justifiable to use ANOVA and follow-up confidence intervals with these data.

Controlling Overall Confidence With Many Confidence Intervals

With g groups, there are $g(g - 1)/2$ pairs of groups to compare. With $g = 3$, for instance, there are $g(g - 1)/2 = 3(2)/2 = 3$ comparisons: Group 1 with Group 2, Group 1 with Group 3, and Group 2 with Group 3.

The confidence interval method just discussed is mainly used when g is small or when only a few comparisons are of main interest. When there are many groups, the number of comparisons can be large. For example, when $g = 10$, there are $g(g - 1)/2 = 45$ pairs of means to compare. If we plan to construct 95% confidence intervals for these comparisons, an error probability of 0.05 applies to *each* comparison. On average, $45(0.05) = 2.25$ of the confidence intervals would *not* contain the true differences of means.

For 95% confidence intervals, the confidence level of 0.95 is the probability that *any particular* confidence interval that we plan to construct will contain the parameter. The probability that *all* the confidence intervals will contain the parameters is considerably smaller than the confidence level for any particular interval. How can we construct the intervals so that the 95% confidence extends to the *entire set* of intervals rather than to *each single* interval? Methods that control the probability that *all* confidence intervals will contain the true differences in means are called **multiple comparison methods**. For these methods, *all* intervals are designed to contain the true parameters *simultaneously* with an overall fixed probability.

Multiple Comparisons for Comparing All Pairs of Means

Multiple comparison methods compare pairs of means with a confidence level that applies simultaneously to the entire set of comparisons rather than to each separate comparison.

The simplest multiple comparison method uses the confidence interval formula from the beginning of this section. However, for each interval it uses a t-score for a more stringent confidence level. This ensures that the overall confidence level is acceptably high. The desired overall error probability is split into equal parts for each comparison. Suppose we want a confidence level of 0.95 that *all* confidence intervals will be simultaneously correct. If we plan five confidence intervals comparing means, then the method uses error probability $0.05/5 = 0.01$ for each one; that is, a 99% confidence level for each separate interval. This approach ensures that the overall confidence level is *at least* 0.95. (It is actually slightly larger.) Called the **Bonferroni** method, it is based on a special case of a probability theorem shown by Italian probabilist, Carlo Bonferroni, in 1936 (Exercise 14.67).

We shall instead use the **Tukey method**. It is designed to give an overall confidence level *very close* to the desired value (such as 0.95), and it has the advantage that its confidence intervals are slightly narrower than the Bonferroni intervals. The Tukey method is more complex, using a sampling distribution pertaining to the difference between the largest and smallest of the g sample means. We do not present its formula, but it is easy to obtain with software.

Recall

John Tukey was responsible for many statistical innovations, including box plots and other methods of exploratory data analysis (EDA). See On the Shoulders of ... John Tukey in Section 2.6 to read more about Tukey and EDA. ◄

Example 6

Tukey method

Number of Good Friends

Picture the Scenario

Example 5 compared the population mean numbers of good friends, for three levels of reported happiness. There, we constructed a *separate* 95% confidence interval for the difference between each pair of means. Table 14.4 displays these three intervals. It also displays the confidence intervals that software reports for the Tukey multiple comparison method.

Table 14.4 Multiple Comparisons of Mean Good Friends for Three Happiness Categories

An asterisk * indicates a significant difference, with the confidence interval not containing 0.

Groups	Difference of means	Separate 95% CIs	Tukey 95% Multiple Comparison CIs
(Very happy, Pretty happy)	$\mu_1 - \mu_2$	(0.7, 5.3)*	(0.3, 5.7)*
(Very happy, Not too happy)	$\mu_1 - \mu_3$	(−1.6,5.8)	(−2.6,6.5)
(Pretty happy, Not too happy)	$\mu_2 - \mu_3$	(−4.4,2.6)	(−5.1,3.3)

Question to Explore

a. Explain how the Tukey multiple comparison confidence intervals differ from the separate confidence intervals in Table 14.4.

b. Summarize results shown for the Tukey multiple comparisons.

Think It Through

a. The Tukey confidence intervals hold with an *overall* confidence level of about 95%. This confidence applies to the entire set of three intervals. The Tukey confidence intervals are wider than the separate 95% confidence intervals because the multiple comparison approach uses a higher confidence level for each separate interval to ensure achieving the overall confidence level of 95% for the entire set of intervals.

b. The Tukey confidence interval for $\mu_1 - \mu_2$ contains only positive values, so we infer that $\mu_1 > \mu_2$. The mean number of good friends is higher, although perhaps barely so, for those who are very happy than for those who are pretty happy. The other two Tukey intervals contain 0, so we cannot infer that those pairs of means differ.

Insight

Figure 14.4 summarizes the three Tukey comparisons from Table 14.4. The intervals have different lengths because the group sample sizes are different.

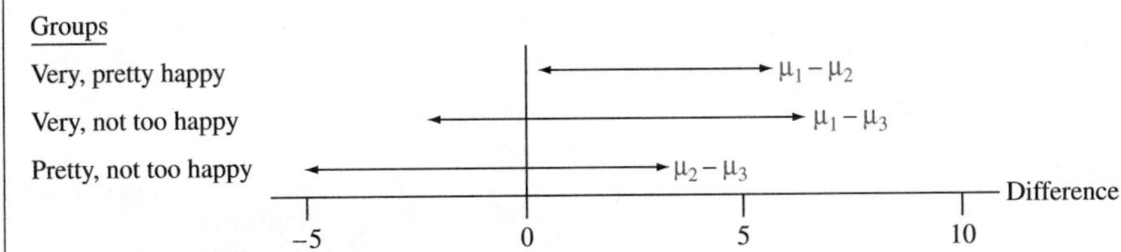

▲ **Figure 14.4** Summary of Tukey Comparisons of Pairs of Means.

Try Exercise 14.15

ANOVA and Regression

ANOVA can be presented as a special case of multiple regression. The factor enters the regression model using *indicator variables.* Each indicator variable takes only two values, 0 and 1, and indicates whether an observation falls in a particular group.

With three groups, we need two indicator variables to indicate the group membership. The first indicator variable is

$$x_1 = 1 \text{ for observations from the first group}$$
$$= 0 \text{ otherwise.}$$

Recall

You can review **indicator variables** in Section 13.5. We used them there to include a categorical explanatory variable in a regression model. ◀

The second indicator variable is

$$x_2 = 1 \text{ for observations from the second group}$$
$$= 0 \text{ otherwise.}$$

The indicator variables identify the group to which an observation belongs as follows:

$$\text{Group 1: } x_1 = 1 \text{ and } x_2 = 0$$
$$\text{Group 2: } x_1 = 0 \text{ and } x_2 = 1$$
$$\text{Group 3: } x_1 = 0 \text{ and } x_2 = 0.$$

We don't need a separate indicator variable for the third group. We know an observation is in that group if $x_1 = 0$ and $x_2 = 0$.

With these indicator variables, the multiple regression equation for the mean of y is

$$\mu_y = \alpha + \beta_1 x_1 + \beta_2 x_2.$$

For observations from the third group, $x_1 = x_2 = 0$, and the equation reduces to

$$\mu_y = \alpha + \beta_1(0) + \beta_2(0) = \alpha.$$

So the parameter α represents the population mean of the response variable y for the last group. For observations from the first group, $x_1 = 1$ and $x_2 = 0$, so

$$\mu_y = \alpha + \beta_1(1) + \beta_2(0) = \alpha + \beta_1$$

equals the population mean μ_1 for that group. Similarly, $\alpha + \beta_2$ equals the population mean μ_2 for the second group (let $x_1 = 0$ and $x_2 = 1$).

Since $\alpha + \beta_1 = \mu_1$ and $\alpha = \mu_3$, the difference between the means

$$\mu_1 - \mu_3 = (\alpha + \beta_1) - \alpha = \beta_1.$$

That is, the coefficient β_1 of the first indicator variable represents the difference between the first mean and the last mean. Likewise, $\beta_2 = \mu_2 - \mu_3$. The beta coefficients of the indicator variables represent differences between the mean of each group and the mean of the last group. Table 14.5 summarizes the parameters of the regression model and their correspondence with the three population means.

Table 14.5 Interpretation of Coefficients of Indicator Variables in Regression Model

The indicator variables represent a categorical predictor with three categories specifying three groups.

	Indicator			
Group	x_1	x_2	**Mean of y**	**Interpretation of β**
1	1	0	$\mu_1 = \alpha + \beta_1$	$\beta_1 = \mu_1 - \mu_3$
2	0	1	$\mu_2 = \alpha + \beta_2$	$\beta_2 = \mu_2 - \mu_3$
3	0	0	$\mu_3 = \alpha$	

Using Regression for the ANOVA Comparison of Means

Recall

Section 13.3 introduced the ***F* test** that all the beta parameters in a multiple regression model equal 0. ◀

For three groups, the null hypothesis for the ANOVA F test is H_0: $\mu_1 = \mu_2 = \mu_3$. If H_0 is true, then $\mu_1 - \mu_3 = 0$ and $\mu_2 - \mu_3 = 0$. In the multiple regression model

$$\mu_y = \alpha + \beta_1 x_1 + \beta_2 x_2$$

with indicator variables, recall that $\mu_1 - \mu_3 = \beta_1$ and $\mu_2 - \mu_3 = \beta_2$. Therefore, the ANOVA null hypothesis H_0: $\mu_1 = \mu_2 = \mu_3$ is equivalent to H_0: $\beta_1 = \beta_2 = 0$ in the regression model. If the beta parameters in the regression model all equal 0, then the mean of the response variable equals α for each group. We can perform the ANOVA test comparing means using the F test of H_0: $\beta_1 = \beta_2 = 0$ for this regression model.

Example 7

Regression analysis

Telephone Holding Times

Picture the Scenario

Let's return to the data we analyzed in Examples 1–4 on telephone holding times for callers to an airline for which the recorded message is an advertisement, Muzak, or classical music.

Questions to Explore

a. Set up indicator variables to use regression to model the mean holding times with the type of recorded message as explanatory variable.

b. Table 14.6 shows a portion of a MINITAB printout for fitting this model. Use it to find the estimated mean holding time for the advertisement recorded message.

c. Use Table 14.6 to conduct the ANOVA F test (which Example 3 had shown).

Table 14.6 Printout for Regression Model $\mu_y = \alpha + \beta_1 x_1 + \beta_2 x_2$ for Telephone Holding Times and Type of Recorded Message

The indicator variables are x_1 for the advertisement and x_2 for Muzak.

Predictor	Coef	SE Coef	T		P
Constant	10.400	1.523	6.83		0.000
x1	−5.000	2.154	−2.32		0.039
x2	−7.600	2.154	−3.53		0.004
Analysis of Variance					
Source	DF	SS	MS	F	P
Regression	2	149.20	74.60	6.43	0.013
Residual Error	12	139.20	11.60		
Total	14	288.40			

Think It Through

a. The factor (type of recorded message) has three categories—advertisement, Muzak, and classical music. We set up indicator variables x_1 and x_2 with

$$x_1 = 1 \text{ for the advertisement (and 0 otherwise)},$$
$$x_2 = 1 \text{ for Muzak (and 0 otherwise)},$$
$$\text{so } x_1 = x_2 = 0 \text{ for classical music.}$$

The regression model for the mean of y = telephone holding time is then

$$\mu_y = \alpha + \beta_1 x_1 + \beta_2 x_2.$$

b. From Table 14.6, the prediction equation is

$$\hat{y} = 10.4 - 5.0x_1 - 7.6x_2.$$

For the advertisement, $x_1 = 1$ and $x_2 = 0$, so the estimated mean is $\hat{y} = 10.4 - 5.0(1) - 7.6(0) = 5.4$. This is the sample mean for the five subjects in that group.

c. From Table 14.6, the *F* test statistic for testing

$$H_0: \beta_1 = \beta_2 = 0$$

is $F = 6.43$, with $df_1 = 2$ and $df_2 = 12$. This null hypothesis is equivalent to

$$H_0: \mu_1 = \mu_2 = \mu_3.$$

Table 14.6 reports a P-value of 0.013. The regression approach provides the same *F* test statistic and P-value as the ANOVA did in Table 14.2 on page 686.

Insight

Testing that the beta coefficients equal zero is a way of testing that the population means are equal. Likewise, confidence intervals for those coefficients give us confidence intervals for differences between means. For instance, since $\beta_1 = \mu_1 - \mu_3$, a confidence interval for β_1 is also a confidence interval comparing μ_1 and μ_3. From Table 14.6, the estimate -5.0 of β_1 has $se = 2.154$. Since $df = 12$ for that interval, $t_{.025} = 2.179$, and a 95% confidence interval for $\mu_1 - \mu_3$ is

$$-5.0 \pm 2.179(2.154), \text{ or } -5.0 \pm 4.7, \text{ which is } (-9.7, -0.3).$$

This agrees with the 95% confidence interval you would obtain using the difference between the sample means and its standard error.

Try Exercise 14.17

Table 14.2: ANOVA Table for Comparing Means

Source	DF	SS	MS	F	P
Group	2	149.2	74.6	6.4	0.013
Error	12	139.2	11.6		
Total	14	288.4			

Table 14.6: ANOVA Table for Regression Model

Source	DF	SS	MS	F	P
Regression	2	149.2	74.6	6.4	.013
Residual	12	139.2	11.6		
Total	14	288.4			

Notice the similarity between the ANOVA table for comparing means (Table 14.2) and the ANOVA table for regression (Table 14.6), both shown again in the margin. The "between-groups sum of squares" for ordinary ANOVA is the "regression sum of squares" for the regression model. This is the variability explained by the indicator variables for the groups. The "error sum of squares" for ordinary ANOVA is the residual error sum of squares for the regression model. This represents the variability within the groups. This sum of squares divided by its degrees of freedom is the mean square error = 11.6 (MS in the error row), which is also the within-groups estimate of the variance of the observations within the groups. The regression mean square is the between-groups estimate = 74.6. The ratio of the regression mean square to the mean square error is the *F* test statistic ($F = 6.4$).

So far, this chapter has shown how to compare groups for a single factor. This is **one-way ANOVA**. Sometimes the groups to compare are the cells of a cross-classification of two or more factors. For example, the four groups (employed men, employed women, unemployed men, unemployed women) result from cross classifying employment status and gender. The next section presents **two-way ANOVA**, the procedure for comparing the mean of a quantitative response variable across categories of each of two factors.

14.2 Practicing the Basics

14.12 House prices and age For the House Selling Prices OR data file on the text CD, the output shows the result of conducting an ANOVA comparing mean house selling prices (in $1000) by Age Category (New = 0 to 24 years old, Medium = 25 to 50 years old, Old = 51 to 74 years old, Very Old = 75 + years old). It also shows a summary table of means and standard deviations of the selling prices, by condition. Consider the New and Medium ages, where 150 of the 200 observations fall.

TRY

a. Using information given in the tables, show how to construct a 95% confidence interval comparing the corresponding population means.

b. Interpret the confidence interval.

Age	N	Mean	StDev
Medium	71	243.6	79.5
New	78	305.8	125.9
Old	37	215.7	85.3
Very Old	14	311.6	188.5

Source	DF	SS	MS	F	P
Age Condition	3	281272	93757	7.70	0.000
Error	196	2387598	12182		
Total	199	2668870			

14.13 Religious importance and educational level An extensive survey by the Pew Forum on Religion & Public Life, conducted in 2007, details views on religion in the United States. Based on interviews with a representative sample of more than 35,000 Americans age 18 and older, the U.S. Religious Landscape Survey found that religious affiliation in the United States is both diverse and dependent on a lot of factors. The table shows results of ANOVA for comparing feelings about the importance of religion in life and the subject's level of education (scale increasing with education: 3 = HS Graduate, 5 = Some College or Associate Degree, 6 = College Graduate, 7 = Post-Graduate Training). It appears that as educational level increases, the importance of religion in daily life decreases. Construct a 95% confidence interval to compare the population mean educational level for the Very Important and Not at All Important religious attitude groups. Interpret the interval.

Summary of ANOVA for mean education level of subjects and the level of importance of religion in daily life

	Very important	Somewhat important	Not too important	Not at all important
Mean	4.55	4.61	4.87	5.12
N	1959	776	309	341

Source	DF	SS	MS	F	P
Educ	8	69.17	8.65	6.65	0.000
Error	3403	4425.25	1.30		
Total	3411	4494.43			

14.14 Comparing telephone holding times Examples 2 and 3 analyzed whether telephone callers to an airline would stay on hold different lengths of time, on average, if they heard (a) an advertisement about the airline, (b) Muzak, or (c) classical music. The sample means were 5.4, 2.8, and 10.4, with $n_1 = n_2 = n_3 = 5$. The ANOVA test had $F = 74.6/11.6 = 6.4$ and a P-value of 0.013.

a. A 95% confidence interval comparing the population mean times that callers are willing to remain on hold for classical music and Muzak is (2.9, 12.3). Interpret this interval.

b. The margin of error was 4.7 for this comparison. Without doing a calculation, explain why the margin of error is 4.7 for comparing *each* pair of means.

c. The 95% confidence intervals are (0.3, 9.7) for $\mu_3 - \mu_1$ and (−2.1, 7.3) for $\mu_1 - \mu_2$. Interpret these two confidence intervals. Using these two intervals and the interval from part a, summarize what the airline company learned from this study.

d. The confidence intervals are wide. In the design of this experiment, what could you change to estimate the differences in means more precisely?

14.15 Tukey holding time comparisons Refer to the previous exercise. We could instead use the Tukey method to construct multiple comparison confidence intervals. The Tukey confidence intervals having *overall* confidence level 95% have margins of error of 5.7, compared to 4.7 for the separate 95% confidence intervals in the previous exercise.

TRY

a. According to this method, which groups are significantly different?

b. Why are the margins of error larger than with the separate 95% intervals?

14.16 REM sleep A psychologist compares the mean amount of time of rapid-eye movement (REM) sleep for subjects under three conditions. She randomly assigns 12 subjects to the three groups, four per group. The sample means for the three groups were 18, 15, and 12. The table shows the ANOVA table from SPSS.

REM sleep

Source	DF	SS	MS	F	P
Group	2	72.00	36.00	0.79	0.481
Error	9	408.00	45.33		
Total	11	480.00			

a. Report and interpret the P-value for the ANOVA *F* test.

b. For the Tukey 95% multiple comparison confidence intervals comparing each pair of means, the margin of error for each interval is 13.3. Is it true or false that since all the confidence intervals contain 0, it is plausible that all three population means equal 0.

c. Would the margin of error for each separate 95% confidence interval be less than 13.3, equal to 13.3, or larger than 13.3? Explain why.

14.17 REM regression Refer to the previous exercise.

TRY

a. Set up indicator variables for a regression model so that an F test for the regression parameters is equivalent to the ANOVA test comparing the three means.

b. Express the null hypothesis both in terms of population means and in terms of regression parameters for the model in part a.

c. The sample means were 18, 15, and 12. Explain why the prediction equation is $\hat{y} = 12 + 6x_1 + 3x_2$. Interpret the three parameter estimates in this equation.

14.18 Outsourcing satisfaction Exercise 14.5 showed an ANOVA for comparing mean customer satisfaction scores for three service centers. The sample means on a scale of 0 to 10 were 7.60 in San Jose, 7.80 in Toronto, and 7.10 in Bangalore. Each sample size = 100, MS error = 0.47, and the F test statistic = 27.6 has P-value < 0.001.

a. Explain why the margin of error for separate 95% confidence intervals is the same for comparing the population means for each pair of cities. Show that this margin of error is 0.19.

b. Find the 95% confidence interval for the difference in population means for each pair of service centers. Interpret.

c. The margin of error for Tukey 95% multiple comparison confidence intervals for comparing the service centers is 0.23. Construct the intervals. Interpret.

d. Why are the confidence intervals different in part b and in part c? What is an advantage of using the Tukey intervals?

14.19 Regression for outsourcing Refer to the previous exercise.

a. Set up indicator variables to represent the three service centers.

b. The prediction equation is $\hat{y} = 7.1 + 0.5x_1 + 0.7x_2$. Show how the terms in this equation relate to the sample means of 7.6 for San Jose, 7.8 for Toronto, and 7.1 for Bangalore.

14.20 Advertising effect on sales Each of 100 restaurants in a fast-food chain is randomly assigned one of four media for an advertising campaign: A = radio, B = TV, C = newspaper, D = mailing. For each restaurant, the observation is the change in sales, defined as the difference between the sales for the month during which the advertising campaign took place and the sales in the same month a year ago (in thousands of dollars).

a. By creating indicator variables, write a regression equation for the analysis to compare mean change in sales for the four media.

b. Explain how you could use the regression model to test the null hypothesis of equal population mean change in sales for the four media.

c. The prediction equation is $\hat{y} = 35 + 5x_1 - 10x_2 + 2x_3$. Estimate the difference in mean change in sales for media (i) A and D, (ii) A and B.

14.21 French ANOVA Refer to Exercise 14.3 about studying French, with data shown again below. Using software,

a. Compare the three pairs of means with separate 95% confidence intervals. Interpret.

b. Compare the three pairs of means with Tukey 95% multiple comparison confidence intervals. Interpret, and explain why the intervals are different than in part a.

Group 1	Group 2	Group 3
4	1	9
6	5	10
8		5

14.3 Two-Way ANOVA

Recall

A **factor** is a categorical explanatory variable. **One-way ANOVA** has a single factor. ◀

One-way ANOVA is a *bivariate* (two-variable) method. It analyzes the relationship between the mean of a quantitative response variable and the groups that are categories of a factor. ANOVA extends to handle two or more factors. With multiple factors, the analysis is *multivariate*. We'll illustrate for the case of two factors. This extension is a **two-way ANOVA**. It enables us to study the effect of one factor at a fixed level of a second factor.

The great British statistician R. A. Fisher (see *On the Shoulders of R. A. Fisher* at the end of Section 8.5) developed ANOVA methods in the 1920s. Agricultural experiments were the source of many of the early ANOVA applications. For instance, ANOVA has often been used to compare the mean yield of a crop for different fertilizers.

Example 8

Two factors

Amounts of Fertilizer and Manure

Did You Know?

A **metric ton** is 1000 kilograms, which is about 2200 pounds. A **hectare** is 10,000 square meters (e.g., 100 meters by 100 meters), which is about 2.5 acres. ◀

Picture the Scenario

This example presents a typical ANOVA application, based on a study at Iowa State University.[2] A large field was portioned into 20 equal-size plots. Each plot was planted with the same amount of corn seed, using a fixed spacing pattern between the seeds. The goal was to study how the yield of corn later harvested from the plots (in metric tons) depended on the levels of use of nitrogen-based fertilizer and manure. Each factor was measured in a binary manner. The fertilizer level was low = 45-kg per hectare or high = 135 kg per hectare. The manure level was low = 84 kg per hectare or high = 168 kg per hectare.

Questions to Explore

a. What are four treatments you can compare with this experiment?

b. What comparisons are relevant when you control for (keep fixed) manure level?

Think It Through

a. Four treatments result from cross-classifying the two binary factors: fertilizer level and manure level. Table 14.7 shows the four treatments, defined for the $2 \times 2 = 4$ combinations of categories of the two factors (fertilizer level and manure level).

Table 14.7 Four Groups for Comparing Mean Corn Yield

These result from the two-way cross classification of fertilizer level with manure level.

	Fertilizer	
Manure	**Low**	**High**
Low		
High		

b. We can compare the mean corn yield for the two levels of fertilizer, controlling for manure level (that is, at a fixed level of manure use). For fields in which manure level was *low*, we can compare the mean yields for the two levels of fertilizer use. These refer to the first row of Table 14.7. Likewise, for fields in which manure level was *high*, we can compare the mean yields for the two levels of fertilizer use. These refer to the second row of Table 14.7.

Insight

Among the questions we'll learn how to answer in this section are: Does the mean corn yield depend significantly on the fertilizer level? Does it depend on the manure level? Does the effect of fertilizer depend on the manure level?

Try Exercise 14.22

[2]Thanks to Dan Nettleton, Iowa State University, for data on which this example is based.

Inference About Effects in Two-Way ANOVA

In two-way ANOVA, a null hypothesis states that the population means are the same in each category of one factor, at each fixed level of the other factor. For example, we could test

H_0: Mean corn yield is equal for plots at the low and high levels of fertilizer, for each fixed level of manure.

Table 14.8a displays a set of population means satisfying this null hypothesis of "no effect of fertilizer level."

Table 14.8 Population Mean Corn Yield Satisfying Null Hypotheses: (a) No Effect of Fertilizer Level, (b) No Effect of Manure Level

		Fertilizer				Fertilizer	
	Manure	Low	High		**Manure**	Low	High
(a)	Low	10	10	**(b)**	Low	10	20
	High	20	20		High	10	20

We could also test

H_0: Mean corn yield is equal for plots at the low and high levels of manure, for each fixed level of fertilizer.

Table 14.8b displays a set of population means satisfying this null hypothesis of "no effect of manure level." The effects of individual factors tested with these two null hypotheses are called **main effects**. We'll discuss a third null hypothesis later in the section.

As in one-way ANOVA, the F tests of hypotheses in two-way ANOVA assume that

- The population distribution for each group is normal.
- The population standard deviations are identical.
- The data result from random sampling or a randomized experiment.

Here, each group refers to a cell in the two-way classification of the two factors. ANOVA procedures still usually work quite well if the population distributions are not normal with identical standard deviations. As in other ANOVA inferences, the quality of the sample is the most important assumption.

The test statistics have complex formulas. We'll rely on software. As in one-way ANOVA, the test for a factor uses two estimates of the variance for each group. These estimates appear in the mean square (MS) column of the ANOVA table.

SUMMARY: *F* Test Statistics in Two-Way ANOVA

For testing the main effect for a factor, the test statistic is the ratio of mean squares,

$$F = \frac{\text{MS for the factor}}{\text{MS error}}.$$

The MS for the factor is a variance estimate based on between-groups variation for that factor. The MS error is a within-groups variance estimate that is always unbiased.

When the null hypothesis of equal population means for the factor is true, the F test statistic values tend to fluctuate around 1. When it is false, they tend to be larger. The P-value is the right-tail probability above the observed F value. That is, it is the probability (presuming H_0 is true) of even more extreme results than we observed in the sample.

Example 9

Testing the main effects

Corn Yield

Picture the Scenario

Let's analyze the relationship between corn yield and the two factors, fertilizer level and manure level. Table 14.9 shows the data and the sample mean and standard deviation for each group.

Table 14.9 Corn Yield by Fertilizer Level and Manure Level

Fertilizer Level	Manure Level	Plot 1	Plot 2	Plot 3	Plot 4	Plot 5	Sample Size	Mean	Std. Dev.
High	High	13.7	15.8	13.9	16.6	15.5	5	15.1	1.3
High	Low	16.4	12.5	14.1	14.4	12.2	5	13.9	1.7
Low	High	15.0	15.1	12.0	15.7	12.2	5	14.0	1.8
Low	Low	12.4	10.6	13.7	8.7	10.9	5	11.3	1.9

Questions to Explore

a. Summarize the factor effects as shown by the sample means.

b. Table 14.10 is an ANOVA table for two-way ANOVA. Specify the two hypotheses tested, give the test statistics and P-values, and interpret.

Table 14.10 Two-Way ANOVA for Corn Yield Data in Table 14.9

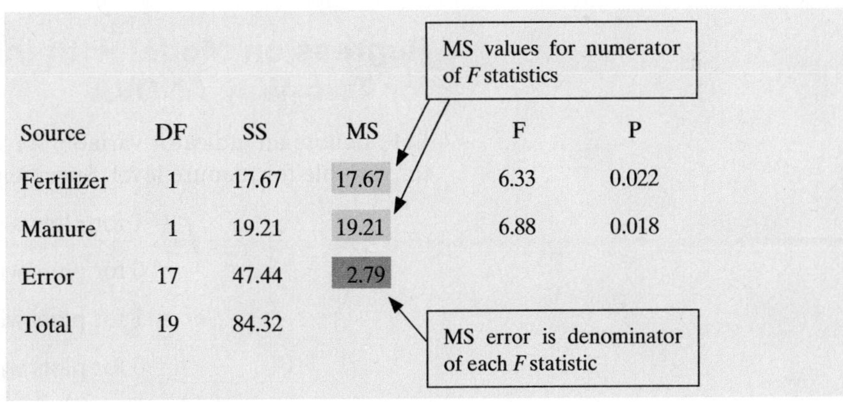

Source	DF	SS	MS	F	P
Fertilizer	1	17.67	17.67	6.33	0.022
Manure	1	19.21	19.21	6.88	0.018
Error	17	47.44	2.79		
Total	19	84.32			

Think It Through

a. Table 14.9 (with means summarized in the margin) shows that for each manure level, the sample mean yield is higher for the plots using more fertilizer. For each fertilizer level, the sample mean yield is higher for the plots using more manure.

Means from Table 14.9

	Fertilizer	
Manure	Low	High
Low	11.3	13.9
High	14.0	15.1

b. First, consider the hypothesis

H_0: Mean corn yield is equal for plots at the low and high levels of fertilizer, for each fixed level of manure.

For the fertilizer main effect, Table 14.10 reports that the between-groups estimate of the variance is 17.67. This is the mean square (MS) for fertilizer in Table 14.10. The within-groups estimate is the MS error, or 2.79. The F test statistic is the ratio,

$$F = 17.67 / 2.79 = 6.33.$$

From Table 14.10, the df values are 1 and 17 for the two estimates. From the F distribution with $df_1 = 1$ and $df_2 = 17$, the P-value is 0.022, also reported in Table 14.10. If the population means were equal at the two levels of fertilizer, the probability of an F test statistic value larger than 6.33 would be only 0.022. There is strong evidence that the mean corn yield depends on fertilizer level.

Next, consider the hypothesis

H_0: Mean corn yield is equal for plots at the low and high levels of manure, for each fixed level of fertilizer.

For the manure main effect, the F test statistic is $F = 19.21 / 2.79 = 6.88$. From Table 14.10, $df_1 = 1$ and $df_2 = 17$, and the P-value is 0.018. There is strong evidence that the mean corn yield also depends on the manure level.

Insight

As with any significance test, the information gain is limited. We do not learn *how large* the fertilizer and manure effects are on the corn yield. We can use confidence intervals to investigate the sizes of the main effects. We'll now learn how to do this by using regression modeling with indicator variables.

Try Exercise 14.27

Regression Model with Indicator Variables for Two-Way ANOVA

Let f denote an indicator variable for fertilizer level and let m denote an indicator variable for manure level. Specifically,

$$\begin{aligned} f &= 1 \text{ for plots with high fertilizer level} \\ &= 0 \text{ for plots with low fertilizer level} \\ m &= 1 \text{ for plots with high manure level} \\ &= 0 \text{ for plots with low manure level.} \end{aligned}$$

The multiple regression model for the mean corn yield with these two indicator variables is

$$\mu_y = \alpha + \beta_1 f + \beta_2 m.$$

To find the population means for the four groups, we substitute the possible combinations of values for the indicator variables. For example, for plots that have high fertilizer level ($f = 1$) and low manure level ($m = 0$), the mean corn yield is

$$\mu_y = \alpha + \beta_1(1) + \beta_2(0) = \alpha + \beta_1.$$

Table 14.11 shows the four means. The difference between the means at the high and low levels of fertilizer equals β_1 for each manure level. That is, the coefficient β_1 of the indicator variable f for fertilizer level equals the difference between the means at its high and low levels, controlling for manure level. The null hypothesis of no fertilizer effect is H_0: $\beta_1 = 0$. Likewise, β_2 is the difference between the means at the high and low levels of manure, for each fertilizer level.

Table 14.11 Population Mean Corn Yield for Fertilizer and Manure Levels

		Indicator Variables		
Fertilizer	**Manure**	f	m	**Mean of** y
High	High	1	1	$\alpha + \beta_1 + \beta_2$
High	Low	1	0	$\alpha + \beta_1$
Low	High	0	1	$\alpha + \beta_2$
Low	Low	0	0	α

We do not need to use regression modeling to conduct the ANOVA F tests. They're easy to do using software. But the modeling approach helps us to focus on estimating the means and the differences among them. We can compare means using an ordinary confidence interval for the regression parameter that equals the difference between those means. The 95% confidence interval has the usual form of

$$\text{parameter estimate} \pm t_{0.025}(se).$$

The *df* for the t-score is the *df* value for the MS error.

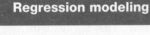

Example 10

Regression modeling

Estimate and Compare Mean Corn Yields

Picture the Scenario

Table 14.12 shows the result of fitting the regression model for predicting corn yield with indicator variables for fertilizer level and manure level.

Table 14.12 Estimates of Regression Parameters for Two-Way ANOVA of the Mean Corn Yield by Fertilizer Level and Manure Level

Predictor	Coef	SE Coef	T	P
Constant	11.6500	0.6470	18.01	0.000
fertilizer	1.8800	0.7471	2.52	0.022
manure	1.9600	0.7471	2.62	0.018

Questions to Explore

a. Find and use the prediction equation to estimate the mean corn yield for each group.

b. Use a parameter estimate from the prediction equation to compare mean corn yields for the high and low levels of fertilizer, at each manure level.

c. Find a 95% confidence interval comparing the mean corn yield at the high and low levels of fertilizer, controlling for manure level. Interpret it.

Think It Through

a. From Table 14.12, the prediction equation is (rounding to one decimal place)

$$\hat{y} = 11.6 + 1.9f + 2.0m.$$

The y-intercept equals 11.6. This is the estimated mean yield (in metric tons per hectare) when both indicator variables equal 0, that is, with fertilizer and manure at the low levels. The estimated means for the other cases result from substituting values for the indicator variables. For instance, at fertilizer level = high and manure level = low, $f = 1$ and $m = 0$, so the estimated mean yield is $\hat{y} = 11.6 + 1.9(1) + 2.0(0) = 13.5$. Doing this for all four groups, we get

	Fertilizer	
Manure	**Low**	**High**
Low	11.6	11.6 + 1.9 = 13.5
High	11.6 + 2.0 = 13.6	11.6 + 1.9 + 2.0 = 15.5

b. The coefficient of the fertilizer indicator variable f is 1.9. This is the estimated difference in mean corn yield between the high and low levels of fertilizer, for each level of manure (for instance, 13.5 − 11.6 = 1.9 when manure level = low).

c. Again, The estimate of the fertilizer effect is 1.9. Its standard error, reported in Table 14.12, is 0.747. From Table 14.10 (see page 703), the df for the MS error is 17. From Table B, $t_{.025} = 2.11$ when $df = 17$. The 95% confidence interval is

$$1.9 \pm 2.11(0.747), \text{ which is } (0.3, 3.5).$$

At each manure level, we estimate that the mean corn yield is between 0.3 and 3.5 metric tons per hectare higher at the high fertilizer level than at the low fertilizer level. The confidence interval contains only positive values (does not contain 0), reflecting the conclusion that the mean yield is significantly higher at the higher level of fertilizer. This agrees with the P-value falling below 0.05 in the test for the fertilizer effect.

Estimated Means

	Fertilizer	
Manure	**Low**	**High**
Low	11.6	13.5
High	13.6	15.5

Table 14.9: Sample Means

	Fertilizer	
Manure	**Low**	**High**
Low	11.3	13.9
High	14.0	15.1

Insight

The *estimated* means are not the same as the *sample* means in Table 14.9. (Both sets are shown again in the margin.) The model *smooths* the sample means so that the difference between the estimated means for two categories of a factor is *exactly* the same at each category of the other factor. For example, from the estimated group means found previously, the fertilizer effect of 1.9 = 13.5 − 11.6 = 15.5 − 13.6.

Try Exercise 14.28

The regression model assumes that the difference between means for the two categories for one factor is the same in each category of the other factor. The next section shows how to check this assumption. When it is reasonable, we can use a single comparison, rather than a separate one at each category of the other variable. In Example 10, we estimated the difference in mean corn yield between the pair of fertilizer levels, a single comparison (with estimate 1.9) holding at each level of manure.

Recall

Section 13.5 introduced the concept of **interaction** between two explanatory variables in their effects on a response variable. In a regression context, **no interaction** implied parallel lines (common slopes). ◀

Exploring Interaction Between Factors in Two-Way ANOVA

Investigating whether **interaction** occurs is important whenever we analyze multivariate relationships. *No interaction* between two factors means that the effect of either factor on the response variable is the same at each category of the other factor. The regression model in Example 10 and the ANOVA tests of main effects assume there is no interaction between the factors. What does this mean in this context?

	Fertilizer	
Manure	Low	High
Low	11.6	13.5
High	13.6	15.5

From Example 10, the estimated mean corn yields for the regression model having an indicator variable for fertilizer level and an indicator variable for manure level are shown in the table in the margin.

What pattern do these show? Let's plot the means for the two fertilizer levels, within each level of manure. Figure 14.5 shows a plot in which the y-axis gives estimated mean corn yields, and points are shown for the four fertilizer–manure combinations. The horizontal axis is not a numerical scale but merely lists the two fertilizer levels. The drawn lines connect the means for the two fertilizer levels, for a given manure level. The absence of interaction is indicated by the *parallel lines.*

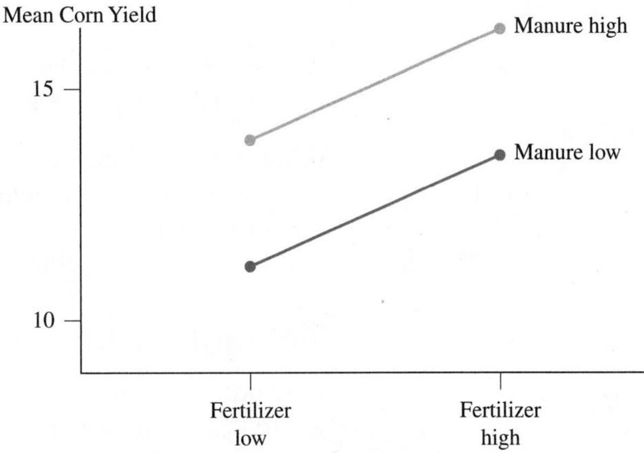

▲ **Figure 14.5 Mean Corn Yield, by Fertilizer and Manure Levels, Showing No Interaction.** The parallel lines reflect an absence of interaction. This implies that the difference in estimated means between the two fertilizer levels is the same for each manure level. **Question** Is it also true that the difference in estimated means between the two manure levels is the same for each fertilizer level?

The parallelism of lines occurs because the difference in the estimated mean corn yield between the high and low levels of fertilizer is the same for each manure level. The difference equals 1.9. Also, the difference between the high and low levels of manure in the estimated mean corn yield is 2.0 for each fertilizer level.

By contrast, Table 14.13 and Figure 14.6 show a set of means for which there is interaction. The difference between the high and low levels of fertilizer in the mean corn yield is $14 - 10 = 4$ for low manure and $12 - 16 = -4$ for high manure. Here, the difference in means depends on the manure level: According to these means, it's better to use a high level of fertilizer when the manure level is low but it's better to use a low level of fertilizer when the manure level is high. Similarly, the manure effect differs at the two fertilizer levels; for the low level, it is $16 - 10 = 6$ whereas at the high level, it is $12 - 14 = -2$. The lines in Figure 14.6 are not parallel.

Table 14.13 Means that Show Interaction Between the Factors in Their Effects on the Response

The effect of fertilizer differs according to whether manure level is low or high. See Figure 14.6.

	Fertilizer	
Manure	Low	High
Low	10	14
High	16	12

In Table 14.13, suppose the numbers of observations are the same for each group. Then the overall mean corn yield, ignoring manure level, is 13 for each fertilizer level (the average of 10 and 16, and the average of 14 and 12). The overall

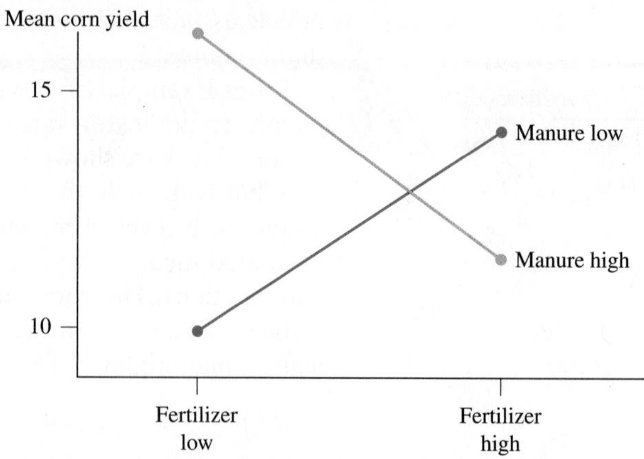

▲ **Figure 14.6** Mean Corn Yield, by Fertilizer and Manure Levels, Displaying Interaction. Question What aspect of the plot reflects the interaction?

difference in means between the two fertilizer levels then equals 0. A one-way analysis of mean corn yield by fertilizer level would conclude that fertilizer level has no effect. However, a two-way analysis that allows for interaction would detect that fertilizer has an effect, but that effect differs according to the manure level.

Testing for Interaction

In conducting a two-way ANOVA, before testing the main effects, it is customary to test a third null hypothesis stating that there is no interaction between the factors in their effects on the response. The test statistic providing the sample evidence of interaction is

$$F = \frac{\text{MS for interaction}}{\text{MS error}}.$$

When H_0 is false, the F statistic tends to be large. Therefore, as usual, the P-value is the right-tail probability.

Example 11

Testing for interaction

Corn Yield Data

Picture the Scenario

Let's return to our analysis of the corn yield data, summarized in the margin table. In Example 10 we analyzed these data assuming no interaction. Let's see if that analysis is valid. Table 14.14 is an ANOVA table for a model that allows interaction in assessing the effects of fertilizer level and manure level on the mean corn yield.

Corn Yields

Fertilizer Level	Manure Level	Mean	Std. Dev.
High	High	15.1	1.3
High	Low	13.9	1.7
Low	High	14.0	1.8
Low	Low	11.3	1.9

Table 14.14 Two-Way ANOVA of Mean Corn Yield by Fertilizer Level and Manure Level, Allowing Interaction

Source	DF	SS	MS	F	P
Fertilizer	1	17.67	17.67	6.37	0.023
Manure	1	19.21	19.21	6.92	0.018
Interaction	1	3.04	3.04	1.10	0.311
Error	16	44.40	2.78		
Total	19	84.32			

F statistic for test of no interaction

Question to Explore

Give the result of the test of H_0: no interaction, and interpret.

Think It Through

The test statistic for H_0: no interaction is

$$F = (\text{MS for interaction})/(\text{MS error}) = 3.04 / 2.78 = 1.10.$$

Based on the F distribution with $df_1 = 1$ and $df_2 = 16$ for these two mean squares, the ANOVA table reports P-value = 0.31. This is not much evidence of interaction. We would not reject H_0 at the usual significance levels, such as 0.05.

Insight

Because there is not much evidence of interaction, we are justified in conducting the simpler two-way ANOVA about main effects. The tests presented previously in Table 14.10 for effects of fertilizer and manure on mean corn yield are valid.

Try Exercise 14.29

It is not meaningful to test the main effects hypotheses when there is interaction. A small P-value in the test of H_0: no interaction suggests that each factor has an effect, but the size of effect for one factor varies according to the category of the other factor. Then, you should investigate the nature of the interaction by plotting the sample cell means, using a plot like Figure 14.6. You should compare categories of one factor separately at different levels of the other factor.

Recall

From the box at the beginning of Section 14.2, the 95% confidence interval comparing two means is

$$(\bar{y}_i - \bar{y}_j) \pm t_{.025}\, se, \text{ where } se = s\sqrt{\frac{1}{n_i} + \frac{1}{n_j}},$$

s is the square root of the MS error, and df for the t distribution is df for MS error. ◀

In Practice Check Interaction Before Main Effects

In practice, in two-way ANOVA you should *first test the hypothesis of no interaction.* If the evidence of interaction is not strong (that is, if the P-value is not small), then test the main effects hypotheses and/or construct confidence intervals for those effects. But if important evidence of interaction exists, plot and compare the cell means for a factor separately at each category of the other factor.

For comparing means in two cells using a confidence interval, use the formula from the box at the beginning of Section 14.2, shown again in the margin. Substitute the cell sample sizes for n_i and n_j and use the MS error for the two-way ANOVA that allows interaction.

Interactions and confidence interval

Example 12

Political Ideology by Gender and Race

Picture the Scenario

In most years, the General Social Survey asks subjects to report their political ideology, measured with seven categories in which 1 = extremely liberal, 4 = moderate, 7 = extremely conservative. Table 14.15 shows results from the 2008 General Social Survey on mean political ideology classified by gender and by race.

Caution

When conducting a two-way ANOVA for samples of different sizes, the analysis (using software) will often have to be performed as a General Linear Model. Notice that the Analysis of Variance table here shows both a Seq SS and an Adj SS column. You should use the Adjusted SS and Adjusted MS for the tests.

Table 14.15 Mean Political Ideology by Gender and by Race

	Race	
Gender	Black	White
Female	4.164 (n = 165)	4.268 (n = 840)
Male	3.819 (n = 116)	4.444 (n = 719)

For the test of H_0: no interaction, software reports an F test statistic of 6.19 with $df_1 = 1$ and $df_2 = 1836$, for a P-value of 0.013. So, in comparing females and males on their mean political ideology, we should do it separately by race. The MS error for the model allowing interaction equals 2.534, so the residual standard deviation is $s = \sqrt{2.534} = 1.592$.

Analysis of Variance for POLVIEWS, using Adjusted SS for Tests

Source	DF	Seq SS	Adj SS	Adj MS	F	P
SEX	1	5.138	1.652	1.652	0.65	0.420
RACE	1	24.786	30.780	30.780	12.15	0.001
SEX*RACE	1	15.693	15.693	15.693	6.19	0.013
Error	1836	4651.981	4651.981	2.534		
Total	1839	4697.598				

Questions to Explore

a. Interpret the significant interaction by comparing sample mean political ideology of females and males for each race descriptively and using 95% confidence intervals.

b. Interpret the confidence intervals derived in part a.

Think It Through

a. The sample means show that for black subjects, females are more conservative (have the higher mean). By contrast, for white subjects, males are more conservative. For a confidence interval comparing mean political ideology for females and males who are black, the standard error (using the sample sizes reported in the table) is

$$se = s\sqrt{\frac{1}{n \text{ for black females}} + \frac{1}{n \text{ for black males}}} =$$

$$1.592\sqrt{\frac{1}{165} + \frac{1}{116}} = 0.193.$$

The 95% confidence interval is

$$(4.164 - 3.819) \pm 1.96(0.193), \text{ which is } 0.345 \pm 0.378, \text{ or } (-0.03, 0.72).$$

Likewise, you can find that the 95% confidence interval comparing mean political ideology for females and males who are white is -0.1768 ± 0.159, or $(-0.33, -0.02)$.

b. Since the confidence interval for black subjects contains zero we cannot infer that there is a difference in the populations. For white subjects, however, all values in the interval are negative. We infer that white females are *less* conservative than white males in the population.

Insight

The confidence interval for white subjects has an endpoint that is close to 0. So, the true gender effect in the population could be small.

Try Exercise 14.33

Why Not Instead Perform Two Separate One-Way ANOVAs?

When you have two factors, rather than performing a two-way ANOVA, why not instead perform two separate one-way ANOVAs? For instance, in Example 12 you could compare the mean political ideology for females and males using a one-way ANOVA, ignoring the information about race. Likewise, you could perform a separate one-way ANOVA to compare the means for blacks and whites, ignoring the information about gender.

The main reason is that you learn more from two-way ANOVA. The two-way ANOVA indicates whether there is interaction. When there is, as in Example 12, it is more informative to compare levels of one factor *separately* at each level of the other factor. This enables us to investigate how the effect depends on that other factor. For instance, a one-way ANOVA of mean political ideology by gender might show *no* gender effect, whereas the two-way ANOVA has shown that there *is* a gender effect but it varies by race.

Similarly, in experimental studies, rather than carrying out one experiment to investigate the effect of one factor and a separate experiment to investigate the effect of a second factor, it is better to carry out a *single* experiment and study both factors at the same time. Besides the advantage of being able to check for interaction, this is more cost effective. If we have funds to experiment with 100 subjects, we can use a sample size of 100 for studying each factor with a two-way ANOVA, rather than use 50 subjects in one experiment about one factor and 50 subjects in a separate experiment about the other factor.

Yet another benefit of a two-way ANOVA is that the residual variability, which affects the MS error and the denominators of the F test statistics, tends to decrease. When we use two factors to predict a response variable, we usually tend to get better predictions (that is, less residual variability) than when we use one factor. With less residual (within-groups) variability, we get larger test statistics, and hence greater power for rejecting false null hypotheses.

Factorial ANOVA

The methods of two-way ANOVA extend to the analysis of several factors. A multifactor ANOVA with observations from all combinations of the factors is called **factorial ANOVA**. For example, with three factors, **three-way ANOVA** considers main effects for those factors as well as possible interactions.

In Practice Use Regression With Categorical and Quantitative Predictors

With several explanatory variables, usually some are categorical and some are quantitative. Then, it is sensible to build a multiple regression model containing both types of predictors. That's what we did in Example 10 in the previous chapter when we modeled house selling prices in terms of house size and condition of the house.

14.3 Practicing the Basics

14.22 Reducing cholesterol An experiment randomly assigns 100 subjects suffering from high cholesterol to one of four groups: low-dose Lipitor, high-dose Lipitor, low-dose Zocor, high-dose Zocor. After three months of treatment, the change in cholesterol level is measured.

TRY

a. Identify the response variable and the two factors.

b. What are four treatments that can be compared with this experiment?

c. What comparisons are relevant when we control for dose level?

14.23 Drug main effects For the previous exercise, show a hypothetical set of population means for the four groups that would have

a. A dose effect but no drug effect.

b. A drug effect but no dose effect.

c. A drug effect and a dose effect.

d. No drug effect and no dose effect.

14.24 Reasons for divorce The 26 students in a statistics class at the University of Florida were surveyed about their attitudes toward divorce. Each received a response score according to how many from a list of 10 possible reasons were regarded as legitimate for a woman to seek a divorce. The higher the score, the more willing the subject is to accept divorce as an option in a variety of circumstances. The students were also asked whether they were fundamentalist or nonfundamentalist in their religious beliefs and whether their church attendance frequency was frequent (more than once a month) or infrequent. The table displays the data and the results of a two-way ANOVA.

a. State the null hypothesis to which the F test statistic in the religion row refers.

b. Show how to use mean squares to construct the F test statistic for the religion main effect, report its P-value, and interpret.

		Religion	
		Fundamentalist	**Non fundamentalist**
Church Attendance	**Frequent**	0, 3, 4, 0, 3, 2, 0, 1, 1, 3	2, 5, 1, 2, 3, 3
	Infrequent	4, 3, 4	6, 8, 6, 4, 6, 3, 7, 4

Source	DF	SS	MS	F	P
Religion	1	11.07	11.07	5.20	0.032
Church_attend	1	36.84	36.84	17.32	0.000
Error	23	48.93	2.13		
Total	25	117.12			

14.25 House prices, age, and bedrooms For the House Selling Prices OR data file on the text CD, the output shows the result of conducting a two-way ANOVA of house selling prices (in thousands) by the number of bedrooms in the house and the age (New, Medium, Old, Very Old—see exercise 14.12) of the houses in Corvallis, Oregon.

a. For testing the main effect of age, report the F test statistic value, and show how it was formed from other values reported in the ANOVA table.

b. Report the P-value for the main effect test for age, and interpret.

Source	DF	SS	MS	F	P
Bedrooms	7	517868	68523	6.79	0.000
Age Condition	3	243063	81021	8.03	0.000
Error	189	1907939	10095		
Total	199	2668870			

14.26 Corn and manure In Example 10, the coefficient of the manure-level indicator variable m is 1.96.

a. Explain why this coefficient is the estimated difference in mean corn yield between the high and low levels of manure, for each level of fertilizer.

b. Explain why the 95% confidence interval for the difference in mean corn yield between the high and low levels of manure is $1.96 \pm 2.11(0.747)$.

14.27 Hang up if message repeated? Example 2 described an experiment in which telephone callers to an airline were put on hold with an advertisement, Muzak, or classical music in the background. Each caller who was chosen was also randomly assigned to a category of a second factor: Whether the message played was five minutes long or ten minutes long. (In each case, it was repeated at the end.) The table shows the data classified by both factors and the results of a two-way ANOVA.

TRY

Telephone holding times by type of recorded message and repeat time

	Repeat Time	
Message	**Ten Minutes**	**Five Minutes**
Advertisement	**8, 11, 2**	**5, 1**
Muzak	**1, 4, 3**	**0, 6**
Classical	**13, 8, 15**	**7, 9**

Source	DF	SS	MS	F	P
Message	2	149.20	74.60	7.09	0.011
Repeat	1	23.51	23.51	2.24	0.163
Error	11	115.69	10.52		
Total	14	288.40			

a. State the null hypothesis to which the F test statistic in the Message row refers.

b. Show how to use mean squares to construct the F test statistic for the Message main effect, report its P-value, and interpret.

c. On what assumptions is this analysis based?

14.28 Regression for telephone holding times Refer to the previous exercise. Let $x_1 = 1$ for the advertisement and 0 otherwise, $x_2 = 1$ for Muzak and 0 otherwise, and $x_1 = x_2 = 0$ for classical music. Likewise, let $x_3 = 1$ for repeating in 10-minute cycles and $x_3 = 0$ for repeating in 5-minute cycles. The display shows results of a regression of the telephone holding times on these indicator variables.

TRY

Regression for telephone holding times

Predictor	Coef	SE Coef	T	P
Constant	8.867	1.776	4.99	0.000
x1	−5.000	2.051	−2.44	0.033
x2	−7.600	2.051	−3.71	0.003
x3	2.556	1.709	1.50	0.163

a. State the prediction equation. Interpret the parameter estimates.

b. Find the estimated means for the six groups in the two-way cross-classification of message type and repeat time.

c. Find the estimated difference between the mean holding times for 10-minute repeats and 5-minute repeats, for a fixed message type. How can you get this estimate from a coefficient of an indicator variable in the prediction equation?

14.29 Interaction between message and repeat time? Refer to
TRY the previous two exercises. When we allow interaction, two of the rows in the new ANOVA table are

```
Source          DF     SS      MS      F      P
Group*Repeat     2   15.02    7.51   0.67  0.535
Error            9  100.67   11.19
```

where Group*Repeat denotes the interaction effect.

a. Show the hypotheses, test statistic, and P-value for the test of H_0: no interaction.

b. What is the implication of the result of this test? For instance, the analyses in previous exercises assumed a lack of interaction. Was it valid to do so?

14.30 Income by gender and job type In 2000, the population mean hourly wage for males was $22 for white-collar jobs, $11 for service jobs, and $14 for blue-collar jobs. For females the means were $15 for white-collar jobs, $8 for service jobs, and $10 for blue-collar jobs.[3]

a. Identify the response variable and the two factors.

b. Show these means in a two-way classification of the two factors.

c. Compare the differences between males and females for (i) white-collar jobs and (ii) blue-collar jobs. Explain why there is interaction, and describe it.

d. Show a set of six population mean incomes that would satisfy H_0: no interaction.

14.31 Ideology by gender and race Refer to Example 12, the sample means from which are shown again below.

Mean political Ideology

	Race	
Gender	**Black**	**White**
Female	4.164	4.2675
Male	3.819	4.4443

a. Explain how to obtain the following interpretation for the interaction from the sample means: "For females there is no race effect on ideology. For males, whites are more conservative by about half an ideology category, on the average."

b. Suppose that instead of the two-way ANOVA, you performed a one-way ANOVA with gender as the predictor and a separate one-way ANOVA with race as the predictor. Suppose the ANOVA for gender does not show a significant effect. Explain how this could happen, even though the two-way ANOVA showed a gender effect for each race. (*Hint:* Will the overall sample means for females and males be more similar than they are for each race?)

c. Refer to part b. Summarize what you would learn about the gender effect from a two-way ANOVA that you would fail to learn from a one-way ANOVA.

14.32 Attractiveness and getting dates The results in the table are from a study of physical attractiveness and subjective well-being (E. Diener et al., *Journal of Personality and Social Psychology*, vol. 69, 1995, pp. 120–129). A panel rated a sample of college students on their physical attractiveness. The table presents the number of dates in the past three months for students rated in the top or bottom quartile of attractiveness.

a. Identify the response variable and the factors.

b. Do these data appear to show interaction? Explain.

c. Based on the results in the table, specify one of the ANOVA assumptions that these data violate. Is this the most important assumption?

Dates and attractiveness

	Number of DATES, MEN			Number of DATES, WOMEN		
ATTRACTIVENESS	**Mean**	**Std. Dev**	***n***	**Mean**	**Std. Dev**	***n***
More	9.7	10.0	35	17.8	14.2	33
Less	9.9	12.6	36	10.4	16.6	27

14.33 Diet and weight gain A randomized experiment[4] measured weight gain (in grams) of male rats under six diets varying by source of protein (beef, cereal, pork) and level of protein (high, low). Ten rats were assigned to each diet. The data are shown in the table that follows and are also available in the Protein and Weight Gain data file on the book's CD.

TRY

a. Conduct a two-way ANOVA that assumes a lack of interaction. Report the *F* test statistic and the P-value for testing the effect of the protein level. Interpret.

b. Now conduct a two-way ANOVA that also considers potential interaction. Report the hypotheses, test statistic and P-value for a test of no interaction. What do you conclude at the 0.05 significance level? Explain.

c. Refer to part b. Allowing interaction, construct a 95% confidence interval to compare the mean weight gain for the two protein levels, for the beef source of protein.

Weight gain by source of protein and by level of protein

	High Protein	Low Protein
Beef	73, 102, 118, 104, 81, 107, 100, 87, 117, 111	90, 76, 90, 64, 86, 51, 72, 90, 95, 78
Cereal	98, 74, 56, 111, 95, 88, 82, 77, 86, 92	107, 95, 97, 80, 98, 74, 74, 67, 89, 58
Pork	94, 79, 96, 98, 102, 102, 108, 91, 120, 105	49, 82, 73, 86, 81, 97, 106, 70, 61, 82

14.34 Regression of weight gain on diet Refer to the previous exercise.

a. Set up indicator variables for protein source and for protein level, and specify a regression model with the effects both of protein level and protein source on weight gain.

[3]*Source*: Data from *The State of Working America 2000–2001*, Economic Policy Institute.

[4]*Source:* Data from G. Snedecor and W. Cochran, *Statistical Methods*, 6th ed. (Iowa State University Press, 1967), p. 347.

b. Fit the model in part a, and explain how to interpret the parameter estimate for the protein level indicator variable.

c. Show how you could test a hypothesis about beta parameters in the model in part a to analyze the effect of protein source on weight gain.

d. Using the fit of the model, find the estimated mean for each of the six diets. Explain what it means when we say that these estimated means do not allow for interaction between protein level and source in their effects on weight loss.

Chapter Review

ANSWERS TO THE CHAPTER FIGURE QUESTIONS

Figure 14.1 No, even if the population means are equal we would expect the sample means to vary due to sample to sample variability. Differences among the sample means is not sufficient evidence to conclude that the population means differ.

Figure 14.2 The data in Figure 14.2b give stronger evidence against H_0 because Figure 14.2b has less variability within each sample than Figure 14.2a.

Figure 14.3 Evidence suggesting the normal with equal standard deviation assumptions are violated would be (1) graphical methods showing extreme skew for the response variable, or (2) the largest group standard deviation is more than about double the smallest group standard deviation when the sample sizes are unequal.

Figure 14.5 Yes. The parallelism of lines implies that no interaction holds no matter which factor we choose for making comparisons of means.

Figure 14.6 The lines are not parallel.

CHAPTER SUMMARY

Analysis of variance (ANOVA) methods compare several groups according to their means on a quantitative response variable. The groups are categories of categorical explanatory variables. A categorical explanatory variable is also called a **factor**.

- The **one-way ANOVA** F test compares means for a single factor. The groups are specified by categories of a single categorical explanatory variable.
- **Two-way ANOVA** methods compare means across categories of each of two factors, at fixed levels of the other factor. When there is **interaction** between the factors in their effects, differences between response means for categories of one factor change according to the category of the other factor.
- **Multiple comparison methods** such as the **Tukey method** compare means for each pair of groups while controlling the overall confidence level.
- Analysis of variance methods can be conducted by using multiple regression models. The regression model uses **indicator variables** as explanatory variables to represent the factors. Each indicator variable equals 1 for observations from a particular category and 0 otherwise.
- The ANOVA methods assume randomization and that the population distribution for each group is normal, with the same standard deviation for each group. In practice, *the randomness assumption is the most important*, and ANOVA methods are *robust* to moderate violations of the other assumptions.

CHAPTER PROBLEMS

Practicing the Basics

14.35 Good friends and marital status Is the number of good friends associated with marital status? For GSS data with marital status measured with the categories (married, widowed, divorced, separated, never married), an ANOVA table reports $F = 0.80$ based on $df_1 = 4, df_2 = 612$.

a. Introduce notation, and specify the null hypothesis and the alternative hypothesis for the ANOVA F test.

b. Based on what you know about the F distribution, would you guess that the test statistic value of 0.80 provides strong evidence against the null hypothesis? Explain.

c. Software reports a P-value of 0.53. Explain how to interpret it.

14.36 Going to bars and having friends Do people who go to bars and pubs a lot tend to have more friends? A recent GSS asked, "How often do you go to a bar or tavern?" The table shows results of ANOVA for comparing the mean number of good friends at three levels of this variable. The very often group reported going to a bar at least several times a week. The occasional group reported going occasionally, but not as often as several times a week.

Summary of ANOVA for mean number of good friends and going to bars

	Very often	Occasional	Never
Mean	12.1	6.4	6.2
Standard dev.	21.3	10.2	14.0
Sample size	41	166	215

Source	DF	SS	MS	F	P
Group	2	1116.8	558.4	3.03	0.049
Error	419	77171.8	184.2		
Total	421	78288.5			

a. State the (i) hypotheses, (ii) test statistic value, and (iii) P-value for the significance test displayed in this table. Interpret the P-value.

b. Based on the assumptions needed to use the method in part a, is there any aspect of the data summarized here that suggests that the ANOVA test and follow-up confidence intervals may not be appropriate? Explain.

14.37 TV watching The 2008 General Social Survey asked 1324 subjects how many hours per day they watched TV, on the average. Are there differences in population means according to the race of the subject (white, black, other)? The sample means were 2.76 for whites ($n = 1014$), 4.38 for blacks ($n = 188$), and 2.70 for other ($n = 122$). In a one-way ANOVA, the between-groups estimate ofthe variance is 215.26 and the within-groups estimate is 6.76.

a. Conduct the ANOVA test and make a decision using 0.05 significance level.

b. The 95% confidence interval comparing the population means is (1.1, 2.2) for blacks and whites, (−0.4, 0.5) for whites and the other category, and (1.0, 2.4) for blacks and the other category. Based on the three confidence intervals, indicate which pairs of means are significantly different. Interpret.

c. Based on the information given, show how to construct the confidence interval that compares the population mean TV watching for blacks and whites.

d. Refer to part c. Would the corresponding interval formed with the Tukey method be wider, or narrower? Explain why.

14.38 Comparing auto bumpers A consumer organization compares the sturdiness of three types of front bumpers. In the study, a particular brand of car is driven into a concrete wall at 15 miles per hour. The response is the amount of damage, as measured by the repair costs, in hundreds of dollars. Due to the potentially large costs, the study conducts only two tests with each bumper type, using six cars. The table shows the data and some ANOVA results. Show the (a) assumptions, (b) hypotheses, (c) test statistic and *df* values, (d) P-value, and (e) interpretation for testing the hypothesis that the true mean repair costs are the same for the three bumper types.

Bumper A	Bumper B	Bumper C
1	2	11
3	4	15

Source	DF	SS	MS	F	P
Bumper	2	148.00	74.00	18.50	0.021
Error	3	12.00	4.00		
Total	5	160.00			

14.39 Compare bumpers Refer to the previous exercise.

a. Find the margin of error for constructing a 95% confidence interval for the difference between any pair of the true means. Interpret by showing which pairs of bumpers (if any) are significantly different in their true mean repair costs.

b. For Tukey 95% multiple comparison confidence intervals, the margin of error is 8.4. Explain the difference between confidence intervals formed with this method and separate confidence intervals formed with the method in part a.

c. Set up indicator variables for a multiple regression model including bumper type.

d. The prediction equation for part c is $\hat{y} = 13 - 11x_1 - 10x_2$. Explain how to interpret the three parameter estimates in this model, and show how these estimates relate to the sample means for the three bumpers.

14.40 Segregation by region Studies of the degree of residential racial segregation often use the *segregation index.* This is the percentage of nonwhites who would have to change the block on which they live to produce a fully nonsegregated city—one in which the percentage of nonwhites living in each block is the same for all blocks in the city. This index can assume values ranging from 0 to 100, with higher values indicating greater segregation. (The national average for large U.S. metropolitan areas in 2009 was 27, down from 33 in 2000.) The table shows the index for a sample of cities for 2005–2009, classified by region.

Segregation index

Northeast	North Central	South	West
Boston: 67	Minneapolis-St. Paul: 56	New Orleans: 64	San Francisco-Oakland: 64
NY, Long Island, Northern NJ: 79	Detroit: 80	Tampa: 58	Dallas-Ft Worth: 57
Philadelphia: 69	Chicago: 78	Miami: 66	Los Angeles: 70
Pittsburgh: 68	Milwaukee: 81	Atlanta: 60	Seattle: 54

Source: Racial and Ethnic Residential Segregation in the United States, 1980–2000, U.S. Census Bureau Series CENSR-3, 2002. www.psc.isr.umich.edu/dis/census/segregation.html.

a. Report the mean and standard deviation for each of the four regions.

b. Define notation, and state the hypotheses for one-way ANOVA.

c. Report the *F* test statistic and its P-value. What do you conclude about the mean segregation indices for the four regions?

d. Suppose we took these data from the Census Bureau report by choosing only the cities in which we know people. Is the ANOVA valid? Explain.

14.41 Compare segregation means Refer to the previous exercise.

a. Using software, find the margin of error that pertains to each comparison using the Tukey method for 95% multiple comparison confidence intervals.

b. Using part a, determine which pairs of means, if any, are significantly different.

14.42 Georgia political ideology The Georgia Student Survey file on the text CD asked students their political party affiliation (1 = Democrat, 2 = Republican, 3 = Independent) and their political ideology (on a scale from 1 = very liberal to 7 = very conservative). The table shows results of an ANOVA, with political ideology as the response variable.

Georgia political ideology and party affiliation

	Political ideology		
Affiliation	**N**	**Mean**	**StDev**
Democrat	8	2.6250	1.0607
Republican	36	5.5000	1.0000
Independent	15	3.4667	0.9155

Source	DF	SS	MS	F	P
PoliticalAff	2	79.629	39.814	40.83	0.000
Error	56	54.608	0.975		
Total	58	134.237			

a. Does the ANOVA assumption of equal population standard deviations seem plausible, or is it so badly violated that ANOVA is inappropriate?

b. Are the population distributions normal? Why or why not? Which is more important, the normality assumption or the assumption that the groups are random samples from the population of interest?

c. The next table shows 95% confidence intervals comparing pairs of means. Interpret the confidence interval comparing Republicans and Democrats.

Affiliations	95% CI		
	Lower	**Center**	**Upper**
Republican–Democrat	2.10	2.87	3.65
Independent–Democrat	−0.02	0.84	1.71
Republican–Independent	1.43	2.03	2.64

d. Explain how you would summarize the results of the ANOVA *F* test and the confidence intervals to someone who has not studied statistics.

14.43 Comparing therapies for anorexia The Anorexia data file on the text CD shows weight change for 72 anorexic teenage girls who were randomly assigned to one of three psychological treatments.

a. Show how to construct 95% confidence intervals to investigate how the population means differ. Interpret them.

b. Report the Tukey 95% multiple comparison confidence intervals. Interpret, and explain why they are wider than the confidence intervals in part a.

14.44 Lot size varies by region? A geographer compares residential lot sizes in four quadrants of a city. To do this, she randomly samples 300 records from a city file on home residences and records the lot sizes (in thousands of square feet) by quadrant. The ANOVA table, shown in the table that follows, refers to a comparison of mean lot sizes for the northeast (NE), northwest (NW), southwest (SW), and southeast (SE) quadrants of the city. Fill in all the blanks in the table.

Lot sizes by quadrant of city

Source	DF	SS	MS	F	P-value
Quadrant	____	____	____	____	0.000
Error	296	1480	____		
Total	____	4180			

14.45 House with garage Refer to the House Selling Price OR data file on the text CD.

a. Set up an indicator variable for whether a house has a garage or not. Using software, put this as the sole predictor of house selling price (in thousands) in a regression model. Report the prediction equation, and interpret the intercept and slope estimates.

b. For the model fitted in part a, conduct the *t* test for the effect of the indicator variable in the regression analysis (that is, test H_0: $\beta = 0$). Interpret.

c. Use software to conduct the *F* test for the analysis of variance comparing the mean selling prices of homes with and without a garage. Interpret.

d. Explain the connection between the value of *t* in part b and the value of *F* in part c.

14.46 Ideal number of kids by gender and race The GSS asks, "What is the ideal number of kids for a family?" When we use a recent GSS to evaluate how the mean response depends on gender and race (black or white), we get the results shown in the ANOVA table.

a. Identify the response variable and the factors.

b. Explain what it would mean if there were no interaction between gender and race in their effects. Show a hypothetical set of population means that would show a strong race effect and a weak gender effect and no interaction.

c. Using the table, specify the null and alternative hypotheses, test statistic, and P-value for the test of no interaction. Interpret the result.

ANOVA of ideal number of kids by gender and race

Source	DF	SS	MS	F	P
Gender	1	0.25	0.25	0.36	0.546
Race	1	16.98	16.98	24.36	0.000
Interaction	1	0.95	0.95	1.36	0.244
Error	1245	867.72	0.70		
Total	1248	886.12			

14.47 Regress kids on gender and race Refer to the previous exercise. Let $f = 1$ for females and 0 for males, and let $r = 1$ for blacks and 0 for whites. The regression model for predicting y = ideal number of kids is $\hat{y} = 2.42 + 0.04f + 0.37r$.

a. Interpret the coefficient of *f*. What is the practical implication of this estimate being so close to 0?

b. Find the estimated mean for each of the four combinations of gender and race.

c. Summarize what you learn about the effects based on the analyses in this and the previous exercise.

14.48 Florida student data For the FL Student Survey data file on the text CD, we use as the response variable sports (the number of weekly hours engaged in sports and other physical exercise). For the explanatory variables, we use gender and whether the student is a vegetarian. The output shows results of a two-way ANOVA.

a. State the hypotheses for the vegetarian main effect.

b. Show how the *F* test statistic for part a was obtained from mean squares reported in the ANOVA table.

c. Report and interpret the P-value of the test for the vegetarian main effect.

Source	DF	SS	MS	F	P
vegetarian	1	7.89	7.89	0.56	0.457
gender	1	64.35	64.35	4.59	0.037
Error	57	799.80	14.03		
Total	59	878.98			

14.49 Regress TV watching on gender and religion When we use a recent GSS and regress y = number of hours per day watching TV on g = gender (1 = male, 0 = female) and religious affiliation ($r_1 = 1$ for Protestant, $r_2 = 1$ for Catholic, $r_3 = 1$ for Jewish, $r_1 = r_2 = r_3 = 0$ for none or other), we get the prediction equation TVHOURS $= 2.65 - 0.186g + 0.666r_1 + 0.373r_2 - 0.270r_3$

a. Interpret the gender effect.

b. Interpret the coefficient of r_1.

c. State a corresponding model for the population, and indicate which parameters would need to equal zero for the response variable to be independent of religious affiliation, for each gender.

14.50 Income, gender, and education According to the U.S. Census Bureau, as of March 2009, the average earnings of full-time workers was estimated to be \$31,666 for females with high school education, \$43,493 for males with high school education, \$60,293 for white females with a bachelor's degree, and \$94,206 for males with a bachelor's degree.

a. Identify the response variable and the two factors.

b. Show these means in a two-way classification of the factors.

c. Compare the differences between males and females for (i) high school graduates and (ii) college graduates. If these are close estimates of the population means, explain why there is interaction. Describe its nature.

14.51 Birth weight, age of mother, and smoking A study on the effects of prenatal exposure to smoke (by J. Nigg and N. Breslau, *Journal of the American Academy of Child & Adolescent Psychiatry*, vol. 46, 2009, pp. 362–369) indicated that mean birth weight was significantly lower for babies born to mothers who smoked during pregnancy. It also suggested that increasing age of the mother resulted in increased effects. Explain how this suggests interaction between smoking status and age of mother in their effects on birth weight of the child.

14.52 TV watching by gender and race When we use the 2008 GSS to evaluate how the mean number of hours a day watching TV depends on gender and race, we get the results shown in the ANOVA table that follows.

a. Identify the response variable and the factors.

b. From the table, specify the test statistic and P-value for the test of no interaction. Interpret the result.

c. Is there a significant (i) gender effect and (ii) race effect? Explain.

d. The sample means were 2.82 for white females, 2.68 for white males, 4.52 for black females, and 4.19 for black males. Explain how these results are compatible with the results of the tests discussed in part c.

Analysis of Variance for TVHOURS, using Adjusted SS for Tests

Source	DF	Seq SS	Adj SS	Adj MS	F	P
RACE	1	419.60	399.12	399.12	58.32	0.000
SEX	1	8.74	8.74	8.74	1.28	0.259
RACE*SEX	1	1.42	1.42	1.42	0.21	0.649
Error	1198	8199.10	8199.10	6.84		
Total	1201	8628.86				

S = 2.61610 R-Sq = 4.98% R-Sq(adj) = 4.74%

14.53 Salary and gender The American Association of University Professors (AAUP) reports yearly on faculty salaries for all types of higher education institutions across the United States. Regard *Salary* as the response variable, *Gender* as the explanatory variable, and *Academic Rank* as the control variable. A regression analysis using these data could include an indicator variable for *Gender* and an indicator variable for *Rank*. The estimated coefficients are −13 (thousands of dollars) for *Gender* ($x_1 = 1$ for female and $x_1 = 0$ for male) and −40 (thousands of dollars) for a lesser assistant *Rank* ($x_2 = 1$ for assistant professor and $x_2 = 0$ for professor).

a. Interpret the coefficient for gender.

b. At particular settings of the other predictors, the estimated mean salary for female professors was 96.2 thousand dollars. Using the estimated coefficients, find the estimated means for (i) male professors and (ii) female assistant professors.

14.54 Political ideology by religion and gender The table shown summarizes responses on political ideology (where 1 = extremely liberal and 7 = extremely conservative) in a General Social Survey by religion and gender. The P-value is 0.414 for testing the null hypothesis of no interaction. Explain what this means in the context of this example. (*Hint*: Is the difference between males and females in sample mean ideology about the same for each religion?)

Political ideology by religion and gender

		Political Ideology				
Religion		**Mean**	**Std Dev.**		**Mean**	**Std. Dev.**
Protestant	Female	4.35	1.44	Male	4.50	1.39
Catholic	Female	3.97	1.29	Male	4.04	1.34
Jewish	Female	2.86	1.49	Male	3.50	1.16
None	Female	3.25	1.31	Male	3.60	1.42

Concepts and Applications

14.55 Number of friends and degree Using the GSS Web site sda.berkeley.edu/GSS, analyze whether the number of good friends (the variable NUMFRIEND) depends on the subject's highest degree (the variable DEGREE). To do so, at this web address look under the Analysis tab to pick comparison of means and then enter the variable names and check the ANOVA stats box. Prepare a short report summarizing your analysis and its interpretation.

14.56 Sketch within- and between-groups variability Sketch a dot plot of data for 10 observations in each of three groups such that

a. You believe the P-value would be very small for a one-way ANOVA. (You do not need to do the ANOVA; merely show points for which you think this would happen.)

b. The P-value in one-way ANOVA would not be especially small.

14.57 A = B and B = C, but A ≠ C? In multiple comparisons following a one-way ANOVA with equal sample sizes, the margin of error with a 95% confidence interval for comparing each pair of means equals 10. Give three sample means illustrating that it is possible that Group A is not significantly different from Group B and Group B is not significantly different from Group C, yet Group A is significantly different from Group C.

14.58 Multiple comparison confidence For four groups, explain carefully the difference between a confidence level of 0.95 for a single comparison of two means and a confidence level of 0.95 for a multiple comparison of all six pairs of means.

14.59 Another Simpson paradox The 25 women faculty members in the humanities division of a college have a mean salary of \$65,000, whereas the five women faculty in the science division have a mean salary of \$72,000. The 20 men in the humanities division have a mean salary of \$64,000, and the 30 men in the science division have a mean salary of \$71,000.

a. Construct a 2 × 2 table of sample mean incomes for the table of gender and division of college. Use weighted averages to find the overall means for men and women.

b. Discuss how the results of a one-way comparison of mean incomes by gender would differ from the results of a two-way comparison of mean incomes by gender, controlling for division of college. (*Note*: This reversal of which gender has the higher mean salary, according to whether one controls division of college, illustrates *Simpson's paradox*.)

14.60 Multiple choice: ANOVA/regression similarities Analysis of variance and multiple regression have many similarities. Which of the following is *not* true?

a. The response variable is quantitative for each.

b. They both have *F* tests for testing that the response variable is statistically independent of the explanatory variable(s).

c. For inferential purposes, they both assume that the response variable *y* is normally distributed with the same standard deviation at all combinations of levels of the explanatory variable(s).

d. They are both appropriate when the main focus is on describing straight-line effects of quantitative explanatory variables.

14.61 Multiple choice: ANOVA variability One-way ANOVA provides relatively more evidence that $H_0: \mu_1 = \cdots = \mu_g$ is false:

a. The smaller the between-groups variation and the larger the within-groups variation.

b. The smaller the between-groups variation and the smaller the within-groups variation.

c. The larger the between-groups variation and the smaller the within-groups variation.

d. The larger the between-groups variation and the larger the within-groups variation.

14.62 Multiple choice: Multiple comparisons For four means, it is planned to construct Tukey 95% multiple comparison confidence intervals for the differences between the six pairs.

a. For each confidence interval, there is a 0.95 chance that the interval will contain the true difference.

b. The probability that all six confidence intervals will contain the true differences is 0.70.

c. The probability that all six confidence intervals will contain the true differences is about 0.95.

d. The probability that all six confidence intervals will contain the true differences is $(0.95)^6$.

14.63 Multiple choice: Interaction There is interaction in a two-way ANOVA model when

a. The two factors are associated.

b. Both factors have significant effects in the model without interaction terms.

c. The difference in true means between two categories of one factor varies among the categories of the other factor.

d. The mean square for interaction is about the same size as the mean square error.

14.64 True or false: Interaction For subjects aged under 50, there is little difference in mean annual medical expenses for smokers and nonsmokers. For subjects aged over 50, there is a large difference. Is it true or false that there is interaction between smoking status and age in their effects on annual medical expenses.

14.65 ◆◆ **What causes large or small *F*?** An experiment used four groups of five individuals each. The overall sample mean was 60.

a. What did the sample means look like if the one-way ANOVA for comparing the means had test statistic $F = 0$? (*Hint:* What would have to happen in order for the between-groups variability to be 0?)

b. What did the data look like in each group if F = infinity? (*Hint*: What would have to happen in order for the within-groups variability to be 0?)

14.66 ◆◆ **Between-subjects estimate** This exercise motivates the formula for the between-subjects estimate of the variance in one-way ANOVA. Suppose each population mean equals μ (that is, H_0 is true) and each sample size equals *n*. Then the sampling distribution of each $\bar{y}_i$ has mean μ and variance σ^2/n, and the sample mean of the $\bar{y}_i$ values is the overall sample mean, $\{\bar{y}\}$.

a. Treating the *g* sample means as *g* observations having sample mean $\bar{y}$, explain why $\Sigma(\bar{y}_i - \bar{y})^2/(g - 1)$

estimates the variance σ^2/n of the distribution of the $\{\bar{y}_i\}$ values.

b. Using part a, explain why the between-groups estimate $\Sigma n(\bar{y}_i - \bar{y})^2/(g - 1)$ estimates σ^2. (For the unequal sample size case, the formula replaces n by n_i.)

14.67 Bonferroni multiple comparisons The Bonferroni theorem states that the probability that *at least* one of a set of events occurs can be no greater than the sum of the separate probabilities of the events. For instance, if the probability of an error for each of five separate confidence intervals equals 0.01, then the probability that *at least* one confidence interval will be in error is no greater than $(0.01 + 0.01 + 0.01 + 0.01 + 0.01) = 0.05$.
◆◆

a. Following Example 10, construct a confidence interval for each factor and guarantee that they both hold with overall confidence level at least 95%. [*Hint*: Each interval should use $t_{.0125} = 2.46$.]

b. Exercise 14.8 referred to a study comparing three groups (smoking status never, former, or current) on various personality scales. The study measured 35 personality scales and reported an F test comparing the three smoking groups for each scale. The researchers mentioned doing a Bonferroni correction for the 35 F tests. If the nominal *overall* probability of Type I error was 0.05 for the 35 tests, how small did the P-value have to be for a particular test to be significant? (*Hint*: What should the Type I error probability be for each of 35 tests in order for the overall Type I error probability to be no more than 0.05?)

14.68 Independent confidence intervals You plan to construct a 95% confidence interval in five different situations with independent data sets.
◆◆

a. Assuming that the results of the confidence intervals are statistically independent, find the probability that *all* five confidence intervals will contain the parameters they are designed to estimate. (*Hint*: Use the binomial distribution.)

b. Which confidence level should you use for each confidence interval so that the probability that all five intervals contain the parameters equals *exactly* 0.95?

14.69 Regression or ANOVA? You want to analyze y = house selling price and x = number of bathrooms (1, 2, or 3) by testing whether x and y are independent.
◆◆

a. You could conduct a test of independence using (i) the ANOVA F test for a multiple regression model with two indicator variables or (ii) a regression t test for the coefficient of the number of bathrooms when it is treated as a quantitative predictor in a straight-line regression model. Explain the difference between these two ways of treating the number of bathrooms in the analysis.

b. What do you think are the advantages and disadvantages of the straight-line regression approach to conducting the test?

c. Give an example of three population means for which the straight-line regression model would be less appropriate than the model with indicator variables.

14.70 Three factors An experiment analyzed how the mean corn yield varied according to three factors: nitrogen-based fertilizer, phosphate-based fertilizer, and potash (potassium chloride)-based fertilizer, each applied at low and at high levels.
◆◆

a. How many groups result from the different combinations of the three factors?

b. Defining indicator variables, state the regression model that corresponds to an ANOVA assuming a lack of interaction.

c. Give possible estimates of the model parameters in part b for which the estimated corn yield would be highest at the high level of nitrogen fertilizer and phosphate fertilizer and the low level of potash.

Student Activities

14.71 Student survey data Refer to the student survey data file that your class created with Activity 3 in Chapter 1. For variables chosen by your instructor, use ANOVA methods and related inferential statistical analyses. Interpret and summarize your findings in a short report, and prepare to discuss your findings in class.

15 Nonparametric Statistics

Example 1

How to Get a Better Tan

Picture the Scenario

Statistics students were asked to design an experiment about a topic of choice, conduct the experiment, and then analyze the data. One student, Allison, decided to compare tanning methods without exposure to the sun to avoid skin cancer risk. She investigated two treatments—a bronze tanning lotion applied twice over a two-day period, or a tanning studio where the person is exposed to ultraviolet (UV) light. We'll refer to these treatments as "tanning lotion" and "tanning studio."

The tanning lotion is much less expensive, but Allison predicted that the tanning studio would give a better tan. To investigate this hypothesis, she recruited five untanned female friends to participate in an experiment. Another student in the class used a random number generator to pick three of the friends to use the tanning lotion. The other two friends used the tanning studio. After three days, Allison evaluated the tans produced. She was blinded to the treatment allocation, not knowing which participants used which tanning method. Allison ranked the friends in terms of the quality of their tans. The ranks went from 1 to 5, with 1 = most natural looking and 5 = least natural-looking.

Questions to Explore

- Once Allison ranked the five tanned participants, how could she summarize the evidence in favor of one treatment over the other?
- How can Allison find a P-value to determine if one treatment is significantly better than the other?

Thinking Ahead

You learned in Sections 10.2 and 10.3 how to compare means for two treatments using *t* tests. The tests assume a *normal* distribution for a quantitative response variable. The *t* tests are *robust*, usually working well even when population distributions are *not* normal. An exception is when the distribution is skewed, the sample sizes are small, and the alternative hypothesis is one-sided.

To use the *t* test, suppose Allison created a quantitative variable by assigning a score between 0 and 10 for each girl to describe the quality of tan. With such small sample sizes (only 2 and 3), she would not be able to assess whether quality of tan is approximately normal. Moreover, her prediction that the studio gives a better tan than the lotion was one-sided. In any case, Allison found it easier to rank the participants than to create a quantitative variable. For these reasons, then, it's not appropriate for her to use a *t* test to compare the tanning methods.

We'll now learn about an alternative way to compare treatments without having to assume a normal distribution for the response variable. **Nonparametric statistical methods** provide statistical inference without this assumption. They use solely the *ranking* of the subjects on the response variable.

Nonparametric Statistical Methods

Nonparametric statistical methods are inferential methods that do not assume a particular form of distribution, such as the normal distribution, for the population distribution.

In Example 2, we'll use the best known nonparametric method, the Wilcoxon test, to analyze the ranking of the tans from Allison's experiment.

Nonparametric methods are especially useful in two cases:

- When the data are ranks for the subjects (as in Example 1) rather than quantitative measurements.
- When it's inappropriate to assume normality, and when the ordinary statistical method is not robust to violations of the normality assumption. For instance, we might prefer not to assume normality because we think the distribution will be skewed. Or, perhaps we have no idea about the distribution shape and the sample size is too small to give us much information about it.

Statisticians developed the primary nonparametric statistical methods starting in the late 1940s, long after most other methods described in this text. Since then, many nonparametric methods have been devised to handle a wide variety of scenarios. This final chapter is designed to show you the idea behind nonparametric methods. In Section 15.1, we'll learn about the most popular nonparametric test, the Wilcoxon test for comparing two groups. Section 15.2 briefly describes other popular nonparametric methods.

15.1 Compare Two Groups by Ranking

Of the five participants in Example 1, three were randomly assigned to use the tanning lotion and two to use a tanning studio. The tans were then ranked from 1 to 5, with 1 representing the most natural-looking tan. Let's consider all the possible outcomes for this experiment. Each possible outcome divides the ranks of 1, 2, 3, 4, 5 into two groups—three ranks for the tanning lotion group and two ranks for the tanning studio group. Table 15.1 shows the possible rankings. For instance, in the first case, the three using the tanning lotion got the three best ranks, (1, 2, 3).

Table 15.1 Possible Rankings of Tanning Quality

Each case shows the three ranks for those using the tanning lotion and the two ranks for those using the tanning studio. It also shows the sample mean ranks and their difference.

Treatment			Ranks		
Lotion	(1, 2, 3)	(1, 2, 4)	(1, 2, 5)	(1, 3, 4)	(1, 3, 5)
Studio	(4, 5)	(3, 5)	(3, 4)	(2, 5)	(2, 4)
Lotion mean rank	2	2.3	2.7	2.7	3
Studio mean rank	4.5	4	3.5	3.5	3
Difference of mean ranks	−2.5	−1.7	−0.8	−0.8	0
Lotion	(1, 4, 5)	(2, 3, 4)	(2, 3, 5)	(2, 4, 5)	(3, 4, 5)
Studio	(2, 3)	(1, 5)	(1, 4)	(1, 3)	(1, 2)
Lotion mean rank	3.3	3	3.3	3.7	4
Studio mean rank	2.5	3	2.5	2	1.5
Difference of mean ranks	0.8	0	0.8	1.7	2.5

We summarize the outcome of the experiment by finding the mean rank for each group and taking their difference. Table 15.1 shows the mean ranks and their differences. For example, in the first case the mean is $(1 + 2 + 3)/3 = 2$ for the tanning lotion and $(4 + 5)/2 = 4.5$ for the tanning studio. The difference is $2 - 4.5 = -2.5$. In other words, for the first case the rank for the lotion tanned participants is on average 2.5 less than the rank for the studio tanned participants.

Allison predicted that the tanning studio would tend to give better tans. When this happens in the sample, the ranks for the tanning studio are smaller than those for the tanning lotion. Then the mean rank is larger for the tanning lotion, and the difference between the mean ranks is positive, as in the last case in the table.

Comparing Mean Ranks: The Wilcoxon Test

For this experiment, the samples were independent random samples—the responses for the participants using the tanning lotion were independent of the responses for the participants using the tanning studio. Suppose that the two treatments have identical effects, in the sense that the quality of tan would be the same regardless of which treatment a participant uses. Then, each of the ten possible outcomes shown in Table 15.1 is equally likely. Each has probability 1/10. Using the ten possible outcomes, we can construct a sampling distribution for the difference between the mean ranks. Table 15.2 shows this sampling distribution. For instance, the difference between the sample mean ranks is 0 for two of the ten samples in Table 15.1, so the probability of this outcome is 2/10.

Table 15.2 Sampling Distribution of Difference Between Sample Mean Ranks

These probabilities apply when the treatments have identical effects. For example, only one of the ten possible samples in Table 15.1 has a difference between the mean ranks equal to −2.5, so this value has probability 1/10.

Difference Between Mean Ranks	Probability
−2.5	1/10
−1.7	1/10
−0.8	2/10
0	2/10
0.8	2/10
1.7	1/10
2.5	1/10

Figure 15.1 displays this sampling distribution. It is symmetric around 0. This is the expected value for the difference between the sample mean ranks if the two treatments truly have identical effects.

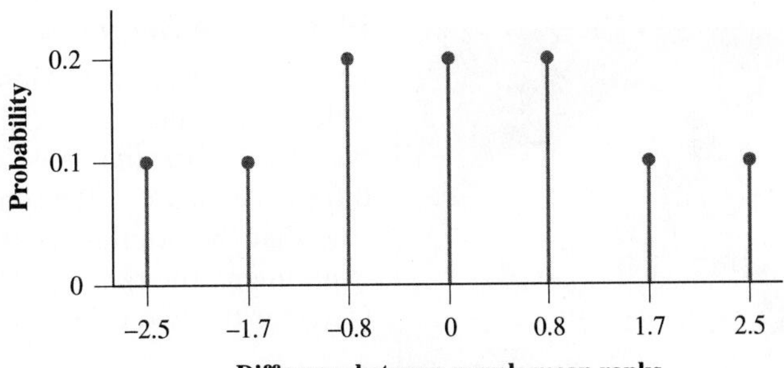

▲ **Figure 15.1 Sampling Distribution of Difference Between Sample Mean Ranks.** This sampling distribution, which is symmetric around 0, applies when the treatments have identical effects. It is used for the significance test of the null hypothesis that the treatments are identical in their tanning quality.

Allison hypothesized that the tanning studio would give a better tan than the tanning lotion. She wanted to test the null hypothesis,

H_0: The treatments are identical in tanning quality,

against the alternative hypothesis

H_a: Better tanning quality results with the tanning studio.

This alternative hypothesis is one-sided. Another way to state H_a is, "The expected value of the sample mean rank is smaller for the tanning studio than for the tanning lotion." On average, we expect the ranks to be smaller (better) for the tanning studio. Thus, we expect the difference between the sample mean rank for the tanning lotion and the sample mean rank for the tanning studio to be positive.

To find the P-value, we presume that H_0 is true. Then, all samples in Table 15.1 are equally likely, and the sampling distribution is the one shown in Table 15.2. The P-value is the probability of a difference between the sample mean rankings like the observed difference or even more extreme, in terms of giving even more evidence in favor of H_a. The test comparing two groups based on the sampling distribution of the difference between the sample mean ranks is called the **Wilcoxon test**. It is named after the chemist-turned-statistician, Frank Wilcoxon, who devised it in 1945.

SUMMARY: Wilcoxon Nonparametric Test for Comparing Two Groups

1. **Assumptions:** Independent random samples from two groups, either from random sampling or a randomized experiment.
2. **Hypotheses:**
   H_0: Identical population distributions for the two groups (this implies equal expected values for the sample mean ranks)
   H_a: Different expected values for the sample mean ranks (two-sided), or
   H_a: Higher expected value for the sample mean rank for a specified group (one-sided)
3. **Test statistic:** Difference between sample mean ranks for the two groups (equivalently, can use sum of ranks for one sample, as discussed after Example 2).
4. **P-value:** One-tail or two-tail probability, depending on H_a, that the difference between the sample mean ranks is as extreme or more extreme than observed.
5. **Conclusion:** Report the P-value and interpret in context. If a decision is needed, reject H_0 if the P-value $\leq$ significance level, such as 0.05.

Example 2

P-value for Wilcoxon test

Tanning Studio Versus Tanning Lotion

Picture the Scenario

Example 1 describes Allison's experiment to determine whether a tanning lotion or a tanning studio produced a better tan. Table 15.1 showed the possible rankings for five tans. Table 15.2 showed the sampling distribution of the difference between the sample mean ranks, presuming the null hypothesis is true that the tanning treatments have identical effects. For Allison's actual experiment, the ranks were (2, 4, 5) for the three using the tanning lotion and (1, 3) for the two using the tanning studio.

Questions to Explore

a. Find and interpret the P-value for comparing the treatments, using the one-sided alternative hypothesis that the tanning studio gives a better tan than the tanning lotion. That is, H_a states that the expected mean rank is larger for the tanning lotion than for the tanning studio, so the difference between the two is positive.

b. What's the smallest possible P-value you could get for this experiment?

Think It Through

a. For the observed sample, the mean ranks are $(2 + 4 + 5)/3 = 3.7$ for the tanning lotion and $(1 + 3)/2 = 2$ for the tanning studio. The test statistic is the difference between the sample mean ranks, $3.7 - 2 = 1.7$. The right tail of the sampling distribution in Figure 15.1 has the large positive differences, for which the ranks tended to be better (lower) with the tanning studio. For the one-sided H_a, the P-value is the probability,

$$\text{P-value} = P(\text{difference between sample mean ranks at least as large as observed}).$$

That is, under the presumption that H_0 is true,

$$\text{P-value} = P(\text{difference between sample mean ranks} \geq 1.7).$$

From Table 15.2 (shown again in the margin) or Figure 15.1 (reproduced below), the probability of a sample mean difference of 1.7 or even larger is $1/10 + 1/10 = 2/10 = 0.20$. This is the P-value. It is not very close to 0. Although there is some evidence that the tanning studio gives a better tan (it *did* have a lower sample mean rank), the evidence is not strong. If the treatments had identical effects, the probability would be 0.20 of getting a sample like we observed or even more extreme.

Recall

Sampling distribution of the difference between mean ranks:

Difference	Probability
−2.5	1/10
−1.7	1/10
−0.8	2/10
0	2/10
0.8	2/10
1.7	1/10
2.5	1/10

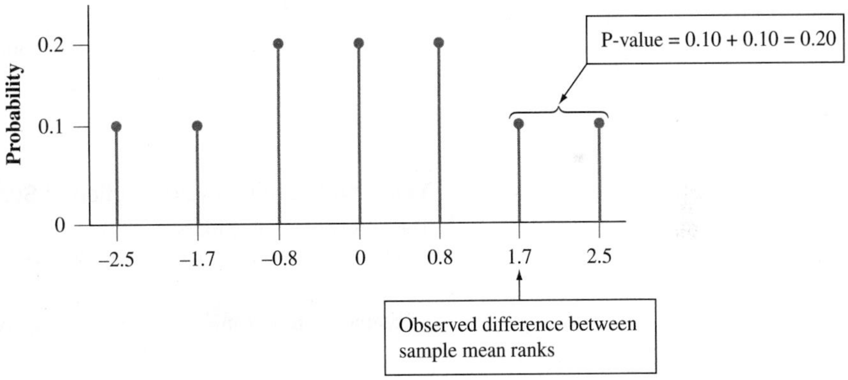

b. In this experiment, suppose the tanning studio gave the two most natural-looking tans. The ranks would then be (1, 2) for the tanning studio and (3, 4, 5) for the tanning lotion. The difference of sample means then equals $4 - 1.5 = 2.5$. It is the most extreme possible sample, and (from Table 15.2 or Figure 15.1) its tail probability is 0.10. This is the smallest possible one-sided P-value.

Insight

With sample sizes of only 2 for one treatment and 3 for the other treatment, it's not possible to get a very small P-value. If Allison wanted to make a decision using a 0.05 significance level, she would never be able to get strong enough evidence to reject the null hypothesis. To get informative results, she'd need to conduct an experiment with larger sample sizes.

Try Exercises 15.1 and 15.2

Suppose Allison instead used a two-sided H_a, namely that there is a treatment difference without specifying which is better. Then the P-value is a two-tail probability. In Example 2, we would get P-value $= 2(0.20) = 0.40$.

The Wilcoxon Rank Sum Statistic The Wilcoxon test can, equivalently, use as the test statistic the sum (or mean) of the ranks in just one of the samples. For example, Table 15.3 shows the possible sum of ranks for the tanning lotion. They have the same probabilities as the differences between the sample mean ranks because the ranks in one sample determine the ranks in the other sample (since the ranks must be the integers 1 through 5) and the mean ranks. For example, the ranks were (2, 4, 5) for the participants using the tanning lotion, for which the sum of ranks is 11. This implies that the ranks were (1, 3) for those using the tanning studio, and it implies that the difference between the mean ranks was $(2 + 4 + 5)/3 - (1 + 3)/2 = 3.7 - 2 = 1.7$. The right-tail probability of the observed rank sum of 11 or a more extreme value is again $1/10 + 1/10 = 0.20$, as seen in the following figure. Some software reports the sum of ranks as the *Wilcoxon rank sum statistic*, sometimes denoted by *W*.

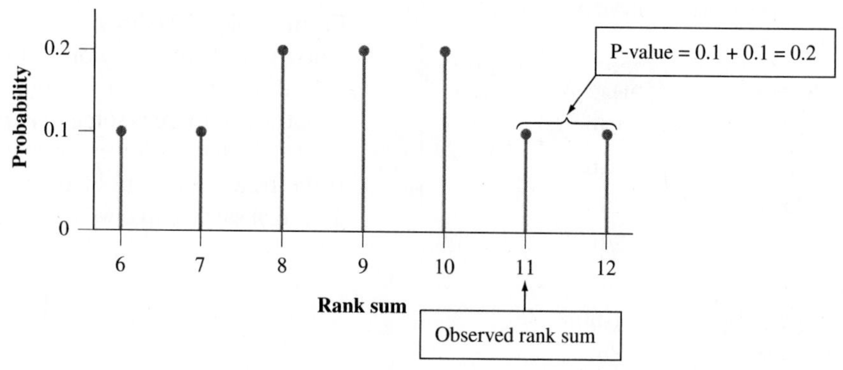

Table 15.3 Sampling Distribution of Sum of Ranks

The observed tanning lotion ranks of (2, 4, 5) have a rank sum of 11. These ranks imply that the tanning studio ranks were (1, 3) and that the difference between the sample mean ranks was 1.7.

Tanning Lotion Ranks	Sum of Tanning Lotion Ranks	Probability
(1, 2, 3)	6	1/10
(1, 2, 4)	7	1/10
(1, 2, 5), (1, 3, 4)	8	2/10
(2, 3, 4), (1, 3, 5)	9	2/10
(2, 3, 5), (1, 4, 5)	10	2/10
(2, 4, 5)	11	1/10
(3, 4, 5)	12	1/10

Large-Sample P-Values Use a Normal Sampling Distribution

With sample sizes of 3 and 2, it was simple to enumerate all the possible outcomes and construct a sampling distribution for the sum of ranks for a sample or for the difference between the mean ranks. With the sample sizes usually used in practice, it is tedious to do this. It's best to use software to get the results. Table 15.4 shows the way SPSS reports results for the Wilcoxon test conducted in Example 2. It shows the P-value of 0.40 for the two-sided alternative hypothesis.

Table 15.4 SPSS Output for Wilcoxon Test with Example 2

The Wilcoxon W is the sum of ranks (1, 3) for the tanning studio. The P-value for the two-sided test is listed opposite "Exact Sig" (where "Sig." is short for significance).

	GROUP	N	Mean Rank	Sum of ranks
	lotion	3	3.67	11.00
	studio	2	2.00	4.00
	Total	5		
Test Statistics				
	Wilcoxon W		4.000	
	Z		−1.155	
	Asymp. Sig. (2-tailed)		.248	
	Exact Sig. [2*(1-tailed Sig.)]		.400	

Two-sided P-value (0.40)

In Words

"Asymp." is short for **asymptotic**, which in statistics means "large-sample" or "as the sample size grows" (mathematically, as it grows toward infinity).

In Example 2 we found the P-value with an exact probability calculation using the actual sampling distribution. With large samples, an alternative approach finds a z test statistic and a P-value based on a normal distribution approximation to the actual sampling distribution. (Some software, such as MINITAB, *only* provides results for the large-sample z test analysis.) For example, in Table 15.4, SPSS reports $z = -1.155$ and Asymp. Sig. (2-tailed) $= 0.248$. These are the results of the large-sample approximate analysis. This analysis is not appropriate for Example 2 because the sample sizes are only 2 and 3 for the two groups.

The large-sample test has a z test statistic because the difference between the sample mean ranks (or the sum of ranks for one sample) has an approximate normal sampling distribution. The z test statistic has the form

$$z = (\text{difference between sample mean ranks})/se.$$

The standard error formula is complex but easily calculated by software. The P-value is then the right-tail, left-tail, or two-tail probability, depending on H_a. *Using the normal distribution for the large-sample test does not mean we are assuming that the response variable has a normal distribution. We are merely using the fact that the sampling distribution for the test statistic is approximately normal.*

Ties Often Occur When We Rank Observations

Often in ranking the observations, some pairs of subjects are tied. They are judged to perform equally well. We then *average the ranks* in assigning them to those subjects. For example, suppose a participant using the tanning studio got the most natural-looking tan (rank 1), two participants using the tanning lotion got the two least natural-looking tans (ranks 4 and 5), but the other participant using the tanning studio and the other participant using the tanning lotion were judged to have equally good tans. Then, those two participants share ranks 2 and 3, and each gets the average rank of 2.5. The ranks are then

Tanning studio: 1, 2.5

Tanning lotion: 2.5, 4, 5

In Practice Software and Implementing the Wilcoxon Test

Software can find tied ranks and then calculate the P-value for the Wilcoxon test. Most software reports only the large-sample approximate P-value when either sample size is larger than some value, typically 20. For example, SPSS reports the small-sample exact P-value only when each sample size is 20 or less.

Using the Wilcoxon Test with a Quantitative Response

In Examples 1 and 2, the response was a rank. When the response variable is quantitative, the Wilcoxon test is applied by converting the observations to ranks. For the combined sample, the observations are ordered from smallest to largest. The smallest observation gets rank 1, the second smallest gets rank 2, and so forth. The test compares the mean ranks for the two samples. Software can implement the test.

Wilcoxon test: finding ranks (large sample)

Example 3

Driving Reaction Times

Picture the Scenario

Example 9 in Chapter 10 discussed an experiment investigating whether or not cell phone use impairs drivers' reaction times. A sample of 64 college students was randomly assigned to a cell phone group or a control group, 32 to each. On a machine that simulated driving situations, participants were instructed to press a "brake button" as soon as possible when they detected a red light. The control group listened to a radio broadcast or to books-on-tape while they performed the simulated driving. The cell phone group carried out a conversation on a cell phone with someone in a separate room.

A subject's reaction time observation is defined to be his or her response time to the red lights (in milliseconds), averaged over all the trials. Figure 15.2 shows box plots of the data for the two groups. Here's some of the data showing the four smallest observations and the four largest observations for each treatment.

Cell phone:	456	468	482	501	....	672	679	688	960
Control:	426	436	444	449	....	626	626	642	648

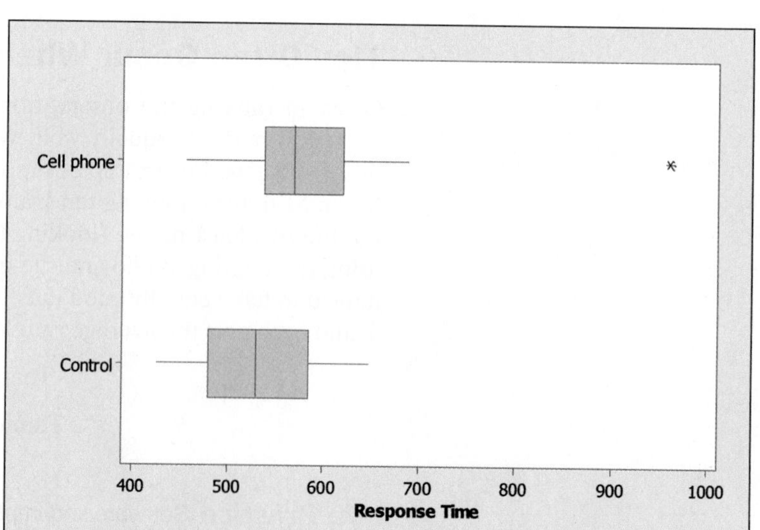

▲ **Figure 15.2 Box Plots of Response Times for Cell Phone Study.** Question Does either box plot show any irregularities that suggest it's safer to use a nonparametric test than a two-sample t test?

The t inferences for comparing the treatment means assume normal population distributions. The box plots do not show any substantial skew, but there is an extreme outlier for the cell phone group. One subject in that group had a very slow mean reaction time, 960 milliseconds.

Questions to Explore

a. Explain how to find the ranks for the Wilcoxon test by showing which of the 64 observations get ranks 1, 2, 63, and 64.

b. Table 15.5 shows the SPSS output for conducting the Wilcoxon test. Report and interpret the mean ranks.

Table 15.5 SPSS Output for Wilcoxon Test with Data from Cell Phone Study

Ranks

	group	N	Mean Rank	Sum of Ranks
TIME	Control	32	27.03	865.00
	Cell phone	32	37.97	1215.00
	Total	64		

Test Statistics

	TIME
Wilcoxon W	865.000
Z	−2.350
Asymp. Sig. (2-tailed)	.019

c. Report the test statistic and the P-value for the two-sided Wilcoxon test. Interpret.

Think It Through

a. Let's look at the smallest and largest observations for each group that were shown above:

Cell phone:	456	468	482	501	....	672	679	688	960
Control:	426	436	444	449	....	626	626	642	648

We give rank 1 to the *smallest* reaction time, so the value 426 gets rank 1. The second smallest observation is 436, which gets rank 2. The largest of the 64 reaction times, which was 960, gets rank 64. The next largest observation, 688, gets rank 63.

b. Table 15.5 reports mean ranks of 27.03 for the control group and 37.97 for the cell phone group. The smaller mean for the control group suggests that that group tends to have smaller ranks, and thus faster reaction times.

c. The z test statistic takes the difference between the sample mean ranks and divides it by a standard error. Table 15.5 reports $z = -2.35$. The P-value of 0.019, reported as "Asymp. Sig. (2-tailed)," is the two-tail probability for the two-sided H_a. It shows strong evidence against the null hypothesis that the distribution of reaction time is identical for the two treatments. Specifically, the sample mean ranks suggest that reaction times tend to be slower for those using cell phones.

Insight

The observation of 960 would get rank 64 if it were *any* number larger than 688 (the second largest value). So, *the Wilcoxon test is not affected by an outlier.* No matter how far the largest observation falls from the next largest, it still gets the same rank. Likewise, no matter how far the smallest observation is below the next smallest, it still gets the rank of 1.

Try Exercise 15.4

Nonparametric Estimation Comparing Two Groups

Throughout this text, we've seen the importance of *estimating* parameters. We learn more from a confidence interval for a parameter than from a significance test about the value of that parameter. Nonparametric estimation methods, like nonparametric tests, do not require the assumption of normal population distributions.

When the response is quantitative, we can compare a measure of center for the two groups. Chapter 10 did this for the mean. Nonparametric methods are often used when the response distribution may be skewed. Then, it can be more informative to summarize each group by the *median.* We then estimate the difference between the population medians for the two groups.

Most software for the Wilcoxon test reports point and interval estimates comparing medians. (Some software, such as MINITAB, refers to the equivalent Mann-Whitney test instead. See Exercise 15.35.) Although this inference does not require a normal population assumption, it *does* require an extra assumption, namely that *the population distributions for the two groups have the same shape.*

Under the extra assumption, here's how software estimates the difference between the population medians. For every possible pair of subjects, one from each group, it takes the difference between the response from the first group and the response from the second group. The point estimate is the median of all those differences. Software also reports a confidence interval for the difference between the population medians. The mechanics of this are beyond the scope of this text.

Example 4

Difference Between Median Reaction Times

Picture the Scenario

Example 3 used the Wilcoxon test to compare reaction time distributions in a simulated driving experiment for subjects using cell phones and for a control group. The MINITAB output in Table 15.6 shows results of comparing the distributions using medians. (It uses the Greek letter name *eta*, which is η, to denote the median.)

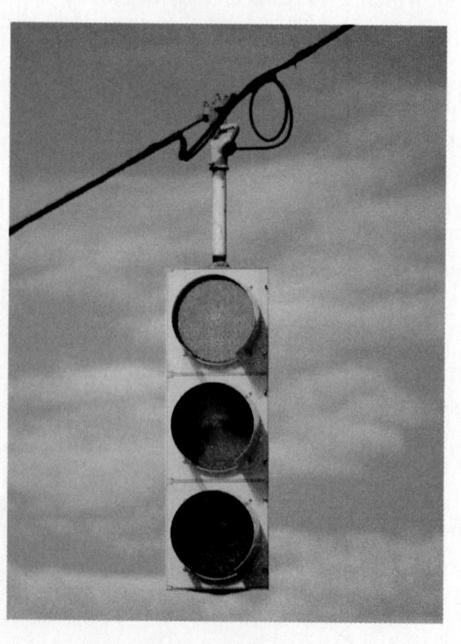

Table 15.6 MINITAB Output for Comparing Medians for Cell Phone Group and Control Group

	N	Median
Cell phone	32	569.00
Control	32	530.00

Point estimate for ETA1-ETA2 is 44.50 ← ETA is MINITAB notation for the median
95.1 Percent CI for ETA1-ETA2 is (8.99, 79.01)
Test of ETA1 = ETA2 vs. ETA1 not = ETA2 is significant at 0.0184 ← P-value

Questions to Explore

a. Report the sample medians and the point estimate of the difference between the population medians.

b. Report and interpret the 95% confidence interval for the difference between the population medians.

Recall

From the discussion before the example, the **point estimate** of the difference between the population medians equals the median of the differences between responses from the two groups. ◀

Think It Through

a. The median reaction times were 569 milliseconds for the cell phone group and 530 milliseconds for the control group. For the cell phone group, for example, half of the reaction times were smaller than 569 milliseconds and half were larger than 569. Table 15.6 reports a point estimate of the difference between the population medians for the two groups of 44.5 milliseconds. (*Note*: This is not the same as the difference between the two sample medians, which is an alternative estimate.)

b. Table 15.6 reports that the 95% confidence interval for the difference between the population medians is (9, 79). Since zero is not contained in the 95% interval, this interval supports that the median reaction times are not the same for the cell phone and control groups. We infer from the interval that the population median reaction time for the cell phone group is between 9 milliseconds and 79 milliseconds larger than for the control group. This inference agrees with the conclusion of the Wilcoxon test that the reaction time distributions differ for the two groups (P-value = 0.02).

Insight

Example 9 in Chapter 10 estimated the difference between the population *means* to be 51.5, with a 95% confidence interval of (12, 91). However, those results were influenced by the outlier of 960 for the cell phone group. When estimation focuses on medians rather than means, outliers do not influence the analysis. *The lack of an influence of outliers is an advantage of the analysis reported here for the medians.*

Try Exercise 15.5

The Proportion of Better Responses for a Group

For comparing two groups, here's another way to summarize the results: Look at all the pairs of observations, such that one observation is from one group and the other observation is from the other group. Then find the proportion of the pairs for which the observation from the first group was better.

We illustrate for the tanning experiment described in Examples 1 and 2. The tanning lotion got ranks (2, 4, 5) and the tanning studio got ranks (1, 3). Figure 15.3 depicts all the pairs of participants (subjects), such that one participant used the tanning lotion and one participant used the tanning studio. The tanning lotion gave a better tan when we pair the participants who got ranks 2 and 3. That is, the participant using the tanning lotion got the better rank, namely, rank 2. However, for all other pairs, the participant who used the tanning studio got the better tan. This is the case when we pair those who got ranks 1 and 2, ranks 1 and 4, ranks 1 and 5, ranks 3 and 4, and ranks 3 and 5. With three participants using the tanning lotion and two participants using the tanning studio, there are $3 \times 2 = 6$ pairs to consider. The tanning studio gave the better tan in five of the six pairs, a sample proportion of 5/6.

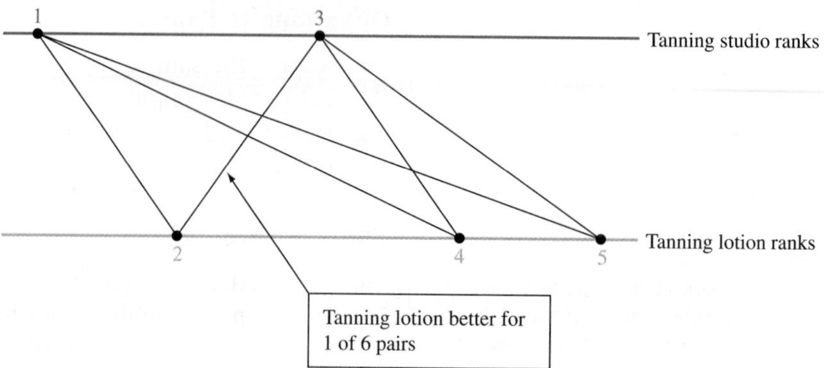

▲ **Figure 15.3 Results of Pairing Participants from Two Treatment Groups.** The tanning lotion got ranks (2, 4, 5) and the tanning studio (with the UV tanning bed) got ranks (1, 3). The tanning studio gave a better tan for five of the six pairs.

This sample proportion is a useful descriptive summary of the effect. "No effect" corresponds to a proportion of 1/2, and greater effects correspond to proportions farther from 1/2. As the proportion gets closer to 0 or to 1, there's a greater difference between the two groups, in the sense that ranks tend to be higher for one group than for the other.

Consider the driving with cell phone experiment analyzed in Examples 3 and 4. For 67% of the pairs the reaction time was greater for the subject using a cell phone, and for 33% of the pairs the reaction time was greater for the subject in the control group. We estimate that the probability is 0.67 of a greater reaction time for the cell phone user.

15.1 Practicing the Basics

15.1 Tanning experiment Suppose the tanning experiment described in Examples 1 and 2 used only four participants, two for each treatment. TRY

a. Show the six possible ways the four ranks could be allocated, two to each treatment, with no ties.

b. For each possible sample, find the mean rank for each treatment and the difference between the mean ranks.

c. Presuming H_0 is true of identical treatment effects, construct the sampling distribution of the difference between the sample mean ranks for the two treatments.

15.2 Test for tanning experiment Refer to the previous exercise. For the actual experiment, suppose the participants using the tanning studio got ranks 1 and 2 and the participants using the tanning lotion got ranks 3 and 4. TRY

a. Find and interpret the P-value for the alternative hypothesis that the tanning studio tends to give better tans than the tanning lotion.

b. Find and interpret the P-value for the alternative hypothesis that the treatments have different effects.

c. Explain why it is a waste of time to conduct this experiment if you plan to use a 0.05 significance level to make a decision.

15.3 Comparing clinical therapies A clinical psychologist wants to choose between two therapies for treating severe mental depression. She selects six patients who are similar in their depressive symptoms and overall quality of health. She randomly selects three patients to receive Therapy 1. The other three receive Therapy 2. After one month of treatment, the improvement in each patient is measured by the change in a score for measuring severity of mental depression—the higher the change score, the better. The improvement scores are

Therapy 1: 25, 40, 45

Therapy 2: 10, 20, 30

a. For the possible samples that could have occurred (with no ties), show the possible ways the six ranks could have been allocated to the two treatments.

b. For each possible allocation of ranks, find the mean rank for each treatment and the difference between the mean ranks.

c. Consider the null hypothesis of identical response distributions for the two treatments. Presuming H_0 is true, construct the sampling distribution of the difference between the sample mean ranks for the two treatments.

d. For the actual data shown above, find and interpret the P-value for the alternative hypothesis that the two treatments have different effects.

15.4 Baby weight and smoking Smoking during pregnancy is a known cause of reduced infant birth weight and other issues that can affect the delivery and mortality. Despite TRY this strong evidence, women who smoke find it difficult to quit even when they are pregnant. Suppose a study is done measuring the birth weight of babies of mothers who smoked during pregnancy and those who did not smoke. The birth weight data in pounds are:

Smokers: 3.1, 4.2, 4.5, 5.0, 6.4, 4.7, 6.0

Nonsmokers: 5.5, 6.5, 7.1, 8.0, 6.8, 7.5, 6.2

a. State the hypotheses for conducting a one-sided (right tailed) Wilcoxon test.

b. Find the ranks for the two groups and their mean ranks.

c. Software reports a small-sample one-sided P-value of 0.022. Interpret.

15.5 Estimating smoking effect Refer to the previous exercise. For these data, MINITAB reports

TRY

```
Point estimate for ETA1-ETA2 is −2.000.
95.9 Percent CI for ETA1-ETA2 is
(−3.300, −0.700).
W = 31.0.
```

Explain how to interpret the reported (a) point estimate and (b) confidence interval.

15.6 Trading volumes In Example 7 in Chapter 8 we compared the number of shares of General Electric stock traded on Mondays and on Fridays during February through April of 2011. The trading volumes (rounded to the nearest million) are as follows:

Mondays: 45, 43, 43, 66, 91, 53, 35, 45, 29, 64, 56

Fridays: 43, 41, 45, 46, 61, 56, 80, 40, 48, 49, 50, 41

Using software,

a. Plot the data. Summarize what the plot shows.

b. State the hypotheses and give the P-value for the Wilcoxon test for comparing the two groups with a two-sided alternative hypothesis.

c. Construct a 95% confidence interval for comparing the population medians. Interpret and explain what (if any) effect the day of the week (Monday versus Friday) has on the median number of shares traded.

d. State the assumptions for the methods in parts b and c.

15.7 Teenage anorexia Example 8 in Section 9.3 and Exercise 10.87 in Chapter 10 described a study that used therapy to treat teenage girls who suffered from anorexia. The girls were randomly assigned to the cognitive behavioral treatment (Group 1) or to the control group (Group 2). The study observed the weight change after a period of time. The output shows results of a nonparametric comparison.

a. Interpret the reported point estimate of the difference between the population medians for the weight changes for the two groups.

b. Interpret the reported confidence interval, and summarize the assumptions on which it is based.

c. Report a P-value for testing the null hypothesis of identical population distributions of weight change. Specify the alternative hypothesis, and interpret the P-value.

MINITAB output comparing weight changes

```
                   N      Median
Cognitive_change   29      1.400
Control_change     26     −0.350
Point estimate for ETA1-ETA2 is 3.05
95.0 Percent CI for ETA1-ETA2 is (−0.60,8.10)
W = 907.0
Test of ETA1 = ETA2 vs ETA1 not = ETA2
is significant at 0.1111
```

15.2 Nonparametric Methods for Several Groups and for Matched Pairs

This final section describes a few other nonparametric methods. We'll first learn about a method that extends the Wilcoxon test to comparisons of mean ranks for *several* groups. We then study two methods for comparing *dependent* samples in the form of *matched pairs* data.

Comparing Mean Ranks of Several Groups: The Kruskal-Wallis Test

Chapter 14 showed that comparisons of means for two groups extend to comparisons for many groups, using the **analysis of variance (ANOVA) *F* test**. Likewise, the Wilcoxon test for comparing mean ranks of two groups extends to a comparison of mean ranks for several groups. This rank test is called the **Kruskal-Wallis test**, named after the statisticians who proposed it in 1952.

As in the Wilcoxon test, the Kruskal-Wallis test assumes that the samples are independent random samples, and the null hypothesis states that the groups have identical population distributions for the response variable. The test determines the ranks for the entire sample and then finds the sample mean rank for each group. The test statistic is based on the between-groups variability in the sample mean ranks. To calculate this test statistic, denote the sample mean rank by $\overline{R}_i$ for group i and by $\overline{R}$ for the combined sample of g groups. The Kruskal-Wallis test statistic is

$$\left(\frac{12}{n(n+1)}\right)\Sigma n_i(\overline{R}_i - \overline{R})^2.$$

The constant $12/(n(n+1))$ is there so that the test statistic values have approximately a chi-squared sampling distribution. Software easily calculates it for us.

The sampling distribution of the test statistic indicates whether the variability among the sample mean ranks is large compared to what's expected under the null hypothesis that the groups have identical population distributions. With g groups, the test statistic has an approximate chi-squared distribution with $g - 1$ degrees of freedom. (Recall that for the chi-squared distribution, the *df* is the mean.) The approximation improves as the sample sizes increase. The larger the differences among the sample mean ranks, the larger the test statistic and the stronger the evidence against H_0. The P-value is the right-tail probability above the observed test statistic value.

When would we use this test? The ANOVA *F* test assumes normal population distributions. The Kruskal-Wallis test does not have this assumption. It's a safer method to use with small samples when not much information is available about the shape of the distributions. It's also useful when the data are merely ranks and we don't have a quantitative measurement of the response variable. Here's a basic summary of the test:

SUMMARY: Kruskal-Wallis Nonparametric Test for Comparing Several Groups

1. **Assumptions:** Independent random samples from several (g) groups, either from random sampling or a randomized experiment
2. **Hypotheses:**
   H_0: Identical population distributions for the g groups
   H_a: Population distributions not all identical
3. **Test statistic:** Uses between-groups variability of sample mean ranks
4. **P-value:** Right-tail probability above observed test statistic value from chi-squared distribution with $df = g - 1$
5. **Conclusion:** Report the P-value and interpret in context.

Example 5

Kruskal-Wallis test

Frequent Dating and College GPA

Picture the Scenario

Tim decided to study whether dating was associated with college GPA. He wondered whether students who date a lot tend to have poorer GPAs. He asked the 17 students in the class to anonymously fill out a short questionnaire in which they were asked to give their college GPA (0 to 4 scale) and to indicate whether during their college careers they had dated regularly, occasionally, or rarely.

Figure 15.4 shows the dot plots that Tim constructed of the GPA data for the three dating groups. Since the dot plots showed evidence of severe skew to the left and since the sample size was small in each group, he felt safer analyzing the data with the Kruskal-Wallis test than with the ordinary ANOVA F test.

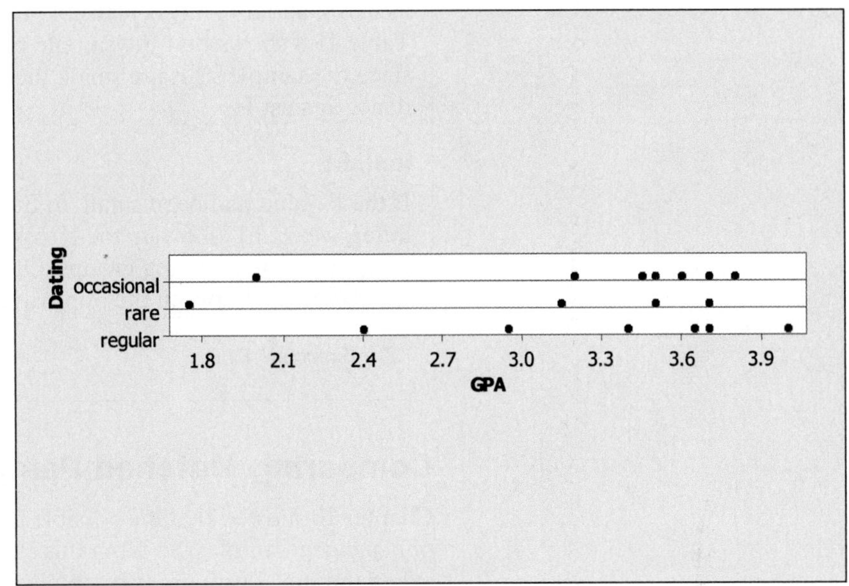

▲ **Figure 15.4 Dot Plots of GPA by Dating Group.** Question Why might we be nervous about using the ordinary ANOVA F test to compare mean GPA for the three dating groups?

Finding ranks for data in Table 15.7

GPA	Rank	
1.75	1	
2.00	2	
2.40	3	
2.95	4	
3.15	5	
3.20	6	
3.40	7	
3.44	8	
3.50	9.5	← average of 9 and 10
3.50	9.5	
3.60	11	
3.67	12	
3.68	13	
3.70	14	
3.71	15	
3.80	16	
4.00	17	

Table 15.7 shows the data, with the college GPA values ordered from smallest to largest for each dating group. The table in the margin shows the combined sample of 17 observations from the three groups and their ranks. Table 15.7 also shows these ranks as well as the mean rank for each group.

Table 15.7 College GPA by Dating Group

Dating Group	GPA Observations	Ranks	Mean Rank
Rare	1.75, 3.15, 3.50, 3.68	1, 5, 9.5, 13	7.1
Occasional	2.00, 3.20, 3.44, 3.50, 3.60, 3.71, 3.80	2, 6, 8, 9.5, 11, 15, 16	9.6
Regular	2.40, 2.95, 3.40, 3.67, 3.70, 4.00	3, 4, 7, 12, 14, 17	9.5

Question to Explore

Table 15.8 shows MINITAB output for the Kruskal-Wallis test. MINITAB denotes the chi-squared test statistic by H. Interpret these results.

Table 15.8 Results of Kruskal-Wallis Test for Data in Table 15.7

Kruskal-Wallis Test: GPA versus Dating

```
Dating           N    Median  AVE. Rank
rare             4    3.325     7.1
occasional       7    3.500     9.6
regular          6    3.535     9.5
H = 0.72    DF = 2   P = 0.696 (adjusted for ties)
```

Think It Through

If H_0: identical population distributions for the three groups were true, the Kruskal-Wallis test statistic would have an approximate chi-squared distribution with $df = 2$. Table 15.8 reports that the test statistic is H = 0.72. The P-value is the right-tail probability above 0.72. Table 15.8 reports this as 0.696, about 0.7. It is plausible that GPA is independent of dating group. Table 15.8 shows that the sample median GPAs are not very different, and since the sample sizes are small, these sample medians do not give much evidence against H_0.

Insight

If the P-value had been small, to find out which pairs of groups significantly differ, we could follow up the Kruskal-Wallis test by a Wilcoxon test to compare each pair of dating groups. Or, we could find a confidence interval for the difference between the population medians for each pair.

Try Exercise 15.8

Comparing Matched Pairs: The Sign Test

Chapter 10 showed that it's possible to compare groups using either *independent* or *dependent* samples. So far in this chapter, the samples have been independent. When the subjects in the two samples are matched, such as when each treatment in an experiment uses the same subjects, the samples are dependent. Then we must use different methods.

For example, the tanning experiment from Examples 1 and 2 could have used a crossover design instead: The participants get a tan using one treatment, and after it wears off they get a tan using the other treatment. The order of using the two treatments is random. For each participant, we observe which treatment gives the better tan. That is, we make comparisons by pairing the two observations for the same participant.

For such a matched pairs experiment, let p denote the population proportion of cases for which a particular treatment does better than the other treatment. Under the null hypothesis of identical treatment effects, $p = 0.50$. That is, each treatment should have the better response outcome about half the time. (We ignore those cases in which each treatment gives the *same* response.) Let n denote the sample number of pairs of observations for which the two responses differ. For large n, we can use the z test statistic to compare the sample proportion $\hat{p}$ to the null hypothesis value of 0.50. (See the margin Recall box.) The P-value is based on the approximate standard normal sampling distribution.

A test that compares matched pairs in this way is called a **sign test**. The name refers to how the method evaluates for each matched pair whether the difference between the first and second response is *positive* or *negative.*

Recall

From Section 9.2, to test H_0: $p = 0.50$ with sample proportion $\hat{p}$ when $n \geq 30$, the test statistic is

$$z = (\hat{p} - 0.50)/se,$$

with $se = \sqrt{(0.50)(0.50)/n}$. ◀

SUMMARY: Sign Test for Matched Pairs

1. **Assumptions:** Random sample of matched pairs for which we can evaluate which observation in a pair has the better response.
2. **Hypotheses:** H_0: Population proportion $p = 0.50$ who have better response for a particular group
   H_a: $p \neq 0.50$ (two-sided) or H_a: $p > 0.50$ or H_a: $p < 0.50$ (one-sided)
3. **Test statistic:** $z = (\hat{p} - 0.50)/se$, as shown in margin recall box.
4. **P-value:** For large samples ($n \geq 30$), use tail probabilities from standard normal. For smaller n, use binomial distribution (discussed in Example 7).
5. **Conclusion:** Report the P-value and interpret in context.

Sign test for matched pairs

Example 6

Time Browsing the Internet or Watching TV

Picture the Scenario

Which do most students spend more time doing—browsing the Internet or watching TV? Let's consider the students surveyed at the University of Georgia whose responses are in the Georgia Student Survey data file. The results for the first three students in the data file (in minutes per day) were

Student	Internet	TV
1	60	120
2	20	120
3	60	90

All three spent more time watching TV. For the entire sample, 35 students spent more time watching TV and 19 students spent more time browsing the Internet. (The analysis ignores the 5 students who reported the same time for each.)

Question to Explore

Let p denote the population proportion who spent more time watching TV. Find the test statistic and P-value for the sign test of H_0: $p = 0.50$ against H_a: $p \neq 0.50$. Interpret.

Think It Through

Here, $n = 35 + 19 = 54$. The sample proportion who spent more time watching TV was $35/54 = 0.648$. For testing that $p = 0.50$, the *se* of the sample proportion is

$$se = \sqrt{(0.50)(0.50)/n} = \sqrt{(0.50)(0.50)/54} = 0.068.$$

The test statistic is

$$z = (\hat{p} - 0.50)/se = (0.648 - 0.50)/0.068 = 2.18.$$

From the normal distribution table (Table A or software), the two-sided P-value is 0.03. This provides considerable evidence that most students spend more time watching TV than browsing the Internet. The conclusion must be tempered by the fact that the data resulted from a convenience sample (students in a class for a statistics course) rather than a random sample of all college students.

Insight

The sign test uses merely the information about *which* response is higher and *how many* responses are higher, not the quantitative information about *how much* higher. This is a disadvantage compared to the corresponding parametric test, the matched pairs t test of Section 10.4, which analyzes the mean of the differences between the two responses. The sign test is most appropriate when we can order the responses but do not have quantitative information, such as in the next example.

Try Exercise 15.10

For small n, we can conduct the sign test by using the *binomial* distribution. The next example illustrates this case as well as a situation in which we can order responses for each pair but do not have quantitative information about how different the responses are.

Sign test for matched pairs (small sample)

Example 7

Crossover Experiment Comparing Tanning Methods

Picture the Scenario

When Allison told another student in the class (Megan) about her planned experiment to compare tanning methods, Megan decided to do a separate tanning experiment. She used a crossover design for a different sample of five untanned female friends. The results of her experiment were that the tanning studio gave a better tan than the tanning lotion for four of the five participants.

Question to Explore

Find and interpret the P-value for testing that the population proportion p of participants for whom the tanning studio gives a better tan than the tanning lotion equals 0.50. Use the alternative hypothesis that this population proportion is larger than 0.50, because Megan predicted that the tanning studio would give better tans.

Recall

Section 6.3 presented the **binomial** distribution. With probability p of success on a trial, out of n independent trials the probability of x successes is

$$P(x) = \frac{n!}{x!(n-x)!}p^x(1-p)^{n-x}.$$ ◀

Think It Through

The null hypothesis is H_0: $p = 0.50$. For H_a: $p > 0.50$, the P-value is the probability of the observed sample outcome or an even larger one. The sample size ($n = 5$) was small, so we use the binomial distribution rather than its normal approximation to find the P-value.

If $p = 0.50$, from the margin Recall box the binomial probability that $x = 4$ of the $n = 5$ participants would get better tans with the tanning studio is

$$P(4) = \frac{5!}{4!(5-4)!}(0.50)^4(0.50)^1 = 0.156.$$

The more extreme result that all five participants would get better tans with the tanning studio has probability $P(5) = (0.50)^5 = 0.031$. The P-value is the right-tail probability of the observed result and the more extreme one, that is, $0.156 + 0.031 = 0.187$. See the margin figure. In summary, the evidence is not strong that more participants get a better tan from the tanning studio than the tanning lotion.

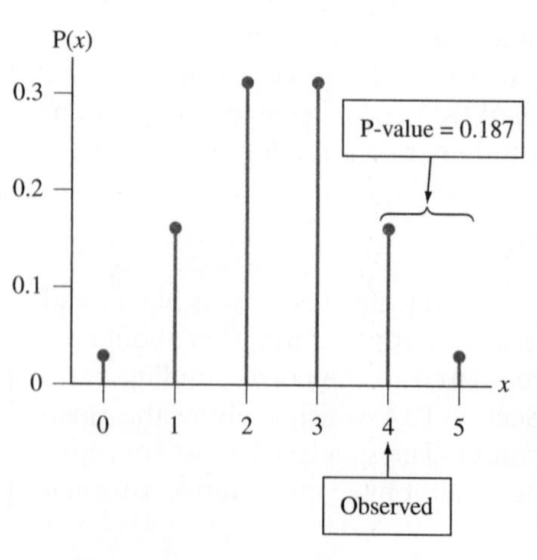

Insight

Megan would instead use the two-sided alternative, H_a: $p \neq 0.50$, if she did not make a prior prediction about which tanning method would be better. The P-value would then be $2(0.19) = 0.38$. With only $n = 5$ observations, the smallest possible two-sided P-value would be $2(0.031) = 0.06$, which occurs when $x = 0$ or when $x = 5$.

Try Exercise 15.11

Ranking Matched Pairs: The Wilcoxon Signed-Ranks Test

With matched pairs data, for each pair the sign test merely observes which treatment does better, but not *how much* better. The **Wilcoxon signed-ranks test** is a nonparametric test designed for cases in which the comparisons of the paired observations can themselves be ranked. For each matched pair of responses, it measures the difference between the responses. It tests the hypothesis,

H_0: population median of difference scores is 0.

The test uses the quantitative information provided by the n difference scores by ranking their magnitudes, in absolute value. (Like the sign test, it ignores observation pairs for which the difference equals 0.) The test statistic is the sum of the ranks for the differences that are positive.

SUMMARY: Wilcoxon Signed-Ranks Test for Matched Pairs

1. **Assumptions:** Random sample of matched pairs for which the differences of observations have a symmetric population distribution and can be ranked.
2. **Hypotheses:**
   H_0: Population median of difference scores is 0.
   H_a: Population median of difference scores is not 0 (one-sided also possible).
3. **Test statistic:** Rank the absolute values of the difference scores for the matched pairs and then find the sum of ranks of the differences that were positive.
4. **P-value:** Software can find a P-value based on all the possible samples with the given absolute differences. (For large samples, it uses an approximate normal sampling distribution, as discussed following.)
5. **Conclusion:** Report the P-value and interpret in context.

Example 8

Wilcoxon signed-rank test

GRE Test Scores

Picture the Scenario

If you want to attend graduate school, taking the Graduate Record Examination (GRE) is usually a requirement. Many graduate schools consider GRE scores for admittance, to qualify for financial aid, to determine fellowships and grants, and for other program research or teaching assignments. The GRE includes three sections designed to test verbal, quantitative, and writing skills.

The verbal and quantitative sections are each scored between 200 and 800. The analytical writing portion of the GRE is given a score between 0 and 6 in half point increments.

In our example, three students volunteered for a study to determine if taking a two-day workshop on GRE preparation improved their GRE analytical writing score from a previous score. Note: The original data was larger ($n = 12$) but a small sample is used in this example to make it easier to explain. The results are shown in the following table:

	Subject		
	1	2	3
Before	2.5	4	1.5
After	3	3.5	3

Recall

How to handle ties is discussed on page 727. ◀

A negative difference (after − before) represents a lower score. A positive difference represents an improved score. Let's test the null hypothesis that the two-day GRE workshop has no effect, in the sense that the population median gained score is 0, against the alternative hypothesis that the population median gained score is positive.

Observations that have a tie score use a decimal point ranking and are ranked the same. The next ranking includes all of the previous rankings.

Questions to Explore

a. For ranks applied to the absolute values of the differences, find the rank sum for the differences that were positive.

b. Consider each of the possible ways that positive and negative signs could be assigned to these three differences. For each case, find the rank sum for the positive differences. Create the sampling distribution of this rank sum that applies if the workshop truly has no effect.

c. Find the P-value for the Wilcoxon signed-ranks test, using the sampling distribution of the rank sum created in part b.

Think It Through

a. The Wilcoxon test begins by calculating the difference and then the absolute value, of each instance. In most applications of the Wilcoxon procedure, the cases in which there is zero difference are eliminated from consideration because they provide no useful information; the remaining absolute differences are then ranked from lowest to highest, with tied ranks included where appropriate.

After sorting the GRE data, the differences have absolute values and ranks as follows:

Subject	Before	After	Difference	Absolute Value	Rank of Absolute Value	Signed Rank
1	2.5	3	0.5	0.5	1.5	1.5
2	4	3.5	−0.5	0.5	1.5	−1.5
3	1.5	3	1.5	1.5	3	3

b. For the difference values of the three subjects, 0.5, − 0.5, and 1.5, Table 15.9 shows all the possible ways the differences could have been positive or negative. For each sample, this table also shows the sum of ranks for the positive differences. The observed data are Sample 1, which had a rank sum of 4.5.

Table 15.9 Possible Samples with Absolute Difference Values of Sample

Subject	1	2	3	4	5	6	7	8	Rank of Absolute Value
1	0.5	0.5	−0.5	0.5	−0.5	0.5	−0.5	−0.5	1.5
2	−0.5	−0.5	−0.5	0.5	−0.5	0.5	0.5	0.5	1.5
3	1.5	−1.5	1.5	1.5	−1.5	−1.5	1.5	−1.5	3
	Sum of Ranks for Positive Differences								
	4.5	1.5	3	6	0	3	4.5	1.5	

If the workshop has no effect, then the eight possible samples in Table 15.9 are equally likely. The table that follows summarizes the sampling distribution of the rank sum for the positive differences, presuming no effect of the workshop. For example, the rank sum was 4.5 for two of the eight samples, so its probability is 2/8.

Sampling Distribution of Rank Sum for the Positive Differences

Rank Sum	Probability
0	1/8
1.5	2/8
3	2/8
4.5	2/8
6	1/8

c. The larger the sum of ranks for the positive differences, the greater the evidence that the workshop has a positive effect. So, the P-value is the probability that this sum of ranks is at least as large as observed. Since three of the eight possible samples had a rank sum for the positive differences of at least 4.5 (the observed value), the P-value is 3/8 = 0.375.

Insight

Suppose we instead used the sign test. We then observe that two of the three differences are positive. For the alternative hypothesis that the workshop has a positive effect, the P-value is the probability that at least two of the three differences are positive, when the chance is 0.50 that any particular difference is positive. Using the binomial distribution, you can find that the P-value is 0.50 (Exercise 15.12).

The sign test ignores the fact that the two positive differences are larger than the negative difference. The Wilcoxon signed-ranks test uses this information. By taking this extra information into account, its P-value of 0.375 is smaller than the P-value of 0.50 from the sign test. However, the P-value is still not small. With only three observations, the one-sided P-value can be no smaller than one-eighth, which is the P-value for the largest possible value (which is 6) for the rank sum of positive differences.

The MINITAB output for the complete data set is shown below. The test of medians between the two groups results in a statistically significant P-value of 0.018.

Wilcoxon Signed Rank CI: Before, After

	N	Estimated Median	Achieved Confidence	Confidence Interval Lower	Confidence Interval Upper
Before	12	3.00	94.5	2.25	3.75
After	12	4.00	94.5	3.00	4.75

Wilcoxon Signed Rank Test: Diff

Test of median = 0.000000 versus median not = 0.000000

	N	Test	Statistic	P	Median
Diff	12	11	60.0	0.018	1.000

Try Exercise 15.12

Caution

For independent samples, rank all of the data combined together, and then compare the sum of the ranks of the two groups. For dependent samples, find the difference between the two groups and then rank the differences. These methods often are confused—especially if the independent samples have equal sample sizes. ◀

Although the Wilcoxon signed-ranks test has the advantage compared to the sign test that it can take into account the *sizes* of the differences and not merely their *sign*, it also has a disadvantage. For the possible samples (such as the eight samples shown in Table 15.9) to be equally likely, it must make an additional assumption: The population distribution of the difference scores must be *symmetric*.

The symmetry assumption for the Wilcoxon signed-ranks test is a bit weaker than the assumption of a normal population distribution that the matched pairs t test of Section 10.4 makes. However, with small samples, there's not much evidence to check this symmetry assumption. Also, recall that the t test is *robust* for violations of the assumption of normality, especially for two-sided inference. The extra assumption of symmetry for the Wilcoxon signed-ranks test weakens the advantage of using a nonparametric method. Because of this, many statisticians prefer to use the matched pairs t test for such data.

Like the Wilcoxon test for comparing mean ranks for two independent samples, the Wilcoxon signed-ranks test can allow for ties and it has a large-sample z test statistic that has an approximate standard normal sampling distribution. For instance, Example 6 used the sign test to compare times spent on Internet browsing and watching TV for 54 students. If we create difference scores

$$\text{Difference} = \text{TV watching time} - \text{Internet browsing time}$$

for the students and analyze them with the Wilcoxon signed-ranks test, MINITAB reports the results shown in Table 15.10.

Table 15.10 Results of Wilcoxon Signed-Ranks Test for Time Differences Between TV Watching and Internet Browsing (from Example 6)

Wilcoxon Signed Ranks Test: Difference

Test of median = 0.000000 versus median not = 0.000000

	N	N for Test	Wilcoxon Statistic	P	Estimated Median
difference	59	54	1012.5	0.020	22.50

The Wilcoxon test statistic is the sum of ranks for the positive differences. The two-sided P-value of 0.020 provides strong evidence that the population median of the differences is not 0. There were 59 subjects, but only 54 were used in the test, because five difference scores were equal to 0. MINITAB also estimates that the population median difference between the time watching TV and the time browsing the Internet was 22.5 minutes. Separately, it provides a 95% confidence interval for the population median of the differences of 2.5 to 40.0 minutes.

Advantages and Limitations of Nonparametric Statistics

Nonparametric methods make weaker assumptions than the parametric methods we've studied in the rest of the text. For comparing two groups, for instance, it's not necessary to assume that the response distribution is normal. This is especially appealing for small samples because then the normality assumption is more important than it is for larger samples, for which the central limit theorem applies.

Statisticians have shown that nonparametric tests are often very nearly as good as parametric tests even in the exact case for which the parametric tests are designed. To illustrate, suppose two population distributions are normal in shape and have the same standard deviation but different means. Then, the Wilcoxon test is very nearly as powerful in detecting this difference as the t test, even though it uses only the ranks of the observations.

Nonetheless, nonparametric methods have some disadvantages. For example, confidence interval methods have not been as thoroughly developed as significance tests. Also, methods have not yet been well developed for multivariate procedures, such as multiple regression. Finally, recall that in many cases (especially two-sided tests and confidence intervals) most parametric methods are *robust*, working well even if assumptions are somewhat violated.

15.2 Practicing the Basics

15.8 How long do you tolerate being put on hold? Examples 1–4 and 7 in Chapter 14 referred to the following randomized experiment: An airline analyzed whether telephone callers to their reservations office would remain on hold longer, on average, if they heard (a) an advertisement about the airline, (b) Muzak, or (c) classical music. For 15 callers randomly assigned to these three conditions, the table shows the data. It also shows the ranks for the 15 observations, as well as the mean rank for each group and some results from using MINITAB to conduct the Kruskal-Wallis test.

TRY

a. State the null and alternative hypotheses for the Kruskal-Wallis test.

b. Identify the value of the test statistic for the Kruskal-Wallis test, and state its approximate sampling distribution, presuming H_0 is true.

c. Report and interpret the P-value shown for the Kruskal-Wallis test.

d. To find out which pairs of groups significantly differ, how could you follow up the Kruskal-Wallis test?

Telephone holding times by type of recorded message

Recorded Message	Holding Time Observations	Ranks	Mean Rank
Muzak	0, 1, 3, 4, 6	1, 2.5, 5, 6, 8	4.5
Advertisement	1, 2, 5, 8, 11	2.5, 4, 7, 10.5, 13	7.4
Classical	7, 8, 9, 13, 15	9, 10.5, 12, 14, 15	12.1

Kruskal-Wallis Test: Holding Time Versus Group

```
Group       N       Median      Ave Rank
Muzak       5       3.000         4.5
advert      5       5.000         7.4
classical   5       9.000        12.1
H = 7.38 DF = 2 P = 0.025 (adjusted for ties)
```

15.9 What's the best way to learn French? Exercise 14.3 gave the data in the table for scores on the first quiz for ninth-grade students in an introductory-level French course. The instructor grouped the students in the course as follows:

Group 1: Never studied foreign language before, but have good English skills

Group 2: Never studied foreign language before; have poor English skills

Group 3: Studied at least one other foreign language

The table also shows results of using MINITAB to perform the Kruskal-Wallis test.

a. Find the rank associated with each observation, and show how to find the mean rank for Group 1.

b. Report and interpret the P-value for the test.

Scores on the quiz

Group 1	Group 2	Group 3
4	1	9
6	5	10
8		5

```
Kruskal-Wallis Test on response
 Group    N   Median    Ave Rank
   1      3   6.000       4.3
   2      2   3.000       2.3
   3      3   9.000       6.2
H = 3.13 DF = 2 P = 0.209 (adjusted for ties)
```

15.10 Sports versus TV Which do students spend more time doing—playing sports or watching TV? For the students surveyed at the University of Florida whose responses are in the FL Student Survey data file, 24 students spent more time playing sports and 30 students spent more time watching TV. Let p denote the corresponding population proportion who spend more time watching TV.

TRY

a. Find the test statistic for the sign test of H_0: $p = 0.50$ against H_a: $p \neq 0.50$.

b. Refer to part a. Find and interpret the P-value.

15.11 Cell phones and reaction times Example 12 in Chapter 10 compared reaction times in a simulated driving test for the same students when they were using a cell phone and when they were not. The table shows data for the first four students. For all 32 students, 26 had faster reaction times when not using the cell phone and 6 had faster reaction times when using it.

TRY

a. Are the observations for the two treatments independent samples, or dependent samples? Explain.

b. Let p denote the population proportion who would have a faster reaction time when not using a cell phone. Estimate p based on this experiment.

c. Using all 32 observations, find the test statistic and the P-value for the sign test of H_0: $p = 1/2$ against H_a: $p > 1/2$. Interpret.

d. What is the parametric method for comparing the scores? What is an advantage of it over the sign test? (*Hint:* Does the sign test use the magnitude of the difference between the two scores, or just its direction?)

Reaction times in cell phone study

Student	Using Cell Phone? No	Yes	Difference
1	604	636	32
2	556	623	67
3	540	615	75
4	522	672	150

15.12 Sign test for GRE scores Consider Example 8, for which the changes in the writing portion GRE scores for the first three people who attended a training workshop were 0.5, −0.5, and 1.5. Show how to use the sign test to test that the probability that the difference is positive equals 0.50 against the alternative hypothesis that it is greater than 0.50.

TRY

15.13 Does exercise help blood pressure? Exercise 10.47 in Chapter 10 discussed a pilot study of people who suffer from abnormally high blood pressure. A medical researcher decides to test her belief that walking briskly for at least half an hour a day has the effect of lowering blood pressure. She randomly samples three of her patients who have high blood pressure. She measures their systolic blood pressure initially and then again a month later after they participate in her exercise program. The table shows the results. Show how to analyze the data with the sign test. State the hypotheses, find the P-value, and interpret.

Subject	Before	After
1	150	130
2	165	140
3	135	120

15.14 More on blood pressure Refer to the previous exercise. The analysis there did not take into account the *size* of the change in blood pressure. Show how to do this with the Wilcoxon signed-ranks test.

a. State the hypotheses for that test, for the relevant one-sided alternative hypothesis.

b. Construct the sampling distribution for the rank sum of the positive differences when you consider the possible samples that have absolute differences in blood pressure of 20, 25, and 15.

c. Using the sampling distribution from part b, find and interpret the P-value. (When every difference is positive, or when every difference is negative, this test and the sign test give the same P-value, for a given alternative hypothesis.)

15.15 More on cell phones Refer to Exercise 15.11. That analysis did not take into account the magnitudes of the differences in reaction times. Show how to do this with the Wilcoxon signed-ranks test, illustrating by using only the four observations shown in the table there.

a. State the hypotheses for the relevant one-sided test.

b. Create the sampling distribution of the sum of ranks for the positive differences.

c. Find the P-value, and interpret.

15.16 Use all data on cell phones Refer to the previous exercise. When we use the data for all 32 subjects, MINITAB reports results in the following for the Wilcoxon signed-ranks test.

Wilcoxon signed-ranks test results

```
Test of median = 0.000000 versus median
not = 0.000000
              N for  Wilcoxon              Estimated
        N     Test   Statistic      P         Median
diff   32      32       490.0    0.000         47.25
```

a. State the null and alternative hypotheses for this test.

b. Explain how MINITAB found the value reported for the Wilcoxon test statistic.

c. Report and interpret the P-value.

d. Report and interpret the estimated median.

Chapter Review

ANSWERS TO THE CHAPTER FIGURE QUESTIONS

Figure 15.2 Yes, there is an extreme outlier for the cell phone group. This may have a large effect on the mean for that group but will not affect the nonparametric test.

Figure 15.4 The dot plots show evidence of severe skew to the left. Because the sample sizes are very small, severe violation of the normality assumption could invalidate the ANOVA *F* test.

CHAPTER SUMMARY

Nonparametric statistical methods provide statistical inference without an assumption of normality about the probability distribution of the response variable.

- The **Wilcoxon test** is a nonparametric test for the null hypothesis of identical population distributions for two groups. It assumes independent random samples and uses the ranks for the combined sample of observations. The P-value is a one- or two-tail probability for the sampling distribution of the difference between the sample mean ranks.
- Nonparametric methods can also estimate differences between groups, for example to construct a confidence interval for the difference between two population **medians**.
- Other popular nonparametric methods include
  - The **Kruskal-Wallis test** for comparing mean ranks of *several* groups with independent random samples.
  - The **sign test** for comparing two groups with *matched pairs* data (dependent samples rather than independent samples) in terms of the proportion of times the response is better for a particular group.
  - The **Wilcoxon signed-ranks test** for comparing two groups with *matched pairs data* when the differences can be ranked.

CHAPTER PROBLEMS

Practicing the Basics

15.17 Car bumper damage An automobile company compares two types of front bumper for their new model by driving sample cars into a concrete wall at 15 miles per hour. The response is the amount of damage to the car, as measured by the repair costs, in dollars. Due to the costs, the study uses only six cars, obtaining results for three bumpers of each type. The results are in the table.

a. Find the ranks and the mean rank for each bumper type.

b. Show that there are 20 possible allocations of ranks to the two bumper types.

c. Explain why the observed ranks for the two groups are one of the two most extreme ways the two groups can differ, for the 20 possible allocations of the ranks.

d. Explain why the P-value for the two-sided test equals 0.10.

Bumper A	Bumper B
2100	1100
2500	1300
2300	1400

15.18 Comparing more bumpers Refer to the previous exercise.

a. Would the results of your analysis change if for the second Bumper A the repair cost was $9000 instead of $2500? What does this illustrate about the analysis?

b. Suppose the company actually wanted to compare *three* bumper types. Which significance test could they use to do this if they did not want to assume that the repair costs have a normal distribution for each bumper type?

15.19 Telephone holding times In Exercise 15.8, the telephone holding times for Muzak and classical music were

Muzak 0, 1, 4, 6, 3
Classical 13, 9, 8, 15, 7

a. For comparing these two groups with the Wilcoxon test, report the ranks and the mean rank for each group.

b. Two groups of size 5 each have 252 possible allocations of rankings. For a two-sided test of H_0: identical distributions with these data, explain why the P-value is $2/252 = 0.008$. Interpret the P-value.

15.20 Treating alcoholics The nonparametric statistics textbook by Hollander and Wolfe (1999) discussed a study on a social skills training program for alcoholics. A sample of male alcoholics was randomly split into two groups. The control group received traditional treatment. The treatment group received the traditional treatment plus a class in social skills training. Every two weeks for a year after the treatment, subjects indicated the quantity of alcohol they consumed during that period. The summary response was the total alcohol intake for the year (in centiliters).

a. Suppose the researchers planned to conduct a one-sided test and believed that the response variable could be highly skewed to the right. Why might they prefer to use the Wilcoxon test rather than a two-sample *t* test to compare the groups?

b. The data were

Controls: 1042, 1617, 1180, 973, 1552, 1251, 1151, 1511, 728, 1079, 951, 1319

Treated: 874, 389, 612, 798, 1152, 893, 541, 741, 1064, 862, 213

Show how to find the ranks and the mean ranks for the two groups.

c. MINITAB reports the results shown in the output below. Report and interpret the P-value.

d. Report and interpret the confidence interval shown in the output.

MINITAB output

```
            N            Median
control     12           1165.5
treated     11            798.0
Point estimate for ETA1-ETA2 is 435.5
95.5 Percent CI for ETA1-ETA2 is (186.0, 713.0)
Test of ETA1 = ETA2 vs ETA1 > ETA2 is
significant at 0.0009
```

15.21 Comparing tans Examples 1 and 2 compared two methods of getting a tan. Suppose Allison conducted an expanded experiment in which nine participants were randomly assigned to one of two brands of tanning lotion or to the tanning studio, three participants to each treatment. The nine were ranked on the quality of tan.

a. Which nonparametric test could be used to compare the three treatments?

b. Give an example of ranks for the three treatments that would have the largest possible test statistic value and the smallest possible P-value for this experiment. (*Hint:* What allocation of ranks would have the greatest between-groups variation in the mean ranks?)

15.22 Comparing therapies for anorexia The Anorexia data file on the text CD shows weight change for 72 anorexic teenage girls who were randomly assigned to one of three psychological treatments. Using software, analyze these data with a nonparametric Kruskal Wallis test to compare the three weight change distributions.

a. State the hypotheses.

b. Report the test statistic and its sampling distribution.

c. Report the P-value and explain how to interpret it.

15.23 Internet versus cell phones For the countries in the Human Development data file on the text CD, in 4 countries a higher percentage of people used the Internet than used cell phones, while in 35 countries a higher percentage of people used cell phones than the Internet.

a. Show how you could use a nonparametric test to compare Internet use and cell phone use in the population of all countries. State the (i) hypotheses, (ii) test statistic value, and (iii) find and interpret the P-value.

b. Is the analysis in part a relevant if the 39 countries in the data file are all the countries of interest to you rather than a random sample of countries? Explain.

15.24 Browsing the Internet Refer to the Georgia Student Survey data file on the text CD. Use a method from this chapter to test whether the amount of time spent browsing the Internet is independent of one's political affiliation. State the (a) hypotheses, (b) test statistic, (c) P-value, and interpret the result.

15.25 GPAs The Georgia Student Survey data file has data on college GPA and high school GPA for 59 University of Georgia students.

a. If you wanted to use a nonparametric test to check your friend's prediction that high school GPAs tend to be higher than college GPAs, which would you use?

b. What would be a reason for using the nonparametric method to do this?

c. Use the test in part a to do this analysis, and interpret the result.

15.26 Sign test about the GRE workshop In Exercise 15.12 on the effect of a GRE training workshop on the writing you used the sign test to evaluate the test differences of 0.5, −0.5, and 1.5. Suppose the test differences were 5.5, −0.5, and 1.5 instead of 0.5, −0.5., and 1.5.

a. State the hypotheses and find the P-value for the one-sided sign test for evaluating the effect of the workshop. Interpret.

b. Compare results to those in Exercise 15.12, and indicate what this tells you about the effect that outliers have on this nonparametric statistical method.

15.27 Wilcoxon signed-ranks test about the GRE workshop Example 8 on the GRE workshop used the Wilcoxon signed-ranks test to evaluate the score differences of 0.5, −0.5, and 1.5. Suppose the test differences were 5.5, −0.5, and 1.5 instead of 0.5 −0.5, and 1.5.

a. State the hypotheses and find the P-value for the one-sided Wilcoxon signed-ranks test for evaluating the effect of the workshop. Interpret.

b. Compare results to those in Example 8 and indicate what this tells you about the effect that outliers have on this nonparametric statistical method.

Concepts and Investigations

15.28 Student survey For the FL Student Survey data file on the text CD, we identify the number of times reading a newspaper as the response variable and gender as the explanatory variable. The observations are as follows:
Females: 5, 3, 6, 3, 7, 1, 1, 3, 0, 4, 7, 2, 2, 7, 3, 0, 5, 0, 4, 4, 5, 14, 3, 1, 2, 1, 7, 2, 5, 3, 7
Males: 0, 3, 7, 4, 3, 2, 1, 12, 1, 6, 2, 2, 7, 7, 5, 3, 14, 3, 7, 6, 5, 5, 2, 3, 5, 5, 2, 3, 3
Using software, analyze these data using methods of this chapter. Write a one-page report summarizing your analyses and conclusions.

15.29 Why nonparametrics? Present a situation for which it's preferable to use a nonparametric method instead of a parametric method, and explain why.

15.30 Why matched pairs? Refer to Example 7. Describe the advantages of an experiment using a crossover design instead of independent samples to compare the tanning methods.

15.31 Complete the analogy The *t* test for comparing two means is to the one-way ANOVA *F* test as the Wilcoxon test is to the ________ test.

15.32 Complete the analogy The *t* test for comparing two means is to the Wilcoxon test (for independent samples) as the matched pairs *t* test is to the ________ (for dependent samples in matched pairs).

15.33 True or false For a one-sided significance test comparing two means with small samples from highly skewed population distributions, it's safer to use a *t* test than a Wilcoxon test. This is because the Wilcoxon test assumes normal population distributions and is not robust if that assumption is violated.

15.34 Multiple choice Nonparametric statistical methods are used

a. Whenever the response variable is known to have a normal distribution.

b. Whenever the assumptions for a parametric method are not *perfectly* satisfied.

c. When the data are ranks for the subjects rather than quantitative measurements or when it's inappropriate to assume normality and the ordinary statistical method is not robust when the normal assumption is violated.

d. Whenever we want to compare two methods for getting a good tan.

15.35 Mann-Whitney statistic ◆◆ For the tanning experiment, Table 15.2 showed the sampling distribution of the difference between the sample mean ranks. Suppose you instead use as a test statistic the sample proportion of pairs of participants for which the tanning studio gave a better tan than the tanning lotion. This is the basis of the **Mann-Whitney** test, devised by two statisticians about the same time as Wilcoxon devised his test.

a. The table shows the proportion of pairs for which the tanning studio gave a better tan, for the possible sample results. Explain how to find these proportions.

b. Using this table, construct the sampling distribution that applies under the null hypothesis for this proportion.

c. For the one-sided alternative hypothesis of a better tan with the tanning studio, find the P-value for the observed sample proportion of 5/6. (*Note*: P-values based on this sample proportion are identical to P-values for the Wilcoxon test based on the mean ranks.)

Sample Proportion of Pairs, One from Each Treatment, for Which Tanning Studio Gave Better Tan Than Tanning Lotion

Treatments	Ranks				
Tanning lotion ranks	(1, 2, 3)	(1, 2, 4)	(1, 2, 5)	(1, 3, 4)	(1, 3, 5)
Tanning studio ranks	(4, 5)	(3, 5)	(3, 4)	(2, 5)	(2, 4)
Proportion	0/6	1/6	2/6	2/6	3/6
Tanning lotion ranks	(2, 3, 4)	(1, 4, 5)	(2, 3, 5)	(2, 4, 5)	(3, 4, 5)
Tanning studio ranks	(1, 5)	(2, 3)	(1, 4)	(1, 3)	(1, 2)
Proportion	3/6	4/6	4/6	5/6	6/6

15.36 **Rank-based correlation** For data on two quantitative ◆◆ variables, x and y, an alternative correlation uses the *rankings* of the data. Let n denote the number of observations on the two variables. You rank the values of the x-variable from 1 to n according to their magnitudes, and you separately rank the values of the y-variable from 1 to n. The correlation computed between the two sets of ranks is called the **Spearman rank correlation**. Like the ordinary correlation, it falls between -1 and $+1$, with values farther from 0 representing stronger association.

a. The ordinary correlation can be strongly affected by a regression outlier. Is this true also for the Spearman rank correlation? Why or why not?

b. If you want to test the null hypothesis of no association, what value for the Spearman rank correlation would go in the null hypothesis?

15.37 **Nonparametric regression** Nonparametric methods ◆◆ have also been devised for regression. Here's a simple way to estimate the slope: For each pair of subjects, the slope of the line connecting their two points is the difference between their y values divided by the difference between their x values. (See the figure.) With n subjects, we can find this slope for each pair of points. (There are $n(n - 1)/2$ pairs of points.) A nonparametric estimate of the slope is the median of all these slopes for the various pairs of points. The ordinary slope (least squares, minimizing the sum of squared residuals) can be strongly affected by a regression outlier. Is this true also for the nonparametric estimate of the slope? Why or why not?

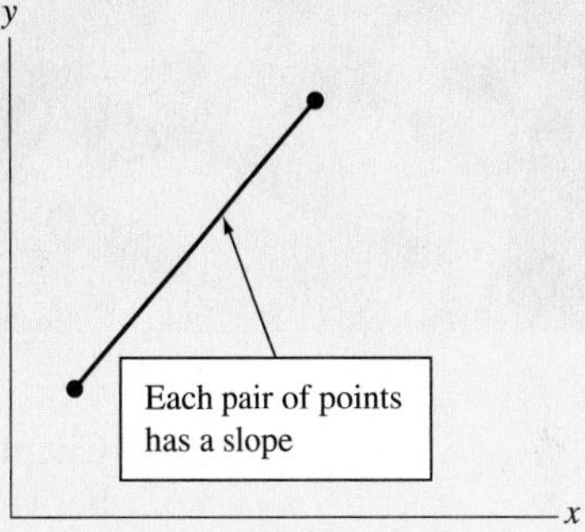

BIBLIOGRAPHY

Hollander, M., and Wolfe, D. (1999). *Nonparametric Statistical Methods*, 2nd edition. Wiley.

Lehmann, E. L. (1975). *Nonparametrics: Statistical Methods Based on Ranks.* Holden-Day.

Part Review

Analyzing Associations

4

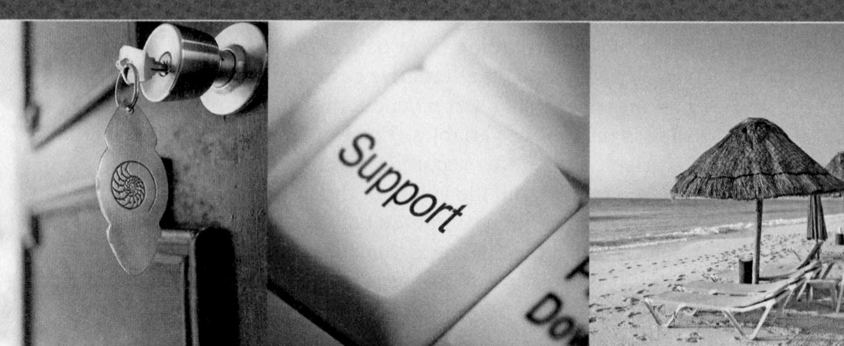

Chapters 11–15 presented the statistical methods for analyzing **associations** between two variables and among many variables. Chapter 11 dealt with associations between two categorical variables, using methods such as the **chi-squared test of independence.** Chapter 12 introduced the **regression model** for two quantitative variables and showed how to conduct inferences about their association. Chapter 13 worked with a **multiple regression model** that can handle several explanatory variables. The **analysis of variance (ANOVA)** methods of Chapter 14 analyzed how the mean of a quantitative response variable depends on one or more categorical explanatory variables. Finally, Chapter 15 used **nonparametric** methods that provide inferences without making an assumption (such as normality) for the shape of the population distribution of the response variable.

This review section gives examples of questions you should be able to answer about the main concepts in Chapters 11–15. The questions are followed by brief summaries as well as references to sections of the text where you can find more detail to strengthen your understanding of these concepts.

Review Questions

- How can you test whether or not there is an *association between two categorical variables?*

The **chi-squared** statistic is used to **test the null hypothesis of independence.** For a contingency table, this statistic compares the observed counts in the cells of the table to expected counts that satisfy H_0: independence, using the formula

$$X^2 = \sum \frac{(\text{observed count} - \text{expected count})^2}{\text{expected count}}.$$

When H_0 is true and the expected cell counts are relatively large (at least about 5), the sampling distribution of X^2 is approximately the **chi-squared distribution.** The **degrees of freedom** for this distribution depend on the number of rows r and the number of columns c in the contingency table, through $df = (r - 1) \times (c - 1)$. The P-value is the right-tail probability above the observed value of X^2. See Section 11.2.

We can describe the pattern of the association by comparing each observed count and expected count using the **standardized residual** (see Section 11.4). A value larger than 3 in absolute value indicates that the cell count is farther from the expected frequency than we'd expect merely by chance, if H_o were true.

- What is the basic *regression model* for two quantitative variables?

  Section 12.1 explained that the **regression model** $\mu_y = \alpha + \beta x$ uses a straight line with *y*-intercept α and slope β to approximate the relationship in the population between *x* and the mean μ_y of the conditional distribution of *y* at the different possible values of *x*. It is designed for a quantitative explanatory variable *x* and a quantitative response variable *y*.

- How can you describe *association* for quantitative variables?

  The **correlation** *r* describes the strength of a linear association between two variables *x* and *y*. It is a standardized slope that represents the value that the slope equals if *x* and *y* have the same standard deviation. The correlation falls between −1 and + 1, with greater absolute values corresponding to stronger association. The squared correlation, r^2, describes the proportional reduction in the sum of squared errors using the prediction equation $\hat{y} = a + bx$ to predict *y* compared to using the sample mean $\bar{y}$ to predict *y*. See Section 12.2.

- How can you *test the null hypothesis of independence* for two quantitative variables?

  For the linear regression model, assuming randomization, a significance test of $H_0: \beta = 0$ for the population slope β tests **statistical independence** of *x* and *y*. It has test statistic

  $$t = (b - 0)/se$$

  for the sample slope *b* and its standard error. You can construct a confidence interval for β to estimate the size of the effect. As Section 12.3 explained, either inference adds to the basic regression model the assumptions of randomization, and a normal conditional distribution of *y* with the same standard deviation at the different possible values of *x*. Inference is robust to violations of the normal assumption, except for one-sided tests with highly skewed distributions.

- How does regression generalize to include more than one explanatory variable?

  The **multiple regression model** relates the mean μ_y of the conditional distribution of a response variable *y* to several explanatory variables. An example of a multiple regression equation, with two predictors, is

  $$\mu_y = \alpha + \beta_1 x_1 + \beta_2 x_2.$$

  Section 13.1 presented this model. Section 13.4 showed that we can check it using **residual plots** by checking whether there is random variability in the residuals when we plot them against explanatory variables.

- How do *correlation* measures generalize when there are multiple predictors?

  Section 13.2 explained that the **multiple correlation** *R* and its square (R^2) describe the predictability of the response variable *y* by the set of explanatory variables. The multiple correlation *R* equals the correlation between the observed values of *y* and the predicted values $\hat{y}$ from the regression equation. Its square, R^2, is the proportional reduction in error from predicting *y* using the prediction equation instead of using $\bar{y}$ (and ignoring the predictors). Both *R* and R^2 fall between 0 and 1, with larger values representing stronger association.

- How can you test the null hypothesis that none of the predictors have an effect on the explanatory variable?

  An ***F* statistic** tests $H_0: \beta_1 = \beta_2 = \ldots = 0$, which states that *y* is independent of all the explanatory variables in the model. The *F* test statistic equals a ratio of mean squares, and it is larger when the sample R^2 is larger. A small P-value suggests that y is associated with at least one of the explanatory variables. See Section 13.3. Table 13.17 in Chapter 13 showed a summary of the basic properties and inference methods for multiple regression (Chapter 13) and bivariate regression (Chapter 12).

- How can you include *categorical explanatory variables* in a regression model?

  Categorical explanatory variables can be included using **indicator variables** (Section 13.5). With two categories (a **binary** predictor), the indicator variable equals 1 when the observation is in the first category and 0 when it is in the second category. If more than two categories exist, the indicator variable equals 1 if the observation falls into the category and 0 otherwise.

- How can you model a *categorical response variable?*

  Section 13.6 showed that the **logistic regression model** describes how the probability that y falls in a particular category depends on the values of explanatory variables. For a quantitative predictor, it has an S shape instead of the usual straight line for ordinary regression.

- What is the purpose of *ANOVA*?

  Analysis of variance (ANOVA) methods compare several groups according to their means for a quantitative response variable. The groups are categories of categorical explanatory variables, called **factors.** The **one-way ANOVA** *F* test (Section 14.1) compares means for a single factor. The groups are categories of a single categorical explanatory variable, and the null hypothesis states that each group has the same population mean. **Two-way ANOVA** methods (Section 14.3) compare means across categories of each of two factors, at fixed levels of the other factor.

- How is *ANOVA* related to *regression*?

  Analysis of variance methods can be conducted using multiple regression models. A regression model uses **indicator variables** as explanatory variables to represent the factors. Each indicator variable equals 1 for observations from a particular category and 0 otherwise. See Sections 14.2 and 14.3. ANOVA also makes the regression inference assumptions of randomization and a normal population distribution for y, with robustness for the normality assumption.

- What are *nonparametric* statistical methods?

  Nonparametric statistical methods provide statistical inference without an assumption of normality about the probability distribution of the response variable. Chapter 15 explained how most nonparametric methods use only the **rankings** of the observations rather than the quantitative values.

- Which *nonparametric* method is used to *compare two groups*?

  The **Wilcoxon test** (Section 15.1) is a nonparametric test for the null hypothesis of identical population distributions for two groups. It assumes independent random samples and uses the ranks for the combined sample of observations, with smaller P-values resulting when the mean ranks for the two samples are farther apart.

A Guide to Choosing a Statistical Method

We congratulate you on getting to the end of this statistics text! We are confident that you now have a better understanding of the analyses that underlie results you hear about from polls and research studies.

At this stage, when you apply statistical methods, whether to exercises such as at the end of this review or to data that you analyze for yourself, you may feel a bit unsure how to know *which* statistical method to use. In the front endpaper of the book, you will see a page with the title "A Guide to Choosing a Statistical Method." This guide is designed to help you think about the factors that determine which method to use.

In practice, there is usually more than one variable in an analysis. So the first step is to distinguish between the *response* variable and the *explanatory* variables. The guide is organized in terms of whether these are categorical or quantitative. For example, if you have a quantitative response variable and want to compare its means for two groups of a categorical explanatory variable, Item 2 in the second part of this guide lists methods that are appropriate. If the explanatory variables are also quantitative, multiple regression methods are appropriate, as mentioned in Item 5 in the second part of the guide.

We suggest that you read through the guide to help refresh your memory about the methods this book presents. Then, check your understanding by trying to answer the questions in the following example and in the Review Exercises.

Example

Review of statistical methods: multiple regression

eBay Selling Prices

Picture the Scenario

Let's analyze a data set of 33 selling prices on eBay of the iPod Touch 3rd Generation (32GB). When we regress y = selling price (in dollars) on x_1 = number of bids placed on the item and x_2 whether the seller offered buyers the buy-it-now option (1 = yes, 0 = no), software (MINITAB) gives us the results shown in the table. The accompanying graph shown is a scatterplot of y verses x_1 with observations identified by whether or not the buy-it-now option was available.

Results of Regression Analysis for eBay Sales

```
The regression equation is
price = 240 - 0.629 bids - 6.77 buynow
Predictor         Coef    SE Coef       T       P
Constant       240.455      5.736   41.92   0.000
bids            -0.629      0.432   -1.45   0.156
buynow          -6.766      7.670   -0.88   0.385
S = 18.6995      R-Sq = 6.8%
Analysis of Variance
Source            DF         SS      MS       F       P
Regression         2      766.8   383.4    1.10   0.347
Residual Error    30    10490.1   349.7
Total             32    11256.9
```

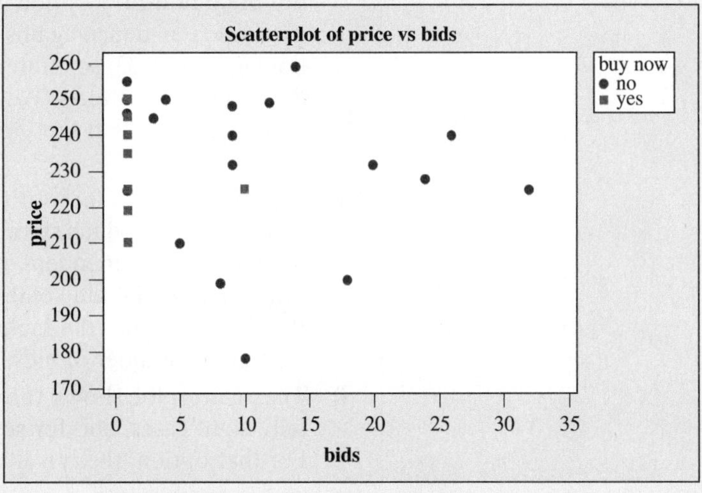

Questions to Explore

a. State the prediction equation, and explain how to interpret the coefficients of the explanatory variables.

b. Find the equation relating the predicted selling price to the number of bids (i) with and (ii) without the buy-it-now option. Interpret.

c. Explain how to interpret the value reported as R-Sq in the printout.

d. Explain the purpose of the *F* statistic in the analysis of variance table, and report and interpret its P-value.

e. Explain the purpose of the *t* statistic for the *buynow* predictor, and report and interpret its P-value.

f. Explain why the scatterplot suggests that you should be very cautious in using this regression model with these data to describe or make inferences about how the selling price depends on the number of bids when the buy-it-now option is used.

Think It Through

a. From the printout, the prediction equation is

$$\hat{y} = 240 - 0.629x_1 - 6.77x_2.$$

For a given option (yes or no for whether it was possible to buy-it-now), the predicted selling price decreases by \$0.63 for each additional bid. For a given number of bids, the predicted selling price is \$6.77 lower when the buy-it-now option is available compared to when it is not available.

b. When the buy-it-now option is available ($x_2 = 1$), the equation is $\hat{y} = 240 - 0.629x_1 - 6.77(1) = 233.2 - 0.629x_1$. Without the buy-it-now option ($x_2 = 0$), the equation is $\hat{y} = 240 - 0.629x_1 - 6.77(0) = 240 - 0.629x_1$. In each case, the slope for the estimated effect of the number of bids is the same (namely, -0.629). This regression model makes the assumption that the effect of a predictor is the same at every level of the other predictors. (This is the "no interaction" assumption discussed in Section 13.4.)

c. R-Sq = 6.8% means that $R^2 = 0.068$. Relatively little variability in selling price (less than 7%) is explained by these predictors. We cannot predict selling price better if we know the number of bids and know whether or not the buy-it-now option was available than if we merely used the sample mean selling price as the predictor.

d. The *F* statistic of 1.10 tests H_0: $\beta_1 = \beta_2 = 0$, which states that the selling price *y* is independent of both the number of bids and whether or not the buy-it-now option was available. The P-value of 0.35 does not give much evidence against this null hypothesis. The sample size was not large, $n = 33$, so inference methods may not have much power for detecting weak effects. We should not make the error of accepting H_0 and concluding that these predictors have no effect for the population of such sales.

e. The *t* statistic of -0.88 for the buynow predictor tests the null hypothesis H_0: $\beta_2 = 0$, which states that for any given number of bids, the selling price *y* is independent of whether or not the buy-it-now option was available. The P-value of 0.38 does not give much evidence against H_0. It is plausible that the mean selling price is the same for each option, for a given number of bids.

f. The scatterplot shows that the buy-it-now option was available for only eight cases, and for seven of those cases there was only one bid. For that option, there is little information about how the selling price depends on the number of bids. We should put little faith in the equation $\hat{y} = 233.2 - 0.629x_1$ from part b for describing the effect of bidding for that option. It is more sensible to fit the model with x_1 as a predictor to the 25 cases that did not have the buy-it-now option, which gives $\hat{y} = 240 - 0.617x_1$. Or we could ignore whether that option was used and fit the model using only x_1 as a predictor with all 33 cases, which gives $\hat{y} = 237 - 0.463x_1$. In either case, we see one observation on the scatterplot, with a selling price of $178, that falls a bit below the overall weak negative trend.

Insight

The scatterplot shows us the danger of using statistical methods without looking first at the data. With the powerful statistical software available to us these days, we can use all sorts of statistical methods whether they are justified or not. It is up to us to check the assumptions and to look at the data to check whether or not it makes sense to use the methods.

Try Exercise R4.14

Part 4 Review Exercises

Practicing the Basics

R4.1 Gender and opinion about abortion The GSS surveys routinely show that in the United States about 40% of males and 40% of females answer yes when asked whether or not a woman should be able to get an abortion if she wants it for any reason (variable ABANY).

a. For males and for females, report the sample conditional distribution for whether or not unrestricted abortion should be legal, using the categories (yes, no).

b. If the population percentage answering yes was exactly 40% for each gender, would gender and opinion about abortion be statistically independent or dependent? Explain.

R4.2 Opinion depends on party? In a July 2011 Gallup poll, 1463 adults were surveyed for their approval rating for President Barack Obama. His approval dropped to a new low in the midst of the debt crisis. Among Democrats, his job approval rating was 72%, while it was 13% with Republicans. Would you characterize the association between political party affiliation and opinion about President Obama's performance as weak, or strong? Explain why.

R4.3 Murders and gender For murders in the United States in 2009 having a single victim and a single offender, the table cross-classifies the sex of the victim by the sex of the offender, using data from www.fbi.gov.

a. Using software, find the test statistic and P-value for testing whether these variables are independent.

b. Estimate a summary measure to describe the association.

	Sex of Victim	
Sex of Offender	Female	Male
Female	182	484
Male	1719	4078

R4.4 Change in opinion The GSS has asked whether you agree or disagree with the statement, "It is much better for everyone involved if the man is the achiever outside the home and the woman takes care of the home and family." The table compares responses in 1977 and in 2008 and shows standardized residuals (in parentheses) for the null hypothesis of independence.

a. The chi-squared statistic equals 236.22. Report the P-value, and interpret.

b. Explain how to interpret the standardized residuals.

	Agree	Disagree	Total
1977	989 (15.37)	514 (−15.37)	1503
2008	481 (−15.37)	827 (15.37)	1308

R4.5 Ma and Pa education For 2008 GSS data, the correlation matrix for subject's number of years of education (EDUC), mother's education (MAEDUC), and father's education (PAEDUC) is as shown in the table. Explain how to interpret this matrix, identifying the pair of variables with the strongest association and giving the implication of the sign of each correlation.

	EDUC	MAEDUC	PAEDUC
EDUC	1.00	0.45	0.47
MAEDUC	0.45	1.00	0.68
PAEDUC	0.47	0.68	1.00

R4.6 Mother's education and yours For 1777 observations from the 2008 GSS on y = number of years of education (EDUC) and x = number of year's of mother's education (MAEDUC), $\hat{y} = 9.61 + 0.358x$ with $se = 0.017$ for the slope estimate.

a. Test the null hypothesis that these variables are independent, and interpret.

b. Find a 95% confidence interval for the population slope. Interpret.

c. What are the assumptions on which the methods in part a and part b are based? In practice, explain how you could use the data to check some of these.

R4.7 Fertility and contraception In the *Human Development Report*, one variable reported for many nations was x = Percentage of adults who use contraceptive methods. The table shows part of a MINITAB printout for a regression analysis using y = fertility (mean number of children per adult woman), for 22 nations listed in that report. For those nations x had a mean of 60.0 and standard deviation of 20.6.

a. Report the prediction equation and find the predicted fertility when (i) $x = 0$ and (ii) $x = 100$. Explain why the difference between these is 100 times the estimated slope.

b. For Belgium, $x = 51.0$ and $y = 1.3$ Find the predicted fertility and the residual. (This is the observation with the largest residual.)

c. The standardized residual for Belgium was −2.97. Interpret it.

Fertility and contraception around the world

```
Predictor        Coef     SE Coef       T       P
Constant        6.663     0.4771     13.97   0.000
CONTRA       -0.06484   0.007539     -8.60   0.000
Analysis of Variance
Source           DF         SS
Regression        1     37.505
Residual Error   20     10.138
Total            21     47.644
```

R4.8 Association between fertility and contraception Refer to the previous exercise.

a. Find r^2 using the SS values listed in the ANOVA table. Interpret its value.

b. Using r^2 and the sign of the slope, find the correlation. Interpret its value.

R4.9 Predicting body fat The accompanying table shows part of a MINITAB printout for a regression of y = percentage of body fat (BF%) on x = body mass index (BMI) for 57 high school female athletes. For these data, BF% has mean = 25.1 and standard deviation = 5.0, and BMI has mean = 22.9 and standard deviation = 3.5.

a. Report and interpret r^2. Is the association strong or weak?

b. Report and interpret the P-value for the test of independence between BF% and BMI. State the conclusion, in context, for the 0.05 significance level.

c. The 95% confidence interval for the true slope is (1.21, 1.46). Explain how this is in agreement with the result of the conclusion of the t test in part b.

d. Report and interpret the residual standard deviation.

Modelling body fat

```
Predictor        Coef     SE Coef      T       P
Constant       -5.556     1.453     -3.82   0.000
BMI           1.33812   0.06264     21.36   0.000
S = 1.6404   R-Sq = 89.2%
Analysis of Variance
Source           DF       SS        MS        F       P
Regression        1   1228.0    1228.0   456.32   0.000
Residual Error   55    148.0       2.7
Total            56   1376.0
```

R4.10 Body weight and lean body mass The College Athletes data set on the text CD comes from a study of University of Georgia Division 1 female athletes from nine sports. SPSS reports the results in the table of regressing TBW = total body weight (in pounds) on LBM = lean body mass, which equals TBW (1 − proportion of body fat):

a. To what does the reported T value of 17.91 refer?

b. Explain how to interpret its associated P-value.

c. Interpret the S and R-Sq values reported on the last line of the table.

Coefficients[a]

Model		Unstandardized Coefficients B	Std. Error	t	Sig.
1	(Constant)	0.415	7.452	0.056	0.956
	LBM	1.225	0.068	17.913	0.000

R = 0.915 $R^2 = 0.838$ S = 6.959

[a]Dependent Variable: TBW

R4.11 Using the Internet For data from several nations, we want to evaluate whether y = percentage Internet penetration (the percentage of adult users) is more strongly associated with x = gross domestic product (GDP, in thousands of dollars per capita) or with x = carbon dioxide emissions per capita.

a. Can we compare the slopes when GDP and carbon dioxide emissions predict Internet use in separate regression equations? Why or why not?

b. Let x = GDP. For recent data on 39 nations (from the UN), for y = percentage Internet penetration, the prediction equation is $\hat{y} = 0.0157x + 0.1239$ whereas for y = percentage using Facebook, the prediction equation is $\hat{y} = 0.0075x + 0.081$. Does it make sense to compare these slopes? Why or why not?

R4.12 Fertility rate and GDP Fitting the exponential regression model to United Nations data with y = fertility rate (mean number of children per adult woman) and x = per capita GSP (in \$10,000 U.S. dollars), we get $\hat{y} = 3.15 \times 0.81^x$.

a. Interpret the coefficients 3.15 and 0.81. Explain why the predicted fertility rate decreases by 19% for a \$10,000 increase in per capita GDP.

b. Predict the fertility rate for nations with (i) $x = 1$ and (ii) $x = 4$.

R4.13 Growth of Wikipedia For the data given in the article "Wikipedia: Modelling Wikipedia's growth" at en.wikipedia.org, between October 2002 and September 2006 the number of English-language articles in

Wikipedia was well approximated by $\hat{y} = 100{,}000 \times 2.1^x$, where x is the time (in years) since January 1, 2003.

a. Interpret the values 100,000 and 2.1.

b. If this equation continues to hold, predict the number of English Wikipedia articles in (i) January 1, 2008 and (ii) January 1, 2013. Do you trust such predictions, or do you see any danger in making them?

R4.14 Distance of college from home The FL Student Survey data file on the text CD gives 60 student responses for a survey given at the University of Florida. Regression y = distance from home (in miles) on x_1 = age and x_2 = gender (male = 0, female = 1) gives the results shown in the table.

TRY

a. State the prediction equation for predicting distance from home (y) from age (x_1) and gender (x_2). Interpret the coefficients of the explanatory variables.

b. Find the equation relating the predicted distance from home to age (i) for females and (ii) for males. Interpret.

c. Interpret the value reported as R-Sq in the printout.

d. Explain the purpose of the F statistic in the analysis of variance table, and report and interpret its P-value.

e. Explain the purpose of the t statistic for the gender predictor, and report and interpret its P-value.

f. Use software to construct a scatterplot for these data. Do you see anything that makes you question the usefulness of the model? Explain.

Regression analysis for Florida Student Survey distance from home, age, and gender

```
The regression equation is
distance_home = 223 + 34.2 age + 25 gender_ind
Predictor Coef    SE Coef     T        P
Constant   222.5  837.5      0.27     0.791
age          34.18  25.98    1.32     0.194
gender       24.6  437.0     0.06     0.955
S = 1685.56  R-Sq = 3.0%
Analysis of Variance
Source       DF       SS          MS        F      P
Regression   2        4925427  2452713  0.87   0.426
Residual
Error        57   161942867  2841103
Total        59   166868294
```

R4.15 Baseball offensive production An analysis of major league baseball team season totals for every team from 1995 through 2004 yielded the prediction equation[1]

$\hat{y}$ = 100.0 + 0.59(1B) + 0.71(2B) + 0.91(3B) + 1.48(HR) + 0.30(BB) + 0.27(SB) − 0.14(OUTS) − 0.20(CS)

where y = number of runs scored in season by the team, 1B = number of singles, 2B = number of doubles, 3B = number of triples, HR = number of home runs, BB = number of bases on balls (walks), SB = number of stolen bases, OUTS = number of outs, and CS = number caught stealing.

a. For which predictor would a one-unit increase have the largest effect on $\hat{y}$? Explain.

b. What does it mean for the effect of SB to be positive but the effect of CS to be negative?

c. Predict the number of runs scored by a team that had 1B = 600, 2B = 100, 3B = 10, HR = 200, BB = 300, SB = 40, OUTS = 4000, CS = 20.

R4.16 Correlates of fertility For data on 145 nations in the World Data for Fertility and Literacy data file, the table shows a correlation matrix for AFR = adolescent fertility rate, ALR = adult literacy rate, and CE = combined enrollment ratio (ratio of enrolled students to all children school age for elementary, middle, and high school). Consider the multiple regression model with y as AFR and predictors ALR and CE.

	AFR	ALR
ALR	−0.661	
CE	−0.576	0.803

a. From looking at the correlation matrix, how is it that you know that R for this model cannot be smaller than 0.661?

b. Suppose the multiple correlation $R = 0.67$ for the model. Interpret.

c. Report and interpret the value of R^2.

R4.17 Predicting pollution Using the Twelve Countries data file, regressing carbon dioxide (CO2) use (a measure of air pollution) on gross domestic product (GDP) gives a correlation of 0.740. With unemployment rate added as a second explanatory variable, the multiple correlation is $R = 0.775$.

a. Interpret the multiple correlation.

b. For predicting, did it help much to add unemployment rate to the model? Explain.

c. Refer to part b. Explain why this does not imply that unemployment rate is itself very weakly correlated with CO_2 emissions.

d. Explain why a large positive standardized residual means that a country has a relatively large value for its GDP and unemployment rate. The two biggest "energy hogs"—the countries with the largest positive standardized residuals—are Australia (5.16) and the United States (3.49). Interpret the standardized residual for Australia.

R4.18 Attitudinal research A study[2] on predicting attitudes toward homosexuality used GSS data to model a response variable with a 4-point scale in which homosexual relations were scaled from 1 = always wrong to 4 = never wrong, with x_1 = education (in years), x_2 = age (in years), x_3 = political conservative (1 = yes, 0 = no), x_4 = religious fundamentalist (1 = yes, 0 = no), and x_5 = whether live in same city as when age = 16 (1 = yes, 0 = no). The model had fit $\hat{y} = 1.53 + 0.09x_1 - 0.01x_2 - 0.49x_3 - 0.39x_4 - 0.15x_5$.

a. Summarize the effect of x_1. Explain what its positive sign represents.

b. Summarize the effect of x_3. Explain what its negative sign represents.

[1]In article by S. Berry, *Chance*, vol. 19, 2006, p. 58.

[2]T. Shackelford and A. Besser, *Individual Differences Research,* vol. 5, 2007.

c. Find the prediction equation relating $\hat{y}$ to x_1 for those who are 20 years old, politically conservative, fundamentalist in religion, but who live in a different city as when 16 years old. For that group, report $\hat{y}$ when $x_1 = 10$ and when $x_1 = 20$, and interpret.

R4.19 Interaction between SES and age in quality of health A study using data from the 2005/2006 National Health Measurement Study (a telephone survey of a nationally representative sample of U.S. adults) noted that health related quality of life (HRQol) measures can be stratified by socioeconomic status (SES) and age.[3] Results indicated that there are SES differences in HRQol in all age groups. Those in the lowest SES groups in the 35–44 age cohort have worse HRQoL than those in higher SES groups in the 65+ age cohort. The HRQol measures tend to be positively correlated with SES (measured by years of education and annual household income). Moreover, the association strengthens with age. For example, the gap in health between low SES and high SES levels tends to be larger at older ages. Represent this by sketching two lines on a graph (one for a low age and one for a high age) relating quality of health to SES, labeling the variables. (This indicates there is *interaction* between SES and age in their effects on health. See the subsection on Interaction in Section 13.4.)

R4.20 Protecting children in car crashes An article[4] analyzed data on nearly 4000 children who had been in car crashes. The response variable was whether the child was injured. Explanatory variables included whether the vehicle was a sports utility vehicle (SUV) or a regular sedan (yes or no), whether the child was using a restraint (yes or no), whether the vehicle rolled over (yes or no), and the weight of the vehicle.

a. Explain why logistic regression is the appropriate type of regression model for this analysis.

b. For this model, explain how each predictor could be measured for the regression equation. (*Hint:* When a predictor is categorical, show how to include it in the model using an indicator variable.)

c. The study concluded that the best action to protect children in a crash is to restrain them properly, especially in SUVs, which were more likely to roll over. If p denotes the probability of injury, then for the way you set up the restraint predictor in part b, indicate whether you would expect its effect to be positive or negative.

R4.21 Political ideology and party affiliation The GSS measures political ideology with scores such as 1 = extremely liberal, 4 = moderate, 7 = extremely conservative. In a recent GSS, when responses were classified by political party affiliation, results were obtained as shown in the table.

Political ideology by party

Group (Party)	n	Mean	Std. Dev.
Democrat	954	3.53	1.32
Independent	279	4.03	1.23
Republican	657	5.00	1.17

a. Identify the response variable and the factor.

b. Providing notation, state the null and alternative hypotheses for conducting a one-way ANOVA with these data.

c. The ANOVA F statistic equals 269.456, and its P-value is 0.000. Explain how to interpret the result of the test.

d. What assumptions are made to use one-way ANOVA? Do they seem plausible for this application?

R4.22 MS error for ideology The ANOVA table (not shown) for the analysis in the previous exercise reported a mean square error of 1.581.

a. If we assume that each party has the same variability in political ideology scores, what is the estimated standard deviation of political ideology for each group?

b. From the table in the previous exercise, explain why the assumption in part a seems reasonable.

c. As a follow-up to the ANOVA, the 95% confidence interval for $\mu_3 - \mu_1$ is (1.35, 1.60). Interpret.

R4.23 Income and education Based on a graph provided by the U.S. Census Bureau in a study of the effect of education on income, in 2005 the mean income (in thousands of dollars) for workers of age 18 and older approximately followed the formula

$$\hat{y} = 20 + 23e + 17s,$$

where $e = 1$ for college graduates and $e = 0$ for high school graduates, and $s = 1$ for men and $s = 0$ for women.

a. Approximate the mean income for women (i) high school graduates and (ii) college graduates.

b. Explain how to interpret the coefficient 23 of e in the prediction equation.

R4.24 Nonparametric rank test Describe a nonparametric method that can compare two groups using independent samples. Specify the hypotheses, and explain what information software uses from the two samples to determine the P-value.

Concepts and Investigations

R4.25 Racial prejudice Several sociologists have reported that racial prejudice varies according to religious group. Examine this issue using the table shown, which summarizes results for white respondents to the 2002 General Social Survey. The variables are Fundamentalism/Liberalism of Respondent's Religion (FUND) and response to the question (RACMAR), "Do you think there should be laws against marriages between blacks and whites?" Analyze these data. Prepare a report, describing your analyses and providing interpretations of the data.

	Favor Laws Against Marriage	
Religious Preference	**Yes**	**No**
Fundamentalist	39	142
Moderate	21	248
Liberal	17	236

[3] S. Robert et. al., *J Gerontol B Psychol Sci Soc Sci*, vol. 64B, 2009, pp. 378–389.

[4] By L. Daly et al. in *Pediatrics*, vol. 117, 2006.

R4.26 GPA and TV watching Using software with the FL Student Survey data file on the text CD, conduct analyses for y = college GPA, using explanatory variables of your choice. Prepare a two-page report, showing descriptive and inferential methods for analyzing the relationship.

R4.27 Analyze your data Refer to the data file you created in Activity 3 at the end of Chapter 1. For variables chosen by your instructor, conduct a statistical analysis. Report both descriptive and inferential statistical analyses, interpreting and summarizing your findings.

R4.28 Review research literature Your instructor will pick a research topic of interest to the class. Find a recent article in a journal or a report on the Internet that gives results of a research study on that topic. Describe the statistical analyses that were used in that article. Prepare a one-page summary of your findings.

R4.29 Predicting college success A study about the factors (such as high school GPA and tests such as the SAT) that affect student performance in college stated that "only 30 percent of the grade variance in college could be explained by the factors that admissions officers examine."[5] Explain how this statement could summarize the result of a regression analysis (with y = college GPA), and explain what the 30 percent refers to.

R4.30 Regression toward mean Explain the concept of "regression toward the mean," illustrating with x = mother's height and y = daughter's height.

R4.31 Why ANOVA? Explain the purpose of the one-way ANOVA F test. Explain how you can follow up that test to learn more about the population means.

R4.32 Variability in ANOVA Explain using sample dot plots of data whether there is *more* evidence or *less* evidence against the null hypothesis in one-way ANOVA when (a) the between-groups variation increases and (b) the within-groups variation increases, other things being equal.

R4.33 Violating regression assumptions The basic regression model assumes that y has a normal conditional distribution, with the same variation at all combinations of predictor values, and with a straight-line trend for the effect of an explanatory variable. Explain what you can use as an alternative model

a. If the trend instead is curved, with an ever-increasing slope, as in modeling population size over time.

b. If y can take only two values, such as "success" and "failure."

R4.34 True or false? For comparing two population distributions, the t test assumes normal population distributions, but the Wilcoxon test does not need this assumption. Moreover, the t test is not robust to violations of the normal assumption when the test is one-sided and the sample size is small. Therefore, for a one-sided significance test comparing two population distributions with small samples from highly skewed population distributions, it's safer to use the Wilcoxon test than the t-test.

[5]See insidehighered.com/news/2007/06/19/admit.

Appendix A

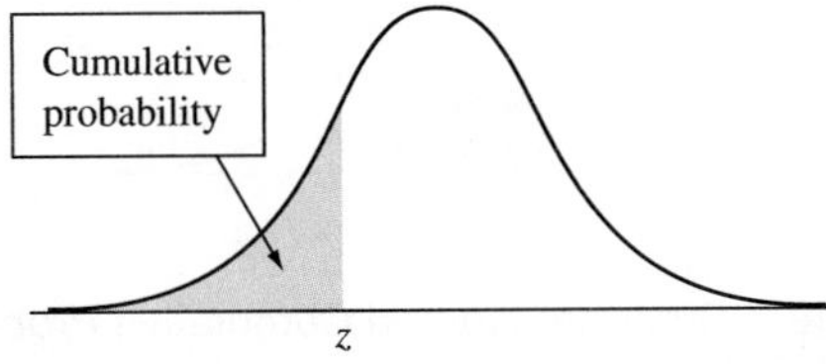

Cumulative probability for z is the area under the standard normal curve to the left of z

z	.00
−5.0	.000000287
−4.5	.00000340
−4.0	.0000317
−3.5	.000233

Table A Standard Normal Cumulative Probabilities

z	.00	.01	.02	.03	.04	.05	.06	.07	.08	.09
−3.4	.0003	.0003	.0003	.0003	.0003	.0003	.0003	.0003	.0003	.0002
−3.3	.0005	.0005	.0005	.0004	.0004	.0004	.0004	.0004	.0004	.0003
−3.2	.0007	.0007	.0006	.0006	.0006	.0006	.0006	.0005	.0005	.0005
−3.1	.0010	.0009	.0009	.0009	.0008	.0008	.0008	.0008	.0007	.0007
−3.0	.0013	.0013	.0013	.0012	.0012	.0011	.0011	.0011	.0010	.0010
−2.9	.0019	.0018	.0018	.0017	.0016	.0016	.0015	.0015	.0014	.0014
−2.8	.0026	.0025	.0024	.0023	.0023	.0022	.0021	.0021	.0020	.0019
−2.7	.0035	.0034	.0033	.0032	.0031	.0030	.0029	.0028	.0027	.0026
−2.6	.0047	.0045	.0044	.0043	.0041	.0040	.0039	.0038	.0037	.0036
−2.5	.0062	.0060	.0059	.0057	.0055	.0054	.0052	.0051	.0049	.0048
−2.4	.0082	.0080	.0078	.0075	.0073	.0071	.0069	.0068	.0066	.0064
−2.3	.0107	.0104	.0102	.0099	.0096	.0094	.0091	.0089	.0087	.0084
−2.2	.0139	.0136	.0132	.0129	.0125	.0122	.0119	.0116	.0113	.0110
−2.1	.0179	.0174	.0170	.0166	.0162	.0158	.0154	.0150	.0146	.0143
−2.0	.0228	.0222	.0217	.0212	.0207	.0202	.0197	.0192	.0188	.0183
−1.9	.0287	.0281	.0274	.0268	.0262	.0256	.0250	.0244	.0239	.0233
−1.8	.0359	.0351	.0344	.0336	.0329	.0322	.0314	.0307	.0301	.0294
−1.7	.0446	.0436	.0427	.0418	.0409	.0401	.0392	.0384	.0375	.0367
−1.6	.0548	.0537	.0526	.0516	.0505	.0495	.0485	.0475	.0465	.0455
−1.5	.0668	.0655	.0643	.0630	.0618	.0606	.0594	.0582	.0571	.0559
−1.4	.0808	.0793	.0778	.0764	.0749	.0735	.0721	.0708	.0694	.0681
−1.3	.0968	.0951	.0934	.0918	.0901	.0885	.0869	.0853	.0838	.0823
−1.2	.1151	.1131	.1112	.1093	.1075	.1056	.1038	.1020	.1003	.0985
−1.1	.1357	.1335	.1314	.1292	.1271	.1251	.1230	.1210	.1190	.1170
−1.0	.1587	.1562	.1539	.1515	.1492	.1469	.1446	.1423	.1401	.1379
−0.9	.1841	.1814	.1788	.1762	.1736	.1711	.1685	.1660	.1635	.1611
−0.8	.2119	.2090	.2061	.2033	.2005	.1977	.1949	.1922	.1894	.1867
−0.7	.2420	.2389	.2358	.2327	.2296	.2266	.2236	.2206	.2177	.2148
−0.6	.2743	.2709	.2676	.2643	.2611	.2578	.2546	.2514	.2483	.2451
−0.5	.3085	.3050	.3015	.2981	.2946	.2912	.2877	.2843	.2810	.2776
−0.4	.3446	.3409	.3372	.3336	.3300	.3264	.3228	.3192	.3156	.3121
−0.3	.3821	.3783	.3745	.3707	.3669	.3632	.3594	.3557	.3520	.3483
−0.2	.4207	.4168	.4129	.4090	.4052	.4013	.3974	.3936	.3897	.3859
−0.1	.4602	.4562	.4522	.4483	.4443	.4404	.4364	.4325	.4286	.4247
−0.0	.5000	.4960	.4920	.4880	.4840	.4801	.4761	.4721	.4681	.4641

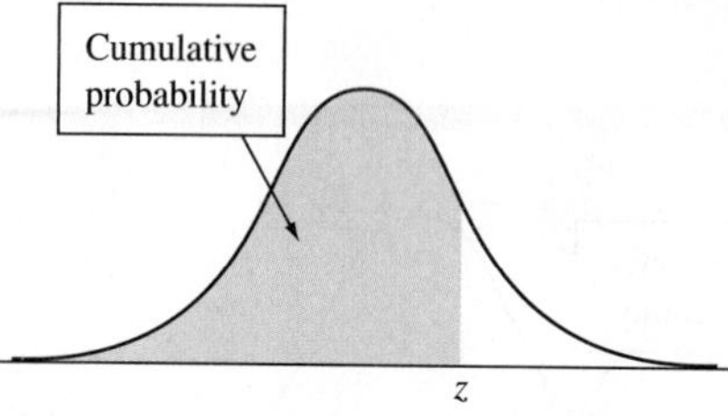

Cumulative probability for z is the area under the standard normal curve to the left of z

Table A Standard Normal Cumulative Probabilities (*continued*)

z	.00	.01	.02	.03	.04	.05	.06	.07	.08	.09
0.0	.5000	.5040	.5080	.5120	.5160	.5199	.5239	.5279	.5319	.5359
0.1	.5398	.5438	.5478	.5517	.5557	.5596	.5636	.5675	.5714	.5753
0.2	.5793	.5832	.5871	.5910	.5948	.5987	.6026	.6064	.6103	.6141
0.3	.6179	.6217	.6255	.6293	.6331	.6368	.6406	.6443	.6480	.6517
0.4	.6554	.6591	.6628	.6664	.6700	.6736	.6772	.6808	.6844	.6879
0.5	.6915	.6950	.6985	.7019	.7054	.7088	.7123	.7157	.7190	.7224
0.6	.7257	.7291	.7324	.7357	.7389	.7422	.7454	.7486	.7517	.7549
0.7	.7580	.7611	.7642	.7673	.7704	.7734	.7764	.7794	.7823	.7852
0.8	.7881	.7910	.7939	.7967	.7995	.8023	.8051	.8078	.8106	.8133
0.9	.8159	.8186	.8212	.8238	.8264	.8289	.8315	.8340	.8365	.8389
1.0	.8413	.8438	.8461	.8485	.8508	.8531	.8554	.8577	.8599	.8621
1.1	.8643	.8665	.8686	.8708	.8729	.8749	.8770	.8790	.8810	.8830
1.2	.8849	.8869	.8888	.8907	.8925	.8944	.8962	.8980	.8997	.9015
1.3	.9032	.9049	.9066	.9082	.9099	.9115	.9131	.9147	.9162	.9177
1.4	.9192	.9207	.9222	.9236	.9251	.9265	.9279	.9292	.9306	.9319
1.5	.9332	.9345	.9357	.9370	.9382	.9394	.9406	.9418	.9429	.9441
1.6	.9452	.9463	.9474	.9484	.9495	.9505	.9515	.9525	.9535	.9545
1.7	.9554	.9564	.9573	.9582	.9591	.9599	.9608	.9616	.9625	.9633
1.8	.9641	.9649	.9656	.9664	.9671	.9678	.9686	.9693	.9699	.9706
1.9	.9713	.9719	.9726	.9732	.9738	.9744	.9750	.9756	.9761	.9767
2.0	.9772	.9778	.9783	.9788	.9793	.9798	.9803	.9808	.9812	.9817
2.1	.9821	.9826	.9830	.9834	.9838	.9842	.9846	.9850	.9854	.9857
2.2	.9861	.9864	.9868	.9871	.9875	.9878	.9881	.9884	.9887	.9890
2.3	.9893	.9896	.9898	.9901	.9904	.9906	.9909	.9911	.9913	.9916
2.4	.9918	.9920	.9922	.9925	.9927	.9929	.9931	.9932	.9934	.9936
2.5	.9938	.9940	.9941	.9943	.9945	.9946	.9948	.9949	.9951	.9952
2.6	.9953	.9955	.9956	.9957	.9959	.9960	.9961	.9962	.9963	.9964
2.7	.9965	.9966	.9967	.9968	.9969	.9970	.9971	.9972	.9973	.9974
2.8	.9974	.9975	.9976	.9977	.9977	.9978	.9979	.9979	.9980	.9981
2.9	.9981	.9982	.9982	.9983	.9984	.9984	.9985	.9985	.9986	.9986
3.0	.9987	.9987	.9987	.9988	.9988	.9989	.9989	.9989	.9990	.9990
3.1	.9990	.9991	.9991	.9991	.9992	.9992	.9992	.9992	.9993	.9993
3.2	.9993	.9993	.9994	.9994	.9994	.9994	.9994	.9995	.9995	.9995
3.3	.9995	.9995	.9995	.9996	.9996	.9996	.9996	.9996	.9996	.9997
3.4	.9997	.9997	.9997	.9997	.9997	.9997	.9997	.9997	.9997	.9998

z	.00
3.5	.999767
4.0	.9999683
4.5	.9999966
5.0	.999999713

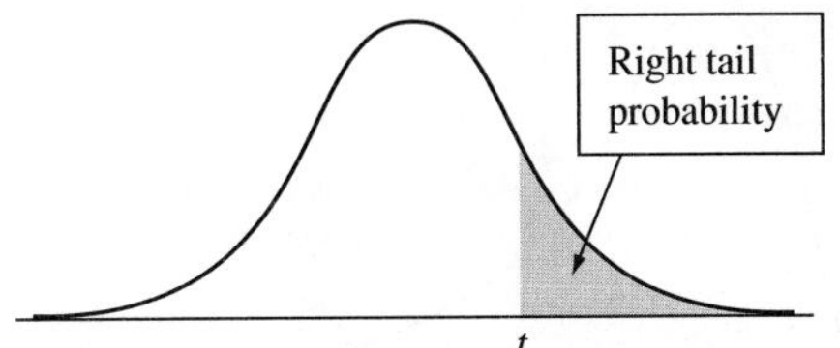

Table B *t* **Distribution Critical Values**

	Confidence Level					
	80%	90%	95%	98%	99%	99.8%
	Right-Tail Probability					
df	$t_{.100}$	$t_{.050}$	$t_{.025}$	$t_{.010}$	$t_{.005}$	$t_{.001}$
1	3.078	6.314	12.706	31.821	63.656	318.289
2	1.886	2.920	4.303	6.965	9.925	22.328
3	1.638	2.353	3.182	4.541	5.841	10.214
4	1.533	2.132	2.776	3.747	4.604	7.173
5	1.476	2.015	2.571	3.365	4.032	5.894
6	1.440	1.943	2.447	3.143	3.707	5.208
7	1.415	1.895	2.365	2.998	3.499	4.785
8	1.397	1.860	2.306	2.896	3.355	4.501
9	1.383	1.833	2.262	2.821	3.250	4.297
10	1.372	1.812	2.228	2.764	3.169	4.144
11	1.363	1.796	2.201	2.718	3.106	4.025
12	1.356	1.782	2.179	2.681	3.055	3.930
13	1.350	1.771	2.160	2.650	3.012	3.852
14	1.345	1.761	2.145	2.624	2.977	3.787
15	1.341	1.753	2.131	2.602	2.947	3.733
16	1.337	1.746	2.120	2.583	2.921	3.686
17	1.333	1.740	2.110	2.567	2.898	3.646
18	1.330	1.734	2.101	2.552	2.878	3.611
19	1.328	1.729	2.093	2.539	2.861	3.579
20	1.325	1.725	2.086	2.528	2.845	3.552
21	1.323	1.721	2.080	2.518	2.831	3.527
22	1.321	1.717	2.074	2.508	2.819	3.505
23	1.319	1.714	2.069	2.500	2.807	3.485
24	1.318	1.711	2.064	2.492	2.797	3.467
25	1.316	1.708	2.060	2.485	2.787	3.450
26	1.315	1.706	2.056	2.479	2.779	3.435
27	1.314	1.703	2.052	2.473	2.771	3.421
28	1.313	1.701	2.048	2.467	2.763	3.408
29	1.311	1.699	2.045	2.462	2.756	3.396
30	1.310	1.697	2.042	2.457	2.750	3.385
40	1.303	1.684	2.021	2.423	2.704	3.307
50	1.299	1.676	2.009	2.403	2.678	3.261
60	1.296	1.671	2.000	2.390	2.660	3.232
80	1.292	1.664	1.990	2.374	2.639	3.195
100	1.290	1.660	1.984	2.364	2.626	3.174
∞	1.282	1.645	1.960	2.326	2.576	3.091

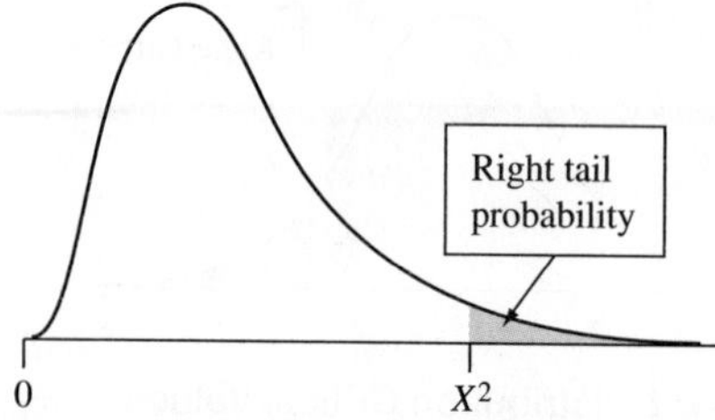

Table C Chi-Squared Distribution for Values of Various Right Tail Probabilities

	Right-Tail Probability						
df	**0.250**	**0.100**	**0.050**	**0.025**	**0.010**	**0.005**	**0.001**
1	1.32	2.71	3.84	5.02	6.63	7.88	10.83
2	2.77	4.61	5.99	7.38	9.21	10.60	13.82
3	4.11	6.25	7.81	9.35	11.34	12.84	16.27
4	5.39	7.78	9.49	11.14	13.28	14.86	18.47
5	6.63	9.24	11.07	12.83	15.09	16.75	20.52
6	7.84	10.64	12.59	14.45	16.81	18.55	22.46
7	9.04	12.02	14.07	16.01	18.48	20.28	24.32
8	10.22	13.36	15.51	17.53	20.09	21.96	26.12
9	11.39	14.68	16.92	19.02	21.67	23.59	27.88
10	12.55	15.99	18.31	20.48	23.21	25.19	29.59
11	13.70	17.28	19.68	21.92	24.72	26.76	31.26
12	14.85	18.55	21.03	23.34	26.22	28.30	32.91
13	15.98	19.81	22.36	24.74	27.69	29.82	34.53
14	17.12	21.06	23.68	26.12	29.14	31.32	36.12
15	18.25	22.31	25.00	27.49	30.58	32.80	37.70
16	19.37	23.54	26.30	28.85	32.00	34.27	39.25
17	20.49	24.77	27.59	30.19	33.41	35.72	40.79
18	21.60	25.99	28.87	31.53	34.81	37.16	42.31
19	22.72	27.20	30.14	32.85	36.19	38.58	43.82
20	23.83	28.41	31.41	34.17	37.57	40.00	45.32
25	29.34	34.38	37.65	40.65	44.31	46.93	52.62
30	34.80	40.26	43.77	46.98	50.89	53.67	59.70
40	45.62	51.80	55.76	59.34	63.69	66.77	73.40
50	56.33	63.17	67.50	71.42	76.15	79.49	86.66
60	66.98	74.40	79.08	83.30	88.38	91.95	99.61
70	77.58	85.53	90.53	95.02	100.4	104.2	112.3
80	88.13	96.58	101.8	106.6	112.3	116.3	124.8
90	98.65	107.6	113.1	118.1	124.1	128.3	137.2
100	109.1	118.5	124.3	129.6	135.8	140.2	149.5

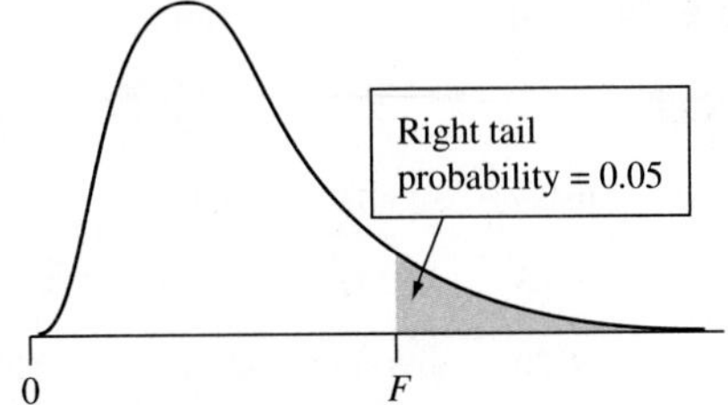

Table D ***F*** **Distribution for Values of Right-Tail Probability = 0.05**

	df_1									
df_2	1	2	3	4	5	6	8	12	24	∞
1	161.45	199.50	215.71	224.58	230.16	233.99	238.88	243.91	249.05	254.31
2	18.51	19.00	19.16	19.25	19.30	19.33	19.37	19.41	19.45	19.50
3	10.13	9.55	9.28	9.12	9.01	8.94	8.85	8.74	8.64	8.53
4	7.71	6.94	6.59	6.39	6.26	6.16	6.04	5.91	5.77	5.63
5	6.61	5.79	5.41	5.19	5.05	4.95	4.82	4.68	4.53	4.37
6	5.99	5.14	4.76	4.53	4.39	4.28	4.15	4.00	3.84	3.67
7	5.59	4.74	4.35	4.12	3.97	3.87	3.73	3.57	3.41	3.23
8	5.32	4.46	4.07	3.84	3.69	3.58	3.44	3.28	3.12	2.93
9	5.12	4.26	3.86	3.63	3.48	3.37	3.23	3.07	2.90	2.71
10	4.96	4.10	3.71	3.48	3.33	3.22	3.07	2.91	2.74	2.54
11	4.84	3.98	3.59	3.36	3.20	3.09	2.95	2.79	2.61	2.40
12	4.75	3.89	3.49	3.26	3.11	3.00	2.85	2.69	2.51	2.30
13	4.67	3.81	3.41	3.18	3.03	2.92	2.77	2.60	2.42	2.21
14	4.60	3.74	3.34	3.11	2.96	2.85	2.70	2.53	2.35	2.13
15	4.54	3.68	3.29	3.06	2.90	2.79	2.64	2.48	2.29	2.07
16	4.49	3.63	3.24	3.01	2.85	2.74	2.59	2.42	2.24	2.01
17	4.45	3.59	3.20	2.96	2.81	2.70	2.55	2.38	2.19	1.96
18	4.41	3.55	3.16	2.93	2.77	2.66	2.51	2.34	2.15	1.92
19	4.38	3.52	3.13	2.90	2.74	2.63	2.48	2.31	2.11	1.88
20	4.35	3.49	3.10	2.87	2.71	2.60	2.45	2.28	2.08	1.84
21	4.32	3.47	3.07	2.84	2.68	2.57	2.42	2.25	2.05	1.81
22	4.30	3.44	3.05	2.82	2.66	2.55	2.40	2.23	2.03	1.78
23	4.28	3.42	3.03	2.80	2.64	2.53	2.37	2.20	2.01	1.76
24	4.26	3.40	3.01	2.78	2.62	2.51	2.36	2.18	1.98	1.73
25	4.24	3.39	2.99	2.76	2.60	2.49	2.34	2.16	1.96	1.71
26	4.23	3.37	2.98	2.74	2.59	2.47	2.32	2.15	1.95	1.69
27	4.21	3.35	2.96	2.73	2.57	2.46	2.31	2.13	1.93	1.67
28	4.20	3.34	2.95	2.71	2.56	2.45	2.29	2.12	1.91	1.65
29	4.18	3.33	2.93	2.70	2.55	2.43	2.28	2.10	1.90	1.64
30	4.17	3.32	2.92	2.69	2.53	2.42	2.27	2.09	1.89	1.62
40	4.08	3.23	2.84	2.61	2.45	2.34	2.18	2.00	1.79	1.51
60	4.00	3.15	2.76	2.53	2.37	2.25	2.10	1.92	1.70	1.39
120	3.92	3.07	2.68	2.45	2.29	2.18	2.02	1.83	1.61	1.25
∞	3.84	3.00	2.60	2.37	2.21	2.10	1.94	1.75	1.52	1.00

Table E Table of Random Numbers

Line/Col.	(1)	(2)	(3)	(4)	(5)	(6)	(7)	(8)
1	10480	15011	01536	02011	81647	91646	69179	14194
2	22368	46573	25595	85393	30995	89198	27982	53402
3	24130	48360	22527	97265	76393	64809	15179	24830
4	42167	93093	06243	61680	07856	16376	39440	53537
5	37570	39975	81837	16656	06121	91782	60468	81305
6	77921	06907	11008	42751	27756	53498	18602	70659
7	99562	72905	56420	69994	98872	31016	71194	18738
8	96301	91977	05463	07972	18876	20922	94595	56869
9	89579	14342	63661	10281	17453	18103	57740	84378
10	85475	36857	53342	53988	53060	59533	38867	62300
11	28918	69578	88231	33276	70997	79936	56865	05859
12	63553	40961	48235	03427	49626	69445	18663	72695
13	09429	93969	52636	92737	88974	33488	36320	17617
14	10365	61129	87529	85689	48237	52267	67689	93394
15	07119	97336	71048	08178	77233	13916	47564	81056
16	51085	12765	51821	51259	77452	16308	60756	92144
17	02368	21382	52404	60268	89368	19885	55322	44819
18	01011	54092	33362	94904	31273	04146	18594	29852
19	52162	53916	46369	58586	23216	14513	83149	98736
20	07056	97628	33787	09998	42698	06691	76988	13602

Answers

Chapter 1

1.1 (a) The 22,000 physicians, the randomization, and the plan to obtain percentages of each group that have heart attacks **(b)** The actual percentages of each group who have heart attacks **(c)** The generalization that aspirin reduces the risk of having a heart attack **1.3** 64.6%; 20.8%; 8.7%; 5.9% **1.5** Responses will differ for each student **1.7 (a)** The 840 respondents to the GSS question **(b)** All adult Americans **(c)** The percentage of the sample with only one good friend, namely 6.1% **1.9 (a)** a new Honda Accord car; **(b)** the new Honda Accords that are chosen for the study; **(c)** all new Honda Accords **1.11 (a)** Descriptive statistics; they summarize data from a population. **(b)** Parameter, since they refer to a population. **1.13 (a)** Very short bars toward the top indicate very few old men and women in 1750. **(b)** Bars are much longer for both men and women in 2010 than in 1750. **(c)** The bars for women in their 70s and 80s in 2010 are longer than those for men. **(d)** From the graph, the longest bars are associated with the 45–49 age group. **1.15 (a)** Yes; **(b)** No. Two random samples may differ by chance. **1.17** The results for this exercise will vary. **1.19** People choose to participate instead of being randomly chosen to participate. The feelings of those who choose to participate usually are not representative of the general population. In general, one should not rely much on the information contained in such samples. **1.21** Answer is data file printed from software. **1.23** Answers will vary. It would be surprising to observe less than 50%. With the applet, most samples fall within 14 points of the true population percentage—from about 56 to 84. **1.25 (a)** All American adults **(b)** The sample data are summarized by a proportion, 0.638. **(c)** The population proportion who had at least one such experience **1.27 (a)** sample data; **(b)** percentages reported; **(c)** reporting margin of error for sample percentage **1.29 (a)** Population: all customers; Sample: 1000 customers **(b)** Descriptive: average sales per person in sample; Inferential: estimate of average sales per person in population. **1.31** (c) **1.33** Will be different for each student. **1.35** You will get a sample proportion as large as 0.638 when the true proportion is 0.20 only *very* rarely.

Chapter 2

2.1 (a) Categorical: observations belong to one of a set of categories. Quantitative: observations are numerical. **(b)** Examples will differ for each student. **2.3 (a)** Quantitative. **(b)** Categorical. **(c)** Categorical. **(d)** Quantitative. **2.5 (a)** Discrete: possible variables are separate values, such as 0, 1, 2,.... Continuous: possible variables form an interval. **(b)** Examples will differ for each student. **2.7 (a)** Continuous. **(b)** Discrete. **(c)** Discrete. **(d)** Continuous. **2.9 (b)** Bar chart. **(c)** We get a better sense of the data when we can see the sizes of the various categories. **2.11 (a)** Categories of a variable. **(b)** Frequency and percentage of weather stations in a given region. **(c)** Bar graph; we can more easily compare the heights of bars. **2.13** The Pareto chart allows us to easily see what outcomes occurred most frequently. **2.15 (b)** Prices ranged from 2195 to 7500, with most selling closer to the lower end of this range. **2.17 (a)** 1|3333445677778899
2|04.

This plot is too compact, making it difficult to visualize where the data fall. **(b)** 1|333344
1|5677778899
2|04.

2.19 (a) 0 mg and 1000 mg. **(b)** 3000, 4000, 11000, 12000, and 14000. **2.21 (a)** −1 to 1, 1 to 3, 3 to 5, 5 to 7, 7 to 9, 9 to 11, 11 to 13, 13 to 15, 15 to 17, and 17 to 19. **(b)** Bimodal; child cereals, on average, have more sugar than adult cereals have. **(c)** The dot and stem-and-leaf plots allow us to see all the individual data points. **(d)** The relative differences among bars would remain the same. **2.23 (a)** Skewed to the right because of some very expensive homes. **(b)** Skewed to the right because of the few faculty who overdraw frequently. **(c)** Symmetric because most would be in the middle, with some higher and some lower. **(d)** Symmetric because most would fall in the middle, with some higher and some lower. **2.25 (a)** Discrete; the value for each person would be a whole number. **(b) (i)** 0; **(ii)** 9; **(iii)** 2; **(iv)** 3. **(c)** Unimodal and somewhat skewed to the right. **2.27 (a)** Somewhat skewed to the left. **(b)** A time plot connects the data points over time to show time trends. **(c)** A histogram shows the number of observations at each level more easily than does the time plot. We also can see the shape of the distribution from the histogram but not from the time plot. **2.29** Overall, the time plot does seem to show a decrease in temperature over time. **2 .31 (a)** Mean: 2601; Median: 1495. **(b)** Some of these nations have much larger populations than others; it is misleading to compare the total emission for a large nation to the total emission for a small nation, because other things being equal, it tends to be higher for the nations with larger populations. **2.33** Both distributions are skewed to the right. When this happens, means are higher than medians. **2.35** The moderate skewness to the left causes the mean to be lower than the median. **2.37 (a)** Mean: 2; median: 0; and mode: 0 **(b)** Mean: 10; median: 0. Outlier affects mean but not median. **2.39 (a)** The most common score (mode), and the middle score (median), will be zero. The few very expensive hospital stays will inflate the mean to be greater than 0. **(b)** Number of times arrested in previous year. **2.41** Skewed to the right because the mean is much higher than the median. **2.43 (a)** 1.7. **(b)** 1.65. **(c)** The mean need not take one of the possible values for the variable. Although the number of children born to each adult woman is a whole number, the mean number of children born per adult woman need not be a whole number. For example, the mean number of children per adult woman is considerably higher in Mexico than in Canada. **2.45 (a)** 0.16; **(b)** 0; **(c)** Median: 0; mean: 1.95. **(d)** The median is the same for both because the median ignores much of the data. **2.47 (a)** Range = 6; six days separate the fewest and most sick days taken. **(b)** $s = 2.38$; the typical distance of an observation from the mean is 2.38. **(c)** Range = 60; standard deviation = 21.06. Both increase when an outlier is added. **2.49** Larger, because the value for Russia is much smaller than the rest of the group, adding additional variability to the data. (In fact, $s = 3.6$, including Russia, compared to 1.0.). **2.51** 12 is most realistic. −10 is negative; 0 indicates *no* variability; 3 is too small and 63 is too large for a typical deviation. **2.53 (a) (i)** The right; **(ii)** Middle. **(b)** The left because it is bell-shaped. **2.55 (a)** 99–167. **(b)** 184. Yes. It is very high; all or nearly all observations will fall within three standard deviations from the mean. **2.57 (a)** 84.0% fall between −0.21 and 0.53; 84.0% fall between −0.58 and 0.90; 99.9% fall between −0.95 and 1.27. **(b)** This is not a bell-shaped distribution but is highly skewed to the right. **2.59 (a)** Skewed to the right. The standard deviation is larger than the mean, the lowest possible value of 0 is less than 1 standard deviation below the mean, and the mean is higher than the median. **(b)** No. The data are not bell-shaped. **2.61 (b)** Standard Deviation = 3.695. This is a typical deviation from the mean unemployment for EU nations. **2.63 (a)** 31; **(b)** 25.5; **(c)** 36; **(d)** 25% of countries have residents who take fewer than 25.5 vacation days, half have residents who take fewer than 31, and 75% of countries have residents who take fewer than 36. **2.65. (a)** One fourth had maximum bench press less than 70 pounds, and one fourth had maximum bench press greater than 90 pounds. **(b)** Symmetric. The mean and median are about the same, and the first and third quartiles are equidistant from the median. **2.67 (a)** The range is more affected by an outlier, and the standard deviation uses all the data. **(b)** IQR is not affected by an outlier. **(c)** The standard deviation takes into account the values of *all* observations.

2.69 (a) Range = 6; there are six days separating the fewest and most sick days taken. **(b)** The range of the middle 50% of scores is 2. **(c)** a. Range: 60; b. IQR: 2. The IQR is least affected by the outlier because it doesn't take the magnitudes of the two extreme scores into account. **2.71** Q1 = median of lower half of sorted data: 3, 3, 3, 4, 4, 4 and 4. It is 4. Q2, = median = average of middle two data values, (4 + 4)/2 = 4. Q3 = median of upper half of sorted data: 4, 4, 4, 4, 5, 5 and 5. It is 4. **2.73 (a)** Skewed left because of the low outliers and the slightly longer left whisker. **(b)** Min: 49, Q1: 52.5, Median = 53.5, Q3 = 55, Max = 57. **2.75 (a)** Minimum: 0; Q1: 0; Median: 0; Q3: 4; Maximum: 10 (part of answer is box plot) **(b)** The middle scores of the whole set of data and of the lower half of the data set are both zero. **(c)** The minimum score also is zero. **2.77 (a)** Outer edges of box: 4.5 and 7.8; ends of whiskers: 3.2, 8.7. **(b)** No; it is only 1.33 standard deviations above the mean. **(c)** 6.3 (mean). **2.79 (a)** −3.1 **(b)** The height is below the mean. **(c)** Yes; it is more than three standard deviations from the mean. **2.81 (a)** The distribution depicted in the box plot (part of answer is box plot) is skewed to the right. Most observations fall between about 0 and 15 but there are a few outliers representing very large values. **(b)** There are several potential outliers; the 1.5(IQR) criterion indicates that observations falling above 20.5 are potential outliers. **2.83 (a)** 22. **(b)** skewed right, indicated by longer whisker for large values and greater distance from median of Q3 than Q1. **(c)** Emissions tend to be much higher in Europe, with Q1 in Europe being above all values in South America. **2.85 (a)** The percentages do not add up to 100, and the Tesco slice seems too large for 27.2%. Also, contiguous colors are similar. **(b)** Bar graph. We would merely have to identify the highest bar. **2.87 (a)** Because the vertical axis does not start at 0, it appears that six times as many people are in the "no, not" column than in the "yes, safer" column. **(b)** With a pie chart, the area of each slice represents the percentage who fall in that category. Therefore, the relative sizes of the slices will always represent the relative percentages in each category. **2.89** The slices do not seem to have the correct sizes, for instance the slice with 16% seems larger than the slice with 19%. **2.91 (a)** Quantitative. **(b)** Quantitative. **(c)** Categorical. **(d)** Categorical. **2.93 (a)** Categorical; add percentages: 13.7, 9.6, 37.6, 6.1, 25.4, 7.6, and 100 for total. **(b)** These values are frequencies for categorical data and so we would use the modal category: Central America. **(c)** The place of birth categories would be organized from highest to lowest percentages. It's easier to make comparisons and to identify the most common outcomes. **2.95 (a)** Optical scanning with a two-column ballot. **(b)** One column with either optical scanning or votomatic. **(c)** Bar for each combination (e.g., optical, one column; optical, two columns) with each bar showing the overall percentage of overvotes. **2.97 (c)** 3.5. **(d)** Slightly skewed to the right. **2.99 (b)** Allows one to see the individual scores. **(c)** The distribution is skewed to the left with one low outlier. **2.101 (a)** Skewed to the right, since most values at 0 but some large values. **(b)** Skewed to the left, since most values at 1 hour or slightly less but some could be quite a bit less. **(c)** Skewed to the right, since some extremely large values. **(d)** Skewed to the left, since most values are high but some very young people die. **2.103** Larger; right skew (some very large values) would pull the mean higher. **2.105 (a)** Skewed to the right. **(b)** 2,090,012 is the mean and $1,646,853 is the median. **2.107 (a)** 1, 2, 4, 6, 7. **(b)** 2, 2, 3, 5, 6. **2.109 (a)** 12 is best; −10 is negative (not possible); 1 indicates almost no variability; 60 is almost as large as the whole range. **(b)** −20. standard deviations must be nonnegative. **2.111 (a)** Skewed to the right; maximum is much farther from the mean than the minimum is, and the lowest possible value of 0 is only 780/506 = 1.54 standard deviations below the mean. **(b)** Skewed to the right; the standard deviation is almost as large as the mean and the smallest possible value is 0. **2.113 (a)** Nearly all heights would fall between 62.2 and 79.6. **(b)** The center for women is about five inches less than the center for men. The variability, however, is very similar. **(c)** 3.07. **2.115 (a)** Numbers are approximate: Min: 0; Q1: 4; Median: 9.5; Q3: 13.5; Max: 18. **(b)** Slightly skewed to the left. **(c)** −1.64. This sugar value falls 1.64 standard deviations below the mean. **2.117 (a)** Symmetric, centered around 41 degrees, range from 30 to 51, with most months between 35 and 45. **(b)** mean = 40.3, std. dev. = 4.2. **(c)** March is colder and varies more than November. **(d)** The box plots make the comparison easier because they are embedded on the same numerical axis. **2.119 (a)** Skewed to the right **(b)** 4 most plausible; −16 is negative; 0 would indicate no variability; 15 is almost as large as the whole range, and 25 is even larger. **2.121 (a)** Range: 14.; IQR: 5.2. **(b)** No, because the minimum is not more than 1.5 IQR below Q1, and the maximum is not more than 1.5 IQR above Q3. **2.123 (a)** The z-score of 1.19 indicates that a person with a blood pressure of 140 falls 1.19 standard deviations above the mean. **(b)** 89 to 153. **2.125 (a)** Neither the minimum nor the maximum score has a z-score more extreme than 3 or −3. **(b)** The maximum is much farther from the mean and median than is the minimum. **(c)** This has a z-score of only −0.62, and so is not unusual. **2.127** The responses will be different for each student depending on the methods used. **2.129** Answers will vary. **2.131 (a)** Sugar values are skewed to the right for adults and skewed to the left for children. **(b)** Adult sodium values vary across a wider range than children. **2.133 (a)**

$$\bar{x} = \frac{69(1) + 240(2) + 221(3) + 740(4) + 268(5) + 327(6) + 68(7)}{1933} = \frac{7950}{1933} = 4.11.$$

(b) 4. **(c)** The median, 4 (representing the Moderate category), is found from the middle score being the 666th score. **2.135** (a). **2.137** (a). **2.139** Standard deviation, much too large. **2.141** Dividing the frequency for a given category by the total number of subjects would give us the proportion. **2.143** $\bar{x} \pm 3s$ would encompass all or nearly all scores. A range of $3s + 3s = 6s$ contains all or nearly all scores, so approximates the range. **2.145 (a)** With greater variability, numbers tend to be further from the mean. Thus, the absolute values of their deviations from the mean would be larger. When we take the average of all these values, the overall MAD is larger than with distributions with less variability. **(b)** The MAD is more resistant than the standard deviation because by squaring the deviations using the standard deviation formula, a large deviation has greater effect.

Chapter 3

3.1 (a) Explanatory: high school GPA; response: college GPA. **(b)** Explanatory: mother's religion; response: number of children **(c)** Explanatory: marital status; response: happiness. **3.3 (a)** The response variable is happiness and the explanatory variable is income. **(b)** Using 2008 data,

Income	Not Too Happy	Pretty Happy	Very Happy	Total	n
Above average	0.06	0.55	0.39	1.00	423
Average	0.13	0.54	0.33	1.00	883
Below average	0.25	0.56	0.19	1.00	687

Those of above average income are most likely to be very happy whereas those of below average income are most likely to be not too happy.
(c) 0.30. **3.5 (a)** Response: binge drinking; explanatory: gender. **(b) (i)** 1908; **(ii)** 2854. **(c)** No; these are not proportions of male and female students; these are counts and there are far more females than males in this study. **(d)**

Gender	Binge Drinker	Non-Binge Drinker	Total	n
Male	0.49	0.51	1.00	3925
Female	0.41	0.59	1.00	6979

(e) It appears that men are more likely than are women to be binge drinkers. **3.7** Either variable could be the outcome of interest. **3.9** Using 2008 data, **(a)** Response: party identification; explanatory: gender **(b) (i)** 0.12; **(ii)** 0.14. **(c) (i)** 0.45; **(ii)** 0.26. **(d)** Conditional proportions suggest that more women than men are Democratic, and that more men than women are Independent or Republican. **3.11 (a)** Positive; as cars age, they tend to have covered more miles. **(b)** Negative; as cars age, they tend to be worth less. **(c)** Positive; older cars tend to have needed more repairs. **(d)** Negative; heavier cars tend to travel fewer miles on a gallon of gas. **3.13 (a)** (box plot is part of answer); skewed to the right. **(b)** Skewed to the right. **(c)** Intuitively, one might expect a strong, positive, linear relationship between the two. **(d)** Joe is confusing the notion of standardization, which involves all *variables* getting divided by *the same* value (the standard deviation), with that of dividing overall GDP by *a given country's* population size to obtain per capita GDP. **3.15 (a)** Internet users and Broadband subscribers. **(b)** Facebook users and Population. **(c)** Because each pair of observations for Internet users and Facebook users are divided by the corresponding nation's population size to obtain the percentages. **3.17 (a)** (answer is scatterplot). **(b)** (4, 13). **(c)** The value of 13 would have to be changed to 10. **3.19** Answer is scatterplot. **3.21 (a) (i)** weight; **(ii)** price. **(b)** Bikes with weights in the middle tend to cost the most (scatterplot is part of answer). **(c)** Negative and fairly small; weight does not appear to affect price strongly in a linear manner. **3.23 (a)** Both box plots indicate that the counts are skewed to the right with few counties in the high ranges of vote counts (two box plots are part of the answer). **(b)** The point close to 3500 on the variable Buchanan is a regression outlier; we were unable to make this comparison from the box plots because there were two separate depictions, one for each candidate (scatterplot is part of the answer). **(c)** About 1000 **(d)** The point close to 3500 on the variable Buchanan is still an outlier (two box plots and a scatterplot are part of the answer). **3.25 (a) (i)** 6.47; **(ii)** 5.75. Connect the points ($x = 10$, $y = 6.47$) and ($x = 40$, $y = 5.75$). **(b)** The y-intercept indicates that when a person cannot do any sit-ups, she/he would be predicted to run the 40-yard dash in 6.71 seconds. The slope indicates that every increase of one sit-up leads to a decrease in running time of 0.024 seconds. **(c)** The slope indicates a negative correlation (the slope and correlation have the same sign). **3.27 (a) (i)** \$2.50; **(ii)** \$122.50. **(b)** For every one point increase in food quality rating, the predicted price of the dinner increases by \$4.00. **(c)** The correlation of 0.53 is a moderate positive correlation; higher costs tend to correspond to higher food quality ratings. **(d)** $b = \left(\frac{s_y}{s_x}\right) = 0.53\left(\frac{20.54}{2.70}\right) = 4.0$. **3.29 (a)** Correlation is positive and so are S_x and S_y. **(b)** 7.92%. **(c)** 5.57%. **3.31 (a)** 506.6. **(b)** 9.5; this means Californians scored about 9.5 points higher on math than what might have been expected based on the reading scores. **(c)** $R^2 = 95.5\%$, so reading is a very good predictor of math. **3.33. (a)** The line fit by MINITAB minimizes the sum of the squared residuals. **(b)** Two cereals had more sodium than we would expect based on their sugar contents. These were Raisin Bran and Rice Krispies. **(c)** Not reliable at all; R^2 is close to zero (0.2%). **3.35 (a)** answer is a scatterplot **(b)** 1.0; $\hat{y} = 4 + 2x$. **(c)** Advertising: 1, 1; Sales: 6, 2. **(d)** $\hat{y} = 4 + 2x$; when there is no advertising, it is predicted that sales will be about \$4,000. For each increase of \$1000 in advertising, predicted sales increase by \$2000. **3.37 (a)** $\hat{y} = 24 + 0.70x$. **(b)** Predicted final exam score = 80. **3.39 (b)** 0.81; strong positive association, with longer study times associated with higher GPAs. **(c)** $\hat{y} = 2.63 + 0.044x$. **(i)** 2.84, **(ii)** 3.72 **3.41 (a)** (answer is a scatterplot). **(b)** From MINITAB: price = 1896 − 40.5 weight. Predicted price decreases by \$40.5 for each pound increase in weight. No, because there were no weights around 0. **(c)** \$681. **3.43 (a)** 1708 for lowa and 1409 for Maine. **(b)** − 142.2; West Virginians scored well below what we would predict based on their participation rate. **(c)** Some states take predominantly the ACT instead of SAT. Those who took the SAT in states with small participation rates are probably strong students planning to apply to good universities nationwide. **3.45 (a)** The observation for 1896 is well below the general trend. **(b)** Value from the regression equation because year looks to be a useful predictor (it can't lead to something less reliable than the sample mean). **(c)** No. In general, unreliable results arise when we extrapolate well outside the range of the data. **3.47 (a)** $\hat{y} = -3.1 + 0.33(15) = 1.85$; $\hat{y} = -3.1 + 0.33(40) = 10.1$. **(b)** $\hat{y} = 8.0 - 0.14(15) = 5.9$; $\hat{y} = 8.0 - 0.14(40) = 2.4$. **(c)** D.C. is a regression outlier because it is well removed from the trend of the rest of the data. **(d)** Because D.C. is so high on both variables, it pulls the line upward on the right and suggests a positive correlation, when the rest of the data (without D.C.) are negatively correlated. The relationship is best summarized after removing D.C. **3.49 (a)** Yes on all three counts. **(b) (1)** the x value (the United States in this case) is relatively low or high compared to the rest of the data; **(2)** the observation (the United States) is a regression outlier, falling quite far from the trend that the rest of the data follow. TV watching in the United States is very high despite the very low birth rate. **(c)** The association is **(i)** very weak without the United States because the six countries, although they vary in birth rates, all have very few televisions, and is; **(ii)** very strong with the United States because the United States is so much higher in number of televisions and so much lower on birth rate that it makes the two variables seem related. **(d)** Because that point has a large effect on pulling the line downward. **3.51 (a)** Perhaps Raisin Bran. **(b)** With Raisin Bran removed, r changed from −0.017 to −0.300, which is a substantial increase. **3.53 (a)** Not likely. **(b)** Values would be higher on both variables as age increases. **(c)** There is an overall positive correlation between height (in inches) and vocabulary (assessed on a scale of 1–10) if we ignore grade; if we look within each grade, we see roughly a horizontal trend and no particular association (scatterplot part of answer). **3.55 (a)** Several possible responses to this exercise (e.g., students who watch TV more may be more likely to see the ads and less likely to spend time with peers who smoke pot). **(b)** Pot smoking (or the lack thereof) might be caused by many variables (e.g., antidrug ads, television watching, regular school attendance, parental influence, neighborhood type). **3.57 (c)** 0.73, **(d) (i)** −0.96, **(ii)** −0.95. **3.59 (a)** Response: eighth grade math scores; explanatory: state **(b)** The third variable is race. Nebraska has the overall higher mean because there is a higher percentage of whites and a lower percentage of blacks in Nebraska than in New Jersey, and overall, whites tended to have higher math scores than blacks. **3.61 (a)** Response: assessed value; explanatory: square feet. **(b)** Response: political party; explanatory: gender. **(c)** Response: income; explanatory: education. **(d)** Response: pounds lost; explanatory: type of diet. **3.63 (a) Opinion about life after death and using 2008 data,**

Gender	Yes	No	Total
Male	76.9	23.1	808
Female	85.2	14.8	979

(b) Women are slightly more likely than are men to respond yes. **3.65 (a)** Income; quantitative **(b)** Degree; categorical. **(c)** A bar graph could have a separate bar for each degree type. **3.67 (a)** Response: gender; explanatory: year. **(b)** For executive, administrative, and managerial

Year	Female	Male	Total
1972	0.197	0.803	1.00
2002	0.459	0.541	1.00

(c) Yes; women made up a larger proportion of the executive work force in 2002 than in 1972. **(d)** Year and type of occupation. **3.69 (a)** Correlation = 0.745 **(b)** $\hat{y} = -48.91 + 0.919x$. Since the y-intercept would correspond to female economic activity = 0, which is well outside of the range of data. **(c)** The predicted value for the United States is −48.91 + 0.919(81) = 25.5 with 15.0 − 25.5 = −10.5 as the corresponding residual. The regression equation overestimates the percentage of women in parliament by 10.5% for the United States. **(d)** $b = 0.56(9.8/7.7) = 0.713$ and $a = 26.5 - 0.713(76.8) = -28.2$. Thus, the prediction equation is given by $\hat{y} = -28.2 + 0.713x$. **3.71 (a)** Slope: 0.56; as urban goes from 0 to 100, predicted crime rate increases by 56. **(b)** Relatively strong, positive relationship. **(c)** $b = r\left(\frac{s_y}{s_x}\right)$; $0.56 = 0.67(28.3/34.0)$. **3.73 (a)** Response: salary; explanatory: height. **(b)** Slope: \$789 per inch. **(c)** An increase of seven inches (from 5 foot 5 to 6 feet) is worth \$789 per inch, or \$5523. **3.75** y-intercept: 0; slope: 1; this means that your predicted college GPA equals your high school GPA. **3.77 (a)** A 1000 increase in x would mean a predicted change in y of −5.2 (poorer mileage). **(b)** 14.04; 2.96. The Hummer gets 2.96 more miles/gallon than one would predict. **3.79 (a)** \$4200. **(b)** $r = 0.79$; **(i)** positive; **(ii)** strong. **3.81 (a)** When economic activity is 0%, predicted birth rate is much higher (36.3) than when economic activity is 100% (6.3). **(b)** women's economic activity; the correlation between birth rate and women's economic activity is larger in magnitude than the correlation between birth rate and GNP. **3.83 (a)** −16,000. **(b)** 3200; **(c)** same (0.50); correlation does not depend on the units used **3.85 (a)** D.C. is the outlier to the far, upper right. This would have an effect on the regression analysis because it is a regression outlier. **(b)** When D.C. is included, the y-intercept decreases and the slope increases. **3.87 (a)** The point with a y value of around 1500 (scatterplot is part of answer). **(b)** Violent crime rate = 2545 − 24.6 high school; for each increase of 1% of people with a high school education, the predicted violent crime rate decreases by 24.6. **(c)** Violent crime rate = 2268 − 21.6 high school; for each increase of 1% of people with a high school education, predicted violent crime rate decreases by 21.6, similar to the slope in part b. **3.89 (a)** Negative relationship (scatterplot is part of answer). **(b)** −0.45; negative association. **(c)** Predicted percentage without health ins = 49.2 − 0.42 HS Grad Rate; for each increase of one percent in the high school graduation rate, the predicted percentage of individuals without health insurance goes down by 0.42; negative relationship. **3.91 (a)** Men tend to be taller and make more money than women. **(b)** When measured, a lurking variable becomes a confounding variable. **3.93 (a)** As people age, they might both sleep more and be more likely to die. **(b)** Subject's age might actually cause people to sleep more and be more likely to die. **3.95** Responses will vary. **3.97** Temp = 119 − 0.029 Year; the regression line indicates a very slight decrease over time, the opposite of what is indicated by the Central Park data. **3.99** San Francisco could be higher than other cities on lots of variables, but that does not mean those variables cause AIDS, as association does not imply correlation. Alternative explanations are that San Francisco has a relatively high gay population or relatively high intravenous drug use, and AIDS is more common among gays and IV drug users. **3.101** Stress level, physical activity, wealth, and social contacts are all possible lurking variables. Any one of these variables may contribute to one's physiological and psychological human health as well as be associated with whether or not a person owns a dog. For example, it may be that people who are more active are more likely to own a dog as well as being physically healthier. **3.103** (b). **3.105** (d). **3.107** (c) **3.109 (a)** If we convert income from British pounds to dollars, we multiply each score by two. Thus, each y-value doubles, the mean of y is now doubled, and the distance of each score from the mean doubles. If this is all so, then the variability (the standard deviation) has now doubled. **(b)** The correlation would not change in value because the correlation is based on standardized versions of the measures—z-scores, rather than raw scores. **3.111 (a)** Algebraically, the formula $a = \bar{y} - b\bar{x}$ becomes $\bar{y} = a + b\bar{x}$. The latter formula is similar to the regression equation, except the generic predicted y, $\hat{y}$, is replaced by the mean of y, $\bar{y}$. Similarly, the generic x is replaced by $\bar{x}$. Thus, a score on any x that is at the mean will predict the mean for y. **(b)** Here are the algebraic steps to go from one formula to the other. **Step 1** Because $a = \bar{y} - b\bar{x}$, we can replace a in the regression equation with the formula $\hat{y} = \bar{y} - b\bar{x} + bx$. **Step 2** We can now subtract $\bar{y}$ from both sides: $\hat{y} - \bar{y} = -b\bar{x} + bx$ or $\hat{y} - \bar{y} = bx - b\bar{x} = b(x - \bar{x})$.

Chapter 4

4.1 (a) Explanatory is whether or not the automated call was placed to the phone on the right ear; response is a specific type of brain activity while the call is being received. **(b)** Experimental; the researchers controlled via randomization to determine whether the call to a given participant would be received during the first PET scan or the second. **4.3 (a)** The response variable is the death rate due to cancer. The explanatory variable is whether one is a Kuna Indian or a resident of mainland Panama. **(b)** Observational study, because the rate of death due to cancer was observed for the two areas over a period of time. The researchers did not assign the subjects to treatments. **(c)** One possible lurking variable is pollution level, which could be much higher in mainland Panama and higher levels of pollution could be associated with higher cancer death rates. **4.5** For example, it is possible that schools that institute drug testing are those in higher crime areas than are those that did not choose to use drug testing. **4.7 (a)** Response: whether woman's infant had a birth defect; explanatory: whether woman was a hairdresser. **(b)** Observational study because women were not randomly assigned to be hairdressers or not. **(c)** No; it's possible that there's a lurking variable, such as socioeconomic status (e.g., people who are less wealthy or educated tending to have poorer health care). **4.9 (a)** Observational study; **(b)** Observational study; **(c)** Experiment **4.11** The seat belt incident might be the exception, rather than what is typical. Death rates are in fact higher for those who do not wear seat belts. **4.13 (a)** Difficulty and expense. **(b)** To have a count of the population size. **(c)** Of the 10 questions on the form, Question 6 pertained to gender and Question 8 pertained to race. **4.15** Using line 10, choose individuals labeled 47 and 53. Using line 18, choose 01 and 54. **4.17** Number the accounts from 01 to 60, then pick 10 random two-digit numbers that fall between 01 and 60. Ignore duplicates. **4.19** Pick five-digit random numbers, ignoring 00000, numbers above 50,000, and duplicates. Keep the first ten numbers that are in range, and find the names associated with those 10 numbers. **4.21 (a)** 10 percentage points; **(b)** 5 percentage points; **(c)** 2.5 percentage points. As n increases, the sample becomes a more accurate reflection of the population, and the margin of error decreases. **4.23 (a)** It provides negative information within the question (e.g., "symbol of past slavery"). **(b)** It has neither explicitly negative nor explicitly positive information. **4.25 (a)** All executive recruiters **(b)** Intended: 1500; actual: 97; percentage of nonresponse: 93.5%. **(c)** Possible sources: nonresponse (those who responded were not representative) and executive recruiters might not be the best source of industries that are hiring. **4.27 (a)** Hong Kong residents. **(b)** Subjects were not assigned to have a particular perception. **(c)** This study used a volunteer sample; it may not be representative of the population. **4.29 (a)** Since the sampling method was not random, not all teenagers are equally likely to respond. **(b)** Some teenagers who have purchased alcohol over the Internet may refuse to respond because they are fearful of getting caught. **(c)** Not all teenagers may have answered the survey question truthfully, particularly if they are fearful of getting in trouble for answering in the affirmative. **4.31 (a)** Not all parts of the population have representation (because they may not have read the paper a long time) **(b)** Instead of random sampling, the newspaper used the 1000 people who subscribed the longest, who were probably the oldest and most conservative in terms of retirement planning. **(c)** Those who take the time to respond might not be representative. A large percentage of people did not respond. **(d)** The question seems to go out of the way to remind readers of volatility of stocks as of late. **4.33 (a)** Experiment; subjects are randomly assigned to treatments **(b)** Among the possible answers: unethical, impossible to ensure that subjects do as assigned and smoke or not smoke according to the assignment, too long to wait for answer **4.35 (a)** Response variable: whether the wart was successfully removed; explanatory variable: type of treatment for removing the wart (duct tape or the placebo). Experimental units: The 103 patients; treatments: duct-tape therapy and the placebo. **(b)** The difference between the number of patients whose warts were successfully removed using the duct-tape method and those using the placebo was not large enough to attribute to the treatment type. The difference in the success rates could be due to random variation. **4.37 (a)** Experimental units are the 878 employees; response variable is smoking cessation status; explanatory variable is treatment status; treatments are info only and info plus incentive. **(b)** Yes; because this is a randomized experiment, we are comfortable with concluding causality and saying the difference was due to the financial incentive. **4.39** Without a placebo or a control comparison group, there is no way to separate the placebo effect from the actual effect of the medication. **4.41** The researchers being blinded prohibits them from deliberately biasing the results, for example by intentionally or unintentionally providing extra support to one of the treatment groups. **4.43 (a)** Recruit volunteers (the experimental units) with a history of high blood pressure; randomly assign to one of two treatments: the new drug or the current drug; explanatory: treatment type; response: blood pressure after experimental period. **(b)** To make the study double-blind, the two drugs would have to look identical so that neither the subjects nor the experimenters know what drug the subjects are taking. **4.45 (a)** Number the students from 1 to 5. Pick one-digit numbers randomly. Select the first female student to have her number picked (1 to 3) and the first male student to have his number picked (4 or 5). Ignore duplicates and digit 0, and digits 6-9. **(b)** Because every sample of size 2 does not have an equal chance of being selected (e.g., samples with two males have probability 0 of being selected.) **(c)** 1 in 3; 1 in 2. **4.47 (a)** Treat each school as a cluster and randomly select clusters (schools). Keep a running tally of students, and continue choosing schools until the tally reaches approximately 20% of all students. **(b)** Answers will vary depending on labeling of schools. **(c)** Answers will vary. **(d)** Stratified random sampling is not possible. **4.49** It is possible that people with lung cancer had a different diet than did those without. These people might have eaten out at restaurants quite a bit, consuming more fat, and smoking socially. It could have been the fat and not the smoking that caused lung cancer. **4.51 (a)** Yes; if we look at enough data, some coincidental relationships are bound to be encountered. **(b)** Anticipated results stated in advance are usually more convincing than results that have already occurred. **(c)** Prospective. **4.53 (a)** The response: blood pressure, heart rate, and reported stress levels; explanatory: whether caffeine taken, with two treatments: caffeine pill or placebo pill; experimental units: 47 regular coffee drinkers **(b)** crossover design because all subjects participate in both treatments on different days **4.55 (a)** No; the researchers are not randomly assigning subjects to living situation. **(b)** Yes; those living with smokers and those not living with smokers. **(c)** Yes; randomization of units to treatments occurs within blocks. **4.57** In an observational study, we observe people in the groups they already are in. In an experiment, we actually assign people to the groups of interest. The major weakness of an observational study is that we can't control (such as by balancing through randomization) other possible factors that might influence the outcome variable. **4.59 (a)** Explanatory is status of protein presence; response is memory status. **(b)** Nonexperimental because the researchers are not assigning any treatments. **(c)** No; either the proteins are present or not; its presence is not something the researchers can control for each patient. **4.61** The study; the story of one person is anecdotal and not as strong evidence as a carefully conducted study with a much larger sample size. **4.63 (a)** Restricting the range would not allow us to determine an association between these variables. **(b)** There are several possible responses. For example, we could select a simple random sample, or we could stratify the sample to get equal numbers of students from each year. **4.65 (a)** Pick three random numbers between 0001 and 3500. **(b)** No; every sample is not equally likely; any possible sample with more than 40 males or fewer than 40 males has probability 0 of being chosen. **(c)** Stratified random sample; it offers the advantage of having the same numbers of men and women in the study, which would be unlikely if the population had a small proportion of one of these, and this is useful for making comparisons. **4.67 (a)** No; we cannot conclude causation from observational data. **(b)** Answers will vary. Possible variables are other chronic health conditions or an unhealthy diet. **4.69 (a)** Explanatory: history of playing video games; response: visual skills. **(b)** Observational study; because the men were not randomly assigned to treatment. **(c)** There are several possible answers (e.g., excellent reaction times might make it easier, and therefore more fun, to play video games, and might also lead to better performance on tasks measuring visual skills). **4.71 (a)** Response: whether have a heart attack; explanatory: treatment group (aspirin or placebo). **(b)** Because physicians were randomly assigned to treatment—either aspirin or placebo. **(c)** Because the experiment is randomized, we can assume that the groups are fairly balanced with respect to exercise. **4.73 (a)** Among those in the aspirin group, 0.493 never smoked, 0.397 smoked in the past, and 0.110 are current smokers. Among those in the placebo group, 0.498 never smoked, 0.391 smoked in the past, and 0.111 currently smoke. These proportions are similar. **(b)** Yes; the treatment groups are balanced in terms of smoking status. It does seem that the randomization process did a good job in achieving balanced treatment groups in terms of smoking status. **4.75 (a)** Nicotine patch only: 6.4 percentage points; bupropion only: 6.4 percentage points; nicotine patch with bupropion: 6.4 percentage points; placebo only: 7.9 percentage points **(b)** Yes; the margin of error for bupropion only indicates that the low end of believable values is 23.9%, whereas the margin of error for placebo only indicates that the high end of believable values is 23.5%. **(c)** No; there is substantial overlap between the ranges indicated by the margins of error. **(d)** The results of the study suggest that two of the treatments, bupropion only and nicotine patch with bupropion, led to higher abstinence percentages than did either of the other two treatments, nicotine only or placebo only. **4.77 (a)** Response: gas mileage; explanatory: brand of gas; treatments: Brands A and B **(b)** 10 cars would be randomly assigned to Brand A and 10 cars to Brand B. **(c)** Each car would be a block. It would first use gas from one brand, and then from the other. **(d)** It would reduce the effects of possible lurking variables because the two groups would be identical. **4.79 (a)** Clusters; **(b)** Because each possible sample does not have an equal chance. For instance, there is no chance of a sample in which just one person from a particular nursing home is sampled. **4.81** This is cluster random sampling. The colleges are the clusters. **4.83** Perhaps any relationship is coincidental, not causal. For example, perhaps those genetically susceptible to acquiring schizophrenia are the same group who especially enjoy marijuana. **4.85** Answers will vary, depending on the study. **4.87** No; due to the volunteer nature of the sample. **4.89** Answers will vary. **4.91 (a)** The sample includes only one cluster. **(b)** Sampling just Fridays is an example of sampling bias (values might be higher than usual because days sampled are at start of weekend). **4.93** This is not a random sampling method. People who approach the street corner are interviewed as they arrive (and as they agree to the interview!), and may not be representative. **4.95 (a)** Such a study would measure how well a treatment works if patients believe in it, rather than how much a treatment works independent of subjects' beliefs about its efficacy. **(b)** Patients might be reluctant to be randomly assigned to one of the treatments because they might perceive it as inferior to another treatment. **(c)** He or she might feel that all patients should get the new treatment. **4.97** Since the disease is rare, this proposed study might not sample anyone that gets it. With the case-control study, one would find a certain number of people who already have the rare disease (the cases), and compare the proportion of these people who had received the vaccine to the proportion in a group of controls who did not have the disease. **4.99** Assign the numbers 01 through 16 to the babies. Choose a random starting point on table. Proceed across table selecting unique two-digit numbers between 01 and 16 until obtaining a group of 8 to show videos in one order. Show the remaining group of 8 videos in the other order. **4.101** (b) **4.103** (b) **4.105** (b) **4.107** (a) **4.109 (a)** 100%. **(b)** 0% or 100%. **(c)** 400. **4.111 (a)** The researchers randomly selected among all possible Standard Metropolitan Statistical Areas (SMSAs) or nonmetropolitan counties. **(b)** The researchers stratified participants based on region, age, and race. **(c)** The GSS Web site notes: "The full-probability GSS samples used since 1975 are designed to give each household an equal probability of inclusion in the sample." **4.113 (a)** $M = 50, n = 125, R = 12$ **(c)** 520.8. **(d)** Column headings: In the census?: Yes (returned form), No (did not return form); Row headings: In the PES?: Yes, No. **4.115** Answers will vary.

Part 1 Answers

R1.1 (a) Sample: the 1245 subjects who responded to the General Social Survey question on astrology. Population: the American adult public. **(b)** Variable: subject's response about whether astrology has scientific truth, which is categorical. **(c)** 651/1245 = 0.52. **R1.3 (a)** Categorical because each person's response falls into one of the four possible categories. **(b)** Statistics because each pertains specifically to the sample of 1077 individuals contacted in the poll. **R1.5** Mean and median cannot be found, because religion is not quantitative. The modal category for religion is Christianity. **R1.7 (a)** modal category = every day, median = a few times a week **(b)** mean = 3.178, less than in 1994. **R1.9 (a)** mean = 80.4, median = 81. **(b)** Yes, mean = 52.1. **(c)** Female economic activity is quantitative and response variable. **R1.11 (a)** Weak. **(b)** As scoring average decreases, the percentage of greens reached in regulation tends to increase. **(c)** Percentage of greens reached in regulation and average number of putts taken on holes for which the green was reached in regulation. **R1.13** Infant mortality rates range from 1.8 to 17. Since the distances from the median to the minimum and lower quartile are less than the distances from the median to the maximum and upper quartile, the distribution of infant mortality rates is skewed to the right. 25% of the nations had infant mortality rates less than 2.775 and 25% of the nations had infant mortality rates above 5.025. The middle 50% of the nations had rates between 2.775 and 5.025. **R1.15 (a)** 300. **(b)** 350. **R1.17 (a)** Skewed to right. **R1.19 (a)** 76.63%. **(b)** 30.45% **(c)** Yes, percentage saying "always wrong" depends considerably on religion. **R1.21 (a)** 5727.25. **(b)** Age could be a lurking variable, but it would not be responsible for the association. This is because age is positively associated with income but negatively associated with years of education. So, as age increases, income tends to go up and education tends to go down, which has a negative influence on the association between income and education. **R1.23 (a)** If 0 were within the range of x-values observed, 22 would represent the predicted child poverty rate for a country with 0% of their gross domestic product spent on social expenditures. If social expenditure

as a percent of gross domestic product increases by 1%, the child poverty rate is predicted to decrease by 1.3%. **(b)** United States: 19.4%, Denmark: 1.2%. **(c)** As social expenditure increases, the child poverty rate tends to decrease. The linear association is strong and negative. **R1.25** The *y*-intercept about 18 and slope about 0.5. **R1.27** As SES increases, mortality tends to decrease and frequency of vacations tends to increase. This may induce a negative association between mortality and vacation time. **R1.29** Response bias. The wording of the last two questions led more respondents to answer affirmatively. **R1.31 (a)** Slope = −0.22. **(b)** Slope = 0.23. **(c)** In part a correlation is negative, and in **(b)** it is positive, because it has the same sign as the slope. **R1.33** Anecdotal evidence is usually not representative of the population. The customers who believed the echinacea was effective were the ones who were more likely to come back in and tell the manager good things. In the randomized experiment, the subjects who received echinacea were randomized so that the groups should tend to be balanced on values of potential lurking variables. The results of such a study are more trustworthy because the results can be attributed to the treatments rather than to lurking variables. **R1.35** Answers will vary but should include the following:
The fitted regression equation for 2006 is $\hat{y} = 11.445 - 0.116x$ where y = WWWHR and x = AGE.

Chapter 5

5.1 The long-run relative frequency definition of probability refers to the probability of a particular outcome as the proportion of times that the outcome would occur in a long run of observations. **5.3** No. In the short run, the proportion of a given outcome can fluctuate a lot. **5.5 (a)** Most likely due to a rounding error. **(b)** It has a relative frequency interpretation. **5.7** No. With a biased sampling design, having a large sample does not remove problems from the sample not being selected to represent the entire population. **5.9** We would be relying on our own judgment rather than objective information such as data, and so would be relying on the subjective definition of probability. **5.11 (b)** About 50. **(c)** The results will depend on the answer to part a. **(d)** 42% of the answers were true. We would expect 50%. They are not necessarily identical, because observed percentage of a given outcome fluctuates in the short run. **(e)** There are some groups of answers that appear nonrandom. For example, there are strings of five trues and eight falses, but this can happen by random variation. **5.13 (a)** Great/in favor, great/opposed, great/no opinion, good/in favor, good/opposed, good/no opinion, fair/in favor, fair/opposed, fair/no opinion, poor/in favor, poor/opposed, poor/no opinion.
(b)

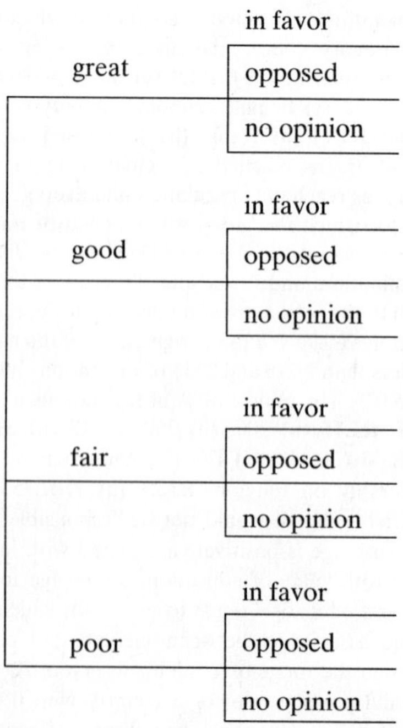

- great
  - in favor
  - opposed
  - no opinion
- good
  - in favor
  - opposed
  - no opinion
- fair
  - in favor
  - opposed
  - no opinion
- poor
  - in favor
  - opposed
  - no opinion

5.15 (a) Answer is tree diagram with 16 possible outcomes. **(b)** The probability of each possible individual outcome is 1/16 = 0.0625. **(c)** 0.3125. **5.17** There is no reason to think that all horses have an equally likely chance of winning. **5.19 (a)** BBB, BBG, BGB, GBB, BGG, GBG, GGB, GGG. **(b)** 1/8. **(c)** 3/8. **(d)** 6/8 or 3/4. **5.21** 20 heads has probability (1/2) to the 20th power, which is 1/1,048,576 = 0.000001. Risk of a one in a million death: 1/1,000,000 = 0.000001. **5.23 (a)** YS; YD; NS; and ND. **(b) (i)** 0.004; **(ii)** 0.284. **(c)** 0.003. **(d)** Answer would have been P(N and D) = P(N) × P(D) = (0.004)(0.284) = 0.001. This indicates that chance of death depends on seat belt use since 0.001 is not equal to 0.003. **5.25 (b)** If A and B were independent events, P(A and B) = P(A) × P(B). Since P(A and B) > P(A)P(B), A and B are not independent. The probability of responding yes on global warming and less on future fuel use is higher than what is predicted by independence. **5.27. (a)** There are eight possible outcomes. Answer is tree diagram. **(b)** 0.488. **(c)** They assumed independence, but that might not be so (e.g., three customers are friends, or three members of the same family). **5.29 (a)** $P(B|S)$. **(b)** $P(B^C|S^C)$. **(c)** $P(S^C|B^C)$. **(d)** $P(S|B)$. **5.31 (a)** 0.7599. **(b)** 0.3299. **(c)** 0.1579. **5.33 (a)** 0.004. **(b) (i)** 0.001. **(ii)** 0.010. **(c)** Neither P(D|wore seat belt) nor P(D|didn't wear seat belt) equals P(D); specifically, 0.001 and 0.010 are different from 0.004. Thus, the events are not independent.
5.35

	Identified as Spam by ASG	
Spam	Yes	No
Yes	7005	835
No	48	

(b) 7005/(7005 + 835) = 0.8935. **(c)** 7005/(7005 + 48) = 0.9932. **5.37 (a)** P(D|NEG) = 6/3927 = 0.0015. **(b)** No; P(NEG|D) = 6/54 = 0.111. **5.39 (a)** 398/969 = 0.41. **(b) (i)** 183/469 = 0.39; **(ii)** 215/500 = 0.43. **(c)** No. **5.41 (a)** 0.30. **(b) (i)** 0.40; **(ii)** 0.40. **(c)** No; the probability that he will make the second shot depends on whether he made the first. **5.43** 0.0416. **5.45 (a)** P(C and A)/P(A) = (1/4)/(1/2) = (1/2). **(b)** No; C implies A. **(c)** P(C and B)/P(B) = (1/4)/(3/4) = (1/3). **(d)** The event "at least one child is female" includes more outcomes than the event "first child is female." **5.47** 0.9329. **5.49** Each student can be matched with 24 other students, for a total of 25(24) pairs. But this considers each pair twice (e.g., student 1 with student 2, and student 2 with student 1), so the answer is 25(24)/2 = 300. **5.51** 0.6954. **5.53** The explanation should discuss the context of the huge number of possible random occurrences that happen in one's life, and the likelihood that at least some will happen (and appear coincidental) just by chance. **5.55 (a)** The probability that the first will go, multiplied by the probability that the second will go, and so on for all 5.4 million, that is (1/5000) taken to the 5.4 millionth power, which is zero to a huge number of decimal places. **(b)** This solution assumes that each person decides independently of all others. This is not realistic because families and friends often make vacation plans together. **5.57 (a)** The first answer is a tree diagram. **(b)** $P(POS) = P(S \text{ and } POS) + P(S^c \text{ and } POS) = 0.0086 + 0.1188 = 0.1274$. **(c)** 0.068. **(d)** They are calculated by multiplying the proportion for each branch by the total number. The proportion of positive tests with breast cancer is 1/(1 + 12) = 0.08. **5.59 (a)** Of the 100,000 who suffer abuse, 40 + 5 = 45 are killed; 5 of those 45 are killed by someone other than the partner. **(b)** 99,955 and 40. **(c)** They differ because the number who are abused is much larger than the number killed, whereas of all those killed, most are killed by the partner. **5.61 (a)** Tree diagram is part of answer. P(Innocent|Match) = P(Innocent and Match)/P(Match) = 0.0000005/0.4950005 = 0.000001. **(b)** Tree diagram is part of answer. P(Innocent|Match) = P(Innocent and Match)/P(Match) = 0.00000099/0.00990099 = 0.0001. When the probability of being innocent is higher, there's a larger probability of being innocent given a match. **(c)** P(Innocent|Match) can be much different from P(Match|Innocent). **5.63 (b)** (0.95) × (0.95) × (0.95) . . . × (0.95) — a grand total of twenty 0.95s (or 0.95^{20}) = 0.36. **5.65 (a)** $7.35. **(b)** It is less; essentially you expect to lose $10 − $7.35 = $2.65 per play on average. **5.67** The gender of a child is independent of the previous children's genders; the chance is still 1/2. **5.69 (a)** There are 18 possible meals [(2 × 3 × 3 × 1) = 18]. Tree diagram is part of answer. **(b)** No; some menu options would be more popular than are others. **5.71 (a)** 0.8142. **(b)** 0.6629. **(c)** That the responses of the two subjects are independent; married couples share many beliefs. **5.73 (a)** 8; tree diagram is part of answer. **(b)** 0.125. **(c)** 0.343. **(d)** Three friends are likely to be similar on many characteristics that might affect performance. **5.75 (a)** 0.84. **(b)** 0.70. **5.77 (a)** The percentages 31% and 1% are conditional probabilities. The 31% is conditioned on the event that the teen says that parents are never present at the parties attend. The 1% is conditioned on the event that the teen says that parents are present at the parties. For both percentages, the event to which the probability refers is a teen reporting that marijuana is available at the parties. **(b)**

Marijuana available	Parents Present Yes	No
Yes	9	133
No	860	295

(c) 860/(860 + 295) = 0.74. **5.79 (a)** (1,1); (1,2); (1,3); (1,4); (1,5); (1,6); (2,1); (2,2); (2,3); (2,4); (2,5); (2,6); (3,1); (3,2); (3,3); (3,4); (3,5); (3,6); (4,1); (4,2); (4,3); (4,4); (4,5); (4,6); (5,1); (5,2); (5,3); (5,4); (5,5); (5,6); (6,1); (6,2); (6,3); (6,4); (6,5); (6,6). **(b)** (1,1); (2,2); (3,3); (4,4); (5,5); and (6,6); 0.167. **(c)** (1,6); (2,5); (3,4); (4,3); (5,2); and (6,1); 0.167. **(d) (i)** 0; **(ii)** 0.333; **(iii)** 0. **(e)** A and B are disjoint. You cannot roll doubles that add up to seven. **5.81** 0.80. **5.83** 0.67. **5.85 (a)** This strategy is a poor one; the chance of an even slot is the same on each spin of the wheel. **(b)** (18/38) to the 26th power, which is essentially 0. **(c)** No; events that seem highly coincidental are often not so unusual when viewed in the context of *all* the possible random occurrences at all times. **5.87** False negatives: test indicates that an adult smoker does not have lung cancer when he or she does have lung cancer. False positives: test indicates the presence of lung cancer when there is none. **5.89 (b)** 0.048; because so few people have this cancer, most of the positive tests will be false. **5.91 (c)** 0.999. **(d)** A positive result is more likely to be in error when the prevalence is lower, as relatively more of the positive results are for people who do not have the condition. Tree diagrams or contingency tables are part of the answer. **5.93** Will vary. **5.95 (a)** The cumulative proportion of heads approaches 0.50 with larger numbers of flips, illustrating the law of large numbers and the long-run relative frequency definition of probability. **(b)** The outcome will be similar to that in part a, with the cumulative proportion of 3 or 4 approaching one third with larger numbers of rolls. **5.97 (a)** 0.545. **(b)** 0.505. **(c)** 0.500; as n increases, the cumulative proportion tends toward 0.50. **5.99 (1)** Specificity: if pregnant, 99% chance of a positive test. **(2)** Sensitivity: if not pregnant, 99% chance of a negative test. **(3)** 99% chance of pregnancy given a positive test. **(4)** 99% chance of not pregnant given a negative test. **5.101 (a)** 2.4%. **(b)** Multiply (1) probability of getting a high school degree by (2) probability of getting a college degree once you had a high school degree by (3) probability of getting a masters' once you had the earlier degrees by (4) probability of getting a Ph.D. once you had the earlier degrees. **(c)** 0.06. **5.103** The event of a person bringing a bomb is independent of the event of any other person bringing a bomb. **5.105** (c) and (d). **5.107** (e). **5.109** (b). **5.111** False. **5.113** Being not guilty is a separate event from the event of matching all the characteristics listed. Suppose there are 100,000 people in the population. Since the probabiity of a match is 0.001, out of the 100,000 people, 100 people would match all the characteristics. Suppose that in the population, 5% of the people are guilty of such a crime, 95% are not guilty. The contingency table is

	Guilty	Not Guilty	**Total**
Match		???	100
No Match			99,900
Total	5000	95,000	100,000

Thus, P(not guilty | match) = ???/100. **5.115** When two events are not independent, P(A and B) = P(A) × P(B | A). If we think about (A and B), as one event, then P[C and (A and B)] = P(A and B) × P(C | A and B). If we replace P(A and B) with its equivalent, P(A) × P(B | A), we see that: P(A and B and C) = P(A) × P(B | A) × P(C | A and B). **5.117 (b)** The simulated probability should be close to 1. **5.119** Answers will vary.

Chapter 6

6.1 (a) Uniform distribution; P(1) = P(2) = P(3) = P(4) = P(5) = P(6) = 1/6. **(b)** There will be a probability for each x from 2 through 12; for example, there are three rolls that add up to four [(1,3), (2,2) (3,1)] ; thus, the probability of four is 3/36 = 0.083. **(c)** $0 \le P(x) \le 1$; and $\sum P(x) = 1$. **6.3 (a)** All are between 0 and 1 and they sum to 1. **(b)** 0.4083. **(c)** Over the course of many at-bats, the average number of bases per at-bat is 0.4083. **6.5 (a)** Not all values are equally likely. **(b)** 0.5. **6.7 (a)** Answers will vary. **(b)** For bet 1, expected winnings is $0.53; for bet 2, expected winnings is $0.53. **6.9 (a)** Females: $\mu = 0(0.01) + 1(0.03) + 2(0.55) + 3(0.31) + 4(0.11) = 2.5$. Males: $\mu = 0(0.02) + 1(0.03) + 2(0.60) + 3(0.28) + 4(0.08) = 2.39$. **(b)** The responses for males tend to be closer to the mean than the responses for the females are.

6.11 (a)

x	P(x)
$90	0.50
$120	0.20
$130	0.30

$\mu = 108$. The expected selling price for the sale of a drill is $108. **(b)** $\mu = 110$. The new strategy gives higher mean profit, in the long run. **6.13 (a)** Someone could watch exactly one hour of TV or 1.8643 hours of TV. **(b)** Each person reported TV watching by rounding to the nearest whole number. **(c)** These curves would approximate the two histograms to represent the approximate distribution if we could measure TV watching in a continuous manner. Then, the area above an interval represents the proportion of people whose TV watching fell in that interval. **6.15** 0.251. **6.17 (a)** 0.9495 − 0.0505 = 0.899, which rounds to 0.90. **(b)** 0.9951 − 0.0049 = 0.9902, which rounds to 0.99. **(c)** 0.7486 − 0.2514 = 0.4972, which rounds to 0.50. **6.19 (a)** Divide by two for amount in each tail, 0.005. Subtract from 1.0 for cumulative probability, 0.995. Look up this probability on Table A to find the z-score of 2.58. **(b)** (a) 1.96. (b) 1.645; graph is part of answer. **6.21 (a)** 0.67. **(b)** 1.645. **6.23 (a)** 1.19. **(b)** 0.12. **(c)** 0.79. **(d)** 141.5. **6.25 (a)** $z = (1000-673)/556 = 0.59$ has a cumulative probability of 0.72. The probability that household electricity use was greater than 1000 kilowatt-hours is 1 − 0.72 or 0.28. **(b)** No, because electricity use = 0 gives a minimum possible z-score of $(0-673)/556 = -1.21$. **6.27 (a) (i)** 0.106; **(ii)** 0.894; **(b)** 137.3. **(c)** 62.7. **6.29 (a)** 5.866, which is extremely high for a normal distribution. **(b)** No; anything more than about 1.2 standard deviations below the mean would be a negative number, and a murder rate cannot be negative. **6.31** $Z_{joe} = 1$, $Z_{kate} = 0.851$; in terms of math scores, Joe is better. **6.33 (a)** Sample space: (SSS, SSF, SFS, SFF, FSS, FSF, FFS, FFF); probability of each is 0.125. The probabilities for $x = 0, 1, 2, 3$ are: P(0) = 0.125, P(1) = 0.375, P(2) = 0.375, P(3) = 0.125. **(b)** Same probabilities as in part a;

e.g., $P(0) = \dfrac{3!}{0!(3-0)!}0.5^0(1-0.5)^{3-0} = 0.125.$

6.35 (d) The graph in part a is symmetric, as is the case anytime p = 0.50. **(e)** The graph in part c is most heavily skewed; the graph with p = 0.01 would be even more skewed. **6.37** P(0) = 0.77, P(1) = 0.22, P(2) = 0.02 (rounded). **6.39 (a)** Same probability of success for each free throw, and results of different free throws are independent. Each trial is binary (make or do not make free throw). **(b)** $n = 10$; $p = 0.90$. **(c) (i)** 0.349; **(ii)** 0.387. **6.41 (a)** $n = 60$; $p = 1/6 = 0.167$. **(b)** $\mu = 10$; $\sigma = 2.89$ **(c)** Yes; 0 is well over 3 standard deviations from the mean.

(d) $P(0) = \dfrac{60!}{0!(60-0)!}0.1667^0(1-0.1667)^{60-0} = (1)(1)(0.0000177) =$ 0.0000177. **6.43 (a)** Yes; binary data (Hispanic or not), same probability of success for each trial (i.e., 0.40), and independent trials (whom you pick for the first juror is not likely to affect whom you pick for the others, and $n < 10\%$ of the population size). $n = 12$; $p = 0.40$. **(b)** 0.002. **(c)** Yes; there is only a 0.2% chance that this would occur if the selection were done randomly. **6.45 (a)** Yes; the sample is less than 10% of the size of the population size. **(b)** No; neither (np) nor [$n(1-p)$] is at least 15. Both are 5. **6.47 (a)** The formula relies on a fixed number of trials, n. **(b)** The binomial applies for X. We only have n for X, not Y. $n = 3$. $p = 0.50$. **6.49 (a)** Discrete; you can only have whole numbers of grandparents. **(b)** They each fall between 0 and 1, and the sum of the probabilities of all possible values is 1. **(c)** 0.50. **6.51 (a)** Probability distribution: P(0) = 0.994, P(45.50) = 0.006. **(b)** Probability distribution: P(0) = 0.999, P(45.50) = 0.001. **(c)** The first; because you have a higher chance of winning. **6.53 (a)** Probability distribution: P(0) = 0.999999, P(100,000) = 0.000001. **(b)** 0.10. **(c)** Because the flyer's return on each $1 spent averages to only $0.10. **6.55 (a)** 1.96. **(b)** 2.58; answer includes graph. **6.57 (a)** 25% between the mean and the positive z-score. Added to 50% below mean, 75% of the normal distribution is below the positive z-score, corresponding to a z-score of 0.67. **(b)** First quartile: 25th percentile; third quartile: 75th percentile. The 75th percentile has a z-score of 0.67. The z-score at the 25th percentile is $z = -0.67$. **(c)** The interquartile range is Q3–Q1 = $(\mu + 0.67\sigma)-(\mu-0.67\sigma) = 2(0.67)\sigma$. **6.59 (a)** 0.076. **(b)** 0.023. **(c)** 0.847. **(d)** 0.006, 0.309, and 0.494. **6.61** 0.018. **6.63** 0.07. **6.65 (a)** 0.04. **(b)** 0.75; each day's take must be independent of every other day's take. **6.67 (a)** 0.997. **(b)** 0.87. **6.69** 0.002; yes, because it would be such a rarely occurring event if they did not exhibit a preference, i.e., just by random chance. **6.71 (a)** Binary data (i.e., yes or no as only options); same probability of success for each call; independent trials (calls). **(b)** No. **(c)** $n = 5$; $p = 0.60$; $\mu = 5(0.60) = 3$. **6.73 (a)** Binary data (death or no death); same probability of success for each person ($p = 0.0001$); independent outcomes for each woman. $n = 1{,}000{,}000$; $p = 0.0001$. **(b)** Mean = 100; standard deviation = 10.0. **(c)** 70 to 130. **(d)** All women would not have same probability (e.g., some drive much more), and outcomes may not be independent (e.g., one woman may usually ride in the same car as another). **6.75 (a)** Binomial. **(b)** Mean = 4.0, standard deviation = 2.0. **(c)** 0.018. **6.77 (a)** The mean increases by one for each doubling. **(b) (i)** 8; **(ii)** 11; **(c)** 4 is a bit more than two standard deviations, and a bell-shaped curve has probability about 0.95 within two standard deviations. **6.79 (a)** The binomial distribution with $n = 190$ and $p = 0.80$ has mean 152 and standard deviation 5.514. Since $np > 15$ and $n(1 - p) > 15$, the normal distribution will approximate this binomial distribution. We expect that the counts within 3 standard deviations of the mean are approximately 135 and 169. The number of seats the plane can handle is 170, just above this range. **(b)** Groups might buy tickets and travel together, violating the assumption of independent trials. **6.81** (d). **6.83 (a)** For group A, 0.16 and for group B, 0.23. **(b)** 0.59. **(c)** The proportion of those not admitted belonging to group B decreases to 0.065; the legislator was correct. **6.85** $\sigma^2 = p^2(1-p) + (1-p)^2p = p(1-p)$. So $\sigma = \sqrt{p(1-p)}$. Since standard deviation of probability distribution equals $\sigma = \sqrt{p(1-p)}$ and since we must divide by $\sqrt{n}$ to get standard error, we have $\sigma/\sqrt{n} = \sqrt{p(1-p)/n}$. **6.87 (a)** The probability of rolling doubles is 1/6. The outcome on the second die has a 1/6 chance of matching the first. To have doubles occur first on the second roll, you'd have to have no match on the first, a 5/6 chance. You'd then have to have a match on the second roll, a 1/6 chance. The probability of both occurring is (5/6)(1/6). For no doubles until the third roll, the first and the second roll would not match, a 5/6 chance for each, followed by doubles on the third, a 1/6 chance. The probability of all of three events is $(5/6)^2(1/6)$. **(b)** By the logic in part a, P(4) = (5/6)(5/6)(5/6)(1/6) or $(5/6)^3(1/6)$. By extension, we could calculate P(x) for any x by $(5/6)^{x-1}(1/6)$. **6.89** Answers will vary.

Chapter 7

7.1 (a) Although the population proportion is 0.53, by random variability it is unlikely that exactly 53 out of 100 polled voters will vote yes. Sample proportions close to 0.53 will be more likely than those further from 0.53. **(b)** The graph of the sample proportions should be bell shaped and centered around 0.53.

(c) Predicted standard deviation $= \sqrt{\dfrac{0.53(1 - 0.53)}{100}} = 0.05$.

(d) The graph should look similar as in part b but shifted so that it is centered around 0.70. Also, standard deviation changes to 0.046. **7.3 (a)** Mean $= p = 0.10$ and standard deviation $= \sqrt{\dfrac{0.10(1 - 0.10)}{4000}} = 0.0047$.

(b) Mean $= p = 0.10$ and standard deviation = 0.0095. **(c)** Mean $= p = 0.10$ and standard deviation = 0.0190. As the sample size gets larger, the standard deviation gets smaller. **7.5 (a)** P(1) = 0.409 and P(0) = 0.591. **(b)** Mean = 0.409, standard deviation = 0.008. **7.7 (a)** Bell shaped, mean = 0.300, standard deviation = 0.020 **(b)** These values are only about a standard deviation from the mean, which is not unusual. **7.9 (a)** P(0 out of 3) = 27/64, P(1 out of 3) = 27/64, P(2 out of 3) = 9/64, P(3 out of 3) = 1/64. **(b)** Mean = 0.25, standard deviation = 0.25. **(c)** For $n = 10$, mean = 0.25, standard deviation = 0.1370; for $n = 100$, mean = 0.25, standard deviation = 0.0433; the mean stays the same and the standard deviation gets smaller as n gets larger. **7.11 (a)** The data distribution consists of 94 1s and 6 0s. **(b)** The population distribution consists of the x values of the 14,201 students, 95.1% of which are 1s and 4.9% of which are 0s. **(c)** Mean = 0.951, standard deviation = 0.0216; it represents the probability distribution of the sample proportion of full-time students in a random sample of 100 students. **7.13 (a)** population distribution. **(b)** The data distribution includes only 0s and 1s. The sampling distribution of the sample proportion refers to proportion values between 0 and 1. For example, with a sample of 100 observations of which 20 were 0s and 80 were 1s, the sample proportion would be 0.8. **7.15 (b) (i)** Mean = 3.50, standard deviation = 1.21. **(ii)** Mean = 3.50, standard deviation = 0.31; As n increases, the sampling distribution becomes more normal in shape and has less variability around the mean. **7.17 (a)** mean = 0.10, standard deviation = 0.10 **(b)** Probability = 0 to many decimal places **7.19** Probability = 0.006. **7.21 (a)** 0.82. **(b)** Larger, because the standard deviation of the sample mean would also be smaller. **7.23 (a)** X is the number of people in a household which consists of both family and nonfamily members; quantitative. **(b)** 4.43 and 2.02; somewhat skewed to the right because of some relatively large households. **(c)** 4.2 and 1.9; shape similar to population distribution. **(d)** 4.43 and 0.135; approximately normal distribution due to Central Limit Theorem. **7.25** Sampling distribution should become more bell shaped and the variability becomes smaller as n increases. **7.27** The sampling distribution is normal even for $n = 2$. If the population distribution is approximately normal, the sampling distribution is approximately normal for all sample sizes. **7.29** Answers will vary from the different simulations. **7.31 (a)** Mean = 0.1667, standard deviation = 0.0373. **(b)** Yes; 8.9 standard deviations. **(c)** Set of 0s and 1s describing whether an American has blue eyes (1) or not (0), 17% 1s and 83% 0s; set of 50 0s and 50 1s describing whether the students in your class have blue eyes or not; probability distribution of the sample proportion, having mean = 0.1667 and standard deviation = 0.0373. **7.33 (a)** Mean = 0.45, standard deviation = 0.144. **(b)** 1.39. **(c)** 3/12 = 0.25 is only 1.39 standard deviations below the mean. **7.35 (a)** $0.65^{1751} \approx 0$. **(b)** 30.7. **7.37 (a)** No; many games have zero or one home runs while a few have three or four or more home runs. The random variable is discrete, and the normal distribution is for continuous random variables. **(b)** Approximately normal with mean = 1 and standard deviation = 0.078. **(c)** Essentially 0. **7.39** 0.9545. **7.41 (a)** No, X is discrete where the standard deviation is as big as the mean and the min value of 0 is only 1 standard deviation below mean, indicating skewness. **(b)** Approximately normal with mean = 1.0 and standard deviation = 0.1. Since $n = 100$, we can expect sampling distribution to be approximately normal even if population distribution is skewed. **7.43 (a)** Target line = mean = 500, control limits are 3 standard deviations from the mean, 494 and 506. **(b)** 0.84, normally distributed population. **7.45** Each time a poll is conducted, a sample proportion is calculated (How many of the 1000 polled Canadians think the prime minister is doing a good job?). The distribution of sample proportions is called a sampling distribution. The sample proportions fluctuate around the population proportion, with the degree of variability around the population proportion being smaller when the sample size is larger. **7.47** This standard deviation tells us how close a typical sample proportion is to the population proportion. **7.49** The shapes are similar in parts a and b. In part c, sampling distribution is bell shaped and displays much less variability than in part b. **7.51** False. **7.53** (c). **7.55 (a)** Sample mean would always be the same as the population mean, no variability of the sample means. As such, the standard deviation would equal zero. **(b)** Exactly the same.

7.57 Standard deviation $= \sqrt{\dfrac{30{,}000 - 300}{30{,}000 - 1}}\dfrac{\sigma}{\sqrt{n}} = 0.995\dfrac{\sigma}{\sqrt{n}}.$

(b) Standard deviation $= \sqrt{\dfrac{30{,}000 - 30{,}000}{30{,}000 - 1}}\dfrac{\sigma}{\sqrt{n}} = 0\dfrac{\sigma}{\sqrt{n}} = 0.$

7.59 (a) P(0) = 0.5, P(1) = 0.5, mean = 0.5. **(b)** 0.4 = proportion of of 0s, 0.6 = proportion of 1s. **(c)** Should be roughly bell shaped around 0.5. **(d)** Mean close to 0.50 and standard deviation close to 0.16, the standard deviation when $p = 0.50$ and $n = 10$.

Part 2 Answers

R2.1 (a) 0.46. **(b)** The probability for the complement of an event. **R2.3 (a)** 0.06. **(b)** The addition rule for the union of two disjoint events. **R2.5 (a)** 0.15. **(b)** 0.72. **R2.7 (a)** 6/49. **(b)** 0.000186. **(c)** 268,920 years. **R2.9** 0.0016. **R2.11 (a) (i)** 0.77. **(ii)** 0.23. **(b)** The distance between the mean and lowest score is much greater than between the mean and highest score, which is only about 1.4 standard deviations above the mean. **(c)** Approximately normal, mean = 591, standard deviation = 14.8. **R2.13 (a)** Standard deviation = 0.014. **(b)** This standard deviation describes how much we can expect the sample proportion to vary from one sample of size 1034 to the next. **R2.15 (a)** Mean = 1000, standard deviation = 300. **(b)** Mean = 880, standard deviation = 276. **(c)** Mean = 1000, standard deviation = 113. This standard deviation describes how much we can expect the sample mean to vary from one sample of seven daily sales to the next. **R2.17** If the population proportion voting for Schumer were 0.50, the sampling distribution of the sample proportion who voted for him would be approximately normal with mean = 0.50 and standard deviation = 0.0119. The z-score associated with a sample proportion value of 0.65 would then be (0.65 − 0.5)/0.0119 = 12.61. A z-score of 12.61 indicates the sample proportion 0.65 is 12.61 standard deviations above the proportion of 0.50. It is highly unlikely to obtain a sample proportion of 0.65 if the actual population proportion is 0.50. Thus, we conclude that the population proportion voting for Schumer is larger than 0.50. **R2.19 (a)** $P(B \mid H) = 0.25$. **(b)** $P(B^C \mid H^C) = 0.95$.

(c)

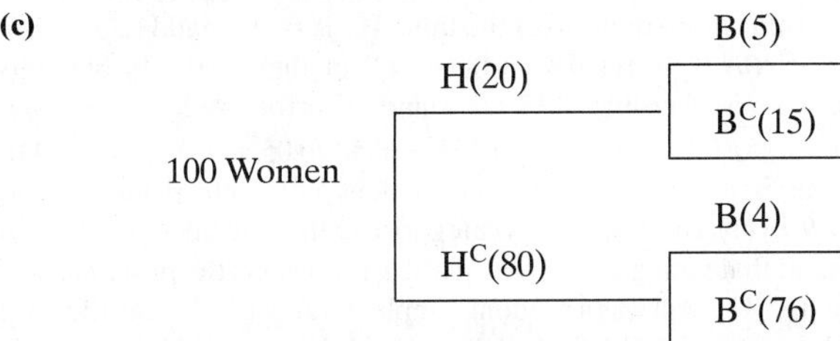

Total, we would expect about 5 + 4 = 9 women to relapse. **R2.21** Since samples are merely representative of the population, we would not expect to obtain the exact same sample mean each time a sample of size n is collected. Rather, we would expect these values to vary about the true population mean. The amount by which the sample means vary is summarized by the sampling distribution of the sample mean which has mean = μ and spread described by its standard deviation = $\sigma/\sqrt{n}$ where μ is the population mean, σ is the population standard deviation and n is the sample size. **R2.23** True. **R2.25** True.

Chapter 8

8.1 (a) Proportion with health insurance in population; mean amount spent on insurance by population. **(b)** Sample proportion and sample mean. **8.3** 0.526 **8.5 (a)** 1.6. **(b)** The sample mean likely falls within 0.7 of the population mean. **8.7 (a)** 0.02. This is the interval containing the most plausible values for the parameter. It is very likely that the population proportion is no more than 0.02 lower or 0.02 higher than the reported sample proportion. **(b)** 0.83 to 0.87; the interval contains the most believable values for the parameter. **8.9** 1.38 to 1.62 is the estimated range that includes the most believable values for the population mean. **8.11** Margins of error typically are based on a 95% confidence interval. They would have multiplied 1.96, the z-score associated with a 95% confidence interval, by the standard error = $\sqrt{0.088(1 - 0.088)/30000} = 0.0016$ giving a margin of error of 1.96 × 0.0016 = 0.003 on the proportion scale, or 0.3%. **8.13 (a)** 0.294. **(b)** 0.025. **(c)** (0.27, 0.32); the most believable values for the population proportion. **(d)** Observations are obtained randomly; number of successes and the number of failures both are greater than 15. Both seem to hold true in this case. **8.15 (a)** Divide the number of those in favor by the total number of respondents, 1263/1902. **(b)** We can be 95% confident that the proportion of the population who are in favor of the death penalty is between 0.6428 and 0.6853, or rounding, (0.64, 0.69). **(c)** If we use this method over and over for numerous random samples, in the long run we make correct inferences (that is, the confidence interval contains the parameter) 95% of the time. **(d)** Yes, all the values in the confidence interval are above 0.50. **8.17 (a)** The "sample p,"1521/2113 = 0.7198, is the proportion of all respondents in the sample who believe stem cell research has merit. We can be 95% confident that the population proportion falls between 0.7007 and 0.7390. **(b)** (0.7390 − 0.7007)/2 = 0.0192. **8.19** We can be 95% confident that the population proportion of individuals who believe in ghosts is between 0.40 and 0.44. **8.21 (a)** We can be 95% confident that the population proportion falls between 0.585 and 0.635. **(b)** The data must be obtained randomly, and the number of successes and the number of failures both must be greater than 15. (They are.) **8.23 (a)** Yes; because 0.50 falls outside of the confidence interval of 0.446 to 0.496. **(b)** No; because 0.50 falls in the confidence interval of 0.437 to 0.505. The more confident we want to be, the wider the confidence interval must be. **8.25** The percentage we'd expect would be 95% and 99%, but the actual values may differ a bit because of sampling variability. **8.27 (a)** 3.22. **(b)** 0.076. **(c)** We're 95% confident the population mean falls between 3.07 and 3.37. **(d)** No; because 2 falls outside the confidence interval. **8.29 (a)** 2.776. **(b)** 2.145. **(c)** 2.977. **8.31 (a)** Data production must have used randomization. The population distribution is assumed to be approximately normal, although this method is robust except in cases such as extreme skew or severe outliers. Here the method is questionable because of an outlier at 80. **(b)** $50 - 2.201 * 11.34/\sqrt{12} = 42.795$ and $50 + 2.201 * 11.34/\sqrt{12} = 57.205$. **(c) (i)** 41.5 − 19.5 = 22 and 54.5 + 19.5 = 74, so 80 is an outlier using 1.5IQR; **(ii)** 50 − 34.02 = 15.98 and 50 + 34.02 = 84.02, so 80 is not an outlier using 3s. **(d)** The new interval is (42.856, 51.689), which is narrower than the other interval and also centered around a smaller value (47.27 instead of 50). **8.33 (a)** The data must be produced randomly, and the population distribution should be approximately normal. **(b)** The population mean hours of TV watching is likely to be between 1.1 and 3.2. **(c)** The confidence interval is wide because of the very small sample size. **8.35 (a)** Mean = 4.14; Standard deviation = 5.08; standard error = 1.92. **(b)** (0.4, 7.9). We are 90% confident that the population mean number of hours per week spent sending and answering e-mail for women of at least age 80 is between 0.4 and 7.9 hours. **(c)** Since there will be many women of age 80 or older who do not use email at all but some who use email a lot, the distribution is likely to be skewed right. Since the confidence interval using the t-distribution is a robust method, the interval in part b should still be valid. **8.37** The most plausible values for the population mean of Monday volume: 40.27 through 63.37; for Friday volume: 42.80 through 57.20. Monday's interval is wider, implying that there is more variation in the number of shares traded on Mondays than on Fridays. **8.39 (a)** The margin of error equals the standard error, 0.0326, multiplied by the t-score for 1932 degrees of freedom and a 95% confidence interval (1.96). Take the sample mean of 4.11 plus and minus the margin of error. **(b)** Yes; because 4.0 falls below the lowest believable value of the confidence interval. **(c) (i)** Wider; **(ii)** wider. **8.41 (i)** 9.8; **(ii)** 4.9; as sample size increases, the margin of error is smaller. **8.43 (a)** No, there is a large proportion of observations at the single value of 0. Very little, because with large random samples the sampling distribution of the sample mean is approximately normal. **(b)** (8.0, 12.0); yes, because $15 is not in the confidence interval. **8.45 (b)** We would expect 5% of the intervals not to contain the true value. **(c)** With a large sample size, the sampling distribution is approximately normal even when the population distribution is not. The assumption of a normal population distribution becomes less important as n gets larger. **8.47** 379. **8.49 (a)** 97. **(b)** 385. **(c)** 664. **(d)** In both cases we need a larger sample size. **8.51 (a)** 246. **(b)** 37.5. **8.53** Diverse populations (medical doctors making a wide range of incomes) have a wider range of observed values, and thus larger standard deviations, than homogeneous populations (entry-level McDonald's workers all making close to minimum wage). Larger standard deviations result in larger standard errors and wider confidence intervals. **8.55 (a)** 1.0. **(b)** 0; no; the sampling distribution is likely to have *some* variability, because the true probability (which determines the exact standard error) is positive. **(c)** (1.0, 1.0); no, because this method works poorly in this case. **(d)** We don't have at least 15 outcomes of each type. Using the small-sample method based on adding 2 outcomes of each type, we can be 95% confident that the proportion is between 0.51 and 1.0 (note, we used 1.0 instead of the calculated value of 1.05, since p cannot exceed 1). **8.57** Yes; 0.10 is not in the confidence interval of 0 to 0.04. **8.59** Sample with replacement from the 10 values, taking 10 observations and find the standard deviation. Do this many, many times. The 95% confidence interval would be the values in the middle 95% of the standard deviation values. **8.61 (a)** Point estimates. **(b)** No; also need to know sample sizes. **8.63** 0.015; 0.018 **8.65** Subtract and add the 1.85% margin of error from/to 81.4%. We are 95% confident that the population percentage who believe in life after death falls between 79.6% and 83.3%. **8.67 (a)** Data were obtained randomly. **(b)** (0.035, 0.045); mainly because of the very large sample size. **(c)** Yes; 10% falls above the highest plausible value in the confidence interval. **8.69** No; because you have data for the entire population; you don't have to estimate it. **8.71 (a)** 496 said "legal." 751 said "not legal." 0.398 and 0.602. **(b)** Minority; 0.50 is above the highest believable value for the percentage who think it should be made legal in the 95% confidence interval of 0.37 to 0.43. **(c)** It appears that the proportion favoring legalization is increasing over time. **8.73** Sample: specific sample of non drinkers. "X," is the number in the sample who said that they have been lifetime abstainers from drinking alcohol, 7380 people. N: sample size of 30,000. Sample p: proportion of total sample, 0.246, who said that they are lifetime abstainers from drinking alcohol. 95.0% CI: 95% confidence interval; we can be 95% confident that the population proportion falls between 0.241 and 0.251. **8.75 (a)** The first result is a 90% confidence interval for the mean hours spent per week sending and answering e-mail for males of at least age 75. The sample mean is 6.38 hours. The sample standard deviation is 6.02. These estimates are based on a sample of size 8. **(b)** The confidence interval is 2.34 to 10.41. We can be 90% confident that the population mean number of hours spent per week sending and answering email for males of at least age 75 is between 2.34 and 10.41 hours. **(c)** Since many men over the age of 75 do not use e-mail but also some use e-mail a large number of hours, this distribution is likely skewed right. Since the t-distribution is robust to violations of the normality assumption, the interval is still valid. **8.77 (a)** 2.262, 2.093, 2.045, 1.96 (the z-score associated with a 95% confidence interval). **(b)** The t distribution approaches the standard normal distribution as the sample size gets larger. **8.79 (a)** ($37,647, $50,021). **(b)** An approximately normal population distribution. **(c)** Not necessarily; the method is robust in terms of the normal distribution assumption. **8.81 (a)** No; the minimum possible value of 0 is barely more than one standard deviation below the mean, and the mean is quite a bit larger than the median. **(b)** Yes, because the normal population assumption is much less important with such a large random sample size. (19.4, 21.2) **8.83** We can be 95% confident that the mean for the population of students is between 0.1 and 2.7. Assumptions: data randomly produced; approximately normal population distribution. **8.85 (a)** We can be 95% confident that the population mean price is between $411 and $845. **(b)** Assumptions: data produced randomly; population distributed normally. The population does not seem to be distributed normally (cluster on the low and high ends, with fewer in the middle). The method is robust with respect to the normal distribution assumption. **8.87 (a)** Box plot suggests that the population distribution is slightly right skewed. There does not seem to be an extreme outlier, so this should not affect the population inferences. **(b)** 90.2; 21.0. **(c)** We can be 95% confident that population mean income is between $8,256 and $9,823. **8.89** We can be 95% confident that the mean number of hours for the population of adult males falls between 2.67 and 3.08. **8.91** You could make inferences using the mean of 13.4 (standard deviation of 3.2) or the proportion of 0.166. For the proportion, $se = 0.0083$ and 95% confidence interval is (0.150, 0.182). **8.93 (a)** $se = 0.029$. **(b)** The lowest possible value of 0 is less than one standard deviation below the mean. **(c)** Because the method is robust with respect to the normal distribution assumption. **8.95 (a)** 385. **(b)** 601; if we can make an educated guess, we can use a smaller sample size. **8.97 (a)** 385; the solution makes the assumption that the standard deviation will be similar now. **(b)** More than $100 because the standard error will be larger than predicted. **(c)** With a larger margin of error, the 95% confidence interval is wider; thus, the probability that the sample mean is within $100 of the population mean is less than 0.95. **8.99** Answers will vary but will explain the logic behind random sampling and how the margin of error decreases as the sample size increases. **8.101** Report should include confidence intervals (17.7, 18.5) for men and (32.2, 33.0) for women. It seems that women do more housework than men, on the average. **8.103** The mean falls in the middle of the confidence interval of (4.0, 5.6). It equals 4.8. **8.105** An extremely large confidence level makes the confidence interval so wide as to have little use. **8.107** (−10.4, 30.4); An outlier can dramatically affect the standard deviation and standard error. **8.109** (a). **8.111** (b) and (e). **8.113 (a)** The confidence interval refers to the population, not the sample, mean. **(b)** The confidence interval is an interval containing possible means, not possible scores. **(c)** $\bar{x}$ is the sample mean; we know exactly what it is. **(d)** If we sampled the entire population even once, we would know the population mean exactly. **8.115** False. **8.117** False. **8.119 (a)** In the long-run, if we took many random samples of this size from the population, intervals based on the samples would capture the population proportion 95% of the time. **(b)** 0.358. **(c)** 19. **(d)** You could increase the confidence level, to 0 for example. **8.121 (a)** $|\hat{p} - p| = 1.96\sqrt{p(1-p)/n}$; $|1 - p| = 1.96\sqrt{p(1-p)/20}$; If we substitute 0.83887 for p, we get 1 − 0.83887 = 0.16113 on both sides of the equation. If we substitute 1 for p, we get 0 on the left and on the right. **(b)** This confidence interval seems more believable because it forms an actual interval and contains more than just the value of 1. **8.123** If the population is normal, the standard error of the median is 1.25 times the standard error of the mean. This means a larger margin of error, and therefore, a wider confidence interval. **8.125** Answers will vary.

Chapter 9

9.1 (a) Null. **(b)** Alternative. **(c)** (a) H_0: $p = 0.50$; H_a: $p \neq 0.50$. (b) H_0: $p = 0.24$; H_a: $p < 0.24$. **9.3** H_0: The pesticide is not harmful; H_a: The pesticide is harmful. **9.5 (a)** Alternative; it has a range of parameter values. **(b)** The relevant parameter is the mean weight change, μ. $H_0: \mu = 0$; this is a null hypothesis. **9.7** Strong evidence against; this sample proportion falls over 3 standard errors from the null hypothesis value. **9.9** Null: psychic will predict the outcome 1/6 of the time; alternative: psychic will predict the outcome more than 1/6 of the time. H_0: $p = 1/6$ and H_a: $p > 1/6$. **9.11 (a)** 0.15. **(b)** 0.30. **(c)** 0.85. **(d)** No; all of them indicate that this null hypothesis is plausible. **9.13 (a)** Standard error $= \sqrt{p_0(1-p_0)/n} = \sqrt{0.5(1-0.5)/100} = 0.05$; $z = (0.35 - 0.50)/0.05 = -3.0$. **(b)** 0.001. **(c)** Yes; this is a very small P-value; it does not seem plausible that $p = 0.50$. **9.15** (1) The response is categorical with outcomes yes or no to the statement that young adults pray daily; p represents the probability of a yes response. The poll was a random sample of 1,679 18–29-year-olds and $np_0 = n(1 - p_0) = 1{,}679(0.5) = 840 \geq 15$. (2) H_0:$p = 0.5$; H_a:p $\neq 0.5$. (3) $\hat{p} = 0.45$ so $z = \dfrac{\hat{p} - p_0}{\sqrt{p_0(1-p_0)/n}} = \dfrac{0.45 - 0.50}{\sqrt{0.5(0.5)/1679}} = -4.10$. The sample proportion is 4.1 standard errors below the null hypothesis value. (4) The P-value ≈ 0 is the probability of obtaining a sample proportion at least as extreme as the one observed, if the null hypothesis is true. (5) Because the P-value is approximately 0, the sample data supports the alternative hypothesis. There is very strong evidence that the percentage of 18–29-year-olds who pray daily is not 50%. **9.17 (a)** p = population proportion of trials guessed correctly; H_0: $p = 0.50$ and H_a: $p > 0.50$. **(b)** $z = -2.10$; sample proportion is a bit more than 2 standard errors less than expected if the null hypothesis were true. **(c)** P-value = 0.98. We cannot conclude that they predict better than by random guessing. **(d)** $np = (130)(0.5) = 65 = n(1-p)$; the sample size was large enough. We also would need to assume randomly selected practitioners and subjects for this to apply to all practitioners and subjects. **9.19 (a)** H_0: $p = 0.60$ and H_a: $p > 0.60$, where p is probability of selecting a male; the company predicts that the proportion of males chosen for management training is the same as in the eligible pool. **(b)** $np = (40)(0.6) = 24$, and $n(1-p) = (40)(0.4) = 16$. Software obtained the test-statistic by subtracting p predicted by the null from the sample p, then dividing by standard error. **(c)** The P-value in the table refers to the alternative hypothesis that $p \neq 0.60$. For the one-sided alternative, it is half this, or about 0.10. **(d)** For a 0.05 significance level, we would not reject the null hypothesis. There is insufficient evidence that there is a greater proportion of male trainees than would be expected. **9.21 (a)** Whether garlic or placebo is more effective; p = population proportion of those for whom garlic is more effective. **(b)** H_0: $p = 0.50$ and H_a: $p \neq 0.50$; there are at least 15 successes (garlic) and failures (placebo). **(c)** $z = 0.98$. **(d)** 0.33; not strong evidence against the null hypothesis that the population proportion is 0.50. **9.23 (a)** Take the difference between sample and null proportions, and divide by standard error. **(b)** The two-tail probability from the standard normal distribution below −1.286 and above 1.286. The P-value of 0.1985 or 0.20 tells us that if the null hypothesis were true, a proportion of 0.20 of samples would fall at least this far from the null hypothesis proportion of 0.50. **(c)** No; it is possible that there is a real difference in the population that we did not detect (perhaps because the sample size was not very large), and we can never accept a null hypothesis. A confidence interval shows that 0.50 is one of many plausible values. **(d)** The 95% confidence interval tells us the range of plausible values, whereas the test merely tells us that 0.50 is plausible. **9.25** This P-value gives strong evidence against the null hypothesis. It would be very unlikely to have a sample proportion of 1.00 if the actual population proportion were 0.50. **9.27 (a)** 2.145. **(b)** 1.762. **(c)** −1.762. **9.29 (a)** Larger; the t-value of 1.20 is less extreme. **(b)** Because larger sample sizes decrease the standard error (by making its denominator larger). **9.31 (a)** The variable is the number of hours worked in the previous week by male workers; the parameter is the population mean work week (in hours) for men. **(b)** H_0: $\mu = 40$; H_a: $\mu > 40$. **(c)** P-value ≈ 0. The P-value is the probability of obtaining a sample with a mean of 45.5 or more hours if the null hypothesis were true. **(d)** Since the P-value is less than the significance level of 0.01, there is sufficient evidence to reject the null hypothesis and to conclude that the population mean work week for men exceeds 40 hours. **9.33 (c)** The P-value of 0.046 is smaller than 0.05, so we have enough evidence to reject the null hypothesis. There is relatively strong evidence that the wastewater limit is being exceeded. **(d)** If it would be unusual to get a sample mean of 2000 if the population mean were 1000, it would be even more unusual to get this sample mean if the population mean were less than 1000. **9.35 (a)** 40, 15, 90, 50, 30, 70, 20, 30, −35, 40, 30, 80, 130; positively skewed; box plot part of answer. **(b) (1)** Assumptions: The data (PEF change scores) are randomly obtained from a normal population distribution. Here, the data are not likely produced using randomization, but are likely a convenience sample. The two-sided test is robust if the population distribution is not normal. **(2)** H_0: $\mu = 0$; H_a: $\mu \neq 0$. **(3)** $t = 4.03$. **(4)** P-value = 0.002. **(5)** strong evidence that PEF levels were lower with salbutamol. **(c)** Assumption of random production of data does not seem valid; convenience sample limits reliability in applying inference to population at large. **9.37 (1)** The data are produced using randomization, from a normal population distribution. Here, the distribution is likely skewed right, but the two-sided test is robust for this assumption. **(2)** H_0: $\mu = 0$; H_a: $\mu \neq 0$. **(3)** $t = 2.37$. **(4)** P-value = 0.04. **(5)** Sufficient evidence to conclude that the coupons led to higher sales than outside posters. **9.39 (a)** Most of the data fall between 4 and 14. The sample size is small so it we cannot tell too much from the plot, but there is no evidence of severe non-normality. **(c) (1)** Data are quantitative, produced randomly; population distribution approximately normal. **(2)** H_0: $\mu = 0$; H_a: $\mu \neq 0$. **(3)** $t = 4.19$. **(4)** P-value = 0.001. **(5)** Strong evidence against the null hypothesis that family therapy has no effect. **9.41** 0.01. If the two-sided test rejects the null hypothesis with significance level 0.01, then the 99% confidence interval does not contain the value in the null hypothesis. **9.43 (a)** 0.05. **(b)** Type I. **9.45 (a)** We rejected the null hypothesis, but therapy actually did not work. **(b)** We failed to reject the null hypothesis, but therapy actually had an effect. **9.47 (a)** If H_0 is rejected, we conclude that the defendant is guilty. **(b)** A Type I error would result in finding the defendant guilty when he or she is actually innocent. **(c)** If we fail to reject H_0, the defendant is found not guilty. **(d)** A Type II error would result in failing to convict a defendant who is actually guilty. **9.49 (a)** We rejected the null hypothesis that there is no disease, but the woman actually does not have cancer. **(b)** We failed to reject the null hypothesis that there is no disease, but the woman actually has cancer. The consequence is failing to detect cancer and treat the cancer when it actually exists. **(c)** More women who do have cancer will have false-negative tests and not receive treatment. **9.51 (a)** Type I; an innocent man or woman is put to death. **(b)** Type II; someone might not receive life-saving treatment when it actually is needed. **9.53 (a)** Test statistic: $t = \dfrac{(498 - 500)}{100/\sqrt{25{,}000}} = -3.16$. **(b)** P-value $= 2P(Z < -3.16) = 0.002$. **(c)** This result is statistically significant because the P-value is very small, but it is not practically significant because the sample mean of 498 is very close to the null hypothesis mean of 500. **9.55** We would expect about 5% of tests to be significant just by chance if the null hypothesis is true, and for 60 tests this is 0.05(60) = 3 tests. **9.57** The following tree diagram is based on 100 studies.

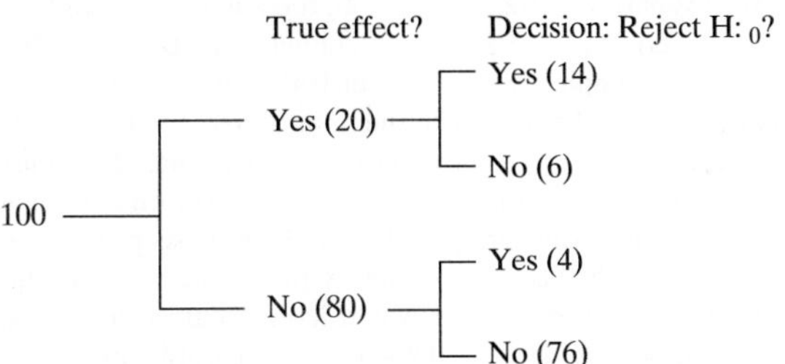

The proportion of actual Type I errors (of cases where the null is rejected) would be about 4/(4+14) = 0.22. **9.59 (a)** Cutoff z-score of 1.645; standard error $= \sqrt{0.5(1 - 0.5)/100} = 0.050$. The value 1.645 standard errors above 0.50 is $0.50 + 1.645(0.050) = 0.582$.

(c) 0.36. **9.61 (a)** Cutoff z-score: 2.33; standard error = 0.0438. The value 2.33 standard errors above 0.333 is $0.333 + 2.33(0.0438) = 0.435$. **(b)** If $p = 0.50$, the z-score for 0.435 in reference to 0.50 is $z = \dfrac{(0.435 - 0.5)}{\sqrt{0.5(1 - 0.5)/116}} = -1.40$;proportion of this curve that is not in the rejection area is 0.08. **9.63 (a)** Standard error $= \sqrt{\dfrac{p_0(1 - p_0)}{n}} = 0.0438$, where $p_0 = 1/3$; for H_a, $z = 1.96$ has a P-value of 0.05. We reject H_0 when $|\hat{p} - 1/3| \geq 1.96(se) = 0.086$, hence we need $\hat{p} \geq 0.086 + 1/3 = 0.419$ or $\hat{p} \leq 1/3 - 0.086 = 0.248$, When H_0 is false, a Type II error occurs if $0.248 < \hat{p} < 0.419$. **(b)** $Z = (0.248 - 0.50)//0.0464 = -5.43$; the probability that $\hat{p}$ is less than this z-score is 0. $z = (0.419 - 0.50)/0.0464 = -1.75$; the probability that $\hat{p}$ is greater than this z-score is 0.96. **(c)** Probability of Type II error is the portion of the curve (when $p = 0.50$) not over the rejection area. This is 0.04. **9.65 (a)** Theory predicts that P(Type II error) will be larger with a p that is closer to that predicted by the null hypothesis. **(b)** Decrease. **(c)** Increase. **9.67 (1)** Data are categorical (correct vs. incorrect guesses) and are obtained randomly. $np = (20)(0.5) < 15$, and $n(1 - p) = (20)(0.5) < 15$ so this test is approximate. **(2)** H_0: $p = 0.50$; H_a: $p > 0.50$. **(3)** $z = 0.89$. **(4)** P-value = 0.19. **(5)** There is not much evidence that the probability of a correct guess is higher than 0.50. **9.69 (a) (1)** Assumptions: The data are categorical (Brown and Whitman) and are obtained randomly; the expected successes and failures are both at least fifteen under H_0; $np = (0.5)(650) \geq 15$, and $n(1 - p) = (0.5)(650) \geq 15$. **(2)** p = population proportion of voters who prefer Brown. Hypotheses: H_0: $p = 0.50$; H_a: $p \neq 0.50$. **(3)** Test statistic: $z = \dfrac{0.554 - 0.50}{\sqrt{0.5(1 - 0.5)/650}} = 2.75$. **(4)** P-value: 0.006. **(5)** Conclusion: We can reject the null hypothesis at a significance level of 0.05; we do have strong evidence that the population proportion of voters who chose Brown is different from 0.50. **(b)** If the sample size had been 50, the test statistic would have been $z = \dfrac{0.56 - 0.50}{\sqrt{0.5(1 - 0.5)/50}} = 0.849$, and the P-value would have been 0.40. We could not have rejected the null hypothesis under these circumstances. **9.71 (1)** Assumptions: The data are categorical (frustrated versus not frustrated); the sample is a random sample of 2505 adults; the expected number of frustrated and not frustrated responses are both at least 15 under the null hypothesis: $np = n(1 - p) = (0.5)(2505) \geq 15$. **(2)** Hypotheses: H_0: $p = 0.50$; H_a: $p \neq 0.50$. **(3)** Test statistic: $z = \frac{0.56 - 0.50}{\sqrt{0.5(1 - 0.5)/2505}} = 6.0$. **(4)** P-value: 0.000. **(5)** Conclusion: We can reject the null hypothesis using a significance level of 0.05 since the P-value is much smaller than 0.05. We have very strong evidence that more than half of the population is frustrated with the federal government. **9.73 (a)** p = proportion of people who would pick Box A; H_0: $p = 0.50$; H_a: $p \neq 0.50$. **(b)** $z = 5.06$; P-value = 0.000. There is very strong evidence to conclude that the population proportion that chooses Box A is not 0.50. Box A seems to be preferred. **9.75** We are testing "p = 0.50 versus not = 0.50." X is 40, the number of people in the sample who preferred the card with the annual cost. 100 is the "N," the size of the whole sample. "Sample p" of 0.40 is the proportion of the sample that preferred the card with the annual cost. "95.0% CI" is the 95% confidence interval. "Z-Value" is the test statistic. "P-Value" tells us that if the null hypothesis were true, the proportion 0.0455 of samples would fall at least this far from the null hypothesis proportion of 0.50. The majority of the customers seem to prefer the card without the annual cost. **9.77 (a)** Type I error would have occurred if we had rejected the null hypothesis, when really women were not being passed over. A Type II error would occur if we had failed to reject the null, but women really were being picked disproportionate to their representation in the jury pool. **(b)** Type I error. **9.79 (b)** Sufficient evidence to reject null and conclude that the population mean is not 0. **(c)** 0.001; we have strong evidence to conclude that the population mean is positive. **(d)** 0.999; insufficient evidence to conclude that the population mean is negative. **9.81 (a)** H_0: $\mu = 40$; H_a: $\mu \neq 40$. **(b) (i)** SE Mean, 0.418, is standard error; **(ii)** $t = 4.76$ is the test statistic; **(iii)** P-value, 0.000, is probability that we'd get a sample mean at least this far from the value in H_0 if the null hypothesis were true. **(c)** Confidence interval supports the conclusion to reject H_0 and support H_a; 40 falls below the range of plausible values. **9.83 (a) (1)** Data are quantitative and have been produced randomly and have an approximate normal population distribution. **(2)** H_0: $\mu = 130$; H_a: $\mu \neq 130$. **(3)** The sample mean is 150.0 and standard deviation is 8.37; $t = 5.85$. **(4)** P-value = 0.002. **(5)** Strong evidence to conclude that the true mean is different from 130. **(b)** We do not know whether the population distribution is normal, but the two-sided test is robust for violations of this assumption. **9.85 (a)** Software indicates a test statistic of −5.5 and a P-value of 0.001. **(b)** For a significance level of 0.05, we would conclude that the process is not in control. **(c)** If we rejected the null hypothesis when it is in fact true, we have made a Type I error and concluded that the process is not in control when it actually is. **9.87 (1)** Data are quantitative and seem to have been produced randomly; also assume approximately normal population distribution. **(2)** H_0: $\mu = 500$; H_a: $\mu < 500$. **(3)** $t = -13.9$. **(4)** P-value = 0.000. **(5)** Extremely strong evidence that the population mean is less than 500. **9.89 (a)** We can reject the null hypothesis. **(b)** It would be a Type I error. **(c)** No, because the P value is less than 0.05; when a value is rejected by a test at the 0.05 significance level, it does not fall in the 95% confidence interval. **9.91 (a)** By choosing a smaller significance level. **(b)** It will be too difficult to reject the null hypothesis, even if the null hypothesis is not true. **9.93 (a)** $\sqrt{0.333(1 - 0.333)/60} = 0.061$. **(b)** When P(Type I error) = significance level = 0.05, $z = 1.645$, the value 1.645 standard errors above 0.333 is 0.433. **(c)** 0.15; Type II error is larger when n is smaller, because a smaller n results in a larger standard error and makes it more difficult to have a sample proportion fall in the rejection region. **9.95** Answers will vary. **9.97 (a)** These data give us a sense of what the probability would look like in the long run. If we look at just a few games, we don't get to see the overall pattern, but when we look at a number of games over time, we start to see the long run probability of the home team winning (in this case 1359/2430). **(b)** $z = \frac{0.5592 - 0.5}{\sqrt{0.5(1 - 0.5)/2430}} = 5.8$. P-value = 0.000. We can add and subtract the result of (1.96)(0.0101), namely the z-score associated with a 95% confidence interval multiplied by the standard error, to the sample proportion to obtain a confidence interval of (0.54, 0.58) for the probability of the home team scoring more runs. The test merely indicates whether $p = 0.50$ is plausible whereas the confidence interval displays the range of plausible values. **9.99 (a)** After seeing the data, we know the direction of results; it is cheating to do a one-tailed test now. **(b)** If there really is no effect, but many studies are conducted, eventually someone will achieve significance, and then the journal will publish a Type I error. **9.101** The subgroups have smaller sample size, so for a particular size of effect will have a smaller test statistic and a larger P-value. **9.103** The studies with the most extreme results will give the smallest P-values and be most likely to be statistically significant. If we could look at how results from all studies vary around a true effect, the most extreme results would be out in a tail, suggesting an effect much larger than it actually is. **9.105** Just because the sample statistic was not extreme enough to conclude that the value at H_0 is unlikely doesn't mean that the value in H_0 is the actual value. A confidence interval would show that there is a whole range of plausible values, not just the null value. **9.107** Statistical significance: strong evidence that the true parameter value is not the value in H_0; practical significance: true parameter is sufficiently different from value in H_0 to be important in practical terms. **9.109** With the probability of a false-positive diagnosis being about 50% over the course of 10 mammograms, it would not be unusual for a woman to receive a false positive over the course of having had many mammograms. Likewise, if you conduct 10 significance tests at the 0.05 significance level in 10 cases in which the null hypothesis is actually true, it would not be surprising if you sometime reject the null hypothesis just by chance. **9.111** Suppose P-value = 0.057. If the significance level were any number above this, we would have enough evidence to reject the null; if it were any number below this, there would not be enough evidence. **9.113** (b). **9.115** (a). **9.117** False. **9.119** False. **9.121** False. **9.123** Sample proportion = 0, standard error = 0, test statistic = infinity, which does not make sense. A significance test is conducted by supposing the null is true, so in finding the test statistic we should substitute the null hypothesis value, giving a more appropriate standard error. **9.125** Answers will vary.

Chapter 10

10.1 (a) The response variable is unemployment rate and the explanatory variable is race. **(b)** The two groups that are the categories of the explanatory variable are white and black individual. **(c)** The samples of white and black individuals were independent. No individuals could be in both samples. **10.3 (a)** The point estimate is 0.21.The proportion of students who reported bingeing at least three times within the past two weeks has apparently decreased between 1999 and 2009. **(b)** The standard error is the standard deviation of the sampling distribution of differences between the sample proportions. $se = 0.0304$. **(c)** (0.15, 0.27) We can be 95% confident that the population mean change in proportion is between 0.15 and 0.27. This confidence interval doe not contain zero; thus, we do have enough evidence to conclude that there was an decrease in the population proportion of UW students who reported binge drinking at least three times in the past two weeks between 1999 and 2009. **(d)** The assumptions are that the data are categorical (reported binge drinking at least three times in the past two weeks versus did not), that the samples are independent and are obtained randomly, and that there are sufficiently large sample sizes. Specifically, each sample should have at least 10 successes and 10 failures. **10.5 (a)** 0.46; 0.63. **(b)** We're 95% confident that the population proportion for females falls between 0.11 higher and 0.22 higher than the population proportion for males; 0 is not in this interval, so we conclude that females are more likely to say that they believe in miracles. Assumptions: categorical data, independent and randomly obtained samples, sufficient sample size. **(c)** The confidence interval has a wide range of plausible values, including some (such as 0.11) that indicate a moderate difference and some (such as 0.22) that indicate a relatively large difference. **10.7 (a)** H_0: $p_1 = p_2$; H_a: $p_1 \neq p_2$. **(b)** If the null were true, we would obtain a difference at least this extreme a proportion 0.14

of the time. **(c)** 0.07. **(d)** This study has smaller samples than the Physicians Health Study did. Therefore, its standard error was larger and its test statistic was smaller. A smaller test statistic has a larger P-value. **10.9 (a)** Assumptions: Each sample must have at least 10 outcomes of each type. The data must be categorical, and the samples must be independent random samples. Notation: p is the probability that someone says that he or she had engaged in unplanned sexual activities because of drinking alcohol. $H_0: p_1 = p_2$; $H_a: p_1 \neq p_2$. **(b)** $\hat{p} = 0.252$; this is the common value of p_1 and p_2, estimated by the proportion of the *total* sample who reported that they had engaged in such activities. **(c)** $se_0 = 0.028$. In this case, the standard error is interpreted as the standard deviation of the estimates $(\hat{p}_1 - \hat{p}_2)$ from different randomized studies using these sample sizes. **(d)** $z = 2.75$. P-value $= 0.006$; if the null hypothesis were true, the probability would be 0.003 of getting a test statistic at least as extreme as the value observed. We have sufficient evidence to reject the null hypothesis; we are able to show that there is a difference in proportions of reports of engaging in unplanned sexual activities because of drinking between 1999 and 2009. **10.11 (a)** Each sample has at least 10 outcomes of each type; data are categorical; independent random samples; p: probability someone developed cancer; H_0: $p_1 = p_2$; H_a: $p_1 \neq p_2$. **(b)** The test statistic is 1.03, and the P-value is 0.30. We do not have strong evidence that there are different results. **(c)** We cannot reject the null hypothesis. **10.13** Independent. Responses from Japan should not be related to responses from the United States. **10.15 (a)** The response variable is the amount of tax the student is willing to add to gasoline in order to encourage drivers to drive less or to drive more fuel-efficient cars; the explanatory variable is whether the student believes that global warming is a serious issue that requires immediate action or not. **(b)** Independent samples; the students were randomly sampled so which group the student falls in (yes or no to second question) should be independent of the other students. **(c)** A 95% confidence interval for the difference in the population mean responses on gasoline taxes for the two groups, $\mu_1 - \mu_2$, is given by $(\bar{x}_1 - \bar{x}_2) \pm t_{.025}(se)$ where $\bar{x}_1$ is the sample mean response on the gasoline tax for the group who responded yes to the second question, $\bar{x}_2$ is the sample mean response on the gasoline tax for the group who responded no to the second question, t is the t-score for a 95% confidence interval and $se = \sqrt{\frac{s_1^2}{n_1} + \frac{s_2^2}{n_2}}$ is the standard error of the difference in mean responses. **10.17 (a)** The margin of error is the t value of 2.576 multiplied by the standard error of 1.20 = 3.1. The bounds are 10.0, and 16.2. **(b)** This interval is wider than the 95% confidence interval because we have chosen a larger confidence level, and thus, the t value associated with it will be higher. To be more confident, we must include a wider range of plausible values. **10.19 (a)** 3.05, 1.92; $n = 605$. **(b)** $(-0.04, 0.38)$. Because 0 is in this interval, there is not evidence of a significant difference in population mean reports of ideal number of children between males and females. **10.21 (a)** $se = 0.97$. **(b)** The 95% CI $(-4.79, -0.81)$ indicates that we can be 95% confident that the population mean difference is between -4.79 and -0.81. Because 0 does not fall in this interval, we can conclude that, on average, the sexually abused students had a lower population mean family cohesion than the non-abused students. **10.23 (a) (i)** Because the mean is very close to 0; **(ii)** Those who reported inhaling had a mean score that was $2.9 - 0.1 = 2.8$ higher. **(b)** Not for the noninhalers. The lowest possible value of 0, which was very common, was only a fraction of a standard deviation below the mean. **(c)** 0.24; it's approximately the standard deviation of the difference between sample means from different studies using these sample sizes. **(d)** Because 0 is not in this interval, we can conclude that there is a difference in population mean HONC scores. **10.25 (a)** 0.36 is the standard deviation of the difference between sample means from different studies using these sample sizes. **(b)** $t = 3.30$; P-value $= 0.001$; very strong evidence that females have higher population mean HONC score. **(c)** No, because in each case the lowest possible value of 0 is less than 1 standard deviation below the mean. This does not affect the validity of our inference greatly because of the robustness of the two-sided test for the assumption of a normal population distribution for each group. **10.27 (a)** H_0: $\mu_1 = \mu_2$; H_a: $\mu_1 \neq \mu_2$. **(b)** P-value $= 0.151$; insufficient evidence to conclude that there is a gender difference in TV watching. **(c)** Yes, because according to the test 0 is a plausible value for the difference between the population means. **(d)** No, because in each case the lowest possible value of 0 is less than 1.2 standard deviations below the mean. This does not affect the validity of our inference greatly because of the robustness of the two-sided test for the assumption of a normal population distribution for each group. Inferences assume randomized study, normal population distributions. **10.29 (a)** Let Group 1 represent the students who planned to go to graduate school and Group 2 represent those who did not. Then, $\bar{x}_1 = 11.67$, $s_1 = 8.34$, $\bar{x}_2 = 9.10$ and $s_2 = 3.70$. The sample mean study time per week was higher for the students who planned to go to graduate school, but the times were also much more variable for this group. **(b)** $se = 2.16$. If further random samples of these sizes were obtained from these populations, the differences between the sample means would vary. The standard deviation of these values would equal about 2.2. **(c)** A 95% confidence interval is $(-1.9, 7.0)$. We are 95% confident that the difference in the mean study time per week between the two groups is between -1.9 and 7.0 hours. Since 0 is contained within this interval, we cannot conclude that the population mean study times differ for the two groups. **10.31 (a)** Let Group 1 represent males and Group 2 represent females. Then, $\bar{x}_1 = 11.9, s_1 = 3.94, \bar{x}_2 = 15.6$ and $s_2 = 8.00$. The sample mean time spent on social networks was higher for females than for males, but notice the apparent outlier for the female group (40). The data were also much more variable for females, but this may also merely reflect the outlier. **(b)** $se = 2.084$. If further random samples of these sizes were obtained from these populations, the differences between the sample means would vary. The standard deviation of these values would equal about 2.1. **(c)** A 90% confidence interval is $(-7.24, -0.17)$. We are 90% confident that the difference in the population mean number of hours spent on social network per week is between -7.2 and -0.17 for males and females. Since 0 is not quite contained within this interval, we can conclude that the population mean time spent on social networks per week differs for males and females. **10.33** With large random samples, the sampling distribution of the difference between two sample means is approximately normal regardless of the shape of the population distributions. Substituting sample standard deviations for unknown population standard deviations then yields an approximate t sampling distribution. With small samples, the sampling distribution is not necessarily bell-shaped if the population distributions are highly non-normal. **10.35 (a)** $t = ((13.2 - 7.3) - 0)/1.78 = 3.31$; P-value: 0.001. **(b)** The assumptions are that the data are quantitative, constitute random samples from two groups, and are from populations with approximately normal distributions. In addition, we assume that the population standard deviations are equal. Given the large standard deviations of the groups, the normality assumption is likely violated, but we're using a two-sided test, so inferences are robust to that assumption. **10.37 (b)** We can be 95% confident that the true population mean difference between change scores is between -1.2 and 41.2. The therapies may not have different means, but if they do the population mean could be much higher for Therapy 1. The confidence interval is wide because the two sample sizes are very small. **(c)** Yes; 0 is no longer in the range of plausible values. **10.39 (a)** The first set of inferences assume equal population standard deviations, but the sample standard deviations suggest this is not plausible. It is more reliable to conduct the second set of inferences, which do not make this assumption. **(b)** Using the second set of results, it appears that the vegetarian students are more liberal than are the nonvegetarian students. **10.41 (a)** The proportion of heart failure for subjects who ate a high intake of fish daily was estimated to be 0.96 times the proportion of heart failure for subjects who ate no fish daily. **(b)** Because the interval contains 1, in the population we can infer that it is plausible the proportion of heart failure is similar for individuals who eat a high intake of fish daily and individuals who eat no fish daily. **10.43** In the population of those living in the Asia-Pacific region, we are 95% confident that the increase in the population death rate for the highly obese group was between 9% and 36%. **10.45** The ratio of the percentages is 14.6%/12.4% = 1.18. The percentage of overweight children in 2008 was estimated to be 1.18 times the percentage of overweight children in 1998. The obesity prevalence among low-income, preschool-aged children increased by 18%. **10.47 (a)** The same patients are in both samples. **(b)** 150; 130; 20; "before" mean minus "after" mean equals mean of the difference scores. **(c)** We can be 95% confident that the difference between the population means is between 7.6 and 32.4. **10.49 (a)** Dependent; the same students are in both samples. **(b)** No, there is quite a bit of variability, but outlying values appear in both directions. **(c)** The 95% confidence interval was obtained by adding and subtracting the margin of error (the *t*-score of 2.262 for $df = 9$ times the standard error of 5.11, which is 11.6) from the mean difference score of 4.0. **(d)** The test statistic was obtained as $t = \frac{\bar{x}_d - 0}{se} = \frac{4.0 - 0}{5.11} = 0.78$; the P-value is 0.45. We cannot conclude that there is a difference in attendance at movies versus sports events. **10.51 (a) (1)** Assumptions: the difference scores are a random sample from a population distribution that is approximately normal. **(2)** H_0: $\mu_1 = \mu_2$. (or population mean of difference scores is 0); H_a: $\mu_1 \neq \mu_2$. **(3)** $t = -1.62$. **(4)** P-value $= 0.14$. **(5)** Insufficient evidence to conclude that the population means are different for movies and parties. **(b)** (−21.1, 3.5). We are 95% confident that the population mean number of times spent attending movies is between 21.1 less and 3.5 higher than the population mean number of times spent attending parties. **10.53 (a)** The standard deviation of the change in weight scores could be much smaller because even though there is a lot of variability among the initial and final weights of the women, most women do not see a large change in weight over the course of the study, so the weight changes would not vary much. **(b)** $se = s_d/\sqrt{n} = 2.0/\sqrt{132} = 0.174$. A 95% confidence interval given by $\bar{X}_d \pm t_{.025}(se)$ has lower endpoint 2.1-(1.98)(0.174)=1.76 and upper endpoint 2.1 + (1.98)(0.174) = 2.44. Thus, the confidence interval is (1.76, 2.44). 15 is not a plausible weight change in the population of freshmen women. The plausible weight change falls in the range from 1.76 to 2.44. **(c)** The data must be quantitative, the sample of difference scores must be a random sample from a population of such difference scores, and the difference scores must have a population distribution that is approximately normal (particularly with samples of size less than 30). **10.55 (1)** Assumptions: the differences in prices are a random sample from a population that is approximately normal.

(2) H_0: $\mu_d = 0$ vs. H_a: $\mu_d \neq 0$. **(3)** $t = \frac{\bar{d} - 0}{s_d/\sqrt{n}} = \frac{4.3}{4.7152/\sqrt{10}} = 2.88$.

(4) P-value=0.02. **(5)** If the null hypothesis is true, the probability of obtaining a difference in sample means as extreme as that observed is 0.02. We would reject the null hypothesis and conclude that there is a significant difference in the mean price of textbooks used at her college between the two sites for $\alpha = 0.05$. **10.57 (a)** They are dependent since matched pairs were formed. **(b)** Let Group 1 represent the juveniles assigned to adult court and Group 2 represent the juveniles assigned to juvenile court. Then $\hat{p}_1 = 673/2097 = 0.32$ and $\hat{p}_2 = 448/2097 = 0.21$.

(c) H_0: $\mu_d = 0$ vs. H_a: $\mu_d \neq 0$, $z = \frac{515 - 290}{\sqrt{515 + 290}} = 7.9$. P-value ≈ 0.

There is extremely strong evidence of a population difference in the re-arrest rates between juveniles assigned to adult court and those assigned to juvenile court. **10.59 (a)** 0.38; 0.43. **(b)** Sample mean: for each subject, their coffee score is 1 if they choose Sanka and 0 if not. The mean of the scores is the proportion using Sanka. The estimated difference of population proportions is difference between the sample means at the two times. **(c)** We can be 95% confident that the population proportion choosing Sanka for the second purchase was between 0.01 and 0.09 higher than the population proportion choosing it for the first purchase. **10.61 (a)** 0.135. **(b) (i)** independent, random samples and sums of two counts at least 30. **(ii)** H_0: $p_1 = p_2$; H_a: $p_1 \neq p_2$; **(iii)** $z = 3.00$; **(iv)** P-value: 0.003; **(v)** Strong evidence that there is a difference between the population proportions. **10.63 (a)** This refers to an analysis of three variables, a response variable, an explanatory variable, and a control variable. The response variable is "whether or not at risk for cardiovascular disease," the explanatory variable is "whether drink alcohol moderately," and the control variables would be socioeconomic status and mental and physical health. **(b)** There is a stronger association between drinking alcohol and its effect on risk for cardiovascular disease for subjects who have a higher socioeconomic status. **10.65 (a)** The proportion of shots made is higher for Barry both for 2-point shots and for 3-point shots, but the proportion of shots made is higher for O'Neal overall. **(b)** O'Neal took almost exclusively 2-point shots, where the chance of success is higher. **10.67 (a)** Higher mean for English-speaking families (1.95) than French-speaking families (1.85). **(b)** In each case, higher mean for French-speaking families: Quebec: French (1.80); English (1.64); other provinces: French (2.14); English (1.97). **(c)** There are relatively more English-speaking families in the "other" provinces, and the "other" provinces have a higher mean regardless of language. This illustrates Simpson's paradox. **10.69** There could be no difference in the prevalence of breast cancer now and in 1900 for women of a given age. Overall, the breast cancer rate would be higher now, because more women live to an old age now, and older people are more likely to have breast cancer. **10.71 (a)** Response: view of the U.S; explanatory: year. **(b)** Independent; the two samples are not linked (e.g., not the same people in both groups). **(c)** The two samples include the same subjects so should be treated as dependent. **10.73 (a)** We are 95% confident that the population proportion of females who have used marijuana is at least 0.01 lower and at most 0.09 lower than the population proportion of males who have used marijuana (when rounded). Because 0 is not in the confidence interval, we can conclude that females and males differ with respect to marijuana use. **(b)** The confidence interval would change only in sign. It would now be (0.01, 0.09) instead of (−0.09,−0.01), rounded. We are 95% confident that the population proportion of males who have used marijuana is at least 0.01 higher and at most 0.09 higher than the population proportion of females who have used marijuana. **10.75 (a)** $se = 0.0228$. **(b)** $z = 5.57$; P-value = 0; we cannot reject the null hypothesis. We reject that the population proportion believing in the afterlife is the same for females and males. **(c)** No, because there actually was a difference between females and males. **(d)** Independent random samples; at least 5 successes and 5 failures in each sample. **10.77 (a)** Box plots are part of answer. Female crabs have a higher median and more variability if they had a mate (right-skewed distribution) than if they did not have a mate (symmetrical). **(b)** 0.5. **(c)** $se = 0.076$. **(d)** We can be 90% confident that the difference between the population mean weights of female crabs with and without a mate is between 0.375 and 0.625 kg. **10.79 (a)** H_0: $\mu_1 = \mu_2$; H_a: $\mu_1 \neq \mu_2$. **(b)** 5.96; 0.000. It would be extremely unusual to obtain a test statistic this large if the null hypothesis were true. **(c)** We can reject the null hypothesis. The population mean is higher for blacks. **(d)** When we reject the null hypothesis with significance level 0.05, the 95% confidence interval does not include 0. **10.81 (a)** The response variable is the number of hours a week spent on the Internet and is quantitative. The explanatory variable is the respondent's gender and is categorical. **(b)** The 99% confidence interval is (0.4, 2.2). We are 99% confident that the population mean number of hours spent on the Internet per week is between 0.4 and 2.2 hours more for males than for females. **(c) (1)** Assumptions: Independent random samples, and number of hours spent on the Internet per week has an approximately normal population distribution for each gender. **(2)** H_0: $\mu_1 = \mu_2$; H_a: $\mu_1 \neq \mu_2$. **(3)** $t = 3.62$. **(4)** P-value $= 0.000$. **(5)** If the null hypothesis is true, the probability of obtaining a difference in sample means as extreme as that observed is close to 0. At $\alpha = 0.01$, we would reject the null hypothesis and conclude that the population mean number of hours a week spent on the Internet differs for males and females. **10.83 (1)** Assumptions: the data are quantitative (child's score); the samples are independent random samples and the population distributions of scores are approximately normal for each group. **(2)** H_0: $\mu_1 = \mu_2$; H_a: $\mu_1 \neq \mu_2$ where Group 1 represents the group with the male tester and Group 2 represents the group with the female tester. **(3)** $se = 0.235$, $t = -1.28$. **(4)** P-value: 0.205. **(5)** If the null hypothesis were true, the probability would be 0.205 of getting a test statistic at least as extreme as the value observed. Since the P-value is quite large, there is not much evidence of a difference in the population mean of the children's scores when the tester is male versus female. **10.85 (a)** The 95% confidence interval is (0.6, 7.4). **(b)** It tells us the P-value = 0.02; we would expect a difference of sample means at least this size only 2% of the time if there were truly no difference between the population means. **10.87 (a)** P-value = 0.10, so there is a probability of 0.10 that we would get a test statistic at least this large if the null hypothesis is true that there is no difference between population mean change scores. **(b)** Data are quantitative; samples are independent and random; population distributions for each group are approximately normal. Based on the box plots (which show outliers and a skew to the right for the cognitive behavioral group), it would not be a good idea to conduct a one-sided test. It is not as robust as the two-sided test to violations of the normal assumption. **(c)** The lowest plausible difference between means is −0.7, a difference of less than 1 pound. The highest plausible difference between means is 7.6. **(d)** We do not reject the null hypothesis that the difference between the population means is 0, and 0 falls in the 95% confidence interval for the difference between the population means. **10.89** We can be 95% confident that the true population mean difference is between −10.8 and 5.2. **10.91** **(a)** We can be 95% confident that the true population mean difference is between −0.7 and 7.6. **(b)** If the null hypothesis were true, there would be probability 0.10 of obtaining a test statistic at least this large (in absolute value). **(c)** P-value = 0.10/2 = 0.05. If the null hypothesis were true, the probability = 0.05 of getting a t statistic of 1.68 or larger. **(d)** Quantitative response variable, independent random samples, and approximately normal population distributions for each group. **10.93** **(a)** The relative risk of 0.68 indicates that, in this study, for those with a family history of prostate cancer and who ate two servings per week of spaghetti/tomato sauce, the sample proportion of prostate cancer was estimated to be 0.68 times the sample proportion of prostate cancer cases for those who ate less than one serving per month of spaghetti/tomato sauce. **(b)** We can be 95% confident that the population mean relative risk value is between 0.31 and 1.51. Because 1.0 does fall in this range, further investigation is required to verify decreased relative risk for prostate cancer. **10.95** **(a)** The ratio of incarceration rates is (1/109)/(1/1563) = 14.3. Males were 14.3 times more likely to be incarcerated than women in 2006. **(b)** The ratio of incarceration rates is (1694/100000)/(252/100000) = 6.7. Black residents were 6.7 times more likely to be incarcerated than white residents in 2006. **10.97** We would explain that there's less than a 5% chance that we'd get a mean at least this much higher after the training course if there were, in fact, no difference. It would have been helpful to have the means, the mean difference, and its standard error (or a confidence interval). **10.99** **(a)** The groups are dependent since they were matched according to age and gender. **(b)** **(1)** Assumptions: the differences in scores are a random sample from a population that is approximately normal. **(2)** H_0: $\mu_d = 0$ versus H_a: $\mu_d \neq 0$. **(3)** $t = 1.95$. **(4)** P-value = 0.057. **(5)** If the null hypothesis is true, the probability of obtaining a difference in sample means at least as extreme as that observed is 0.057. This is some, but not strong, evidence that there is a difference in the mean scores between children of alcoholics versus children of nonalcoholics. **(c)** We assume that the population of differences is approximately normal and that our sample is a random sample from this distribution. **10.101** **(a)** The samples were dependent since the same women were sampled before and after their surgeries. **(b)** No. In order to find the t statistic, we need to know the standard deviation of the differences. **10.103** **(a)** Dependent; the same people are answering both questions. **(b)** We can be 95% confident that the true population mean difference between ratings of the influence of TV and rock music is between −0.3 and 1.3. **(c)** The P-value indicates that there is a probability of 0.21 that we would obtain a test statistic at least this large if the null hypothesis were true. **10.105** **(a)** High dose: 69/86 = 0.802; low dose: 61/86 =0.709. **(b)** There is a probability of 0.10 that the test statistic $z = 1.63$ would be at least this extreme if the null hypothesis were true. **10.107** Relative risk =3. The proportion of obesity for those having lots of fast food and TV watching is three times the proportion of obesity for those who do not. **10.109** **(a)** Ignore victim's race: proportion 0.11 of white defendants receive the death penalty, proportion 0.08 of black defendants receive the death penalty. When we take victim's race into account, the direction of the association changes, with black defendant more likely to get the death penalty. Specifically, when victim was white: white defendants: 0.11; black defendants: 0.23. When victim was black: White defendants: 0.00; black defendants: 0.03. **(b)** Death penalty imposed more when victim was white, and white victims were more common when defendant was white. **10.111** **(a)** If most of the female faculty had been hired recently, they would be fewer years from their degree and would have lower incomes. So, overall, mean could be lower for females. **(b)** If more women seek positions in low-salary colleges and more men in high-salary colleges, there could be no difference in means between males and females in each college yet overall a higher mean for men. **10.113** Answers will vary. **10.115** **(a)** Sample standard deviations and sample sizes for the two groups. **(b)** No; if we have the information on the entire population, there's no need to make inferences. **10.117** **(a)** Whether obese (yes or no) and wage. **(b)** Education level; the women could be paired according to education level and then compared in obesity rates. **10.119** (d). **10.121** False. **10.123** False. **10.125** We can be 95% confident that the difference between the population mean is between 9.6 and 17.0. **10.127** **(a)** **(i)** $\hat{p}_1 = \hat{p}_2 = 0$ because there are no successes in either group (i.e., 0/10 = 0); **(ii)** $se = 0$ because there is no variability in either group if all responses are the same (and can also see this from formula for se); **(iii)** The 95% confidence interval would be (0,0) because we'd be adding 0 to 0 (se multiplied by z would always be 0 with se of 0, regardless of the confidence level). **(b)** The new confidence interval, (−0.22, 0.22), is far more plausible than (0, 0). **10.129** Answers will vary.

Part 3 Answers

R3.1 **(a)** $se = \sqrt{\hat{p}(1 - \hat{p})/n} = \sqrt{0.20(1 - 0.20)/2303} = 0.0083$. **(b)** The standard error would be twice as large ($\sqrt{4} = 2$). In order to increase the precision of estimates, the standard error must get smaller, which happens when the sample size increases. Since the sample size is in the denominator through its square root, it must quadruple for the standard error to be half as large. **R3.3** **(a)** 0.54; (0.51, 0.57). **(b)** For H_0: $p_1 = p_2$ and H_a: $p_1 \neq p_2$, test statistic $z = 10.2$, P-value ≈ 0. We reject the null hypothesis and conclude that the proportion of Floridians who say sexual relations between two adults of the same sex is always wrong has decreased from years 1988 to 2006. **R3.5** **(a)** No; let p be the proportion of adult Floridians who favor raising taxes to handle the problem. Then, the test statistic for testing: H_0: $p = 0.5$ versus H_a: $p \neq 0.5$ is $z = 1.39$ with a P-value of 0.16. There is not much evidence about whether a majority or minority of Floridians favored raising taxes to handle the government's problem of not having enough money to pay for all of its services. Assumptions: the variable is categorical, random sample, $np_0 \geq 15$ and $n(1 - p_0) \geq 15$. **(b)** 1066. **R3.7** **(a)** The hypotheses are H_0: $p = 0.5$ versus H_a: $p \neq 0.5$. The assumptions are as follows: **(i)** the variable, whether gays should be allowed to serve in the military, is categorical; **(ii)** the sample is a random sample; **(iii)** the sample size, 1029, is large enough to ensure that the sampling distribution of the sample proportion is approximately normal. (Check that the number of successes and failures are both greater than 15.) **(b)** Test statistic: $z = \dfrac{\hat{p} - p}{\sqrt{p_0(1 - p_0)/n}} = 12.83$. The sample proportion, 0.70, lies 12.83 standard errors above the hypothesized value of 0.5. **(c)** P-value = $2P(z > 12.83) = 0.000$. If the null is true, the probability of obtaining a sample result at least as extreme as that observed is about 0.000. **(d)** Since the P-value is negligible, there is strong evidence that in 2010 a majority of U.S. adults thought that gays should be allowed to serve in the military. **R3.9** **(a)** The standard error was found by $s/\sqrt{n} = 8.37/\sqrt{166} = 0.65$. **(b)** We are 95% confident that in 2008 the population mean number of male sex partners for females between the ages of 20 and 29 was between 3.7 and 6.3 (actually reported as between 4 and 6 male partners). **(c)** No, probably highly skewed to right. Since the sample size is quite large, the confidence interval is still valid because the sampling distribution will still be bell-shaped by the central limit theorem. **(d)** Median since it is more appropriate for highly skewed distributions. **R3.11** **(a)** Independent; the respondents differ from one year to the next. **(b)** The percentage favoring legalization showed a noteworthy increase in the 1970s but dipped back down in the 1980s. This percentage has been increasing fairly steadily since 1991. **R3.13** Dependent since the same 20 people were observed watching both of the films. **R3.15** **(a)** We are 95% confident that the population proportion of black youth who listen to rap music every day is between 0.09 and 0.17 higher than the population proportion of Hispanic youths who listen to rap music every day. **(b)** 0. **R3.17** **(a)** Let p_1 denote the proportion of fundamentalists who answered definitely not true and p_2 denote the proportion of liberals who an-swered definitely not true; H_0: $p_1 = p_2$; H_a: $p_1 \neq p_2$. **(b)** $z = 10.1$. **(c)** P-value $\approx$ 0. Presuming H_o is true, there is an approximate 0 probability of obtaining the test statistic 10.1 or higher. **(d)** Since the P-value < 0.05, we conclude that the population proportion who responded definitely not true is different for those who classify themselves as religious fundamentalists than for those who classify themselves as liberal in their religious beliefs. **R3.19** (0.15, 0.23). We are 95% confident that the population proportion of men who reported spending no time on cooking and washing up during a typical day is between 0.15 and 0.23 higher than the population proportion of women who responded the same. **R3.21 (a) (1)** Assumptions: the response is categorical for both groups; the groups are independent and the samples are random; the sample sizes are large enough so that there are at least five successes and five failures for each group. **(2)** Let Group 1 be the sample of women and Group 2 the sample of men. Then the hypotheses are H_0: $p_1 = p_2$ versus H_a: $p_1 \neq p_2$. **(3)** $z = 0.48$. **(4)** P-value = 0.629. **(5)** The P-value is quite large indicating that the test statistic observed is not unusual under the null hypothesis. **(b)** (−0.015, 0.025). Since 0 is contained in the interval, there may be no difference in the population proportions of males and females who are compulsive buyers. **R3.23 (a)** The 95% confidence interval is given by (21.1, 26.8). We are 95% confident that the mean percentage expectation of time spent on chores is between 21.1% and 26.8% more for women than for men. Since 0 does not fall within this interval, we can conclude that the mean percentage expectation time spent on chores is higher for women than for men. **(b)** The assumptions are that the samples are random and independent and that the number percentage expectation of time spent on chores is approximately normally distributed for the two groups. **R3.25 (a)** Males: $\bar{x}_1 = 1.28$; Females: $\bar{x}_2 = 1.67$. **(b)** (1.51, 1.83). **(c)** (−0.6, −0.2). We can be 95% confident that the population mean number of days that males felt lonely is between 0.6 and 0.2 days less than women felt lonely. **R3.27 (a)** The P-value, 0.000, is the probability of obtaining a test statistic at least as extreme as that observed if the null hypothesis is true. There is extremely strong evidence that the population mean number of sex partners differs for males and females. **(b)** We are 95% confident that the population mean number of sex partners over the past year is between 0.16 and 0.52 more for males than for females. **(c)** We assume that the samples are independent, random samples from distributions that are approximately normal. **R3.29 (a)** The difference in the proportion of women who suffer from anorexia versus those who suffer from binge eating is −0.025. **(b)** The proportion of women who suffer from binge eating is 3.5 times the proportion of women who suffer from anorexia. **R3.31 (a) (i)** 7; **(ii)** 7. The two methods give identical answers. **(b)** P-value = 0.002. The probability of obtaining a test statistic at least as extreme as that observed, when the null hypothesis is true, is 0.002. **(c)** (4.0, 10.0). We are 90% confident that the population mean difference in scores after and before taking the training course is between 4.0 and 10.0. **(d)** The 90% confidence interval does not contain 0. Likewise, the significance test has P-value below 0.10. Each inference suggests that 0 is not a plausible value for the population mean difference. **R3.33** Answers will vary but could include a significance test comparing the population means ($t = -0.82$, P-value = 0.42) and/or a 95% CI for the difference in population means for females and males, which is (−2.2, 0.9). **R3.35** Descriptive: the proportion of people who said that they had not discussed matters of importance with anyone over the past six months was 0.089 in 1985 and 0.25 in 2004. Inferential: 95 CI = (−0.19, −0.13); the population proportion of people who said that they had not discussed matters of importance with anyone over the past six months was less in 1985 than in 2004. **R3.37** By the sample size formula $\frac{\sigma^2 z^2}{m^2}$, the needed sample size is proportional to the squared standard deviation. To estimate the mean income for all lawyers in the United States, we would need a large sample because of the wide range of salaries (large standard deviation) due to differences in specialty, experience, location, etc. To estimate the mean income for all entry-level employees at Burger King restaurants in the United States, pay is likely to be fairly homogeneous (small standard deviation) so that a small sample would suffice. **R3.39** If we assume no difference in mean weight gain for the population of freshman men and women, the probability of obtaining a sample difference as large as or larger than that observed would be quite small. This probability is called the P-value and when it is small enough, it contradicts the statement of no difference, providing us with sufficient evidence to reject this statement and conclude the two groups have differing mean weight gains. **R3.41** Yes, an increase of 12% in the percentage of teenagers who are overweight seems practically significant as well. Statistical significance means that the sample results were unusual enough to reject the null hypothesis of no difference. The results are practically significant if the difference is large in practical terms. **R3.43** If sample sizes were large and similiar, (1.9, 16.1). We are 95% confident that the difference in population mean prose literacy scores for Canada and the United States is between 1.9 and 16.1. **R3.45(a)** Margin of error $= z\sqrt{\hat{p}(1-\hat{p})/n} = 1.96\sqrt{0.6(0.4)/1000} = 0.03$. **(b)** By the formula in part a, the margin of error changes for different sample proportions. It tends to get smaller as the sample proportion moves toward 0 or 1. **(c)** The numerators, $\hat{p}(1-\hat{p})$ and $(1-\hat{p})\hat{p}$, are the same.

Chapter 11

11.1 (a) Response: political party identification; explanatory: gender. **(b)**

	Political Party Identification				
Gender	**Democrat**	**Independent**	**Republican**	**Total**	***n***
Female	39.2%	35.4%	25.4%	100%	1076
Male	33.4%	40.7%	25.9%	100%	896

Women are more likely than are men to be Democrats, whereas men are more likely than are women to be Independents. **(c)** Distributions should show percentages in the party categories that are the same for men and women, such as (36%, 38%, 26%). **11.3 (a)** These distributions refer to those of *y* at given categories of *x*.

	Type of Student	
Gender of Student	**Undergraduate**	**Graduate**
Female	78%	22%
Male	84%	16%

(b) x and y are dependent because the probability of a randomly chosen student being an undergraduate or graduate changes according to gender. **11.5 (a)**

	Happiness Of Marriage			
Income	**Very Happy**	**Pretty Happy**	**Not Too Happy**	***n***
Below	123	105	7	235
Average	291	151	17	459
Above	172	83	6	261

(b)

	Happiness Of Marriage			
Income	**Very Happy**	**Pretty Happy**	**Not Too Happy**	***n***
Below	52.3%	44.7%	3.0%	235
Average	63.4%	32.9%	3.7%	459
Above	65.9%	31.8%	2.3%	261

Average-income people tend to have slightly less happy marriages than those in the other two brackets; high-income people tend to have slightly happier marriages. **(c)** Regardless of income brackets, marital happiness tends to be higher than general happiness for the Very Happy and the Not Too Happy categories. **11.7** Answers will vary depending on the column variable selected. **11.9 (a)** H_0: Gender and happiness are independent. H_a: Gender and happiness are dependent. **(b)** The P-value is not especially small. If the null hypothesis were true, the probability would be 0.79 of getting a test statistic at least as extreme as the value observed. So, there is not strong evidence against the null hypothesis, and it is plausible that gender and happiness are independent.

11.11 (a)

Gender	Believe	Don't Believe	Total
Male	620	189	808
Female	835	143	979

(b)

Gender	Believe	Don't Believe	Total
Male	658.7	150.3	808
Female	796.3	181.7	979

There are more women who believe than expected and more men who do not believe than expected. **(c)** $X^2 = (620 - 658.7)^2/658.7 + \cdots + (143 - 181.7)^2/181.7 = 22.362$. **11.13 (a)** (61.1%, 38.9%) and (5.9%, 94.1%); suggests marijuana use much more common for those who have smoked cigarettes than for those who have not. **(b) (1)** Assumptions: two categorical variables (cigarette use and marijuana use), randomization, expected count at least five in all cells. **(2)** H_0: marijuana use independent of cigarette use; H_a: dependence. **(3)** $X^2 = 642.0$. **(4)** P-value: 0.000. **(5)** Extremely strong evidence that marijuana use and cigarette use are associated. **11.15 (a)** H_0: opinion about the environment independent of sex; H_a: dependence. **(b)** $r = 2$; $c = 5$; $df = 4$. **(c)** The P-value is 0.09. **(i)** No; **(ii)** Yes. **(d) (i)** Fail to reject the null hypothesis, and conclude that variables may be independent; **(ii)** Reject the null hypothesis, there is evidence to suggest that opinion depends on gender. **11.17 (a)**

	H_a	No H_a
placebo	28	656
aspirin	18	658

(b) (1) 2 categorical variables, randomization, expected count at least 5 in all cells. **(2)** H_0: whether have heart attack independent of treatment; H_a: dependence. **(3)** $X^2 = 2.1$. **(4)** P-value: 0.14. **(5)** It is plausible that treatment and heart attack incidence are independent. **11.19 (a)** Data show a strong association between happiness and belief in life after death. **(b)** 0.03 (as reported by GSS); results would be unusual if variables were independent. **(c)** There is strong evidence that happiness depends on opinion about the afterlife. **11.21 (a)** H_0: $p = 0.75$. **(b)** $X^2 = 3.46$; $df = 1$. **(c)** P-value $= 0.06$. The probability of obtaining a test statistic at least as extreme as that observed, assuming the null hypothesis is true, is 0.06. Some evidence against the null, but not very strong. **11.23 (a) (1)** 2 categorical variables, randomization, expected count at least 5 in all cells. **(2)** H_0: party identification independent of race; H_a: dependence. **(3)** $X^2 = 177.312$. **(4)** P-value: 0.000. **(5)** Very strong evidence that party identification and race are associated. **(b)** 0.395; strong; it is quite far from 0.00, the difference corresponding to no association. **11.25 (a)** Reject the null hypothesis of independence, because P-value is less than 0.05. The data suggests happiness depends of independence on highest degree. **(b)** No; with a large sample size, even a weak association could be statistically significant. **(c)** 0.21; in the sample, happiness is higher among those who completed college than among those who did not complete high school. **(d)** 3.3; college graduates are 3.3 times as likely to be happy as students who did not complete high school. **11.27 (a)** The proportion who had used alcohol is 0.33 higher for cigarette users versus non-cigarette users. **(b)** The proportion who had used cigarettes is 0.60 higher for alcohol users than for those who had not used alcohol. **(c)** Alcohol users are 5.3 times as likely to have smoked as are non-alcohol users. **11.29 (a)** 949/100,000 $=$ 0.00949 of men were incarcerated, whereas 67/100,000 $=$ 0.00067 for women. The relative risk of being incarcerated is 0.00949/0.00067 $=$ 14.2. Men were 14.2 times as likely as women were to be incarcerated. **(b)** The difference of proportions being incarcerated is $0.00949 - 0.00067 = 0.009$. The proportion of men who are incarcerated is 0.009 higher than the proportion of women who are incarcerated. **(c)** The response in part a is more appropriate because it shows there is a substantial gender effect, which the difference does not show when both proportions are close to 0. **11.31 (a)** False; larger chi-squared might be due to a larger sample size rather than a stronger association. **(b)** Yes; evidence suggests race is more strongly associated. **11.33 (a)** The observed count falls 4.681 standard errors above the expected count. **(b)** They designate conditions in which the observed counts are much higher than the expected counts, relative to what we'd expect due to sampling variability. **(c)** They designate conditions in which the observed counts are much lower than the expected counts, relative to what we'd expect due to sampling variability. **11.35 (a)** Observed count for this cell is only 0.1 standard errors below the expected count. **(b)** People who are not happy in their marriage are more likely to be not happy in general than would be expected if the variables were independent. People who are pretty happy in their marriage are more likely to be pretty happy overall than would be expected if the variables were independent. Lastly, people who are very happy in their marriage are more likely to be very happy overall than would be expected if the variables were independent. **11.37** Females tend to be Democrat more often and males tend to be Democrat less often than one would expect if party ID were independent of gender.

11.39 (a)

	S	F	Total
control	2	6	8
treatment	12	0	12

(b) (1) There are two binary categorical variables, and randomization was used. **(2)** H_0: $p_1 = p_2$; H_a: $p_1 \neq p_2$. **(3)** Test statistic: 2. **(4)** P-value: 0.001. **(5)** Strong evidence that care and diet are associated with ability to solve a task. **(c)** Because the expected cell counts are less than 5 for at least some cells. **11.41 (a)** 0.24; it is plausible that nervousness and treatment are independent. **(b)** No, because two cells have an expected count of less than five. **11.43 (a)**

	Y	N
male	0.94	0.06
female	0.94	0.06

(b) Yes; percentages of men and women who would vote for a qualified woman may be the same. **11.45 (1)** Two categorical variables, randomization, expected count at least 5 in all cells. **(2)** H_0: blood test result independent of Down syndrome status; H_a: dependence. **(3)** $X^2 = 114.4$, $df = 1$. **(4)** P-value $= 0.000$. **(5)** Very strong evidence of an association between test result and actual status. **11.47 (a)** The response variable is level of common cold symptoms and the explanatory variable is treatment (placebo versus Immumax). **(b)** I would explain that if the decrease of common cold symptoms did not depend on whether one took Immumax or placebo, then it would be quite unusual to observe the results actually obtained. This provides relatively strong evidence of decreased cold symptoms for those taking Immumax. **11.49 (1)** Two categorical variables, randomization, expected count at least 5 in all cells. **(2)** H_0: political party independent of sex; H_a: dependence. **(3)** $X^2 = 8.294$; $df = 2$. **(4)** P-value $= 0.016$. **(5)** strong evidence that party identification and gender are associated. **11.51 (a) (i)** The assumptions are that there are two categorical variables (group and cardiovascular event), that randomization was used to obtain the data, and that the expected count was at least five in all cells; **(ii)** H_0: Group and cardiovascular event are independent. H_a: Group and cardiovascular event are dependent; **(iii)** $X^2 = 10.7$; $df = 2$; **(iv)** P-value: 0.005; **(v)** The P-value is very small. If the null hypothesis were true, the probability would be close to 0 of getting a test statistic at least as extreme as the value observed. We have very strong evidence that there is an association between cardiovascular event and group. **(b)** The proportion of those on placebo who had a stroke is 0.013. The proportion of those on aspirin who had a stroke is 0.011. Thus, the relative risk is 0.013/0.011 $=$ 1.2. Those on placebo are 1.2 times as likely as those on aspirin to have stroke. **11.53 (a)** The proportion who were injured is 0.06 higher for those who did not wear a seat belt than for those who wore a seat belt. **(b)** People were 1.95 times as likely to be injured if they were not wearing a seat belt than if they were wearing a seat belt. **11.55** Those with no partners or two or more partners are less likely to be very happy, and those with one partner are more likely to be very happy, than would be expected if these variables were independent. **11.57 (a)** H_0: $p_1 = p_2$; H_a: $p_1 \neq p_2$; p_1: population proportion of those who are aggressive in the group that watches less than one hour of TV per day; p_2: population proportion of those who are aggressive in the group that watches more than one hour of TV per day. **(b)** 0.0001; very strong evidence that TV watching and aggression are associated. **11.59** Report could include: Contingency table of summary counts and Pearson chi-squared $= 21.4$, $df = 6$, P-Value $= 0.002$. **11.61** Answers will vary.

11.63 (a)

	Has Young Children	
Gray Hair	**Yes**	**No**
Yes	0	4
No	5	0

(b) Yes. **(c)** There often are third factors, such as age in this case, that influence an association. **11.65 (a)** Difference of proportions with U.S. as Group 1: 0.000061; Britain as Group 1: −0.000061. **(b)** U.S. as Group 1: 48; Britain as Group 1: 0.021. **(c)** Relative risk; the difference between proportions might be very small, even when one is many times larger than the other. **11.67** False. **11.69** True. **11.71 (a)** The chi-squared value for a right-tail probability of 0.05 and df = 1s 3.84, which is the z value for a two-tail probability of 0.05 squared: (1.96)(1.96) = 3.84. **(b)** The chi-squared value for P-value of 0.01 and df = 1 is 6.64. This is the square of the z value for a two-tail P-value of 0.01, which is 2.58. **11.73** The two observed values in a given row (or column) must add up to the same total as the two expected values in that same row (or column). **11.75 (a)**

F1, F2, F3	F1, F2, M1	F1, F2, M2	F1, F2, M3
F1, F3, M1	F1, F3, M2	F1, F3, M3	F1, M1, M2
F1, M1, M3	F1, M2, M3	F2, F3, M1	F2, F3, M2
F2, F3, M3	F2, M1, M2	F2, M1, M3	F2, M2, M3
F3, M1, M2	F3, M1, M3	F3, M2, M3	M1, M2, M3

This contingency table shows that two males were chosen (M1 and M3) and one was not (M2). It also shows that one female was chosen (F2) and two were not (F1 and F3). **(b)** Ten samples have a difference greater than or equal to 1/3. A difference greater than or equal to 1/3 indicates more men than women in the sample.

Chapter 12

12.1 (a) Response: mileage; explanatory: weight. **(b)** $\hat{y} = 45.6 - 0.0052x$; y-intercept is 45.6; slope is −0.0052. **(c)** For each 1000-pound increase, the predicted mileage will decrease by 5.2 miles per gallon. **(d)** No; the y-intercept is the predicted miles per gallon for a car that weighs 0 pounds. **12.3 (a)** 322.67 kg. **(b)** 117.5 kg. **(c)** The y-intercept indicates that for male athletes who cannot perform any repetitions for a fatigue bench press, the predicted max bench press is 117.5 kg. As RTFBP increases from 0 to 35, the predicted 1RMBP increases from 117.5 to 322.6 kg. **12.5** For a given x-value, there will not be merely one y value—not every elementary schoolgirl in your town who is a given height will weigh the same. **12.7 (a)** There appears to be a positive association between GPA and study time. **(b)** Predicted GPA = 2.63 + 0.0439 Study time. For every one-hour increase in study time per week, GPA is predicted to increase by about .04 points. **(c)** 3.73. **(d)** −0.13. The observed GPA for Student 2, who studies an average of 25 hours per week, is 3.6 which is 0.13 points below the predicted GPA of 3.73. **12.9 (a)** Response: college GPA; explanatory: high school GPA; answer includes scatterplot. There is a ceiling effect that limits our ability to predict college GPA. **(b)** From software: predicted CGPA = 1.19 + 0.637 HSGPA. **(i)** 3.10; **(ii)** 3.74; predicted college GPA increases 0.64 (the slope) for one-unit increase in high school GPA. **12.11** 0.0015. **12.13 (a)** For a $1000 increase in GDP, the predicted percentage using the Internet increases by 1.6%, and the predicted percentage using Facebook increases by less than 1% (0.75%). **(b)** Because the slope of GDP to Internet use is larger (over double) than is the slope relating GDP to Facebook use, an increase in GDP would have a slightly greater impact on the percentage using the Internet than on the percentage using Facebook. **12.15 (a) (i)** 6.46; **(ii)** 5.73; the difference, 0.73, equals the slope multiplied by the difference in numbers of sit-ups: (0.024346)(30) = 0.73.[should be negative] **(b)** $r = b(\frac{s_x}{s_y}) = -0.024346(\frac{6.887}{0.365}) = -0.46$. **12.17 (a)** Height and weight correlation for females was 0.553 and slightly lower for males at 0.457. Height and body fat percentages correlation for females was 0.216 and slightly higher for males at 0.232. Weight and body fat percentages correlation for females was 0.871 and almost the same for males at 0.883. Correlations for both male and female athletes are very close in values. **(b) (i)** Percentage of body fat and weight have the strongest association; **(ii)** Height and percentage of body fat have the weakest association. There is a fairly strong, positive association between height and weight. As one goes up, the other tends to go up. **12.19 (a)** 650. **(b)** x = 800 is 3 standard deviations above the mean, so the predicted y value is 0.5(3) = 1.5 standard deviations above the mean. **12.21 (a)** r = 0.81. There is a fairly strong, positive, linear association between GPA and study time. **(b)** r^2 = 0.656. The error using $\hat{y}$ to predict y is 66% smaller than the error using $\bar{y}$ to predict y. **12.23 (a)** 270. **(b)** No; this decrease could occur merely because of regression to the mean. **12.25** Regression to the mean; stocks that are relatively very high one year will, on the average, move toward the mean at a later time. **12.27** There is a 75% reduction in error in predicting a car's mileage based on knowing the weight, compared to predicting by the mean mileage. This relatively large value means that we can predict a car's mileage quite well if we know its weight. **12.29** University of Connecticut; Yale would have a restricted range of high school GPA values, with nearly all its students clustered very close to the top, and the correlation is weaker when the range of predictor values is restricted. **12.31 (a)** From software, the correlation for females between number of bench presses before fatigue (BRTF(60)) and maximum bench press (1RMBENCH) is 0.80, a strong positive association. The correlation for males is 0.91, stronger than the female correlation. **(b) (i)** The median for females is 10. Using only the x-values below 10, the correlation is 0.48. The median for males is 17. Using only the x-values below 17, the correlation is 0.93; **(ii)** Using only the x-values above the median of 10 for females, the correlation is 0.67. Using only the x-values above the median of 17 for males, the correlation is 0.57. They are so different because the correlation usually is smaller in absolute value when the range of predictor values is restricted. **12.33 (a) (i)** Assume randomization, linear trend with normal conditional distribution for y and the same standard deviation at different values of x; **(ii)** H_0: $\beta = 0$; H_a: $\beta \neq 0$; **(iii)** t = 11.6; **(iv)** P-value =0.000; **(v)** Very strong evidence that an association exists between the size and price of houses; this is extremely unlikely to be due to random variation. **(b)** $b \pm t_{.025}(se) = 77.008 \pm 1.985(6.626)$, which is (64, 90). **(c)** An increase of $100 is outside the confidence interval and so is an implausible value. **12.35 (a) (i)** Assume randomization, linear trend with normal conditional distribution for y and the same standard deviation at different values of x. These data were not gathered using randomization, and so inferences are highly tentative; **(ii)** H_0: $\beta = 0$; H_a: $\beta \neq 0$; **(iii)** t = 9.6; **(iv)** P-value = 0.000; **(v)** Very strong evidence that an association exists between these variables. **(b)** (4.2, 6.4); it gives us a range of plausible values for the slope of the straight line describing the population. **12.37 (a)** Having more daughters is good. **(b) (i)** Assume that there was roughly a linear relationship between variables, and that the data were gathered using randomization and that the population y-values at each x-value follow a normal distribution, with roughly the same standard deviation at each x-value; **(ii)** H_0: $\beta = 0$; H_a: $\beta \neq 0$; **(iii)** t = 1.52; **(iv)** P-value = 0.13. v) It is plausible that there is no association. **(c)** (−0.1, 1.0). Zero is a plausible value for this slope. **12.39 (a)** Means: advertising is 2, sales is 7. Standard deviations: advertising is 2.16, sales is 2.16. **(b)** $\hat{y} = 5.286 + 0.857x$. **(c)** For H_0: $\beta = 0$; H_a: $\beta > 0$, t = 2.35, P-value =0.14, it is plausible that there is no association between advertising and sales. **12.41** (−0.11,−0.05). On average, GPA decreases by between 0.05 and 0.11 points for every additional class that is skipped. **12.43 (a)** 15.0 = number of 60-pound bench presses and 105.00 = maximum bench press for athlete 10, 85.90 = predicted maximum bench press, 19.10 = difference between the actual and predicted maximums, 2.41 =residual divided by the standard error that describes the sampling variability of the residuals. **(b)** No; we would expect about 5% of standardized residuals to have an absolute value above 2.0. **12.45 (a)** Distribution of standardized residuals and hence conditional distribution of maximum bench press. **(b)** The conditional distribution seems to be approximately normal. **12.47 (a)** 448 = predicted amount spent on clothes in past year for those in 12th grade of school. **(b)** (427, 469): range of plausible values for the population mean of dollars spent on clothes for 12th grade students in the school. **(c)** (101, 795): range of plausible values for the individual observations (dollars spent on clothes) for all the 12th grade students at the school. **12.49 (a)** The square root of 1303.72 is 36.1. This is the estimated standard deviation of maximum leg presses for female athletes who can do a

fixed number of 200-pound leg presses. **(b)** Standard deviation = 36.1; (277.6, 422.0). **12.51 (a)** Total SS = residual SS + regression SS, where residual SS = error in using the regression line to predict y, regression SS = how much less error there is in predicting y using the regression line compared to using $\bar{y}$. **(b)** Sum of squares around mean divided by $n - 1$ is 192,787/56 = 3442.6, and its square root is 58.7. This estimates the overall standard deviation of y-values whereas the residual s estimates the standard deviation of y-values at a fixed value of x. **(c)** The F test statistic is 92.87; its square root is the t statistic of 9.64. **12.53 (a)** MS values: 200,000, 22,580.6; F: 8.86. **(b)** H_0: $\beta = 0$ against H_a: $\beta \neq 0$.
12.55

Source	DF	SS	MS	F	P
Regression	1	1.9577	1.9577	19.52	0.000
Residual Error	57	5.7158	0.1003		
Total	58	7.6736			

(a) Total = 7.6736 = 1.9577 + 5.7158, where residual SS =5.7158 = error in using the regression line to predict y, regression SS =1.9577 = how much less error there is in predicting y using the regression line compared to using $\bar{y}$. **(b)** 0.32; estimates standard deviation of y at fixed value of x, and describes typical size of the residuals about the regression line. **(c)** 0.36; this describes variability about the overall mean of $\bar{y}$ at all values of x, not just those at a specific value of x. **12.57 (a)** $(1.072)^{10} = 2.0$. **(b)** $(1.10)^{10} = 2.59$; the effect here is multiplicative, not additive. **12.59 (a)** $1.15^5 = 2.0$. **(b)** $1.15^{10} = 4.0$. **(c)** $1.15^{20} = 16.4$. **12.61 (a)** Scatterplot shows the relation between variables is curvilinear. **(b)** −16.8; no, a weight cannot be negative. **(c)** Yes. **(d) (i)** 80.6; **(ii)** 1.3. **(e)** Indicates that predicted weight multiplies by 0.813 each week. **12.63 (a)** At a fixed values of x there is variability in the values of y so we can't specify individual y values using x but we can try to specify the mean of those values and how that mean changes as x changes. **(b)** Since y values vary at a fixed value of x, the model has a σ parameter to describe the variability of the conditional distribution of those y values at each fixed x. **12.65** Response: height of children; explanatory: height of parents. Children tend to be short but not as short as parents. Because of regression toward the mean, the value of y tends (on the average) to be not so far from its mean as the x value is from its mean. **12.67 (a)** 210,991. This house sold for $210,991 more than would have been predicted. **(b)** This observation is 4.02 standard errors higher than predicted. **12.69 (a)** residual standard deviation of y: variability of the y-values at a particular x-value; standard deviation: variability of all of the y-values **(b)** Variability of y-values at a given x is about the same as variability of all y observations $r^2 = 0.13$. **12.71 (a)** The plausible values range from 338 to 365 for the mean of y values for all female high school athletes having $x = 80$. **(b)** For all female high school athletes with a maximum bench press of 80, we predict that 95% of them have maximum leg press between about 248 and 455 pounds. The 95% PI is for a single observation y, whereas the confidence interval is for the mean of y. **12.73 (a)** $1000 \times 2^5 = 32{,}000$. **(b)** $1000 \times 2^{10} = 1{,}024{,}000$. **(c)** 1000×2^x. **12.75 (a)** 1900: $\hat{y} = 1.424 \times 1.014^0 = 1.42$ billion; 2010: $\hat{y} = 1.424 \times 1.014^{100} = 6.57$ billion. **(b)** The fit of the model corresponds to a rate of growth of 1.4% per year because multiplying by 1.014 adds an additional 1.4% each year. **(c) (i)** The predicted population size doubles after 50 years because $1.014^{50} = 2.0$, the number by which we'd multiply the original population size; **(ii)** It quadruples after 100 years: $1.014^{100} = 4.0$. **(d)** The exponential regression model is more appropriate for these data because the log of the population size and the year number are more highly correlated ($r = 0.99$) than are the population size and the year number. **12.77 (a)** The 3 outlying points represent outliers—values more than $1.5 \times IQR$ beyond either Q1 or Q3. **(b)** From software: Diff = −9.125 + 1.178 Run; difference is positive when $-9.125 + 1.178(\text{runs}) > 0$, which is equivalent to runs $> 9.125/1.178 = 7.7$. **(c)** Runs, hits, and difference are positively associated with one another. Errors are negatively associated with those three variables.

	Run	Hits	Errors
Hits	0.819		
Errors	−0.259	−0.154	
Difference	0.818	0.657	−0.501

(d) From software, the P-value of 0.000 for testing that the slope equals 0 provides extremely strong evidence that DIFF and RUNS are associated. **12.79** Report would interpret results from:

high_sch_GPA = 3.44 − 0.0183 TV
S = 0.446707 R-Sq = 7.2% R-Sq(adj) = 5.6%

Source	DF	SS	MS	F	P
Regression	1	0.8921	0.8921	4.47	0.039
Residual Error	58	11.5737	0.1995		
Total	59	12.4658			

12.81 (a) In this case, the predictions should exactly match the observations, i.e., $y = x$ which translates to the true y-intercept equaling 0 and the true slope equaling 1. **(b)** No. The P-value for testing that the true y-intercept is 0 is large, so that we are unable to conclude that the y-intercept differs from 0. **12.83** Explain how players or teams that had a particularly good or bad year tended to have results in the following year that were not so extreme (i.e., regression toward the mean). **12.85** As the range of values reflected by each sample is restricted, the correlation tends to decrease when we consider just students of a restricted range of ages. **12.87 (a)** The slope would be two times the original slope. **(b)** The correlation would not change because it is independent of units. **(c)** The t statistic would not change because although the slope doubles, so does its standard error. (The result of a test should not depend on the units we use.) **12.89 (a)** There are two parameters α and β, and so $df = n - 2$. **(b)** There is only one parameter, and therefore, $df = n - 1$. **12.91 (a)** The percentage would likely fluctuate over time, and this would not be a linear relationship. **(b)** Annual medical expenses would likely be quite high at low ages, then lower in the middle, then high again, forming a parabolic, rather than linear, relationship. **(c)** The relation between these variables is likely curvilinear. Life expectancy increases for awhile as per capita income increases, then gradually levels off. **12.93 (a)** The statement is referring to additive growth, but this is multiplicative growth. There is an exponential relation between these variables. **(b)** $\hat{y} = \$175{,}000 \times 0.966^{10} = \$123{,}825$; the percentage decline for the decade is about 29.2%. **12.95** (b). **12.97** (d). **12.99** The best response is (b). **12.101 (a)** At 0 impact velocity, there would be 0 putting distance. The line would pass through the point having coordinates (0, 0). **(b)** If x doubles, then x^2 (and hence the mean of y) quadruples; e.g., if x goes from 2 to 4, then x^2 goes from 4 to 16. **12.103** Because $r^2 = \dfrac{\Sigma(y - \bar{y})^2 - \Sigma(y - \hat{y})^2}{\Sigma(y - \bar{y})^2}$, and dividing each term by approximately n (actually, $n - 1$ and $n - 2$) gives the variance estimates, it represents the relative difference between the quantity used to summarize the overall variability of the y values and the quantity used to summarize the residual variability. **12.105 (a)** Error is calculated by subtracting the mean from the actual score, y. If this difference is positive, then the observation must fall above the mean. **(b)** $\epsilon = 0$ when the observation falls exactly at the mean. **(c)** Since the residual $e = y - \hat{y}$, we have $y = \hat{y} + e = a + bx + e$. As $\hat{y}$ is an estimate of the population mean, e is an estimate of ϵ. **(d)** It does not make sense to use the simpler model, $y = \alpha + \beta x$ that does not have an error term because it is improbable that every observation will fall exactly on the regression line.

Chapter 13

13.1 (a) $\hat{y} = 134.3$. **(b)** −19.3. The actual total body weight is 19.3 pounds lower than predicted. **13.3 (a) (i)** 3.80; **(ii)** 1.60; **(b)** $\hat{y} = 0.20 + 0.50x_1 + 0.002(500) = 0.20 + 0.50x_1 + 1 = 1.20 + 0.50x_1$. **(c)** $\hat{y} = 0.20 + 0.50x_1 + 0.002(600) = 0.20 + 0.50x_1 + 1.2 = 1.40 + 0.50x_1$. **13.5 (a) (i)** 18.3; **(ii)** 12.5; **(b)** When education goes up $10 = 80 - 70$, predicted crime rate changes by 10 multiplied by the slope, $10(-0.58) = -5.8$. **(c) (i)** $\hat{y} = 59.12 - 0.5834x_1 + 0.6825(0) = 59.12 - 0.5834x_1$; **(ii)** $\hat{y} = 59.12 - 0.5834x_1 + 0.6825(50) = 93.2 - 0.5834x_1$; **(iii)** $\hat{y} = 59.12 - 0.5834x_1 + 0.6825(100) = 127.4 - 0.5834x_1$; **(d)** The line passing through the points having urbanization = 50 has a negative slope. The line passing through all the data points has a positive slope. Simpson's paradox occurs because the association between crime rate and education is positive overall but is negative at each fixed value of urbanization. It happens because urbanization is positively associated with crime rate and with education. As urbanization increases, crime rate and education tend to increase, giving an overall positive association between crime rate and education. **13.7 (a)** $\hat{y}$ = \$26,417,000 + \$168,300(GIR) + \$33,859(SS) − \$19,784,000(AvePutt) − \$44,725(Events). **(b)** The average of putts after reaching the green (AvePutt) will increase a golfer's score and therefore decrease his or her earnings. The number of PGA events entered (Events) is a bit harder to explain. One would think that the more events a golfer participated in, the more money he or she would earn. However, the negative sign implies the opposite. An excellent golfer may enter only select events on the tour. If he or she is able to win enough by August, the golfer would be hesitant to participate in more events instead of resting or participating in other promotions. A worse golfer has to play in more events to make a living. **(c)** $\hat{y}$ = \$26,417,000 + \$168,300(60) + \$33,859(50) − \$19,784,000(1.5) − \$44,725(20) = \$7,637,450. **13.9** We don't need to control x_2 if it's not associated with x_1. Changes in x_2 will not have an impact on the effect of x_1 on y. **13.11 (a)** $R^2 = \frac{\Sigma(y-\bar{y})^2 - \Sigma(y-\hat{y})^2}{\Sigma(y-\bar{y})^2}$ = (4354684931 − 1993805006)/4354684931 = 0.542. **(b)** Using these variables together to predict attendance reduces the prediction error by 54%, relative to using $\bar{y}$ alone to predict attendance. Yes, the prediction is at least somewhat better. **(c)** $R = \sqrt{R^2} = \sqrt{0.542} = 0.736$; there is a moderately strong association between the observed attendance and the predicted attendance. **13.13** Height is not strongly correlated with either body fat or age. **13.15** Because x_1 and x_2 are themselves highly correlated, once one of them is in the model, the remaining one does not help much in adding to the predictive power. **13.17 (a)** The correlation between predicted house selling price and actual house selling price is 0.72. **(b)** If selling price is measured in thousands of dollars, each y value would be divided by 1000. For example, \$145,000 would become 145 thousands of dollars. Each slope also would be divided by 1000 (e.g., a slope of 63 for house size on selling price in dollars corresponds to 0.063 in thousands of dollars for $\hat{y}$). **(c)** The multiple correlation would not change because it is not dependent on units. **13.19 (a)** It means that, in the population of all students, high school GPA doesn't predict college GPA for students having any given value for study time. **(b) (1)** Assumptions: We assume a random sample and that the model holds (each explanatory variable has a straight-line relation with μ_y, controlling for the other predictors, with the same slope for all combinations of values of other predictors in model, and there is a normal distribution for y with the same standard deviation at each combination of values of the predictors). Here, the 59 students were a convenience sample, not a random sample, so inferences are highly tentative. **(2)** Hypotheses: H_0: $\beta_1 = 0$; H_a: $\beta_1 \neq 0$. **(3)** Test statistic: $t = (b_1 - 0)/se = 0.6434/0.1458 = 4.41$. **(4)** P-value: 0.000. **(5)** Conclusion: The P-value of 0.000 gives strong evidence against the null hypothesis that $\beta_1 = 0$. At common significance levels, such as 0.05, we can reject H_0. **13.21 (a)** 0.32 describes the typical size of the residuals and also estimates the standard deviation of y at fixed values of the predictors. **(b)** Between 2.97 and 4.25. **13.23** The first test analyzes the effect of LP_200 at any given fixed value of BP_60, whereas the second test describes the overall effect of LP_200 ignoring other variables. These are different effects, so one can exist when the other does not. In this case, it is likely that LP_200 and BP_60 are strongly associated with one another, and the effect of LP_200 is weaker once we control for BP_60. **13.25 (a)** H_0: $\beta_1 = \beta_2 = 0$; the null hypothesis states that neither of the two explanatory variables have an effect on the response variable y. **(b)** 3.17. **(c)** 51.39 with a P-value of 0.000. Extremely strong evidence against the null, and we can reject H_0. At least one of the two explanatory variables has an effect on BP. **13.27 (a)** (0.04, 0.17). **(b)** The confidence interval gives plausible values for the slope, the amount that mean mental impairment will increase when life events score increases by one, when controlling for SES. For an increase of 100 units in life events, we multiply each endpoint of the confidence interval by 100, giving (4, 17). **13.29 (a)** H_0: $\beta_1 = \beta_2 = \beta_3 = 0$ means that house selling price is independent of size of home, number of bedrooms, and age. **(b)** The large F-value and small P-value provide strong evidence that at least one of the three explanatory variables has an effect on selling price. **(c)** The results of the t tests at the 5% significance level tell us that house size and number of bedrooms are statistically significant predictors when controlling for the other explanatory variables. Age is not statistically significant. **13.31 (a)** Conditional distribution. **(b)** That the distribution may be skewed to the right rather than normal. **13.33** One might think this suggests less variability at low levels and even less at high levels of BP_60, but this may merely reflect fewer points in those regions. Overall, it seems OK. **13.35 (a)** U-shaped scatterplot. **(b)** U-shaped pattern, with residuals above 0 for small age and large age and residuals below 0 for medium-size age values. **13.37 (a)** Bottom left and bottom middle plots; strong, positive associations. **(b)** $\hat{y}$ = 55.012 + 0.16676LBM + 1.6575REP_BP; 1.66 is the amount that maximum bench press changes for a one-unit increase in number of repetitions, controlling for lean body mass. **(c)** 0.832; using these variables together to predict BP reduces the prediction error by 83%, relative to using $\bar{y}$ alone to predict BP. **(d)** 0.91; there is a strong association between the observed BPs and the predicted BPs. **(e)** F = (7641.5/50.7) = 150.75; P-value is 0.000; strong evidence that BP is not independent of these two predictors. **(f) (1)** Assumptions: The 64 athletes were a random sample (actually a convenience sample, so inferences are highly tentative). We must assume that the model holds. **(2)** Hypotheses: H_0: $\beta_1 = 0$; H_a: $\beta_1 \neq 0$. **(3)** Test statistic: $t = 2.22$. **(4)** P-value: 0.030. **(5)** Conclusion: The P-value of 0.030 gives relatively strong evidence against the null hypothesis that $\beta_1 = 0$. **(g)** The histogram suggests that the residuals are roughly bell-shaped about 0. They fall between about −3 and +3. The shape suggests that the conditional distribution of the response variable is roughly normal. **(h)** The plot of residuals against values of REP_BP describes the degree to which the response variable is linearly related to this particular explanatory variable. It suggests that the residuals are less variable at smaller values of REP_BP than at larger values of REP_BP. **(i)** The individual with REP_BP around 32 and standardized residual around −3 had a BP value considerably lower than expected. **13.39** For large values of lot size, residuals would be negative and have decreasing trend. **13.41 (a)** For each increase in one pound, predicted price of bike decreased by \$53.75. **(b)** When suspension type is front end the predicted price is \$643.60 cheaper. **13.43 (a) (i)** Minimum: 61.3 + 0.35(12) = 65.5; **(ii)** Maximum: 61.2 + 0.35(54) = 80.2. **(b)** When controlling for region, an increase of one hour leads to a decrease in predicted defects of 0.78 per 100 cars. Japanese facilities had 36 fewer predicted defects, on average, than did other facilities when controlling for time. **(c)** Because the direction of the association between time and defects changed with the variable of whether facility is Japanese was added. **(d)** Because, overall, Japanese facilities have fewer defects and take less time; other facilities have more defects and take more time. When the data are looked at together, this leads to an overall positive association between defects and time. (Student's answer will include a scatterplot.) **13.45 (a)** From software: price in thousands = 64.7 + 0.0674 House_Size + 40.3 Garage. For homes with a garage, price in thousands = 105 + 0.0674 House_Size. For homes without a garage, price in thousands = 64.7 + 0.0674 House_Size. **(b)** The coefficient, 40.3, indicates that the predicted selling price for houses with a garage is \$40,300 higher than for houses without.

Predictor	Coef	SE Coef	T	P
Constant	64.66	17.07	3.79	0.000
House Size	0.067431	0.004596	14.67	0.000
Garage	40.29	13.39	3.01	0.003

13.47 (a) Interaction. **(b)** Males: $\hat{y}$ = 1.98 + 0.0055(380) = 4.07; Females: $\hat{y}$ = 1.60 + 0.0065(380) = 4.07. **(c)** No; it is dangerous to extrapolate beyond our existing data. We do not know whether trends will change in the future.

13.49 $\hat{p} = \frac{e^{-3.52+0.105x}}{1+e^{-3.52+0.105x}} = \frac{e^{-3.52+0.105(25)}}{1+e^{-3.52+0.105(25)}} = (0.41/1.41) = 0.29.$

13.51 (a) Q1: 0.48; Q3: 0.81. **(b)** 0.81 − 0.48 = 0.33. **13.53 (a)** 0.31. **(b)** 0.73; probability of voting Republican increases as income increases. **13.55 (a)** As family income increases, people are more likely to vote Republican. As number of years of education increases, people are more likely to vote Republican. Men are more likely to vote Republican than are women. **(b)** (i) $\hat{p}$ = 0.47; (ii) $\hat{p}$ = 0.42. **13.57 (a)** White. **(b)** Black defendants who had white victims.

13.59 (a) The plots that pertain to selling price as a response variable are those across the top row. The highly discrete nature of x_2 and x_3 limits the number of values these variables can take on. This is reflected in the plots, particularly the plot for bedrooms by baths. **(b)** From software: Pred Price = 39001 + 53.21 House Size − 7885 Bedrooms + 57796 Bath. When number of bedrooms and number of bathrooms are fixed, an increase of one in size of home leads to an increase of 53.2 in predicted selling price. **(c)** $R^2 = 0.603$; this indicates that predictions are about 60% better when using the prediction equation instead of using the sample mean $\bar{y}$ to predict y. **(d)** The multiple correlation, 0.77, is the square root of R^2. It is the correlation between the observed y values and the predicted $\hat{y}$ values. **(e)** **(i)** Assumptions: multiple regression equation holds, data gathered randomly, normal distribution for y with same standard deviation at each combination of predictors; **(ii)** Hypotheses: H_0: $\beta_1 = \beta_2 = \beta_3 = 0$; H_a: At least one β parameter differs from 0; **(iii)** Test statistic: $F = 99.14$; **(iv)** P-value: for df (3,196); 0.000; **(v)** Conclusion: If the null hypothesis were true, the probability would be close to 0 of getting a test statistic at least as extreme as the value observed. We have very strong evidence that at least one explanatory variable has an effect on y; **(f)** The t statistic is −1.29 with a one-sided P-value of 0.200/2 = 0.100. If the null hypothesis were true, the probability would be 0.100 of getting a test statistic at least as extreme as the value observed. At a significance level of 0.05, we cannot reject the null. It is plausible that the number of bedrooms does not have an effect on selling price. This is likely not significant because it is correlated with the other explanatory variables in this model. It might be associated with selling price on its own, but might not provide additional predictive information over and above the other explanatory variables. **(g)** This histogram describes the shape of the conditional distribution of y at given values of the explanatory variables. It suggests that the distribution may be slightly skewed to the right. **(h)** This plot depicts the size of the residuals for the different house sizes observed in this sample. It indicates possibly greater residual variability (and hence, greater variability in selling price) as house size increases. **13.61 (a)** *Difference* = −5.00 + 0.934 Hits − 1.61 Errors; for each increase of one hit, the predicted difference increases by 0.93, and for each increase of one error, the predicted difference decreases by 1.61. **(b)** Six. **13.63** Slope with x_3 in the model: effect of poverty when controlling for percentage of single-parent families, as well as percent living in urban areas; slope without x_3 in the model: effect of poverty when controlling only for percent living in urban areas. **13.65 (a)** $F = 61791/1119 = 55.2$; P-value: 0.000; if the null hypothesis were true, the probability would be close to 0 of getting a test statistic at least as extreme as the value observed. We have very strong evidence that at least one of these explanatory variables predicts y better than the sample mean does. **(b)** The significance test would not be relevant if we were not interested in nations beyond those in the study; if this were the case, the group of nations that we studied would be a population and not a sample. **13.67** Since the effect of education on income changes depending on gender, the explanatory variables, education and gender, are said to interact. **13.69 (a)** Negative; if an individual used AZT, the probability of developing symptoms was lower. **b)** Black/yes: $\hat{p}$ = 0.14; black/no: $\hat{p}$ = 0.26 **(c) (1)** Assumptions: The data were generated randomly. The response variable is binary. **(2)** Hypotheses: H_0: $\beta_1 = 0$; H_a: $\beta_1 \neq 0$; **(3)** Test statistic: −2.58 **(4)** P-value: 0.010 **(5)** Conclusion: The P-value of 0.010 gives sufficient evidence against the null hypothesis that $\beta_1 = 0$. We can reject H_0. **13.71** The reports can include information such as regression equation: college_GPA = 2.83 + 0.203 high_sch_GPA − 0.0092 sports. **13.73** The two-paragraph summary report will be different for each student. It should indicate that two of the noncategorical explanatory variables—years of experience and merit rating—are both positively associated with salary. The rankings of a nontenure track instructor is negatively associated. In addition, the influence of the increased factors have a greater affect on increasing salary. One other noncategorical explanatory variable of market influence is negatively associated indicating that OSU salaries are better than comparable salaries at other institutions. For the categorical variables, predicted salary is higher among those with a full professor ranking and in the business discipline; professors in nontenure track positions are also negatively associated with salary. **13.75** (d). **13.77 (a)** False; the multiple correlation is at least as large as the ordinary correlations. **(b)** False; it falls between 0 and 1. **(c)** False; R^2 describes how well you can predict y using a set of explanatory variables together in a multiple regression model. **(d)**True. **13.79 (a)** True; the slope for this variable is positive in the bivariate equation. **(b)** False; such a change occurs only when we ignore x_2. **(c)** True; the slope for x_2 is 0.003 when controlling for x_1. 0.003. multiplied by 100 is 0.30. **13.81** y = math achievement score, x_1 = height, x_2 = age for a sample of children from all the different grades in a school system. **13.83** When there are many explanatory variables, doing the F test first provides protection from doing lots of t tests and having one of them be significant merely by random variation when, in fact, there truly are no effects in the population. **13.85** When $x = 0$, you likely can't afford a home no matter what. Similarly, when $x = 50{,}000$, you likely can afford a home no matter what. But at $x = 500$ the extra income may have an effect on whether you feel you can afford a home. This suggests an S-shaped curve for the relationship and thus a logistic regression model. **13.87** When you add a predictor, if it has no effect its coefficient is 0. Then the prediction equation is exactly the same as with the simpler model without that variable and R will be exactly the same as before. If having a nonzero coefficient results in better predictions overall, then R will increase. **13.89** Answer is scatterplot showing two lines with positive slopes, the one for Florida falling below the one for Louisiana. The points for Florida should scatter about its line but tend to have higher age values than the points for Louisiana. **13.91 (a)** When $p = 0.5$, $p(1 - p) = 0.5(1 - 0.5) = 0.25$; 0.25 multiplied by β is the same as $\beta/4$. **(b)** $0.1(1 - 0.1) = 0.09$; $0.3(1 - 0.3) = 0.21$; $0.7(1 - 0.7) = 0.21$; $0.9(1 - 0.9) = 0.09$; as p gets closer and closer to 1, the slope approaches 0. (Student's answer will include a sketched curve.) **13.93** The responses will be different for each class.

Chapter 14

14.1 (a) Response variable: performance gap; factor: which hotel the guest stayed in; categories: five hotels. **(b)** H_0: $\mu_1 = \mu_2 = \mu_3 = \mu_4 = \mu_5$; H_a: at least two of the population means are unequal. **(c)** $df_1 = g - 1 = 5 - 1 = 4$; $df_2 = N - g = 125 - 5 = 120$. **(d)** From a table or software, $F = 2.45$. **14.3 (a) (i)** Assumptions: Independent random samples, normal population distributions with equal standard deviations; **(ii)** Hypotheses: H_0: $\mu_1 = \mu_2 = \mu_3$; H_a: at least two population means are unequal; **(iii)** Test statistic: $F = 2.50$ ($df_1 = 2, df_2 = 5$); **(iv)** P-value = 0.18; **(v)** Conclusion: It is plausible that H_0 is true. **(b)** The sample sizes are very small. **(c)** Observational; a lurking variable might be school GPA. Perhaps higher GPA students are more likely to have previously studied a language and higher GPA students also tend to do better on quizzes than other students. **14.5 (a)** H_0: $\mu_1 = \mu_2 = \mu_3$; μ_1 represents the population mean satisfaction rating for San Jose, μ_2 for Toronto, and μ_3 for Bangalore. **(b)** 27.6 = 13.00/0.47; $df_1 = 2$ and $f_2 = 297$. **(c)** Very small P-value of 0.000 gives strong evidence against the null; we would reject the null and conclude at least two population means differ. **14.7 (a)** μ_1 represents the population mean ideal number of kids for Protestant; μ_2 represents the population mean for Catholic; μ_3 for Jewish; μ_4 for those of Another Religion; H_0: $\mu_1 = \mu_2 = \mu_3 = \mu_4$. **(b)** Independent random samples; normal population distributions with equal standard deviations. **(c)** $F = 5.48$; P-value = 0.001. Very strong evidence that at least two of the population means differ. **(d)** No; ANOVA tests only whether at least two population means are different.

14.9 (a)

```
Variable gender  N   Mean   StDev
sports   f      31  4.129  3.640
         m      29  6.310  3.828
```

(b) Hypotheses: H_0: $\mu_1 = \mu_2$; H_a: $\mu_1 \neq \mu_2$. $F = 5.12$; P-value = 0.027. Strong evidence that the population means differ. **(c)** The t statistic would be the square root of the F statistic, and the P-values would be identical.

14.11 (a)

```
Level        N    Mean    StDev
famchange   17   7.265   7.157
cogchange   29   3.007   7.309
conchange   26  -0.450   7.989
```

The means of these groups are somewhat different. The standard deviations are similar. **(b)** $F = 5.42$; P-value = 0.006; strong evidence that at least two population means are different. **(c)** The assumptions are that there are independent random samples and normal population distributions with equal standard deviations. There is evidence of skew, but the test is robust with respect to this assumption. The subjects were randomly assigned to treatments. Note that since this was not a random sample of subjects suffering from anorexia, the scope of inference to the population may be limited. **14.13** $(\bar{y}_1 - \bar{y}_2)$ $t_{0.25}\, s\sqrt{\frac{1}{n_1} + \frac{1}{n_2}} = (4.55 - 5.12) \pm 1.960(1.140)\sqrt{\frac{1}{1959} + \frac{1}{341}} = -0.57 \pm$ 0.13; which is (−0.70, − 0.44). Because 0 does not fall in this confidence interval, we can infer at the 95% confidence level that the population means are different (higher for the not important group than for the very important group). **14.15 (a)** Classical music and Muzak. **(b)** Because the Tukey method uses an overall confidence level of 95% for the entire set of intervals. **14.17 (a)** $x_1 = 1$ for observations from the first group and =0 otherwise; $x_2 = 1$ for observations from the second group and = 0 otherwise. **(b)** H_0: $\mu_1 = \mu_2 = \mu_3$; H_0: $\beta_1 = \beta_2 = 0$. **(c)** The intercept estimate is the mean for the third group, which is 12. The first estimated regression coefficient is the difference between the means for the first and third groups, or 18 − 12 = 6, and the second is the difference between the means for the second and third groups, or 15 − 12 = 3. **14.19 (a)** $x_1 = 1$ for observations from San Jose and = 0 otherwise; $x_2 = 1$ for observations from Toronto and = 0 otherwise. **(b)** 7.1 is the sample mean for Bangalore, 0.5 is the difference between the sample means for San Jose and Bangalore, and 0.7 is the difference between the sample means for Toronto and Bangalore. **14.21 (a)** Group 2 – Group 1: (−8.7, 2.7), Group 3 – Group 1: (−3.1, 7.1) Group 3 – Group 2: (−0.7, 10.7) Because 0 falls in all three confidence intervals, we cannot infer that any of the pairs of population means are different. **(b)** Note: Statistical software such as MINITAB is needed to complete this solution. Group 2 – Group 1: (−10.3, 4.3) Group 3 – Group 1: (−4.5, 8.5), Group 3 – Group 2: (−2.3, 12.3). Again, we cannot infer that any of the pairs of population means are different. The intervals are wider because we are now using a 95% confidence level for the overall set of intervals.

14.23 (a)

	Lipitor	Zocor
Low	10	10
High	20	20

(b)

	Lipitor	Zocor
Low	10	20
High	10	20

(c)

	Lipitor	Zocor
Low	10	20
High	20	30

(d)

	Lipitor	Zocor
Low	10	10
High	10	10

14.25 (a) $F = 81021/10095 = 8.03$. **(b)** The small P-value of 0.000 provides strong evidence that the population mean house selling price depends on the age of the house. If the null hypothesis were true, the probability would be 0.000 of getting a test statistic at least as extreme as the value observed. **14.27 (a)** H_0: Population mean holding time is equal for the three types of messages, for each fixed level of repeat time. **(b)** $F = 74.60/10.52 = 7.09$; the small P-value of 0.011 provides strong evidence that the population mean holding time depends on the type of message. **(c)** Population distribution for each group is normal, the population standard deviations are identical, and the data result from a random sample or randomized experiment. **14.29 (a)** H_0: no interaction; $F = 0.67$; P-value = 0.535. **(b)** Large P-value; it is plausible that there is no interaction. This lends validity to the previous analyses that assumed a lack of interaction. **14.31 (a)** The means for women of each race are almost the same, whereas the mean for white males is about 0.6 higher than the mean for black males. **(b)** The gender effect was such that women had the higher mean for blacks but men had the higher mean for whites. Overall, ignoring race, the means may be quite similar for men and for women, and the one-way ANOVA for testing the gender effect may not be significant. **(c)** From a two-way ANOVA, we learn that the effect of gender differs based on race (as described previously), but we do not learn this from the one-way ANOVA. **14.33 (a)** $F = 13.90$; P-value =0.000; very strong evidence that the mean weight gain depends on the protein level. **(b)** H_0: no interaction; $F = 2.75$; P-value: 0.07, so we cannot reject the null hypothesis. No interaction is plausible. **(c)** (7.7, 33.9). **14.35 (a)** Population means, μ_1 for married, μ_2 for widowed, μ_3 for divorced, μ_4 for separated, and μ_5 for never married. H_0: $\mu_1 = \mu_2 = \mu_3 = \mu_4 = \mu_5$. H_a: at least two of the population means are different. **(b)** No; large values of F contradict the null, and when the null is true the expected value of the F statistic is approximately 1. **(c)** The null hypothesis is plausible. **14.37 (a)** $F = 215.26/6.76 = 31.84$; this test statistic has a P-value far smaller than 0.05; we can reject the null hypothesis. **(b)** The mean differences between blacks and whites and between blacks and other we can conclude are statistically significant; it appears that blacks watch more hours of TV per day than either whites or those in the category other. **(c)** $(4.38 - 2.76) \pm (1.962)(2.6)\sqrt{\frac{1}{188} + \frac{1}{1014}} = 1.62 \pm 0.4051 = (1.22, 2.03)$. **(d)** Wider; it uses a 95% confidence level for the overall set of intervals. **14.39 (a)** $3.182(2.00)\sqrt{\frac{1}{2} + \frac{1}{2}} = 6.4$. **(b)** The confidence interval formed using the Tukey 95% multiple comparison uses a 95% confidence level for the overall set of intervals. **(c)** Let $x_1 = 1$ for Bumper A and 0 otherwise, $x_2 = 1$ for Bumper B and 0 otherwise, and $x_1 = x_2 = 0$ for Bumper C. **(d)** 13 is the estimated mean damage cost for Bumper C, −11 is the difference between the estimated mean damage costs between Bumpers A and C, and −10 is the difference between the estimated mean damage costs between Bumpers B and C. **14.41 (a)** From software: 16.2. **(b)** All intervals contain 0; therefore, no pair is significantly different. **14.43 (a)** Software gives the following 95% confidence intervals: control and cog: (−7.5, 0.6), family and cog: (−8.8, 0.3), family and control: (−12.4, −3.0). Only the interval for family and control does not include 0. We can infer that the population mean weight change is greater among those who receive family therapy than among those in the control group. **(b)** Software gives the following Tukey 95% multiple comparison confidence intervals: Difference between control and cognitive: (−8.3, 1.4). Difference between family and cognitive: (−9.8, 1.3). Difference between family and control: (−13.3, −2.1). The interpretations are the same as for the intervals in part a. The intervals are wider because the 95% confidence level is for the entire set of intervals rather than for each interval separately. **14.45 (a)** Let $x_1 = 1$ for having a garage and 0 otherwise. From software: House price in thousands = $247 + 26.3x_1$. The intercept, 247, is the estimated mean selling price (in thousands) when a house does not have a garage, and 26.3 (in thousands) is the difference in the estimated mean selling price between a house with and without a garage. **(b)** From software:

```
Predictor     Coef    SE Coef     T       P
Garage       26.28     19.27    1.36   0.174
```

If the null hypothesis were true, the probability would be 0.174 of getting a test statistic at least as extreme as the value observed. We do not have sufficient evidence that the population mean house selling price is significantly higher with a garage than without a garage. **(c)** From software:

```
Analysis of Variance
Source            DF       SS      MS     F      P
Regression         1    24830   24830  1.86  0.174
Residual Error   198  2644040   13354
Total            199  2668870
```

As with the regression model, the P-value is 0.174; this is the same result. **(d)** The value of t in part b is the square root of the value of F in part c. **14.47 (a)** The coefficient of f (0.04) is the difference between the estimated population means for men and women for each level of race. The fact that it is close to 0 indicates that there is a small estimated difference between the population means for men and women, given race. **(b)** Black female: 2.83, white female: 2.46; black male: 2.79; white male: 2.42. **(c)** From the P-value of 0.000 for race, we can conclude that on the average blacks report a higher ideal number of children than whites do (by 0.37). **14.49 (a)** The mean estimate for men is 0.19 lower than the mean estimate for women at each fixed level of religion. **(b)** The mean estimate for Protestants is 0.67 higher than the mean estimate for other religions at each fixed level of gender. **(c)** $\mu_y = \alpha + \beta_1 g + \beta_2 r_1 + \beta_3 r_2 + \beta_4 r_3$; for the response variable to be independent of religion, for each gender, β_2, β_3, and β_4 would need to equal 0. **14.51** This suggests an interaction since smoking status has a different impact at different levels of age. There is a bigger mean difference between smokers and non-smokers among older women than among younger women. **14.53 (a)** The coefficient for gender, −13 (thousands of dollars), indicates that at fixed levels of race, men have higher estimated mean salaries than women. Women salaries are reduced by 13 (thousands of dollars). **(b) (i)** 109.2 (calculation: 96.2 + 13); **(ii)** 56.2 (calculation: 96.2 − 40). **14.55** The short report will be different for each student, but can include and interpret the *F* test statistic of 0.81 and the P-value of 0.52 for number of friends and degree of education. **14.57** Means 10, 17, 24 for A, B, C. **14.59 (a)**

	Female	Male
Humanities	65,000	64,000
Science	72,000	71,000

The overall mean for women is [25(65,000) + 5(72,000)]/30 = 66,167. The overall mean for men is [20(64,000) + 30(71,000)]/50 = 68,200. **(b)** A one-way comparison of mean income by gender would reveal that men have a higher mean income than women do. A two-way comparison of mean incomes by gender, however, would show that at fixed levels of university divisions, women have the higher mean. **14.61** (c). **14.63** (c). **14.65 (a)** Identical sample means. **(b)** No variability (standard deviation of 0) within each sample. **14.67 (a)** *f*: 1.88 ± 2.46(0.7471), which is (0.04, 3.7), *m*: 1.96 ±2.46(0.7471), which is (0.1, 3.8). **(b)** 0.05/35 = 0.0014. **14.69 (a) (i)** number of bathrooms would be treated as a categorical variable; **(ii)** number of bathrooms would be a treated as a quantitative variable and we'd be assuming a linear trend. **(b)** The straight-line regression approach would allow us to know whether increasing numbers of bathrooms led to an increasing mean house selling price. The ANOVA would only let us know that the mean house price was different at each of the three categories. Moreover, we could not use the ANOVA as a prediction tool for other numbers of bathrooms. **(c)** Mean selling price $150,000 for one bathroom, $100,000 for two bathrooms, $200,000 for three bathrooms. (There is not an increasing or decreasing overall trend.) **14.71** The short reports will be different for each class.

Chapter 15

15.1 (a) Lotion	(1,2)	(1,3)	(1,4)	(2,3)	(2,4)	(3,4)
(b) Studio	(3,4)	(2,4)	(2,3)	(1,4)	(1,3)	(1,2)
Lotion mean rank	1.5	2.0	2.5	2.5	3.0	3.5
Studio mean rank	3.5	3.0	2.5	2.5	2.0	1.5
Difference of mean ranks	−2.0	−1.0	0.0	0.0	1.0	2.0

(c) P(−2.0) = 1/6, P(−1.0) = 1/6, P(0.0) = 2/6, P(1.0) = 1/6, P(2.0) = 1/6. **15.3** Parts a and b together: example given for seven combinations only; there would be 20.

Treatment	Ranks						
Therapy 1	(1,2,3)	(1,2,4)	(1,2,5)	(1,2,6)	(1,3,4)	(1,3,5)	(1,3,6)
Therapy 2	(4,5,6)	(3,5,6)	(3,4,6)	(3,4,5)	(2,5,6)	(2,4,6)	(2,4,5)
Therapy 1 mean rank	2.0	2.33	2.67	3.0	2.67	3.0	3.33
Therapy 2 mean rank	5.0	4.67	4.33	4.0	4.33	4.0	3.67
Difference of mean ranks	−3.0	−2.33	−1.67	−1.0	−1.67	−1.0	−0.33

(c) Differences between mean ranks followed by probabilities: (−3.00, 1/20) (−2.33, 1/20) (−1.67, 2/20) (−1.00, 3/20) (−0.33, 3/20) (0.33, 3/20) (1.00, 3/20) (1.67, 2/20) (2.33, 1/20) (3.00, 1/20). **(d)** The P-value = 0.20; if the treatments had identical effects, the probability would be 0.10 of getting a sample like we observed, or even more extreme, in either direction. **15.5 (a)** The point estimate of −2.000 is an estimate of the difference between the population median birth weight for the smokers group and the population median birth weight for the nonsmokers group. **(b)** The confidence interval of (−3.300, −0.700) estimates that the population median birth weight for the smokers group is between 3.3 and 0.7 pounds below the population median birth weight for the nonsmoking group. Because 0 does not fall in the confidence interval, there is evidence that there is a difference between the population medians for the two groups. **15.7 (a)** The estimated difference between the population median weight change for the cognitive behavioral treatment group and the population median weight change for the control group is 3.05. **(b)** The confidence interval estimates that the population median weight change for the cognitive-behavioral group is between 0.6 below and 8.1 above the population median weight change for the control group. The assumption needed is that the girls were randomly assigned to the three treatment groups. **(c)** P-value = 0.11 for testing against the alternative hypothesis of different expected median ranks; there is not much evidence against the null hypothesis. It is plausible that the population distributions are identical. **15.9 (a)** Group 1 ranks: 2, 5, 6; Group 2 ranks: 1, 3.5; Group 3 ranks: 3.5, 7, 8. The mean rank for Group 1 = (2 + 5 + 6)/3 = 4.33. **(b)** P-value = 0.209; it is plausible that the population median quiz score is the same for each group. **15.11 (a)** Dependent; all students receive both treatments. **(b)** 0.81. **(c)** $z = 3.54$; P-value = 0.0002. Strong evidence that the population proportion of drivers who would have a faster reaction time when not using a cell phone is greater than 1/2. **(d)** Matched-pairs *t* test; the sign test uses merely the information about *which* response is higher and how many, not the quantitative information about *how much* higher. **15.13** H_0: $p = 0.50$. H_a: $p > 0.50$; $P(3) = (0.50)^3 = 0.125$. The evidence is not strong that walking lowers blood pressure. **15.15 (a)** H_0: population median of difference scores is 0; H_a: population median of difference scores is > 0. **(c)** P-value = 1/16 = 0.06; there is some, but not strong, evidence that cell phones tend to impair reaction times. **15.17 (a)** Bumper A: ranks are 4, 6, 5; mean is 5. Bumper B: ranks are 1, 2, 3; mean is 2.

(b)

Treatment	Ranks
Bumper A	(1, 2, 3) (1,2,4) (1,2,5) (1,2,6) (1,3,4) (1,3,5) (1,3,6) (1,4,5) (1,4,6) (1,5,6)
Bumper B	(4,5,6) (3,5,6) (3,4,6) (3,4,5) (2,5,6) (2,4,6) (2,4,5) (2,3,6) (2,3,5) (2,3,4)

Treatment	Ranks
Bumper A	(2,3,4) (2,3,5) (2,3,6) (2,4,5) (2,4,6) (2,5,6) (3,4,5) (3,4,6) (3,5,6) (4,5,6)
Bumper B	(1,5,6) (1,4,6) (1,4,5) (1,3,6) (1,3,5) (1,3,4) (1,2,6) (1,2,5) (1,2,4) (1,2,3)

(c) There are only two ways in which the ranks are as extreme as in this sample: Bumper A with 1, 2, 3 and Bumper B with 4, 5, 6, or Bumper A with 4, 5, 6 and Bumper B with 1, 2, 3. **(d)** The P-value is 0.10 because out of 20 possibilities, only two are this extreme. 2/20 = 0.10.

15.19 (a)

Group	Ranks	Mean Rank
Muzak	1,2,4,5,3	3.0
Classical	9,8,7,10,6	8.0

(b) There are only two cases this extreme (i.e., Muzak has ranks 1–5 or Muzak has ranks 6–10). Thus, the P-value is the probability that one of these two cases would occur out of the 252 possible allocations of rankings. The probability would be 0.008 of getting a sample like we observed or even more extreme. **15.21 (a)** Kruskal-Wallis test. **(b)** All examples would have one group with ranks 1–3, one with ranks 4–6, and one with ranks 7–9. **15.23 (a) (i)** H_0: Population proportion $p = 0.50$ who use the cell phone more than Internet; H_a: $p \neq 0.50$; **(ii)** $se = \sqrt{(0.50)(0.50)/n} = \sqrt{(0.50)(0.50)/39} = 0.08$, $z = (\hat{p} - 0.50)/se = (0.897 - 0.50)/0.08 = 4.96$; **(iii)** P-value = 0.000; if the null hypothesis were true, the probability would be near 0 of getting a test statistic at least as extreme as the value observed. We have extremely strong evidence that a majority of countries have more cell phone usage than Internet usage. **(b)** This would not be relevant if the data file were comprised only of countries of interest to us. We would know the population parameters so inference would not be relevant. **15.25 (a)** We could use the sign test for matched pairs or the Wilcoxon signed-ranks test. **(b)** One reason for using a nonparametric method is if we suspected that the population distributions were not normal, for example, possibly highly skewed, because we have a one-sided alternative and parametric methods are not then robust. **(c)** From software (used CGPA-HSGPA):

```
Test of median = 0.000000 versus median < 0.000000
                 N
                for      Wilcoxon              Estimated
        N       Test     Statistic     P         Median
C20    59       55       268.5       0.000    -0.1850
```

The P-value of 0.000 gives very strong evidence that population median high school GPA is higher than population median college GPA. **15.27 (a)** H_0: population median of difference scores is 0; H_a: population median of difference scores > 0.

Possible Samples with Absolute Difference Values of Sample

Subject	1	2	3	4	5	6	7	8	Rank of Absolute Value
1	5.5	5.5	−5.5	5.5	−5.5	5.5	−5.5	−5.5	3
2	−0.5	−0.5	−0.5	0.5	−0.5	0.5	0.5	0.5	1
3	1.5	−1.5	1.5	1.5	−1.5	−1.5	1.5	−1.5	2
	Sum of Ranks for Positive Differences								
	5	3	2	6	0	4	3	1	

The rank sum is 5 one-eighth of the time, and is more extreme (i.e., 6) one-eighth of the time (values that were greater than the observed). Thus, the P-value = 2/8 = 0.25. If the null hypothesis were true, the probability would be 0.25 of getting a test statistic at least as extreme as the value observed. It is plausible that the null hypothesis is correct and that the population median of difference scores is not positive. **(b)** The p-value is smaller than in Example 8. Outliers do not have an effect on this nonparametric statistical method. **15.29** One example is when the population distribution is highly skewed and the researcher wants to use a one-sided test. **15.31** Kruskal-Wallis. **15.33** False. **15.35 (a)** The proportions are calculated by pairing up the subjects in every possible way, and then counting the number of pairs for which the tanning studio gave a better tan. **(b)**

Proportion	Probability
0/6	1/10
1/6	1/10
2/6	2/10
3/6	2/10
4/6	2/10
5/6	1/10
6/6	1/10

(c) The P-value would be 2/10. The probability of an observed sample proportion of 5/6 or more extreme (i.e., 6/6) is 2/10 = 0.20. **15.37** No; we are taking the median of all slopes, and the median is not susceptible to outliers.

Part 4: Review Exercises

R4.1 (a)

Unrestricted abortion should be legal?

Gender	Yes	No
Male	0.40	0.60
Female	0.40	0.60

(b) Independent since the opinion of the respondent is the same regardless of the respondent's gender. **R4.3 (a)** $X^2 = 1.557$; P-value 0.212. **(b)** The difference in the proportions of male versus female offenders when the victim was female is 1719/1901 = 0.904 versus 182/1901 = 0.096, which = 0.808. The difference in the proportions of male versus female offenders when the victim was male is 4078/4562 = 0.894 versus 484/4562 = 0.106, which is = 0.788. Among female offenders, 182/666 = 0.273 had female victims. Among male offenders, 1719/5797 = 0.297 had female victims. Male offenders were 1.09 times more likely to have a female victim than were female offenders. **R4.5** Mother's education and father's education have the strongest linear association with a correlation of 0.68. All of the linear associations are positive (i.e., one variable tends to increase as the other increases) since all of the correlations are positive. **R4.7 (a)** $\hat{y} = 6.663–0.06484x$; **(i)** $\hat{y} = 6.663$; **(ii)** $\hat{y} = 0.179$. The difference is the change in x (100 − 0) times the estimated slope (−0.0648). **(b)** $\hat{y} = 3.4$; $y - \hat{y} = -2.1$. **(c)** The observation for Belgium was 2.97 standard errors below the value predicted from the regression line. **R4.9 (a)** r^2 is 0.89; strong. **(b)** P-value ≈ 0; if $\beta = 0$, it would be extremely unusual to get a sample slope as far from 0 as $b = 1.338$. **(c)** Since 0 is not included in the interval, it is not a plausible value for the slope. This agrees with part b. **(d)** 1.64. The residual standard deviation estimates the standard deviation of y-values at a fixed value of x. **R4.11 (a)** No, the slopes are in different units. **(b)** Yes, the slopes measure the change in the percentage of those using either the Internet or Facebook for a one-unit ($1000) change in GDP. Because the slope of GDP to Internet use is larger (over double) than is the relation of GDP to Facebook use, an increase in GDP would have a slightly greater impact on the percentage using the Internet than on the percentage using Facebook. **R4.13 (a)** 100,000 represents the predicted number of English-language articles in Wikipedia as of January 1, 2003. 2.1 represents the estimated multiplicative effect on the mean of y for each one-year change in x. **(b)** **(i)** $\hat{y} = 4{,}084{,}101$. **(ii)** $\hat{y} = 166{,}798{,}810$. Do not trust predictions. Since posting articles on the Internet is a relatively new practice, it is possible that the number of articles posted in a given time period will level off, thereby changing the relationship between x and y. **R4.15 (a)** HR because its coefficient has the greatest magnitude. **(b)** As the number of stolen bases increases, the predicted number of runs scored will increase. As the number caught stealing increases, the predicted number of runs scored will decrease. **(c)** $\hat{y} = 367$. **R4.17 (a)** $R = 0.775$ is the correlation between the observed CO2 values and those predicted using GDP and unemployment rate as predictors. **(b)** No. The multiple correlation increased only very slightly. **(c)** It is possible that unemployment rate and GDP are closely associated so that once one of these variables is in the model, adding the other does not do much to increase the predictive power of the model. **(d)** A large positive standardized residual means that the observed CO_2 value is significantly larger than the value predicted by the multiple regression model. Australia's observed CO2 level was 5.16 standard deviations above the value predicted by the multiple regression model. **R4.19** The line for the age = high group should have a greater positive slope. **R4.21 (a)** Response: political ideology score; factor: respondent's political party. **(b)** H_0: $\mu_1 = \mu_2 = \mu_3$; H_a: at least two of the population means are unequal; μ_1 denotes the population mean political ideology score for Democrats; μ_2 denotes the population mean political ideology score for Independents; and μ_3 indicates the population mean political ideology score for Republicans. **(c)** Since the P-value ≈ 0, there is very strong evidence against the null hypothesis. **(d)** Independent random samples from normal population distributions with equal standard deviations, which seems plausible. **R4.23 (a)** **(i)** $\hat{y}$ = $20,000 for women high school graduates; **(ii)** $\hat{y}$ = $43,000 for women college graduates. **(b)** The estimated population mean income in 2008 for college graduates is $23,000 higher than for high school graduates when gender is held constant. **R4.25** Answers should include a chi-squared test of independence ($X^2 = 28.2$, P-value = 0.000) and follow-up residual analysis and differences of proportions for comparing pairs of rows on the proportion answering yes. **R4.27** Answers will vary. **R4.29** This could summarize the result of a regression analysis using college GPA as the response and the various factors considered by the admissions officers as the explanatory variables. The 30% refers to R^2 and gives the proportional reduction in error using $\hat{y}$ to predict y rather than using $\bar{y}$ to predict y. **R4.31** One-way ANOVA is a method used to compare the population means of several groups. The null hypothesis is that all of the groups come from populations with equal means and the alternative is that at least two of these population means differ. If the null hypothesis is rejected, we can compute confidence intervals for pairs of means to determine which population means are different as well as how different they are. **R4.33 (a)** Exponential regression model. **(b)** Logistic regression model.

Index

D

E

F

G

H

I

K

L

P

Q

R

S

T

Index of Applications

A

B

C

E

F

H

I

L

M

N

P

R

S

T

W

Photo Credits

PREFACE **p. xxiv** photo of Christine Franklin by Dale Green

CHAPTER 1 **p. 1** (from left to right) Skip Odonnell/iStockphoto; Alexander/Fotolia; GK/Fotolia; Corepics/Fotolia **p. 2** Skip Odonnell/iStockphoto **p. 3** Colin Cuthbert/Photo Researchers, Inc. **p. 9** S44/Zuma Press/Newscom **p. 11** Richard Levine/Alamy **p. 17** 1000 Words/Shutterstock

CHAPTER 2 **p. 23** Alexander/Fotolia **p. 24** Light & Magic Photography/Shutterstock **p. 27** Davidk79/Fotolia **p. 29** Jurie Maree/Shutterstock **p. 31** Branislav Senic/iStockphoto **p. 41** Alija/iStockphoto **p. 50** Wrangler/Shutterstock **p. 52** Alexandr Shebanov/iStockphoto **p. 58** Andy Dean Photography/Shutterstock **p. 65** AGE Fotostock/SuperStock **p. 71** Digital Vision/Thinkstock **p. 78** Time & Life Pictures/Getty Images

CHAPTER 3 **p. 89** GK/Fotolia **p. 90** Kevers/Dreamstime **p. 92** WB/Shutterstock **p. 94** Asharkyu/Shutterstock **p. 99** NetPhotos/Alamy **p. 103** Gary I. Rothstein/AP Images **p. 111** Bumihills/Shutterstock **p. 114** Bill Florence/Shutterstock **p. 117** Allstar Picture Library/Alamy **p. 120** Imago Sportfotodienst/Newscom **p. 125** 4x6/iStockphoto **p. 133** Jiri Foltyn/Shutterstock **p. 134** Kevers/Dreamstime

CHAPTER 4 **p. 149** Corepics/Fotolia **p. 150** Tryfonov/Fotolia **p. 153** Monkeybusiness/Dreamstime **p. 156** Roger L. Wollenberg/UPI/Newscom **p. 160** Zerkalka/Dreamstime **p. 164** Forestpath/Shutterstock **p. 167** Pearson Education, Inc. **p. 173** Paul Matthew Photography/Shutterstock **p. 180** Ryan McVay/Thinkstock **p. 181** Mihai Simonia/Shutterstock **p. 182** Rob Marmion/Shutterstock **p. 184** Spike Mafford/Thinkstock

PART 1 REVIEW **p. 198** (left to right) Skip Odonnell/iStockphoto; Alexander/Fotolia; GK/Fotolia; Corepics/Fotolia **p. 201** Reinhold Foeger/Shutterstock

CHAPTER 5 **p. 207** (from left to right) Anna Minkevich/iStockphoto; Toddtaulman/Dreamstime; Mark Evans/iStockphoto **p. 208** Anna Minkevich/iStockphoto **p. 209** Chepko Danil Vitalevich/Shutterstock **p. 213** Tomo/Shutterstock **p. 219** Hemera Technologies/Thinkstock **p. 220** Jill Fromer/iStockphoto **p. 222** Michelle Milano/Shutterstock **p. 232** Libby Chapman/iStockphoto **p. 234** Vlad Mereuta/Shutterstock **p. 235** Zentilia/Shutterstock **p. 243** Mark Stout/Fotolia **p. 245** Mikephotos/Dreamstime **p. 247** F1online digitale Bildagentur GmbH/Alamy **p. 250** Janet Wall/Fotolia

CHAPTER 6 **p. 263** Toddtaulman/Dreamstime **p. 264** Jeff Greenberg/Alamy **p. 266** Doug James/Icon SMI/Newscom **p. 269** Nikada/iStockphoto **p. 270** Noam Fein/Dreamstime **p. 273** Sonya Etchison/Shutterstock **p. 281** Jaimie Duplass/Shutterstock **p. 283** Lightpoet/Fotolia **p. 289** Peter Atkins/Fotolia **p. 291** Pressmaster/Fotolia **p. 295** Gabe Palmer/Alamy **p. 303** Jeff Zelevansky/Icon SMI

CHAPTER 7 **p. 305** Mark Evans/iStockphoto **p. 306** Billy E. Barnes/PhotoEdit Inc. **p. 312** Laura A. Oda/MCT/Newscom **p. 319** Photosani/Shutterstock **p. 322** Andrejs Zemdega/iStockphoto **p. 325** Steve Mason/Shutterstock

PART 2 REVIEW **p. 338** (from left to right) Anna Minkevich/iStockphoto; Toddtaulman/Dreamstime; Mark Evans/iStockphoto **p. 341** Pressmaster/Dreamstime

CHAPTER 8 **p. 347** (from left to right) Jose AS Reyes/Shutterstock; Goran Bogicevic/Shutterstock; Pitb_1/Fotolia **p. 348** Jose AS Reyes/Shutterstock **p. 349** Lisafx/Dreamstime **p. 354** Buccina Studios/Thinkstock **p. 357** JustASC/Shutterstock **p. 359** Copestello/Shutterstock **p. 361** Yuri Arcurs/Dreamstime **p. 368** Scyther5/Shutterstock **p. 372** Mike Flippo/Shutterstock **p. 376** (from left to right) Copyright © 2005 ISI Newsletter Volume 29, No. 2 (86) 2005, International Statistical Institute (ISI), Den Haag, The Netherlands; Trevorb687/Dreamstime **p. 379** Rolle Rod/SIPA/Newscom **p. 383** Tim Graham/Alamy **p. 385** Rob Mattingley/iStockphoto **p. 389** Medioimages/Photodisc/Thinkstock **p. 390** A. Barrington Brown/Photo Researchers, Inc.

CHAPTER 9 **p. 400** Goran Bogicevic/Shutterstock **p. 401** Jovan Nikolic/Shutterstock **p. 403** Jupiterimages/Thinkstock **p. 406** Jovan Nikolic/Shutterstock **p. 410** Cameron Cross/Shutterstock **p. 415** Dean Mitchell/Shutterstock **p. 422** Ryan McVay/Thinkstock **p. 432** Taiga/Shutterstock **p. 438** Junial Enterprises/Shutterstock **p. 441** (from top to bottom) Kurniawan1972/Dreamstime; Christopher Halloran/Shutterstock **p. 444** Yuri Arcurs/Shutterstock **p. 445** (from top to bottom) Copyright © 1959 G. Paul Bishop; With kind permission of the Biometrika Trustees

CHAPTER 10 **p. 460** Pitb_1/Fotolia **p. 461** Tanyae/Dreamstime **p. 462** Copyright © 2009 Jim Borgman. Distributed by Universal Uclick. Reprinted with permission. All rights reserved. **p. 463** Tom Bochsler Photography Limited/Pearson Education, Inc. **p. 470** CandyBoxPhoto/Shutterstock **p. 475** Naluwan/Shutterstock **p. 479** Mark William Richardson/Shutterstock **p. 482** Lucidwaters/Dreamstime **p. 489** Capifrutta/Shutterstock **p. 501** Igor Zh./Shutterstock **p. 509** Robert J. Daveant/Shutterstock

PART 3 REVIEW **p. 524** (from left to right) Jose AS Reyes/Shutterstock; Goran Bogicevic/Shutterstock; Pitb_1/Fotolia **p. 527** Lincoln Rogers/Shutterstock

CHAPTER 11 **p. 535** (from left to right) R. Gino Santa Maria/Shutterstock; Malota/Dreamstime; Amana Productions/Photolibrary New York; Sodafish bvba/iStockphoto; Byron Moore/iStockphoto **p. 536** R. Gino Santa Maria/Shutterstock **p. 537** Oliveromg/Shutterstock **p. 544** Andresr/Shutterstock **p. 549** Comstock/Shutterstock **p. 558** Getty Images **p. 560** Magdalena Bujak/Shutterstock **p. 567** Piyachok Thawornmat/Shutterstock

CHAPTER 12 **p. 576** Malota/Dreamstime **p. 577** Valueline/Photodisc/Thinkstock **p. 584** Hxdbzxy/Shutterstock **p. 587** Victoria/Fotolia **p. 590** Photodisc **p. 596** AKG Images/Newscom **p. 601** Jack Hollingsworth/Photodisc Red/Getty Images **p. 607** Andresr/Dreamstime **p. 616** Adisa/Shutterstock

CHAPTER 13 **p. 629** Amana Productions/Photolibrary New York **p. 630** Mike Flippo/Shutterstock **p. 632** StockLite/Shutterstock **p. 643** Muzsy/Shutterstock **p. 662** Jupiterimages/Thinkstock **p. 666** Olli Wang/Fotolia

CHAPTER 14 **p. 679** Sodafish bvba/iStockphoto **p. 680** Yuri Arcurs/Shutterstock **p. 682** George Doyle/Thinkstock **p. 692** De Visu/Shutterstock **p. 701** Maksud/Shutterstock **p. 705** Auremar/Shutterstock

CHAPTER 15 **p. 720** Byron Moore/iStockphoto **p. 721** Kuleczka/Shutterstock **p. 724** Sandra Cunningham/Shutterstock **p. 730** Susan Montgomery/Shutterstock

PART 4 REVIEW **p. 748** (from left to right) R. Gino Santa Maria/Shutterstock; Malota/Dreamstime; Amana Productions/Photolibrary New York; Sodafish bvba/iStockphoto; Byron Moore/iStockphoto **p. 751** Goldenangel/Shutterstock

A Guide to Choosing a Statistical Method

CATEGORICAL RESPONSE VARIABLE (ANALYZING PROPORTIONS)

1. If there is only one categorical response variable, use
   - Descriptive methods of Chapter 2 (Sections 2.1 and 2.2)
   - Inferential methods of Section 8.2 (confidence interval) and Section 9.2 (significance test) for proportions
2. To compare proportions of a categorical response variable for two or more groups of a categorical explanatory variable, use
   - Descriptive methods of Chapter 3 (Sections 3.1 and 3.4)
   - Inferential methods of Sections 10.1 and 10.4 for comparing proportions between two groups
   - Inferential methods of Chapter 11 for comparing two or more proportions or testing the independence of two categorical variables
3. If working with a binary response variable with quantitative and/or categorical explanatory variables (predictors), use logistic regression methods of Section 13.6

QUANTITATIVE RESPONSE VARIABLE (ANALYZING MEANS)

1. If there is only one quantitative response variable, use
   - Descriptive methods of Chapter 2
   - Inferential methods of Section 8.3 (confidence interval) and Section 9.3 (significance test) for a mean
2. To compare means of a quantitative response variable for two groups of a categorical explanatory variable, use
   - Descriptive methods of Chapter 2
   - Inferential methods of Sections 10.2 and 10.3 for independent samples
   - Inferential methods of Section 10.4 for dependent samples
   - Nonparametric tests in Section 15.1 for independent samples or Section 15.2 for dependent samples
3. To compare several means of a quantitative response variable for two or more groups of a categorical explanatory variable, use
   - ANOVA methods of Chapter 14 for independent samples, which are equivalent to regression methods with indicator variables for categorical predictors
   - Nonparametric methods from Section 15.2
4. To analyze the association of a quantitative response variable and quantitative explanatory variable, use regression and correlation
   - Descriptive methods of Chapters 3 and 12
   - Inferential methods of Chapter 12 (Sections 12.3 and 12.4)
5. To analyze the association of a quantitative response variable and several explanatory variables (predictors), use
   - Multiple regression methods of Chapter 13 (Sections 13.1–13.4) for several quantitative predictors
   - Multiple regression methods of Section 13.5 with indicator variables for categorical predictors

Data Files and Applet Usage

DATASETS	EXAMPLES	EXERCISES	
		Required	Referenced
al_team_statistics	3-9		
anorexia	9-8	8.30, 9.39-40, 14.11, 14.43, 15.22	9.34, 9.45, 9.46, 10.87-88, 10.91, R3.44
baseballs_hr_hitters		2.126	
buchanan_and_the_butterfly_ballot	3-6, 3-10	3.23	
car_weight_and_mileage		3.77	12.1, 12.27
central_park_yearly_temps	2-9, 3-12	2.117	2.27, 2.73, 2.108
cereal	2-4, 2-5, 2-7, 2-10, 2-16, 2-17	2.22, 3.33, 3.51, 3.84	2.14, 2.19-21, 2.35, 2.115, 2.124, 2.131, 3.34
cigarette_tax		2.114	2.56, 2.68
college_atheletes	13-4-7	2.55, 13.37	2.66, 13.1, 13.12, 13.31, R4.10
credit_card_ and_income	13-12, 13-13		13.49
energy_and_EU	2-18		2.78
european_union_unemployment		2.61	2.64, 2.77
fl_crime	3-14	13.5	8.38, 12.66
fl_student_survey		1.21, 1.37, 2.81, 3.14, 9.94, 10.28, 10.112, R3.33, 11.59, 12.79, 13.71, 14.9, 14.48, 15.10, 15.28, R4.14, R4.26	
georgia_student_survey	12-13, 15-6	2.127, 3.96, 12.9, 12.10, 12.20, 12.42, 12.54, 12.70, 13.2, 13.18, 15.24, 15.25	12.44, 13.19, 14.42
heads_of_household		7.58	
heights	2-15	2.113	2.52
high_jump	3-11, 13-11	3.90	13.40, 13.47
high_school_female_athletes	12-1-12-3	2.54, 2.65, 12.80, 13.84	12.45, 12.71, R4.9
house_selling_prices_FL			12.33, 12.34, 12.46, 12.67-69
house_selling_prices_OR	13-1 – 13-3, 13-8 – 13-10	13.10, 13.30, 13.38, 13.45, 13.59, 14.45	13.8, 13.17, 13.29, 13.39, 13.42, 13.46, 14.12, 14.25
hs_graduation_rates		3.89	2.121
human_development		3.78, 3.80, 15.23	
income_in_public_housing		8.87	
internet_use	3-4, 3-5, 3-7	3.12, 3.30	3.15, 3.29, 12.13
long_jump			3.45
male_athlete_strength		12.31	12.3, 12.17
mental_health		13.27, 13.28	
mountain_bike		3.21, 3.41, 3.42, 8.85, 13.41	
newnan_ga_temps		2.29, 3.97	
NL_team_statistics		3.95	3.38
protein_and_weight_gain		14.33	14.34
quality_and_productivity		13.43	
sat2010		3.31, 3.43	
sharks	2-2		2.13
softball		12.77, 13.61	13.16
tv_europe		3.52	
twelve_countries	12-5		13.14, R4.17
us_statewide_crime	3-13	3.50, 3.86, 3.88, 12.30	3.44, 3.47, 3.72, 3.85
us_temperatures		3.46	
whooping_cough		3.66	2.28
world_data_for_fertility_and_literacy			13.64, R4.16

APPLETS*	ACTIVITY	EXAMPLE	EXERCISES
Sample from a population	1-2		1.22, 1.23, 1.35, 4.115
Sampling distributions	7-1, 7-2		7.1, 7.2, 7.25-7.27, 7.29, 7.44, 7.49, 8.26, 8.60
Random numbers	6-1		4.16-17, 5.63, 5.117-120, 6.89
Long-run probability demonstrations	5-1	5-2	4.100, 5.10, 5.11, 5.12, 5.95
Mean versus median	2-1		2.38, 2.148
Standard deviation			2.62
Confidence intervals for a proportion	8-2		8.25, 8.26
Confidence intervals for a mean			8.45
Hypothesis tests for a proportion	9-2		
Correlation by eye			3.114-115
Regression by eye			3.114-115

*See page x for descriptions of the applets.